Mielke (Ed.)
Analysis, Modeling and Simulation of Multiscale Problems

Alexander Mielke
Editor

Analysis, Modeling and Simulation of Multiscale Problems

With 167 Figures, 32 Colored Plates, and 11 Tables

Editor

Alexander Mielke

Weierstraß-Institut für Angewandte
Analysis und Stochastik
Mohrenstraße 39
10117 Berlin, Germany

and

Institut für Mathematik
Humboldt-Universität zu Berlin
Rudower Chaussee 25
12489 Berlin, Germany

E-mail: mielke@wias-berlin.de

Library of Congress Control Number: 2006932292

Mathematics Subject Classification (2000): 35-XX, 49-XX, 70-XX, 74-XX, 81-XX, 92-XX

ISBN-10 3-540-35656-8 Springer Berlin Heidelberg New York
ISBN-13 978-3-540-35656-1 Springer Berlin Heidelberg New York

Springer is a part of Springer Science+Business Media

springer.com

Typesetting by the editor using a Springer TEX macro package
Production: LE-TEX Jelonek, Schmidt & Vöckler GbR, Leipzig
Cover design: WMXDesign GmbH, Heidelberg

Printed on acid-free paper 46/3100/YL - 5 4 3 2 1 0

Preface

L'étude approfondie de la nature
est la source la plus féconde
des découvertes mathématiques.
J.B.J. Fourier (1768–1830)

Recent technological advances allow us to study and manipulate matter on the atomic scale. Thus, the traditional borders between mechanics, physics and chemistry seem to disappear and new applications in biology emanate. However, modeling matter on the atomistic scale ab initio, i.e., starting from the quantum level, is only possible for very small, isolated molecules. Moreover, the study of mesoscopic properties of an elastic solid modeled by 10^{20} atoms treated as point particles is still out of reach for modern computers. Hence, the derivation of coarse grained models from well accepted fine-scale models is one of the most challenging fields. A proper understanding of the interaction of effects on different spatial and temporal scales is of fundamental importance for the effective description of such structures. The central question arises as to which information from the small scales is needed to describe the large-scale effects correctly.

Based on existing research efforts in the German mathematical community we proposed to the Deutsche Forschungsgemeinschaft (DFG) to strengthen the mathematical basis for attacking such problems. In May 1999 the DFG decided to establish the

DFG Priority Program (SPP 1095)
Analysis, Modeling and Simulation of Multiscale Problems.

After another reviewing process the official start in September 2000 involved about 25 research groups all over Germany, see

`http://www.mathematik.uni-stuttgart.de/~mehrskalen`

for information about these groups and their activities. In the sequel, this program has inspired a number of multiscale initiatives worldwide, for example, the *Center for Integrative Multiscale Modeling and Simulation* (CIMMS) at CalTech, the *Bath Institute for Complex Systems* and the European Network *Multi-scale modeling and characterization of phase transformations in advanced materials* (MULTIMAT).

The aim of our priority program has been to combine different expertise in the mathematical community and to enhance the development of mathematical methods and concepts for the study of multiscale problems. On the one hand, these problems are driven by needs in applications in physics, chemistry, or biology. On the other hand, the challenge to bridge theories at different scales leads to deep and fundamental questions within mathematics. This book tries to span this variety of the subject by surveying and highlighting the work done in the groups of the priority program. In the vast mathematical area of multiscale problems the program has focused its efforts on several specific fields that can be characterized by the following subjects:

- weak-convergence methods, relaxations, and Gamma convergence
- Young measures, gradient Young measures, Wigner measures
- homogenization, averaging, and adiabatic evolution
- singular perturbations, boundary layers, interfaces, and point defects
- formation, stability, and dynamics of sharp interfaces
- microstructures in elastic solids, thin films or rods, and in micro-magnetism
- derivation of continuous models from spatially discrete models
- quantum-chemical and quantum-mechanical models
- quantum and semi-classical models for semi-conductors and their coupling
- numerical algorithms for microstructures in solids
- exponential integrators for Hamiltonian systems with fast oscillations
- almost invariant sets and metastability

Most of the work reported in this book has been supported substantially by the DFG through the priority program. The major part of the funding went into positions for PhD students and post-doctoral researchers. In fact, this program has been an ideal way to educate a new generation of young scientists to cope with the future challenges in mathematical modeling, analysis and numerics. We are glad to see that about half of the contributors to this book are such young scientists. The additional support for travel and the guest program has proved to be crucial for the interaction and mutual stimulation of the participating groups, in particular the annual colloquia as wells as a large series of specialized workshops.

Bernhard Nunner and Frank Kiefer from the DFG made our life as easy as possible concerning the administrative duties while pushing our scientific achievements to new frontiers. In this goal, they could rely on the steady support and expertise of the referees Wolfgang Dahmen, Gero Friesecke, Thomas Y. Hou, Bernhard Kawohl, Volker Mehrmann, Umberto Mosco, Marek Niezgódka, Tomaš Roubíček, Stefan Sauter, Valery P. Smyshlyaev, and Gabriel Wittum. These referees have been valuable advisors for the projects in this program and played an essential role in shaping our view of the mathematics for multiscale problems.

Finally, Alexander Mielke would like to thank Stefanie Siegert. Her dedication and precise management was often hidden in the background, but constituted an essential part of the smooth coordination of the program.

Of course, such a priority program on multiscale problems cannot be successful without the creative work of the involved researchers and without their openness for collaboration. We, as the initiators of this program, are grateful to all these companions in this endeavor and hope that the achievements within the program will prove to be rewarding for all of them.

München *Folkmar Bornemann*
Leipzig *Stephan Luckhaus*
Berlin *Alexander Mielke* (coordinator)
Leipzig *Stefan Müller*
May 2006

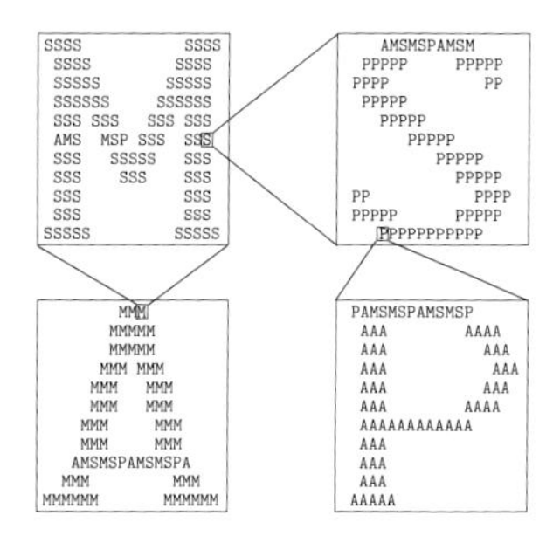

Contents

List of Contributors

Anton Arnold
Institut für Analysis und Scientific Computing
Technische Universität Wien
Wiedner Hauptstraße 8
1040 Wien, Austria

Steffen Arnrich
Fakultät für Mathematik und Informatik
Universität Leipzig
Augustusplatz 10/11
04103 Leipzig

Lev Balykov
Crystal Growth group
Research center caesar
Ludwig-Erhard-Allee 2
53175 Bonn

Sören Bartels
Institut für Mathematik
Humboldt-Universität zu Berlin
Rudower Chaussee 25
12489 Berlin

Sebastian Bauer
Fachbereich Mathematik
Universität Duisburg-Essen
45117 Essen

Thomas Blesgen
Max-Planck-Institut für Mathematik in den Naturwissenschaften
Inselstraße 22–26
04103 Leipzig

Folkmar Bornemann
Zentrum für Mathematik
Technische Universität München
Boltzmannstraße 3
85747 Garching bei München

Carsten Carstensen
Institut für Mathematik
Humboldt-Universität zu Berlin
Rudower Chaussee 25
12489 Berlin

Vladimir Chalupecky
Department of Mathematics
Faculty of Nuclear Sciences and Physical Engineering
Czech Technical University
Trojanova 13
120 00 Prague 2, Czech Republic

David Cohen
Mathematisches Institut
Universität Tübingen
72076 Tübingen

Sergio Conti
Universität Duisburg-Essen
Campus Duisburg
Fachbereich Mathematik
Lotharstraße 65
47057 Duisburg

Michael Dellnitz
Institut für Mathematik
Universität Paderborn
33095 Paderborn

Georg Dolzmann
Mathematics Department
University of Maryland
College Park
MD 20742-4015, U.S.A.

Wolfgang Dreyer
Weierstraß-Institut für Angewandte
Analysis und Stochastik
Mohrenstraße 39
10117 Berlin

Christof Eck
Institute for Applied Mathematics
Universität Erlangen-Nürnberg
Martensstraße 3
91058 Erlangen

Heike Emmerich
Center for Computational Engineering Science
Institute of Minerals Engineering
RWTH Aachen
Mauerstraße 5
52064 Aachen

Heinz-Jürgen Flad
Institut für Informatik und Praktische Mathematik
Universität zu Kiel
Christian-Albrechts-Platz 4
24098 Kiel
and
Max-Planck-Institut für Mathematik
in den Naturwissenschaften
Inselstraße 22–26
04103 Leipzig

Jürgen Fuhrmann
Weierstraß-Institut für Angewandte
Analysis und Stochastik
Mohrenstraße 39
10117 Berlin

Harald Garcke
NWF I – Mathematik
Universität Regensburg
93040 Regensburg

Johannes Giannoulis
Weierstraß-Institut für Angewandte
Analysis und Stochastik
Mohrenstraße 39
10117 Berlin

Wolfgang Hackbusch
Max-Planck-Institut für Mathematik
in den Naturwissenschaften
Inselstraße 22–26
04103 Leipzig

Klaus Hackl
Institut für Mechanik
Ruhr-Universität Bochum
Universitätsstraße 150
44801 Bochum

Michael Herrmann
Humboldt-Universität zu Berlin
Institut für Mathematik
Unter den Linden 6
10099 Berlin

Mirko Hessel-von Molo
Institut für Mathematik
Universität Paderborn
33095 Paderborn

Ronald H.W. Hoppe
Institut für Mathematik
Universität Augsburg
86159 Augsburg
and
Department of Mathematics
University of Houston
Houston, TX 77204–3008

Ulrich Hoppe
Institut für Mechanik
Ruhr-Universität Bochum
Universitätsstraße 150
44801 Bochum

Tobias Jahnke
BioComputing Group
Institut für Mathematik II
Freie Universität Berlin
Arnimallee 2–6
14195 Berlin

Ansgar Jüngel
Institut für Mathematik
Universität Mainz
Staudingerweg 9
55099 Mainz

Hans-Christoph Kaiser
Weierstraß-Institut für Angewandte Analysis und Stochastik
Mohrenstraße 39
10117 Berlin

Dietmar Kolb
Institut für Physik
Universität Kassel
Heinrich-Plett-Straße 40
34132 Kassel

Thomas Koprucki
Weierstraß-Institut für Angewandte Analysis und Stochastik
Mohrenstraße 39
10117 Berlin

Ganeshram Krishnamoorthy
Center for Computational Engineering Science
Institute of Minerals Engineering
RWTH Aachen
Mauerstr. 5
52064 Aachen

Markus Kunze
Fachbereich Mathematik
Universität Duisburg-Essen
45117 Essen

Matthias Kurzke
Institute for Mathematics and its Applications
University of Minnesota
207 Church Street SE
Minneapolis, MN 55455, U.S.A.

David Jung Chul Kwak
NWF I – Mathematik
Universität Regensburg
93040 Regensburg

Caroline Lasser
Fachbereich Mathematik und Informatik
Freie Universität Berlin
Arnimallee 2–6
14195 Berlin

Martin Lenz
Institut für Numerische Simulation
Universität Bonn
Nußallee 15
53115 Bonn

Katina Lorenz
Mathematisches Institut
Universität Tübingen
72076 Tübingen

Christian Lubich
Mathematisches Institut
Universität Tübingen
72076 Tübingen

Stephan Luckhaus
Fakultät für Mathematik und Informatik
Universität Leipzig
Augustusplatz 10/11
04103 Leipzig

Hongjun Luo
Institut für Physik
Universität Kassel
Heinrich-Plett-Straße 40
34132 Kassel

Karsten Matthies
Department of Mathematical Sciences
University of Bath
Bath BA2 7AY, United Kingdom

Christof Melcher
Department of Mathematics
Humboldt-Universität zu Berlin
Unter den Linden 6
10099 Berlin

Philipp Metzner
Institut für Mathematik II
Freie Universität Berlin
Arnimallee 2-6
14195 Berlin

Alexander Mielke
Weierstraß-Institut für Angewandte Analysis und Stochastik
Mohrenstraße 39
10117 Berlin
and
Institut für Mathematik
Humboldt-Universität zu Berlin
Rudower Chaussee 25
12489 Berlin

Roger Moser
Department of Mathematical Sciences
University of Bath
Bath BA2 7AY, United Kingdom

Luca Mugnai
Fakultät für Mathematik und Informatik
Universität Leipzig
Augustusplatz 10/11
04103 Leipzig

Britta Nestler
Fachbereich Informatik
Hochschule Karlsruhe
Moltkestraße 30
76133 Karlsruhe

Barbara Niethammer
Institut für Mathematik
Humboldt-Universität zu Berlin
Rudower Chaussee 25
12489 Berlin

Antonio Orlando
Institut für Mathematik
Humboldt-Universität zu Berlin
Rudower Chaussee 25
12489 Berlin
and
School of Engineering
University of Wales Swansea
Singleton Park
SA2 8PP Swansea UK

Felix Otto
Institut für Angewandte Mathematik
Rheinische Friedrich-Wilhelms-Universität Bonn
Wegelerstraße 10
53115 Bonn

Gianluca Panati
Zentrum Mathematik
Technische Universität München
Boltzmannstraße 3
85747 Garching bei München

Svetozara I. Petrova
Institut für Mathematik
Universität Augsburg
86159 Augsburg
and
Institute for Parallel Processing
Bulgarian Academy of Science
1113 Sofia, Bulgaria

Robert Preis
Institut für Mathematik
Universität Paderborn
33095 Paderborn

Jens D.M. Radmeacher
Weierstraß-Institut für Angewandte Analysis und Stochastik
Mohrenstraße 39
10117 Berlin

Andreas Rätz
Crystal Growth group
Research center caesar
Ludwig-Erhard-Allee 2
53175 Bonn

Martin Rumpf
Institut für Numerische Simulation
Universität Bonn
Nußallee 15
53115 Bonn

Christof Schütte
Institut für Mathematik II
Freie Universität Berlin
Arnimallee 2-6
14195 Berlin

Herbert Spohn
Zentrum Mathematik
Technische Universität München
Boltzmannstraße 3
85747 Garching bei München

Börn Stinner
NWF I – Mathematik
Universität Regensburg
93040 Regensburg

Torben Swart
Fachbereich Mathematik und Informatik
Freie Universität Berlin
Arnimallee 2–6
14195 Berlin

Stefan Teufel
Mathematisches Institut
Universität Tübingen
Auf der Morgenstelle 10
72076 Tübingen

Juan J.L. Velázquez
Departamento de Matemática Aplicada
Facultad de Matemáticas
Universidad Complutense
28040 Madrid, Spain

Axel Voigt
Crystal Growth group
Research center caesar
Ludwig-Erhard-Allee 2
53175 Bonn
and
Institut für Angewandte Mathematik
Universität Bonn
Wegelerstraße 10
53115 Bonn

Jessika Walter
Department of Mathematics
Ecole Polytechnique Fédérale de Lausanne
1015 Lausanne, Switzerland

Ulrich Weikard
Fachbereich Mathematik
Universität Duisburg
Lotharstraße 63/65
47048 Duisburg

Frank Wendler
Institut für Angewandte Forschung
Hochschule Karlsruhe
Moltkestraße 30
76133 Karlsruhe

Exponential Estimates in Averaging and Homogenisation

Karsten Matthies

University of Bath, Department of Mathematical Sciences, Bath, BA2 7AY, United Kingdom. K.Matthies@maths.bath.ac.uk

Summary. Many partial differential equations with rapid spatial or temporal scales have effective descriptions which can be derived by homogenisation or averaging. In this article we deal with examples, where quantitative estimates of the error is possible for higher order homogenisation and averaging. In particular, we provide theorems, which allow homogenisation and averaging beyond all orders by giving exponential estimates of appropriately averaged and homogenised descriptions. Methods include iterated averaging transformations, optimal truncation of asymptotic expansions and highly regular solutions (Gevrey regularity). Prototypical examples are reaction-diffusion equations with heterogeneous reaction terms or rapid external forcing, nonlinear Schrödinger equations describing dispersion management, and second-order linear elliptic equations.

1 Introduction

Many classical multiscale problems are modelled by (partial) differential equations which are heterogeneous in space (explicitly depending on the space variable x) or nonautonomous (explicitly dependent on time t). The multiscale character is introduced, if these dependencies are rapid, i.e. they are on a fast scale. Typically, this is achieved by introducing a small parameter ε and considering dependencies of the form x/ε or t/ε. Application areas, where such descriptions are used, include continuum mechanics, chemical reactions, nonlinear optics, ecology, and celestial mechanics among others. The x/ε describes e.g. the underlying varying microstructure of the medium, whereas an t/ε dependency is used for rapid external excitations.

Then there are two main approaches to such problems when comparing the heterogeneous/ nonautonomous partial differential equations with their homogeneous/ autonomous counterparts. Firstly one can identify effects that are created by the multiscale structure. Secondly one can try to find effective descriptions by homogeneous and autonomous equations without any explicit multiscale structure. This process is usually called homogenisation for space-dependent problems and averaging for time-dependent problems. Of course, these two approaches are related, as effective descriptions can be used to estimate the size of effects.

There are many different methods to derive the homogenised or averaged descriptions, these can be grouped roughly into two groups. The first one is using weak convergence methods (see e.g. [Tar79, Tar86, Bor98, GM*97, JKO94]), which are applicable to many problems but which usually do not provide a quantitative bound on the effectiveness, i.e. on the error of approximation. The next group of methods is based on asymptotic expansions in the small parameter ε. When ruling out purely formal methods, then the rigorous asymptotic techniques can be also used to derive quantitative error bounds on the approximation error. These will be then in some finite or even exponential order in ε. Some classical methods can be found in [AKN97, BLP78, BP84, JKO94, LM88]. To derive such estimates, the finer analysis often also requires additional assumptions on the structure of the differential equation.

Besides certain regularity assumptions on the solutions, we will mainly consider cases where the underlying microstructure and external excitation is either periodic or quasiperiodic. When expanding a rigorous finite order procedure into a series, even in a very benign example, like periodic averaging of analytic ordinary differential equations

$$\dot{x} = f(x, t/\varepsilon),$$

convergence of the expansion cannot be expected in general. Nevertheless, beyond a finite asymptotic expansion, there are exponential estimates in several aspects, early examples are by Nekhoroshev, and Neishtadt; see, e.g., [Nek79, Nei84, LM88]. These then yield upper estimates on all kinds of effects created by the periodic nonautonomous structure.

Here, the purpose of this article is to describe a number of asymptotic techniques which provide effective descriptions up to exponentially small errors for wide classes of multiscale problems. The remainder of the article has the following structure. First we will provide a list of prototypical partial differential equations and interesting solutions involving multiple scales in Sect. 2. Then we will describe several finite order methods like classical averaging, homogenisation and their relation to normal forms in the dynamical systems literature (Sect. 3). Detailed results on techniques for upper exponential estimates for the earlier examples will be given in Sect. 4. In Sect. 5, we will use these results to estimate several possible effects. We conclude with a discussion, where we are e.g. describing some situations, where lower estimates are possible, Sect. 6.

2 Examples

We provide some typical examples for which effective descriptions beyond every order will be possible. For larger classes of examples and more general assumptions, we refer to the relevant papers [Mat01, MS03, Mat04, KMS06].

2.1 Partial differential equations with rapid time dependence

The basic example of nonautonomous differential equations is the periodic ordinary differential equation

$$\dot{u} = f(u, t/\varepsilon) \text{ with } f(., \tau) = f(., \tau + 1) \text{ for all } \tau \in \mathbf{R} \tag{2.1}$$

To describe more general dependencies than the purely periodic one, we introduce the phase $\phi \in T^m = (\mathbf{R}/\mathbf{Z})^m, m \in \mathbf{N}$. Then we rewrite the autonomous equation as

$$\begin{aligned} \dot{u} &= f(u, \phi) \\ \dot{\phi} &= \frac{1}{\varepsilon}\Omega(u, \phi) \end{aligned} \tag{2.2}$$

with $\Omega(u, \phi) \geq c_0 > 0$. Then $m = 1, \Omega(u, \phi) = 1$ recovers the periodic case. Results on averaging in this situation, under certain assumptions on Ω, can be found in [BM61, SV85]. Exponential estimates are due to Neishtadt [Nei84] for $m = 1$ and to Simó [Sim94] for $m > 1$ and

$$\begin{aligned} &\Omega(u, \phi) = \omega \in \mathbf{R}^m \text{ with Diophantine conditions on } \omega \\ &|(\ell, \omega)| \geq \gamma |\ell|^{-\tau} \text{ for some } \gamma > 0, \tau > m - 1 \text{ and all } \ell \in \mathbf{Z}^m \end{aligned} \tag{2.3}$$

Partial differential equations with such a structure can be found in the context of systems of reaction-diffusion equations

$$\begin{aligned} u_t &= D\Delta u + f(u) + g(u, \phi) \\ \dot{\phi} &= \frac{1}{\varepsilon}\Omega(u, \phi) \\ (u(0), \phi(0)) &= (u_0, \phi_0) \in X \times T^m \end{aligned} \tag{2.4}$$

with $D = \text{diag}(d_1, \ldots, d_n)$. We will consider the functiion u on $[0, 1]^d$ with periodic boundary conditions. Initial conditions are in the phase space $X = H^s_{per}([0, 1]^d, \mathbf{R}^n)$ with $s > d/2$ to ensure the embedding of X into C^0 and differentiability of the nonlinearities in X. Reaction diffusion equations are the prime example for pattern formation, for a review see [FS03]. An external forcing can be introduced in light sensitive reactors by periodic changes, see e.g. [SSW99, RM*03] for a framework and an example.

Another example are nonlinear Schrödinger type equations, which e.g. describe the evolution of pulses in optical fibres [NM92]. The evolution then describes the changes of the pulse while propagating along the fibre, so changes in the material due to dispersion management and localised amplification are described as

$$\begin{aligned} \mathrm{i}u_t &= d(\phi)u_{xx} + C(\phi, |u|)u \\ \dot{\phi} &= \frac{1}{\varepsilon}\Omega(u) + g(u,\phi) \end{aligned} \tag{2.5}$$

As a phase-space we use as in (2.4) some Sobolev space, denoted again by X. Variants describing general interaction between fast oscillations and pulses can be also described in the setting of Hamiltonian PDE. For the necessary frame work and the description of non-adiabatic coupling, we refer the reader to [MS03].

2.2 Partial differential equations with rapid space dependence

First we consider a heterogeneous version of (2.4), where the heterogeneity is in the reaction term, examples of these can be found in several modelling areas [BHR05, Kee00, KS98]:

$$u_t = \Delta_{x,y}u + f(u, x/\varepsilon) \tag{2.6}$$

We will be in particular interested in the behaviour in infinite cylinders, i.e. $x \in \mathbf{R}$ and $y \in \Sigma$ with Σ a bounded cross-section. In particular, let $\Sigma = [0,1]^d$ with periodic boundary conditions in y. When looking for stationary solutions, we obtain

$$\Delta_{x,y}u + f(u, x/\varepsilon) = 0. \tag{2.7}$$

We rewrite the equation and use the idea of spatial dynamics. It is a way to construct special solutions to PDE on unbounded domains. For this we let

$$U = \begin{pmatrix} u \\ u_x \end{pmatrix}; \quad A = \begin{pmatrix} 0 & I \\ -\Delta_y & 0 \end{pmatrix}; \quad F(U, x/\varepsilon) = \begin{pmatrix} 0 \\ -f(u, x/\varepsilon) \end{pmatrix}$$

Renaming x as time t, we again obtain an equation

$$U_t = AU + F(U, t/\varepsilon), \tag{2.8}$$

which has the form of a rapidly forced evolution equation. The phase space X is a function space on the cross-section Σ like $X = H^{s+1}(\Sigma, \mathbf{R}^n) \times H^s(\Sigma, \mathbf{R}^n)$. Even if the Cauchy problem is not well-posed, this method of spatial dynamics has a long history, see [Kir82] and further work (see, e.g.[IM91, AM95, FS03] and the references therein).

When considering travelling waves in heterogeneous media one is using the ansatz

$$u(x, y, t) = v(x - ct, y, x/\varepsilon),$$

i.e. the profile v of the travelling wave is changing periodicly while moving through the periodic medium. This can be also formulated as a spatial dynamics problem, for details see [MSU06].

Variants also include heterogeneities in the main part like in classical homogenisation theory. The homogenisation of heterogeneous second-order elliptic equations is appearing in many stationary problems, consider

$$-\nabla \cdot (\mathcal{A}(x/\varepsilon)\nabla u)(x)) = f(x). \tag{2.9}$$

The matrix $\mathcal{A} \in L^\infty(T^d)$ on $T^d = (\mathbf{R}/\mathbf{Z})^d$ is assumed to be symmetric and uniformly elliptic. Furthermore we assume boundary conditions for x in some bounded domain, here again periodic boundary conditions.

3 Finite order estimates and normal forms

The basic idea of this approach is to transform the equation to derive an effective, simpler version of the differential equation. The method proved to be very successful in the analysis of finite dimensional dynamical systems [LM88, SV85]. Now we consider problems, that can be written as an evolution equation with external forcing like (2.4,2.8) and under further assumption also (2.5) (see [Mat04]). More examples and references can be found in [Mat04, Ver05, Ver06], including other variants of explicitly time dependent partial differential equations and other near-identity transformations to obtain the form in (3.1). We consider

$$\begin{aligned} u_t &= Au + f(u,\phi) \\ \dot{\phi} &= \frac{1}{\varepsilon}\Omega(u,\phi), \end{aligned} \tag{3.1}$$

where $u \in X$ for some appropriate phase space X. Then a near-identity transformation on a ball $B_R(X)$ can be written in the form

$$u = v + \varepsilon W(v,\phi) \tag{3.2}$$

The transformed equation can be derived from

$$\partial_t u = \partial_t v + \varepsilon \partial_v W(v,\phi)\partial_t v + \varepsilon \partial_\phi W(v,\phi)\partial_t \phi$$

then

$$\begin{aligned} \partial_t v =& (I + \varepsilon \partial_v W(v,\phi))^{-1}\{A(v + \varepsilon W(v,\phi)) + f(v + \varepsilon W(v,\phi),\phi) \\ & - \varepsilon \partial_\phi W(v,\phi)\frac{1}{\varepsilon}\Omega(v,\phi)\}. \end{aligned}$$

So depending on the form of Ω, one can try to reduce the externally forced term f by an appropriate choice of W. It is notational convenient to split f such that

$$f(v,\phi) = \hat{f}(v) + g(v,\phi) \text{ such that}$$
$$\langle g\rangle = \int_{T^m} g(v,\psi)\mathrm{d}\psi = 0.$$

In the simplest case of periodic external forcing, i.e. $m = 1, \Omega \equiv 1$, we let

$$W(v,\phi) = \int_0^\phi g(v,\psi)\mathrm{d}\psi,$$

then the transformed equation has the form

$$\begin{aligned}\partial_t v =& (I + \varepsilon\partial_v W(v,\phi))^{-1}\big\{A(v+\varepsilon W(v,\phi)) + \hat{f}(v+\varepsilon W(v,\phi)) \\ &+ g(v+\varepsilon W(v,\phi),\phi) - g(v,\phi)\big\}.\end{aligned}$$

This removes the lowest order nonautonomous terms. A problem is to estimate the remainder r, if we rewrite the equation as

$$v_t = Av + \bar{f}(v;\varepsilon) + r(v,\phi;\varepsilon). \tag{3.3}$$

The remainder involves terms depending on the unbounded operator A, such that r is only formally small. A more promising variant is to use some bounded Galerkin type approximation to perform the estimates, where the Galerkin approximation v_N is chosen depending on ε. For all our examples, there exists a sequence of (Galerkin) projections $(P_N)_{N\in\mathbf{N}}$ which satisfy

1. the sequence of projections converges strongly to the identity on the phase space X,
$$\lim_{N\to\infty} P_N u = u \text{ in } X \text{ for all } u \in X; \tag{3.4}$$
2. the projections P_N commute with A on its domain of definition
$$P_N A u = A P_N u \text{ for all } u \in D(A); \tag{3.5}$$
3. the operator A is bounded on $\operatorname{Rg} P_N$,
$$|AP_N u|_X \le N|P_N u|_X \text{ for all } u \in X. \tag{3.6}$$

The equation with projection P_N on the approximation space is

$$\partial_t u_N = Au_N + P_N f(u_N,\phi). \tag{3.7}$$

Then the transformed equation is

$$\begin{aligned}\partial_t v_N = &(I + \varepsilon\partial_{v_N} W(v_N,\phi))^{-1}\big\{A(v_N+\varepsilon W(v_N,\phi)) + f(v_N+\varepsilon W(v_N,\phi),\phi) \\ &-\varepsilon\partial_\phi W(v_N,\phi)\frac{1}{\varepsilon}\Omega(v_N,\phi)\big\}.\end{aligned} \tag{3.8}$$

Regrouping again yields

$$\partial_t v_N = A v_N + \bar{f}_N(v_N; \varepsilon) + r_1(v_N, \phi; \varepsilon) \tag{3.9}$$

then the remainder term is of order

$$\sup_{\|v_N\| \leq R} \varepsilon \|A v_N\| \|g(v_N, \phi)\|,$$

for V in a large ball $B_R(X)$. The remainder is small for an appropriate choice of $N(\varepsilon)$. E.g. when we are choosing $N(\varepsilon)$ such that $\|A v_N\| \sim \varepsilon^{-1/2}$, we obtain a rigorous estimate on $r_1 \in \mathcal{O}(\varepsilon^{1/2})$ in the approximation space $P_N X$, this is uniform for $\varepsilon \to 0, N(\varepsilon) \to \infty$.

The important property is now, that (3.9) has still the form of the original equation (3.7). So the equation can be transformed again to obtain a new remainder term r_2. This is of order $\sup_{\|v_N\| \leq R} \varepsilon \|A v_n\| \|r_1(v_N, \phi)\|$, such that $r_2 \in \mathcal{O}(\varepsilon)$ for the same choice of $N(\varepsilon)$ with $\varepsilon \to 0, N(\varepsilon) \to \infty$. The transformed equation has again the same form. Hence this procedure can be iterated to obtain arbitrary finite order estimate $\mathcal{O}(\varepsilon^k)$, provided we can ensure enough regularity of the nonlinearity.

In a detailed analysis taking into account all parts of the remainder term one can see, that the constant in the $\mathcal{O}(\varepsilon^k)$ remainder will become large with k. So this procedure will not lead to a convergent asymptotic expansion. But keeping track of the constants in the remainder estimates depending on k and ε is crucial for later exponential estimates.

Here of course, the error of the Galerkin approximation is still to be estimated. When extending the transformation back to the full space by

$$u = v + \varepsilon P_N W(P_N v, \phi),$$

this leads to

$$\partial_t v = A v + \bar{f}(v; \varepsilon) + r(v, \phi; \varepsilon)$$

with additional terms in $\bar{f}$ and r due to the Galerkin approximation. The additional terms are all of the form $G(v - P_N v)$.

Another situation with an explicit choice of W is for $m = 1$ and $\Omega(v, \phi) = \bar{\Omega}(v) + \varepsilon \beta(v, \phi)$. The phase ϕ will also be transformed. We choose the explicit change of coordinate $(u_N, \phi) = (v_N, \psi) + \varepsilon W(v_N, \psi)$

$$W(v_N, \psi; \varepsilon) = \begin{pmatrix} W^1(v_N, \psi; \varepsilon) \\ W^2(v_N, \psi; \varepsilon) \end{pmatrix}, \tag{3.10}$$

with

$$W^1(v_N, \psi; \varepsilon) = \frac{1}{\bar{\Omega}(v_N)} \int_0^{\phi} g(v_N, \tau) \mathrm{d}\tau$$

and

$$\begin{aligned} W^2(v_N, \psi; \varepsilon) = \frac{1}{\bar{\Omega}(v_N)} \int_0^{\phi} & \beta(v_N, \tau) \\ & + \partial_{v_N} \bar{\Omega}(v_N)(W^1_{k+1}(v_N, \tau) - \langle W^1(v_N, .) \rangle) \mathrm{d}\tau, \end{aligned}$$

where $\langle \, . \, \rangle$ again denotes the T^1-average.

The third situation is the quasiperiodic case, i.e. $m > 1$ and $\Omega(v, \phi) = \omega$ with $\omega = (\omega_1, \dots, \omega_m)$ fulfilling Diophantine conditions: there are constants $\gamma > 0$ and $\tau > m - 1$ such that for all $\ell \in \mathbf{Z}^p$

$$|(\ell, \omega)| \geq \gamma |\ell|^{-\tau} \tag{3.11}$$

as for the finite dimensional result (2.3). Then the transformation is given terms of the Fourier expansion of the phase dependent term. Letting

$$g(u_N, \phi; \varepsilon) = \sum_{\ell \in \mathbf{Z}^p} g_\ell(u_N; \varepsilon) \exp(\mathrm{i}2\pi(\ell, \phi))$$

then we transform $u_N = v_N + \varepsilon W(v_N, \phi; \varepsilon)$ with

$$W(v_N, \phi; \varepsilon) = \sum_{\ell \in \mathbf{Z}^p} \frac{g_\ell(v_N; \varepsilon)}{2\pi\mathrm{i}(\ell, \omega)} \exp(\mathrm{i}2\pi(\ell, \phi)) \tag{3.12}$$

Classical homogenisation theory of (2.9) gives an asymptotic expansion (cf. e.g. [BP84]):

$$u^{\varepsilon,N}(x) = \sum_{n=0}^{N+2} \varepsilon^m u^{(n)}(x, x/\varepsilon), \tag{3.13}$$

where the functions $u^{(l)}(x, y)$ are required to be periodic in the fast variable y. For this problem one can construct in this way a full asymptotic expansion with $u^{(l)}$ adopting the following form (see e.g. [BP84], [CS04]):

$$u^{(n)}(x, y) = \sum_{l=0}^{n} \sum_{|k|=l} N_k(y) D_x^k v_{n-l}(x), \tag{3.14}$$

where $N_0(y) \equiv 1$ and $N_k(y)$ are periodic solutions of the main ($|k| = 1$) and higher order ($|k| > 1$) unit cell problems in y. The functions $v_s(x)$, $s \geq 0$, solve certain recurrent systems of equations in x, see [BP84, KMS06]. This cannot be easily rephrased as an iterative procedure but the idea of tracking the dependence of the constants and taking expansion depending on ε can also be used in the context of (2.9). For other quantitative estimates on homogenisation involving quasiperiodic terms see [FV01].

The method of iterative transformation procedures are encountered widely in dynamical systems theory as the concept of normal forms, see e.g. [AKN97, Van89]. An important question is the description of the behaviour near an equilibrium of

$$\dot{u} = Au + f(u), u \in \mathbf{R}^n,$$

where f is of higher order in u. A particular simple form is sought for f. If A has purely imaginary spectrum and is semisimple (i.e. there are no Jordan blocks of size 2 or bigger in its complex Jordan normal form), then we obtain a normal form transformation by averaging, see [Van89][Sec.2.4]. When we assume that $\exp(At)$ is periodic with period T, then all terms can be removed by appropriate coordinate changes, except those which are invariant under the averaging operator

$$\pi f(u) = \frac{1}{T} \int_0^T \exp(-At) f(\exp(At)u) \mathrm{d}t$$

for all $u \in \mathbf{R}$. But while removing terms, which are not invariant under π, one also changes certain higher order invariant terms, such that pure averaging will not give a correct normal form.

The relation of averaging and purely imaginary spectrum has also been used in the context of wave equations and similar partial differential equations, see [Bam03a, Bam03b, Bam06, Kr89, Ver05, Ver06]. Here also the reductions to Galerkin approximations were used depending on a small parameter, which is introduced via scaling $u \mapsto \varepsilon u$ to obtain

$$\partial_t u = Au + \varepsilon \tilde{f}(u, \varepsilon),$$

with $\tilde{f}(u, \varepsilon) = 1/\varepsilon^2 f(\varepsilon u) \in \mathcal{O}(1)$. The equation is then simplified by averaging the semigroup $\exp(At)$. Here the name 'Galerkin averaging' was introduced. The development of 'Exponential averaging' in [Mat01, MS03] was independent of this.

4 Exponential estimates

We will now collect methods to move from finite order normal form and averaging transformations to exponential estimates in the framework described above. In particular, we obtain results to all orders $\mathcal{O}(\varepsilon^k)$ and beyond. In the last section, we derived iterative estimates on the Galerkin approximation space $P_N X$, and obtained error terms due to the Galerkin approximation. We will control both errors at the same time. There are three variables to choose to minimise the error. Firstly there is ε, which is given, then we have the choice of the number of normal form steps and there is also the index N of the Galerkin approximation.

The crucial ingredient, which dictates the optimal coupling, is the dependence of the error on N, which is in our examples a question of regularity. As it is a question about the decay of spatial Fourier coefficients for periodic boundary conditions or the decay of the Fourier transform for problems on $\mathbf{R}^d$. A class of function spaces is introduced to capture this. We define the Gevrey space in the following way: Assuming that there exists a closed, densely defined, boundedly invertible operator $\Gamma_{\sigma,p}$ with domain of definition

$$\mathcal{G}_{\sigma,\nu} := D(\Gamma_{\sigma,\nu}) \subset D(A), \tag{4.1}$$

such that $\operatorname{Rg} P_N \subset \mathcal{G}_{\sigma,\nu}$, $\Gamma_{\sigma,\nu}(\operatorname{Rg} P_N) = \operatorname{Rg} P_N$ for all N, and

$$\Gamma_{\sigma,\nu} A u = A \Gamma_{\sigma,\nu} u \text{ for all } u \in \operatorname{Rg} P_N.$$

We equip the Gevrey spaces $\mathcal{G}_{\sigma,\nu}$ with the graph norm

$$|u|_{\mathcal{G}_{\sigma,p}} = |u|_X + |\Gamma_{\sigma,\nu} u|_X. \tag{4.2}$$

For the theorem, we will assume that Gevrey-regular functions $u \in G_{\sigma,\nu}$ are exponentially well approximated by the Galerkin projections P_N, i.e.

$$|\Gamma_{\sigma,\nu}^{-1}(I - P_N)| \leq C_0 \exp(-\sigma/N^\nu), \tag{4.3}$$

for N-independent constants $C_0(\sigma, \nu)$. Now we are in a position to formulate the following theorems about exponential estimates.

Theorem 4.1 (Exponential averaging of parabolic equation under periodic forcing [Mat01]). *Let the nonlineairties f, g be entire functions on $\mathbf{R}^n$, let g be continuous on T^1 and let $\Omega \equiv 1$ in equation (2.4). Then the equation can be transformed on bounded sets in X by a real analytic and time-periodic change of coordinates for $0 < \varepsilon < \varepsilon_0$*

$$u = v + \varepsilon W(v, t/\varepsilon; \varepsilon) \tag{4.4}$$

with W bounded on any ball of radius R in X. The transformed nonautonomous terms r are exponentially small after a short transient, but the equation may contain nonlocal terms $\bar{f}$:

$$\frac{\partial}{\partial t} v(x,t) = D\Delta v(x,t) + f(v(x,t)) + \bar{f}(v(t);\varepsilon)(x) + r(v(t), t/\varepsilon; \varepsilon)(x), \tag{4.5}$$

with $v(0) = u_0$, $|v(0)|_X < R$ and $t \in (0, t^*)$

$$\sup_{|v(0)|_X < R} |\alpha(v(t))|_X \leq C\varepsilon \exp(-\min(t,c)\varepsilon^{-1/3}),$$

$$\sup_{|v(0)|_X < R} |\bar{f}(V(t))|_X \leq C\varepsilon + C\exp(-\min(t,c)\varepsilon^{-1/3})$$

where C, c, ε_0 do not depend on u_0.

Proof. For a detailed proof see [Mat01]. We will sketch the proof using the transformation of (3.7) in the previous section. Adapting the results of [Nei84] and a coupling

$$N(\varepsilon)\varepsilon^\alpha = 1$$

we can estimate the remainder term in the approximation space $P_N X$. For this we use a complex extension and a Cauchy estimate. Performing $k = [\varepsilon^{-1+\alpha}]$ transformation steps as in (3.9) and proving that $|r_{j+1}| \leq |r_j|/2$ for $j = 1, \ldots, k$ yields that the remainder term on the approximation space is

$$|r_k| \leq C 2^{-k} \leq C \exp(-c\varepsilon^{-1+\alpha})$$

for an appropriate choice of C and c. The autonomous correction term is at most of order $\mathcal{O}(\varepsilon)$. To estimate the effect of the transformation on the full space X, we use a regularity result for equation (2.4). Letting $A = \Delta$ and

$$\Gamma_{\sigma,\nu} = \exp(-\sigma|A|^{\nu}) \tag{4.6}$$

we obtain that

$$u(t) \in \begin{cases} \mathcal{G}_{t,1/2} & \text{for } t \in [0, t^*] \\ \mathcal{G}_{t^*,1/2} & \text{for } t > t^* \end{cases}$$

as long as $|u(t)|_X$ remains bounded. This and similar results can be found in [Pro91, TB*96, FT98], adaptions to equations with nonlocal operators as they appear in the transformed equations can be found in [Mat01]. Using this regularity result and (4.3), we can estimate the additional error terms due the Galerkin approximation of the form $G(v - P_N v)$. Thus the overall remainder can be bounded by

$$\begin{aligned} & |r_{k(\varepsilon)}|_X + |G(v(t) - P_{N(\varepsilon)}v(t))|_X \\ \leq\ & C\exp(-c\varepsilon^{-1+\alpha}) + C_0\exp(-\min(t,t^*)/N(\varepsilon)^{1/2}) \\ =\ & C\exp(-c\varepsilon^{-1+\alpha}) + C_0\exp(-\min(t,t^*)\varepsilon^{-\alpha/2}). \end{aligned}$$

Choosing $\alpha = 2/3$ yields the desired result for an appropriate choice of C and c. The other results including the estimate on $\bar{f}$ are direct consequences of the detailed analysis in [Mat01].

So far the analysis was about the equation, now we compare the solutions of (4.5) with solutions of the truncated equation,

$$\partial_t u = \Delta u + f(u) + \bar{f}(u;\varepsilon) \tag{4.7}$$

Corollary 4.2 (Gronwall estimates with Gevrey initial data). *Let the assumptions of Theorem 4.1 hold, and additionally assume that (2.4) has only globally bounded solutions. Fix $R > 0$, the maximal amplitude of the solution. Then for any $t_0 > 0$ there are constants $\varepsilon_0(t_0) > 0$, and $C'(t_0), c'(t_0) > 0$ such that the following holds.*

Let $u(t)$ be a solution to the truncated equation (4.7) with norm bounded by R in the Gevrey space $\mathcal{G}_{t^*,1/2}$, for a time interval $0 \leq t \leq t_0 < \infty$.

Then there exists a unique solution $v(t)$ on $0 \leq t \leq t_0$ to (4.5) with initial value $u(0)$. Moreover, the solutions are exponentially close in $\varepsilon < \varepsilon_0$,

$$|v(t) - u(t)|_X \leq C'\exp\left(-c'\varepsilon^{-1/3}\right),$$

for all $0 \leq t \leq t_0$.

Proof. The difference $w(t) = v(t) - u(t)$ satisfies the equation

$$\partial_t w = A(t)w + r(t),$$

where

$$A(t) = A + \int_0^1 (\partial_u f + \partial_u \bar{f})(\tau u + (1-\tau)v)\mathrm{d}\tau$$

and

$$r(t) = r(v(t)) - r(u(t)),$$

where for the sake of notation, we omitted the arguments t and ε. Since f, $\bar{f}$, and r possess bounded derivatives on bounded sets of $\mathcal{G}_{t^*,1/2}$, the result is an immediate consequence of a standard Gronwall lemma. Note that Gevrey initial data will stay in the Gevrey space $\mathcal{G}_{t^*,1/2}$, such that there is no transient in the exponential estimate. We also note that the estimate on the remainder r is in the X-topology, only, such that the closeness result only holds in this topology. Furthermore when starting with arbitrary initial data, the transient in the exponential estimate would destroy an exponential estimate.

In the next theorem, we use smooth initial data in $\mathcal{G}_{\sigma,1/2}$, which is defined in the same way as in (4.1) with the same function $\Gamma_{\sigma,\nu}$ as in (4.6). We denote a ball of radius R in a Banach space Y by $B_R(Y)$.

Theorem 4.3 (Exponential averaging of Gevrey regular solutions of nonlinear Schrödinger equations [MS03]). *Consider equation (2.5) with analytic nonlinearities C, g. Assume $\Omega(u) \geq c_0 > 0$ for all $u \in X, \phi \in T^1$. Consider initial data in $\mathcal{G}_{\sigma,1/2}$. There exists a near-identity transformation $I + \varepsilon W$, defined on the ball $B_R(X) \times T^1$, which eliminates adiabatically the fast phase, up to an exponentially small non-adiabatic effect and which is analytic on $B_R(\mathcal{G}_{\sigma,1/2}) \times T^1$. In the new variables (v, ψ), the evolution equation reads*

$$\begin{aligned} i\partial_t v &= \Delta v + \bar{C}(v;\varepsilon) + r^1(v,\psi;\varepsilon), \\ \partial_t \psi &= \frac{1}{\varepsilon}(\Omega(v) + \tilde{\Omega}(v;\varepsilon)) + r^2(v,\psi;\varepsilon). \end{aligned} \tag{4.8}$$

The transformed nonlinearities r^1, r^2, $\bar{C}$ and $\tilde{\Omega}$ are bounded on the ball $B_R(X)$ and $B_R(\mathcal{G}_{\sigma,1/2})$ respectively, uniformly in $0 < \varepsilon < \varepsilon_0$.

The non-adiabatic interaction terms r^1 and r^2 are exponentially small in ε, i.e., there exist constants $c, C > 0$ such that

$$|r^1(v,\psi;\varepsilon)|_X + |r^2(v,\psi;\varepsilon)| < C\exp\left(-c\varepsilon^{-1/3}\right), \tag{4.9}$$

for all $v \in B_R(\mathcal{G}_{\sigma,1/2})$, and all $\psi \in T^1$. The adiabatic corrections $\bar{C}$ and $\tilde{\Omega}$ are small in Gevrey spaces,

$$|\tilde{C}(v;\varepsilon)|_{\mathcal{G}_{\sigma,1/2}} \leq C\varepsilon^{1/3}, \qquad |\tilde{\Omega}(v;\varepsilon)| \leq C\varepsilon,$$

for $v \in B_R(\mathcal{G}_{\sigma,1/2})$.

Proof. A detailed proof for general evolution equation or Hamiltonian partial differential equations can be found in [MS03]. The proof is similar to the proof of theorem 4.1. The formal transformations we consider are given in (3.10). Estimates on the remainder term in approximation space can be derived in the same way as above. For the analysis of the error due to the Galerkin space, we use that the initial data are in $\mathcal{G}_{\sigma,1/2}$ and that (2.5) is well-posed on this space for finite times. Then the estimates on the non-adiabatic remainder in $P_N B_R$ and the error of the Galerkin approximation can be balanced again to obtain the desired exponential estimate.

A comparable corollary of Gronwall type also holds.

Theorem 4.4 (Averaging of Gevrey regular solutions of parabolic equations under quasiperiodic forcing [Mat04]). *Consider the reaction-diffusion equation (2.4) with analytic nonlinearities and $\Omega(u,\phi)=\omega$ with Diophantine conditions (3.11). Then, for any ball of radius R in $\mathcal{G}_{\sigma,\nu}$ there exists an $\varepsilon_0>0$, such that for $0<\varepsilon<\varepsilon_0$ there exists a near identity transformation of both $\mathcal{G}_{\sigma,1/2}$ and X to*

$$\begin{aligned} \partial_t v &= \Delta v + f(v) + \bar{g}(v,\varepsilon) + r(v,\psi,\varepsilon) \\ \partial_t \psi &= \frac{1}{\varepsilon}\omega \end{aligned} \tag{4.10}$$

with initial conditions $v(0)=u_0; \theta(0)=\theta_0$ and with $\bar{g}$ and r both bounded on balls in X, furthermore the remainder term is exponentially small on balls of the Gevrey space.

$$\begin{aligned} |\bar{g}(v,\varepsilon)|_X &\le C\varepsilon^{(\tau+1)/(\tau+3)} \\ |r(v,\theta,\varepsilon)|_X &\le C(|v|_{\mathcal{G}_{\sigma,\nu}})\exp(-c\varepsilon^{-1/(\tau+3)}). \end{aligned}$$

Details, variants and extensions are given in [Mat04].

Proof. A major part is already the estimate on the approximation space, see [Sim94]. The remainder on this space is $\mathcal{O}(\exp(-c\varepsilon^{(\varepsilon N(\varepsilon))^{1/(1+\tau)}}))$. The error estimate of the Galerkin approximation is as above of order $\mathcal{O}(\exp(-cN(\varepsilon)^{1/2})$, such that the optimal coupling is $N(\varepsilon)=\varepsilon^{-2/(3+\tau)}$, which yields the exponential estimate.

Theorem 4.5 (Homogenisation via spatial dynamics [Mat05]). *Consider (2.7) with analytic nonlinearity $f(.,.)$, which is periodic in the second component. Then there exist $\varepsilon_0>0$, $c,C>0$ and a t-periodic transformation of (2.8) on a ball $B_R(X)$ for $0<\varepsilon<\varepsilon_0$ to*

$$V_t = AV + F(V) + \bar{F}(V,\varepsilon) + r(V,t/\varepsilon,\varepsilon), \tag{4.11}$$

where $\bar{F},\alpha$ are differentiable for $V\in B_R(X)$, nonlinear and nonlocal in the cross-section, but local in t. When considering bounded solutions $V(.)\in$

$BC(\mathbf{R}, X)$ of the original equation (2.8) then the influence of the fast scale on $V(.)$ is exponentially small, uniformly on balls in $BC(\mathbf{R}, X)$:

$$\|r(V(.), ./\varepsilon, \varepsilon)\|_{BC(\mathbf{R},X)} \leq C \exp(-c\varepsilon^{-1/2}).$$

The correction term can be estimated on $B_R(X)$ by

$$\sup_{V(.)\in B_X(R)} \|\bar{F}(V(.), \varepsilon)\|_{BC(\mathbf{R},X)} \leq C\varepsilon.$$

Proof. A complete proof of this homogenisation and of variants can be found in [Mat05, MW06]. The estimates on the approximation space are similar to the estimates in theorem 4.1. Using again the coupling

$$N(\varepsilon)\varepsilon^{\alpha} = 1.$$

Then we use again a regularity result. We let A as in (2.8) and use

$$\varGamma_{\sigma,\nu} = \exp(-\sigma|A|^{\nu})$$

in the definition of the Gevrey norm in (4.1). Then we obtain that the globally bounded solutions are in fact highly regular as functions on the cross-section Ω

$$V(t) \in \mathcal{G}_{\sigma_*,1}$$

for some $\sigma_* > 0$ with estimates uniform in t. The set of all such solutions is sometimes called the attractor of the spatial dynamics equation. Then the overall remainder on bounded sets within the attractor can be bounded by

$$\begin{aligned} & |r_{k(\varepsilon)}|_X + |G(V(t) - P_{N(\varepsilon)}V(t))|_X \\ \leq\ & C\exp(-c\varepsilon^{-1+\alpha}) + C_0\exp(-\sigma_*/N(\varepsilon)) \\ =\ & C\exp(-c\varepsilon^{-1+\alpha}) + C_0\exp(-\sigma_*\varepsilon^{-\alpha}). \end{aligned}$$

Choosing $\alpha = 1/2$ yields the result for appropriate C and c. The results hold in the same way for any function $V(.)$, which is smooth enough, e.g. by being a solution of a similar averaged equation.

Theorem 4.6 (Homogenisation of elliptic operators [KMS06]). *Suppose $\mathcal{A} \in L^{\infty}(T^d)$, $f \in \mathcal{G}_{\sigma,1/2}$ (as defined above) and $\int_{T^d} f = 0$ in equation (2.9). Let u^{ε} be the solution. Then there exist ε-independent constants $C > 0$, $c > 0$, $\kappa > 0$, such that for any $N \sim \kappa\varepsilon^{-1}$ the approximation* (3.13) *has the error bound:*

$$\|u^{\varepsilon,N} - u^{\varepsilon}\ ;\ H^1(\mathbb{T})\| \leq C\exp(-c\varepsilon^{-1}).$$

The proof in [KMS06] is based on a careful analysis of the remainder term $u^{\varepsilon,k} - u^{\varepsilon}$, then the error can be estimated in an exponential way for $k = N \sim \kappa\varepsilon^{-1}$, where only regularity of f is needed, but not on the regularity of the matrix $\mathcal{A}$.

5 Effects

In this section we discuss some effects due to the heterogeneous or nonautonomous structure of the equation.

5.1 Splitting of homoclinic orbits

An important difference in the dynamics between autonomous and non-autonomous differential equation are the existence of transversal homoclinic orbits. So consider a parameter dependent version of (2.4) without rapid forcing

$$\partial_t u = \Delta u + f(u, \lambda) \tag{5.1}$$

with $\lambda \in \mathbf{R}$. A homoclinic orbit is a solution, which is biasymptotic for $t \to \pm\infty$ to some hyperbolic equilibrium u_0, i.e. $\Delta u_0 + f(u_0, \lambda) = 0$ and the operator $\Delta + Df(u_0, \lambda)$ has only spectrum away from the imaginary axis. Under some non-degeneracy conditions, this homoclinic orbit will only exist for special values λ_0. Whereas in the externally forced equation

$$\partial_t u = \Delta u + f(u, \lambda) + g(u, t/\varepsilon) \tag{5.2}$$

the stable and unstable manifold will intersect transversally, creating rich dynamics nearby. The main assertion of theorem 4.1 is, that the nonautonomous dynamics of (5.2) can be described by the exponentially close autonomous equation as in (4.7). It is possible to show by some further analysis that the equilibrium persists as a hyperbolic periodic orbit, exponentially close to an equilibrium of the truncated equation after the transformation of theorem 4.1. Then also the phase-portraits with unstable and parts of the stable manifolds will be exponentially close, see [Mat01]. From this, it is easy to see, that transversality effects can only occur in a small parameter interval $(\lambda^-(\varepsilon), \lambda^+(\varepsilon))$ with

$$|\lambda^+(\varepsilon) - \lambda^-(\varepsilon)| \leq C \exp(-c\varepsilon^{-1/3}),$$

such that these effects were called "invisible chaos" [FS96].

Using functional analytic methods as in [Mat03a] one obtains better results for large classes of parabolic equations, by analysing the problem in complex time. A more detailed analysis is possible for ordinary differential equations including lower estimates, see e.g. [Gel99, HMS88].

5.2 Pinning

Pinning describes a phenomenon in equations like (2.6), where a travelling wave does not propagate due to heterogeneous structures in the medium. Thus travelling waves will be instead standing waves, i.e. they are solutions to (2.7). A pinned wave is then homoclinic or heteroclinic orbit in the spatial dynamics setting (2.8). Pinning occurs when there a transversal intersection of stable and unstable manifolds. The general idea is similar to the analysis of the transversal intersection above. Here it is more convenient to find solutions nearby a pulse U_0 as zeros of

$$I(V) = \partial_t(U_0 + V) - A(U_0 + V) - F(U_0 + V) \in BC(\mathbf{R}, X)$$

Using Fredholm properties or exponential dichotomies, the problem is reduced to a low-dimensional problem, where the effects of the heterogeneous terms can be estimated to be of order $\mathcal{O}(\exp(-c\varepsilon^{-1/2}))$, see [MW06].

6 Discussion

In this article we have given an overview on results about averaging and homogenisation for partial differential equations beyond every order. The first step of this analysis are finite-order averaging estimates, which were iterated, to obtain exponential estimates on the transformed equation. With Gronwall-type estimates, we can extend this to estimates about the finite time behaviour. For particular solutions, equilibria, stable and unstable manifolds and connecting orbits the analysis can be extended to infinite times too. Due to the general nature of the procedure there remain several problem, which require a finer analysis.

The construction of the transformation and the remainder terms are iterative so they are not easily computable, but they coincide to finite order with their finite order counter-parts. Therefore, it is possible to show by direct calculations, that the transformed nonlinearities are typically not-local in x, even if the original nonlinearities are local, see [Mat04].

A very subtle point are lower estimates for both averaging procedures and particular effects. For the general averaging procedure, there is an example in [MS03], showing that the averaging results cannot be improved by any other "averaging-type" transformation. In particular there is difference between ordinary differential equations and infinite dimensional systems in what kind of exponents can be achieved by exponential averaging. In [KMS06], a particular simple choice of the heterogeneous matrix allows explicit calculations and shows, that the truncation of the asymptotic expansion is optimal.

For particular effects like the splitting of homoclinic orbits, there are several ways to obtain lower estimates in the finite dimensional case, see [HMS88, Gel99]. A crucial part is extending the analysis to complex time and analysing the time-singularities in the complex plane. This idea is also used in other cases of exponential analysis [BT05, CM05] and references therein. Some ways of introducing exponential asymptotics into numerical analysis can be found in [MBS00, Mat03b]. The relation of exponential averaging and integrable systems is still open, but see [Bam99, Pös99] for related results.

The analysis here was restricted to cases to some partial differential equations with periodic boundary conditions, the needed abstract properties, such that the theorems hold for large classes of evolution equations are given e.g. in [MS03, Mat04]. Then these results will hold e.g. for problems on the domain $\mathbf{R}^d$. The effect of boundary conditions and of boundary layers [Neu00] on exponential homogenisation still remains to be analysed.

Acknowlegdments. This work has been supported by the DFG Priority Program 1095 "Analysis, Modelling and Simulation of Multiscale Problems" under Fi 441/12. The author is grateful to Bernold Fiedler for support and encouragement during the work on this project at Freie Universiät Berlin.

References

[AM95] A. Afendikov, and A. Mielke. Bifurcations of Poiseuille flow between parallel plates: three-dimensional solutions with large spanwise wavelength. *Arch. Rational Mech. Anal.*, 129(2):101–127, 1995.

[AKN97] V. Arnold, V. Kozlov, and A. Neishtadt. *Mathematical Aspects of Classical and Celestial Mechanics.* Springer Verlag, Berlin, 1997.

[BP84] N.S. Bakhvalov, and G.P. Panasenko. *Homogenization: Averaging Processes in Periodic Media.* Nauka, Moscow 1984. (in Russian). English translation in: Mathematics and Its Applications (Soviet Series) **36**, Kluwer Academic Publishers, Dordrecht-Boston-London 1989.

[Bam99] D. Bambusi. Nekhoroshev theorem for small amplitude solutions in nonlinear Schrödinger equations. *Math. Z.*, 230:345–387, 1999.

[Bam03a] D. Bambusi. An averaging theorem for quasilinear Hamiltonian PDEs. *Ann. Henri Poincaré* 4(4):685–712, 2003.

[Bam03b] D. Bambusi. Birkhoff normal form for some nonlinear PDEs. *Comm. Math. Phys.* 234(2):253–285, 2003.

[Bam06] D. Bambusi. Galerkin averaging method and Poincaré normal form for some quasilinear PDEs. *Ann. Scuola. Norm. Sup. Pisa* to appear, 2006.

[BLP78] A. Bensoussan, J.-L. Lions, and G.-C. Papanicolaou. *Asymptotic analysis for periodic structures.* North Holland, Amsterdam, 1978.

[BT05] V. Betz, and S. Teufel. Precise coupling terms in adiabatic quantum evolution: the generic case *Comm. Math. Phys.*, 260:481–509, 2005.

[BHR05] H. Berestycki, F. Hamel, and L. Roques. Analysis of the periodically fragmented environment model. II. Biological invasions and pulsating travelling fronts. *J. Math. Pures Appl. (9)*, 84:1101–1146, 2005.

[BM61] N. Bogoliubov, and Y. Mitroploski. *Asymptotic Methods in the theory of nonlinear oscillations*, Gordon and Breach, New York, 1961.

[Bor98] F. Bornemann. *Homogenization in time of singularly perturbed mechanical systems.* Lecture Notes in Mathematics 1687, Springer-Verlag, Berlin, 1998.

[CS04] K.D. Cherednichenko, and V.P. Smyshlyaev. On full two-scale expansion of the solutions of nonlinear periodic rapidly oscillating problems and higher-order homogenized variational problems. *Arch. Ration. Mech. Anal.*, 174:385–442, 2004.

[CM05] S.J. Chapman, and D.B. Mortimer. Exponential asymptotics and Stokes lines in a partial differential equation. *Proc. R. Soc. Lond. Ser. A Math. Phys. Eng. Sci.* 461:2385–2421, 2005.

[FT98] A.B. Ferrari, and E.S. Titi. Gevrey regularity for nonlinear analytic parabolic equations. *Comm. Part. Diff. Eq.*, 23(1-2):1–16, 1998.

[FS03] B. Fiedler, and A. Scheel. Spatio-temporal dynamics of reaction-diffusion patterns. in Kirkilionis, Markus (ed.) et al., Trends in nonlinear analysis. On the occasion of the 60th birthday of Willi Jäger. Berlin: Springer. 23–152, 411–417, 2003.

[FS96] B. Fiedler, and J. Scheurle. Discretization of homoclinic orbits, rapid forcing and "invisible" chaos. *Mem. Amer. Math. Soc.*, 119, 1996.

[FV01] B. Fiedler, and M. Vishik. Quantitative homogenization of analytic semigroups and reaction diffusion equations with diophantine spatial frequencies. *Adv. Differ. Equ.*, 6:1377–1408, 2001.

[Gel99] V. Gelfreich. A Proof of the Exponentially Small Transversality of the Separatrices for the Standard Map. *Comm. Math. Phys.*, 201: 155–216, 2000.

[GM*97] P. Gérard, P. Markowich, N.J. Mauser, and F. Poupaud. Homogenization limits and Wigner transforms. *Comm. Pure Appl. Math.*, 50: 323–379, 1997.

[HMS88] P. Holmes, J. Marsden and J. Scheurle. Exponentially small splittings of separatrices with applications to KAM theory and degenerate bifurcations. Hamiltonian dynamical systems (Boulder, CO, 1987), *Contemp. Math.*, 81: 213–244, Amer. Math. Soc., Providence, RI, 1988.

[IM91] G. Iooss, and A. Mielke. Bifurcating time–periodic solutions of Navier–Stokes equations in infinite cylinders. *J. Nonlinear Science*, 1:107–146, 1991.

[JKO94] V.V. Jikov, S.M. Kozlov, and O.A. Oleinik. *Homogenization of Differential Operators and Integral Functionals.* Springer-Verlag, Berlin, 1994.

[KMS06] V. Kamotski, K. Matthies, and V. Smyshlyaev. Exponential homogenization of linear second order elliptic problems with periodic coefficients BICS Preprint 2/2006.

[Kee00] J.P. Keener. Propagation of waves in an excitable medium with discrete release sites. *SIAM J. Appl. Math.*, 61(1):317–334, 2000.

[KS98] J.P. Keener, and J. Sneyd. *Mathematical physiology.* Springer, New York, 1998.

[Kir82] K. Kirchgässner. Wave solutions of reversible systems and applications. *J. Diff. Eq.*, 45:113–127, 1982.

[Kr89] M.S. Krol. On a Galerkin-averaging method for weakly nonlinear wave equations. *Math. Methods Appl. Sci.*, 11:649–664, 1989.

[LM88] P. Lochak, and C. Meunier. *Multiphase Averaging for Classical Systems*, Appl. Math. Sc. **72**, Springer-Verlag, New York 1988.

[MBS00] A.M. Matache, I. Babuška, and C. Schwab. Generalized p-FEM in homogenization. *Numer. Math.* 86:319–375, 2000.

[Mat01] K. Matthies. Time-averaging under fast periodic forcing of parabolic partial differential equations: exponential estimates. *J. Differential Equations*, 174(1):133–180, 2001.

[Mat03a] K. Matthies. Exponentially small splitting of homoclinic orbits of parabolic differential equations under periodic forcing. *Discrete Contin. Dyn. Syst.*, 9(3):585–602, 2003.

[Mat03b] K. Matthies. Backward error analysis of a full discretisation scheme for a class ofparabolic partial differential equations *Nonlinear Anal.*, 52(3):805–826, 2003.

[Mat04] K. Matthies. Exponential Averaging of Rapid Quasiperiodic Forcing, preprint, 2004.

[Mat05] K. Matthies. Homogenization of exponential order for elliptic systems in infinite cylinders. *Asymptot. Anal.* 43:205–232, 2005.

[MS03] K. Matthies, and A. Scheel. Exponential Averaging of Hamiltonian Evolution Equations, *Trans. Amer. Math. Soc.*, 355, 747–773, 2003.

[MW06] K. Matthies, and C.E. Wayne. Wave pinning in strips. *Proc.Roy. Soc. Edinburgh A*, to appear 2006.

[MSU06] K. Matthies, G. Schneider, and H. Uecker. Exponential averaging and traveling waves in rapidly varying periodic media *Mathematische Nachrichten*, to appear 2006.

[Nei84] A.I. Neishtadt. The separation of motions in systems with rapidly rotating phase. *J. Appl. Math. Mech.*, 48:133–139, 1984.

[Nek79] N. N. Nekhorošev, An exponential estimate of the time of stability of nearly integrable Hamiltonian systems, (Russian) *Uspehi Mat. Nauk*, 32:5–66, 1977.

[Neu00] A. Neuss-Radu. A result on the decay of the boundary layers in the homogenization theory. *Asymptot. Anal.* 23:313–328, 2000.

[NM92] A. Newell, J. Moloney. *Nonlinear Optics*, Advanced Topics in the Interdisciplinary Sciences, Addison-Wesley, Redwood City, 1992.

[Pös99] J. Pöschel. On Nekhoroshev estimates for a nonlinear Schrödinger equation and a theorem by Bambusi, *Nonlinearity* 12:1587–1600, 1999.

[Pro91] K. Promislow. Time analyticity and Gevrey regularity for solutions of a class of dissipative partial differential equations. *Nonlinear Anal.*, 16(11):959–980, 1991.

[RM*03] S. Rüdiger, D.G. Miguez, A.P. Munuzuri, F. Saguesa, and J. Casademunt. Dynamics of Turing patterns under spatio-temporal forcing. *Phys. Rev. Lett.*, 90(12):128301, 2003.

[Sim94] C. Simó. Averaging under fast quasiperiodic forcing, in *Hamiltonian Mechanics* (Torun, 1993), *NATO Adv. Sci. Inst.Ser. B Phys.* 331:13–34, 1994.

[SV85] J.A. Sanders, and F. Verhulst. *Averaging methods in nonlinear dynamical systems.* Applied Mathematical Sciences 59, Springer, New York, 1985.

[SSW99] B. Sandstede, A. Scheel, and C. Wulff. Bifurcations and dynamics of spiral waves. *J. Nonlinear Science*, 9:439–478, 1999.

[TB*96] P. Takáč, P. Bollerman, A. Doelman, A. van Harten, and E. S. Titi. Analyticity of essentially bounded solutions to semilinear parabolic systems and validity of the Ginzburg-Landau equation. *SIAM J. Math. Anal.*, 27(2):424–448, 1996.

[Tar79] L. Tartar. Compensated compactness and applications to partial differential equations. in: Nonlinear analysis and mechanics: Heriot-Watt Symp., Vol. 4, Edinburgh 1979, *Res. Notes Math.*, 39: 136-212, 1979.

[Tar86] L. Tartar. Oscillations in nonlinear partial differential equations: compensated compactness and homogenization, in: Nolinear systems of partial differential equations in applied mathematics, *Lectures in Applied Mathematics* 74, AMS, Providence, 243–266, 1986.

[Van89] A. Vanderbauwhede. Centre manifolds, normal forms and elementary bifurcations. in:*Dynamics reported*, Vol. 2, Wiley, Chichester, 89–169, 1989.

[Ver05] F. Verhulst. *Methods and applications of singular perturbations. Boundary layers and multiple timescale dynamics.* Texts in Applied Mathematics, 50. Springer, New York, 2005.

[Ver06] F. Verhulst. On averaging methods for partial differential equations preprint, 2006.

[Xin00] J. Xin. Front propagation in heterogeneous media. *SIAM Rev.*, 42(2):161–230, 2000.

Multiscale Problems in Solidification Processes

Christof Eck[1], Harald Garcke[2], and Björn Stinner[2]

[1] Institute for Applied Mathematics, University Erlangen-Nürnberg, Martensstraße 3, 91058 Erlangen. eck@am.uni-erlangen.de
[2] NWF I – Mathematik, University of Regensburg, 93040 Regensburg. harald.garcke@mathematik.uni-regensburg.de, bjoern.stinner@mathematik.uni-regensburg.de

Summary. Our objective is to describe solidification phenomena in alloy systems. In the classical approach, balance equations in the phases are coupled to conditions on the phase boundaries which are modelled as moving hypersurfaces. The Gibbs-Thomson condition ensures that the evolution is consistent with thermodynamics. We present a derivation of that condition by defining the motion via a localized gradient flow of the entropy. Another general framework for modelling solidification of alloys with multiple phases and components is based on the phase field approach. The phase boundary motion is then given by a system of Allen-Cahn type equations for order parameters. In the sharp interface limit, i.e., if the smallest length scale δ related to the thickness of the diffuse phase boundaries converges to zero, a model with moving boundaries is recovered. In the case of two phases it can even be shown that the approximation of the sharp interface model by the phase field model is of second order in δ. Nowadays it is not possible to simulate the microstructure evolution in a whole workpiece. We present a two-scale model derived by homogenization methods including a mathematical justification by an estimate of the model error.

1 Introduction

Solidification of alloys based on iron, aluminum, copper, zinc, nickel, and other materials which are of importance in industrial applications involves the occurrence of structures on an intermediate length scale of some μm between the atomic scale of the crystal lattice and the typical size of the workpiece. This so-called microstructure consists of regions (in the following labelled phases) differing in the crystalline structure, in the composition or only in the orientation of the crystal lattice, and it is responsible for a broad range of material properties and, hence, for the quality and durability of the material.

Being a result of the solidification process the microstructure is not in thermodynamic equilibrium. Its formation is classically modelled using moving hypersurfaces for the phase boundaries. The Gibbs-Thomson condition couples the form and the motion of the interface to its surface energy and to the local thermodynamic potentials of the adjacent phases. In addition, balance equations for the internal energy and the concentrations of the components have to be taken into account. This leads to diffusion equations in the phases and jump conditions on the moving phase boundaries.

In the last years, the phase field approach has emerged as a powerful tool to simulate microstructure formation. Phase field variables are introduced standing for the presence of related phases. Instead of jumping across the phase boundaries, the phase field variables and all the thermodynamic quantities change smoothly but rapidly within a narrow transition layer. It scales with a new length scale δ smaller than the typical scale of the microstructure to be described. This leads to the notion of a diffuse interface in contrast to the sharp interface model with the moving phase boundaries. The Gibbs-Thomson condition is replaced by a diffuse version which can be viewed as a gradient flow of an appropriate Ginzburg-Landau energy. The balance equations for the conserved quantities are reformulated in terms of the new variables where the jump conditions enter in a natural way. As a main advantage, numerically simulating microstructure formation is restricted to solving a system of parabolic differential equations, and explicit tracking of the phase boundaries in the sharp interface model is avoided.

The limit of vanishing diffuse interface thickness, i.e., the limit as $\delta \searrow 0$, is of particular interest. The first question is whether a related sharp interface model is obtained in the following sense: given solutions to the diffuse interface model for every δ, is there a limit of the solutions, and which equations do the limiting fields fulfill? This question is related to the calibrations problem when quantitatively investigating a certain alloy. Usually, material parameters such as latent heats, surface tensions, and several mobility and diffusion coefficients entering the sharp interface model are measured in experiments, and the question is how they should enter the diffuse interface model.

Problems involving multiple length scales not only result from the modelling approach but are also inherent in the physical problems itself. Diffusion of the temperature is much faster than mass diffusion. Because of the boundary conditions – solidifying workpieces are usually cooled from outside – and the release of latent heat the temperature field is expected to suffer changes over a scale proportional to the size of the the workpiece. On the other hand, the concentrations of the components should exhibit strong gradients only near the solidification front. The available numerical techniques and computational power only allow for the simulation of small domains in acceptable computation time, the direct computation of the microstructure of a whole workpiece is not feasible. For the latter purpose, macroscopic models involving heuristic assumptions on the distribution of the solidified parts and the released latent heat have been in use. Newer mathematical methods are based on a two-scale approach and allow for effective, homogenized equations for the temperature distribution but also for taking the microstructure evolution into account.

The structure of the present article is as follows. The first section is dedicated to models for alloy solidification. First, the governing equations from the classical approach for modelling alloy solidification are presented. In particular, the Gibbs-Thomson condition is derived by locally varying the entropy. After, the phase field approach is presented. In the second section, the relation between the sharp and the diffuse approaches is elucidated. Comments on the calibration problem are given including appropriate potentials for the phase field model with good calibration properties. Exemplary, a model for a binary alloy is derived. In the third section, a mathematically rigorous approach to the derivation of homogenized models for phase transitions with equiaxed dendritic microstructure is given. Asymptotic expansions are used to derive a macroscopic heat equation coupled to microscopic cell problems for the dendritic growth. A mathematical justification is carried out, i.e., an estimate is established comparing the solution to the two-scale model with that to the original model.

Acknowlegdments. This work has been supported by the German Research Foundation (DFG) through the Priority Program 1095 "Analysis, Modeling and Simulation of Multiscale Problems". We thank Britta Nestler and Frank Wendler for the inspiring discussions, mainly emerging from numerical simulations based on the presented phase field model (see also their contribution on page 113).

2 Models for alloy solidification

The production of certain microstructural morphologies is often achieved by imposing appropriate conditions before and during the solidification process. In order to get a deeper understanding of the process, the scientific challenge is to describe the microstructure formation with a mathematical model where the imposed conditions enter the equations governing the evolution as initial and boundary values or as additional forces and parameters.

A framework for continuum modelling of alloy solidification can be derived from thermodynamic principles for irreversible processes (cf. [Mu01]). Balancing the conserved quantities energy and mass respectively concentrations of the components yields diffusion equations in the bulk phases as well as continuity and jump conditions on the moving phase boundaries. A coupling of the phase boundary motion to the thermodynamic quantities of the adjacent phases, the Gibbs-Thomson condition, is derived by localizing an appropriate gradient flow of the entropy. The balance equations and the Gibbs-Thomson condition, together with certain angle conditions in junctions where several phases meet and which are due to local force balance, enable to show that the local entropy production is non-negative and to prove an entropy inequality.

An entropy functional involving bulk and surface contributions plays a central role also in non-equilibrium thermodynamics. In the phase field approach, the interfacial entropy (or energy) is modelled with the help of a Ginzburg-Landau type functional. Evolution equations for the phase fields can then be derived as gradient flows (see [FP90]) or within the theory of rational thermodynamics (see [AP96, Ha06]). A small length scale is involved which is related to the thickness of the interfacial layers.

We proceed as follows. First, the classical approach to model alloy solidification, namely with moving phase boundaries, is presented. In the second subsection, the Gibbs-Thomson condition is derived. After, the phase field variables are defined, and the phase field approach is presented. As an example, a model for non-isothermal solidification of a binary alloy involving two phases is derived. Finally we briefly comment on the solvability of the differential equations of the phase field model.

For general informations on the theory and models of phase transitions we refer to the books [BS96, Vi96]. In this section, partial derivatives sometimes are denoted by subscripts after a comma. For example, $s_{,e}$ is the partial derivative of the function $s = s(e, \hat{c})$ with respect to the variable e.

2.1 Classical approach with moving hypersurfaces

An alloy of $N \in \mathbb{N}$ components occupying an open domain $\Omega \in \mathbb{R}^d$, $d = 1, 2, 3$, during some time interval $I_{\mathcal{T}} = (0, \mathcal{T})$ is considered. Changes in volume or pressure are neglected (cf. [Ha94], Sect. 5.1). Moreover, the mass density is assumed to be constant (only concentrations will be considered). The only transport mechanism is diffusion, and there are no forces present leading to flows or deformations. Such effects can strongly influence the growing structures (cf. [Da01]). The applicability of the model presented in the following is therefore restricted to cases where such effects can be neglected.

Let $M \in \mathbb{N}$ be the number of possible phases. The domain Ω is decomposed into sub-domains $\Omega_1(t), \dots, \Omega_M(t)$, $t \in I_{\mathcal{T}}$, which are called phases. The phase boundaries $\Gamma_{\alpha\beta}(t) := \overline{\Omega_\alpha(t)} \cap \overline{\Omega_\beta(t)}$, $1 \leq \alpha \neq \beta \leq M$, are supposed to be piecewise smooth evolving points, curves, or hypersurfaces, depending on the dimension (cf. Def. A.1 in the Appendix). The unit normal on $\Gamma_{\alpha\beta}$ pointing into phase Ω_β is denoted by $\nu_{\alpha\beta}$. If $d \geq 2$ the intersections of the curves or hypersurfaces are denoted by $T_{\alpha\beta\delta}(t) := \overline{\Omega_\alpha(t)} \cap \overline{\Omega_\beta(t)} \cap \overline{\Omega_\delta(t)}$ for pairwise different $\alpha, \beta, \delta \in \{1, \dots, M\}$, and the points where the phase boundaries hits the external boundary by $T_{\alpha\beta,ext}(t) := \overline{\Omega_\alpha(t)} \cap \overline{\Omega_\beta(t)} \cap \partial\Omega$. If $d = 2$ then $T_{\alpha\beta\delta}$ is a set of triple junctions, i.e., piecewise smooth evolving points. If $d = 3$ triple lines can appear which are piecewise smooth evolving curves.

During evolution, it may happen that one of the phases vanishes, namely if the adjoining phase boundaries coalesce. It is also possible that a piece of a phase boundary vanishes so that one of the sets $T_{\alpha\beta\delta}$ includes a quadruple point or line. Typically, the latter configuration is not in mechanical equilibrium and will instantaneously split up forming new phase boundaries (see [GNS99, BGN06]). It is supposed that such singularities only occur at finitely many times $t \in I_{\mathcal{T}}$ during the evolution. This is why only piecewise smooth evolution is assumed. The evolution equations stated in the following are only valid for times at which no such singularity occurs.

Before proceeding let us introduce some notation. For $K \in \mathbb{N}$ define the sets

$$\mathrm{H}\Sigma^K := \Big\{ v \in \mathbb{R}^K \,:\, \sum_{i=1}^{K} v_i = 1 \Big\}, \quad \Sigma^K := \Big\{ v \in \mathrm{H}\Sigma^K \,:\, v_i \geq 0 \quad \forall i \Big\}. \tag{2.1}$$

The tangent space on $\mathrm{H}\Sigma^K$ can be naturally identified in every point $v \in \mathrm{H}\Sigma^K$ with the subspace

$$\mathrm{T}_v \mathrm{H}\Sigma^K \cong \mathrm{T}\Sigma^K := \Big\{ w \in \mathbb{R}^K \,:\, \sum_{i=1}^{K} w_i = 0 \Big\}. \tag{2.2}$$

The map $\mathcal{P}^K : \mathbb{R}^K \to \mathrm{T}\Sigma^K$ is the orthogonal projection given by

$$\mathcal{P}^K w = \Big(w_k - \frac{1}{K} \sum_{l=1}^{K} w_l \Big)_{k=1}^{K} = \Big(\mathrm{Id}_K - \frac{1}{K} \mathbf{1}_K \otimes \mathbf{1}_K \Big) w$$

where $\mathbf{1}_K = (1, \ldots, 1) \in \mathbb{R}^K$ and Id_K is the identity on $\mathbb{R}^K$.

The following bulk fields are considered in the phases Ω_α, $1 \leq \alpha \leq M$:

$$\begin{aligned}
c_i^\alpha &: \text{concentration of component } i,\ 1 \leq i \leq N, \\
c_0^\alpha := e^\alpha &: \text{internal energy density}, \\
f^\alpha &: \text{(Helmholtz) free energy density}, \\
\mu_i^\alpha &: \text{chemical potential of component } i,\ 1 \leq i \leq N, \\
T^\alpha &: \text{temperature}, \\
s^\alpha &: \text{entropy density}, \\
u_0^\alpha := \tfrac{-1}{T^\alpha} &: \text{inverse negative temperature}, \\
u_i^\alpha := \tfrac{\overline{\mu}_i^\alpha}{T^\alpha} &: \text{reduced chemical potential difference of component } i,\ 1 \leq i \leq N.
\end{aligned}$$

On the interfaces $\Gamma_{\alpha\beta}$, $1 \leq \alpha \neq \beta \leq M$, there are the following surface fields:

$$\begin{aligned}
\nu_{\alpha\beta} &: \text{unit normal pointing into } \Omega_\beta, \\
\sigma_{\alpha\beta}(\nu_{\alpha\beta}) &: \text{surface tension}, \\
\gamma_{\alpha\beta}(\nu_{\alpha\beta}) &: \text{capillarity coefficient}, \\
m_{\alpha\beta}(\nu_{\alpha\beta}) &: \text{mobility coefficient}, \\
v_{\alpha\beta} &: \text{normal velocity towards } \nu_{\alpha\beta}, \\
\kappa_{\alpha\beta} &: \text{curvature}.
\end{aligned}$$

The concentrations fulfill the constraint $\hat{c}^\alpha = (c_1^\alpha, \ldots, c_N^\alpha) \in \Sigma^N$. Following [Mu01], Sect. 11.2, the evolution in the phases is governed by balance equations for the conserved quantities, i.e.,

$$\partial_t c_i^\alpha = -\nabla \cdot J_i^\alpha = \nabla \cdot \left(\sum_{j=0}^{N} L_{ij}^\alpha \nabla u_j^\alpha \right), \quad 0 \leq i \leq N. \tag{2.3}$$

Let us briefly comment on the fluxes J_i^α. In thermodynamics of irreversible processes the relations between the fields are based on the principle of local thermodynamic equilibrium. In the present situation the entropy density s^α is a function of the conserved quantities. Its derivatives are the inverse temperature and the chemical potential difference reduced by the temperature, i.e.,

$$s^\alpha = s^\alpha(e^\alpha, \hat{c}^\alpha) \quad \text{and} \quad ds^\alpha = \frac{1}{T^\alpha} de^\alpha + \frac{-\overline{\mu}^\alpha}{T^\alpha} \cdot d\hat{c}^\alpha = -u^\alpha \cdot c^\alpha.$$

In the above equation the identity $\overline{\mu}^\alpha = \mathcal{P}^N \mu^\alpha$ was used, where $\mu^\alpha = (\mu_1^\alpha, \ldots, \mu_N^\alpha)^T$. The fluxes are postulated to be linear combinations of the thermodynamic forces ∇u_j^α, $0 \leq j \leq N$, with coefficients L_{ij}^α which may depend on the thermodynamic potentials u_j^α or on the conserved quantities c_i^α. This phenomenological theory was already introduced in [On31]. It is assumed that the matrix $L = (L_{ij}^\alpha)_{i,j=0}^N$ is positive semi-definite. To fulfill the constraint $\hat{c}^\alpha \in \Sigma^N$ it is required that $\sum_{i=1}^N L_{ij}^\alpha = 0$, $1 \leq j \leq N$, which also means that $\sum_{i=1}^N J_i^\alpha = 0$.

Onsager's law of reciprocity states the symmetry of L and can be proven and experimentally observed if the fluxes and forces are independent (cf. [KY87], Sect. 3.8). The above fluxes are not independent. But even in the present case Onsager's law can be shown to hold by a certain choice of the coefficients (see [KY87], Sect. 4.2, and the reference therein; there the calculation is performed for the isothermal case, but another additional independent force can be taken into account without any problem). We remark that, considering $J_i - J_N$, the definition of the fluxes as above is equivalent to the definition in [Mu01], Sect. 11.2.

On the phase boundaries $\Gamma_{\alpha\beta}$ the continuity conditions

$$[u_i]_\alpha^\beta = 0, \quad 0 \leq i \leq N, \tag{2.4}$$

have to be satisfied. Mass and energy balance imply furthermore the jump conditions

$$[c_i]_\alpha^\beta v_{\alpha\beta} = [J_i]_\alpha^\beta \cdot \nu_{\alpha\beta}, \quad 0 \leq i \leq N. \tag{2.5}$$

Here, $[\cdot]_\alpha^\beta$ denotes the jump of the quantity in brackets across $\Gamma_{\alpha\beta}$, e.g., $[e]_\alpha^\beta = e^\beta - e^\alpha$.

The matrix of surface tensions $(\sigma_{\alpha\beta}(\nu))_{\alpha,\beta}$ is symmetric for every unit vector ν (the diagonal entries are not of interest and may be set to zero). The relation between surface tension and capillarity coefficient is given by

$$\gamma_{\alpha\beta}(\nu_{\alpha\beta}) = \frac{\sigma_{\alpha\beta}(\nu_{\alpha\beta})}{T_{ref}}$$

with some reference temperature T_{ref}. The surface tensions are one-homogeneous in their argument while the mobility coefficients $m_{\alpha\beta}(\nu_{\alpha\beta})$ are zero-homogeneous in their argument.

The evolution of the phase boundaries is coupled to the thermodynamic fields by the Gibbs-Thomson condition

$$m_{\alpha\beta}(\nu_{\alpha\beta})v_{\alpha\beta} = -\nabla_\Gamma \cdot D\gamma_{\alpha\beta}(\nu_{\alpha\beta}) + \Big[- u_0 f(T,\hat{c}) + \sum_{i=1}^{N} u_i c_i \Big]_\alpha^\beta \tag{2.6}$$

which is derived in the following subsection. By $\nabla_\Gamma \cdot$ the surface divergence is denoted. In the case of an isotropic surface entropy, i.e., $\gamma_{\alpha\beta}(\nu) = \overline{\gamma}_{\alpha\beta}|\nu|$ with some constant $\overline{\gamma}_{\alpha\beta}$ independent of the direction, there is the identity $-\nabla_\Gamma \cdot D\gamma_{\alpha\beta}(\nu) = \overline{\gamma}_{\alpha\beta}\kappa_{\alpha\beta}$ where $\kappa_{\alpha\beta}$ is the mean curvature.

To avoid wetting effects (cf. [Ha94], Sect. 3.4, for a discussion and references) the surface tensions are assumed to fulfill the constraints

$$\sigma_{\alpha\beta} + \sigma_{\beta\delta} > \sigma_{\alpha\delta}. \tag{2.7}$$

Capillary forces acting on $\Gamma_{\alpha\beta}$ are related to the vectors (cf. [CH74, WM97])

$$\xi_{\alpha\beta}(\nu_{\alpha\beta}) := D\sigma_{\alpha\beta}(\nu_{\alpha\beta}) = \sigma_{\alpha\beta}(\nu_{\alpha\beta})\nu_{\alpha\beta} + D_{S^{d-1}}\sigma_{\alpha\beta}(\nu_{\alpha\beta}) \tag{2.8}$$

where $D_{S^{d-1}}$ is the surface gradient on the sphere S^{d-1}. The identity $D = D_{S^{d-1}} + \nu_{\alpha\beta}(\nu_{\alpha\beta} \cdot D)$ was used as well as the fact that $\sigma_{\alpha\beta}$ is one-homogeneous implying $D\sigma_{\alpha\beta}(\nu_{\alpha\beta}) \cdot \nu_{\alpha\beta} = \sigma_{\alpha\beta}(\nu_{\alpha\beta})$.

In points x belonging to $T_{\alpha\beta\delta}$ forces are in equilibrium. In the three-dimensional case $T_{\alpha\beta\delta}$ consists of triple lines that can be oriented with a unit tangent vector $\tau_{\alpha\beta\delta}(x)$. If the whole space is cut with the plane orthogonal to $\tau_{\alpha\beta\delta}(x)$ through x then the picture in Fig. 2.1 is obtained. Due to the surface tension $\Gamma_{\alpha\beta}$ exerts a force on x which is given by $\xi_{\alpha\beta}(\nu_{\alpha\beta}(x)) \times \tau_{\alpha\beta\delta}(x)$, whence equilibrium of forces means that

$$0 = \sum_{(i,j)\in\mathcal{A}} \xi_{ij}(\nu_{ij}(x)) \times \tau_{\alpha\beta\delta}(x) \tag{2.9}$$

where $\mathcal{A} := \{(\alpha,\beta),(\beta,\delta),(\delta,\alpha)\}$. A short calculation shows that in the situation of Fig. 2.1

$$\xi_{\alpha\beta}(\nu_{\alpha\beta}) \times \tau_{\alpha\beta\delta} = (\nabla\sigma_{\alpha\beta}(\nu_{\alpha\beta}) \cdot \tau_{\alpha\beta})(-\nu_{\alpha\beta}) + \sigma_{\alpha\beta}(\nu_{\alpha\beta})\tau_{\alpha\beta}.$$

Similarly, if $x \in T_{\alpha\beta,ext}$ there is a unit tangent vector $\tau_{\alpha\beta,ext}(x)$, and the force acting on x is given by $\xi_{\alpha\beta}(\nu_{\alpha\beta}(x)) \times \tau_{\alpha\beta,ext}(x)$. Force balance in x implies that this force is not tangential to $\partial\Omega$. Since it is already orthogonal to $\tau_{\alpha\beta,ext}(x)$ by definition this is true if and only if

$$\xi_{\alpha\beta}(\nu_{\alpha\beta}(x)) \cdot \nu_{ext}(x) = 0. \tag{2.10}$$

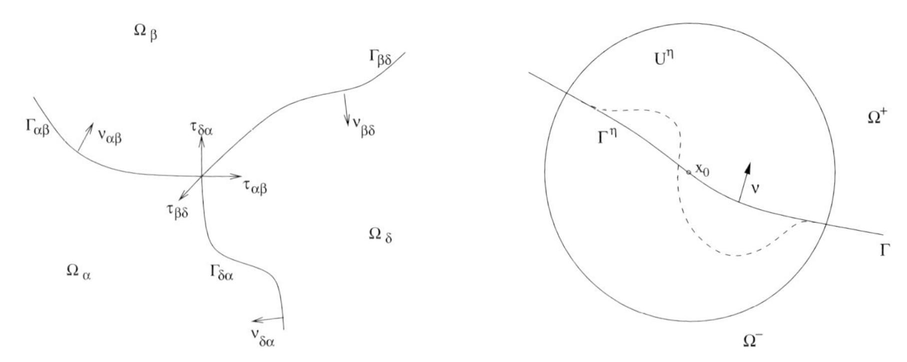

Fig. 2.1. On the left: triple junction x with orientations of the forming curves; such a picture is also obtained in the 3D-case by cutting the space with the plane spanned by $\nu_{\alpha\beta}(x)$, $\tau_{\alpha\beta}(x)$. On the right: local situation around a point x_0 on a phase boundary for the derivation of the Gibbs-Thomson condition; a local deformation is indicated by the dashed line.

In particular, angle conditions in $T_{\alpha\beta\delta}$ and $T_{\alpha\beta,ext}$ are due to the above force balance conditions (2.9) and (2.10).

To obtain a well-posed problem the governing equations (2.3)–(2.6), (2.9), and (2.10) must be provided with initial conditions for the fields and the moving boundaries and boundary conditions. If not otherwise stated, the isolated case

$$J_i^\alpha \cdot \nu_{ext} = 0 \text{ on } \partial\Omega, \quad 0 \leq i \leq N,\, 1 \leq \alpha \leq M, \tag{2.11}$$

is considered.

The total entropy of the system being given by

$$S(t) = \sum_{\alpha=1}^{M} \int_{\Omega_\alpha(t)} s^\alpha(c^\alpha) d\mathcal{L}^d - \sum_{\alpha<\beta,\, \alpha,\beta=1}^{M} \int_{\Gamma_{\alpha\beta}(t)} \gamma_{\alpha\beta}(\nu_{\alpha\beta})\, d\mathcal{H}^{d-1} \tag{2.12}$$

it can be shown that the evolution equations (2.3)–(2.11) imply non-negative entropy production:

Theorem 2.1. *The entropy (2.12) satisfies*

$$\begin{aligned} \frac{d}{dt} S(t) \quad &= \quad \sum_{1\leq\alpha\leq M} \int_{\Omega_\alpha(t)} \sum_{i,j=0}^{N} \nabla u_i^\alpha \cdot L_{ij}^\alpha \nabla u_j^\alpha \, d\mathcal{L}^d \\ &\quad + \sum_{1\leq\alpha<\beta\leq M} \int_{\Gamma_{\alpha\beta}(t)} m_{\alpha\beta}(v_{\alpha\beta})^2 \, d\mathcal{H}^{d-1}. \end{aligned}$$

The proof can be found in the Appendix of [GNS04].

2.2 Derivation of the Gibbs-Thomson condition

In this section a physical motivation of the Gibbs-Thomson condition (2.6) based on thermodynamic principles is given. The idea is to define the motion of the phase boundaries as a gradient flow of the entropy. On the set of admissible surfaces the tangent space of a surface is defined by the smooth real valued functions f on the surface supplied with a weighted L^2-product. A variation of the surface entropy in the direction f is then the rate of change of the entropy when deforming the surface towards its normal with a strength given by f. Such a deformation of a phase boundary usually changes the volumes of the adjacent phases. Thanks to this fact the bulk fields can enter the Gibbs-Thomson condition. But changes in the conserved quantities must be counterbalanced. Since (2.6) is a local motion law, only local deformations of an η-ball around a point x_0 on a phase boundary are considered. Conservation of energy and mass is ensured by taking a non-local Lagrange multiplier into account. In the limit as $\eta \to 0$ all terms become local after appropriate scaling so that the desired equation is obtained.

For keeping the presentation simple we do not consider the general situation as in the previous subsection but the one depicted in Fig. 2.1. There, Γ is a smooth compactly embedded $d-1$-dimensional hypersurface separating two phases Ω^+ and Ω^- with unit normal ν pointing into Ω^+. Such a surface respectively configuration is said to be admissible.

Definition 2.2. *Let G be the set of the admissible surfaces. The tangent space is defined by $T_\Gamma G := C^1(\Gamma, \mathbb{R})$. A Riemannian structure on $T_\Gamma G$ is defined by the weighted L^2 product*

$$(v, \xi)_\Gamma := \int_\Gamma m(\nu) v \xi \, d\mathcal{H}^{d-1} \quad \forall v, \xi \in T_\Gamma G$$

where $m(\nu)$ is a non-negative mobility function.

The bulk fields for energy density and concentrations, here denoted by c^0, are allowed to suffer jump discontinuities across Γ, but the potentials $s_{,c} = -u$ are supposed to be Lipschitz continuous. Within the phases Ω^+ and Ω^- all fields are smooth.

Variations of the entropy are based on local deformations of the domain. Let $x_0 \in \Gamma$ and consider the family of open balls $\{U^\eta\}_{\eta>0}$ centered in x_0 with radius η. Given arbitrary functions $\xi^\eta \in C_0^1(U^\eta)$ it can be shown that that there are vector fields

$$\boldsymbol{\xi}^\eta \in C_0^1(U^\eta, \mathbb{R}^d) \text{ with } \boldsymbol{\xi}^\eta = \xi^\eta \nu \text{ on } \Gamma^\eta := \Gamma \cap U^\eta. \tag{2.13}$$

The solution $\theta^\eta : U^\eta \to U^\eta$ to

$$\theta^\eta(0, y) = y, \quad \theta^\eta_{,\delta}(\delta, y) = \boldsymbol{\xi}^\eta(\theta^\eta(-\delta, y)) \text{ for } \delta \in [-\delta_0^\eta, \delta_0^\eta],$$

yields a local deformation of U^η. The restriction of δ is such that $\Gamma^\eta := U^\eta \cap \Gamma$ remains a smooth surface imbedded into U^η, i.e., the sets

$$\Gamma_\delta^\eta = \{\theta^\eta(\delta, x) \, : \, x \in \Gamma^\eta\}, \quad \delta \in [-\delta_0^\eta, \delta_0^\eta],$$

define an evolving $(d-1)$-dimensional surface in U^η in the sense of Def. A.1.

A short calculation yields the identity

$$\frac{d}{d\delta} \det \theta_{,x}^\eta(\delta, x) = \nabla \cdot \boldsymbol{\xi}^\eta(\theta^\eta(\delta, x)) \det \theta_{,x}^\eta(\delta, x). \tag{2.14}$$

The functional mapping L^1-functions on U^η onto their mean value is denoted by $\mathcal{M}^\eta$, i.e.,

$$\mathcal{M}^\eta : L^1(U^\eta) \to \mathbb{R}^m, \quad \mathcal{M}^\eta(f) := \frac{1}{|U^\eta|} \int_{U^\eta} f(x) \, dx = \fint_{U^\eta} f(x) \, dx$$

where $|U^\eta| = \mathcal{L}^d(U^\eta)$ with the d-dimensional Lebesgue measure $\mathcal{L}^d$.

Definition 2.3. *Under the local deformation θ^η of U^η the densities of the conserved quantities are*

$$c(\delta, y) := c^0(\theta^\eta(-\delta, y)) - \mathcal{M}^\eta\big(c^0(\theta^\eta(-\delta, \cdot)) - c^0(\cdot)\big), \quad y \in U^\eta. \tag{2.15}$$

The local entropy consists of the bulk part

$$S_B^\eta(\delta) := \int_{U^\eta} s(c(\delta, y)) \, dy \tag{2.16}$$

and the surface part

$$S_S^\eta(\delta) := -\int_{\Gamma_\delta^\eta} \gamma(\nu(\delta)) \, d\mathcal{H}^{d-1}. \tag{2.17}$$

The Lagrange multiplier $\mathcal{M}^\eta\big(c^0(\theta^\eta(-\delta, \cdot)) - c^0(\cdot)\big)$ in (2.15) ensures that energy and mass are conserved under the deformation.

Lemma 2.4. *The derivative of the bulk entropy with respect to δ in $\delta = 0$ is*

$$\frac{d}{d\delta} S_B^\eta(0) = \int_{U^\eta} \Big(s(c^0) + \mathcal{M}^\eta(u) \cdot c^0\Big) \nabla \cdot \boldsymbol{\xi}^\eta \, dx.$$

Proof. By definition (2.15), the bulk entropy (2.16) is

$$\int_{U^\eta} s\Big(c^0(\theta^\eta(-\delta, y)) - \mathcal{M}^\eta\big(c^0(\theta^\eta(-\delta, \cdot)) - c^0\big)\Big) \, dy$$
$$= \int_{U^\eta} s\Big(c^0(x) - \mathcal{M}^\eta\big(c^0(\theta^\eta(-\delta, \cdot)) - c^0\big)\Big) \det \theta_{,x}(\delta, x) \, dx$$

where the transformation $y = \theta^\eta(\delta, x)$ was used. The equation (2.14) yields together with $\theta^\eta(0,x) = x$ and $\det(\theta^\eta_{,x}(0,x)) = \det \mathrm{Id} = 1$

$$\frac{d}{d\delta}\int_{U^\eta} c^0(\theta^\eta(-(\cdot),z))\,dz\Big|_{\delta=0} = \frac{d}{d\delta}\int_{U^\eta} c^0(x)\det\theta^\eta_{,x}(\delta,x)\,dx\Big|_{\delta=0}$$
$$= \int_{U^\eta} c^0(x)\nabla\cdot\boldsymbol{\xi}^\eta(x)\,dx.$$

With $s_{,c} = -u$ the desired identity can be shown as follows:

$$\frac{d}{d\delta}S^\eta_B(0) = \int_{U^\eta} s\Big(c^0(x) - \mathcal{M}^\eta\big(c^0(\theta^\eta(0,\cdot)) - c^0\big)\Big)\nabla\cdot\boldsymbol{\xi}^\eta(x)\,dx$$
$$- \int_{U^\eta} s_{,c}(c^0(x))\cdot\frac{d}{d\delta}\frac{1}{|U^\eta|}\int_{U^\eta} c^0(\theta^\eta(-(\cdot),z))\,dz\Big|_{\delta=0}\,dx$$
$$= \int_{U^\eta} s(c^0(x))\nabla\cdot\boldsymbol{\xi}^\eta(x)\,dx + \frac{1}{|U^\eta|}\int_{U^\eta} u(x)\,dx\cdot\int_{U^\eta} c^0(x)\nabla\cdot\boldsymbol{\xi}^\eta(x)\,dx$$
$$= \int_{U^\eta}\Big(s(c^0) + \mathcal{M}^\eta(u)\cdot c^0\Big)\nabla\cdot\boldsymbol{\xi}^\eta(x)\,dx.$$

Lemma 2.5. *The derivative of the surface entropy with respect to δ in $\delta = 0$ is*

$$\frac{d}{d\delta}S^\eta_S(0) = -\int_{\Gamma^\eta}\nabla_\Gamma\cdot D\gamma(\nu)\,\xi^\eta\,d\mathcal{H}^{d-1}.$$

Proof. Interpreting $\{\Gamma^\eta_\delta\}_\delta$ as evolving surface, the scalar normal velocity is ξ^η and the vectorial normal velocity is $\boldsymbol{\xi}^\eta = \xi^\eta\nu$. The scalar curvature is denoted by κ_Γ. Applying Th. A.4 from the Appendix yields (observe that the boundary integrals over $\partial\Gamma^\eta$ vanish since the velocity $\boldsymbol{\xi}^\eta$ has a compact support in U^η)

$$\frac{d}{d\delta}S^\eta_S(0) = -\int_{\Gamma^\eta}\big(\partial^\circ\gamma(\nu) - \gamma(\nu)\,\boldsymbol{\xi}^\eta\cdot\boldsymbol{\kappa}_\Gamma\big)d\mathcal{H}^{d-1}$$

which is using (A.3), (A.2), (A.4), and the one-homogeneity of γ

$$= \int_{\Gamma^\eta}\big(\nabla\gamma(\nu)\cdot\nabla_\Gamma\xi^\eta + \nabla\gamma(\nu)\cdot\nu\,\kappa_\Gamma\,\xi^\eta\big)d\mathcal{H}^{d-1}.$$

Applying Th. A.3 to $\boldsymbol{\varphi} = \nabla\gamma(\nu)\xi^\eta$ (again the boundary integral vanishes) and again (A.2) on the last term it follows the desired identity:

$$\ldots = \int_{\Gamma^\eta}\big(-\nabla_\Gamma\cdot\nabla\gamma(\nu)\,\xi^\eta - \boldsymbol{\kappa}_\Gamma\cdot\nabla\gamma(\nu)\,\xi^\eta + \nabla\gamma(\nu)\cdot\boldsymbol{\kappa}_\Gamma\,\xi^\eta\big)d\mathcal{H}^{d-1}$$
$$= -\int_{\Gamma^\eta}\big(\nabla_\Gamma\cdot\nabla\gamma(\nu)\,\xi^\eta\big)d\mathcal{H}^{d-1}.$$

As stated at the beginning of this section, the goal is to define the motion as a localized version of a gradient flow. This is realized in the following definition. Let $|\Gamma^\eta| := \mathcal{H}^{d-1}(\Gamma^\eta)$.

Definition 2.6. *The motion of the phase boundary Γ is defined as follows: In each point $x_0 \in \Gamma$ the identity*

$$\lim_{\eta\to 0} \frac{1}{|\Gamma^\eta|}(v, \xi^\eta)_\Gamma = \lim_{\eta \to 0} \frac{1}{|\Gamma^\eta|} \frac{d}{d\delta}(S_B^\eta + S_S^\eta)(0) \tag{2.18}$$

holds for all families of functions $\xi^\eta \in C_0^1(U^\eta)$ where $S_B^\eta(\delta)$ and $S_S^\eta(\delta)$ are defined by (2.16) *and* (2.17) *respectively.*

Theorem 2.7. *The localized gradient flow* (2.18) *yields the Gibbs-Thomson condition* (2.6).

To prove the theorem the following lemma is useful:

Lemma 2.8. *Let $g \in L^\infty(U^\eta)$ with $g \in C^1(\overline{\Omega^+ \cap U^\eta})$ and $g \in C^1(\overline{\Omega^- \cap U^\eta})$, and let $z \in \mathbb{R}$ be given. There is a family of functions $\{\xi^\eta\}_{\eta>0} \subset C^1(U^\eta)$ with $\xi^\eta(x_0) = z$ for all η such that*

$$\begin{aligned} \frac{1}{|\Gamma^\eta|} \int_{U^\eta} g \nabla \cdot \boldsymbol{\xi}^\eta \, dx &= - \fint_{\Gamma^\eta} [g]_-^+ \xi^\eta \, d\mathcal{H}^{d-1} - \frac{1}{|\Gamma^\eta|} \int_{U^\eta} \nabla g \cdot \boldsymbol{\xi}^\eta \, dx \\ &\to -[g(x)]_-^+ z \qquad \text{as } \eta \to 0 \end{aligned}$$

where the functions $\boldsymbol{\xi}^\eta$ are uniformly bounded and satisfy condition (2.13). *By g^+ the limit of g in $x \in \Gamma$ when approximated from the side Ω^+ is denoted. Analogously g^- is defined when approximating $x \in \Gamma$ from Ω^-, and $[g]_-^+ = g^+ - g^-$ is the difference.*

Proof. The first identity follows from the divergence theorem applied to the two parts $U^\eta \cap \Omega^+$ and $U^\eta \cap \Omega^-$ of U^η using that $\boldsymbol{\xi}^\eta$ vanishes on the external boundary ∂U^η. For the limiting behavior consider the functions

$$\tilde{\xi}^\eta := \begin{cases} z & \text{on } U^{\eta - \eta^2}, \\ 0 & \text{on } U^\eta \backslash U^{\eta-\eta^2}. \end{cases}$$

Let ζ be a smooth function with compact support on the unit ball $U^1(0) \subset \mathbb{R}^d$ such that $\int_{\mathbb{R}^d} \zeta = 1$ and define ξ^η by the convolution of $\tilde{\xi}^\eta$ with $\eta^{-3d}\zeta(\cdot/\eta^3)$, i.e.,

$$\xi^\eta(x) := \left(\eta^{-3d}\zeta(\tfrac{\cdot}{\eta^3}) * \tilde{\xi}^\eta\right)(x).$$

Then for η small enough $\xi^\eta = z$ on $\Gamma \cap U^{\eta - 2\eta^2} =: \tilde{\Gamma}^\eta$.

Observe that thanks to the smoothness of Γ the $\mathcal{H}^{d-1}$-measure of $\Gamma^\eta \backslash \tilde{\Gamma}^\eta$ is of order η^d whence $|\Gamma^\eta \backslash \tilde{\Gamma}^\eta| / |\Gamma^\eta| = O(\eta)$ as $\eta \to 0$. By assumption, the function $f = [g]_-^+$ is Lipschitz continuous on Γ. Thanks to the special choice of ξ^η it can easily be derived that

$$\fint_{\Gamma^\eta} f\xi^\eta \, d\mathcal{H}^{d-1} = \fint_{\Gamma^\eta} f z \, d\mathcal{H}^{d-1} + \fint_{\Gamma^\eta} f(\xi^\eta - z) \, d\mathcal{H}^{d-1} \to f(x_0)z$$

as $\eta \to 0$. As moreover the $\mathcal{L}^d$-measure of U^η is of order η^d but the $\mathcal{H}^{d-1}$-measure of Γ^η is of order η^{d-1} and since $|\nabla g \cdot \boldsymbol{\xi}^\eta|$ is bounded in U^η the assertion follows.

Proof. (Th. 2.7) First, observe that $\mathcal{M}^\eta(u) \to u(x_0)$ as $\eta \to 0$ since u is Lipschitz continuous. Choose some arbitrary $z \in \mathbb{R}$ and a family of functions $\{\xi^\eta\}_{\eta>0}$ as in Lemma 2.8 and let $\{\boldsymbol{\xi}^\eta\}_{\eta>0}$ be the corresponding vector fields. Then

$$\begin{aligned}\frac{1}{|\Gamma^\eta|} \int_{U^\eta} \mathcal{M}^\eta(u) \cdot c^0(x) \nabla \cdot \boldsymbol{\xi}^\eta(x) \, dx &= \mathcal{M}^\eta(u) \cdot \frac{1}{|\Gamma^\eta|} \int_{U^\eta} c^0(x) \nabla \cdot \boldsymbol{\xi}^\eta(x) \, dx \\ &\to \quad u(x_0) \cdot [c^0(x_0)]_-^+ z = [u \cdot c^0]_-^+(x_0) z.\end{aligned}$$

The limit of the right hand side of (2.18) is, using the Lemmata 2.4, 2.5, and 2.8,

$$\begin{aligned}&\frac{1}{|\Gamma^\eta|} \frac{d}{d\delta}(S_B^\eta + S_S^\eta)(0) \\ &= \frac{1}{|\Gamma^\eta|} \int_{U^\eta} \Big(s(c^0) + \mathcal{M}^\eta(u) \cdot c^0 \Big) \nabla \cdot \boldsymbol{\xi}^\eta \, dx - \fint_{\Gamma^\eta} \nabla_\Gamma \cdot \nabla\gamma(\nu) \, d\mathcal{H}^{d-1} \\ &\to \left(-[s(c^0)]_-^+(x_0) + \left[\frac{e^0}{T}\right]_-^+(x_0) + \left[\frac{-\bar{\mu} \cdot \hat{c}^0}{T}\right]_-^+(x_0) - \nabla_\Gamma \cdot \nabla\gamma(\nu(x_0)) \right) z \\ &= \left(\left[\frac{f(T, \hat{c}^0) - \bar{\mu} \cdot \hat{c}^0}{T}\right]_-^+(x_0) - \nabla_\Gamma \cdot \nabla\gamma(\nu(x_0)) \right) z.\end{aligned}$$

For the last two lines the identities $c^0 = (e^0, \hat{c}^0)$, $u_0 = -\frac{1}{T}$, $(u_1, \ldots, u_N)^T = \frac{\bar{\mu}}{T}$, and the thermodynamic relation $e = f + sT$ were applied. The left hand side of (2.18) yields in the limit as $\eta \to 0$

$$\frac{1}{|\Gamma^\eta|}(v, \xi^\eta)_\Gamma = \fint_{\Gamma^\eta} m(\nu) v \xi^\eta \, d\mathcal{H}^{d-1} \to m(\nu(x_0)) v(x_0) z.$$

Since $z \in \mathbb{R}$ can be chosen arbitrarily the condition (2.6) follows in x_0.

2.3 Phase field approach

In phase field models, the individual phases are distinguished by phase field variables. In different phases they attain different values, and interfaces are modelled by a diffuse interface layer, i.e., the phase fields and all other thermodynamic quantities change smoothly on a thin transition layer (the diffuse interface) instead of suffering discontinuous transitions.

Let $\phi = (\phi_\alpha)_{\alpha=1}^M$ where each variable ϕ_α describes the local fraction of a corresponding phase α. The vector of these phase field variables is required to fulfill the constraint $\phi \in \Sigma^M$. The interfacial contribution in (2.12) is replaced by a Ginzburg-Landau type functional (cf. [LG50]) of the form

$$-\int_\Omega \left(\delta a(\phi, \nabla\phi) + \frac{1}{\delta} w(\phi)\right) dx. \tag{2.19}$$

The function $a : \Sigma^M \times (\mathrm{T}\Sigma^M)^d \to \mathbb{R}$ is a gradient energy density which is assumed to be smooth and to satisfy

$$a(\phi, X) \geq 0 \text{ and } a(\phi, \eta X) = \eta^2 a(\phi, X) \; \forall (\phi, X, \eta) \in \Sigma^M \times (\mathrm{T}\Sigma^M)^d \times \mathbb{R}^+.$$

The function $w : \Sigma^M \to \mathbb{R}$ is smooth and has exactly M global minima at the points $e_\beta = (\delta_{\alpha\beta})_{\alpha=1}^M$, $1 \leq \beta \leq M$, with $w(e_\beta) = 0$, i.e.,

$$w(\phi) \geq 0, \quad \text{and } w(\phi) = 0 \Leftrightarrow \phi = e_\beta \text{ for some } \beta \in \{1, \dots, M\}.$$

Possible choices for a and w will be given later. We also refer to the article of Nestler and Wendler on page 113.

The surface contribution to the entropy is described above. Let us now comment on the bulk entropy contribution and its dependence on the phase field variables. The (Helmholtz) free energy of the system can be defined as an appropriate interpolation of the free energies $\{f^\alpha(T, \hat{c})\}_\alpha$ of the possible phases, i.e.,

$$f(T, \hat{c}, \phi) = \sum_{\alpha=1}^M f^\alpha(T, \hat{c}) h(\phi_\alpha) \tag{2.20}$$

with an interpolation function $h : [0, 1] \to [0, 1]$ satisfying $h(0) = 0$ and $h(1) = 1$. By the thermodynamic relations $s = -f_{,T}$ and $e = f + Ts$ the entropy and the internal energy can be expressed in terms of $(T, \hat{c}, \phi)$. By appropriate assumptions on f, inversely, the temperature can be expressed as a function in $(e, \hat{c}, \phi) = (c, \phi)$ whence also the entropy, $s(c, \phi) = -f_{,T}(T(c, \phi), \hat{c}, \phi)$. Short calculations taking the change of variables into account yield

$$s_{,c}(c, \phi) = -u(c, \phi), \quad s_{,\phi}(c, \phi) = -\frac{f_{,\phi}(T(c, \phi), \hat{c}, \phi)}{T(c, \phi)}.$$

The total entropy of the system is now

$$S(c, \phi) = \int_\Omega \left(s(c, \phi) - \left(\delta a(\phi, \nabla\phi) + \frac{1}{\delta} w(\phi)\right)\right) dx.$$

The evolution of the system is determined by a gradient flow of the entropy for the phase field variables coupled to balance equations for the conserved variables such that the second law of thermodynamics is fulfilled. To allow for anisotropy in the mobility of the phase boundaries, again a weighted L^2-product is used. Given a smooth field $\phi : \Omega \to \Sigma^M$ let

$$(w, v)_{\omega,\phi} := \int_\Omega \delta\, \omega(\phi, \nabla\phi)\, w \cdot v \, dx \quad \forall w, v \in C^\infty(\Omega; \mathrm{T}\Sigma^M).$$

The function ω is supposed to be smooth, positive, and homogeneous of degree zero in the second variable, i.e.,

$$\omega(\phi, X) \geq 0 \text{ and } \omega(\phi, \eta X) = \omega(\phi, X) \quad \forall(\phi, X, \eta) \in \Sigma^M \times \mathbb{R}^{d\times M} \times \mathbb{R}^+.$$

The evolution of the system is defined by

$$(\partial_t \phi, v)_{\omega,\phi} = \left\langle \frac{\delta S}{\delta \phi}(c, \phi), v \right\rangle \quad \forall v \in C^\infty(\Omega, \mathrm{T}\Sigma^M).$$

Taking the boundary condition

$$a_{,\nabla\phi_\alpha}(\phi, \nabla\phi) \cdot \nu_{ext} = 0, \quad 1 \leq \alpha \leq M, \tag{2.21}$$

into account this means that for all $\alpha \in \{1, \dots, M\}$

$$\delta\omega(\phi, \nabla\phi)\partial_t \phi_\alpha = \delta\nabla \cdot a_{,\nabla\phi_\alpha}(\phi, \nabla\phi) - \delta a_{,\phi_\alpha}(\phi, \nabla\phi) - \frac{1}{\delta} w_{,\phi_\alpha}(\phi) + s_{,\phi_\alpha}(c, \phi) - \lambda \tag{2.22}$$

with the Lagrange factor (due to the constraint $\sum_\alpha \phi_\alpha = 1$)

$$\lambda = \frac{1}{M} \sum_{\alpha=1}^{M} \left(\delta\nabla \cdot a_{,\nabla\phi_\alpha}(\phi, \nabla\phi) - \delta a_{,\phi_\alpha}(\phi, \nabla\phi) - \frac{1}{\delta} w_{,\phi_\alpha}(\phi) + s_{,\phi_\alpha}(c, \phi) \right).$$

It is also possible to consider multi-well potentials of obstacle type (cf. [BE91]). Then the differential equation (2.22) becomes a variational inequality.

The balance equations for the conserved quantities read

$$\partial_t c_i = -\nabla \cdot J_i(c, \phi, \nabla u(c, \phi)) = \nabla \cdot \left(\sum_{j=0}^{N} L_{ij}(c, \phi) \nabla u_j(c, \phi) \right). \tag{2.23}$$

The fact that the Onsager coefficients $L_{ij}(c, \phi)$ can differ in the different phases may be modelled by interpolating the coefficients $\{L_{ij}^\alpha\}_\alpha$ of the pure phases analogously as done for the free energy. The matrix $L = (L_{ij})_{i,j=0}^N$ then remains symmetric and positive semi-definite. Moreover, the condition $\sum_{i=1}^N L_{ij}(c, \phi) = 0$, $1 \leq j \leq N$, remains satisfied. In addition to initial conditions boundary conditions are imposed which, in the isolated case, are of the form

$$J_i(c, \phi, \nabla u(c, \phi)) \cdot \nu_{ext} = 0, \quad 0 \leq i \leq N. \tag{2.24}$$

In [GNS04] the following entropy inequality is shown:

Theorem 2.9. *If the system under consideration evolves following* (2.22) *and* (2.23) *then it holds that*

$$\frac{d}{dt} s(c, \phi) \geq -\nabla \cdot \left(\sum_{i=0}^{N} (-u_i) J_i - \delta \sum_{\alpha=1}^{M} a_{,\nabla\phi_\alpha} \partial_t \phi_\alpha \right).$$

If the boundary conditions (2.21) *and* (2.24) *hold then* $\frac{d}{dt} S(c, \phi) \geq 0$.

2.4 Example for calibration: binary alloy, two phases

The framework for phase field modelling of alloy solidification presented in the previous subsection generalizes earlier models that have successfully been applied to describe phenomena like dendritic and eutectic growth. By postulating appropriate free energies f, surface terms a and w, Onsager coefficients L_{ij}, and a kinetic mobility function ω, for example, the models used in [Ca89, PF90, WMB92] can be derived (see [GNS04, St05b]). In the following we will exemplify the choices to model non-isothermal solidification of a binary alloy involving a solid and as liquid phase. For more complex cases of multiple phases and components we refer to the article of Nestler and Wendler.

Let $M = 2$ and $N = 2$. According to the model of an ideal solution, the free energy density of the liquid phase is defined by

$$f^{(l)}(T, \hat{c}) := \sum_{i=1}^{2} -\frac{L_i}{2}\frac{T - T_i}{T_i} c_i + \frac{R_g}{v_m} T \sum_{i=1}^{2} c_i \log(c_i) - c_p T \log(\frac{T}{T_{ref}}),$$

and the free energy of the solid phase by

$$f^{(s)}(T, \hat{c}) := \sum_{i=1}^{2} \frac{L_i}{2}\frac{T - T_i}{T_i} c_i + \frac{R_g}{v_m} T \sum_{i=1}^{2} c_i \log(c_i) - c_p T \log(\frac{T}{T_{ref}}).$$

The quantities L_A and L_B are the latent heats of the pure substances $A = 1$ and $B = 2$, T_A and T_B are the melting temperatures, R_g is the gas constant, v_m the molar volume (supposed to be constant), c_p the specific heat, and T_{ref} some reference temperature, e.g., the mean value of the melting temperatures. In the following, the entropy differences s_A and s_B between the phases will appear. They are defined by $s_i := L_i/T_i$, $i = A, B$. Moreover let $R := R_g/v_m$. For simplicity assume that $L_A = s_A T_A = L_B = s_B T_B =: 2L$.

To simplify the presentation further we now consider dimensionless equations. Whenever thermodynamic quantities appear in the following, we will use the same letters but they are thought to be appropriately rescaled. In particular we are able to set $c_p = 1$ and $T_{ref} = 1$. Interpolating the free energies of the pure phases with the interpolation function $h(\phi) = \phi$ in the sense of (2.20) yields

$$f(T, c, \phi) := \Big(c_1 \frac{s_A}{2}(T_A - T) + c_2 \frac{s_B}{2}(T_B - T)\Big)(\phi_1 - \phi_2) + RT \sum_{i=1}^{2} c_i \log(c_i) - T \log(T).$$

Since $\phi_1 + \phi_2 = 1$ and $c_1 + c_2 = 1$ it is sufficient to consider $\Phi = \phi_1 - \phi_2$ and $C = c_1$ in order to distinguish the phases and to describe the alloy composition. We then have $\Phi = 1$ in the liquid phase, $\Phi = -1$ in the solid phase, and C is the concentration of component A. The free energy density can then be written in the form

$$\begin{aligned}\tilde{f}(T,C,\Phi) &:= f(T,C,1-C,\tfrac{1+\Phi}{2},\tfrac{1-\Phi}{2}) \\ &= \tfrac{1}{2}\big(Cs_A(T_A-T)+(1-C)s_B(T_B-T)\big)\Phi \\ &\quad + RT\big(C\log(C)+(1-C)\log(1-C)\big) - T\log(T) \qquad (2.25)\end{aligned}$$

resulting in the internal energy density

$$\tilde{e}(T,C,\Phi) = \tfrac{1}{2}\big(CL_A+(1-C)L_B\big)\Phi + T =: L\Phi + T.$$

Setting $L_{0i} = L_{i0} := 0$ for $i = 1,2$ and $L_{00} := K(\Phi)T^2$ the energy flux becomes

$$-L_{00}\nabla u_0 = -K(\Phi)T^2\nabla\frac{-1}{T} = -K(\Phi)\nabla T$$

whence the balance equation for the energy reads

$$\partial_t\tilde{e} = \partial_t T + L\partial_t\Phi = \nabla\cdot\big(K(\Phi)\nabla T\big). \qquad (2.26)$$

Since $\overline{\mu}_1 = f_{,c_1} - \frac{1}{2}(f_{,c_1}+f_{,c_2}) = \frac{1}{2}(f_{,c_1}-f_{,c_2}) = \frac{1}{2}\tilde{f}_{,C}$ we have

$$-u_2 = u_1 = \frac{\overline{\mu}_1}{T} = \frac{1}{2}\frac{s_B-s_A}{2}\Phi + \frac{R}{2}\big(\log(C)-\log(1-C)\big)$$

whence

$$-\nabla u_2 = \nabla u_1 = \frac{1}{2}\Big(\frac{s_B-s_A}{2}\nabla\Phi + R\frac{1}{C(1-C)}\nabla C\Big).$$

Choosing $\tilde{D}(\Phi)C(1-C) =: L_{11} = -L_{12} = -L_{21} = L_{22}$ with some diffusivity coefficient $\tilde{D}(\Phi)$ a short calculation gives

$$\begin{aligned}-\partial_t c_2 = \partial_t c_1 = \partial_t C &= \nabla\cdot\big(\tilde{D}(\Phi)R\nabla C\big) \\ &\quad + \nabla\cdot\Big(\tilde{D}(\Phi)C(1-C)\frac{s_B-s_A}{2}\nabla\Phi\Big).\end{aligned} \qquad (2.27)$$

Subtracting the equations for the two phase field variables ϕ_1 and ϕ_2 yields

$$\delta^2\omega\partial_t\Phi = \delta^2\big(\nabla\cdot(a_{,\nabla\phi_1}-a_{,\nabla\phi_2})-(a_{,\phi_1}-a_{,\phi_2})\big)-(w_{,\phi_1}-w_{,\phi_2})-\frac{\delta}{T}(f_{,\phi_1}-f_{,\phi_2}).$$

The standard double-well potential $w(\phi) := 9\gamma\phi_1^2\phi_2^2$ for some $\gamma > 0$ related to the surface tension (see below) gives

$$(w_{,\phi_1}-w_{,\phi_2})\big(\tfrac{1+\Phi}{2},\tfrac{1-\Phi}{2}\big) = \tfrac{9}{4}\gamma p'(\Phi) \quad \text{where } p(\Phi) = \tfrac{1}{2}(\Phi^2-1)^2.$$

Moreover it holds that

$$-\tfrac{\delta}{T}(f_{,\phi_1}-f_{,\phi_2}) = -\tfrac{\delta}{T}2\tilde{f}_{,\Phi} = -\tfrac{\delta}{T}\big(Cs_A(T_A-T)+(1-C)s_B(T_B-T)\big).$$

The surface gradient term is set to $a(\phi, \nabla\phi) := \gamma|\phi_1\nabla\phi_2 - \phi_2\nabla\phi_2|^2 = \gamma|\frac{1}{4}\nabla\Phi|^2$. Short calculations give

$$a_{,\phi_1} - a_{,\phi_2} = 0, \quad a_{,\nabla\phi_1} - a_{,\nabla\phi_2} = 2\gamma(\phi_1\nabla\phi_2 - \phi_2\nabla\phi_2)(\phi_1 - \phi_2) = \gamma\nabla\Phi.$$

Finally, let $\xi := \frac{2}{3}\delta$, $\alpha := \frac{\omega}{\gamma}$, and replace the surface energy $T\gamma =: \sigma$ by a temperature independent constant (i.e., replace T in that term by some reference temperature T_{ref} and assume that variations of σ in the temperature can be neglected). Then the evolution of the phase field variable is governed by

$$\xi^2\alpha\partial_t\Phi = \xi^2\Delta\Phi - p'(\Phi) - \frac{2\xi}{3\sigma}\big(Cs_A(T_A - T) + (1 - C)s_B(T_B - T)\big). \quad (2.28)$$

The model consisting of equations (2.26)–(2.28) and some additional conditions will be used in the following section to sketch the method of relating a phase field model to a sharp interface model and in the last section to describe dendritic solidification.

2.5 Some remarks on the solvability of the phase field model

The reduced grand canonical potential is defined to be the Legendre transform of the negative entropy with respect to the conserved quantities,

$$\psi(u, \phi) = (-s)^*(c(u, \phi), \phi).$$

With its help it is possible to reformulate the differential equations using (u, ϕ) as variables (cf. [St05b]). The parabolic system then has the structure

$$\partial_t\psi_{,u_i}(u, \phi) = \nabla \cdot \left(\sum_{j=0}^{N} L_{ij}(\psi_{,u}(u, \phi), \phi)\nabla u_j\right),$$

$$\omega(\phi, \nabla\phi)\partial_t\phi_\alpha = \nabla \cdot a_{,\nabla\phi_\alpha}(\phi, \nabla\phi) - a_{,\phi_\alpha}(\phi, \nabla\phi) - w_{,\phi_\alpha}(\phi) + \psi_{,\phi_\alpha}(u, \phi) - \lambda$$

where $0 \le i \le N$, $1 \le \alpha \le M$. When rigorously analyzing these equations the main difficulties arise from the growth properties of ψ in u and the nonlinearities involving $\nabla\phi$.

An ideal solution formulation of the free energy density has the structure

$$f(T, c) = T\log(T) + T\sum_i c_i\log(c_i) + \ldots$$

As a result, in ψ a term $-\log(-u_0)$ appears. In particular, when solving the differential equations it must be shown that $u_0 < 0$ almost everywhere. Moreover, ψ is only of at most linear growth in the u_i, $1 \le i \le N$. A control of terms involving $\psi_{,u}$ obtained by standard estimates for parabolic equations do not provide much information of u itself any more. These difficulties have been independently tackled in [AP92] and [LV83] respectively.

Based on those results, the above system including the phase field variables is analyzed in [St05b] by approximating ψ with a perturbed potential of quadratic growth in u. The main task is to derive suitable estimates and, based on the estimates, to develop and apply appropriate compactness arguments in order to go to the limit as the perturbation vanishes. It is assumed that the matrix of Onsager coefficients $L = (L_{ij})_{ij}$ is positive (on a certain subspace) uniformly in their arguments. If a degenerating coefficient matrix is considered as in the previous subsection it may be better to switch to $(T, \hat{c})$ or $(e, \hat{c})$ as variables, e.g. cf. [Ec04a].

Managing the phase field variables is kept simple in [St05b] by appropriate assumptions on a, w, and ω. The interesting case of a involving the terms $\phi_\alpha \nabla \phi_\beta - \phi_\beta \nabla \phi_\alpha$ (which give a good approximation of the direction of $\nu_{\alpha\beta}$) is still open. Non-local models have been considered by multiple authors (for instance, we refer to [BS96, SZ03, KRS05]). There, the energy is the only conserved quantity, and the difficulties with the logarithmic term in u_0 are tackled by performing a Moser type iteration to get L^∞-bounds for u_0 and $1/u_0$.

3 Relation between the approaches and calibration

The relation between the phase field model and the free boundary problem presented in the previous chapter can be established using the method of matched asymptotic expansions. Generalizing methods developed in [CF88, Ca89, BGS98, GNS98] this has been done in [GNS04]. The procedure is as follows: It is assumed that the solution to the phase field model can be expanded in δ-series in the bulk regions occupied by the phases (outer expansions) and, using rescaled coordinates, in the interfacial regions (inner expansion). Given suitable relations between the functions and parameters of the phase field model on the one hand and the parameters in the free boundary problem on the other hand the functions to leading order of the δ-series solve the governing equations of the free boundary problem. It should be remarked that this procedure is a formal method in the sense that it is not rigorously shown that the assumed expansions in fact exist and converge. But in some cases this ansatz could be verified (cf. [DS95, St96, CC98, Di04]).

If the phase field model is considered as an approximation of the free boundary problem fast convergence with respect to δ is desired. An improvement of the approximation was obtained in [KR98] in the context of thin interface asymptotics. The analysis led to a positive correction term in the kinetic coefficient of the phase field equation balancing undesired terms of order δ in the Gibbs-Thomson condition and raising the stability bound of explicit numerical methods. Besides, the better approximation allows for larger values of δ and, therefore, for coarser grids. In particular, it is possible to consider the limit of vanishing kinetic undercooling which is important in applications.

Numerical simulations of appropriate test problems reveal an enormous gain in efficiency thanks to a better approximation.

In [Al99] the analysis was extended to the case of different diffusivities in the phases and both classical and thin interface asymptotics were discussed. By choosing different interpolation functions for the free energy density and the internal energy density (the function h in (2.20)) an approximation of second order could still be achieved but the gradient structure of the model and thermodynamic consistency were lost. Based on those ideas it was shown in [An02] that even an approximation of third order is possible by using high order polynomials for the interpolation. In [MWA00] an approach based on an energy and an entropy functional was used providing more degrees of freedom to tackle the difficulties with unequal diffusivities in the phases while avoiding the loss of the thermodynamic consistency. Both classical and thin asymptotics are discussed in that article as well as the limit of vanishing kinetic undercooling. In a more recent analysis in [RB*04], a binary alloy also involving different diffusivities in the phases was considered and a better approximation was obtained by adding a small additional term to the mass flux (anti-trapping mass current, the ideas stem from [Ka01]).

We have shown in [GS06] that, for two-phase multi-component systems with arbitrary phase diagrams, there is a correction term to the kinetic coefficient such that the model with moving boundaries is approximated to second order in the small parameter δ. A new feature compared to the existing results is that, in general, this correction term depends on temperature and chemical potentials. Indeed, up to some numerical constants, the latent heat appears in the correction term obtained by Karma and Rappel [KR98]. Analogously, the equilibrium jump in the concentrations enters the correction term when investigating an isothermal binary alloy. But from realistic phase diagrams it is obvious that this jump depends on the temperature leading to a temperature dependent correction term in the non-isothermal case.

In this chapter, the procedure to get an second order approximation will be sketched for a simple model describing solidification of a pure substance. The model is based on the model in Sect. 2.4. There, the small quantity $\xi = \frac{2}{3}\delta$ was introduced and will be used instead of δ. In addition to the free boundary problem which appears as problem to leading order a correction problem to the next order is derived by continuing the asymptotic analysis. The goal is to obtain that fields identically zero solve the correction problem. It turns out that the above mentioned correction term to the kinetic coefficient is necessary to allow for this solution. The model equations including assumptions, asymptotic expansions, and matching conditions are listed in the following subsection. After, the asymptotic analysis is performed. Finally, the leading order problem and the correction problem are stated. In [GS06], numerical tests have been performed to show that a better approximation of the free boundary problem thanks to the kinetic correction term is really obtained.

3.1 The simplified model and assumptions

In order to present the main ideas to obtain a second order approximation a simple model for solidification of a pure substance is considered, namely, the model in Sect. 2.4 where we set $C \equiv 1$. In the definition of the free energy density (2.25) Φ is replaced by a term $h(\Phi)$ with an interpolation function $h : [-1,1] \to [0,1]$ which is symmetric, i.e., $h(-\Phi) = -h(\Phi)$, and fulfills $h'(\pm 1) = 0$. For s_A we simply write s, and T_A is replaced by T_m. The kinetic coefficient splits into a main part and a positive correction term of order ξ, i.e., $\alpha = \alpha_0 + \xi\alpha_1$. The correction term will later be determined and turn out to be crucial to get a higher order approximation of the related free boundary problem. The heat diffusivity K is assumed to be independent of the phase field variable. The governing equations then have the form

$$\xi^2(\alpha_0 + \xi\alpha_1)\partial_t\Phi = \xi^2\Delta\Phi - p'(\Phi) - \frac{2\xi}{3\sigma}\big(s(T_m - T)\big)h'(\Phi), \tag{3.1}$$

$$\partial_t T + L\partial_t h(\Phi) = K\Delta T. \tag{3.2}$$

To obtain a well-posed problem initial and boundary conditions have to be imposed. Consider a domain $\Omega \subset \mathbb{R}^2$ and a time interval $I_\mathcal{T} := (0, \mathcal{T})$. For $\xi > 0$ let $(T(t,x;\xi), \Phi(t,x;\xi))$, $x \in \Omega$, $t \in I_\mathcal{T}$, denote smooth solutions to (3.1)–(3.2) given the same initial and boundary conditions. We suppose that, for all times, there exist two phases separated by a diffuse interfacial layer which is bounded away from the boundary of the domain Ω. Here, we do not carry out the asymptotic analysis for the initial and boundary conditions but only give some remarks. That analysis is carried out in [St05b], Sect. 3.2, and [GS06].

The following procedure of matching asymptotic expansions is outlined with great care in [FP95, DW05]. Here, only the main ideas for the two-dimensional case are sketched. The family

$$\Gamma(t;\xi) := \{x \in \Omega : \Phi(t,x;\xi) = 0\}, \quad \xi > 0, t \in I_\mathcal{T}, \tag{3.3}$$

is supposed to be a set of smooth curves in Ω. They are demanded to be uniformly bounded away from $\partial\Omega$ and to depend smoothly on (ξ, t) such that, if $\xi \searrow 0$, some limiting curve $\Gamma(t;0)$ is obtained. With $\Omega^l(t;\xi)$ and $\Omega^s(t;\xi)$ we denote the regions occupied by the liquid phase (where $\Phi(t,x;\xi) > 0$) and the solid phase (where $\Phi(t,x;\xi) < 0$) respectively.

Let $\gamma(t,s;0)$ be a parameterization of $\Gamma(t;0)$ by arc-length s for every $t \in I_\mathcal{T}$. The vector $\nu(t,s;0)$ denotes the unit normal on $\Gamma(t;0)$ pointing into $\Omega^l(t;0)$, and $\tau(t,s;0) := \partial_s\gamma(t,s;0)$ denotes the unit tangential vector. For ξ small enough the curves $\Gamma(t;\xi)$ can be parametrized over $\Gamma(t;0)$ using some distance function $d(t,s;\xi)$,

$$\gamma(t,s;\xi) := \gamma(t,s;0) + d(t,s;\xi)\nu(t,s;0).$$

Close to $\xi = 0$ we assume that there is the expansion $d(t,s;\xi) = \xi^1 d_1(t,s) + \xi^2 d_2(t,s) + O(\xi^3)$. Also the curvature $\kappa(t,s;\xi)$ and the normal velocity $v(t,s;\xi)$

of $\Gamma(t;\xi)$ are smooth and can be expanded in ξ-series (cf. the Appendix of [GS06]):

$$\kappa(t,s;\xi) = \kappa(t,s;0) + \xi\big(\kappa(t,s;0)^2 d_1(t,s) + \partial_{ss} d_1(t,s)\big) + O(\xi^2), \tag{3.4}$$

$$v(t,s;\xi) = \partial_t \gamma(t,s;\xi)\cdot\nu(t,s;\xi) = v(t,s;0) + \xi\,\partial^\circ d_1(t,s) + O(\xi^2). \tag{3.5}$$

Here, ∂° denotes the normal time derivative, see (A.1) for a definition.

We suppose that in each domain $E \subset \mathbb{R}^2$ such that $\overline{E} \subset \Omega\backslash\Gamma(t;0)$ the solution can be expanded in a series close to $\xi = 0$ (**outer expansion**):

$$\begin{aligned} T(t,x;\xi) &= \sum_{k=0}^{K} \xi^k \theta_k(t,x) + O(\xi^{K+1}), \\ \Phi(t,x;\xi) &= \sum_{k=0}^{K} \xi^k \varphi_k(t,x) + O(\xi^{K+1}). \end{aligned} \tag{3.6}$$

Let z be the $\frac{1}{\xi}$-scaled signed distance of x from $\Gamma(t;0)$. Hence, in a neighborhood of $\Gamma(t;0)$ we can write for $z \neq 0$

$$\hat{T}(t,s,z;\xi) := T(t,x(t,s,z);\xi), \quad \hat{\Phi}(t,s,z;\xi) := \Phi(t,x(t,s,z);\xi).$$

An essential assumption is now that $\hat{T}$ and $\hat{\Phi}$ can be expanded in these new variables (**inner expansion**),

$$\hat{T}(t,s,z;\xi) = \sum_{k=0}^{K} \xi^k T_k(t,s,z) + O(\xi^{K+1}), \tag{3.7}$$

$$\hat{\Phi}(t,s,z;\xi) = \sum_{k=0}^{K} \xi^k \Phi_k(t,s,z) + O(\xi^{K+1}), \tag{3.8}$$

and that these expansions are valid for $z \in \mathbb{R}$. The notion is that, since the interfacial thickness scales with ξ, one can expect a meaningful convergence behavior when rescaling the space with $1/\xi$ in the normal direction.

Given $x \notin \Gamma(t;0)$ clearly $z(t,x) = \mathrm{dist}(x,\Gamma(t;0))/\xi \to \pm\infty$ as $\xi \searrow 0$. On the other hand, in that limit x is located in one of the two phases, and the closer it lies to the interface $\Gamma(t;0)$ the better the series of the functions $\theta_k(t,x)$ approximates the value of the temperature on the interface. These facts are reflected by the following matching conditions relating the outer and inner expansions (see [St05b], Sect. 3.1, and [GS06] for the derivation): As $z \to \pm\infty$

$$T_0(z) \sim \theta_0(0^\pm), \tag{3.9}$$

$$T_1(z) \sim \theta_1(0^\pm) + (\nabla\theta_0(0^\pm)\cdot\nu)z, \tag{3.10}$$

$$\partial_z T_1(z) \sim \nabla\theta_0(0^\pm)\cdot\nu, \tag{3.11}$$

$$\partial_z T_2(z) \sim \nabla\theta_1(0^\pm)\cdot\nu + \big((\nu\cdot\nabla)(\nu\cdot\nabla)\theta_0(0^\pm)\big)z \tag{3.12}$$

and analogously for Φ. Here, for a function $g(t,x) = \hat{g}(t,s,r)$ with the signed distance $r = \mathrm{dist}(x, \Gamma(t;0))$

$$g(0^+) := \lim_{r \searrow 0} \hat{g}(t,s,r), \quad g(0^-) := \lim_{r \nearrow 0} \hat{g}(t,s,r).$$

3.2 Outer solutions

Away from $\Gamma(t;0)$, i.e., in domains $E \subset \mathbb{R}^2$ with $\overline{E} \subset \Omega \backslash \Gamma(t;0)$, the expansions (3.6) are plugged into the differential equations. All terms are expanded in ξ-series.

To leading order ξ^0 equation (3.1) yields the identity $0 = -p'(\varphi_0)$. The only stable solutions are the minima of p, hence $\varphi_0 \equiv \pm 1$. These values distinguish the two phases because, since the result is independent of ξ, necessarily $\varphi_0 = 1$ in Ω^l and $\varphi_0 = -1$ in Ω^s.

To the next order ξ^1 the identity

$$0 = -p''(\varphi_0)\varphi_1 - \frac{2}{3\sigma} s(T_m - \theta_0) h'(\varphi_0)$$

follows. By $h'(\pm 1) = 0$ and $p''(\pm 1) = 4$ we obtain $\varphi_1 \equiv 0$ as the only solution.

The energy balance equation (3.2) yields the heat equation, to leading order for θ_0 and to the next order for θ_1:

$$\partial_t \theta_k = K \Delta \theta_k, \quad k = 0, 1.$$

Observe that it is possible to replace θ_0 by the internal energies $e^{(l)}(\theta_0) = \theta_0 + L$ of the liquid phase or $e^{(s)} = \theta_0 - L$ of the solid phase.

The initial conditions and boundary conditions on $\partial\Omega$ are independent of ξ and, hence, only enter θ_0 and φ_0 respectively. The higher order corrections fulfill homogeneous initial and boundary conditions. Boundary conditions on $\Gamma(t;0)$ will be obtained by matching the expansions with the expansions in the interfacial region.

3.3 Inner solutions

Derivatives with respect to (t,x) transform into derivatives with respect to (t,s,z) as follows:

$$\begin{aligned}
\frac{d}{dt} &= -\tfrac{1}{\xi} v \partial_z + \partial^\circ - (\partial^\circ d_1)\partial_z + O(\xi), \\
\Delta_x &= \tfrac{1}{\xi^2}\partial_{zz} - \tfrac{1}{\xi}\kappa\partial_z \\
&\quad + (\partial_s d_1)^2 \partial_{zz} - 2\partial_s d_1 \partial_{sz} - (\kappa^2(z + d_1) - \partial_{ss} d_1)\partial_z + \partial_{ss} + O(\xi).
\end{aligned}$$

The phase field equation first yields

$$0 = \partial_{zz}\Phi_0 - p'(\Phi_0). \tag{3.13}$$

By (3.3) and the assumption that (3.8) holds true for $\xi = 0$ we have $\Phi_0(z = 0) = 0$. The matching conditions (3.9) imply

$$\Phi_0(t, s, z) \to \varphi_0(t, s; 0^\pm) = \pm 1 \text{ as } z \to \pm\infty.$$

Hence, the solution to (3.13) is $\Phi_0(t, s, z) = \tanh(z)$ and only depends on z.

For the conserved variable we get $0 = K\partial_{zz}T_0$ to order ξ^{-2}. By the matching conditions (3.9) T_0 has to be bounded as $z \to \pm\infty$, hence we see that T_0 must be constant with respect to z which means $T_0 = T_0(t, s)$. The matching condition (3.9) furthermore implies that $T_0(t, s)$ is exactly the value of θ_0 in the point $\gamma(t, s; 0) \in \Gamma(t; 0)$ from both sides of the interface. In particular,

$$\theta_0 \text{ is continuous across the interface } \Gamma(t; 0).$$

To order ξ^1 equation (3.1) yields

$$-\alpha_0 v \partial_z \Phi_0 = \partial_{zz}\Phi_1 - \kappa\partial_z\Phi_0 - p''(\Phi_0)\Phi_1 - \frac{2}{3\sigma}s(T_m - T_0)h'(\Phi_0). \qquad (3.14)$$

From the outer solutions we have $\varphi_1(t, s, 0^\pm) = 0$ and $\nabla\varphi_0(t, s, 0^\pm) \cdot \nu = 0$. Due to the matching condition (3.10) we conclude $\Phi_1(t, s, z) \to 0$ as $z \to \pm\infty$. The operator $\mathcal{L}(\Phi_0) = \partial_{zz} - w''(\Phi_0)$ is self-adjoint. Differentiating (3.13) with respect to z we obtain that $\partial_z\Phi_0$ lies in the kernel of $\mathcal{L}(\Phi_0)$. Since $\Phi_0(-z) = -\Phi_0(z)$, $\partial_z\Phi_0$ and $h'(\Phi_0)$ are even, (3.14) allows for an even solution. In the following we will assume that Φ_1 indeed is even.

A solvability condition can be deduced by multiplying the equation with $\partial_z\Phi_0$ and integrating over $\mathbb{R}$ with respect to z:

$$\begin{aligned} 0 &= \int_{\mathbb{R}} \Big((\kappa - \alpha_0 v)(\partial_z\Phi_0(z))^2 + \frac{2}{3\sigma}s(T_m - T_0)h'(\Phi_0(z))\partial_z\Phi_0(z) \Big)\, dz \\ &= \frac{4}{3}(\kappa - \alpha_0 v) + \frac{4}{3\sigma}s(T_m - \theta_0) \end{aligned} \qquad (3.15)$$

where we used that $\int_{\mathbb{R}}(\partial_z\Phi_0)^2 dz = \frac{4}{3}$. Up to the factor $\frac{4}{3}$ this is the Gibbs-Thomson condition (2.6).

The system (3.2) becomes to the order ξ^{-1}

$$-v\partial_z(T_0 + Lh(\Phi_0)) = K\partial_{zz}T_1.$$

Integrating two times with respect to z furnishes

$$\begin{aligned} T_1 &= -\frac{1}{K}\Big(vL\int_0^z h(\Phi_0)dz' + (vT_0 - A)z \Big) + \bar{\tau} \qquad (3.16) \\ &\sim -\frac{1}{K}\Big((v(T_0 + L) - A)z - vLH \Big) + \bar{\tau} \text{ as } z \to \infty \\ &\sim -\frac{1}{K}\Big((v(T_0 - L) - A)z - vLH \Big) + \bar{\tau} \text{ as } z \to -\infty \end{aligned}$$

where A and $\bar{\tau}$ are integration constants and

$$H := \int_0^\infty (1 - h(\Phi_0(z')))dz' = \int_{-\infty}^0 (1 + h(\Phi_0(z')))dz'.$$

Here, we used the fact that Φ_0 converges to constants exponentially fast, so that the integral $\int_0^z$ has been replaced by $\int_0^\infty$ while the linear terms remain. By (3.10)

$$\theta_1(0^\pm) = \bar{\tau} + \frac{v}{K} LH \tag{3.17}$$

which means, in particular, that

$$\theta_1 \text{ is continuous across } \Gamma(t;0). \tag{3.18}$$

With (3.11) and $T_0 = \theta_0(0^\pm)$ the following jump condition is obtained on $\Gamma(t;0)$:

$$[-K\nabla\theta_0]_s^l \cdot \nu = \big(v(T_0 + L) - A\big) - \big(v(T_0 - L) - A\big) = v[e(\theta_0(0))]_s^l. \tag{3.19}$$

Since Φ_0 only depends on z the phase field equation to order ξ^2 gives

$$\begin{aligned}
&- \alpha_0 v \partial_z \Phi_1 - \alpha_1 v \partial_z \Phi_0 - \alpha_0 (\partial^\circ d_1) \partial_z \Phi_0 \\
= \; & \partial_{zz}\Phi_2 - p''(\Phi_0)\Phi_2 + (\partial_s d_1)^2 \partial_{zz}\Phi_0 - (\kappa^2(z + d_1) + \partial_{ss} d_1)\partial_z \Phi_0 \\
&- \kappa \partial_z \Phi_1 - \frac{1}{2} p'''(\Phi_0)(\Phi_1)^2 + \frac{2}{3\sigma} s(T_m - T_0) h''(\Phi_0)\Phi_1 + \frac{2}{3\sigma} s T_1 h'(\Phi_0).
\end{aligned}$$

To guarantee that Φ_2 exists there is again a solvability condition which is obtained by multiplying with $\partial_z \Phi_0$ and integrating over $\mathbb{R}$ with respect to z. The terms involving Φ_1 vanish. For this purpose, equation (3.14) and the assumption that Φ_1 is even is used. Let

$$\begin{aligned}
J :&= \int_0^\infty \partial_z (h \circ \Phi_0)(z) \int_0^z (1 - (h \circ \Phi_0)(z'))dz'dz \\
&= \int_{-\infty}^0 \partial_z (h \circ \Phi_0)(z) \int_z^0 (1 + (h \circ \Phi_0)(z'))dz'dz.
\end{aligned}$$

Using (3.16) to replace T_1 and, after, (3.17) to replace $\bar{\tau}$ a short calculation shows that the solvability condition becomes (remember that $2L = sT_m$)

$$\begin{aligned}
0 = \; & \sigma(-\alpha_0 \partial^\circ + \partial_{ss} + \kappa^2) d_1 - s\theta_1 \\
& + v\Big(- \sigma\alpha_1 + (H + J)\frac{1}{K} T_m s^2 \Big).
\end{aligned} \tag{3.20}$$

We remark that $\partial^\circ d_1$ and $(\partial_{ss} + \kappa^2) d_1$ are the first order corrections of the normal velocity and the curvature of $\Gamma(t, s; \xi)$ (see (3.5) and (3.4) respectively). Indeed, when inserting the expansions for $T = \theta_0 + \xi\theta_1 + \ldots$ and the interface distance $d = \xi d_1 + \ldots$ into the Gibbs-Thomson condition $\sigma\alpha v = \sigma\kappa + s(T_m - T)$

then, to leading order, we get (3.15), and the first line of (3.20) is the equation to first order in ξ.

The goal is to obtain that $\theta_1 \equiv 0$ and $d_1 \equiv 0$ are solutions to the equations they have to fulfill. For this purpose, the second line of (3.20) must vanish. But by suitable choice of the additional correction term α_1 in the kinetic coefficient, namely

$$\alpha_1 = (H + J)\frac{\sigma}{K} T_m s^2, \tag{3.21}$$

this in indeed ensured.

Analogously to the above correction to the Gibbs-Thomson condition we are interested in deriving a first order correction to the jump condition (3.19). The equation (3.2) yields to order ξ^0

$$\begin{aligned} &- v\partial_z(T_1 + Lh'(\Phi_0)\Phi_1) + (\partial^\circ - (\partial^\circ d_1)\partial_z)(T_0 + Lh(\Phi_0)) \\ &\quad = K\left(\partial_{zz}T_2 - \kappa\partial_z T_1 + \partial_{ss}T_0\right). \end{aligned}$$

Integrating once with respect to z leads to

$$\begin{aligned} - K\partial_z T_2 = & \underbrace{v(T_1 + Lh'(\Phi_0)\Phi_1) - B}_{(i)} \\ & + \underbrace{\int_0^z (-\partial^\circ + (\partial^\circ d_1)\partial_z)(T_0 + Lh(\Phi_0))dz'}_{(ii)} - \underbrace{\kappa K T_1}_{(iii)} + K\partial_{ss}T_0 z \end{aligned}$$

where B is an integration constant. We need to collect the terms contributing to $\nabla\theta_1 \cdot \nu$. In view of (3.12) this means that the terms linear in z are not of interest. Applying (3.10) to Φ_1, T_1 and by the assumption $h'(0) = h'(1) = 0$ it holds that

$$(i) \sim v\theta_1 - B + (\dots)z \text{ as } z \to \pm\infty.$$

Furthermore, since $\partial^\circ \Phi_0 = 0$,

$$\begin{aligned} (ii) &= -\partial^\circ(T_0 \pm L)z + (\partial^\circ d_1)L(h(\Phi_0))\big|_0^z \\ &\sim -\partial^\circ e^{(l)} + (\partial^\circ d_1)L \qquad \text{as } z \to \infty, \\ &\sim -\partial^\circ e^{(l)} - (\partial^\circ d_1)L \qquad \text{as } z \to -\infty. \end{aligned}$$

By (3.10) and (3.18) we get $(iii) = \kappa K\theta_1 + (\dots)z$ as $z \to \pm\infty$. Finally, the first order correction of the jump condition (3.19) at the interface is

$$[-K\nabla\theta_1]_s^l \cdot \nu = v\theta_1 + 2L(\partial^\circ d_1).$$

3.4 Summary of assumptions and stated problems

Let us now collect the equations. First, the problem to leading order is stated:

(LOP) Find a function $\theta_0 : I_T \times \Omega \to \mathbb{R}$ and a family of curves $\{\Gamma(t;0)\}_{t \in I_T}$ separating Ω into two domains $\Omega^l(t;0)$ and $\Omega^s(t;0)$ such that

$$\partial_t e^{(p)}(\theta_0) = K \Delta \theta_0, \qquad \text{in } \Omega^p(t;0),\ t \in I_T,\ p = s, l,$$

and such that on $\Gamma(t;0)$ there holds for all $t \in I_T$:

$$\begin{aligned} \theta_0 \text{ is continuous,} \\ [-K\nabla\theta_0]_s^l \cdot \nu = v[e(\theta_0)]_s^l, \\ \sigma\alpha_0 v = \sigma\kappa + s(T_m - \theta_0). \end{aligned}$$

If we define α_1 as in (3.21) then the correction problem reads as follows:

(CP) Let $(\theta_0, \{\Gamma(t;0)\}_t)$ be a solution to (LOP). Let $l(t)$ be the length of $\Gamma(t;0)$ and set $S_{I_T} := \{(t,s) : t \in I_T, s \in [0, l(t))\}$. Then find functions $\theta_1 : I_T \times \Omega \to \mathbb{R}$ and $d_1 : S_{I_T} \to \mathbb{R}$ such that

$$\partial_t \theta_1 = K \Delta \theta_1, \qquad \text{in } \Omega^p(t;0),\ t \in I_T,\ p = s, l,$$

and such that on $\Gamma(t;0)$ there holds for all $t \in I_T$:

$$\begin{aligned} \theta_1 \text{ is continuous,} \\ [-K\nabla\theta_1]_s^l \cdot \nu = v\theta_1 + (\partial^\circ d_1)\,[e(\theta_0)]_s^l \\ \sigma\alpha_0\,\partial^\circ d_1 = \sigma(\partial_{ss} + \kappa^2)d_1 - s\theta_1. \end{aligned}$$

Obviously, $(\theta_1, d_1) \equiv 0$ is a solution to the correction problem (as previously remarked, the boundary conditions on $\partial\Omega$ are homogeneous). If this solution is unique then the leading order problem is approximated to second order in ξ by the phase field model. Problem (CP) is in fact the linearization of (LOP), i.e., the problem resulting from (LOP) when inserting the expansions $T = \theta_0 + \xi\theta_1 + \ldots$ and $d = \xi d_1 + \ldots$. We point out again that the choice (3.21) is crucial in order to guarantee that the undesired terms in (3.20) vanish.

3.5 Numerical example

In [GS06] several numerical tests have been performed revealing that the free boundary problem can indeed be better approximated by the phase field model with the correction term. Fig. 3.1 shows the results for an undercooled binary alloy (the potentials, physical parameters, and initial values are precisely stated in [GS06], Sect. 4.3). A planar solid-liquid front moves into the liquid phase. On the right the figure shows the profiles of the concentration of one component during the solidification. The position of the interface, i.e,

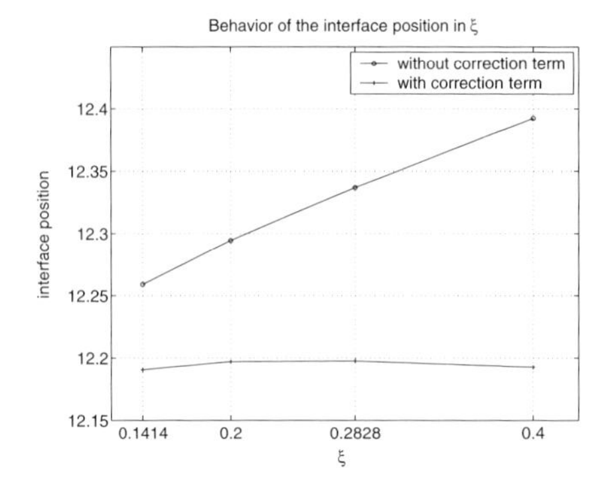

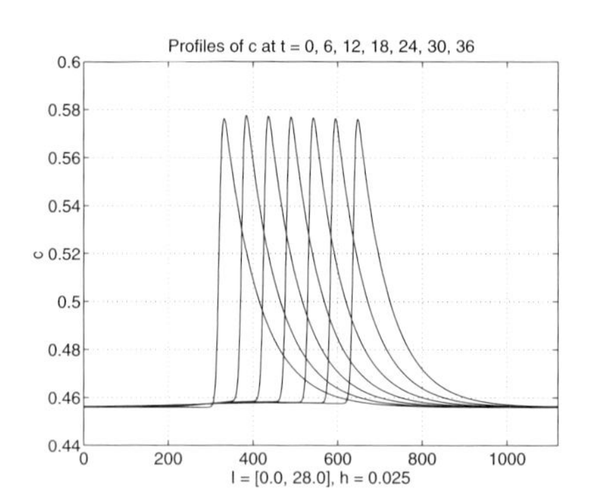

Fig. 3.1. Numerical test of the correction term. On the left: the position of the interface depicted over ξ. On the right: profiles of the concentration c during evolution.

the point where $\Phi(t, x; \xi) = 0$, is depicted on the left for several values of ξ, the other parameters being fixed.

Simulating with the correction term (3.21) in the phase field equation and varying ξ the changes in the interface position turned out to be of about 10^{-3} which is smaller than the grid spacing $\Delta x = 0.02$. In contrast, if the correction term was not taken into account changes of several grid points were observed. This behavior in ξ indicates that the approximation of the sharp interface solution is improved thanks to the correction term.

3.6 Remarks on the multi-phase case

When multiple phases are present the asymptotic analysis leads to a leading order problem consisting of the equations (2.3)–(2.6), (2.9), and (2.10) (cf. [GNS98, GNS04]). Indeed, the procedure presented in the previous subsections yields the equations (2.3)–(2.6). To obtain the force balance (2.9) (and, analogously, (2.10)) it is assumed that, away from the triple junction on a diffuse phase boundary, the situation is just as in the case of two phases.

Aiming for a second order approximation of the force balance we observed that, in general, in the interfacial regions not only the phase field variables of the adjacent phases are present but also phase fields corresponding to other phases appear. It turned out that these artificial third phase contributions do not trouble the first order asymptotic analysis but a second order analysis. As a first step we therefore developed and analyzed suitable multi-well potentials w that avoid the third phase contributions (cf. [St05a, GHS06]), smooth potentials as well as potentials of obstacle type.

As an additional feature, the calibration of the phase field model with respect to given surface energies $\sigma_{\alpha\beta}(\nu)$ and mobility coefficients $m_{\alpha\beta}(\nu)$ becomes much simpler. It is shown in [BBR05] that the Γ–limit of (2.19) as $\delta \to 0$

has the form of the surface contribution in (2.12), and a relation between the $\sigma_{\alpha\beta}$ and the functions a and w is derived. Using matched asymptotic expansions, [St91] for the isotropic case and [GNS98] for the general case proposed the simpler relation

$$\sigma_{\alpha\beta}(\nu) = \inf_p \Big\{ \int_{-1}^{1} \sqrt{w(p)a(p, p' \otimes \nu)}dy, \\ p \in C^{0,1}([-1,1]; \Sigma^M),\, p(-1) = e_\alpha,\, p(1) = e_\beta \Big\}. \quad (3.22)$$

Using numerical simulations they got evidence that this formula seems to hold true for a large class of anisotropies.

The new potentials w are such that solutions to (3.22) exist with $p_i \not\equiv 0$ only if $i = \alpha, \beta$. Moreover, it is possible to adapt coefficients in w and calibration functions in a such that the integral in (3.22) becomes a given surface energy. Similarly, the relation between the $m_{\alpha\beta}(\nu)$ and $\omega(\phi, \nabla\phi)$ becomes much simpler thanks to the new potentials.

4 A homogenized two-scale model for a binary mixture

In this section we apply the theory of homogenization to a simplified physical situation with periodic equiaxed dendritic microstructure which is described by a phase field model for a binary alloy. The resulting model will be a *two-scale model* that consists of a macroscopic heat equation and of microscopic cell problems that describe the evolution of the phases and the microscopic solute transport at each point of the macroscopic domain. In order to justify the formal asymptotic expansion, an estimate is established that compares the solution of the two-scale model to that of the original model.

The phase transition problem to be considered is given by equations (2.26)–(2.28), i.e.,

$$\partial_t T + L\partial_t \Phi - \nabla \cdot (K(\Phi)\nabla T) = 0, \quad (4.1)$$

$$\partial_t C - \nabla \cdot (D_1(\Phi)\nabla C) - \nabla \cdot (D_2(C,\Phi)\nabla \Phi) = 0, \quad (4.2)$$

$$\alpha\xi^2\partial_t \Phi - \xi^2 \Delta\Phi + p'(\Phi) + q(T,C,\Phi) = 0, \quad (4.3)$$

to be solved in the time-space cylinder $Q_{T\Omega} := I_T \times \Omega$ with time interval $I_T := [0,T]$ and domain $\Omega \subset \mathbb{R}^d$. The diffusion tensors are assumed to be Lipschitz-functions of the phase field Φ, they shall be symmetric, $K_{ij} = K_{ji}$, $D_{1,ij} = D_{1,ji}$ for $i,j = 1,\dots,d$, as well as elliptic and bounded,

$$k_0|z|^2 \le K_{ij}z_iz_j \le k_1|z|^2, \quad d_0|z|^2 \le D_{1,ij}z_iz_j \le d_1|z|^2 \quad (4.4)$$

for all $z \in \mathbb{R}^d$ with positive constants $k_0 \le k_1$ and $d_0 \le d_1$ independent of Φ. Here and in the sequel, the sum convention is used. The function D_2 :

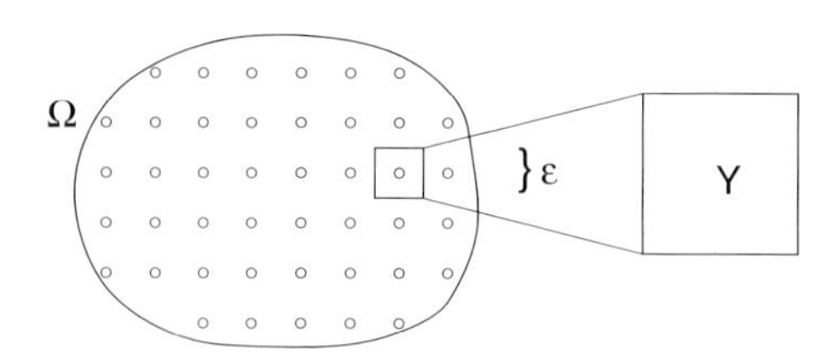

Fig. 4.1. Periodic microstructure

$\mathbb{R}^2 \to \mathbb{R}^{d,d}$ is Lipschitz and bounded. The function p represents the double-well potential $p(\Phi) = \frac{1}{2}(\Phi^2 - 1)^2$, and $q : \mathbb{R}^3 \to \mathbb{R}$ is a Lipschitz function. The differential equations are supplemented by Dirichlet conditions for the temperature and homogeneous Neumann conditions for concentration and phase field,

$$T = T_{ibc}, \quad \big(D_1(\Phi)\nabla C + D_2(C,\Phi)\nabla\Phi\big) \cdot \nu_{ext} = 0, \quad \nabla\Phi \cdot \nu_{ext} = 0 \tag{4.5}$$

on $S_{T\Omega} := I_T \times \partial\Omega$, and by initial conditions

$$T(0,\cdot) = T_{ibc}, \quad C(0,\cdot) = C_{ic} \text{ and } \Phi(0,\cdot) = \Phi_{ic} \tag{4.6}$$

on Ω. For simplicity of the notation the Dirichlet condition and the initial condition of the temperature are given by the same function T_{ibc} that is defined on $Q_{T\Omega}$.

Let us introduce some notation for function spaces. Spaces of functions with continuous derivatives of order β are denoted by $C^\beta(Q)$, $L_r(Q)$ is the Lebesgue space of functions whose r-th power has an integral, $W_r^k(Q)$ is the Sobolev space of functions with derivatives of order k whose r-th power is integrable, and $H^\beta(Q) = W_2^\beta(Q)$. In anisotropic spaces of the type $C^{k,\ell}(I \times Q)$ or $W_r^{k,\ell}(I \times Q)$ with time interval I, the index k refers to the time variable and ℓ to the space variables.

4.1 Asymptotic expansion and the two-scale model

To construct a model that is suitable for a very small scale of the evolving dendritic structures, we consider a sequence of problems of varying scale $\varepsilon > 0$, study the limit $\varepsilon \to 0$ of their solutions, and construct a limit problem that is valid for the limit of these solutions. This limit problem may be used as an approximation for situations with small but non-vanishing scale ε.

This procedure is done for an idealized equiaxed dendritic microstructure that consists of equiaxed crystals growing at the nodes of a uniform grid with edge length given by the scale parameter ε, see Fig. 4.1. This situation is generated by the initial data

$$T_{ibc}^{(\varepsilon)}(x) = T_{ibc}^{(0)}(x), \quad C_{ic}^{(\varepsilon)}(x) = C_{ic}^{(0)}\left(x, \tfrac{x}{\varepsilon}\right) \text{ and } \Phi_{ic}^{(\varepsilon)}(x) = \Phi_{ic}^{(0)}\left(x, \tfrac{x}{\varepsilon}\right) \tag{4.7}$$

with functions $T_{ibc}^{(0)} \in L_2(\Omega)$, $C_{ic}^{(0)}, \Phi_{ic}^{(0)} \in L_2\big(\Omega; C_\#(Y)\big)$. The domain Y is a *unit cell*, by definition this is a bounded, simply connected Lipschitz domain with the property that $\mathbb{R}^d$ can be represented as union of shifted copies of $\overline{Y}$ with no intersection of their interiors. For simplicity of the presentation, the volume of the unit cell is scaled to one. The standard example for Y is the unit cube $Y = [0,1]^d$. The set $C_\#(Y)$ contains all periodic continuous functions in $\mathbb{R}^d$ with periodicity cell Y, the subscript # indicates periodic boundary conditions with respect to $y \in Y$. Condition (4.7) describes instantaneous nucleation at time $t = 0$ of a periodic distribution of nuclei. In order to obtain a well-defined asymptotic limit for $\varepsilon \to 0$, it is necessary to scale some given data in dependence of ε. Here it is assumed

$$\xi = \varepsilon \xi_0, \quad \alpha = \varepsilon^{-2}\alpha_0, \quad \text{and } D_\ell = \varepsilon^2 D_\ell^{(0)}, \ \ell = 1,2. \tag{4.8}$$

The scaling of ξ is obvious: if the size of a solid crystal is proportional to ε, and if we model this crystal by a diffuse interface model, then the width of the diffuse interface must be bounded by const $\cdot\, \varepsilon$ with a constant that is small compared to the size of the crystal. Hence ξ_0 is a small phase field parameter that is fixed in the asymptotic expansion. The relaxation parameter α is scaled such that the total relaxation factor $\alpha\xi^2$ in the phase field equation remains constant. The scaling of the solute diffusivity is motivated by the fact that dendritic structures are created by a competition between a diffusional instability and surface energy. At least one of the diffusivities K or D_1, D_2 has to be scaled in dependence of ε. Since solute diffusivity is usually smaller than heat conductivity, it is natural to scale D_1 and D_2. The fact that D_ℓ and ξ are both scaled proportional to ε^2 does not indicate that they are of similar size: in fact we expect $D_\ell^{(0)}$ to be of the size 1 and ξ_0 to be small compared to $D_\ell^{(0)}$, but the relation D_ℓ/ξ is kept fixed.

In order to study the limit $\varepsilon \to 0$, the existence of an asymptotic expansion

$$u_\varepsilon(t,x) = u_0\big(t,x,\tfrac{x}{\varepsilon}\big) + \varepsilon\, u_1\big(t,x,\tfrac{x}{\varepsilon}\big) + \varepsilon^2 u_2\big(t,x,\tfrac{x}{\varepsilon}\big) + \cdots \quad \text{for } u = T, C, \Phi \tag{4.9}$$

is assumed. The existence of such an asymptotic expansion is not guaranteed. The result of the calculation will be justified in the next section. The gradient of a function $x \mapsto u(x, \frac{x}{\varepsilon})$ is given by $\nabla u = \nabla_x u + \frac{1}{\varepsilon}\nabla_y u$, where ∇_x and ∇_y denote the gradients with respect to the first and second variables of u, respectively. The asymptotic expansions (4.9) and the formal relation $\nabla = \nabla_x + \frac{1}{\varepsilon}\nabla_y$ are used in the differential equations (4.1)–(4.3). Then the coefficients of different powers of ε are compared, starting from the lowest order. For the Φ-dependent conductivities we use a Taylor expansion that is abbreviated by $K_\varepsilon = K_0 + \varepsilon K_1 + \varepsilon^2 K_2 + \cdots$ with $K_0 = K(\Phi_0)$ and analogous expansions for $D_1^{(0)}(\Phi)$, $D_2^{(0)}(C,\Phi)$. The validity of these expansions with a remainder of order ε^β requires $K, D_1^{(0)} \in C^\beta\big(\mathbb{R}; \mathbb{R}^{d,d}\big)$ and $D_2^{(0)} \in C^\beta\big(\mathbb{R}^2; \mathbb{R}^{d,d}\big)$.

The **problem of 1st order** consists of the terms of order ε^{-2} in the heat equation (4.1); these are

$$\begin{aligned} -\nabla_y \cdot \big(K_0 \nabla_y T_0\big) &= 0 \text{ in } Q_{T\Omega Y} := I_T \times \Omega \times Y, \\ T_0 \text{ is } Y&\text{-periodic with respect to } y. \end{aligned}$$

The solutions of this problem are constant with respect to y, hence $T_0(t,x,y) = T_0(t,x)$ is independent of y.

The **problem of 2nd order** is given by the terms of order ε^{-1} in the heat equation,

$$\begin{aligned} -\nabla_y \cdot \big(K_0(\nabla_y T_1 + \nabla_x T_0)\big) &= 0 \text{ in } Q_{T\Omega Y}, \\ T_1 \text{ is } Y&\text{-periodic with respect to } y. \end{aligned}$$

This is a linear elliptic equation for T_1 with right hand side defined in terms of T_0. Its solution can be represented by

$$T_1(t,x,y) = \sum_{j=1}^{d} H_j(t,x,y)\, \partial_{x_j} T_0(t,x)$$

with the solutions H_j of the local cell problem

$$-\nabla_y \cdot \big(K_0 \nabla_y H_j\big) = \nabla_y \cdot \big(K_0\, e_j\big), \quad H_j \text{ is } Y\text{-periodic},$$

where e_j is the j-th unit vector of $\mathbb{R}^d$. Both K_0 and H_j depend on Φ_0.

The **problem of 3rd order** consists of the terms of order ε^0 in the heat equation, the diffusion equation and the phase field equation,

$$\begin{aligned} \partial_t T_0 + L\partial_t \Phi_0 - \nabla_y \cdot \big(K_0(\nabla_y T_2 + \nabla_x T_1) + K_1(\nabla_y T_1 + \nabla_x T_0)\big) \\ -\nabla_x \cdot \big(K_0(\nabla_y T_1 + \nabla_x T_0)\big) = 0, \end{aligned} \tag{4.10}$$

$$\partial_t C_0 - \nabla_y \cdot \big(D_1^{(0)}(\Phi_0)\nabla_y C_0\big) - \nabla_y \cdot \big(D_2^{(0)}(C_0,\Phi_0)\nabla_y \Phi_0\big) = 0, \tag{4.11}$$

$$\alpha_0 \xi_0^2 \partial_t \Phi_0 - \xi_0^2 \Delta_y \Phi_0 + p'(\Phi_0) + q(T_0, C_0, \Phi_0) = 0 \tag{4.12}$$

on $Q_{T\Omega Y}$, supplemented by periodic boundary conditions on ∂Y for T_2, C_0 and Φ_0. Equations (4.11) and (4.12) do not contain any derivatives with respect to x. Hence they can be interpreted as a set of differential equations defined on $Q_{TY} := I_T \times Y$ for every parameter $x \in \Omega$. Equation (4.10) is transformed into a macroscopic equation for $T_0 = T_0(t,x)$ by integration with respect to $y \in Y$. Due to the periodic boundary conditions the ∇_y-term disappears and the homogenized heat equation is obtained,

$$\partial_t T_0 + L\partial_t \overline{\Phi}_0 - \nabla \cdot \big(K^*(\Phi_0)\nabla T_0\big) = 0$$

with solid volume fraction $\overline{\Phi}_0(t,x) := \int_Y \Phi_0(t,x,y)\, dy$ and the effective heat conductivity

$$K_{ij}^*(\Phi_0) := \int_Y \left(K_{ij}(\Phi_0) + \sum_{k=1}^{d} K_{ik}(\Phi_0)\partial_{y_k} H_j(\Phi_0) \right) dy.$$

The effective heat conductivity K^*_{ij} is symmetric, elliptic and bounded with the same constants k_0 and k_1 as the original matrix K, see e.g. [JKO94] or [Ho97].

Let us sum up the obtained two-scale model. It consists of

- The macroscopic heat equation

$$\partial_t\big(T_0 + L\overline{\Phi}_0\big) - \nabla\cdot\big(K^*(\Phi_0)\nabla T_0\big) = 0 \quad \text{in } Q_{T\Omega} = I_T \times \Omega \tag{4.13}$$

with boundary conditions and initial conditions

$$T_0 = T^{(0)}_{ibc} \quad \text{on } S_{T\Omega} = I_T \times \partial\Omega \quad \text{and} \quad T_0(0,\cdot) = T^{(0)}_{ibc} \quad \text{in } \Omega.$$

- The definition of the averaged phase field $\overline{\Phi}_0(t,x) = \int_Y \Phi_0(t,x,y)\,dy$ and the effective heat conductivity

$$K^*_{ij}(\Phi_0) = \int_Y K_{ik}(\Phi_0)\big(\delta_{jk} + \partial_{y_k} H_j(\Phi_0)\big)\,dy \tag{4.14}$$

with the Kronecker symbol δ_{jk} via the solutions $H_j = H_j(\Phi_0)$ of the local cell problems

$$-\nabla_y\cdot\big(K(\Phi_0)(\nabla_y H_j + e_j)\big) = 0 \quad \text{in } Y \tag{4.15}$$

with periodic boundary conditions.
- The microscopic problems

$$\partial_t C_0 - \nabla_y\cdot\big(D_1^{(0)}(\Phi_0)\nabla_y C_0\big) - \nabla_y\cdot\big(D_2^{(0)}(C_0,\Phi_0)\nabla_y\Phi_0\big) = 0, \tag{4.16}$$

$$\alpha_0\xi_0^2\partial_t\Phi_0 - \xi_0^2\Delta_y\Phi_0 + p'(\Phi_0) + q(T_0,C_0,\Phi_0) = 0 \tag{4.17}$$

in $Q_{TY} = I_T \times Y$ with periodic boundary conditions and initial data

$$C_0(0,x,y) = C^{(0)}_{ic}(x,y), \quad \Phi_0(0,x,y) = \Phi^{(0)}_{ic}(x,y) \text{ for } y \in Y.$$

These equations must be solved for every point $x \in \Omega$ of the macroscopic domain.

4.2 Analysis of the two-scale model

The existence of weak solutions to the two-scale model is proved in [Ec04c], Theorem 3.3, by a fixed point approach. Uniqueness of the solution is also proved in [Ec04c], Theorems 3.4 and 3.5. The results can be summed up as:

Theorem 4.1. *Let $\Omega \subset \mathbb{R}^d$ be a C^2-smooth domain of dimension $d = 2$ or $d = 3$, $Y \subset \mathbb{R}^d$ be a unit cell, let $K, D_1 : \mathbb{R} \to \mathbb{R}^{d,d}$ be Lipschitz, symmetric and satisfy the condition (4.4), let $T_{ibc} \in W^{1,2}_r(Q_{T\Omega}) \cap H^1\big(I_T; W^1_s(\Omega)\big)$ with $r > d$, $s > 1$ for $d = 2$ and $s > 6/5$ for $d = 3$, $C_{ic}, \Phi_{ic} \in L_\infty\big(\Omega; W^{2-2/\ell}_{\ell\#}(Y)\big) \cap W^1_r\big(\Omega; L_2(Y)\big)$ with $\ell > 1 + d/2$, $0 \le C_{ic} \le 1$, suppose $D_2 \in C^{0,1}\big(\mathbb{R}^2; \mathbb{R}^{d,d}\big)$ with $D_2(C,\Phi) = 0$ for $C \notin [0,1]$, $p(\Phi) = \frac{1}{2}\big(\Phi^2 - 1\big)^2$, $q : \mathbb{R}^3 \to \mathbb{R}$ is Lipschitz and satisfies the growth condition $|q(T,C,\Phi)| \le \mathrm{const}(1 + |T| + |C| + |\Phi|)$, and let L, ξ, α be positive constants. Then there exists a unique weak solution (T,C,Φ) of the two-scale model (4.13)–(4.17).*

An estimate for the model error is derived in [Ec04b] under appropriate assumptions concerning the regularity of the solutions for both the original model and the two-scale model. Let $(T_\varepsilon, C_\varepsilon, \Phi_\varepsilon)$ denote the solutions of the original model (4.1)–(4.3), (4.5), (4.7) with the scaling (4.8) of the parameters and (T_0, C_0, Φ_0) be the solutions of the two-scale model (4.13)–(4.17) with initial data $T_{ibc}^{(0)}, C_{ic}^{(0)}, \Phi_{ic}^{(0)}$. The error estimate is done in terms of *macroscopic reconstructions of scale* ε for the solutions of the two-scale model:

$$u_0^\varepsilon(t,x) := u_0(t,x,x/\varepsilon) \quad \text{for } u \in \{T, C, \Phi\}.$$

The required regularity for the solutions of the original model is

$$\begin{aligned} &\|T_\varepsilon\|_{H^{1/2,1}(Q_{T\Omega})} + \|T_\varepsilon\|_{L_\infty(I_T;L_2(\Omega))} + \varepsilon\|C_\varepsilon\|_{H^{1/2,1}(Q_{T\Omega})} \\ &\quad + \varepsilon\|\Phi_\varepsilon\|_{H^{1/2,1}(Q_{T\Omega})} + \|C_\varepsilon\|_{L_\infty(Q_{T\Omega})} + \|\Phi_\varepsilon\|_{L_\infty(Q_{T\Omega})} \le \text{const}_1 \end{aligned} \tag{4.18}$$

with a constant const_1 independent of ε. The solution of the two-scale model is supposed to satisfy

$$\begin{aligned} &T_0 \in W_r^{1,2}(Q_{T\Omega}) \cap H^{1/2+\beta}\big(I_T; H^1(\Omega)\big), \\ &C_0, \Phi_0 \in L_\infty\big(\Omega; C^{1,2}(Q_{TY})\big), \quad \nabla_x C_0, \nabla_x \Phi_0 \in L_\infty\big(\Omega; W_s^{1,2}(Q_{TY})\big) \end{aligned} \tag{4.19}$$

with parameters $r > d+2$, $s > d$ and $\beta > 0$.

Theorem 4.2. *Let $\Omega \subset \mathbb{R}^d$ be a bounded Lipschitz domain and Y be a unit cell, let $K, D_1^{(0)} \in C^2\big(\mathbb{R}; \mathbb{R}^{d,d}\big)$ be bounded and elliptic as described in* (4.4), *$D_2^{(0)} \in C^{0,1}\big(\mathbb{R}^2; \mathbb{R}^{d,d}\big)$ be bounded, $q : \mathbb{R}^3 \to \mathbb{R}$ be globally Lipschitz and $p(\Phi) = \frac{1}{2}\big(\Phi^2 - 1\big)^2$. The solutions of the original model and the two-scale model satisfy the regularity properties* (4.18) *and* (4.19)*. Let*

$$T_1^\varepsilon(t,x) := T_0(t,x) + \varepsilon H_j\big(t,x,\tfrac{x}{\varepsilon}\big)\, \partial_{x_j} T_0(t,x)$$

be the first order term in the asymptotic expansion for the temperature. Then

$$\begin{aligned} &\|T_\varepsilon - T_0\|_{L_\infty(I_T;L_2(\Omega))} + \|C_\varepsilon - C_0^\varepsilon\|_{L_\infty(I_T;L_2(\Omega))} + \|\Phi_\varepsilon - \Phi_0^\varepsilon\|_{L_\infty(I_T;L_2(\Omega))} \\ &\quad + \|T_\varepsilon - T_1^\varepsilon\|_{L_2(I_T;H^1(\Omega))} \le \text{const}\, \varepsilon^{1/2} \end{aligned}$$

with const independent of ε.

This theorem guarantees the order of approximation $\varepsilon^{1/2}$ for the two-scale model. The exponent of ε is limited to $1/2$, because the two-scale model does not approximate the original model of scale ε close to the boundary of the domain. It must be expected that the domain of an equiaxed dendritic crystal growing close to the boundary is not a full shifted copy of εY, but a subdomain obtained by intersecting with Ω. This generates an additional error of order $\varepsilon^{1/2}$.

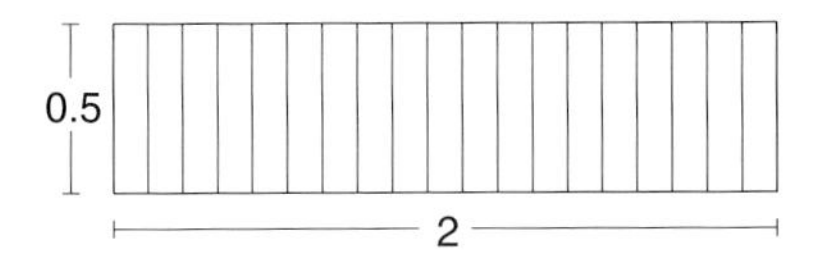

Fig. 4.2. Macroscopic domain of the numerical example

4.3 Numerical example

In order to illustrate the two-scale model we present the results of numerical computations for two space dimensions. The computations are done with constant heat conductivity $K = 1$ – hence no elliptic cell problem must be solved, – constant solute diffusivity $D_1(\Phi) = 1$, $D_2(C, \Phi) = -0.05$ and latent heat $2L = 1$. The function q in the phase field model is given by

$$q(\nabla_y \Phi, T, c, \Phi) = \left(1 - \Phi^2\right) \cdot 1.2 \cdot \arctan\left(\tfrac{\xi}{1.2 \cdot \sigma(\nabla_y \Phi)} \left(T + 10 \cdot C - \tfrac{1}{2}\Phi - 2\right)\right).$$

The quantity σ here is correlated with the surface tension for the sharp interface limit $\xi \to 0$. Its dependence on $\nabla_y \Phi$ is introduced in order to describe the dependence of the surface tension on the orientation of the surface. The problem can be reformulated in terms of the function $\mu = 10 \cdot C - \frac{1}{2}\Phi - 2$ that plays the role of a chemical potential; the diffusion equation then takes the form

$$\partial_t \left(\mu + \tfrac{1}{2}\Phi\right) - \Delta_y \mu = 0,$$

and the constitutive function q in the phase field equation is

$$q(\nabla_y \Phi, T, \mu, \Phi) = \left(1 - \Phi^2\right) \cdot 1.2 \cdot \arctan\left(\tfrac{\xi}{1.2 \cdot \sigma(\nabla_y \Phi)} (\mu + T)\right).$$

The precise form of $\sigma(\nabla \Phi)$ is

$$\sigma(\nabla \Phi) = \sigma_0 \left(1 - (m^2 - 1)\sigma_1 \cos(m(\Theta(\nabla \Phi) - \Theta_0))\right),$$

where σ_0 describes the average value, σ_1 is the strength of the anisotropy, m describes the symmetry pattern of the dendrites, $\Theta(\nabla \Phi)$ is the angle between $\nabla \Phi$ and the x_1-axis and Θ_0 is an offset angle. The special choice of q involving the arctan function is chosen in order to ensure that the minima of the potential for fixed $T, \mu, \nabla_y \Phi$ are kept at $\Phi = \pm 1$, even for large deviations from the equilibrium melting point; following the proposition of Kobayashi [Ko93].

Both the global heat equation and the microscopic problems are discretized by bilinear finite elements on uniform rectangular grids. The equations are decoupled by the time discretization in the following way: first a partially linearized version of the phase field equation is solved with temperature, concentration and $\nabla_y \Phi$ taken from the previous time step, then the diffusion equation is solved. This is done for every grid point of the macroscopic grid, then the global heat equation is solved. The decoupled linear equations are discretized with respect to time by the Crank-Nicolson scheme. This gives

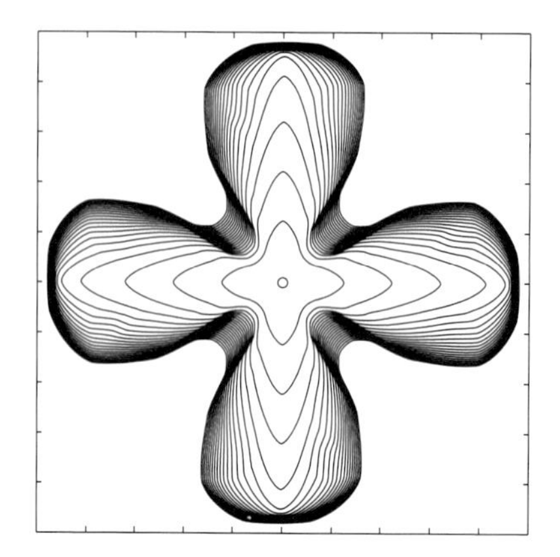
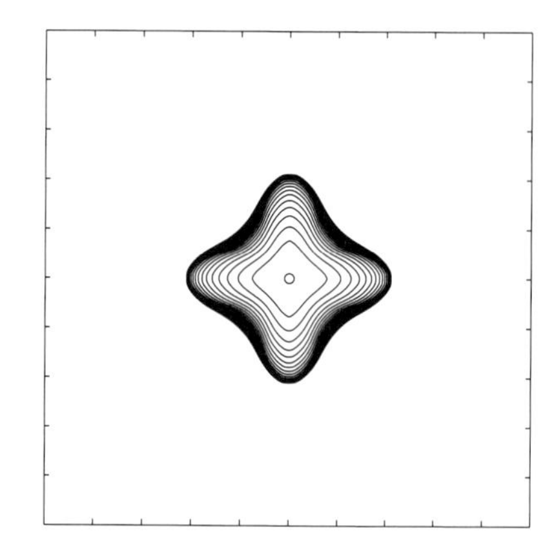

Fig. 4.3. Evolution of left and right crystal for $\Theta_0 = 0$

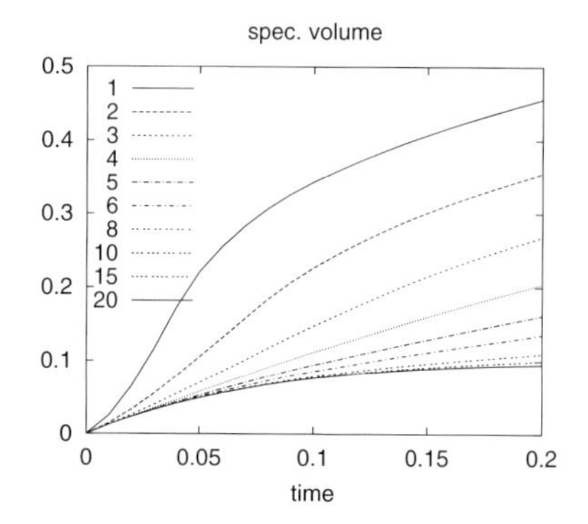

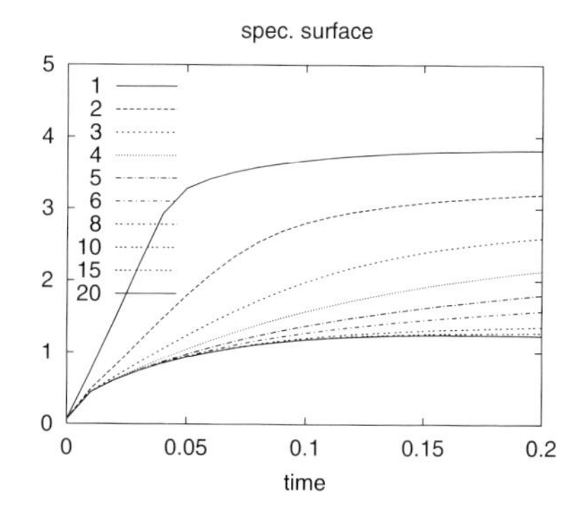

Fig. 4.4. Evolution of specific data for $\Theta_0 = 0$

a semi-implicit time-discretization of the two-scale model, with an implicit discretization of the main parts of the differential operators.

The examples to be presented are computed for $\sigma_0 = 0.0002$, $m = 4$, $\xi = 0.005$, $\alpha = 5$ and $\sigma_1 = 0.05$. The initial conditions are $T = -0.1$ and $\mu(T, c, \Phi) = -0.1$, this adds up to a total initial undercooling of -0.2. The unit cell for the microscopic problem is $Y = [0, 1]^2$, the initial solid nucleus is a sphere of radius $r = 0.05$ located at the midpoint $(0.5, 0.5)$ of Y. The boundary conditions are periodic boundary conditions for the microscopic problems and given heat fluxes for the macroscopic equations. The macroscopic domain is $\Omega = [0, 2] \times [0, 0.5]$; we prescribe homogeneous heat fluxes $\nabla T \cdot \nu_{ext} = 0$ on $[0, 2] \times \{0\}$, $\{2\} \times [0, 0.5]$ and $[0, 2] \times \{0.5\}$, on the remaining part $\{0\} \times [0, 0.5]$ of the boundary we prescribe the heat flux $\nabla T \cdot \nu_{ext} = -1$. The macroscopic equation is discretized by a uniform rectangular grid with 19×1 elements; it is essentially one-dimensional, the crystals evolving at the same x_1-position are equal. The microscopic problems are solved with uniform rectangular grids

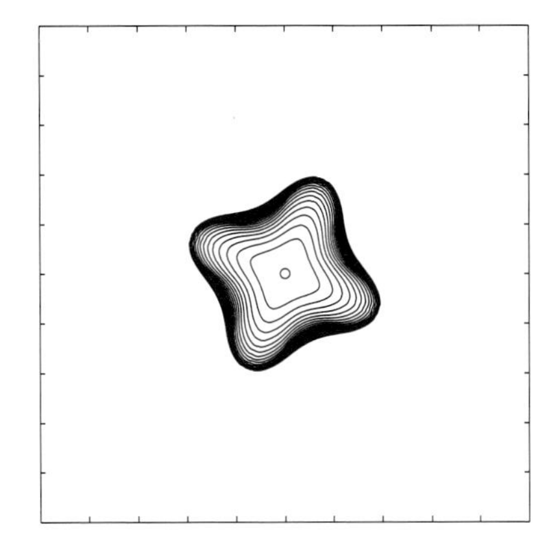

Fig. 4.5. Evolution of left and right crystal for $\Theta_0 = 0.4$

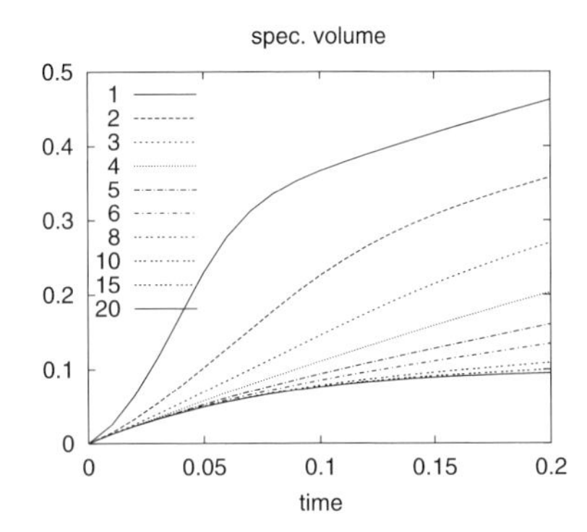

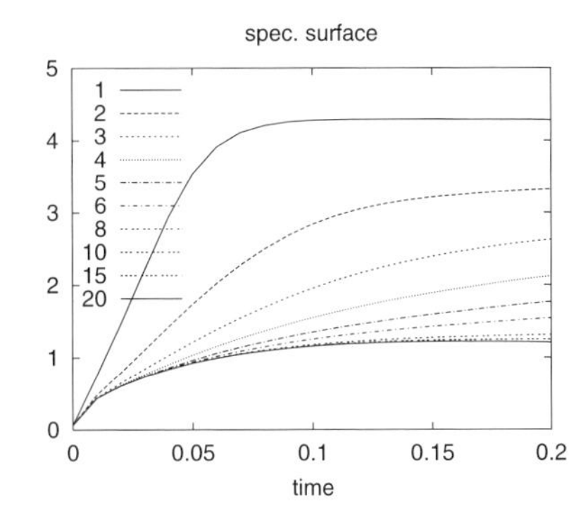

Fig. 4.6. Evolution of specific data for $\Theta_0 = 0.4$

of 300×300 elements. The time step is $\Delta t = 2 \cdot 10^{-5}$, the final time of all computations is $t = 0.2$.

The figures show the results for three different orientations $\Theta_0 = 0$, $\Theta_0 = 0.4$ and $\Theta_0 = \pi/4$ for the anisotropy of the surface tension. Figures 4.3, 4.5 and 4.7 show the evolution of the left and right crystals from the initial time $t = 0$ to $t = 0.2$ in twenty steps. The left crystal is that growing at $x_1 = 0$, the right that at $x_1 = 2$. Due to the boundary cooling at $x_1 = 0$ the left crystal grows quickly, whereas the right one evolves rather slowly; its driving force is limited to the initial undercooling. For the left crystal, the offset angle $\Theta_0 = 0$ leads to shorter dendrites than the other angles, here the interaction of neighboring crystals happens earlier than in the cases $\Theta_0 = 0.4$ and $\Theta_0 = \pi/4$. This effect is not visible for the right crystals which are in an early stage of their evolution. In Fig. 4.4, 4.6 and 4.8 the evolution of the specific data (specific volume and specific surface) is depicted for selected crystals in a row in x_1-direction, the number corresponds to the position of the crystal, starting with position 1 at $x_1 = 0$. Further examples are presented in [Ec04a] and [Ec04b].

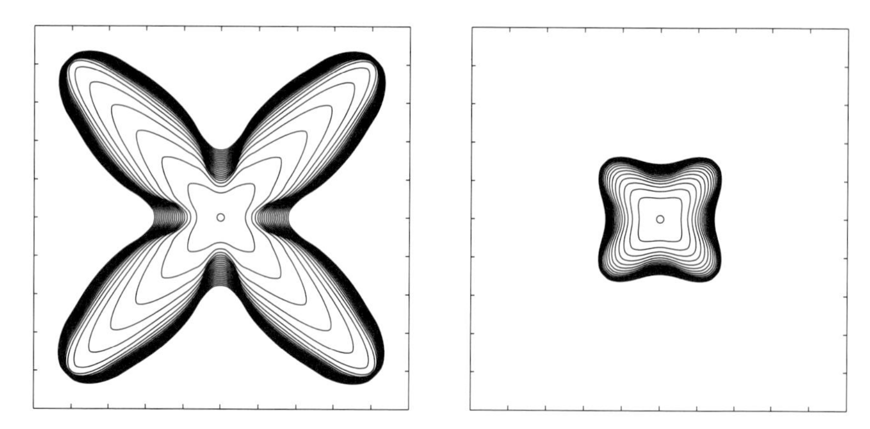

Fig. 4.7. Evolution of left and right crystal for $\Theta_0 = \pi/4$

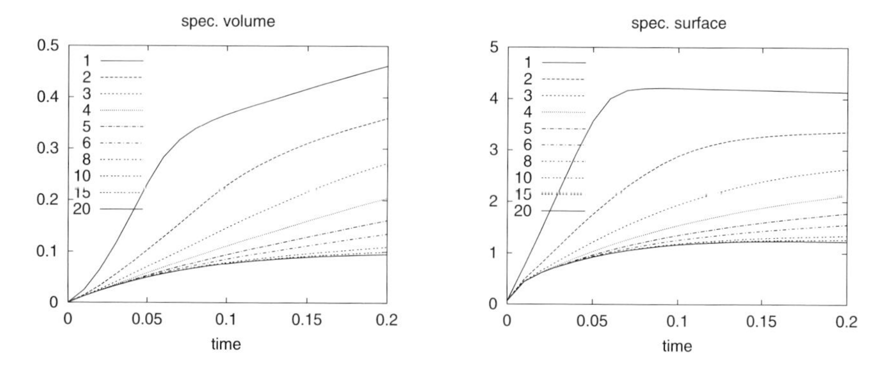

Fig. 4.8. Evolution of specific data for $\Theta_0 = \pi/4$

4.4 Some remarks on the numerical analysis

Error estimates for simple finite element discretizations of both the original model for scale ε and the two-scale model are derived and compared in [Ec02]. For linear or bilinear finite elements on a grid with mesh size h and a discretization with respect to time by an implicit Euler scheme with time step Δt, the error for the original model of scale ε is

$$\text{const}_1 \left(\left(\frac{h}{\varepsilon} \right)^2 + \Delta t \right).$$

This estimate reveals the typical convergence properties of the chosen discretization for parabolic equations: convergence of second order with respect to the space variables and of first order with respect to time. The factor $1/\varepsilon$ of the mesh size h accounts for the obvious fact that the microstructure starts to be properly resolved for $h \ll \varepsilon$ only. The discretization of the two-scale model uses a global grid for the macroscopic heat equation defined on Ω, at each node of this global grid the local cell problem defined on Y is solved with a local grid for Y. The mesh size of both grids is related to h_0, the time step is again Δt. Then the error estimate for the two-scale model is

$$\mathrm{const}_2 \left(h_0^2 + \Delta t \right) .$$

Obviously no dependence on ε is present here. In order to have comparable computational complexity, the mesh sizes h_0 for the two-scale model and h for the original model should scale according to $h_0 \sim \sqrt{h}$. Respecting also the model error of order $\varepsilon^{1/2}$ for the two-scale model we conclude that the two-scale model is superior, if the mesh size used for the original model is larger than the threshold $h_\varepsilon = \mathrm{const}_3 \varepsilon^{5/4}$ with a suitable const_3.

4.5 Conclusion

The presented two-scale model is an approximate model for a problem with scale ε of the microstructure, with increasing accuracy for decreasing ε. Numerical computations with this model are valid for a whole range of microscale parameters $\varepsilon \in (0, \varepsilon_0]$ with the appropriate diffusivities. The model is suitable for material with *fast* heat diffusion and *slow* solute diffusion, where the temperature is assumed to be essentially constant on the microscopic scale, while the solute transport is neglected on the macroscopic scale.

Extensions of the presented two-scale model may be possible for more complex physical phenomena, for example phase transitions with convection, and physically more realistic situations, in particular for non-periodic microstructures. The extension to models with convection is probably possible by the application of techniques similar to those presented here to available phase field models that include convection, see e.g. [AMW00], [BD*99], [NW*00]. For phase transitions with density differences between solid and liquid – where convection cannot be avoided – it may be necessary to use a unit cell that is fixed in Lagrangian coordinates but moves and deforms with the flow in an Eulerian description. Non-periodic microstructures can be described by a probabilistic description of the initial conditions, then it is possible to apply techniques of random homogenization of the type described in [JKO94]. A corresponding stochastic version of the two-scale model can be found in [Ec04a].

A Facts on evolving surfaces and transport identities

Let $I_\mathcal{T} = (0, \mathcal{T}) \subset \mathbb{R}$ be a time interval and let $m, d \in \mathbb{N}$ with $m \leq d$.

Definition A.1. *$(\Sigma_t)_{t \in I_\mathcal{T}}$ is an evolving m-dimensional surface in $\mathbb{R}^d$ if*

1. *for each $t \in I_\mathcal{T}$, the surface Σ_t can be parameterized over a fixed smooth orientable submanifold $U \subset \mathbb{R}^{m+1}$,*

2. *the set* $\Sigma' := \{x' = (t,x) : t \in I_T, x \in \Sigma_t\} \subset \mathbb{R} \times \mathbb{R}^d$ *is a smooth* $m+1$*-dimensional surface,*
3. *the tangent space* $\mathrm{T}_{x'}\Sigma'$ *is nowhere purely spatial, i.e.,* $\mathrm{T}_{x'}\Sigma' \neq \{0\} \times V$ *with* $V \cong \mathbb{R}^m$.

The spatial tangent space of dimension m in $x \in \Sigma_t$ is denoted by $\mathrm{T}_x\Sigma_t$, the spatial normal space of dimension $d-m$ by $\mathrm{N}_x\Sigma_t := (\mathrm{T}_x\Sigma_t)^\perp$. There is a unique vector field $\mathbf{v}_\Sigma : \Sigma' \to \mathbb{R}^{d+1}$ such that $(1, \mathbf{v}_\Sigma(t,x)) \in \mathrm{T}_{x'}\Sigma'$ and $\mathbf{v}_\Sigma(t,x) \in \mathrm{N}_x\Sigma_t$; $\mathbf{v}_\Sigma(t,x)$ is the vectorial normal velocity of the evolving surface. It can be verified that

$$\mathrm{T}_{x'}\Sigma' = \{(s, s\mathbf{v}_\Sigma(x')) + (0,\tau) : s \in \mathbb{R}, \tau \in \mathrm{T}_x\Sigma_t\},$$
$$\mathrm{N}_{x'}\Sigma' = \{(-\mathbf{v}_\Sigma(x') \cdot \nu, \nu) : \nu \in \mathrm{N}_x\Sigma_t\}.$$

Let φ be a smooth scalar field on Σ'. The derivative

$$\partial^\circ \varphi(x') := \partial_{(1,\mathbf{v}_\Sigma(x'))}\varphi(x') \text{ in } x' = (t,x) \in \Sigma', \tag{A.1}$$

is the normal time derivative of φ in x' and describes the variation of φ when following the curve $\delta \mapsto c(\delta) \in \Sigma_{t+\delta}$ defined by $c(0) = x$ and $\partial_\delta c(\delta) = \mathbf{v}_\Sigma(t+\delta, c(\delta))$, $\delta \in (t-\delta_0, t+\delta_0)$ with some small $\delta_0 > 0$.

Let $(\tau_k(t,x))_{k=1}^m$ be an orthonormal basis of $\mathrm{T}_x\Sigma_t$. By $\partial_{\tau_k}\varphi(x)$ the differential of φ into direction $(0,\tau_k) \in \mathrm{T}_{x'}\Sigma'$ is denoted. The surface gradient of φ in x' is defined by $\nabla_\Sigma\varphi(x') := \sum_{k=1}^m \partial_{\tau_k}\varphi(x')\tau_k \in \mathrm{T}_x\Sigma_t$. Let $\boldsymbol{\varphi}$ be a smooth vector field on Σ'. The surface divergence of $\boldsymbol{\varphi}$ in x' is defined by $\nabla_\Sigma \cdot \boldsymbol{\varphi}(x') := \sum_{k=1}^m \partial_{\tau_k}\boldsymbol{\varphi}(x') \cdot \tau_k$.

If $m = d-1$ the normal space $\mathrm{N}_x\Sigma_t$ has dimension one, and Σ' is orientable. Then there is a smooth vector field ν_Σ of unit normals, $\nu_\Sigma(x') \in \mathrm{N}_x\Sigma_t$, $|\nu_\Sigma(x')|_2 = 1$. The (scalar) curvature and the curvature vector then are defined by

$$\kappa_\Sigma := -\nabla_\Sigma \cdot \nu_\Sigma, \qquad \boldsymbol{\kappa}_\Sigma := \kappa_\Sigma\nu_\Sigma. \tag{A.2}$$

Moreover, the (scalar) normal velocity then is defined by

$$v_\Sigma = \mathbf{v}_\Sigma \cdot \nu_\Sigma, \tag{A.3}$$

and the following relation, derived in [Gu00], Chapter 15b, holds:

$$\partial^\circ \nu_\Sigma = -\nabla_\Sigma v_\Sigma. \tag{A.4}$$

Definition A.2. $\Gamma' := (\Gamma_t)_t$ *is an evolving* m*-dimensional subsurface of* Σ' *if*

1. *the set* Γ_t *is a relatively open connected subset of* Σ_t *for each* $t \in I_T$,
2. *the boundary* $\partial\Gamma' := (\partial\Gamma_t)_t$ *consists of a finite number of evolving* $m-1$*-dimensional surfaces such that, locally for each* $t \in I_T$, $\partial\Gamma_t$ *is the graph of a Lipschitz continuous map.*

A vectorial normal velocity $\mathbf{v}_{\partial\Gamma}$ can be assigned to the pieces of $\partial\Gamma'$ while Γ' obviously has the same vectorial normal velocity as Σ', namely $\mathbf{v}_\Sigma$.

In some point $x \in \partial\Gamma_t$ the tangent cone on Γ_t is denoted by $\mathrm{T}_x\Gamma_t$. If x is in the interior of one of the pieces the cone is a half-space of $\mathrm{T}_x\Sigma_t$. Besides then the boundary of $\mathrm{T}_x\Gamma_t$ in $\mathrm{T}_x\Sigma_t$ coincides with the tangent space of the boundary $\partial\Gamma_t$, i.e., $\partial\mathrm{T}_x\Gamma_t = \mathrm{T}_x\partial\Gamma_t$. In such points x there is a unique unit vector $\tau_\Gamma \in \mathrm{T}_x\Sigma_t \cap \mathrm{N}_x\partial\Gamma_t$ with $\tau_\Gamma \cdot \tilde{\tau} \leq 0$ for all $\tilde{\tau} \in \mathrm{T}_x\Gamma_t$. This vector τ_Γ is said to be the external unit normal of Γ_t with respect to Σ_t.

Let $m = d-1$ and $d \leq 3$. First, a divergence theorem is stated for a smooth surface with piecewise smooth Lipschitz boundary like Γ_t as in Definition A.2:

Theorem A.3. ([Be86], Corollary 4) *In the above situation there is the following identity:*

$$\int_{\Gamma_t} (\nabla_\Sigma \cdot \boldsymbol{\varphi} + \boldsymbol{\kappa}_\Sigma \cdot \boldsymbol{\varphi})\, d\mathcal{H}^m(x) = \int_{\partial\Gamma_t} \boldsymbol{\varphi} \cdot \tau_\Gamma d\mathcal{H}^{m-1}.$$

If $\boldsymbol{\varphi}$ is a tangent vector field then $\boldsymbol{\kappa}_\Sigma \cdot \boldsymbol{\varphi} = 0$ so that one gets the usual divergence theorem on surfaces. It should be remarked that the proof in [Be86] is performed for smooth $\partial\Gamma_t$ but there is a brief note on the above case of a piecewise smooth boundary at the end of Sect. II(2). Finally, a transport identity is stated:

Theorem A.4. ([Be86], Theorem 1) *In the above situation it holds for every $t \in I_T$ that*

$$\frac{d}{dt}\left(\int_{\Gamma_t} \varphi\, d\mathcal{H}^m\right)\bigg|_t = \int_{\Gamma_t} (\partial^\circ\varphi - \varphi\mathbf{v}_\Sigma \cdot \boldsymbol{\kappa}_\Sigma)\, d\mathcal{H}^m + \int_{\partial\Gamma_t} (\varphi\mathbf{v}_{\partial\Gamma} \cdot \tau_\Gamma)\, d\mathcal{H}^{m-1}.$$

Remark A.5. If $\mathbf{v}_\Sigma = 0$ and $\boldsymbol{\kappa}_\Sigma = 0$ then Γ_t is flat, ∂° reduces to ∂_t and $\mathbf{v}_{\partial\Gamma}$ is tangential. Altogether, the Reynold's transport theorem is obtained.

References

[Al99] Almgren, R.F.: Second-order phase field asymptotics for unequal conductivities. SIAM J. Appl. Math. **59**, 2086–2107 (1999)

[AP92] Alt, H.W., Pawlow, I.: Existence of solutions for non-isothermal phase separation. Adv. Math. Sci. Appl. **1**, 319–409 (1992)

[AP96] Alt, H.W., Pawlow, I.: On the entropy principle of phase transition models with a conserved order parameter. Adv. Math. Sci. Appl. **6** 291–376 (1996)

[An02] Andersson, C.: Third order asymptotics of a phase-field model. TRITA-NA-0217, Dep. of Num. Anal. and Comp. Sc., Royal Inst. of Technology, Stockholm (2002)

[AMW00] Anderson, D.M., McFadden, G.B., Wheeler, A.A: A phase field model for solidification with convection. Physica D **135**, 175–194 (2000)

[BGN06] Barrett, J.W., Garcke, H., Nürnberg, R.: On the Variational Approximation of Combined Second and Fourth Order Geometric Evolution Equations. Preprint No. 07/2006, Faculty of Mathematics, University of Regensburg.

[BBR05] Bellettini, G., Braides, A., Riey, G.: Variational Approximation of Anisotropic Functionals on Partitions. Ann. Mat. **184**, 75–93 (2002)

[BD*99] Beckermann, C., Diepers, H.-J., Steinbach, I., Karma, A., Tong, X.: Modeling melt convection in phase field simulations of solidification. J. Comput. Phys. **154**, 468–496 (1999)

[Be86] Betounes, D.E.: Kinematics of submanifolds and the mean curvature normal. Arch. Rat. Mech. Anal. **96**, 1–27 (1986)

[BE91] Blowey, J.F., Elliott, C.M.: The Cahn-Hilliard gradient theory for phase separation with nonsmooth free energy. I. Mathematical analysis. Europ. J. Appl. Math. **2**, 233–280 (1991)

[BS96] Brokate, M, Sprekels, J.: Hysteresis and Phase Transitions. Appl. Math. Sc. **121**, Springer (1996)

[BGS98] Bronsard, L., Garcke, H., Stoth, B.: A multi-phase Mullins–Sekerka system: Matched asymptotic expansions and an implicit time discretization for the geometric evolution problem. Proc. Roy. Soc. Edinburgh Sect. A **128**, 481–506 (1998)

[CC98] Caginalp, G., Chen, X.: Convergence of the phase field model to its sharp interface limits. Europ. J. Appl. Math. **9**, 417–445 (1998)

[CF88] Caginalp, G., Fife, P.C.: Dynamics of layered interfaces arising from phase boundaries. SIAM J. Appl. Math. **48**, 506–518 (1988)

[Ca89] Caginalp, G.: Stefan and Hele Shaw type models as asymptotic limits of the phase field equations. Phys. Rev. A **39**, 5887–5896 (1989)

[CH74] Cahn, J.W., Hoffmann, D.W.. A vector thermodynamics for anisotropic surfaces II. Curved and faceted surfaces. Acta Metall. Mater., **22**, 1205–1214 (1974)

[Da01] Davis, S.H.: Theory of solidification. Cambridge University Press (2001)

[DS95] De Mottoni, P., Schatzman, M.: Geometrical evolution of developed interfaces. Trans. Amer. Math. Soc. **347**, 1533–1589 (1995)

[Di04] Dirr, N.: A Stefan problem with surface tension as the sharp interface limit of a nonlocal system of phase-field type. J. Statist. Phys. **114**, 1085–1113 (2004)

[DW05] Dreyer, W., Wagner, B.: Sharp-interface model for eutectic alloys, Part I: Concentration dependent surface tension. Interf. Free Bound. **7**, 199–227 (2005)

[Ec02] Eck, C.: Finite element error estimates for a two–scale phase field model describing liquid–solid phase transitions with equiaxed dendritic microstructure. SPP 1095 Mehrskalenprobleme, Preprint 67 (2002) http://www1.am.uni-erlangen.de/ẽck/papers/pfhee.pdf

[Ec04a] Eck, C.: A two-scale phase field model for liquid-solid phase transitions of binary mixtures with dendritic microstructure. Habilitation Thesis, Universität Erlangen (2004)

[Ec04b] Eck, C.: Homogenization of a phase field model for binary mixtures. Multiscale Modeling and Simulation **3**(1), 1-27 (2004)

[Ec04c] Eck, C.: Analysis of a two-scale phase field model for liquid-solid phase transitions with equiaxed dendritic microstructures. Multiscale Modeling and Simulation **3**(1), 28-49 (2004)

[FP90] Fife, P.C., Penrose, O.: Thermodynamically consistent models of phase-field type for the kinetics of phase transitions. Physica D **43**, 44–62 (1990)

[FP95] Fife, P.C., Penrose, O.: Interfacial dynamics for thermodynamically consistent phase-field models with nonconserved order parameter. EJDE 1995 **16**, 1–49 (1995)

[GHS06] Garcke, H. Haas, R., Stinner, B.: On Ginzburg-Landau type free energies for multi-phase systems. In preparation.

[GNS04] Garcke, H., Nestler, B., Stinner, B.: A diffuse interface model for alloys with multiple components and phases. SIAM J. Appl. Math., **64**, 775–799 (2004)

[GNS98] Garcke, H., Nestler, B., Stoth, B.: On anisotropic order parameter models for multi-phase systems and their sharp interface limits. Phys. D **115**, 87–108 (1998)

[GNS99] Garcke, H., Nestler, B., Stoth, B.: A multi phase field concept: numerical simulations of moving phase boundaries and multiple junctions. SIAM J. Appl. Math. **60**, 295–315 (1999)

[GS06] Garcke, H., Stinner, B.: Second order phase field asymptotics for multi-component systems. To appear in Interf. Free Bound. (2006)

[Gi77] Giusti, E.: Minimal surfaces and functions of bounded variation. Notes on Pure Math. **10**, Austr. Nat. Univ. Canberra (1977)

[Gu00] Gurtin, M.E.: Configurational forces as basic concepts of continuum physics. Appl. Math. Sc. **137**, Springer (2000)

[Ha94] Haasen, P.: Physikalische Metallkunde. 3. ed., Springer (1994)

[Ha06] Haas, R.: Modelling and analysis for general non-isothermal convective phase field systems. In preparation.

[Ho97] Hornung, U. (ed.): Homogenization and Porous Media. Springer, New York (1997)

[JKO94] Jikov, V.V, Kozlov, S.M., Oleinik, O.A: Homogenization of Differential Operators and Integral Functionals, Springer, Berlin - Heidelberg (1994)

[KR98] Karma, A., Rappel, J.-W.: Quantitative phase-field modeling of dendritic growth in two and three dimensions. Phys. Rev. E **57**, 4323–4349 (1998)

[KY87] Kirkaldy, J.S., Young, D.J.: Diffusion in the condensed state. The Institute of Metals, London (1987)

[Ka01] Karma, A.: Phase-field formulation for quantitative modeling of alloy solidification. Phys. Rev. Lett. **87**, 115701–1-4 (2001)

[Ko93] Kobayashi, R.: Modeling and numerical simulation of dendritic crystal growth. Physica D **63** 410–423 (1993)

[KRS05] Krejči, P., Rocca, E., Sprekels, J.: Nonlocal temperature-dependent phase-field models for non-isothermal phase transitions. WIAS Preprint No. 1006 (2005)

[LG50] Landau, L.D., Ginzburg, V.I.: K teorii sverkhrovodimosti. Zh. Eksp. Teor. Fiz **20**, 1064–1082 (1950), english translation: On the theory of superconductivity, Collected Papers of L.D. Landau, D. ter Haar (ed.), Pergamon, Oxford, UK, 626–633 (1965)

[LV83] Luckhaus, S., Visintin, A.: Phase transition in multicomponent systems. Man. Math. **43**, 261–288 (1983)

[MWA00] McFadden, G.B., Wheeler, A.A., Anderson, D.M.: Thin interface asymptotics for an energy/entropy approach to phase-field models with unequal conductivities. Phys. D **144**, 154–168 (2000)

[Mu01] Müller, I.: Grundzüge der Thermodynamik. 3. ed., Springer (2001)

[NW*00] Nestler, B., Wheeler, A.A., Ratke, L., Stöcker, C.: Phase-field model for solidification of a monotectic alloy with convection. Physica D **141**, 133–154 (2000)

[On31] Onsager, L.: Reciprocal relations in irreversible processes I. Phys. Rev., **37**, 405–426 (1931)

[PF90] Penrose, O., Fife, P.C.: Thermodynamically consistent models of phase field type for the kinetics of phase transition. Phys. D **43**, 44–62 (1990)

[RB*04] Ramirez, J.C., Beckermann, C., Karma, A., Diepers, H.-J.: Phase-field modeling of binary alloy solidification with coupled heat and solute diffusion. Phys. Rev. E **69**, 51607–1-16 (2004)

[SZ03] Sprekels, J., Zheng, S.: Global existence and asymptotic behaviour for a nonlocal phase-field model for non-isothermal phase transitions. J. Math. Anal. Appl. **279**, 97–110 (2003)

[St91] Sternberg, P.: Vector-valued local minimizers of nonconvex variational problems. Rocky Mountain J. Math. **21**, 799–807 (1991)

[St05a] Stinner, B.: Surface energies in multi phase systems with diffuse phase boundaries. To appear in the Proc. of the conference on Free Boundary Problems in Coimbra, Portugal (2005)

[St05b] Stinner, B.: Derivation and analysis of a phase field model for alloy solidification. Doctoral Thesis, Universität Regensburg (2005).

[St96] Stoth, B.: A sharp interface limit of the phase field equations, one-dimensional and axisymmetric. Europ. J. Appl. Math. **7**, 603–633 (1996)

[Vi96] Visintin, A.: Models of phase transitions. Progr. Nonlin. Diff. Eq. Appl. **28**, Birkhäuser, Boston (1996)

[WMB92] Wheeler, A.A., McFadden, G.B., Boettinger, W.J.: Phase-field model for isothermal phase transitions in binary alloys, Phys. Rev. A **45**, 7424–7439 (1992)

[WM97] Wheeler, A.A., McFadden, G.B.: On the notion of a ξ-vector and a stress tensor for a general class of anisotropic diffuse interface models. Proc. Roy. Soc. London Ser. A **453**, 1611–1630 (1997)

Multiscale Modeling of Epitaxial Growth: From Discrete-Continuum to Continuum Equations

Lev Balykov[4], Vladimir Chalupecky[3], Christof Eck[1], Heike Emmerich[2], Ganeshram Krishnamoorthy[2], Andreas Rätz[4], and Axel Voigt[4,5]

[1] Institut für Angewandte Mathematik, Universität Erlangen-Nürnberg, Martensstraße 3, 91058 Erlangen. eck@am.uni-erlangen.de
[2] Computational Materials Engineering, Center for Computational Engineering Science and Institute of Minerals Engineering, RWTH Aachen. emmerich@ghi.rwth-aachen.de, ganesh@ghi.rwth-aachen.de
[3] Department of Mathematics, Faculty of Nuclear Sciences and Physical Engineering, Czech Technical University. chalupec@kmlinux.fjfi.cvut.cz
[4] Crystal Growth group, research center caesar, Ludwig-Erhard-Allee 2, 53175 Bonn. balykov@caesar.de, raetz@caesar.de, voigt@caesar.de
[5] Institut für Angewandte Mathematik, Universität Bonn, Wegelerstraße 10, 53115 Bonn.

Summary. Imposed by the crystal lattice, at the surface of a crystal, there exist atomic steps, which separate exposed lattice planes that differ in height by a single lattice spacing. These steps are long-living lattice defects, which make them suitable as a basis for the description of surface morphology on a mesoscopic length scale and thus are an ideal approach to overcome the different length scales, which range from several atoms in lateral direction to micrometers in horizontal direction. This paper summerizes an approach how the thermodynamics and kinetics of atomic steps can be coarse grained to continuum models for the evolving surface. We discuss phase-field approximations to the step dynamics model and apply them to various growth procedures.

1 Introduction

A deep understanding of the fundamental physics underlying epitaxial growth techniques is a prerequisite for any significant success in semiconductor nanotechnology. The frontiers in developing novel electronic devices lie in the realm of the length scales from nanometers to microns. Controlling the surface morphology during epitaxial growth at such scales is a challenging task. In this regime kinetic effects on an atomistic scale strongly influence the morphology. Thus a continuum theory cannot describe the growth process appropriately. On the other hand the applicability of discrete models on an atomistic scale is limited due to its high computational cost if time and length scales related to device applications are considered.

As first pointed out by Burton, Cabrera and Frank [7] an intermediate regime between atomistic and continuum modeling can be used to combine the appropriate atomistic properties with the computational ease of continuum models. At crystalline surfaces the evolution of the steps, their advancement, nucleation and annihilation, can be used to describe the surface morphology of the growing film. This allows to break the model in a $2+1$-dimensional model, continuous in the lateral directions but discrete in height. The evolution of steps is implicitly connected to atomistic motion via attachment and detachment at step edges and therefore allows to incorporate essential kinetic effects on an atomistic scale. Various step flow models have been developed which incorporate more and more of the relevant atomistic effects. These models have proven to be extremely powerful in understanding surface morphologies and the formation of patterns at the nanometer and micrometer scales, see [25, 37] for recent reviews. During the last years powerful numerical techniques for the solution of these problems have been developed and make step dynamics an attractive alternative to atomistic modeling approaches in epitaxial crystal growth. Level-set based methods [9, 35, 39], front-tracking type approaches [3, 4], phase-field methods [30, 23, 36, 34, 53] and geometry-based simulations [29] have been developed.

In this contribution we will mainly concentrate on phase-field approximations for the discrete-continuum model. Thereby the discrete height is approximated by a continuous phase-field variable. With this interpretation, the phase-field approximation can be viewed as a coarse grained continuous model for the underlying discrete-continuum step flow model. The contribution is organized as follows: In Section 2 we introduce the Burton-Cabrera-Frank model and discuss some of its limitations. Sections 3 and 4 are devoted to two different growth methods, Liquid Phase Epitaxy (LPE) and Molecular Beam Epitaxy (MBE), respectively. In the case of LPE, the film grows from a solution and transport mechanisms in the solution can influence the surface evolution and vice versa, and thus have to be considered in addition. In MBE the film grows from a vapour phase, here transport in the vapour does not need to be taken into account. For both situations we discuss a phase-field approximation and apply it to simulate epitaxial growth in a spiral growth model.

2 Burton-Cabrera-Frank model

In this section, we present a general BCF-like model for step-flow growth or island growth of homoepitaxial films.

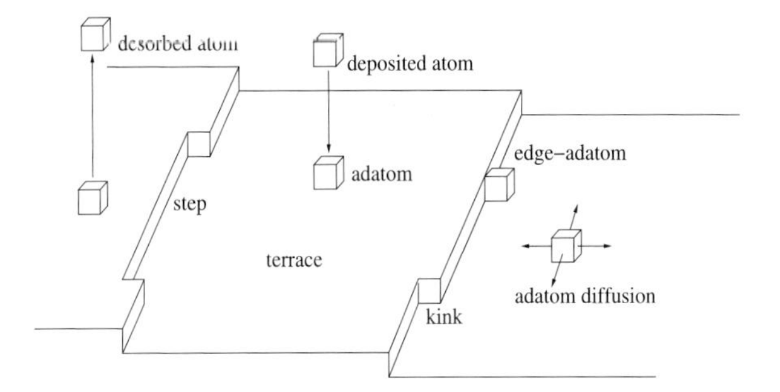

Fig. 2.1. Microscopic processes in epitaxial growth of thin films.

2.1 The BCF framework

The Burton-Cabrera-Frank (BCF) equations [7] serve as a prototype of discrete-continuous step flow models, discrete in the height resolving the atomic layers in the growth direction, but coarse grained in the lateral direction. The steps, separating terraces of different height, are assumed to be smooth curves and serve as free boundaries for an adatom diffusion on terraces. Mathematically such a model is a free boundary problem that consists of a diffusion equation for the adatom density on the terraces, boundary conditions at the steps and a velocity law for the motion of the steps. For a schematic picture of an epitaxial growing surface see Fig. 2.1.

We denote by $\Omega \subset \mathbb{R}^2$ the projected domain of a film surface in a two-dimensional Cartesian coordinate system, and assume that Ω is independent of time t. In addition we denote by $\Omega_i = \Omega_i(t) \subset \mathbb{R}^2$, $i = 0, 1, \ldots$, islands or terraces of discrete height i. We denote further the corresponding steps (or terrace boundaries) by

$$\Gamma_i(t) = \overline{\Omega_i(t)} \cap \overline{\Omega_{i-1}(t)}, \qquad i = 1, 2, \ldots.$$

Denote by $c_i = c_i(x,t)$ the adatom concentration on terrace $\Omega_i(t)$ at time t. The adatom diffusion on a terrace is described by the diffusion equation for the adatom density

$$\partial_t c_i - \nabla \cdot (D_{T,i} \nabla c_i) = -\tau^{-1} c_i + F_T - M_T \qquad \text{in } \Omega_i(t), \tag{2.1}$$

where $D_{T,i}$ is the adatom diffusion coefficient on a terrace Ω_i, τ^{-1} is the desorption rate, F_T is the deposition flux rate onto the terrace and M_T is the loss due to nucleation of adatom islands. In this paper we will not address the nucleation of adatom islands M_T. For a detailed description on nucleation and a way how to include it into the step flow model we refer to [38]. The fluxes of adatoms to the steps are given by

$$-D_{T,i} \nabla c_i \cdot \mathbf{n}_i - v_i c_i = f_{i,+}, \tag{2.2}$$

$$D_{T,i-1} \nabla c_{i-1} \cdot \mathbf{n}_i + v_i c_{i-1} = f_{i,-}, \tag{2.3}$$

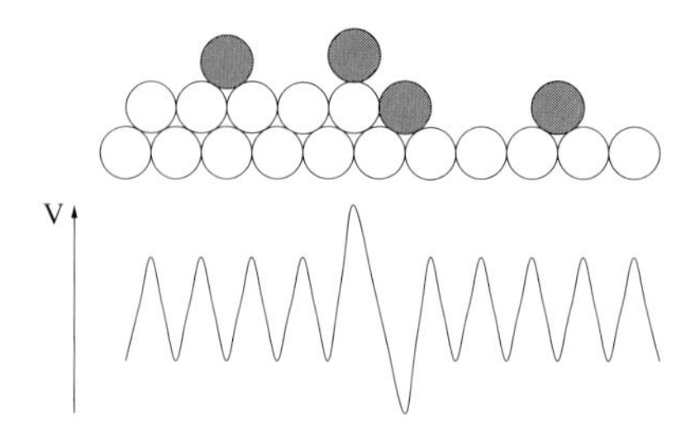

Fig. 2.2. The Ehrlich-Schwoebel barrier.

where $f_{i,+}$ is the net flux from the upper terrace $\Omega_i(t)$ and $f_{i,-}$ the net flux from the lower terrace $\Omega_{i-1}(t)$ to the boundary $\Gamma_i(t)$, $\mathbf{n}_i$ is the unit normal of the step $\Gamma_i(t)$ pointing from the upper to the lower terrace, and v_i is the normal velocity of the step $\Gamma_i(t)$ with the convention that $v_i > 0$ if the movement of $\Gamma_i(t)$ is in the direction of $\mathbf{n}_i$. The terms $v_i c_i$ describe the convection of adatoms due to the motion of the step and are needed for mass conservation. The fluxes are now related to the deviation of the adatom density at the step from its equilibrium value

$$f_{i,+} = k_+(c_i - c_i^{eq}), \tag{2.4}$$
$$f_{i,-} = k_-(c_{i-1} - c_i^{eq}), \tag{2.5}$$

where $k_+ = k_+(\theta)$ and $k_- = k_-(\theta)$ are the kinetic attachment rates from the upper and lower terrace to the step $\Gamma_i(t)$, respectively; θ defines the angle between the normal and a low index crystallographic direction. For finite $k_\pm$ the adatom density is discontinuous at the steps and for $k_+ \neq k_-$ growth instabilities can occur, see [25] for a recent review. Step meandering is known for $k_+ > k_-$ (step Ehrlich-Schwoebel barrier) and step bunching occurs for $k_+ < k_-$ (inverse step Ehrlich-Schwoebel barrier). It is observed in many material systems that, in order to stick to a step from an upper terrace, an adatom must overcome an additional energy barrier (see Fig. 2.2), the Ehrlich-Schwoebel (ES) barrier [17, 43, 44]. Its inverse counterpart is much less understood.

In the limit $k_+, k_- \to \infty$ the adatom density becomes continuous and the boundary condition reduces to $c_i = c_{i-1} = c_i^{eq}$, where the equilibrium adatom density c_i^{eq} is given by

$$c_i^{eq} = c^* \left(1 + \frac{\Omega_a \tilde{\gamma}}{k_B T} \kappa_i\right), \tag{2.6}$$

with $\tilde{\gamma} = \tilde{\gamma}(\theta)$ the step stiffness, κ_i the curvature of $\Gamma_i(t)$, c^* the equilibrium adatom density of a straight step, k_B the Boltzmann constant, T the temperature and Ω_a the area of an adsorption site. For the motion of the moving boundaries, we assume the following law for the normal velocity v_i of the step $\Gamma_i(t)$:

$$\frac{v_i}{\Omega_a} = f_{i,+} + f_{i,-} + \partial_s(\nu\partial_s(\Omega_a\tilde{\gamma}\kappa_i)) + \partial_s((f_{i,+} + f_{i,-})\beta), \tag{2.7}$$

where $\nu = \nu(\theta)$ is the mobility for diffusion along the steps, $\beta = \beta(\theta)$ is an effective anisotropy to model the effect of a kink Ehrlich-Schwoebel barrier. The kink Ehrlich-Schwoebel barrier results from an additional energy barrier for the diffusion of an edge-adatom around a kink or a corner. The presence of such a barrier affects the nucleation of kinks at the steps and generates a nonlinear current, which can also contribute to a meandering instability. The last term in (2.7) is purely kinetic and has been introduced by [20]. ∂_s denotes the tangential derivative along the steps. The term $\partial_s(\nu\partial_s(\tilde{\gamma}\kappa_i))$ represents diffusion along steps.

All the introduced models assume implicitly a close-to-equilibrium situation, which is not fulfilled in many applications related to epitaxy. This results from the linear dependency of the flux $f_{i,\pm}$ on the equilibrium adatom density c_i^{eq} in the boundary conditions (2.4) and (2.5). Step flow models without this assumption are rare [8, 1, 2]. The idea behind these models is a mean field approach, in which an atomistic description of the steps is incorporated into the discrete-continuum model by allowing for two additional quantities along the step: step adatoms and kinks.

However, for the applications presented in the following several simplifications are possible: i) we can neglect nucleation M_T, ii) we can neglect desorption τ by assuming the average time for an adatom to reach a step is much smaller than the time to desorb, which is fullfilled for small distances between steps, iii) we assume that there is one diffusion coefficient D_T for all terraces. Thus the system reduces to

$$\partial_t c_i - \nabla \cdot (D_T \nabla c_i) = F_T \tag{2.8}$$

and

$$-D_T \nabla c_i \cdot \mathbf{n}_i = f_{i,+}, \tag{2.9}$$

$$D_T \nabla c_{i-1} \cdot \mathbf{n}_i = f_{i,-}. \tag{2.10}$$

We further consider the case where the adatom concentration is continuous at the boundary. Thus the boundary conditions at the steps read

$$c_i = c_{i-1} = c^* \left(1 + \frac{\Omega_a\tilde{\gamma}}{k_B T}\kappa_i\right) + \alpha\frac{v_i}{\Omega_a}, \tag{2.11}$$

where we have included a term αv_i accounting for kinetic attachment effects. In the velocity law we neglect diffusion along edges and the effect of the kink Ehrlich Schwoebel barrier. Thus we obtain

$$\frac{v_i}{\Omega_a} = f_{i,+} + f_{i,-}. \tag{2.12}$$

2.2 Thermodynamic consistency

As pointed out in [11] the above model is not necessarily consistent with the second law of thermodynamics. To ensure consistency restrictions on the constitutive relations (2.4) and (2.5) are required. [11] provide an alternative set of boundary conditions, derived from a configurational force balance, for which the second law holds, which corresponding to our situation reads

$$f_{i,+} = l_+(\mu_i - [[\omega_i]]), \tag{2.13}$$
$$f_{i,-} = l_-(\mu_{i-1} - [[\omega_i]]), \tag{2.14}$$

with μ_i the terrace chemical potential, defined through $\mu_i = \partial_{c_i}\Psi_i$, with $\Psi_i(c_i)$ the terrace adatom free energy. Here $\omega_i = \Psi_i - \mu_i c_i$ is the grand canonical potential and $[[\omega_i]] = \omega_i - \omega_{i-1}$. The jump in the grand canonical potential results from the discontinuity of the adatom concentration and might have significant influence on the stability of steps. As in the case of an ideal lattice gas $\Psi_i(c_i)$ is convex. If we further assume $\Psi_i(0) = 0$ and define the adatom equilibrium density c_i^{eq} such that $\mu_i(c_i^{eq}) = \mu_{i-1}(c_i^{eq}) = 0$ we can expand μ_i and ω_i close to c_i^{eq} to obtain in leading order

$$\mu_i(c_i) = \mu_i'(c_i^{eq})(c_i - c_i^{eq}), \tag{2.15}$$
$$\omega_i(c_i) = \omega_i(c_i^{eq}) + c_i^{eq}\mu_i'(c_i^{eq})(c_i - c_i^{eq}), \tag{2.16}$$

which inserted into (2.13) and (2.14) yield

$$f_{i,+} = k_+(c_i - c_i^{eq} - c_i^{eq}[[c_i]]), \tag{2.17}$$
$$f_{i,-} = k_-(c_{i-1} - c_i^{eq} - c_i^{eq}[[c_i]]), \tag{2.18}$$

with $k_\pm = l_\pm \mu_i'(c_i^{eq})$. The original form (2.4) and (2.5) is thus only obtained as a first-order approximation under the additional assumption that the adatoms are sufficiently rarified ($c_i \ll c_i^{eq}$) such that the term $c_i^{eq}[[c_i]]$ can be neglected.

2.3 Spiral growth

We will apply the BCF model in the following in its classical form to two important growth technologies and numerically investigate a spiral growth mode. Spiral growth remains to be an important growth mechanism for various materials, including compound semiconductors, see e.g. [48]. The growth mode is initiated from a screw dislocation from which a single step originates (see Fig. 2.3). Thus the growth is not limited by the nucleation rate and therefore allows growth also under low supersaturations [6]. Instead of the independent layer structure discussed so far, the crystal surface is now just one layer, which overlaps itself helicoidally. The crystal grows now by simply turning the screw, through the advancement of the spiral emanating from the disclocation [31], see also Fig. 2.4.

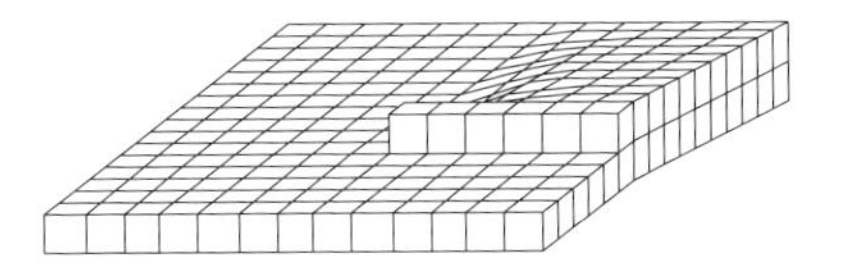

Fig. 2.3. Dislocation at a flat surface. This dislocation might trigger spiral growth.

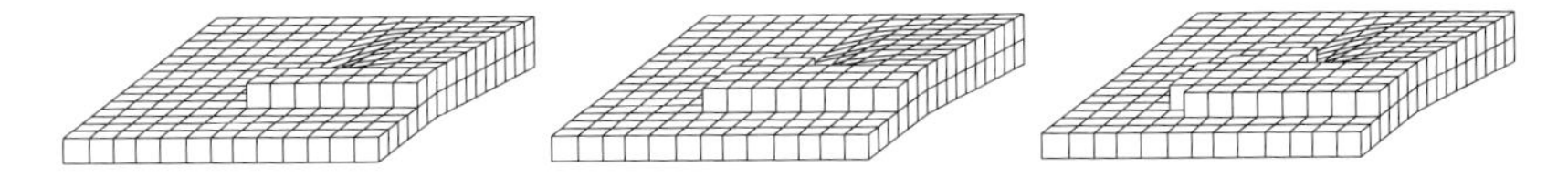

Fig. 2.4. Step formation due to the pinning at a screw dislocation.

A basic question arising in spiral growth is the prediction of the final surface slope, which is influenced by the interplay of diffusion processes on the terraces and along the steps, attachment kinetics at the steps and presumably also from mass transport in the bulk.

2.4 Phase-field approximation

A phase-field model for this situation was considered in [41], where formal matched asymptotic analysis was used to show the convergence for vanishing interfacial thickness and kinetic parameter to the sharp interface model (2.8)–(2.12). The phase-field model reads

$$\frac{1}{\Omega_a}\partial_t\phi + \partial_t c = \nabla \cdot (D_T \nabla c) + F \tag{2.19}$$

$$\frac{\alpha}{\Omega_a}\varepsilon^2\partial_t\phi = \frac{c^*\Omega_a\tilde{\gamma}}{k_B T}(\varepsilon^2\Delta\phi - G'(\phi)) + \varepsilon(c - c^*), \tag{2.20}$$

where ϕ is the phase-field variable, ε a small parameter determining the width of the diffuse interface and α a kinetic parameter. For applications in step flow or island growth $G(\phi)$ is a multiwell potential with minima at the discrete heights of the terraces. A more general phase-field model incorporating finite kinetic coefficients k_+, k_- has been introduced in [34]. An approximation in which the effect of edge diffusion is considered was discussed in [42] and a combination of both and thus the general case described in Sect. 2 is considered in a combined phase-field level-set approach in [21].

In order to adapt the phase-field model to the special situation of a spiral growth mode we follow the approach in [23]. We define the multiwell potential as

$$G(\phi) = H(\phi - \phi_0)$$

with ϕ_0 representing the screw dislocation and defining an initial value; $2\pi\phi_0$ defines the angle of (x, y) in the $x - y$ plane. The phase transition happens in radial direction and ϕ measures the continuous height of the spiral. H is now defined through

$$H(\psi) = (k+1-\psi)^2(k-\psi)^2, \qquad \psi \in [k, k+1], \quad k \in \mathbb{Z}.$$

We use no-flux conditions on the boundary of the domain $\Omega = (-L/2, L/2)^2$

$$\partial_n c = \partial_n \phi = 0.$$

3 Application in Liquid Phase Epitaxy

Liquid Phase Epitaxy [46] is the name for an epitaxial technique, which combines solution growth and epitaxy taking advantage of the potentialities of both. Its development was given impetus from (a) technical requirements for thin (1 to 5 μm) garnet films, and (b) its ability to grow pure semiconductor compounds and their alloys. Later it was developed further to manufacture thin-film devices using compound semiconducting materials. It proved tremendous superiority to the early attempts to fabricate such thin films by means as e.g. polishing bulk crystals. Since then it has been successfully applied in many laboratories around the world. Moreover, it allowed to obtain fundamental new information concerning the generic processes of crystal growth.

Despite its large fundamental and technological impact, models of LPE so far are restricted to one-dimensional models predicting the evolution of the integrated growth rate $h(t)$ of the epitaxial surface. First results were obtained neglecting hydrodynamic convection, which is a good approximation only for quiescent liquids. For the case of first-order surface kinetics, one can reduce this problem to various perturbation series [32, 27, 19, 49].

For LPE systems, where the assumption of a quiescent liquid is no longer satisfied, hydrodynamics effects have to be taken into account. Usually one imposes a forced flow to the substrate via a "rotating-disk" configuration. A rotating disk tends to draw material towards its surface in the normal direction, and then expells it radially in the manner of a centrifugal fan. If one applies this picture to LPE one has to take into account that the fluid-velocity's normal component cannot vanish at the disk's surface because mass conservation requires it to be proportional to the crystal's growth rate [51]. So far, hydrodynamics have adressed this problem when the amount of "suction" is indepentent of time. In this case the fluid flow problem is effectively decoupled from the mass-transfer problem [47, 52]. Numerical simulations of the coupled, time dependent problem are still lacking.

Moreover, two-dimensional studies of the problem, which do not only resolve the integrated growth rate but also the two-dimensional microscopic morphology of the substrate's surface, are still an open issue. LPE is particularly rich in morphological features. It is tempting to apply [24, 12] the theory of Mullins and Sekerka [33] for an analysis of the stability of these morphologies. But technically this is incorrect, because the unperturbed state is almost always time dependent. Thus numerical simulations of the full time dependent problem are essential to adress the question which processing parameters trigger which final stage morphology.

A previous approach to LPE developed by two of the authors [18] treats this problem assuming a quiescent solution, i.e. negecting fluid flow, but taking into account the evolution of the microscopic surface morphology. It is based on a phase-field model which allows us to consider precisely the physics of the underlying phase-diagram for the heteroepitaxial LPE growth process. Moreover, it applies to the situation, where the main driving force of the concentration field results from its gradients along the surface.

In the following subsection we will focus on an extension of this previous approach, which takes into account also (a) adatom concentration gradients perpendicular to the surface and (b) hydrodynamic transport in the solution.

3.1 A model for Liquid Phase Epitaxy

We consider the growth of a thin epitaxial layer at the bottom of a domain $Q \subset \mathbb{R}^3$ filled with a liquid solution that contains the adatoms as depicted in Fig. 3.1. The adatoms in the liquid solution are transported to the solid layer by diffusion and convection, there they may deposit on the epitaxial surface. At time t, the domain Q is partitioned into a domain $Q_L(t)$ filled by the liquid solution and a domain $Q_S(t)$ filled by the solid phase. The interface $S = S(t)$ between Q_L and Q_S is represented by the graph of a function h over the bottom surface S_0 of Q,

$$S(t) = \{\mathbf{x} + h(t, \mathbf{x})\mathbf{e}_3 \,|\, \mathbf{x} \in S_0\}, \tag{3.1}$$

where $\mathbf{e}_3$ is the 3rd unit vector. Fluid flow and adatom transport in Q_L are described by a Navier-Stokes system and a convection-diffusion equation,

$$\begin{aligned} \nabla \cdot \mathbf{v} &= 0 \\ \partial_t \mathbf{v} + (\mathbf{v} \cdot \nabla)\mathbf{v} - \eta \Delta \mathbf{v} &= \nabla p, \end{aligned} \tag{3.2}$$

$$\partial_t c^V + \mathbf{v} \cdot \nabla c^V - D_V \Delta c^V = 0 \tag{3.3}$$

with fluid velocity $\mathbf{v}$, pressure p, viscosity η, mass specific concentration c^V and diffusion constant D_V. The epitaxial growth is governed by a BCF-model,

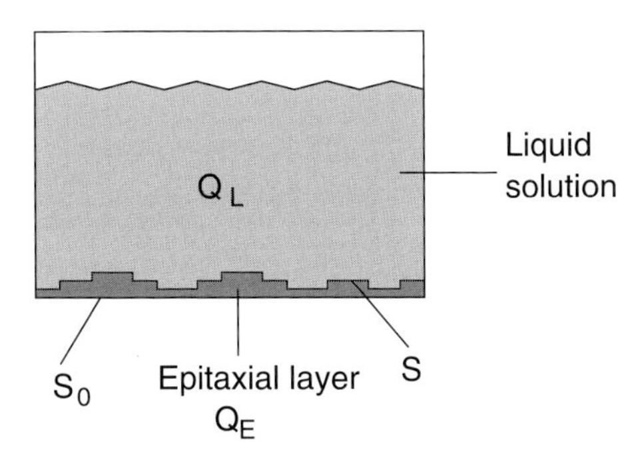

Fig. 3.1. Liquid Phase Epitaxy

$$\partial_t c^S = D_S \Delta c^S + \frac{c^V}{\tau_V} - \frac{c^S}{\tau_S} \quad \text{in } S_0 \setminus \Gamma(t), \tag{3.4}$$

$$\left.\begin{aligned} c^S &= c^*\big(1 + \kappa \Omega_a \gamma/(k_B T)\big), \\ v_\Gamma &= D_S \Omega_a \left[\tfrac{\partial c^S}{\partial n}\right] \end{aligned}\right\} \quad \text{on } \Gamma(t). \tag{3.5}$$

Here c^S is the surface density of adatoms, D_S is a surface diffusion coefficient, τ_V is the mean time for the deposition of adatoms from the solution to the surface, τ_S is the mean time for the desorption of adatoms from the surface to the solution, the curve Γ describes the monoatomar steps, v_Γ is the velocity of the steps, and $\left[\frac{\partial c^S}{\partial n}\right]$ is the difference of the normal derivatives on both sides of the interface.

The volume equations (3.2), (3.3) and the surface equation (3.4) are coupled by boundary conditions that model the conservation of total mass and mass of the adatoms. If ϱ_V and ϱ_E denote the densities of the liquid and solid adatom phase, m_a is the mass of a single atom and J_S is the density of the surface measure for the surface S, parametrized over S_0, then these conditions are

$$v_n = \frac{m_a}{J_S}\left(\frac{1}{\varrho_V} - \frac{1}{\varrho_E}\right)\left(\frac{c^V}{\tau_V} - \frac{c^S}{\tau_S}\right) \quad \text{and} \tag{3.6}$$

$$D_V \frac{\partial c^V}{\partial n} = \frac{m_a}{J_S}\left(1 - c^V\right)\left(\frac{c^S}{\tau_S} - \frac{c^V}{\tau_V}\right). \tag{3.7}$$

The tangential fluid velocity at the interface S vanishes, $\mathbf{v}_t(t, \mathbf{x}) = 0$ for $\mathbf{x} \in S(t)$. The movement of the interface S is given by the condition

$$\partial_t h = \frac{m_a}{\varrho_E}\left(\frac{c^V}{\tau_V} - \frac{c^S}{\tau_S}\right). \tag{3.8}$$

The model is completed by initial conditions for $\mathbf{v}$, c^V and c^S, by an initial partition of Q into a liquid domain $Q_L(0)$ and a solid domain $Q_S(0)$ with interface $S(0)$ given by an initial condition for $h(0, \mathbf{x})$, by initial steps $\Gamma(0)$ on the solid layer S_0 and by boundary conditions for (3.2), (3.3), (3.4).

3.2 Homogenization and two-scale model

Liquid Phase Epitaxy involves various different length scales: the epitaxial layer is measured in atom diameters, while the continuum models for diffusion and fluid flow, and also the model for the adatom diffusion on the liquid-solid interface, are valid for a much larger scale. The liquid-solid interface often exhibits a specific microstructure, whose scale is larger than that of an atom diameter but much smaller than that of the diffusion and fluid flow processes in the liquid solution. In this section we use homogenization techniques to derive a model that is suitable for such situations. It is necessary to choose suitable scaling properties of some coefficients in dependence of a scale parameter η that represents the size of the microstructure. The size of the adatoms is scaled proportional to η, this leads to

$$\Omega_a = \eta^2 \Omega_a^0, \quad h_a = \eta\, h_a^0 \ \text{ and } \ m_a = \eta^3 m_a^0. \tag{3.9}$$

The epitaxial microstructure of scale η requires the relations

$$D_S = \eta^2 D_S^0, \quad \tau_V = \eta^3 \tau_V^0, \quad c^* = \eta^{-2} c_0^* \ \text{ and } \ \gamma = \eta^{-1}\gamma_0. \tag{3.10}$$

In the limit $\eta \to 0$ two phenomena must be described: small scale oscillations in the epitaxial microstructure, and boundary layers for the solutions of (3.2), (3.3) close to the interface. These phenomena require a combination of the standard homogenization technique for the oscillations and the technique of matched asymptotic expansions for the computation of the boundary layer.

The homogenization is done for a simple periodic setting with periodicity cell $Y \subset \mathbb{R}^2$ of area 1. A periodicity cell is a bounded Lipschitz domain with the property that $\mathbb{R}^2$ can be represented as union of shifted copies of Y with empty intersection; the simplest example is the unit square $Y = [0,1]^2$. The corresponding initial conditions for the concentrations are

$$\begin{aligned} c_\eta^{S0}(\mathbf{x}) &= \eta^{-2} c^{S0}(\mathbf{x}, \mathbf{x}/\eta) \ \text{ for } \mathbf{x} \in S_0 \text{ and} \\ c_\eta^{V0}(\mathbf{x}) &= \eta\, c^{V0}(\mathbf{x}) \ \text{ for } \mathbf{x} \in Q_L(0). \end{aligned}$$

The function $c^{S0} = c^{S0}(\mathbf{x}, \mathbf{y})$ is Y-periodic in its second variable $\mathbf{y} \in Y$. The scaling of c^S by η^{-2} makes sense, because c^S measures the number of atoms per unit area, and the area of single atoms is proportional to η^2; the scaling by η^{-2} keeps the area covered by adatoms constant. The scaling of c^V is motivated by the exchange of matter between solid and liquid phase: the thickness of the solid phase is proportional to η, therefore the total mass of adatoms that deposit on the surface is also proportional to η. The scaling by η accounts for this rate of mass exchange; it is reasonable for a *dilute* solution. The adatom density c^S is expanded into the power series

$$c_\eta^S(t, \mathbf{x}) = \eta^{-2} c_0^S(t, \mathbf{x}, \mathbf{x}/\eta) + \eta^{-1} c_1^S(t, \mathbf{x}, \mathbf{x}/\eta) + \cdots. \tag{3.11}$$

The surface $\Gamma = \Gamma_\eta(t)$, its curvature $\kappa = \kappa_\eta$ and its velocity $v_\Gamma = v_{\Gamma\eta}$ are also expanded in an appropriate *two-scale* sense, for details we refer to [16].

The outer expansions for the volume fields $\mathbf{v}$, p and c^V are valid far away from the liquid-solid interface $S_\eta(t)$. They are given by the power series

$$\begin{aligned} \mathbf{v}_\eta(t, \mathbf{x}) &= \mathbf{v}_0(t, \mathbf{x}) + \eta\, \mathbf{v}_1(t, \mathbf{x}) + \cdots, \\ p_\eta(t, \mathbf{x}) &= p_0(t, \mathbf{x}) + \eta\, p_1(t, \mathbf{x}) + \cdots, \\ c_\eta^V(t, \mathbf{x}) &= \eta\, c_0^V(t, \mathbf{x}) + \eta^2 c_1^V(t, \mathbf{x}) + \cdots. \end{aligned} \tag{3.12}$$

The inner expansion for $\mathbf{v}, p, c^V$ is valid close to $S_\eta(t)$. Since this interface is a priori unknown, the inner expansion is done after the transform of variables $\tilde{\mathbf{x}} = \mathbf{x} - h_\eta(t, \mathbf{x})\mathbf{e}_3$. Then $\mathbf{v}, p, c^V$ are expanded in power series

$$\begin{aligned}
\mathbf{v}_\eta(t, \mathbf{x}) &= \mathbf{v}_0(t, \mathbf{x}, \mathbf{x}/\eta) + \eta\, \mathbf{v}_1(t, \mathbf{x}, \mathbf{x}/\eta) + \cdots, \\
p_\eta(t, \mathbf{x}) &= p_0(t, \mathbf{x}, \mathbf{x}/\eta) + \eta\, p_1(t, \mathbf{x}, \mathbf{x}/\eta) + \cdots, \\
c_\eta^V(t, \mathbf{x}) &= \eta\, c_0^V(t, \mathbf{x}, \mathbf{x}/\eta) + \eta^2 c_1^V(t, \mathbf{x}, \mathbf{x}/\eta) + \cdots.
\end{aligned}$$

This expansion combines both the homogenization by formal asymptotics and the inner expansion for the boundary layer. The former is realized by Y-periodic oscillations of the components (y_1, y_2) in $u^V(t, \mathbf{x}, \mathbf{y})$ for $u \in \{\mathbf{v}, p, c^V\}$, the latter via the component y_3. As a consequence, the functions $\mathbf{y} \to u_j(t, \mathbf{x}, \mathbf{y})$ for $u \in \{\mathbf{v}, p, c^V\}$, $j = 0, 1, 2, \ldots$, are defined in the domain $Y \times (0, +\infty)$, with periodic boundary conditions at the lateral surface $(y_1, y_2) \in \partial Y$. The condition for $y_3 \to +\infty$ is derived by matching the inner and outer expansion and the condition for $y_3 = 0$ is given by the coupling conditions (3.6), (3.7). The height function $h = h_\eta$ is also expanded into a series,

$$h_\eta(t, \mathbf{x}) = \eta\, h_0(t, \mathbf{x}, \mathbf{x}/\eta) + \eta^2 h_1(t, \mathbf{x}, \mathbf{x}/\eta) + \cdots. \tag{3.13}$$

Using these formal expansions in (3.2)–(3.8) it is possible to derive a two-scale model that consists of the following relations:

- Macroscopic equations for fluid flow and solute diffusion in Q:

$$\partial_t c_0^V + \mathbf{v}_0 \cdot \nabla c_0^V - D_V \Delta c_0^V = 0, \tag{3.14}$$

$$\begin{aligned}
\nabla \cdot \mathbf{v}_0 &= 0, \\
\partial_t \mathbf{v}_0 + (\mathbf{v}_0 \cdot \nabla)\mathbf{v}_0 - \eta \Delta \mathbf{v}_0 + \nabla p_0 &= 0.
\end{aligned} \tag{3.15}$$

- A microscopic BCF-model in Y to be solved for every $\mathbf{x} \in S_0$:

$$\partial_t c_0^S - D_S^0 \Delta_{\mathbf{y}} c_0^S = \frac{1}{\tau_V^0} c_0^V - \frac{1}{\tau_S} c_0^S \quad \text{for } \mathbf{y} \in Y, \tag{3.16}$$

$$\left.\begin{aligned}
c_0^S &= c_0^*\left(1 + \kappa_0 \Omega_a^0 \gamma_0 / (k_B T)\right), \\
v_{\Gamma 0} &= D_S^0 \Omega_a^0 \left[\frac{\partial c_0^S}{\partial n}\right]
\end{aligned}\right\} \text{ for } \mathbf{y} \in \Gamma_0(t, \mathbf{x}). \tag{3.17}$$

- Coupling conditions on S_0

$$D_V \frac{\partial c_0^V}{\partial n} = m_a^0 \left(\frac{\bar{c}_0^S}{\tau_S} - \frac{c_0^V}{\tau_V^0}\right), \tag{3.18}$$

$$\mathbf{v} = 0 \tag{3.19}$$

with the *microscopic mean value* $\bar{c}_0^S(t, \mathbf{x}) = \int_Y c_0^S(t, \mathbf{x}, \mathbf{y})\, dy$. These equations serve as boundary conditions for (3.14), (3.15) on S_0.

The details of the derivation can be found in [16]. The equations for fluid flow here decouple from the other equations. The fluid velocity can be computed in a pre-processing step, and the main computation consists in solving the remaining equations with given fluid velocity.

3.3 Analysis of the two-scale model

An analysis is available for a slightly modified version of the two-scale model, where the sharp interface version (3.16), (3.17) of the BCF-model is replaced by a the following adaption of the phase field model (2.19), (2.20) to liquid-phase epitaxy:

$$\partial_t c_0^S - D_S^0 \Delta_{\mathbf{y}} c_0^S + (\Omega_a^0)^{-1} \partial_t \phi_0 = \frac{c_0^V}{\tau_V^0} - \frac{c_0^S}{\tau_S}, \tag{3.20}$$

$$\alpha_0 \varepsilon_0^2 \partial_t \phi_0 - \varepsilon_0^2 \Delta_{\mathbf{y}} \phi_0 + G'(\phi_0) + p_0(c_0^S, \phi_0) = 0. \tag{3.21}$$

The function p_0 here represents a possibly generalized version of the term $\frac{\varepsilon k_B T}{c^* \tilde{\gamma} \Omega_a}(c - c^*)$ in a rescaled variant of (2.20). The lower indices 0 indicate solutions of the homogenized problem. A basic assumption for the analysis is that G' and p are Lipschitz functions with at most linear growth, $|G'(\phi)| \leq C(1 + |\phi|)$ and $p_0(c, \phi) \leq C(1 + |c| + |\phi|)$. For the modified model (3.14), (3.15), (3.20), (3.21), (3.18), (3.19), the existence and uniqueness of its solution can be proved, see [16], Theorem 1. The main tool of the existence proof is the Schauder fixed point theorem, applied to a composition of the solution operators for the family of microscopic problems on one hand and for the macroscopic problem on the other hand.

For the phase field versions of both the original and the two-scale model, an estimate of the model error is avaliable. The original model of scale η is given by the equations (3.2), (3.3), (3.6)–(3.8) and the phase field version of the BCF-model

$$\partial_t c^S - D_S \Delta c^S + \Omega_a^{-1} \partial_t \phi = \frac{c^V}{\tau_V} - \frac{c^S}{\tau_S}, \tag{3.22}$$

$$\alpha \varepsilon^2 \partial_t \phi - \varepsilon^2 \Delta \phi + G'(\phi) + p(c^S, \phi) = 0. \tag{3.23}$$

The scaling of the phase field variables ε and α is $\varepsilon = \eta \varepsilon_0$ and $\alpha = \eta^{-2} \varepsilon_0$, the relation between p and p_0 is

$$p(c, \phi) = p_0(\eta^2 c, \phi).$$

This scaling property is in harmony with the scaling of c^S in (3.11).

Let $(\mathbf{v}_\eta, p_\eta, c_\eta^V, c_\eta^S, \phi_\eta)$ be the solution of the described original problem for scale η and let $(\mathbf{v}_0, p_0, c_0^V, c_0^S, \phi_0)$ be the solution of the described two-scale model. Due to the scaling relations (3.11), (3.12) the comparison of models is done with the rescaled functions

$$\tilde{c}_\eta^V = \eta^{-1} c_\eta^V \quad \text{and} \quad \tilde{c}_\eta^S = \eta^2 c_\eta^S.$$

In order to compare the functions $\tilde{c}_\eta^S$, ϕ_η of variables $(t, \mathbf{x})$ and the functions c_0^S, ϕ_0 of variables $(t, \mathbf{x}, \mathbf{y})$ we employ *macroscopic reconstructions* of two-scale type

$$u^\eta(t, \mathbf{x}) = u_0(t, \mathbf{x}, \mathbf{x}/\eta) \quad \text{for } u = c_0^S, \phi.$$

Let $Q_\eta(t)$ be the liquid domain and $S_\eta(t)$ be the liquid-solid interface for the problem of scale η. The boundary conditions for both the original model of scale η and the two-scale model on $\partial Q_\eta(t) \setminus S_\eta(t)$ are Neumann conditions for c_η^V, c_0^V and Dirichlet conditions for $\mathbf{v}_\eta$, $\mathbf{v}_0$; the given boundary data for both models are identical. Further boundary conditions are homogeneous Neumann conditions for c_η^S, ϕ_η on ∂S_0, for c_0^S and ϕ_0 there are no boundary conditions on ∂S_0, because there are no $\mathbf{x}$-derivatives in the corresponding differential equations. The initial conditions for both models are compatible in the sense $\tilde{c}_\eta^V(0, \mathbf{x}) = c_0^V(0, \mathbf{x})$, $\mathbf{v}_\eta(0, \mathbf{x}) = \mathbf{v}_0(0, \mathbf{x})$ and

$$\tilde{c}_\eta^S(0, \mathbf{x}) = c_0^S(0, \mathbf{x}, \mathbf{x}/\eta), \quad \phi_\eta(0, \mathbf{x}) = \phi_0(0, \mathbf{x}, \mathbf{x}/\eta).$$

Then under some moderate additional assumptions on the regularity of the solutions the following estimate for the model error is availabe, see [16]:

$$\begin{aligned} &\left\| \tilde{c}_\eta^S - c_0^{S\eta} \right\|_{L_\infty(I; L_2(S_0))} + \left\| \Phi_\eta - \Phi_0^\eta \right\|_{H^1(I; L_2(S_0))} \\ &\quad + \sup_{t \in I} \Big(\left\| (\tilde{c}_\eta^V - c_0^V)(t) \right\|_{L_2(Q_\eta(t))} + \| (\mathbf{v}_\eta - \mathbf{v}_0)(t) \|_{L_2(Q_\eta(t))} \Big) \\ &\le C \eta^{1/2}. \end{aligned} \tag{3.24}$$

The presented estimate reveals an approximation order of $\eta^{1/2}$ that is lower than the order η as could be expected from the asymptotic expansion. This deterioration has two reasons. First, the macroscopic fluid flow and convection-diffusion processes happen in the whole domain Q for the two-scale model and in the liquid domain $Q_\eta(t)$ for the original model. These domains differ by a volume of size proportional to η, which leads to an error of order $\eta^{1/2}$ in L_2-type estimates. Second, in a general setting it must be expected that the microstructure evolving at the boundary of S_0 does not occupy a full shifted periodicity cell ηY but some "truncated" domain; this generates an additional error of order $\eta^{1/2}$ in the $L_2(S_0)$-norm.

3.4 Numerical approach

The scheme we employ for the simulations in the following section is based on the method of lines in time. After discretizing the problem by finite differences in space, we solve the resulting ODE system by the standard Runge-Kutta method of fourth order with fixed time step and impose a constant velocity field defined as

$$\mathbf{v}_{ijk} = \left(v_0 \sqrt{k h_z^V}, 0, 0 \right),$$

where the ijk denote the three spatial directions and h_z^V the mesh size in the z direction in the volume. Moreover, in the following simulations we have constant inflow at the left hand side of the simulated region and free outflow at the right hand side.

3.5 Simulation results

First simulations allow to visualize e.g. how the variation of the macroscopic flow field as depicted in Fig. 3.2 is accompanied by a variation of the concentration field in the melt which in turn results in a systematic variation of microstructure evolution as depicted in Fig. 3.3. The parameters for the simulations are given in Table 3.1. Systematic parameter studies to show the limitations of previous perturbative studies [24, 12] of microstructure evolution in LPE are still open, just as their analysis to extract relations between processing parameters and morphological stability of the material sample which promise to allow for a desired more efficient parameter control in the sensitive crystallization process of LPE grown material systems [13].

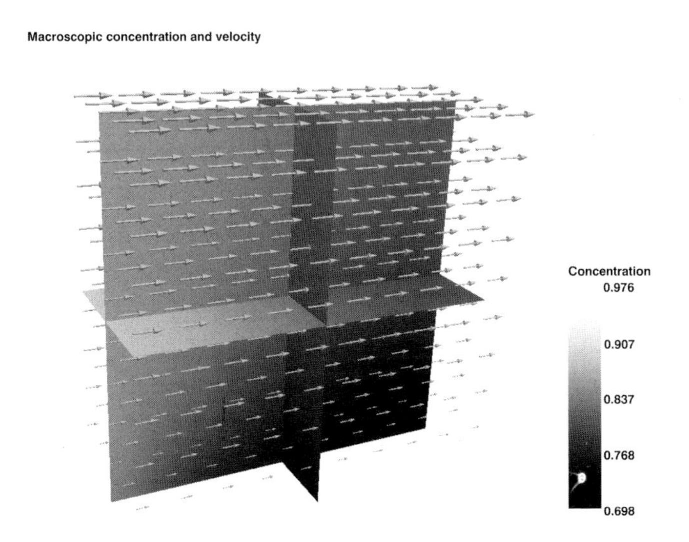

Fig. 3.2. Dynamics at the macroscale for the parameters given in Table 3.1.

Table 3.1. Simulation parameters for simulations depicted in Fig. 3.2 and 3.3

α	1	λ_1	19.591	τ_S	5	Ω_a	2	m_a	0.009
ε	1	D_S	10	τ_V	2	D_V	10	v_0	0.1

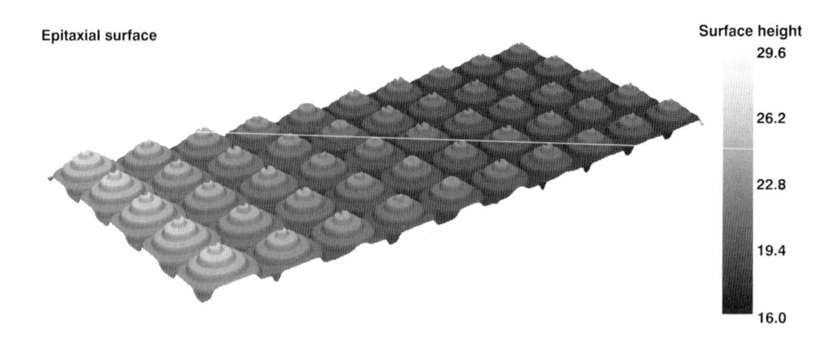

Fig. 3.3. Dynamics at the microscale for the parameters given in Table 3.1.

4 Application in Molecular Beam Epitaxy

Here we will consider Molecular Beam Epitaxy (MBE), which allows a decoupling from macroscopic processes in the growth furnace and the microscopic description on the surface of the growing film. Thus the model in Sect. 2 is directly applicable.

4.1 Phase-field approximation

Now we are concerned with the phase-field model (2.19)–(2.20). Measuring the length in atomic distances (i.e. $\Omega_a = 1$) the equations read

$$\partial_t \phi + \partial_t c = \nabla \cdot (D_T \nabla c) + F \tag{4.1}$$

$$\alpha \varepsilon^2 \partial_t \phi = \frac{c^* \tilde{\gamma}}{k_B T} (\varepsilon^2 \Delta \phi - G'(\phi)) + \varepsilon (c - c^*). \tag{4.2}$$

We consider the initial conditions

$$\phi(\mathbf{x}, 0) = \phi_0(\mathbf{x}), \quad c(\mathbf{x}, 0) = c^*,$$

where the function ϕ_0 introduced in Sect. 2.4 represents a screw dislocation and c^* is the equilibrium concentration for straight steps. Furthermore we use no-flux conditions on the boundary of the domain $\Omega = (-L/2, L/2)^2$

$$\partial_n c = \partial_n \phi = 0.$$

4.2 Numerical approach

Following the numerical approach for a viscous Cahn-Hilliard equation described in [40] equations (2.19)–(2.20) are discretized in space by linear finite elements and in time in a semi-implicit way. Here the derivative of the double well $G'(\phi)$ is treated explicity. The resulting linear equations are solved as a system.

4.3 Simulation results

The algorithm is implemented in the adaptive finite element toolbox AMDiS [50]. In the simulations the following parameters are used: $D_T = 10.0$, $c^* = 0.1$, $\frac{\tilde{\gamma}}{k_B T} = 1.0$ and $\alpha = 1.0$. Fig. 4.1 shows the phase-field variable and adatom density at a given time instant after the growth of several monolayers. A more quantitative picture is given in Fig. 4.2, which shows the development towards a stationary profile. For $t = 1500$ the width of the terraces, the stepspacing l, is constant. We can now investigate how the stepspacing l depends on the deposition flux F. Before we analyse this numerically let us consider a theoretical approach used in [23]. The stepspacing l can be computed by solving the cubic equation

$$19c^* \frac{\tilde{\gamma}}{k_B T} = l^3 \frac{F}{4D_T} \left(1 + \frac{1}{l} \frac{4D_T c^* \tilde{\gamma} \alpha}{k_B T} \right).$$

A comparison of the numerical results with this theory is given in Table 4.1. A second interesting quantity is the surface width ω, which is a common measure

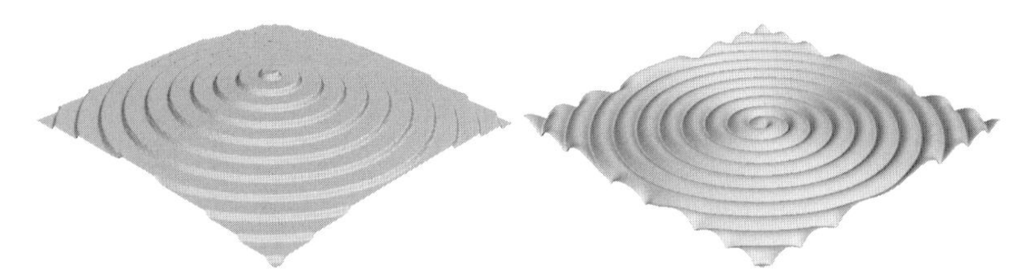

Fig. 4.1. Phase-field variable ϕ (left) and adatom concentration c (right) at a given time instant.

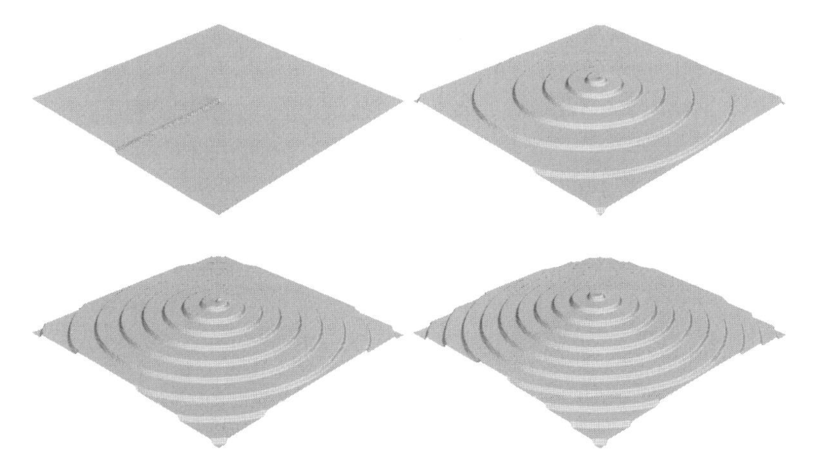

Fig. 4.2. Phase-field variable ϕ at different time steps $t = 0$, $t = 100$, $t = 300$ and $t = 1500$. The deposition flux used is $F = 0.2$. One only sees the height profile measured from the minimal terrace height.

Table 4.1. Comparison of theoretically and numerically obtained step spacing in stationary profile.

F	0.2	0.1	0.05	0.025
l_{theory}	6.1	8.0	10.3	13.3
$l_{\mathrm{simulation}}$	6.6	8.5	11.3	15.0

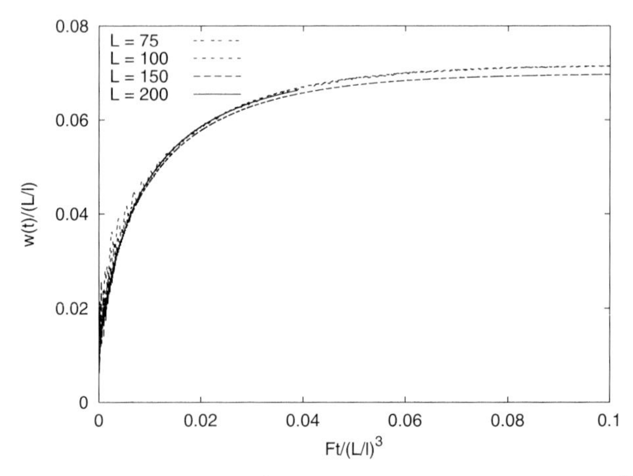

Fig. 4.3. Rescaled surface width $\omega\frac{l}{L}$ as a function of rescaled time $t\frac{Fl^3}{L^3}$ for various values of L.

of film roughness and is defined as [23]

$$\omega(t) = \frac{1}{2}\left(\frac{1}{|\Omega|}\int_\Omega \phi^2 - \overline{\phi}^2 \, dx\right)^{1/2}$$

with the mean value $\overline{\phi} = |\Omega|^{-1}\int_\Omega \phi \, dx$. In the limit $t \to \infty$ one obtains

$$\lim_{t\to\infty} \omega(t) \sim \frac{L}{l}.$$

Fig. 4.3 shows the data collaps by rescaling the surface width and time using this ansatz. Thus the time needed to reach the steady state solution scales as $(L/l)^3$.

A similar approach with the incorporation of the kinetic boundary conditions (2.4), (2.5) is under investigation [26].

Acknowlegdments. This work has been supported by the DFG Priority Program 1095 "Analysis, Modeling and Simulation of Multiscale Problems" under Ec 151/5-1, Em 68/13-1, and Vo 899/3-1.

References

1. L. Balykov, A. Voigt. A kinetic model for step flow growth of [100] steps. *Phys. Rev. E*, 72(2):022601, 2005.
2. L. Balykov, A. Voigt. A $2+1$ dimensional terrace-step-kink model for epitaxial growth far from equilibrium. *Multiscale Model. Sim.*, 5(1):45–61, 2006.
3. E. Bänsch, F. Haußer, O. Lakkis, B. Li, A. Voigt. Finite element method for epitaxial growth with attachment-detachment kinetics. *J. Comput. Phys.*, 194:409–434, 2004.
4. E. Bänsch, F. Haußer, A. Voigt. Finite element method for epitaxial growth with thermodynamic boundary conditions. *SIAM J. Sci. Comp.*, 26:2029–2046, 2005.
5. E. Bauser. Atomic mechanisms in semiconductor liquid phase epitaxy. *Handbook of Crystal Growth*, Vol. 3, ed. D. T. J. Hurle, North-Holland, Amsterdam, 1994.
6. W. K. Burton, N. Cabrera, F. C. Frank. Role of dislocations in crystal growth. *Nature*, 163(4141):398–399, 1949.
7. W. K. Burton, N. Cabrera, F. C. Frank. The growth of crystals and the equilibrium of their surfaces. *Phil. Trans. Roy. Soc. London Ser. A*, 243(866):299–358, 1951.
8. R. E. Caflisch, W. E, M. F. Gyure, B. Merriman, C. Ratsch. Kinetic model for a step edge in epitaxial growth. *Phys. Rev. E*, 59(6):6879– 6887, 1999.
9. R. E. Caflisch, M. F. Gyure, B. Merriman, S. Osher, C. Ratsch, D. Vvedensky, J. Zink. Island dynamics and the level set method for epitaxial growth. *Applied Math Letters*, 12:13–22, 1999.
10. G. Caginalp. An analysis of a phase field model of a free boundary. *Arch. Ration. Mech. Anal.*, 92:205–245, 1986.
11. P. Cermelli, M. Jabbour. Multispecies epitaxial growth on vicinal surfaces with chemical reactions and diffusion. *Proc. Royal Soc. A-Math. Phy.*, 461(2063):3483–3504, 2005.
12. A. A. Chernov, T. Nishinaga. *Growth shapes and stability, in Morphology of Crystals*, ed. I. Sungawa (Terra Scientific Publ. Co.,), p. 270, 1987.
13. W. Dorsch, S. Christiansen, M. Albrecht, P. O. Hansson, E. Bauser, H. P. Strunk. Early growth stages of $Ge_{0.85}Si_{0.15}$ on Si(001) from Bi solution *Surf. Sci.*, 896:331, 1994.
14. Ch. Eck. Analysis of a two–scale phase field model for liquid–solid phase transitions with equiaxed dendritic microstructures. *Multiscale Model. Sim.*, 3(1):28–49, 2004.
15. Ch. Eck, H. Emmerich. Models for liquid phase epitaxy. Preprint 146, DFG SPP 1095 "Mehrskalenprobleme", 2004.
16. Ch. Eck, H. Emmerich. A two–scale model for liquid phase epitaxy. Preprint 196, DFG SPP 1095 "Mehrskalenprobleme", 2006.
17. G. Ehrlich, F. G. Hudda. Atomic view of surface diffusion: tungsten on tungsten. *J. Chem. Phys.*, 44:1036–1099, 1966.
18. H. Emmerich, Ch. Eck. Morphology- transitions at heteroepitaxial surfaces. *Cont. Mech. Thermodynamics*, 17:373, 2006.
19. R. Ghez. Expansions in time for the solution of one-dimensional Stefan problems of crystal growth. *Int. J. Heat Mass Transfer*, 23:425, 1980.
20. F. Haußer, B. Li, A. Voigt. Step dynamics with kink Ehrlich-Schwoebel effect. In preparation.

21. F. Haußer, A. Rätz, A. Voigt. A level-set phase-field approach to step flow. In preparation.
22. V. V Jikov, S. M. Kozlov, O. A Oleinik. *Homogenization of Differential Operators and Integral Functionals*, Springer, Berlin - Heidelberg, 1994.
23. A. Karma, M. Plapp. Spiral surface growth without desorption. *Phys. Rev. Lett.*, 81:4444–4447, 1998.
24. L. D. Khutoryanskii, P. P. Petrov. *Sov. Phys. Crystallogr.*, 23:571, 1978.
25. J. Krug. Introduction to step dynamics and step instabilities. In A. Voigt, editor, *Multiscale modeling of epitaxial growth*, volume 149 of *ISNM*, pages 59–95. Birkhäuser, Basel, 2005.
26. J. Krug, T. Michely, A. Rätz, A. Voigt. In preparation.
27. F. P. J. Kuijpers, G. F. M. Beenker. The exact solution of the Stefan problem describing the growth rate of binary III-V compounds for LPE with linear cooling. *J. Cryst. Growth*, 48:411, 1979.
28. O. A. Ladyženskaja, V. A. Solonnikov, N. N. Ural'ceva. *Linear and Quasilinear Equations of Parabolic Type.* AMS Transl. Math. Monographs Vol. 23, Providence, Rhode Island, 1968.
29. M. Z. Li, J. W. Evans. Modeling of island formation during submonolayer deposition: A stochastic geometry- based simulation approach. *Multiscale Model. Sim.*, 3(3):629–657, 2005.
30. F. Liu, H. Metiu. Stability and kinetics of step motion on crystal surfaces. *Phys. Rev. E*, 49:2601–2616, 1997.
31. T. Michely, J. Krug. *Islands, Mounds, and Atoms: Patterns and Processes in Crystal Growth Far from Equilibrium.* Springer, 2004.
32. H. Müller-Krumbhaar. Diffusion theory for crystal growth at arbitrary solute concentration. *J. Chem. Phys.*, 63:5131, 1975.
33. W. W. Mullins, R. F. Sekerka. Stability of a planar interface during solidification of a dilute binary alloy. *J. Appl. Phys.*, 35:444, 1964.
34. F. Otto, P. Penzler, A. Rätz, T. Rump, A. Voigt. A diffuse interface approximation for step flow in epitaxial growth. *Nonlinearity*, 17:477–491, 2004.
35. M. Petersen, C. Ratsch, R. E. Caflisch, A. Zangwill. Level set approach to reversible epitaxial growth. *Phys. Rev. E*, 64(6):061602, 2001.
36. O. Pierre-Louis. Phase field models for step flow. *Phys. Rev. E*, 68(2):021604, 2003.
37. O. Pierre-Louis. Dynamics of crystal steps. *C. R. Phys.*, 6(1), 2005.
38. P. Politi, C. Castellano. Process of irreversible nucleation in multilayer growth. i. failure of the mean-field approach. *Phys. Rev. E*, 66(3):031605, 2002.
39. C. Ratsch, M. F. Gyure, R. E. Caflisch, F. Gibou, M. Petersen, M. Kang, J. Garcia, D. D. Vvedensky. Level-set method for island dynamics in epitaxial growth. *Phys. Rev. B*, 65(19):195403, 2002.
40. A. Rätz, A. Ribalta, A. Voigt. Surface evolution of elastically stressed films under deposition by a diffuse interface model. *J. Comput. Phys.*, 214(1):187–208, 2006.
41. A. Rätz, A. Voigt. Phase-field models for island dynamics in epitaxial growth. *Applicable Analysis*, 83:1015–1025, 2004.
42. A. Rätz, A. Voigt. A diffuse step-flow model with edge-diffusion. In A. Voigt, editor, *Multiscale modeling of epitaxial growth*, volume 149 of *ISNM*, pages 115–126. Birkhäuser, Basel, 2005.
43. R. L. Schwoebel. Step motion on crystal surfaces II. *J. Appl. Phys.*, 40:614–618, 1969.

44. R. L. Schwoebel, E. J. Shipsey. Step motion on crystal surfaces. *J. Appl. Phys.*, 37:3682–3686, 1966.
45. M. B. Small, E. Ghez, E. Giess. Liquid Phase Epitaxy. *Handbook of Crystal Growth*, Vol. 3, ed. D. T. J. Hurle, North–Holland, Amsterdam, 1994.
46. M. B. Small, E. Ghez, E. Giess. *Handbook of Crystal Growth*, Vol. 3, ed. D. T. J. Hurle, North-Holland, Amsterdam 1994.
47. E. M. Sparrow, J. L. Gregg. Mass transfer, flow and heat transfer about a rotating disk. *Trans. ASME J. Heat Transfer*, 82C:294, 1960.
48. G. Springholz, A. Y. Ueta, N. Frank, G. Bauer. Spiral growth and threading dislocations for molecular beam epitaxy of pbte on $BaF_2(111)$ studied by scanning tunneling microscopy. *Appl. Phys. Lett.*, 69(19):2822–2824, 1996.
49. N. Tokuda. A solution to a crystal growth Stefan problem by Lagrange-Bürmann expansions. *J. Cryst. Growth*, 67:358–369, 1984.
50. S. Vey, A. Voigt. AMDiS - adaptive multidimensional simulations. *Comput. Vis. Sci.*, to appear.
51. W. R. Wilcox. Crystallization flow. *J. Cryst. Growth*, 12:93, 1972.
52. L. O. Wilson, N. L. Schryer. Flow between a stationary and a rotating disk with suction. *J. Fluid Mech.*, 85:479, 1978.
53. Y. M. Yu, B. G. Liu. Phase-field model of island growth in epitaxy. *Phys. Rev. E*, 69(2):021601, 2004.

On Asymptotic Limits of Cahn-Hilliard Systems with Elastic Misfit

Harald Garcke and David Jung Chul Kwak

NWF I – Mathematik, Universität Regensburg, 93040 Regensburg.
harald.garcke@mathematik.uni-regensburg.de,
david.kwak@mathematik.uni-regensburg.de

1 Introduction

The aim of this work is to study the sharp interface limit of the Cahn-Hilliard equation in situations in which elastic stresses appear. The Cahn-Hilliard equation is a phase field model in the sense that interfaces are diffuse, i.e. across an interface an order parameter representing the phases changes its state rapidly, but in a smooth way. If elastic stresses are present, the Cahn-Hilliard equation has to be coupled to an elasticity system and this extended set of equations is called the Cahn-Larché system. For the Cahn-Hilliard equation it is well known that if the interfacial thickness $\varepsilon > 0$ tends to zero, the Mullins-Sekerka model is recovered. The Mullins-Sekerka model is a sharp interface model and can be formulated as a classical free boundary model. Also the sharp interface model can be extended to include elastic effects and it is the goal of this paper to discuss recent attempts to relate the Cahn-Larché system and the elastically modified Mullins-Sekerka model. We refer to the article by Garcke et al. [GL*06] in this book for more information on phase separation and Ostwald ripening which are both phenomena that can be modelled with the help of the Cahn-Larché system and the extended Mullins-Sekerka model. We also refer to [GL*06] for a discussion of situations where the two models can be reasonably used to recover the above phenomena.

Some work has been done already to study the sharp interface limit of the Cahn-Larché system. Fried and Gurtin [FG94] and Leo, Lowengrub and Jou [LLJ98] used the method of formally matched asymptotic expansions to relate the two models. Using this technique one has to assume that a smooth solution of the sharp interface model exists and fulfills certain smoothness properties, but to our knowledge there are no rigorous results known so far for the asymptotic limit of the Cahn-Larché system.

We present three results which relate the Cahn-Larché model to the sharp interface model. In Sect. 4 we will first show that the Cahn-Larché free energy, which is a Ginzburg-Landau type energy supplemented by contributions from elasticity, has a Γ-limit for ε tending to zero. The Γ-limit contains the classical surface energy together with elastic terms. Furthermore we will show in Section 4 that one can pass to the limit in the Euler-Lagrange equation for minimizers of the Cahn-Larché energy in order to obtain an elastically modified Gibbs-Thomson equation. This result generalizes a result of Luckhaus and Modica [LM89] to the Cahn-Larché system.

For general solutions we are going to use arguments and techniques from geometric measure theory to get rigorous results. Here one uses a priori estimates and compactness arguments to show convergence of the concentration, the chemical potential and the deformation vector. The main part is then to derive the Gibbs-Thomson law from the Cahn-Larché system. We are going to use methods introduced by Ilmanen [Ilm93], Soner [Son95] and Chen [Che96] in order to perform the limiting process in the context of the theory of varifolds. The analysis for the Cahn-Larché system is more complicated due to the fact that elastic terms appear in the Gibbs-Thomson equation through the so-called Eshelby tensor.

The outline of this work is as follows. After introducing the governing models in Sect. 2 we review basic knowledge on geometric measure theory and present related work in Sect. 3. In Sect. 4 we consider the stationary case and in Sect. 5 we discuss the general case in the context of geometric measure theory. Sect. 5 is part of the ongoing PhD thesis of the second author and we refer to the thesis [Kwa06] for more details.

2 The models

We start reviewing elasticity theory and the models which our analysis is based on.

2.1 Introduction to mechanics

We shortly introduce the basic concepts of linear elasticity, for a detailed introduction we refer to [Gur72], [Cia88] and [Bra91]. Denoting by a bounded region $\Omega \subset \mathbb{R}^n$ the reference state, we introduce the *deformation vector* $\mathbf{u}\colon \Omega \to \mathbb{R}^n$. Since in the applications we have in mind only small deformations appear, we consider a theory which is based on the *linearized strain tensor*

$$\mathcal{E}(\mathbf{u}) = 1/2(\nabla \mathbf{u} + \nabla \mathbf{u}^T).$$

The *elastic energy density* W is typically of quadratic form

$$W(c, \mathcal{E}) = \tfrac{1}{2}\big(\mathcal{E} - \mathcal{E}^*(c)\big) : \mathcal{C}(c)\big(\mathcal{E} - \mathcal{E}^*(c)\big) \tag{2.1}$$

with a symmetric and positive definite *elasticity tensor* $\mathcal{C}(c)$. We call $\mathcal{E}^*(c) = \mathcal{E}^* c$ the *eigenstrain* corresponding to c which describes the energetically favorable strain at concentration c. If $\mathcal{C}(c) = \mathcal{C}$ does not depend on the concentration, we speak of *homogeneous elasticity*, otherwise we use the term *inhomogeneous elasticity*. For the theory we are going to present in this work we will make the assumption that for a suitable constant $C > 0$ the following properties of W hold

$$\begin{aligned} W \in C^1(\mathbb{R}\times\mathbb{R}^{n\times n}, \mathbb{R}) \quad &\text{such that} \\ |W(c,\mathcal{E})| &\leq C(1+|c|^2+|\mathcal{E}|^2), \\ |W_{,\mathcal{E}}(c,\mathcal{E})| &\leq C(1+|c|^2+|\mathcal{E}|), \\ |W_{,c}(c,\mathcal{E})| &\leq C(1+|c|+|\mathcal{E}|). \end{aligned} \tag{2.2}$$

We assume in addition that $W(c,\mathcal{E})$ only depends on the symmetric part of $\mathcal{E} \in \mathbb{R}^{n\times n}$ and $W_{,\mathcal{E}}$ is strongly monotone, i.e. there exists a constant $c_1 > 0$ such that

$$(W_{,\mathcal{E}}(c,\mathcal{E}_1) - W_{,\mathcal{E}}(c,\mathcal{E}_2)) : (\mathcal{E}_2 - \mathcal{E}_1) \geq c_1|\mathcal{E}_2 - \mathcal{E}_1|^2. \tag{2.3}$$

We remark that an elasticity energy W according to equation (2.1) with $\mathcal{E}^*(c) = \mathcal{E}^* c$ does not fulfill (2.2), if the elasticity tensor $\mathcal{C}$ depends on the concentration c.

The mechanical equilibrium is attained on a much faster time scale compared to concentration changing by diffusion. This is why we assume that the mechanical equilibrium is attained instantaneously, so that the equation for the mechanics (2.4) does not involve any time derivatives and we hence consider at each time $t > 0$ the *quasi-stationary* system:

$$\operatorname{div} S = \operatorname{div} W_{,\mathcal{E}}(c, \mathcal{E}(\mathbf{u})) = 0 \tag{2.4}$$

where $S = S(c,\mathcal{E}) = W_{,\mathcal{E}}(c,\mathcal{E})$ is the *stress tensor*.

For definiteness we demand the deformation vector $\mathbf{u}$ to be in $\mathrm{X}_{\mathrm{ird}}^{\perp}$ with

$$\mathrm{X}_{\mathrm{ird}} := \{\mathbf{u} \in H^{1,2}(\Omega,\mathbb{R}^n) \mid \text{there exist } b \in \mathbb{R}^n \text{ and a skew symmetric } A \in R^{n\times n} \text{ such that } \mathbf{u}(x) = b + Ax\}$$

and $\mathrm{X}_{\mathrm{ird}}^{\perp}$ is the space perpendicular to $\mathrm{X}_{\mathrm{ird}}$ where perpendicular is meant with respect to the $H^{1,2}$-inner product. We remark that the energies of both the phase field and sharp interface models depend on $\mathbf{u}$ only through $\mathcal{E}(\mathbf{u})$ and hence the infinitesimal rigid part of $\mathbf{u}$ has no influence on the evolution of c. We have the *Korn inequality*

$$\|\mathbf{u}\|_{H^{1,2}(\Omega)} \leq \tilde{C}\|\mathcal{E}(\mathbf{u})\|_{L^2(\Omega)}$$

for all $\mathbf{u} \in \mathrm{X}_{\mathrm{ird}}^{\perp}$ for some constant $\tilde{C}$ (see Zeidler [Zei88]). In particular we obtain using (2.3) and an energy argument that $\mathbf{u} \in \mathrm{X}_{\mathrm{ird}}^{\perp}$ is uniquely determined by (2.4) and a stress-free boundary condition.

2.2 Phase field model

The Cahn-Larché model is based on the Ginzburg-Landau energy

$$\mathbf{E}_{\mathrm{pf}}^{\varepsilon}(c,\mathbf{u}) = \int_{\Omega} \left(\frac{\varepsilon}{2}|\nabla c|^2 + \frac{1}{\varepsilon}\Psi(c) + W(c,\mathcal{E}(\mathbf{u})) \right) \tag{2.5}$$

where $\varepsilon > 0$ is a small parameter related to the thickness of the diffuse interface, c is a scaled concentration difference, Ψ is a polynomial double well potential which we take to be

$$\Psi(c) = \frac{1}{4}(c^2-1)^2. \tag{2.6}$$

In the diffuse interface model the evolution problem related to (2.5) is the Cahn-Larché system

$$\partial_t c = \Delta w, \tag{2.7}$$

$$w = \frac{\delta \mathbf{E}_{\mathrm{pf}}^{\varepsilon}}{\delta c} = -\varepsilon \Delta c + \tfrac{1}{\varepsilon}\Psi'(c) + W_{,c}(c,\mathcal{E}(\mathbf{u})), \tag{2.8}$$

$$\operatorname{div} S = \operatorname{div} \frac{\delta \mathbf{E}_{\mathrm{pf}}^{\varepsilon}}{\delta \mathbf{u}} = 0 \tag{2.9}$$

where w is the chemical potential. We can view this system as the H^{-1} gradient flow of the energy functional (2.5), see [GL*06]. This structure will lead to crucial energy estimates of the Cahn-Larché system. The existence of solutions to this phase field system has been shown in [Gar00] and [Gar03]. The results are cited in Subsection 3.2.

2.3 Sharp interface model

The energy for the sharp interface limit is given by

$$\mathbf{E}_{\mathrm{si}} = \int_{\Gamma} 2\sigma \, d\mathcal{H}^{n-1} + \sum_{k=+,-} \int_{\Omega_k} W_k(\mathcal{E}(\mathbf{u}))\, dx \tag{2.10}$$

where $\sigma > 0$ is a surface energy constant and Γ is the interface (a hypersurface). The notation $\int_{\Gamma} . \, d\mathcal{H}^{n-1}$ denotes the integration with respect to the $(n-1)$-dimensional surface measure (the Hausdorff measure) and Ω_-, Ω_+ are the distinct regions occupied by the two phases with the corresponding elastic energy densities

$$W_-(\mathcal{E}) := W(-1,\mathcal{E}), \quad W_+ := W(+1,\mathcal{E}).$$

To simplify notation we set $W_+ = 0$ in Ω_- and vice versa, since then we can write $\sum_k \int_{\Omega_k} W_k = \int_{\Omega} \sum_k W_k$. Furthermore the surface energy density is

$$\sigma = \tfrac{1}{2} \int_{-\infty}^{\infty} (\tfrac{1}{2}(z'(y))^2 + \Psi(z(y)))dy,$$

where z is the solution of

$$-z'' + \Psi'(z) = 0 \qquad \text{with } z(-\infty) = -1 \text{ and } z(\infty) = 1.$$

One can easily compute

$$\sigma = \int_{-\infty}^{\infty} z'(y)\sqrt{\Psi(z(y))/2}\, dy = \int_{-1}^{1} \sqrt{\Psi(y)/2}\, dy.$$

The evolution problem related to the sharp interface energy is a modified Mullins-Sekerka problem

$$\Delta w = 0 \qquad \text{in } \Omega_-(t) \text{ and } \Omega_+(t), \tag{2.11}$$

$$V = -\tfrac{1}{2}[\nabla w]_-^+ \cdot \nu \qquad \text{on } \Gamma(t), \tag{2.12}$$

$$w = \sigma\kappa + \tfrac{1}{2}\nu^T[W\mathrm{Id} - (\nabla \mathbf{u})^T S]_-^+ \nu \qquad \text{on } \Gamma(t), \tag{2.13}$$

$$\operatorname{div} S = 0 \qquad \text{in } \Omega_-(t) \text{ and } \Omega_+(t), \tag{2.14}$$

$$[S\nu]_-^+ = [\mathbf{u}]_-^+ = 0,\ [w]_-^+ = 0 \qquad \text{on } \Gamma(t)$$

where $\Omega_-(t)$ and $\Omega_+(t)$ are the regions occupied by the phases at time t, $\Gamma(t)$ is the interface separating these regions, ν is the unit normal along the interface pointing towards Ω_+, V is the normal velocity of the interface and $[\,.\,]_-^+$ denotes the jump of the quantity in the brackets across the interface, e.g. $[w]_-^+ = w^+ - w^-$. κ is the mean curvature of $\Gamma(t)$ with the sign convention that κ is positive, if $\Gamma(t)$ is curved in the direction of ν. In contrast to its standard definition the mean curvature is taken here to be the sum of the principle curvatures. The first two equations are classical laws describing quasi-static diffusion driven by a chemical potential w. The third equation is the modified Gibbs-Thomson equation stating that the system is in local thermodynamical equilibrium.

Since we want to restrict our analysis to closed systems, we take homogeneous Neumann boundary conditions. In the phase field model this means

$$\nabla c \cdot \nu_\Omega = \nabla w \cdot \nu_\Omega = 0, \quad S\nu_\Omega = 0,$$

where ν_Ω denotes the outer unit normal of Ω. In the sharp interface model the condition for the concentration changes to an angle condition for the interface, so altogether the boundary conditions for the sharp interface model are

$$\angle(\Gamma(t), \partial\Omega) = 90^\circ, \quad \nabla w \cdot \nu_\Omega = 0, \quad S\nu_\Omega = 0.$$

3 Preliminaries

We introduce notations and recall some known facts about measures and varifolds (see also [EG92], [Fed69] and [Sim83]). We end this section by precisely stating the problems we want to analyze in this paper and with a discussion of related work.

3.1 Geometric measure theory

First we recall the definition of a Radon measure μ on an open set $\Omega \subset \mathbb{R}^n$ as a Borel regular measure that is finite on compact sets. To a measure μ we introduce the notion of densities on Ω for $x \in \Omega$

$$\theta^{*n-1}(\mu, x) = \limsup_{\rho \to 0} \frac{\mu(\Omega \cap B_\rho(x))}{\omega_{n-1}\rho^{n-1}},$$

$$\theta_*^{n-1}(\mu, x) = \liminf_{\rho \to 0} \frac{\mu(\Omega \cap B_\rho(x))}{\omega_{n-1}\rho^{n-1}}.$$

Here ω_{n-1} is the volume of the $(n-1)$-dimensional unit ball. If $\theta^{*n-1}(\mu, x)$ and $\theta_*^{n-1}(\mu, x)$ coincide, this common value will be denoted by $\theta^{n-1}(\mu, x)$.

Now we look on the set of $(n-1)$-dimensional subspaces

$$\mathbb{P}^{n-1} := \{P \mid P \text{ is a } (n-1)\text{-dimensional subspace in } \mathbb{R}^n\} = \mathbb{S}^{n-1}/\{\pm 1\}.$$

We will use the same notation P for the orthogonal projection onto the subspace P. On $\mathbb{P}^{n-1}$ we use the metric induced by endomorphisms:

$$d(P, Q) := \|P - Q\|_{End}.$$

This enables us to define a varifold:

Definition 3.1. *A* varifold V *is a Radon measure on the* Grassmanian

$$G(\Omega) := \Omega \times \mathbb{P}^{n-1}.$$

Remark 3.2.

- Such varifolds are in fact $(n-1)$-varifolds. We use such varifolds, since we want to describe interfaces. One can see them to give spatial and tangential information independently of each other.
  Defining varifolds simply as Radon measures on $\Omega \times \mathbb{P}^{n-1}$, we have weakened the usual view that the tangential information is solely given by the spatial information (of a neighborhood).
- For a C^1-hypersurface $\mathcal{M}$, we can introduce a corresponding varifold V by setting
$$dV(x, P) = d\mathcal{H}^{n-1}\lfloor_{\mathcal{M}}(x)\, \delta_{T_x\mathcal{M}}(P).$$

Finally we introduce the mass measure of a varifold.

Definition 3.3. *The* mass measure *of a varifold is defined by*

$$\mu_V(A) := \int_{A \times \mathbb{P}^{n-1}} dV(x, P) \quad \textit{for } A \subset \Omega.$$

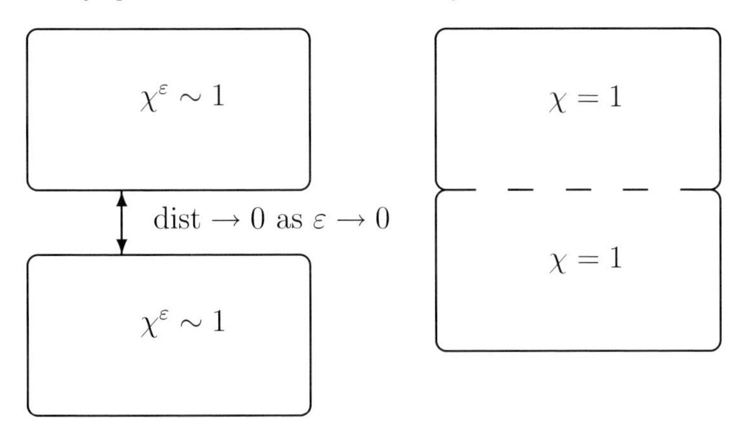

Fig. 3.1. An example where phase field interfaces lead to a varifold in the sharp interface limit.

The motivation to use varifolds is that the limiting interface will not provide sufficient smoothness to fulfill some kind of Gibbs-Thomson law in the classical sharp interface sense. In fact Schätzle has shown in [Sch97] that even the BV-formulation of the Gibbs-Thomson law breaks down when two interfaces touch each other. Introducing the notion of varifolds enable us to come up with a formulation which extends the model beyond the time of topological changes.

Remark 3.4. Bronsard and Stoth studied the related Allen-Cahn equation and proved that in the limit there exist interfaces with arbitrary high multiplicity, also called *phantom interfaces*, see [BS96]. Figure 3.1 gives an illustration of a time-independent example. Assume that the two regions of approximations χ^ε merge to one when letting $\varepsilon \to 0$. Then the dashed line is a phantom interface. Such phantom interfaces are not captured when using only characteristic functions.

First Variation of a varifold

In the smooth classical sense the Gibbs-Thomson law incorporates the mean curvature κ. Actually the curvature term occurs through the first variation of the area. For varifolds one has to use the first variation formula derived in Allard [All72] and Simon [Sim83].

As it can be found in the aforementioned works of Allard and Simon, the first variation of a varifold is given by

$$\delta V(X) = \int_{G(\Omega)} DX(x) \cdot P \, dV(x, P) \quad \text{for } X \in C_0^1(\Omega, \mathbb{R}^n) \tag{3.1}$$

where $DX(x) \cdot P$ is defined to be the inner product between linear mappings and $DX(x) \cdot P$ turns out to be the divergence of X with respect to the linear subspace P.

In fact, this coincides with the mean curvature in the smooth case. Using the Gauss theorem on a C^2-hypersurface $\mathcal{M}$

$$\int_{\mathcal{M}} X \cdot \nu_{\mathcal{M}} \kappa_{\mathcal{M}} \, d\mathcal{H}^{n-1} = \int_{\mathcal{M}} \operatorname{div}_{\mathcal{M}} X \, d\mathcal{H}^{n-1}$$

with $\nu_{\mathcal{M}}$ an arbitrary unit normal to $\mathcal{M}$ and $\kappa_{\mathcal{M}}$ the mean curvature of $\mathcal{M}$ with the sign according to $\nu_{\mathcal{M}}$. One notices that for $X \in C_0^1(\Omega, \mathbb{R}^n)$ the variation of the area can thus be read as the surface divergence of the vector field, i.e. the full divergence minus the normal part of DX.

In the case that the varifold is less smooth, but still has locally bounded first variation, one gets the following decomposition:

If $\|\delta V\|$ is a Radon measure, i.e.

$$\forall K \subset\subset \Omega \; \exists c_K > 0 \colon \quad |\delta V(X)| < c_K \|X\|_\infty \quad \forall X \in C_0^1(K, \mathbb{R}^n),$$

the first variation of V can be seen as a bounded operator on $C^0(\Omega, \mathbb{R}^n)$ and one has a $\|\delta V\|$-measurable function $\nu \colon \Omega \to \mathbb{P}^{n-1}$ such that

$$\delta V(X) = -\int_\Omega X \cdot \nu \, d\|\delta V\|.$$

We now take the Lebesgue decomposition of $\|\delta V\|$ with respect to μ_V:

$$\delta V(X) = \int_\Omega X \cdot \nu \, d\|\delta V\| = \int_\Omega X \cdot \mathbf{H}_V \, d\mu_V + \int_Z X \cdot \nu \, d\sigma \tag{3.2}$$

where $\mathbf{H}_V$ is the Radon-Nikodym derivative of $\|\delta V\|$ with respect to μ_V multiplied with the normal function ν:

$$\mathbf{H}_V(x) = \nu(x) D_{\mu_V} \|\delta V\|(x).$$

$\mathbf{H}_V$ is called *generalized mean curvature vector*. The set of singularities $Z := \{x \in \mathbb{R}^n \mid D_{\mu_V} \|\delta V\|(x) = \infty\}$ is the *generalized boundary* of V with *generalized boundary measure* σ, *generalized unit co-normal* $\nu|_Z$ and $\mu_V(Z) = 0$.

Rectifiability

For a $(n-1)$-rectifiable set $M \subset \Omega$ there exists for $\mathcal{H}^{n-1}$-a.e. $x \in M$ the approximate tangent plane to M, denoted by $T_x^{\text{app}} M$ (see [Sim83] for details). To such a set M one can associate a varifold V_M by setting

$$V_M(A) := \mathcal{H}^{n-1}\big(\{x \in \Omega \mid (x, T_x^{\text{app}} M) \in A\}\big) \quad \text{for } A \subset G(\Omega).$$

Definition 3.5. *A varifold V is* rectifiable, *if there exist $\theta_i > 0$ and $(n-1)$-rectifiable sets $M_i \subset \Omega$ for $i \in \mathbb{N}$ such that*

$$V = \sum_{i \in \mathbb{N}} \theta_i \, V_{M_i}.$$

Since varifolds represent an abstract concept, one goal is to confirm rectifiability, if not even *integrality*, which is the case when all θ_i are integers in the above identity.

The relation between rectifiability and the first variation is stated in the following theorem by Allard (see [All72]).

Theorem 3.6 (Allard). *Suppose a varifold V has locally bounded first variation in Ω and $\theta^{n-1}(\mu_V, x) > 0$ for μ_V-a.e. $x \in \Omega$, then V is already a rectifiable varifold.*

Remark 3.7. Especially for a varifold V with locally bounded first variation in Ω the restriction of V onto $\{x \mid \theta^{*n-1}(\mu_V, x) > 0\} \times \mathbb{P}^{n-1}$ is rectifiable.

The next theorem by Schätzle shows that the structure of the first variation can lead to the desired rectifiability (see [Sch01]).

Theorem 3.8 (Schätzle). *Let W be a varifold in $\Omega \subset \mathbb{R}^n$, $w \in H^{1,p}(\Omega)$, $n/2 < p < n$, $F \subset \Omega$ such that the characteristic function χ_F lies in $BV(\Omega)$. Furthermore we suppose*

1. $\delta W(\eta) = \int_\Omega \operatorname{div}(w\eta)\chi_F \quad \forall \eta \in C_0^1(\Omega, \mathbb{R}^n)$,
2. $|\nabla \chi_F| \leq \mu_W$ *and*
3. $\|w\|_{H^{1,p}(\Omega)} + \mu_W(\Omega) \leq \Lambda$ *for some $\Lambda \in \mathbb{R}$.*

Then W is rectifiable and has locally bounded first variation satisfying

$$\|\mathbf{H}_W\|_{L^s(\mu_W \lfloor B(x_0,r))} \leq C_{n,p}(r)\Lambda^{1+1/s} \qquad \forall B(x_0, 2r) \subset \Omega,$$

where $s \in \mathbb{R}$ such that $\frac{n-1}{s} = \frac{n}{p} - 1$.

The main part of the proof is to show a particular monotonicity formula for the density of the mass measure:

Lemma 3.9 (Monotonicity Formula). *For a varifold W which fulfills the assumptions of Theorem 3.8 the function*

$$\rho \mapsto \rho^{-(n-1)}\mu_W(B_\rho(x_0)) + C_{n,p} \min(1, d)^{-1} \Lambda \rho^\alpha \quad \forall x_0 \in \Omega, 0 < \rho < d$$

is non-decreasing for $\alpha = 1 - \frac{n-1}{s} \in (0,1)$ with $d = \operatorname{dist}(x_0, \partial\Omega)$.

Once this monotonicity formula is verified, one can use the following theorem by Ziemer.

Theorem 3.10 (Ziemer). *Let μ be a Radon measure on $\mathbb{R}^n$. Then the following statements are equivalent:*

1. $\mathcal{H}^{n-1}(A) = 0$ *implies that* $\mu(A) = 0$ *for all Borel sets* $A \subset \mathbb{R}^n$ *and there is a constant* $\bar{C}$ *such that* $\left|\int \phi d\mu\right| \leq \bar{C}\|\phi\|_{BV(\mathbb{R}^n)}$ *for all* $\phi \in BV(\mathbb{R}^n)$.
2. *There is a constant* $\bar{C}$ *such that* $\mu(B(x,r)) \leq \bar{C}r^{n-1}$.

By the theorem of Ziemer we obtain from Lemma 3.9 local bounds for the measure μ_W, i.e. for all $\phi \in BV(\Omega)$ and $B_\rho(x_0) \subset \Omega$

$$\left|\int_\Omega \phi \chi_{B_\rho(x_0)} d\mu_W\right| \leq \bar{C}\|\phi\|_{BV(\mathbb{R}^n)}.$$

Now, we choose $\phi = |w|^s$, which is in $H^{1,1}(\Omega)$ by imbedding theorems, and the first variation of the varifold W can be therefore estimated by

$$|\delta W(\eta)| \leq \left|\int (w\eta) d\mu_W\right| \leq \|w\|_{L^s(\mu_W)}\|\eta\|_{L^{s^*}(\mu_W)}.$$

By this estimate the first variation can be interpreted as a Radon measure and the above inequality leads to rectifiability of the varifold through the theorem of Allard.

3.2 Assumptions and notations

We start with solutions of the Cahn-Larché systems fulfilling the following assumptions (see also [Gar00] and [Gar03]) for $\Omega \subset \mathbb{R}^n$ open and bounded with smooth boundary. We consider for all $\varepsilon > 0$

$$\begin{aligned}
c^\varepsilon &\in L^2_{\text{loc}}(0,\infty; H^{2,2}(\Omega)) \cap H^{1,2}_{\text{loc}}(0,\infty; H^{-1,2}(\Omega)),\\
w^\varepsilon &\in L^2_{\text{loc}}(0,\infty; H^{1,2}(\Omega)),\\
\mathbf{u}^\varepsilon &\in L^2_{\text{loc}}(0,\infty; H^{2,2}(\Omega)^n)
\end{aligned}$$

such that the following weak formulation is fulfilled for all $T > 0$

$$\int_0^T \langle \partial_t c^\varepsilon, \zeta_1\rangle dt = \int_0^T\int_\Omega \nabla w^\varepsilon \cdot \nabla\zeta_1 \, dxdt, \tag{3.3}$$

$$\int_0^T\int_\Omega w^\varepsilon \zeta_2 \, dxdt = \int_0^T\int_\Omega \varepsilon\nabla c^\varepsilon \cdot \nabla\zeta_2 + \tfrac{1}{\varepsilon}\Psi'(c^\varepsilon)\zeta_2 + W_{,c}(c^\varepsilon, \mathbf{u}^\varepsilon)\zeta_2 \, dxdt, \tag{3.4}$$

$$0 = \int_0^T\int_\Omega S : D\zeta_3 \, dxdt \tag{3.5}$$

for all $\zeta_1 \in L^2(0,T; H^{1,2}(\Omega))$, $\zeta_2 \in L^2(0,T; H^{1,2}(\Omega)) \cap L^\infty(\Omega \times [0,T])$ and $\zeta_3 \in L^2(0,T; H^{1,2}(\Omega)^n)$. Here, the notation of $c^\varepsilon \in L^2_{\text{loc}}(0,\infty; H^{1,2}(\Omega))$ means that for all times $T > 0$ one has $c^\varepsilon \in L^2(0,T; H^{1,2}(\Omega))$ and $\langle .,.\rangle$ is the duality pairing between $H^{-1,2}(\Omega)$ and $H^{1,2}(\Omega)$. In contrast to other notations we define $H^{-1,2}(\Omega)$ as the dual of $\{c \in H^{1,2}(\Omega) \mid \int_\Omega c = 0\}$.

As initial conditions we assume that for all $\varepsilon > 0$

1. the initial energy is bounded: $\mathbf{E}^{\varepsilon}_{\mathrm{pf}}(0) \leq \mathbf{E}_0$ and
2. the integral of the initial concentration does not depend on ε, i.e. there exists a constant $m_0 \in (-1, 1)$ such that $\int_\Omega c_0^\varepsilon = m_0|\Omega|$.

Remark 3.11. The existence of weak solutions of the Cahn-Larché system has been shown in [Gar00] and [Gar03]. But so far it has not been verified in general, if the concentration and deformation vector are indeed in $H^{2,2}(\Omega)$ for almost all t.

In the case that W is of the quadratic form (2.1) with constant elasticity tensor $\mathcal{C}$, i.e. in the homogeneous case, the equation determining $\mathbf{u}$ can be read as an elliptic system with constant coefficients, where only the right-hand side depends on the concentration:

$$\operatorname{div} S = 0 \quad \Longleftrightarrow \quad \operatorname{div}[\mathcal{C}\mathcal{E}(\mathbf{u}^\varepsilon)] = \operatorname{div}[\mathcal{C}\mathcal{E}^*(c^\varepsilon)].$$

Since c^ε is in $H^{1,2}(\Omega)$, the right-hand side is in $L^2(\Omega)$, which leads $\mathbf{u}^\varepsilon$ to be in $H^{2,2}(\Omega)$ by elliptic regularity theory.

On the other side, $W_{,c} = \mathcal{E}^* : \mathcal{C}[\mathcal{E}(\mathbf{u}^\varepsilon) - \mathcal{E}^* c^\varepsilon]$ is in $L^2(\Omega)$. Now, equation (3.4) can be read as an elliptic equation for c^ε and again elliptic regularity theory can be used.

If one considers inhomogeneous elasticity, the elasticity system (2.4) contains possibly non-continuous coefficients $\mathcal{C}(c^\varepsilon)$ so that one cannot argue as in the homogeneous case. Though, in low dimensions due to Sobolev imbedding theorems the concentration functions c^ε are continuous and therefore elliptic regularity theory for smooth coefficients can be used.

Nevertheless, for the general case we are presenting in this work we have to include this assumptions in order to get the correct Gibbs-Thomson law, see Sect. 5.4.

One important first observation for the limiting process $\varepsilon \to 0$ is to identify

$$e^\varepsilon(c^\varepsilon) := \tfrac{\varepsilon}{2}|\nabla c^\varepsilon|^2 + \tfrac{1}{\varepsilon}\Psi(c^\varepsilon)$$

as the *interfacial energy density* in the phase field model. Heuristically, this is exactly the quantity one observes to carry the interfacial energy of the phase field model, and the goal is to show convergence to a quantity that will be understood up to a factor as the $\mathcal{H}^{n-1}$-measure of the interface.

The second important function is the so-called *discrepancy measure*

$$\xi^\varepsilon(c^\varepsilon) := \tfrac{\varepsilon}{2}|\nabla c^\varepsilon|^2 - \tfrac{1}{\varepsilon}\Psi(c^\varepsilon). \tag{3.6}$$

As it is stated in Theorem 5.7, in the limit $\varepsilon \to 0$ the discrepancy measure will be non-positive, which means that the Ψ-part is larger than the $|\nabla c^\varepsilon|^2$-part.

3.3 Related works

One important source of this work is the paper by Chen [Che96]. He has studied the asymptotic limit of the Cahn-Hilliard model. Chen showed for arbitrary spatial dimensions that solutions of the Cahn-Hilliard system converge globally in time to some generalized sharp-interface solution. He did not show that the limit varifold is rectifiable, but in the case $p = 2, n = 3$ one can use the Theorem 3.8 by Schätzle to deduce rectifiability for the limit varifold for the Cahn-Hilliard systems without elasticity, see Remark 5.3.

There is one significant difference to results for the related Allen-Cahn models which are proposed to describe motion of phase boundaries driven by surface tension:

$$\varepsilon \frac{\partial c}{\partial t} = \varepsilon \Delta c - \varepsilon^{-1} \Psi'(c).$$

Ilmanen [Ilm93] has studied the limiting behavior of the Allen-Cahn equation towards the mean curvature flow in the sense of Brakke [Bra78] and confirmed that one gets in the limit

$$\xi = 0.$$

This is also known as *equipartition of energy*. It is quite interesting to note that the interface energy is asymptotically equally distributed between the $|\nabla c^\varepsilon|^2$- and the $\Psi(c^\varepsilon)$-part. Moreover this result can be used for further results, namely it is easier to deduce the fact that the resulting interface varifold is rectifiable.

After Ilmanen [Ilm93] first used geometric measure theory to prove such convergence in $\Omega = \mathbb{R}^n$, Soner [Son95] improved the result for more general settings. Hutchinson and Tonegawa studied in [HT00] the asymptotic behavior of critical, not necessarily minimal points of the Cahn-Hilliard energy functional. In their work they also used geometric measure theory and derived local estimates for the discrepancy measure (3.6). By that, they gained convergence results for bounded domains. In their (time-independent) setting the limit varifold turns out to be integral, i.e. the interface has indeed integer multiplicity modulo a surface constant almost everywhere. Moreover local minimizers of the Cahn-Hilliard energy functional converge to a local area minimizer subject to a volume constraint. Later Tonegawa extended with similar estimates the results by Ilmanen and showed that time-dependent solutions of the Allen-Cahn equation converge to an integral varifold, cf. [Ton03].

4 The stationary case

Before we study the evolution problem, we consider the sharp interface limit of the Ginzburg-Landau energy $\mathbf{E}^\varepsilon_{\text{pf}}(c, \mathbf{u})$ in the limit ε tending to zero. As the Cahn-Larché system conserves the integral of the concentration c, we will consider $\mathbf{E}^\varepsilon_{\text{pf}}$ subject to an integral constraint on c. In fact in this case one can show that $\mathbf{E}_{\text{si}}$ is the Γ-limit of $\mathbf{E}^\varepsilon_{\text{pf}}$, even if we take the constraint into account. Furthermore, we present a result stating that the Lagrange multipliers related to minimizers of $\mathbf{E}_{\mathrm{pf}}^{\varepsilon}$ will converge to a Lagrange multiplier related to a minimizer of $\mathbf{E}_{\mathrm{si}}$ subject to a volume constraint. The results we present will generalize results of Modica [Mod87] and Luckhaus and Modica [LM89] to the case including elastic effects.

4.1 The Γ-limit of the Cahn-Larché energy

In this subsection we study solutions of the variational problems:

($\mathbf{P}^{\varepsilon}$) Find a minimizer $(c, \mathbf{u}) \in H^{1,2}(\Omega) \times X_{ird}^{\perp}$ of $\mathbf{E}_{\mathrm{pf}}^{\varepsilon}$ subject to the constraint $\frac{1}{|\Omega|} \int_{\Omega} c = m_0$, where $m_0 \in (-1, 1)$ is a given constant.

We will now present a result stating that solutions to ($\mathbf{P}^{\varepsilon}$) converge along subsequences to a minimizer of the functional

$$\mathbf{E}^0 \colon L^1(\Omega) \times \mathrm{X}_{\mathrm{ird}}^{\perp} \to \mathbb{R} \cup \{\infty\}$$

where

$$\mathbf{E}^0(c, \mathbf{u}) = \begin{cases} 2\sigma \mathcal{H}^{n-1}(\partial\{c = 1\} \cap \Omega) + \int_{\Omega} W(c, \mathcal{E}(\mathbf{u})) & \text{if } c \in BV(\Omega, \{-1, 1\}) \\ & \text{and } \frac{1}{|\Omega|} \int_{\Omega} c = m_0, \\ \infty & \text{otherwise.} \end{cases}$$

The following theorem now states that $\mathbf{E}^0$ is the Γ-limit of $\mathbf{E}_{\mathrm{pf}}^{\varepsilon}$. We also obtain that minimizers of $\mathbf{E}_{\mathrm{pf}}^{\varepsilon}$ approximate minimizers of $\mathbf{E}_{\mathrm{si}}$, if we take a volume constraint into account. The limiting variational problem is a partitioning problem taking interfacial energy and elastic effects into account.

The following theorem has been shown in [Gar00].

Theorem 4.1. *Assume that the assumptions of Ψ and W as stated above hold and let Ω be a bounded domain with Lipschitz boundary. Then it holds:*

1. *For all $(c^{\varepsilon_k}, \mathbf{u}^{\varepsilon_k})_{k \in \mathbb{N}} \in H^{1,2}(\Omega) \times \mathrm{X}_{ird}^{\perp}$ with $c^{\varepsilon_k} \to c$ in $L^1(\Omega)$ and $\mathbf{u}^{\varepsilon_k} \to \mathbf{u}$ in $L^2(\Omega, \mathbb{R}^n)$ as ε_k tends to zero, it holds*

$$\mathbf{E}^0(c, \mathbf{u}) \le \liminf_{k \to \infty} \mathbf{E}_{pf}^{\varepsilon_k}(c^{\varepsilon_k}, \mathbf{u}^{\varepsilon_k}).$$

2. *For any $(c, \mathbf{u}) \in L^1(\Omega) \times \mathrm{X}_{ird}^{\perp}$ and any sequence $\varepsilon_k \to 0, k \in \mathbb{N}$, there exists a sequence $(c^{\varepsilon_k}, \mathbf{u}^{\varepsilon_k})_{k \in \mathbb{N}} \in H^{1,2}(\Omega) \times \mathrm{X}_{ird}^{\perp}$ with $c^{\varepsilon_k} \to c$ in $L^1(\Omega)$ and $\mathbf{u}^{\varepsilon_k} \to \mathbf{u}$ in $L^2(\Omega, \mathbb{R}^n)$ as $\varepsilon_k \to 0$ such that*

$$\mathbf{E}^0(c, \mathbf{u}) \ge \limsup_{k \to \infty} \mathbf{E}_{pf}^{\varepsilon_k}(c^{\varepsilon_k}, \mathbf{u}^{\varepsilon_k}).$$

3. *Let* $(c^\varepsilon, \mathbf{u}^\varepsilon)$ *be solutions of problem* $(\mathbf{P}^\varepsilon)$. *Then there exists a sequence* $\varepsilon_k \to 0, k \in \mathbb{N}$ *and* $(c, \mathbf{u}) \in L^1(\Omega) \times \mathrm{X}_{ird}^\perp$ *such that*

$$\begin{aligned} c^{\varepsilon_k} &\to c \quad \text{in } L^1(\Omega), \\ \mathbf{u}^{\varepsilon_k} &\to \mathbf{u} \quad \text{in } H^{1,2}(\Omega, \mathbb{R}^n) \end{aligned}$$

and $(c, \mathbf{u})$ *is a global minimizer of* $\mathbf{E}^0$.

For the proof and for a generalization to the situation of more than two phases we refer to [Gar00].

4.2 Convergence of the Lagrange multipliers

For a minimizer $(c, \mathbf{u})$ of $\mathbf{E}^0$ it can be shown that a constant Lagrange multiplier ω exists such that

$$2\sigma\kappa + \nu \cdot \left[W\mathrm{Id} - (\nabla\mathbf{u})^T W_{,\mathcal{E}}\right] \nu = 2\omega. \tag{4.1}$$

A minimizer of $\mathbf{E}^0$ minimizes $\mathbf{E}_{\mathrm{si}}$ subject to a volume constraint and ω is the Lagrange multiplier related to this constraint.

In the case that no elastic effects are present, we obtain that the mean curvature is constant and the term $\nu \cdot \left[W\mathrm{Id} - (\nabla\mathbf{u})^T W_{,\mathcal{E}}\right] \nu$ modifies this law. In particular the mean curvature can be inhomogeneous along the interface. The identity (4.1) and its non-equilibrium analogue (2.13) can be interpreted as a generalized Gibbs-Thomson equation.

Absolute minimizers $(c^\varepsilon, \mathbf{u}^\varepsilon)$ of $\mathbf{E}_{\mathrm{pf}}^\varepsilon$ have a constant Lagrange ω^ε which fulfills in a distributional sense (see [Gar00])

$$-\varepsilon\Delta c^\varepsilon + \tfrac{1}{\varepsilon}\Psi'(c^\varepsilon) + W_{,c}(c^\varepsilon, \mathcal{E}(\mathbf{u}^\varepsilon)) = \omega^\varepsilon. \tag{4.2}$$

In [Gar00] it is shown that the Lagrange multipliers of $(c^\varepsilon, \mathbf{u}^\varepsilon)$ converge (along subsequences) to a Lagrange multiplier ω of the sharp interface variational problem. Here we state the result in detail.

Theorem 4.2. *Let* Ω *be a domain with a* C^1*-boundary and assume that* Ψ *and* W *fulfill the conditions stated above. Furthermore let* $(c^\varepsilon, \mathbf{u}^\varepsilon) \in H^{1,2}(\Omega) \times \mathrm{X}_{ird}^\perp$ *be a solution of the variational problem* $(\mathbf{P}^\varepsilon)$ *with Lagrange multipliers* ω^ε. *Then for each sequence* $(\varepsilon_k)_{k\in\mathbb{N}} \to 0$ *such that*

$$\begin{aligned} c^{\varepsilon_k} &\to c \quad \text{in } L^1(\Omega), \\ \mathbf{u}^{\varepsilon_k} &\to \mathbf{u} \quad \text{in } H^{1,2}(\Omega, \mathbb{R}^n) \end{aligned}$$

it holds

$$\omega^{\varepsilon_k} \to \omega,$$

where ω *is the Lagrange multiplier for the absolute minimizer* $(c, \mathbf{u})$ *of* $\mathbf{E}^0$, *compare* (4.1).

For a proof we refer to [Gar00]. We remark that although the method of Luckhaus and Modica [LM89] for the case without elasticity is used in the proof, one cannot follow their arguments in a straightforward way. This is due to the fact that not enough regularity is known for the minimizer $(c^\varepsilon, \mathbf{u}^\varepsilon)$ of $\mathbf{E}^\varepsilon_{\mathrm{pf}}$. In the proof of Theorem 4.2 one uses variations of $\mathbf{E}^\varepsilon_{\mathrm{pf}}$ with respect to the independent variables and shows that the resulting Lagrange multiplier is related to the Lagrange multiplier from (4.2), which is the first variation with respect to the dependent variables.

Remark 4.3. We also note that a minimizer $(c, \mathbf{u})$ of $\mathbf{E}^0$ also fulfills

$$\int_\Omega 2\sigma(\nabla \cdot \xi - \nu \cdot \nabla\xi\nu)|\nabla\chi_{\{c=-1\}}| + \int_\Omega (W\mathrm{Id} - (\nabla\mathbf{u})^T W_{,\mathcal{E}}) : \nabla\xi = \int_\Omega \lambda c \nabla \cdot \xi$$

for all $\xi \in C^\infty(\bar{\Omega}, \mathbb{R}^n)$ with $\xi \cdot \nu_\Omega = 0$ on $\partial\Omega$. Here, $\nu = -\frac{\nabla\chi_{\{c=-1\}}}{|\nabla\chi_{\{c=-1\}}|}$ is the generalized outer unit normal to $\{c = -1\}$ which is a $|\nabla\chi_{\{c=-1\}}|$-measurable function. The above identity is a weak formulation of the modified Gibbs-Thomson equation (4.1) (see [Gar00]).

5 The time-dependent case

For the evolutionary system we start with a suitable weak formulation of the sharp interface problem. Through the limiting process one cannot expect that the resulting limit objects are smooth enough such that equations (2.12)–(2.14) can be verified in a classical way. Besides concentration, the chemical potential and deformation vector which converge quite straightforward in the limiting process, to formulate a Gibbs-Thomson law we need both a characteristic function and a varifold, which represents the interface as motivated in Sect. 3.1 including possible phantom interfaces.

After stating the theorem we give an overview on the proof, not giving all the details due to the limited space and refer to [Kwa06] for a full treatment.

5.1 Statement of the main theorem

First we specify the notion of a generalized solution of the sharp interface model.

Definition 5.1 (Generalized solution). *$(M, V, w, \mathbf{u})$ is said to be a* generalized solution of the modified Mullins-Sekerka problem, *if*

$M \subset \Omega \times [0, \infty)$, $w \in L^2_{loc}(0, \infty; H^{1,2}(\Omega))$, $\mathbf{u} \in L^2_{loc}(0, \infty; H^{1,2}(\Omega)^n)$

V is a Radon measure on $\Omega \times \mathbb{P}^{n-1} \times (0, \infty)$.

Moreover $\chi_M \in C^0([0, \infty); L^1(\Omega)) \cap L^\infty(0, \infty; BV(\Omega))$ and

V^t is a varifold on Ω for all $t > 0$

such that for all $T > 0$, for almost every $0 < \tau < t < T$ and for all test functions $\zeta \in C_0^1(\bar{\Omega} \times [0,T))$, $\mathbf{Y} \in C_0^1(\Omega, \mathbb{R}^n)$ and $\mathbf{X} \in L_0^2(0,T; H^{1,2}(\Omega, \mathbb{R}^n))$ the following holds:

1. $\int_0^T \int_\Omega [-2\chi_{M^t}\partial_t\zeta + \nabla w \nabla \zeta] = \int_\Omega 2\chi_{M^0}\zeta(.,0)$,
2. $2\int_\Omega \chi_{M^t} \operatorname{div}(w\mathbf{Y}) = \langle \partial V^t, \mathbf{Y}\rangle + \sum_{k=+,-} \int_\Omega (W_k^t Id - (\nabla \mathbf{u})^T S_k^t) : D\mathbf{Y}$,
3. $dV^t(x,P) = \sum_i \rho_i^t(x)\delta_{\nu_i^t(x)}(P) d\mu^t(x) dP$,
4. $d\mu^t(x) \geq 2\sigma |D\chi_{M^t}|(x)dx$,
5. $\mu^t(\Omega) + \sum_k \int_\Omega W_k^t + \int_\tau^t \int_\Omega |\nabla w|^2 \leq \mu^\tau(\Omega) + \sum_k \int_\Omega W_k^\tau$,
6. $\int_0^T \int_\Omega S : D\mathbf{X}\, dxdt = 0$

where $\rho_i^t \in [0,1], \sum_i \rho_i^t \geq 1, \sum_i \nu_i^t \otimes \nu_i^t = Id$ and μ^t is a Radon measure on $\bar{\Omega}$. An upper index $\{.\}^t$ denotes the time.

Remark 5.2. Let us discuss the definition in more detail. The first equation is the weak formulation of the diffusion equations (2.11) and (2.12). In the bulk the chemical potential will be harmonic. Equation 2 is the Gibbs-Thomson law (2.13) in a weak formulation (cf. Remark 4.3 and [Gar00]). Equations 3 and 4 describe properties of the varifold. Inequality 4 allows that the varifold can possibly see phantom interfaces. Equation 5 states the dissipation of the free energy and equation 6 states in a weak form that the stress is divergence free in the bulk, cf. (2.14), and at the same time one obtains that the normal jump of the stress is zero across the interface.

One should notice that the Gibbs-Thomson law has two terms which represent the interface and vanish in the bulk, but the elastic term stays a volume integral. The reason for this is that the elastic energy is a non-local volume energy. So, one has to be aware in the limiting process that both $\frac{\varepsilon}{2}|\nabla c^\varepsilon|^2$ and $\Psi(c^\varepsilon)$ converge to a $(n-1)$-dimensional measure while W stays n-dimensional.

Remark 5.3. In the case of Cahn-Hilliard systems, i.e. without any elastic terms, equation 2 in Definition 5.1 becomes of the same form as in the Theorem 3.8 by Schätzle. This means that one can deduce rectifiability of the varifold in the case without elasticity, at least for the case $p = 2, n = 3$.

Theorem 5.4 (Main). *Let the assumptions mentioned in Sect. 3.2 hold. Then there is a sequence $\varepsilon_i \to 0$ and a generalized solution $(M, V, w, \mathbf{u})$ as in Definition 5.1 such that for all $T > 0$*

1. $c^{\varepsilon_i} \to -1 + 2\chi_M$ *in* $C^{1/9}([0,T]; L^2(\Omega))$ *and almost everywhere,*
2. $w^{\varepsilon_i} \to w$ *weakly in* $L^2(0,T; H^{1,2}(\Omega))$,
3. $\mathbf{u}^{\varepsilon_i} \to \mathbf{u}$ *in* $L^2(0,T; H^{1,2}(\Omega)^n)$.

More precisely the varifold is obtained in the following way.

Proposition 5.5. *For the sequence of Theorem 5.4 it further holds:*

1. *There exist Radon measures μ, μ_{kl} on $\bar{\Omega} \times [0, \infty)$ such that*

$$e^{\varepsilon_i}(c^{\varepsilon_i})dxdt \to d\mu(x,t), \tag{5.1}$$
$$\varepsilon_i c^{\varepsilon_i}_{x_k} c^{\varepsilon_i}_{x_l} dxdt \to d\mu_{kl}(x,t) \tag{5.2}$$

both as Radon measures on $\bar{\Omega} \times [0,T]$ for all $T > 0$.

2. *For all $\mathbf{Y} \in C^1_0(\Omega \times [0,T], \mathbb{R}^n)$ it holds:*

$$\int_0^T \langle \partial V^t, \mathbf{Y} \rangle = \int_0^T \int_\Omega \nabla \mathbf{Y} : [d\mu(x,t) Id - (d\mu_{ij}(x,t))_{ij}].$$

Remark 5.6. The first part of the proposition follows easily by the energy estimates and using compactness properties of Radon measures. So it is left to show that the measures μ and μ_{ij} can be indeed identified as a varifold. This is essentially done by proving Theorem 5.7.

We define for $\varepsilon > 0$ the set

$$\mathcal{K}_\varepsilon := \{(c,v) \in H^{2,2}(\Omega) \times L^2(\Omega) \mid -\varepsilon \Delta c + \varepsilon^{-1} \Psi'(c) = v \text{ in } \Omega \text{ and } \partial_\nu c = 0 \text{ on } \partial\Omega\}.$$

Theorem 5.7. *There exist a constant $\eta_0 \in (0,1]$ and continuous and non-increasing functions $M_1(\eta), M_2(\eta) \colon (0, \eta_0] \to (0, \infty)$ such that for every $\eta \in (0, \eta_0]$, every $\varepsilon \in (0, M_1(\eta)^{-1}]$ and every $(c,v) \in \mathcal{K}_\varepsilon$ it holds*

$$\int_\Omega (\xi^\varepsilon(c))^+ \leq \eta \int_\Omega e^\varepsilon(c) + \varepsilon M_2(\eta) \int_\Omega v^2. \tag{5.3}$$

Remark 5.8. In the application of Theorem 5.7, v will be the sum

$$v = w^\varepsilon - W_{,c}(c^\varepsilon, \mathcal{E}(\mathbf{u}^\varepsilon)).$$

5.2 Convergence of concentration and chemical potential

One crucial a priori estimate is due to the H^{-1} gradient flow property of the Cahn-Larché system with respect to the energy functional (2.5) which we already mentioned in Sect. 2.2. For more details see [Gar03].

Lemma 5.9. *For all $\varepsilon > 0$ and $0 < \tau < t$ it holds*

$$\mathbf{E}^\varepsilon_{pf}(t) + \int_\tau^t \int_\Omega |\nabla w^\varepsilon|^2 \leq \mathbf{E}^\varepsilon_{pf}(\tau).$$

From equations (3.3) and (3.4) one easily gets the following a priori estimates:

Lemma 5.10. *For all $\varepsilon > 0$ and almost all $t > 0$ it holds*

1. $\frac{1}{|\Omega|}\int_\Omega c^\varepsilon(.,t) = m_0$,
2. $\int_\Omega |c^\varepsilon|^4 \leq C(1+\mathbf{E}_0)$,
3. $\int_\Omega (|c^\varepsilon|-1)^2 \leq C\varepsilon\mathbf{E}_0$.

Remark 5.11. The first equation describes one feature of the phase field model: *conservation of mass* over time. This is essentially due to the diffusion which is driven by a potential and the Neumann boundary conditions.

We introduce the auxiliary function

$$\tilde{c}^\varepsilon(x,t) := \int_{-1}^{c^\varepsilon(x,t)} \sqrt{\tilde{\Psi}(s)/2}ds,$$

which is also known as the *Modica ansatz*. Here $\tilde{\Psi}(s) := \min(\Psi(s), 1+|s|^2)$ is used, so one has

$$C_1|s_1 - s_2|^2 \leq |\tilde{c}(s_1) - \tilde{c}(s_2)| \leq C_2|s_1 - s_2|(1+|s_1|+|s_2|)$$

for some $C_1, C_2 > 0$. Using this auxiliary function it is possible to obtain bounds in $BV(\Omega)$.

Lemma 5.12. *For solutions to the Cahn-Larché system the Modica ansatz leads to*

$$\|\tilde{c}^\varepsilon\|_{L^\infty(0,\infty;H^{1,1}(\Omega))} + \|\tilde{c}^\varepsilon\|_{C^{1/8}([0,\infty);L^1(\Omega))} + \|c^\varepsilon\|_{C^{1/8}([0,\infty);L^2(\Omega))} \leq C. \quad (5.4)$$

With these uniform bounds one can pass to the limit $\varepsilon \to 0$ along a sequence and together with Lemma 5.10 identify a set $M \subset \Omega \times [0,\infty)$ such that we have the following lemma:

Lemma 5.13. *For solutions of the Cahn-Larché system there exists a sequence $\varepsilon_j \to 0$ such that*

- $\tilde{c}^{\varepsilon_j}(x,t) \to 2\sigma\chi_M$ *in* $C^{1/9}([0,T];L^1(\Omega))$,
- $c^{\varepsilon_j}(x,t) \to -1 + 2\chi_M$ *in* $C^{1/9}([0,T];L^2(\Omega))$ *and almost everywhere*

for all $T > 0$.

This set M then defines $\Omega_-(t)$ for all $t > 0$.

This proves the first convergence statement of the main theorem. For the chemical potential we observe that the following Poincaré type inequality holds:

Lemma 5.14. *For the solutions of the Cahn-Larché system we obtain*

$$\|w^\varepsilon(.,t)\|_{H^{1,2}(\Omega)} \leq C(\mathbf{E}^\varepsilon_{pf}(t) + \|\nabla w^\varepsilon(.,t)\|_{L^2(\Omega)}). \quad (5.5)$$

To prove this lemma we test equation (3.4) with $\mathbf{X}\cdot\nabla c^\varepsilon$ for $\mathbf{X} \in C_0^1(\Omega, \mathbb{R}^n)$ to get

$$\int w^\varepsilon \mathbf{X}\cdot\nabla c^\varepsilon = \int \varepsilon\nabla c^\varepsilon \cdot \nabla(\mathbf{X}\cdot\nabla c^\varepsilon) + \tfrac{1}{\varepsilon}\Psi'(c^\varepsilon)\mathbf{X}\cdot\nabla c^\varepsilon + W_{,c}(c^\varepsilon, \mathcal{E}(\mathbf{u}^\varepsilon))\mathbf{X}\cdot\nabla c^\varepsilon$$
$$= \int \varepsilon\left(\nabla c^\varepsilon \cdot D\mathbf{X}\nabla c^\varepsilon - \tfrac{1}{2}\operatorname{div}\mathbf{X}\,|\nabla c^\varepsilon|^2\right) + (\tfrac{1}{\varepsilon}\Psi' + W_{,c})\mathbf{X}\cdot\nabla c^\varepsilon.$$

Now we see that via partial integration

$$\int D\mathbf{X} : (\Psi\mathrm{Id}) = \int \operatorname{div}\mathbf{X}\,\Psi = -\int \mathbf{X}\cdot\nabla c^\varepsilon\,\Psi', \tag{5.6}$$
$$\int D\mathbf{X} : (W\mathrm{Id}) = \int \operatorname{div}\mathbf{X}\,W = -\int \mathbf{X}\cdot\nabla c^\varepsilon\,W_{,c} + \mathbf{X}_i W_{,\mathcal{E}_{kl}}\partial_i\partial_k \mathbf{u}_l^\varepsilon \tag{5.7}$$
$$= -\int \mathbf{X}\cdot\nabla c^\varepsilon W_{,c} - (\partial_k\mathbf{X}_i)W_{,\mathcal{E}_{kl}}\partial_i\mathbf{u}_l^\varepsilon, \tag{5.8}$$

where we used equation (3.5). With $W_{,\mathcal{E}_{kl}} = S_{kl}$ we obtain

$$\int \operatorname{div}(w^\varepsilon\mathbf{X})c^\varepsilon = \int D\mathbf{X} : \left[e^\varepsilon(c^\varepsilon)\mathrm{Id} - \varepsilon\nabla c^\varepsilon\otimes\nabla c^\varepsilon + W\mathrm{Id} - (\nabla\mathbf{u}^\varepsilon)^T S\right]. \tag{5.9}$$

We now introduce the mean value of w^ε as $\bar{w}^\varepsilon$ and use integration by parts to obtain

$$\int_\Omega w^\varepsilon\mathbf{X}\cdot\nabla c^\varepsilon = -\int_\Omega \nabla w^\varepsilon\cdot\mathbf{X}c^\varepsilon - \int_\Omega (w^\varepsilon - \bar{w}^\varepsilon)c^\varepsilon\operatorname{div}\mathbf{X} - \bar{w}^\varepsilon\int_\Omega c^\varepsilon\operatorname{div}\mathbf{X}. \tag{5.10}$$

Combining equation (5.9) and (5.10), one arrives at

$$\bar{w}^\varepsilon = \frac{1}{\int_\Omega c^\varepsilon\operatorname{div}\mathbf{X}}\int_\Omega D\mathbf{X} : \left[(e^\varepsilon(c^\varepsilon) + W(c^\varepsilon, \mathbf{u}^\varepsilon))\mathrm{Id} - \varepsilon\nabla c^\varepsilon\otimes\nabla c^\varepsilon - (\nabla\mathbf{u}^\varepsilon)^T S\right]$$
$$- \nabla w^\varepsilon\cdot\mathbf{X}c^\varepsilon - (w^\varepsilon - \bar{w}^\varepsilon)c^\varepsilon\operatorname{div}\mathbf{X}\,dx,$$

where we choose a smooth $\mathbf{X}$ such that $\int_\Omega c^\varepsilon\operatorname{div}\mathbf{X} \neq 0$. Using elliptic regularity theory, the lemma can be verified.

With this bound we conclude to weak convergence of the chemical potential.

Corollary 5.15. *There exist constants $C, \varepsilon_0 > 0$ such that for all $\varepsilon \in (0, \varepsilon_0]$ and all $T > 0$ it holds*

$$\int_T^{T+1} \|w^\varepsilon(.,t)\|_{H^{1,2}(\Omega)} \leq C. \tag{5.11}$$

Therefore, for a sequence $\varepsilon_j \to 0$ there exists a function $w \in L^2(0, T; H^{1,2}(\Omega))$ such that

$$w^{\varepsilon_j} \to w \qquad \text{weakly in } L^2(0, T; H^{1,2}(\Omega)).$$

5.3 Convergence of deformation

Using the monotonicity of $W_{,\mathcal{E}}$, see (2.3), we obtain that the elastic energy density fulfills

$$W(c,\mathcal{E}) \geq C_0|\mathcal{E}|^2 - C_1(|c|^2+1)$$

for some constants $C_0, C_1 > 0$. Therefore we have for solutions $(c^\varepsilon, \mathbf{u}^\varepsilon)$

$$\int_\Omega |\mathcal{E}(\mathbf{u}^\varepsilon)|^2\,dx \leq C\left(1 + \int_\Omega W(c^\varepsilon, \mathcal{E}(\mathbf{u}^\varepsilon))\,dx + \int_\Omega |c^\varepsilon|^2\,dx\right).$$

Since the W-term is bounded by the total energy $\mathbf{E}^\varepsilon_{\mathrm{pf}}$ and the c^ε-term by the a priori estimate in Lemma 5.10, we have that $\|\mathcal{E}(\mathbf{u}^\varepsilon)\|_{L^2(\Omega)}$ is bounded uniformly in t and ε. By Korn's inequality we can also control the deformation vector $\mathbf{u}^\varepsilon$ in $H^{1,2}(\Omega)^n$.

Since $L^2(0,T;H^{1,2}(\Omega)^n)$ is a reflexive space, we have weak compactness of the deformation vector, i.e. for all sequences $(\varepsilon_j)_{j\in\mathbb{N}}$ there exists a subsequence $(\varepsilon_{j_k})_{k\in\mathbb{N}}$ such that

$$\mathbf{u}^{\varepsilon_{j_k}} \to \mathbf{u} \quad \text{weakly in } L^2_{\mathrm{loc}}(0,\infty;H^{1,2}(\Omega)^n)$$

for some $\mathbf{u} \in L^2_{\mathrm{loc}}(0,\infty;H^{1,2}(\Omega)^n)$.

Now we use again the monotonicity of $W_{,\mathcal{E}}$ to get

$$\begin{aligned} &c_1\|\mathcal{E}(\mathbf{u}^\varepsilon - \mathbf{u})\|^2_{L^2(\Omega\times(0,T))} \\ &\leq \int_{\Omega\times(0,T)} \big(W_{,\mathcal{E}}(c^\varepsilon, \mathcal{E}(\mathbf{u}^\varepsilon)) - W_{,\mathcal{E}}(c^\varepsilon, \mathcal{E}(\mathbf{u}))\big) : \mathcal{E}(\mathbf{u}^\varepsilon - \mathbf{u}) \\ &= -\int_{\Omega\times(0,T)} W_{,\mathcal{E}}(c^\varepsilon, \mathcal{E}(\mathbf{u})) : \mathcal{E}(\mathbf{u}^\varepsilon - \mathbf{u}). \end{aligned} \tag{5.12}$$

The last equality is due to the divergence free stress tensor, cf. (2.4). One should notice that only $W_{,\mathcal{E}}(c^\varepsilon, \mathcal{E}(\mathbf{u}^\varepsilon))$, but not $W_{,\mathcal{E}}(c^\varepsilon, \mathcal{E}(\mathbf{u}))$ is divergence free, since only in the former term the respective deformation function $\mathbf{u}^\varepsilon$ meets the condition (2.4).

By the strong convergence of the concentration function and the weak convergence of the deformation field, the right hand side of equation (5.12) goes to zero, i.e. we obtain strong convergence of the strain tensor for the sequence $(\varepsilon_{j_k})_{k\in\mathbb{N}}$. By Korn's inequality we have that the deformation vector converges strongly in $L^2_{\mathrm{loc}}(0,\infty;H^{1,2}(\Omega)^n)$. Then for almost all t we have that $\nabla\mathbf{u}^{\varepsilon_{j_k}}(t)$ converges strongly to $\nabla\mathbf{u}(t)$ in $L^2(\Omega)$.

This verifies the third convergence statement of the main theorem. So far, we have shown the convergences as stated in the theorem, but we still have to verify, if the limit functions do represent a generalized solution according to Definition 5.1. Indeed the first diffusion equation immediately follows from equation (3.3) and the convergences of the concentration function and potential. The identity 6 in Definition 5.1 follows from (3.5) in the limit $\varepsilon \to 0$, as $\nabla\mathbf{u}^\varepsilon$ and c^ε converge strongly. The other conditions require the specification of the varifold.

5.4 The limit varifold and the Gibbs-Thomson law

This part deals with the limit varifold V. It is mainly derived from the convergence mentioned in Proposition 5.5 and we show that using Theorem 5.7 we verify the remaining conditions of Definition 5.1.

The energy density $e^\varepsilon(c^\varepsilon) := \frac{\varepsilon}{2}|\nabla c^\varepsilon|^2 + \frac{1}{\varepsilon}\Psi(c^\varepsilon)$ and $\varepsilon \nabla c^\varepsilon \otimes \nabla c^\varepsilon$ are bounded by the initial energy:

$$\int_0^T \int_\Omega e^\varepsilon(c^\varepsilon)\, dx\, dt \leq \int_0^T \mathbf{E}^\varepsilon_{pf}(t)\, dt \leq T\mathbf{E}_0$$

$$\int_0^T \int_\Omega \varepsilon\, |(\nabla c^\varepsilon)_i (\nabla c^\varepsilon)_j|\, dx\, dt \leq \int_0^T \int_\Omega e^\varepsilon(c^\varepsilon)\, dx\, dt \leq T\mathbf{E}_0.$$

By compactness there exist Radon measures μ and μ_{ij} according to (5.1) and (5.2). But since we also have energy estimates for all times t, we can split the measures $d\mu(x,t)$ and $d\mu_{ij}(x,t)$ into a spatial and time component, both being still Radon measures:

$$d\mu(x,t) = d\mu^t(x)dt, \qquad d\mu_{ij}(x,t) = d\mu^t_{ij}(x)dt.$$

The energy estimates in Lemma 5.9 show that the energies of the phase field solutions are non-increasing. This feature carries through the limit ε going to zero:

Lemma 5.16. *For a sequence $\varepsilon_k \to 0$ there exists a non-increasing function $\mathbf{E}\colon [0,\infty) \to [0,\infty)$ such that for all $t > 0$*

$$\mathbf{E}^{\varepsilon_k}_{pf}(t) \to \mathbf{E}(t).$$

One has to verify that this function $\mathbf{E}$ is indeed the energy of the sharp interface model. As mentioned above the interfacial energy converges to a Radon measure μ. Together with the strong convergence of the deformation vector $\mathbf{u}$, the function $\mathbf{E}$ can be identified as the energy of the limiting system:

$$\mathbf{E}(t) = \mu^t(\Omega) + \int_\Omega \sum_{k=+,-} W_k(\mathcal{E}(\mathbf{u}^t)).$$

This shows that part 5 of Definition 5.1 is fulfilled in the limit $\varepsilon \to 0$.

Equation (5.9) gives in the limit $\varepsilon \to 0$

$$\int 2\chi_{\Omega_-} \operatorname{div}(w\mathbf{X}) = \int D\mathbf{X} : [d\mu \mathrm{Id} - (d\mu_{ij})_{ij}] + \int D\mathbf{X} : (W\mathrm{Id} - (\nabla \mathbf{u})^T S).$$

Remark 5.17. The claim is now that $\int D\mathbf{X} : [d\mu \mathrm{Id} - (d\mu_{ij})_{ij}]$ can be seen as the first variation of a varifold. This will prove Proposition 5.5. Hence, part 2 of Definition 5.1 will be verified.

Proof (Proposition 5.5). For $\mathbf{Y}, \mathbf{Z} \in C^0(\bar{\Omega} \times [0,T], \mathbb{R}^n)$ one gets

$$\int_0^T \int_\Omega \mathbf{Y}^T \cdot (\varepsilon_k \nabla c^k \otimes \nabla c^k)\mathbf{Z} \leq \int_0^T \int_\Omega |\mathbf{Y}||\mathbf{Z}| e^{\varepsilon_k}(c^{\varepsilon_k}) + \int_0^T \int_\Omega |\mathbf{Y}||\mathbf{Z}| \xi^{\varepsilon_k}(c^{\varepsilon_k}).$$

This means that in the limit $\varepsilon \to 0$, the last integral is non-positive and one gets the inequality

$$\int_0^T \int_\Omega \mathbf{Y}^T \cdot (d\mu_{ij})_{ij} \mathbf{Z} \leq \int_0^T \int_\Omega |\mathbf{Y}||\mathbf{Z}| d\mu$$

which means that the measures μ_{ij} are absolutely continuous with respect to μ. Then there exist μ-measurable functions ν_{ij} such that $d\mu_{ij}(x,t) = \nu_{ij}(x,t) d\mu(x,t)$ and we get

$$0 \leq \int_0^T \int_\Omega \mathbf{Y} \cdot \big(\mathrm{Id} - (\nu_{ij})_{ij}\big) \mathbf{Z}\, d\mu(x,t). \tag{5.13}$$

Since the matrix $(\nu_{ij})_{ij}$ inherits the symmetry from (5.2), the matrix is positive semi-definite and by (5.13) it further holds $0 \leq (\nu_{ij})_{ij} \leq \mathrm{Id}$. This means one can write this matrix as $(\nu_{ij})_{ij} = \sum_{i=1}^n \tilde{\rho}_i \boldsymbol{\nu}_i \otimes \boldsymbol{\nu}_i$ where $\tilde{\rho}_i \in [0,1], \sum_i \boldsymbol{\nu}_i \otimes \boldsymbol{\nu}_i = \mathrm{Id}$. Moreover $\sum_i \tilde{\rho}_i \leq 1$, since actually for $y \in C^0(\Omega \times [0,T])$

$$\int_0^T \int_\Omega y\, \varepsilon_k \underbrace{\mathrm{tr}(\nabla c^k \otimes \nabla c^k)}_{=|\nabla c^{\varepsilon_k}|^2} = \int_0^T \int_\Omega y\, (e^{\varepsilon_k}(c^{\varepsilon_k}) + \xi^{\varepsilon_k}(c^{\varepsilon_k}))$$

and $\lim_{k\to\infty} \varepsilon_k \,\mathrm{tr}(\nabla c^k \otimes \nabla c^k) = \sum_i (\nu_{ii}) d\mu$. Recall that the trace of a matrix is the sum of its eigenvalues.

Setting $\rho_i := \tilde{\rho}_i + \frac{1}{n-1}\left(1 - \sum_{j=1}^n \tilde{\rho}_j\right) \in [0,1]$ we get

$$\mathrm{Id} - (\nu_{ij})_{ij} = \mathrm{Id} - \sum_i \tilde{\rho}_i \boldsymbol{\nu}_i \otimes \boldsymbol{\nu}_i = \sum_i \rho_i \,(\mathrm{Id} - \boldsymbol{\nu}_i \otimes \boldsymbol{\nu}_i)\,.$$

Thus we can see the limiting varifold as

$$dV(x,P) = \sum_i \rho_i(x) d\mu(x) \delta_{\boldsymbol{\nu}_i}(P)$$

where $\delta_{\boldsymbol{\nu}_i}$ is the projection onto the hyperplane normal to $\boldsymbol{\nu}_i$.

5.5 Control of discrepancy measure

In the case of homogeneous elasticity we have

$$|W_{,c}(c, \mathcal{E}(\mathbf{u}))| \leq C(1 + |c| + |\mathcal{E}(\mathbf{u})|)$$

which leads $W_{,c}(c^\varepsilon, \mathcal{E}(\mathbf{u}^\varepsilon))$ to be in $L^2(\Omega)$ for almost all times $t > 0$. So we can follow the proof of Chen in [Che96] for the estimation of the discrepancy measure.

The proof is based on a blow-up technique for which we need some preparatory lemmas.

Lemma 5.18. *For each $\eta > 0$ there is a constant $R(\eta) > 2$ such that for all $R > R(\eta)$ the following holds:*

If

$$\hat{\Omega} = \{(x', x_n) \in B_R \mid x_n > Y(x')\}$$

is a domain in $\mathbb{R}^n$, $Y \colon \mathbb{R}^{n-1} \to \mathbb{R}$ satisfying

$$Y(0') \leq 0, \quad \nabla_{x'} Y(0') = 0', \quad \|D^2_{x'} Y\|_{C^0(B'_R)} \leq R^{-3} \tag{5.14}$$

and if $(\mathbf{c}, \mathbf{v}) \in H^{1,2}(\hat{\Omega}) \times L^2(\hat{\Omega})$ with

$$-\Delta \mathbf{c} + \Psi'(\mathbf{c}) = \mathbf{v} \qquad \text{in } \hat{\Omega}, \tag{5.15}$$

$$\frac{\partial}{\partial \nu} \mathbf{c} = 0 \qquad \text{on } \{(x', x_n) \in B_R \mid x_n = Y(x')\}, \tag{5.16}$$

$$\|\mathbf{v}\|_{L^2(B_R \cap \hat{\Omega})} \leq R^{-1} \tag{5.17}$$

then the following inequality holds

$$\int_{B_1 \cap \hat{\Omega}} \left(|\nabla \mathbf{c}|^2 - 2\Psi(\mathbf{c})\right)^+ \leq \eta \int_{B_2 \cap \hat{\Omega}} \left(|\nabla \mathbf{c}|^2 + \Psi'(\mathbf{c})^2 + \Psi(\mathbf{c}) + \mathbf{v}^2\right) + \int_{\{x \in B_1 \cap \hat{\Omega} \mid |\mathbf{c}| \geq 1 - \eta\}} |\nabla \mathbf{c}|^2. \tag{5.18}$$

Proof. For the proof, which we roughly sketch, one studies the interfacial region:

$$\hat{\Omega}_1 := \{x \in B_1 \cap \hat{\Omega} \mid |\mathbf{c}| \leq 1 - \eta\}.$$

For the case that $|\hat{\Omega}_1|$ is sufficiently small, one gets by Hölder inequality

$$\|\nabla \mathbf{c}\|_{L^2(\hat{\Omega}_1)} \leq |\hat{\Omega}_1|^{m^*} \|\nabla \mathbf{c}\|_{L^{2^*}(\hat{\Omega}_1)} \leq C\eta \|\nabla \mathbf{c}\|_{H^{1,2}(B_1 \cap \hat{\Omega})}$$

where $m^* = \frac{2\,2^*}{2-2^*}$ with $2^* = \frac{2n}{n-2}$ for $n > 2$ and $2^* = 7$ otherwise. One can notice that one η appears on the right hand side, which finally leads to the statement. For the other case $|\hat{\Omega}_1|$ being large one can use a contradiction argument. Through this assumption the homogeneous equation $\Delta \mathbf{c} = \Psi'(\mathbf{c})$ is recovered. Here one can use elliptic regularity theory to get smoothness of the function $\mathbf{c}$. Comparison with viscosity functions yields that $\mathbf{c}$ would be in fact bounded in $[-1, 1]$. Results by Modica [Mod85] then finally finish the proof.

Now we need a control on the bulk energy of the interface. This is shown in the following lemma.

Lemma 5.19. *There exist positive constants C_0 and η_0 such that for every $\eta \in (0, \eta_0]$, every $\varepsilon \in (0, 1]$ and every $(c, v) \in \mathcal{K}_\varepsilon$ the following holds*

$$\int_{\{x \in \Omega \mid |c| \geq 1 - \eta\}} \left(e^\varepsilon(c) + \varepsilon^{-1} \Psi'(c)^2\right) \leq C_0 \eta \int_{\{x \in \Omega \mid |c| \leq 1 - \eta\}} \varepsilon |\nabla c|^2 + C_0 \varepsilon \int_\Omega v^2. \tag{5.19}$$

The proof of this lemma is based on the convexity property of Ψ'' for values $|c| \geq 1-\eta$. One combines both

$$\int_\Omega v\psi = \int_\Omega \varepsilon\psi'(c)|\nabla c|^2 + \varepsilon^{-1}\Psi'(c)\psi$$

from the equation (3.4) and the Young inequality

$$\int_\Omega v\psi \leq \int_\Omega \left(\tfrac{\varepsilon}{2}v^2 + \tfrac{1}{2\varepsilon}\psi^2\right)$$

where $\psi = \Psi'$ except in the bulk, so that one has bounds for ψ' in $\{|c| \geq 1-\eta\}$.

Proof (Theorem 5.7). We give a simple sketch of the proof for Theorem 5.7. As already mentioned above we use a blow-up technique. The set Ω is covered by balls $B_{R\varepsilon}(x_j)$ where R is as in Lemma 5.18. Changing variables to $y \to x_j + \varepsilon y$ and rescaling $\mathbf{v}^j(y) = \varepsilon v^\varepsilon(x_j + \varepsilon y)$, one gets the equation

$$-\Delta_y \mathbf{c}^j + \Psi'(\mathbf{c}^j) = \mathbf{v}.$$

By this blow-up process the right-hand side $\mathbf{v}$ can be decreased so much that Lemma 5.18 is applicable. Using Lemma 5.19 this ends the proof. If $\mathbf{v}$ does not fulfill the assumptions of Lemma 5.18, other elliptic estimates can be used. The careful choice of the covering then ensures that by assembling the covering the desired estimate is attained.

References

[All72] W. K. Allard. *On the first variation of a varifold.* Ann. of Math., **95** (1972), pp. 417–491.

[Bra91] D. Braess. *Finite Elemente.* Springer-Verlag, Berlin-Heidelberg-New York, 1991.

[Bra78] K. A. Brakke. *The motion of a surface by its mean curvature.* Mathematical Notes, Princeton University Press, 1978.

[BS96] L. Bronsard and B. Stoth. *On the existence of high multiplicity interfaces.* Math. Res. Let., **3** (1996), pp. 41–50.

[Che96] X. Chen. *Global asymptotic limit of solutions of the Cahn-Hilliard equation.* J. Differential Geom., **44** (1996), pp. 262–311.

[Cia88] P. G. Ciarlet. *Elasticity Theory, Volume I: Three-dimensional Elasticity.* North-Holland, Amsterdam, 1988.

[EG92] L. C. Evans and R. F. Gariepy. *Measure Theory and Fine Properties of Functions.* CRC Press, Boca Raton, 1992.

[Fed69] H. Federer. *Geometric Measure Theory.* Springer-Verlag, Berlin-Heidelberg-New York, 1969.

[FPL95] P. Fratzl, O. Penrose, and J. L. Lebowitz. *Modelling of phase separation in alloys with coherent elastic misfit.* J. Stat. Physics, **95** (1999), nos. 5/6, pp. 1429–1503.

[FG94] E. Fried and M. E. Gurtin. *Dynamic solid-solid transitions with phase characterized by an order parameter.* Physica D, **72** (1994), pp. 287–308.

[Gar00] H. Garcke. *On mathematical models for phase separation in elastically stressed solids.* Habilitation thesis, University Bonn, 2000.

[Gar03] H. Garcke. *On Cahn-Hilliard systems with elasticity.* Proc. Roy. Soc. Edinburgh Sect. A, **133** (2003), no. 2, pp. 307–331.

[GL*06] H. Garcke, M. Lenz, B. Niethammer, M. Rumpf, U. Weikard. *Multiple scales in phase separating systems with elastic misfit.* In "Analysis, Modeling and Simulation of Multiscale Problems, A. Mielke (edr), Springer-Verlag, 2006."

[Gur72] M. E. Gurtin. *The Linear Theory of Elasticity.* Handbuch der Physik, Vol. VIa/2, Springer, S. Flügge and C. Truesdell (eds.), Berlin, 1972.

[HT00] J. Hutchinson and Y. Tonegawa. *Convergence of phase interfaces in the van der Waals-Cahn-Hilliard theory.* Calc. Var. and Part. Diff. Equat., **10** (2000), no. 1, pp. 49–84.

[Ilm93] T. Ilmanen. *Convergence of the Allen-Cahn equation to Brakke's motion by mean curvature.* J. Differential Geom., **38** (1993), no. 2, pp. 417–461.

[Kwa06] D. Kwak. PhD thesis, University of Regensburg, in preparation.

[LLJ98] P. H. Leo, J. S. Lowengrub and H. J. Jou. *A diffuse interface model for microstructural evolution in elastically stressed solids.* Acta Mater., **46** (1998), pp. 2113–2130.

[LM89] S. Luckhaus, L. Modica. *The Gibbs-Thomson relation within the gradient theory of phase transitions.* Arch. Rat. Mech. Anal., **107** (1989), pp. 71–83.

[Mod85] L. Modica. *A gradient bound and a Liouville theorem for nonlinear Poisson equations.* Comm. Pure Appl. Math., **38** (1985), pp. 679–684.

[Mod87] L. Modica. *The gradient theory of phase transitions and the minimal interface criterion.* Arch. Rat. Mech. Anal., **98** (1987), pp. 123–142.

[Sim83] L. Simon. *Lectures on Geometric Measure Theory.* Proceedings of the Centre for Mathematical Analysis, Australian National University, Vol. 3, 1983.

[Sch97] R. Schätzle. *A counterexample for an approximation of the Gibbs-Thomson law.* Adv. Math. Sci. Appl., **7** (1997), no. 1, pp. 25–36.

[Sch01] R. Schätzle. *Hypersurfaces with mean curvature given by an ambient Sobolev function.* J. Differential Geom., **58** (2001), pp. 371–420.

[Son95] H. M. Soner. *Convergence of the phase field equations to the Mullins-Sekerka problem with a kinetic undercooling.* Arch. Rat. Mech. Anal., **131** (1995), pp. 139–197.

[Ton03] Y. Tonegawa. *Integrality of varifolds in the singular limit of reaction-diffusion equations.* Hiroshima Math. J., **33** (2003), no. 3, pp. 323–341.

[Zei88] E. Zeidler. *Nonlinear Functional Analysis and its Applications IV.* Applications to Mathematical Physics, Springer, Berlin-Heidelberg-New York, 1988.

Simulations of Complex Microstructure Formations

Britta Nestler[1] and Frank Wendler[2]

[1] Faculty of Computer Science, Karlsruhe University of Applied Sciences.
britta.nestler@hs-karlsruhe.de
[2] Institute of Applied Research, Karlsruhe University of Applied Sciences.
frank.wendler@hs-karlsruhe.de

Summary. In this article, we review the progress on phase-field modelling, sharp interface asymptotics, numerical simulations and applications to microstructure evolution and pattern formation in materials science. Model formulations and computations of pure substances and of binary alloys are discussed. Furthermore, a thermodynamically consistent class of non-isothermal phase-field models for crystal growth and solidification in complex alloy systems is presented. Explicit expressions for the different energy density contributions are proposed. Multicomponent diffusion in the bulk phases including interdiffusion coefficients as well as diffusion in the interfacial regions are discussed. Anisotropy of both, the surface energies and the kinetic coefficients is incorporated in the model formulation. A 3D parallel simulator based on a finite difference discretization is introduced illustrating the capability of the model to simultaneously describe the diffusion processes of multiple components, the phase transitions between multiple phases and the development of the temperature field. The numerical solving method contains parallelization and adaptive strategies for optimization of memory usage and computing time. Applying the computational methods, we show a variety of simulated microstructure formations in complex multicomponent alloy systems occuring on different time and length scales. In particular, we present 2D and 3D simulation results of dendritic and eutectic solidification in pure substances and binary and ternary alloys. Another field of application is the modelling of competing polycrystalline grain structure formation and grain growth.

1 Introduction

Materials science plays a tremendous role in modern engineering and technology, since it is the basis of the entire microelectronics and foundry industry, as well as many other industries. The manufacture of almost every man-made object and material involves phase transformations and solidification at some stage. Metallic alloys are the most widely-used group of materials in industrial applications. During the manufacture of castings, solidification of metallic melts occurs involving many different phases and hence, various kinds of phase

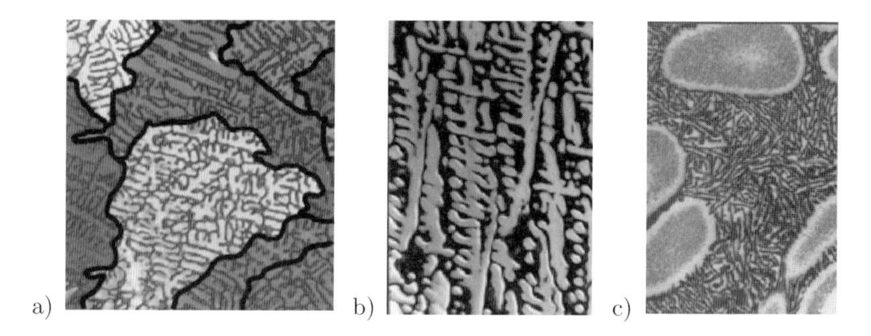

Fig. 1.1. Experimental micrographs of $Al - Si$ alloy samples: a) Grain structure with different crystal orientations, b) network of primary Al dendrites and c) interdendritic eutectic microstructure of two distinguished solid phases in the regions between the primary phase dendrites. (See page 683 for a colored version of the figure.)

transitions [KF92]. The solidification is accompanied by a variety of different pattern formations and complex microstructure evolutions. Depending on the process conditions and on the material parameters, different growth morphologies can be observed, significantly determining the material properties and the quality of the castings. For improving the properties of materials in industrial production, the detailed understanding of the dynamical evolution of grain and phase boundaries is of great importance. Since numerical simulations provide valuable information of the microstructure formation and give access for predicting characteristics of the morphologies, it is a key to understanding and controlling the processes and to sustaining continuous progress in the field of optimizing and developing materials.

The solidification process involves growth phenomena on different length and time scales. For theoretical investigations of microstructure formation it is essential to take these multiscale effects as well as their interaction into consideration. The experimental photographs in Fig. 1.1 give an illustration of the complex network of different length scales that exist in solidification microstructures of alloys.

The first image (Fig. 1.1 a)) shows a polycrystalline Al-Si grain structure after an electrolytical etching preparation. The grain structure contains grain boundary triple junctions which themselves shown an individual physical behaviour. The coarsening by grain boundary motion takes place on a long time scale. If the magnification is enlarged, a dendritic substructure in the interior of each grain can be resolved (Fig. 1.1 b)). Each orientational variant of the polycrystalline structure consists of a dendritic array in which all dendrites of a specific grain have the same crystallographic orientation. The third image in Fig. 1.1 c) displays the interdendritic eutectic structure at a higher resolution, where eutectic lamellae solidify between the primary dendritic phase. In a eutectic solidification of this kind, two distinct solid phases S_1 and S_2

grow into an undercooled melt if the temperature is below the critical eutectic temperature. Within the interdendritic eutectic lamellae, a phase boundary triple junction of the two solid phases and the liquid occurs. The dendrites and the eutectic lamellae grow into the melt on a micrometer scale and during a short time period. Once the dendrites and the eutectic lamellae impinge one another, grain boundaries are formed.

Alloy systems with multiple components are an important class of materials, in particular for technical applications and processes. The microstructure formation of a material plays a central role for a broad range of mechanical properties and, hence, for the quality and the durability of the material. Aiming for a continuous optimization of materials properties, the study of pattern formation during alloy solidification has been a focus of many experimental and, recently, also of computational work. Since the microstructure characteristics are a result of the process conditions used during production, the analysis of the fundamental correlation between the processing pathway and the microstructure is of fundamental importance. Multiple components in alloys are combined with the appearance of multiple phases leading to complex phase diagrams, various phase transformations and different types of solidification. Modelling and numerical simulations aim to predict microstructure evolution in multicomponent alloys in order to virtually design materials. However, the great number of material parameters and of physical variables involved in systems yields a complexity that remains a big challenge for future work. In particular, the gain of statistically meaningful data from computations requires simulations in sufficiently large domains with a tremendous need of memory and computing time resources. To treat complex systems, high performance computing, parallelization and optimized algorithms including adaptive mesh generators are mandatory.

The phase-field method has become an important tool for tackling free boundary problems such as grain boundary motion [Sea96, GNS99a], and for simulating crystal growth, solidification and pattern formation phenomena in alloys [WMB96, TN*98, BS98, KWC00, PK00, NW00, JGD01, SPG01, BW*02, GBP02, AB*02, KO*03, RB*04]. The advantage of the phase-field method lies in the formulation of diffuse interfaces of a finite thickness. Explicit front tracking is avoided by using smooth continuous variables locating the grain and phase boundaries. By asymptotic expansions for vanishing interface thickness, it can be shown that classical sharp interface models including physical laws at interfaces and multiple junctions are recovered [Cag89, GNS98]. Since phase-field models can be derived on the basis of classical irreversible thermodynamics [PF90, WS*93, GNS04], they can be applied to processes close to thermodynamical equilibrium, i.e. at relatively small driving forces. Extensions of the phase-field approach to describe strongly nonequilibrium solidification, solute trapping and solute drag effects at large driving forces are discussed in [AW*98, Gal01]. The scaling problem of quantitatively modelling the low growth rate regime where the microstructure is typically several orders of magnitude larger than the microscopic capillary length has been overcome by a so-called thin interface approximation of the phase-field model [KR96, Al99, KKS99, MWA00, Kav01].

The purpose of this article is to extend the advances of the phase-field approach to model general multiphase solidification in situations close to local thermodynamical equilibrium. The underlying formulation of an entropy density functional is given in Sect. 2 and explicit expressions for the free energy densities of the bulk phases and of the interfaces are discussed. A method is described how the artifical appearance of a foreign phase contributions at a two-phase boundary can be avoided. Formulations defining the bulk, interdiffusion and interfacial diffusion coefficients as well as different types of surface energies and kinetic anisotropies are presented. In particular, an expression of crystalline (facetted) anisotropy is given that can be used for modelling general crystal shapes with an arbitrary number of edges and corners in three spatial dimensions. The essential ingredients of the phase-field model are summarized, the numerical method for solving the governing equations is briefly explained in Sect. 3. Examples of possible applications to numerically simulate dendritic growth, eutectic solidification in binary and ternary alloys, moving grain and phase boundaries and the formation of polycrystalline grain structures are given in Sect. 4. The simulation results are intended to illustratively show the potential of the phase-field model in computing and numerically analyzing complex pattern formation.

2 Phase-field Model for Multiphase and Polycrystalline Systems

The phase-field model allowing for an arbitrary number of phases (or grains) and components is derived from an entropy density functional in a thermodynamically consistent way, [GNS04, NGS05]. The formulation is defined solely via the bulk free energies of the individual phases, the surface energy densities (surface entropy densities, respectively) of the interfaces, the diffusion and mobility coefficients. Thus, the full set of phase-field evolution equations is defined by quantities which can be measured. Since the bulk free energies determine the phase diagrams (see, e.g., Chalmers [Cha77], Haasen [Haa94]), the phase-field model can be used to describe phase transitions, in principal, for arbitrary phase diagrams. For a number of binary and ternary alloys, the bulk free energies are provided by thermodynamical data base programmes and can be directly used in simulation applications.

The phase-field model for a general class of multicomponent and multiphase (or polycrystalline) alloy systems is formulated consisting of K components and N different phases (or grains) in a domain $\Omega \subset \mathbb{R}^3$. The domain Ω is separated in phase regions $\Omega_1, \ldots, \Omega_N$ occupied by the N phases as schematically illustrated in the left image of Fig. 2.1 for the situation $N = 4$. The middle and right images show examples of an Al-Si grain structure with

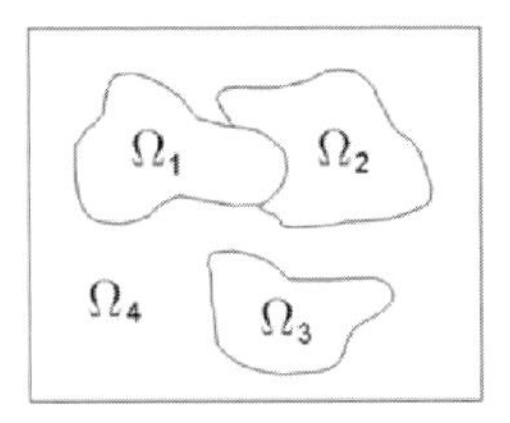

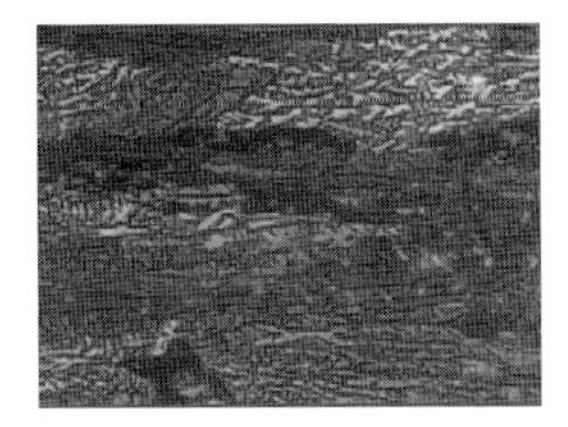

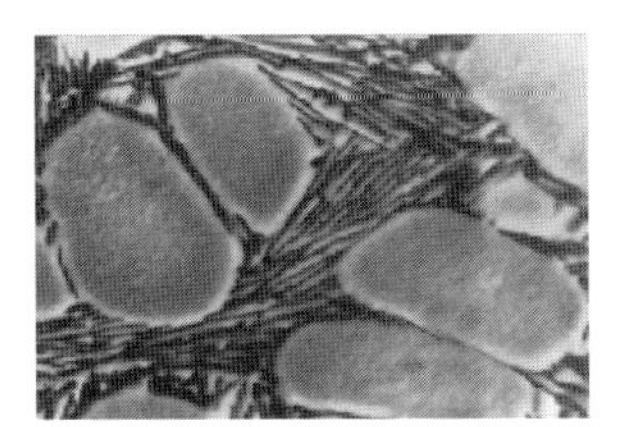

Fig. 2.1. Left image: Schematic drawing of a domain separation by four different phase regions; Middle image: Polycrystalline grain structure in Al-Si; Right image: Multiphase solification microstructure with dendrites and a eutectic structure. (See page 683 for a colored version of the figure.)

grains of different crystallographic orientations and of a real multiphase structure with primary dendrites and an interdendritic eutectic substructure.

The concentrations of the components are represented by a vector $\boldsymbol{c}(\mathbf{x}, t) = (c_1(\mathbf{x}, t), \ldots, c_K(\mathbf{x}, t))$. Similarly, the phase fractions are described by a vector-valued order parameter $\boldsymbol{\phi}(\mathbf{x}, t) = (\phi_1(\mathbf{x}, t), \ldots, \phi_N(\mathbf{x}, t))$. The variable $\phi_\alpha(\mathbf{x}, t)$ denotes the local fraction of phase of grain α. The phase-field model is based on an entropy functional of the form

$$S(e, \boldsymbol{c}, \boldsymbol{\phi}) = \int_\Omega \left(s(e, \boldsymbol{c}, \boldsymbol{\phi}) - \left(\varepsilon a(\boldsymbol{\phi}, \nabla \boldsymbol{\phi}) + \frac{1}{\varepsilon} w(\boldsymbol{\phi}) \right) \right) dx.$$

We assume that the bulk entropy density s depends on the internal energy density e, the concentrations $\boldsymbol{c}$ and the phase-field variable $\boldsymbol{\phi}$. We require that the concentrations of the components and of the phase-field variables fulfill the constraints

$$\sum_{i=1}^{K} c_i = 1, \quad \sum_{\alpha=1}^{N} \phi_\alpha = 1. \tag{2.1}$$

2.1 Entropy density contributions

It will be convenient to use the free energy $f(T, \boldsymbol{c}, \boldsymbol{\phi})$ as a thermodynamical potential. We therefore postulate the Gibbs relation

$$df = -sdT + \sum_{i=1}^{K} \mu_i dc_i + \sum_{\alpha=1}^{N} r_\alpha d\phi_\alpha.$$

Here, T is the temperature, $\mu_i = f_{,c_i}$ are the chemical potentials, and $r_\alpha = f_{,\phi_\alpha}$ are potentials due to the appearance of different phases. We set

$$e = f + sT,$$

and hence

$$de = Tds + \sum_i \mu_i dc_i + \sum_\alpha r_\alpha d\phi_\alpha,$$
$$ds = \tfrac{1}{T}de - \sum_i \tfrac{\mu_i}{T} dc_i - \sum_\alpha \tfrac{r_\alpha}{T} d\phi_\alpha.$$

If we interpret s as a function of $(e, \mathbf{c}, \boldsymbol{\phi})$, then we have

$$s_{,e} = \frac{1}{T}, \quad s_{,c_i} = \frac{-\mu_i}{T}, \quad s_{,\phi_\alpha} = \frac{-r_\alpha}{T}.$$

We note that given the free energy densities of the pure phases $f_\alpha(T, \mathbf{c})$, we obtain the total free energy $f(T, \mathbf{c}, \boldsymbol{\phi})$ as a suitable interpolation of the free energies f_α of the individual phases in the system. By inserting the free energy f into the phase-field method enables to model systems with a very general class of phase diagrams. In the way it is formulated, the model can describe systems with concave entropies $s_\alpha(e, \mathbf{c})$ in the pure phases. This corresponds to free energies $f_\alpha(T, \mathbf{c})$ which are concave in T and convex in $\mathbf{c}$. In the case where $f(T, \mathbf{c})$ is not convex in the variable $\mathbf{c}$, the free energy needs to contain gradients of the concentrations (as in the Cahn-Hilliard model).

Choosing the liquid phase to be the last component ϕ_N of the phase-field vector $\boldsymbol{\phi}$, an ideal solution formulation of the bulk free energy density reads

$$f_{id}(T,\mathbf{c},\boldsymbol{\phi}) = \sum_{\alpha=1}^{N} \sum_{i=1}^{K} \left(c_i L_i^\alpha \tfrac{T - T_i^\alpha}{T_i^\alpha} h(\phi_\alpha) \right) + \sum_{i=1}^{K} \left(\tfrac{R_g}{v_m} T c_i \ln(c_i) \right) - c_v T \ln(\tfrac{T}{T_M}),$$

with $L_i^N = 0$ and L_i^α, $i = 1, \ldots, K$, $\alpha = 1, \ldots, N-1$, being the latent heat per unit volume of the phase transition from phase α to the liquid phase and of pure component i. Furthermore, T_i^α, $i = 1, \ldots, K$, $\alpha = 1, \ldots, N-1$ is the melting temperature of the i-th component in phase α, T_M is a reference temperature. c_v, the specific heat and v_m, the molar volume are assumed to be constant, R_g is the gas constant. With a suitable choice of the function $h(\boldsymbol{\phi})$ satisfying $h(0) = 0$ and $h(1) = 1$, e.g. $h(\phi_\alpha) = \phi_\alpha$ or $h(\phi_\alpha) = \phi_\alpha^2(3 - 2\phi_\alpha)$, the free energy density f is an interpolation of the individual free energy densities f_α. We can calculate the entropy density

$$s = -f_{,T} = -\sum_{\alpha=1}^{N} \sum_{i=1}^{K} \left(c_i \frac{L_i^\alpha}{T_i^\alpha} h(\phi_\alpha) \right) - \sum_{i=1}^{K} \left(\frac{R}{v_m} c_i \ln(c_i) \right) + c_v \ln(T),$$

so that

$$e = f + Ts = -\sum_{\alpha=1}^{N} \sum_{i=1}^{K} (c_i L_i^\alpha h(\phi_\alpha)) + c_v T.$$

We note that if $L_i^\alpha = L^\alpha$ for all components i, then e does not depend on c. The chemical potentials $\mu_i(T, \mathbf{c}, \boldsymbol{\phi}) = f_{,c_i}(T, \mathbf{c}, \boldsymbol{\phi})$ are given by

$$\mu_i(T, \mathbf{c}, \boldsymbol{\phi}) = \sum_{\alpha=1}^{N} \left(L_i^\alpha \frac{T - T_i^\alpha}{T_i^\alpha} h(\phi_\alpha) \right) + \frac{R}{v_m} T(\ln(c_i) + 1). \tag{2.2}$$

A more general expression for the bulk free energy density of alloys is the Redlich-Kister-Muggianu model of subregular solution

$$f_{sr} = f_{id} + \sum_{i=1}^{K}\sum_{j=1}^{K} c_i c_j \sum_{\nu=0}^{M} M_{ij}^{(\nu)} (c_i - c_j)^{\nu},$$

with binary interaction coefficients $M_{ij}^{(\nu)}$ depending on the parameter ν. For $M = 0$, the Redlich-Kister-Muggianu ansatz takes the form of a regular solution model. In most applications, in particular to metallic systems, M takes a maximum value of two. A ternary term $\sim c_i c_j c_k$ can be added to describe the excess free enthalpy.

The thermodynamics of the interfaces gives additional contributions to the entropy given by a Ginzburg-Landau functional of the form

$$-\int_{\Omega} \left(\varepsilon a(\phi, \nabla\phi) + \frac{1}{\varepsilon} w(\phi) \right) dx.$$

Here, $a(\phi, \nabla\phi)$ is the gradient entropy density which is assumed to be homogeneous of degree two in the second variable; i.e., $a(\phi, \eta\nabla\phi) = \eta^2 a(\phi, \nabla\phi), \forall \eta \in \mathbb{R}^+$. The simplest form of the gradient entropy density is

$$a(\phi, \nabla\phi) = \sum_{\alpha=1}^{N} |\nabla\phi_\alpha|^2.$$

However, it has been shown [GNS98, GNS99b] that gradient entropies of the form

$$a(\phi, \nabla\phi) = \sum_{\alpha<\beta} A_{\alpha\beta}(\phi_\alpha \nabla\phi_\beta - \phi_\beta \nabla\phi_\alpha), \tag{2.3}$$

where $A_{\alpha\beta}$ are convex, homogeneous degree two functions, are more convenient with respect to the calibration of parameters in the phase-field model to the surface terms in the sharp interface model.

In liquid-solid and solid-solid phase transitions the anisotropy has a large effect on the pattern formation. A choice of the gradient entropies that leads to anisotropic surface terms is

$$a(\phi, \nabla\phi) = \sum_{\alpha<\beta} \gamma_{\alpha\beta}\, a_c(q_{\alpha\beta})^2\, |q_{\alpha\beta}|^2, \tag{2.4}$$

where $\gamma_{\alpha\beta}$ represent surface entropy densities and $q_{\alpha\beta} = (\phi_\alpha \nabla\phi_\beta - \phi_\beta \nabla\phi_\alpha)$ are generalized gradient vectors. The formulation using the generalized gradient vectors $q_{\alpha\beta}$ allows to distinguish the physics of each phase (or grain) boundary by providing enough degrees of freedom. Anisotropy of the surface entropy density is modeled by the factor $a_c(q_{\alpha\beta})^2$ depending on the orientation of the interface. Isotropic phase boundaries are realized by $a_c(q_{\alpha\beta}) = 1$. Weakly anisotropic crystals with an underlying cubic symmetry in 3D can be modeled by the expression

$$a_c(q_{\alpha\beta}) = 1 - \delta_{\alpha\beta}\Big(3 - 4\frac{|q_{\alpha\beta}|_4^4}{|q_{\alpha\beta}|^4}\Big), \tag{2.5}$$

with $\delta_{\alpha\beta}$ being the strength of the anisotropy of the α-β interface. The norms are given by $|q_{\alpha\beta}|_4^4 = \sum_{i=1}^3 (q_i^4)$ and $|q_{\alpha\beta}|^4 = \Big(\sum_{i=1}^3 (q_i^2)\Big)^2$ with $q_i = (\phi_\alpha \frac{\partial}{\partial x_i}\phi_\beta - \phi_\beta \frac{\partial}{\partial x_i}\phi_\alpha)$.

The function in Eq. (2.5) for the cubic crystal symmetry is plotted in Fig. 2.2 a) for an anisotropic strength $\delta = 0.15$. The equilibrium shape of a crystal can be determined by finding the inner envelope of the planes perpendicular to all radius vectors from the center to the surface depicted in Fig. 2.2 (Wulff-construction). In the directions of the depressions, curved faces evolve during growth.

For a strongly anisotropic crystal of facetted type, we define

$$a_c(q_{\alpha\beta}) = \max_{1\le k\le n_{\alpha\beta}} \Big\{\frac{q_{\alpha\beta}}{|q_{\alpha\beta}|} \cdot \eta_{\alpha\beta}^k\Big\}, \tag{2.6}$$

where $\eta_{\alpha\beta}^k, k = 1, \ldots, n_{\alpha\beta}$ are the $n_{\alpha\beta}$ corners of the Wulff shape of the α-β transition leading to flat crystal faces with sharp edges. These evolve in the direction of the cusps. In principal, Eq. (2.6) allows to model arbitrary crystal shapes with $n_{\alpha\beta}$ corners. For a comparison with the expression for smooth anisotropy in Eq. (2.5), we display in Fig. 2.2 b) the function $a_c(q_{\alpha\beta})$ in Eq. (2.6) also for a cubic crystal symmetry. In Fig. c) the simulation results of two crystals with 45° rotated orientation growing from adjacent nuclei are shown. Each grain develops its minimum energy surfaces in contact with the melt and at their interface.

The interfacial entropy density contribution $w(\phi)$ is a nonconvex function with N global minima corresponding to the N phases or grains in the system. As an extension of the standard double-well potential, one may take the standard multi-well potential

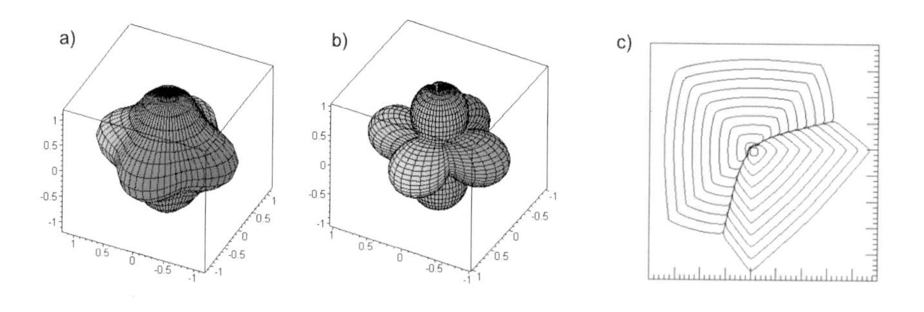

Fig. 2.2. 3D surface plot of a) a smooth and b) a facetted cubic anisotropy. c) contour plots of two adjacent growing, 45° misoriented cubic crystals applying the smooth anisotropy formulation in Eq. (2.5) with $\delta = 0.2$. (See page 684 for a colored version of the figure.)

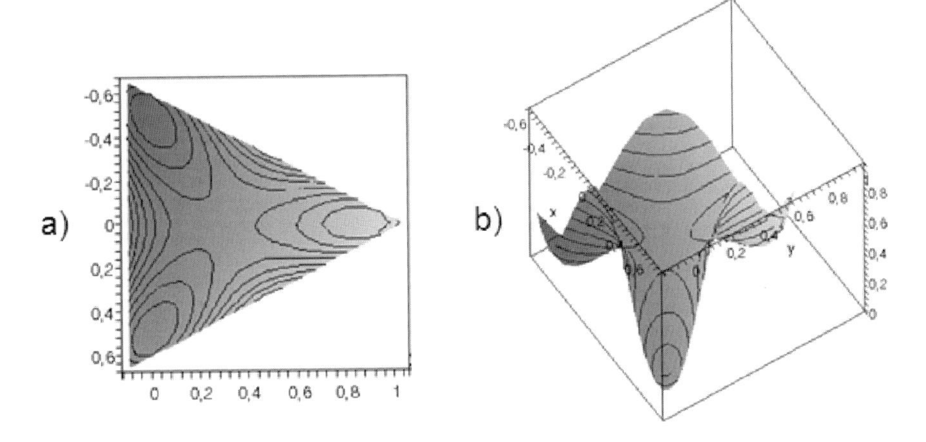

Fig. 2.3. Plot of the multi-well potential $w_{st}(\boldsymbol{\phi})$ for $N = 3$ and equal surface entropy densities $\gamma_{\alpha\beta}$. (See page 684 for a colored version of the figure.)

$$w_{st}(\boldsymbol{\phi}) = 9 \sum_{\alpha<\beta} \gamma_{\alpha\beta} \phi_\alpha^2 \phi_\beta^2 \tag{2.7}$$

or a higher order variant

$$\tilde{w}_{st}(\boldsymbol{\phi}) = w_{st}(\boldsymbol{\phi}) + \sum_{\alpha<\beta<\delta} \gamma_{\alpha\beta\delta} \phi_\alpha^2 \phi_\beta^2 \phi_\delta^2. \tag{2.8}$$

For practical computations the multi-obstacle potential yields good calibration properties. It is defined by

$$w_{ob}(\boldsymbol{\phi}) = \frac{16}{\pi^2} \sum_{\alpha<\beta} \gamma_{\alpha\beta} \phi_\alpha \phi_\beta, \tag{2.9}$$

with a higher order variant

$$\tilde{w}_{ob}(\boldsymbol{\phi}) = w_{ob}(\boldsymbol{\phi}) + \sum_{\alpha<\beta<\delta} \gamma_{\alpha\beta\delta} \phi_\alpha \phi_\beta \phi_\delta, \tag{2.10}$$

where w_{ob} and $\tilde{w}_{ob}$ are defined to be infinity whenever $\boldsymbol{\phi}$ is not on the Gibbs simplex. In Figs. 2.3 and 2.4, we show a plot of both expressions, the multi-well and the multi-obstacle potential for the case of three phases ($N = 3$).

We refer to [GNS99a] and [GNS99b] for a further discussion of the properties of the surface terms $w_{st}, \tilde{w}_{st}, w_{ob}$ and $\tilde{w}_{ob}$. We assume for simplicity that $a(\boldsymbol{\phi}, \nabla\boldsymbol{\phi})$ and $w(\boldsymbol{\phi})$ and, hence, the interfacial contributions to the entropy, do not depend on $(T, \boldsymbol{c})$.

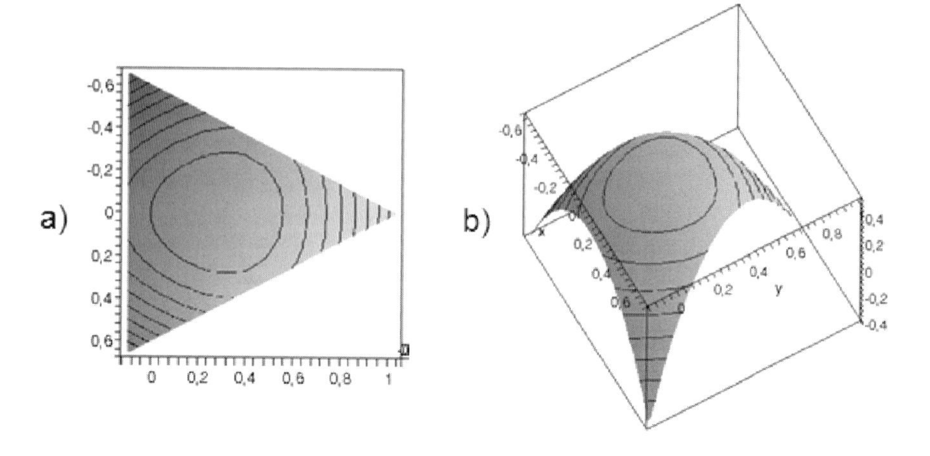

Fig. 2.4. Plot of the multi-obstacle potential $w_{ob}(\phi)$ for $N = 3$ and equal surface entropy densities $\gamma_{\alpha\beta}$. (See page 684 for a colored version of the figure.)

2.2 Evolution equations

The energy and mass balance equations can be derived from the energy flux J_0 and from the fluxes of the components $J_1, \ldots, J_K$ by

$$\frac{\partial e}{\partial t} = -\nabla \cdot J_0 \qquad \text{energy balance,} \tag{2.11}$$

$$\frac{\partial c_i}{\partial t} = -\nabla \cdot J_i \qquad \text{mass balances} \tag{2.12}$$

and are coupled to a set of phase-field equations

$$\tau\varepsilon \frac{\partial \phi_\alpha}{\partial t} = \frac{\delta S}{\delta \phi_\alpha} - \lambda \qquad \text{phase-field equations,} \tag{2.13}$$

in such a way that the second law of thermodynamics is fulfilled in an appropriate local version.

In order to derive the expressions for the fluxes $J_0, \ldots, J_K$, we use the generalized thermodynamic potentials $\frac{\delta S}{\delta e} = \frac{1}{T}$ and $\frac{\delta S}{\delta c_i} = \left(\frac{-\mu_i}{T}\right)$, which will drive the transition. Now, we appeal to nonequilibrium thermodynamics and postulate that the fluxes are linear functions of the thermodynamic driving forces $\nabla \frac{\delta S}{\delta e}, \nabla \frac{\delta S}{\delta c_1}, \ldots, \nabla \frac{\delta S}{\delta c_K}$ to obtain

$$\begin{aligned} J_0 &= L_{00}(T,c,\phi)\nabla \tfrac{\delta S}{\delta e} + \textstyle\sum_{j=1}^N L_{0j}(T,c,\phi)\nabla \tfrac{\delta S}{\delta c_j} \\ &= L_{00}(T,c,\phi)\nabla \tfrac{1}{T} + \textstyle\sum_{j=1}^N L_{0j}(T,c,\phi)\nabla \tfrac{-\mu_j}{T}, \end{aligned}$$

$$\begin{aligned} J_i &= L_{i0}(T,c,\phi)\nabla \tfrac{\delta S}{\delta e} + \textstyle\sum_{j=1}^N L_{ij}(T,c,\phi)\nabla \tfrac{\delta S}{\delta c_j} \\ &= L_{i0}(T,c,\phi)\nabla \tfrac{1}{T} + \textstyle\sum_{j=1}^N L_{ij}(T,c,\phi)\nabla \tfrac{-\mu_j}{T}, \end{aligned}$$

with mobility coefficients $(L_{ij})_{i,j=0,\ldots,K}$. To fulfill the constraint $\sum_{i=1}^K c_i = 1$ in Eq. (2.1) during the evolution, we assume

$$\sum_{i=1}^{K} L_{ij} = 0, \qquad j = 0, \ldots, K,$$

which implies $\sum_{i=1}^K J_i = 0$, and, hence, $\partial_t(\sum_{i=1}^K c_i) = \nabla \cdot (\sum_{i=1}^K J_i) = 0$. We further assume that L is symmetric (Onsager relations) and positive semidefinite; i.e.,

$$\sum_{i,j=0}^{K} L_{ij}\xi_i\xi_j \geq 0 \qquad \forall \xi = (\xi_0, \ldots, \xi_K) \in \mathbf{R}^{K+1}.$$

This condition ensures that an entropy inequality is satisfied. Cross effects between mass and energy diffusion are included in the model. One can neglect them by setting $L_{i0} = 0$ and $L_{0j} = 0$ for all $i, j \in \{1, \ldots, K\}$. In general, the mobility coefficients $(L_{ij})_{i,j=0,,\ldots,K}$ are allowed to depend on T, $\boldsymbol{c}$ and $\boldsymbol{\phi}$. Given some heat and mass diffusion coefficients, $k = k(T, \boldsymbol{c}, \boldsymbol{\phi})$ and $D_i = D_i(T, \boldsymbol{c}, \boldsymbol{\phi})$, the mobility coefficients L_{ij} read

$$L_{ji} = L_{ij} = \frac{v_m}{R_g} D_i c_i \left(\delta_{ij} - \frac{D_j c_j}{\sum_{k=1}^K D_k c_k} \right), \tag{2.15}$$

for $i, j = 1, \ldots, K$ and then recursively define

$$L_{0j} = -\frac{v_m}{R_g} \sum_{\alpha=1}^{N} \sum_{i=1}^{K} L_{ji} h(\phi_\alpha) L_i^\alpha, \tag{2.16}$$

$$L_{00} = kT^2 + \frac{v_m}{R_g} \sum_{\alpha,\beta}^{N,N} \sum_{i,j}^{K,K} h(\phi_\alpha) L_i^\alpha L_{ji} h(\phi_\beta) L_j^\beta, \tag{2.17}$$

where δ_{ij} denotes the Kronecker delta and L_i^α are the latent heats of fusion.

The formulation in Eqs. (2.15)-(2.17) takes bulk diffusion effects including interdiffusion coefficients into account. The dependence of the mass and heat diffusion coefficients on $\boldsymbol{\phi}$ can be realized by e.g. linear expansions. To also consider enhanced diffusion in the interfacial region of phase or grain boundaries, additional terms proportional to $\phi_\alpha\phi_\beta$ with interfacial diffusion coefficients $D_i^{\alpha\beta}(T, \boldsymbol{c}, q_{\alpha\beta})$ need to be added. Altogether, we suggest for mass and heat diffusion

$$D_i = \sum_{\alpha=1}^{N} D_i^\alpha(T, \boldsymbol{c})\phi_\alpha + \frac{1}{\varepsilon} \sum_{\alpha<\beta} D_i^{\alpha\beta}(T, \boldsymbol{c}, q_{\alpha\beta})\phi_\alpha\phi_\beta, \tag{2.18}$$

$$k = \sum_{\alpha=1}^{N} k^\alpha(T, \boldsymbol{c})\phi_\alpha, \tag{2.19}$$

i.e. in particular that the diffusion coefficients in Eq. (2.15) and (2.17) can be anisotropic.

For the nonconserved phase-field variables $\phi_1, \dots, \phi_N$, we assume that the evolution is such that the system locally tends to maximize entropy conserving concentration and energy at the same time. Therefore, we postulate

$$\tau\varepsilon\partial_t\phi_\alpha = \varepsilon\Big(\nabla\cdot a_{,\nabla\phi_\alpha}(\boldsymbol{\phi},\nabla\boldsymbol{\phi}) - a_{,\phi_\alpha}(\boldsymbol{\phi},\nabla\boldsymbol{\phi})\Big) - \frac{w_{,\phi_\alpha}(\boldsymbol{\phi})}{\varepsilon} - \frac{f_{,\phi_\alpha}(T,\mathbf{c},\boldsymbol{\phi})}{T} - \lambda, \tag{2.20}$$

where we denote with $a_{,\phi_\alpha}$, $w_{,\phi_\alpha}$, $f_{,\phi_\alpha}$ and $a_{,\nabla\phi_\alpha}$ the derivative with respect to the variables corresponding to ϕ_α and $\nabla\phi_\alpha$, respectively.

For material systems with anisotropic kinetics, the kinetic coefficient τ may depend on the generalized gradient vectors $q_{\alpha\beta}$ in a similar way as the gradient energies $a(\boldsymbol{\phi},\nabla\boldsymbol{\phi})$ in Eq. (2.5). The quantity $\tau = \tau(\boldsymbol{\phi},\nabla\boldsymbol{\phi})$ in Eq. (2.13) models an anisotropic kinetic coefficient of the form

$$\tau(\boldsymbol{\phi},\nabla\boldsymbol{\phi}) = \tau_0 + \sum_{\alpha<\beta} B_{\alpha\beta}(q_{\alpha\beta})$$

with $B_{\alpha\beta}(q_{\alpha\beta}) = 0$ if $q_{\alpha\beta} = 0$. Possible choices are

$$B_{\alpha\beta} = \tau^0_{\alpha\beta}\left(1 + \zeta_{\alpha\beta}\left(3 \pm 4\frac{|q_{\alpha\beta}|_4^4}{|q_{\alpha\beta}|^4}\right)\right) - \tau_0 \quad \text{or} \tag{2.21}$$

$$B_{\alpha\beta} = \tau^0_{\alpha\beta} \max_{1\le k\le r_{\alpha\beta}} \left\{\frac{q_{\alpha\beta}}{|q_{\alpha\beta}|}\cdot \xi^k_{\alpha\beta}\right\} - \tau_0, \tag{2.22}$$

if $q_{\alpha\beta} \neq 0$ for weakly cubic (Eq. (2.21)) or strongly facetted (Eq. (2.22)) kinetic anisotropies with $r_{\alpha\beta}$ corners $\xi^k_{\alpha\beta}$. $\zeta_{\alpha\beta}$ determines the strength of the kinetic anisotropy similar to $\delta_{\alpha\beta}$ in Eq. (2.5) for the surface energy anisotropy. Systems with isotropic kinetics are realized by setting $\zeta_{\alpha\beta} = 0$.

λ is an appropriate Lagrange multiplier such that the constraint $\sum_1^N \phi_\alpha = 1$ in Eq. (2.1) is satisfied; i.e.,

$$\lambda = \frac{1}{N}\sum_{\alpha=1}^{N}\left[\varepsilon\Big(\nabla\cdot a_{,\nabla\phi_\alpha}(\boldsymbol{\phi},\nabla\boldsymbol{\phi}) - a_{,\phi_\alpha}(\boldsymbol{\phi},\nabla\boldsymbol{\phi})\Big) - \frac{1}{\varepsilon}w_{,\phi_\alpha}(\boldsymbol{\phi}) - \frac{f_{,\phi_\alpha}(T,\mathbf{c},\boldsymbol{\phi})}{T}\right]$$

It has been shown in [GNS04] as a result that inequality

$$\begin{aligned}\partial_t(\text{entropy}) &= \partial_t\left(s(e,\mathbf{c},\boldsymbol{\phi}) - \varepsilon a(\boldsymbol{\phi},\nabla\boldsymbol{\phi}) - \frac{1}{\varepsilon}w(\boldsymbol{\phi})\right)\\ &\ge -\nabla\cdot\left(\sum_{i=0}^{K}\frac{-\mu_i}{T}J_i - \varepsilon\sum_{\alpha=1}^{N} a_{,\nabla\phi_\alpha}\partial_t\phi_\alpha\right)\end{aligned}$$

holds and therefor the derived phase-field equation ensures a positive local entropy production.

2.3 Non-dimensionalization

A dimensionless form of the system of governing equations was necessary on account of computational efficiency and accuracy. To non-dimensionalize the system with respect to time, space, temperature and inner energy density, we assume

$$t[s] = \tau^* \cdot \tilde{t}, \qquad x[m] = \ell \cdot \tilde{x}, \qquad T[K] = \vartheta \cdot \tilde{T}, \qquad e\left[\frac{J}{m^3}\right] = (c_v\vartheta) \cdot \tilde{e},$$

where $\tilde{t}, \tilde{x}, \tilde{T}, \tilde{e}$ are dimensionless and $\tau^*[s], \ell[m], \vartheta[K], c_v[J/m^3K]$ are reference quantities. The notation $\tilde{}$ indicates dimensionless quantities. Introducing dimensionless latent heats $\tilde{L_i^\alpha}$ and a rescaled gas constant $\tilde{R}_v$, the chemical potential μ_j (Eq. (2.2)) follows as

$$\mu_j\left[\frac{J}{m^3}\right] = (c_v\vartheta)\cdot\tilde{\mu}_j \quad \text{with} \quad L_i^\alpha\left[\frac{JK}{m^3K}\right] = (c_v\vartheta)\cdot\tilde{L_i^\alpha}, \quad R\left[\frac{J}{molK}\right] = v_m c_v \tilde{R}_v.$$

With the dimensionless mobility coefficients (Eqs. (2.15)-(2.17)) $\tilde{L}_{00}$, $\tilde{L}_{0j}$, $\tilde{L}_{i0}$, $\tilde{L}_{ij}$ of the form

$$L_{00}\left[\frac{JK}{sm}\right] = \frac{\ell^2 c_v \vartheta^2}{\tau^*}\tilde{L}_{00}, \qquad L_{0j}\left[\frac{m^2K}{s}\right] = \frac{\ell^2\vartheta}{\tau^*}\tilde{L}_{0j},$$

$$L_{i0}\left[\frac{m^2K}{s}\right] = \frac{\ell^2\vartheta}{\tau^*}\tilde{L}_{i0}, \qquad L_{ij}\left[\frac{m^5K}{Js}\right] = \frac{\ell^2}{c_v\tau^*}\tilde{L}_{ij}.$$

and with the dimensionless mass and heat diffusivities

$$D_i^\alpha\left[\frac{m^2}{s}\right] = \frac{\ell^2}{\tau^*}\tilde{D}_i^\alpha \qquad \text{and} \qquad k^\alpha\left[\frac{J}{msK}\right] = \frac{\ell^2 c_v}{\tau^*}\tilde{k}^\alpha,$$

the energy and mass diffusion equations (Eqs. (2.11) and (2.12)) can directly be used with dimensionless quantities.

The phase-field equations with the surface contributions $a(\boldsymbol{\phi}, \nabla\boldsymbol{\phi})$ (Eq. 2.4) and $w(\boldsymbol{\phi})$ (Eqs. (2.7) - (2.10)) are treated as follows: The surface entropy densities $\gamma_{\alpha\beta}$, the kinetic coefficients τ and the interface width ε are scaled as

$$\gamma_{\alpha\beta}\left[\frac{J}{m^2K}\right] = \gamma_0 \cdot \tilde{\gamma}_{\alpha\beta}, \quad \tau = \tau_0 \cdot \tilde{\tau} \quad \text{and} \quad \varepsilon[m] = \tilde{\varepsilon} \cdot \ell.$$

Inserting these quantities into the phase-field equation (Eq. (2.13)) (both sides with the dimension of an entropy density $[J/m^3K]$) and division by c_v gives

$$\tilde{\tau}\tilde{\varepsilon}\frac{\tau_0\ell}{\tau^*\, c_v}\,\partial_{\tilde{t}}\phi_\alpha = \tilde{\varepsilon}\frac{\gamma_0}{\ell c_v}\left(\tilde{\nabla}\cdot\tilde{a}_{,\tilde{\nabla}\phi_\alpha}(\boldsymbol{\phi},\tilde{\nabla}\boldsymbol{\phi}) - \tilde{a}_{,\phi_\alpha}(\boldsymbol{\phi},\tilde{\nabla}\boldsymbol{\phi})\right)$$
$$-\frac{1}{\tilde{\varepsilon}}\frac{\gamma_0}{\ell c_v}\tilde{w}_{,\phi_\alpha}(\boldsymbol{\phi}) - \frac{\tilde{f}_{,\phi_\alpha}(T,\boldsymbol{c},\boldsymbol{\phi})}{\tilde{T}} - \tilde{\lambda}, \quad \alpha = 1,\ldots,N.$$

Table 2.1. Data set for pure Ni with dimensional and nondimensionalized values

Parameter	label	unit	dim. value	nondim. value
melting temperature	T_m	K	1728	1.0
latent heat	L	J/m^3	$2.35 \cdot 10^9$	0.251
specific heat	c_p	J/m^3K	$5.42 \cdot 10^6$	1.0
average surface tension	σ_0	J/m^2	0.37	$1.58 \cdot 10^{-4}$
thermal diffusivity	k_T	m^2/s	$1.55 \cdot 10^{-5}$	$1.991 \cdot 10^{-2}$
surface energy anisotropy	δ_c			0.018
kinetic anisotropy	δ_k			0.13

When making the choice $\gamma_0 = \ell c_v$ and $\tau = \frac{\ell \tau_0}{c_v}$, the phase-field equation retains its original form (Eq. 2.13) with dimensionless quantities instead of the dimensional ones. Finally, if we consider the Gibbs-Thomson equation describing the motion of a sharp interface with the curvature κ (square brackets indicate a jump of the respective value)

$$\beta_{\alpha\beta}\, v \left[\frac{m}{s}\right] = \gamma_{\alpha\beta} \left[\frac{J}{m^2 K}\right] \kappa \left[\frac{1}{m}\right] + \frac{([f]_\alpha^\beta - \sum_i \bar{\mu}_i \, [c_i]_\alpha^\beta)}{T} \left[\frac{J}{m^3 K}\right], \tag{2.23}$$

it can easily be seen that the mobility coefficients $\beta_{\alpha\beta}$ are equal to the kinetic coefficient τ in the phase-field equation (Eq. (2.13)).

The length scale parameter ℓ can be related to the size of the domain resolved with N_x grid points. In the case of the pure Ni system, the size of a thermal dendrite is $Nx \cdot \ell \cdot \Delta\tilde{x} = 2 \cdot 10^{-5} m$. With a feasible number of gridpoints $N_x = 500$ and a dimensionless cell spacing $\Delta\tilde{x} = 1.0$, we have $\ell = 4.0 \cdot 10^{-8} m$. Using thermophysical data for pure Ni from [BK*02], we obtain the nondimensional data set listed in Table 2.1.

3 Numerical Methods

In the previous section, a general multi-component multi-phase-field model was formulated leading to a system of coupled partial differential equations (Eqs. (2.11) - (2.13)). In the general case, there will be one energy equation, K concentration equations and N phase-field equations to be solved numerically.

Our numerical algorithm for solving the general system of equations is based on a finite difference scheme with explicit time update on a regular grid. Despite its disadvantages concerning stability and limited time step size, this straight forward approach is justified by the applicability of various sophisticated techniques to reduce the computational effort and to save memory. This becomes especially important when treating problems with a high number of phases N as in the case of grain growth. Further, the coupling of the numerical scheme with finite difference solvers for fluid flow or elasticity is greatly facilitated (the approach of coupling to a Navier-Stokes solver is specified in [SNW05]). In the following, the discretization of the balance equations (Eqs.

(2.11) and (2.12)) and of the phase-field equation (Eq. (2.13)) is considered in detail.

The phase-field variable ϕ_α defines the smallest length scale and the largest spatial gradients due to its rapid change from 0 to 1 over the width of the interface. When using a regular grid, it must be chosen fine enough to adequately resolve the diffuse interface layer of the phase-field variables. Depending on the kind of spatial discretization, it should extend over at least 5 to 10 grid points. The diffusion lengths for the temperature T and for the concentrations c_i are much greater than the interface width, so that one could consider the application of lower grid resolutions for these conserved fields. We anticipate this option to keep a higher accuracy, since over the diffuse layer in general, both, the phase-field as well as the diffusion fields may change rapidly.

3.1 FD discretization and staggered grid

The complete set of evolution equations of the model (Eqs. (2.11) - (2.13)) can be treated in the following simplified divergence form, where u_l, u_m stands for the respective field quantity (l, m indicative for e, c_i, ϕ_α):

$$\frac{\partial u_l}{\partial t} = rhs = \nabla \cdot \mathbf{f}\Big(u_m, \frac{\partial u_m}{\partial x_i}\Big), \tag{3.1}$$

All terms including field variables and their spatial derivatives are arguments of the vector valued function $\mathbf{f}$. In case of the phase-field equation (Eq. (2.13)) the right hand side (rhs) contains additional source terms, which in general are small by a value compared to the divergence term, so that the general character of Eq. (3.1) is kept preserved. Concerning the time discretization, an explicit forward Euler scheme is applied with a time update according to ${u_l}^{n+1} = {u_l}^n + \Delta t \cdot rhs^n$ (time steps are superscripted). This explicit scheme requires a control of the temporal step width Δt for each individual equation. For the case of a 3D simulation and an identical grid step width Δx in each space dimension, the criterion for stability would suggest a stepwidth of

$$\Delta t \leq min \left\{ \frac{\Delta x^2}{6k_{max}}, \frac{\Delta x^2}{6D_{max}}, \frac{\Delta x^2}{6\frac{\gamma_{\alpha\beta}}{\tau_0}} \right\},$$

where k_{max} and D_{max} are the maximum values for all heat and mass diffusion coefficients, $\gamma_{\alpha\beta}$ is the maximum surface entropy coefficient among all appearing phases and τ_0 is the kinetic coefficient.

Since the right hand sides of the conserved order parameter equations (Eqs. (2.11) and (2.13)) as well as the first term in Eq. (2.13) consist of a divergence term, a two-step algorithm was developed: First, the vector flux quantities are calculated using right-sided finite differences and they are stored in a memory buffer for multiple access. This buffer holds the flux values of three adjacent 2D-layers during the layerwise calculation, shifted through the 3D grid along the z-direction. In a second step, the divergence is evaluated using left-sided differences. This results in an extremely memory saving numerical scheme without any redundant calculations, since for each field variable only a single 3D array must be stored in memory ([WN06]). The spatial discretization of the phase-field and of the balance equations differs and is described in the following paragraph.

Balance equations:

In the nonlinear energy and mass diffusion equations, the physical diffusion coefficients are incorporated in the Onsager coefficients L_{ij}, which may depend on $\boldsymbol{\phi}$, $\boldsymbol{c}$ and T. The discretization on a regular grid with spatial indices (i, j, k) for the following simplified energy equation (mass diffusion cross terms have been omitted for clarity):

$$\frac{\partial e}{\partial t} = -\nabla \cdot \mathbf{J}_0 = -\nabla \cdot \left(L_{00}(T, \boldsymbol{c}, \boldsymbol{\phi}) \nabla \left(\frac{1}{T} \right) \right) \tag{3.2}$$

is accomplished in the *FTCS* scheme (forward in time, centered in space). First, to compute the divergence on the right hand side of Eq. (3.2), all components of the energy flux vector are assembled. For a grid cell with indices (i, j, k) the spatial derivatives are approximated with right sided finite differences, and the Onsager coefficients L_{00} are evaluated at the respective intermediate grid positions, in the center boundary of two adjacent grid cells (see Fig. 3.1 b)). For example, the x component of the energy flux reads:

$$L_{00} \frac{\partial}{\partial x} \left(\frac{1}{T} \right) \Big|_{i,j,k} \simeq L_{00} \Big|_{i+\frac{1}{2},j,k} D_x^+ \left(\frac{1}{T} \right) \Big|_{i,j,k} \tag{3.3}$$

with

$$L_{00} \Big|_{i+\frac{1}{2},j,k} = \frac{1}{2} \left(L_{00} \big|_{i+1,j,k} + L_{00} \big|_{i,j,k} \right)$$

and

$$D_x^+ \left(\frac{1}{T} \right) \Big|_{i,j,k} = \frac{1}{\Delta x} \left(\frac{1}{T} \Big|_{i+1,j,k} - \frac{1}{T} \Big|_{i,j,k} \right),$$

where D_x^+ indicates the forward difference operator in x direction. As a next step, the divergence operation is carried out using left sided finite differences of the flux components in Eq. (3.3). The resulting spatial-temporal scheme has an accuracy of convergence of order $(\Delta x)^2$ in space and of order Δt in time.

Phase-field equations:

For the phase-field equation (Eq. (2.20)), the correct treatment of anisotropy in the gradient entropy density $a(\boldsymbol{\phi}, \nabla\boldsymbol{\phi})$ is important, especially when reducing the numerical interface to a desirable low number of grid points.

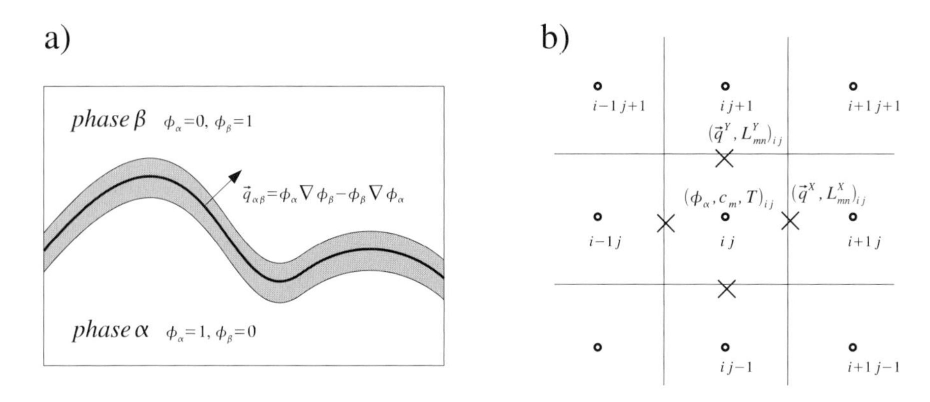

Fig. 3.1. a) Schematical view of a two-phase region with diffuse interface (shaded) and generalized gradient vector $\mathbf{q}_{\alpha\beta}$. b) 2D sketch of the finite difference grid with the field variables $\boldsymbol{\phi}, \mathbf{c}, T$ at central positions (∘), $\mathbf{q}$-vectors and with transport coefficients L_{mn} at staggered positions (×).

It is convenient to keep the formulation using generalized gradient vectors $q_{\alpha\beta} = \phi_\alpha \nabla \phi_\beta - \phi_\beta \nabla \phi_\alpha$ (Fig. 3.1 a)) in the solution algorithm. We take $a(\boldsymbol{\phi}, \nabla \boldsymbol{\phi})$ in the form of Eqs. (2.3) and (2.4) and carry out the variational derivatives with respect to $\nabla \phi_\alpha$ and ϕ_α to get the two anisotropic *rhs* terms of the phase-field equation, namely:

$$a_{,\nabla\phi_\alpha} = \sum_{\beta\neq\alpha} \frac{\partial A_{\alpha\beta}}{\partial q_{\alpha\beta}} (-\phi_\beta) \tag{3.4}$$

$$a_{,\phi_\alpha} = \sum_{\beta\neq\alpha} \frac{\partial A_{\alpha\beta}}{\partial q_{\alpha\beta}} \nabla \phi_\beta \tag{3.5}$$

$$\text{with} \quad \frac{\partial A_{\alpha\beta}}{\partial q_{\alpha\beta}} = 2\,\gamma_{\alpha\beta} \left(a_c(q_{\alpha\beta}) \frac{\partial a_c}{\partial q_{\alpha\beta}} |q_{\alpha\beta}|^2 + a_c^2(q_{\alpha\beta})\, q_{\alpha\beta} \right). \tag{3.6}$$

The entropy flux term $a_{,\nabla\phi_\alpha}$ in Eq. (3.4) needs a special attention due to the influence of the anisotropy function $a_c(q_{\alpha\beta})$. Three different vectors $q^x_{\alpha\beta}$, $q^y_{\alpha\beta}$ and $q^z_{\alpha\beta}$ are computed for each grid point, evaluated at the upper side x, y and z boundaries of the cell volume. The spatial derivatives of ϕ_α on these staggered grid positions include a combination of right-sided and central differences of the neighboured grid positions (see Fig. 3.1). To compute each spatial component of the vector valued function $a_{,\nabla\phi_\alpha}$, the respective vector $q^{x,y,z}_{\alpha\beta}$ is used. The divergence of $a_{,\nabla\phi_\alpha}$ appearing in Eq. (2.20) is calculated subsequently with left sided differences, taking advantage of the same 3-layer buffer mechanism as described above. For the second anisotropic term, $a_{,\phi_\alpha}$, the gradient vectors $q_{\alpha\beta}$ are evaluated at central positions using central differences of ϕ_α.

We find that by this treatment (equivalent to a 16-point stencil) a second order accuracy in space is obtained. For the 2D case the discretization is equivalent to the scheme published in [And00]. If no anisotropy is present, $q_{\alpha\beta}$ can be calculated exclusively with right-sided differences of ϕ_α within a 7-point stencil, in this case sufficient to guarantee second order accuracy.

3.2 Optimization of the Computational Algorithm

The phase-field model has been implemented in its complete generality for simulations in 2D and 3D describing phase transitions with energy and mass diffusion in alloy systems for an arbitrary number of components and phases. An extensive program package (*PACE3D*: Parallel Algorithms for Crystal Evolution in 3D) was developed in the programming language C due to performance reasons, but uses an object oriented approach; a data base is provided to choose different formulations of free energies, anisotropies, potentials, mass and heat diffusion functions. Moreover, artificially generated noise of various distributions can be added to the phase-field variable (non-conservative) or to energy or concentration fields (conservative noise). Several **adaptive strategies** reduce the computational effort:

- The equations for the phase fields are exclusively solved in their respective diffuse interface area by the use of an activation flag field. This field is set in each time step by an optimized gradient test routine.
- A dynamical memory concept reduces the memory costs especially for a high number of phase-field variables: For the majority of the grid points, ϕ_α assumes values of 1.0 or 0.0 in all regions of pure bulk phases and can be referenced by pointers to constant unit vectors.

Since both strategies are especially effective with a narrow interface profile, we favorize the use of the multi-obstacle potential in Eq. (2.9). This potential has a strong separating character and reduces the number of interface grid points (with $0 < \phi_{i,j,k} < 1$).

For most alloy systems, there are great differences in the thermal and mass diffusivities leading to differences in the evolutional time scale. If the Eqs. (2.11) and (2.12) are solved conjointly, the stability criterion demands for the use of the small time step width which is usually the heat diffusion scale. To reduce the computational effort, different step widths, integer multiples of the smallest (heat diffusion) scale, are used to solve the three kinds of Eqs. (2.11),(2.12) and (2.13).

3.3 Parallelization

Two parallelization concepts of the finite difference algorithm are integrated in the source code and can be chosen aside or in combination: For high performance computing on Linux clusters, **distributed computing** is realized via the (LAM-) Message Passing Interface (MPI) routines. The approach splits the three-dimensional simulation grid into multiple sub-grids, so that each available node gets assigned a specific part of the simulation space. To avoid unnecessary complexity of the code and to reduce boundary data exchange, the simulation is exclusively subdivided along the z-direction. In this way, we have minimal interdependencies between neigbouring simulation subdomains and a maximum speedup factor, a result of extensive software tests. By mapping the number N_z of grid points in z-direction to its value n_z in each node subdomain, we are able to use the same code base as well as simulation description for serial as well as distributed simulations, effectively decoupling the simulation from the hardware executing it. An MPI-based boundary exchange mechanism ensures that before each simulation time step commences, the outer planes of each node are copied into the excess space, provided by its immediate neighbour. A more extensive description of the parallelization can be found in [NW*05]

An additional approach to exploit the power of multiprocessor workstations and supercomputers was realized by **shared memory parallelization** via OpenMP: the execution of the spacial loops is subdivided into different threads running on different processors of a single node. Therefore, only slight modifications of the code are necessary, e.g. the introduction of indexed loop variables.

An important feature for parallel phase-field simulations is the implementation of an appropriate load balancing mechanism. Due to the advancing fronts, which require more calculations, the demand for computational power is locally non-uniform. In our case, this was achieved by performing statistics of the load of each node and of the data transfer times. A redistribution of the simulation area after an optimal number of time steps is initiated. Additionally, to adapt the code to networks with lower capabilities requiring data compression, we evaluate run-length and quadtree encoding of the exchanged boundary data.

4 Applications

To give an impression about the applicability of the model and simulation tools, a selection of different microstructures examined in more detail is given below. Here our focus lays on solidification processes in pure (Ni), in binary alloys (Ni/Cu) and on the description of grain growth phenomena. In the simulations, the complete binary coded data for temperature, concentrations and phase-field parameters is stored in floating point accuracy and analyzed a posteriori. To tackle the problem of 3D data analysis, a complete software toolbox was developed, including means for visualization, determination of front velocity, temperature and gradients, calculating contour lines, one and two dimensional intersection profiles and growth rates.

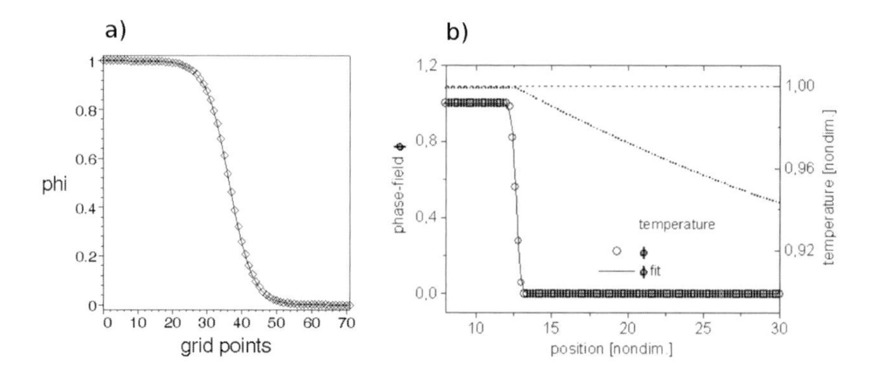

Fig. 4.1. a) Phase-field profile of a planar front growth in an undercooled melt (pure Ni) for high grid resolution (double well potential). b) Phase-field and temperature profiles with reduced resolution ensuring sufficient accuracy.

4.1 Parameter Tests

To verify the accuracy of the numerical method, different grid sizes and parameter tests were carried out for the Ni and NiCu data sets. The predictions of the Gibbs-Thomson equation (Eq. (2.23)) for the growth of a planar front resp. a spherical shape with radius R were examined. For these simple cases, where the curvature κ is equal to 0 resp. to $1/R$, the tests gave the predicted growth velocities v by Eq. (2.23) very closely. Finite size effects of the numerical grid were verified over a long simulation period by evaluating the shape deformations of a spherical solid in equilibrium with the surrounding melt. With the lateral and diagonal radii R_{10} and R_{11} from the simulation the grid anisotropy $\delta_{grid} = \frac{R_{10}-R_{11}}{R_{10}+R_{11}}$ was evaluated. Grid anisotropy was found to be always much below the anisotropy strength δ for sphere diameters greater than 20 grid points with the Ni dataset (Table 2.1). 1D simulations of planar front solidification with a diffusion field were used to optimize the values of the grid spacing $\Delta\tilde{x}$ and of the diffuse interface width ε with respect to an agreement with the analytical solution.

The calculated phase-field profile corresponds well with the expected $\frac{1}{2}(1 - arctan(x - x_0)/\varepsilon)$ shape (Fig. 4.1 a)) and the temperature profile follows a $erfc(x-x_0)$ form (Fig. 4.1 b)) as expected from literature (cp. [Dav01]). x_0 denotes the front location. As a result the diffuse interface is adequately resolved with about 5 to 8 grid points to produce correct results.

4.2 Dendritic Growth

The 2D and 3D simulations of dendritic solidification in pure Ni were accomplished with the dimensionless dataset of Table 2.1. In the simulation runs, a

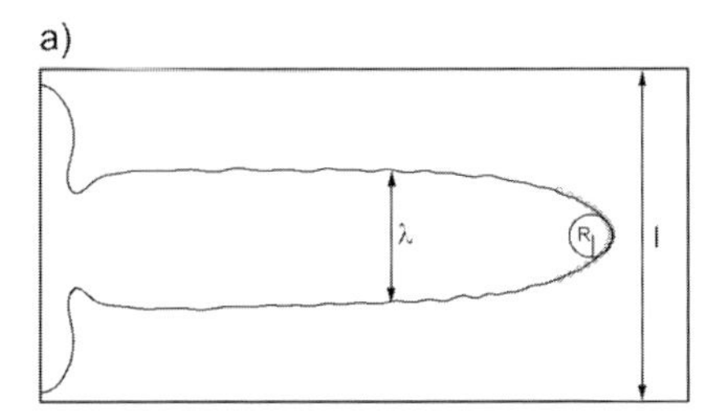

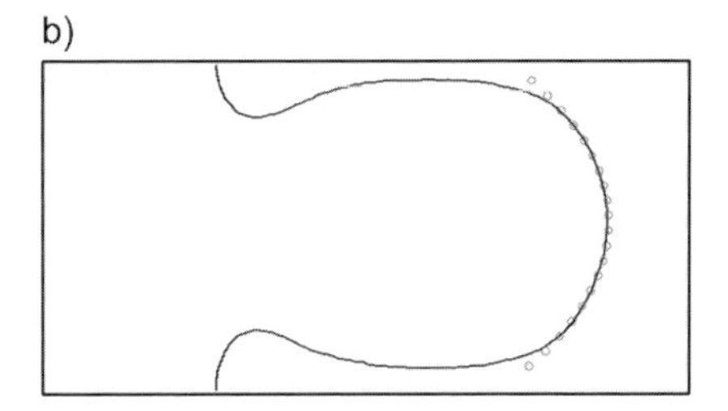

Fig. 4.2. a) A dendrite in a channel of width l with tip radius R and with a finger width λ ($l = 15\ \mu m$); b) liquid finger growth for a narrow channel with $l = 1\ \mu m$.

small amplitude of gaussian distributed noise was added to the front, strong enough to initiate the evolution of dendritic side arms. An example for this pattern in 3D, an equiaxial Ni dendrite, is given in Fig. 4.5 c).

Real alloy microstructures are determined by the competitive growth with a high number of initial dendrite tips, often starting from a planar configuration. A quick selection of the tip spacing due to the retarding effect of the emitted latent heat at the front follows, so that only a few dendritic fingers survive. 2D simulations to study the influence of the dendritic spacing l, of the undercooling $\Delta = c_p(T_m - T)/L$ and of the crystal orientation were carried out. In order to mimic the situation of an array of dendrites, a single nucleus was confined in a long simulation domain with isolation boundary conditions at the bottom and periodic conditions at the long sides. The calculations were executed until the tip velocity approximated an asymptotic value, i.e a steady state was reached. The microscopic solvability theory predicts the operating conditions for a steady state dendrite tip with radius R (inversely proportional to the tip curvature) and tip velocity v_{tip} [BM91]. In Fig. 4.2 a) the geometric parameters of a channel dendrite: the tip radius R, the finger width λ and the channel spacing l are shown. The tip radius was determined by fitting a parabola to the tip region, as indicated by small circles. In a first series the spacing was varied over more than one order of magnitude at a constant undercooling $\Delta = 0.4$. The grid resolution therefore was adapted to avoid finite size effects for narrow channels. The relation of the finger width and of the channel spacing λ/l shows a clear increase from an asymptotic value of $\lambda/l = 0.4$ for large spacings to $\lambda/l = 0.85$ at $l = 1.0\ \mu m$ (cp. Fig. 4.3 a)), which can be explained with the increasing surface tension in a narrow channel. This can also be seen in Fig. 4.3 b), where the tip radius R deviates at the left side from the strictly logarithmic characteristics for wider channels. Also we observe for narrow channels an unsteady evolution of the tip and a viscious finger like behaviour of the interface shape (Fig. 4.2 b)), in accordance with an earlier phase-field study on a model data set [SPG01].

The orientation dependency, an important question when considering situations with more than one nucleus, is essential for selection and overgrowth phenomena during grain formation processes in alloy systems. The effect on

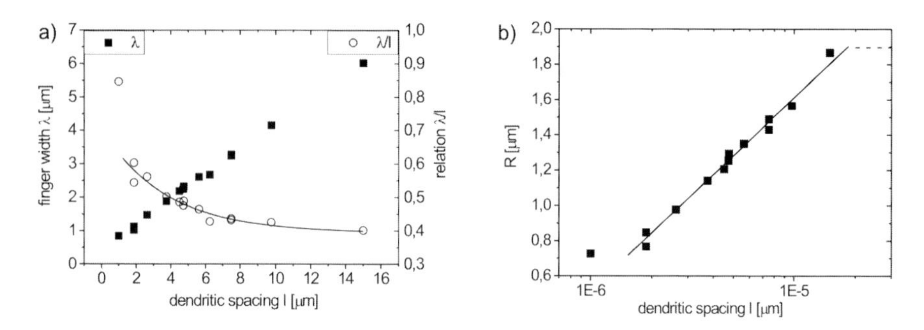

Fig. 4.3. Variation of the dendritic spacing l. a) dendritic finger width λ and fraction λ/l, b) tip radius R (dashed line: value for an equiaxial dendrite).

the growth velocity v for an increasing orientation angle between the dendrite and the long channel direction is given in Fig. 4.4 a). The profile of the dendrite becomes asymmetric with side arms starting to evolve exclusively in the front direction. At an angle of more than 40° two dendritic tips start to compete, so that in a narrow channel a transition to cellular growth takes place (Fig. 4.4 b and c).

Some examples for dendritic array simulations in 3D are shown in Fig. 4.5. A single dendrite in a 7.5 μm channel (a) develops an approximately quadratic cross section. Chart b) presents an array of dendrites emerging from a rough front with a 15° inclined orientation with respect to the surface normal. Due to the compact growth pattern, only a minor side arm formation appears as compared to the equiaxed single nucleus in chart c) (domain: 15 μm).

For comparison, the case of solutal growth in a $Ni_{0.59}Cu_{0.61}$ melt with an isothermal undercooling of 20 K below solidus is shown in Fig. 4.6. The dendrite evolves with a pronounced side arm formation. The dependence of

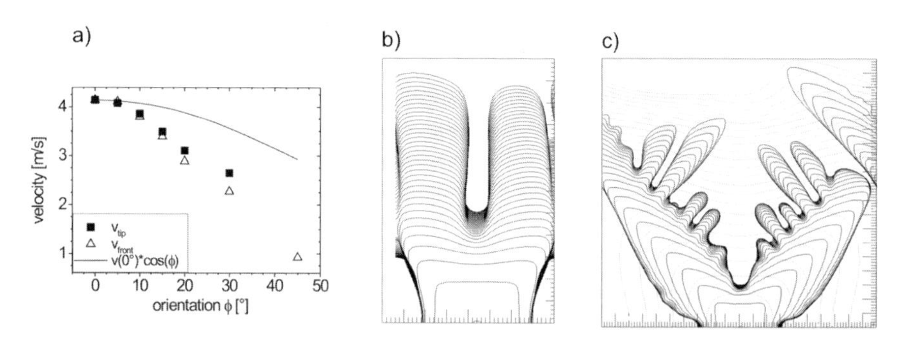

Fig. 4.4. Variation of the orientation angle ϕ. a) tip v_{tip} and front velocity $v_{front} = v_{tip}\cos(\phi)$. Contour plots of growing dendrites with an orientation of 45° in a narrow channel (b)) and in a wide domain (c)).

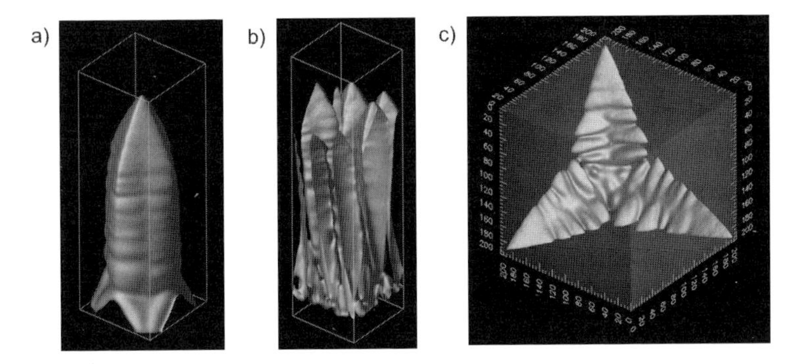

Fig. 4.5. 3D Ni dendritic growth at an undercooling of $\Delta = 0.6$. a) single channel dendrite, b) dendritic array with an orientation inclination of 15° with rsp. to normal, c) equiaxial dendrite. (See page 685 for a colored version of the figure.)

the tip velocity on the channel width is negligible when exceeding a critical width, since the solutal diffusion coefficient is three to four orders of magnitude lower than the thermal coefficient in pure Ni. However, for NiCu a strong dependence on the growth orientation ϕ (anisotropy direction) was found. The plot of the tip velocity v_{tip} in Fig. 4.6 b) indicates the occurance of a minimum at an angle of 15°, whereas the actual front velocity $v_{front} = v_{tip}\, cos(\phi)$ shows only minor changes within a range of 15° to 35°.

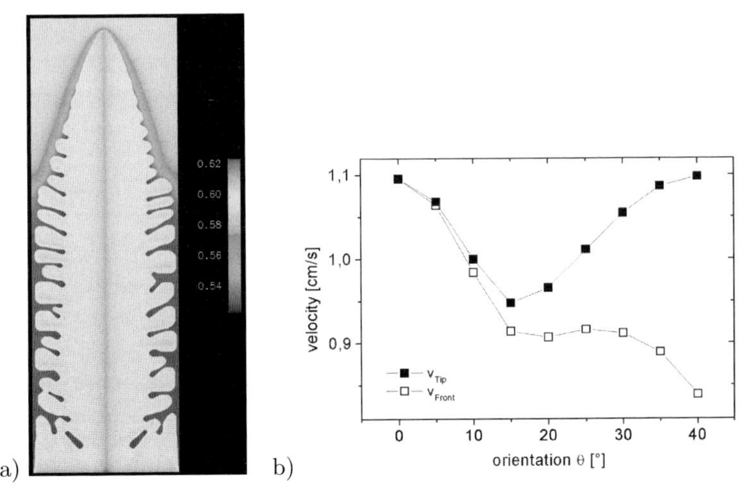

Fig. 4.6. Channel growth for the $Ni_{0.59}Cu_{0.61}$ system. a) Single dendrite arm (unrotated) with Ni concentration map. b) Dependancy on orientation: tip v_{tip} and front velocity $v_{front} = v_{tip} \cos(\phi)$.

4.3 Eutectic Growth

In the following, microstructure simulations of binary and ternary phase transformation processes are shown to illustrate the wide variety of realistic growth structures and morphologies in multicomponent multiphase systems that can be described and investigated by the phase-field model. Fields of applications are eutectic grain boundary formations and structure evolutions in ternary systems [RMC03] as well as eutectic colony growth involving ternary impurity effects [AF00] which will be shown by the following examples.

To perform the simulations in Figs. 4.7-4.13, we considered a ternary eutectic model alloy of three components A, B and C ($i = 1, \dots, 3$), three solid phases α, β, γ ($\alpha = 1, \dots, 3$) and one liquid phase L ($\alpha = 4$). We non-dimensionalized the model equations (Eqs. (2.11-2.13)) and, for initialization of the computations, chose the following parameter set: Equal grid spacings for the two/three coordinates at a value $\Delta x = 0.01$, a diffuse interface thickness $\epsilon = 0.05$, surface entropy densities $\gamma_{\alpha\beta} = 0.001$, an isotropic kinetic coefficient $\tau = 0.2$, zero diffusion in the solid phases $D^{\alpha,\beta,\gamma}_{i=1,\dots,3} = 0.0$ and diffusion coefficients in the liquid $D^L_{i=1,\dots,3} = 0.01$. Further, we constructed a completely symmetric phase diagram with dimensionless data for the latent heats of fusion L_i^α and for the melting temperatures T_i^α:

$$(L_i^\alpha)_{\substack{i=1,\dots,3\\ \alpha=1,\dots,4}} = \begin{pmatrix} 1.47 & 1.00 & 1.00 & 0.00 \\ 1.00 & 1.47 & 1.00 & 0.00 \\ 1.00 & 1.00 & 1.47 & 0.00 \end{pmatrix} \tag{4.1}$$

$$(T_i^\alpha)_{\substack{i=1,\dots,3\\ \alpha=1,\dots,4}} = \begin{pmatrix} 1.50 & 0.50 & 0.50 & 0.00 \\ 0.50 & 1.50 & 0.50 & 0.00 \\ 0.50 & 0.50 & 1.50 & 0.00 \end{pmatrix}, \tag{4.2}$$

where $\alpha = 4$ is assumed to be the liquid phase L. As a result, the phase fractions of the three solid phases at the ternary eutectic temperature are equal. Further, we considered the solidification process under the condition of isothermally undercooled melts.

In Fig. 4.7, the formation of two eutectic grains in the binary $A-B$ 'edge' system of initial composition $(c_A, c_B, c_C) = (0.5, 0.5, 0.0)$ has been simulated in a 2D domain of 270×540 grid points. The simulation involves pattern formation on different length scales. On a larger scale, grains with different orientations due to anisotropy of the surface entropy densities $\gamma_{\alpha\beta}$ grow and form a eutectic grain boundary. To include anisotropic effects, we used the facetted formulation of Eq. (2.6) for a cubic crystal symmetry and defined two sets of four corners for the upper and for the lower grain, whereas the corners of the lower grain are rotated by $10°$ with respect to the growth direction. On a smaller scale, a lamellar eutectic substructure solidifies: Below a critical eutectic temperature T_e (here $T_e = 1.0$), a parent liquid phase L transforms into two solid phases α and β in a binary eutectic reaction: $L \to \alpha + \beta$. The white and light grey colored regions as well as the black and dark grey

 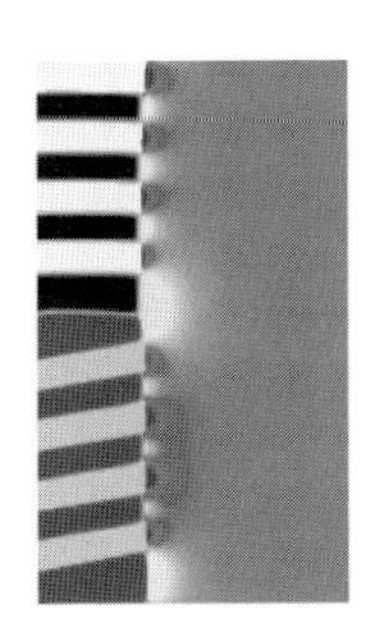 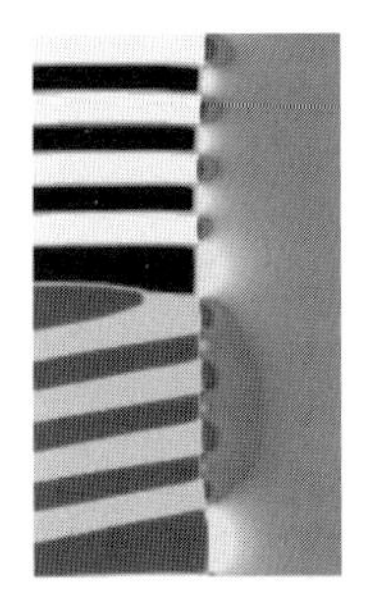 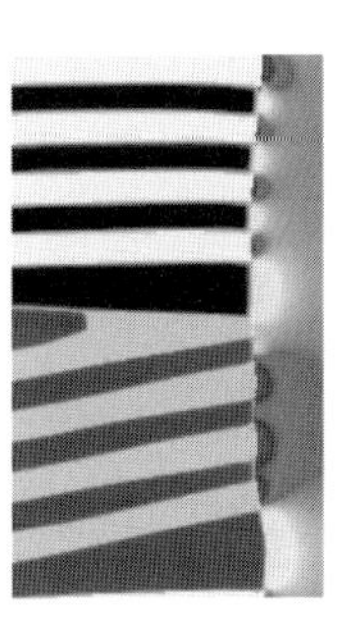

Fig. 4.7. Growth of two eutectic grains (white/black and light/dark-grey) of a binary A-B alloy with different crystal orientations into an isothermally undercooled melt (continuous grey scale) at four time steps.

colored regions represent the same solid phases, namely α and β, with just a different orientation. The results illustrates the capability of the model to distinguish several phases and grains at the same time. The images visualize the phase evolution and the concentration profile of the alloy component B in the liquid ahead of the growing solid phases at different time steps. Concentration depleted zones occur in dark grey and concentration enriched zones appear in light grey.

Fig. 4.8 shows in a) 2D and in b) 3D simulations results of regular eutectic lamellae growing from an undercooled melt with initial composition $(c_A, c_B, c_C) = (0.5, 0.5, 0.0)$. In a symmetric phase system, the two solid phases grow with equal phase fractions. At off-eutectic composition $c = 0.4$, the volume fractions of the two solid phases are different. The proportions of the phase fractions in our simulations can well be related to the classical lever rule. Next, we investigate the widely observed phenomenon of regular oscillations along the solid-solid interface driven by the motion of the triple junctions, see Figs. 4.9 and 4.10. A transition from stable lamellar growth to an oscillatory structure is found for varying initial phase fractions at the eutectic composition. A characteristic amplitude and wave length of the oscillation is established during solidification. The 3D microstructure performs an alternating topological change from α-solid rods embedded in a β-solid matrix to β-solid rods embedded in an α solid matrix and so on.

Depending on the position in the phase diagram, ternary alloy solidification may involve phase changes of four different phases and diffusion of three alloy components A, B and C. At the ternary eutectic composition, three solid phases grow into an undercooled melt via the reaction $L \rightarrow \alpha+\beta+\gamma$. While simultaneously growing, the solid phases mutually enhance each other's growth conditions as they reject opposite components of the alloy into the liquid. We have set an equal initial composition vector of $(c_A, c_B, c_C) = (0.\overline{3}, 0.\overline{3}, 0.\overline{3})$. For isotropic phases, this leads to very regular lamellar structures as those in Fig. 4.11 a) and b). The three images in a) display the concentration fields of

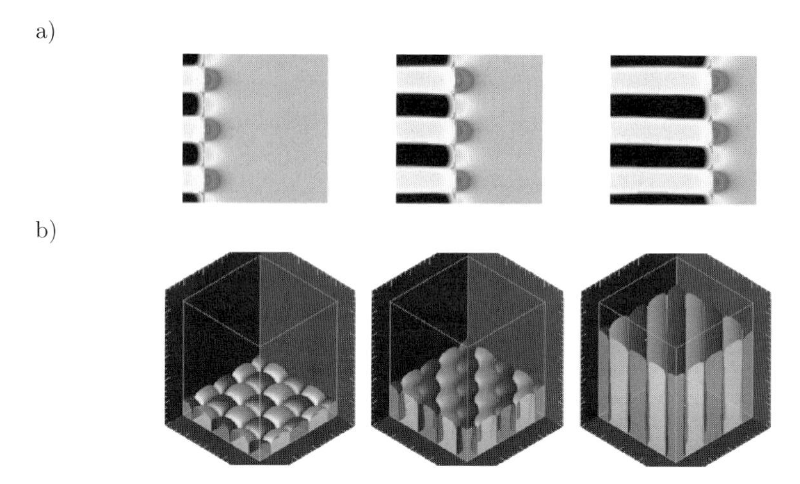

Fig. 4.8. Establishment of regular lamellar solidification at the eutectic composition in 2D (a) and 3D (b). (See page 685 for a colored version of the figure.)

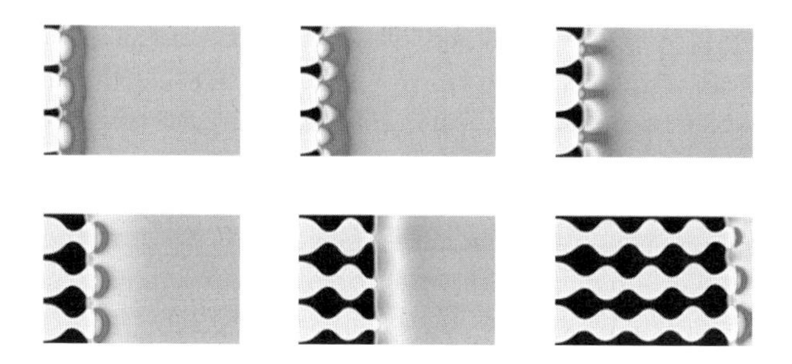

Fig. 4.9. Regular oscillations along the solid-solid interface driven by the motion of the triple junction/triple line in 2D. (See page 685 for a colored version of the figure.)

the three components A, B and C in front of the growing eutectic lamellae with a phase sequence of $\alpha|\beta|\alpha|\gamma|\alpha|\ldots$ at the same intermediate time step. It can be observed that the white α phase consumes component A from the melt and pushes components B and C into the melt. The respective process happens for the two other solids β and γ. For comparison of the diffusion fields, Fig. 4.11 b) shows the concentration of C for a phase sequence $\alpha|\beta|\gamma|\alpha|\ldots$ By performing phase-field simulations, the stability of different phase sequences for varying solidification conditions can be investigated. The diffusion processes of the three components are illustrated in a 2D domain of size 200×200.

Fig. 4.12 shows a time sequence of a 3D simulation of ternary eutectic solidification in a computational domain $60 \times 90 \times 90$. The computation was initialized with cubic crystal shapes. During the evolution, a regular hexagonal structure of the three isotropic solid phases with $120°$ angles between the solid

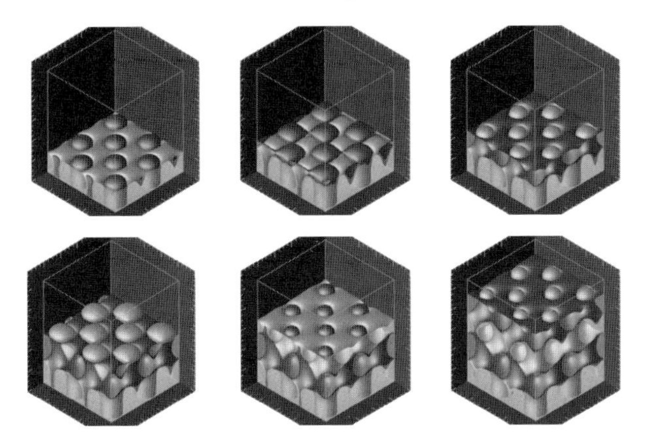

Fig. 4.10. Topological change of the microstructure due to oscillations along the solid-solid interface in 3D. (See page 686 for a colored version of the figure.)

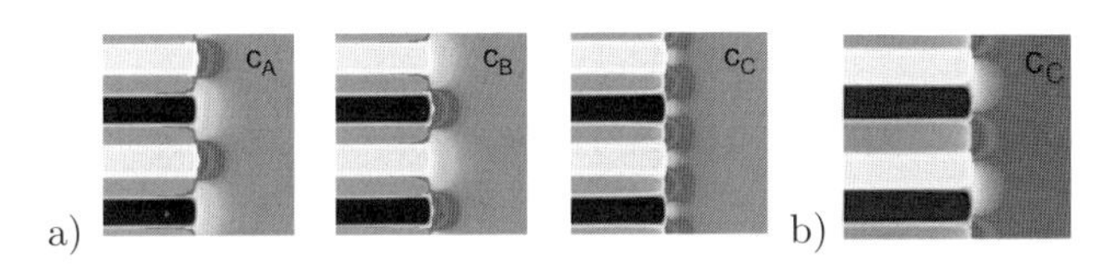

Fig. 4.11. a) Concentration fields c_A, c_B and c_C of a ternary eutectic lamellar solidification front with a solid phase configuration $(\alpha|\beta|\alpha|\gamma|\ldots)$. b) Concentration field c_C for a phase sequence $(\alpha|\beta|\gamma|\ldots)$.

Fig. 4.12. Formation of a 3D hexagonal rod-like structure in a ternary eutectic system with isotropic surface energies and three different solid phases α, β and γ. (See page 686 for a colored version of the figure.)

phases is established as steady growth configuration in 3D in analogy to the lamellar structure in 2D. This symmetry breaks if anisotropy is included.

The simulation in Fig. 4.13 was conducted with an initial composition vector of $(c_A, c_B, c_C) = (0.47, 0.47, 0.06)$ so that the concentration component c_C acts as a ternary impurity of minor amount. As can be seen in the first two images, the solid phase α in white color is formed by using up the concentration c_A whereas solid phase β rejects A atoms. If a γ solid phase containing c_C as its major composition is introduced, it is instable and immediately dissolves for these concentration proportions. Neither the α phase nor the β

Fig. 4.13. Simulation of lamellar eutectic growth in a ternary system with an impurity component c_C: The concentration profile of the main component c_A in melt is shown in the left and centered images for two time steps. The ternary impurity c_C is pushed ahead of the growing eutectic front so that concentration enriched zones of component c_C can be observed at the solid-liquid interface in the right image.

phase engulfs the concentration c_C so that it increases all along the solid-liquid interface. The simulated evolution process recovers the experimentally observed effect that the impurity becomes enriched ahead of the solidifying lamellae and builds up. At larger computational domains, we expect the effect of cell/colony formation to occur.

4.4 From Solidification to Grain Coarsening

A problem with a high relevance to technical applications is the evolution of polycrystalline alloy microstructures consisting of many grains with individual orientations. Solidification as well as grain boundary and concentration reorganization are successive steps in casting respectively heat treatment.

The influence of crystal anisotropy is very important when examining non-equilibrium structures such as dendrites or the interaction of different crystal grains in a polycrystal. In our model, we treat each grain as an individual phase with its own orientation, given by three angles of rotation with respect to the coordinate axes. This approach is not the computationally most effective one as compared to two-phase models with an additional misorientation variable (cp. [GP*04]), but offers benefits with respect to accuracy and parameter choice: if the angular dependence of the surface free energy is given, correct surface energy contributions are incorporated and no additional unknown model parameters, such as the mobility of reorientation, are necessary.

Concerning the numerical realisation in the model with N phases, an $N \times N$ matrix is automatically generated at simulation start and keeps the information of the orientation relations of each $\alpha - \beta$ interface. More precisely, a special anisotropy is not related to a phase with a special orientation, but rather to a phase boundary. This has the great advantage, that each α-β phase boundary can be related to an individual surface energy as well as to an inclination dependence. Both parameters: the misorientation of two grains and the inclination of the grain boundary with respect to a coordinate system fixed in one of the grains, affect the surface energy (see Kazaryan et. al. [KW*00]). In the phase-field equation, only two additional matrix multiplications, one for

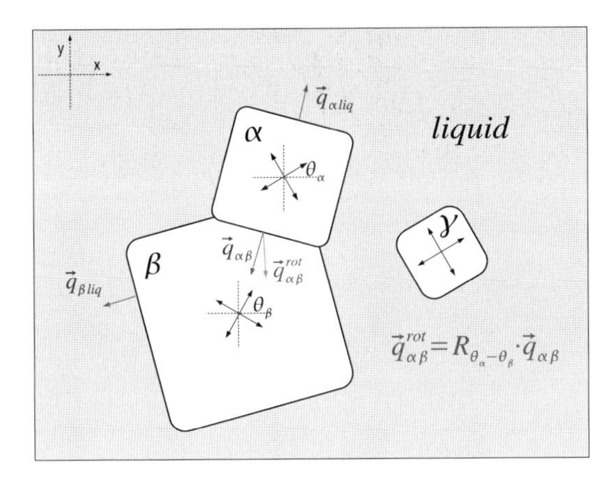

Fig. 4.14. Treatment of multi-phase anisotropy: For three solid phases α, β and γ, the anisotropy with respect to the liquid as well to each solid phase must be defined. Numerically, the $q_{\alpha\beta}$ vectors are rotated by the misorientation angle $\theta_\alpha - \theta_\beta$ and then used in the phase-field equation.

rotating the $q_{\alpha\beta}$-vector and another for back-rotating the resulting entropy flux vector $a_{,\nabla\phi_\alpha}$ (Eq. (3.4)) enter the program sequence. Fig. 4.14 gives an overview of this relationship.

For the simulations presented in the following, the NiCu data set from Sect. 4.2 with identical initial conditions (melt composition: $Ni_{0.59}Cu_{41}$, 20 K below solidus) were taken. First, the selection of dendritic grains with 10

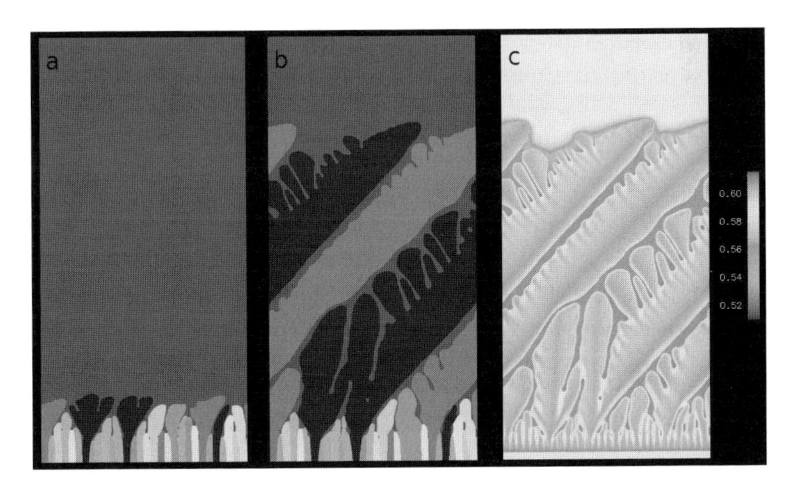

Fig. 4.15. Selection process in a polycrystalline dendritic front: The colours in a) and b) indicate the orientations of the dendrites for two different time steps, whereas in c) the Ni concentration is shown. (See page 686 for a colored version of the figure.)

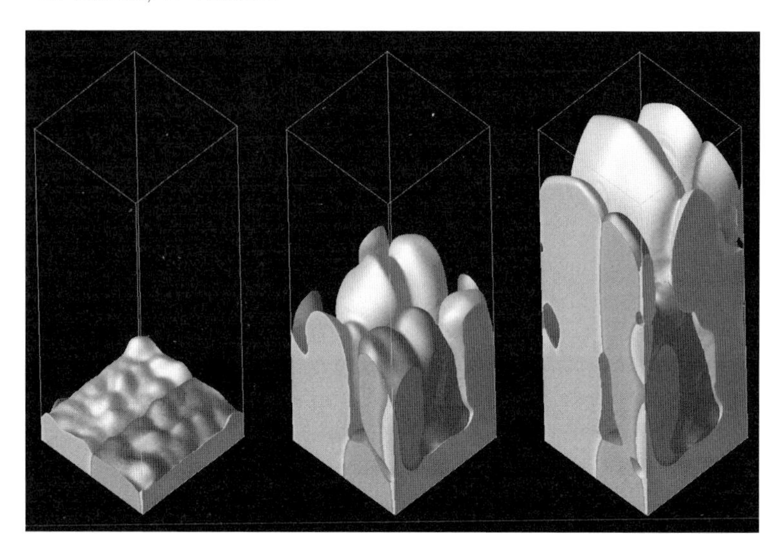

Fig. 4.16. Three time steps of two misaligned NiCu grains in 3D starting from a rough planar initial state (lateral periodic boundary conditions). (See page 687 for a colored version of the figure.)

different orientation orientations over the range of $[-45°, +45°]$ starting from the bottom wall with periodic horizontal boundary conditions is depicted in Fig. 4.15. A rapid overgrowth of the more vertically aligned orientations takes place. The result is a nearly 45° misaligned growth direction of two surviving dendrite branches. A similar process with two differently oriented, initially planar and rough grains is shown in Fig. 4.16, where the grain with a stronger inclination with respect to the vertical axis (lighter grey) overgrows the second.

As a final example, equiaxial grain growth processes for $Ni_{0.59}Cu_{0.41}$ polycrystals and for a random initial distribution of nuclei are discussed. The evolution of the grain structure combines the growth of many differently aligned crystal seeds into the same melt reservoir and the final interaction of the fully solidified individual grains. In Fig. 4.17, three time steps of the solidification process in 2D are shown on a computational domain of 800×800 grid points. The release of Cu in front of the interface slows the phase transformation down. The fully solidified crystals show a low Ni concentration at the grain boundaries, frozen in due to the very slow solid-solid mass diffusion. The analog situation in 3D is given in Fig. 4.18 a) and b) with about 30 separated grains. Here the level sets $\phi_\alpha = 0.5$ are displayed for 9 grains.

When raising the temperature of this grain structure in a subsequent process (Fig. 4.18 c)), but keeping it below the solidus temperature of the equilibrium composition, the structure starts to melt along the edges separating three or more grains. When continuing this processing, about 30 % of the grains shrink and vanish, whereas the rest grows and the remolten liquid

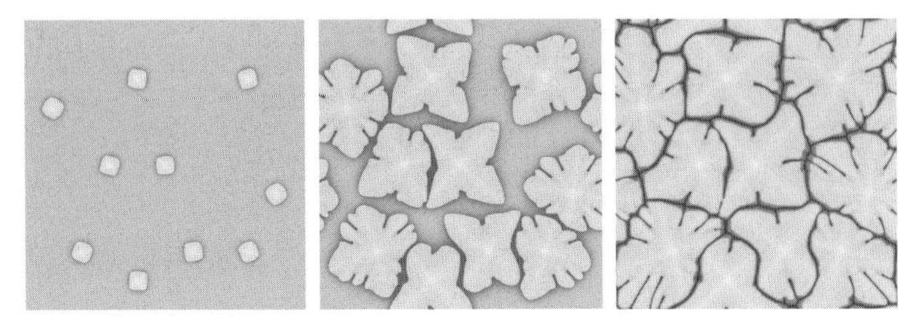

Fig. 4.17. Growth of dendritic NiCu grains into a 20 K undercooled melt illustrated by the Ni concentration (range: 0.41–0.62). The complete solidification (right image) is reached after further reducing the temperature by 15 K in a second step. (See page 687 for a colored version of the figure.)

fraction gradually disappears. A reduction of the interface curvature and a minimization of grain boundary energy is the driving force of the process.

In addition to the visualization, the diffuse phase-field data can be used for various analysis purposes of the material properties. An application useful for the extraction of morphological information which is motivated by crystallographic texture analysis is presented. The surfaces of a grain in contact with neighbours of different orientation evolve due to local curvature, kinetic and surface energy anisotropy. Since the surface normals can be computed by $\nabla\phi_\alpha/|\phi_\alpha|$, a stereographic projection on the equatorial plane reveals information about prefered faces or shape distributions. In Fig. 4.19 an originally spherical grain in a matrix of a second phase with strong cubic anisotropy develops an octahedral shape with distinct faces. This corresponds to the formation of maxima (bright spots in Fig. 4.19 a)–c)) at these directions from an originally homogeneous distribution. The method can also be useful to get average values on a complete set of grains, if one backrotates the surface

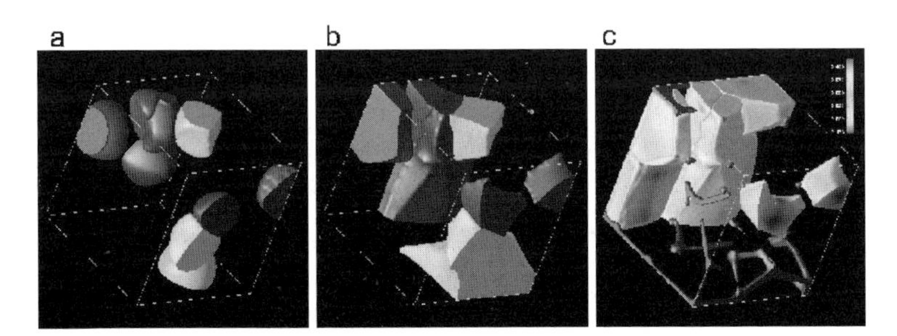

Fig. 4.18. a) and b): Growth of a polycrystalline NiCu structure with 30 grains on a domain of size $100 \times 100 \times 100$. The isosurfaces of selected grains for two time steps are displayed. c): Heat treatment with partial melting along the grain vertices (from [WN06]). (See page 688 for a colored version of the figure.)

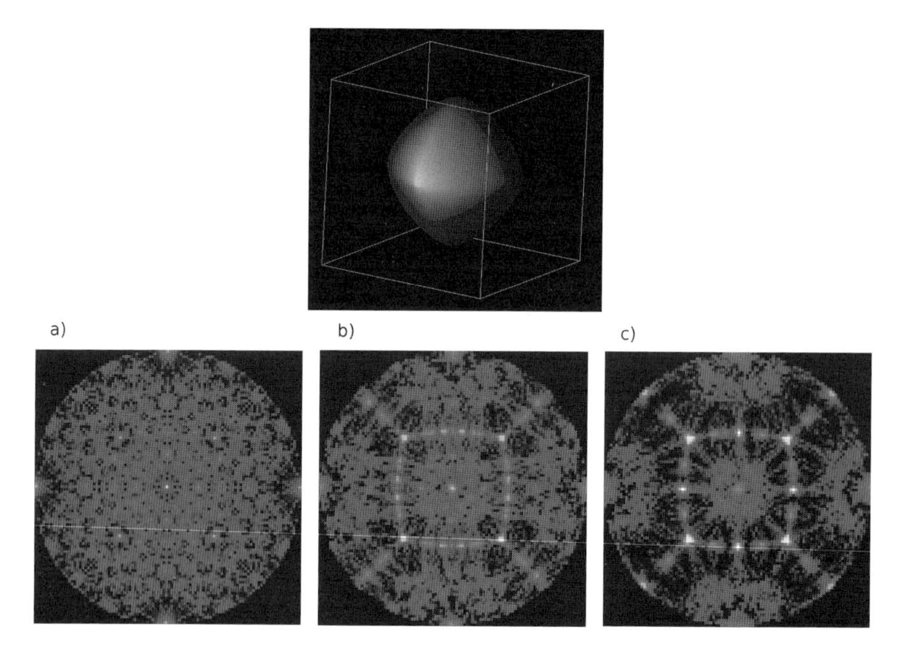

Fig. 4.19. Analysis of surface orientations for an evolving grain (top) using a stereographic projection of the surface normals for three time steps in a)–c). A spherical grain turns into an facetted shape with distinct octahedral faces due to surface energy anisotropy (triangle spots on diagonals in image c)).

normals according to the respective grain orientation before performing the stereographic projection.

4.5 Grain Structures in Geological Materials

As shown in the previous subsection, a numerical scheme which efficiently handles a large number of phase-field parameters can be used to simulate grain growth. We applied our model to two basic geological topics in microstructure evolution: first, grain growth with melt inclusions and second, the crack-seal mechanism in rock veins. Both processes are characterized by a strong interdependence between triple point movements on a small length scale and the resulting large-scale grain morphology.

Geological Microstructure Evolution with Melt Inclusions

Rock formation takes place in deep inner layers of the earth's crust under high temperatures and pressures, where microscopic rock grains are in equilibrium with a low fraction of melt. Fig. 4.20 shows a thin section of an experimental model grain system, where some special local geometries are marked: fluid

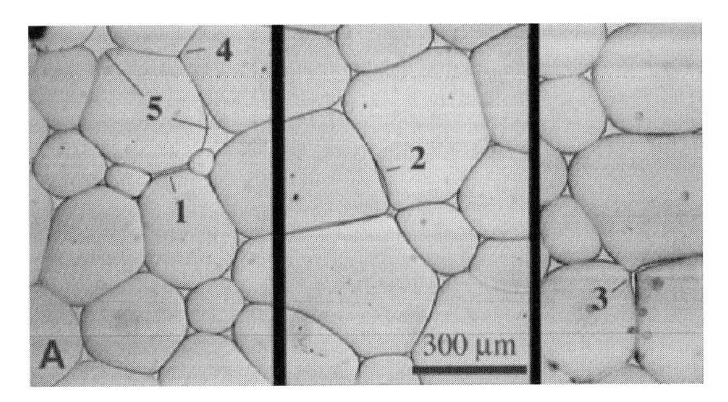

Fig. 4.20. Thin section of an experimental geological model system with grains and intermediate melt pockets (from [BB05]). Typical morphologies of liquid inclusions at grain boundaries and triple junctions are marked (see text).

boundaries (1), grain boundaries with melt lenses (2), edge shaped melt inclusions (3), 'dry' triple junctions (4) and large melt pools (5).

The grain structure is expected to be porous with a high permeability: the melt inclusions are thought to be connected in the third (hidden) dimension allowing for an unlimited exchange of liquid between them. For the simulations, the driving forces $f_{\phi_\alpha}(T, \mathbf{c}, \boldsymbol{\phi})$ for the phase transition in the phase-field Eq. (2.20) were set to zero. Since the total amount of liquid phase is conserved, a new mechanism for the preservation of phase volume was realized in the framework of the multi-phase-field model. Volume preservation of an individual phase α in the domain Ω requires

$$\int_\Omega \phi_\alpha = const. \quad \Leftrightarrow \quad \partial_t \int_\Omega \phi_\alpha = 0.$$

This can be numerically fulfilled by a redistribution of the changed phase volume after each time step at the diffuse interface, weighted with the local kinetic coefficient $\tau(\phi, \nabla\phi)$. To add the missing phase volume fraction exclusively to the interface area, the first derivative of the interpolation function $h(\phi)$ is used, which differs from 0 only for $0 < \phi < 1$. The nonlocal correction term produced in this way was shown to be equivalent to an additional driving force $-\mu_\alpha h(\phi_\alpha)$ on the right hand side of Eq. (2.20), where μ_α represents the strength of this artificial force [WSN06].

2D Simulations were carried out for a various number of solid grains and one fluid phase with preserved volume (see [WZ*06]). To exclude the influence of the domain boundaries, periodic boundary conditions were applied. In a first step, wetting angles and local dynamics of the fluid inclusions were examined. Therefore, a special periodic geometry consisting of 4 hexagonal grains in Fig. 4.21 was designed. This configuration reveals 120° angles between the grains and proved to be stable under calculation for identical solid-solid and solid-liquid surface energies, $\gamma_{ss} = \gamma_{sl}$. Melt inclusions were inserted for different ratio of the surface energies γ_{ss}/γ_{sl} and at different positions along the grain

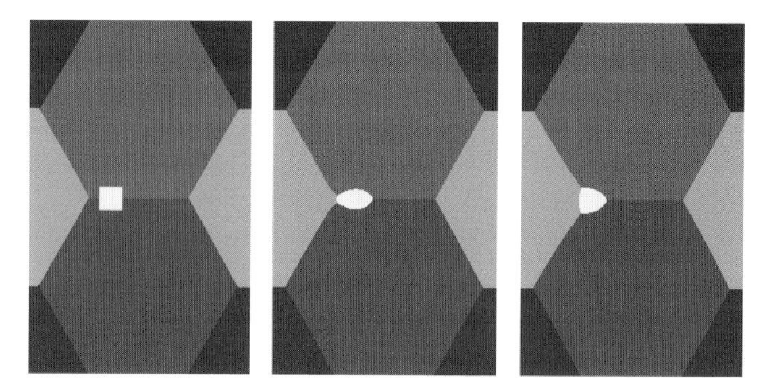

Fig. 4.21. Periodic test geometry with an inserted melt inclusion (light grey) at the horizontal boundary, shown for three time steps (from left to right). (See page 688 for a colored version of the figure.)

boundaries. Fig. 4.21 shows the evolution of a square shaped inclusion, which slowly moves into the nearby triple point. At this location, the total curvature is minimized and the fluid phase reveals stable angles according to the relationship $2cos(\theta/2) = \gamma_{ss}/\gamma_{sl}$ [BWT79]. If the liquid phase is distributed at different positions along a grain boundary, a coarsening process starts in the course of which grains with higher local curvature shrink faster until only one melt pocket survives. This is also initiated for identical shaped inclusions due to a small amount of artificial noise. On a larger scale, the evolution of grain ensembles was studied. Using a Voronoi partitioning algorithm, a random grain structure was produced and circular melt inclusions were placed at the triple points. Besides the grain growth, a coarsening also of the liquid pockets occurs associated with an enlargement of their average diameter. In Fig. 4.22, two steps of the evolution can be seen: Within a small time period, the melt inclusions at the triple points shrink away and on a comparatively longer time scale the inclusions adjacent to four or more grain phases (quadruple points, right picture) perform a ripening process. This is in agreement with

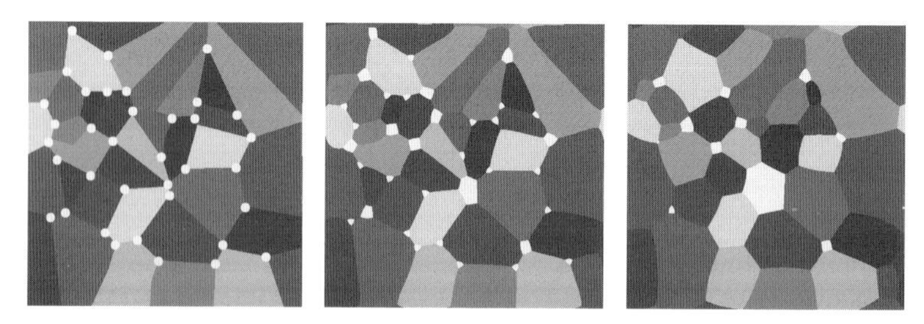

Fig. 4.22. Coarsening of a grain structure with fluid inclusions (light grey) at the triple junctions for three time steps and for a ratio of the surface energies $\gamma_{ss}/\gamma_{sl} = 1.2$. (See page 688 for a colored version of the figure.)

the Neumann-Mullins law which anticipates an area shrinkage rate $\frac{d}{dt}A(t)$ of $\frac{d}{dt}A(t) \propto (n-6)$, with n being the number of edges of the evolving grain [Mul56]. Finally most of the melt is collected in the central hexagonal area.

The Crack-Seal Mechanism in Rock Vein Formation

Rock veins are clearly observable separated regions in a rock matrix consisting of a different polycrystalline mineral. They were formed in solutal growth from a hydrothermal solution which infiltrates a macroscopic crack in the rock. In the veins, a various number of block shaped or fibroid textures were found, an example is shown in Fig. 4.23 a). Essential for the formation of these elongated fibrous crystals is a repeated opening of the crack and a resealing with the mineral phase. In the resulting morphology, the history of the process is stored, which takes place on a long geological time scale, whereas the sealing of an individual crack opening happens quickly. For the 2D simulations, inner boundaries were introduced to represent the outline of the crack (see [NSW06]). As shown in the scheme in Fig. 4.23 b) the rock matrix is modeled via two barrier areas on top and bottom, not linked to memory and containing no phase-field information. The barrier cells serve as markers for the boundary cells next to the gap in between, which contains the liquid phase. As for the outer domain boundary, any boundary condition can be chosen at the barrier walls. To initiate the polycrystalline growth process, a large number of seeds of 10 different orientations is distributed along the upper crack boundary. After the complete solidification of the liquid gap, the lower part of the barrier is shifted a definite number of cells in y- and x-direction to model a crack opening combined with a shear. At the x borders of the simulation domain, periodic conditions were assumed. The additional non-barrier cells

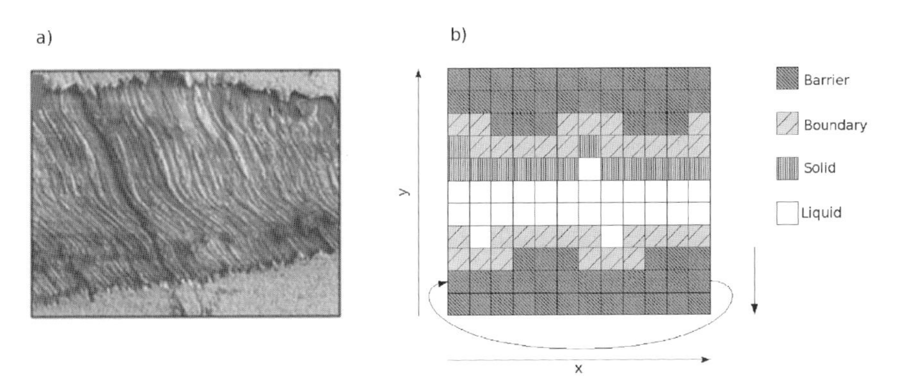

Fig. 4.23. a) Fibroid structure in a rock vein (taken from [Hil00]). b) Schematic view of the simulation setup consisting of the rock matrix (barrier cells), of the inner boundary, solid starting grains and the liquid gap area.

in the simulation grid are initilized as liquid phase. This process is repeated until the lower boundary of the simulation domain is reached. For the first results we desist from introducing a diffusion field or treating the advective flow into the opened crack. A driving force is established only between the solid phases with different orientations and the liquid phase. To amplify the role of the solid-liquid phase transition, solid-solid kinetic coefficients were chosen as $\tau_{ss}^0 = 10\,\tau_{sl}^0$ (cp. Eq. (2.21)). Various anisotropies and crack geometries were studied. As an example, three time steps of two different, randomly initiated simulation runs are depicted in Fig. 4.24 for facetted anisotropies (Eqn. 2.6) of the grains. Contrary to the case of weaker anisotropies, always grain orientations with optimal alignment along the opening-shear direction of the crack survive the selection process (dark and light grey regions). The morphology of the resulting grain boundaries is dominated by the geometry and the opening and shear rate of the crack. The dependence of the growth morphology on the shear rate (the horizontal crack movement) is demonstrated in Fig. (4.25a) and b)) for two simulation runs. A cubic surface energy anisotropy in the form of Eqn. (2.5) with $\delta_{sl} = 0.08$ was chosen. In the case of a vertical opening, also grains with strongly misaligned growth directions (medium grey) tend to be selected and form vertical grain boundaries. An increased horizontal shear leads to a wave-like pattern with increasing frequency and a selection of the optimal growth directions (dark and light grey).

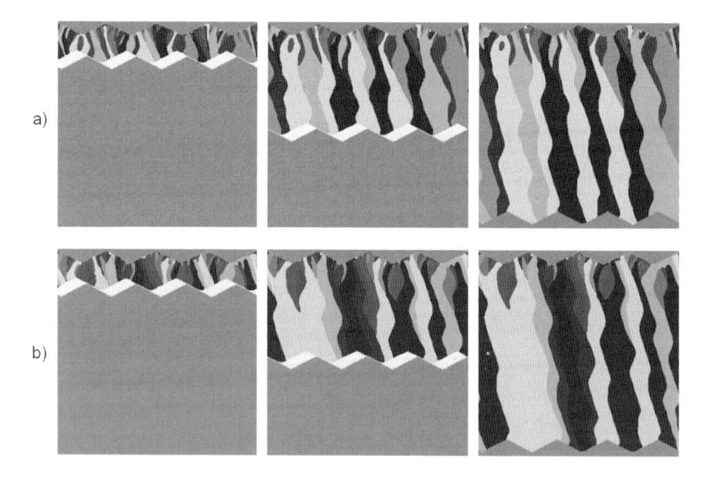

Fig. 4.24. Polycrystalline grain growth in a crack-seal process with facetted anisotropy of the surface energies of the grain boundaries. a) and b) show three time steps of two simulation runs with different starting grain distribution. (See page 689 for a colored version of the figure.)

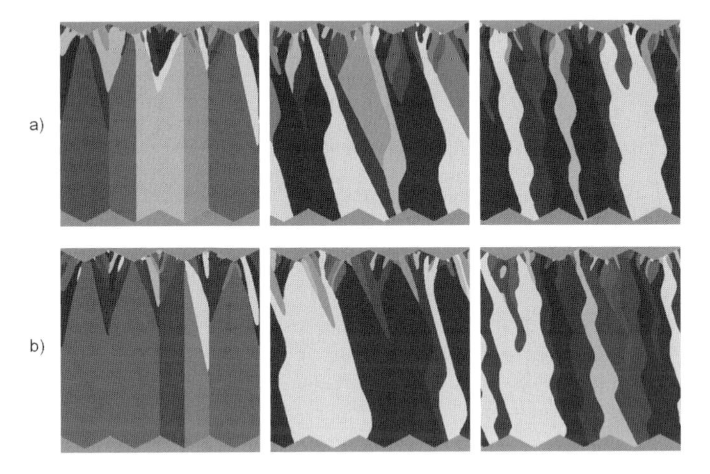

Fig. 4.25. Effect of different shear rates on the resulting morpholgy. From left to right: no shear, $\Delta x = 10$ cells, $\Delta x = 20$ cells. Two simulation runs a) and b) with different grain distributions are displayed. (See page 689 for a colored version of the figure.)

5 Conclusion

A phase-field model for multiphase solidification has been formulated in a general and thermodynamically consistent way. Explicit expressions for different energy density contributions are given, which are appropriate to the specific physical process. Multicomponent diffusion in the bulk phases including interdiffusion coefficients as well as diffusion in the interfacial regions are discussed. Anisotropy of both, the surface energies and the kinetic coefficients is incorporated in the model formulation. Simulation results of dendritic, eutectic and grain growth show the capability of the model to describe phase transitions and complex multiscale microstructure formation. Building upon results on modelling rapid solidification [AW*98, Gal01] and on deriving a thin interface analysis [KR96, Al99, KKS99, MWA00, Kav01], challenges for future developments are the extension of the presented model to strong nonequilibrium solidification and to small undercooling situations. The effective numerical multi-phase algorithm can be utilized in a straight way to incorporate fluid flow (see [SNW05]) and elasticity (in preparation).

Acknowlegdments. The authors gratefully acknowledge the financial support of the German Research Foundation, Grant Nos. Ne 882/1–1 to 1–3.

References

[AW*98] Ahmad, N.A., Wheeler, A.A., Boettinger, W.J., McFadden, G.B.: Solute trapping and solute drag in a phase-field model of rapid solidification. Phys. Rev. E **58**, 3436–3450 (1998)

[AF00] Akamatsu S., Faivre, G.: Traveling waves, two-phase fingers, and eutectic colonies in thin-sample directional solidification of a ternary eutectic alloy. Phys. Rev. E **61**, 3757–3770 (2000)
[Al99] Almgren, R.F.: Second-order phase field asymptotics for unequal conductivities. SIAM J. Appl. Math. **59** 2086–2107 (1999)
[And00] Andersson, C.: Computation of dendrites on parallel distributed memory architectures. In: Engquist, B., Johnsson, L., Hammill, M., Short, F. (eds.), Lecture Notes in Computational Science and Engineering, Vol. 13, Springer Verlag (2000)
[AB*02] Apel, M., Boettger, B., Diepers, H.-J., Steinbach, I.: 2D and 3D phase–field simulations of lamellae and fibrous eutectic growth. J. Cryst. Growth, **237 - 239**, 154–158 (2002)
[BS98] Bi, Zh. and Sekerka, R.F.: Phase-field model of solidification of a binary alloy. Physica A **261**, 95–106 (1998)
[BW*02] Boettinger, W.J., Warren, J.A., Beckermann, C., Karma, A.: Phase–field simulation of solidification. Annu. Rev. Mater. Res. **32**, 163–194 (2002)
[BB05] Bons, P.D., Becker, J.K.: personal communication
[BK*02] Bragard, J., Karma, A., Lee, Y.H., Plapp, M.: Linking phase-field and atomistic simulations to model dendritic solidification in highly undercooled melts. Interface Science **10**, 121–136 (2002)
[BM91] Brener, E.A., Mel'nikov, V.I.: Pattern selection in two-dimensional dendritic growth. Adv. Phys. **40**, 53–97 (1991)
[BWT79] Bulau, J.R., Waff, H.S., Tyburczy, J.A.: Mechanical and thermodynamic constraints on fluid distribution in partial melts. J. Geophys. Res. **84**, 6102–6108 (1979)
[Cag89] Caginalp, G.: Stefan and Hele Shaw type models as asymptotic limits of the phase field equations. Phys. Rev. A **39**, 5887–5896 (1989)
[Cha77] Chalmers, B.: Principles of solidification. Krieger, Melbourne, FL (1977)
[Dav01] Davis, S.H.: Theory of solidification. Cambridge University Press (2001), chapt. 2.1, pp. 7
[Gal01] Galenko, P.: Phase-field model with relaxation of the diffusion flux in nonequilibrium solidification of a binary system. Phys. Lett. A **287**, 190–197 (2001)
[GHS06] Garcke, H., Haas, R., Stinner, B., in preparation
[GNS98] Garcke, H., Nestler, B., Stoth, B.: On anisotropic order parameter models for multi-phase systems and their sharp interface limits. Physica D **115**, 87–108 (1998)
[GNS99a] Garcke, H., Nestler, B., Stoth, B.: A multi phase field concept: numerical simulations of moving phase boundaries and multiple junctions. SIAM J. Appl. Math. **60**, 295–315 (1999)
[GNS99b] Garcke, H., Nestler, B., Stoth, B.: Anisotropy in multi-phase sytems: a phase-field approach. J. Interfaces and Free Boundary Problems, **1**, 175–198 (1999)
[GNS04] Garcke, H., Nestler, B., Stinner, B.: A diffuse interface model for alloys with multiple components and phases. SIAM J. Appl. Math. **64**, 775–799 (2004)
[GBP02] Gránásy, L., Börzsönyi, T., Pusztai, T.: Nucleation and bulk crystallization in binary phase field theory. Phys. Rev. Lett. **88**, 206105-1 – 206105-4 (2002); Crystal nucleation and growth in binary phase-field theory. J. Cryst. Growth **237 - 239**, 1813–1817 (2002)

[GP*04] Gránásy, L., Pusztai, T., Börzsönyi, T., Warren, J.A., Douglas, J.F.: A general mechanism of polycrystalline growth. Nature Materials **3**, 645–650 (2004)

[Haa94] Haasen, P.: Physikalische Metallkunde. 3rd ed., Springer, Berlin (1994)

[Hil00] Hilgers, Ch.: Vein growth in fractures - experimental, numerical and real rock studies. Doctoral thesis, RWTH Aachen (2000)

[JGD01] Jeong, J.-H., Goldenfeld, N., Dantzig, J.A.: Phase field model for three–dimensional dendritic growth with fluid flow. Phys. Rev. E, **64**, 041602-1 – 041602-14 (2001)

[Kav01] Karma, A.: Phase-field formulation for quantitative modeling of alloy solidification. Phys. Rev. Lett. **87**, 115701–115704 (2001)

[KR96] Karma A., Rappel, W.-J.: Phase-field method for computationally efficient modeling of solidification with arbitrary interface kinetics. Phys. Rev. E **53** (4), R3017–R3020 (1996); Quantitative phase-field modeling of dendritic growth in two and three dimensions. Phys. Rev. E **57**, 4323–4349 (1998)

[KW*00] Kazaryan, A., Wang, Y., Dregia, S.A., Patton, B.R.: Generalized phase-field model for computer simulation of grain growth in anisotropic systems. Phys. Rev. B **61**, 14275–14278 (2000)

[KKS99] Kim, S.G., Kim, W.T., Suzuki, T.: Phase-field model for binary alloys. Phys. Rev. E **60**, 7186–7197 (1999)

[KO*03] Kobayashi, H., Ode, M., Kim, S.G., Kim, W.T., Suzuki, T.: Phase-field model for solidification of ternary alloys coupled with thermodynamic database. Scripta Mat. **48**, 689–694 (2003)

[KWC00] Kobayashi, R., Warren, J.A., Carter, W.C.: A continuum model of grain boundaries. Physica D **140**, 141–150 (2000)

[KF92] Kurz, W., Fischer, D.J.: Fundamentals of Solidification. 3rd ed. Trans. Tech. Publications, Aedermannsdorf, Switzerland, 1992

[MWA00] McFadden, G.B., Wheeler, A.A., Anderson, D.M.: Thin interface asymptotics for an energy/entropy approach to phase-field models with unequal conductivities. Physica D **144** (1-2), 154–168 (2000)

[Mul56] Mullins, W.W.: Two–dimensional motion of idealized grain boundaries. J. Appl. Phys. **27**, 900–904 (1956)

[NGS05] Nestler, B., Garcke, H., Stinner, B.: Multicomponent alloy solidification: Phase-field modeling and simulations. Phys. Rev. E **71**, 041609–041614 (2005)

[NSW06] Nestler, B., Selzer, M., Wendler, F.: Ein Kontinuumsmodell zur Beschreibung der Mikrostrukturausbildung bei der Versiegelung von Brüchen in Gesteinsadern. To appear in 'Berichte aus der Umweltinformatik', Proc. of the workshop 'Simulation in den Umwelt- und Geowissenschaften, Medizin und Biologie'. Shaker-Verlag Aachen, in print (2006)

[NW00] Nestler, B., Wheeler, A.A.: Anisotropic phase-field model: interfaces and junctions Phys. Rev. E **57**, 2602–2609 (1998); A multi-phase-field model of eutectic and peritectic alloys: numerical simulation of growth structures. Physica D **138** (1-2), 114–133 (2000)

[NW*05] Nestler, B., Wendler, F., Frodl, T., Schabunow, D.: Ein paralleler 3D Simulator zum mikrostrukturbasierten Materialdesign. 239–258 In: Spath, D., Haasis, K., Klumpp, D. (eds.), Aktuelle Trends in der Softwareforschung. Proc. of the doIT Software Research Day, Fraunhofer IRB Verlag, Stuttgart, ISBN 3-8167-6955-1, 239 - 258 (2005)

[PF90] Penrose O., Fife, P.C.: Thermodynamically consistent models of phase-field type for the kinetics of phase transitions. Physica D **43**, 44–62 (1990)

[PK00] M. Plapp and A. Karma, Phys. Rev. Lett. **84** (8), 1740–1743 (2000); Akamatsu, S., Plapp, M., Faivre, G., Karma, A., Phys. Rev. E **66** 030501(R) (2002)

[RB*04] Ramirez, J.C., Beckermann, C., Karma, A., Diepers, H.-J.: Phase-field modeling of binary alloy solidification with coupled heat and solute diffusion. Phys. Rev. E **69**, 051607 (2004)

[RMC03] Rios, C., Milenkovic, S., Caram, R.: A novel ternary eutectic in the Nb-Al-Ni system. Scripta Mat. **48**, 1495–1500 (2003)

[SPG01] Sabouri-Ghomi, M., Provatas, N., Grant, M.: Solidification of a supercooled liquid in a narrow channel. Phys. Rev. Lett. **86**, 5084–5087 (2001)

[SNW05] Selzer, M., Nestler, B., Wendler, F.: A coupled 3D simulator for solidification microstructures with fluid flow. 124–130, In: Hülsemann, F., Kowarschik, M., Rüde, U. (eds.), Frontiers in Simulation. Proc. of the 18th Symposium on Simulationstechnique in Erlangen, SCS Publishing House e. V., Erlangen ISBN 3-936150-41-9, 124 - 130 (2005)

[Sea96] Steinbach, I., Pezzolla, F., Nestler, B., Seesselberg, M., Prieler, R., Schmitz, G.J., Rezende, J.L.L.: A phase field concept for multiphase systems. Physica D **94**, 135–147 (1996)

[TN*98] Tiaden, J., Nestler, B., Diepers, H.–J., Steinbach, I.: The multiphase-field model with an integrated concept for modelling solute diffusion. Physica D **115**, 73–86 (1998)

[WS*93] Wang, S.-L., Sekerka, R.F., Wheeler, A.A., Murray, B.T., Coriell, S.R., Braun, R.J., McFadden, G.B.: Thermodynamically-consistent phase-field models for solidification. Physica D **69** (1-2), 189–200 (1993)

[WN06] Wendler, F., Nestler, B.: 3D phase-field simulations of polycrystalline and multiphase microstructures, In: Proc. of the conference of Casting, Welding and Advanced Solidification Processes XI (MCWASP), Opio, France (2006)

[WSN06] Wendler, F., Stinner, B., Nestler, B.: in preparation

[WZ*06] Wendler, F., Zamora-Morschhäuser, M., Nestler, B., Selzer, M.: Phasenfeldsimulation der Korngrenzenbewegung und des Kornwachstums in geologischen Materialien. To appear in 'Berichte aus der Umweltinformatik', Proc. of the workshop 'Simulation in den Umwelt- und Geowissenschaften, Medizin und Biologie'. Leipzig (2006)

[WMB96] Wheeler, A.A., McFadden, G.B., Boettinger, W.J.: Phase-field model for solidification of a eutectic alloy. Proc. Roy. Soc. Ser. A **452**, 495–525 (1996)

Multiple Scales in Phase Separating Systems with Elastic Misfit

Harald Garcke[1], Martin Lenz[2], Barbara Niethammer[3], Martin Rumpf[2], and Ulrich Weikard[4]

[1] NWF I – Mathematik, Universität Regensburg, 93040 Regensburg. harald.garcke@mathematik.uni-regensburg.de
[2] Institut für Numerische Simulation, Rheinische Friedrich-Wilhelms-Universität Bonn, Nußallee 15, 53115 Bonn. martin.lenz@ins.uni-bonn.de, martin.rumpf@ins.uni-bonn.de,
[3] Institut für Mathematik, Humboldt-Universität zu Berlin, Unter den Linden 6, 10099 Berlin. niethamm@mathematik.hu-berlin.de
[4] Fachbereich Mathematik, Gerhard-Mercator-Universität Duisburg, Lotharstr. 63/65, 47048 Duisburg. weikard@math.uni-duisburg.de

1 Introduction

In this article we review recent attempts to understand the interaction of different length and time scales in phase separating systems with elastic misfit. Phase separation occurs for example if an alloy is quenched below a critical temperature, where a homogeneous mixture of the alloy components is not stable. The early stage of the separation process, where different phases, characterized by the respective concentrations of the alloy components, appear is called spinodal decomposition.

The Cahn-Hilliard model [CH58] and its extension with elasticity, the Cahn-Larché model [CL82, CL73], have originally been introduced to model spinodal decomposition. Later numerical simulations (see e.g. [Ell89]) and formally matched asymptotic expansions (see [Pe89]) showed that the Cahn-Hilliard equation can also describe a process on a slower intermediate time scale in which the regions occupied by the phases rearrange in order to decrease their free energy. In the case that elastic contributions can be neglected, the free energy is essentially given by the surface energy and the evolution leads to nearly spherical disjoint components, called particles (see Fig. 1.1). If anisotropic elastic effects are present the shapes resemble the anisotropy of the elastic energy (see Fig. 1.2).

In the late stage, when the system has already minimized its energy locally, interactions between particles become important. In the case that no elastic energy is relevant, small particles shrink, while larger ones grow, a coarsening process known as Ostwald Ripening. The influence of elastic in-

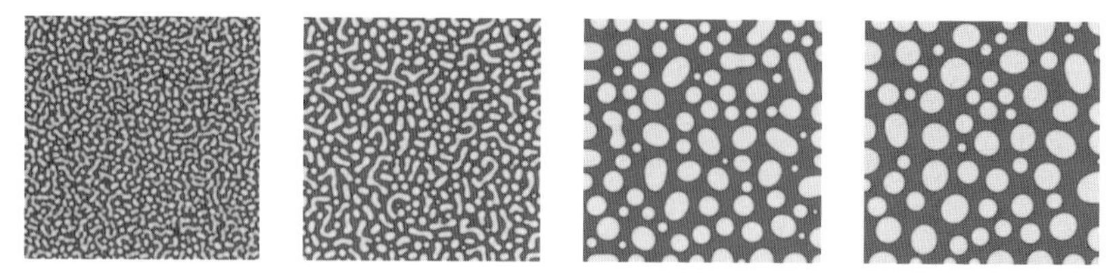

Fig. 1.1. Evolution starting from a perturbation of a uniform state. (See page 690 for a colored version of the figure.)

Fig. 1.2. Alignment of interfaces driven by homogeneous, anisotropic elasticity. (See page 690 for a colored version of the figure.)

teractions, e.g. through an elastic misfit due to different lattice constants, can drastically influence the coarsening process. The shape of the particles changes from spherical to cuboidal or plate shape, particles can align or even split. In particular on the large time scale the elastic energy which scales like a volume becomes comparable to the surface energy and it might be possible to stabilize the coarsening process ("inverse coarsening"). For a review on the modelling of phase separation in alloys with elastic misfit we refer to [FPL99]. To model the late stage regime, often so called sharp interface models are used, which also appear as singular limits of the Cahn-Hilliard equation (see [GK06]). In contrast to the latter, the boundary between different phases is given by a hypersurface.

In this overview we will discuss both the Cahn-Hilliard equation with elasticity (the Cahn-Larché system) and a Mullins-Sekerka type model with anisotropic and inhomogeneous elasticity. Although we will also discuss some aspects of modelling and mathematical analysis, our main focus will be on computational aspects.

First, we will introduce the governing models and their interpretation as a gradient flow in Sect. 2. The latter will be relevant for the set up of a reduced model to simulate large particle ensembles. In Sect. 3 we study the effect of elastic contributions on spinodal decomposition within the Cahn-Larché model. In Sect. 4 we will explain how the Cahn-Larché system can be solved efficiently and present computational results for the Cahn-Hilliard equation and the Cahn-Larché system. In Sect. 5 we study coarsening rates for a large system of particles. Here we observe a transient coarsening behaviour in the Cahn-Hilliard model without elasticity and we will also see effects of elasticity on the ripening process.

In Sect. 6 we introduce a boundary integral formulation of the Mullins-Sekerka evolution and a corresponding boundary integral method. Simulations for systems with a few particles will be presented, which in particular show typical particle shapes and display when a certain pattern such as alignment of particles appears. Finally, in Sect. 7, we will use the gradient-flow perspective for the Mullins-Sekerka evolution to derive a reduced model, in which particle shapes are extremely simple. With this approach we can efficiently simulate larger particle systems.

2 The models

2.1 The Cahn-Larché model

We consider the case of a binary alloy, i.e. two alloy components are present with concentrations c_1 and c_2. We choose the concentration difference $c = c_1 - c_2$ as variable which due to the constraint $c_1 + c_2 = 1$ determines the concentrations. The deformation field is denoted by u and since we consider models that are based on linearized elasticity we introduce the linearized strain tensor

$$\varepsilon(F) := \frac{1}{2}(F + F^T), \text{ with } F = \nabla u.$$

The free energy of the system is then given by

$$\mathcal{E}[c,u] = \int_\Omega \{\frac{\gamma}{2}|\nabla c|^2 + \psi(c) + W(c, \nabla u)\}\, dx \tag{2.1}$$

where $\Omega \subset \mathbf{R}^d$ is a bounded domain, $\gamma > 0$ is a small interfacial parameter, $\psi : \mathbf{R} \to \mathbf{R}$ is the non-convex free energy density and $W : \mathbf{R} \times \mathbf{R}^{d\times d} \to \mathbf{R}$ is the elastic energy density. A homogeneous free energy density ψ for a mean field model at a fixed absolute temperature is

$$\psi(c) = \tfrac{R\theta}{2}\{(1+c)\ln(1+c) + (1-c)\ln(1-c)\} + \tfrac{R\theta_c}{2}(1-c^2)\,. \tag{2.2}$$

Here θ_c is the critical temperature and R is the gas constant scaled by the (constant) molar volume. For θ below the critical temperature θ_c the energy density ψ has two global minima c_-, c_+ and hence a non-convex form. For shallow quenches, i.e. $0 \ll \theta < \theta_c$ one usually takes a smooth approximation to (2.2) of the form

$$\psi(c) = b(c^2 - a^2)^2\,, \quad 0 < a < 1,\ b > 0. \tag{2.3}$$

As elastic energy density W we take a quadratic function in the strain tensor ε and set

$$W(c, \nabla u) = \frac{1}{2}(\varepsilon(\nabla u) - \bar{\varepsilon}(c)) : C(c)(\varepsilon(\nabla u) - \bar{\varepsilon}(c)).$$

Here $\overline{\varepsilon}(c)$ is the symmetric misfit strain (also called eigenstrain), $C(c)$ is the fourth rank elasticity tensor and $A : B := tr(A^T B)$ for linear mappings A and B. As the elasticity tensor C is assumed to be symmetric and positive definite we obtain that $\overline{\varepsilon}(c)$ is the energetically favourable and hence stress free strain at concentration c. Typically $\overline{\varepsilon}$ is affine linear, i.e.

$$\overline{\varepsilon}(c) = \varepsilon^1 + \varepsilon^* c$$

where $\varepsilon^1, \varepsilon^* \in \mathbf{R}^{d\times d}$ are symmetric. We allow for an elasticity tensor that can be different for the two components and hence C can depend on the concentration c.

For an isotropic material we obtain

$$C(c)\varepsilon = 2\mu(c)\varepsilon + \lambda(c)\, tr(\varepsilon)\mathrm{Id}$$

where the Lamé moduli μ and λ depend on the concentration c.

For a material with cubic symmetry we have

$$C(c)\varepsilon = 2\mu(c)\varepsilon + \lambda(c)\, tr\, \varepsilon \mathrm{Id} + \mu'(c)\mathrm{diag}\, \varepsilon$$

where $\mathrm{diag}\, \varepsilon$ is the matrix that one obtains, if all off-diagonal entries are set to zero. In general C is an arbitrary fourth rank tensor $C(c) = (C_{ij\, i'j'}(c))$ and using the symmetry conditions

$$C_{ij\, i'j'} = C_{ij\, j'i'} = C_{ji\, i'j'} = C_{i'j'\, ij}$$

one can compute that for $d = 3$ there are 21 degrees of freedom in C which of course in general will be restricted by crystal symmetry.

For example in a cubic system we obtain that $C_{1111} = C_{2222} = C_{3333}$, $C_{iijj} = C_{ii\kappa\kappa}$ (for i, j, κ mutually different), $C_{2323} = C_{3131} = C_{1212}$ and all other entries in C either follow from the above by symmetry or they are zero. Sometimes a fourth rank tensor in $\mathbf{R}^3$ is denoted by C_{ij} (Voigt notation). In this case the indices i, j take values $1, 2, 3, 4, 5, 6$ and they stand for the pairs $11, 22, 33, 23, 31, 12$ in the original notation. This means in a cubic system we only need to specify C_{11}, C_{12} and C_{44}. All other parameters are determined by symmetry. For a discussion of other symmetry classes we refer to Gurtin [Gu72]. We will also always assume that $C(c)$ is positive definite and bounded uniformly in c.

Taking mechanical effects in the Cahn-Hilliard model into account we obtain the system

$$\partial_t c = \Delta w, \tag{2.4}$$

$$w = \frac{\delta \mathcal{E}}{\delta c} = -\gamma \Delta c + \psi'(c) + W_{,c}(c, \nabla u), \tag{2.5}$$

$$0 = \frac{\delta \mathcal{E}}{\delta u} = -\nabla \cdot W_{,F}(c, \varepsilon(\nabla u)), \tag{2.6}$$

which we sometimes also call the Cahn-Larché system (see [CL82, CL73]). Here $\frac{\delta E}{\delta c}$ denotes the first variation of E with respect to c and $W_{,c}$ is the partial derivative with respect to c (the same notation holds with respect to u). We remark that for simplicity in (2.4) the mobility is taken to be 1. The chemical potential w is the diffusion potential and is given by the first variation of energy with respect to concentration. The quantity $S = W_{,F}$ with $F = \nabla u$ is the stress and hence (2.6) are the mechanical equilibrium equations from the theory of elasticity.

The set of equations then has to be completed by appropriate boundary conditions which can be e.g. periodic boundary conditions or Neumann boundary conditions for w and c and a prescribed normal stress at the boundary for the u-equation.

2.2 The Cahn-Larché system as a gradient flow

The Cahn-Larché system can be viewed as a gradient flow. A gradient flow is the flow in the direction of steepest descent in an energy landscape. This framework requires a differentiable manifold $\mathcal{M}$, and a vector field f, which attaches a tangent vector $f(x) \in T_x\mathcal{M}$ to every point $x \in \mathcal{M}$. The vector field f defines a dynamical system $\dot{x} = f(x)$. A gradient flow is a dynamical system where f is the negative gradient $-\operatorname{grad}\mathcal{E}$ of a function $\mathcal{E}$ on $\mathcal{M}$. The notion of a gradient requires a Riemannian structure, that is, a metric tensor g on $\mathcal{M}$. Then, the precise formulation of $\dot{x} = -\operatorname{grad}\mathcal{E}_x$ is

$$g_{x(t)}(\dot{x}(t), y) + \langle \operatorname{diff}\mathcal{E}_{x(t)}, y\rangle = 0 \quad \text{for all } y \in T_{x(t)}\mathcal{M} \text{ and for all } t. \tag{2.7}$$

If we choose $y = \dot{x}(t)$ we observe that the value of $\mathcal{E}$ decreases along trajectories.

We now give two possibilities to view the Cahn-Larché system as a gradient flow. First we choose

$$\mathcal{M} := \left\{ c : \Omega \to \mathbf{R} \,\middle|\, \int_\Omega c\,dx = \int_\Omega c_0\,dx \right\},$$

where $c_0 : \Omega \to \mathbf{R}$ is the concentration at time zero. The tangent space is then given as

$$T_c\mathcal{M} := \left\{ v : \Omega \to \mathbf{R} \,\middle|\, \int_\Omega v\,dx = 0 \right\}$$

and the metric tensor on $T_c\mathcal{M}$ is given by the H^{-1} norm, that is

$$g_c(v, \tilde{v}) := \int_\Omega \nabla\mu_v \cdot \nabla\mu_{\tilde{v}}\,dx$$

where μ_v (respectively $\mu_{\tilde{v}}$) has mean value zero and fulfills

$$\int_\Omega \nabla\mu_v \cdot \nabla\xi\,dx = \int_\Omega v\,\xi\,dx \quad \text{for all} \quad \xi \in H^1(\Omega).$$

We remark that

$$g_c(v, \tilde{v}) = \int_\Omega \mu_v \tilde{v}\, dx.$$

In what follows we will write $\mu_v = (-\Delta)^{-1} v$.

We define

$$\mathcal{E}(c) = \int_\Omega \{\frac{\gamma}{2}|\nabla c|^2 + \psi(c)\}\, dx + \min_u \int_\Omega W(c, \nabla u)\, dx \tag{2.8}$$

and claim that

$$\langle \operatorname{diff} \mathcal{E}(c), \tilde{v}\rangle = \int_\Omega \{-\gamma \Delta c + \psi'(c) + W_{,c}(c, \nabla u_c)\}\tilde{v}\, dx$$

where u_c solves (2.6) for given c. It should be remarked that the last term in (2.8) can be written as $\int_\Omega W(c, \nabla u_c)\, dx$ which means that also u depends on c. Since (2.6) holds it can be computed that this dependence gives no contribution to the differential. We now obtain that

$$\langle \operatorname{diff} \mathcal{E}, \tilde{v}\rangle = g_c(\partial_t c, \tilde{v}) = \int_\Omega (-\Delta)^{-1}\partial_t c\, \tilde{v}\, dx$$

is equivalent to (2.4)–(2.5) if we set $w := (-\Delta)^{-1}\partial_t c$.

Another gradient flow perspective for the Cahn-Larché system uses the energy (2.1) and uses the manifold

$$\mathcal{M} := \left\{(c, u) : \Omega \to \mathbf{R} \times \mathbf{R}^d \,\middle|\, \int_\Omega c = \int_\Omega c_0 + \text{ boundary conditions for } u\right\}$$

with a corresponding tangent space $T_{(c,u)}\mathcal{M}$. The metric tensor is then chosen to be degenerate with respect to u. In fact we choose

$$g_{(c,u)}((v, w), (\tilde{v}, \tilde{w})) = \int_\Omega \nabla \mu_v \cdot \nabla \mu_{\tilde{v}}\, dx$$

with μ_v and $\mu_{\tilde{v}}$ as above.

We remark that the gradient flow property has been used in Garcke [Ga03b] to show existence of solutions to the Cahn-Larché system.

2.3 The Mullins-Sekerka evolution

In the Mullins-Sekerka model the interface between two phases is described by the boundary $\partial\{\chi = 1\}$, where χ is the characteristic function of one of the phases. We restrict our presentation to the case $\Omega = \mathbf{R}^d$.

The evolution is driven by the reduction of an energy, which is given by

$$\mathcal{E}[\chi, u] := \int_{\mathbf{R}^d} |\nabla \chi| + \int_{\mathbf{R}^d} W(\chi, \nabla u)\, dx,$$

where $\int_{\mathbf{R}^d} |\nabla\chi|$ denotes the perimeter of the set $\{\chi = 0\}$ in $\mathbf{R}^d$. That is, the energy is the sum of interfacial area, which is due to surface tension, and an elastic part, which depends on χ and the deformation field u. In the following we consider linearized elasticity, that is we take

$$\begin{aligned} W(\chi, F) :=& \chi W_1(F) + (1-\chi)W_0(F)\,, \\ W_\alpha(F) :=& \frac{1}{2}C^\alpha(\varepsilon(F) - \bar{\varepsilon}_\alpha) : (\varepsilon(F) - \bar{\varepsilon}_\alpha)\,, \end{aligned}$$

in particular, we allow as above that the elasticity tensor is anisotropic as well as inhomogeneous, i.e. different in each phase; and we allow for a misfit between the two phases, $\bar{\epsilon}_1 \neq \bar{\epsilon}_0$. The misfit may also be anisotropic, i.e. it is not necessarily a multiple of the identity.

The evolution of the interface is driven by the gradient of the chemical potential μ, that is the normal velocity v is given by

$$v = [\partial_\nu \mu] \qquad \text{on the interface } \Gamma := \partial\{\chi = 1\}, \tag{2.9}$$

where ν is the outer normal on $\partial\{\chi = 1\}$, and

$$[\partial_\nu \mu] := \lim_{x\to\Gamma, x\in\{\chi=0\}} \partial_\nu\mu - \lim_{x\to\Gamma, x\in\{\chi=1\}} \partial_\nu\mu$$

denotes the jump of the normal component of the gradient across the interface. The chemical potential μ is determined for each time t via

$$-\Delta\mu = 0 \qquad \text{in the bulk } \mathbf{R}^d\backslash\Gamma, \tag{2.10}$$

$$\mu = \kappa + \nu\cdot[E(u)]\nu \qquad \text{on } \Gamma\,, \tag{2.11}$$

where the jump of the Eshelby tensor

$$E(\chi, F) := W(\chi, F)\mathbf{1} - F^T\frac{\partial W}{\partial F}(\chi, F) \tag{2.12}$$

can be computed from the solution of the elastic equation (see below in (2.13), (2.14)).

We assume that the mechanical fields relax at each time t instantaneously to equilibrium, which yields

$$\operatorname{div}\sigma = 0, \qquad \text{in } \mathbf{R}^d\backslash\Gamma, \tag{2.13}$$

$$[\sigma\cdot\nu] = 0, \qquad \text{on } \Gamma. \tag{2.14}$$

Here, σ denotes the stress tensor, which is given by $\sigma = \frac{\partial W}{\partial F}(\chi, \nabla u)$.

2.4 The Mullins-Sekerka evolution as a gradient flow

We now argue that also the Mullins-Sekerka free boundary problem formally fits into the gradient flow framework: $\mathcal{M}$ has to be chosen as the manifold of all sets, representing the particle phase, with fixed volume, i.e.

$$\mathcal{M} := \left\{ \chi : \mathbf{R}^d \to \{0,1\} \,\middle|\, \int_{R^d} \chi \, dx = \mathcal{V} \, , \, \mathrm{supp}\chi \subset\subset \mathbf{R}^d \right\} .$$

The tangent space $T_\chi \mathcal{M}$ can be described by all admissible normal velocities of Γ, that is

$$T_\chi \mathcal{M} := \left\{ v : \Gamma \to \mathbf{R} \,\middle|\, \int_\Gamma v \, d\mathcal{H}^{d-1} = 0 \right\} .$$

The metric tensor is given here by the H^{-1} norm in the bulk. More precisely

$$g_\chi(v, \tilde{v}) := \int_{\mathbf{R}^d} \nabla \mu_v \cdot \nabla \mu_{\tilde{v}} \, dx \, ,$$

where μ_v (respectively $\mu_{\tilde{v}}$) is the bounded solution of the elliptic problem

$$-\Delta \mu_v = 0 \quad \text{in} \quad \mathbf{R}^d \backslash \Gamma \, , \tag{2.15}$$

$$[\nabla \mu_v \cdot \nu] = v \quad \text{on } \Gamma \, . \tag{2.16}$$

After an integration by parts we obtain

$$g_\chi(v, \tilde{v}) = \int_\Gamma -\mu_v \tilde{v} \, d\mathcal{H}^{d-1}.$$

Our assumption in the previous chapter was that we have a clear separation of time scales such that the mechanical fields can be assumed to relax instantaneously to equilibrium given a phase distribution χ. Thus, we replace our energy by

$$\mathcal{E}(\chi) = \int |\nabla \chi| + \min_u \int_{\mathbf{R}^d} W(\chi, \nabla u) \, dx \, .$$

Indeed, it follows that $\min_u \int_{\mathbf{R}^d} W(\chi, \nabla u) \, dx = \int_{\mathbf{R}^d} W(\chi, \nabla u_\chi) \, dx$, where u_χ solves (2.13), (2.14) for given χ.

We have now all the ingredients for a gradient flow evolution at hand. In order to compute it explicitly we have to calculate the differential of $\mathcal{E}$.

First, we recall the well-known result that the first variation of surface area is the mean curvature, that is for $\tilde{\mathcal{E}}(\chi) := \int |\nabla \chi|$ we have

$$\langle \mathrm{diff} \tilde{\mathcal{E}}, \tilde{v} \rangle = \int_\Gamma \kappa \, \tilde{v} \, d\mathcal{H}^{d-1}$$

for all $\tilde{v} \in T_\chi \mathcal{M}$. The differential of the elastic part of the energy is (compare [Ga03a])

$$\begin{aligned} \langle \mathrm{diff} \hat{\mathcal{E}}, \tilde{v} \rangle &= \frac{d}{d\delta} \hat{\mathcal{E}}[\chi_\delta]|_{\delta=0} \\ &= \int_\Gamma \left(W(\chi, \nabla u) \mathbf{1} - (\nabla u)^T \frac{\partial W}{\partial F}(\chi, \nabla u) \right) \nu \cdot \nu \tilde{v} \, d\mathcal{H}^{d-1} \, . \end{aligned}$$

The part in brackets is again the Eshelby tensor $E(\chi, \nabla u)$ (2.12), and the jump of its normal part across the interface is the contribution of the elastic energy to the evolution of the particles.

Therefore we have for the gradient flow, evaluating (2.7), that

$$0 = g_\chi(v, \tilde{v}) + \langle \text{diff}\mathcal{E}, \tilde{v}\rangle = \int_\Gamma (\kappa + \nu \cdot [E(\chi, \nabla u)]\nu - \mu_v)\, \tilde{v}\, d\mathcal{H}^{d-1} \qquad (2.17)$$

for all $\tilde{v} \in T_\chi \mathcal{M}$. We see in fact that for the direction of steepest descent the corresponding potential satisfies – up to an irrelevant additive constant – the Gibbs-Thomson law with elasticity (2.11).

3 Spinodal decomposition

At high temperatures the free energy ψ is convex and hence a homogeneous state is stable. If now the system is quenched below the critical temperature θ_c the homogeneous state becomes unstable and different phases form which can be distinguished by a different chemical concentration. This process happens on a very short time scale and the regions with different phases have sizes which are given by a small length scale. If the elasticity tensor or the eigenstrains are anisotropic, one will observe that the phase regions orientate themself in certain directions (see [GMW03],[GRW01]) for numerical simulations). We will now describe how one can make these observations quantitative. We first solve the linearized Cahn-Larché system with the help of Fourier transformation (see Khachaturyan [Kha83]). Then a method developed by Maier-Paape and Wanner allows to show that one will see certain patterns after spinodal decomposition with a probability close to one for the nonlinear evolution (we refer to Garcke, Maier-Paape and Weikard [GMW03] for details). We will assume here that C does not depend on c which is the elastically homogeneous case.

Linearization of the Cahn-Larché system around a constant stationary state $(c, u) = (c_m, 0)$, where $c_m \in \mathbf{R}$ is constant, gives

$$\partial_t c = (-\Delta)(\gamma\Delta c - \psi''(c_m)c + \varepsilon^* : S), \qquad (3.1)$$

$$\nabla \cdot S = 0, \qquad (3.2)$$

$$S = C(\varepsilon(\nabla u) - \varepsilon^* c). \qquad (3.3)$$

We consider the system (3.1)–(3.3) on $\Omega = (0, 2\pi) \times \cdots \times (0, 2\pi)$ with periodic boundary conditions. For a given c we can compute u from (3.2)–(3.3) by Fourier transformation and we can express $\varepsilon^* : S$ as a function in c. The result will be denoted as

$$\mathcal{L}(c) = \varepsilon^* : S.$$

For

$$\varphi_\kappa(x) = \mathrm{e}^{\mathrm{i}\kappa\cdot x}, \quad \mathbf{i} \text{ being the imaginary unit,}$$

with

$$\kappa = (\kappa_1, \ldots, \kappa_d) \in \mathbf{Z}^d$$

one obtains (see [Kha83, GMW03])

$$\mathcal{L}(\varphi_\kappa) = L(\kappa)\varphi_\kappa$$

with

$$L(\kappa) = \varepsilon^* : (C[Z(\kappa)S^*\kappa\kappa^T] - S^*)$$

where $S^* := C[\varepsilon^*]$ and $Z(\kappa)$ is the inverse of

$$Z^{-1}(\kappa) = \left(\sum_{j,m}^{d} C_{ijmn}\kappa_j\kappa_m \right)_{i,n=1,\ldots,d} .$$

An important observation is that L is homogeneous of degree 0 which implies that $\mathcal{L}$ is a pseudo-differential operator of order 0. The function L can be computed more explicitly in certain cases, e.g. if C is isotropic or has a cubic symmetry (see [GMW03]). In the particular case of cubic symmetry one obtains that certain directions $\kappa \in \mathbf{Z}^d$ are stronger amplified by L than others. This has important consequences for (3.1)–(3.3). If we consider solutions to (3.1)–(3.3) of the separation of variables form

$$c(x,t) = f(t)\mathrm{e}^{\mathrm{i}\kappa\cdot x},$$

we obtain

$$f(t) = \alpha \mathrm{e}^{\lambda_{\kappa,\gamma} t}, \ \alpha \in \mathbf{R},$$

with

$$\lambda_{\kappa,\gamma} = |\kappa|^2(-\gamma|\kappa|^2 - \psi''(c_m) + L(\kappa)).$$

If c_m is such that $\psi''(c_m) < 0$, one obtains in the case without elasticity that all κ with a certain wave length are amplified the most. Now in case of anisotropic elasticity also the direction of κ plays an important role when we want to determine the most unstable waves. It turns out (see [Kha83, GMW03] and the references therein) that in case of cubic anisotropy either directions parallel to the coordinate axes or directions parallel to the diagonals of the coordinate axes are amplified more by the influence of elastic interactions. Which of the two cases occur depends on the parameter $\Delta C := C_{11} - C_{12} - 2C_{44}$. One speaks of positive anisotropy if $\Delta C > 0$ and of negative anisotropy if $\Delta C < 0$.

We will demonstrate this for the case of negative anisotropy. In Fig. 3.1 we show the most amplified eigenmodes and a typical function which is a linear combination of basis functions with these eigenmodes. In Fig. 3.2 we show a typical solution of the Cahn-Larché system after spinodal decomposition. We show the modulus of the Fourier coefficients and the sign of the concentration difference in the case of cubic negative anisotropy. One clearly sees the cubic anisotropy which is in contrast to the isotropic case where patterns do not

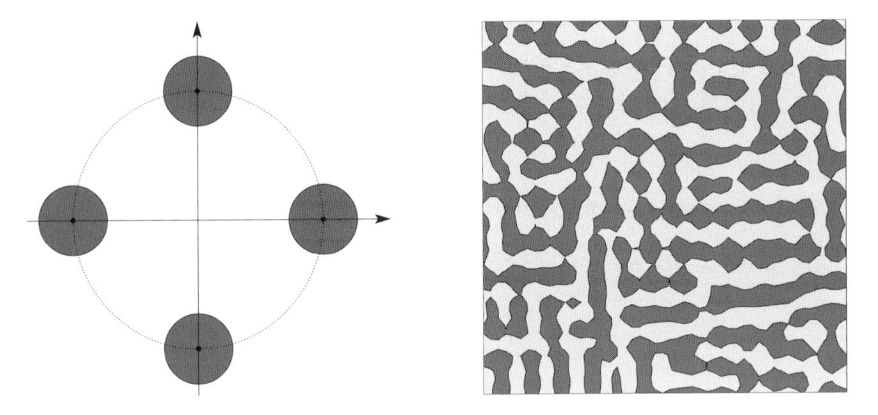

Fig. 3.1. The most amplified eigenmodes in the (κ_1, κ_2)-plane (left) and a typical pattern (right) for negative anisotropy ($\Delta C < 0$)

Fig. 3.2. Cubic anisotropy of the elasticity tensor; modulus of the Fourier coefficient (left) and sign of the concentration difference c (right)

follow a direction. In fact typical solutions look like in Fig. 1.1 to the left (see [GMW03] for more details). We also refer to [GMW03] for the proof of a theorem which roughly speaking says that with a probability close to one, the evolution to initial data which are randomly chosen out of a neighborhood of a uniform state will be dominated by an invariant manifold which is tangential to the most unstable eigenfunctions of the linearized operator.

4 Numerical approximation of the Cahn-Larché system

The Cahn-Larché system has a variational structure and hence it is natural to use a finite element method for the discretization. The formulation of the

Cahn-Larché system (2.4)–(2.6) is of second order in space and we will use continuous piecewise affine elements to approximate c, w and u.

For a polyhedral domain Ω we choose a quasi-uniform family $\{\mathcal{T}^h\}_{h>0}$ of partitionings of Ω into disjoint simplices with maximal element size $h := \max_{s\in\mathcal{T}^h}\{\operatorname{diam} s\}$, so that $\overline{\Omega} = \bigcup_{s\in\mathcal{T}^h} \overline{s}$. Associated to $\mathcal{T}^h$ is the finite element space of continuous piecewise affine elements

$$S^h := \{\varphi \in C^0(\overline{\Omega}) \mid \varphi_{|s} \text{ is linear for all } s \in \mathcal{T}^h\} \subset H^1(\Omega).$$

To formulate a finite element discretization we introduce the lumped mass scalar product $(\cdot,\cdot)^h$ instead of the L^2 scalar product $(\cdot,\cdot)$ as follows: For $v_1, v_2 \in C^0(\overline{\Omega})$ let

$$(v_1, v_2)^h := \int_\Omega \pi^h(v_1 v_2)$$

where $\pi^h : C^0(\overline{\Omega}) \to S^h$ is the interpolation operator, such that $(\pi^h\eta)(p) = \eta(p)$ for all nodes of $\mathcal{T}^h$.

Then a semi-implicit scheme for (2.4)–(2.6) reads as follows.

We search for $c^h, w^h : [0,T] \to S^h$ and $u^h : [0,T] \to (S^h)^d$ such that

$$(\partial_t c^h, \varphi^h)^h = -(\nabla w^h, \nabla\varphi^h), \tag{4.1}$$

$$(w^h, \varphi^h)^h = \gamma(\nabla c^h, \nabla\varphi^h) + (\psi'(c^h), \varphi^h)^h + (W_{,c}(c^h, \nabla u^h), \varphi^h), \tag{4.2}$$

$$0 = (\varepsilon(\nabla u^h) - \overline{\varepsilon}(c^h), C(c^h)\varepsilon(\nabla\xi^h)) \tag{4.3}$$

holds for all $\varphi^h \in S^h, \xi^h \in (S^h)^d$ and all $t \in [0,T]$.

In order to obtain a fully discrete scheme one needs to introduce a time discretization. The simplest implicit time discretization is the implicit Euler scheme in which the time derivative in (4.1) is discretized in the following way

$$(\partial_t c^h, \varphi^h)^h \rightsquigarrow \left(\frac{c_n^h - c_{n-1}^h}{\tau_n}, \varphi^h\right)^h.$$

Here we divided the time interval $[0,T]$ into N steps with length τ_n and set $t_n := \sum_{i=1}^n \tau_i$. The discrete solution at time t_n is denoted by (c_n^h, w_n^h, u_n^h). The resulting numerical scheme has been analyzed in [GRW01, GW05]. In [GRW01] optimal error estimates have been shown in the case that C does not depend on the concentration (homogeneous elasticity). In the case of inhomogeneous elasticity a convergence proof has been given in [GW05].

The fully discrete scheme has the properties that mass is conserved and that the total discrete free energy decreases (see [GRW01, GW05]). The last observation is a consequence of the fact that the discrete problem reflects the gradient flow property of the continuous problem and this is an important fact in the analysis of the scheme (see [GW05]).

It turns out that the so-called Θ-scheme [BGP87, MU94] leads to a more efficient but hard to analyze time discretization. All the computations presented in the following are with the help of the Θ-scheme, but we made sure that computations with the implicit Euler scheme lead to qualitatively similar results

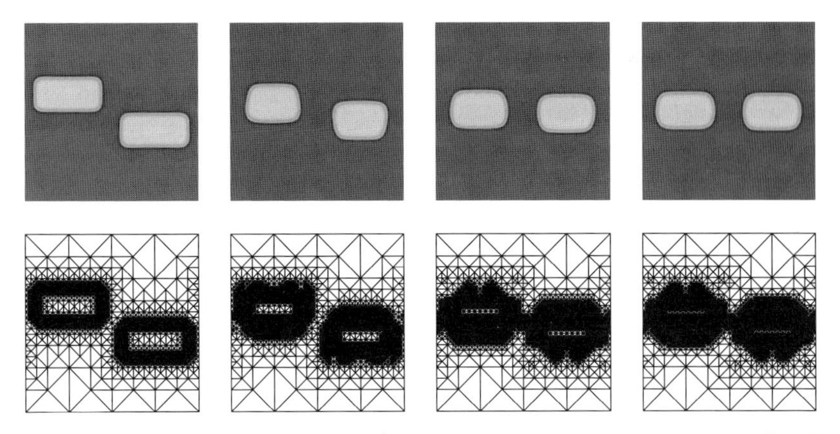

Fig. 4.1. Alignment of two particles (anisotropic inhomogeneous elasticity), adaptive computational grids

although with higher computational effort. We consider adaptive triangular grids in space and a corresponding a posteriori error control [GRW01, GW05]. The discrete linear systems were solved with the help of the BICG and GMRES algorithms and for the nonlinear discrete problem we used Newton's method (see [GRW01, GW05] for more details).

Another approach to solve the Cahn-Larché system numerically uses spectral methods. We refer e.g. to the work of Dreyer and Müller [DM00] and Leo, Lowengrub and Jou [LLJ98] and the references therein. Due to the nonlinear structure of the Cahn-Larché system, approaches based on spectral methods loose their efficiency. This is in particular true in the case where the elastic constants are different in the two phases (inhomogeneous elasticity).

To conclude this section we report on some numerical simulations with the above robust and efficient numerical method. We have studied various qualitative effects of the Cahn-Larché model including homogeneous elasticity. We observe e.g. the following (see also [LLJ98])

- particles align their faces to the elastically soft directions of the material (see Fig. 1.2),
- particles align in rows (see Fig. 4.1),
- always the harder phase forms particles in the softer phase independent of the volume fraction (see Fig. 4.2),
- in the case of inhomogeneous elasticity one observes that particles do not merge when close to each other but instead repel each other (see Fig. 4.3).

The numerical approach for the Cahn-Larché model turns out to be efficient for ensembles ranging from a couple of particles to a few thousand particles and has been applied to derive experimental results on growth laws (see the following section).

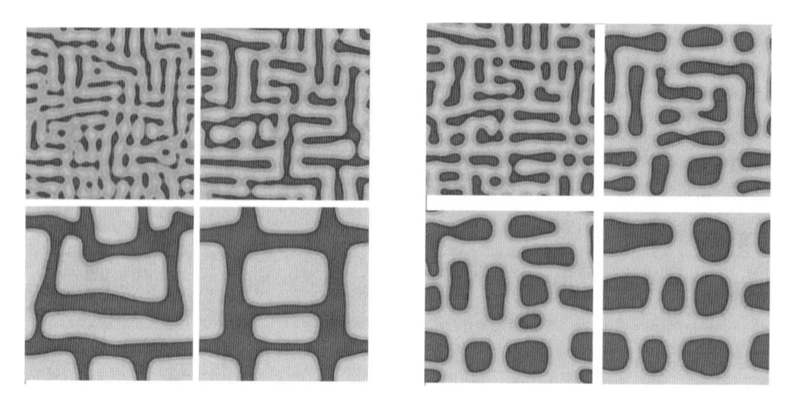

Fig. 4.2. Effects of inhomogeneous elasticity: On the left side the green phase is the elastically harder one, the blue phase is softer. On the right side it is vice versa. The volume fraction of both phases are the same. (See page 690 for a colored version of the figure.)

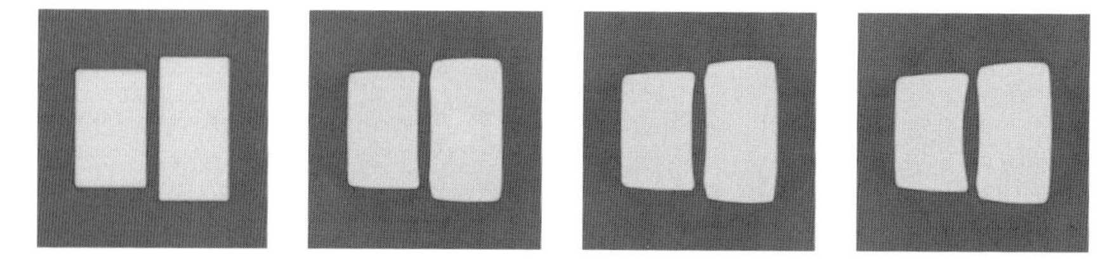

Fig. 4.3. Repulsion of two particles due to anisotropic elasticity

5 Ostwald ripening within the Cahn-Hilliard and Cahn-Larché models

A relevant issue in coarsening systems is an estimate of the coarsening rate of the system. The latter can be expressed by the rate of growth of mean particle size or by the rate of decrease of surface energy. Dimensional arguments give, that the coarsening rate in diffusion controlled coarsening, as described by the Cahn-Hilliard equation or the Mullins-Sekerka evolution, is proportional to $t^{1/3}$. Weak time-averaged upper estimates of this coarsening rate have been established in [KO02] for the Cahn-Hilliard model without elasticity. It turns out that the proof goes through without any difference for the Cahn-Hilliard equation with elasticity. Whether or not this estimate is sharp, say, for generic data, in the case with elasticity is however not clear.

Using the adaptive finite element method described in Sect. 4 we made an attempt to study coarsening rates for large particle systems. We first considered the Cahn-Hilliard model without elasticity. Due to the adaptive grids (see Fig. 4.1) and the time discretization based on the Θ-scheme simulations with about 4000 particles after the initial phase of the particle formation

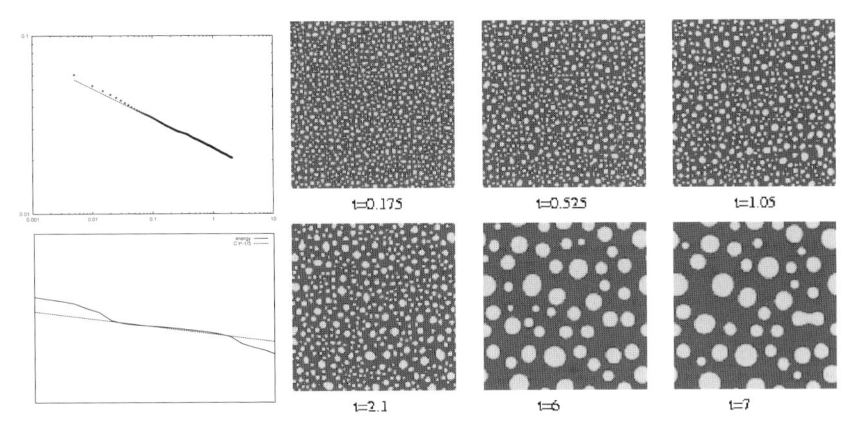

Fig. 5.1. Graph of the energy at an early and a very late stage of the evolution (two graphs on the left side), different time steps of the evolution (on the right side). (See page 691 for a colored version of the figure.)

have been feasible. Let us give a brief summary of the results for the original Cahn-Hilliard model without elasticity (details can be found in [GN*03]):

- The observed decay rates for the energy and the growth of the averaged particle size are in correspondence with the basic LSW theory. There in 2D one has a decay of the energy like:

$$E = Ct^{-\frac{1}{3}}.$$

- Depending on the initial data (arbitrary distributed small particles (cf. Küpper, Masbaum [MK94]), a slightly perturbed homogeneous mixture, a homogeneous mixture with arbitrary positioned localized seeds for particles) we observe a rather long intermediate behaviour with energy decay and particle growth rates different from the expectations.

Figure 5.1 shows results obtained by our extensive numerical tests. On the left the energy is plotted in double logarithmic scale over time. In this representation the expected polynomial decay should turn out as a straight line. This is the case albeit with the unexpected exponent of $-\frac{1}{6}$. In fact, we made the observation that after spinodal decomposition the system settles for a wrong exponent for quite long time. Depending on the volume fractions of the two phases the exponents observed range from $-\frac{1}{6}$ to the expected $-\frac{1}{3}$ in the case where we start with equal volume fractions (see Figure 5.1 upper left graph). However, after a long time the speed of the energy decay changes and we see a behaviour in line with the expectations. In the example shown the energy decay at times $t > 7$ differs significantly from the behaviour at earlier times. We observe a graph like in the left lower part of Fig. 5.1. Here we see time phases where the energy goes according to

$$E \approx Ct^{-\frac{1}{3}}$$

which are intersected by short periods, when the energy decays faster. In these short periods one sees particles vanishing, whereas between these steeper declines particles are just growing and shrinking with the number of particles constant.

In the case of the Cahn-Larché model with elasticity we have observed so far, that the coarsening rates are affected by the presence or absence of elasticity as well as by the homogeneity of the elasticity. Anisotropy seems to play a minor role (cf. Fig. 5.2). However, it may be the case that the coarsening rates change at later times as in the standard Cahn-Hilliard model.

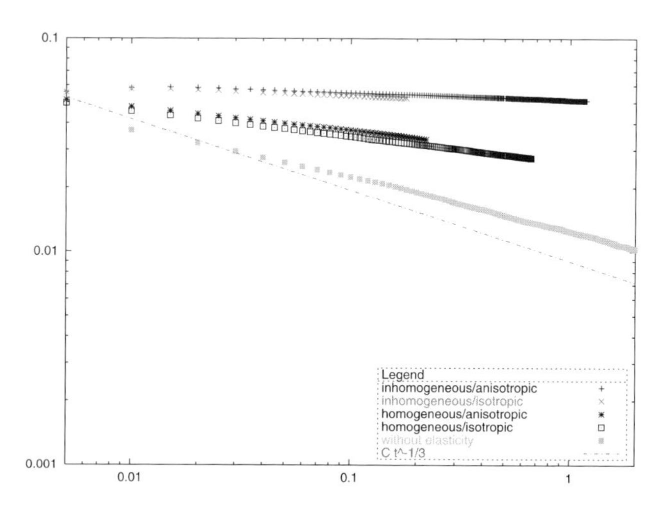

Fig. 5.2. Graph of the nonelastic part of the energy. (See page 691 for a colored version of the figure.)

6 Simulation of the sharp interface model

Different from the diffuse interface model which allows a straightforward discretization via finite elements (cf. Sect. 4) the interface propagation in the Mullins-Sekerka sharp interface model (2.9)–(2.14) and the computation of the corresponding energy contributions and their variations has preferably been implemented using the boundary element approach. Hence, the linear elliptic subproblems for the chemical potential (2.10),(2.11) and the elastic displacement (2.13),(2.14) are transformed into integral equations on the interface between the different phases. They are then dicretized based on a collocation-type ansatz. Thus, for $d = 2$ the interfaces are resolved by polygonal lines whose vertex positions are updated in the actual evolution. The two dimensional, evolving phase domains have not to be meshed and adapted in each time step of the evolution. Indeed the interface geometry enters the formulation via appropriate Poisson type kernel functions and kernel functions for linear, anisotropic elasticity to be integrated in the collocation ansatz on the polygonal lines [Ha95]. In each time step only the vertex positions representing the interface have to be updated. This discretization approach has among others already been successfully applied by Voorhees, Lowengrub and coworkers [ATV01, JLL97, TAV04a, TAV04b, VMJ92].

Let us first depict this transformation for the chemical potential subproblem. Let $\psi_{x_0}(x) := -\frac{1}{2\pi}\ln|x-x_0|$ be the fundamental solution for the Laplacian in $\mathbf{R}^2$, i.e. $\Delta\psi_{x_0}(x) = \delta(x-x_0)$ in the sense of distributions. Applying Greens formula we obtain for points x_0 on a smooth interface Γ:

$$\mu(x_0) = \int_\Gamma \{[\mu](x)\partial_\nu\psi_{x_0}(x) - \psi_{x_0}(x)[\partial_\nu\mu](x)\}\,dx \\ + \int_{\partial B_R} \{\mu(x)\partial_{\tilde\nu}\psi_{x_0}(x) - \psi_{x_0}(x)\partial_{\tilde\nu}\mu(x)\}\,dx\,.$$

Here, μ is the chemical potential, $[\cdot]$ the usual jump operator and B_R is a large ball containing all particles. Recalling that μ is continuous across the interface (2.11) the jump of the chemical potential $[\mu]$ vanishes. Furthermore, for $R\to\infty$ the integral over ∂B_R converges to a constant $c(t)$ solely depending on time. This additional degree of freedom reflects the conservation of the overall particle volume. Finally taking into account the governing equation for the normal velocity of the interface (2.9) the Mullins-Sekerka problem can be rewritten in the following form:

Let $\Gamma(t)$ be the interface with normal velocity v, κ its curvature, and $E = E(\chi,\nabla u)$ the Eshelby tensor (2.12), then at time t

$$\kappa(x_0,t) + [E(x_0,t)]\nu(x_0,t)\cdot\nu(x_0,t) + c(t) = \int_{\Gamma(t)} \psi_{x_0}(x)v(x,t)\,d\mathcal{H}^1 \quad (6.1)$$

for every x_0 on $\Gamma(t)$ and the velocity field fulfills the constraint $0 = \int_{\Gamma(t)} v\,d\mathcal{H}^1$.

The solution of the quasi stationary elastic subproblem is required for the evaluation of the Eshelby tensor E on Γ. Let C_α be the elasticity tensor, where the index α indicates either the matrix or the particle phase, and denote by $\bar\epsilon_\alpha$ the misfit. Let u be the displacement on the interface and τ the normal stress defined as the difference between actual strain and eigenstrain in normal direction: $\tau = \sigma\nu = C(\epsilon(\nabla u) - \bar\epsilon_\alpha)\nu$. We recall from (2.14) that τ is continuous across the interface. Now we consider the matrix valued fundamental solutions $\psi^\alpha_{x_0}$ for linear, anisotropic elasticity from [CR78, Cle87] and obtain again by Greens formula an integral equation

$$\frac{1}{2}u(x_0) = \int_\Gamma \partial_{C_\alpha\nu}\psi^\alpha_{x_0}(x)u(x) - \psi^\alpha_{x_0}(x)\left(\tau(x) + C_\alpha\bar\epsilon_\alpha\nu\right)\,d\mathcal{H}^1\,.$$

Let us remark that in case x_0 coincides with a vertex on a polygonal interface, a matrix $c(x_0)$ depending on the direction of the two edges at x_0 is applied to the displacement $u(x_0)$ on the left hand side replacing the factor $\frac{1}{2}$. Given the above integral equation for the matrix and for the particle phase, the displacement u and the normal stress τ are up to a constant displacement uniquely determined. The computation of the Eshelby tensor requires the evaluation of the full displacement gradient ∇u. For given u and τ this gradient can be computed differentiating the above integral equation with respect to x_0. This differentiation applies to the integral kernels, thus increasing the order of singularity. In particular, for the kernel $\partial_{C_\alpha \nu} \psi_{x_0}^\alpha$ a hypersingular integral has to be evaluated.

In the actual spatial discretization the integral equations are assumed to be fulfilled at appropriate collocation points on a polygonal interface and the displacement, the normal stresses, the chemical potential, and the interface velocity are approximated in a corresponding discrete space. Two particular useful choices are either piecewise constant ansatz functions on the polygon segments and segment centers as collocation points, or a piecewise linear functions and vertices as collocation points. For the elastic subproblem, piecewise linear ansatz functions are the appropriate choice, since piecewise constant ansatz functions do not make sense with respect to the above sketched evaluation of the deformation gradient on the interface.

For the notion of a discrete curvature on the polygonal interface we refer to [Dz91] and define on vertex x_i a curvature vector

$$\kappa_i \nu_i := - \frac{\frac{x_{i+1}-x_i}{\|x_{i+1}-x_i\|} - \frac{x_i - x_{i-1}}{\|x_i - x_{i-1}\|}}{\frac{\|x_{i+1}-x_i\| + \|x_i - x_{i-1}\|}{2}}, \tag{6.2}$$

where ν_i represents a unit length vector and κ_i the discrete, scalar curvature required for a spatially discrete Mullins-Sekerka model. Finally, a suitable time discretization for (6.1) has to be considered. An explicit treatment of the discrete curvatures κ_i would result in severe time step restrictions. Thus, we evaluate the normal direction – according to the above equation – at the old time step and redefine a semi-implicit scalar curvature as the scalar product of this time explicit normal field with a semi-implicit curvature vector. For the latter, we again follow [Dz91] and consider time implicit vertex positions but a time explicit edge length in the above formula (6.2).

Let us depict two types of particle interaction in the presence of inhomogeneous and anisotropic elasticity. Figure 6.1 shows the attraction of particles in case of a strongly inhomogeneous elasticity with a hard particle phase and a considerably softer matrix phase. Figure 6.2 renders the alignment of particles, which can be observed in the presence of strongly anisotropic elasticity.

Fig. 6.1. Three time steps of a discrete Mullins-Sekerka evolution showing the attraction of two soft particles in case of isotropic but inhomogeneous elasticity. The matrix phase is four times harder than the particle phase

Fig. 6.2. Three time steps of a discrete Mullins-Sekerka evolution with particles lining up

7 Reduced sharp interface model for larger systems

In order to make simulations for large particle systems feasible, we now set up a reduced model of the Mullins-Sekerka evolution with elasticity. The reduction is based on the observation that in the case of a cubic anisotropy in the elasticity, particles become quickly rectangular, whereas the long-time behavior is dominated by long-range interactions. This motivates to reduce the gradient flow of the Mullins-Sekerka evolution to the submanifold of rectangular particles. We will see, that such a reduction is in very good agreement with the full evolution for a small set of particles. We will then also present first results for larger particle ensembles.

We restrict our dynamical system to the submanifold $\mathcal{N} \subset \mathcal{M}$ which consists of sets which are the union of disjoint rectangular particles aligned with the coordinate axes.

To define $\mathcal{N}$ we first need to introduce some notation. As indicated in Fig. 7.1 each particle will be identified by the two points

$$p = (p^-, p^+) = ((p_x^-, p_y^-), (p_x^+, p_y^+)) \in \mathbf{R}^2 \times \mathbf{R}^2.$$

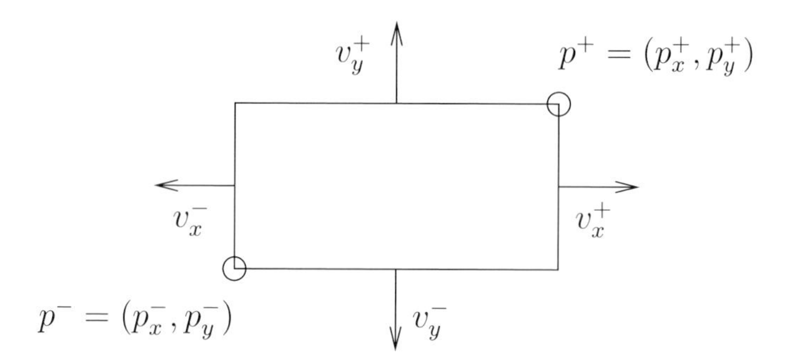

Fig. 7.1. Configuration of one rectangular, axis-aligned particle

We denote the edges perpendicular to the x-axes by b_x^- and b_x^+, the ones perpendicular to the y-axes by b_y^- and b_y^+, more precisely $b_x^- := \{p_x^-\} \times [p_y^-, p_y^+]$, $b_x^+ := \{p_x^+\} \times [p_y^-, p_y^+]$ and $b_y^- := [p_x^-, p_x^+] \times \{p_y^-\}$, $b_y^+ := [p_x^-, p_x^+] \times \{p_y^+\}$. Consequently the volume of a particle p is given by $a_p = |b_x^-| \cdot |b_y^-|$ and the boundary length by $l_p := |b_x^-| + |b_y^-| + |b_x^+| + |b_y^+|$.

The normal velocities of the sides are given by

$$v_p = (v_x^-, v_x^+, v_y^-, v_y^+) \in \mathbf{R}^4 .$$

Now our submanifold $\mathcal{N}$ can be identified with the space

$$\mathcal{N} := \left\{ \mathbf{P} = \{p_i\}_i , i = 1, \ldots, N \,\middle|\, \sum_{p_i} a_{p_i} = \mathcal{V} \right\} \subset \mathbf{R}^{4N},$$

where N is the number of particles, and the tangent space with the hyperplane

$$T_{\mathbf{P}}\mathcal{N} := \left\{ \mathbf{V} = \{v_i\}_i , i = 1, \ldots, N \,\middle|\, \begin{array}{l} \sum_i \left(|b_x^-| ((v_i)_y^- + (v_i)_y^+) \right. \\ \qquad \left. + |b_y^-| ((v_i)_x^- + (v_i)_x^+) \right) = 0 \end{array} \right\} .$$

The surface energy $\tilde{\mathcal{E}}$ can be expressed as

$$\tilde{\mathcal{E}} = \int |\nabla \chi| = 2 \sum_i \left(|(b_i)_x^-| + |(b_i)_y^+| \right),$$

such that the variation of $\tilde{\mathcal{E}}$ with respect to $\tilde{v} \in T_p\mathcal{N}$ is given by

$$\langle \mathrm{diff} \tilde{\mathcal{E}}, \tilde{v} \rangle = 2 \sum_i \left((\tilde{v}_i)_x^- + (\tilde{v}_i)_x^+ + (\tilde{v}_i)_y^- + (\tilde{v}_i)_y^+ \right) . \tag{7.1}$$

We are now going to consider what interfacial condition is satisfied for the direction of steepest descent as given in (2.17). For that notice that we can for any vector $w \in \mathbf{R}^{4n}$ construct an element $\tilde{v} \in T_p\mathcal{N}$ by setting

$$\tilde{v}_i = w_i - \frac{\sum_j (w_j b_j)_x^- + (w_j b_j)_x^+ + (w_j b_j)_y^- + (w_j b_j)_y^+}{\sum_j l_{p_j}} =: w_i - \bar{w}.$$

Using (2.17) and (7.1) we easily find that there is a constant C such that for all $i \in 1, \ldots, N$ we have

$$\fint_{(b_i)_\alpha^\beta} \mu \, d\mathcal{H}^1 = \frac{2}{|(b_i)_\alpha^\beta|} + \fint_{(b_i)_\alpha^\beta} [E_{(b_i)_\alpha^\beta}] \, d\mathcal{H}^1 + C \tag{7.2}$$

where $\alpha = x, y$, $\beta = +, -$, $i = 1, \ldots, N$, $[E_b] := [E(\chi, \nabla u)] : \nu_b \nu_b^T$ and $\fint_b := \frac{1}{|b|} \int_b$, where $E(\chi, \nabla u)$ is the Eshelby tensor from (2.12). The term $\frac{2}{|(b_i)_\alpha^\beta|}$ can be interpreted as a crystalline curvature which also appears for surface energies with cubic crystalline anisotropy (see e.g. Taylor [Tay78] and Gurtin [Gu93]).

We summarize the above to define the evolution in the restricted setting. This will also motivate the order in which the equations are evaluated in the numerical algorithm.

For a given particle configuration $\mathbf{P} = \{p_i : i = 1, \ldots, N\}$ we compute in view of (2.13), (2.14) the elastic deformation u from

$$\begin{aligned} \operatorname{div} \frac{\partial W}{\partial F}(\chi, \nabla u) &= 0 && \text{in } \mathbf{R}^2 \backslash \Gamma, \\ [u] = \left[\frac{\partial W}{\partial F}(\chi, \nabla u) \cdot \nu \right] &= 0 && \text{on } \Gamma := \bigcup_{i,\alpha,\beta} (b_i)_\alpha^\beta \end{aligned}$$

for $\alpha = x, y$, $\beta = +, -$, $i = 1, \ldots, N$. The chemical potential μ is given (cf. (2.15) and (7.2)) by

$$\begin{aligned} \Delta \mu &= 0 && \text{in } \mathbf{R}^2 \backslash \Gamma, \\ \fint_{(b_i)_\alpha^\beta} \mu d\mathcal{H}^1 &= \frac{2}{|(b_i)_\alpha^\beta|} + \fint_{(b_i)_\alpha^\beta} [E_{(b_i)_\alpha^\beta}] \, d\mathcal{H}^1 + C \end{aligned}$$

for $\alpha = x, y$, $\beta = +, -$ and $i = 1, \ldots, N$, so that the velocities can be derived from (2.16) via

$$(v_i)_\alpha^\beta = -[\partial_{\nu_{(b_i)_\alpha^\beta}} \mu] \qquad \text{for } \alpha = x, y, \ \beta = +, -, \ i = 1, \ldots, N.$$

This evolution is well-defined until the side of a particle shrinks to zero. Then we remove this particle and continue with the remaining particles.

To compare the reduced model to the full Mullins-Sekerka evolution, we simulate the interaction of a group of particles. Indeed, from a start configuration for the full model, we compute a couple of small time steps to allow the particle shapes to relax to their preferred form (which happens rather quickly). For this configuration of nearly rectangular particles we construct a

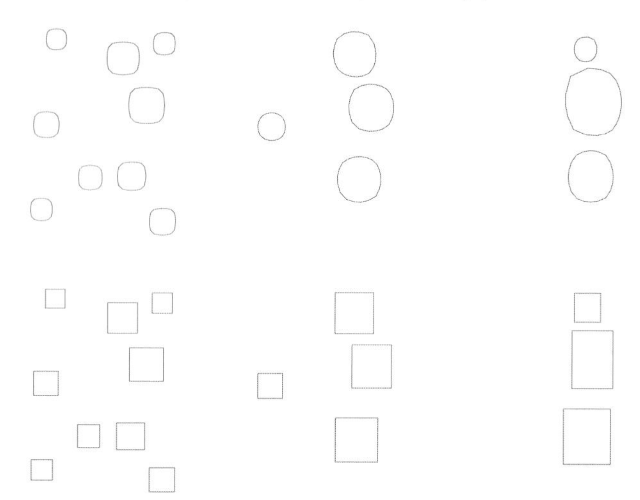

Fig. 7.2. Evolution of the interface for the full Mullins-Sekerka model (top) and for the reduced model (bottom). In both cases the times $t = 0; 0.0011; 0.004$ are depicted.

matching starting configuration for the reduced model. This configuration is then considered as the initial data both for the reduced and the full model. In both models particles below a certain small diameter are deleted completely. Figure 7.2 shows computational results for both models at different time steps of the evolution. The corresponding evolution of the interfacial energy over time is depicted in Fig. 7.3. Furthermore, plots of the elastic strain, stress and the energy density are compared in Fig. 7.4. Indeed, one observes a striking qualitative similarity and basically the same temporal behaviour.

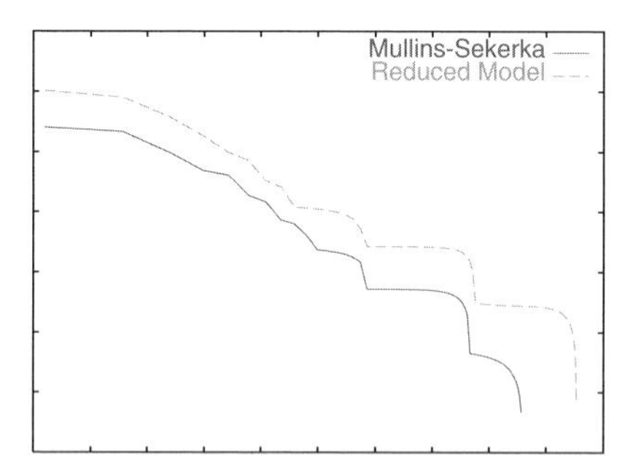

Fig. 7.3. The evolution of the interfacial energy is rendered for both models. Time and energy axis are logarithmic

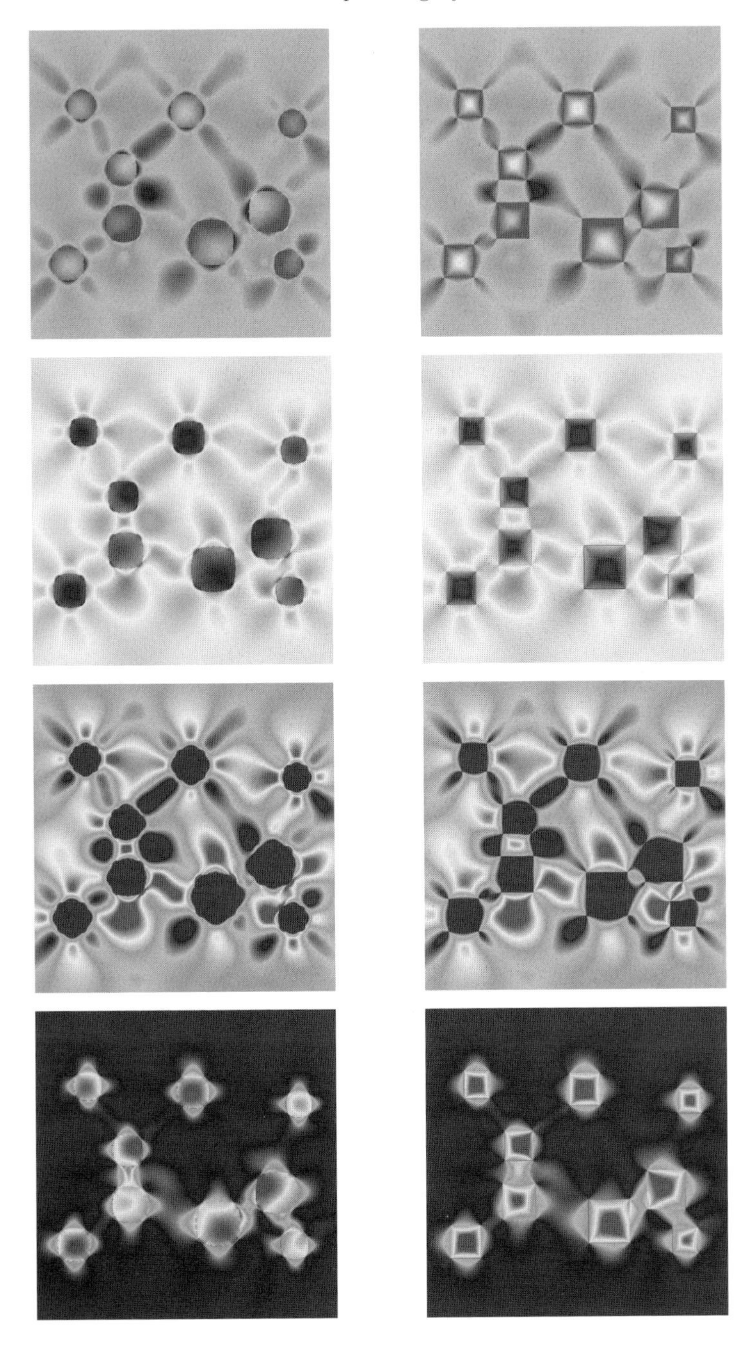

Fig. 7.4. The trace of the elastic strain (both top rows, compared to zero strain and eigenstrain), the trace of the stress (middle) and energy density (bottom) for the Mullins-Sekerka (left) and the reduced model (right), in the initial configuration. (See page 692 for a colored version of the figure.)

Finally, time steps from the evolution of a moderately large particle ensemble with about one thousand particles and homogeneous anisotropic elasticity are shown in Fig. 7.5.

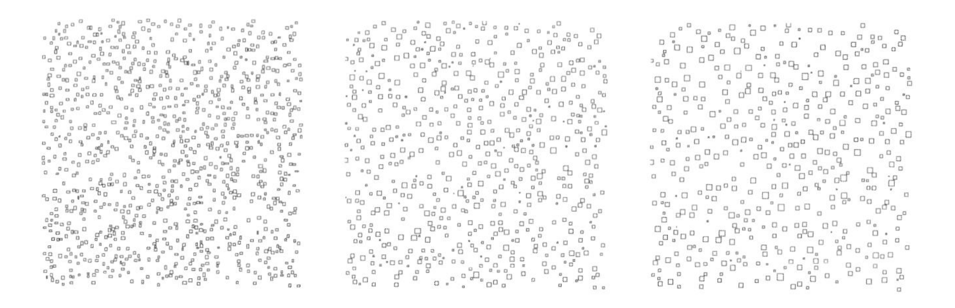

Fig. 7.5. Coarsening of a moderately large particle ensemble

References

[ATV01] N. Akaiwa, K. Thornton, and P.W. Voorhees. Large–Scale Simulations of Microstructural Evolution in Elastically Stressed Solids. *J. Comp. Phys.*, 173, 61–86, 2001.

[BGP87] M.O. Bristeau, R. Glowinski, and J. Periaux. Numerical methods for the Navier-Stokes equations: Applications to the simulation of compressible and incompressible viscous flows. In: Computer Physics Report, Research Report UH/MD-4. University of Houston, 1987.

[CH58] J.W. Cahn and J.E. Hilliard. Free energy of a nonuniform system. I. Interfacial free energy. *Journal of Chemical Physics*, 28, 258–267, 1958.

[CL73] J.W. Cahn and F.C. Larché. A linear theory of thermomechanical equilibrium of solids under stress. *Acta Metall.*, 21, 1051–1063, 1973.

[CL82] J.W. Cahn and F.C. Larché. Surface stress and the chemical equilibrium of single crystals II: solid particles imbedded in a solid matrix. *Acta Metall.*, 30, 51–56, 1982.

[Cle87] D.L. Clements. Green's Functions for the Boundary Element Method. *Boundary Elements IX: Proceedings of the 9th International Conference on Boundary Elements*, 1987.

[CR78] D.L. Clements and F.J. Rizzo. A Method for the Numerical Solution of Boundary Value Problems Governed by Second-Order Elliptic Systems. *J. Inst. Math. Appl.*, 22, 197–202, 1978.

[DM00] W. Dreyer and W.H. Müller. A study of the coarsening in tin/lead solders. *Int. J. Solids Struct.*, 37, 381–3871, 2000.

[Dz91] G. Dziuk. An algorithm for evolutionary surfaces. *Numer. Math.*, 58, 603–611, 1991.

[Ell89] C.M. Elliott The Cahn-Hilliard model for the kinetics of phase transitions, in Mathematical Models for Phase Change Problems, J.F. Rodrigues (ed.). *International Series of Numerical Mathematics*, 88, Birkhäuser-Verlag, Basel, 35–73, 1989.

[FPL99] P. Fratzl, O. Penrose, and J.L. Lebowitz. Modelling of phase separation in alloys with coherent elastic misfit. *J. Stat. Physics*, 95 5/6, 1429–1503, 1999.

[Ga03a] H. Garcke. On Mathematical Models for Phase Separation in Elastically Stressed Solids. Habilitation Thesis 2003

[Ga03b] H. Garcke. On Cahn-Hilliard systems with elasticity. *Proc. Roy. Soc. Edinburgh*, 133 A, 307–331, 2003.

[GK06] H. Garcke and D.J.C. Kwak. On asymptotic limits of Cahn-Hilliard systems with elastic misfit. In "Analysis, Modeling and Simulation of Multiscale Problems, A. Mielke (edr), Springer-Verlag, 2006."

[GMW03] H. Garcke, S. Maier-Paape, and U. Weikard. Spinodal decomposition in the presence of elastic interactions. In: S. Hildebrand, H. Karcher (Eds.), *Geometric Analysis and Nonlinear Partial Differential Equations,* Springer, 2003.

[GN*03] H. Garcke, B. Niethammer, M. Rumpf, and U. Weikard. Transient coarsening behaviour in the Cahn-Hilliard model *Acta Mat.*, 51, 2823–2830, 2003.

[GRW01] H. Garcke, M. Rumpf, and U. Weikard, The Cahn-Hilliard equation with elasticity: Finite element approximation and qualitative studies. *Interfaces and Free Boundaries*, 3, 101–118, 2001.

[GW05] H. Garcke and U. Weikard. Numerical Approximation of the Cahn-Larché equation. *Numer. Math.*, 100, Number 4, 639–662, 2005.

[Gu72] M.E. Gurtin. The Linear Theory of Elasticity. *Handbuch der Physik*, Vol. VIa/2, Springer, S. Flügge and C. Truesdell (eds.), Berlin, 1972.

[Gu93] M.E. Gurtin. *Thermomechanics of evolving phase boundaries in the plane.* Oxford Mathematical Monographs. New York, 1993.

[Ha95] W. Hackbusch. *Integral Equations. Theory and Numerical Treatment.* Birkhäuser 1995.

[JLL97] H.-J. Jou, P.H. Leo, and J.S. Lowengrub. Microstructural Evolution in Inhomogeneous Elastic Media. *J. Comp. Phys.*, 132, 109–148, 1997.

[Kha83] A.G. Khachaturyan. *Theory of Structural Transformations in Solids.* Wiley, New York, 1983.

[KO02] R.V. Kohn and F. Otto. Upper bounds for coarsening rates. *Comm. Math. Phys.*, 229, 375–395, 2002.

[LLJ98] P.H. Leo, J.S. Lowengrub, and H.J. Jou. A diffuse interface model for microstructural evolution in elastically stressed solids. *Acta Mater.*, 46, 2113–2130, 1998.

[MK94] N. Masbaum and T. Küpper. Simulation of particle growth and Ostwald ripening via the Cahn-Hilliard equation. *Acta Metallurgica et Materialia*, 42, No.6, 1847–1858, 1994.

[MU94] S. Müller-Urbaniak. Eine Analyse des Zwischenschritt-θ-Verfahrens zur Lösung der instationären Navier-Stokes-Gleichungen. Preprint 94-01 des SF 359, 1994.

[Pe89] R.L. Pego. Front migration in the nonlinear Cahn-Hilliard equation. *Proc. Roy. Soc. London*, Ser. A 422, no. 1863, 261–278, 1989.

[Tay78] J.E. Taylor. Crystalline variational problems. *Bull. Amer. Math. Soc.*, 84 (4), 568–588, 1978.

[TAV04a] K. Thornton, N. Akaiwa, and P.W. Voorhees. Large-scale simulations of Ostwald ripening in elastically stressed solids. I. Development of Microstructure. *Acta Materialia*, 52, 1353–1364, 2004.

[TAV04b] K. Thornton, N. Akaiwa, and P.W. Voorhees. Large-scale simulations of Ostwald ripening in elastically stressed solids. II. Coarsening kinetics and particle size distribution. *Acta Materialia*, 52, 1365–1378, 2004.

[VMJ92] P.W. Voorhees, G.B. McFadden, and W.C. Johnson. On the morphological development of second phase particles in elastically stressed solids. *Acta Metall.*, 40, 2979–2992, 1992.

[Wei02] U. Weikard. Numerische Lösungen der Cahn-Hilliard-Gleichung und der Cahn-Larché-Gleichung. PhD thesis, Bonn, 2002.

A General Theory for Elastic Phase Transitions in Crystals

Steffen Arnrich[1], Thomas Blesgen[2], and Stephan Luckhaus[1]

[1] Mathematisches Institut, Universität Leipzig, Augustusplatz 10–11, 04103 Leipzig. arnrich@mathematik.uni-leipzig.de, luckhaus@mathematik.uni-leipzig.de

[2] Max-Planck-Institute for Mathematics in the Sciences, Inselstrasse 22–26, 04103 Leipzig. blesgen@mis.mpg.de,

Summary. We derive a general theory for elastic phase transitions in solids subject to diffusion under possibly large deformations. After stating the physical model, we derive an existence result for measure-valued solutions that relies on a new approximation result for cylinder functions in infinite settings.

1 Introduction

The general idea of the present work is to derive a satisfying mathematical theory of phase transitions in single-crystals where the elastic properties of the material are described by nonlinear laws, see [Ogd97], [Hol00]. This generalises common existing models with linear stress-strain laws, starting with pioneering works by Khattchaturyan, [Kha83], and makes the model applicable to a somewhat broader class of materials.

Beside prescribing the elastic behaviour of the material, the other fundamental assumption is that all occuring phase transitions are reconstitutive, i.e. that there are no plastic deformations of the crystal.

For the proof of existence of solutions to our new model, the experiences with the Stefan problem, [Vis96] and [LS95], were found to be very valuable. But as it turns out, due to the nonlinear aspects of elasticity, a formulation within the framework of Sobolev functions is not general enough. This leads in a natural way to the formulation of the solution as Young measures presented in Sect. 4.1.

Historically, there is a variety of approaches to phase transitions in solids and it is not possible to give here a comprehensive overview. Instead, we refer the interested reader to the monographs and survey articles [Vis96, WS*93, Mül98, BS96, SR86] and [GHM01]. After these classifying comments we give now a general survey of this article.

We consider the elastic theory of single crystals at constant temperature where the free energy density depends on the local concentration of one or more species of particles in such a way that for a given local concentration vector certain lattice geometries (phases) are preferred.

The local concentration of the molecules may change due to diffusion. The time scales typical of diffusion and of elastic deformation are usually significantly different and in good approximation it is admissible to assume that the deformation adjusts infinitely fast to the local situation. In the developed model there is surface energy contributing to the free energy of the crystal and the model allows for m different coexisting macroscopic phases. We will assume that the crystal does not possess interstitials and that the time-evolution of the boundary of the domain is known.

After deriving the physical model with the above properties we will discuss the existence of solutions to this model by means of an implicit time discretisation and will show that in the limit of vanishing time step the time-discrete solutions converge in the sense of Young measures on suitable Banach spaces with separable dual. To achieve this goal, we will derive a general approximation result for cylinder functions in the infinite setting. Finally we can proof an energy inequality for the limit solution.

2 Model and implicit time discretisation

In this section we present our model. In the first subsection the mathematical equations are derived whereas the second subsection presents a reasonable solution strategy. This section is only heuristic and shall make the reader familiar with the general ideas. A deep mathematical treatment is done later.

2.1 Derivation of the model

To describe the physical phenomenon presented in the introduction, we make use of non-equilibrium thermodynamics, see [dGM84], [KP98], and of continuum mechanics, see [Gur81], [Cia88]. We neglect the atomistic structure of the crystal and disregard possible effects of the microstructure. The model is based upon the following basic considerations: The diffusion is caused by the gradients of the chemical potentials. The diffusive flux causes a local change of the free energy of the crystal. The free energy shall depend on particle density, elasticity of the crystal and phase parameter only with a term representing the surface energy of the boundary layers.

In good approximation we can assume that the system is in mechanical equilibrium. For the analytical treatment, we will assume in this work that deformation and phase parameters are global minimisers of the free energy with respect to the present particle densities.

At starting time $t = 0$ the crystal is described by a non-empty, bounded Lipschitz domain $\Omega \subset \mathbb{R}^3$. Let $\mathbb{R}_+ := [0, \infty[$. The evolution of Ω is given by a family of C^2-diffeomorphisms $\{\Psi_t : \mathbb{R}^3 \to \mathbb{R}^3 : t \in \mathbb{R}_+\}$ with

$$\Psi_0(x) = x \text{ for all } x \in \Omega, \tag{2.1}$$

$$\left(\mathbb{R}_+ \times \mathbb{R}^3 \ni (t,x) \mapsto \Psi_t(x) \in \mathbb{R}^3\right) \in C^2(\mathbb{R}_+ \times \mathbb{R}^3, \mathbb{R}^3), \tag{2.2}$$

$$\left(\mathbb{R}_+ \times \mathbb{R}^3 \ni (t,x) \mapsto \Psi_t^{-1}(x) \in \mathbb{R}^3\right) \in C^2(\mathbb{R}_+ \times \mathbb{R}^3, \mathbb{R}^3). \tag{2.3}$$

The domain occupied by the crystal at time $t \geq 0$ is denoted by $\Omega_t := \Psi_t(\Omega)$.

The mechanical deformation is given by a family of mappings $\{\Phi_t : \mathbb{R}_+ \times \Omega \to \Omega_t : t \in \mathbb{R}_+\}$ that satisfy for all $t \in \mathbb{R}_+$ and for all $x \in \Omega$

$$\Phi_0(x) = x, \tag{2.4}$$

$$\Phi_t \in W^{1,3+\delta}(\Omega, \mathbb{R}^3) \text{ and } \Phi_t^{-1} \in W^{1,3+\delta}(\Omega_t, \mathbb{R}^3) \text{ exists}, \tag{2.5}$$

$$\det \nabla \Phi_t > 0 \text{ a.e. in } \Omega. \tag{2.6}$$

Here, $\delta > 0$ can be arbitrary. The conditions (2.5) and (2.6) ensure that Φ_t are deformations. The condition $\delta > 0$ guarantees the integrability of the functional determinant and therefore that the volume is finite. Condition (2.4) reflects the fact that the initial state is undeformed.

The space- and time-dependent particle densities of the $n \in \mathbb{N}$ different species of molecules are described by $\rho_i : \mathbb{R}_+ \times \mathbb{R}^3 \to \mathbb{R}$, $i = 1, \ldots, n$. The following natural conditions are postulated for $t \in \mathbb{R}_+$ and $1 \leq i \leq n$:

$$\rho_i(t) \in L^1(\Omega_t), \tag{2.7}$$

$$\rho_i(t) \geq 0 \quad \text{a.e. in } \Omega_t, \tag{2.8}$$

$$\int_{\Omega_t} \rho_i(t,x)\, dx = \int_{\Omega} \rho_{0_i}(x)\, dx > 0, \tag{2.9}$$

$$\sum_{i=1}^{n} \rho_i(t) \circ \Phi_t \det \nabla \Phi_t \leq 1 \qquad \text{a.e. in } \Omega. \tag{2.10}$$

The functions ρ_{0_i} are given initial values with

$$\rho_{0_i} \in L^1(\Omega), \quad 0 < \int_{\Omega} \rho_{0_i}(x)\, dx, \quad 0 \leq \rho_{0_i} \quad \text{a.e. in } \Omega, \tag{2.11}$$

$$\sum_{i=1}^{n} \rho_i \leq 1 \text{ a.e. in } \Omega, \quad \int_{\Omega} \sum_{i=1}^{n} \rho_{0_i}(x)\, dx < |\Omega|. \tag{2.12}$$

Equation (2.9) ensures the conservation of mass, (2.10) is due to the fact that the crystal does not possess interstitials and that the number of lattice positions in a volume element is a uniform constant. We assume $(2.12)_2$ with strict inequality as we assume a non-vanishing vacancy density.

Let $m \in \mathbb{N}$ denote the number of different possible phases. The phases are described by a family of phase vectors $\left\{\chi_t := (\chi_{j t})_{j=1}^{m} : \Omega_t \to \mathbb{R} : t \in \mathbb{R}_+\right\}$, where the initial value χ_0 is given. Here, $\chi_{j_t}(x)$ determines whether the material point x at time $t \in \mathbb{R}_+$ is in phase j, $1 \leq j \leq m$, i.e. χ_{j_t} are characteristic functions. As any point in the crystal belongs to exactly one phase, the functions χ_{j_t} fulfil for any $1 \leq j \leq m$ and $t \in \mathbb{R}_+$

$$\chi_{j_t}(1-\chi_{j_t}) = 0, \tag{2.13}$$

$$\sum_{j=1}^{m} \chi_{j_t} = 1, \tag{2.14}$$

$$\chi_{j_t} \in \mathrm{BV}(\,\Omega_t\,). \tag{2.15}$$

With (2.14) we may write the surface energy between phase i and j at time t as

$$S_t : \mathrm{L}^1(\Omega_t) \times \mathrm{L}^1(\Omega_t) \to \bar{\mathbb{R}}_+ := \mathbb{R}_+ \cup \{+\infty\},$$

$$(p_1, p_2) \mapsto \begin{cases} \frac{1}{2} \int\limits_{\Omega_t} \left(|\nabla p_1| + |\nabla p_2| - |\nabla(p_1+p_2)|\right), & \text{if } \int\limits_{\Omega_t} |\nabla p_i| < \infty, i = 1,2, \\ \infty, \text{ otherwise.} \end{cases} \tag{2.16}$$

Assuming further that the densities of the surface energy ς_{ij} on the interface between phase i and phase j are positive constants with $\varsigma_{ij} = \varsigma_{ji}$, the surface energy $F_t^s(\chi_t)$ at time $t \in \mathbb{R}_+$ can be introduced by

$$F_t^s : \mathrm{L}^1\left(\Omega_t, \mathbb{R}^m\right) \to \mathbb{R}_+, \quad p = (p_k)_{k=1}^m \mapsto \sum_{i=1}^m \sum_{j=1}^m \sigma_{ij} S_t(p_i, p_j) \tag{2.17}$$

with $\sigma_{ij} := \frac{\varsigma_{ij}}{2}$, $1 \leq i, j \leq m$. Additionally we postulate for $1 \leq i, j, k \leq m$

$$\sigma_{ij} = \sigma_{ji}, \tag{2.18}$$

$$(i \neq j \text{ and } k \notin \{i,j\}) \Rightarrow \sigma_{ij} \leq \sigma_{ik} + \sigma_{kj}. \tag{2.19}$$

So, energetically it is not favourable to add a third phase between two other existing phases.

We assume that the volume density of phase j depends on the particle concentrations and the gradients of the deformations. It is given by a measurable function (see p.9 in [Bau92]) $f_j : \mathbb{R}^n \times \mathsf{M}^3 \to \bar{\mathbb{R}}_+$, $\mathsf{M}^k := \mathrm{M}(k \times k, \mathbb{R})$, $k \in \mathbb{N}$.

Let for $t \in \mathbb{R}_+$

$$B_t := \mathrm{L}^1\left(\Omega_t, \mathbb{R}^n\right) \times \mathrm{W}^{1,3+\delta}\left(\Omega_t, \mathbb{R}^3\right) \times \mathrm{L}^1\left(\Omega_t, \mathbb{R}^m\right), \tag{2.20}$$

$$\mathrm{Def}_t := \{(r,d,p) \in B_t : \det\nabla d \neq 0 \text{ a.e. in } \Omega_t\}, \tag{2.21}$$

$$F_t^v : \mathrm{Def}_t \to \bar{R} := \mathbb{R} \cup \{-\infty, +\infty\},$$

$$(r, d, p = (p_k)_{k=1}^m) \mapsto \sum_{j=1}^m \int\limits_{\Omega_t} p_j(x) f_j\left(r(x), (\nabla d)^{-1}(x)\right) dx. \tag{2.22}$$

By (2.22) the volumetric free energy $F_t^v\left(\rho_t, \Phi_t^{-1}, \chi_t\right)$ of the crystal at time $t \in \mathbb{R}_+$ is introduced. For the total free energy of the system at time $t \in \mathbb{R}_+$ we write $F_t(\rho(t), \Phi_t^{-1}, \chi_t)$ with

$$F_t : \mathrm{Def}_t \to \bar{\mathbb{R}}, \ (r, d, p) \mapsto F_t^v(r, d, p) + F_t^s(p). \tag{2.23}$$

For the subsequent formal derivation we assume that all functions are sufficiently smooth. We extend the density vector $\rho(t)$ by 0 to a function on the whole of $\mathbb{R}^n$. We use the notation $\partial v := (\partial v_l)_{l=1}^k$, where ∂ is an arbitrary differential operator.

The evolution in time of the particle densities is described by the continuity equation

$$\partial_t \rho(t) = -\mathrm{div}\mathcal{F}_t \text{ in } \Omega_t, \tag{2.24}$$

where $\mathcal{F}_t := \left(\mathcal{F}_{i_t}\right)_{i=1}^n$, and $\mathcal{F}_{i_t}$ is the particle flux of species i, $1 \leq i \leq n$, at time $t \in \mathbb{R}_+$. In our case, $\mathcal{F}_{i_t}$ consists of two components, the diffusive flux $\tilde{J}_{i_t}$, and the mechanical flux M_{i_t}.

We write

$$\mathcal{F}_t = -J_t + M_t, \ \left(J_t := (-\tilde{J}_{i_t})_{i=1}^n, M_t := (M_{i_t})_{i=1}^n\right), \quad t \in \mathbb{R}_+. \tag{2.25}$$

For the mechanical flux we easily find

$$M_{i_t} = \rho_i(t)\partial_t \Phi_t \circ \Phi_t^{-1} \text{ in } \Omega_t, \ t \in \mathbb{R}_+, 1 \leq i \leq n. \tag{2.26}$$

We introduce the notations $\rho(t)\partial_t\Phi_t \circ \Phi_t^{-1} := \left(\rho_i(t)\partial_t\Phi_t \circ \Phi_t^{-1}\right)_{i=1}^n$ and write (2.24) as

$$\partial_t \rho(t) = \mathrm{div}\left(J_t - \rho(t)\partial_t \Phi_t \circ \Phi_t^{-1}\right) \text{ in } \Omega_t, \ t \in \mathbb{R}_+, \tag{2.27}$$

or equivalently

$$\partial_t \left(\rho(t) \circ \Phi_t \mathrm{det}\nabla\Phi_t\right) = \mathrm{div} J_t \circ \Phi_t \mathrm{det}\nabla\Phi_t \text{ in } \Omega, \ t \in \mathbb{R}_+. \tag{2.28}$$

At fixed constant temperature the diffusive fluxes are caused by the negative gradients of the chemical potentials which are the thermodynamic forces, [KK93], [dGM84]. According to Onsager's postulate, [KY87], [Ons31a], [Ons31b], [dGM84], every thermodynamic flux is a linear combination of the thermodynamic forces. So we set

$$J_{i_t} = \sum_{k=1}^{n} L_{ik}\nabla\mu_{k_t} \text{ in } \Omega_t, \quad 1 \leq i \leq n, \quad t \in \mathbb{R}^+, \tag{2.29}$$

$$\text{or } J_t = L\nabla\mu_t \text{ in } \Omega_t, \quad \left(\mu_t = (\mu_{k_t})_{k=1}^n\right), \quad t \in \mathbb{R}_+ \tag{2.30}$$

with a symmetric and positive definite matrix $L := (L_{ik})_{i,k=1}^n$ and the chemical potential μ_i of species i, where the symmetry and positive definiteness of L comes from Onsager's reciprocity relation, [Ons31a], [Ons31b], [dGM84].

According to the definition of the chemical potential one has in Ω_t for any $t \in \mathbb{R}_+$ and $i = 1, \ldots, n$

$$\mu_{i_t} = \sum_{j=1}^{m} \chi_{j_t} \partial_{r_i} f_j \left(\rho(t), \nabla \Phi_t \circ \Phi_t^{-1}\right) = \sum_{j=1}^{m} \chi_{j_t} \partial_{r_i} f_j(\rho(t), (\nabla \Phi_t^{-1})^{-1}). \quad (2.31)$$

Now we formulate the aforementioned minimality condition on the free energy. Considering the time-evolution of the deformation of a representative volume element and keeping in mind that the number of particles only changes due to diffusion, we find for two possible deformations Φ_t^1, Φ_t^2 the relation

$$\rho^1(t) \circ \Phi_t^1 \mathrm{det} \nabla \Phi_t^1 = \rho^2(t) \circ \Phi_t^2 \mathrm{det} \nabla \Phi_t^2 \text{ in } \Omega, \quad (2.32)$$

where $\rho^1(t), \rho^2(t)$ are the densities corresponding to Φ_t^1, Φ_t^2. So the minimality condition for any $t \in \mathbb{R}_+$ reads

$$F_t\left(\rho(t), \Phi_t^{-1}, \chi_t\right) = \min_{\tilde{\Phi} \in D_t, \tilde{\chi} \in P_t} F_t\left(\hat{\rho}(t) \circ \tilde{\Phi}^{-1} \det \nabla \tilde{\Phi}^{-1}, \tilde{\Phi}^{-1}, \tilde{\chi}\right), \quad (2.33)$$

with

$$D_t := \left\{\Phi \in \mathcal{W}^\delta : \tilde{\Phi}(\Omega) = \Omega_t, \exists \Phi^{-1} \in \mathcal{W}_t^\delta, \mathrm{det} \nabla \Phi > 0 \text{ a.e. in } \Omega\right\}, \quad (2.34)$$

$$P_t := \left\{\chi \in \mathcal{BV} : \sum_{j=1}^{m} \chi_j = 1, \chi_j(1 - \chi_j) = 0, j = 1, \ldots, m\right\}, \quad (2.35)$$

$$\hat{\rho}(t) := \rho(t) \circ \Phi_t \mathrm{det} \nabla \Phi_t \quad (2.36)$$

for $t \in \mathbb{R}_+$ and the setting $\mathcal{BV} := \mathrm{BV}(\Omega_t, \mathbb{R}^m)$, $\mathcal{W}^\delta := \mathrm{W}^{1,3+\delta}(\Omega, \mathbb{R}^3)$, $\mathcal{W}_t^\delta := \mathrm{W}^{1,3+\delta}(\Omega_t, \mathbb{R}^3)$.

To conclude, our model consists of the equations (2.4)–(2.6), (2.7)–(2.10), (2.13)–(2.15), (2.27) or (2.28), (2.29) or (2.30), (2.31) and (2.33).

2.2 Solution strategy - implicit time discretisation

The objective is to solve the model equations. To this end we discretise the equations implicitly in time. The ansatz is the same as in [Vis96], [Luc94], [LS95].

The following argument is only heuristic. We exploit the minimality condition on the free energy (2.33) and choose a suitable approximation of (2.28).

If we formally consider the time derivative of $F = F(\rho, \Phi, \chi)$, we find (omitting the dependence on t)

$$\mathrm{d}_t F = \partial_\rho F \partial_t \rho + \partial_\Phi F \partial_t \Phi + \partial_\chi F \partial_t \chi.$$

From (2.33) it follows $\partial_\chi F \partial_t \chi = 0$, $\partial_\rho F \partial_\Phi \rho + \partial_\Phi F = 0$. Consequently,

$$\mathrm{d}_t F = \partial_\rho F(\partial_t \rho - \partial_\Phi \rho \partial_t \Phi). \tag{2.37}$$

Now we compute $\partial_\Phi \rho \partial_t \Phi$ from (2.32) to obtain

$$\partial_\Phi \rho \partial_t \Phi = -\rho \mathrm{Tr}(\nabla \Phi^{-1} \nabla \partial_t \Phi \circ \Phi^{-1}) - \langle \nabla \rho, \partial_t \Phi \circ \Phi^{-1} \rangle \tag{2.38}$$

$$= -\mathrm{div}(\rho \partial_t \Phi \circ \Phi^{-1}) = -\mathrm{div} M, \tag{2.39}$$

where $\mathrm{Tr}(A) := \sum_{k=1}^{3} A_{kk}$ is the trace of $A \in \mathsf{M}^3$. If we plug (2.27) and (2.39) into (2.37) we find

$$d_t F = \partial_\rho F(\mathrm{div} J - \mathrm{div} M + \mathrm{div} M) = \partial_\rho F \mathrm{div} J.$$

Assuming that the surface terms do not depend on ρ, it follows, see [BB92] p.58,

$$d_t F(t) = \sum_{i=1}^{n} \int_{\Omega_t} \sum_{j=1}^{m} \chi_{j_t}(x) \partial_{r_i} f_j \left(\rho_t(x), \nabla \Phi_t \circ \Phi_t^{-1}(x)\right) \mathrm{div} J_{i_t}(x)\, dx$$

$$= \sum_{i=1}^{n} \int_{\Omega_t} \mu_{i_t}(x) \mathrm{div} J_{i_t}(x)\, dx, \quad t \in \mathbb{R}_+. \tag{2.40}$$

Assuming further that the normal component of $J_{i,t}$ vanishes on $\partial\Omega_t$, which follows from (2.28) and (2.9), we get with the inner product $\langle a, b \rangle := \sum_{k=1}^{3} a_k b_k$ in $\mathbb{R}^3$

$$\sum_{i=1}^{n} \mu_{i_t} \mathrm{div} J_{i_t} = \sum_{i=1}^{n} \mathrm{div}(\mu_{i_t} J_{i_t}) - \langle \nabla \mu_{i_t}, J_{i_t} \rangle, \ t \in \mathbb{R}_+.$$

With the divergence theorem and (2.30) we find

$$d_t F(t) = -\int_{\Omega_t} \sum_{i=1}^{n} \langle \nabla \mu_{i_t}(x), J_{i_t}(x) \rangle\, dx = -2\mathrm{Q}_t(J_t) = -2\mathrm{Q}_t^*(\nabla \mu_t) \tag{2.41}$$

$$= -\mathrm{Q}_t(J_t) - \mathrm{Q}_t^*(\nabla \mu_t), \tag{2.42}$$

where for $t \in \mathbb{R}_+$

$$\mathrm{Q}_t : \left(\mathrm{L}^2(\Omega_t, \mathbb{R}^3)\right)^n \to \mathbb{R}, \ G \mapsto \frac{1}{2} \int_{\Omega_t} (L^{-1} G, G)\, dx, \tag{2.43}$$

$$Q_t^* : \left(\mathrm{L}^2(\Omega_t, \mathbb{R}^3)\right)^n \to \mathbb{R}, \ G \mapsto \frac{1}{2} \int_{\Omega_t} (G, LG)\, dx. \tag{2.44}$$

Here we introduced the symbol

$$(\cdot,\cdot) : \left(\mathbb{R}^3\right)^n \to \mathbb{R}, \ (a,b) \mapsto \sum_{k=1}^{n} \langle a_k, b_k \rangle. \tag{2.45}$$

In (2.41) and (2.42), Q_t^* denotes the Fenchel conjugate to Q_t. We call (2.42) the $Q - Q^*$-formulation of the problem. In general, every system of equations originating from non-equilibrium thermodynamics can be written in the form

$$\mathrm{d}_t F + Q + Q^* + G + G^* \leq 0,$$

where Q and G are certain convex functionals and Q^*, G^* are their convex conjugates. Therefore, Eq. (2.41) can be written in the form

$$\int_{\Omega_t} \langle \nabla_r f(x), \mathrm{div} J_t(x) \rangle \, dx + \partial_J \mathrm{Q}_t(J_t)(J_t) = 0, \quad t \in \mathbb{R}_+, \tag{2.46}$$

with $\langle \nabla_r f, \mathrm{div} J_t \rangle := \sum\limits_{i=1}^{n} \sum\limits_{j=1}^{m} \chi_{j_t} \partial_{r_i} f_j \left(\rho_t, \left(\nabla \Phi_t^{-1}\right)^{-1} \right) \mathrm{div} J_{i_t}$.

Now we approximate (2.28) for given discrete step size $h > 0$ by

$$\rho(t+h) = \hat{\rho}(t) \circ \Phi_{t+h}^{-1} \mathrm{det} \nabla \Phi_{t+h}^{-1} + h \mathrm{div} J_{t+h}. \tag{2.47}$$

We see that for known $(\rho(t), \Phi_t, \chi_t)$, a minimiser (Φ, χ, J) of the functional

$$(\tilde{\Phi}, \tilde{\chi}, \tilde{J}) \mapsto E_{t+h}(\tilde{\Phi}, \tilde{\chi}, \tilde{J}) := F_{t+h} \left(\hat{\rho}(t) \circ \tilde{\Phi}^{-1} \mathrm{det} \nabla \tilde{\Phi}^{-1}, \tilde{\Phi}^{-1}, \tilde{\chi} \right) + h \mathrm{Q}_t(\tilde{J}) \tag{2.48}$$

satisfies (2.46). Additionally, due to $\rho := \hat{\rho}(t) \circ \tilde{\Phi}^{-1} \mathrm{det} \nabla \tilde{\Phi}^{-1} + h \mathrm{div} J$, every minimiser fulfils the equation

$$\begin{aligned} 0 = \partial_J E_{t+h}(\Phi, \chi, J)(\delta J) &= h \int_{\Omega_{t+h}} \langle \nabla_\rho f(\rho, \nabla \Phi, \chi), \mathrm{div} \delta J \rangle \, dx + h \partial_J Q(J)(\delta J) \\ &= -h \int_{\Omega_{t+h}} (\nabla \mu, \delta J) \, dx + h \int_{\Omega_{t+h}} (L^{-1} J, \delta J) \, dx \end{aligned}$$

and it holds $J = L \nabla \mu$ and (2.33) is fulfilled.

Motivated by these considerations we arrive at the following implicit time discrete version of our original problem:

Let $\rho(t)$, Φ_t, χ_t be the solutions of the problem at time t. Then $\rho(t+h)$, Φ_{t+h} and χ_{t+h} are given by $\rho(t+h) := \hat{\rho}(t) \circ \tilde{\Phi}^{-1} \mathrm{det} \nabla \tilde{\Phi}^{-1} + h \mathrm{div} J$, $\Phi_t := \Phi$, $\chi_t := \chi$, where (Φ, χ, J) is a minimiser of (2.48).

3 The time-discrete system

In this paragraph we show the existence of minimisers of (2.48) in a suitable function space. To this end it is necessary to make (2.48) precise. We make further assumptions on the structure of $f_1, \ldots, f_m$, which are motivated by the direct method in the calculus of variations, see [BB92], [Dac89] and (2.6), (2.10).

First we want to extend Def_t on B_t (see (2.20), (2.21)) and want to ensure that the domain where F_t is finite is closed. The difficulty here is that in the definition of F_t the inverse of the gradients of the deformations occur. The following ansatz solves this problem, taking (2.6) and (2.10) into account.

For $A \in \mathsf{M}^3$ let

$$\mathrm{cof}A := \begin{pmatrix} A_{22}A_{33} - A_{23}A_{32} & A_{13}A_{32} - A_{12}A_{33} & A_{12}A_{23} - A_{22}A_{23} \\ A_{23}A_{31} - A_{21}A_{33} & A_{11}A_{33} - A_{13}A_{31} & A_{13}A_{21} - A_{11}A_{23} \\ A_{21}A_{32} - A_{22}A_{32} & A_{12}A_{31} - A_{11}A_{32} & A_{11}A_{22} - A_{12}A_{21} \end{pmatrix}.$$

With this definition we have $A\,\mathrm{cof}A = (\mathrm{cof}A)A = (\det A)E_3 := (\det A)(\delta_{kl})_{k,l=1}^3$. It is important to notice that for invertible $A \in \mathsf{M}^3$ it holds $\frac{1}{\det A}\mathrm{cof}A = A^{-1}$. Therefore we make the following conditions on $f_1, \ldots, f_m$:

$$\begin{aligned} &\text{There exist } g_j : \mathbb{R}^n \times \mathsf{M}^3 \times \mathsf{M}^3 \times \mathbb{R} \to \bar{\mathbb{R}}_+ \text{ such that for all } (r, A) \in \mathbb{R}^n \times \mathsf{M}^3 \\ &\qquad f_j(r, A^{-1}) = g_j(r, A, \mathrm{cof}A, \det A), \text{ if } \det A \neq 0. \end{aligned} \tag{3.1}$$

Furthermore we demand that for all $(r, A, B, d) \in \mathbb{R}^n \times \mathsf{M}^3 \times \mathsf{M}^3 \times \mathbb{R}$

$$g_j(r, A, B, d) \in \mathbb{R}_+ \text{ iff } (r, A, B, d) \in \mathrm{Z}, \quad 1 \leq j \leq m, \tag{3.2}$$

with the admissible set

$$\mathrm{Z} := \left\{ (r, A, B, d) \in \mathbb{R}^n \times \mathsf{M}^3 \times \mathsf{M}^3 \times \mathbb{R} : d \geq 0, \sum_{i=1}^n r_i d \leq 1 \text{ and } r \geq 0 \right\}. \tag{3.3}$$

Here, the condition $r \geq 0$ for $r \in \mathbb{R}^n$ has to be understood componentwise.

As given in (3.1), g_j denotes the argument of F_t^v in the position of f_j. Now we define for $t \in \mathbb{R}_+$ with the abbreviation $\mathbf{d} := (\nabla d, \mathrm{cof}\nabla d, \det\nabla d)$, $d \in \mathcal{W}_t^\delta$,

$$\begin{aligned} &F_t^v : \mathrm{B}_t \to \bar{\mathbb{R}},\ (r, d, p) \mapsto \\ &\begin{cases} \sum_{j=1}^m \int_{\Omega_t} p_j(x) g_j(r(x), \mathbf{d}(x)) dx, & \text{if } \sup_{j=1}^m \int_{\Omega_t} |p_j(x)|\, g_j(r(x), \mathbf{d}(x))\, dx < \infty, \\ +\infty, & \text{otherwise,} \end{cases} \end{aligned} \tag{3.4}$$

$$F_t : \mathrm{B}_t \to \bar{\mathbb{R}}, \quad (r, d, p) \mapsto F_t^s(p) + F_t^v(r, d, p). \tag{3.5}$$

Since Q_t is defined on $\left(\mathrm{L}^2(\Omega_t, \mathbb{R}^3)\right)^n$, it remains to define the divergence on $\mathrm{L}^2(\Omega_t, \mathbb{R}^3)$. This is done in a way adapted to the equations such that the conservation of mass (2.9) and the implication (2.40)⇒(2.41) holds.

$$\begin{aligned} \mathrm{div}_t : L^2(\Omega_t, \mathbb{R}^3) &\to W^{1,-2}(\Omega_t), \\ \mathrm{div}_t j(\xi) := -\int_{\hat{\Omega}} \langle j, \nabla\xi \rangle\, dx, &\quad \xi \in W^{1,2}(\hat{\Omega}, \mathbb{R}^n). \end{aligned} \tag{3.6}$$

In the following we write div instead of div_t.

Remark. *As is shown in [Anr06], for a lower semicontinuous convex function $g : \mathbb{R}^n \to \bar{\mathbb{R}}$ that satisfies $C_1 g(x) + C_2 \geq \|x\|^q$ for all $x \in \mathbb{R}^n$ and positive constants C_1, C_2 and $q > 1$, one can define $\int_\Omega g(\mathrm{div}J)$ in a natural way by setting*

$$\int_\Omega g(\mathrm{div}J) = +\infty, \quad \textit{if } \mathrm{div}J \notin \mathrm{L}^q(\Omega, \mathbb{R}^n).$$

Now we are in the position to formulate (2.48) appropriately. Let $t \in \mathbb{R}_+$, $h > 0$, $\rho(t)$ and Φ_t be given such that (2.4)–(2.6) and (2.7)–(2.10) are satisfied. Then introduce

$$\begin{aligned} E_{t+h} &: \mathrm{B}_t \times \left(\mathrm{L}^2(\Omega_t, \mathbb{R}^3)\right)^n \mapsto \bar{\mathbb{R}}, \\ &(d, p, G) \mapsto F_{t+h}(\hat{\rho}(t) \circ d\det\nabla d + h\mathrm{div}G, d, p) + h\mathrm{Q}_t(G). \end{aligned} \tag{3.7}$$

We call the variational problem $E_{t+h} \to \min$ the *time-discrete* system. The existence of minimisers is ensured by the following theorem.

Theorem 3.1. *Let $t \in \mathbb{R}_+$ and $h > 0$. In addition to the earlier assumptions, let $f_1, \ldots, f_m$ be lower semicontinuous and convex, satisfying*

$$c_1^j f_j(r, A, B, d) + c_2^j \geq \|A\|^{3+\delta}, \tag{3.8}$$

$$c_3^j f_j(r, A, B, d) + c_4^j \geq \frac{\|B\|^{3+\delta}}{|d|^{2+\delta}}, \tag{3.9}$$

$$c_5^j f_j(r, A, B, d) + c_6^j \geq |d|^{1+\frac{\delta}{3}} \tag{3.10}$$

for all $(r, A, B, d) \in \mathbb{R}^n \times \mathsf{M}^3 \times \mathsf{M}^3 \times \mathbb{R}$ with non-negative constants c_k^j, $k = 1, \ldots, 6$. Then E_{t+h} possesses a minimiser (Φ, χ, J) that fulfils (2.5), (2.6), (2.7)–(2.10) and (2.13)–(2.15) with $\rho := \hat{\rho}(t) \circ \tilde{\Phi}^{-1}\det\nabla\tilde{\Phi}^{-1} + h\mathrm{div}J$.

The proof of this theorem can be found in [Anr06].

Idea of proof: We use the direct method in the calculus of variations. First we show the lower semicontinuity of the functionals in a suitable topology (weak topology for J and Φ, strong topology for χ; a new proof for the lower semicontinuity of F_t^s with methods from elementary convex algebra is given in [Anr04]). For a minimising sequence $(\Phi_k, \chi_k, J_k)_{k\in\mathbb{N}}$, Eq. (3.8) yields the boundedness of $\|\Phi_k^{-1}\|_{\mathcal{W}_t^\delta}$, (3.9) gives the boundedness of $\|\Phi_k\|_{\mathcal{W}^\delta}$, (3.10) gives the boundedness of $\|\det\nabla\Phi_k^{-1}\|_{\mathrm{L}^{1+\frac{\delta}{3}}}$ which implies the boundedness of $\|\mathrm{div}J_k\|_{\mathrm{L}^{1+\frac{\delta}{3}}}$.

The norm $\|\chi_k\|_{\mathrm{BV}}$ can be estimated by F_t^s, see [Anr04], which guarantees the compactness in L^1, whereas the L^2-norm of J_k is estimated by Q_t.

Exploiting the differentiability properties of functions in $\mathcal{W}_t^\delta$, see [GMS98], and properties of the weak convergence in these spaces, see [Cia88], the existence of minimisers with the stated properties can be proved.

Corollary 3.2. *If for* $1 \leq i \leq n$, $1 \leq j \leq m$ *the derivatives* $\partial_{r_i} f_j$ *exist in* $\overset{\circ}{\mathrm{Z}}$ *and if* $\lim_{k\to\infty} |\partial_{r_i} f_j(r_k, A_k, B_k, d_k)| = +\infty$ *for every sequence* $(r_k, A_k, B_k, d_k)_{k\in\mathbb{N}}$ *with* $\lim_{k\to\infty} (r_k, A_k, B_k, d_k) = (r, A, B, d) \in \partial\mathrm{Z}$, *then for a minimiser of Thm. 3.1 it holds* $J = \nabla\mu$ *with* $\mu_i = \sum_{j=1}^{m} \chi_j \partial_{r_i} f_j(\rho, \mathbf{\Phi})$, $1 \leq i \leq n$.

The complete proof of this statement is given in [Anr06].
Idea of proof: Since the formal method presented at the end of Sect. 2.2 cannot be applied, we approximate f_j by suitable smooth and convex functions f_j^k from below and solve the corresponding variational problem $E_{t+h}^k \to \min$ for which we have $J^k = \nabla\mu^k$, $\mu^k = \nabla_r f^k(\rho^k, \mathbf{\Phi}^k) \in \mathrm{W}^{1,2}(\Omega_t)$.

Using convexity arguments and elementary measure-theoretic results one can then show with the Poincaré inequality the existence of a subsequence with $J^{k_l} \to J$, $\nabla\mu^{k_l} \to \nabla\mu$ and $\nabla_r f^{k_l}(\rho^{k_l}, \mathbf{\Phi}^{k_l}) \to \nabla_r f(\rho, \mathbf{\Phi})$ as $l \to \infty$.

4 The continuous system

According to Thm. 3.1 we can construct for given $h > 0$ a sequence of time-discrete solutions. We discuss the limit $h \to 0$ and show that the discrete solutions converge in a sense that has yet to be specified to a solution of the original problem. As we have at most weak convergence, due to the non-linearity of F_t we cannot expect that the weak limit satisfies (2.33). Additionally, the equations do not provide a condition on Φ_t and χ_t in time. If we consider the problem on a fixed given time interval $[0, T]$ for $T > 0$ and regard ρ, Φ, χ, J, μ as mappings from $[0, T]$ to a certain topological space X, we notice the analogy to Young measures that yield solutions to our problem in case X is finite-dimensional (or locally compact), see [Eva91], [Mül04], [Ped99].

4.1 Formulation of the problem with Young measures

As the domain Ω_t is time dependent, so is the function space X_t containing $\rho(t), \Phi_t, \chi_t, J_t, \mu_t$. The space $X := \bigcup_{t\in[0,T]} X_t$ has no 'nice' topological properties. Therefore we consider the quantities $\hat{\rho}(t) := \rho(t) \circ \Phi_t \det\nabla\Phi_t, \hat{\Phi}_t := \Psi_t^{-1} \circ \Phi_t, \hat{\chi}_t := \chi_t \circ \Psi_t, \hat{J}_t := J_t \circ \Psi_t, \hat{\mu}_t := \mu_t \circ \Psi_t$ on the reference domain Ω to formulate the equations. So we transform with Φ_t respectively Ψ_t for $t \in \mathbb{R}_+$. The corresponding time-discrete solutions exist according to Thm. 3.1 since Φ_t possesses the transformation property, see [GMS98]. In the following, $\hat{\alpha}$ always denotes the transform of α. In analogy to the common weak formulation, the measure-valued formulation reads:

Let $\hat{X} := \mathrm{L}^2(\Omega, \mathbb{R}^n) \times \mathcal{W}^\delta \times \mathrm{L}^{1+\frac{3}{\delta}}(\Omega, \mathbb{R}^m) \times \left(\mathrm{L}^2(\Omega, \mathbb{R}^3)\right)^n \times \mathrm{W}^{1,2}(\Omega, \mathbb{R}^n)$ be equipped with the product topology of the weak topology in coordinates 2, 4, 5 and the strong topology in coordinates 1 and 3. We look for a mapping

$\mathcal{P} : [0,T] \to \mathcal{R}(\hat{X})$, $t \mapsto \mathcal{P}_t$, where $\mathcal{R}(\hat{X})$ is the space of signed Radon measures over $\hat{X}$ with $\mathcal{P}_t \geq 0$ and $\mathcal{P}_t(\hat{X}) = 1$ for almost all $t \in [0,T]$ such that (with $\hat{x} := (\hat{\rho}, \hat{\phi}, \hat{\chi}, \hat{J}, \hat{\mu})$)

$$\int_0^T \vartheta'(t) \int_{\hat{X}} \int_{\Omega} \hat{\rho}(y)\hat{\xi}(y)\, dy\, d\mathcal{P}_t(\hat{x})\, dt$$
$$= \int_0^T \vartheta(t) \int_{\hat{X}} \int_{\Omega} \hat{J}(y) \nabla(\hat{\xi} \circ \hat{\Phi}^{-1} \circ \Psi_t^{-1}) \circ \Psi_t(y) \det \nabla \Psi_t(y)\, dy\, d\mathcal{P}_t(\hat{x})\, dt, \quad (4.1)$$

$$\int_0^T \vartheta(t) \int_{\hat{X}} \int_{\Omega} \hat{\xi}(y) L^{-1} \hat{J}(y) \det \nabla \Psi_t(y)\, dy\, d\mathcal{P}_t(\hat{x})\, dt$$
$$= -\int_0^T \vartheta(t) \int_{\hat{X}} \int_{\Omega} \hat{\mu}(y) \mathrm{div}(\hat{\xi} \circ \Psi_t^{-1}) \circ \Psi_t(y) \det \nabla \Psi_t(y)\, dy\, d\mathcal{P}_t(\hat{x})\, dt, \quad (4.2)$$

$$\int_0^T \vartheta(t) \int_{\hat{X}} \int_{\Omega} \hat{\mu}(y)\hat{\xi}(y) \det \nabla \Psi_t(y)\, dy\, d\mathcal{P}_t(\hat{x})\, dt$$
$$= \int_0^T \vartheta(t) \int_{\hat{X}} \int_{\Omega} \sum_{j=1}^m \hat{\chi}_j \circ \hat{\Phi}(y) \partial_r \hat{f}_j \left(\hat{\rho}(y), \hat{\mathbf{\Phi}}(y)\right) \hat{\xi}(y)\, dy\, d\mathcal{P}_t(\hat{x})\, dt \quad (4.3)$$

for all $\vartheta \in C_0^\infty([0,T])$, $\hat{\xi} \in C_0^\infty(\bar{\Omega}, \mathbb{R}^n)$ and

$$\mathrm{supp}\, \mathcal{P}_t \subset \Big\{ (\hat{\rho}, \hat{\phi}, \hat{\chi}, \hat{J}, \hat{\mu}) \in \hat{X} : \hat{\rho} \in \mathrm{supp}\, \mathcal{P}_t\big|_{\hat{X}_1},$$
$$\hat{F}(\hat{\rho}, \hat{\phi}, \hat{\chi}) = \min_{(\tilde{\Phi}, \tilde{\chi}) \in \hat{X}_2 \times \hat{X}_3} F(\hat{\rho}, \tilde{\Phi}, \tilde{\chi}) \Big\} \quad (4.4)$$

for almost all $t \in [0,T]$. Here, $\hat{X}_l$ denotes the l-th component of $\hat{X}$, $1 \leq l \leq 5$. Furthermore, let the conditions analogous to (2.4)–(2.6), (2.7)–(2.10), (2.13)–(2.15) be fulfilled for almost all $t \in [0,T]$ on $\mathrm{supp}\, \mathcal{P}_t$.

4.2 Construction of the Young measures

The key to the proof of existence of measure-valued solutions to the continuous problem is given by the following theorem.

Theorem 4.1. *(Analogon to Young measures in the infinite setting)*
Let $I_T := [0,T] \subset \mathbb{R}$ for $T > 0$ and λ_T be the Lebesgue measure on I_T, $(\nu_i)_{i \in \mathbb{N}}$

be a sequence of positive Radon measures on I_T with $\nu_i \overset{*}{\to} \lambda_T$ for $i \to \infty$, let X_1, X_2 be Banach spaces with separable X_1^*, X_2^*, define

$$(X, \|\cdot\|_X) := (X_1 \times X_2, \|\cdot\|_{X_1} + \|\cdot\|_{X_2}), \quad (X, \tau) := (X_1 \times X_2, \|\cdot\|_{X_1} \times w_{X_2}),$$

and $(\gamma_i : [0,T] \to X)_{i\in\mathbb{N}}$ be a mapping.

If $K_1^n \subset\subset (X_1, \|\cdot\|_{X_1})$, $K_2^n \subset\subset (X_2, w_{X_2})$ and

$$\nu_i\left(M_i^n := \{t \in I_T : \gamma_i(t) \notin K_1^n \times K_2^n\}\right) < \frac{1}{n} \quad \text{for } i \in \mathbb{N}, \tag{4.5}$$

then there exists a subsequence $(\gamma_{i_k})_{k\in\mathbb{N}}$ and a mapping $\mathcal{P} : I_T \to \mathcal{R}(X)$ with

$$\mathcal{P}_t \geq 0, \quad \mathcal{P}_t(X) = 1 \ \text{ for almost all } t \in I_T \tag{4.6}$$

and

$$\left([0,T] \ni t \mapsto \int_X f(t,x)\, d\mathcal{P}_t(x) \in \mathbb{R}\right) \in \mathrm{L}^\infty([0,T]), \tag{4.7}$$

$$\lim_{k\to\infty} \int_0^T f(t, \gamma_{i_k}(t))\, d\nu_{i_k}(t) = \int_0^T \int_X f(t,x)\, d\mathcal{P}_t(x)\, dt \tag{4.8}$$

for all $f \in \mathrm{C}_b([0,T] \times X)$.

Corollary 4.2. *Additional to the assumptions of Thm. 4.1 let there exist a $q \geq 0$ such that for all $i \in \mathbb{N}$*

$$\|\gamma_i\|_X \in \mathrm{L}^1(I_T) \text{ and } \int_{M_i^n} \|\gamma_i\|_X^q\, d\nu_i(t) < \frac{1}{n}. \tag{4.9}$$

Let $f : (I_T \times X, |\cdot| \times \tau) \to \mathbb{R}$ fulfil for a constant $C > 0$

$$f(t,x) \leq C\,(1 + \|x\|^q) \quad \text{for all } (t,x) \in I_T \times X. \tag{4.10}$$

Let f be bounded from below and let f be either lower semi-continuous or lower semicontinuous with respect to the second argument and satisfy a uniform continuity in time, i.e. for any $n \in \mathbb{N}$ and given $\epsilon > 0$ there exists a $\delta(n,\epsilon)$ with $|f(x,t) - f(x,t')| < \epsilon$ for $|t - t'| < \delta(n,\epsilon)$, $t, t' \in I_T$ and all $x \in K_1^n \times K_2^n$. If one of these two conditions is met, it follows

$$f(t,\cdot) \in \mathrm{L}^1\left(X, \mathcal{B}_X, \mathcal{P}_t\right) \ \text{for almost all } t \in I_T, \tag{4.11}$$

$$\left([0,T] \ni t \mapsto \int_X f(t,x)\, d\mathcal{P}_t(x)\, dt \in \mathbb{R}\right) \in \mathrm{L}^1([0,T]), \tag{4.12}$$

$$\liminf_{k\to\infty} \int_0^T f\left(t, \gamma_{i_k}(t)\right) d\nu_{i_k}(t) \geq \int_0^T \int_X f(t,x)\, d\mathcal{P}_t(x)\, dt. \tag{4.13}$$

A continuous function f that is not necessarily bounded from below and satisfies (4.10) fulfils

$$\lim_{k\to\infty} \int_0^T f\left(t, \gamma_{i_k}(t)\right) d\nu_{i_k}(t) = \int_0^T \int_X f(t,x)\, d\mathcal{P}_t(x)\, dt. \tag{4.14}$$

We remind that a cylinder function is defined as follows:

Definition 4.3. *Let X be a topological vector space. A function $f : X \to \mathbb{R}$ is called a cylinder function on X if for some $p \in \mathbb{N}$ there exists a $\alpha \in X^{*^p}$ and a $g \in \mathrm{C}_b(\mathbb{R}^p)$ such that f has the representation $f = g \circ \alpha$. The symbol Z_X denotes the set of all cylinder functions on X.*

Corollary 4.4. *From the assumptions of Thm. 4.1 it follows*

$$\mathcal{P}_t\left(X \setminus \bigcup_{n=1}^{\infty} K_1^n \times K_2^n\right) = 0 \text{ for almost all } t \in I_T. \tag{4.15}$$

The longer, technical proofs of this statement can be found in [Anr06]. Crucial is the following Lemma that is also proved in [Anr06].

Lemma 4.5. *(Approximation Lemma.) Let X be a Banach space, $f : X \to \mathbb{R}$ be strongly continuous, $g : X \to \mathbb{R}$ be weakly continuous, $K \subset X$ strongly compact and $L \subset X$ weakly compact. Then for any $\epsilon > 0$ there exist cylinder functions $\Gamma_\epsilon : X \to \mathbb{R}$ and $\Theta_\epsilon : X \to \mathbb{R}$ with*

$$\max_{x\in K} |f(x) - \Gamma_\epsilon(x)| < \epsilon \text{ and } \max_{x\in L} |g(x) - \Theta_\epsilon(x)| < \epsilon.$$

In addition it holds $\max_{x\in X} |\Gamma_\epsilon(x)| \le \max_{x\in K} |f(x)|$ and $\max_{x\in X} |\Theta_\epsilon(x)| \le \max_{x\in L} |g(x)|$

Remark. *The approximating cylinder functions can be chosen as elements of a fixed countable set.*

Idea of proof: The strategy to prove Thm. 4.1 is to first show the statements in the finite-dimensional case using slicing theorems, [Ped04],, and then to approximate with cylinder functions. Essentially, the extension to the infinite-dimensional case is an application of Riesz' representation theorem for positive linear forms on the space of continuous functions on completely regular spaces.

The proofs of the corollaries rely on the fact that lower continuous functions can be approximated on compact sets from below by continuous functions, see [Anr06].

4.3 Measure valued solutions

In earlier sections we have established the mathematical tools needed to formulate the final theorem. The proof of the next theorem is contained in [Anr06].

Theorem 4.6. *The continuous system (2.4)-(2.6), (2.7)-(2.10), (2.13)-(2.15), (2.27) or (2.28), (2.29) or (2.30), (2.31) and (2.33) possesses a solution in the sense of Sect. 4.1 if the requirements of Thm. 3.1 and Corollary 3.2 are met. There exists $\hat{\rho} \in \mathrm{L}^2([0,T] \times \Omega, \mathbb{R}^n)$ with $\mathcal{P}_t = \delta_{\hat{\rho}(t)} \times \tilde{\mathcal{P}}_t$ for almost all $t \in [0,T]$ and the following energy inequality holds:*

$$\mathsf{E}(\tau_2) - \mathsf{E}(\tau_1) \leq - \int_{\tau_1}^{\tau_2} \mathsf{Q}(t) + \mathsf{Q}^*(t)\, dt \text{ for almost all } \tau_1, \tau_2 \in [0,T], \tag{4.16}$$

where

$$\mathsf{E} : [0,T] \to \mathbb{R},\ t \mapsto \int_{\hat{X}} \hat{F}(\hat{\rho}, \hat{\Phi}, \hat{\chi}, t)\, d\mathcal{P}_t(\hat{x}), \tag{4.17}$$

$$\mathsf{Q} : [0,T] \to \mathbb{R},\ t \mapsto \int_{\hat{X}} \hat{Q}(\hat{J}, t)\, d\mathcal{P}_t(\hat{x}), \tag{4.18}$$

$$\mathsf{Q}^* : [0,T] \to \mathbb{R},\ t \mapsto \int_{\hat{X}} \hat{Q}^*(\nabla \hat{\mu}, t)\, d\mathcal{P}_t(\hat{x}). \tag{4.19}$$

Idea of proof: (All subsequences are labelled as the original sequence; we use the abbreviation $\hat{x} := (\hat{\rho}, \hat{\Phi}, \hat{\chi}, \hat{J}, \hat{\mu})$.)

For $i \in \mathbb{N}$ let $h(i) := \frac{T}{2^i}$. Then, according to Thm. 3.1 and Corollary 3.2, we can construct recursively a finite sequence $\left(\hat{x}_{lh(i)}\right)_{l=0}^{2^i}$ which solves the time-discrete problem. Now we define $\hat{x}_i$ for $i \in \mathbb{N}$ as the step function corresponding to the sequence which is continuous from left. One can show that $\left(\gamma_i := (\hat{\rho}_i, \hat{\Phi}_i, \hat{\chi}_i, \hat{J}_i, \hat{\mu}_i)\right)_{i \in \mathbb{N}}$ satisfies the assumptions of Thm. 4.1. In the next step one proves with the help of Kolmogoroff's compactness criterium, see [Wlo82], the existence of a subsequence with the property

For a given smooth Dirac sequence $(\phi_j)_{j \in \mathbb{N}}$ there exists for every $j \in \mathbb{N}$ a $\hat{\rho}^j \in \mathrm{L}^2([0,T] \times \Omega, \mathbb{R}^n)$ such that

$$\lim_{k \to \infty} \int_0^T \int_\Omega \left| \hat{\rho}_i^j(t,x) - \hat{\rho}^j(t,x) \right|^2 dx\, dt = 0.$$

Furthermore we need:

Lemma 4.7. *Let X be a T3a-space, $\nu \in \mathcal{PM}(X)$ with $\nu(X) = 1$ and $g \in \mathrm{C}(X,X)$. If for every $f \in \mathrm{C_b}(X)$*

$$\int_X f^2(g(x))\, d\nu(x) - \left(\int_X f(g(x))\, d\nu(x) \right)^2 = 0, \tag{4.20}$$

then there exists $x_0 \in X$ with $g(x) = x_0$ for all $x \in \operatorname{supp} \nu$.

Applying Thm. 4.1 and the above Lemma on the subsequence $(\gamma_i)_{i\in\mathbb{N}}$, we find the existence of $\hat{\rho}$ with $\mathcal{P}_t = \delta_{\hat{\rho}(t)} \times \tilde{\mathcal{P}}_t$ for almost all $t \in [0,T]$.

With the exception of (4.2) and (4.4), the remaining equations follow essentially from Thm. 4.1 and the Corollaries 4.2 and 4.4.

For the proof of the minimality condition (4.4), we show with (4.1) that

$$\mathsf{E}(t) = \min_{(\Phi,\chi)\in\hat{X}_2\times\hat{X}_3} \hat{F}(t,\hat{\rho}(t),\Phi,\chi) \quad \text{for } t\in[0,T].$$

The validity of (4.2) relies on the subgradient-inequality

$$\int_{\hat{X}}\int_{\Omega} \mu(\tilde{\rho}-\rho)\,dy\,d\mathcal{P}_t(\hat{x}) \le \int_{\hat{X}}\int_{\Omega} f(\tilde{\rho}) - f(\rho)\,dy\,d\mathcal{P}_t(\hat{x}),$$

where (4.4) is used.

The proof of the energy inequality is based on the following considerations. Define for $i\in\mathbb{N}$, $1\le k\le 2^i$ and $t := kh(i)$

$$J_{t_i}^{\inf} := \operatorname*{argmin}_{J\in L^2(\Omega_t,\mathbb{R}^3)^n} \left[F_t\left(\rho_i(t)+h(i)\mathrm{div}J,\Phi_{t_i},\chi_{t_i}\right)+h(i)Q_t(J)\right],$$

$$\rho_i^{\inf}(t) := \rho_i(t) + h\mathrm{div}J_{t_i}^{\inf}$$

and the corresponding continuation for $t\neq kh(i)$. Then it holds

$$h(i)\mu_{j_i}^{\inf}\mathrm{div}J_i^{\inf} = \mu_{j_i}^{\inf}\left(\rho_i^{\inf}-\rho_i\right) \ge f_j\left(\rho_i^{\inf},\mathbf{\Phi}_i\right) - f_j\left(\rho_i,\mathbf{\Phi}_i\right)$$

and (Young's inequality)

$$\begin{aligned} Q_t\left(J_{t_i}^{\inf}\right)+Q_t^*\left(\nabla\mu_{t_i}^{\inf},t\right) &= \int_{\Omega_t}\left\langle J_{t_i}^{\inf},\nabla\mu_{t_i}^{\inf}\right\rangle dx \\ &= -\sum_{j=1}^{m}\int_{\Omega_t}\chi_{j_{t_i}}\mu_{j_{t_i}}^{\inf}\mathrm{div}J_{t_i}^{\inf}\,dx \end{aligned}$$

for $i\in\mathbb{N}$, $1\le j\le m$, $t\in[0,T]$ and $x\in\Omega_t$. This yields

$$\begin{aligned} &F_t\left(\rho_i^{\inf}(t),\Phi_{t_i},\chi_{t_i}\right) - F\left(\rho_i(t),\Phi_{t_i},\chi_{t_i}\right) \\ &\qquad \le -h(i)\left[Q_t\left(J_{t_i}^{\inf}\right)+Q_t^*\left(\nabla\mu_{t_i}^{\inf}\right)\right] \end{aligned}$$

which can be rewritten as

$$\begin{aligned} &F_t\left(\rho_i^{\inf}(t),\Phi_{t_i},\chi_{t_i}\right)+h(i)Q_t\left(J_{t_i}^{\inf}\right) \\ &\qquad \le F_t\left(\rho_i(t),\Phi_{t_i},\chi_{t_i}\right) - h(i)Q_t^*\left(\nabla\mu_{t_i}^{\inf}\right). \end{aligned}$$

Therefore, due to

$$F_{t+h(i)}\left(\rho_i(t+h(i)), \Phi_{(t+h(i))_i}, \chi_{(t+h(i))_i}\right) + h(i)Q_{t+h(i)}\left(J_{(t+h(i))_i}\right) \\ \leq F_t\left(\rho_i^{\mathrm{inf}}(t), \Phi_{t_i}, \chi_{t_i}\right) + h(i)Q_t\left(J_{t_i}^{\mathrm{inf}}\right),$$

we obtain the estimate

$$F_{t+h(i)}\left(\rho_i(t+h(i)), \Phi_{(t+h(i))_i}, \chi_{(t+h(i))_i}\right) - F_t\left(\rho_i^{\mathrm{inf}}(t), \Phi_{t_i}, \chi_{t_i}\right) \\ \leq -h(i)\left[Q_{t+h(i)}\left(J_{(t+h(i))_i}\right) + Q_t^*\left(\nabla\mu_{t_i}^{\mathrm{inf}}\right)\right]. \quad (4.21)$$

Next, the inequality (4.21) is rewritten in terms of $\hat{F}$, $\hat{Q}$, $\hat{Q}_{\mathrm{inf}}$. Then we consider the two sequences

$$\left(\gamma_i^1 := (\hat{\rho}_i, \hat{\Phi}_i, \hat{\chi}_i, \hat{J}_i, \hat{\mu}_i, \hat{\mu}_{\mathrm{inf}_i})\right)_{i\in\mathbb{N}} \text{ and } \left(\gamma_i^2 := (\hat{\rho}_{\mathrm{inf}_i}, \hat{\Phi}_i, \hat{\chi}_i, \hat{J}_i, \hat{\mu}_i, \hat{\mu}_{\mathrm{inf}_i})\right)_{i\in\mathbb{N}}$$

and show that they generate the same measure $\breve{\mathcal{P}}_t = \delta_{\hat{\rho}(t)} \times \bar{\mathcal{P}}_t$ for almost all $t \in [0,T]$. From (4.21) it follows (4.16) with the corresponding $\mathsf{Q}_{\mathrm{inf}}^*$. Estimating the subgradient inequality we can finally show $\mathsf{Q}_{\mathrm{inf}}^* = \mathsf{Q}^*$.

References

[Anr04] Arnrich, S.: Lower Semicontinuity of the Surface Energy Functional-An Alternative Proof. Preprint **148**, DFG Priority Programme 1095 Analysis, Modelling and Simulation of Multiscale Problems, 2004.

[Anr06] Arnrich, S.: Ein allgemeines maßwertiges Modell für Phasenübergänge in Kristallen, Ph.D. Thesis, University of Leipzig (2006)

[Bau92] Bauer, H: Maß- und Integrationstheorie. De Gruyter, Berlin; New York (1992).

[BB92] Blanchard, P, Brüning, E.: Variational Methods in Mathematical Physics. A Unified Approach. Springer Berlin (1992)

[BS96] Brokate, M, Sprekels, J: Hysteresis and Phase Transitions, Springer, Berlin (1996)

[Cia88] Ciarlet, P.G.: Mathematical Elasticity. North Holland, Amsterdam (1988)

[Dac89] Dacorogna B.: Direct Methods in the Calculus of Variations. Springer Berlin, Heidelberg (1989)

[EG92] Evans, L.C., Gariepy, R.F.: Measure Theory and Fine Property of Functions. CRC Press, London (1992)

[Eva91] Evans, L.C.: Weak Convergence Methods for Nonlinear Partial Differential Equations (Regional Conference Series in Mathematics, No 74) CBMS/74. American Mathematical Society, USA (1991)

[GMS98] Giaquinta, M., Modica, G., Soucek, J.: Cartesian currents in the calculus of variations. Springer, Berlin, Heidelberg (1998)

[GHM01] Georgi, H.O., Häggström, O., Maess, C.: The random geometry of equilibrium phases, *Phase Transitions and Critical Phenomena*, **18**, (C. Domb and J.L. Lebowitz, eds.), Academic Press, London, 1–142 (2001)

[dGM84] de Groot, S.R., Mazur, P.: Non-Equillibrium Thermodynamics. Dover Publications, New York (1984)

[Gur81] Gurtin, M.E.: An Introduction to Continuum Mechanics. Academic Press, San Diego (California), (1981)

[Hol00] Holzapfel, G.A.: Nonlinear Solid Mechanics, Wiley, New York (2000)
[Kha83] Khachaturyan, A: Theory of Structural Transformation in Solids, *manuscripta mathematica*, **43**, 261–288 (1983)
[KY87] Kirkaldy, J.S., Young, D.J.: Diffusion in the Condensed State. The Institute of Metals, London (1987)
[KK93] Kittel, C., Krömer, H.: Physik der Wärme. 4. Edition. R. Oldenbourg Verlag GmbH, München (1993)
[KP98] Kondepudi, D., Prigogine, I.: Modern Thermodynamics. John Wiley & Sons Ltd, Cichester(England) (1998)
[Luc94] Luckhaus, S.: Solidification of Alloys and the Gibbs-Thomson Law. Preprint of SFB 256 no. **335** (1994)
[LS95] Luckhaus, S., Sturzenhecker, T.: Implicit time discretization for the mean curvature flow equation. *Calc. Var.*, **3**, 253–271 (1995),
[Mül98] Müller, S.: Variational models for microstructure and phase transitions. Lecture notes no.: 2, Max-Planck-Institute for Mathematics in the Sciences, Leipzig (1998)
[Mül04] Müller, S.: Weak Convergence Methods for Partial Differential Equations, Lecture 2004/05, Leipzig, 2004.
[Ogd97] Ogden, R.W.: Non-linear Elastic Deformations, Dover Pub., Dover (1997)
[Ons31a] Onsager, L: Reciprocal relations in irreversible processes I. *Phys. Rev.*, **37**, 405–426 (1931)
[Ons31b] Onsager, L: Reciprocal relations in irreversible processes II. *Phys. Rev.*, **38**, 2265–2279 (1931)
[Ped99] Pedregal, P.: Optimization, relaxatian and Young measures. *BULLETIN (New Series) OF THE American Mathematical Society*, **36/1**, 27–58 (1999)
[Ped04] Pedregal, P: Γ-convergence through Young measures. *SIAM J. Math. Anal.*, **36** (2004), 423–440 (2004)
[SR86] Slemrod, M., Royburd, V.: Measure-valued solutions to a problem in dynamic phase transitions, Arch. Rat. Mech. Anal., **93**, 61–79 (1986)
[Vis96] Visintin, A.: Models of Phase Transitions. Birkhäuser, Boston (1996)
[WS*93] Wang, S., Sekerka, R., Wheeler, A., Murray, B., Coriell, C., Braun, R., Mc Fadden, G: Thermodynamically consistent phase field models for solid solidification, *Physica D*, **69**, 189–200 (1993)
[Wlo82] Wloka, J.: Partielle Differentialgleichungen. Teubner, Stuttgart (1982)
[Zie89] Ziemer, W.P.: Weakly Differentiable Functions. Sobolev Spaces and Functions of Bounded Variations. Springer, New York (1989)

Relaxation and the Computation of Effective Energies and Microstructures in Solid Mechanics

Sören Bartels[1], Carsten Carstensen[1], Sergio Conti[2], Klaus Hackl[3], Ulrich Hoppe[3], and Antonio Orlando[1,4]

[1] Institut für Mathematik, Humboldt-Universität zu Berlin.
{sba,cc,ao}@math.hu-berlin.de
[2] Fachbereich Mathematik, Universität Duisburg-Essen, Campus Duisburg.
conti@math.uni-duisburg.de
[3] Institut für Mechanik, Ruhr-Universität Bochum.
{klaus.hackl,ulrich.hoppe}@rub.de
[4] School of Engineering, University of Wales Swansea, UK

Summary. We address the numerical analysis of relaxed formulations for scalar and vectorial nonconvex variational problems originating from models for solid-solid phase transitions and crystal plasticity. We discuss algorithms for the approximation of the quasiconvex envelope using laminates, rank-one convexity, and polyconvexity, and present some numerical applications to benchmarks problems, and to a model for single-slip crystal plasticity.

1 Introduction

Variational models based on nonlinear elasticity, and their mathematical analysis, have proved useful for the study of phase transitions and microstructures in elastic solids, starting with the seminal work of Ball & James [BJ87, BJ92]. The methods of relaxation have in some cases lead to new understandings on the mesoscopic phase diagram [DSD02], on the microscopic origin of complex domain patterns [KM94], and on geometrical conditions relevant for the design of new devices and materials [Bha03].

Mathematically, one minimizes the functional

$$E(u) = \int_\Omega W(Du)\,dx \tag{1.1}$$

over the set of admissible deformations $u : \Omega \subset \mathbb{R}^n \to \mathbb{R}^m$, with $u \in W^{1,p}(\Omega;\mathbb{R}^m)$, $u = u_D$ on $\partial\Omega$, and W the energy density of the crystal. Here $W^{1,p}(\Omega;\mathbb{R}^m)$ denotes standard Sobolev spaces with $p \in (1,\infty)$ related to the growth of W, and (1.1) may include lower-order terms representing external forces.

One says that the functional E predicts a microstructure if a minimum does not exist, and gradients of infimizing sequences exhibit oscillations on finer and finer scales. Objective of the research is the analysis and numerical simulation of those infimizing sequences and/or their most relevant features.

The determination of low-energy states of such functionals E by standard finite element methods will typically yield mesh dependent results with oscillations in Du_h on a length scale comparable with the mesh size. Further, the computations can be very sensitive to mesh orientation and miss completely the description of the real microstructural configuration. For instance, the precise characterization of the minimizers of a non-convex problem in [BP04] shows that they develop complicated branching structures and are therefore difficult to detect numerically. One is therefore interested in alternative approaches, which do not attempt a direct numerical minimization, and is lead to the concept of relaxation.

From a physical point of view, relaxation focuses on macroscopic features and on the average material behaviour, rather than on the details of the microstructure. This means that one operates a separation of scales, and tries to extract from the microscale all information that is relevant for the macroscale, and no more. The macroscopic deformation is then determined by studying a problem which contains an effective energy density, which automatically accounts for the optimal local microstructure.

From a mathematical point of view, relaxation theory aims to replace an ill posed problem with a well posed one (at least as far as existence is concerned), preserving the essential features of the original problem. This can be achieved following basically two approaches. The first option is to enlarge the class of the competing functions, allowing for measure-valued solutions [You80, Ped97, Rou96]. The second one is to focus on the weak limits of infimizing sequences. Weak convergence, which qualitatively corresponds to convergence of averages, eliminates the fine-scale oscillations and gives a limit which only contains information on the macroscopic scale. The idea is, therefore, to study the behaviour of infimizing sequences by characterizing their limit points as minimizers of a new functional [But89, Dac89, Mue99].

Lack of strong convergence of infimizing sequences, and lack of existence of a minimizer, is strictly related to the lack of weak lower semicontinuity of the functional (1.1) on the space $W^{1,p}(\Omega;\mathbb{R}^m)$, which in turn is equivalent to quasiconvexity of W, under suitable continuity and growth conditions [Dac89, Mue99]. Precisely, if W is coercive then weak sequential lower semicontinuity (and hence quasiconvexity) is a sufficient condition for the existence of minimizers. If instead W is not quasiconvex, then one must expect fine-scale oscillations in the gradients of infimizing sequences. The relaxation of $E(u)$ is achieved in this case by replacing W with its quasiconvex envelope W^{qc}, the largest quasiconvex function bounded from above by W. Knowledge of W^{qc} would permit an accurate simulation of the macroscopic features of E.

Unfortunately, analytical formulas for the quasiconvex envelope are known only for very few energy densities W. Consequently, one is interested to numerical relaxation, which aims at an efficient approximation of W^{qc}. This will be illustrated and discussed below, considering model examples of microstructures in phase transitions and in elastoplasticity described by scalar and vector nonconvex variational problems. We call the minimization of (1.1) scalar if $n \wedge m := \min\{n, m\} = 1$, vector otherwise.

The remaining part of the paper is organized as follows. Sect. 2 analyzes generalized and relaxed formulations of a scalar nonconvex minimization problem for a two-well energy density. Sect. 3 deals, instead, with nonconvex vector variational problems by introducing the notions of quasiconvexity, rank-one convexity, and polyconvexity. Sect. 4 describes numerical algorithms for the evaluation of the rank-one convex and polyconvex envelope as approximation of the quasiconvex envelope. Applications to models for microstructure in phase transitions and plasticity are given in Sect. 5. Finally, Sect. 6 concludes the paper with some observations.

2 The scalar double-well problem and its relaxation

We consider in this section the anti-plane shear simplification of the Ericksen-James energy density

$$W(F) := |F - F_1|^2 |F - F_2|^2 \quad \text{for } F \in \mathbb{R}^2, \tag{2.1}$$

with F_1, $F_2 \in \mathbb{R}^2$, $F = Du$, and $u : \Omega \subset \mathbb{R}^2 \to \mathbb{R}$, as a typical example of a scalar nonconvex minimization problem. This reads as follows.

Problem 2.1. Seek $u \in \mathcal{A}$ that minimizes

$$E(u) = \int_\Omega W(Du)\, dx + \alpha \int_\Omega |u - f|^2\, dx, \tag{P}$$

over the set of admissible functions $\mathcal{A} := u_D + W_0^{1,4}(\Omega; \mathbb{R})$, with $u_D \in W^{1,4}(\Omega; \mathbb{R})$ prescribed, $\alpha \geq 0$, and $f \in L^2(\Omega; \mathbb{R})$.

As discussed in the Introduction, direct minimization of (P) is difficult [Lus96]. The rest of this section discusses alternative approaches. Precisely, in Sect. 2.1 we introduce the concept of Young measures, and in Subsections 2.2 and 2.3 generalizations of (P) with Young measures (problem (GP)) and by convexification (problem (CP)) are discussed. Convergence of adaptive mesh refinement algorithms is discussed in the Sect. 2.4, whereas Sect. 2.5 summarizes the main results for the formulations (CP) and (GP) for an *ad hoc* extension to 2D of the broken Tartar problem [NW93] developed in [CJ03].

2.1 Young measures capture oscillations

Infimizing sequences (u_ℓ) for (P) are typically weakly but not strongly convergent in $W^{1,p}(\Omega;\mathbb{R})$ and the corresponding weak limits are in general not solutions of (P), because of the lack of weak lower semicontinuity. Young measures provide the mathematical tool for representing the weak-$*$ limit, whenever it exists, of sequences $(f(u_\ell)) \subset L^\infty(\Omega;\mathbb{R})$ with $f \in C_0(\mathbb{R};\mathbb{R})$. Here $C_0(\mathbb{R}^m;\mathbb{R})$ for $m \geq 1$ is the space of the functions $f \in C(\mathbb{R}^m;\mathbb{R})$ such that $\lim_{|x|\to\infty} f(x) = 0$. Definitions and properties of Young measures are given next in relation to their use for sequences $(u_\ell) \subset L^\infty(\Omega \subset \mathbb{R}^n;\mathbb{R}^m)$.

Definition 2.2 ([Mue99]). *Denote with $\mathcal{M}(\mathbb{R}^m)$ the set of all finite signed Radon measures supported in $\mathbb{R}^m$, and with $L^\infty_w(\Omega;\mathcal{M}(\mathbb{R}^m))$ the space of functions $\nu = (\nu_x)_{x\in\Omega}$ defined in $\Omega \subset \mathbb{R}^n$ and with values in $\mathcal{M}(\mathbb{R}^m)$ such that*

$$\langle \nu; g\rangle : \Omega \to \mathbb{R}, \quad x \mapsto \langle \nu_x; g\rangle := \int_{\mathbb{R}^m} g \, d\nu_x$$

are measurable for all $g \in C_0(\mathbb{R}^m;\mathbb{R})$. Let $YM(\Omega;\mathbb{R}^m)$ be the set of all $\nu \in L^\infty_w(\Omega;\mathcal{M}(\mathbb{R}^m))$ which are probability measures, i.e. $\nu_x \geq 0$ and $\nu_x(\mathbb{R}^m)=1$ for almost all $x \in \Omega$. The elements of $YM(\Omega;\mathbb{R}^m)$ are called Young measures.

Theorem 2.3 (Existence theorem for Young measures [Mue99]). *Assume the sequence (u_ℓ) bounded in $L^\infty(\Omega;\mathbb{R}^m)$. Then there exists a compact set $K \subset \mathbb{R}^m$, a subsequence $(u_{\ell_j}) \subset (u_\ell)$, and a Young measure $\nu = (\nu_x)_{x\in\Omega} \in L^\infty_w(\Omega;\mathcal{M}(\mathbb{R}^m))$ such that:*

(i) supp $\nu_x \subseteq K$ a.e. in Ω

(ii) for each $f \in C_0(\mathbb{R}^m;\mathbb{R})$ we have

$$\int_\Omega f(u_{\ell_j}) h \, dx \to \int_\Omega \bar{f} h \, dx \text{ for every } h \in L^1(\Omega;\mathbb{R}), \tag{2.2}$$

where

$$\bar{f}(x) = \langle \nu_x; f\rangle := \int_{\mathbb{R}^m} f d\nu_x \text{ for a.e. } x \in \Omega. \tag{2.3}$$

Definition 2.4. *We call $\nu = (\nu_x)_{x\in\Omega}$ in Thm. 2.3 the Young measure associated with (or generated by) the sequence (u_{ℓ_j}).*

Remark 2.5. (i) From Thm. 2.3 one obtains a criterion for strong convergence, and consequently a criterion to decide on the occurrence or not of oscillations. Given $u_\ell \overset{*}{\rightharpoonup} u$ in $L^\infty(\Omega;\mathbb{R}^m)$, then $u_\ell \to u$ strongly in $L^p(\Omega;\mathbb{R}^m)$ with $p < \infty$ if and only if $\nu_x = \delta_{u(x)}$ a.e. in Ω [Mue99].
(ii) By making specific choices for f, we can read off certain information regarding the structure of the Young measures. For instance, if $u_\ell \overset{*}{\rightharpoonup} u$ in $L^\infty(\Omega;\mathbb{R}^m)$ and $f = \text{id}$ in a neighbourhood of supp ν, then

$$u(x) = \int_{\mathbb{R}^m} \lambda d\nu_x(\lambda)$$

where $(\nu_x)_{x\in\Omega}$ is the Young measure associated with (u_ℓ).

Since microstructures are associated with oscillations in the gradients of infimizing sequences, one is mainly interested in understanding what are the Young measures associated with the sequence (Du_ℓ). These are called gradient Young measures.

Definition 2.6 ([Mue99]). *An element $\nu \in L^\infty_w(\Omega; \mathcal{M}(\mathbb{R}^m))$ is called a $W^{1,\infty}$-gradient Young measure generated by (u_ℓ) if it is the Young measure generated by the sequence of gradients (Du_ℓ) with (u_ℓ) weakly-$*$ convergent in $W^{1,\infty}(\Omega; \mathbb{R}^m)$.*

Remark 2.7. (i) From eq. (2.2) with $f = \mathrm{id}$ in a neighbourhood of supp ν, the weak-$*$ limit Du of a sequence of gradients (Du_ℓ) for an infimizing sequence (u_ℓ) for (P) is related to the gradient Young measure generated by (u_ℓ) by

$$Du(x) = \int_{\mathbb{R}^2} F d\nu_x(F) = \langle \nu_x; \mathrm{id}\rangle \text{ a.e. in } \Omega. \tag{2.4}$$

Therefore the gradient Young measure ν permits to compute the macroscopic strain Du. Analogously, specifying $f = DW$ around supp ν in (2.2), if $(DW(Du_\ell))$ is weakly-$*$ convergent to some σ, one obtains

$$\sigma(x) = \int_{\mathbb{R}^2} DW(F) d\nu_x(F) = \langle \nu_x; DW\rangle \text{ a.e. in } \Omega. \tag{2.5}$$

(ii) Specifying then $f = W$ around supp ν in (2.2) one has

$$\lim_{\ell\to\infty} \int_\Omega W(Du_\ell)\, dx = \int_\Omega \langle \nu_x; W\rangle\, dx. \tag{2.6}$$

2.2 Relaxation with Young measures and their numerical approximation

Equation (2.6) along with (2.4) motivate the following generalized problem.

Problem 2.8. Seek a minimizer $(u, \nu) \in \mathcal{B}$ of

$$GE(u,\nu) := \int_\Omega \langle \nu_x, W\rangle\, dx + \alpha \int_\Omega |u - f|^2\, dx \tag{GP}$$

over $\mathcal{B} := \{(u,\nu) \in \mathcal{A} \times YM(\Omega; \mathbb{R}^2) : Du(x) = \langle \nu_x; \mathrm{id}\rangle \text{ for a.e. } x \in \Omega\}$.

The relevance of problem (GP) follows from relaxation theory.

Theorem 2.9 ([Rou96, Thm. 5.2.1][Ped97, Thm. 4.4]).
Problem (GP) *has a solution and there holds*

$$\mathit{inf}_{u\in\mathcal{A}} E(u) = \mathit{min}_{(u,\nu)\in\mathcal{B}} GE(u,\nu).$$

Moreover, if (u_ℓ) *is a weakly convergent infimizing sequence for* (P), *with weak limit* u, *that generates the gradient Young measure* ν, *then* (u,ν) *is a minimizer for* (GP). *Vice versa, if* (u,ν) *is a solution of* (GP) *then there is a weakly convergent infimizing sequence* (u_ℓ) *such that its weak limit is* u *and* ν *is the Young measured generated by* (u_ℓ).

Remark 2.10. Given $\nu \in L^\infty_w(\Omega;\mathcal{M}(\mathbb{R}^2)$, the compatibility condition $\langle \nu_x;\mathrm{id}\rangle = Du(x)$ with $u\in W^{1,\infty}(\Omega;\mathbb{R})$, and $\operatorname{supp}\nu_x \subseteq K$ compact subset of $\mathbb{R}^2$, characterize in the scalar case completely the gradient Young measures associated with sequences.

Numerical approximations of (GP) have been proposed in [NW93, CR98, Rou96a, Ped95, KMR05]. Those involve a discretization of the admissible set $\mathcal{A}\times YM(\Omega;\mathbb{R}^2)$ and care of the differential constraint.

Within a finite element scheme, denote with $\mathcal{T}$ a regular triangulation of Ω, and by $\mathcal{E}$ and $\mathcal{N}$ the set of all edges and vertices, respectively. Then introduce the following finite dimensional spaces

$$\begin{aligned}\mathcal{S}^1(\mathcal{T}) &:= \{v_h \in C(\bar\Omega) : \forall T\in\mathcal{T},\ v_h|_T \text{ is affine}\},\\ \mathcal{S}^1_0(\mathcal{T}) &:= \{v_h \in \mathcal{S}^1(\mathcal{T}) : v_h = 0 \text{ on } \partial\Omega\}.\end{aligned}$$

Let $\mathcal{K} := \mathcal{N}\cap\Omega$ denote the set of free nodes, the Dirichlet boundary conditions u_D are discretized by nodal interpolation, i.e. $u_{D,h}\in\mathcal{S}^1(\mathcal{T})$ with

$$u_{D,h}(z) = u_D(z) \text{ if } z\in\mathcal{K} \text{ and } u_{D,h}(z) = 0 \text{ if } z\in\mathcal{N}\setminus\mathcal{K}.$$

A conforming finite element method of (GP) is obtained by replacing the space $\mathcal{A}$ with $\mathcal{A}_h := u_{D,h} + \mathcal{S}^1_0(\mathcal{T})$ whereas the set of Young measures $YM(\Omega;\mathbb{R}^2)$ is approximated by element-wise constant measures, i.e. homogeneous Young measures ν_T expressed as a convex combination of Dirac measures supported at the nodes of a triangulation τ of a convex polygonal domain $\omega\subset\mathbb{R}^2$ with mesh size d. That is, denote by $\mathcal{N}_d(\omega)$ the set of nodes of the triangulation τ of ω, we assume

$$\nu_{T,d} = \sum_{F_{T,j}\in\mathcal{N}_d(\omega)} a_{F_{T,j}}\delta_{F_{T,j}} \tag{2.7}$$

with known atoms $F_{T,j}\in\mathcal{N}_d(\omega)$, and unknown coefficients $a_{T,j}$. We denote this set with $\mathcal{L}^0(\mathcal{T};\mathcal{PM}_{h,d})$ where $\mathcal{PM}_{h,d}$ is the set of probability measures expressed as in (2.7). Consider the set

$$\mathcal{B}_{h,d} = \{(v_h,\mu_{h,d})\in\mathcal{A}_h\times\mathcal{L}^0(\mathcal{T};\mathcal{PM}_{h,d}) : \forall T\in\mathcal{T},\ Dv_h|_T = \langle\mu_{h,d}|_T;\mathrm{id}\rangle\}, \tag{2.8}$$

the discrete generalized problem reads

Problem 2.11. Seek $(u_h, \nu_{h,d}) \in \mathcal{B}_{h,d}$ such that

$$\text{Minimize } GE(u_h, \nu_{h,d}) \text{ over } \mathcal{B}_{h,d}. \qquad (\text{GP}_{h,d})$$

An existence result for $(\text{GP}_{h,d})$ follows as for (GP). Let $\mathcal{L}^0(\mathcal{T}; \mathbb{R}^2)$ be the set of piecewise constant functions on $\mathcal{T}$ with values in $\mathbb{R}^2$, W^{**} the convex envelope of W (defined in Sect. 2.3) and $W_d^c = (\mathcal{P}_\tau W)^{**}$ the convex envelope of $\mathcal{P}_\tau W$, the nodal interpolation of W associated with the triangulation τ of ω. Let $\sigma = DW^{**}(Du)$ for a solution $u \in \mathcal{A}$ of (CP) (see Problem 2.13), then we have the following a-priori and a-posteriori error bounds

Theorem 2.12 ([Bar04, Thm. 4.6 & Thm. 4.8]). *Assume $u \in \mathcal{A}$ solution of (CP), $(u_h, \nu_{h,d}) \in \mathcal{B}_{h,d}$ solution of $(\text{GP}_{h,d})$, and $\lambda_{h,d} \in \mathcal{L}^0(\mathcal{T}; \mathbb{R}^2)$ the Lagrange multiplier associated with the constraint $Dv_h|_T = \langle \mu_{h,d}|_T; \text{id}\rangle$. Then, there holds*

$$\begin{aligned} \|\sigma - \lambda_{h,d}\| &\le C \inf_{v_h \in \mathcal{A}_h} (\|\nabla(u - v_h)\| + \alpha\|u - v_h\|) \\ &\quad + C\|\partial W_d^c - DW^{**}\|_{L^\infty(\omega)}\,; \end{aligned} \tag{2.9}$$

$$\begin{aligned} \|\sigma - \lambda_{h,d}\|^2 \le C\Big\{ &(\sum_{T\in\mathcal{T}} h_T^2 \|f + \text{div}\lambda_{h,d} + 2\alpha(f - u_h)\|)^{1/2} \\ &+ (\sum_{E\in\mathcal{E}} h_E \|[\lambda_{h,d}\cdot n_E]\|^2)^{1/2} + \|\partial W_d^c - DW^{**}\|_{L^\infty(\omega)} \\ &+ \|h_{\mathcal{E}}^{3/2} \partial_{\mathcal{E}}^2 u_D / \partial s^2\|_{L^2(\Gamma_D)} \Big\}. \end{aligned} \tag{2.10}$$

Since $\|\partial W_d^c - DW^{**}\|$ can be bounded from above in terms of grid size d and D^2W^{**}, together with the density of the finite element spaces in $\mathcal{A}$, from (2.9) one proves $\lambda_{h,d} \to \sigma$ in L^2 as h, $d \to 0$, whereas (2.10) represents a basic ingredient of the multilevel adaptive scheme for the definition of the support of the Young measures developed by [Bar04].

2.3 Relaxation via convex envelopes

By minimizing the two contributions in (GP) separately one obtains another relaxation of (P). For fixed $F = Du$ one can find a probability measure $\nu = (\nu_x)_{x\in\Omega}$ such that ν minimizes the expression $\langle \mu; W\rangle$ among all the probability measures μ satisfying $\langle \mu; \text{id}\rangle = F$. In some cases, it is also possible to obtain an explicit expression for the convex hull of W, defined by

$$W^{**}(F) = \min_{\substack{\mu \in YM(\Omega;\mathbb{R}^2) \\ \langle \mu; \text{id}\rangle = F}} \langle \mu, W\rangle. \tag{2.11}$$

The notation is motivated by the fact that for continuous W the convex envelope coincides with the bipolar function. We recall that, since we are in the scalar case, convexity and quasiconvexity coincide.

Problem (GP) and (2.11) motivate to consider the convexified problem.

Problem 2.13. Seek $u \in \mathcal{A}$ that minimizes

$$E^c(u) := \int_\Omega W^{**}(Du)\,dx + \alpha \int_\Omega |u - f|^2\,dx. \tag{CP}$$

Likewise problem (GP), the importance of problem (CP) follows from relaxation theory.

Theorem 2.14 ([Dac89]). *Problem* (CP) *has a solution and there holds*

$$\inf_{u \in \mathcal{A}} E(u) = \min_{u \in \mathcal{A}} E^c(u). \tag{2.12}$$

Moreover, if (u_ℓ) *is a weakly convergent minimizing sequence of* (P) *and* u *is its weak limit, then* u *is a solution of* (CP)*. Vice versa, if* u *is a solution of* (CP) *then there exists a weakly convergent minimizing sequence of* (P) *having* u *as weak limit. The stress field* $\sigma = DW^{**}(u)$ *is unique and independent of* u *among the solutions of* (CP) *[Fri94].*

Remark 2.15. Whenever $\alpha > 0$ in (P), problem (CP) admits a unique solution u. If $\alpha = 0$, however, the numerical treatment of (CP) requires the introduction of a perturbation in E^c, usually in the form of a strictly convex functional scaled by a small quantity. The introduction of the stabilized term finds its justification in the need of including some kind of 'selection mechanism' in the model which *(i)* ensures uniqueness on discrete level, *(ii)* is necessary for the design of convergent iterative solvers [BCHH04], and *(iii)* with some stabilizations terms for standard low-order finite element methods yields strong convergence of the gradients [BCPP04]. The same happens when dealing with quasiconvex envelopes in the vectorial case, see, e.g., [CDD02].

2.4 Adaptive finite element methods for relaxed formulations

An h-finite element adaptive algorithm consists of successive loops of the form

$$\text{SOLVE} \rightarrow \text{ESTIMATE} \rightarrow \text{MARK} \rightarrow \text{REFINE} \tag{2.13}$$

designed to produce with less computational effort more efficient meshes by targeted local refinements. The use of such algorithms for the direct finite element minimization, however, does not always lead to an improved convergence rate in the stress error and also unclear is its convergence. For degenerately convex problems with C^1 energy density W characterized for some constants p, r, s by the conditions

$$\begin{aligned} |DW(A) - DW(B)|^r &\le c(1 + |A|^s + |B|^s)(W(B) - W(A) \\ &\qquad - DW(A)\cdot(B - A)), \\ c_l(|A|^p - 1) \le W(A) &\le c_u(|A|^p + 1), \end{aligned} \tag{2.14}$$

to hold for all $A, B \in \mathbb{R}^n$, [Car06] proves the convergence of the stress fields $\sigma_0, \sigma_1, \dots$ produced by (2.13) to $\sigma = DW^{**}(Du)$ in $L^{r/(1+s/p)}(\Omega; \mathbb{R}^2)$. In the algorithm (2.13), the step `MARK` is realized by the criterion introduced by [Dor96] where one marks the edges $E \in \mathcal{M} \subset \mathcal{E}$ such that $\Theta \sum_{E\in\mathcal{E}} \eta_E^2 \leq \sum_{E\in\mathcal{M}} \eta_E^2$ with η_E the edge contribution to the global error estimator. In the step `REFINE`, on the other hand, one refines each triangle T with an edge in $\mathcal{M}$ such that an inner node is created, with possible further refinements that guarantee that $\|h_j Df\|_{L^2(\Omega;\mathbb{R}^2)}$ tends to zero as $j \to \infty$ and the resulting triangulation is regular.

2.5 A 2D scalar benchmark problem

In this section we report on the analysis of (P) in the particular case of $\alpha = 1$, f and u_D given in [CJ03], $\Omega = (0,1) \times (0,3/2)$, and the two wells $F_1 := -(3,2)/\sqrt{13}$ and $F_2 = -F_1$. The convex envelope W^{**} was computed in [CP97], and is

$$W^{**}(F) = ((|F|^2 - 1)_+)^2 + 4(|F|^2 - ((3,2)\cdot F)^2) \tag{2.15}$$

with $(\cdot)_+ := \max\{0,\cdot\}$ and the symbol $\cdot$ the inner product in $\mathbb{R}^2$. From relaxation theory, we have the following result.

Theorem 2.16 ([CJ03]). *The problem* (CP) *has a unique solution* $u \in \mathcal{A}$

$$\inf_{v\in\mathcal{A}} E(v) = \min_{v\in\mathcal{A}} E^{**}(v) = E^{**}(u), \tag{2.16}$$

characterized as the solution of the Euler-Lagrange equation

$$\int_\Omega \sigma \cdot Dv\,dx + 2\int_\Omega (u-f)v\,dx = 0 \textit{ for all } v \in W_0^{1,4}(\Omega;\mathbb{R}), \tag{2.17}$$

where $\sigma := DW^{**}(Du)$. *Furthermore, any infimizing sequence* (u_ℓ) *of* (P) *is bounded in* $W^{1,4}(\Omega;\mathbb{R})$ *and generates a sequence of stresses* $\sigma_\ell := DW(Du_\ell)$ *convergent in measure toward* $\sigma = DW^{**}(Du)$.

For this problem, moreover, one obtains an analytical expression for the gradient Young measure which is unique and is given by

$$\nu_x = \lambda(F)\delta_{S_+(F)} + (1-\lambda(F))\delta_{S_-(F)}, \tag{2.18}$$

with $F = Du$ and

$$\lambda(F) = \frac{1}{2}(1 + F_2\cdot F(1-|\mathbb{P}F|^2)^{-1/2}) \in [0,1], \tag{2.19}$$

$$S_\pm(F) = \begin{cases} \mathbb{P}F \pm F_2(1-|\mathbb{P}F|^2)^{-1/2} & \text{for } |F|<1, \\ F & \text{for } 1<|F|, \end{cases} \tag{2.20}$$

where $\mathbb{P} = \mathbb{I} - F_2 \otimes F_2$ (with $\otimes$ tensor product of vectors of $\mathbb{R}^2$).

Remark 2.17. Since σ_ℓ converges toward σ, from (2.5) the stress field $\sigma = DW^{**}(Du)$ can then be represented as

$$\sigma(x) = \int_{\mathbb{R}^2} DW \, d\nu_x, \tag{2.21}$$

with ν given in (2.18) [Fri94].

With the notation of Sect. 2.2 the Galerkin discretization of (2.17) reads

Problem 2.18. Seek $u_h \in \mathcal{A}_h$ such that

$$\int_\Omega \sigma_h \cdot Dv_h \, dx + 2 \int_\Omega (u_h - f) v_h \, dx = 0 \text{ for all } v_h \in \mathcal{S}_0^1(\mathcal{T}) \tag{CP$_h$}$$

with $\sigma_h := DW^{**}(Du_h)$

Strong convergence in $L^{4/3}(\Omega;\mathbb{R}^2)$ of the stress fields σ_h results from the a priori error estimate [CP97]

$$\|\sigma - \sigma_h\|_{L^{4/3}(\Omega;\mathbb{R}^2)} \leq c_1 \inf_{v_h \in \mathcal{A}_h} \|u - v_h\|_{W^{1,4}(\Omega;\mathbb{R})}. \tag{2.22}$$

This is obtained using the condition

$$|DW^{**}(A) - DW^{**}(B)|^2 \leq c(1+|A|^2+|B|^2)(DW^{**}(B) - DW^{**}(A)) : (B - A) \tag{2.23}$$

that holds for any $A, B \in \mathbb{R}^2$ together with some $p = 4$ and $q = 3$ growth conditions on W and on DW^{**}, respectively. Another application of (2.23) also shows the reliability of residual and averaging based error estimates whereas the efficiency follows from standard arguments; that is, one has also

$$c_2 \eta_M - h.o.t. \leq \|\sigma - \sigma_h\|_{L^{4/3}(\Omega;\mathbb{R}^2)} \leq c_2 \eta_M^{1/2} + h.o.t. \tag{2.24}$$

The minimal averaging error estimator η_M that enters (2.24) is defined by

$$\eta_M = \Big(\sum_{T \in \mathcal{T}} \eta_T^{4/3} \Big)^{3/4} \text{ for } \eta_T = \|\sigma_h - \sigma_h^*\|_{L^{4/3}(T;\mathbb{R}^2)},$$

with $\sigma^* \in \mathcal{S}^1(\mathcal{T})^2$ that minimizes

$$\|\sigma_h - \tau_h\|_{L^{4/3}(\Omega;\mathbb{R}^2)} \text{ among } \tau_h \in \mathcal{S}^1(\mathcal{T})^2.$$

Figure 2.1 displays experimental convergence rates for $\|\sigma - \sigma_h\|_{L^{4/3}(\Omega;\mathbb{R}^2)}$ and the error estimators η_M and $\eta_M^{1/2}$ for uniform and adaptive mesh refinement. The adaptive refinement strategy leads to significantly reduced error and improved experimental convergence rates.

Remark 2.19. The two-sided estimates (2.24) shows that lower bounds are no valid upper bounds and vice versa, due to the different exponents for η_M in the reliability and efficiency estimate. This miss balance is referred to as reliability-efficiency gap [CJ03].

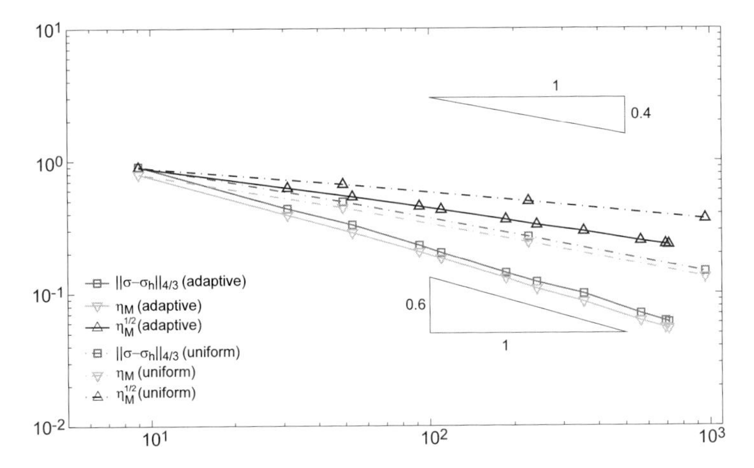

Fig. 2.1. The 2D benchmark problem. Experimental convergence rates for $\|\sigma - \sigma_h\|_{L^{4/3}(\Omega;\mathbb{R}^2)}$ and the error estimators η_M and $\eta_M^{1/2}$ plotted against degrees of freedom N with a logarithmic scale for uniform and adaptive mesh refinement.

3 Nonconvex vector variational problems

For scalar nonconvex variational problems the convexity of $W(Du)$ with respect to $F = Du$ ensures the weak (weak-$*$) sequential lower semicontinuity of the functional $E(u) = \int_\Omega W(Du)$ on $W^{1,p}(\Omega;\mathbb{R}^m)$ for $1 \leq p < \infty$ (resp. $p = \infty$). Along with suitable growth conditions on W, one can prove the existence of minimizers using the direct method of the calculus of variations. In the vectorial case a weaker condition is sufficient, namely, quasiconvexity.

3.1 Quasiconvexity and effective energy density

Quasiconvexity was introduced by Morrey in 1952 as a condition on the energy density W which is equivalent, under appropriate growth conditions, to weak sequential lower semicontinuity of the functional E [Mor52].

Definition 3.1. *Given a function $W : \mathbb{R}^{3\times 3} \to \mathbb{R}$, we say that W is quasiconvex at $F \in \mathbb{R}^{3\times 3}$ if for every open and bounded set $\omega \subseteq \mathbb{R}^3$ one has*

$$\int_\omega W(F + Dy(x))\,dx \geq \int_\omega W(F)\,dx = |\omega| W(F) \text{ for each } y \in W_0^{1,\infty}(\omega;\mathbb{R}^3). \tag{3.1}$$

Quasiconvexity lies at the heart of the relaxation theory for functionals of the type $E(u) = \int_\Omega W(Du)\,dx$. An instrumental role is played by the quasiconvex envelope of W defined as the pointwise supremum of the quasiconvex functions that are bounded from above by W, i.e., for each $F \in \mathbb{R}^{3\times 3}$,

$$W^{qc}(F) = \sup\{f(F) : f \leq W \text{ with } f \text{ quasiconvex}\}. \tag{3.2}$$

Under suitable growth conditions on W, the weakly (weakly-$*$ for $p = \infty$) sequentially lower semicontinuous envelope of the functional

$$E(u) = \int_{\Omega} W(Du)\,dx + \mathcal{L}(u), \tag{3.3}$$

has the following integral representation

$$E^{qc}(u) = \int_{\Omega} W^{qc}(Du)\,dx + \mathcal{L}(u). \tag{3.4}$$

Here $\mathcal{L}(u)$ is a linear term representing external forces.

The link between the minimization of (3.3) and (3.4) is given by relaxation theory.

Theorem 3.2 ([Dac89, Mue99]). *Let $u_D \in W^{1,p}(\Omega;\mathbb{R}^3)$ be fixed, $\mathcal{A} = u_D + W_0^{1,p}(\Omega;\mathbb{R}^3)$, and assume W to have p-growth and be p-coercive. Then the relaxed problem*

$$\textit{Minimize } E^{qc}(u) \textit{ amongst } u \in \mathcal{A}, \tag{QP}$$

has a solution and there holds

$$\min_{u\in\mathcal{A}} E^{qc}(u) = \inf_{u\in\mathcal{A}} E(u). \tag{3.5}$$

Furthermore, any solution u of (QP) is the weak limit of an infimizing sequence for (3.3).

The quasiconvex envelope of W at F can be characterized equivalently as [Dac89, Mue99]

$$W^{qc}(F) = \inf_{\substack{y\in W^{1,\infty}(\omega;\mathbb{R}^3)\\ y=Fx \text{ on } \partial\omega}} \frac{1}{|\omega|}\int_{\omega} W(Dy(x))\,dx. \tag{3.6}$$

Inequality (3.1) states that the deformation $u(x) = Fx$ is a minimizer of $\int_{\Omega} W(Dy)\,dx$ subject to its own boundary values. As such $W^{qc}(F)$ represents the infimum of the average energy taken over all possible microstructures $y = y(x)$ that satisfy the boundary condition $y(x) = Fx$ on $\partial\Omega$, with the least energy achieved by the deformation $y = Fx$ itself.

Remark 3.3. (*i*) For $n \geq 2$, $m \geq 3$ it has been shown in [Kri99] that there does not exist a local characterization of (3.1), that is, there is no set of inequalities on W and its derivatives at an arbitrary matrix F which is necessary and sufficient for W to be quasiconvex. As a result, quasiconvexity is a very difficult property to verify in practice. Only few examples of analytical expressions of quasiconvex envelopes of particular functions are known with notable examples reported in [KS86, Koh91, DSD02, CT05, CO05].

(*ii*) The generalized formulation with gradient Young measures does not circumvent the quasiconvexification. The set of admissible gradient Young measures, besides the conditions listed in Remark 2.10, is characterized by the fact that Jensen's inequality should hold for any quasiconvex function, i.e. [KP91]

$$f(\langle \nu_x; \mathrm{id}\rangle) \leq \langle \nu_x; f\rangle \text{ a.e. } x \in \Omega, \text{ for all quasiconvex functions } f. \tag{3.7}$$

For the constructive characterization and evaluation of W^{qc} for general W one is, therefore, faced with a direct minimization of a nonconvex functional with linear boundary conditions and no lower order terms on an arbitrary domain ω, whose solution may be, however, very difficult to tackle with. Necessary or sufficient conditions for quasiconvexity have been, therefore, introduced providing some insight for the analysis of microstructures.

3.2 Rank-one convexity and laminated microstructures

A necessary condition for quasiconvexity is rank-one convexity, stating convexity of the function W along all rank-one directions.

Definition 3.4. *A function* $W : \mathbb{R}^{3\times 3} \to \mathbb{R}$ *is rank-one convex if for all* $A, B \in \mathbb{R}^{3\times 3}$ *such that* $\mathrm{rank}(A - B) \leq 1$ *and all* $\lambda \in (0, 1)$,

$$W(\lambda A + (1-\lambda)B) \leq \lambda W(A) + (1-\lambda)W(B). \tag{3.8}$$

Equivalently, W *is rank-one convex if for all* $A \in \mathbb{R}^{3\times 3}$ *and all* a, $n \in \mathbb{R}^3$ *the function* $\lambda \mapsto W(A + \lambda a \otimes n)$ *is convex on* $\mathbb{R}$. *This is in turn equivalent to*

$$W(A + \lambda a \otimes n) \leq \lambda W(A + a \otimes n) + (1-\lambda)W(A), \tag{3.9}$$

for all $\lambda \in (0, 1)$, *and all* A, a *and* n.

The following considerations illustrate the relevance of rank-one convexity in the analysis of microstructures. By letting $y \in W^{1,\infty}(\Omega; \mathbb{R}^3)$ with $y(x) = Fx$ on $\partial\omega$, $W^{qc}(F)$ provides a macroscopic description of all possible microstructures with average deformation F. In the evaluation of the infimum (3.6) it may be convenient to restrict $y = y(x)$ to a subclass of $W^{1,\infty}(\Omega; \mathbb{R}^3)$ corresponding only to certain microstructure patterns. For example, one can consider the deformations $y_\ell = y_\ell(x)$ describing first order laminates, and with $y_\ell(x) = Fx$ on $\partial\omega$. The corresponding sequence of gradients will, therefore, oscillate between two phases

$$F_0 = F + (1-\lambda)a \otimes n \text{ and } F_1 = F - \lambda a \otimes n \tag{3.10}$$

with some $a, n \in \mathbb{R}^{3\times 3}$, $\lambda \in (0, 1)$, and $F_0 - F_1 = a \otimes n$. The gradient Young measure associated with (y_ℓ) will be homogeneous and equal to

$$\nu = \lambda\delta_{F_0} + (1-\lambda)\delta_{F_1}.$$

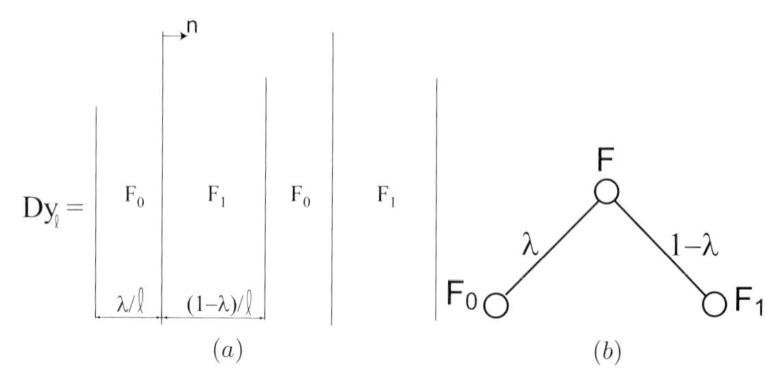

Fig. 3.1. (a) Microstructural patterns in first order laminates. (b) Graph representation.

For such infimizing sequences (y_ℓ) one has [Dac89, Mue99]

$$\lim_{\ell\to\infty}\frac{1}{|\omega|}\int_\omega W(Dy_\ell(x))\,dx = \lambda W(F_0)+(1-\lambda)W(F_1). \tag{3.11}$$

In the class of the first order laminates defined by (3.10), those that realize the lowest energetic content will therefore be solution of the problem

$$\begin{aligned} R^{(1)}W(F) = \inf\{\lambda W(\underbrace{F+(1-\lambda)a\otimes n}_{F_0}) + (1-\lambda)W(\underbrace{F-\lambda a\otimes n}_{F_1}) \;: \\ 0\le\lambda\le 1 \text{ and } a,n\in\mathbb{R}^3\}. \end{aligned} \tag{3.12}$$

If $\lambda = 0$ or $\lambda = 1$ then no microstructure will occur. The graphical interpretation of condition (3.10) and the corresponding microstructure pattern are depicted in Fig. 3.1.

For F_0 and F_1 given as above, consider the convex combination

$$F_0 = \lambda_0 F_{00} + (1-\lambda_0)F_{01} \text{ and } F_1 = \lambda_1 F_{10} + (1-\lambda_1)F_{11}, \tag{3.13}$$

with

$$F_{00}-F_{01} = a_0\otimes n_0 \text{ and } F_{10}-F_{11} = a_1\otimes n_1. \tag{3.14}$$

By replacing (3.13) into (3.10) one obtains

$$F = \lambda\lambda_0 F_{00} + \lambda(1-\lambda_0)F_{01} + (1-\lambda)\lambda_1 F_{10} + (1-\lambda)(1-\lambda_1)F_{11}. \tag{3.15}$$

The graphical interpretation of this decomposition and the corresponding microstructure pattern are shown in Fig. 3.2.

Microstructures defined by (3.15) are called second order laminates. One can therefore inquire on the second order laminates (if they exist) that minimize $1/|\omega|\int_\omega W(Du)\,dx$. Those will be solution of the following global nonlinear optimization problem

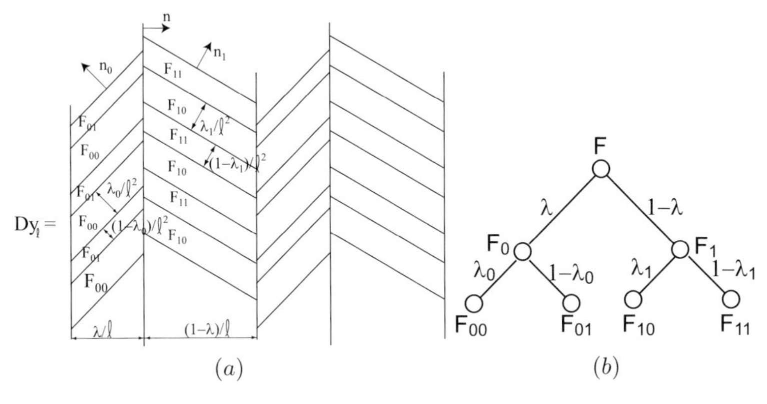

Fig. 3.2. (a) Microstructural patterns in second order laminates. (b) Graph representation.

$$R^{(2)}W(F) = \inf\Big\{ \lambda\lambda_0 W(F_{00}) + \lambda(1-\lambda_0)W(F_{01}) + (1-\lambda)\lambda_1 W(F_{10}) + (1-\lambda)(1-\lambda_1)W(F_{11}) \ :\ 0 \le \lambda, \lambda_0, \lambda_1 \le 1, \text{ and } a, n, a_0, n_0, a_1, n_1 \in \mathbb{R}^3 \Big\}. \tag{3.16}$$

It can be shown that it is also

$$R^{(2)}W(F) = \inf\{ \lambda R^{(1)}W(F_0) + (1-\lambda)R^{(1)}W(F_1) \ :\ 0 \le \lambda \le 1,\ a, n \in \mathbb{R}^3 \} \tag{3.17}$$

with F_0, F_1 defined as in (3.10). The iteration of (3.17) produces laminates of order $k \in \mathbb{N}$ such that [KS86]

$$W^{rc}(F) = \lim_{k\to\infty} R^{(k)}W(F), \tag{3.18}$$

and there holds

$$W^{qc} \le W^{rc} \le \cdots \le R^{(k)}W \le \cdots \le R^{(2)}W \le R^{(1)}W \le W. \tag{3.19}$$

In (3.18), W^{rc} denotes the rank-one convex envelope of W defined by (3.2) with rank-one convex functions.

3.3 A lower bound to W^{qc}: polyconvex envelope

Polyconvexity was introduced by Ball in [Bal77] as a structural condition on W compatible with some physical requirements that simple convexity would violate, and that was sufficient to ensure existence of minimizers for nonlinear finite strain elasticity. Both polyconvexity and convexity provide sufficient conditions for quasiconvexity.

Definition 3.5 ([Bal77]). *The function $W : \mathbb{R}^{3\times3} \to \mathbb{R}$ is polyconvex if there exists a convex function $g : \mathbb{R}^{3\times3} \times \mathbb{R}^{3\times3} \times \mathbb{R} \to \mathbb{R}$ such that*

$$W(F) = g(T(F)) \text{ for each } F \in \mathbb{R}^{3\times3}. \tag{3.20}$$

Here

$$T : F \in \mathbb{R}^{3\times3} \to T(F) = (F, \mathit{cof}\ F, \mathit{det}\ F) \in \mathbb{R}^{3\times3} \times \mathbb{R}^{3\times3} \times \mathbb{R}. \tag{3.21}$$

The function g is not defined uniquely from W. Using Carathéodory theorem it can be shown that one possible choice is [Dac89]

$$g(T(F)) = \inf_{\substack{A_i \in \mathbb{R}^{3\times3} \\ \lambda_i \in \mathbb{R}}} \{\sum_{i=1}^{19} \lambda_i W(A_i) : \lambda_i \geq 0, \sum_{i=1}^{19} \lambda_i = 1, \sum_{i=1}^{19} \lambda_i T(A_i) = T(F)\}. \tag{3.22}$$

The value of the polyconvex envelope W^{pc} at $F \in \mathbb{R}^{3\times3}$, can be, therefore, characterized equivalently as solution of the following minimization problem.

$$W^{pc}(F) = \inf_{\substack{A_i \in \mathbb{R}^{3\times3} \\ \lambda_i \in \mathbb{R}}} \{\sum_{i=1}^{19} \lambda_i W(A_i) : \lambda_i \geq 0, \sum_{i=1}^{19} \lambda_i = 1, \sum_{i=1}^{19} \lambda_i T(A_i) = T(F)\}. \tag{3.23}$$

The semiconvex notions introduced so far reduce to convexity in the scalar case, whereas in the vector case, their relation is represented in the following diagram

$$W \text{ convex } \Rightarrow W \text{ polyconvex } \Rightarrow W \text{ quasiconvex} \Rightarrow W \text{ rank-one convex}, \tag{3.24}$$

with the converse not holding in general [Mue99]. In view of (3.24) one has

$$W^c \leq W^{pc} \leq W^{qc} \leq W^{rc} \leq \cdots \leq R^{(k)}W \leq \cdots \leq R^{(2)}W \leq R^{(1)}W \leq W, \tag{3.25}$$

with W^{pc} and W^{rc} providing lower and upper bound to W^{qc}, respectively.

From Thm. 3.2 it follows

$$\min_{u\in\mathcal{A}} \int_\Omega W^{qc}(Du)\,dx = \inf_{u\in\mathcal{A}} \int_\Omega W^{rc}(Du)\,dx = \cdots = \inf_{u\in\mathcal{A}} \int_\Omega R^{(k)}W(Du)\,dx$$
$$\cdots = \inf_{u\in\mathcal{A}} \int_\Omega R^{(1)}W(Du)\,dx = \inf_{u\in\mathcal{A}} \int_\Omega W(Du)\,dx. \tag{3.26}$$

4 Numerical relaxation

The evaluation of rank–one convex and polyconvex envelopes for a characterization of the quasiconvex hull is an extremely complex task since the energy density W is defined on four– or nine–dimensional matrix spaces in the space dimension $n = m = 2$ or $n = m = 3$ but can be reduced using invariance under rotations. Furthermore, the definition of an envelope is typically not local, that is, the value at a given $F \in \mathbb{R}^{n\times n}$ depends, in general, on the values of W on the whole space $\mathbb{R}^{n\times n}$ and not just on a bounded neighborhood of F.

In view of the difficulty involved in checking analytically the previous notions, one tries to resort to efficient numerical algorithms for the approximation to rank–one convex and polyconvex envelope, referred to as numerical relaxation, exploiting growth conditions and qualitative properties of W.

4.1 Numerical polyconvexification

For $F \in \mathbb{R}^{3\times 3}$ the value of the polyconvex envelope at F, $W^{pc}(F)$, given by eq. (3.23) involves a nonlinear optimization problem whose solution may be very difficult. Given a finite set of nodes $\mathcal{N}_{\delta,r} = \delta\mathbb{Z}^{3\times3} \cap \overline{B_r(0)} \subseteq \mathbb{R}^{3\times 3}$, δ mesh size such that $0 \le \delta \le r$ and r large enough so that $F \in \text{co}\, \mathcal{N}_{\delta,r}$, convex hull of $\mathcal{N}_{\delta,r}$, an approximation to $W^{pc}(F)$ can be obtained by solving the following linear optimization problem over the space $\mathbb{R}^{\#\mathcal{N}_{\delta,r}}$ with $\#\mathcal{N}_{\delta,r}$ the cardinality of the discrete set $\mathcal{N}_{\delta,r}$.

$$W^{pc}_{\delta,r}(F) = \inf_{\theta_A \in \mathbb{R}^{\#\mathcal{N}_{\delta,r}}} \Big\{ \sum_{A\in\mathcal{N}_{\delta,r}} \theta_A W(A) : \theta_A \ge 0, \sum_{A\in\mathcal{N}_{\delta,r}} \theta_A = 1, \sum_{A\in\mathcal{N}_{\delta,r}} \theta_A T(A) = T(F) \Big\}. \tag{4.1}$$

Under the assumption that $W \in C^{1,\alpha}_{loc}(\mathbb{R}^{3\times3};\mathbb{R})$ with $\alpha \in [0,1]$ [Bar04a] shows that there exists $r' < r$ such that the following estimates holds

$$|W^{pc}_{\delta,r}(F) - W^{pc}(F)| \le c\delta^{1+\alpha} |W|_{C^{1,\alpha}_{loc}(B_{r'}(0))} \tag{4.2}$$

obtained by constructing a continuous piecewise multilinear approximation to W^{pc}. Furthermore, let $\lambda^F_{\delta,r} \in \mathbb{R}^{19}$ denote the Lagrangian multiplier associated with the constraints

$$\sum \theta_A A = F, \quad \sum \theta_A \text{cof} A = \text{cof} F, \text{ and } \sum \theta_A \det A = \det F. \tag{4.3}$$

If additionally $\alpha > 0$ and $W^{pc} \in C^{1,\alpha}_{loc}(\mathbb{R}^{3\times3};\mathbb{R})$ then an approximation to $\sigma := DW^{pc}(F)$ is given by $\lambda^F_{\delta,r} \circ DT(F)$, where $DT(F)$ is the Gateaux derivative of T, and $\circ$ denotes the composition operator between $\lambda^F_{\delta,r} \in \mathcal{L}(\mathbb{R}^{19};\mathbb{R})$ and $DT \in \mathcal{L}(\mathbb{R}^9;\mathbb{R}^{19})$ (with $\mathcal{L}(\mathbb{R}^m;\mathbb{R}^n)$ space of linear operators of $\mathbb{R}^m$ into $\mathbb{R}^n$).

The solution of (4.1) involves a large number of unknowns equal to the cardinality of the discrete set $\mathcal{N}_{\delta,r}$. The combination of an active set strategy with local grid refinement and coarsening to avoid checking a Weierstrass-type maximum principle in all the nodes of $\mathcal{N}_{\delta,r}$ leads to a very efficient but still reliable algorithm that computes $W^{pc}_{\delta,r}(F)$ [Bar04a].

4.2 Numerical finite lamination

Approximations to the rank-one convex envelope W^{rc} can be realized by $R^{(k)}W$ by iterating the construction described in Sect. 3 and motivated by the condition (3.18). The algorithm proposed by [Dol99, Dol03], on the other hand, performs convexification along rank-one directions until the function is stable under this operation. A pseudo-algorithm for the approximation of W^{rc} would therefore have the following main ingredients:

Algorithm 4.1 (Numerical lamination)
(a) $k = 0$; $R^{(k)}W = W$.
(b) For certain F, *and for* $a, n \in \mathbb{R}^3$, $g(t) =$*convexify* $R^{(k)}W(F + ta \otimes n)$.
(c) $R^{(k+1)}W(F) = g(0)$ *and compare with* $R^{(k)}W(F)$ *to stop, otherwise set* $k = k + 1$ *and go to (b).*

Approximations are, therefore, introduced in step (a), by restricting the space $\mathbb{R}^{3\times3}$ where to evaluate W, and in step (b) where only discrete set of rank-one directions will be considered.

With the notation of Sect. 4.1, introduce the discrete set of rank-one directions

$$\mathcal{R}^1_\delta = \{\delta R \in \mathbb{R}^{3\times3} : R = a \otimes n, \text{ with } a, n \in \mathbb{Z}^3\},$$

and for $R \in \mathcal{R}^1_\delta$ the following set $\ell_{R,\delta} := \{\ell \in \mathbb{Z} : F + \ell\delta R \in \overline{\text{co}\mathcal{N}_{\delta,r}}\}$. For assigned $R \in \mathcal{R}^1_\delta$, the elements of $\ell_{R,\delta}$ identify the intersection of the grid $\overline{\text{co}\mathcal{N}_{\delta,r}}$ with the direction $F + tR$. In step (a) of the Algorithm 4.1, one set $R^{(0)}W = I_{\delta,r}W$ as nodal interpolation of W in $\text{co}\mathcal{N}_{\delta,r}$ whereas at step (c) one solves the following optimization problem

$$R^{(k+1)}_{\delta,r}W(F) = \inf_{R\in\mathcal{R}^1_\delta} \inf_{\theta\in\mathbb{R}^{\#\ell_{R,\delta}}} \Big\{ \sum_{\ell\in\ell_{R,\delta}} \theta_\ell R^{(k)}_{\delta,r}W(F+\delta\ell R) : \theta_\ell \geq 0, \sum_{\ell\in\ell_{R,\delta}} \theta_\ell = 1 \Big\},$$

with $R^{(k)}_{\delta,r}W := \infty$ in $\mathbb{R}^{3\times3}$ and nodal interpolation of $R^{(k)}_{\delta,r}W$ in $\overline{\text{co}\mathcal{N}_{\delta,r}}$ at variance of the algorithm proposed in [Dol99].

Assuming $W \in C^{1,1}(\mathbb{R}^{3\times3};\mathbb{R})$ and equal to W^{rc} in $\mathbb{R}^{3\times3} \setminus B_r(0)$ with some r, bounds on a and b in the definition of $\mathcal{R}^1_\delta$, and that there exists a lamination level L such that $R^{(L)}_{\delta,r}W = W^{rc}$, [Bar04b] improves the estimate of [DW00]

$$\|R^{(k+1)}_{\delta,r}W - W^{rc}\|_{L^\infty(\text{co}\mathcal{N}_{\delta,r};\mathbb{R})} \leq c\delta\,. \tag{4.4}$$

Even if one does not know L and r, $R^{(k+1)}_{\delta,r}W$ provides, however, an upper bound to W^{rc} for all $k \geq 0$, $r \geq \delta > 0$ and $F \in \text{co}\mathcal{N}_{\delta,r}$.

5 Phase transitions and plasticity as vector nonconvex minimization problems

This section discusses the numerical analysis and approximation of relaxed formulations for two types of nonconvex vector stored energy densities. In the first example the quasiconvex envelope is known whereas for the other one no analytical expression of any semiconvex envelope is available. In the latter case, therefore, we proceed to numerical relaxation by computing the polyconvex and lamination convex envelope.

5.1 Compatible phase transitions in elastic solids

We consider a solid with two phases, whose energy density takes the form

$$W(F) = \min \{W_1(F), W_2(F)\}. \tag{5.1}$$

In a geometrically linear context, the energy of each phase is

$$W_j = \frac{1}{2}\mathbb{C}(F - F_j) : (F - F_j)\,, \tag{5.2}$$

where $\mathbb{C}$ is the linear elasticity tensor, the symbol $:$ the inner product in $\mathbb{R}^{n\times n}$, $n = 2, 3$, and F_j the stress-free configuration of phase j.

Since W is not rank-one convex, thus neither quasiconvex, the functional E in (3.3) is not sequentially weakly lower semicontinuous. Assuming that the two wells F_1 and F_2 are rank-one connected, then there exists an affine function that equals W at the two wells and is elsewhere a strict lower bound of W and, therefore, there is no attainment of minimizer.

The quasiconvex envelope of W is given by [Koh91]

$$W^{qc} = \begin{cases} W_2(F) & \text{if } W_2(F) + \gamma \le W_1(F), \\ \frac{1}{2}(W_1(F) + W_2(F)) & \\ -\frac{1}{4\gamma}(W_2(F) - W_1(F))^2 - \frac{\gamma}{4} & \text{if } |W_2(F) - W_1(F)| \le \gamma, \\ W_1(F) & \text{if } W_1(F) + \gamma \le W_2(F), \end{cases} \tag{5.3}$$

with $\gamma = 1/2\langle F_2 - F_1, \mathbb{C}(F_2 - F_1)\rangle$ for rank-one connected wells. In this case, W^{qc} belongs to $C^1(\mathbb{R}^{n\times n}; \mathbb{R})$ and is convex. Further, from a result of [CP00], one can show that the following conditions hold true for W^{qc} and are, in fact, equivalent [HL93, CHO06]

$$|DW^{qc}(E) - DW^{qc}(F)| \le L|E - F|\,, \tag{5.4}$$

$$\frac{1}{L}|DW^{qc}(E) - DW^{qc}(F)|^2 \le (DW^{qc}(E) - DW^{qc}(F)) : (E - F)\,, \tag{5.5}$$

$$\frac{1}{2L}|DW^{qc}(E) - DW^{qc}(F)|^2 \le W^{qc}(E) - W^{qc}(F) - DW^{qc}(F) : (E - F)\,, \tag{5.6}$$

for any $E, F \in \mathbb{R}^{n\times n}$. Given the functional

$$\mathcal{H}(u) := \int_\Omega W^{qc}(\varepsilon(u))\, dx + \int_\Omega f u\, dx + \|u\|^2_{L^2(\Omega;\mathbb{R}^n)} \tag{5.7}$$

with $\varepsilon(u) = \operatorname{sym} Du$, using (5.4)–(5.6), and the following condition

$$W^{qc}(E) - W^{qc}(F) - DW^{qc}(F) : (E-F) \le (DW^{qc}(F) - DW^{qc}(E)) : (F-E) \tag{5.8}$$

that holds for any $E, F \in \mathbb{R}^{n\times n}$ for the convexity of W^{qc}, [CHO06] prove the convergence of (2.13) for the minimization of (5.7) over $\mathcal{A} := u_D + W^{1,2}_0(\Omega;\mathbb{R}^n)$ and the preasymptotic convergence rate of the energy. More precisely, let $\delta_h := \mathcal{H}(u_h) - \mathcal{H}(u)$, with u and u_h minimizers of $\mathcal{H}$ over $\mathcal{A}$ and $\mathcal{A}_h$, respectively. Then, there holds

$$\delta_\ell + \|\sigma - \sigma_\ell\|^2_{L^2(\Omega;\mathbb{R}^{n\times n})} + \|u - u_\ell\|^2_{L^2(\Omega;\mathbb{R}^n)} \le C\big((\delta_\ell - \delta_{\ell+1})^{1/2} + \operatorname{osc}_\ell\big) \tag{5.9}$$

with $\sigma := DW^{qc}(\varepsilon(u))$, $C > 0$ depending on the mesh regularity and material parameters, and osc_ℓ a node-patchwise definition of the data oscillations. The observation that $(\mathcal{H}_\ell)$ is a Cauchy sequence yields, finally, that

$$\sigma_h \to \sigma \text{ in } L^2(\Omega;\mathbb{R}^{n\times n}), \quad \text{and} \quad u_h \to u \text{ in } L^2(\Omega;\mathbb{R}^n), \tag{5.10}$$

provided that one controls also the data oscillations.

5.2 Single-slip elastoplasticity

We consider here a simplified model for plastic deformation in ductile single crystals. We focus on two spatial dimensions, and on the case that only a single slip system is active, which is described by an orthonormal pair of vectors s and n, with $s \in \mathbb{S}^1$ (where $\mathbb{S}^1 = \{x \in \mathbb{R}^2 : |x| = 1\}$) the slip direction on the slip plane and $n \in \mathbb{S}^1$ the normal to the slip plane. In a geometrically nonlinear context, we assume the multiplicative decomposition of the deformation gradient $F = F_e F_p$ with $F_p = I + \gamma s \otimes n$, where $\gamma \in \mathbb{R}$ is referred to as plastic slip. Hardening is included through a single internal variable $p \in \mathbb{R}$. Within the framework of rate-independent processes [Mie03, Mie04a, Mie04b, Mie05], we consider monotonic loading, or equivalently the first time step in a time-discrete scheme, and, set equal to zero the initial values of the internal variables (γ, p). Minimizing out locally the internal variables leads to a variational formulation analogous to (1.1), which can again be analysed using the discussed methods of the calculus of variations. The analogy with the study of martensitic microstructures via continuum models based on nonlinear elasticity, and the study via a variational problem expressed only in terms of the deformation gradient $F = D\phi$, was advanced for the first time in [OR99].

The constitutive behaviour of the single crystal can be described in terms of two potentials: the free energy density $W(F_e, p)$ and the dissipation potential

$J(\gamma, p)$. The free energy density is sum of an elastic and a plastic contribution as follows

$$W(F_e, p) = W_e(F_e) + W_p(p), \tag{5.11}$$

with

$$W_e(F_e) = U(F_e) + \frac{\mu}{2}(|F_e|^2 - 2), \quad W_p(p) = \frac{h}{2}p^2, \tag{5.12}$$

and $U(F_e)$ a polyconvex function defining a Neo-Hookian material, such as

$$U(F_e) = \begin{cases} \frac{\kappa}{4}((\det F_e)^2 - 1) - \frac{\kappa + 2\mu}{2}\log(\det F_e) & \text{if } \det F_e > 0 \\ +\infty & \text{else,} \end{cases} \tag{5.13}$$

with μ, κ material constants and h the hardening moduli. The dissipation potential $J(\gamma, p)$ is

$$J(\gamma, p) = \begin{cases} \tau_{cr}|\gamma| & \text{if } |\gamma| + p \leq 0 \\ \infty & \text{else}, \end{cases} \tag{5.14}$$

with τ_{cr} the critical shear stress. This is the same model considered in [CHM02, BCHH04, MLG04].

For this particular example, by minimizing with respect to the internal variables (γ, p), we obtain a closed form of the condensed energy $W_{\text{cond}}(F)$ as

$$W_{\text{cond}}(F) = U(F) + \frac{\mu}{2}(|F|^2 - 2) - \frac{1}{2}\frac{(max(0, \mu|Cs \cdot n| - \tau_{cr}))^2}{\mu Cs \cdot s + h}, \tag{5.15}$$

with $C = F^T F$. The energy density (5.15) is not rank-one convex and, hence, not quasiconvex. As a result, one may expect non attainment of minimizers for the corresponding functional, and developments of oscillations in the gradients of low-energy deformations. For the case under consideration, the occurrence of such microstructures can be shown by a direct finite element simulation using representative volume elements under periodic boundary conditions, cf. [HH02]. Figure 5.1 shows two typical results of these simulations: Oscillations in the plastic slip field γ, forming first and second order laminates. These oscillations are highly mesh-dependent with the number of oscillations growing towards infinity when the mesh becomes finer and finer.

The macroscopic material behaviour can be, however, understood by minimizing out locally the possible microstructures and defining the quasiconvex envelope of W_{cond}. Unfortunately, a closed form for condensed energies of the kind of W_{cond} is known only in few simplified cases [Con03, CT05, CO05]. We therefore resort to an approximation to the rank-one convex envelope $W^{rc}_{\text{cond}}(F)$ (Sect. 4) based on laminates.

Let $a, b \in \mathbb{S}^1$ with $a = (\cos\alpha, \sin\alpha)$ and $b = (\cos\beta, \sin\beta)$, then all the rank one matrices can be expressed in $\mathbb{R}^{2\times 2}$ as $\rho a \otimes b$ for α, β, $\rho \in \mathbb{R}$. Considering first order laminates, the average energy is given by

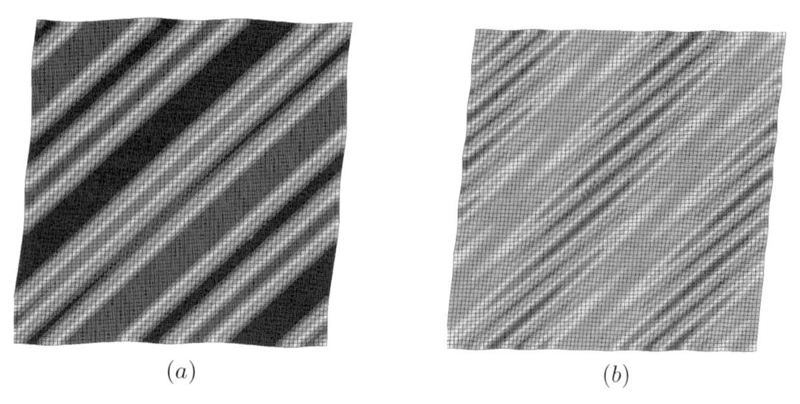

(a) (b)

Fig. 5.1. Single-slip plasticity (a) First order laminates and (b) Second order laminates as from (3.12) and (3.16) respectively, assuming periodic boundary conditions. (See page 693 for a colored version of the figure.)

$$E^{lc(1)}(F;\alpha,\beta,\lambda,\rho) = \lambda W_{\text{cond}}(F+(1-\lambda)\rho a\otimes b) \\ +(1-\lambda)W_{\text{cond}}(F-\lambda\rho a\otimes b) \tag{5.16}$$

with the microscopic energy $W_{\text{cond}}(F)$ defined in (5.15). Let $q=(\alpha,\beta,\lambda,\rho)$ and introduce the feasible set

$$\Sigma = \{q\in\mathbb{R}^4 : \ \alpha,\beta,\rho\in\mathbb{R}, \quad \lambda\in[0,\,1]\}\,,$$

the first order laminate envelope is obtained by solving the following global optimization problem

$$R^{(1)}W_{\text{cond}}(F) = \min_{q\in\Sigma} E^{lc(1)}(F;q)\,, \tag{5.17}$$

under the constraints

$$\det(F+(1-\lambda)\rho a\otimes b)>0\,,\quad \det(F-\lambda\rho a\otimes b)>0\,. \tag{5.18}$$

Following the definitions in (3.15) corresponding minimization problems can be set up for higher order laminates. The growing number of optimization variables, however, strongly limits a practical application. Already for low order laminates the numerical search for the minimizer of (5.17) turns out to be difficult, because the objective function may present an exponential number of nearby optimal local minima [Car01].

Within the techniques of global optimization for the solution of (5.17), probabilistic global search procedures are the one commonly adopted. Applying a local search several times starting from randomly chosen sampling points leads, however, to an inefficient global search, because the same local minimum may be identified over and over. As an improvement, clustering methods attempt to avoid this inefficiency by carefully selecting points at which the local search is initiated.

Algorithm 5.1 (Clustering method)
Input F, initial population $q_i \in \Sigma$ of n starting points, tolerance ε.
(a) (Sampling and reduction): Sample the objective function $E^{lc(1)}$ at q_i and reduce population taking the m best points giving the least value.
(b) (Clustering): Identify clusters, such that the points inside a cluster are 'close' to each other, and the clusters are 'separated' from each other.
(c) (Center of attraction): Identify a center of attraction in each cluster.
(d) (Local search): Start a local search from the center of attraction and stop when a minimum is reached within the tolerance ε.
Output the value of $R^{(1)}W_{cond}(F)$.

The final local search step is done by using sequential quadratic programming methods with simple bounds [NW99]. Since in a finite element framework the above algorithm has to be performed at every material point (e.g. Gauss point), for real applications it is important to develop fast techniques for the numerical relaxation. In the literature the computational effort related to the global search is usually reduced by fixing some laminate related parameters on the basis of conjectures motivated by physical considerations [ORS00, AFO03, ML03, MLG04].

Mixed analytical-numerical relaxation

A different approach to the relaxation of W_{cond} over laminates has been pursued in [CCO06]. Rather than attacking the global minimization by a brute-force global optimization algorithm that is anyway computationally very expensive, [CCO06] exploit the structure of the problem both to achieve a fundamental understanding on the optimal microstructure and, in parallel, to design an efficient numerical relaxation scheme. Inspired by results based on the global optimization [BCHH04] and on analytical relaxation in the case of rigid elasticity and no self-hardening [CT05], we determine analytically a second order laminate which has 'good' energy and furnishes an upper bound to the relaxed energy.

We consider first an elastically rigid problem where the elastic part of the deformation is assumed to be a rotation, and only the contribution from the plastic free energy is considered, i.e., dissipation is neglected. The condensed energy for this case is then given by

$$W'(F) = \begin{cases} \frac{h}{2}\gamma^2 \text{ if } F = Q(I + \gamma s \otimes m) \quad Q \in SO(2)\,, \\ \infty \quad \text{else}\,, \end{cases} \tag{5.19}$$

with the quasiconvex envelope obtained as follows

Theorem 5.1 ([Con03, Con05]). *The quasiconvex, rank-one convex, and polyconvex envelope of $W'(F)$ are equal and given by*

$$W'_{qc}(F) = \begin{cases} \frac{h}{2}(|Fm|^2 - 1) \ \textit{if} \ \det F = 1 \ \textit{and} \ |Fs| \le 1\,, \\ \infty \qquad \textit{else}\,. \end{cases} \tag{5.20}$$

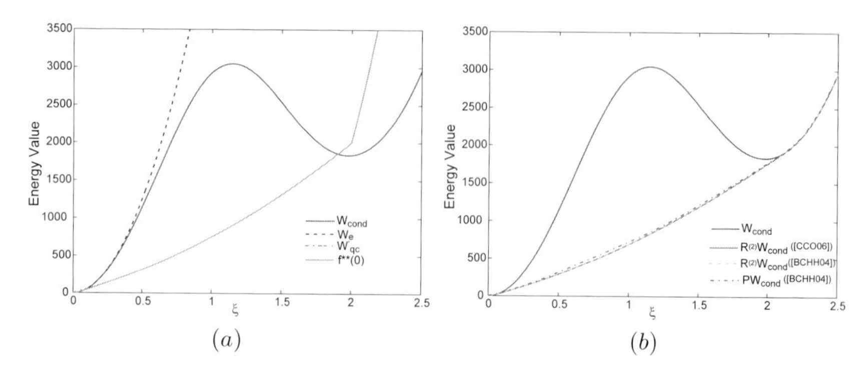

Fig. 5.2. (a) Bounds to the quasiconvex envelope of the condensed energy for zero dissipation; (b) Polyconvex and second-order laminate envelope for the condensed energy density in single-slip plasticity.

The optimal energy is given by a first-order laminate, which is supported on two matrices which have plastic deformation γ of the same magnitude and opposite sign.

We then construct a more refined model by assuming the microstructure to have the form of a laminate of second order, which is supported either on rigid-plastic deformations or on purely elastic ones. In this case, assuming volume-preserving deformations, the relaxation is reduced to a global minimization of a function of only one variable which defines the orientation of the laminate. Using this solution and the splitting of F_p from Thm. 5.1, we obtain an approximate second order laminate. The latter is then used as a starting point for the local minimization of the full energy density, including dissipation, and removing the kinematic constraint.

Figure 5.2(a) depicts the condensed energy W_{cond} (see eq. (5.15)) together with W_e (see eq (5.12)), W'_{qc} (see eq (5.20)) and the value of the energy over the approximate second order laminate (which we denote by $f^{**}(0)$) for the case of a pure shear strain $F = I + \xi r \otimes r^\perp$ with $r = (1,0)$, $r^\perp = (0,1)$ and for the material constants $\mu = 1.0 \cdot 10^4$MPa, $\kappa = 1.5 \cdot 10^4$ MPa, $h = 1.0 \cdot 10^3$ MPa and $\tau_{cr} = 10$ MPa.

Figure 5.2(b) shows a very good quantitative agreement for the values of the relaxed energy with those in [BCHH04] which had required a significantly higher numerical effort and compares approximations of the polyconvex hull $W^{pc}_{\delta,r}(F)$, realized with the procedure described in Sect. 4.1. A finer analysis at small deformations reveals however some differences, which will be discussed elsewhere [CCO06].

Figure 5.3 depicts finally the value of the volume fractions λ and λ_1 whereas $\lambda_0 = 1$. Initially, the material is in a homogeneous elastic state. Then an elastic state and a mixture of two opposite–slip plastic states appears. The volume fraction of the elastic phase starts at 100% and then decreases contin-

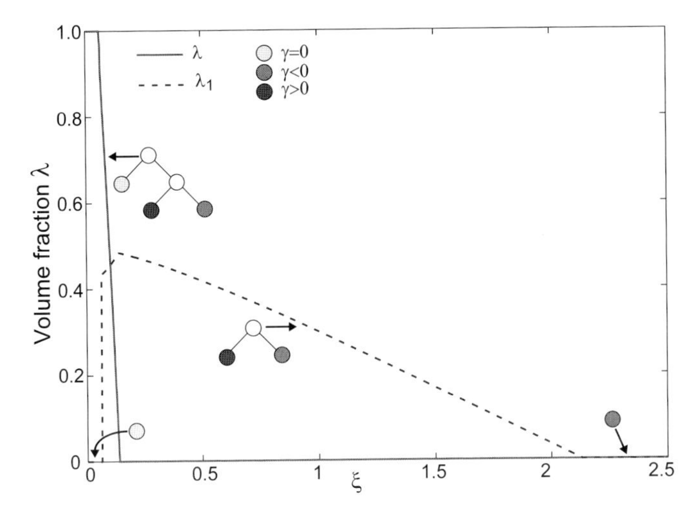

Fig. 5.3. Volume fractions λ and λ_1 for different values of ξ

uously until it vanishes at a shear $\xi = 0.13$. Both plastic phases then progress with slowly varying volume fractions until the homogeneous phase F is stable.

6 Conclusions

In this paper we have considered the numerical analysis of relaxed formulations for variational formulations lacking lower semicontinuity, and discussed algorithms for the approximation of the quasiconvex envelope of energy densities, in cases where it is not known in closed form. Relaxed solutions convey important information on the macroscopic behaviour of the microstructure, summarized by the relaxation theory. A resulting benefit is that the approximation of macroscopic quantities does not pose severe difficulties and classical algorithms for numerical optimization can be efficiently employed.

Acknowlegdments. This work has been supported by the DFG Priority Program 1095 'Analysis, Modeling and Simulation of Multiscale Problems' under projects Ca 151/13 and Co 304/1.

References

[AFO03] Aubry, S., Fago, M., Ortiz, M.: A constrained sequential-lamination algorithm for the simulation of sub-grid microstructure in martensitic materials. Comput. Methods Appl. Mech. Engrg., **192**, 2823–2843 (2003)

[Bal77] Ball, J.M.: Convexity conditions and existence theorems in nonlinear elasticity. Arch. Rat. Mech. Anal., **63**, 337–403 (1977)

[BJ87] Ball, J.M., James, R.D.: Fine phase mixtures as minimizer of energy. Arch. Rat. Mech. Anal., **100**, 13–52 (1987)

[BJ92] Ball, J.M., James, R.D.: Proposed experimental tests for the theory of fine microstructures and the two–well problem. Phil. Trans R. Soc. Lond. A, **338**, 389–450 (1992)

[Bar04] Bartels, S.: Adaptive approximation of Young measure solutions in scalar nonconvex variational problems. SIAM J. Numer. Anal., **42**, 505–529 (2004)

[Bar04a] Bartels, S.: Reliable and efficient approximation of polyconvex envelopes. SIAM J. Numer. Anal., **43**, 363–385 (2004)

[Bar04b] Bartels, S.: Linear convergence in the approximation of rank-one convex envelopes. M2AN Math. Model. Numer. Anal., **38**, 811–820 (2004)

[BCPP04] Bartels S., Carstensen, C., Plecháč, P., Prohl, A.: Convergence for stabilisation of degenerately convex minimisation problems. Interfaces and Free Boundary, **6**, 253–269 (2004)

[BCHH04] Bartels, S., Carstensen, C., Hackl, K., Hoppe, U.: Effective relaxation for microstructure simulations: algorithms and applications. Comput. Methods Appl. Mech. Engrg., **193**, 5143–5175 (2004)

[BP04] Bartels, S., Prohl, A.: Multiscale resolution in the computation of crystalline microstructure. Numer. Math., **96**, 641–660 (2004)

[Bha03] Bhattacharya, K.: Microstructure of Martensite. Oxford University Press, New York (2003)

[But89] Buttazzo, G.: Semicontinuity, Relaxation, and Integral Representation in the Calculus of Variations. Pitman Research Notes in Mathematics Series, **207**, Longman Scientific & Technical, New York (1989)

[Car01] Carstensen, C.: Numerical analysis of microstructure. In: Blowey, J.F., Coleman, J.P., Craig, A.W. (eds) Theory and Numerics of Differential Equations, Durham 2000. Springer, New York (2001)

[Car06] Carstensen, C.: On the convergence of adaptive FEM for convex minimization problems. In preparation.

[CCO06] Carstensen, C., Conti, S., Orlando, A.: Mixed analytical-numerical relaxation in single-slip crystal plasticity. In preparation.

[CHM02] Carstensen, C., Hackl, K., Mielke, A.: A variational formulation of incremental problems in finite elasto–plasticity. Proc. R. Soc. Lond. A, **458**, 299–317 (2002)

[CHO06] Carstensen, C., Hu Jun, Orlando, A.: A convergent adaptive finite elmenet method for compatible phase transitions in elastic solids. In preparation.

[CJ03] Carstensen, C., Jochimsen K.: Adaptive finite element methods for microstructures? Numerical experiments for a two-well benchmark. Computing, **71**, 175–204 (2003)

[CP97] Carstensen, C., Plecháč, P.: Numerical solution of the scalar double–well problem allowing microstructure. Math. Comp., **66**, 997–1026 (1997)

[CP00] Carstensen, C., Plecháč, P.: Numerical analysis of compatible phase transitions in elastic solids. SIAM J. Numer. Anal., **37**, 2061–2081 (2000)

[CR98] Carstensen, C., Roubíček, T.: Numerical approximation of Young measures in nonconvex variational problems. Numer. Math., **84**, 395–415 (1998)

[Con03] Conti, S.: Relaxation of single-slip single-crystal plasticity with linear selfhardening. Preprint (2003)

[Con05] Conti, S.: Microstructure and relaxation in single-crystal plasticity. Oberwolfach Reports, **2**, 3000-3003 (2005)

[CDD02] Conti, S., DeSimone, A., Dolzmann, G.: Soft elastic response of stretched sheets of nematic elastomers: a numerical study. J. Mech. Phys. Solids, **50**, 1431—1451 (2002)

[CO05] Conti, S., Ortiz, M.: Dislocation microstructures and the effective behavior of single crystals. Arch. Rational Mech. Anal., **176**, 103–147 (2005)

[CT05] Conti, S., Theil, F.: Single-slip elastoplastic microstructures Arch. Rational Mech. Anal., **178**, 125–148 (2005)

[Dac89] Dacorogna, B.: Direct Methods in the Calculus of Variations. Springer, Berlin (1989)

[DSD02] DeSimone, A., Dolzmann, G.: Macroscopic response of nematic elastomers via relaxation of a class of $SO(3)$-invariant energies. Arch. Rat. Mech. Anal., **161**, 181–204 (2002)

[Dol99] Dolzmann, G.: Numerical computation of rank-one convex envelopes. SIAM J. Numer. Anal., **36**, 1621–1635 (1999)

[Dol03] Dolzmann, G.: Variational Methods for Crystalline Microstructure. Analysis and Computation. Lecture Notes in Mathematics, **1803**, Springer, Berlin (2003)

[DW00] Dolzmann, G., Walkington, N.J.: Estimates for numerical approximations of rank one convex envelopes. Numer. Math., **85**, 647–663 (2000)

[Dor96] Dörfler, W.: A convergent adaptive algorithm for Poisson's equation. SIAM J. Numer. Anal., **33**, 1106-1124 (1996)

[Fri94] Friesecke, G.: A necessary and sufficient condition for nonattainment and formation of microstructure almost everywhere in scalar variational problems. Proc. R. Soc. Edinburgh, **124A**, 437–471 (1994)

[HH02] Hackl, K., Hoppe, U.: On the calculation of microstructures for inelastic materials using relaxed energies. In: Miehe, C. (ed) IUTAM Symposium on Computational Mechanics of Solid Materials at Large Strains. Kluwer, Dordrecht (2002)

[HL93] Hiriart-Urruty, J.-B., Lemaréchal, C.: Fundamentals of Convex Analysis. Springer Verlag, Heidelberg (1993)

[KP91] Kinderlehrer, D., Pedregal, P.: Characterizations of Young measures generated by gradients. Arch. Rat. Mech. Anal., **115**, 329–365 (1991)

[Koh91] Kohn, R.V.: The relaxation of a double-well energy. Cont. Mech. Thermodynam., **3**, 193–236 (1991)

[KM94] Kohn, R.V., Müller, S.: Surface energy and microstructure in coherent phase transitions. Comm. Pure Appl. Math., **47**, 405–435 (1994)

[KMR05] Kružík, M., Mielke, A., Roubíček, T.: Modelling of microstructure and its evolution in shape-memory-alloy single-crystals, in particular in CuAlNi. Meccanica, **40**, 389–418 (2005)

[KS86] Kohn, R.V., Strang, G.: Optimal design and relaxation of variational problems I, II, III. Comm. Pure Appl. Math., **39**, 113–137, 139–182, 353–377 (1986)

[Kri99] Kristensen, J.: On the non-locality of quasiconvexity. Ann. Inst. H. Poincaré Anal. Non-Lin., **16** 1–13 (1999)

[Lus96] Luskin, M.: Computation of crystalline microstructure. Acta Numerica, **5**, 191–257 (1996)

[ML03] Miehe, C., Lambrecht, M.: Analysis of microstructure development in shearbands by energy relaxation of incremental stress potentials: large strain theory for standard dissipative solids. Int. J. Numer. Meth. Engng., **58**, 1–41 (2003)

[MLG04] Miehe, C., Lambrecht, M., Gürses, E.: Analysis of material instabilities in inelastic solids by incremental energy minimization and relaxation methods: evolving deformation microstructures in finite plasticity. J. Mech. Phy. Solids, **52**, 2725–2769 (2004)

[Mie03] Mielke, A.: Energetic formulation of multiplicative elasto–plasticity using dissipation distances. Cont. Mech. Thermodynamics, **15**, 351–382 (2003)

[Mie04a] Mielke, A.: Existence of minimizers in incremental elasto–plasticity with finite strains. SIAM J. Math. Analysis, **36**, 384-404 (2004)

[Mie04b] Mielke, A.: Deriving new evolution equations for microstructures via relaxation of variational incremental problems. Comput. Methods Appl. Mech. Engrg., **193**, 5095–5127 (2004)

[Mie05] Mielke, A.: Evolution in rate-independent systems (ch. 6). In: Dafermos, C., Feireisl, E. (eds) Handbook of Differential Equations: Evolutionary Equations. Vol 2, Elsevier, Amsterdam (2005), pp. 461–559

[Mor52] Morrey, C.B.: Quasi-convexity and the lower semicontinuity of multiple integrals. Pacific J. Math., **2**, 25-53 (1952)

[Mue99] Müller, S.: Variational models for microstructure and phase transitions. In: Hildebrandt, S., Struwe, M. (eds) Calculus of Variations and Geometric Evolution Problems. Lecture Notes in Mathematics, **1713**, Springer, Berlin (1999)

[NW93] Nicolaides, R.A., Walkington, N.: Computation of microstructure utilizing Young measure representations. J. Intelligent Material Systems Structures, **5**, 457–462 (1993)

[NW99] Nocedal, J., Wright, S.J.: Numerical Optimization. Springer, New York (1999)

[OR99] Ortiz, M., Repetto, E.A.: Nonconvex energy minimization and dislocation structures in ductile single crystals. J. Mech. Phys. Solids, **47**, 397–462 (1999)

[ORS00] Ortiz, M., Repetto, E.A., Stainier, L.: A theory of subgrain dislocation structures. J. Mech. Phys. Solids, **48**, 2077–2114 (2000)

[Ped95] Pedregal, P.: Numerical approximation of parametrized measures. Numer. Funct. Anal. and Optimiz., **16**, 1049–1066 (1995)

[Ped97] Pedregal, P.: Parametrized Measures and Variational Principles. Birkhäuser, Basel (1997)

[Rou96] Roubíček, T.: Relaxation in Optimization Theory and Variational Calculus. Walter de Gruyter, Berlin (1996)

[Rou96a] Roubíček, T.: Numerical approximation of relaxed variational problems. J. Convex Analysis, **3**, 329–347 (1996)

[You80] Young, L.C.: Lectures on the Calculus of Variations and Optimal Control Theory. AMS Chelsea Publishing, New York (1980)

Derivation of Elastic Theories for Thin Sheets and the Constraint of Incompressibility

Sergio Conti[1] and Georg Dolzmann[2]

[1] Fachbereich Mathematik, Universität Duisburg-Essen, Lotharstraße 65, 47057 Duisburg. conti@math.uni-duisburg.de
[2] Mathematics Department, University of Maryland, College Park, MD 20742-4015, U.S.A. dolzmann@math.umd.edu

Summary. We discuss the derivation of two-dimensional models for thin elastic sheets as Γ-limits of three-dimensional nonlinear elasticity. We briefly review recent results and present an extension of the derivation of a membrane theory, first obtained by LeDret and Raoult in 1993, to the case of incompressible materials. The main difficulty is the construction of a recovery sequence which satisfies pointwise the nonlinear constraint of incompressibility.

1 Introduction

The formulation and the study of reduced theories for the elastic properties of thin films is a traditional field of (heuristic) multiscale analysis, dating back to Euler. A rigorous derivation of such models from three-dimensional elasticity theory is instead a recent development, and is based on the mathematical tools of Γ-convergence, which permit passage to the limit for variational functionals.

The starting point is three-dimensional nonlinear elasticity. Precisely, let the cylindrical domain $\Omega_h = \omega \times (-h/2, h/2)$ represent the reference configuration of the sheet, where $\omega \subset \mathbb{R}^2$ is the cross-section and h the (small) height. The deformation field $u : \Omega_h \to \mathbb{R}^3$ is determined by minimizing the elastic energy, which after scaling with the film thickness h takes the form

$$E_h[u, f] = \frac{1}{h} \int_{\Omega_h} \big(W_{3\mathrm{D}}(\nabla u) - f_h \cdot u\big)\, \mathrm{d}x\,, \tag{1.1}$$

where f_h is a given vector field representing the applied forces. The deformation u can be additionally subject to boundary conditions on all or part of the lateral boundary $(\partial\omega) \times (-h/2, h/2)$. Different scalings of f_h with h lead to different scalings of the optimal energy, and correspondingly to different limiting functionals. In the relevant cases the term $f_h \cdot u$ turns out to be a continuous perturbation of the term $W_{3\mathrm{D}}(\nabla u)$.

The history of the theory of thin sheets and shells is paved with a number of simplifications of the general elasticity functional (1.1) aiming at a direct description of the behavior of the sheet as a two-dimensional object, without resolving explicitly the structure in the third dimension. Different assumptions on the energy density W_{3D} and on the scaling of the forces (and hence of the energy) lead to drastically different limiting functionals. Mathematically, for different energy scalings one deals with problems from relaxation, from variational convergence of singularly perturbed problems, and from the geometry of isometric immersions.

A number of heuristic derivations led to the formulation of the most relevant limiting theories in the field; the study was then made systematic by means of asymptotic expansions, which rely on a-priori assumptions on the structure or at least on the regularity of the minimizers, see [Cia97]. Only recently it has become possible to derive in a rigorous way two-dimensional reduced models from the variational problem of elasticity (1.1), by using the concept of Γ-convergence, as introduced by De Giorgi and his school [DGF75], see also [Dal93, Bra02]. We focus here on the derivation of a nonlinear membrane theory for incompressible materials, and generalize the result obtained for finite-valued energy densities by LeDret and Raoult [LR93, LR95] to the case where the nonlinear unit-determinant constraint is imposed. Essentially the same result was independently obtained by Trabelsi [Tra05, Tra06].

2 Rigorous derivation of elastic theories for thin films

2.1 General framework

Γ-convergence corresponds to convergence of energy along minimizing sequences for a family of functionals and all continuous perturbations [DGF75, Dal93, Bra02]. Precisely, a family of functionals $\mathcal{F}_\varepsilon : X \to \mathbb{R} \cup \{\infty\}$ Γ-converges to $\mathcal{F}_0 : X \to \mathbb{R} \cup \{\infty\}$ if the following properties hold:

(i) For every pair of sequences $\varepsilon_j \to 0$, $u_j \to u$, one has

$$\mathcal{F}_0[u] \leq \liminf_{j\to\infty} \mathcal{F}_{\varepsilon_j}[u_j] .$$

(ii) For every $u \in X$ and every sequence $\varepsilon_j \to 0$ there is a sequence $u_j \to u$ with

$$\mathcal{F}_0[u] = \lim_{j\to\infty} \mathcal{F}_{\varepsilon_j}[u_j] .$$

The notion of Γ-convergence depends on the topology of the space X (which in this discussion is assumed to be metrizable). The most natural choices are those for which the family $\mathcal{F}_\varepsilon$ is equicoercive, i.e.,

$$\text{for any } t \in \mathbb{R} \text{ the set } \bigcup_\varepsilon \{u \in X : \mathcal{F}_\varepsilon[u] \leq t\} \text{ is precompact.}$$

Indeed, if $\Gamma - \lim \mathcal{F}_\varepsilon = \mathcal{F}_0$ and the $\mathcal{F}_\varepsilon$ are equicoercive, then the set of minimizers of $\mathcal{F}_0$ coincides with the set of accumulation points of minimizing sequences of the family $\mathcal{F}_\varepsilon$. We also recall that functionals defined on subspaces $Y \subset X$ can be trivially extended setting them to be ∞ on $X \setminus Y$.

Appropriate notions for the convergence of a family of deformations $u_h : \Omega_h \to \mathbb{R}^3$ to a limit $v : \omega \to \mathbb{R}^3$ are best defined after rescaling to a single, fixed domain. We define the rescaling operator $\mathcal{T}_h$ by

$$(\mathcal{T}_h u)(x', x_3) = u(x', hx_3) \qquad \text{for } x' \in \omega \text{ and } x_3 \in \left(-\frac{1}{2}, \frac{1}{2}\right). \tag{2.1}$$

For the limiting map $v : \omega \to \mathbb{R}^3$ the rescaling reduces to $(\mathcal{T}_0 v)(x', x_3) = v(x')$. The relevant convergences are weak and strong convergence in $W^{1,p}(\Omega_1; \mathbb{R}^3)/\mathbb{R}^3$. The quotient means that an additive constant has been factored out, i.e., we say that $\mathcal{T}_h u_h \to \mathcal{T}_0 v$ in $W^{1,p}(\Omega_1; \mathbb{R}^3)/\mathbb{R}^3$ if there are $b_h \in \mathbb{R}^3$ such that $\mathcal{T}_h u_h - b_h \to \mathcal{T}_0 v$ in $W^{1,p}$ (weakly or strongly, respectively).

Given a Borel measurable function $W_{3\mathrm{D}} : \mathbb{M}^{3\times 3} \to [0, \infty]$, we shall consider the family of functionals $I_h : W^{1,p}(\Omega_h; \mathbb{R}^3) \to [0, \infty]$ defined by

$$I_h[u] = \frac{1}{h} \int_{\Omega_h} W_{3\mathrm{D}}(\nabla u) \,\mathrm{d}x \,. \tag{2.2}$$

2.2 Smooth energy densities

Γ-convergence of the functionals I_h to a membrane theory was obtained by LeDret and Raoult in 1993-1995 [LR93, LR95], for energy densities with p growth (see (2.6)). The limit functional $J : W^{1,p}(\omega; \mathbb{R}^3) \to [0, \infty)$ is

$$J[u] = \int_\omega W_{2\mathrm{D}}^{\mathrm{qc}}(\nabla u) \,\mathrm{d}x' \,. \tag{2.3}$$

The energy density $W_{2\mathrm{D}}^{\mathrm{qc}}$ is obtained by first optimizing over all possible directions of the normal gradient b, i.e., setting

$$W_{2\mathrm{D}}(F') = \inf \left\{ W_{3\mathrm{D}}(F'|b) : b \in \mathbb{R}^3 \right\}, \tag{2.4}$$

and then replacing $W_{2\mathrm{D}}$ by its quasiconvex envelope [Dac89, Mül99],

$$W_{2\mathrm{D}}^{\mathrm{qc}}(F') = \inf \left\{ \int_{(0,1)^2} W_{2\mathrm{D}}(F' + \nabla\psi) \,\mathrm{d}x' : \psi \in W_0^{1,\infty}((0,1)^2; \mathbb{R}^3) \right\}. \tag{2.5}$$

Theorem 2.1 (From [LR93, LR95]). *Let ω be a bounded and connected Lipschitz domain in $\mathbb{R}^2$, and suppose that $W_{3\mathrm{D}} : \mathbb{M}^{3\times 3} \to \mathbb{R}$ satisfies*

$$\frac{1}{c}|F|^p - c \le W_{3\mathrm{D}}(F) \le c|F|^p + c, \qquad \textit{for some } p \in (1, \infty). \tag{2.6}$$

Then the sequence $I_h \circ \mathcal{T}_h^{-1}$ is weak-$W^{1,p}(\Omega_1; \mathbb{R}^3)/\mathbb{R}^3$ equicoercive, and

$$\Gamma - \lim_{h\to 0} I_h \circ \mathcal{T}_h^{-1} = J \circ \mathcal{T}_0^{-1}$$

with respect to the weak-$W^{1,p}(\Omega_1;\mathbb{R}^3)$ topology.

The functional $J \circ \mathcal{T}_0^{-1}$ is defined by (2.3) on $\mathcal{T}_0 W^{1,p}(\omega;\mathbb{R}^3)$, and extended by ∞ to the rest of $W^{1,p}(\Omega_1;\mathbb{R}^3)$. By the compactness result, the statement is equivalent to Γ-convergence with respect to the strong L^1 topology.

If the energy tends to zero with h, one is interested in considering rescaled functionals, such as $h^{-\alpha} I_h$ for $\alpha > 0$. The limit can be finite only on the null set of J. To analyze this regime we assume that

$$W_{\mathrm{3D}}(QF) = W_{\mathrm{3D}}(F) \text{ for all } Q \in \mathrm{SO}(3) \text{ and all } F \in \mathbb{M}^{3\times 3}, \tag{2.7}$$

$$W_{\mathrm{3D}}(\mathrm{Id}) = 0, \qquad W_{\mathrm{3D}}(F) \geq C \mathrm{dist}^2(F, \mathrm{SO}(3)), \qquad \text{and} \tag{2.8}$$

$$W_{\mathrm{3D}} \text{ is } C^2 \text{ smooth in a neighborhood of Id.} \tag{2.9}$$

Then J is minimized by short maps, i.e., maps $v : \omega \to \mathbb{R}^3$ such that $|v(x') - v(y')| \leq |x' - y'|$ whenever the segment $[x', y']$ is contained in ω.

Theorem 2.2 (From [CM05]). *Let ω be a bounded Lipschitz domain in $\mathbb{R}^2$, $\alpha \in (0, 5/3)$, and let W_{3D} obey (2.7-2.9). Then*

$$\Gamma - \lim_{h\to 0} \frac{1}{h^\alpha} I_h \circ \mathcal{T}_h^{-1} = I_0 \circ \mathcal{T}^{-1}, \qquad I_0[v] = \begin{cases} 0 & \text{if } v \text{ is short,} \\ \infty & \text{else,} \end{cases}$$

with respect to the weak $W^{1,p}$ topology.

A different behavior is instead obtained if Dirichlet boundary conditions are imposed, both in the compressive case [BC*00, JS01, BC*02, JS02, Con03] and applying external forces to a membrane whose boundary is kept fixed [CMM06]. The latter work has in particular permitted to derive by Γ-convergence a relaxed version of the classical membrane theory by Föppl.

For $\alpha = 2$ one obtains a plate theory, which describes the bending-dominated regime. A rigorous derivation, confirming the expression first obtained by Kirchhoff in 1850, was recently obtained by Friesecke, James and Müller [FJM02] (see also [Pan03]).

Theorem 2.3 (From [FJM02]). *Let ω be a bounded Lipschitz domain in $\mathbb{R}^2$, and suppose that W_{3D} satisfies (2.7-2.9). Then the sequence $I_h \circ \mathcal{T}_h^{-1}$ is strong-$W^{1,2}(\Omega_1;\mathbb{R}^3)/\mathbb{R}^3$ equicoercive, and with respect to the same topology*

$$\Gamma - \lim_{h\to 0} \frac{1}{h^2} I_h \circ \mathcal{T}_h^{-1} = J_{\mathrm{plate}} \circ \mathcal{T}_0^{-1}.$$

The limit J_{plate} is defined by

$$J_{\mathrm{plate}}[v] = \begin{cases} \displaystyle\frac{1}{24} \int_\omega Q_2(\mathrm{II}_v)\,\mathrm{d}x' & \text{if } v \in W^{2,2}(\omega;\mathbb{R}^3), \text{ and } \nabla v \in \mathrm{O}(2,3) \text{ a.e.,} \\ \infty & \text{else,} \end{cases}$$

$\mathrm{II}_v = (\nabla v)^T \nabla(\partial_1 v \wedge \partial_2 v)$ *and*

$$Q_2(F'') = \inf\left\{Q_3(F'' + d \otimes e_3 + e_3 \otimes d) : d \in \mathbb{R}^3\right\}, \qquad Q_3 = \frac{\partial^2 W_{3\mathrm{D}}}{\partial F^2}(\mathrm{Id}).$$

2.3 Incompressible materials

The foregoing results assume that the free energy density is finite (or even smooth) on an open set in the space of deformation gradients. This assumption excludes the important class of rubber-like materials for which the energy density incorporates an incompressibility constraint, in the sense that the energy is infinite for all deformation gradients F that do not satisfy the nonlinear condition $\det F = 1$. This situation arises, for example, in the analysis of soft elasticity in thin sheets of liquid crystal elastomers [WT96, DD02, CDD02, WT03, ACD06].

We focus here on the derivation of a nonlinear membrane theory for incompressible materials. Theorem 2.1 concerns finite-valued energy densities; it was later extended by Ben Belgacem to energy densities which are infinite if the determinant of the deformation gradient is negative [Bel96b, Bel96a]. We characterize here the Γ-limit for incompressible materials and prove that the mechanism by which the limiting energy density is obtained is similar. It is a remarkable feature that the incompressibility constraint is lost in the Γ-limit. Essentially the same result was independently obtained by Trabelsi [Tra05, Tra06]. We give here a full proof, including a discussion of the relevant ideas from differential geometry. At variance with [Tra05, Tra06], we do not need to take the rank-one convex envelope of $W_{2\mathrm{D}}$ first.

Theorem 2.4. *Let ω be a bounded and connected Lipschitz domain in $\mathbb{R}^2$, and suppose that $W_{3\mathrm{D}} : \mathbb{M}^{3\times 3} \to \mathbb{R} \cup \{\infty\}$ has the form*

$$W_{3\mathrm{D}}(F) = \begin{cases} W_0(F) & \text{if } \det F = 1, \\ \infty & \text{else,} \end{cases} \tag{2.10}$$

where $W_0 : \mathbb{M}^{3\times 3} \to \mathbb{R}$ is continuous, nonnegative, and satisfies

$$\frac{1}{c}|F|^p - c \le W_0(F) \le c|F|^p + c, \qquad \text{for some } p \in (1, \infty). \tag{2.11}$$

Then the sequence $I_h \circ \mathcal{T}_h^{-1}$ is weak-$W^{1,p}(\Omega_1;\mathbb{R}^3)/\mathbb{R}^3$ equicoercive, and

$$\Gamma - \lim_{h\to 0} I_h \circ \mathcal{T}_h^{-1} = J \circ \mathcal{T}_0^{-1}$$

with respect to the weak-$W^{1,p}(\Omega_1;\mathbb{R}^3)$ topology. Equivalently:

(i) (Compactness) *For every sequence $h_j \to 0$, and every sequence $u_j : \Omega_{h_j} \to \mathbb{R}^3$ such that $I_{h_j}[u_j] < C < \infty$, there exists a $v \in W^{1,p}(\omega;\mathbb{R}^3)$ and a subsequence such that*

$$\mathcal{T}_{h_{j_k}} u_{j_k} - \frac{1}{|\omega|}\int_{\Omega_1} \mathcal{T}_{h_{j_k}} u_{j_k}\, \mathrm{d}x\, \mathrm{d}x_3 \rightharpoonup \mathcal{T}_0 v \qquad \text{in } W^{1,p}(\Omega_1;\mathbb{R}^3).$$

(ii) (Lower bound) *If additionally* $\mathcal{T}_{h_j} u_j \rightharpoonup \mathcal{T}_0 v$ *in* $W^{1,p}(\Omega_1;\mathbb{R}^3)$, *then*

$$\liminf_{j\to\infty} I_{h_j}[u_j] \geq J[v]\,.$$

(iii) (Upper bound) *For any* $u \in W^{1,p}(\omega;\mathbb{R}^3)$ *and any sequence* $h_j \to 0$ *there is a sequence* $u_j \in C^\infty(\Omega_{h_j};\mathbb{R}^3)$ *such that* $\mathcal{T}_{h_j} u_j \rightharpoonup \mathcal{T}_0 v$ *in* $W^{1,p}(\Omega_1;\mathbb{R}^3)$ *and*

$$\limsup_{j\to\infty} I_{h_j}[u_j] \leq J[v]\,.$$

The proof of this theorem involves two main parts: (i) the verification of the compactness statement together with the liminf inequality, and (ii) the construction of a recovery sequence. The definition of the limiting energy W^*_{2D} is based on the notion of quasiconvexity in the sense of Morrey, which naturally relates the lower semicontinuity arguments needed in part (i) with the explicit constructions needed in part (ii).

The main difficulty is here the construction of the recovery sequence for the upper bound. This requires several steps that are carried out in Sect. 4–6. The naïve idea is to try to extend a given function $u : \omega \to \mathbb{R}^3$ in such a way that the gradient of the extension u_h is approximately

$$F(x', x_3) = \left(\partial_1 u(x'), \partial_2 u(x'), \frac{\partial_1 u \wedge \partial_2 u}{|\partial_1 u \wedge \partial_2 u|^2}(x')\right).$$

This expression satisfies the condition $\det F(x, x_3) = 1$, however F is not, in general, a gradient field. This construction method can in fact be made precise if $|\partial_1 u \wedge \partial_2 u| \neq 0$ for all x in ω, see Sect. 5 and 6 for details. One crucial step in the construction of the recovery sequence is therefore to show the existence of a small perturbation of u in $W^{1,p}$ with the property that the wedge product does not vanish in ω. This is also a crucial step in the proof of Whitney's embedding theorem [Whi43, Whi44a, Whi44b], where Whitney proves the existence of an approximation in C^0. Indeed, Whitney's arguments can also be used in the situation at hand, and permit to obtain the desired $W^{1,p}$ approximation as well. We summarize the main steps of his argument in Sect. 4, for the special case of interest here, focussing on the few additional estimates and corrections needed for this application. Whitney's construction does not, however, provide a good energy estimate. In particular, in order to control the energy of the modified function we need a good estimate for the L^p norm of the inverse of the wedge product, i.e., for $1/|\partial_1 u \wedge \partial_2 u|$. This is achieved by means of local perturbations with smooth oscillations in the spirit of the work of Nash and Kuiper [Nas54, Kui55].

Also a plate theory can be obtained for incompressible materials. The following result was announced in [CD05], its proof will appear elsewhere.

Theorem 2.5. *Let* ω *be a bounded, convex Lipschitz domain in* $\mathbb{R}^2$, *let* W_{3D} *be as in (2.10), and* W_0 *obey (2.11). Let* $Q_3^{\rm inc} = \partial^2 W_0/\partial F^2(\mathrm{Id})$, *and*

$$Q_2^{\rm inc}(F'') = \inf \left\{ Q_3(F'' + d \otimes e_3 + e_3 \otimes d) : d \in \mathbb{R}^3 \,, \;\; \mathrm{Tr}(G|d) = 0 \right\} .$$

Then all assertions of Theorem 2.3 hold, with $J_{\rm plate}$ replaced by

$$J^{\rm inc}_{\rm plate}[v] = \begin{cases} \dfrac{1}{24} \displaystyle\int_\omega Q_2^{\rm inc}(\mathrm{II}_v)\,\mathrm{d}x' & \text{if } v \in W^{2,2}(\omega;\mathbb{R}^3)\,, \text{ and } \nabla v \in \mathrm{O}(2,3) \text{ a.e.,} \\ \infty & \text{else.} \end{cases}$$

3 Incompressible membranes: lower bound

We now start the proof of the equicoercivity and the lower bound. For simplicity we drop in the following the prime on 3×2 matrices. We first observe that due to the continuity and coercivity of W_0, (2.4) reduces to

$$W_{\rm 2D}(F) = \begin{cases} \min\limits_{b \in \mathbb{R}^3} W(F|b) & \text{if } \operatorname{rank} F = 2, \\ \infty & \text{else.} \end{cases} \tag{3.1}$$

The growth condition (2.11) implies, for all $F \in \mathbb{M}^{3\times 2}$ with $\operatorname{rank} F = 2$,

$$\frac{1}{c}|F|^p + \frac{1}{c}\frac{1}{|F_1 \wedge F_2|^p} - c \le W_{\rm 2D}(F) \le c|F|^p + c\frac{1}{|F_1 \wedge F_2|^p} + c\,. \tag{3.2}$$

We first study some properties of the relaxed energy $W_{\rm 2D}^*$. It is convenient to work with piecewise affine and continuous functions. We denote by P_c^1 the space of Lipschitz functions whose gradient is constant on each of at most countably many disjoint open triangles T_j, whose union is, up to a null set, equal to ω. The properties of $W_{\rm 2D}^*$ listed in the next lemma follow immediately using the results and the techniques of [BM84, Fon88, Dac89].

Lemma 3.1. *The function $W_{\rm 2D}^*$ has the following properties:*

(i) the definition does not depend on the choice of the domain ω, as long as it is open, bounded, and $|\partial\omega| = 0$;
(ii) for any $\delta > 0$ there is $\varphi \in P_c^1 \cap W_0^{1,p}(\omega;\mathbb{R}^3)$ such that

$$\frac{1}{|\omega|}\int_\omega W_{\rm 2D}(F + \nabla\varphi)\,\mathrm{d}x \le W_{\rm 2D}^*(F) + \delta \qquad \text{and} \qquad \int_\omega |\varphi|^p\,\mathrm{d}x \le \delta\,;$$

(iii) there is a constant c such that

$$\frac{1}{c}|F|^p - c \le W_{\rm 2D}^*(F) \le c|F|^p + c\,;$$

(iv) $W_{\rm 2D}^*$ is quasiconvex and Lipschitz continuous on bounded sets.

We are now ready to prove the compactness and the lower bound. This part of our result follows from the characterization given above and the results by LeDret and Raoult. For completeness we give a short self-contained proof.

Proof of Theorem 2.4 (i) and (ii). We work with the rescaled functions $U_j = \mathcal{T}_{h_j} u_j$. By the growth condition (2.11) we have

$$\int_{\Omega} |\partial_1 U_j|^p + |\partial_2 U_j|^p + \frac{1}{h_j^p} |\partial_3 U_j|^p \, \mathrm{d}x \, \mathrm{d}x_3 \leq c \,. \tag{3.3}$$

After subtracting the average value, the sequence U_j is uniformly bounded in $W^{1,p}(\Omega; \mathbb{R}^3)$. We may choose a subsequence (not relabeled) that converges weakly in $W^{1,p}$. Further, $\partial_3 U_j \to 0$ strongly in L^p by (3.3), hence the limit does not depend on x_3. This concludes the proof of the compactness part.

The lower bound is proved by lower semicontinuity (see [LR95]). Consider the function $Z : \mathbb{M}^{3\times 3} \to \mathbb{R}$ defined by $Z(F) = W_{2\mathrm{D}}^*(\widehat{F})$ where $\widehat{F} = (F_1|F_2)$. By Lemma 3.1 this function is quasiconvex, continuous, and has p-growth from above. Therefore the functional $\widetilde{I}[u] = \int_\Omega Z(\nabla u) \, \mathrm{d}x \, \mathrm{d}x_3$ is lower semicontinous with respect to weak $W^{1,p}$ convergence [AF84]. Since $W(F) \geq W_{2\mathrm{D}}(\widehat{F}) \geq W_{2\mathrm{D}}^*(\widehat{F}) = Z(F)$ we obtain

$$\begin{aligned} \liminf_{j\to\infty} \int_{\Omega} W \left(\partial_1 U_j | \partial_2 U_j | \frac{1}{h_j} \partial_3 U_j \right) \mathrm{d}x \, \mathrm{d}x_3 \\ \geq \int_{\Omega} Z \left(\partial_1 V | \partial_2 V \right) \mathrm{d}x \, \mathrm{d}x_3 = \int_{\omega} W_{2\mathrm{D}}^* \left(\partial_1 v | \partial_2 v \right) \mathrm{d}x \,, \end{aligned}$$

with $V = \mathcal{T}_0 v$. This concludes the proof of the liminf inequality.

4 Approximation with regular maps

The key ingredient in the construction of a recovery sequence is an approximation result for functions in the Sobolev space $W^{1,p}(\omega; \mathbb{R}^3)$ by smooth and regular functions, i.e., by smooth functions u such that $\mathrm{rank}(\nabla u) = 2$ everywhere, as discussed above. This approximation theorem follows essentially from Whitney's [Whi43, Whi44a, Whi44b] work on the nature of singularities of maps $u : \mathbb{R}^n \to \mathbb{R}^{2n-1}$. Moreover, it is crucial to ensure the lower bound (4.2) below. We shall obtain it by using methods developed by Nash [Nas54] and Kuiper [Kui55]. For the convenience of the reader we sketch below the construction for the two-dimensional case of interest here.

The statement in the following proposition shows (i) that any smooth map $u : \omega \to \mathbb{R}^3$ can be uniformly approximated by regular maps, i.e., maps which fulfill $\mathrm{rank}\, \nabla u = 2$ everywhere; (ii) that if the map u is regular away from a connected set $\Gamma \subset \omega$ containing $\partial\omega$, then the regular maps can be taken to differ from u only in a neighbourhood of Γ, and (iii) that in a (smaller) neighbourhood of Γ the stronger condition $|\partial_1 u \wedge \partial_2 u| > c$ can be enforced, for some universal constant c (i.e. the approximating maps are order one away from being degenerate).

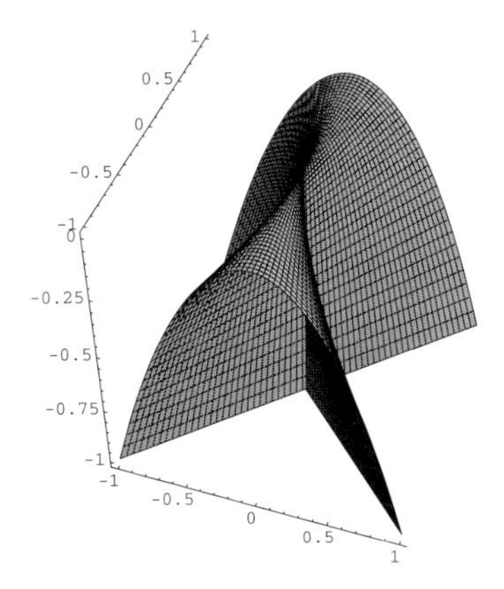

Fig. 4.1. The fundamental singularity of mappings from $\mathbb{R}^2$ into $\mathbb{R}^3$: "Whitney's umbrella", given by the equations in (4.4) (up to a rigid rotation).

Proposition 4.1. *Let $\omega \subset \mathbb{R}^2$ be a Lipschitz domain, $\Gamma \subset \bar{\omega}$ a connected set which contains $\partial\omega$, and define for $\eta > 0$ the set $\Gamma_\eta = \{x \in \bar{\omega} : \operatorname{dist}(x, \Gamma) < \eta\}$. If $u \in C^\infty(\bar{\omega}; \mathbb{R}^3)$ and $\operatorname{rank} \nabla u = 2$ on $\omega \setminus \Gamma_\eta$, then for any $\delta > 0$ there exists a $w \in C^\infty(\bar{\omega}; \mathbb{R}^3)$ such that $\|u - w\|_{C^0(\omega)} \leq \delta$,*

$$|\nabla w| \leq c\,(|\nabla u| + 1)\,, \qquad |\partial_1 w \wedge \partial_2 w| \geq c|\partial_1 u \wedge \partial_2 u| \quad \text{on } \omega \tag{4.1}$$

and

$$w = u \text{ on } \bar{\omega} \setminus \Gamma_{2\eta}\,, \qquad |\partial_1 w \wedge \partial_2 w| \geq c \text{ on } \Gamma_\eta\,. \tag{4.2}$$

All constants are absolute constants.

Proof. We split the argument into several steps in which we use small perturbations to show that we can accomplish the following goals: 1. The rank of ∇u is greater than or equal to one everywhere. 2. The rank of ∇u is equal to two everywhere except at finitely many points which correspond to the canonical singularity of mappings from $\mathbb{R}^2$ into $\mathbb{R}^3$, called "Whitney umbrella". This means that for each $p \in \omega$ where ∇u does not have full rank, there is a direction $e = e(p) \in S^1$ with

$$\nabla_e u = 0\,, \qquad \operatorname{span}\left(\nabla_{e^\perp} u, \nabla^2_{ee^\perp} u, \nabla^2_{ee} u\right) = \mathbb{R}^3 \tag{4.3}$$

(here $\nabla_e u = \nabla u \cdot e$, and so on). Equivalently, u can be locally represented, after a smooth change of coordinates in the domain and in the target, as

$$(x_1, x_2) \to (x_1^2, x_2, x_1 x_2) \tag{4.4}$$

(the equivalence is shown in [Whi43, Th. 1]). 3. The rank of ∇u is equal to two everywhere. 4. The lower bound on the angle between $\partial_1 u$ and $\partial_2 u$ (see

(4.2)) holds. The Steps 1.-3. follow Whitney's original work, and the final step is based on ideas of Nash and Kuiper.

Step 1: For all $\delta > 0$ there exists a smooth function u_δ such that $u = u_\delta$ on $\omega \setminus \Gamma_{2\eta}$, $|\nabla u_\delta - \nabla u| \leq \delta$, $|\partial_1 u_\delta \wedge \partial_2 u_\delta| \geq \frac{1}{2}|\partial_1 u \wedge \partial_2 u|$ on $\Gamma_{2\eta} \setminus \Gamma_\eta$ and $\partial_2 u_\delta \neq 0$ everywhere.

Let ψ_η be a smooth cutoff function supported in $\Gamma_{2\eta}$ and such that $\psi_\eta = 1$ on Γ_η. We define, for $\alpha \in \mathbb{R}^3$, the function $u^{(\alpha)}(x) = u(x) + \psi_\eta(x) x_2\, \alpha$. For α sufficiently small we get $|\nabla(\psi_\eta(x)\alpha x_2)| \leq \delta$. We assert that we can choose α arbitrarily small so that $u^{(\alpha)}_{,2}(x) \neq 0$ everywhere, and $\operatorname{rank} \nabla u^{(\alpha)} = 2$ outside Γ_η. Consider first points in Γ_η, where $u^{(\alpha)}_{,2} = u_{,2} + \alpha$. The set $u_{,2}(\bar{\omega}) \subset \mathbb{R}^3$ has dimension two, hence there are arbitrarily small α such that $-\alpha \notin u_{,2}(\bar{\omega})$, or equivalently $u_{,2}(x) + \alpha \neq 0$ for all $x \in \bar{\omega}$. Now consider points in the compact set $\bar{\Gamma}_{2\eta} \setminus \Gamma_\eta = \bar{\Gamma}_{2\eta} \setminus \{x \in \mathbb{R}^2 : \operatorname{dist}(x, \Gamma) < \eta\}$. Here

$$\nabla u^{(\alpha)}(x) = \nabla u(x) + \psi_\eta(x) \alpha \otimes e_2 + x_2\, \alpha \otimes \nabla \psi_\eta\,,$$

hence the distance to ∇u is controlled by α times constants depending on Γ and η. Since $\operatorname{rank} \nabla u = 2$ on $\bar{\Gamma}_{2\eta} \setminus \Gamma_\eta$, the function $|\partial_1 u \wedge \partial_2 u|$ has a positive minimum on the same set, hence for $|\alpha|$ sufficiently small we have $|\partial_1 u^{(\alpha)} \wedge \partial_2 u^{(\alpha)}| \geq \frac{1}{2}|\partial_1 u \wedge \partial_2 u|$. Finally, outside of $\Gamma_{2\eta}$ the two functions coincide, hence $u_\delta = u^{(\alpha)}$ has the desired properties for $|\alpha|$ small enough. For simplicity we write in the following u for u_δ.

Step 2: Suppose that u is a function with the properties established in Step 1. Then for all $\delta > 0$ there exists a smooth function u_δ such that $u = u_\delta$ on $\omega \setminus \Gamma_{2\eta}$, $|\nabla u_\delta - \nabla u| \leq \delta$ on ω, $|\partial_1 u_\delta \wedge \partial_2 u_\delta| \geq \frac{1}{4}|\partial_1 u \wedge \partial_2 u|$ on $\Gamma_{2\eta} \setminus \Gamma_\eta$, and $\operatorname{rank}(\nabla u_\delta) = 2$ except at finitely many points in Γ_η which are of the Whitney umbrella type (4.3).

For $\beta, \gamma \in \mathbb{M}^{3\times 2}$ and ψ_η as in Step 1, let $u_{\beta,\gamma}(x) = u(x) + \psi_\eta(x)\,[\beta x + x_1 \gamma x]$. The fact that x_1 appears explicitly in the perturbation is related to the choice of having $u_{,2} \neq 0$ in Step 1. As above, the function u has been modified only inside $\Gamma_{2\eta}$. For sufficiently small β and γ the gradient of the correction is uniformly small, hence the bounds for $|\nabla u_\delta - \nabla u|$ and $|\partial_1 u_\delta \wedge \partial_2 u_\delta|$ follow. At the same time the set of $(\beta, \gamma) \in \mathbb{R}^{12}$ for which $u_{\beta,\gamma}$ has (in Γ_η) singular points not of the type (4.3) has dimension at most 11 [Whi43, Th. 2]. Therefore one can choose arbitrarily small β and γ so that all singular points of $u_{\beta,\gamma}$ are of that class. Finally, singularities of the type (4.3) are necessarily isolated, and therefore $u_{\beta,\gamma}$ can have at most finitely many singularities in the compact set $\bar{\omega}$. As before, we are going to call $u_{\beta,\gamma}$ again u.

Step 3: Suppose that u is a function with the properties established in Step 2. Then for all $\delta > 0$ there exists a smooth function u_δ such that $u = u_\delta$ on $\omega \setminus \Gamma_{2\eta}$, $|\nabla u_\delta| \leq C(|\nabla u| + 1)$ and $|u - u_\delta| \leq \delta$ on $\bar{\omega}$, $|\partial_1 u_\delta \wedge \partial_2 u_\delta| \geq \frac{1}{2}|\partial_1 u \wedge \partial_2 u|$ on $\Gamma_{2\eta} \setminus \Gamma_\eta$, and $\operatorname{rank}(\nabla u_\delta) = 2$ everywhere.

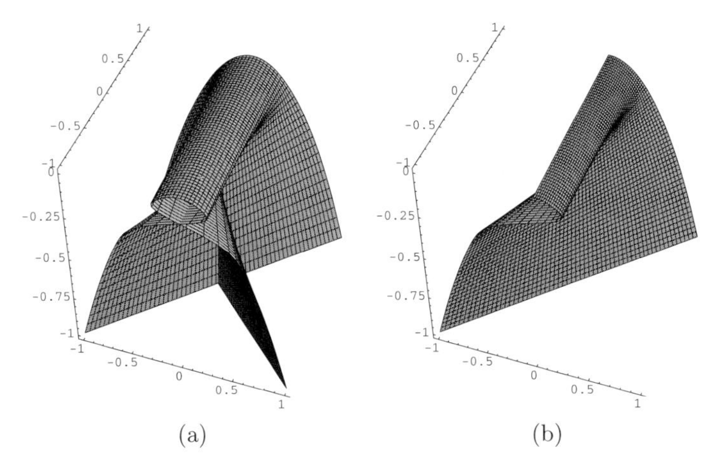

Fig. 4.2. (a) Modification of the map in Figure 4.1 in order to remove the singularity. (b) Plot of one half of surface in (a). This view illustrates that the modification is a smooth surface in $\mathbb{R}^3$ for which the tangent plane in each point has full rank. (See page 693 for a colored version of the figure.)

This step cannot be accomplished with purely local perturbations. Indeed, Whitney derived a sum rule which relates the total charge of the singularities to the behavior of the map at the boundary [Whi44b]. This implies that they cannot be eliminated without modifying the boundary values. The proof of Whitney's embedding builds upon the fact that singularities can be moved to the boundary by means of C^0 perturbations of the given map (this is, however, in general impossible with C^1 perturbations). We sketch the construction and show in particular that one does not need to modify the map outside $\Gamma_{2\eta}$, and that the stated bounds can be satisfied.

Let $p_1, \ldots, p_k \in \Gamma_\eta$ be the finitely many points where ∇u has rank one. We first assert that there are smooth arcs A_i of finite length in Γ_η which are closed, disjoint, and such that A_i starts in x_i and ends on $\partial\omega$. To construct these arcs, we first choose k distinct points $q_1, \ldots, q_k \in \partial\omega$. Each p_i can be connected to q_i by an arc of finite length contained in Γ_η. Indeed, Γ is connected and can be covered by finitely many balls of radius η; we can therefore choose an arc that has length at most 3η inside each of the balls and therefore finite total length. By a standard perturbation argument the arcs can be modified (the pairings of the points p_i and q_j may also change) so that the new arcs are disjoint, and the assertion is proven. It follows that the distance $\mathrm{dist}(A_i, A_j)$, $i \neq j$, is bounded from below by a constant $\varepsilon > 0$, and we can also assume $\varepsilon < \eta$.

In a neighbourhood of each x_i the map u is, in suitable coordinates, of the form (4.3) (see Fig. 4.1). We modify the surface as illustrated in Fig. 4.2, inserting a thin cylinder which (in the original coordinates) follows the path A_i from the singularity to the boundary of ω. If each of these corrections is supported on an $\varepsilon/3$-neighbourhood of the arc A_i then the ones constructed for different singularities do not interfere, and do not change u outside of $\Gamma_{2\eta}$. Notice that this construction may generate large curvature in the cylinders, but the increase in C^1 norm is bounded. From now on we denote by u the function obtained after these modifications.

Step 4: Suppose that u is a function with the properties established in Step 3. For any $\delta > 0$, there is a smooth function w such that

$$|\partial_1 w \wedge \partial_2 w| > \frac{1}{2} \text{ on } \Gamma_\eta, \qquad |\partial_1 w \wedge \partial_2 w| \geq \frac{1}{8}|\partial_1 u \wedge \partial_2 u| \text{ on } \Gamma_{2\eta} \setminus \Gamma_\eta$$

with $w = u$ on $\omega \setminus \Gamma_{2\eta}$; $|\nabla w| \leq C(|\nabla u| + 1)$ and $|w - u| \leq \delta$ on $\bar{\omega}$.

In this step we need to modify u so that $|w_{,1} \wedge w_{,2}|$ has a lower bound of order one. We use hereafter for brevity the notation $w_{,i} = \partial_i w$. Our approach follows that used by Nash and Kuiper [Nas54, Kui55], with the difference that here we only need two iteration steps. The basic idea is to insert locally oscillations to increase the appropriate components of the gradient.

We first sketch the construction for the simpler case of one spatial dimension. The corresponding statement is that any smooth parameterized curve $\gamma : [0,1] \to \mathbb{R}^2$ with $\gamma' \neq 0$ can be modified so that $|\gamma'| \geq 1$. The key idea is to insert fine-scale oscillations normal to γ', and to multiply them by a cutoff function to localize the correction. Precisely, one writes $\gamma_\varepsilon(t) = \gamma(t) + \varepsilon\psi(t)\Phi(t/\varepsilon)$, with Φ a periodic oscillating vector field, and ψ a smooth cutoff function, which is equal to one on the regions where γ' is small. Taking the derivative one sees that the term $\psi\Phi'$ is of order one, whereas $\varepsilon\psi'\Phi$ is of order ε, hence negligible for small ε.

Assume now that γ parameterizes a line segment, $\gamma(t) = (\gamma_1(t), 0)$, with $\gamma_1' > 0$. This is the generic case, if one works in a coordinate system that moves with the curve (here one needs $\gamma' \neq 0$, and additional smoothness of γ). Since the normal component Φ_2 oscillates on a scale ε, its derivative is often zero. Hence a uniform control on $|\gamma_\varepsilon'|$ from below cannot be gained via the normal derivative alone. We therefore seek a bound from below for the other component, $|\gamma_1' + \psi\Phi_1'|$. In the region where γ_1' is large, $\psi = 0$ and there is no correction. In the region where γ_1' is small, $\psi = 1$ and the term $\psi\Phi_1'$ dominates. Hence there are points where the two terms in $|\gamma_1' + \psi\Phi_1'|$ are equal, and one needs to ensure that no cancellation can occur.

The argument above illustrates that one cannot use simple oscillations, like $\Phi(t) = (\sin t, \cos t)$, because the map $t \to \Phi'(t)$ would move around the unit circle without any sign condition. One resorts instead to a more elaborated oscillation pattern, where the map $t \to \Phi'(t)$ does not move around a circle, but rather oscillates forward and backward on an arc of a circle, with non-uniform speed so that its average vanishes (and the primitive Φ can be periodic). The precise definition of $\varphi = \Phi'$ is given in (4.5) below.

We now give the proof for the two-dimensional case. Our starting point is the map $u \in C^\infty(\bar{\omega}; \mathbb{R}^3)$ constructed in Step 3. Since $u_{,1} \wedge u_{,2} \neq 0$ everywhere,

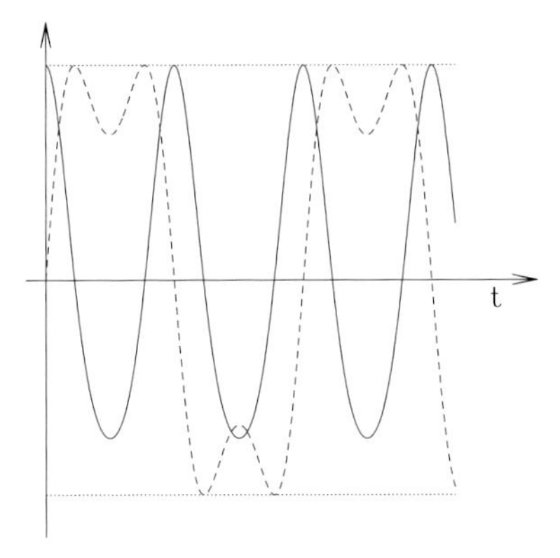

Fig. 4.3. The curves $\varphi_1(t)$ (full curve) and $\varphi_2(t)$ (dashed curve) given in (4.5), as functions of t. Note that φ_1 becomes negative only in a region where φ_2 is well separated from zero. The arc covered by the point (φ_1, φ_2) is illustrated in Figure 4.4.

there exists a $\rho > 0$ such that $|u_{,1} \wedge u_{,2}| \geq \rho > 0$ everywhere, and the normal vector $\nu = u_{,1} \wedge u_{,2}/|u_{,1} \wedge u_{,2}|$ is smooth on $\bar{\omega}$. Let ψ_η be again a smooth, nonnegative cutoff function such that $\psi_\eta = 1$ on Γ_η and $\psi_\eta = 0$ outside $\Gamma_{2\eta}$. We set, for $\varepsilon > 0$ to be chosen later,

$$v(x) = u(x) + \varepsilon\psi_\eta(x)\left[\Phi_1\left(\frac{x_1}{\varepsilon}\right)\mu + \Phi_2\left(\frac{x_1}{\varepsilon}\right)\nu\right].$$

Here $\mu = u_{,1}/|u_{,1}|$ and $\Phi_{1,2}$ are 2π-periodic primitives of

$$\varphi_1(t) = \frac{\cos(\alpha \sin t)}{\sin \alpha}, \qquad \varphi_2(t) = \frac{\sin(\alpha \sin t)}{\sin \alpha} \tag{4.5}$$

with $\Phi_{1,2}(0) = 0$. The number $\alpha \in (0, \pi)$ is chosen so that φ_1 has average zero over its period $(0, 2\pi)$ (the existence of α follows from the mean-value theorem; numerically, $\alpha \simeq 2.4$, see also Fig. 4.3). We find that

$$\nabla v = \nabla u + (\psi_\eta\varphi_1\mu + \psi_\eta\varphi_2\nu) \otimes e_1 + \mathcal{O}(\varepsilon). \tag{4.6}$$

Here and below, we denote by $\mathcal{O}(\varepsilon)$ quantities which are uniformly bounded by $C_{u,\eta}\varepsilon$, where we write $C_{u,\eta}$ for generic constants which can depend on u (specifically, on $\|u\|_{C^2}$ and $\min|u_{,1} \wedge u_{,2}|$), on η, on Γ and on ω, but not on ε.

Consider now the derivative $v_{,1}$. Up to $\mathcal{O}(\varepsilon)$ terms, it coincides with

$$z(x,t) = (|u_{,1}|(x) + \psi_\eta(x)\varphi_1(t))\,\mu(x) + \psi_\eta(x)\varphi_2(t)\nu(x) \tag{4.7}$$

evaluated at $t = x_1/\varepsilon$. Considering z as a function of t for fixed x, we see that

$$|z| \geq \max(\psi_\eta, |u_{,1}| \sin \alpha) \tag{4.8}$$

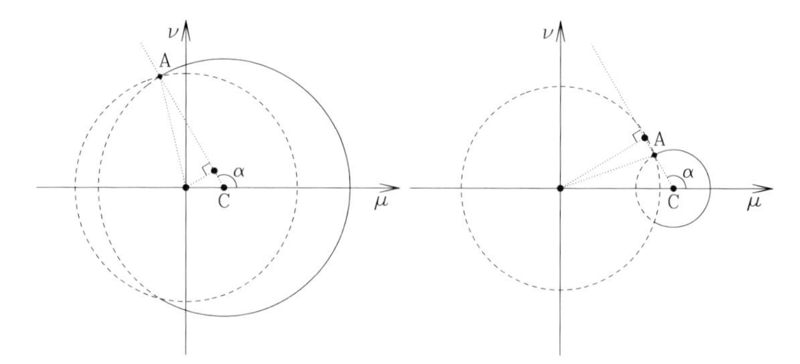

Fig. 4.4. For fixed x, the values of $t \to z(x,t)$ (see (4.7)) are located on an arc of a circle in the μ, ν plane. The radius is $\psi/\sin\alpha$, the center $C = (m, 0)$, where $m = |u_{,1}| \geq 0$. The closest point to the origin is $A = (m - \psi \cot\alpha, \psi)$ (since two distinct circles intersect in at most two points), and it corresponds to $t = \pi/2$. Its distance $|A|$ from the origin $O = (0,0)$ is greater than or equal to the distance d of O to the orthogonal projection of O onto the line through A and C. This distance is given by $d = m\sin(\pi - \alpha) = m \sin\alpha$. At the same time, $|A| \geq |A_2| = \psi$.

(see Fig. 4.4). In particular, in the region where $\psi_\eta = 1$ (covering Γ_η) $|v_{,1}| \geq 1 + \mathcal{O}(\varepsilon)$, whereas on the rest $|v_{,1}| \geq (\sin\alpha)|u_{,1}| + \mathcal{O}(\varepsilon)$. We finally verify that the wedge product $v_{,1} \wedge v_{,2}$ does not vanish (the uniform lower bound on its length will be obtained only after superposition of additional oscillations in the next step). A direct computation shows that

$$v_{,1} \wedge v_{,2} = |u_{,1} \wedge u_{,2}| \left(1 + \frac{\psi_\eta}{|u_{,1}|}\varphi_1\right) \nu + |u_{,2}|\psi_\eta \varphi_2 \, \mu' + \mathcal{O}(\varepsilon)$$

where $\mu' = \nu \wedge u_{,2}/|u_{,2}|$. The two unit vectors ν and μ' are orthogonal. Further, $|u_{,1} \wedge u_{,2}|/|u_{,1}| \leq |u_{,2}|$. Therefore

$$|v_{,1} \wedge v_{,2}| \geq |u_{,1} \wedge u_{,2}| \left| \left(1 + \frac{\psi_\eta}{|u_{,1}|}\varphi_1\right) \nu + \frac{\psi_\eta}{|u_{,1}|}\varphi_2 \, \mu' \right| + \mathcal{O}(\varepsilon) .$$

Reasoning as in (4.7-4.8) we get

$$|v_{,1} \wedge v_{,2}| \geq |u_{,1} \wedge u_{,2}| \max\left(\frac{\psi_\eta}{|u_{,1}|}, \sin\alpha\right) + \mathcal{O}(\varepsilon) .$$

Now we choose $\varepsilon > 0$ (depending on $C_{u,\eta}$) so small that all the $\mathcal{O}(\varepsilon)$ terms in the foregoing equations are less than $1/4$ of the leading-order terms. The function v defined in this way belongs to $C^2(\bar{\omega})$ and satisfies $\|v\|_{C^2} \leq C_{u,\eta}$ and $|v_{,1} \wedge v_{,2}| \geq |u_{,1} \wedge u_{,2}|/2$. Moreover, $|v_{,1}| \geq 3/4$ in Γ_η.

We finally insert oscillations in the orthogonal direction by defining

$$w(x) = v(x) + \widetilde{\varepsilon}\psi_\eta(x) \left[\Phi_1\left(\frac{x_2}{\widetilde{\varepsilon}}\right)\widetilde{\mu} + \Phi_2\left(\frac{x_2}{\widetilde{\varepsilon}}\right)\widetilde{\nu}\right] .$$

Here $\Phi_{1,2}$ are the same functions as above, and

$$\widetilde{\nu} = \frac{v_{,1} \wedge v_{,2}}{|v_{,1} \wedge v_{,2}|}, \qquad \widetilde{\mu} = \frac{\widetilde{\nu} \wedge v_{,1}}{|\widetilde{\nu} \wedge v_{,1}|} = \frac{\widetilde{\nu} \wedge v_{,1}}{|v_{,1}|}.$$

Note that $(v_{,1}/|v_{,1}|, \widetilde{\nu}, \widetilde{\mu})$ form an orthonormal basis. We find

$$\nabla w = \nabla v + \psi_\eta \left(\varphi_1 \widetilde{\mu} + \varphi_2 \widetilde{\nu}\right) \otimes e_2 + \mathcal{O}(\widetilde{\varepsilon}), \tag{4.9}$$

and by the definition of the basis vectors,

$$\begin{aligned} w_{,1} \wedge w_{,2} &= v_{,1} \wedge \left(v_{,2} + \psi_\eta \left(\varphi_1 \widetilde{\mu} + \varphi_2 \widetilde{\nu}\right)\right) + \mathcal{O}(\widetilde{\varepsilon}) \\ &= v_{,1} \wedge \left((\widetilde{\mu} \cdot v_{,2} + \psi_\eta \varphi_1)\widetilde{\mu} + \psi_\eta \varphi_2 \widetilde{\nu}\right) + \mathcal{O}(\widetilde{\varepsilon}) \\ &= (\widetilde{\mu} \cdot v_{,2} + \psi_\eta \varphi_1)\, v_{,1} \wedge \widetilde{\mu} + \psi_\eta \varphi_2\, v_{,1} \wedge \widetilde{\nu} + \mathcal{O}(\widetilde{\varepsilon}). \end{aligned}$$

The vectors $v_{,1} \wedge \widetilde{\mu}$ and $v_{,1} \wedge \widetilde{\nu}$ are orthogonal, both have length $|v_{,1}|$, and

$$\widetilde{\mu} \cdot v_{,2} = \frac{(\widetilde{\nu} \wedge v_{,1}) \cdot v_{,2}}{|v_{,1}|} = \frac{(v_{,1} \wedge v_{,2}) \cdot \widetilde{\nu}}{|v_{,1}|} = \frac{|v_{,1} \wedge v_{,2}|^2}{|v_{,1}|\,|v_{,1} \wedge v_{,2}|} = \frac{|v_{,1} \wedge v_{,2}|}{|v_{,1}|}.$$

Reasoning as in (4.7-4.8) above (see also Fig. 4.4) we obtain

$$|w_{,1} \wedge w_{,2}| \geq |v_{,1}| \max\left(\psi_\eta, \frac{|v_{,1} \wedge v_{,2}|}{|v_{,1}|} \sin\alpha\right) + \mathcal{O}(\widetilde{\varepsilon}).$$

We choose $\widetilde{\varepsilon}$ so that the last correction is smaller than

$$\min\left(\frac{1}{6}, \frac{1}{2} \min_{\bar{\omega}} |v_{,1} \wedge v_{,2}| \sin\alpha\right).$$

On Γ_η we get $|w_{,1} \wedge w_{,2}| \geq |v_{,1}| - 1/6 \geq 1/2$, and globally

$$|w_{,1} \wedge w_{,2}| \geq |v_{,1} \wedge v_{,2}| \sin\alpha + \mathcal{O}(\widetilde{\varepsilon}) \geq \frac{1}{2} |v_{,1} \wedge v_{,2}| \sin\alpha.$$

The estimate for $|\nabla w|$ follows from (4.6) and (4.9) and concludes the proof.

5 An extension subject to the incompressibility constraint

The next important step in the construction of a recovery sequence is the extension of a regular map $w : \omega \to \mathbb{R}^3$ to a map $u_h : \Omega_h \to \mathbb{R}^3$ with $\det \nabla u_h = 1$ everywhere. In view of the previous results we may assume that $\operatorname{rank}(\nabla w) = 2$ everywhere. Heuristically, one can think of extending w by setting $u(x, x_3) = w(x) + x_3 \nu(x)$, where

$$\nu = \frac{\partial_1 w \wedge \partial_2 w}{|\partial_1 w \wedge \partial_2 w|^2} \tag{5.1}$$

(ν here does not represent a unit vector, but rather a scaled normal). In the case that w is itself affine, the deformation gradient turns out to be $\nabla u = (\nabla w, \nu)$, which has determinant one. However, if the function w is not affine one obtains additional terms from the x_1 and x_2 derivatives of ν. These will be compensated for via a nonlinear change of coordinates.

Consider now the control of the energy. We focus here on the situation that $W_{2\mathrm{D}}(\nabla w)$ is small, and seek u_h so that $W_{3\mathrm{D}}(\nabla u_h)$ is correspondingly small. The additional step of passing from $W_{2\mathrm{D}}^*$ to $W_{2\mathrm{D}}$ will be dealt with in Lemma 6.2 below. The definition of $W_{2\mathrm{D}}$ in terms of $W_{3\mathrm{D}}$ shows that, at least on large parts of the domain, the x_3 derivative of u_h should not be ν, but rather b as obtained from the minimization in (3.1). This b depends in a discontinuous way on ∇w (generically), hence a suitable smoothing-interpolation scheme will be needed (see Proof of Theorem 2.4(iii) in Sect. 6). Therefore in the next proposition we assume only that ν satisfies the constraint

$$(\partial_1 w \wedge \partial_2 w) \cdot \nu = 1 \quad \text{in } \omega\,. \tag{5.2}$$

Proposition 5.1. *Let $w, \nu \in C^\infty(\bar\omega; \mathbb{R}^3)$ satisfy (5.2) on $\bar\omega$. Then there exist an $h_0 > 0$ and an extension $v \in C^\infty(\bar\omega \times (-h_0, h_0); \mathbb{R}^3)$ such that $v(x, 0) = w(x)$ and* $\det \nabla v = 1$ *everywhere. Moreover, for all $x_3 \in (-h_0, h_0)$ the pointwise bound*

$$|\nabla v - (\nabla w|\nu)| \le C x_3 \tag{5.3}$$

holds, where C can depend on w and ν.

Remark: If we define v_h to be the restriction of v to Ω_h, for $h < h_0$, then (5.3) implies that $\mathcal{T}_h v_h \to \mathcal{T}_0 w$ strongly in $W^{1,p}(\Omega; \mathbb{R}^3)$ as $h \to 0$.

Proof. We define $u \in C^\infty(\bar\Omega; \mathbb{R}^3)$ by $u(x, x_3) = w(x) + x_3 \nu(x)$ and compute

$$\det \nabla u = \det\left(\nabla w + x_3 \nabla \nu \,|\nu\right) = 1 + x_3 P(x) + x_3^2 Q(x) \tag{5.4}$$

where P and Q are combinations of the components of ν, $\nabla\nu$, and ∇w, which are all uniformly bounded in $\bar\omega$. Hence for small x_3 the determinant is close to one, and in particular $|\det \nabla u - 1| \le C x_3$ (C depends on w and ν). We define $v = u \circ \Phi$, with $\Phi(x, x_3) = (x, \varphi(x, x_3))$, and construct φ so that $\det \nabla v = (\det \nabla u \circ \Phi)\, \partial_3 \varphi = 1$. Therefore φ has to be a solution of the ODE

$$\partial_3 \varphi(x, x_3) = \frac{1}{\det \nabla u(x, \varphi(x, x_3))}$$

subject to $\varphi(x, 0) = 0$. It follows from standard results in the theory of parameter-dependent ODEs that φ is smooth, and the previous bound on $\det \nabla u$ implies that $|\varphi(x, x_3) - x_3| \le C x_3^2$. Hence there is $h_0 > 0$ such that v is smooth on $\bar\omega \times (-h_0, h_0)$, with $\det \nabla v = 1$.

The next step is the estimate for the gradient of v. By the chain rule, this requires a good control of the gradient of φ. To this end we rewrite the ODE as an integral equation and differentiate. Let F be a primitive of $\det \nabla u$ with respect to x_3, i.e.

$$F(x,t) = \int_0^t \det \nabla u(x,s)\, \mathrm{d}s\,.$$

Then the ODE above is equivalent to the implicit equation $F(x, \varphi(x, x_3)) = x_3$. Differentiating with respect to x_1 we obtain

$$\int_0^{\varphi(x,x_3)} \frac{\partial}{\partial x_1} \det \nabla u(x,s)\, \mathrm{d}s + \det \nabla u\,(x, \varphi(x,x_3))\, \frac{\partial}{\partial x_1} \varphi(x,x_3) = 0\,.$$

The expansion in (5.4) shows that for $x_3 < h_0$ the determinant $\det \nabla u$ takes values between 1/2 and 2. Taking the x_1 derivative of the same expression, we see that also $\partial_1 \det \nabla u$ is bounded by Cx_3 (C depends here on ν and on the first and second gradients of w and ν). It follows that

$$|\partial_1 \varphi| \le 2 \int_0^{\varphi} |\partial_1 \det \nabla u|\, \mathrm{d}s \le C x_3^2\,.$$

The estimate for $\partial_2 \varphi$ is analogous. We finally observe that $v(x, x_3) = w(x) + \varphi(x, x_3)\nu(x)$, and hence

$$\nabla v = (\partial_1 w + \partial_1 \varphi\, \nu + \varphi\, \partial_1 \nu | \partial_2 w + \partial_2 \varphi\, \nu + \varphi\, \partial_2 \nu | \partial_3 \varphi\, \nu)\,.$$

Since $|\partial_1 \varphi| + |\partial_2 \varphi| \le C x_3^2$, and $|\partial_3 \varphi - 1| \le C x_3$, we obtain (5.3), with C depending on w and ν.

6 Construction of the recovery sequence

For any function $u \in W^{1,p}(\omega; \mathbb{R}^3)$ we need to construct functions $u_h \in W^{1,p}(\Omega_h; \mathbb{R}^3)$ such that $\mathcal{T}_h u_h$ converges weakly in $W^{1,p}$ to $\mathcal{T}_0 u$ and

$$\limsup_{h \to 0} I_h[u_h] \le J[u].$$

In doing so we need to correctly account for the two relaxation steps, which lead from W_{3D} to W_{2D} and then to W^*_{2D}. It is natural to proceed in the reverse order, and start by constructing a sequence u_j such that the integral of $W_{\mathrm{2D}}(\nabla u_j)$ converges to the integral $W^*_{\mathrm{2D}}(\nabla u)$. We first reduce to the case of piecewise affine functions u, and then insert a suitable test function in each affine piece.

Lemma 6.1. *For any $u \in W^{1,p}(\omega; \mathbb{R}^3)$ there is a sequence $u_j \in P^1_c \cap W^{1,p}(\omega; \mathbb{R}^3)$ such that $u_j \to u$ strongly in $W^{1,p}$ and*

$$\lim_{j \to \infty} \int_\omega W^*_{\mathrm{2D}}(\nabla u_j)\, \mathrm{d}x = \int_\omega W^*_{\mathrm{2D}}(\nabla u)\, \mathrm{d}x\,.$$

Proof. By density of $P_c^1 \cap W^{1,p}(\omega;\mathbb{R}^3)$ in $W^{1,p}(\omega;\mathbb{R}^3)$ there is a sequence $u_j \in P_c^1 \cap W^{1,p}$ converging to u in $W^{1,p}$. The energy converges since W_{2D}^* is continuous and has p-growth from above (Lemma 3.1, (iii) and (iv)). Indeed, it suffices to apply Fatou's lemma first to $W_{2D}^*(\nabla u_j)$ and then to $C|\nabla u_j|^p - W_{2D}^*(\nabla u_j)$. The triangulation underlying u_j is denoted by $\mathcal{T}^{(j)}$.

Lemma 6.2. *For any $u \in W^{1,p}(\omega;\mathbb{R}^3)$ there is a sequence of functions $z_j \in P_c^1 \cap W^{1,p}(\omega;\mathbb{R}^3)$ such that $z_j \rightharpoonup u$ weakly in $W^{1,p}(\omega;\mathbb{R}^3)$ and*

$$\limsup_{j\to\infty} \int_\omega W_{2D}(\nabla z_j)\,dx \le \int_\omega W_{2D}^*(\nabla u)\,dx\,.$$

Proof. Let u_j be the sequence obtained in Lemma 6.1, and fix $\eta_j \to 0$. By definition, ∇u_j is constant in each of the at most countably many triangles $T_n^{(j)}$. We shall modify it separately in each of them. For any $T_n^{(j)}$ we can find by (ii) of Lemma 3.1 a function $\varphi_n^{(j)} \in P_c^1 \cap W_0^{1,p}(T_n^{(j)})$ such that

$$\frac{1}{|T_n^{(j)}|}\int_{T_n^{(j)}} W_{2D}(\nabla u_j + \nabla\varphi_n^{(j)})\,dx \le W_{2D}^*(\nabla u_j) + \eta_j\,,$$

and $\int_{T_n^{(j)}} |\varphi_n^{(j)}|^p\,dx \le \eta_j |T_n^{(j)}|$. We define $z_j = u_j + \varphi_n^{(j)}$ in each $T_n^{(j)}$. We have

$$\int_\omega W_{2D}(\nabla z_j)\,dx \le \int_\omega W_{2D}^*(\nabla u_j)\,dx + \eta_j|\omega| \quad \text{and} \quad \int_\omega |z_j - u_j|^p\,dx \le \eta_j|\omega|\,.$$

Using the coercivity of the energy, this implies that z_j is uniformly bounded in $W^{1,p}(\omega;\mathbb{R}^3)$. Since $z_j \to u$ in L^p, by the uniqueness of the weak limit, z_j converges weakly in $W^{1,p}(\omega;\mathbb{R}^3)$ to u.

In order to use the results from Sect. 4 we need to pass to smooth functions. This can be accomplished by convolution. However, we need to ensure that the smoothing does not modify the energy significantly. This is not entirely trivial, since we are now dealing with the non-convex, unbounded energy W_{2D}. In this smoothing step we make again use of the fact that the sequence is composed of piecewise affine functions. Precisely, we show that the smoothing does not change z_j away from the "jump set" Γ_j, which is defined as the union of the boundaries of the countably many triangles on which z_j is affine. This argument is for a single z_j, for uniformity we keep the index j.

Lemma 6.3. *Let $z_j \in P_c^1 \cap W^{1,p}(\omega;\mathbb{R}^3)$. Then for any $\eta > 0$ there exists $a_{j,\eta} \in C^\infty(\bar\omega;\mathbb{R}^3)$ such that*

$$a_{j,\eta} = z_j \quad \text{on } \omega\setminus\Gamma_{j,\eta}\,, \qquad \text{and} \qquad \lim_{\eta\to 0}\int_{\Gamma_{j,\eta}} \left(1+|\nabla a_{j,\eta}|^p\right)dx = 0\,.$$

Here $\Gamma_j = \partial\omega \cup \bigcup_n \partial T_n^{(j)}$, where $\{T_n^{(j)}\}_{n\in\mathbb{N}}$ are the triangles on which z_j is affine, and

$$\Gamma_{j,\eta} = \{x \in \omega : \operatorname{dist}(x,\Gamma_j) < \eta\}\,.$$

This implies immediately that $a_{j,\eta} \to z_j$ in $W^{1,p}(\omega;\mathbb{R}^3)$ as $\eta \to 0$.

Proof. It is east to see that

$$\lim_{\eta\to 0} |\Gamma_{j,\eta}| = 0 \quad \text{and} \quad \lim_{\eta\to 0} \int_{\Gamma_{j,\eta}} |\nabla z_j|^p \,\mathrm{d}x = 0\,. \tag{6.1}$$

To construct the functions $a_{j,\eta}$ we choose an extension $z_j \in W^{1,p}(\mathbb{R}^2;\mathbb{R}^3)$ (we use the same letter to denote this extension). We then mollify z_j with a kernel supported on a ball of radius $\eta/2$, and restrict the result to $\bar{\omega}$. Since z_j is locally affine on $\omega \setminus \Gamma_j$, we have $a_{j,\eta} = z_j$ on $\omega \setminus \Gamma_{j,\eta}$. Further,

$$\int_{\Gamma_{j,\eta}} |\nabla a_{j,\eta}|^p \,\mathrm{d}x \le \int_{\Gamma_{j,2\eta}} |\nabla z_j|^p \,\mathrm{d}x + \int_{(\partial\omega)_{2\eta}} |\nabla z_j|^p \,\mathrm{d}x\,,$$

where $(\partial\omega)_{2\eta} = \{x \in \mathbb{R}^2 : \mathrm{dist}(x,\partial\omega) < 2\eta\}$. Both terms tend to zero as $\eta \to 0$, hence $a_{j,\eta} \to z_j$ in $W^{1,p}$.

The following proposition summarizes what we have accomplished so far in this section, and incorporates the result of Sect. 4.

Proposition 6.4. *Let ω be a bounded Lipschitz domain in $\mathbb{R}^2$, and suppose that $W_{3\mathrm{D}}$ satisfies (2.10–2.11). Then for any $u \in W^{1,p}(\omega;\mathbb{R}^3)$ there exists a sequence of functions $u_j \in C^\infty(\bar{\omega};\mathbb{R}^3)$ such that $u_j \rightharpoonup u$ weakly in $W^{1,p}(\omega;\mathbb{R}^3)$,* rank $\nabla u_j = 2$ *everywhere, and*

$$\limsup_{j\to\infty} \int_\omega W_{2\mathrm{D}}(\nabla u_j)\,\mathrm{d}x \le \int_\omega W_{2\mathrm{D}}^*(\nabla u)\,\mathrm{d}x\,.$$

Moreover, each of the functions u_j is affine on finitely many triangles $T_n^{(j)} \subset \omega$, $n = 1,\ldots,N(j)$, where we may choose $N(j)$ such that the complement of these triangles is small in the sense that

$$\limsup_{j\to\infty} \int_{\omega_j} \big(W_{2\mathrm{D}}(\nabla u_j) + 1\big)\,\mathrm{d}x = 0 \qquad \textit{with } \omega_j = \omega \setminus \bigcup_{n=1}^{N(j)} T_n^{(j)}\,. \tag{6.2}$$

Proof. Let z_j be the sequence given in Lemma 6.2. For each fixed j, and each $\eta > 0$, let $a_{j,\eta}$ be as in Lemma 6.3. We shall now approximate the latter with regular functions. In order to do this, we apply Proposition 4.1 to the functions $a_{j,\eta}$, and denote the result by $b_{j,\eta}$ (if ω is not connected, it suffices to treat each connected component separately). On $\Gamma_{2\eta}$ we estimate the energy by means of the upper bound in the growth condition (3.2), giving

$$\int_{\Gamma_{2\eta}} W_{2\mathrm{D}}(\nabla b_{j,\eta})\,\mathrm{d}x \le C \int_{\Gamma_{2\eta}} \Big(|\nabla b_{j,\eta}|^p + \frac{1}{|\partial_1 b_{j,\eta} \wedge \partial_2 b_{j,\eta}|^p} + 1\Big)\,\mathrm{d}x.$$

The term $|\nabla b_{j,\eta}|^p + 1$ is by (4.1) bounded by a constant times $|\nabla a_{j,\eta}|^p + 1$. The second term in the integrand is bounded on Γ_η by a universal constant by (4.2), and on $\Gamma_{2\eta} \setminus \Gamma_\eta$, again in view of (4.1) and (3.2), by

$$\frac{1}{|\partial_1 b_{j,\eta} \wedge \partial_2 b_{j,\eta}|^p} \le \frac{c}{|\partial_1 a_{j,\eta} \wedge \partial_2 a_{j,\eta}|^p} \le cW_{2D}(\nabla a_{j,\eta}) + c = cW_{2D}(\nabla z_j) + c\,,$$

where in the last step we used that $\nabla a_{j,\eta} = \nabla z_j$ outside Γ_η. We obtain

$$\int_{\Gamma_{2\eta}} W_{2D}(\nabla b_{j,\eta})\, dx \le c\int_{\Gamma_{2\eta}} \left(W_{2D}(\nabla z_j) + |\nabla a_{j,\eta}|^p\right) dx + c|\Gamma_{2\eta}|.$$

Both quantities converge to zero as $\eta \to 0$. Since by construction $b_{j,\eta} = a_{j,\eta}$ on $\omega \setminus \Gamma_{2\eta}$, we conclude

$$\lim_{\eta\to 0} \int_\omega W_{2D}(\nabla b_{j,\eta})\, dx \le \int_\omega W^*_{2D}(\nabla z_j)\, dx.$$

Moreover,

$$\int_\omega |\nabla b_{j,\eta} - \nabla z_j|^p\, dx \le C\int_{\Gamma_{2\eta}} \left(|\nabla b_{j,\eta}|^p + |\nabla z_j|^p\right) dx \le C\int_{\Gamma_{2\eta}} \left(|\nabla z_j|^p + 1\right) dx$$

tends to zero as $\eta \to 0$, hence $b_{j,\eta} \to z_j$ strongly in $W^{1,p}(\omega;\mathbb{R}^3)$. Taking a suitable diagonal subsequence we can choose $u_j = b_{j,\eta(j)}$ such that u_j converges weakly in $W^{1,p}(\omega;\mathbb{R}^3)$ to u. This concludes the proof.

Proof of Theorem 2.4 (iii). Given $u \in W^{1,p}(\omega;\mathbb{R}^3)$ and a sequence $h_j \to 0$, we need to construct $u_j \in C^\infty(\Omega_{h_j};\mathbb{R}^3)$ such that

$$\limsup_{j\to\infty} I_{h_j}[u_j] \le J[u]$$

and $\mathcal{T}_{h_j} u_j \rightharpoonup \mathcal{T}_0 u$ weakly in $W^{1,p}(\Omega;\mathbb{R}^3)$. By Proposition 6.4 we can find $u_j \in C^\infty(\bar\omega;\mathbb{R}^3)$ such that $u_j \rightharpoonup u$ in $W^{1,p}$, $\operatorname{rank} \nabla u_j = 2$ everywhere, and

$$\limsup_{j\to\infty} \int_\omega W_{2D}(\nabla u_j)\, dx \le \int_\omega W^*_{2D}(\nabla u)\, dx\,. \tag{6.3}$$

For each j there are finitely many disjoint triangles $T_n^{(j)} \subset \omega$, $n = 1, \dots, N(j)$, such that u_j is affine on each of these triangles, and such that $\omega_j = \omega \setminus \cup_{n=1}^{N(j)} T_n^{(j)}$ is small, in the sense of (6.2). We now use Proposition 5.1 to construct the extensions to the sets $\omega \times (0, h_j)$. For simplicity we drop in the following argument the index j from all quantities except u_j. We first choose the field ν, which will determine the derivative of u_j in the x_3 direction. We set (see (5.1)) $\nu_0 = (\partial_1 u_j \wedge \partial_2 u_j)/\left|\partial_1 u_j \wedge \partial_2 u_j\right|^2$ and then modify ν_0 inside each of the finitely many triangles on which u_j is affine. Let T_n be one of them, $F_n = \nabla u_j$ on T_n, and $b_n \in \mathbb{R}^3$ be such that

$$W_{3D}(F_n|b_n) = W_{2D}(F_n) = \min_{a\in\mathbb{R}^3} W_{3D}\left(F_n|a\right).$$

By the growth condition $|b_n| \le c(|F_n| + |\nu_0| + 1)$, with c depending only on W_0. We define a smooth vector field ν by

$$\nu(x) = \begin{cases} \nu_0(x) + \psi_n(x)(b_n - \nu_0(x)) & \text{if } x \in T_n\,, \\ \nu_0(x) & \text{else.} \end{cases}$$

Here $\psi_n \in C_0^\infty(T_n;[0,1])$ is a cutoff function such that $\psi_n = 1$ on a subset T_n^η of T_n of area at least $(1-\eta)|T_n|$. Since on T_n

$$0 = \det(F|b) - \det(F|\nu_0) = (F_1 \wedge F_2) \cdot (b - \nu_0)$$

(we dropped here the index n as well) we obtain

$$(\partial_1 u_j \wedge \partial_2 u_j) \cdot \nu = (F_1 \wedge F_2) \cdot \nu_0 + \psi\,(F_1 \wedge F_2) \cdot (b - \nu_0) = 1\,,$$

i.e. (5.2) is satisfied. Finally, consider the enlarged exceptional set $\Gamma_j^{(\eta)} = \omega \setminus \cup T_n^\eta$. From the above estimates it follows that

$$\lim_{j\to\infty} \limsup_{\eta\to 0} \int_{\Gamma_j^{(\eta)}} \big(W_{2\mathrm{D}}(\nabla u_j) + 1\big)\, \mathrm{d}x = 0\,. \tag{6.4}$$

We now apply Proposition 5.1 to the pair (u_j, ν), to obtain a function $v_j \in C^\infty(\bar{\Omega}_{h_j};\mathbb{R}^3)$, with $\det \nabla v_j = 1$ everywhere. By the uniform estimate (5.3) and the continuity of W_0 we can estimate the energy, for $h < h_j$, by

$$I_h[v_j] \le \frac{1}{h} \int_{\Omega_h} W_0(\nabla v_j)\, \mathrm{d}x\ \mathrm{d}x_3 \le \int_\omega W_0(\nabla u_j|\nu)\, \mathrm{d}x + Ch\,.$$

The constant C can depend on u_j and ν. We focus on the first term. In most of the area, u_j is affine and ν equals the corresponding b_n. On the remainder, the energy is small by the growth condition. Precisely,

$$\begin{aligned} \int_\omega W_0(\nabla u_j|\nu)\, \mathrm{d}x &\le \int_{\cup T_\eta^n} W_0(F_n|b_n)\, \mathrm{d}x + C \int_{\Gamma_j^{(\eta)}} \big(|\nabla u_j|^p + |\nu|^p + 1\big)\, \mathrm{d}x \\ &\le \int_\omega W_{2\mathrm{D}}(\nabla u_j)\, \mathrm{d}x + C \int_{\Gamma_j^{(\eta)}} \big(W_{2\mathrm{D}}(\nabla u_j) + 1\big)\, \mathrm{d}x\,. \end{aligned}$$

Therefore

$$\limsup_{h\to 0} I_h[v_j] \le \int_\omega W_{2\mathrm{D}}(\nabla u_j)\, \mathrm{d}x + C \int_{\Gamma_j^{(\eta)}} \big(W_{2\mathrm{D}}(\nabla u_j) + 1\big)\, \mathrm{d}x\,.$$

By (6.3) and (6.4) we obtain

$$\limsup_{j\to\infty} \limsup_{\eta\to 0} \limsup_{h\to 0} I_h[v_j] \le J[u]$$

hence taking a diagonal subsequence concludes the proof.

Acknowlegdments. We thank Stefan Müller for bringing [Bel96b] to our attention, and Hervé LeDret for making [Bel96a] available to us. Part of this work was done while SC was affiliated with, and GD visiting the Max Planck Institute for Mathematics in the Sciences in Leipzig, and was supported by the DFG through project Con 304/1 within SPP 1095. The work of GD was also partially supported by the NSF through grants DMS0405853 and DMS0104118.

References

[ACD06] J. Adams, S. Conti, and A. DeSimone. Soft elasticity and microstructure in smectic C elastomers. *Preprint*, 2006.

[AF84] E. Acerbi and N. Fusco. Semicontinuity problems in the calculus of variations. *Arch. Rat. Mech. Anal.*, 86, 125–145, 1984.

[BC*00] H. Ben Belgacem, S. Conti, A. DeSimone, and S. Müller. Rigorous bounds for the Föppl-von Kármán theory of isotropically compressed plates. *J. Nonlinear Sci.*, 10, 661–683, 2000.

[BC*02] H. Ben Belgacem, S. Conti, A. DeSimone, and S. Müller. Energy scaling of compressed elastic films. *Arch. Rat. Mech. Anal.*, 164, 1–37, 2002.

[Bel96a] H. B. Belgacem. *Modélisation de structures minces en élasticité non linéaire.* PhD thesis, Univ. Paris 6, 1996.

[Bel96b] H. B. Belgacem. Une méthode de Γ-convergence pour un modèle de membrane non linéaire. *C. R. Acad. Sci. Paris*, 323, 845–849, 1996.

[BM84] J. M. Ball and F. Murat. $W^{1,p}$-quasiconvexity and variational problems for multiple integrals. *J. Funct. Anal.*, 58, 225–253, 1984.

[Bra02] A. Braides. *Γ-convergence for beginners.* Oxford University Press, Oxford, 2002.

[CD05] S. Conti and G. Dolzmann. Derivation of a plate theory for incompressible materials. *Preprint*, 2005.

[CDD02] S. Conti, A. DeSimone, and G. Dolzmann. Semi-soft elasticity and director reorientation in stretched sheets of nematic elastomers. *Phys. Rev. E*, 66, 061710, 2002.

[Cia97] P. G. Ciarlet. *Theory of plates*, volume II of *Mathematical elasticity.* Elsevier, Amsterdam, 1997.

[CM05] S. Conti and F. Maggi. Confining thin elastic sheets and folding paper. *Preprint*, 2005.

[CMM06] S. Conti, F. Maggi, and S. Müller. Rigorous derivation of Föppl's theory for clamped elastic membranes leads to relaxation. *Preprint, to appear in SIAM J. Math. Anal.*.

[Con03] S. Conti. Low-energy deformations of thin elastic sheets: isometric embeddings and branching patterns. Habilitation thesis, Universität Leipzig, 2003.

[Dac89] B. Dacorogna. *Direct methods in the calculus of variations.* Springer-Verlag, New York, 1989.

[Dal93] G. Dal Maso. *An introduction to Γ-convergence.* Birkhäuser, Boston, 1993.

[DD02] A. DeSimone and G. Dolzmann. Macroscopic response of nematic elastomers via relaxation of a class of $SO(3)$-invariant energies. *Arch. Rat. Mech. Anal.*, 161, 181–204, 2002.

[FJM02] G. Friesecke, R. James, and S. Müller. A theorem on geometric rigidity and the derivation of nonlinear plate theory from three dimensional elasticity. *Comm. Pure Appl. Math*, 55, 1461–1506, 2002.

[Fon88] I. Fonseca. The lower quasiconvex envelope of the stored energy function for an elastic crystal. *J. Math. pures et appl.*, 67, 175–195, 1988.

[DGF75] E. De Giorgi and T. Franzoni. Su un tipo di convergenza variazionale. *Atti Accad. Naz. Lincei Rend. Cl. Sci. Mat. (8)*, 58, 842–850, 1975.

[JS01] W. Jin and P. Sternberg. Energy estimates of the von Kármán model of thin-film blistering. *J. Math. Phys.*, 42, 192–199, 2001.

[JS02] W. Jin and P. Sternberg. In-plane displacements in thin-film blistering. *Proc. R. Soc. Edin. A*, 132A, 911–930, 2002.

[Kui55] N. Kuiper. On C^1 isometric imbeddings I and II. *Proc. Kon. Acad. Wet. Amsterdam A*, 58, 545–556 and 683–689, 1955.

[LR93] H. LeDret and A. Raoult. Le modèle de membrane nonlinéaire comme limite variationelle de l'élasticité non linéaire tridimensionelle. *C. R. Acad. Sci. Paris*, 317, 221–226, 1993.

[LR95] H. LeDret and A. Raoult. The nonlinear membrane model as a variational limit of nonlinear three-dimensional elasticity. *J. Math. Pures Appl.*, 73, 549–578, 1995.

[Mül99] S. Müller. Variational models for microstructure and phase transitions. In F. Bethuel et al., editors, *in: Calculus of variations and geometric evolution problems*, Springer Lecture Notes in Math. 1713, pages 85–210. Springer-Verlag, 1999.

[Nas54] J. Nash. C^1 isometric imbeddings. *Ann. Math.*, 60, 383–396, 1954.

[Pan03] O. Pantz. On the justification of the nonlinear inextensional plate model. *Arch. Rat. Mech. Anal.*, 167, 179–209, 2003.

[Tra05] K. Trabelsi. Incompressible nonlinearly elastic thin membranes. *C. R. Acad. Sci. Paris, Ser. I*, 340, 75–80, 2005.

[Tra06] K. Trabelsi. Modeling of a nonlinear membrane plate for incompressible materials via Gamma-convergence. *To appear in Anal. Appl. (Singap.).*

[Whi43] H. Whitney. The general type of singularity of a set of $2n-1$ smooth functions of n variables. *Duke Math. J.*, 10, 161–172, 1943.

[Whi44a] H. Whitney. The self-intersections of a smooth n-manifold in $2n$-space. *Ann. Math.*, 45, 220–246, 1944.

[Whi44b] H. Whitney. The singularities of a smooth n-manifold in $2n-1$-space. *Ann. Math.*, 45, 247–293, 1944.

[WT96] M. Warner and E. M. Terentjev. Nematic elastomers - a new state of matter? *Prog. Polym. Sci.*, 21, 853–891, 1996.

[WT03] M. Warner and E. M. Terentjev. *Liquid Crystal Elastomers.* Oxford Univ. Press, 2003.

Domain Walls and Vortices in Thin Ferromagnetic Films

Matthias Kurzke[1], Christof Melcher[2], and Roger Moser[3]

[1] Institute for Mathematics and its Applications, University of Minnesota, 207 Church Street SE, Minneapolis, MN 55455, USA. kurzke@ima.umn.edu
[2] Department of Mathematics, Humboldt-Universität zu Berlin, Unter den Linden 6, 10099 Berlin, Germany. melcher@mathematik.hu-berlin.de
[3] Department of Mathematical Sciences, University of Bath, Bath BA2 7AY, United Kingdom. r.moser@maths.bath.ac.uk

1 Introduction

1.1 Multiscale problems in micromagnetics

Ferromagnetic materials show a fascinating variety of magnetization patterns on scales ranging from a few nanometers to hundreds of microns. The formation and evolution of these patterns is at the heart of numerous magnetic devices, including the ubiquitous magnetic storage media. Somewhat surprisingly, this large variety of patterns can be understood as (local) minimizers of a simple, yet subtle, functional, the micromagnetic energy. Their dynamics is described by an associated evolution equation, the Landau-Lifshitz (or Landau-Lifschitz-Gilbert) equation, which combines Hamilitonian and dissipative aspects.

Until recently the micromagnetic energy was mostly analyzed in one of two ways. The first approach is to consider special ansatz functions (inspired by physical intuition) with a few free parameters and then to optimize over these parameters. While this approach has lead to valuable insights, it is also limited in its scope. In particular one cannot detect something which has not been put in the ansatz. The second approach is large scale computation. This has been successful for answering specific questions for submicron devices. Due to the wide separation of the relevant scales, direct numerical simulation is, however, restricted to the smallest scales and cannot cover the full picture. Perhaps even more importantly, it answers specific questions, but provides little insight in general principles and understanding.

In the last decade a new approach to micromagnetics has emerged, and the SPP 1095 has had an important impact in shaping it. This approach is based on two ideas. First, considerable insight can be gained by the identification of optimal scaling laws involving the natural parameters, such as material constants or geometric quantities, and the corresponding magnetization patterns. This amounts to establishing upper and lower bounds on the micromagnetic energy. While for the former one can often rely on intuition and test functions developed in the physics community, the latter requires in general new mathematical ideas. The second approach is to derive simplified theories in certain limiting parameter regimes, e.g., for thin films. These theories reduce the complexity of the magnetic energy landscape (and the dynamics in that landscape) and allow one to get a better insight into the essential structures.

In this paper we focus on the mathematical analysis of the statics and dynamics of magnetization structures in thin films. For a broader review of recent developments, written for a more general science audience, we refer to [DK*05].

1.2 The micromagnetic energy and associated variational problems

We first discuss the functional from which all our results are ultimately derived, the micromagnetic energy. This energy is a sum of terms of various types. Depending on certain material parameters and on the shape and size of the ferromagnetic sample, any of these terms can play a dominant role, or an interplay between several of them can take place. This explains the multitude of different patterns derived from this theory.

We consider an open domain $\Omega \subset \mathbb{R}^3$ which represents the ferromagnetic body that we study. The magnetization of this body is given by a vector field $\mathbf{m} : \Omega \to \mathbb{R}^3$. Below the Curie temperature, the magnetization is saturated, which means that $\mathbf{m}$ is of constant length. We use a normalization such that $\mathbf{m}$ has values in the unit sphere $\mathbb{S}^2$. Now we consider several energies associated to $\mathbf{m}$.

The so-called exchange energy models the tendency towards parallel alignment of neighboring magnetization vectors in the underlying atomic lattice. It is given by the functional

$$\frac{d^2}{2} \int_\Omega |\nabla \mathbf{m}|^2 \, d\mathbf{x}.$$

Here d is a material constant, called the exchange length.

The ferromagnetic material may have crystalline anisotropies which prefer certain directions of $\mathbf{m}$. An integral of the form

$$Q \int_\Omega \phi(\mathbf{m}) \, d\mathbf{x}$$

represents such anisotropies. Here $\phi : \mathbb{S}^2 \to [0, \infty)$ is a fixed function and Q is another material constant. Usually ϕ is assumed to be smooth, often even a polynomial.

The magnetization induces a magnetic field, often called the stray field or demagnetizing field, which obeys the static Maxwell equations. It can be represented by a potential $u : \mathbb{R}^3 \to \mathbb{R}$ which solves the equation

$$\Delta u = \operatorname{div}(\chi_\Omega \mathbf{m}) \quad \text{in } \mathbb{R}^3.$$

Here χ_Ω is the characteristic function of Ω (in other words, we extend $\mathbf{m}$ by 0 outside of Ω). If Ω is bounded, the saturation condition guarantees that $\mathbf{m} \in L^p(\Omega, \mathbb{R}^3)$ for every $p \in [1, \infty]$. Hence there exists a unique solution of this equation in the Sobolev space $H^1(\mathbb{R}^3)$. The induced field is then represented by $-\nabla u$. Its energy is called the magnetostatic energy and given by

$$\frac{1}{2} \int_{\mathbb{R}^3} |\nabla u|^2 \, d\mathbf{x}.$$

The interaction with an external field $h : \mathbb{R}^3 \to \mathbb{R}^3$ induces an energy term

$$- \int_\Omega \mathbf{h} \cdot \mathbf{m} \, d\mathbf{x}$$

that prefers alignment of the magnetization with the external field. In this paper, applied fields appear as driving forces in the context of moving domain walls.

The micromagnetic energy is the sum of all four energies, that is,

$$E(\mathbf{m}) = \frac{d^2}{2} \int_\Omega |\nabla \mathbf{m}|^2 \, d\mathbf{x} + Q \int_\Omega \phi(\mathbf{m}) \, d\mathbf{x} + \frac{1}{2} \int_{\mathbb{R}^3} |\nabla u|^2 \, d\mathbf{x} - \int_\Omega \mathbf{h} \cdot \mathbf{m} \, d\mathbf{x}. \tag{1.1}$$

The exchange term is of leading order in this functional, but the constant in front of it is typically small. The other three terms are of order 0, but one of them (the magnetostatic energy) involves the non-local pseudo-differential operator $\nabla \Delta^{-1} \operatorname{div}$. In some situations, its behavior is quite different from the behavior of the other terms. Under certain conditions, these energies may be in competition with one another. For instance, the exchange energy favors constant magnetizations, whereas the magnetostatic energy prefers vector fields which are divergence free (in $\mathbb{R}^3$). Because of the jump at the boundary, the two conditions cannot be satisfied simultaneously. The anisotropy term, on the other hand, may not penalize a varying vector field in principle, at least not if the function ϕ has several minima (which is usually the case), but it favors rapid transitions between different states – unlike the exchange energy. An analysis of such interplays can explain some of the observed patterns in ferromagnets.

We study two types of variational problems associated to the micromagnetic energy. Minimizers of E, or more generally, local minimizers and (stable) critical points represent the stable magnetization patterns of our ferromagnet. If we write

$$\nabla_{L^2} E(\mathbf{m}) = -d^2 \Delta \mathbf{m} + Q \nabla \phi(\mathbf{m}) + \nabla u - \mathbf{h} \tag{1.2}$$

for the L^2-gradient of E (without the saturation constraint), then these variational problems give rise to the Euler-Lagrange equation

$$(\mathbf{1} - \mathbf{m} \otimes \mathbf{m})\nabla_{L^2} E(\mathbf{m}) = \nabla_{L^2} E(\mathbf{m}) - (\mathbf{m} \cdot \nabla_{L^2} E(\mathbf{m}))\mathbf{m} = 0 \quad \text{in } \Omega.$$

(Here $\mathbf{1}$ denotes the identity (3×3)-matrix.) That is, the projection of $\nabla_{L^2} E(\mathbf{m})$ onto the tangent space $T_{\mathbf{m}}\mathbb{S}^2$ vanishes. This equation can also be expressed in the form

$$d^2(\Delta \mathbf{m} + |\nabla \mathbf{m}|^2 \mathbf{m}) - Q\nabla\phi(\mathbf{m}) + (\mathbf{1} - \mathbf{m} \otimes \mathbf{m})(\mathbf{h} - \nabla u) = 0 \quad \text{in } \Omega.$$

Moreover, we have homogeneous Neumann boundary conditions

$$\frac{\partial \mathbf{m}}{\partial \nu} = 0 \quad \text{on } \partial\Omega.$$

A model for the dynamical behavior of the magnetization is given by the Landau-Lifshitz equation

$$\frac{\partial \mathbf{m}}{\partial t} + \gamma \mathbf{m} \wedge \nabla_{L^2} E(\mathbf{m}) + \alpha \mathbf{m} \wedge \frac{\partial \mathbf{m}}{\partial t} = 0, \tag{1.3}$$

also called the Landau-Lifshitz-Gilbert equation. Here $\wedge$ denotes the vector product in $\mathbb{R}^3$. Both γ and α are fixed constants, and we require that $\alpha\gamma > 0$. Another common way to write the equation is

$$\frac{\partial \mathbf{m}}{\partial t} = \hat{\gamma} \mathbf{m} \wedge \nabla_{L^2} E(\mathbf{m}) + \hat{\alpha} \mathbf{m} \wedge (\mathbf{m} \wedge \nabla_{L^2} E(\mathbf{m})). \tag{1.4}$$

The two versions are equivalent for

$$\hat{\alpha} = \frac{\alpha\gamma}{\alpha^2 + 1} \quad \text{and} \quad \hat{\gamma} = -\frac{\gamma}{\alpha^2 + 1}.$$

The terms in (1.3) and (1.4) with coefficients γ and $\hat{\gamma}$, respectively, describe a magnetic precession. We call them the gyromagnetic terms. The terms with coefficients α and $\hat{\alpha}$, respectively, are damping terms (hence the sign condition on $\alpha\gamma$, giving rise to the condition $\hat{\alpha} > 0$). From the mathematical point of view, the damping terms are the more important ones, because they make the problem parabolic. Without them, the equations would be of the type of a nonlinear Schrödinger equation.

Taking the vector product with $\mathbf{m}$ in all terms of (1.3), we obtain a third equivalent version of the equation,

$$\tilde{\alpha} \frac{\partial \mathbf{m}}{\partial t} + \tilde{\gamma} \mathbf{m} \wedge \frac{\partial \mathbf{m}}{\partial t} = (\mathbf{m} \otimes \mathbf{m} - \mathbf{1})\nabla_{L^2} E(\mathbf{m}), \tag{1.5}$$

where

$$\tilde{\alpha} = \frac{\alpha}{\gamma} \quad \text{and} \quad \tilde{\gamma} = -\frac{1}{\gamma}.$$

Representing the equation in this form is convenient because it underlines the similarity to the negative L^2-gradient flow for the functional E subject to the constraint $|\mathbf{m}| = 1$. (In fact this gradient flow is (1.5) for $\tilde{\alpha} = 1$ and $\tilde{\gamma} = 0$.) We will normally use the Landau-Lifshitz equation in the form (1.5).

It is natural to impose a homogeneous Neumann boundary condition also for the Landau-Lifshitz equation.

Apart from the obvious quantities d, Q, ϕ, and $\mathbf{h}$, the qualitative behavior of E and of solutions of the above variational problems also depends on the shape and the size of the sample Ω. Some idea of the dependence on the size can be gained by studying the scaling properties of the four terms which contribute to the micromagnetic energy. Suppose for a number $\lambda > 0$, we replace Ω by $\lambda\Omega$ and the vector field $\mathbf{m}$ by $\mathbf{m}(x/\lambda)$ (and similarly $\mathbf{h}$ by $\mathbf{h}(x/\lambda)$). Then the exchange energy is multiplied by the factor λ, whereas the other energy terms are multiplied by λ^3. Thus it is to be expected that for a very small sample, the exchange energy determines the behavior of $\mathbf{m}$ to a large extent; that is, a minimizer of E is nearly constant. For a very large sample, on the other hand, the exchange energy is insignificant, and the behavior of $\mathbf{m}$ is ruled by the other terms.

On the other hand, we can use rescalings to eliminate one of the parameters in our problem. Replacing Ω by $\lambda\Omega$, and replacing simultaneously d by λd, we obtain a functional whose energy landscape differs only by a constant. This way we can normalize the problem such that either d or Q become 1, or such that Ω has unit size.

1.3 Thin films

The results in this paper are concerned with ferromagnetic bodies in the shape of thin films. That is, we consider domains of the form

$$\Omega_\delta = \Omega \times (0, \delta),$$

where Ω is now a two-dimensional domain and $\delta > 0$ is small compared with the size of Ω. We either study the limit behavior of the micromagnetic energy and its variational problems as $\delta \searrow 0$, or we use the thinness of Ω_δ as a justification for working directly in two dimensions and with an energy that approximates the micromagnetic energy for thin films. In both cases, the projection of $\mathbf{m}$ onto the plane $\mathbb{R}^2 \times \{0\}$ and the third component of $\mathbf{m}$ play different roles. It is therefore convenient to use the notation $\mathbf{m} = (m, m_3)$ for the magnetization vector field, where $m = (m_1, m_2)$. Similarly we often write $\mathbf{x} = (x, x_3) = (x_1, x_2, x_3)$ for a generic point in $\mathbb{R}^3$. Sometimes, however, it is more convenient to use coordinates (x, y, z) in $\mathbb{R}^3$.

The reduction to two dimensions – whether by a rigorous asymptotic analysis or formally – decreases the complexity of the problems that we study. Nevertheless, a rich variety of patterns can still be observed, and there exist different asymptotic regimes for the thin-film limit which give rise to different reduced theories. These regimes are determined by certain relations between the parameters involved in the problem, of which we have now four (under the assumption that the shape of Ω is fixed, but the size can still be varied by scaling): In addition to the material constants d and Q, we have the thickness δ and a length scale L of the cross-section Ω. When we are not interested in the behavior of $\mathbf{m}$ near $\partial\Omega$, we may assume $\Omega = \mathbb{R}^2$, and then L need not be considered. If we choose to neglect certain terms of the micromagnetic energy, this may of course reduce the number of parameters further. For instance, if we consider only the exchange energy and the magnetostatic energy (which we do in a substantial part of this paper), then the asymptotic regime depends on the behavior of the ratio d/L as we let the aspect ratio δ/L converge to 0. Some asymptotic regimes for this thin-film limit have been studied by Gioia and James [GJ97]; Carbou [Car01]; DeSimone, Kohn, Müller, and Otto [DK*02]; and Kohn and Slastikov [KS05a, KS05b]. Another regime is discussed in this paper, first through a simplified model in two dimensions, then by an asymptotic analysis for the micromagnetic energy on Ω_δ for $\delta \searrow 0$. This theory also establishes a link to the theory of Ginzburg-Landau vortices, which were first studied by Bethuel, Brezis, and Hélein [BBH93, BBH94].

A further dimensional reduction is made in the context of parametrized domain wall models, that we investigate in detail. Such models represent the basic building blocks within larger domain patterns or more complex domain wall structures. Of particular interest is the regime of weakly anisotropic (soft) thin films, where such transition layers significantly differ from those more common in phase transitions.

2 Domain Walls: Internal Structure and Dynamics

The primary phenomenon that one associates with magnetic pattern formation is the decomposition of a magnetic body into almost uniformly magnetized regions. The so-called magnetic domains are separated by thin transition layers, called domain walls, that interact in a complex network. The structure of such domain walls is among the central concerns of micromagnetic theory. While the analysis of domain walls is mathematically an interesting matter of its own, the physical relevance relies in the resulting mutual interaction having large impact on the global magnetic microstructure, especially when nonlocal effects dominate. In reduced thin-film theories, domain walls often emerge as line singularities while fine structural properties no longer have any effect. Breaking the resulting degeneracy by means of transparent selection principles rising from higher order contributions remains a major challenge. On the other hand, magnetic domain walls can exhibit internal substructures themselves or can be made up as a complex composite, such as the *cross-tie wall*, cf. [HS98] pp. 240-241.
The simplest domain wall patterns are one-dimensional and appear as extreme cases in a hierarchy of domain wall models that emerge in diverse parameter

regimes: Within a *Bloch wall*, the magnetization vector performs a rotation perpendicular to the transition axis. The main feature is the avoidance of magnetic volume charges, so that this wall type is energetically favorable in bulk situations and essentially equivalent to the transition problem arising from Cahn-Hilliard models. Such models exhibit sharply localized and rapidly decaying transition profiles. Our analysis shows that this behavior can largely change when nonlocal interactions dominate and internal length scales fail to be determined by dimensional analysis. The *Néel wall*, where the transition proceeds in-plane, is preferred in suitable thin-film regimes and characterized by the avoidance of magnetic surface charge. The presence of three energy components with different scaling behavior gives rise to multiple length scales. The typical feature of a Néel wall is the very long logarithmic tail of transition profiles. Such behavior has been predicted by heuristic arguments and numerical simulation (cf. [RS71, Gra99]) in order to explain long-range interaction of Néel walls, when neighboring tails overlap. Here we demonstrate rigorously how the main analytical feature of the variational principle, a critical regularity property, gives rise to the typical logarithmic decay behavior [Mel02, Mel03, Mel04a]. This global approach served in addition to resolve the spatial scaling laws in terms of all involved parameters and to derive a somewhat universal limiting profile that reflects the decay.
The evolution of magnetic patterns in the presence of applied fields is closely related to the motion of domain walls. Gyrotropic domain wall motion is based on the Landau-Lifshitz-Gilbert (LLG) equations, that describe a damped precession of the magnetization vector about the effective field, i.e. mathematically a hybrid heat and Schrödinger flow for the free energy. An appropriate local description relies on the concept of moving fronts that propagate with constant speed. Traveling wave solutions for the associate LLG dynamics represent a natural dynamic counterpart to static domain walls. As it turns out they provide valuable insight into the mechanisms and properties of domain wall dynamics, where besides energetics and spatial structures, kinematic quantities as *wall mobility* and *wall mass* come into play.
Whereas in the equilibrium case the magnetization path is dictated by energetics and particularly stray-field interactions, a second mechanism effects the shape of a moving domain wall: the precession dynamics as prescribed by LLG pushes the magnetization vector away from its optimal path, taking into account a gain in stray-field energy. Many interesting effects originate from this balance of energetic and dynamic forces, especially when enhanced by strong shape anisotropy in the regime of thin films. The bulk situation, however, can surprisingly be solved explicitly by means of a famous construction by Walker, i.e. a tilted version of the Landau and Lifshitz solution for the standard Bloch wall, cf. [Wal63]. Again all spatial and temporal scales can be read off by dimensional reasoning. The natural question on whether such a construction can be perturbed to the regime of finite layers has been answered in the affirmative [Mel04b]. Indeed, a suitable choice of canonical coordinates transforms the associate LLG system into a weakly coupled Schrödinger/reaction-

diffusion system and makes it accessible for spectral methods. The analysis also demonstrates that the finite layer perturbation is indeed a singular one and how this relates to slow decay.
In the regime of thin films the competition between stray field and precession is singular, that is to say, the asymptotically hard constraint of in-plane magnetization is geometrically incompatible with LLG. In order to derive an effective evolution equation, the change of spatial scaling has to be accompanied with a change of times scale. An effective thin-film limit for LLG with finite Gilbert damping has been carried out in [EG01, KS05a] and in Thm. 4.9, where the gyromagnetic precession term effectively turns into a large damping term as well. While the overall relaxation dynamics is captured correctly, oscillatory phenomena, such as spin waves or domain wall resonances, are suppressed in such a limit. In order to account for these effects we consider the complementary regime where Gilbert damping is comparable to the relative thickness. We show that in this regime LLG keeps its oscillatory features and turns into a damped nonlocal wave map equation [CMO06]. In the context of domain wall motion it provides a mechanical analogy and sheds some new light on the notion of wall mass. For small applied fields, the traveling wave problem, modeled on this wave-type dynamics, reduces to the question of linear stability for stationary Néel walls. Then the implicit function theorem provides existence and determines the mobility of traveling Néel walls.

2.1 Mathematical framework for planar domain walls

Let us consider an infinitely extended uniaxial magnetic film that is represented by $\Omega_\delta = \mathbb{R}^2 \times (0, \delta)$ and oriented by the anisotropy (easy) axis $\mathbb{R}\,\hat{\mathbf{e}}_2$. We consider parameterized transitions along $\mathbb{R}\,\hat{\mathbf{e}}_1$ (that we call transition axis) that connect antipodal states on the easy axis, i.e.

$$\mathbf{m} : \mathbb{R} \to \mathbb{S}^2 \quad \text{with} \quad \mathbf{m}(\pm\infty) = (0, \pm 1, 0).$$

In the following we denote the transition parameter by x and the vertical coordinate by z, i.e. we set $x_1 = x$ and $x_3 = z$. Under the hypothesis that, within the film, the magnetization varies only along the transition axis, we identify the associated global magnetization field $\mathbf{m}(\mathbf{x}) = \mathbf{m}(x)\chi_{(0,\delta)}(z)$ that is defined for $\mathbf{x} \in \mathbb{R}^3$. Then $\mathbf{m} = \mathbf{m}(\mathbf{x})$ induces the stray field ∇u determined by the potential equation $\Delta u = \nabla \cdot \mathbf{m}$ in $\mathcal{D}'(\mathbb{R}^3)$. We observe that $u = u(x, z)$ where the z dependence only stems from shape anisotropy. Thus the micromagnetic energy induces the following averaged domain wall energy per unit length:

$$E(\mathbf{m}) = \frac{1}{2} \fint_0^\delta \left\{ d^2 \int |\mathbf{m}'|^2 \, dx + \int \nabla u \cdot \mathbf{m} \, dx + Q \int (1 - m_2^2) \, dx \right\} \, dz. \quad (2.1)$$

The vertical average in (2.1) is redundant for the *exchange* and *anisotropy* portion. In order to perform a dimensional reduction for the *stray field*, we introduce a reduced stray-field operator

$$\mathcal{S}_\delta : \mathbf{m} \mapsto \fint_0^\delta \nabla u \, dz \quad \text{where} \quad \Delta u = \nabla \cdot \mathbf{m} \quad \text{in} \quad \mathcal{D}'(\mathbb{R}^3).$$

Changing the order of integration, the averaged stray-field energy can be expressed as

$$E_{\text{stray}}(\mathbf{m}) = \frac{1}{2} \fint_0^\delta \int \nabla u \cdot \mathbf{m} \, dx \, dz = \frac{1}{2} \int_{\mathbb{R}} \mathcal{S}_\delta(\mathbf{m}) \cdot \mathbf{m} \, dx.$$

A straightforward calculation shows that the operator $\mathcal{S}_\delta$ has an interpretation in terms of Fourier multiplication operators. Indeed, we have

$$\mathcal{S}_\delta(\mathbf{m}) = \Big[\sigma(\delta D)\, m_1, 0\,, \big(1 - \sigma(\delta D)\big)\, m_3\Big] : \mathbb{R} \to \mathbb{R}^3,$$

where $\sigma(D) f = \mathcal{F}^* \left(\xi \mapsto \sigma(\xi) \hat{f}(\xi)\right)$. The basic Fourier multiplier $\sigma(\xi)$ is given by

$$\sigma(\xi) = \left(1 - \frac{1 - \exp(-|\xi|)}{|\xi|}\right) \sim \begin{cases} \frac{1}{2}|\xi| & \text{for low frequencies } \xi \\ 1 & \text{for high frequencies } \xi. \end{cases} \tag{2.2}$$

The reduced stray-field operator can equivalently be described by means of convolution kernel, cf. [Gra99], [Mel03] for a derivation and a detailed discussion. Accordingly, the reduced stray-field energy can be written as

$$E_{\text{stray}}(\mathbf{m}) = \frac{1}{2} \int \sigma(\delta\xi)\, |\hat{m}_1(\xi)|^2 \, d\xi + \frac{1}{2} \int \big(1 - \sigma(\delta\xi)\big)\, |\hat{m}_3(\xi)|^2 \, d\xi. \tag{2.3}$$

The advantage of the Fourier representation is that one can easily read off the asymptotic form of interaction from the asymptotic behavior of Fourier multipliers. From (2.3) one can separate a local contribution and a nonlocal one that vanishes in the bulk regime

$$E_{\text{stray}}(\mathbf{m}) = \frac{1}{2} \, \|m_1\|_{L^2}^2 + \frac{1}{2} \int \big(1 - \sigma(\delta\xi)\big) \Big\{ |\hat{m}_3(\xi)|^2 - |\hat{m}_1(\xi)|^2 \Big\} \, d\xi. \tag{2.4}$$

Indeed, from (2.2) we deduce that $\sigma(\delta\,\xi) \to 1$ in the regime when $\delta|\xi| \to \infty$. Thus, for a corresponding family of transitions $\mathbf{m}$ so that m_1 and m_3 are uniformly bounded in $L^2(\mathbb{R})$, we infer that

$$E_{\text{stray}}(\mathbf{m}) = \frac{1}{2} \, \|m_1\|_{L^2}^2 + o(1). \tag{2.5}$$

We observe that in the bulk regime, the stray-field interaction reduces to a local contribution having the form of an additional anisotropy term that penalizes magnetizations that point along the transition axis. In the complementary thin-film regime when $\delta|\xi| \to 0$ we have $\sigma(\delta\,\xi) \to \delta|\xi|$ and, for m_1 and m_3 uniformly bounded in $H^1(\mathbb{R})$, an asymptotic expansion

$$E_{\text{stray}}(\mathbf{m}) = \frac{1}{2} \left\| m_3 \right\|_{L^2}^2 + \frac{\delta}{4} \left\| m_1 \right\|_{\dot{H}^{1/2}}^2 + o(\delta), \tag{2.6}$$

where $\|f\|_{\dot{H}^{1/2}}^2 = \int |\xi||\hat{f}(\xi)|^2 \, d\xi$ denotes the homogeneous $H^{1/2}$-norm. The zero order contribution can be interpreted as the residual surface charge interaction having the form of an additional anisotropy that penalizes vertical magnetizations. The first order term corresponds to residual volume charge interaction.

From a variational point of view the leading order stray-field contribution in (2.5) and (2.6), respectively, determines asymptotically a geodesic magnetization path. Whereas in the bulk situation the stray field interaction can be eliminated completely by choosing a path perpendicular to the transition axis, i.e. $m_1 = 0$ (Bloch walls), the penalty on the vertical component as $\delta \to 0$ enforces in-plane rotations, i.e. $m_3 = 0$ (Néel walls), taking into account internal stray fields that typically appear to the leading order, see Fig. 2.1.

Complete elimination of stray-field interaction cannot be achieved by means of one-dimensional transition modes. In somewhat thicker films, however, the symmetric Néel wall would lead to comparatively large stray-field contribution. But for an attempt to construct a stray-field free transition layer one has to abandon the symmetry assumption and to permit variations in the vertical direction. Such an object, referred to as **asymmetric Bloch wall**, has been discovered by Hubert, cf. [HS98] pp. 245-249, where volume charges are avoided by a vortex construction in the wall center. At the same time numerical simulations have confirmed a dramatic decrease of energy by breaking the wall symmetry. Recently, a rigorous verification based on an ansatz-free interpolation argument has been provided by Otto in [Ott02].

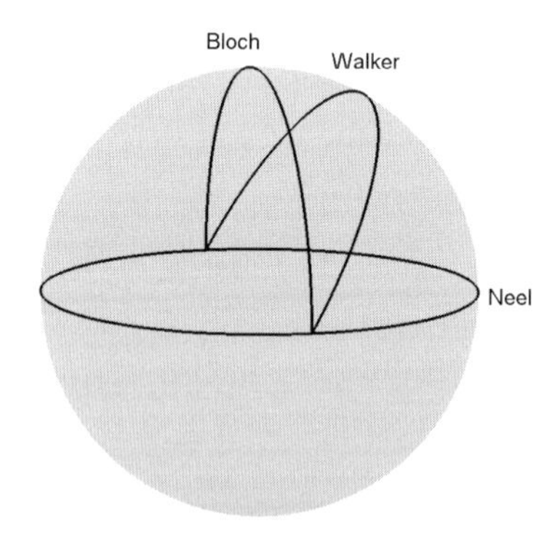

Fig. 2.1. The figure shows the **Bloch** and the **Néel** wall path as perpendicular geodesic connections of antipodal states. It also shows an intermediate geodesic that corresponds to a moving domain wall in the bulk regime where dynamic forces lead to an inclination towards the Néel wall path, the so-called **Walker** path. We will refer to the polar angle φ between the Bloch wall and the Walker path as the Walker angle.

Bloch walls versus Néel walls

The infinite Bloch wall in bulk samples has been the first micromagnetic object proposed and calculated in the seminal work by Landau and Lifshitz, cf. [LL35]. Once the stray-field energy is fully eliminated by choosing an appropriate path, the corresponding optimal profile and minimal energy can be found by nowadays standard variational methods. Indeed, from (2.1) we get for $\mathbf{m} = (0, m_2, m_3) : \mathbb{R} \to \mathbb{S}^2$ with $m_2(\pm\infty) = \pm 1$,

$$E(\mathbf{m}) = \frac{d^2}{2} \int |\mathbf{m}'|^2 \, dx + \frac{Q}{2} \int (1 - m_2^2) \, dx.$$

The length scale $w = \sqrt{d^2/Q}$ defines the typical Bloch wall domain width. Rescaling by w and renormalizing the energy by the factor $4\sqrt{d^2 Q}$ yields

$$E(\mathbf{m}) = \frac{1}{4} \int |\mathbf{m}'|^2 \, dx + \frac{1}{4} \int (1 - m_2^2) \, dx.$$

Using the identity $|\mathbf{m}'|^2 = (m_2')^2/(1 - m_2^2)$ we deduce the optimality relation

$$m_2' = 1 - m_2^2 \quad \text{with} \quad m_2(\pm\infty) = \pm 1 \tag{2.7}$$

that is uniquely solved by $m_2(x) = \tanh(x)$. Thus in the original scaling

$$\mathbf{m}(x) = \big(0, \tanh(x/w), \operatorname{sech}(x/w)\big). \tag{2.8}$$

Moreover we deduce from (2.7) and Young's inequality that

$$E(\mathbf{m}) \geq \frac{1}{2} \int |m_2'| \, dx \geq 1,$$

that is attained under equipartition for $m_2(x) = \tanh(x)$. Thus we recover from scaling that the Bloch wall energy per unit length is given by $e_0 = 4\sqrt{d^2 Q}$. One may wonder whether the Bloch wall path is indeed optimal; but this is a simply consequence of $|\mathbf{m}'|^2 \geq (m_2')^2/(1 - m_2^2)$ that holds true for any $\mathbf{m} : \mathbb{R} \to \mathbb{S}^2 \in H^1$ and Young's inequality that imply the same lower energy bound.

Twenty years later, Louis Néel realized that in a regime where the film thickness becomes comparable to the Bloch wall width, a transition mode within the film plane can lower the total energy decisively, cf. [Née55]. New ideas, however, had to be developed in order to provide a satisfactory analysis of this multiscale object. Indeed, for an in-plane rotation $m : \mathbb{R} \to \mathbb{S}^1$ the thin-film approximation of (2.1) yields

$$E(\mathbf{m}) = \frac{d^2}{2} \int |m'|^2 \, dx + \frac{\delta}{4} \int |\xi| |\hat{m}_1(\xi)|^2 \, d\xi + \frac{Q}{2} \int |m_1|^2 \, dx. \tag{2.9}$$

Unlike the Bloch wall problem where only two energy components remain that can be balanced by a single length scale, the Néel wall problem incorporates two characteristic length scales. Those are connected with the competition of two energy components, respectively: In order to highlight the competition between *stray field* and *anisotropy* we rescale by the tail width $w = \delta/(2Q)$. With the small aspect ratio $\mathcal{Q} = 4\,\kappa^2\, Q$ where $\kappa = d/\delta \simeq 1$, we obtain the following singular perturbation problem

$$E_{\mathcal{Q}}(m) = \frac{\mathcal{Q}}{2}\,\|m\|^2_{\dot{H}^1} + \frac{1}{2}\|u\|^2_{\dot{H}^{1/2}} + \frac{1}{2}\|u\|^2_{L^2} \to \min \qquad (2.10)$$
$$m = (u,v) : \mathbb{R} \to \mathbb{S}^1 \quad \text{with} \quad u(0) = 1$$

that captures the logarithmic decay behavior as we will see below. There is a second characteristic length scale that is smaller than the tail width and related to the width of the core in the very center of the Néel wall. In the above regime it coincides merely with the exchange length d, and rescaling yields an expression

$$\frac{1}{2}\|m\|^2_{\dot{H}^1} + \frac{1}{4\kappa}\|u\|^2_{\dot{H}^{1/2}} + \frac{Q}{2}\|u\|^2_{L^2}$$

that, as Q tends to zero, highlights the competition between *exchange* and reduced *stray-field* energy.

2.2 The logarithmic tail of Néel walls

The main analytical feature of the variational problem (2.10) is that the energy gives only uniform control of the $H^{1/2}$-norm as $\mathcal{Q}$ tends to zero. Since the $H^{1/2}(\mathbb{R})$ norm just fails to control the modulus of continuity, the pointwise constraint $u(0) = 1$ is delicate and one might expect a logarithmic singularity in a renormalized setting. Logarithmic tails of Néel walls have indeed been predicted by heuristic arguments and the resulting very long range interaction between different Néel walls has important consequences, cf. [HS98] pp. 242-245, [RS71] with some extensions in [Gra99]. Logarithmic scaling for the energy has recently been established in [Gra99] and the following refined version is announced in [DK*99]:

Theorem 2.1. *As $\mathcal{Q}$ tends to zero the minimal energies behave like*

$$\inf E_{\mathcal{Q}} = \frac{\pi}{2}\big(1 + o(1)\big)\ln(1/\mathcal{Q})^{-1}$$

where the infimum is taken over transitions that are admissible according to (2.10).

Similar scaling laws have been derived in the case of periodic Néel wall arrays, where tails are confined by those of neighboring walls, cf. [DK*03]. They were used to heuristically quantify their mutual repulsive force, that is particularly interesting in the context of optimal spacing for the cross-tie wall. It is remarkable that, in case of finite Néel walls, the above energy asymptotics in Thm. 2.1 holds true when the infimum is taken over y-periodic transitions $m = m(x, y)$, cf. [DKO05]. Here, the quality factor Q is replaced by the aspect ratio δ/w, i.e. film thickness by tail width, so that $\mathcal{Q} = 4\,\kappa^2\,\frac{\delta}{w}$. The proof is based on a dynamic system argument and a sharp interpolation inequality between L^∞ and BV. The result proves in particular (nonlinear) stability of the one-dimensional Néel wall with respect to two-dimensional variations in the plane, a result yet unknown for infinite Néel walls.

The proof of Thm. 2.1 is based on a perturbation argument; its shows that minimal energies exhibit the same asymptotics as the minimal energies for the relaxed problem (2.12) to be introduced below. It turns out to provide an pointwise logarithmic lower bound as well and motivates the main result [Mel03, Mel04a]:

Theorem 2.2. *Let u_Q be a minimizing profile for the variational principle (2.10). Then u_Q is symmetric-decreasing and exhibits a logarithmic tail in the sense that*

$$u_{\mathcal{Q}}(x) \simeq \frac{\ln(1/x)}{\ln(1/\mathcal{Q})} \quad \text{for all } \mathcal{Q} \lesssim x \lesssim 1 \text{ and } 0 < \mathcal{Q} < 1/4.$$

The notations $a \lesssim b$ and $a \simeq b$ mean that, for some universal constant $0 < c < \infty$ we have $a \le c\,b$ and $\frac{1}{c} b \le a \le c\,b$, respectively. It can also be shown that beyond the logarithmic tail a Néel wall profile decays only quadratically $\ln(1/\mathcal{Q})\,u_{\mathcal{Q}}(x) \simeq x^{-2}$ as $|x| \to \infty$, cf. [Mel02], a fact that is related to the limited regularity of associated Fourier multipliers. Renormalization yields in addition a universal limiting profile that captures the essential decay behavior:

Theorem 2.3. *For any sequence $\mathcal{Q} \to 0$ so that the corresponding sequence of renormalized profiles $U_{\mathcal{Q}} = \ln(1/\mathcal{Q})\,u_{\mathcal{Q}}$ converges in the sense of distributions, the weak limit U_0 is a multiple of the fundamental solution of the operator $(-\Delta)^{1/2} + 1$, and the convergence is strong in L^2_{loc}.*

For fractional derivatives of order $s > 0$ we use the notation $(-\Delta)^{s/2} f = \mathcal{F}^*(\xi \mapsto |\xi|^s \hat{f}(\xi))$, where, for $s = 2$, we have $(-\Delta) f = -\frac{d^2 f}{dx^2}$. The fundamental solution $G \in L^2(\mathbb{R})$ with $(-\Delta)^{1/2} G + G = \delta_0$ is well known as a Fourier integral. It is smooth away from the origin, symmetrically decreasing and has the following expansion:

$$G(x) = \frac{1}{\pi} \begin{cases} \ln(1/|x|) - \gamma + \mathcal{O}(x) & \text{as } |x| \to 0 \\ x^{-2} + \mathcal{O}(x^{-4}) & \text{as } |x| \to \infty, \end{cases} \tag{2.11}$$

where γ denotes Euler's constant.

Logarithmic lower bounds

We introduce a linear comparison problem arising from relaxation that can be solved explicitly. It is shown in [Mel04a] that relaxation of (2.10) leads to

$$E_{\mathcal{Q}}^*(u) = \frac{\mathcal{Q}}{2}\|u\|_{\dot{H}^1}^2 + \frac{1}{2}\|u\|_{\dot{H}^{1/2}}^2 + \frac{1}{2}\|u\|_{L^2}^2 \to \min \quad \text{in} \quad \{u(0)=1\}. \tag{2.12}$$

A standard convexity argument implies the existence of a unique minimizer that satisfier the Euler-Lagrange equation

$$\mathcal{Q}(-\Delta)u^* + (-\Delta)^{1/2}u^* + u^* = \Lambda(\mathcal{Q})\,\delta_0 \quad \text{in} \quad \mathcal{D}'(\mathbb{R}). \tag{2.13}$$

Expanding the associate Fourier multiplier into partial fractions, (2.13) can be solved in terms of the fundamental solution (2.11), so that the following properties can be read off:

Proposition 2.4. *The unique solutions $u_{\mathcal{Q}}^*$ of the relaxed variational principles (2.12) exhibit logarithmic tails in the sense that for $0 < \mathcal{Q} < 1/4$:*

$$u_{\mathcal{Q}}^*(x) = \frac{\Lambda(\mathcal{Q})}{\pi}\big(1+o(1)\big)\big[\ln(1/x) + r_{\mathcal{Q}}(x)\big] \quad \textit{for} \quad \mathcal{Q} < x < 1/e,$$

where the functions $r_{\mathcal{Q}}(x)$ are uniformly bounded in the above regime. The Lagrange multiplier $\Lambda(\mathcal{Q})$ agrees with twice the minimal energy and has the asymptotic behavior (cf. Thm. 2.1)

$$\Lambda(\mathcal{Q}) = \pi\big(1+o(1)\big)\ln(1/\mathcal{Q})^{-1}.$$

Surprisingly, the relaxed variational principle not only provides an energetic but also a pointwise lower bound, thus, in view of Proposition 2.4, a logarithmic lower bound.

Proposition 2.5. *Let $u_{\mathcal{Q}}$ be a minimizing profile for variational principle (2.10). Then the solution $u_{\mathcal{Q}}^*$ of the relaxed variational principle (2.12) is a pointwise lower bound.*

Proof. The idea is to derive suitable pseudo-differential inequalities for the profiles to be compared. We observe that $|m'|^2 = |u'|^2/(1-u^2)$ and deduce the following Euler-Lagrange equation

$$\mathcal{Q}\left\{-\frac{d}{dx}\left(\frac{u'}{1-u^2}\right) + \left(\frac{u'}{1-u^2}\right)^2 u\right\} + (-\Delta)^{1/2}u + u = 0 \tag{2.14}$$

that holds true for a Néel wall profile u in $\{u \neq 1\}$. By Proposition 2.8 this equation holds true in $\mathbb{R}\setminus\{0\}$. Now the essential ingredients are symmetric convexity of comparison profiles stated in Proposition 2.4 and the following global maximum principle for the nonlocal field operator $(-\Delta)^{1/2}$:

Lemma 2.6. *Suppose that the function $u \in H^1(\mathbb{R})$ is smooth in $\mathbb{R}\setminus\{0\}$ and that u attains a global maximum at $x_0 \neq 0$. Then $(-\Delta)^{1/2}u$ is smooth in a neighborhood of x_0 and $(-\Delta)^{1/2}u(x_0) \geq 0$.*

We consider $w = u^* - u \in H^1(\mathbb{R})$ with $w(0) = 0$. From (2.13) and (2.14) we deduce, with the positive coefficient $a(x) = \mathcal{Q}/(1 - u^2(x))$, that

$$a(x)(-\Delta)w + (-\Delta)^{1/2}w + w \leq 0 \quad \text{in} \quad \mathbb{R} \setminus \{0\}.$$

Since w is smooth away from the origin, the Lemma 2.6 applies and excludes a global maximum in $\mathbb{R} \setminus \{0\}$. But from Propositions 2.4 and 2.8 we infer that $w(x)$ decays as $|x| \to \infty$, and the proof is complete.

Logarithmic upper bounds

The key observation is that logarithmic upper bounds are captured by sharp elliptic regularity bounds that are uniform in $\mathcal{Q}$. For magnetizations $m_\mathcal{Q}$ of bounded Néel wall energy (2.10) we have $\|u_\mathcal{Q}\|^2_{H^{1/2}} \leq E_\mathcal{Q}(m_\mathcal{Q})$. Then Sobolev embedding implies, for any $p \in (2, \infty)$, a bound $\|u_\mathcal{Q}\|^2_{L^p} \leq c(p) E_\mathcal{Q}(m_\mathcal{Q})$. A PDE argument, however, shows that for any such p the purely energetic argument misses the optimal scaling by a full factor $E_\mathcal{Q}(m_\mathcal{Q})$ and provides in addition an estimate on the growth of optimal constants. Qualitatively, the same is true for fractional Sobolev norms $H^{1/2}_q$ that are strictly weaker than $H^{1/2}$.

Proposition 2.7. *For a critical point $m_\mathcal{Q} = (u_\mathcal{Q}, v_\mathcal{Q})$ of (2.10) we have*

$$\|u_\mathcal{Q}\|_{L^p} \leq c\, p\, E_\mathcal{Q}(m_\mathcal{Q}) \quad \textit{for each} \quad p \in (1, \infty) \tag{2.15}$$

for some universal constant $c > 0$ and

$$\|u_\mathcal{Q}\|_{H^{1/2}_q} \leq c(q)\, E_\mathcal{Q}(m_\mathcal{Q}) \quad \textit{for each} \quad q \in (1, 2), \tag{2.16}$$

where the constant $c(q) > 0$ only depends on q.

Proof. We outline the main steps: Projection of the Euler-Lagrange system

$$\nabla E_\mathcal{Q}(m) = \nabla E_\mathcal{Q}(m)\, m \otimes m \tag{2.17}$$

onto its first component equation yields, after a suitable decomposition of the non-linearity, an equation of the form

$$\mathcal{Q}\,(-\Delta)u + (-\Delta)^{1/2}u + u = e_\mathcal{Q}[m]\, u + r[u], \tag{2.18}$$

where $e_\mathcal{Q}[m] = \mathcal{Q}\,|m'|^2 + |(-\Delta)^{1/4}u|^2 + |u|^2$ is twice the energy density and $r[u]$ is a defect distribution arising from the in compatibility of nonlocal interaction and the geometric constraint $|m| = 1$. In fact, for any test function φ, we have

$$\langle r[u], \varphi \rangle = \Big\langle (-\Delta)^{1/4}u,\, \big[(-\Delta)^{1/4}, (u\,\varphi)\big] u \Big\rangle.$$

The operator on the left hand side of (2.18) is uniformly first-order elliptic, while the right hand side is essentially L^1-bounded by the energy, and in that case the claim would follow from a simple Fourier argument. Commutator estimates, however, show that uniform bounds for $r[u]$ are slightly weaker than L^1 and rather distributional, i.e in $H_q^{-1/2}$ for $q \in (1,2)$. By means of elliptic regularity theory we get a uniform bounds in $H_q^{1/2}$, and the claim follows from asymptotic inequalities for fractional integration.

From strict rearrangement inequalities that are valid for fractional Sobolev norms and a simple bootstrap argument based on the Euler-Lagrange system (2.17), we deduce the following symmetry and smoothness result for optimal profiles $u = u_Q$.

Proposition 2.8. *A Néel wall profile is smooth and symmetrically decreasing.*

Remark 2.9. The above proposition is strict in the sense that a Néel wall profile $m = e^{i\theta}$ cannot have a plateau at 0, and the associate phase function θ is strictly increasing, cf. Lemma 2.18.

Proof of Thm. 2.2

In view of Propositions 2.4 and 2.5 it remains to prove the logarithmic upper bound. Let $u = u_Q$ be a Néel wall profile. Proposition 2.8 implies that the pointwise values are below the local averages. Thus Hölder's inequality and Proposition 2.7 yield

$$0 \le u(x) \le \fint_0^x u\, dy \le \left(\fint_0^x |u|^p\, dy \right)^{1/p} \le c\, p \left(\frac{1}{x} \right)^{1/p} \inf_{\mathcal{M}} E_Q$$

that is a family of upper bounds parameterized by p. The pointwise optimal choice of p, given by $p(x) = \ln(1/x)$, and Thm. 2.1 yield the logarithmic upper bound.

Proof of Thm. 2.3

From Thm. 2.1 and Proposition 2.7 we deduce that U_Q is uniformly bounded in, say, $H^{1/2}_{3/2}(\mathbb{R})$, so we can assume that $U_Q \rightharpoonup U_0$ weakly in $L^2(\mathbb{R})$ and strongly on bounded intervals. From the profile equation (2.18) we get

$$\mathcal{Q}(-\Delta)U_Q + (-\Delta)^{1/2}U_Q + U_Q = \ln(1/\mathcal{Q})\Big(e_Q[m_Q]\, u_Q + r[u_Q]\Big).$$

Obviously, the left hand side converges to $(-\Delta)^{1/2}U_0 + U_0$ in the sense of distributions. Thus it remains to show that

(i) the distribution $(-\Delta)^{1/2}U_0 + U_0$ is supported at the origin
(ii) the distribution $\ln(1/\mathcal{Q})r[u_Q]$ converges to a finite measure.

Claim (i) can be deduced from (2.14) using the uniform pointwise convergence of Néel wall profiles to zero away from the origin. Claim (ii) follows from an iteration of the arguments in of Proposition 2.7. Indeed, for $u = u_Q$ we have a decomposition of the remainder distribution $\langle r[u], \varphi\rangle = \langle f[u], \varphi\rangle + \langle g[u], \varphi\rangle$, where

$$\langle f[u], \varphi\rangle = \langle (-\Delta)^{1/4} u, [(-\Delta)^{1/4}, u^2]\varphi\rangle,$$
$$\langle g[u], \varphi\rangle = \langle (-\Delta)^{1/4} u, u^2 (-\Delta)^{1/4}\varphi\rangle.$$

Product and commutator estimates for fractional derivatives (cf. [Hof98, KP88]) and the energy estimate in Thm. 2.1 show that

$$|\langle f[u], \varphi\rangle| \le c\, \|u\|_{L^\infty} \|u\|^2_{H^{1/2}} \|\varphi\|_{C^0} \le c \ln(1/\mathcal{Q})^{-1} \|\varphi\|_{C^0}.$$

Thus we find that the contribution coming from $\ln(\mathcal{Q})\, f[u]$ is asymptotically a finite measure. On the other hand, according to the $H^{1/2}_q$ estimate in Proposition 2.7,

$$|\langle g[u], \varphi\rangle| \le c\, \|u\|_{L^\infty} \|u\|_{L^6} \|u\|_{H^{1/2}_{3/2}} \|\varphi\|_{H^{1/2}_6} \le c\, \ln(1/\mathcal{Q})^{-2} \|\varphi\|_{H^{1/2}_6},$$

so that the contribution from $\ln(\mathcal{Q}) g[u]$ vanishes as a distribution as $\mathcal{Q}$ tends to 0.

2.3 Domain wall motion in finite layers

When an external magnetic field $\mathbf{h} = H\,\hat{\mathbf{e}}_2$ is applied that points towards the easy axis, the end-states are no longer equally preferred. Consequently, one expects the domain wall to become unstable and start to move. In gyrotropic domain wall models the evolution of magnetization distributions is characterized by the Landau-Lifshitz-Gilbert equation

$$\begin{aligned} &\mathbf{m} \wedge \partial_t \mathbf{m} + \alpha\, \partial_t \mathbf{m} + \gamma\,(1 - \mathbf{m}\otimes\mathbf{m}) \nabla E(\mathbf{m}) = (1 - \mathbf{m}\otimes\mathbf{m})\,\mathbf{h} \qquad (2.19)\\ &\mathbf{m} : \mathbb{R}\times(0,\infty) \to \mathbb{S}^2 \quad \text{with} \quad \mathbf{m}(\pm\infty, t) = (0, \pm 1, 0) \quad \text{for} \quad t \in (0,\infty). \end{aligned}$$

where $E(\mathbf{m})$ is the internal domain wall energy. We introduce the aspect ratio $\kappa = d/\delta$ and assume for simplicity that $Q = 1$. Renormalizing space and energy by the exchange length d, we get from (2.1) and (2.4) an (internal) domain wall energy of the form

$$E_\kappa(\mathbf{m}) = \frac{1}{2}\int |\mathbf{m}'|^2 + \frac{1}{2}\int (1 - m_2^2)\, dx + \frac{1}{2}\int m_1^2\, dx + G_\kappa(\mathbf{m}), \qquad (2.20)$$

where the nonlocal portion of the stray-field energy is given by

$$G_\kappa(\mathbf{m}) = \frac{1}{2}\int \big(1 - \sigma(\kappa\xi)\big)\Big(|\hat{m}_3(\xi)|^2 - |\hat{m}_1(\xi)|^2\Big)\, d\xi. \qquad (2.21)$$

Regarding the dynamic problem, a special class of solutions to constant coefficient systems are traveling wave solutions, i.e. solutions of the form $\mathbf{m} = \mathbf{m}(x + c\,t)$, that describe a motion of constant speed c. In the case $\kappa = 0$ the traveling wave ansatz turns (2.19) into a constrained nonlinear system of ordinary differential equations. Surprisingly, these equations can be solved explicitly. The solutions are referred to as Walker's exact solutions, see [Wal63, HS98]. Our goal is to show that this situation is indeed generic and can be perturbed to layers of large but finite diameter. As this corresponds to the case of small nonlocal interaction, the nonlocal character of the equations will play a minor role. We will show that, in suitable coordinates, domain wall motion according to Landau-Lifshitz dynamics fits into the context of nonlocal, weakly coupled reaction-diffusion systems. But first we review Walker's construction.

Walker's exact solutions

For $\kappa = 0$ we have a transition energy

$$E_0(\mathbf{m}) = \frac{1}{2}\int |\mathbf{m}'|^2\,dx + \frac{1}{2}\int (1 - m_2^2)\,dx + \frac{1}{2}\int m_1^2\,dx.$$

The traveling wave ansatz $\mathbf{m} = \mathbf{m}(x + c\,\gamma\,t)$ yields the system

$$c\,\alpha\,\mathbf{m}' + c\,\mathbf{m} \wedge \mathbf{m}' + \big(1 - \mathbf{m} \otimes \mathbf{m}\big)\nabla E_0(\mathbf{m}) = \big(1 - \mathbf{m} \otimes \mathbf{m}\big) \cdot \mathbf{h}. \qquad (2.22)$$

The above system can be solved explicitly, a calculation that has been first carried out by Walker. The original calculations become more transparent, when the equation is considered in the canonical orthogonal frame $\{\mathbf{m}', \mathbf{m} \wedge \mathbf{m}'\}$ on the tangent space of $\mathbb{S}^2$ along $\mathbf{m}$. It turns out that the following assumptions can be met:

(I) Under the assumption that dissipation compensates the driving force, the system decomposes into three equations,

$$c\,\alpha\,|\mathbf{m}'|^2 = H\,m_2', \quad \nabla E_0(\mathbf{m}) \cdot \mathbf{m}' = 0, \quad \text{and} \quad \nabla E_0(\mathbf{m}) \cdot \mathbf{m} \wedge \mathbf{m}' = c\,|\mathbf{m}'|^2.$$

(II) Under the assumption that the wall moves with constant polar inclination angle φ, i.e. $\mathbf{m} \wedge \mathbf{m}' = |\mathbf{m}'|\nu$ for some constant unit vector ν, the energy is the Bloch wall energy with increased anisotropy $Q(\varphi) = 1 + \sin^2\varphi$,

$$E_0(\mathbf{m}) = \frac{1}{2}\int |\mathbf{m}'|^2\,dx + \frac{Q(\varphi)}{2}\int (1 - m_2^2)\,dx.$$

Moreover $|\mathbf{m}'|^2 = (m_2')^2/(1 - m_2^2)$ and $\frac{\partial \mathbf{m}}{\partial \varphi} = |\mathbf{m}'|\nu$ holds for such $\mathbf{m}$.

We deduce that, up to scaling, all equations in (I) have the form $|\mathbf{m}'|^2 = 1 - m_2^2$, i.e. $m_2' = 1 - m_2^2$, and can be solved jointly. Matching parameters gives the transition profile, inclination angle, and propagation speed

$$m_2(x) = \tanh\left[\sqrt{1+\sin^2\varphi}\,x\right], \ \sin(2\varphi) = \frac{H}{\alpha}, \ \text{and } c = \frac{\sin(2\varphi)}{2\sqrt{1+\sin^2\varphi}}. \quad (2.23)$$

Obviously a peak velocity ~ 0.4 (i.e. about $\sim 100\frac{m}{sec}$ for a typical garnet material) is reached at for finite field-strength H beyond which the construction breaks down.

Walker's construction shows that the dynamics of domain walls in bulks samples is accompanied by a decrease of domain wall width by a factor $(1+\sin^2\varphi)^{-\frac{1}{2}}$ and an increase of domain wall energy by the inverse factor. From Walker's construction we deduce that the first order correction for domain wall energies vanishes at small velocities:

$$e_H = e_0\,(1+\sin^2\varphi)^{\frac{1}{2}} = e_0 + \frac{1}{2}\,M\,c^2 + o\left(c^2\right).$$

The second order correction can be viewed as a kinetic energy contribution. Thus, the factor $M = \partial_c^2 e_H$ at $c=0$ is referred to as the wall mass, a notion that has been introduced by Döring, cf. [Dör48]. In the original scaling the wall mass is given by $M = e_0/(2d^2)$. We will encounter the wall mass again later in the context of traveling Néel walls rising naturally from a wave-type interpretation of Landau-Lifshitz-Gilbert dynamics in thin films.

Stability and perturbation of Walker's construction

Theorem 2.10. *For sufficiently small field strength H there is a threshold $\kappa(H) > 0$ such that, whenever $\kappa < \kappa(H)$, there is a traveling wave for the Landau-Lifshitz-Gilbert dynamics that connects antipodal states.*

Proof. We perform a stability analysis based on a suitable choice of canonical coordinates that transforms (2.19) into a weakly coupled 2×2 system of reaction-diffusion type. For this purpose we combine standard stereographic coordinates with a polar rotation by the Walker angle φ that maps the Walker path into the Bloch wall path. In these coordinates $\mathbb{C} \ni z \mapsto \mathbf{m}[z] \in \mathbb{S}^2$, the Walker path is given by the line segment $z_0 : \mathbb{R} \to \{0\}\times[-1,1]$. For $\mathbf{m} = \mathbf{m}[z]$, functional gradients transform according to $\lambda^{-2}(z)\nabla_z E_\kappa(\mathbf{m}) = \nabla E_\kappa(\mathbf{m})$, where $\lambda(z)$ is the conformal factor. Moreover, (2.19) becomes a damped Schrödinger equation

$$\alpha\,\partial_t z + i\,\partial_t z - \gamma\,D_x\partial_x z + \gamma\left(f(H,z) + b(\kappa,z)\right) = 0, \qquad (2.24)$$

where $z(\cdot,t) : \mathbb{R} \to \mathbb{C}$ with $z(\pm\infty,t) = (0,\pm 1)$. The mapping

$$f(H,z) = \lambda^{-2}(z)\,\nabla_z \int \left[\frac{1}{2}(m_1^2 - m_2^2) - H\,m_2\right] dx \quad \text{for} \quad \mathbf{m} = \mathbf{m}[z]$$

involves anisotropy, applied field, and the limiting (local) portion of stray-field interaction. For small enough H, it is bi-stable in its second component. With the notation in (2.21), the map $b(\kappa,z) = \lambda^{-2}(z)\,\nabla_z G_\kappa(\mathbf{m})$ is a nonlocal perturbation from stray-field interaction that is continuous in κ with $b(0,\cdot)$=0.

Remark 2.11. An important observation is that the mapping $\kappa \mapsto b(\kappa, \cdot)$, considered as a family of nonlinear operators on suitable function spaces, is not differentiable at $\kappa = 0$. Indeed,

$$\frac{d}{d\kappa} E_{\text{stray}}(\mathbf{m})\big|_{\kappa=0} = \frac{1}{2}\|m_1\|^2_{\dot{H}^{-1/2}} - \frac{1}{2}\|m_3\|^2_{\dot{H}^{-1/2}}$$

with singular behavior at low frequencies that conflicts with slow decay properties in the presence of internal stray fields. The perturbation at $\kappa \sim 0$ is therefore singular, and only continuous versions of the implicit function theorem are at our disposal.

Finally,

$$D_x \partial_x z = \partial_x^2 z + \Gamma(z)\langle \partial_x z, \partial_x z \rangle$$

denotes the second covariant derivative. Surprisingly, the form of Walker solutions remains almost unchanged:

$$z_0(x) = \left(0, \tanh\left[\frac{1}{2}\left(1 + \sin^2\varphi\right)^{1/2} x\right]\right).$$

Recall that the associate propagation speed c_0 inherits the Walker angle φ, so for each H we identify the Walker solution with the pair (z_0, c_0). Introducing a moving frame $x \mapsto x + c\gamma t$, (2.24) reads like

$$G\big((z,c),\kappa\big) = -D_x\partial_x z + c(\alpha + i)\partial_x z + f(H,z) + b(\kappa, z) = 0.$$

It turns out that our choice of canonical stereographic coordinates provides an almost triangulation for the linearized problem. Its spectral properties can be summarized as follows:

Proposition 2.12. *For sufficiently small field strength H the linearization at the Walker solution (z_0, c_0), has the form*

$$\frac{\partial G}{\partial z}\big((z_0, c_0), 0\big) = \begin{bmatrix} L_1 & M_2 \\ M_1 & L_2 \end{bmatrix} : H^2(\mathbb{R};\mathbb{C}) \to L^2(\mathbb{R};\mathbb{C}),$$

where $L_1 : H^2(\mathbb{R}) \to L^2(\mathbb{R})$ has a bounded inverse while L_2 and its L^2-adjoint have zero as a simple eigenvalue with eigenfunctions $v_0' = \operatorname{Im} z_0'$ and ψ_0, respectively, so that the integral $\int v_0\, \psi_0\, dx > 0$ exists. Moreover, $\|M_1\|$ can be made arbitrarily small by choosing H small.

Since $z_0'(x) = \frac{d}{d\lambda}|_{\lambda=0} z_0(x+\lambda)$ can be seen as the infinitesimal generator of translation symmetry, the proposition suggests that degeneracy only stems from translation invariance. Thus we introduce the extended functional equation

$$\mathcal{G}\big((z,c),\kappa\big) = \Big[G\big((z,c),\kappa\big), \operatorname{Im} z(0)\Big] = (0,0). \tag{2.25}$$

Its linearization with respect to (z, c) at the Walker solution (z_0, c_0) has the form

$$\mathcal{L}_0 = \begin{bmatrix} L_1 & M_2 & -v_0' \\ M_1 & L_2 & \alpha\, v_0' \\ 0 & \delta_0 & 0 \end{bmatrix}.$$

In view of Proposition 2.12, the invertibility of $\mathcal{L}_0 : H^2(\mathbb{R};\mathbb{C})\times\mathbb{R} \to L^2(\mathbb{R};\mathbb{C})\times\mathbb{R}$ for sufficiently small H would follow from a Schur-type argument once we have shown invertibility of the 2×2 matrix on the lower right. But this follows from a standard Fredholm argument (cf. the proof of Thm. 2.15), taking into account Proposition 2.12 and the positivity of v_0'. Now the continuous version of the implicit function theorem implies the solvability of (2.25) for sufficiently small $\kappa > 0$.

2.4 Domain wall motion in thin films

A wave-type limit for Landau-Lifshitz-Gilbert

Gyromagnetic precession is geometrically incompatible with the asymptotic constraint of in-plane magnetization that is imposed by stray-filed interaction, in other words, the competition between energetic and dynamic forces becomes singular in a thin-film limit. Thus domain wall motion in thin films should be governed by a suitable effective limit for LLG

$$\partial_t \mathbf{m} + \gamma\, \mathbf{m} \wedge \nabla E(\mathbf{m}) + \alpha\, \mathbf{m} \wedge \partial_t \mathbf{m} = 0 \tag{2.26}$$

as the relative thickness δ/d tends to zero. We recall that Gilbert damping α is a small parameter as well, that is to say, precession proceeds much faster than relaxation. Prior work on thin-film reductions for LLG, leading to enhanced dissipation, cf. [EG01, KS05a] and Thm. 4.9, consider the regime when $\delta/d \ll \alpha$. In order to preserve the oscillatory features of LLG dynamics we take into account small Gilbert damping as well. As it turns out, the effective dynamics depends on asymptotic regime as α and the relative thickness δ/d tend to zero. Rescaling space by the tail width $w = \delta/(2Q)$ and energy by the quality factor, we get, for 3-dimensional transitions $\mathbf{m} = (m, m_3) : \mathbb{R} \to \mathbb{S}^2$, a domain wall energy of the form

$$E_\varepsilon(\mathbf{m}) = E_0(m) + \frac{\mathcal{Q}}{2} \int |m_3'|^2\, dx + \frac{1}{2\varepsilon^2} \|m_3\|_{L^2}^2, \tag{2.27}$$

where $\mathcal{Q} = 4\,\kappa^2 Q$ and $\varepsilon^2 = Q$. The in-plane portion of the energy $E_0(m)$ is given by

$$E_0(m) = \frac{\mathcal{Q}}{2} \int |m'|^2 + \frac{1}{2} \left\|m_1\right\|_{\dot{H}^{1/2}}^2 + \frac{1}{2} \int (1 - m_2^2)\, dx. \tag{2.28}$$

For in-plane magnetizations, it agrees with the standard Néel wall energy that we have considered before. We investigate the regime when $\varepsilon \to 0$ while $\mathcal{Q}$ is uniformly bounded from above and below; in other words $\varepsilon \sim \delta/d$ can be considered as a relative thickness. Let us consider the associated LLG equation in the asymptotic regime when $\alpha(\varepsilon)/\varepsilon \to \nu$. Rescaling time by ε/γ, the system (2.26) becomes

$$\mathbf{m}_t + \varepsilon\, \mathbf{m} \wedge \nabla E_\varepsilon(\mathbf{m}) + \alpha\, \mathbf{m} \wedge \mathbf{m}_t = 0. \tag{2.29}$$

Theorem 2.13. *Let* $\mathbf{m}_\varepsilon : \mathbb{R} \times (0,\infty) \to \mathbb{S}^2$ *be a family of global solutions of (2.29) with uniformly bounded initial energy* $E_\varepsilon(\mathbf{m}_\varepsilon(0)) \le c$. *Suppose that* $\alpha(\varepsilon)/\varepsilon \to \nu$ *and the in-plane components* $m_\varepsilon \rightharpoonup m$ *converge locally in* L^2. *Then*

$$\left[\partial_t^2 m + \nu\, \partial_t m + \nabla E_0(m)\right] \wedge m = 0. \tag{2.30}$$

Proof. We let $\mathbf{m} = (m, \varepsilon v)$, i.e. we blow-up the vertical component. Then the energy can be written as $E_\varepsilon(\mathbf{m}) = E_0(m) + G_\varepsilon(v)$, where $G_\varepsilon(v) = \frac{1}{2}\int \varepsilon^2 |v'|^2 + |v|^2\, dx$. For $\nu_\varepsilon = \alpha(\varepsilon)/\varepsilon$, the Landau-Lifshitz-Gilbert system (2.29) can be written as

$$\partial_t \begin{pmatrix} m \\ v \end{pmatrix} + \begin{bmatrix} 0 & -\varepsilon^2 v & m_2 \\ \varepsilon^2 v & 0 & -m_1 \\ -m_2 & m_1 & 0 \end{bmatrix} \begin{pmatrix} \nabla E(m) + \nu_\varepsilon\, \partial_t m \\ \nabla G_\varepsilon(v) + \varepsilon^2 \nu_\varepsilon\, \partial_t v \end{pmatrix} = 0.$$

The energy inequality implies the requisite a priori estimates

$$\nu_\varepsilon \int_0^T \|\partial_t m_\varepsilon\|_{L^2}^2 + E_0(m_\varepsilon(T)) \le E_\varepsilon(\mathbf{m}_\varepsilon(0)),$$

$$\varepsilon^2 \nu_\varepsilon \int_0^T \|\partial_t v_\varepsilon\|_{L^2}^2 + G_\varepsilon(v_\varepsilon(T)) \le E_\varepsilon(\mathbf{m}_\varepsilon(0)),$$

and passing to the limit yields the following set of equations

$$\begin{aligned} \partial_t m - v\, m^\perp &= 0, \\ \partial_t v + m^\perp \cdot \left[\nabla E(m) + \nu\, \partial_t m\right] &= 0. \end{aligned}$$

From that system the vertical blow-up function v can be eliminated. Indeed, the first equation is equivalent to $\partial_t v = \partial_t^2 m \cdot m^\perp$. Substitution into the second equation yields the result.

Remark 2.14. Under further regularity assumptions, especially the validity of the energy inequality, the asymptotic limit holds true in higher dimensional situations as well.

Traveling waves and kinematic properties of Néel walls

The latter asymptotic limit suggests the following dynamic model for the evolution of Néel walls in thin films under the influence of a constant applied field $h = H\hat{\mathbf{e}}_2$ that points towards one of the end states determined by anisotropy:

$$\Big(\partial_t^2 m + \nu\,\partial_t m + \nabla E_0(m)\Big) \wedge m = h \wedge m, \tag{2.31}$$
$$m : \mathbb{R} \times [0,T) \to \mathbb{S}^1 \text{ with } m(\pm\infty, t) = (0, \pm 1).$$

Representing the transition vector in polar coordinates $m = e^{i\theta}$, the transition energy becomes

$$\mathcal{E}(\theta) = \frac{\mathcal{Q}}{2}\int |\theta'|^2 \, dx + \frac{1}{2}\|\cos\theta\|^2_{H^{1/2}} \quad \text{for} \quad \theta(\pm\infty) = \pm\frac{\pi}{2},$$

where $\|f\|^2_{H^{1/2}} = \int \big(1+|\xi|\big)|\hat{f}(\xi)|^2 \, d\xi$ denotes the full $H^{1/2}$ norm incorporating anisotropy and stray-field interaction. Then the reduced dynamic equation (2.31) reads

$$\partial_t^2\theta + \nu\,\partial_t\theta + \nabla\mathcal{E}(\theta) = h \cdot ie^{i\theta}, \tag{2.32}$$
$$\theta : \mathbb{R} \times [0,T) \to \mathbb{R} \text{ with } \theta(\pm\infty, t) = (0, \pm\pi/2).$$

The latter damped wave dynamics invites for a kinematic interpretation for the wall center as a point mass with constant force and dynamic friction. The argument will be rather informal; asymptotically, however, the kinematic findings will be justified rigorously in the context of the traveling wave result below. Indeed, if H is assumed to be suitably small, the moving phase $\theta = \theta(x,t)$ is presumably close to the stationary phase profile θ_0 shifted by $q(t)$, the center of the wall at time t. Hence we make an ansatz $\theta(x,t) = \theta_0(x,t) + \theta_1(x,t)$ where, with a slight abuse of notation, $\theta_0(x,t) = \theta_0(x + q(t))$ and $\theta_1(x,t)$ is assumed to be a small perturbation. Then we approximate

$$\nabla\mathcal{E}(\theta) = \nabla\mathcal{E}(\theta_0) + L_0\theta_1, \quad \text{where} \quad L_0 = D\nabla\mathcal{E}(\theta_0),$$

and $\cos(\theta) = \cos(\theta_0) - \sin(\theta_0)\theta_1$. Now if θ_1 is a solution of the linearized problem $\partial_t^2\theta_1 + \nu\partial_t\theta_1 + L_0\theta_1 + H\sin(\theta_q)\,\theta_1 = 0$, then

$$\theta_0''\,|\dot{q}|^2 + \theta_0'\,\ddot{q} + \nu\,\theta_0'\,\dot{q} = H\,\cos\theta_0.$$

Thus, the associate momentum $p(t) = M\,\dot{q}(t)$, where $M = \frac{1}{2}\int |\theta_0'|^2 \, dx$ can be interpreted as the wall mass, satisfies the equation $\dot{p} + \nu\,p = H$ with terminal momentum $p^* = H/\nu$. We observe that M is consistent with the wall mass we encountered in section 2.3 rising from the infinitesimal increase of energy. Accordingly, the mobility, i.e. the rate of change of propagation speed with respect to H, is given by $\beta = 1/(M\nu)$, consistent with our rigorous perturbation result:

Theorem 2.15. *For sufficiently small field strength H there is a traveling wave for the reduced Landau-Lifshitz-Gilbert dynamics*

$$c^2\,\theta'' + c\,\nu\,\theta' + \nabla\mathcal{E}(\theta) = H\cos\theta$$

that connects antipodal states at infinity $\theta(\pm\infty) = \pm\pi/2$. Moreover, the propagation speed has an expansion $c = \beta H + o(H)$ where the wall mobility $\beta = 1/(M\nu)$ is related to the wall mass $M = \frac{1}{2}\int |\theta_0'|^2 \, dx$ taken from a stationary Néel wall θ_0.

Proposition 2.16. *Suppose that* θ_0 *is a critical point of* $\mathcal{E}(\theta)$ *subjected to center and boundary conditions* $\theta(0) = 0$ *and* $\theta(\pm\infty) = \pm\pi/2$. *Then the Hessian*

$$D^2\mathcal{E}(\theta_0)\langle\varphi,\varphi\rangle \geq \int |\theta'|^2|\varphi|^2 \, dx + \left\|\varphi \sin\theta\right\|^2_{H^{1/2}} \geq 0$$

is non-negative for any admissible variation φ.

Proof. Let $\langle f,g\rangle_{H^{1/2}} = \mathrm{Re} \int \big(1+|\xi|\big) \hat{f}(\xi)\, \bar{\hat{g}}(\xi)\, d\xi$ be the $H^{1/2}$ inner product. Then

$$D^2\mathcal{E}(\theta_0)\langle\varphi,\varphi\rangle = \mathcal{Q}\int |\varphi'|^2 - \big\langle \cos\theta_0, \varphi^2\cos\theta_0\big\rangle_{H^{1/2}} + \left\|\varphi\sin\theta\right\|^2_{H^{1/2}}.$$

In order to estimate the middle term, we deduce from the Euler-Lagrange equation

$$\mathcal{Q}\int \theta_0' \left(\varphi^2 \cot\theta_0\right)' \, dx = \big\langle \cos\theta_0, \varphi^2\cos\theta_0\big\rangle_{H^{1/2}}.$$

Recalling that $-d(\cot\theta)/d\theta = 1+\cot^2\theta$, then the claim follows immediately.

The Proposition states in particular that any critical point of $\mathcal{E}$ is in fact the phase of a minimizing Néel wall.

Corollary 2.17. *For centered Néel walls* θ_0, *the linearization* $L_0 = D\nabla\mathcal{E}(\theta_0)$ *extends to a self-adjoint operator on* $L^2(\mathbb{R})$ *having zero as a simple eigenvalue with eigenspace spanned by* θ_0'.

We need the following refinement of Proposition 2.8 that in particular rules out a plateau of Néel wall profiles:

Lemma 2.18. *The phase* θ_0 *of a stationary Néel wall is strictly increasing.*

Now the proof of Thm. 2.15 follows closely the one carried out in [BF*97]. We let $G\big((\theta,c),H\big) = c^2\,\theta'' + c\,\nu\,\theta' + \nabla\mathcal{E}(\theta) - H\cos\theta$ and consider the extended functional equation

$$\mathcal{G}\big((\theta,c),H\big) = \big[G\big((\theta,c),H\big),\theta(0)\big] = (0,0).$$

Observe that for a stationary Néel wall θ_0 and $\theta = \theta_0 + \delta\theta$, the mapping $\big((\theta,c),H\big) \mapsto \mathcal{G}\big((\theta,c),H\big)$ is smooth. The linearization with respect to the first two components (θ,c) at the stationary Néel wall $(\theta_0,0)$ reads like

$$\mathcal{L}_0 = \begin{bmatrix} L_0 & \nu\,\theta_0' \\ 0 & \delta_0 \end{bmatrix}.$$

As a mapping $\mathcal{L}_0 : H^2(\mathbb{R})\times\mathbb{R} \to L^2(\mathbb{R})\times\mathbb{R}$ it has a bounded inverse, i.e. for every $(f,b) \in L^2(\mathbb{R})\times\mathbb{R}$ there is $(\phi,c) \in H^2(\mathbb{R})\times\mathbb{R}$ so that $L_0\psi + \nu\,\theta_0'\,c = f$ and $\phi(0) = b$. According to the Fredholm alternative and Corollary 2.17 the first equation is solvable provided $c\,\nu\int|\theta_0'|^2\,dx = \int f\,\theta_0'\,dx$ that fixes c and determines ϕ up to a multiple of θ_0'. But in view of Lemma 2.18, the second equation provides uniqueness. Now the implicit function theorem implies the existence of a differentiable branch $H \mapsto (\theta(H),c(H))$, so that $c'(0)\,\nu\int|\theta_0'|^2\,dx = 2H$, and the claim follows.

3 Boundary vortices in a 2D model

In this section, we present results on a specific thin-film limit of the magnetic energy for a special regime of rather small films. We will analyze a limit of a two-dimensional functional derived by Kohn and Slastikov [KS05b]. They considered *soft* magnetic films without an external field, which corresponds to a functional that uses only the exchange and magnetostatic terms. They studied the asymptotic behavior of the corresponding version of (1.1) in a thin film $\Omega_\delta = \Omega \times (0, \delta)$ with $\operatorname{diam} \Omega = 1$ for $\delta \to 0$ and $\frac{d^2}{\delta \log \frac{1}{\delta}} \to \frac{\varepsilon}{2\pi} \in (0, \infty)$. The energy divided by $\frac{2\pi\delta^2}{\varepsilon d^2}$ then Γ-converges to the limit functional

$$E_{KS}^\varepsilon(m) = \frac{1}{2} \int_\Omega |\nabla m|^2 + \frac{1}{2\varepsilon} \int_{\partial\Omega} (m \cdot \nu)^2 \tag{3.1}$$

defined on maps $m \in H^1(\Omega, S^1)$ (so $m_3 = 0$). Here ν is a unit normal to $\partial\Omega$. The Kohn-Slastikov theorem shows that for this special scaling, the nonlocal contribution arising as the energy of the induced field reduces to a *local* term charging the boundary. The reason for this simplification of the functional lies in a separation of scales between the energy contribution of volume and surface charges to the field energy.

In the following, we present results of Kurzke [Kur06, Kur04, Kur05] on the limit of (3.1) as $\varepsilon \to 0$, for a simply connected domain Ω. The results share some features with those of Moser [Mos03, Mos04] that are presented in Section 4. In particular, sequences of minimizers develop vortices on the boundary.

Due the two-scale process of first letting $\delta \to 0$ to obtain (3.1), a two-dimensional problem, and then letting $\varepsilon \to 0$, our approach can be seen as a simplified model for the boundary vortices in Section 4. Since our functional here is local, the results are more detailed, especially for the asymptotic dynamics.

The sequence of functionals E_{KS}^ε is rather similar to the Ginzburg-Landau functional for superconductivity of [BBH94]. With $m_0 = \tau$ being a continuous unit tangent field to $\partial\Omega$, we are (after rescaling and renaming variables) considering the variational problem for $m : \Omega \to \mathbb{R}^2$: Minimize

$$\frac{1}{2} \int_\Omega |\nabla m|^2 + \frac{1}{2\varepsilon} \int_{\partial\Omega} (1 - (m \cdot m_0)^2) d\mathcal{H}^1$$

subject to $|m| = 1$ in Ω as $\varepsilon \to 0$. This problem has an interior constraint and a boundary penalty.

Bethuel, Brezis and Hélein [BBH94] studied the behavior as $\varepsilon \to 0$ of

$$\frac{1}{2} \int_\Omega |\nabla m|^2 + \frac{1}{4\varepsilon^2} \int_\Omega (1 - |m|^2)^2$$

subject to $m = m_0$ on $\partial\Omega$, so this problem has a boundary constraint and an interior penalty.

Common to both problems is that, as long as m_0 has nonzero topological degree, there is no map in $H^1(\Omega;\mathbb{R}^2)$ that satisfies the constraint and makes the penalty term zero. This is due to the fact that a continuous map $w : \partial\Omega \to S^1$ can be extended to a continuous map $\overline{w} : \overline{\Omega} \to S^1$ if and only if $\deg(w) = 0$. Although H^1 maps need not be continuous, the argument still carries through to show that there is not even an extension of finite H^1 energy. Both problems are thus forced to develop singularities as $\varepsilon \to 0$, and the minimum energy will become unbounded.

We call the singularities of both problems *vortices*, since minimizers converge as $\varepsilon \to 0$ to maps that have the form $\frac{z-a_i}{|z-a_i|}$ near the singularities a_i. In the Ginzburg-Landau case, these vortices are interior and each carries a topological degree of 1; in the Kohn-Slastikov case, the singularities lie on the boundary, and we only see one half of the vortex. Each "boundary vortex" corresponds to a transition from m_0 to $-m_0$ or vice versa, and can be viewed as carrying $\frac{1}{2}$ topological charge.

It was shown in [Kur06, Kur04] that a single boundary vortex carries an energy of $\frac{\pi}{2} \log \frac{1}{\varepsilon}$ (see Thm. 3.3), and that the interaction of these vortices is governed by the next order term in the energy expansion, a renormalized energy that can be calculated by the solution of a linear boundary problem (see 3.6). In [Kur05] it was shown that this renormalized energy actually governs the motion of the vortices in the natural time scaling, when time is accelerated by a factor of $\log \frac{1}{\varepsilon}$ (see Thm. 3.19).

In the following, we will explain these results in more detail, and in the proper frameworks of two orders of Γ-convergence and Γ-convergence of gradient flows.

A major advantage of the simplified energy (3.1) is that the problem can be made *scalar* since $m \in S^1$ can be written as $m = e^{iv}$. The energy functional can then be rewritten as

$$\mathcal{E}_\varepsilon(v) = \frac{1}{2}\int_\Omega |\nabla v|^2 + \frac{1}{2\varepsilon}\int_{\partial\Omega} \sin^2(v-g), \tag{3.2}$$

where g is a function with $ie^{ig} = \nu$. Since Ω is simply connected, the degree of ν as a map from $\partial\Omega$ (which is homeomorphic to S^1) to S^1 is 1, and so g needs to have a jump of height -2π, but can otherwise be chosen as smooth as $\partial\Omega$. As $\varepsilon \to 0$, minimizers v_ε of $\mathcal{E}_\varepsilon$ will now satisfy $\sin^2(v_\varepsilon - g) \approx 0$ on most of $\partial\Omega$, but due to the jump of g, this will not be possible everywhere, and so singularities will develop that correspond precisely to the fast transition from $m \approx \tau$ to $m \approx -\tau$.

We can obviously generalize this to g having a jump of height $-2\pi D$, $D \in \mathbb{Z}$, corresponding to $\deg(e^{ig}) = D$, which will we do in the following. Without loss of generality, we will assume $D \geq 0$. The magnetic case of the Kohn-Slastikov functional is given by $D = 1$.

3.1 Highest order asymptotics

The following calculation gives an upper bound for the energy of a single boundary vortex. We will use a localized energy $\mathcal{E}_\varepsilon(v;B)$ that is defined by

$$\mathcal{E}_\varepsilon(v;B) = \frac{1}{2}\int_{\Omega\cap B} |\nabla v|^2 + \frac{1}{2\varepsilon}\int_{\partial\Omega\cap\overline{B}} \sin^2(v-g). \tag{3.3}$$

To simplify matters we will assume that $\Omega \cap B_R(0) = B_R^+(0)$ is a half-ball and that $g = 0$, corresponding to the constant tangent. We set $\Gamma_R = \partial\Omega \cap \overline{B_R(0)}$ for the flat boundary which we assume to be part of the x-axis in $z = x + iy$-plane.

Proposition 3.1. *There is a sequence $v_\varepsilon \in H^1(B_R(0))$ with $v_\varepsilon|_{\Gamma_R} \to v_* = \pi\chi_{x<0}$ in all $L^p(\Gamma_R)$ for $1 \le p < \infty$ and*

$$\mathcal{E}_\varepsilon(v_\varepsilon; B_R) \le \frac{\pi}{2}\log\frac{R}{\varepsilon} + C. \tag{3.4}$$

Proof. Set $v_\varepsilon = \arg(z)$ in $B_R^+(0) \setminus B_\varepsilon^+(0)$. Choose any H^1 continuation w with $0 \le w \le \pi$ of $\arg|_{\partial B_1^+(0)}$ to $B_1^+(0)$ and set $v_\varepsilon(z) = w(\varepsilon z)$ inside $B_\varepsilon^+(0)$. This sequence obviously satisfies the claims.

This shows that for $R = O(1)$, a typical vortex has an energy of approximately $\frac{\pi}{2}\log\frac{1}{\varepsilon} + O(1)$. A combination of $2D$ such vortices to counter the jump of g leads to the following upper bound:

Proposition 3.2. *Minimizers v_ε of $\mathcal{E}_\varepsilon$ satisfy*

$$\mathcal{E}_\varepsilon(v_\varepsilon) \le \pi D \log\frac{1}{\varepsilon} + C(\Omega). \tag{3.5}$$

A different interpretation of "every vortex carries an energy of $\frac{\pi}{2}\log\frac{1}{\varepsilon}$" is given by the following Γ-convergence theorem:

Theorem 3.3. *Assume (v_ε) is a sequence of functions with $\mathcal{E}_\varepsilon(v_\varepsilon) \le M\log\frac{1}{\varepsilon}$. Then there exists a sequence of $a_\varepsilon \in 2\pi\mathbb{Z}$ such that the boundary traces $w_\varepsilon = (v_\varepsilon - a_\varepsilon)|_{\partial\Omega}$ are bounded in an Orlicz space of type e^L, and in particular, $\|w_\varepsilon\|_{L^p(\partial\Omega)} \le C(M)$.*

The sequence w_ε is then precompact in the strong topology of $L^1(\partial\Omega)$, and every cluster point w satisfies $w - g \in \mathrm{BV}(\partial\Omega;\pi\mathbb{Z})$. In addition, we have the lower bound inequality

$$\liminf_{\varepsilon\to 0}\frac{1}{2\log\frac{1}{\varepsilon}}\int_\Omega |\nabla v_\varepsilon|^2 \ge \frac{1}{2}\int_{\partial\Omega}|D(w-g)|. \tag{3.6}$$

Conversely, for every u with $u - g \in \mathrm{BV}(\partial\Omega;\pi\mathbb{Z})$ there exists a sequence $u_\varepsilon \in H^1(\Omega)$ such that the trace satisfies $u_\varepsilon \to u$ in $L^1(\partial\Omega)$ and with

$$\lim_{\varepsilon\to 0}\frac{1}{2\log\frac{1}{\varepsilon}}\int_\Omega |\nabla u_\varepsilon|^2 = \frac{1}{2}\int_{\partial\Omega}|D(u-g)|. \tag{3.7}$$

We will not prove 3.3 here. A proof based on a nonlocal representation using the $H^{1/2}$ seminorm and rearrangement inequalities is given in [Kur06], and an extension to higher dimensions via slicing that utilizes the Orlicz bound is shown in [Kur05]. Both proofs are based on ideas of Alberti-Bouchitté-Seppecher [ABS94, ABS98] for similar functionals with a coercive instead of periodic potential.

Remark 3.4. Theorem 3.3 shows that, since the BV type limit functional has a lower bound of πD, the energy of minimizers of $\mathcal{E}_\varepsilon$ is $\pi D \log \frac{1}{\varepsilon} + o(\log \frac{1}{\varepsilon})$. For other converging sequences, we obtain just the number of jump points with multiplicities, but the limit functional is independent of the position of these jump points.

3.2 Separation of vortices and renormalized energy

We will relate the dependence of the energy on the position of the singularities to a renormalized energy given as follows:

Definition 3.5 (Possible limit functions). *Let $d_i \in \mathbb{Z}$ with $\sum_i d_i = 2D$ and $a_i \in \partial\Omega$ be distinct points. We define the canonical limit function $v_* = v_*(a_i, d_i)$ to be a harmonic function with $\sin^2(v_* - g) = 0$ such that its trace on $\partial\Omega$ jumps by $-\pi d_i$ at the point a_i.*

The renormalized energy *is defined to be*

$$W(a_i, d_i) = \frac{1}{2} \lim_{\rho \to 0} \left(\int_{\Omega \setminus \bigcup_i B_\rho(a_i)} |\nabla v_*|^2 - \pi \sum_i d_i^2 \log \frac{1}{\rho} \right). \tag{3.8}$$

The renormalized energy can be expressed via the solution of a linear boundary value problem for the Laplacian, see [Kur04].

With the renormalized energy, we can formulate the following second-order Γ-convergence type theorem. For minimizers, we will have $N = 2D$ as above.

Theorem 3.6. *If (v_ε) is a sequence of functions with $\mathcal{E}_\varepsilon(v_\varepsilon) \leq M \log \frac{1}{\varepsilon}$ and $v_\varepsilon \to v_*(a_i, d_i)$ in $L^2(\partial\Omega)$, with $d_i \in \{\pm 1\}$ and a_i distinct, $i = 1, \dots, N$, then*

$$\liminf_{\varepsilon \to 0} \left(\mathcal{E}_\varepsilon(v_\varepsilon) - \frac{\pi N}{2} \left(\log \frac{1}{\varepsilon} + 1 - \log 2 \right) \right) \geq W(a_i, d_i). \tag{3.9}$$

If additionally (v_ε) are stationary points of $\mathcal{E}_\varepsilon$, then $v_\varepsilon \to v_*$ in $W^{1,p}(\Omega)$ for $p < 2$ and in H^1_{loc} away from the a_i, and (3.9) holds with equality.

Furthermore, for any $d_i \in \{\pm 1\}$ and a_i distinct, $i = 1, \dots, N$, there exists a sequence of functions w_ε such that $w_\varepsilon \to v_*(a_i, d_i)$ and

$$\lim_{\varepsilon \to 0} \left(\mathcal{E}_\varepsilon(v_\varepsilon) - \frac{\pi N}{2} \left(\log \frac{1}{\varepsilon} + 1 - \log 2 \right) \right) = W(a_i, d_i). \tag{3.10}$$

We will prove this theorem only partially, for the special case of minimizers, where the bounds and the convergence follow from a comparison argument. The general case can be shown by means of some extra PDE estimates and a regularization technique of Yosida type, replacing the sequence by an improved sequence that minimizes a modified functional, see [Kur05].

Proposition 3.7 (Euler-Lagrange equations). *Stationary points v_ε of $\mathcal{E}_\varepsilon$ satisfy the equations*

$$\Delta v_\varepsilon = 0 \quad \text{in } \Omega, \tag{3.11}$$

$$\frac{\partial v_\varepsilon}{\partial \nu} = -\frac{1}{2\varepsilon} \sin 2(v_\varepsilon - g) \quad \text{on } \partial\Omega. \tag{3.12}$$

Lemma 3.8 (Rellich-Pohožaev identity). *For a Lipschitz domain G and a harmonic function $v \in H^2(G)$, there holds*

$$\int_{\partial G} \frac{\partial v}{\partial \nu}(z \cdot \nabla v) = \frac{1}{2} \int_{\partial G} z \cdot \nu |\nabla v|^2. \tag{3.13}$$

Proof. This is easily seen using by testing $\Delta v = 0$ with $z \cdot \nabla v$.

An easy consequence is

Lemma 3.9. *For a starshaped Lipschitz domain G, there exists constants such that every harmonic function $v \in H^2(G)$ satisfies*

$$c \int_{\partial G} \left| \frac{\partial v}{\partial \tau} \right|^2 \le \int_{\partial G} \left| \frac{\partial v}{\partial \nu} \right|^2 \le C \int_{\partial G} \left| \frac{\partial v}{\partial \tau} \right|^2. \tag{3.14}$$

Following ideas of [BBH94] and [Str94], we relate the penalty term $\frac{1}{2\varepsilon} \int_{\partial\Omega} \sin^2(v-g)$ to a radial derivative of the energy:

Definition 3.10. *For $z_0 \in \partial\Omega$ and $v \in H^2(\Omega)$ we set*

$$A(\rho) = A_{v,\varepsilon,z_0}(\rho) = \rho \int_{\partial B_\rho(z_0) \cap \Omega} |\nabla v|^2 d\mathcal{H}^1 + \frac{\rho}{\varepsilon} \int_{\partial B_\rho(z_0) \cap \partial\Omega} \sin^2(v-g) d\mathcal{H}^0. \tag{3.15}$$

For stationary points of the energy, A can be used to bound the penalty term:

Proposition 3.11. *There exists $\varepsilon_0 > 0$ and $C > 0$ such that for all $\varepsilon < \varepsilon_0$, $\rho < \varepsilon^{3/4}$, any stationary point v of $\mathcal{E}_\varepsilon$, and any $z_0 \in \partial\Omega$ there holds*

$$\frac{1}{2\varepsilon} \int_{\Gamma_\rho(z_0)} \sin^2(v-g) \le A(\rho) + C\sqrt{\varepsilon}. \tag{3.16}$$

Proof. For simplicity, we show this only for $g = 0$ and a flat boundary. We use $z_0 = 0$ and apply (3.13) on the domain $\omega_\rho = \Omega \cap B_\rho(0)$, which shows

$$\frac{1}{2}\int_{\partial\omega_\rho} z\cdot\nu|\nabla v|^2 = \rho\int_{\partial B_\rho\cap\Omega}\left|\frac{\partial v}{\partial\nu}\right|^2 + \int_{\Gamma_\rho}\frac{\partial v}{\partial\nu} z\cdot\nabla v. \tag{3.17}$$

Using the Euler-Lagrange equations, we obtain

$$\frac{\rho}{2}\int_{\partial B_\rho\cap\Omega}|\nabla v|^2 = \rho\int_{\partial B_\rho\cap\Omega}\left|\frac{\partial v}{\partial\nu}\right|^2 - \frac{1}{2\varepsilon}\int_{\Gamma_\rho}\sin 2v(z\cdot\nabla v). \tag{3.18}$$

Integrating by parts, we see that

$$\frac{1}{2\varepsilon}\int_{\Gamma_\rho}\sin^2(v) = \frac{\rho}{2\varepsilon}\int_{\partial B_\rho\cap\partial\Omega}\sin^2(v)d\mathcal{H}^0 - \frac{1}{2\varepsilon}\int_{\Gamma_\rho}\sin 2v(z\cdot\nabla v) \tag{3.19}$$

$$= \frac{\rho}{2\varepsilon}\int_{\partial B_\rho\cap\partial\Omega}\sin^2(v)d\mathcal{H}^0 + \frac{\rho}{2}\int_{\partial B_\rho\cap\Omega}|\nabla v|^2 - \rho\int_{\partial B_\rho\cap\Omega}\left|\frac{\partial v}{\partial\nu}\right|^2 \tag{3.20}$$

$$\leq \frac{1}{2}A(\rho). \tag{3.21}$$

We obtain the following criterion for vortex-free parts of the boundary:

Proposition 3.12. *There exist constants $\gamma > 0$ and $C > 0$ such that for every $z_0 \in \partial\Omega$, $\varepsilon < \varepsilon_0$, $\rho < \varepsilon^{3/4}$ and every stationary point v of $\mathcal{E}_\varepsilon$ with $A(\rho) < \gamma$ there holds*

$$\sup_{\Gamma_{\rho/2}}\sin^2(v-g) < \frac{1}{4} \tag{3.22}$$

and

$$\frac{1}{2\varepsilon}\int_{\Gamma_{\rho/2}}\sin^2(v-g) \leq C. \tag{3.23}$$

Proof. By Lemma 3.9, we can estimate

$$\int_{\Gamma_\rho}\left|\frac{\partial v}{\partial\tau}\right|^2 \leq C\int_{\partial\omega_\rho}\left|\frac{\partial v}{\partial\nu}\right|^2 \leq C\int_{\partial B_\rho\cap\Omega}|\nabla v|^2 + C\int_{\Gamma_\rho}\left|\frac{\partial v}{\partial\nu}\right|^2. \tag{3.24}$$

We thus can estimate, using Sobolev embedding in one dimension

$$\begin{aligned}[v]^2_{C^{0,1/2}(\Gamma_\rho)} &\leq C\int_{\Gamma_\rho}\left|\frac{\partial v}{\partial\tau}\right|^2 \\ &\leq C\left(\frac{1}{\rho}A(\rho) + \frac{1}{\varepsilon^2}\int_{\Gamma_\rho}\sin^2(v-g)\right) \leq \frac{C}{\varepsilon}(2\gamma + C\sqrt{\varepsilon_0}).\end{aligned}$$

Assuming now that $\sin^2(u(z)-g(z)) \geq \frac{1}{4}$ for some $z \in \Gamma_{\rho/2}$ and choosing γ and ε_0 sufficiently small then leads to a contradiction.

Disintegrating the energy radially, we see

Lemma 3.13. *Let (v_ε) be a sequence of stationary points of $\mathcal{E}_\varepsilon$ satisfying the logarithmic energy bound $\mathcal{E}_\varepsilon(v_\varepsilon) \leq M \log \frac{1}{\varepsilon}$. Then for any $z_0 \in \partial\Omega$, the function $A(\rho) = A_{v_\varepsilon,\varepsilon,z_0}(\rho)$ defined above satisfies*

$$\inf_{\varepsilon^{6/7} \leq \rho \leq \varepsilon^{5/6}} A(\rho) \leq \frac{84}{\log \frac{1}{\varepsilon}} \mathcal{E}_\varepsilon(v_\varepsilon; \Omega \cap B_{\varepsilon^{5/6}}(z_0)) \leq 84M \tag{3.25}$$

and

$$\inf_{5\varepsilon^{5/6} \leq \rho \leq 5\varepsilon^{4/5}} A(\rho) \leq 60M. \tag{3.26}$$

Using Vitali's covering lemma, we can use this and (3.16) to show a local upper bound on the penalty term near an almost singularity and in a second step a covering of the set of almost singularities. This leads to

Proposition 3.14. *There is a constant $N = N(g, \Omega, M)$ such that for any sequence of stationary points v_ε satisfying the energy bound $E_\varepsilon(v_\varepsilon) \leq M \log \frac{1}{\varepsilon}$, the approximate vortex set $S_\varepsilon = \{z \in \partial\Omega : \sin^2(v_\varepsilon(z) - g(z)) \geq \frac{1}{4}\}$ can be covered by at most N balls of radius ε, such that the $\varepsilon/5$ balls around the same centers are disjoint.*

For comparison arguments we shall need the following lower bound for the energy on half-annuli whenever $v - g$ is in different wells of $\sin^2$ on both parts of the boundary:

Proposition 3.15. *Let $0 < \rho < R \leq R_0$, R_0 sufficiently small, $z_0 \in \partial\Omega$, w.l.o.g. $z_0 = 0$. We examine the "half-annulus" $D_{R,\rho} = (B_R \setminus \overline{B_\rho}) \cap \Omega$, which can be described by choosing functions $\vartheta_1(r), \vartheta_2(r)$ as $\{re^{i\vartheta} : \vartheta_1(r) < \vartheta < \vartheta_2(r), \rho < r < R\}$ with $|\vartheta_2(r) - \vartheta_1(r) - \pi| \leq Cr$. Assume also that for $j = 1, 2$ there holds $(v - g)(re^{i\vartheta_j(r)}) \in (k_j\pi - \delta, k_j\pi + \delta)$ for some $k_j \in \mathbb{Z}$ and some small δ. Then*

$$\mathcal{E}_\varepsilon(v; D_{R,\rho}) \geq \frac{\pi}{2}(k_2 - k_1)^2 \log \frac{R}{\rho} - C(k_2 - k_1)^2 (R + \frac{\varepsilon}{\rho}). \tag{3.27}$$

Proof. For simplicity, we will assume $g = 0$, $\vartheta_1 = 0$ and $\vartheta_2 = \pi$, corresponding to a flat boundary. We will set $v_j(r) = v(re^{i\vartheta_j})$ for the function on the two boundary components. We also assume w.l.o.g. $k_1 = k$ and $k_2 = 0$. Using polar coordinates, disregarding the radial derivative and by use of Hölder's inequality, we calculate

$$\begin{aligned}
\int_{D_{R,\rho}} |\nabla v|^2 &\geq \int_\rho^R \frac{1}{r} \int_0^\pi \left| \frac{\partial u}{\partial \vartheta} \right|^2 d\vartheta dr \\
&\geq \frac{1}{\pi} \int_\rho^R \left(\int_0^\pi \left| \frac{\partial v}{\partial \vartheta} \right| \right)^2 \geq \frac{1}{\pi} \int_\rho^R \frac{(v_1 - v_2)^2}{r} dr.
\end{aligned}$$

We rewrite $v_1 - v_2 = k\pi - (v_1 - k\pi) - v_2$. Using the lower bound $\sin^2(t - k_i\pi) \geq \sigma t^2$ valid for $|t| < \delta$ with some $\sigma = \sigma(\delta)$, we can estimate

$$\mathcal{E}_\varepsilon(v; D_{R,\rho}) \geq \frac{1}{2}\int_\rho^R \frac{1}{\pi r}\left(k\pi - ((v_1 - k\pi) - v_2)\right)^2 + \frac{\sigma}{\varepsilon}\left(v_1^2 + v_2^2\right)^2 dr.$$

On the last term, we use the inequality $v_1^2 + v_2^2 \geq \frac{1}{2}(v_1 - k\pi + v_2)^2$. Then we use the inequality $\alpha(A-B)^2 + \beta B^2 \geq \frac{1}{\frac{1}{\alpha}+\frac{1}{\beta}} A^2$ on $A = k\pi$ and $B = (v_1 - k\pi - v_2)$. The claim then follows by integration.

We now recall that v_ε are minimizers of $\mathcal{E}_\varepsilon$ satisfying the upper bound

$$\mathcal{E}_\varepsilon(v_\varepsilon) \leq \pi D \log\frac{1}{\varepsilon} + C_0 \tag{3.28}$$

for some constant C_0, where D is the degree of e^{ig}. We will use an appropriate lower bound for the energy away from the vortex set to show convergence by a comparison argument. The same arguments also hold for stationary points satisfying (3.28).

By Proposition 3.14, there exist $a_j^\varepsilon \in \partial\Omega$, $1 \leq j \leq N_\varepsilon \leq N$ such that the approximate vortex set S_ε satisfies $S_\varepsilon \subset \bigcup_{1\leq j\leq N_\varepsilon} B_\varepsilon(a_j^\varepsilon)$. Passing to a subsequence of $\varepsilon \to 0$, we can assume that $N_\varepsilon = N_0$ is constant and $a_j^\varepsilon \to a_j^0$ as $\varepsilon \to 0$. Note that the a_j^0 need not be distinct. We define for $0 < \sigma < \frac{1}{2}\min_{a_j^0 \neq a_{j'}^0} \operatorname{dist}(a_j^0, a_{j'}^0)$ the sets $\Omega_\sigma^\varepsilon = \Omega \setminus \bigcup_j B_\sigma(a_j^\varepsilon)$ and $\Omega_\sigma^0 = \Omega\setminus\bigcup_j B_\sigma(a_j^0)$. With this setup (and this subsequence), the lower bound of Proposition 3.15 can be combined with the arguments of Struwe [Str94] to show

Proposition 3.16. *There is a constant $C = C(g, \Omega, C_0)$ such that $\mathcal{E}_\varepsilon(v_\varepsilon; \Omega_\sigma^\varepsilon) \leq \pi D \log\frac{1}{\sigma} + C$.*

We obtain convergence to the canonical harmonic function:

Proposition 3.17. *Let (v_ε) be a sequence of critical points satisfying the energy bound*

$$\mathcal{E}_\varepsilon(v_\varepsilon) \leq \pi D \log\frac{1}{\varepsilon} + C_0.$$

Then there is a subsequence and $N = 2D$ points $a_1, \ldots, a_N \in \partial\Omega$ such that

$$\int_{\Omega'} |\nabla v_\varepsilon|^2 \leq M(\Omega') < \infty \tag{3.29}$$

for all open Ω' with $\overline{\Omega'} \subset \overline{\Omega}\setminus\{a_1, \ldots, a_N\}$. Additionally, there hold the bounds

$$\int_\Omega |\nabla v_\varepsilon|^p \leq C(p) \tag{3.30}$$

uniformly in ε for all $1 \leq p < 2$. In particular, after adding a suitable $z_\varepsilon \in 2\pi\mathbb{Z}$, a subsequence of (v_ε) converges weakly in H^1_{loc} and $W^{1,p}$, $p < 2$, to a harmonic function v_*. The limit has the properties that $(v_* - g)$ is piecewise constant on $\partial\Omega \setminus \{a_1, \ldots, a_N\}$, with values in $\pi\mathbb{Z}$, and jumps by $-\pi$ at the points a_j.

Proof. We use the setup described above, in particular, we use the points a_j^0 as defined there. Note that for $\varepsilon < \varepsilon_0(\sigma)$, there holds $\Omega_\sigma^0 \subset \Omega_{\sigma/2}^\varepsilon$ and so by Proposition 3.16,

$$\int_{\Omega_\sigma^0} |\nabla v_\varepsilon|^2 \le 2\mathcal{E}_\varepsilon(v_\varepsilon; \Omega_{\sigma/2}^\varepsilon) \le 2\pi D \log \frac{2}{\sigma} + C, \tag{3.31}$$

which proves (3.29). To obtain the L^p bounds (3.30), fix a $\sigma > 0$ and $1 \le p < 2$. Then by Hölder's inequality and Proposition 3.16,

$$\begin{aligned}
\int_\Omega |\nabla v_\varepsilon|^p &\le \int_{\Omega_\sigma^\varepsilon} |\nabla v_\varepsilon|^p + \sum_{\ell=1}^\infty \int_{\Omega_{2^{-\ell}\sigma}^\varepsilon \setminus \Omega_{2^{-\ell+1}\sigma}^\varepsilon} |\nabla v_\varepsilon|^p \\
&\le C + \sum_{\ell=1}^\infty |\Omega_{2^{-\ell}\sigma}^\varepsilon \setminus \Omega_{2^{-\ell+1}\sigma}^\varepsilon|^{1-p/2} \left(\int_{\Omega_{2^{-\ell}\sigma}^\varepsilon} |\nabla v_\varepsilon|^2 \right)^{p/2} \\
&\le C + c\sum_{\ell=1}^\infty 2^{-(1-p/2)\ell} \left(2\pi D \log \frac{1}{2^\ell \sigma} + C \right)^{p/2} \\
&\le C.
\end{aligned}$$

From this L^p gradient bound, we obtain the weak compactness up to translation. The weak limit v_* is harmonic since $\int_\Omega \nabla v_* \cdot \nabla \varphi = \lim_{\varepsilon \to 0} \int_\Omega \nabla v_\varepsilon \cdot \nabla \varphi = 0$ for all $\varphi \in C_c^\infty(\Omega)$. That the boundary values satisfy $v_* - g \in \pi\mathbb{Z}$ with possible jumps at the a_i follows from $\int_{\partial\Omega} \sin^2(v_\varepsilon - g) \to 0$ and $(v_\varepsilon - g)$ being close to $\pi\mathbb{Z}$ outside the approximate vortex set S_ε.

That the vortices are indeed single and $N = 2D$ can then be shown by some refined arguments which prove that higher-order vortices must have far higher energy.

The energy of these limit functions v_* away from a_i is the renormalized energy of (3.8), and the energy of v_ε on Ω_ρ converges to that of v_*. To prove equality in (3.9), we thus need to calculate the energy of v_ε close to a_i. This is done by an ε-scale blowup, which leads to a half-space problem. The solutions of the half-space problem are explicitly known and essentially unique (Toland [Tol97], see also Cabré and Sola-Morales [CS04] for a more general uniqueness theorem). Comparing v_ε with the rescaled half-space solution and some estimates (see [Kur04]) then show the rest of Thm. 3.6. One can even show

Proposition 3.18. *For a sequence v_ε of stationary points of $\mathcal{E}_\varepsilon$, the configuration of vortex points (a_i) is stationary for the renormalized energy $W(a_i, d_i)$. For minimizers v_ε, it is minimizing.*

3.3 Motion of vortices

Theorem 3.6 and the previous proposition show that the renormalized energy W governs the interaction of the vortices on an energetic level. It can be shown that also the motion of the vortices by the gradient flow (which corresponds to the LLG flow since we restrict possible magnetizations to a plane) is given by the renormalized energy:

Theorem 3.19. *Let $0 < T \leq \infty$ and let (v_ε) be a sequence of solutions of*

$$\lambda_\varepsilon \partial_t v_\varepsilon = \Delta v_\varepsilon \quad \text{in } \Omega \times (0,T) \tag{3.32}$$

$$\frac{\partial v_\varepsilon}{\partial \nu} = -\frac{1}{2\varepsilon} \sin 2(v_\varepsilon - g) \quad \text{on } \partial\Omega \times (0,\infty). \tag{3.33}$$

For the initial conditions we assume that $v_\varepsilon(0) \to v_*(a_i, d_i)$ for $d_i \in \{\pm 1\}$ and distinct a_i. Furthermore, v_ε is supposed to be initially well-prepared, *meaning that*

$$\mathcal{E}_\varepsilon(v_\varepsilon(0)) - \frac{\pi N}{2} \log \frac{1}{\varepsilon} - \frac{\pi N}{2}(1 - \log 2) \leq W(a_i, d_i) + o(1) \tag{3.34}$$

as $\varepsilon \to 0$.

Depending on the asymptotic behavior of λ_ε, we then have:

(i) If $\lambda_\varepsilon = \frac{1}{\log \frac{1}{\varepsilon}}$, then there exists a time $T^* > 0$ such that for all $t \in [0, T^*)$, there holds $v_\varepsilon(t) \to v_*(a_i(t), d_i(0)))$. Furthermore, the $a_i(t)$ satisfy the motion law

$$\frac{da_i}{dt} = -\frac{2}{\pi} \frac{\partial}{\partial a_i} W(a_i(t), d_i(0)) \tag{3.35}$$

in the tangent space at a_i to $\partial\Omega$. If $T^* < T$ is the maximal time with these properties, then as $t \to T^*$, there exist $i \neq j$ such that $a_i(t)$ and $a_j(t)$ converge to the same point.

The energy of $v_\varepsilon(t)$ satisfies the expansion

$$\mathcal{E}_\varepsilon(v_\varepsilon(t)) = \frac{\pi N}{2} \log \frac{1}{\varepsilon} + \frac{\pi N}{2}(1 - \log 2) + W(\mathbf{a}(t), \mathbf{d}) + o(1) \tag{3.36}$$

as $\varepsilon \to 0$.

(ii) If $\lambda_\varepsilon \log \frac{1}{\varepsilon} \to 0$ as $\varepsilon \to 0$, then for almost every $t \in [0,T)$ we have $v_\varepsilon(t) \to v_*(a_i(0), d_i(0))$, so there is no motion.

(iii) If $\lambda_\varepsilon \log \frac{1}{\varepsilon} \to \infty$ as $\varepsilon \to 0$, then for almost every $t \in [0,\infty)$ we have $v_\varepsilon(t) \to v_*(b_i, d_i)$ with $\nabla W(b_i, d_i) = 0$, so the system instantaneously jumps into a critical point.

Again, there are strong similarities between this result for the motion of boundary vortices and those in the theory of gradient flow motion of interior Ginzburg-Landau type vortices as studied by Jerrard and Soner [JS98] and Lin [Lin96a, Lin96b].

The proof in [Kur05] is based on the technique of Γ-convergence of gradient flows of Sandier and Serfaty [SS04a], applied to the functionals

$$\mathcal{F}^\varepsilon(u) = \mathcal{E}^\varepsilon(u) - \frac{\pi N}{2}(\log \frac{1}{\varepsilon} + 1 - \log 2) \tag{3.37}$$

and the limit functional

$$\mathcal{F}(a_i) = W(a_i, d_i). \tag{3.38}$$

We need some additional definitions:

Definition 3.20. *We say that functionals $\mathcal{F}^\varepsilon$ Γ-converge to $\mathcal{F}$ along the trajectory $u_\varepsilon(t)$ with respect to the convergence "$\stackrel{S}{\rightharpoonup}$" if there exist $u(t)$ and a subsequence such that for all t, $u_\varepsilon(t) \stackrel{S}{\rightharpoonup} u(t)$ and*

$$\liminf_{\varepsilon \to 0} \mathcal{F}^\varepsilon(u_\varepsilon(t)) \geq \mathcal{F}(u(t)). \tag{3.39}$$

The energy excess *$D_\varepsilon(t)$ and the* limiting energy excess *$D(t)$ for a sequence $u_\varepsilon(t)$ are defined via*

$$D_\varepsilon(t) = \mathcal{E}^\varepsilon(u_\varepsilon(t)) - \mathcal{E}(u(t)), \qquad D(t) = \limsup_{\varepsilon \to 0} D_\varepsilon(t). \tag{3.40}$$

If $u_\varepsilon(t)$ are solutions to the gradient flow for $\mathcal{E}^\varepsilon$ that satisfy $D(0) = 0$, they are said to be initially well-prepared.

The proof of Thm. 3.19 relies on the following version of Sandier and Serfaty's theorem on the Γ-convergence of gradient flows:

Theorem 3.21 (Sandier-Serfaty [SS04a]). *Assume $\mathcal{F}^\varepsilon \in C^1(\mathcal{M})$ and $\mathcal{F} \in C^1(\mathcal{N})$. Let u_ε be a sequence of solutions of the gradient flow for $\mathcal{F}^\varepsilon$ on $[0,T)$ with respect to the metric structure X_ε that satisfy*

$$\mathcal{F}^\varepsilon(u_\varepsilon(0)) - \mathcal{F}^\varepsilon(u_\varepsilon(t)) = \int_0^t \|\partial_t u_\varepsilon(s)\|_{X_\varepsilon}^2 \, ds. \tag{3.41}$$

Assume $u_\varepsilon(0) \stackrel{S}{\rightharpoonup} u_0$, that $\mathcal{F}^\varepsilon$ Γ-converges to $\mathcal{F}$ along the trajectory $u_\varepsilon(t)$, and that (u_ε) is initially well-prepared. Furthermore, assume that (LB) and (CON) hold:

(LB) For a subsequence such that $u_\varepsilon(t) \stackrel{S}{\rightharpoonup} u(t)$, we have $u \in H^1((0,T);\mathcal{N})$ and there exists $f \in L^1(0,T)$ such that for every $s \in [0,T)$ there holds

$$\liminf_{\varepsilon \to 0} \int_0^s \|\partial_t u_\varepsilon(t)\|_{X_\varepsilon}^2 \, dt \geq \int_0^s \left(\|\partial_t u\|_{T_{u(t)}\mathcal{N}}^2 - f(t) D(t) \right) dt. \tag{3.42}$$

(CON) If $u_\varepsilon(t) \stackrel{S}{\rightharpoonup} u(t)$, there exists a locally bounded function g on $[0,T)$ such that for any $t_0 \in [0,T)$ and any v defined in a neighborhood of t_0 that satisfies $v(t_0) = u(t_0)$ and $\partial_t v(t_0) = -\nabla_{T_{u(t_0)}\mathcal{N}} \mathcal{E}(u(t_0))$, there exists a sequence $v_\varepsilon(t)$ such that $v_\varepsilon(t_0) = u_\varepsilon(t_0)$ and the following inequalities hold:

$$\limsup_{\varepsilon \to 0} \|\partial_t v_\varepsilon(t_0)\|_{X_\varepsilon}^2 \leq \|\partial_t v(t_0)\|_{T_{v(t_0)}\mathcal{N}}^2 + g(t_0) D(t_0) \tag{3.43}$$

$$\liminf_{\varepsilon \to 0} \left(-\frac{d}{dt}\Big|_{t=0} \mathcal{F}^\varepsilon(v_\varepsilon(t)) \right) \geq -\frac{d}{dt}\Big|_{t=0} \mathcal{F}(v(t)) - g(t_0) D(t_0). \tag{3.44}$$

Then $u_\varepsilon(t) \stackrel{S}{\rightharpoonup} u(t)$ *which is the solution of the gradient flow for* $\mathcal{E}$ *with respect to the structure of* $T\mathcal{N}$.

This is applicable to our case since (3.32) with the nonlinear boundary condition (3.33) is the gradient flow of $\mathcal{F}^\varepsilon$ with respect to the norm $\sqrt{\lambda_\varepsilon}\,\|\cdot\|_{L^2}$, which we will use as the spaces X_ε in the terminology of the theorem above. The functionals $\mathcal{F}^\varepsilon$ are defined on $\mathcal{M} = H^1(\Omega)$. As the sense of convergence, we use $v_\varepsilon \stackrel{S}{\rightharpoonup} (a_i)$ if $v_\varepsilon \to v_*(a_i, d_i)$ in $L^2(\partial\Omega)$. The necessary Γ-convergence for $\stackrel{S}{\rightharpoonup}$ follows from (3.9).

The limit functional is defined on $\mathcal{N} = \{(a_i)_{i=1,\dots,N} : a_i \neq a_j \text{ for } i \neq j\}$, which is an open subset of the (flat) Riemannian manifold $(\partial\Omega)^N$. The approach of [SS04a] for Euclidean limit spaces carries over to this situation without changes. As the limiting norm on the tangent spaces $T_a\mathcal{N}$ which are identified with $\mathbb{R}^N$ we use the constant Riemannian metric $\sqrt{\frac{\pi}{2}}\,\|\cdot\|_{\mathbb{R}^N}$.

Theorem 3.21 allows us to break up the proof of Theorem 3.19 into two separate parts, a lower bound and a construction. The proof of the lower bound relies on an anisotropic version of (3.6) in higher dimensions. This leads to a product estimate like the one of Sandier-Serfaty [SS04b], which can then be used to separate space- and time-derivatives to show (3.42).

The construction used to show the upper bound inequalities is done by taking a well-prepared sequence and "pushing" the vortices along the boundary with the flow of a vector field that is conformal close to the vortex. The conformality ensures that the highest order of the energy does not change by the flow. With some more detailed local estimates related to (3.9), the estimates (3.43) and (3.44) then follow, as is detailed in [Kur05].

4 Boundary vortices in a refined model

In this section we discuss a thin-film regime that is related to the theory of the the previous section; indeed the theory of Section 3 can be regarded as a simplified version of what is to follow (but, as mentioned earlier, it can also be seen as an asymptotic analysis of a model arising in the thin-film theory of Kohn and Slastikov [KS05b]). We now examine the development of boundary vortices in thin films for a model that is closer to the actual micromagnetic model than the one discussed previously, although we still use some simplifications. In particular we consider now domains that are three-dimensional (but thin in one dimension) and magnetization vector fields with values in $\mathbb{S}^2$ (not $\mathbb{S}^1$). We continue to neglect the anisotropy term in the micromagnetic energy functional and the external magnetic field, but we consider the exchange energy and the magnetostatic energy in the form that they have in the functional E.

Naturally, the problem becomes more difficult when we drop some of the simplifications. It is not surprising, therefore, that we need more assumptions to obtain less information about the asymptotic behavior of the magnetization. But the results we find for this model are consistent with those for the more simplified model, which shows that the latter does indeed describe the significant features of the thin-film limit in the asymptotic regime we study.

We consider the family of domains

$$\Omega_\delta = \Omega \times (0, \delta)$$

for $\delta > 0$, where $\Omega \subset \mathbb{R}^2$ is open and bounded. We also assume that Ω is simply connected and that its boundary is smooth. The outer normal vector on $\partial\Omega$ is denoted by ν. The energy functional (without the anisotropy term and the external field) is then

$$E(\mathbf{m}) = \frac{d^2}{2} \int_{\Omega_\delta} |\nabla \mathbf{m}|^2 \, d\mathbf{x} + \frac{1}{2} \int_{\mathbb{R}^3} |\nabla u|^2 \, d\mathbf{x},$$

where, as usual, the function $u \in H^1(\mathbb{R}^3)$ is determined by the condition

$$\Delta u = \operatorname{div}(\chi_{\Omega_\delta} \mathbf{m}) \quad \text{in } \mathbb{R}^3.$$

We assume for the moment that the shape of Ω is fixed, whereas its size can still be varied by scaling. The problem then involves three length scales: the exchange length d, the thickness δ, and the length scale of the cross-section, measured, e.g., by $L = \operatorname{diam} \Omega$. The asymptotic regime we consider is characterized by the condition that d^2 is of the same magnitude as $L\delta$. For simplicity, we assume $d^2 = L\delta$. Rescaling Ω allows us to normalize $L = 1$, which gives rise to the relation $d^2 = \delta$ between the exchange length and the thickness (and which means that Ω is fixed henceforth).

Since we study the asymptotic behavior of variational problems associated to the micromagnetic energy as $\delta \searrow 0$, we now denote a generic magnetization vector field in Ω_δ by $\mathbf{m}^\delta = (m^\delta, m_3^\delta) \in H^1(\Omega_\delta, \mathbb{S}^2)$. The corresponding potential for the induced magnetic field is then the unique solution $u^\delta \in H^1(\mathbb{R}^3)$ of

$$\Delta u^\delta = \operatorname{div}(\chi_{\Omega_\delta} \mathbf{m}^\delta) \quad \text{in } \mathbb{R}^3. \tag{4.1}$$

We also consider the maps

$$\overline{\mathbf{m}}^\delta = \frac{1}{\delta} \int_0^\delta \mathbf{m}^\delta(x, s) \, ds, \quad x \in \Omega,$$

so that we can pass to a limit in certain spaces of functions on Ω (usually Sobolev spaces). The limit will then be a map $\mathbf{m} = (m, m_3) : \Omega \to \mathbb{S}^2$. If we consider the functions $\delta^{-1} u^\delta$ and apply equation (4.1) to a test function $\phi \in C_0^\infty(\mathbb{R}^3)$, we can formally pass to the limit. We obtain (formally) a limit function $u : \mathbb{R}^3 \to \mathbb{R}$ with

$$\int_{\mathbb{R}^3} \nabla u \cdot \nabla \phi \, d\mathbf{x} = \int_\Omega \mathbf{m}(x) \cdot \nabla \phi(x, 0) \, dx \tag{4.2}$$

for all $\phi \in C_0^\infty(\mathbb{R}^3)$. For the problems we consider here, we have typically $m_3 = 0$ in Ω and $m \cdot \nu = 0$ on $\partial\Omega$. If we have furthermore $m \in W^{1,4/3}(\Omega, \mathbb{S}^1)$, then equation (4.2) does in fact determine a function $u \in H^1(\mathbb{R}^3)$ uniquely. We will see that this function describes in part the limit of the magnetostatic energy. Also important is its trace on $\Omega \times \{0\}$. We denote this trace by u_0.

It is convenient to divide the micromagnetic energy by δ^2. That is, we consider the family of functionals

$$E_\delta(\mathbf{m}^\delta) = \frac{1}{2\delta} \int_{\Omega_\delta} |\nabla \mathbf{m}^\delta|^2 \, d\mathbf{x} + \frac{1}{2\delta^2} \int_{\mathbb{R}^3} |\nabla u^\delta|^2 \, d\mathbf{x}.$$

Critical points of E_δ satisfy the Euler-Lagrange equation

$$\delta(\Delta \mathbf{m}^\delta + |\nabla \mathbf{m}^\delta|^2 \mathbf{m}^\delta) - (\mathbf{1} - \mathbf{m}^\delta \otimes \mathbf{m}^\delta) \nabla u^\delta = 0 \quad \text{in } \Omega_\delta. \tag{4.3}$$

It is natural to impose homogeneous Neumann boundary conditions, i.e.,

$$\frac{\partial \mathbf{m}^\delta}{\partial x_3} = 0 \quad \text{in } \Omega \times \{0, \delta\}, \tag{4.4}$$

$$\frac{\partial \mathbf{m}^\delta}{\partial \nu} = 0 \quad \text{on } \partial\Omega \times (0, \delta). \tag{4.5}$$

Stable critical points also satisfy

$$\left. \frac{d^2}{ds^2} \right|_{s=0} E_\delta \left(\frac{m + s\psi}{|m + s\psi|} \right) \geq 0$$

for all $\psi \in C^\infty(\overline{\Omega_\delta}, \mathbb{R}^3)$. Standard calculations transform this inequality into

$$0 \leq \int_{\Omega_\delta} (|\nabla \dot{\mathbf{m}}^\delta|^2 - |\dot{\mathbf{m}}^\delta|^2 (|\nabla \mathbf{m}^\delta|^2 + \delta^{-1} \mathbf{m}^\delta \cdot \nabla u^\delta) \, d\mathbf{x} + \frac{1}{\delta} \int_{\mathbb{R}^3} |\nabla u^\delta|^2 \, d\mathbf{x}, \tag{4.6}$$

where $\dot{\mathbf{m}}^\delta = (\mathbf{1} - \mathbf{m}^\delta \otimes \mathbf{m}^\delta)\psi$.

For the Landau-Lifshitz equation, there exist two interesting time scales, similarly as in the previous section. The first one gives rise to the equation

$$\frac{\partial \mathbf{m}^\delta}{\partial t} = -\hat{\gamma} \mathbf{m}^\delta \wedge (\Delta \mathbf{m}^\delta - \delta^{-1} \nabla u^\delta) - \hat{\alpha} \mathbf{m}^\delta \wedge (\mathbf{m}^\delta \wedge (\Delta \mathbf{m}^\delta - \delta^{-1} \nabla u^\delta)) \tag{4.7}$$

in $\Omega_\delta \times (0, T)$, which is equivalent to

$$R_{\mathbf{m}^\delta} \frac{\partial \mathbf{m}^\delta}{\partial t} = \Delta \mathbf{m}^\delta + |\nabla \mathbf{m}^\delta|^2 \mathbf{m}^\delta - \frac{1}{\delta} (\mathbf{1} - \mathbf{m}^\delta \otimes \mathbf{m}^\delta) \nabla u^\delta. \tag{4.8}$$

Here we use the abbreviations

$$R_{\mathbf{m}^\delta} X = \tilde{\alpha} X + \tilde{\gamma} \mathbf{m}^\delta \wedge X \quad \text{and} \quad \tilde{\alpha} = \frac{\hat{\alpha}}{\hat{\alpha}^2 + \hat{\gamma}^2}, \quad \tilde{\gamma} = \frac{\hat{\gamma}}{\hat{\alpha}^2 + \hat{\gamma}^2}.$$

We normally use the form (4.8) of the equation. We impose homogeneous Neumann boundary data again, that is,

$$\frac{\partial \mathbf{m}^\delta}{\partial x_3} = 0 \quad \text{in } \Omega \times \{0, \delta\} \times [0, T), \tag{4.9}$$

$$\frac{\partial \mathbf{m}^\delta}{\partial \nu} = 0 \quad \text{on } \partial\Omega \times (0, \delta) \times [0, T). \tag{4.10}$$

This is the time scale where we expect the development of stationary boundary vortices in the limit $\delta \searrow 0$, in analogy to the results of the previous section. To study the dynamical behavior of the vortices, on the other hand, we need to rescale the time axis by the factor $\log\log\frac{1}{\delta}$ (accelerating the time by this factor). The equation then becomes

$$\frac{R_{\mathbf{m}^\delta}\frac{\partial \mathbf{m}^\delta}{\partial t}}{\log\log\frac{1}{\delta}} = \Delta\mathbf{m}^\delta + |\nabla\mathbf{m}^\delta|^2\mathbf{m}^\delta - \frac{1}{\delta}(\mathbf{1} - \mathbf{m}^\delta \otimes \mathbf{m}^\delta)\nabla u^\delta \quad \text{in } \Omega_\delta \times (0, T). \tag{4.11}$$

The boundary conditions remain of the form (4.9), (4.10).

We want to reproduce the asymptotic theory of the previous section for the model given by the energy E_δ and the equations (4.3), (4.8), and (4.11). There are several additional difficulties here, however, that we have to overcome. First, the target space for our maps is now $\mathbb{S}^2$, not $\mathbb{S}^1$, which means that $\mathbf{m}^\delta$ can no longer be represented by a single phase function. The curvature of $\mathbb{S}^2$ also has the consequence that we have to consider equations with nonlinear terms involving first derivatives of $\mathbf{m}^\delta$. Together with the fact that our domains Ω_δ are now three-dimensional, this means that we must expect solutions of the equations with singularities. To simplify the presentation of the results, we always assume here that we have smooth solutions; but without this assumption, regularity is an issue that requires extra care.

The most important new aspect of this model, however, is the nonlocal operator appearing in the magnetostatic energy. This is at first a major impediment to using the methods from the theory of Ginzburg-Landau vortices, for these methods require pointwise comparisons between integrands of the lower order energy terms. To overcome this difficulty, we compare E_δ with another functional that has only local terms, namely

$$F_\delta(\mathbf{m}^\delta) = \frac{1}{2\delta}\int_{\Omega_\delta}\left(|\nabla\mathbf{m}^\delta|^2 + \frac{(m_3^\delta)^2}{\delta}\right)d\mathbf{x} + \frac{\log\frac{1}{\delta}}{2\delta}\int_{\partial\Omega\times(0,\delta)}(m^\delta\cdot\nu)^2\,d\mathcal{H}^2.$$

Here $\mathcal{H}^k$ denotes the k-dimensional Hausdorff measure. The functional F_δ can be thought of as the three-dimensional equivalent of

$$G_\delta(\mathbf{m}^\delta) = \frac{1}{2}\int_{\Omega}\left(|\nabla\mathbf{m}^\delta|^2 + \frac{(m_3^\delta)^2}{\delta}\right)dx + \frac{1}{2}\log\frac{1}{\delta}\int_{\partial\Omega}(m\cdot\nu)^2\,d\mathcal{H}^1,$$

where for the latter functional, we consider $\mathbf{m}^\delta \in H^1(\Omega, \mathbb{S}^2)$.

The connection to the theory discussed earlier is obvious. The connection to the theory of Ginzburg-Landau vortices becomes even more apparent when one observes that $(m_3^\delta)^2 = 1 - |m^\delta|^2$ (since $\mathbf{m}^\delta$ has values in the unit sphere), recovering thus an integrand in the first integral of G_δ that is similar to the one used in most works on Ginzburg-Landau vortices. Less obvious is the connection between E_δ and F_δ. Before we give any rigorous arguments, we look at this question heuristically. The magnetostatic energy seeks to minimize the divergence of $\chi_{\Omega_\delta}\mathbf{m}^\delta$, which consists of two parts: the divergence of $\mathbf{m}^\delta$ in the interior of Ω_δ on the one hand, and the distribution given by the perpendicular part of $\mathbf{m}^\delta$ on $\partial\Omega_\delta$ on the other hand. The latter further splits into two parts according to the natural decomposition of the boundary into $\Omega \times \{0, \delta\}$ and $\partial\Omega \times (0, \delta)$. The part coming from $\operatorname{div} \mathbf{m}^\delta$ now gives a contribution to the magnetostatic energy which is of the same order as the exchange energy. Both of the other parts correspond to one of the terms in F_δ.

The next few lemmas give a more precise description of the relation between these functionals.

Lemma 4.1. *For $\delta \in (0, e^{-e}]$ and $\mathbf{m}^\delta \in H^1(\Omega_\delta, \mathbb{R}^3)$, the inequality*

$$\|\nabla u^\delta\|^2_{L^2(\mathbb{R}^3)} \le C\sqrt{\delta}\|\nabla \mathbf{m}^\delta\|^2_{L^{4/3}(\Omega_\delta)} + C\|m_3^\delta(\,\cdot\,, 0)\|^2_{L_{4/3}(\Omega)} + C\int_{\partial\Omega} \|\chi_{(0,\delta)} m^\delta(x, \,\cdot\,) \cdot \nu(x)\|^2_{H^{-1/2}(\mathbb{R})} d\mathcal{H}^1(x)$$

holds for a constant C that depends only on Ω. Here $u^\delta \in H^1(\mathbb{R}^3)$ is the function determined by (4.1).

Proof. We have

$$\begin{aligned}\int_{\mathbb{R}^3} |\nabla u^\delta|^2 \, d\mathbf{x} &= \int_{\Omega_\delta} \nabla u^\delta \cdot \mathbf{m}^\delta \, d\mathbf{x} \\ &= \left(\int_{\Omega \times \{\delta\}} - \int_{\Omega\times\{0\}}\right) u^\delta m_3^\delta \, dx + \int_{\partial\Omega}\int_0^\delta u^\delta m^\delta \cdot \nu \, dx_3 \, d\mathcal{H}^1 \\ &\quad - \int_0^\delta \int_\Omega u^\delta \operatorname{div} \mathbf{m}^\delta \, dx \, dx_3.\end{aligned}$$

Now we use the continuous trace operators $H^1(\mathbb{R}^3) \to L^4(\Omega)$ for every slice $\Omega \times \{x_3\}$ and the continuous trace operator $H^1(\Omega) \to L^2(\partial\Omega, H^{1/2}(-1, 1))$ to estimate the traces of u in these spaces. Furthermore, an integration of $\frac{\partial m_3^\delta}{\partial x_3}$ along vertical lines gives

$$\|m_3^\delta(\,\cdot\,, 0) - m_3^\delta(\,\cdot\,, \delta)\|^2_{L^{4/3}(\Omega)} \le \sqrt{\delta}\|\nabla \mathbf{m}\|^2_{L^{4/3}(\Omega_\delta)},$$

and the desired estimate then follows from the Hölder inequality.

Lemma 4.2. *For $\delta \in (0, e^{-e}]$, let $m^\delta \in H^1(\Omega, \mathbb{S}^1)$ and $\mathbf{m}^\delta(x, x_3) = (m^\delta(x), 0)$. Then*

$$\|\nabla u^\delta\|_{L^2(\mathbb{R}^3)} \leq C\delta^2 \left(\|\nabla m^\delta\|^2_{L^{4/3}(\Omega)} + \log \frac{1}{\delta} \|m^\delta \cdot \nu\|^2_{L^2(\partial\Omega)} \right)$$

for a constant C that depends only on Ω.

Proof. A direct computation shows that the characteristic function $\chi_{(0,\delta)}$ of the interval $(0, \delta)$ satisfies

$$\|\chi_{(0,\delta)}\|_{H^{-1/2}(\mathbb{R})} \leq c\delta^2 \left(1 + \log \frac{1}{\delta}\right)$$

for a certain constant c which is independent of δ. The claim now follows directly from Lemma 4.1.

Proposition 4.3. *There exists a constant C, dependent only on Ω, such that*

$$\inf_{H^1(\Omega_\delta, \mathbb{S}^2)} E_\delta \leq \pi \log\log \frac{1}{\delta} + C$$

for $\delta \in (0, e^{-e}]$.

Proof. We construct a map $m^\delta \in H^1(\Omega, \mathbb{S}^1)$ with two standard vortices centered at two different points on the boundary, similarly as in the proof of Proposition 3.1 (with $\epsilon = 1/\log \frac{1}{\delta}$). For $\mathbf{m}^\delta(x, x_3) = (m^\delta(x), 0)$, the estimates of Proposition 3.1, together with Lemma 4.2, give a bound for $E_\delta(\mathbf{m}^\delta)$ of the desired form.

In fact the quantity $\pi \log\log \frac{1}{\delta}$ gives also a lower bound for the infimum of E_δ in $H^1(\Omega_\delta, \mathbb{S}^2)$ up to a constant. That is, it determines the asymptotic behavior of this infimum. The proof of the lower estimate is technically more involved; we therefore give only a sketch of the proof here.

Lemma 4.4. *For $\delta \in (0, e^{-e}]$ and $\mathbf{m}^\delta \in H^1(\Omega_\delta, \mathbb{S}^2)$, the inequality*

$$\int_{\Omega_\delta} (m_3^\delta)^2 \, d\mathbf{x} \leq C \left[\delta \int_{\Omega_\delta} |\nabla m_3^\delta|^2 \, d\mathbf{x} + \int_{\mathbb{R}^3} |\nabla u^\delta|^2 \, d\mathbf{x} + \delta^2 \right] \tag{4.12}$$

holds for a constant C that depends only on Ω.

Sketch of the proof. We test (4.1) with a function $\phi \in C^{0,1}(\mathbb{R}^3)$ which is defined on $\Omega \times \mathbb{R}$ by

$$\phi(x, x_3) = \begin{cases} 0, & \text{if } x_3 \leq 0 \text{ or } x > 2\delta, \\ \int_0^{x_3} m_3(x, s)\, ds & \text{if } 0 < x_3 \leq \delta, \\ (2 - x_3/\delta) \int_0^\delta m_3(x, s)\, ds & \text{if } \delta < x_3 \leq 2\delta, \end{cases}$$

and extended suitably to $\mathbb{R}^3$. We recover the left-hand side of (4.12) as one of the terms in the resulting equation (after an integration by parts). All other terms can then be estimated with standard methods.

For $s \geq 0$ we now define the sets

$$V_s = \{x \in \Omega : \ \mathrm{dist}(x, \partial\Omega) < s\}, \quad V_s^\delta = V_s \times (0, \delta),$$
$$\Gamma_s = \{x \in \Omega : \ \mathrm{dist}(x, \partial\Omega) = s\}, \quad \Gamma_s^\delta = \Gamma_s \times (0, \delta).$$

We fix $s_0 > 0$ such that Γ_s is a smooth curve for every $s \in (0, 2s_0]$. Moreover, we define

$$\kappa(\delta) = \frac{1}{\log \frac{1}{\delta}}.$$

Lemma 4.5. *There exists a constant C, depending only on Ω, such that for every $\delta \in (0, e^{-e}]$ and every $s \in [0, s_0]$ with $s \leq \kappa(\delta)$, the inequality*

$$\log \frac{1}{\delta} \int_{\Gamma_s^\delta} (m^\delta \cdot \nu)^2 \, d\mathcal{H}^2 \leq C \int_{V^\delta_{\kappa(\delta)}} \left(|\nabla \mathbf{m}^\delta|^2 + \frac{(m_3^\delta)^2}{\delta} \right) d\mathbf{x} + \frac{C}{\delta} \int_{\mathbb{R}^3} |\nabla u^\delta|^2 \, d\mathbf{x} + C\delta \tag{4.13}$$

is satisfied for any $\mathbf{m}^\delta \in H^1(\Omega_\delta, \mathbb{S}^2)$.

Sketch of the proof. The idea is to test (4.1) with a suitably constructed function ϕ satisfying $\phi = m^\delta \cdot \nu$ on Γ_0^δ and supported on a $\kappa(\delta)$-neighborhood of Γ_0^δ. An integration by parts on one side of the resulting equation then yields, among other terms, the left-hand side of (4.13) for $s = 0$. A careful estimate of the other terms gives the required inequality for Γ_0^δ. To obtain the corresponding inequality for other values of s, integrate the derivative of $(m^\delta \cdot \nu)^2$ along rays in the direction of $-\nu$.

Proposition 4.6. *For any $K \in \mathbb{R}$ there exists a constant C, depending only on Ω and K, such that the following holds. Suppose $\delta \in (0, e^{-e}]$ and $s \in [0, s_0]$. If $\mathbf{m}^\delta \in H^1(\Omega_\delta, \mathbb{S}^2)$ satisfies*

$$E_\delta(\mathbf{m}^\delta) \leq \pi \log \log \frac{1}{\delta} + K, \tag{4.14}$$

then

$$\frac{1}{\delta} \int_{\Omega_\delta} \left(|\nabla m_3^\delta|^2 + \left| \frac{\partial \mathbf{m}^\delta}{\partial x_3} \right|^2 + \frac{(m_3^\delta)^2}{\delta} \right) d\mathbf{x} + \frac{1}{\delta} \int_{V^\delta_{\kappa(\delta)}} |\nabla \mathbf{m}^\delta|^2 \, d\mathbf{x} + \frac{1}{\delta^2} \int_{\mathbb{R}^3} |\nabla u^\delta|^2 \, d\mathbf{x} + \frac{\log \frac{1}{\delta}}{\delta} \int_{\Gamma_s^\delta} (m^\delta \cdot \nu)^2 \, d\mathcal{H}^2 \leq C \tag{4.15}$$

Sketch of the proof. With the arguments from Section 3, combined with similar arguments from the theory of Ginzburg-Landau vortices, applied to slices of the form $\Omega \backslash V_s \times \{t\}$, we obtain the estimate

$$\frac{1}{2\delta}\int_{\Omega_\delta\setminus V_s^\delta}\left(\left|\frac{\partial m^\delta}{\partial x_1}\right|^2+\left|\frac{\partial m^\delta}{\partial x_2}\right|^2+\frac{(m_3^\delta)^2}{\delta}\right)d\mathbf{x}$$
$$+\frac{\log\frac{1}{\delta}}{\delta}\int_{\Gamma_s^\delta}(m^\delta\cdot\nu)^2\,d\mathcal{H}^2\geq\pi\log\log\frac{1}{\delta}-C_1 \quad (4.16)$$

for a certain constant C_1 that depends only on Ω. Combining this with Lemma 4.4, Lemma 4.5, and (4.14), we obtain the desired inequality.

Proposition 4.7. *There exists a constant C, dependent only on Ω, such that*

$$\inf_{H^1(\Omega_\delta,\mathbb{S}^2)} E_\delta\geq\pi\log\log\frac{1}{\delta}-C$$

for $\delta\in(0,e^{-e}]$.

Proof. Choose a minimizer $\mathbf{m}^\delta$ of E_δ in $H^1(\Omega_\delta,\mathbb{S}^2)$, then (4.14) holds for a certain constant K by Proposition 4.3. Thus $\mathbf{m}^\delta$ satisfies (4.15) and (4.16), and the claim follows.

Proposition 4.3 and Proposition 4.6 together describe the asymptotic behavior of the minimal energy as $\delta\searrow 0$ up to an additive constant, and it is the behavior we expect also for the functionals F_δ or G_δ. Moreover, if $\mathbf{m}^\delta$ is independent of x_3, we can estimate each term in E_δ by a combination of terms in G_δ, and vice versa, according to Lemmas 4.1–4.5. This already relates the asymptotic regime studied here to the model used in Section 3. We discover more similarities, however, when we study the asymptotic behavior of critical points of E_δ or solutions of the Landau-Lifshitz equations (4.8) and (4.11)

If we have a family of solutions $\mathbf{m}^\delta$ of one of the variational problems associated to E_δ, it is natural to apply variants of the usual arguments from the theory of Ginzburg-Landau vortices to the slices $\Omega\times\{s\}$ with $0<s<\delta$. More precisely, we use arguments like those discussed in the previous section for the behavior near $\partial\Omega$, and arguments from the theory of Bethuel, Brezis, and Hélein [BBH93, BBH94] for the behavior in the interior of Ω. This is the key element in the proofs of each of the results that follow. We omit a detailed presentation of these proofs (since similar arguments have been discussed earlier), but we give a brief discussion of some additional arguments that are needed in each case. For the complete proofs, the reader is referred to [Mos03, Mos04, Mos05].

Theorem 4.8. *For $\delta\in(0,e^{-e}]$, suppose $\mathbf{m}^\delta\in C^\infty(\Omega,\mathbb{S}^2)$ are stable critical points of E_δ, i.e., solutions of (4.3) satisfying the boundary conditions (4.4) and (4.5), such that (4.6) holds for every $\psi\in C^\infty(\overline{\Omega}_\delta,\mathbb{R}^3)$. Suppose further that there exists a number K such that*

$$E_\delta(\mathbf{m}^\delta)\leq\pi\log\log\frac{1}{\delta}+K$$

for every δ. *Then there exist a sequence* $\delta_k \searrow 0$, *two distinct points* $x^1, x^2 \in \partial\Omega$, *and a map* $\mathbf{m} = (m, 0) \in W^{1,1}(\Omega, \mathbb{S}^1 \times \{0\})$ *with* $m \cdot \nu = 0$ *on* $\partial\Omega$, *such that the following holds.*

(i) For any $p < 2$, *the sequence* $\{\overline{\mathbf{m}}_{\delta_k}\}$ *converges weakly in* $W^{1,p}(\Omega, \mathbb{R}^3)$ *to* $\mathbf{m}$. *The convergence also holds weakly in* $H^1(\Omega', \mathbb{R}^3)$ *for every* $\Omega' \subset \Omega$ *with* $\overline{\Omega'} \subset \overline{\Omega} \backslash \{x^1, x^2\}$.

(ii) The limit map m *satisfies*

$$\Delta m + |\nabla m|^2 m - \nabla u_0 + (m \cdot \nabla u_0) = 0 \quad \text{in } \Omega, \tag{4.17}$$

where u_0 *is the trace on* $\Omega \times \{0\}$ *of the function determined by (4.2).*

(iii) If $\mathbb{R}^2$ *is identified with the complex plane* $\mathbb{C}$ *by* $z = x_1 + ix_2$ *(and similarly* $z^1 = x_1^1 + ix_2^1$ *and* $z^2 = x_1^2 + ix_2^2$ *for* $x^1 = (x_1^1, x_2^1)$ *and* $x^2 = (x_1^2, x_2^2)$*), then* m *has the representation*

$$m(z) = \frac{z - z^1}{|z - z^1|} \frac{z - z^2}{|z - z^2|} e^{i\theta(z)}$$

for a function $\theta \in C^0(\overline{\Omega})$ *which solves*

$$\Delta\theta = m_1 \frac{\partial u_0}{\partial x_2} - m_2 \frac{\partial u_0}{\partial x_1} \quad \text{in } \Omega. \tag{4.18}$$

Thus at least at the lowest possible energy level, we observe the development of two boundary vortices in the limit. Note also that the limit equation (4.17) is formally the Euler-Lagrange equation for the (formal) functional

$$\frac{1}{2} \int_{\Omega} |\nabla m|^2 \, dx + \frac{1}{2} \int_{\mathbb{R}^3} |\nabla u|^2 \, d\mathbf{x},$$

where u denotes the function in $H^1(\mathbb{R}^3)$ defined by (4.2). It turns out, however, that this quantity is identically infinite. It can be replaced by a functional involving the Dirichlet energy of θ, and then (4.17) truly becomes an Euler-Lagrange equation, but we omit the details here.

The stability condition (4.6) is needed in the proof of this theorem in order to estimate the Dirichlet energy of $\mathbf{m}^\delta$ in small cylinders of the form

$$(B_{\mu\sqrt{\delta}}(x) \cap \Omega) \times (0, \delta),$$

where $\mu > 0$ is a fixed constant. This allows to use estimates from the regularity theory of harmonic maps and to conclude that

$$|\nabla \mathbf{m}^\delta| \leq \frac{C}{\sqrt{\delta}}$$

for a certain constant C which is independent of δ. Apart from the fact that such a gradient estimate is normally used in the theory of Ginzburg-Landau vortices, it is in this context also important for another reason: It means that $\mathbf{m}^\delta$ varies very little in the third direction if δ is small, for the thickness of Ω_δ is small compared with $|\nabla \mathbf{m}^\delta|$. This permits to work on a suitable slice $\Omega \times \{s\}$ and pretend that the domain is two-dimensional for much of the proof. The previously mentioned arguments then give (i) and the representation of m in (iii).

To derive the limit equation (4.17), we first take the vector product with $\mathbf{m}^\delta$ on both sides of (4.3), which gives

$$\delta \operatorname{div}(\mathbf{m}^\delta \wedge \nabla \mathbf{m}^\delta) = \mathbf{m}^\delta \wedge \nabla u^\delta = 0 \quad \text{in } \Omega_\delta.$$

This form of the equation has the advantage that it does no longer explicitly contain the term $|\nabla \mathbf{m}^\delta|^2 \mathbf{m}^\delta$ (which would be difficult to handle with the weak convergence that we have). In particular it is then possible to pass to the limit and to show that m satisfies

$$\operatorname{div}(m_1 \nabla m_2 - m_2 \nabla m_1) = m_1 \frac{\partial u_0}{\partial x_2} - m_2 \frac{\partial u_0}{\partial x_1} \quad \text{in } \Omega.$$

This equation is exactly (4.18) if m is represented by θ as in (iii). Finally, the equation is also equivalent to (4.17).

Theorem 4.9. *For $T \in (0,\infty]$ and $\delta \in (0, e^{-e}]$, suppose $\mathbf{m}^\delta \in C^\infty(\Omega_\delta \times [0,T), \mathbb{S}^2)$ satisfy the Landau-Lifshitz equations (4.11) with boundary conditions (4.9), (4.10). Also suppose that the initial data*

$$\hat{\mathbf{m}}^\delta(\mathbf{x}) = \mathbf{m}^\delta(\mathbf{x}, 0)$$

satisfy

$$E_\delta(\hat{\mathbf{m}}^\delta) \le \pi \log\log \frac{1}{\delta} + K$$

and

$$|\nabla \hat{\mathbf{m}}^\delta| \le \frac{K}{\sqrt{\delta}}$$

in Ω_δ for a constant K that is independent of δ. Then there exist a sequence $\delta_k \searrow 0$, a map $\mathbf{m} = (m, 0) \in L^\infty([0,T), W^{1,1}(\Omega, \mathbb{S}^1 \times \{0\}))$ with $m \cdot \nu = 0$ on $\partial\Omega \times [0,T)$, and two distinct points $x^1, x^2 \in \partial\Omega$, such that the following holds.

(i) The sequence $\{\overline{\mathbf{m}}^{\delta_k}\}$ converges weakly in $L^\infty([0,T), W^{1,p}(\Omega, \mathbb{R}^3))$ to $\mathbf{m}$ for any $p < 2$, and also weakly* in $L^\infty([0,T), H^1(\Omega', \mathbb{R}^3))$ for any $\Omega' \subset \Omega$ with $\overline{\Omega'} \subset \overline{\Omega} \backslash \{x^1, x^2\}$.*

(ii) The limit map solves

$$\tilde{\alpha} \frac{\partial m}{\partial t} = \Delta m + |\nabla m|^2 m - \nabla u_0 + (m \cdot \nabla u_0) m \quad \text{in } \Omega \times (0,T), \qquad (4.19)$$

where $u_0(\,\cdot\,, t)$ is the trace on $\Omega \times \{0\}$ of the function determined by (4.2) for almost every fixed $t \in [0,T)$.

(iii) It is of the form

$$m(z,t) = \frac{z - z^1}{|z - z^1|} \frac{z - z^2}{|z - z^2|} e^{i\theta(z,t)},$$

where $\theta \in C^0(\overline{\Omega} \times [0,T))$ *is a solution of*

$$\tilde{\alpha} \frac{\partial \theta}{\partial t} = \Delta\theta - m_1 \frac{\partial u_0}{\partial x_2} + m_2 \frac{\partial u_0}{\partial x_1} \quad \text{in } \Omega \times (0,T).$$

(Here we use an identification of $\mathbb{R}^2$ *with* $\mathbb{C}$ *as in the previous theorem.)*

This is the time scale where we have stationary boundary vortices. The limit equation (4.19) is formally the L^2-gradient flow for the formal functional mentioned earlier (up to a constant). It is also the true gradient flow for a related functional.

It is interesting here to compare the limit equation (4.19) with the original Landau-Lifshitz equation, especially if the latter is in the form (4.7). We had originally a gyromagnetic term with coefficient $\hat{\gamma}$ and a damping term with coefficient $\hat{\alpha}$. The gyromagnetic term has vanished in the thin-film limit (as it is to be expected when $\mathbf{m}$ remains in the plane $\mathbb{R}^2 \times \{0\}$). We still have a damping term, but the damping coefficient is now

$$\frac{1}{\tilde{\alpha}} = \hat{\alpha} + \frac{\hat{\gamma}^2}{\hat{\alpha}}.$$

Thinking of $\hat{\gamma}$ as a fixed constant and of $\hat{\alpha}$ as small in comparison, we are in the seemingly paradox situation that decreasing the damping coefficient $\hat{\alpha}$ accelerates the dynamics in the thin-film limit. This phenomenon has already been discovered by formal computations by W. E and C. García-Cervera [EG01].

With the gradient estimate that we impose on the initial data, we can use the same methods as in the proof of Thm. 4.8 to obtain the same development of boundary vortices for $\hat{\mathbf{m}}^\delta$ that we have found for stable critical points. To prove this for times $t > 0$, we need slightly different arguments. Here we calculate how the energy develops locally with time, that is, for a function $\eta \in C_0^\infty(\mathbb{R}^3)$ with $\eta = 0$ in a neighborhood of $\{x^1, x^2\} \times [0,\delta]$, we calculate

$$\frac{d}{dt}\left(\int_{\Omega_\delta} \eta |\nabla \mathbf{m}^\delta|^2 \, d\mathbf{x} + \frac{1}{\delta} \int_{\mathbb{R}^3} \eta |\nabla u^\delta|^2 \, d\mathbf{x}\right). \tag{4.20}$$

We find that away from the vortex center points x^1, x^2, the energy increases at most by a constant that is independent of δ in bounded time intervals. We can then again use arguments from the theory of Ginzburg-Landau vortices for fixed times $t > 0$.

Theorem 4.10. *Under the conditions of Thm. 4.9, but with (4.8) replaced by (4.11), there exist a sequence* $\delta_k \searrow 0$, *two curves* $x^1, x^2 \in C^{0,1/2}([0,T), \partial\Omega)$, *and a map*

$$\mathbf{m} = (m, 0) \in \bigcap_{p<2} L^\infty([0,T), W^{1,p}(\Omega, S^1 \times \{0\}))$$

with $m \cdot \nu = 0$ *on* $\partial\Omega \times [0,T)$, *such that the following holds.*

(i) For any $p < 2$ *and any* $q < \infty$, *the sequence* $\{\overline{\mathbf{m}}^{\delta_k}\}$ *converges weakly in* $L^q([0,T), W^{1,p}(\Omega, \mathbb{R}^3))$ *to* $\mathbf{m}$.

(ii) For almost every $t \in [0,T)$ *and every* $\Omega' \subset \Omega$ *with* $\overline{\Omega'} \subset \overline{\Omega} \backslash \{x^1(t), x^2(t)\}$, *the map* $m(\cdot, t)$ *belongs to* $H^1(\Omega', \mathbb{S}^1)$.

(iii) The equation

$$\Delta m + |\nabla m|^2 m - \nabla u_0 + (m \cdot \nabla u_0) m = 0 \quad \text{in } \Omega \times (0,T)$$

holds, where $u_0(\cdot, t)$ *is the trace on* $\Omega \times \{0\}$ *of the function determined by (4.2) for almost every fixed* $t \in [0,T)$.

(iv) The map m *is of the form*

$$m(z,t) = \frac{z - z^1(t)}{|z - z^1(t)|} \frac{z - z^2(t)}{|z - z^2(t)|} e^{i\theta(z,t)},$$

where $\theta \in C^0(\overline{\Omega} \times [0,T])$ *is a solution of*

$$\Delta\theta = m_1 \frac{\partial u_0}{\partial x_1} - m_2 \frac{\partial u_0}{\partial x_2} \quad \text{in } \Omega \times (0,T).$$

In contrast to the situation of Thm. 4.9, we now have moving boundary vortices. This means in particular that when we calculate the evolution of the localized energy in (4.20), the vortex centers may enter the support of η after some time. For this reason, the estimates we obtain are not quite as good as before, and the type of convergence we find is weaker. On the other hand, an analysis of the energy increase over a fixed time interval permits to estimate the distance that the vortex centers have moved in this time (since most of the micromagnetic energy is concentrated in the vortex centers). This way we obtain the Hölder continuity of x^1 and x^2.

Finally, comparing Thm. 4.10 with the results of Section 3, especially Thm. 3.19, we see that one statement is missing here: We do not have any information about the law that governs the motion of the vortices. There is no obvious reason why the model discussed here should have a significantly different behavior in this respect, but the technical difficulties mentioned earlier make it hard to carry over the arguments from the simpler model. This problem thus remains open.

Acknowlegdments. The authors are grateful for the support of SPP 1095 through the grants De 772/1-1, De 772/1-2 and Mu 1067/6-3. They would like to thank A. DeSimone and S. Müller for many stimulating discussions at the MPI for Mathematics in the Sciences, and the latter in particular for his valuable suggestions and comments on this work.

References

[ABS94] G. Alberti, G. Bouchitté, and P. Seppecher, *Un résultat de perturbations singulières avec la norme $H^{1/2}$*, C. R. Acad. Sci. Paris Sér. I Math. **319** (1994), no. 4, 333–338.

[ABS98] G. Alberti, G. Bouchitté, and P. Seppecher, *Phase transition with the line-tension effect*, Arch. Rational Mech. Anal. **144** (1998), no. 1, 1–46.

[BF*97] P. W. Bates, P. C. Fife, X. Ren, and X. Wang, *Traveling Waves in a Convolution Model for Phase Transitions*, Arch. Rational Mech. Anal. **138** (1997), no. 2, 105–136.

[BBH93] F. Bethuel, H. Brezis, and F. Hélein, *Asymptotics for the minimization of a Ginzburg-Landau functional*, Calc. Var. Partial Differential Equations **1** (1993), no. 2, 123–148.

[BBH94] F. Bethuel, H. Brezis, F. Hélein, *Ginzburg-Landau vortices*, Progress in Nonlinear Differential Equations and their Applications, 13, Birkhäuser, Boston, 1994.

[CS04] X. Cabré and J. Solà-Morales, *Layer solutions in a half-space for boundary reactions*, Comm. Pure Appl. Math. **58** (2005), no. 12, 1678–1732.

[CMO06] A. Cappella-Kort, C. Melcher, and F. Otto, in preparation

[Car01] G. Carbou, *Thin layers in micromagnetism*, Math. Models Methods Appl. Sci. **11** (2001), no. 9, 1529–1546.

[DKO05] A. DeSimone, H. Knüpfer, and F. Otto, *2-d stability of the Néel wall*, SFB 611 Preprint **224**, 2005.

[DK*99] A. DeSimone, R. V. Kohn, S. Müller, and F. Otto, *Magnetic microstructure - a paradigm of multiscale problems*, Proceedings of the ICIAM 99 (Edinburgh), Eds. J. M. Ball and J. C. R. Hunt, (1999), 175–199.

[DK*02] A. DeSimone, R. V. Kohn, S. Müller, and F. Otto, *A reduced theory for thin-film micromagnetics*, Comm. Pure Appl. Math. **55** (2002), no. 11, 1408–1460.

[DK*03] A. DeSimone, R. V. Kohn, S. Müller, and F. Otto, *Repulsive interaction of Néel walls, and the internal length scale of the Cross-tie wall*, SIAM Multiscale Model. Simul. **1** (2003), 57–104.

[DK*05] A. DeSimone, R. V. Kohn, S. Müller, and F. Otto, *Recent analytical developments in micromagnetics*, in: The science of hysteresis (G. Bertotti, I. Mayergoyz, eds.), Vol. II, pp. 269–381, Elsevier, 2005.

[Dör48] W. Döring, *Über die Trägheit der Wände zwischen Weißschen Bezirken*, Z. Naturforschung **3a** (1948), 373–379.

[EG01] W. E and C. J. García-Cervera, *Effective dynamics in ferromagnetic thin films*, J. Appl. Phys. **90** (2001), no. 1, 370–374.

[Gra99] C. J. Gracía-Cervera, *Magnetic domains and magnetic domain walls*, Ph.D. thesis, New York University, 1999.

[GJ97] G. Gioia and R. D. James, *Micromagnetics of very thin films*, Proc. R. Soc. Lond. Ser. A **453** (1997), 213–223.

[Hof98] S. Hofmann, *An Off-Diagonal $T(1)$ Theorem and Applications*, J. Funct. Analysis **160** (1998), 581–622.

[HS98] A. Hubert and R. Schäfer, *Magnetic Domains, The Analysis of Magnetic Microstructures*, Springer-Verlag, Berlin-Heidelberg-New York, 1998.

[JS98] R. L. Jerrard and H. M. Soner, *Dynamics of Ginzburg-Landau vortices*, Arch. Rational Mech. Anal. **142** (1998), no. 2, 99–125.

[KP88] T. Kato and G. Ponce, *Commutator estimates and the Euler and Navier-Stokes equation*, Comm. Pure Appl. Math **16** (1988), no. 7, 891–907.

[KS05a] R. V. Kohn and V. V. Slastikov, *Effective dynamics for ferromagnetic thin films: a rigorous justification*, Proc. R. Soc. Lond. Ser. A **461** (2005), 143–154.

[KS05b] R. V. Kohn and V. V. Slastikov, *Another thin-film limit of micromagnetics*, Arch. Ration. Mech. Anal. **178** (2005), no. 2, 227–245.

[Kur04] M. Kurzke, *Boundary vortices in thin magnetic films*, Preprint 14, Max Planck Institute for Mathematics in the Sciences, 2004, To appear in Calc. Var. PDE.

[Kur05] M. Kurzke, *The gradient flow motion of boundary vortices*, Preprint 2030, IMA, 2005, To appear in Ann. IHP Analyse nonlineaire.

[Kur06] M. Kurzke, *A nonlocal singular perturbation problem with periodic well potential*, ESAIM:COCV **12** (2006), 52–63.

[LL35] L. D. Landau and E. Lifshitz, *On the theory of the dispersion of magnetic permeability in ferromagnetic bodies*, Phys. Z. Sovietunion **8** (1935), 153–169.

[Lin96a] F. H. Lin, *Some dynamical properties of Ginzburg-Landau vortices*, Comm. Pure Appl. Math. **49** (1996), no. 4, 323–359.

[Lin96b] F. H. Lin, *A remark on the previous paper: "Some dynamical properties of Ginzburg-Landau vortices"*, Comm. Pure Appl. Math. **49** (1996), no. 4, 361–364.

[Mel02] C. Melcher, *Néel walls and regularity in thin film micromagnetics*, Dissertation, Universität Leipzig, 2002.

[Mel03] C. Melcher, *The logarithmic tail of Néel walls*, Arch. Rational Mech. Anal. **168** (2003), 83–113.

[Mel04a] C. Melcher, *Logarithmic lower bounds for Néel walls*, Calc. Var. Partial Differential Equations **168** (2004), 209–219.

[Mel04b] C. Melcher, *Domain wall motion in ferromagnetic layers*, Physica D **192** (2004), 249–264.

[Mos03] R. Moser, *Ginzburg-Landau vortices for thin ferromagnetic films*, AMRX Appl. Math. Res. Express **2003**, no. 1, 1–32.

[Mos04] R. Moser, *Boundary vortices for thin ferromagnetic films*, Arch. Ration. Mech. Anal. **174** (2004), no. 2, 267–300.

[Mos05] R. Moser, *Moving boundary vortices for a thin-film limit in micromagnetics*, Comm. Pure Appl. Math. **58** (2005), no. 5, 701–721.

[Née55] L. Néel, *Energie des parois de Bloch dans les couches minces*, C. R. Acad. Sci. Paris **241** (1955), 533–536.

[Ott02] F. Otto, *Cross-over in scaling laws: a simple example in micromagnetics*, Proceedings of the International Congress of Mathematicians, Volume III (Beijing, 2002), 829–838, Higher Ed. Press, Beijing, 2002.

[RS71] H. Riedel and A. Seeger, *Micromagnetic Treatment of Néel Walls*, Phys. Stat. Sol. (B) **46** (1971), 377–384.

[SS04a] E. Sandier and S. Serfaty, *Gamma-convergence of gradient flows with applications to Ginzburg-Landau*, Comm. Pure Appl. Math. **57** (2004), no. 12, 1627–1672.

[SS04b] E. Sandier and S. Serfaty, *A product-estimate for Ginzburg-Landau and corollaries*, J. Funct. Anal. **211** (2004), no. 1, 219–244.

[Str94] M. Struwe, *On the asymptotic behavior of minimizers of the Ginzburg-Landau model in 2 dimensions*, Differential Integral Equations **7** (1994), no. 5-6, 1613–1624.

[Tol97] J. F. Toland, *The Peierls-Nabarro and Benjamin-Ono equations*, J. Funct. Anal. **145** (1997), no. 1, 136–150.

[Wal63] L. R. Walker, unpublished. Reported in J. F. Dillon Jr., *Domains and Domain Walls*, in Magnetism, Eds. G. R. Rado and H. Suhl, Academic Press, New York, 1963.

Wavelet-Based Multiscale Methods for Electronic Structure Calculations

Heinz-Jürgen Flad[1,2], Wolfgang Hackbusch[1], Hongjun Luo[3], and Dietmar Kolb[3]

[1] Max-Planck-Institut für Mathematik in den Naturwissenschaften, Inselstraße 22–26, 04103 Leipzig. wh.mis.mpg.de
[2] Institut für Informatik und Praktische Mathematik, Christian-Albrechts-Universität zu Kiel, Christian-Albrechts-Platz 4, 24098 Kiel. hjf@numerik.uni-kiel.de
[3] Institut für Physik, Universität Kassel, Heinrich-Plett-Straße 40, 34132 Kassel. luo@physik.uni-kassel.de, kolb@physik.uni-kassel.de

Summary. In order to treat multiple energy- and length-scales in electronic structure calculations for extended systems, we have studied a wavelet based multiresolution analysis of electron correlations. Wavelets provide hierarchical basis sets that can be locally adapted to the length- and energy-scales of physical phenomena. The inherently high dimensional many-body problem can be kept tractable by using the sparse grid method for the construction of multivariate wavelets. These so called "hyperbolic" wavelets provide sparse representations for correlated wavefunctions and can be combined with diagrammatic techniques from quantum many-particle theory into a diagrammatic multiresolution analysis. Using sparsity features originating from the hierarchical structure and vanishing moments property of wavelet bases, this leads to many-particle methods with almost linear computational complexity for the treatment of electron correlations.

We are aiming towards applications in semiconductor physics where quasi two-dimensional many-particle systems provide challenging computational problems. Such kind of systems are metallic slabs and interacting excitons confined in quantum wells of semiconductor heterostructures. As a first step we developed a multiresolution Hartree-Fock method suitable for quasi two-dimensional extended systems. Special emphasis has been laid on low rank tensor product decompositions of orbitals, which take into account the strongly anisotropic character of these systems in one direction.

1 Introduction

Quantum many-particle theory provides a general framework for the description of microscopic processes appearing in solid state physics and chemistry. Some of the fundamental difficulties encountered for accurate calculations of these processes are inherently related with their multiscale character. The huge diversity of methods and techniques in many-particle theory partly emerged from efforts to create tailormade models which describe the essential physics at certain energy or length-scales. However, synergetic effects, caused by couplings between different scales, often prevent their separate treatment. Due to a lack of *a priori* knowledge of the strength of these couplings, it might become necessary to use many-particle models which provide an accurate description of the system over a whole range of energy- or length-scales. In virtue of the multiscale character of such models, it is tempting to study possible applications of some recent mathematical developments in the field of multiscale analysis. In particular wavelet based multiresolution analysis [Dau92, Mal98, Mey92] proved to be a useful tool, both from the analytical and computational point of view [Dah97, DeV98]. Wavelets represent stable multiscale basis sets that can be locally adapted according to the length- and energy-scales of physical processes. Depending on the specific application, wavelets can be equiped with a variety of useful properties including compact support, (bi)orthogonality and vanishing moments [Dau92]. Sparse approximations exist for functions containing local singularities [Mal98] and for a large class of differential as well as (singular) integral operators [Dah97]. Furthermore, because of the hierarchical structure of wavelet bases, it becomes feasible to apply sparse grid methods [BG04] for high-dimensional problems.

We have studied possible applications of multiresolution analysis for many-electron systems which can be considered as a paradigm for fermionic many-particle systems in condensed matter physics. The antisymmetry of fermionic wavefunctions introduces, via Pauli's exclusion principle, a multiscale structure into these systems. Their multiscale character expresses itself in the energy- and length-scales of physical processes extending over several orders of magnitude. Typical examples, ranging from high (short) to low (large) energy- (length-) scales, are processes inside atomic cores, covalent chemical bonding, van der Waals interactions and magnetic couplings between unpaired electron spins in molecules and solids. Besides conventional many-electron systems, we want to consider quasiparticles, like electron hole systems, which are of fundamental interest in semiconductor physics. These quasiparticles combine to excitonic systems similar to molecules. Of special interest are semiconductor heterostructures [Chu95] where excitons are confined in the direction perpendicular to the semiconductor layers. For nanostructured materials, the diameter of a free exciton becomes larger than the thickness of a layer. These excitons, therefore, provide interesting two-scale problems with properties that lie somewhere between that of free 2d- and 3d-excitons.

During the last few years considerable interest in numerical techniques from multiresolution analysis has been emerged within the field of electronic structure calculations. Most of this work has been done in the context of *density functional theory* (DFT) and the *Hartree-Fock* (HF) method [Ari99, Goe98, HF*04]. Recently published papers discuss applications to solids [EA02, TBJ03] and possible extensions of multiresolution analysis to the density-matrix approach [NTR02] which avoids the cumbersome calculation of eigenvalues and provides linear scaling with respect to the size of the system. Another interesting application are semi-classical calculations of polarons and bipolarons solvated in a liquid [CF*05]. Wavelets were used for the quantum description of the polaron and bipolaron, whereas the solvent has been treated classically by an integral equation.

In DFT, the many-particle problem is mapped onto a system of noninteracting particles [Kohn99], resulting in a significant computational simplification. All many-particle aspects are represented by an approximate exchange-correlation potential in the *Kohn-Sham* (KS) equation. The fundamental problem of this approach is that the mapping can be done only approximately and systematic ways for improvement are presently still missing. Multiresolution analysis can therefore just be applied to multiscale aspects of independent-particle systems. For such systems, the most important multiscale features are variations of the electron density between atomic core and valence regions in molecules and solids. The conventional numerical approach to a simultaneous treatment of core and valence electrons in KS or HF equations uses *Gaussian type orbital* (GTO) bases [HJO99], consisting of atomic centered Gaussian basis functions with variable exponents, multiplied by harmonic polynomials. GTO bases have already reached a high level of sophistication in quantum chemistry and seem to be almost optimal for such kind of problems. However, in contrast to wavelets, GTO bases are not stable in a mathematically rigorous sense and might become ill behaved if very high accuracies have to be achieved. There exists a variety of many-particle models in solid state physics where GTO bases are not really appropriate due to certain geometrical constraints imposed on the system. This is the case e.g. for quasi two-dimensional many-particle systems already mentioned above, like excitons in semiconductor heterostructures. For such kind of systems wavelets seem to be an interesting alternative. We proposed a multiresolution HF method for quasi two-dimensional systems in Ref. [FH*05b], which will be briefly discussed in Sect. 5.

Beside DFT, there exists another kind of approach where a direct solution of the many-particle problem is tried to achieve. Starting from an effective noninteracting particle model, like HF, the solution of the interacting many-particle system can be approached via *configuration interaction* (CI) methods, *many-body perturbation theory* (MBPT) or *coupled cluster* (CC) methods. All effects which are not taken into account by the HF solution are summarized under the phrase of electron correlations. We refer to the monographs [Ful93, HJO99] for a thorough discussion of this subject. The original formulations of these post HF methods rely heavily on the eigenfunctions of the HF Hamiltonian. This turns out to be an obstacle for a direct application of multiresolution analysis. In order to circumvent this problem, we have adopted ideas from local correlation methods, which enable an efficient treatment of electron correlations in extended systems. Here we have to mention the local ansatz of Stollhoff and Fulde [Ful93, SF80], the increment method of Stoll [Sto80] and the local correlation methods of Pulay and Saebø [SP93], which have been further developed into linear scaling methods by Werner and collaborators [SHW99]. Conventional numerical approaches for these methods use GTO bases to represent electron correlations. It is common knowledge, however, that GTO bases are not well adapted for this purpose [HJO99]. Within our approach, GTO bases were replaced by wavelets, which provide stable multi-scale bases that can be locally adapted for the description of electron correlations. In Sect. 2, we discuss wavelet approximations of correlated wavefunctions, cf. Refs. [FH*02, LK*02], and provide some simple examples in order to illustrate the underlying concepts. These concepts have been set into a broader context in Sect. 4, where we briefly outline various many-particle methods and discuss their computational complexity using a diagrammatic multiresolution analysis, see Refs. [FH*05a, LK*06] for further details.

The main part of the paper focus on wavelet methods and the nonrelativistic many-particle Schrödinger equation. There are however many other interesting topics concerning multiscale problems in electronic structure calculations. Two such topics related to recent work done by the group in Kassel are briefly outlined at the end of the paper. Appendix B summarizes recent developments in multi-grid methods for KS and HF equations, whereas Appendix C provides a short glimpse into the field of relativistic electronic structure theory.

2 Wavelet approximation of correlated wavefunctions

The nonrelativistic Schrödinger equation within the Born-Oppenheimer approximation provides a firm basis for electronic structure calculations in quantum chemistry and solid state physics. We are focusing on solutions of the stationary Schrödinger equation

$$H\,\Psi\left(\mathbf{r}_1, \mathbf{r}_2, \ldots, \mathbf{r}_N\right) = E\,\Psi\left(\mathbf{r}_1, \mathbf{r}_2, \ldots, \mathbf{r}_N\right), \tag{2.1}$$

where the Hamiltonian

$$H = \sum_{i=1}^{N} \left(-\frac{1}{2}\Delta_i + V_{\mathrm{ext}}(\mathbf{r}_i) \right) + \sum_{i<j} \frac{1}{|\mathbf{r}_i - \mathbf{r}_j|}$$

includes Coulomb interactions between the electrons and an external potential due to the nuclei. Atomic units have been used throughout this paper. Recently, interesting results concerning the regularity of the solutions of the many-particle Schrödinger equation (2.1) in certain Sobolev spaces of mixed derivatives have been obtained by Yserentant [Yse04, Yse05]. These regularity results open the possibility for a multiscale approximation of the total wavefunction Ψ. Closely related to this approach is the sparse grid method of Griebel an coworkers [GG00]. We pursue a more restricted approach by studying multiscale aspects within conventional many-particle theories. This has the significant advantage that we can rely on a wealth of experiences in physics and chemistry. A natural framework for the representation of many-electron wavefunctions Ψ is the product ansatz

$$\Psi(\mathbf{r}_1, \mathbf{r}_2, \ldots, \mathbf{r}_N) = \mathcal{F}\,\Phi(\mathbf{r}_1, \mathbf{r}_2, \ldots, \mathbf{r}_N), \tag{2.2}$$

where the correlation operator $\mathcal{F}$ acts on a mean-field solution Φ. In general, $\mathcal{F}$ has to be understood as a linear operator, who's specific properties must be derived for example from MBPT or CC theory. We have adopted a simplified ansatz, where the correlation factor $\mathcal{F}(\mathbf{r}_1, \mathbf{r}_2, \ldots, \mathbf{r}_N)$ represents a symmetric function of the electron coordinates, which corrects for the inadequacies of a given approximate wavefunction Φ. Typically, Φ is taken from the HF approximation of the original Schrödinger equation (2.1). The HF wavefunction is given in terms of a Slater determinant which corresponds to an antisymmetrized product of orbitals. These orbitals are single-particle wavefunctions obtained by solving the HF equation. In general it is not possible to represent the exact many-particle wavefunction via a correlation factor $\mathcal{F}(\mathbf{r}_1, \ldots, \mathbf{r}_N)$. We have to deal with a constraint variational problem where the expectation value of the energy

$$E[\mathcal{F}] = \frac{\int d^3r_1 \ldots d^3r_N\, \mathcal{F}\Phi(\mathbf{r}_1, \mathbf{r}_2, \ldots, \mathbf{r}_N)\, H\, \mathcal{F}\Phi(\mathbf{r}_1, \mathbf{r}_2, \ldots, \mathbf{r}_N)}{\int d^3r_1 \ldots d^3r_N\, \mathcal{F}\Phi(\mathbf{r}_1, \mathbf{r}_2, \ldots, \mathbf{r}_N)\, \mathcal{F}\Phi(\mathbf{r}_1, \mathbf{r}_2, \ldots, \mathbf{r}_N)}, \tag{2.3}$$

is minimized only with respect to the correlation factor $\mathcal{F}$ instead of the full wavefunction Ψ. The latter would be equivalent to solving the Schrödinger equation (2.1). Within our previous work, we have studied wavelet approximations of correlation factors for simple atoms [FH*02], exactly solvable many-particle models [LK*02] and the homogeneous electron gas [FH*05a]. This can be done using either a linear ansatz

$$\mathcal{F}(\mathbf{r}_1, \mathbf{r}_2, \ldots, \mathbf{r}_N) = \sum_p \sum_{\mathbf{J}}{}' \sum_{\mathbf{A}} f^{(p)}_{\mathbf{J},\mathbf{A}}\; \mathcal{F}^{(p)}_{\mathbf{J},\mathbf{A}}(\mathbf{r}_1, \mathbf{r}_2, \ldots, \mathbf{r}_N), \tag{2.4}$$

or an exponential ansatz, usually called Jastrow factor [Cla79],

$$\mathcal{F}(\mathbf{r}_1, \mathbf{r}_2, \ldots, \mathbf{r}_N) = \exp\left[\sum_p \sum_{\mathbf{J}}{}' \sum_{\mathbf{A}} f^{(p)}_{\mathbf{J},\mathbf{A}}\; \mathcal{F}^{(p)}_{\mathbf{J},\mathbf{A}}(\mathbf{r}_1, \mathbf{r}_2, \ldots, \mathbf{r}_N)\right], \tag{2.5}$$

where symmetrized wavelet tensor products

$$\begin{aligned} \mathcal{F}^{(1)}_{j,\mathbf{a}}(\mathbf{r}_1, \ldots, \mathbf{r}_N) &= \gamma^{(s)}_{j,\mathbf{a}}(\mathbf{r}_1) + \gamma^{(s)}_{j,\mathbf{a}}(\mathbf{r}_2) + \cdots + \gamma^{(s)}_{j,\mathbf{a}}(\mathbf{r}_N), \\ \mathcal{F}^{(2)}_{\mathbf{J},\mathbf{A}}(\mathbf{r}_1, \ldots, \mathbf{r}_N) &= \gamma^{(s_1)}_{j_1,\mathbf{a}_1}(\mathbf{r}_1)\, \gamma^{(s_2)}_{j_2,\mathbf{a}_2}(\mathbf{r}_2) + \cdots + \gamma^{(s_1)}_{j_1,\mathbf{a}_1}(\mathbf{r}_{N-1})\, \gamma^{(s_2)}_{j_2,\mathbf{a}_2}(\mathbf{r}_N), \\ &\;\;\vdots \end{aligned} \tag{2.6}$$

are formed from isotropic 3d-wavelets $\gamma^{(s)}_{j,\mathbf{a}}(\mathbf{r})$. These wavelets are constructed by taking mixed tensor products of univariate wavelets $\psi_{j,a}$ and scaling functions $\varphi_{j,a}$ on the same level j

$$\begin{aligned}
\gamma_{j,\mathbf{a}}^{(0)}(\mathbf{r}) &= \varphi_{j,a_x}(x)\,\varphi_{j,a_y}(y)\,\varphi_{j,a_z}(z), \\
\gamma_{j,\mathbf{a}}^{(1)}(\mathbf{r}) &= \psi_{j,a_x}(x)\,\varphi_{j,a_y}(y)\,\varphi_{j,a_z}(z), \\
&\vdots \\
\gamma_{j,\mathbf{a}}^{(4)}(\mathbf{r}) &= \psi_{j,a_x}(x)\,\psi_{j,a_y}(y)\,\varphi_{j,a_z}(z), \\
&\vdots \\
\gamma_{j,\mathbf{a}}^{(7)}(\mathbf{r}) &= \psi_{j,a_x}(x)\,\psi_{j,a_y}(y)\,\psi_{j,a_z}(z).
\end{aligned} \tag{2.7}$$

For the sake of notational simplicity we have used the notation $\gamma_{j,\mathbf{a}}^{(0)}$ for 3d-scaling functions. A brief introduction into univariate wavelets has been given in Appendix A. The isotropic 3d-wavelets belong to well defined levels j, however, there are seven different types of them, according to the various combinations of univariate wavelets and scaling functions. In those cases where details concerning type, level and location of wavelets are not relevant, we use single Greek wavelet indices (γ_α) to simplify our notation. Capital Latin and Greek multi-indices are used to denote arrays of indices of the same kind like in the case of the tensor products (2.6).

Isotropic wavelet constructions become impracticable beyond three dimensions. For the approximation of correlation factors, we, therefore, have to switch to sparse grids [BG04]. This construction became known in the wavelet community as hyperbolic wavelets [DKT98]. The concept of hyperbolic wavelets is based on a special kind of hierarchical ordering and truncation scheme for standard tensor product wavelets

$$\gamma_{j_1,\mathbf{a}_1}^{(s_1)}(\mathbf{r}_1)\,\gamma_{j_2,\mathbf{a}_2}^{(s_2)}(\mathbf{r}_2)\ldots\gamma_{j_p,\mathbf{a}_p}^{(s_p)}(\mathbf{r}_p), \tag{2.8}$$

which appear in the symmetrized products (2.6). Due to their anisotropic character, we cannot assign a unique level to these tensor products. Instead, we have to take their level sums in order to get a hierarchical ordering. The sparse grid condition for the multilevel-index $\mathbf{J}$ can be expressed as a constraint on the shifted sum of wavelet levels

$$|\mathbf{J}| := \sum_{i=1}^{p} (j_i - l_0 + 1) \leq \tilde{Q}, \tag{2.9}$$

where p is the number of 3d-wavelets in the tensor products (2.8) and j_i are their corresponding levels. According to relation (2.9), the sparse grid threshold parameter $\tilde{Q}$ determines the possible combinations of 3d-wavelet levels that are taken into account in the tensor product expansions (2.4) and (2.5), starting at the coarsest level l_0 which appears in the isotropic 3d-wavelet basis. As a consequence of the sparse grid condition, the number of p-electron tensor products (2.6) is of $O(M \log(M)^{p-1})$ with respect to the cardinality $M \sim O(2^{3\tilde{Q}})$ of the underlying 3d-wavelet basis. For small p, the growth is almost linear in M, however, it still increases exponentially with respect to p. As a consequence we can achieve only modest values of p in the expansions (2.4) and (2.5). A prime on the sum with respect to the multilevel-index $\mathbf{J}$ indicates that only those tensor products are taken into account, which either satisfy the sparse grid condition (2.9) or belong to a specific adaptive refinement.

Recently, best N-term approximation theory [DeV98] has been extended by Nitsche to hyperbolic wavelets [Nit03]. Taking into account the asymptotic behaviour of Jastrow factors near coalescence points of electrons it is possible to derive an estimate for the approximation error of the pair-correlation function in the H^1 Sobolev space [FHS06a]

$$\inf_{\#\{\mathcal{F}_{\Omega}^{(2)}\}\leq N} \|\mathcal{F}^{(2)} - \sum_{\Omega} f_{\Omega}^{(2)} \mathcal{F}_{\Omega}^{(2)}\|_{H^1} \lesssim N^{-\frac{1}{2}}, \tag{2.10}$$

where $\#\{\mathcal{F}_{\Omega}^{(2)}\}$ denotes the number of two-electron tensor products (2.6) in the expansion (2.5). In best N-term approximations these tensor products have to be chosen in order to minimize the approximation error in H^1. The estimate (2.10) is sharp for all sufficiently regular wavelets with a minimum number of vanishing moments. In the case of hyperbolic 6d-wavelets based on isotropic 3d-wavelets γ_α, this estimate requires at least three vanishing moments for the underlying univariate wavelet ψ. For comparison, a direct construction of hyperbolic 6d-wavelets from a univariate wavelet basis requires only two vanishing moments for the same estimate. It is essentially the inter-electron cusp at coalescence points of electrons that restricts the order of approximation. The hyperbolic tensor products from isotropic 3d-wavelets are especially convenient for an adaptive refinement in the neighbourhood of a cusp [FH*02, FH*05a]. Numerical studies for an exactly solvable many-particle model [LK*02] and the homogeneous electron gas [FH*05a] demonstrate that a hyperbolic tensor product construction according to the sparse grid criteria (2.9) with additional diagonal refinement turns out to be very close to best N-term approximation.

3 Variational calculations for the linear ansatz

The Rayleigh-Ritz variational principal can be applied without further approximations to the expectation value of the energy (2.3) for the linear expansion of the correlation factor (2.4). It yields a generalized eigenvalue problem of the form

$$\mathbf{H}\,\mathbf{f} = E\,\mathbf{M}\,\mathbf{f}, \tag{3.1}$$

for the variational parameters f_Λ, with matrix elements

$$H_{\Omega\Lambda} = \int d^3r_1, \ldots, d^3r_N \; \mathcal{F}_\Omega \Phi \; H \; \mathcal{F}_\Lambda \Phi \tag{3.2}$$

$$M_{\Omega\Lambda} = \int d^3r_1, \ldots, d^3r_N \, \mathcal{F}_\Omega \Phi \, \mathcal{F}_\Lambda \Phi. \tag{3.3}$$

This direct approach avoids any uncontrolled approximations. We want to mention however, that in order to develop methods suitable for large scale applications it becomes necessary to switch from the linear to the exponential ansatz (2.5) in order to keep the computational complexity within reasonable bounds. This topic is further discussed in Sect. 3.2.

Before we start to review some applications of the linear ansatz, a few remarks are in order concerning the computation of the matrix elements (3.2) and (3.3). It turns out that this step is the bottleneck of our approach and requires a careful analysis concerning the computational complexity. Due to the tensor product ansatz for the correlation factor $\mathcal{F}$, the Coulomb interaction part of the matrix elements $H_{\Omega\Lambda}$ factors into the standard one-electron

$$\langle \eta_{\mathbf{A}} | \frac{1}{r_{\mathbf{C}}} \rangle := \int d^3r \, \frac{\eta_{\mathbf{A}}(\mathbf{r})}{|\, \mathbf{r} - \mathbf{C} \,|} \tag{3.4}$$

and two-electron

$$\langle \eta_{\mathbf{A}} | \frac{1}{r_{12}} | \eta_{\mathbf{B}} \rangle := \int d^3r_1 d^3r_2 \, \eta_{\mathbf{A}}(\mathbf{r}_1) \, \frac{1}{|\, \mathbf{r}_1 - \mathbf{r}_2 \,|} \, \eta_{\mathbf{B}}(\mathbf{r}_2) \tag{3.5}$$

Coulomb integrals. These integrals have to be calculated for various products of orbitals ϕ_i and isotropic 3d-wavelets

$$\eta_{\mathbf{A}}(\mathbf{r}) = \begin{cases} \phi_i(\mathbf{r})\phi_j(\mathbf{r}) \\ \phi_i(\mathbf{r})\phi_j(\mathbf{r})\gamma_\alpha(\mathbf{r}) \\ \phi_i(\mathbf{r})\phi_j(\mathbf{r})\gamma_\alpha(\mathbf{r})\gamma_\beta(\mathbf{r}). \end{cases}$$

The factorization of matrix elements (3.2) and (3.3) into lower dimensional integrals is already a complex task for systems containing more than two particles. We have to rely on diagrammatic techniques from many-particle theory in order to derive a recurrence scheme for the evaluation of these matrix elements from the basic integrals (3.4) and (3.5). In Sect. 4.1, we present a brief outline of the recurrence scheme and of the underlying diagrammatic techniques.

A new numerical feature arises from the products of wavelets and orbitals which appear in the integrals (3.4) and (3.5). Estimates for wavelet products, based on their Sobolev regularity, have been derived in Ref. [FH*02], indicating that a sufficiently high regularity of the wavelet basis is required for an efficient numerical treatment of these integrals. In an intermediate step, we first calculate the auxiliary function

$$\mathcal{R}(\mathbf{r_1}) := \int d^3r_2 \, \frac{1}{|\, \mathbf{r}_1 - \mathbf{r}_2 \,|} \, \eta_{\mathbf{B}}(\mathbf{r_2})$$

by performing a matrix times vector multiplication in the nonstandard representation [BCR91, Bey92] of the Coulomb interaction and product function

$\eta_{\mathbf{B}}$. For wavelets with several vanishing moments, the Coulomb interaction matrix elements decay very fast with increasing distance from the diagonal. If wavelets appear in the product function $\eta_{\mathbf{B}}$, it has compact support and the corresponding wavelet expansion is strongly peaked around the wavelet belonging to the finest scale in the product. The auxiliary function $\mathcal{R}$ therefore has a sparse representation in the dual wavelet basis

$$\mathcal{R}(\mathbf{r}) \approx \sum_{\alpha} \mathcal{R}_{\alpha} \, \tilde{\gamma}_{\alpha}(\mathbf{r}).$$

In order to calculate the two-electron Coulomb integrals (3.5) it remains to perform a scalar product with the product function $\eta_{\mathbf{A}}$

$$\langle \eta_{\mathbf{A}} \mid \frac{1}{r_{12}} \mid \eta_{\mathbf{B}} \rangle \approx \sum_{\alpha} \langle \eta_{\mathbf{A}} \mid \tilde{\gamma}^{\alpha} \rangle \mathcal{R}_{\alpha}$$

Both vectors involved in the scalar product have sparse representations in the wavelet basis.

3.1 Results for the helium atom and a many-particle model

We have demonstrated the feasibility of our method by applying it to the helium atom and isoelectronic Li^+ ion. Actually this is a standard model for electron correlations in quantum chemistry, which shows many characteristic features of large systems. From a technical point of view, it already required the full machinery for calculating one- and two-electron integrals within the wavelet basis. For our calculations we have taken the biorthogonal SDD6 wavelets with six vanishing moments [Swe96]. Further details concerning the wavelet basis are given in Appendix A. Our results obtained so far are only of preliminary character, due to the fact that the isotropic 3d-wavelet basis γ_{α} was only crudely adapted to the size of the atoms. For full technical details we refer to Ref. [FH*02]. Nevertheless our wavelet approximations recovered 98% and 97% of the correlation energy [4] for He and Li^+, respectively. The correlation energies for various levels of refinement are shown in Fig. 3.1. As a reference, the wavelet level $j = 0$ corresponds to a grid separation of 1 bohr. In the case of the He atom, the dominant contribution comes from the level $j = -2$, which already recovered 83% of the correlation energy. Compared with it, the contribution of the next coarser level $j = -3$ is almost negligible (< 0.1 mhartree). At finer scales, contributions to the correlation energy decrease quite fast. A definite statement concerning the asymptotic convergence behaviour cannot yet be drawn from our calculations. According to the estimate (2.10), we expect an asymptotic convergence of $O(2^{-3j})$ where we assumed diagonal dominance due to the inter-electron cusp. Our results show that rather accurate energies can be achieved already with a rather

[4] The correlation energy corresponds to the difference between exact and HF energy

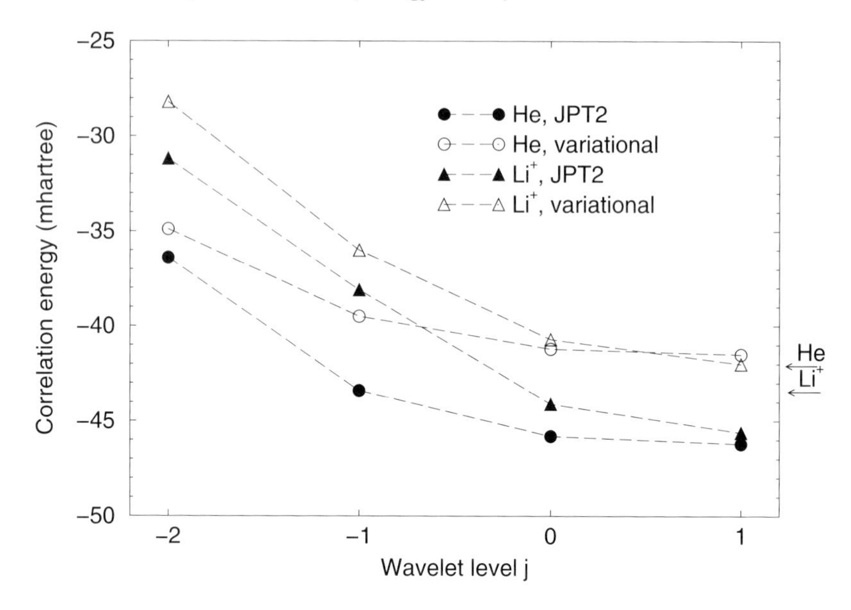

Fig. 3.1. Correlation energies of He and Li^+ from variational and JPT2 calculations for various levels of refinement. The level $j = 0$ corresponds to a resolution length of 1 bohr. Calculations have been performed with the SDD6 wavelet.

small number of wavelets. This is encouraging in view of the fact that we have used regular Cartesian grids, which were not especially adapted to spherical symmetry, except that we have placed the nucleus at the origin.

Exactly solvable many-particle models provide an interesting playground for method development. Especially convenient for our purposes was a one dimensional bosonic model suggested by Koprucki and Wagner [KW00]. The Hamiltonian of this model is given by

$$H_{\text{model}} = -\frac{1}{2}\sum_{i=1}^{N}\frac{\partial^2}{\partial x_i^2} + \frac{1}{2}\sum_{i=1}^{N} x_i^2 + \sum_{i<j}\left(\delta(x_i - x_j) - \frac{1}{2}|x_i - x_j|\right), \quad (3.6)$$

which corresponds to a system of coupled harmonic oscillators with repulsive contact and long-range interactions. The bosonic ground state wavefunctions of this model

$$\Psi_0(x_1, x_2, \ldots, x_N) = \mathcal{N}\prod_{i<j}\exp\left[\frac{1}{2}|x_i - x_j|\right]\prod_{i=1}^{N}\exp\left[-\frac{1}{2}x_i^2\right],$$

have a simple product structure with a cusp similar to electrons. We have studied the convergence behaviour of various approximation schemes [LK*02] in L^∞, L^2 and H^1 norms and for the energy. Because of the simplicity of the model, we were able to consider fully adaptive wavelet grids where tensor

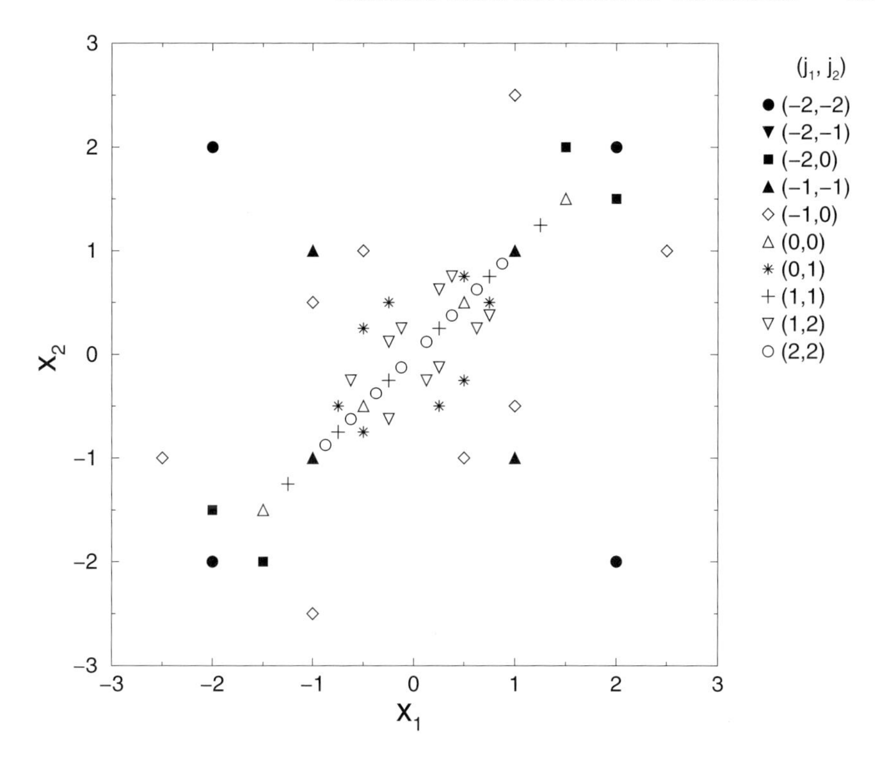

Fig. 3.2. Fully adaptive wavelet grid for the two-particle model (3.6), based on the contribution of tensor products $\psi_{j_1,a_1}(x_1)\ \psi_{j_2,a_2}(x_2)$ to the energy.

products have been selected according to their contribution to the energy. These grids, depicted in Fig. 3.2 for the two-particle case, show pronounced diagonal dominance for wavelets on fine levels. It turned out that hyperbolic wavelets with appropriate diagonal refinement at the cusps have comparable computational complexity as a fully adaptive treatment based on the energy contribution of individual wavelet tensor products. For both schemes, the energy shows an asymptotic convergence rate of $O(2^{-j})$ with respect to the level of refinement. This roughly corresponds to $O(N^{-1})$ convergence with respect to number of tensor product wavelets.

3.2 Size-consistency error for the linear ansatz

The linear hyperbolic wavelet expansion of the correlation factor (2.4) enables a strictly variational treatment of Schrödinger's equation via the generalized eigenvalue problem (3.1). For typical applications to extended systems, size-consistency [5] is required, which means that expectation values of operators

[5] In the terminology of quantum chemistry this property is often called size-extensivity.

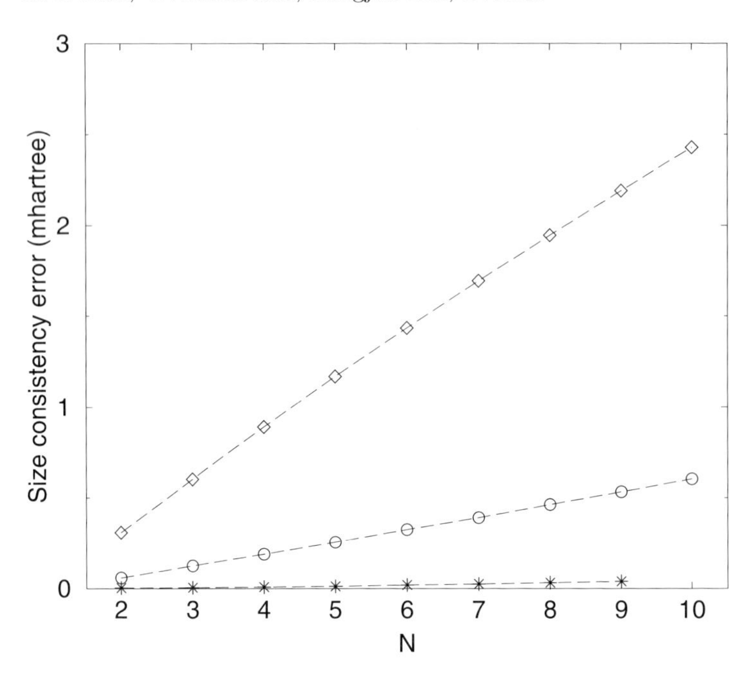

Fig. 3.3. Size consistency error per atom $\Delta E = E_{\mathrm{He}_N}/N - E_{\mathrm{He}}$ (mhartree) of hyperbolic wavelet and SDCI expansions for systems of noninteracting helium atoms. The wavelet approximation for the correlation factor of a single helium atom has been taken from Ref. [FH*02] (E_{He} = -2.89904 hartree, two levels $j = -2, -1$). Two different sparse grid constraints $\tilde{Q} = 4$ ($\circ$) and $\tilde{Q} = 6$ ($*$) have been imposed to the product wavefunction of the helium clusters. For comparison SDCI calculations ($\diamond$) have been performed with a s, p VQZ basis set (E_{He} = -2.90015 hartree).

that correspond to extensive thermodynamic properties, like the energy, are proportional to the system size. Within the linear ansatz, size-consistency is hard to achieve. Not only an inflationary number of variational parameters, but even more severe, an increasing complexity of the matrix elements (3.2) and (3.3) makes the linear hyperbolic wavelet expansion to a challenging problem for large systems. It has been already demonstrated by one of us [Hac01], that in order to keep the computational complexity under control an additional constraint has to be imposed. Besides the sparse grid condition, we have to truncate the first sum in the expansion (2.4) at $p \leq p_{\max} \ll N$, where $p_{\max}$ may slightly vary with the size of the system. This resembles to truncation schemes for the CI method in quantum chemistry, where typically only *single and double excitations* (SDCI) are taken into account.

In order to illustrate this problem, we have studied the size-consistency error of hyperbolic wavelet expansions for a standard benchmark problem in quantum chemistry, namely systems of N noninteracting helium atoms. Such kind of model recovers essential physical aspects of the problem [HJO99].

Due to Pauli's principle, molecules can roughly be described by interacting electron pairs. The dominant intra-pair interactions are taken into account by this model, whereas inter-pair correlation effects are excluded. The latter are more subtle and not easily accessible to analysis. For the sake of computational simplicity, we consider a standard wavelet tensor product expansion for a single helium atom

$$\Psi_{\mathrm{He}}(\mathbf{r}_1,\mathbf{r}_2) = \left[\sum_{p=0}^{2}\sum_{\mathbf{J}}\sum_{\mathbf{A}} f_{\mathbf{J},\mathbf{A}}^{(p)}\ \mathcal{F}_{\mathbf{J},\mathbf{A}}^{(p)}(\mathbf{r}_1,\mathbf{r}_2)\right]\phi(\mathbf{r}_1)\,\phi(\mathbf{r}_2), \tag{3.7}$$

where the sparse grid condition has not been applied. The corresponding wavefunction for a cluster of N noninteracting helium atoms is given by the product

$$\Psi_{\mathrm{He}_N}(\mathbf{r}_1\ldots\mathbf{r}_{2N}) = \prod_{i=1}^{N}\Psi_{\mathrm{He}}(\mathbf{r}_{2i-1},\mathbf{r}_{2i})\,. \tag{3.8}$$

A special permutational symmetry with respect to an interchange of variables among different helium atoms is not required due to the underlying assumption of a large spatial separation of the atoms. Obviously, the total energy scales linearly with the number of helium atoms $E_{\mathrm{He}_N} = NE_{\mathrm{He}}$. After some reordering of the product wavefunction (3.8), the wavelet tensor products for the cluster can be arranged like in the linear ansatz (2.4)

$$\overline{\mathcal{F}}_{\mathbf{J},\mathbf{A}}^{(p)}(\mathbf{r}_1\ldots\mathbf{r}_{2N}) = \prod_{i=1}^{N}\mathcal{F}_{\mathbf{J}_i,\mathbf{A}_i}^{(p_i)}(\mathbf{r}_{2i-1},\mathbf{r}_{2i})\,,\ \text{with}\ \ p=\sum_i p_i.$$

Applying the hyperbolic wavelet approximation to the reordered wavefunction

$$\Psi_{\mathrm{He}_N}(\mathbf{r}_1\ldots\mathbf{r}_{2N}) \approx \left[\sum_{p}\sum_{|\mathbf{J}|\le\tilde{Q}}\sum_{\mathbf{A}}\overline{f}_{\mathbf{J},\mathbf{A}}^{(p)}\ \overline{\mathcal{F}}_{\mathbf{J},\mathbf{A}}^{(p)}(\mathbf{r}_1\ldots\mathbf{r}_{2N})\right]\prod_{i=1}^{N}\phi(\mathbf{r}_{2i-1})\,\phi(\mathbf{r}_{2i}),$$

only those tensor products $\overline{\mathcal{F}}_{\mathbf{J},\mathbf{A}}^{(p)}$ are taken into account for which the sparse grid condition (2.9) is satisfied. On each truncation level $\tilde{Q}$, the coefficients $\overline{f}_{\mathbf{J},\mathbf{A}}^{(p)}$ have been reoptimized with respect to the energy. In order to limit the computational effort, we have contracted the atomic wavelet coefficients on each of the tensor product levels $\mathbf{J}$. This did not introduce any noticeable error into our calculations. In Fig. 3.3 we have shown the size consistency error per atom $\Delta E = E_{\mathrm{He}_N}/N - E_{\mathrm{He}}$ in the case of a two-scale wavelet expansion for a single helium atom. We refer to Ref. [FH*02] for further details concerning this wavefunction. For both thresholds $\tilde{Q} = 4, 6$, a single helium atom can afford all tensor products in the wavefunction (3.7), however, for clusters of several helium atoms, restrictions with respect to the full product wavefunction (3.8) are imposed. The hyperbolic wavelet approximation is less stringent than SDCI, in a sense that certain multicenter tensor products are still possible. For illustration we compare in Fig. 3.3 with SDCI calculations in a *s, p valence quadruple zeta* (VQZ) basis set. It can be seen that the size consistency error is much smaller already for $\tilde{Q} = 4$ compared to the SDCI method. Increasing the sparse grid parameter to $\tilde{Q} = 6$ results in a very small size consistency error, which seems to be acceptable for systems with up to twenty electrons. In principle it appears to be possible, to adjust the parameter $\tilde{Q}$ to the system size in order to achieve a given accuracy. However in view of the difficulties concerning the evaluation of matrix elements, this becomes impracticable beyond a certain size of the system. For comparison, the exponential Jastrow ansatz (2.5) preserves the product structure of the wavefunction (3.8) if only one- and two-particle tensor products (2.6) are taken into account.

4 Jastrow perturbation theory and the local ansatz

A rigorous treatment of the exponential ansatz (2.5), based on Rayleigh-Ritz's variational principle, is almost unfeasible due to the highly nonlinear character of the variational problem. Instead it is possible to derive various approximation schemes which preserve the size-consistency of the exponential ansatz. For Jastrow-type wavefunctions, such kind of methods are the local ansatz of Stollhoff and Fulde [Ful93, SF80] and Krotscheck's *Fermi hypernetted chain* (FHNC) method [CKP92, Kro85]. Without additional symmetries, the nonlinear system of FHNC equations is very challenging from a computational point of view and we refrain from a numerical treatment within multiresolution analysis. We have adapted, however, the diagrammatic representation of matrix elements, which we have slightly modified by introducing additional elements from multiresolution analysis. Instead of the nonlinear FHNC equations, we have taken the local ansatz as a guideline for our work. The local ansatz can be considered as a power series expansion of the exponential ansatz [SF80] which leads to a linear system of equations

$$\sum_{\Lambda} \Big(\langle \mathcal{F}_{\Omega}^{\dagger} \, H \, \mathcal{F}_{\Lambda} \rangle - \langle \mathcal{F}_{\Omega}^{\dagger} \, \mathcal{F}_{\Lambda} \rangle \langle H \rangle \Big) \, f_{\Lambda} = -\langle \mathcal{F}_{\Omega}^{\dagger} H_1 \rangle \tag{4.1}$$

where the correlation energy is given by the expression

$$E_{\text{corr}} = \sum_{\Lambda} \langle H_1 \, \mathcal{F}_{\Lambda} \rangle \, f_{\Lambda}. \tag{4.2}$$

To simplify our notation we have introduced a short-hand notation $\langle \cdots \rangle := \langle \Phi_{\text{HF}} | \cdots | \Phi_{\text{HF}} \rangle$ for expectation values with respect to the HF wavefunction. The residual interaction $H_1 := H - H_{\text{SCF}}$ is defined as the difference between the exact and HF Hamiltonian. Furthermore, we have assumed $\langle \mathcal{F}_{\Lambda} \rangle = 0$, which can be imposed in a natural way using the formalism of second quantization discussed below. The considerable simplification due to linearization is at the expense of an accurate treatment of long-range correlations. For our envisaged applications in quantum chemistry and solid state physics, where electrons or excitons are confined by an external potential, this is perfectly justified. Whereas long-range correlations become important for metallic systems [Ful93] where they cause a divergence in the energy for finite order perturbation expansions. Guided by the local ansatz, we have studied a perturbative approach for Jastrow factors [LK*06]. Such kind of approach seems to be natural from a physical point of view, however, to the best of our knowledge, it has not been described in the literature. This is likely due to the failure of finite order perturbation theory for most of the standard applications of Jastrow factors in condensed matter physics, where long-range correlations are important.

From a formal point of view, *Jastrow perturbation theory* (JPT) is closely related to *coupled cluster perturbation theory* (CCPT) presented in [HJO99]. The symmetrized tensor products (2.6) are conveniently expressed in terms of second quantization, which introduces additional flexibility into the perturbation analysis. This can be used to remove redundancies from the many-particle basis and to reduce computational complexity which is essential for practical applications. In order to illustrate our assertions, we consider an arbitrary two-particle basis function as an operator in second quantization

$$\mathcal{F}^{(2)}_{\alpha,\beta} \equiv \hat{\mathcal{F}}^{(2)}_{\alpha,\beta} := \frac{1}{2} \sum_{pqst} \langle pq|\mathcal{F}^{(2)}_{\alpha,\beta}|st\rangle c^{\dagger}_{p} c^{\dagger}_{q} c_t c_s, \tag{4.3}$$

with

$$\langle pq|\mathcal{F}^{(2)}_{\alpha,\beta}|st\rangle := \int\int d^3x_1 d^3x_2\, \phi_p(\mathbf{x}_1)\phi_q(\mathbf{x}_2)\gamma_\alpha(\mathbf{r}_1)\gamma_\beta(\mathbf{r}_2)\phi_s(\mathbf{x}_1)\phi_t(\mathbf{x}_2)$$
$$+\ \alpha \longleftrightarrow \beta$$

where $\mathbf{x}_i := (\mathbf{r}_i, \sigma_i)$ denotes the combined spatial and spin coordinate of a particle. The creation $c^{\dagger}_p$ and annihilation c_s operators refer to the single particle orbitals ϕ_q of H_{SCF}. In the following, we denote virtual orbitals by $a, b, \cdots$, occupied orbitals by $i, j, \cdots$ and arbitrary orbitals by $p, q, \cdots$. We want to mention that virtual orbitals $\{\phi_a\}$ are introduced for purely formal reasons and that for actual calculations only occupied orbitals $\{\phi_i\}$ are required. This is due to the fact that the underlying tensor product basis (2.6) consists of simple functions which allows us to use the identity

$$\sum_a \phi_a(\mathbf{x}_1)\phi^*_a(\mathbf{x}_2) = \delta(\mathbf{r}_1 - \mathbf{r}_2)\, \delta_{\sigma_1,\sigma_2} - \sum_i \phi_i(\mathbf{x}_1)\phi^*_i(\mathbf{x}_2), \tag{4.4}$$

we refer to Ref. [FH*05a] for further details. An obvious choice, in the spirit of CC theory, for simplified two-particle operators (4.3) are cluster-type operators

$$\hat{\mathcal{F}}^{(2,c)}_{\alpha,\beta} := \frac{1}{2} \sum_{abij} \langle ab | \mathcal{F}^{(2)}_{\alpha,\beta} | ij \rangle c^\dagger_a c^\dagger_b c_j c_i, \tag{4.5}$$

which preserve commutativity of the many-particle basis. This property turns out to be essential for the exponential perturbation scheme. On a first glance, the second quantized Jastrow ansatz (4.5) resembles closely to CC theory. There is however an essential difference inasmuch as the underlying function basis (2.6) does not guaranty convergence of the product ansatz (2.2) to the exact wavefunction in the complete basis set limit. For that reason we take the variational energy (2.3) as the starting point for our perturbation analysis instead of the CC energy and projection equations. The latter assume an exact ansatz for the wavefunction and provide the basis for the CCPT approach [HJO99].

We have performed a perturbation analysis for the linear and exponential representation of the correlation factor [LK*06]. For cluster-type operators, we obtain in the linear case a natural truncation scheme with respect to the degree p of the many-particle basis $\mathcal{F}^{(p)}$ that preserves size-consistency for the energy. It requires e.g. only two-particle operators at second-order but four-particle operators already for third-order energies. In order to avoid such an increase of computational complexity, we switched to the exponential Jastrow-type ansatz, where size-consistency can be achieved by restricting to two-particle operators at all orders of perturbation theory. The first order approximation to the correlation operator $\hat{\mathcal{F}}_1 = \sum_\Lambda \hat{\mathcal{F}}_\Lambda f_{\Lambda,1}$ given by

$$\sum_\Lambda \Big(\langle \hat{\mathcal{F}}^\dagger_\Omega H_{\mathrm{SCF}} \hat{\mathcal{F}}_\Lambda \rangle - \langle \hat{\mathcal{F}}^\dagger_\Omega \hat{\mathcal{F}}_\Lambda \rangle \langle H_{\mathrm{SCF}} \rangle \Big) f_{\Lambda,1} = -\langle \hat{\mathcal{F}}^\dagger_\Omega H_1 \rangle, \tag{4.6}$$

yields the second order (JPT2) correlation energy

$$E_2 = \sum_\Lambda \langle H_1 \hat{\mathcal{F}}_\Lambda \rangle f_{\Lambda,1}.$$

By comparison with the local ansatz (4.1) and (4.2), we observe a close relationship between both approaches. Actually it turns out that the local ansatz is almost equivalent to third-order perturbation theory. Concerning computational complexity, the JPT2 method is considerably simpler than the local ansatz. The Hamiltonian on the left hand side of Eq. (4.6) contains only one-particle operators in contrast to the local ansatz (4.1) where the full Hamiltonian appears.

In order to demonstrate the accuracy of the JPT2 method, we compare in Fig. 3.1 the correlation energies for He and Li^+, at various levels of refinement, with those obtained from variational calculations. It can be seen that JMP2 overestimates the correlation energy by $\approx 10\%$ for these two-electron systems. This has to be compared with standard second order Møller-Plesset perturbation theory which underestimates the correlation energy for He by a similar amount. Another property of interest is the behaviour of the wavefunction near the inter-electron cusp. We have plotted in Fig. 4.1 the angular

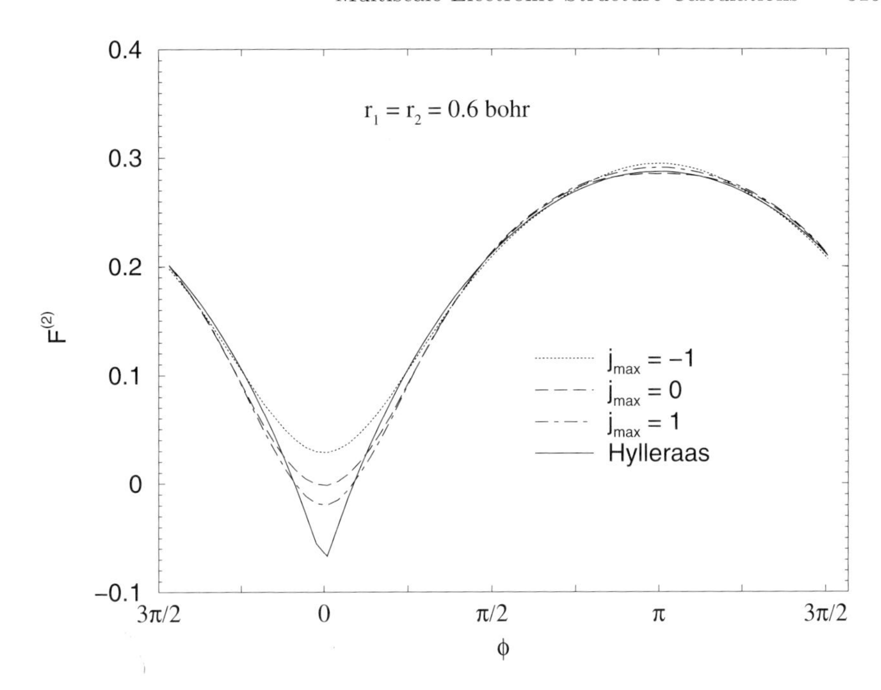

Fig. 4.1. Angular dependence of the correlation function $\mathcal{F}^{(1)} + \mathcal{F}^{(2)}$ of the He atom at various levels of refinement. The wavelet level $j = 0$ corresponds to a resolution length of 1 bohr. Both electrons are at a distance of 0.6 bohr with respect to the nucleus. Calculations have been performed with the SDD6 wavelet. For comparison, the corresponding correlation function obtained from a highly accurate Hylleraas CI calculation is shown.

dependence of the correlation function $\mathcal{F}^{(1)} + \mathcal{F}^{(2)}$ of the He atom at various levels of refinement. Both electrons were kept fixed at a distance of 0.6 bohr with respect to the nucleus. The accuracy of the wavelet approximations can be judged by comparing them with a correlation function obtained from a very accurate Hylleraas CI calculation [FHM84]. Despite of the rather crude approximation, our finest level corresponds to a resolution length of 0.5 bohr, the JMP2 correlation function provides a fairly accurate description of the correlation hole.

4.1 Diagrammatic multiresolution analysis

Straightforward application of diagrammatic techniques from many-particle theory [LM86] enables a formal evaluation of the matrix elements which appear in the local ansatz (4.1) or JPT2 method (4.6). The resulting Goldstone diagrams can be further transformed into FHNC diagrams [CKP92] using the identity (4.4). We have studied the computational complexity for the evaluation of these diagrams in detail for the local ansatz [FH*05a]. An important

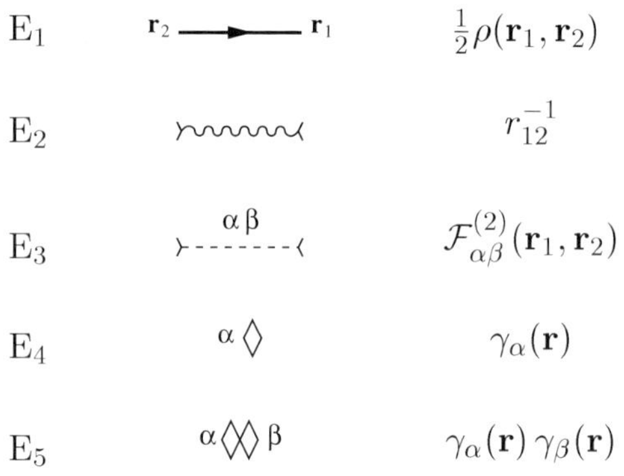

E$_1$	$\mathbf{r}_2$ ——→—— $\mathbf{r}_1$	$\frac{1}{2}\rho(\mathbf{r}_1, \mathbf{r}_2)$
E$_2$	(wavy line)	r_{12}^{-1}
E$_3$	α β (dashed line)	$\mathcal{F}_{\alpha\beta}^{(2)}(\mathbf{r}_1, \mathbf{r}_2)$
E$_4$	α ◊	$\gamma_\alpha(\mathbf{r})$
E$_5$	α ◊◊ β	$\gamma_\alpha(\mathbf{r})\,\gamma_\beta(\mathbf{r})$

Fig. 4.2. Diagrammatic elements of multiresolution analysis.

aspect of this analysis is the multiscale structure of the underlying wavelet basis. This leads to a diagrammatic multiresolution analysis where diagrams can be classified according to their contribution on specific energy- and length-scales. The diagrammatic elements of our multiresolution analysis, depicted in Fig. 4.2, have been adopted from FHNC theory [CKP92]. By a slight abuse of our notation (2.6), we have introduced the pair-correlation basis functions

$$\mathcal{F}_{\alpha\beta}^{(2)}(\mathbf{r}_1, \mathbf{r}_2) = \gamma_\alpha(\mathbf{r}_1)\,\gamma_\beta(\mathbf{r}_2) + \gamma_\alpha(\mathbf{r}_2)\,\gamma_\beta(\mathbf{r}_1). \tag{4.7}$$

Each vertex in a diagram corresponds to an integral $\int d^3r$. Spinless HF density matrices $\rho(\mathbf{r}_1, \mathbf{r}_2)$ are represented in terms of spatial orbitals

$$\rho(\mathbf{r}_1, \mathbf{r}_2) = \sum_i n_i\, \phi_i(\mathbf{r}_1)\, \phi_i(\mathbf{r}_2)^*,$$

where n_i is the occupation number of the i'th orbital.

Because of the tensor product structure of pair-correlation basis functions (4.7), it becomes possible to build up all the diagrams in a recurrent manner from a few types of basic diagrams. At the onset of our recurrence scheme we consider those parts of the diagrams which are directly linked to the Coulomb interaction. These basic Coulomb diagrams, shown in Fig. 4.3, determine the computational complexity of our approach. Various types of pointwise products and wavelet tensor products appear in the basic Coulomb diagrams. The cardinalities of the corresponding sets with respect to the size $M = O(2^{3\tilde{Q}})$ of the underlying isotropic 3d-wavelet basis (2.7) are listed in Table 4.1. In order to derive these cardinality estimates, we have taken into account the compact support and hierarchical structure of the 3d-wavelets as well as the sparse grid condition (2.9) for hyperbolic tensor products.

The number of basic Coulomb diagrams increases almost quadratically with respect to the isotropic 3d-wavelet basis. In order to achieve further reductions to almost linear computational complexity, we considered the effects

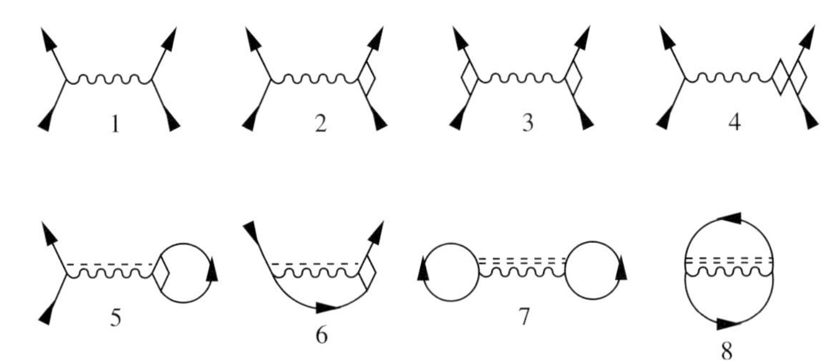

Fig. 4.3. Basic Coulomb diagrams of the recurrence scheme.

Table 4.1. Cardinalities of certain sets of pointwise and tensor products for isotropic 3d-wavelets which appear in the diagrammatic multiresolution analysis.

Wavelet (tensor) product	Diagrammatic element	Cardinality
$\gamma_\alpha(\mathbf{r})$	◇	M
$\gamma_\alpha(\mathbf{r})\,\gamma_\beta(\mathbf{r})$	◇◇	$O(M\log(M))$
$\mathcal{F}_{\alpha\beta}\,(\mathbf{r}_1,\mathbf{r}_2)$	--------	$O(M\log(M))$
$\gamma_\alpha(\mathbf{r}_1)\,\mathcal{F}_{\nu\mu}\,(\mathbf{r}_1,\mathbf{r}_2)$	◇-------	$O(M^2)$
$\mathcal{F}_{\alpha\beta}\,(\mathbf{r}_1,\mathbf{r}_2)\;\mathcal{F}_{\nu\mu}\,(\mathbf{r}_2,\mathbf{r}_3)$	----V---	$O(M^2\log(M))$
$\mathcal{F}_{\alpha\beta}\,(\mathbf{r}_1,\mathbf{r}_2)\;\mathcal{F}_{\nu\mu}\,(\mathbf{r}_1,\mathbf{r}_2)$	========	$O(M^2)$

of the vanishing moment property of wavelets on various sets of diagrams. This requires further assumptions concerning the smoothness of the density matrix. Obviously, nuclear cusps or a highly oscillating behaviour of the density matrix within the support of a wavelet spoils this useful property. If such kind of behaviour is restricted to atomic core regions, which correspond to a comparatively small portion of the total volume, it can be handled by adaptive local refinements of the 3d-wavelet basis. For our initial studies we have neglected such kind of complications and essentially considered valence electron systems. A prominent example is the homogeneous electron gas, where oscillations are bounded by the Fermi momentum. This led us to a simplified model for the sets of basic type 3, 5 and 6 diagrams

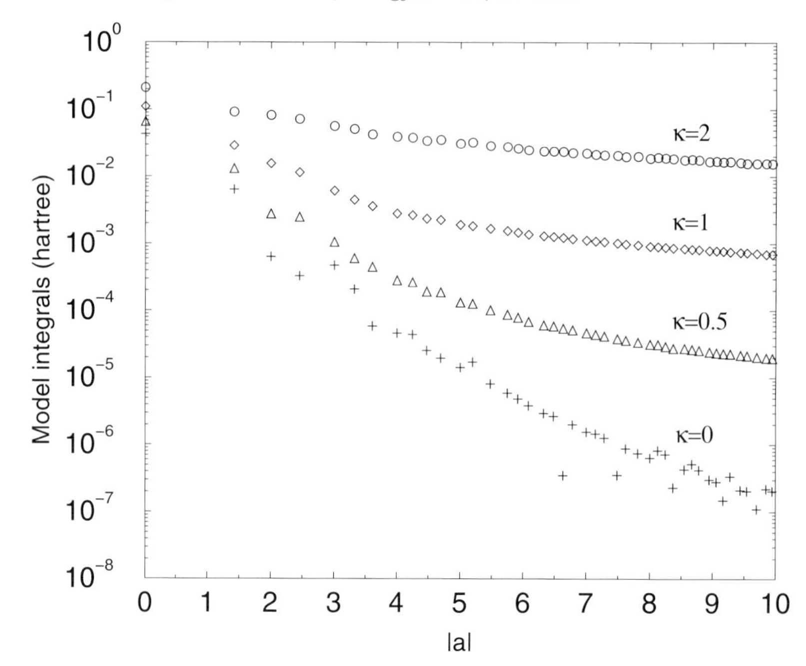

Fig. 4.4. a) Model integrals $I_{j,l}^{(s,t)}(\mathbf{a},\boldsymbol{\kappa},\boldsymbol{\kappa}')$ for the basic type 3, 5, and 6 diagrams. Absolute values of the integrals $I_{0,0}^{(1,0)}(\mathbf{a},\boldsymbol{\kappa},\boldsymbol{\kappa})$ are plotted on a logarithmic scale versus the distance $|\mathbf{a}|$. The plane wave parameters were chosen to be isotropic $\kappa_1 = \kappa_2 = \kappa_3$ with values $\kappa_i = 2$ (○). Isotropic 3d-wavelets and scaling functions were generated from the univariate SDD6 wavelet basis.

$$I_{j,l}^{(s,t)}(\mathbf{a},\boldsymbol{\kappa},\boldsymbol{\kappa}') = \int d^3r_1 \int d^3r_2 \, e^{-i\boldsymbol{\kappa}\mathbf{r}_1} \, \gamma_{j,\mathbf{a}}^{(s)}(\mathbf{r}_1) \, \frac{1}{|\,\mathbf{r}_1 - \mathbf{r}_2\,|} \, e^{i\boldsymbol{\kappa}'\mathbf{r}_2} \, \gamma_{l,\mathbf{0}}^{(t)}(\mathbf{r}_2), \tag{4.8}$$

where only wavelets and the Coulomb interaction are left. All effects of the remainder of the diagrams on the vertices are represented by plane waves, through which we modulated the oscillations by varying the momenta $\boldsymbol{\kappa},\boldsymbol{\kappa}'$. The presence of oscillatory functions in the integrand has important consequences for its large distance behaviour i.e. $|\mathbf{a}| \to \infty$. Our numerical studies for the model integrals (4.8), shown in Fig. 4.4, demonstrate that the vanishing moment property only provides additional sparsity if $2^{-j}|\kappa_i|$ is smaller than a certain critical value. Such kind of sparsity can be related to the behaviour of the derivatives of the Fourier transform of the mother wavelet near the origin.

Pointwise wavelet products on the vertices of diagrams also destroy the vanishing moment property. This partially affects the sets of basic type 5 and 6 diagrams and is most severe for the sets of basic type 7 and 8 diagrams, where wavelet products appear on both vertices of the Coulomb interaction. Since the cardinalities of these sets of diagrams increase quadratically with the size

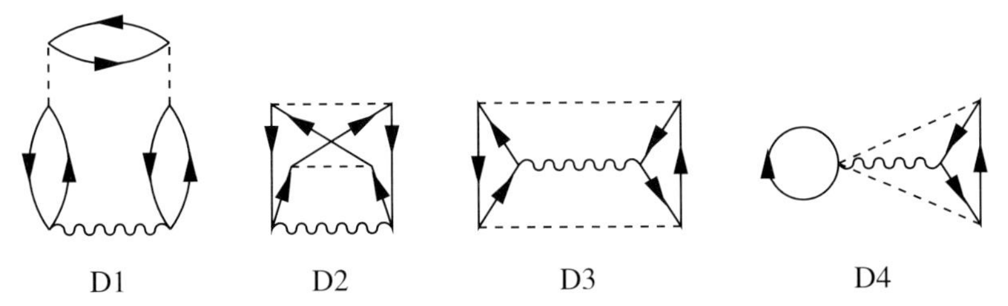

Fig. 4.5. Selected diagrams for numerical studies with FHNC//0 pair-correlation function for a homogeneous electron gas.

of the 3d-wavelet basis it is essential to achieve additional sparsity through the vanishing moment property. A possible resolution to this problem are wavelet expansions of pointwise wavelet products on the vertices of the diagrams. This results in an almost linear complexity for all of the basic Coulomb diagrams. Similar arguments apply to the remaining parts of the recurrence scheme.

We have studied the convergence behaviour of wavelet expansions for selected diagrams, shown in Fig. 4.5, in the case of a homogeneous electron gas for which "optimal" pair-correlation functions $\mathcal{F}^{(2)}_{\text{FHNC}}(\mathbf{r}_1, \mathbf{r}_2)$ have been obtained from FHNC//0 calculations [KKQ85]. For our calculations we have chosen an electron density of $r_s = 2.07$ ($\kappa_F = 0.927\,\text{bohr}^{-1}$). This corresponds to the average valence electron density in aluminium. The cubic supercell with edge length 12 bohr contains 54 electrons. The FHNC//0 pair-correlation function was expanded in terms of hyperbolic wavelet tensor products with additional diagonal refinement

$$\mathcal{F}^{(2)}_{\text{FHNC}}(\mathbf{r}_1, \mathbf{r}_2) = \sum_{\Lambda} f_\Lambda \mathcal{F}^{(2)}_\Lambda(\mathbf{r}_1, \mathbf{r}_2).$$

The approximation error for the diagrams D1 to D4 provides some insight into the performance of our approach for realistic problems. It can be seen from Fig. 4.6 that the overall convergence of these diagrams is rather similar. Fairly high accuracies have been already achieved at rather coarse levels. This is due to relatively weak couplings between pair-correlation functions and the Coulomb potential in the diagrams D1 to D4.

5 Multiresolution Hartree-Fock for quasi two-dimensional extended systems

Our further development of diagrammatic multiresolution analysis is aiming towards applications in the field of semiconductor physics, especially for strongly anisotropic systems like semiconductor heterostructures. A common feature of our envisaged applications is that they are closely related to molecular systems. Traditional methods in quantum chemistry based on Gaussian

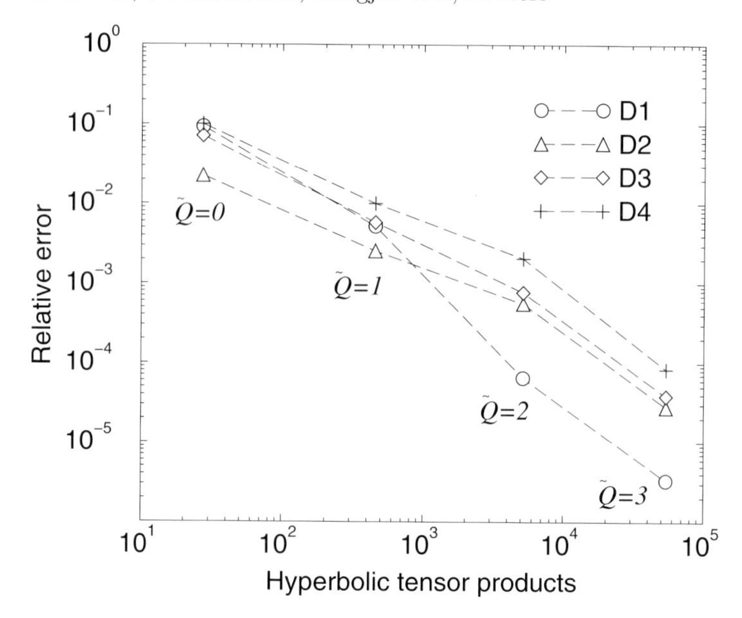

Fig. 4.6. Convergence of selected diagrams for hyperbolic wavelet expansions with diagonal refinement of a FHNC//0 pair-correlation function in the case of a homogeneous electron gas at $r_s = 2.07$. Isotropic 3d-wavelets have been obtained from the univariate SDD6 wavelet basis. Translational symmetry has been only taken into account with respect to the coarsest wavelet level $l_0 = -2$ in the expansion.

type basis functions, however, cannot be easily applied due to the pronounced two-scale character of these systems. In contrast to this, wavelet bases can be adapted in a straightforward manner to highly anisotropic problems. As a preparatory step for the implementation of the local ansatz or JPT methods, we have developed a multiresolution HF method for quasi two-dimensional extended many-particle systems [FH*05b]. Most prominent examples are jellium models for metallic surfaces and slabs. Within these models, the positively charged atomic cores are approximated by a constant background charge density [NP01]. Such kind of models are homogeneous in two directions and show a strongly inhomogeneous behaviour in the third direction, which is perpendicular to the surface of the slab. Due to the homogeneity of jellium models in two dimensions, HF orbitals

$$\phi_n^{\boldsymbol{\kappa}}(\mathbf{r}) = \zeta_n^{\boldsymbol{\kappa}}(x)\, e^{\imath \boldsymbol{\kappa}_{\|} \mathbf{r}_{\|}} \tag{5.1}$$

factor into products of two dimensional plane waves and functions $\zeta_n^{\boldsymbol{\kappa}}(x)$ that describe the behaviour of the orbitals perpendicular to the surface. Inside the slab these functions behave like plane waves and decay exponentially at the surface. Amazingly there has been no exact HF calculations for jellium surfaces or slabs reported in the literature. This is due to the nonlocal exchange operator present in the HF equations, which makes the perpendicular part of the orbitals $\zeta_n^{\boldsymbol{\kappa}}(x)$ depending on the momenta parallel to the slab. However, approximate HF calculations, following the pioneering work of Bardeen [Bar36], have been published [SKG77, SM80]. Within these calculations, the nonlocal exchange operator was approximated by a local potential. As a consequence, the perpendicular part $\zeta_n^{\kappa_x}(x)$ does no longer depend on the momenta in the parallel directions, which greatly simplifies the numerical treatment. The same argument applies to DFT, where only local potentials appear in the KS equation.

Starting with Lang and Kohn's seminal paper [LK71], jellium surfaces have been extensively studied for various kinds of density functionals [HJ74, LP75, PE01]. They proved to be of considerable significance for the development of new density functionals. Therefore a strong need for accurate many-particle benchmark calculations exists. However, due to the presence of periodic boundary conditions and the absence of localized orbitals, these systems pose great technical difficulties for traditional quantum many-particle methods. There are essentially only two such methods which can deal with jellium surfaces, *quantum Monte Carlo* (QMC) [AC96, LN*92] and the FHNC method [KKQ85]. It exists an ongoing controversy concerning the accuracy of these calculations [NP01]. Significant deviations with respect to results obtained from DFT calculations exist and presently the reason for these discrepancies has not been settled.

In order to solve the HF equation with periodic boundary conditions, we have chosen a supercell approach, similar to the setting used for QMC calculations [AC96, LN*92]. This seems to be an appropriate choice with respect to intended extensions to many-particle methods like the local ansatz. A further significant advantage of this approach is that it can handle impurities, which becomes important for applications where electrons or excitons are confined in quantum dots embedded into a semiconductor heterostructure. The supercell approach is based on the standard many-particle Schrödinger equation (2.1), where the Coulomb interaction between electrons has been replaced by the Ewald potential which takes into account the Coulomb interaction between electrons inside the supercell as well as with all their images in periodic copies of the supercell. Due to the quasi two-dimensional character of the systems, we expect that the coupling between the perpendicular and parallel directions in the Hamiltonian remains rather weak. The orbitals are therefore expressed as linear combinations

$$\phi_n^{\boldsymbol{\kappa}}(\mathbf{r}) = \sum_{i=1}^{K} \left(\sum_{\alpha} C_{n,i,\alpha}^{\boldsymbol{\kappa}} \, \zeta_{\alpha,i}^{\boldsymbol{\kappa}}(\mathbf{r}) \right) \tag{5.2}$$

of tensor product basis functions

$$\zeta_{\alpha,i}^{\boldsymbol{\kappa}}(\mathbf{r}) = \psi_{\alpha}^{\kappa_1}(x) \Phi_i^{\boldsymbol{\kappa}_\parallel}(y,z),$$

where univariate supercell wavelets $\psi_{\alpha}^{\kappa_1}$ are taken in the perpendicular direction and contracted isotropic 2d-supercell wavelets $\gamma_{\alpha}^{\boldsymbol{\kappa}_\parallel}$ in the parallel directions

$$\Phi_i^{\boldsymbol{K}_\|}(y,z) = \sum_\alpha d_{i,\alpha}^{\boldsymbol{K}_\|} \gamma_\alpha^{\boldsymbol{K}_\|}(y,z). \tag{5.3}$$

For a detailed presentation of our approach we refer to Ref. [FH*05b]. In view of Eq. (5.1) for the jellium case, it is reasonable to assume that the Kronecker rank K of the tensor product decomposition for HF orbitals (5.2) can be kept small and almost independent of the size of the supercell through an appropriate choice of the contracted basis functions (5.3). A possible choice for these basis functions are HF orbitals of the corresponding two dimensional homogeneous system or systematic basis functions like plane waves.

Acknowlegdments. The authors gratefully acknowledge Dr. T. Koprucki (Berlin) and Prof. R. Schneider (Kiel) for useful discussions. This work has been supported by the DFG Priority Program 1095 "Analysis, Modeling and Simulation of Multiscale Problems".

Appendices

A Basic notions of multiresolution analysis

The purpose of this appendix is to provide some basic definitions and properties of wavelets which are required for this paper. For a complete exposition of this subject, we refer to the excellent monographs of Daubechies [Dau92], Meyer [Mey92] and Mallat [Mal98].

In one dimension, multiresolution analysis provides a partition of the Hilbert space $L^2(\mathbb{R})$ into an infinite sequence of ascending subspaces $\cdots \subset V_{j-1} \subset V_j \subset V_{j+1} \subset \cdots$, where the index j runs over all integers. The union of these subspaces $\bigcup_j V_j$ is dense in $L^2(\mathbb{R})$. On each subspace V_j, the scaling function $\varphi(x)$ generates a basis

$$\varphi_{j,a}(x) := 2^{j/2}\varphi(2^j x - a), \tag{A.1}$$

via the operations of dilation and translation. The dilation factor 2^j scales the size of the basis functions, which means that with increasing j, the $\varphi_{j,a}$ provide a finer resolution in $L^2(\mathbb{R})$. An explicit embedding of V_j into the larger space V_{j+1} is given by the refinement relation

$$\varphi(x) = 2\sum_a h_a\, \varphi(2x-a),$$

where the number of nonzero filter coefficients h_a is finite for the scaling functions used in our applications. Wavelet spaces W_j are defined as complements of V_j in V_{j+1}. The corresponding wavelet basis is generated from a mother wavelet $\psi(x)$ analogous to Eq. (A.1)

$$\psi_{j,a}(x) := 2^{j/2}\psi(2^j x - a).$$

This construction leads to a hierarchical decomposition of $L^2(\mathbb{R}) = \bigoplus_{j\in\mathbb{Z}} W_j$ into wavelet subspaces W_j [Dau92]. In a biorthogonal wavelet basis there exists a sequence of dual spaces $\tilde{V}_j$, $\tilde{W}_j$, which satisfy the orthogonality relations $\tilde{W}_j \perp V_j$ and $\tilde{V}_j \perp W_j$. The corresponding dual wavelets $\tilde{\psi}_{j,a} := 2^{j/2}\tilde{\psi}(2^j x - a)$ and scaling functions $\tilde{\varphi}_{j,a} := 2^{j/2}\tilde{\varphi}(2^j x - a)$ generate a biorthogonal basis in $L^2(\mathbb{R})$

$$\langle \varphi_{j,a} | \tilde{\varphi}_{j,b} \rangle = \delta_{a,b}, \qquad \langle \psi_{j,a} | \tilde{\psi}_{l,b} \rangle = \delta_{j,l}\, \delta_{a,b}.$$

An arbitrary function in $L^2(\mathbb{R})$ can be expanded in a biorthogonal wavelet basis

$$f(x) = \underbrace{\sum_a \langle \tilde{\varphi}_{l_0,a} | f \rangle \, \varphi_{l_0,a}(x)}_{V_{l_0}} + \underbrace{\sum_{j=l_0}^{\infty} \sum_a \langle \tilde{\psi}_{j,a} | f \rangle \, \psi_{j,a}(x)}_{\bigoplus_{l_0 \le j} W_j}, \tag{A.2}$$

where the scaling function and wavelet coefficients are given by scalar products with respect to the dual basis. The multiscale approximation of smooth functions reveals an important sparsity feature due to the vanishing moments property of wavelets. Depending on the specific choice of the wavelet basis, a certain number of moments vanish

$$\int dx\, x^k \, \tilde{\psi}(x) = 0, \ \text{for} \ \ k = 0, \ldots, n-1.$$

This property has a significant effect on the magnitude of wavelet coefficients, as can be seen from local Taylor series expansions

$$f(x) = c_0 + \cdots + c_{n-1}(x - 2^{-j}a)^{n-1} + R_{n-1}(x)(x - 2^{-j}a)^n, \tag{A.3}$$

around the centers of wavelets $\tilde{\psi}_{j,a}(x)$. Inserting the Taylor series expansions (A.3) into the scalar products yields the following estimate for the wavelet coefficients

$$\begin{aligned} \left| \langle \tilde{\psi}_{j,a} | f \rangle \right| &:= \left| \int dx\, f(x)\, \tilde{\psi}_{j,a}(x) \right| \\ &\le \sup_{\mathrm{supp}\{\tilde{\psi}_{j,a}\}} |R_{n-1}(x)| \; 2^{-j(n+1/2)} \int dx \, \left| x^n \tilde{\psi}(x) \right|, \end{aligned}$$

where the supremum of the remainder $|R_{n-1}(x)|$ has to be taken with respect to the support of the wavelet $\tilde{\psi}_{j,a}(x)$. For functions with rapidly converging local Taylor series, the corresponding wavelet expansions (A.2), therefore, converge very fast with respect to the level j. This leads to sparse wavelet representations for these functions.

For our numerical studies, we have used the univariate symmetric biorthogonal wavelets with six vanishing moments of Sweldens [Swe96] and the corresponding univariate scaling function of Deslauriers and Dubuc [DD89]. In the text, we refer to this basis as the SDD6 wavelet basis.

B Multi-grid methods for Kohn-Sham and Hartree-Fock equations

An alternative approach to correlated many-electron systems is the density functional method [Kohn99]. There, multiscale aspects arise from the use of *multi-grid* (MG) methods for solving the discretized KS equation and from the different spatial scales defined by the physics of the KS orbitals, in particular if heavy atoms are involved. The spatial extension of core and valence orbitals may differ by more than two orders of magnitude. This has to be compared with the high accuracy requirements of quantum chemical calculations. A relative accuracy of $10^{-8 \text{ to } -9}$ in total energies is required in oder to obtain three significant digits for the binding energy of a molecule with heavy atoms. From these considerations it is obvious that measures have to be taken in order to keep the computational effort under control.

The *finite element method* (FEM) can be used in order to discretize the non-relativistic energy density functional

$$E_{tot}^{S} = \sum_{i=1}^{N} \langle \phi_i \mid \hat{h} \mid \phi_i \rangle - \frac{1}{2} \int \rho(\mathbf{r}) V_H(\mathbf{r}) d^3r \qquad \text{(B.1)}$$
$$- \frac{1}{4} \int \rho(\mathbf{r}) V_{exc}(\mathbf{r}) d^3r + E_{nuc},$$

where the one-particle Hamiltonian

$$\hat{h} = -\frac{1}{2}\Delta + V_H + V_{exc} + V_{nuc},$$

contains the Hartree potential V_H, the exchange-correlation potential V_{exc}, and the electron-nucleus external potential V_{nuc}. Actually, $E_{tot}^S = E_{tot}^S[\rho]$ can be considered as a functional of the single particle density $\rho = \sum_i |\phi_i|^2$. Variation of the functional (B.1) with respect to the orbitals ϕ_i, under conservation of their norms, leads to the KS eigenvalue equation

$$\hat{h}\phi_i(\mathbf{r}) = \varepsilon_i \phi_i(\mathbf{r}). \qquad \text{(B.2)}$$

The KS equation was first solved, using a combination of FEM and MG, for diatomic molecules which enable a reduction to two dimensional problems due to rotational symmetry. For a better analytic adaptation to the Coulomb singularities, two successive transformations of the coordinates have been performed. Namely first to elliptic hyperbolic coordinates and then by a further transformation to intrinsic coordinates (see e.g. Ref. [KK01]) where the ansatz functions are polynomials and the hierarchical grids are thus easily established. In cartesian coordinates this corresponds to complicated transcendental approximating functions on rotated triangular domains with curved boundaries. The MG method allowed a fast and comprehensive investigation of many different types of density functionals (LDA, GGA and the CS orbital functional) for dimers of the first long period chemical elements ranging from He_2 to F_2.

Further extensions to a non-linear MG scheme for solids were considered in Ref. [WW*05], where pseudo potentials have been used to get rid of the core electrons. Therefore it was possible to treat the KS equations in three dimensions on a finite difference grid. The non-linear MG proved to be quite favorable in order to cut down the number of updates of the potentials. On the other hand it has already been found in 2d molecular calculations, which used a linear MG, that the self-consistent-field iterations could be favorably coupled with the MG iterations and thus the gain by a non-linear MG would not have been dramatic.

Another application of MG methods concerns the HF equation [BHK06] which differs from Eq. (B.2) in such a way that the single particle Hamiltonian is now orbital dependent via the non-local exchange potential which replaces the local V_{exc} of the density functional. Otherwise things work rather similarly, but of course the non-local HF equation requires a considerably more complicated data handling in a MG scheme. The convergence properties are better than in density functional calculations, because the HF exchange potential does not have the $\rho^{1/3}$ behaviour which is non-analytic for $\rho \rightarrow 0$ which happens exponentially for $r \rightarrow \infty$.

C Relativistic aspects of electronic structure calculations

The non-relativistic treatment of many-electron systems misses essential physics the heavier the atoms become. Therefore an extension to a relativistic density functional description was done and pursued in the traditional 4-spinor approach using FEM, after a relativistic *linear combination of atomic orbitals* (LCAO) method had already been used since the early 80ties. Many problems arose, unknown from non- relativistic approaches; they were finally traced to the fact that the relativistic problem always has as solutions not only scattering states in the positive continuum of the spectrum and the discrete bound states, but also positronic states in the negative continuum below $-mc^2$.

The Dirac equation, which properly describes relativistic physics, can be solved rather optimal for atoms on a logarithmic mesh ranging from some small radius r_0 to some r_{max}, big enough not to cut off too much from the exponential tails of the solutions. Usually, the innermost region from r=0 to r_0 is treated analytically with a Taylor series approximation plus singular factors, in order to account for the singular behaviour for point nuclei. The wavefunction with total angular momentum $j = 1/2$ (generally for $j = (2n + 1)/2$ its $n^{\rm th}$ derivative) diverges with non-integer powers at nuclei, unlike the non-relativistic case where only cusps appear. Alternatively, nuclei can be treated as extended objects which, however, leads to other kinds of difficulties concerning the regularity of the wavefunction at the boundary of the nucleus.

If more than one atom is involved (molecules, cluster, solids, plasmas, etc.) one is faced with the problem of different scales defined by different atoms. In a stationary problem this may be handled by different scales at the known nuclear sites which can be done, but requires high effort. However in a non-stationary case these centers are moving and this has to be accounted for by a changing map of scales. This is a rather hopeless task except that physics allows a separation into localized and non-localized contributions where the localized terms have to be moved with the nuclei whereas for the non-localized parts a simple multiscale treatment operating on a fixed multi-level space with time evolving expansion coefficients suffices. Such a scheme consists of the combination of analytical manipulations with fully numerical ones, which in a rather simple two atomic case, using the Dirac-Fock Slater density functional, has already been realized in a "stationary description", see Ref. [DHK98] for further details.

The Dirac equation results from a variation of the relativistic energy functional under the contraint of norm conservation of the wavefunction. Unfortunately there exists no simple way to separate minimum from stationary solutions. In fact they always mix to some extend and these admixtures may be identified as spurious contributions of the negative (positronic) continuum to the electronic states one is interested in. This used to be an almost unsurmountable problem for approximate solutions of the Dirac equation, except for atoms where the asymptotic analytical behaviour of bound states is known. For molecules, the recently developed relativistic minimax functionals [ES99, DES00, ZKK04, ZK*05, KKR04, ZKK06] provide a loophole from this dilemma. It was a big step ahead that one could prove the boundedness of the minimax functional from below for any sufficiently smooth square integrable (i.e. normalizable) function [DE*00] and thus may obtain the best solution in an approximation space by minimization.

For the construction of the eigenvalues of an operator $\hat{H}$ which has a gap (here $-mc^2$ till $+mc^2$) in its continuous spectrum and is unbounded from above and below, the minimax principle can be expressed through the Raleigh quotient

$$\lambda_k = \inf_{\substack{dim G = k \\ G \text{ subspace of } F_+}} \sup_{\substack{\phi \neq 0 \\ \phi \in (G \oplus F_-)}} \frac{\langle \phi \mid \hat{H} \mid \phi \rangle}{\langle \phi \mid \phi \rangle},$$

where $F_+ \oplus F_-$ is an orthogonal decomposition of a well-chosen space of smooth square integrable functions. It has been proven in Refs. [ES99, DES00, DE*00, DES03] that the sequence of minimax energies λ_k equals the sequence of eigenvalues of $\hat{H}$ in the interval $(-mc^2, +mc^2)$. There is however a price to pay, namely the equations become non-linear in the eigenvalues. This causes a severe completeness problem of the spectrum, as there is usually not known a priori how many solutions are below a given energy. In the worst case one has to construct one solution after the other starting from the lowest energy solution. This at least can be formulated in a stable and absolutely safe way

[ZKK04]. However it turned out to be possible to derive a very accurate linear approximation (LARM) [ZK*05] by doubling the variational parameter space.

References

[AC96] P. H. Acioli, and D. M. Ceperley. Diffusion Monte Carlo study of jellium surfaces: Electronic densities and pair correlation functions. *Phys. Rev. B*, 54, 17199–17207, 1996.

[Ari99] T. A. Arias. Multiresolution analysis of electronic structure: Semicardinal and wavelet bases. *Rev. Mod. Phys.*, 71, 267–312, 1999.

[Bar36] J. Bardeen. Theory of the work function II. The surface double layer. *Phys. Rev.*, 49, 653–663, 1936.

[BHK06] O. Beck, D. Heinemann, and D. Kolb. Fast and accurate molecular Hartree-Fock with a finite element multigrid method. to be published.

[Bey92] G. Beylkin. On the representation of operators in bases of compactly supported wavelets. *SIAM J. Numer. Anal.*, 6, 1716–1740, 1992.

[BCR91] G. Beylkin, R. R. Coifman, and V. Rokhlin. Fast wavelet transforms and numerical algorithms I. *Commun. Pure Appl. Math.*, 44, 141–183, 1991.

[BG04] H.-J. Bungartz, and M. Griebel. Sparse grids. *Acta Numerica*, 13, 147–269, 2004.

[CKP92] C. E. Campbell, E. Krotscheck and T. Pang. Electron correlations in atomic systems. *Physics Reports*, 223, 1–42, 1992.

[CF*05] C. N. Chuev, M. V. Fedorov, H. Luo, D. Kolb, E. G. Timoshenko. 3D wavelet treatment of solvated bipolaron and polaron. *J. Theo. & Comp. Chem.*, 4, 751–767, 2005.

[Chu95] S. L. Chuang. *Physics of Optoelectronic Devices* Wiley, New York, 1995.

[Cla79] J. W. Clark. Variational theory of nuclear matter. in D. H. Wilkinson, editor, *Progress in Nuclear and Particle Physics* Vol. 2, pages 89–199. Pergamon, Oxford, 1979.

[Dah97] W. Dahmen. Wavelet and multiscale methods for operator equations. *Acta Numerica*, 6, 55–228, 1997.

[Dau92] I. Daubechies. *Ten Lectures on Wavelets*, Vol. 61 of *CBMS-NSF Regional Conference Series in Applied Mathematics*. SIAM, Philadelphia PA, 1992.

[DD89] G. Deslauriers and S. Dubuc. Symmetric iterative interpolation processes. *Constr. Approx.*, 5, 49–68, 1989.

[DeV98] R. A. DeVore. Nonlinear approximation. *Acta Numerica*, 7, 51–150, 1998.

[DKT98] R. A. DeVore, S. V. Konyagin, and V. N. Temlyakov. Hyperbolic wavelet approximation. *Constr. Approx.*, 14, 1–26, 1998.

[DES00] J. Dolbeault, M. J. Esteban, E. Séré. Variational characterization for eigenvalues of Dirac operators. *Calc. Var.* , 10, 321–347, 2000. .

[DES03] J. Dolbeault, M. J. Esteban, and E. Séré. A variational method for relativistic computations in atomic and molecular physics. *Int. J. Quantum Chem.* ,93, 149–155, 2003.

[DE*00] J. Dolbeault, M. J. Esteban, E. Séré, and M. Vanbreugel. Minimization methods for the one-particle Dirac equation. *Phys. Rev. Lett.*, 85, 4020–4023, 2000.

[DHK98] C. Düsterhöft, D. Heinemann, and D. Kolb. Dirac-Fock-Slater calculations for diatomic molecules with a finite element defect correction method (FEM-DKM). *Chem. Phys. Lett.*, 296, 77–83, 1998.

[EA02] T. D. Engeness, and T. A. Arias. Multiresolution analysis for efficient, high precision all-electron density-functional calculations. *Phys. Rev. B*, 65, 165106 (10 pages), 2002.

[ES99] M. J. Esteban, and E. Séré. Solutions of the Dirac-Fock equations for atoms and molecules. *Commun. Math. Phys.*, 203, 499–530, 1999.

[FHS06a] H.-J. Flad, W. Hackbusch, and R. Schneider. Best N-term approximation in electronic structure calculations. I. One-electron reduced density matrix. *ESAIM: M2AN*, 40, 49–61, 2006.

[FHS06a] H.-J. Flad, W. Hackbusch, and R. Schneider. Best N-term approximation in electronic structure calculations. II. Jastrow factors. MPI-MIS Preprint No. 80 (2005).

[FH*02] H.-J. Flad, W. Hackbusch, D. Kolb and R. Schneider. Wavelet approximation of correlated wavefunctions. I. Basics. *J. Chem. Phys.*, 116, 9641–9657, 2002.

[FH*05a] H.-J. Flad, W. Hackbusch, H. Luo and D. Kolb. Diagrammatic multiresolution analysis for electron correlations. *Phys. Rev. B.*, 71, 125115 (18 pages), 2005.

[FH*05b] H.-J. Flad, W. Hackbusch, H. Luo and D. Kolb. Wavelet Approach to quasi two-dimensional extended many-particle systems. I. Supercell Hartree-Fock method. *J. Comp. Phys.*, 205, 540–566, 2005.

[FHM84] D. E. Freund, B. D. Huxtable, and J. D. Morgan III. Variational calculations on the helium isoelectronic sequence. *Phys. Rev. A*, 29, 980–982, 1984.

[Ful93] P. Fulde. *Electron Correlations in Molecules and Solids, 2nd ed.* Springer, Berlin, 1993.

[GG00] J. Garcke, and M. Griebel. On the computation of the eigenproblems of hydrogen and helium in strong magnetic and electric fields with the sparse grid combination technique. *J. Comp. Phys.*, 165, 694–716, 2000.

[Goe98] S. Goedecker. *Wavelets and their Application for the Solution of Differential Equations.* Presses Polytechniques Universitaires et Romandes, Lausanne, 1998.

[Hac01] W. Hackbusch. The efficient computation of certain determinants arising in the treatment of Schrödinger's equation. *Computing*, 67, 35–56, 2001.

[HJ74] J. Harris, and R. O. Jones. The surface energy of a bounded electron gas. *J. Phys. F: Metal Phys.*, 4, 1170–1186, 1974.

[HF*04] R. J. Harrison, G. I. Fann, T. Yanai, Z. Gan, and G. Beylkin. Multiresolution quantum chemistry: Basic theory and initial applications. *J. Chem. Phys.*, 121, 11587–11598, 2004.

[HJO99] T. Helgaker, P. Jørgensen and J. Olsen. *Molecular Electronic-Structure Theory.* Wiley, New York, 1999.

[Kohn99] W. Kohn. Nobel lecture: Electronic structure of matter-wave functions and density functionals. *Rev. Mod. Phys.*, 71, 1253–1266, 1999.

[KW00] T. Koprucki and H.-J. Wagner. New exact ground states for one-dimensional quantum many-body systems. *J. Stat. Phys.*, 100, 779–790, 2000.

[Kro85] E. Krotscheck. Theory of inhomogeneous quantum systems. III. Variational wave functions for Fermi fluids. *Phys. Rev. B*, 31, 4267–4278, 1985.

[KKQ85] E. Krotscheck, W. Kohn and G.-X. Qian. Theory of inhomogeneous quantum systems. IV. Variational calculations of metal surfaces. *Phys. Rev. B*, 32, 5693–5712, 1985.

[KK01] O. Kullie, and D. Kolb. High accuracy Dirac-finite-element (FEM) calculations for H_2^+ and Th_2^{179+}. *Eur. Phys. J.*, D17, 167–173, 2001.

[KKR04] O. Kullie, D. Kolb, and A. Rutkowki. Two-spinor fully relativistic Finite-Element (FEM) solution of the two-center Coulombic problem. *Chem. Phys. Lett.*, 383, 215–221, 2004.

[ZKK06] O. Kullie, H. Zhang, J. Kolb, and D. Kolb. Relativistic density functional calculations using two-spinor Minimax FEM and LCAO for ZnO, CdO, HgO, UubO and Cu_2, Ag_2, Au_2, Uuu_2. submitted to *J. Chem. Phys.*

[LK71] N. D. Lang, and W. Kohn. Theory of metal surfaces: Work functions. *Phys. Rev. B*, 3, 1215–1223, 1971.

[LP75] D. C. Langreth, and J. P. Perdew. The exchange-correlation energy of a metallic surface. *Solid State Commun.*, 17, 1425–1429, 1975.

[LN*92] X.-P. Li, R. J. Needs, R. M. Martin, and D. M. Ceperley. Green's-function quantum Monte Carlo study of a jellium surface. *Phys. Rev. B*, 45, 6124–6130, 1992.

[LM86] I. Lindgren and J. Morrison. *Atomic Many-Body Theory.* Springer, Berlin, 1986.

[LK*02] H. Luo, D. Kolb, H.-J. Flad, W. Hackbusch and T. Koprucki. Wavelet approximation of correlated wavefunctions. II. Hyperbolic wavelets and adaptive approximation schemes. *J. Chem. Phys.*, 117, 3625–3638, 2002.

[LK*06] H. Luo, D. Kolb, H.-J. Flad, and W. Hackbusch. Perturbative calculation of Jastrow factors. MPI-MIS Preprint 2006.

[Mal98] S. Mallat. *A Wavelet Tour of Signal Processing* Academic Press, San Diego, 1998.

[Mey92] Y. Meyer. *Wavelets and Operators.* Cambridge University Press, 1992.

[NP01] M. Nekovee and J. M. Pitarke. Recent progress in the computational many-body theory of metal surfaces. *Computer Phys. Comm.*, 137, 123–142, 2001.

[NTR02] A. M. N. Niklasson, C. J. Tymczak, and H. Röder Multiresolution density-matrix approach to electronic structure calculations. *Phys. Rev. B*, 66, 155120 (15 pages), 2002.

[Nit03] P.-A. Nitsche. Best N term approximation spaces for sparse grids. Research Report No. 2003-11, Seminar für Angewandte Mathematik, ETH Zürich.

[PE01] J. M. Pitarke, and A. G. Eguiluz. Jellium surface energy beyond the local-density approximation: Self-consistent-field calculations. *Phys. Rev. B*, 63, 045116 (10 pages), 2001.

[SP93] S. Saebø, and P. Pulay. Local treatment of electron correlation. *Annu. Rev. Phys. Chem.*, 44, 213–236, 1993.

[SKG77] V. Sahni, J. B. Krieger, and J. Gruenebaum. Metal surface properties in the linear potential approximation. *Phys. Rev. B*, 15, 1941–1949, 1977.

[SM80] V. Sahni, and C. Q. Ma Hartree-Fock theory of the inhomogeneous electron gas at a jellium metal surface: Rigorous upper bounds to the surface energy and accurate work functions. *Phys. Rev. B*, 22, 5987–5996, 1980.

[SHW99] M. Schütz, G. Hetzer, and H.-J. Werner. Low-order scaling local electron correlation methods. I. Linear scaling local MP2. *J. Chem. Phys.*, 111, 5691–5705, 1999.

[Sto80] H. Stoll. On the correlation energy of graphite. *J. Chem. Phys.*, 97, 8449–8454, 1992.

[SF80] G. Stollhoff, and P. Fulde. On the computation of electronic correlation energies within the local approach. *J. Chem. Phys.*, 73, 4548–4561, 1980.

[Swe96] W. Sweldens. The lifting scheme: A custom-design construction of biorthogonal wavelets. *Appl. Comp. Harm. Anal.*, 3, 186–200, 1996.

[TBJ03] K. S. Thygesen, M. V. Bollinger, and K. W. Jacobsen. Conductance calculations with a wavelet basis set. *Phys. Rev. B*, 67, 115404 (11 pages), 2003.

[WW*05] J. Wang, Y. Wang, S. Yu, and D. Kolb. Nonlinear algorithm for the solution of the Kohn-Sham equations in solids. *J. Phys.: Condens. Matter*, 17, 3701–3715, 2005.

[Yse04] H. Yserentant. On the regularity of the electronic Schrödinger equation in Hilbert spaces of mixed derivatives. *Numer. Math.*, 98, 731–759, 2004.

[Yse05] H. Yserentant. Sparse grid spaces for the numerical solution of the electronic Schrödinger equation. *Numer. Math.*, 101, 381–389, 2005.

[ZKK04] H. Zhang, O. Kullie, and D. Kolb. Minimax LCAO approach to the relativistic two center Coulomb problem and its finite element (FEM) spectrum. *J. Phys. B: At. Mol. Opt. Phys.*, 37, 905–916, 2004.

[ZK*05] H. Zhang, O. Kullie, J. Kolb, H. Luo, and D. Kolb. Linear approximation to relativistic minimax (LARM) applied to a LCAO description of the two-center Coulomb problem. *J. Phys. B: At. Mol. Opt. Phys.*, 38, 2955-2963, 2005.

Multi-Scale Modeling of Quantum Semiconductor Devices

Anton Arnold[1] and Ansgar Jüngel[2]

[1] Institut für Analysis und Scientific Computing, Technische Universität Wien, Wiedner Hauptstr. 8, A-1040 Wien, Austria. anton.arnold@tuwien.ac.at
[2] Institut für Mathematik, Universität Mainz, Staudingerweg 9, 55099 Mainz. juengel@mathematik.uni-mainz.de

Summary. This review is concerned with three classes of quantum semiconductor equations: Schrödinger models, Wigner models, and fluid-type models. For each of these classes, some phenomena on various time and length scales are presented and the connections between micro-scale and macro-scale models are explained. We discuss Schrödinger-Poisson systems for the simulation of quantum waveguides and illustrate the importance of using open boundary conditions. We present Wigner-based semiconductor models and sketch their mathematical analysis. In particular we discuss the Wigner-Poisson-Focker-Planck system, which is the starting point of deriving subsequently the viscous quantum hydrodynamic model. Furthermore, a unified approach to derive macroscopic quantum equations is presented. Two classes of models are derived from a Wigner equation with elastic and inelastic collisions: quantum hydrodynamic equations and their variants, as well as quantum diffusion models.

1 Introduction

The modern computer and telecommunication industry relies heavily on the use of semiconductor devices. A very important fact of the success of these devices is that their size can be very small compared to previous electronic devices (like the tube transistor). While the characteristic length of the first semiconductor device (a germanium transistor) built by Bardeen, Brattain, and Shockley in 1947 was 20μm, the characteristic size has been decreased up to now to some deca-nanometers only. With decreasing device length quantum mechanical effects are becoming more and more important in actual devices. In fact, there are devices, for instance tunneling diodes or quantum wave guides, whose function is based on quantum effects. The development of such devices is usually supported by computer simulations to optimize the desired operating features. Now, in order to perform the numerical simulations, mathematical equations are needed that are both physically accurate and numerically solvable with low computational cost.

We wish to model the flow of electrons in a semiconductor crystal with the goal to predict macroscopically measurable quantities by means of computer simulations. Although the physical process is always the transport of charged particles in a solid crystal, we need to devise different mathematical models because of the wide range of operating conditions and the desired need of accuracy. Moreover, since in some cases we are not interested in all the available physical information, we also need simpler models which help to reduce the computational cost in the numerical simulations.

We shall discuss three model classes: Schrödinger, Wigner, and fluid-type models. Schrödinger models describe the purely ballistic transport of electrons and holes, and they are employed for simulations of quantum waveguides and nano-scale semiconductor heterostructures. As soon as scattering mechanisms (between electrons and phonons, e.g.) become important, one has to resort to Wigner function or the equivalent density matrix formalism. For practical applications Wigner functions have the advantage to allow for a rather simple, intuitive formulation of boundary conditions at device contacts or interfaces. On the other hand, the Wigner equation is posed in a high dimensional phase space which makes its numerical solution extremely costly. As a compromise, fluid-type models can provide a reasonable approximation and is hence often used. Since one only computes the measurable physical quantities in these fluid models, they are computationable cheap. Moreover, classical boundary conditions can also be employed here.

The multi-scale character in semiconductor device modeling becomes manifest in a hierarchy of models that differ in mathematical and numerical complexity and incorporate physical phenomena on various time and length scales. The microscopic models clearly include the highest amount of information, but they involve the highly oscilatory Schrödinger and Wigner functions. However, the macroscopic variables of interest for practitioners (like particle and current densities) are typically much smoother. Hence, it is very attractive (particularly with respect to reduce numerical costs) to settle for simplified macroscopic quantum transport models. Scaling limits allow to relate these models and to obtain important information for the range of validity (and the limitations) of the reduced macro-scale models. Starting from Wigner-Boltzmann-type equations it is indeed possible to obtain a unified derivation of quantum hydrodynamic and quantum diffusion models.

2 Microscopic picture I: Schrödinger models

This section is concerned with Schrödinger-Poisson models for semiconductor device simulations. Such models are only applicable in the ballistic regime, i.e. to devices or subregions of devices, where the quantum mechanical transport is the dominant phenomenon and scattering plays only a minor role. As particular examples we name interferences in quantum waveguides, the tunneling through nano-scale semiconductor heterostructures (in a resonant tunneling diode, e.g.), and the ballistic transport along nano-size channels of MOSFETs. In all of these examples we are dealing with *open quantum systems*, refering to a model on a finite geometry along with *open boundaries* (this contrasts with the situation in Sect. 3.1, where we shall consider collisional open quantum systems). Here, the transport model (the Schrödinger equation, e.g.) is posed on a finite domain $\Omega \subset \mathbb{R}^d$ (d being the space dimension). At the device contacts or interfaces *open boundary conditions* are specified, such that an incoming current can be prescribed and outgoing electron waves will not be reflected at such boundaries.

2.1 Quantum waveguide simulations

In this subsection we discuss simulation models for quantum waveguides. These are novel electronic switches of nanoscale dimensions. They are made of several different layers of semiconductor materials such that the electron flow is confined to small channels or waveguides. Due to their sandwiched structure the relevant geometry for the electron current is essentially two dimensional. Figure 2.1 shows the example of a T-shaped *quantum interference transistor*. The actual structure can be realized as an etched layer of GaAs between two layers of doped AlGaAs (cf. [Ram02]). Applying an external potential at the gate (i.e. above the shaded portion of the stub, the "allowed region" for the electrons, and hence the geometry (in particular the stub length) can be modified. This allows to control the current flow through such an electronic device. It makes it a switch, which resambles a transistor – but on a nano-scale. Such a device shows sharp peaks in conductance that are due to the presence of bound states in the stub (see Fig. 2.2, 2.3). It is expected that these novel devices will operate at low power and high speed.

The electron transport through a quantum waveguide can be modeled in good approximation by a two dimensional, time dependent Schrödinger-Poisson system for the wave functions $\psi_\lambda(x,t)$, indexed by the energy variable $\lambda \in \Lambda \subset \mathbb{R}$. The (possibly time-dependent) spatial domain $\Omega \subset \mathbb{R}^2$ consists of (very long) leads and the active switching region (e.g. T-shaped as in Fig. 2.1). In typical applications electrons are continuously fed into the leads as plane waves ψ_λ^{pw}. The Schrödinger model now reads

$$i\frac{\partial \psi_\lambda}{\partial t} = -\frac{1}{2}\Delta\psi_\lambda + V(x,t)\psi_\lambda, \quad x \in \Omega,\ \lambda \in \Lambda,\ t > 0. \tag{2.1}$$

The potential $V = V_e + V_s$ consists of an external, applied potential V_e and the selfconsistent potential satisfying the Poisson equation with Dirichlet boundary conditions:

$$\begin{aligned} -\Delta V_s(x,t) = n(x,t) &= \int_\Lambda |\psi_\lambda(x,t)|^2 g(\lambda)\, d\lambda, \quad x \in \Omega, \\ V_s = 0, &\quad \text{on } \partial\Omega. \end{aligned} \tag{2.2}$$

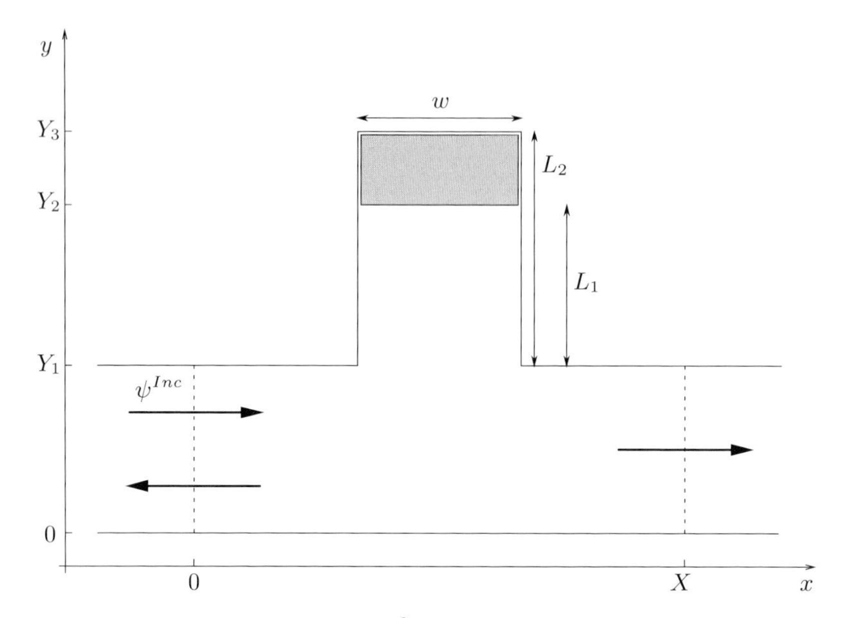

Fig. 2.1. T-shaped geometry $\Omega \subset \mathbb{R}^2$ of a quantum interference transistor with source and drain contacts to the left and right of the channel. Applying a gate voltage above the stub allows to modify the stub length from L_1 to L_2 and hence to switch the transistor between the on- and off-states.

Here, n is the spatial electron density and $g(\lambda)$ denotes the prescribed statistics of the injected waves (Fermi-Dirac, e.g.). In the simplest case (i.e. a 1D approximation) open or "transparent" boundary conditions at the contacts or interfaces take the form

$$\frac{\partial}{\partial \eta}(\psi_\lambda - \psi_\lambda^{pw}) = -e^{-i\pi/4}\sqrt{\partial_t}\,(\psi_\lambda - \psi_\lambda^{pw}),\ \lambda \in \Lambda, \tag{2.3}$$

where η denotes the unit outward normal vector at each interface. $\sqrt{\partial_t}$ is the fractional time derivative of order $\frac{1}{2}$, and it can be rewritten as a time-convolution of the boundary data with the kernel $t^{-3/2}$. For the derivation of the 2D-variant of such transparent boundary conditions and the mathematical analysis of this coupled model (2.1)-(2.3) we refer to [BMP05, Ar01].

The discretization of such a model poses several big numerical challenges, both for stationary and for transient simulations: Firstly, the wave function is highly oscillatory for larger energies, while the macroscopic variables of interest (particle density n and potential V) are rather smooth. Secondly, solutions to (2.1)-(2.3) can exhibit sharp peaks in the curve of conductance versus injection energy (both in quantum waveguides and in resonant tunneling diodes). To cope with these two problems, WKB-type discretization schemes for the 1D stationary analogue of the above model and adaptive energy grids (for $\lambda \in \Lambda$) were devised in [BP06].

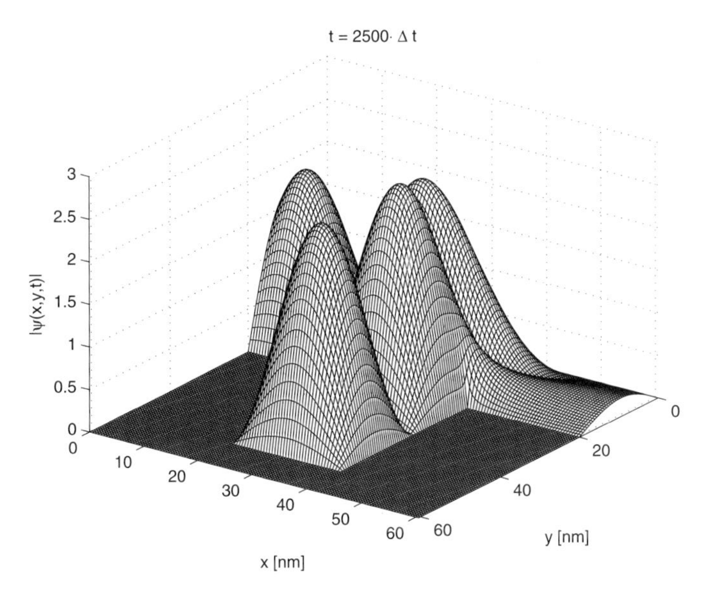

Fig. 2.2. Stationary Schrödinger wave function for T-shaped waveguide with short stub (i.e. $L_1 = 32$ nm) – "off state"

Thirdly, the numerical discretization of the transparent boundary condition (2.3) is very delicate in the time dependent case, as it may easily render the initial-boundary value problem unstable and introduce hugh spurious wave reflections. Based on a Crank-Nicolson finite difference discretization of the Schrödinger equation, unconditionally stable *discrete transparent boundary conditions* were developed in [Arn98] for the one-dimensional and in [AES03, AS06] for the two-dimensional problem. An extension of such discrete open boundary conditions for (multiband) kp-Schrödinger equations appearing in the simulation of quantum heterostructures were developed in [ZA*05].

To close this subsection we present some first simulations of the electron flow through the T-shaped waveguide from Fig. 2.1 with the dimensions $X =$ 60 nm, $Y_1 = 20$ nm. These calculations are based on the linear Schrödinger equation for one wave function with $V \equiv 0$ and the injection of a mono-energetic plane wave with 130 meV, entering in the transparent boundary condition (2.3). The simulation was based on a compact forth order finite difference scheme ("Numerov scheme") and a Crank-Nicolson discretization in time [AS06].

Important device data for practitioners are the current-voltage (I-V) characteristics, the ratio between the on- and the (residual) off-current as well as the switching time between these two stationary states. Depending on the size and shape of the stub, the electron current is either reflected (off-state of the device, see Fig. 2.2) or it can flow through the device (on-state, see Fig. 2.3). Starting from the stationary state in Fig. 2.2, the swiching of the device was

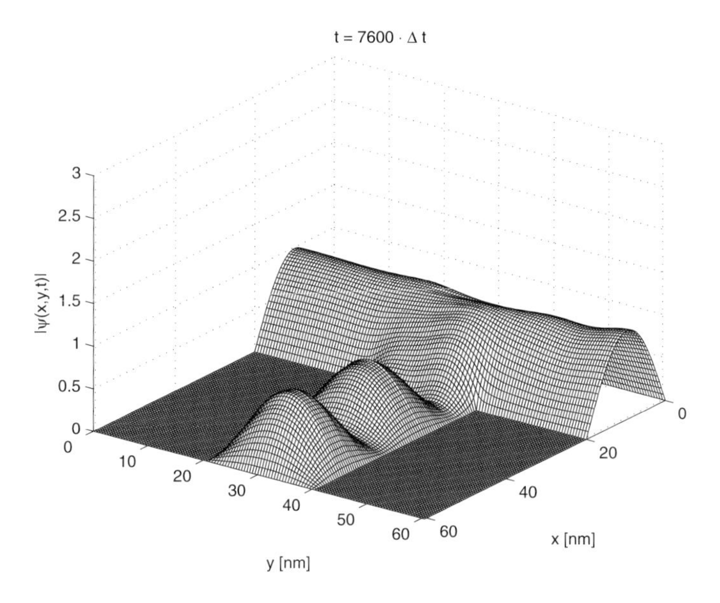

Fig. 2.3. Stationary Schrödinger wave function for T-shaped waveguide with long stub (i.e. $L_2 = 40.5$ nm) – "on state"

realized by an instantaneous extension of the stub length from $L_1 = 32$ nm to $L_2 = 40.5$ nm. After a transient phase the new steady state Fig. 2.3 is reached after about 4 ps.

3 Microscopic picture II: Wigner models

In this section we shall present and discuss semiconductor models that are based on the quantum-kinetic Wigner formalism. As mentioned before there are two main reasons for using this framework in applications (indeed, mostly for time dependent problems): In contrast to Schrödinger models the Wigner picture allows to include the modeling of scattering phenomena in the form of a *quantum Boltzmann equation*. Secondly, this quantum-kinetic framework makes is easier to formulate reasonable boundary conditions at the device contacts, using guidance and inspiration from classical kinetic theory [Fre87]. This approach makes indeed sense, as quantum effects are not important close to the (typically highly doped) contact regions.

The Wigner function $f = f(x, v, t)$ is one of several equivalent formalisms to describe the (mixed) state of a physical quantum system (cf. [Wig32]). It is a real-valued quasi-distribution function in the position-momentum (x, p) phase space at time t. In collision-free regimes, the quantum equivalent of the Liouville equation of classical kinetic theory governs the time evolution of f. In the d-dimensional whole space it reads

$$\partial_t f + p \cdot \nabla_x f + \theta[V]f = 0, \quad t > 0, \quad (x,p) \in \mathbb{R}^{2d}, \tag{3.1}$$

where the (real-valued) potential $V = V(x)$ enters through the pseudo-differential operator $\theta[V]$ defined by

$$\begin{aligned}(\theta[V]f)(x,p,t) &= \frac{i}{(2\pi)^{d/2}} \int_{\mathbb{R}^d} \delta V(x,\eta) \mathcal{F}_{p\to\eta} f(x,\eta,t) e^{ip\cdot\eta}\, d\eta \\ &= \frac{i}{(2\pi)^d} \int_{\mathbb{R}^d} \int_{\mathbb{R}^d} \delta V(x,\eta) f(x,p',t) e^{i(p-p')\cdot\eta}\, dp'\, d\eta. \end{aligned} \tag{3.2}$$

Here, $\delta V(x,\eta) := \frac{1}{\varepsilon}\left[V(x+\frac{\varepsilon\eta}{2}) - V(x-\frac{\varepsilon\eta}{2})\right]$, and $\mathcal{F}_{p\to\eta} f$ denotes the Fourier transform of f with respect to p, and $\varepsilon > 0$ is the scaled Planck constant.

For semiconductor device simulations it is crucial to include the self-consistent potential. The electrostatic potential $V(x,t)$ is hence time-dependent and obtained as a solution to the Poisson equation

$$\Delta_x V(x,t) = n(x,t) - C(x), \tag{3.3}$$

with the particle density

$$n(x,t) = \int_{\mathbb{R}^d} f(x,p,t)\, dp. \tag{3.4}$$

Here, $C(x)$ denotes the time-independent doping profile of the device, i.e. the spatial density of the doping ions implanted into the semiconductor crystal.

3.1 Wigner models for open quantum systems

Until now we only considered the ballistic and hence reversible quantum transport of the electrons in a one-particle (Hartree) approximation. Such purely ballistic models (either in the Wigner or Schrödinger framework - cf. Sect. 2) are useful semiconductor models for device lengths in the order of the electrons' mean free path. For larger devices, however, scattering phenomena between electrons and phonons (i.e. thermal vibrations of the crystal lattice) or among the electrons have to be taken into account. In this case an appropriate collision operator $Q(f)$ has to be added as a right hand side of (3.1). In contrast to Sect. 2 the term *open quantum system* refers here to the interaction of our considered electron ensemble with some 'environment' (an external phonon bath, e.g.) and not to the influence of the boundaries or contacts.

For the classical semiconductor Boltzmann equation excellent models for the most important collisional mechanisms have been derived (cf. [MRS90]) and are incorporated into today's commercial simulation tools. In quantum kinetic theory, however, realistic and numerically usable collison models are much less developed. In contrast to classical kinetic theory, quantum collision operators are actually non-local in time (cf. the Levinson equation [Lev70] as one possible model). However, since most of the current numerical simulations involve only local in time approximations, we shall confine our discussion to such collision operators. The three most used models are firstly relaxation time approximations of the form

$$Q(f) = \frac{f_0 - f}{\tau}, \tag{3.5}$$

with some appropriate steady state f_0 and the relaxation time $\tau > 0$ [KK*89, Arn95a]. Secondly, many applications (cf. [Str86, GG*93, RA05]) use quantum Fokker-Planck models:

$$Q(f) = \beta \mathrm{div}_p(pf) + \sigma \Delta_p f + 2\gamma \mathrm{div}_p(\nabla_x f) + \alpha \Delta_x f, \tag{3.6}$$

(cf. [CL83, CE*00] for a derivation) with the friction parameter $\beta \geq 0$. The non-negative coefficients α, γ, σ constitute the phase-space diffusion matrix of the system. We remark that one would have $\alpha = \gamma = 0$ in the Fokker-Planck equation of classical mechanics [Ris84].

As a third option, the Wigner equation (3.1) is often augmented by a semi-classical Boltzmann operator [KN05]. However, since this model is quantum mechanically not consistent, we shall not discuss it any further. Finally, we mention the quantum-BGK type models [DR03] that were recently introduced for deriving quantum hydrodynamics (cf. Sect. 4.2 for details).

At the end of this section we briefly list the numerical methods developed so far for Wigner function-based device simulations. Virtually all simulations were carried out for one dimensional resonant tunneling diodes. The earliest approaches were based on finite difference schemes for the relaxation-time Wigner-Poisson system [Fre87, KK*89]. Spectral collocation methods were designed as an efficient alternative to discretize the non-local pseudo-differential operator $\theta[V]$ (cf. [Rin90]). In [AR95] this approach was combined with an operator splitting between the transport term $p \cdot \nabla_x$ and $\theta[V]$ which has also been common practice for Boltzmann type equations. This mixed operator splitting/spectral collocation technique was recently extended to the Wigner-Fokker-Planck system in [DA06]. In [KN05] the classical Monte Carlo method was extended to Wigner models, and it has the potential to make multi-dimensional simulations feasible. Since the Wigner function takes both positive and negative values, novel algorithms for particle creation and annihilation had to be developed within this Monte Carlo approach.

3.2 Open quantum systems in Lindblad form

Since the Wigner function takes also negative values, it is a-priori not clear why the macroscopic particle density satisfies $n(x,t) := \int f(x,p,t)\,dp \geq 0\ \forall x, t$. This physically important non-negativity is a consequence of the non-negativity of the density matrix (operator) that is associated with a Wigner function: Let $\widehat{\rho}$ be an operator on $L^2(\mathbb{R}^d)$ with integral kernel $\rho(x,x') = \rho \in L^2(\mathbb{R}^{2d})$, i.e.

$$(\widehat{\rho}\phi)(x) = \int_{\mathbb{R}^d} \rho(x,x')\phi(x')\,dx' \quad \forall \phi \in L^2(\mathbb{R}^d). \tag{3.7}$$

The Wigner transform of the density matrix $\widehat{\rho}$ is now defined as the following Wigner function f (cf. [Wig32, LP93]):

$$W(\widehat{\rho})(x,p) := f(x,p) = \frac{1}{(2\pi)^d} \int_{\mathbb{R}^d} \rho(x + \frac{\varepsilon}{2}\eta, x - \frac{\varepsilon}{2}\eta) e^{i\eta\cdot p}\,d\eta. \tag{3.8}$$

In terms of density matrices, a mixed quantum states is described as a positive trace class operator on $L^2(\mathbb{R}^d)$ (i.e. $\widehat{\rho} \in \mathcal{I}_1$, $\widehat{\rho} \geq 0$), mostly with the normalization $\mathrm{Tr}\widehat{\rho} = 1$, where Tr denotes the operator trace. The positivity of $\widehat{\rho}$ as an operator then implies pointwise positivity of the particle density

$$n(x) := \rho(x,x) \geq 0, \quad x \in \mathbb{R}^d, \tag{3.9}$$

and of the corresponding kinetic energy

$$E_{kin} := \frac{1}{2}\int_{\mathbb{R}^{2d}} |p|^2 f(x,p)\,dxdp = \frac{\varepsilon^2}{2}\mathrm{Tr}(-\Delta\widehat{\rho}) = \frac{\varepsilon^2}{2}\mathrm{Tr}(\sqrt{-\Delta}\widehat{\rho}\sqrt{-\Delta}) \geq 0. \tag{3.10}$$

The time evolution of a density matrix is given by the Heisenberg-von Neumann equation, obtained by applying the Wigner transform (3.7), (3.8) to the Wigner equation (3.1). It reads

$$\begin{aligned} &i\varepsilon\widehat{\rho}_t = H(t)\widehat{\rho} - \widehat{\rho}H(t), \quad t > 0, \\ &\widehat{\rho}(t=0) = \widehat{\rho}_I, \end{aligned} \tag{3.11}$$

with the (possibly time dependent) Hamiltonian $H(t) = -\frac{\varepsilon^2}{2}\Delta + V(x,t)$. For an open quantum system the right hand side of (3.11) has to be augmented by a non-Hamiltonian term $iA(\widehat{\rho})$. It is well known from [Lin76] that such quantum evolution equations preserve the positivity of $\widehat{\rho}(t)$ (more precisely, it is actually the complete positivity) if and only if the dissipative term $A(\widehat{\rho})$ satisfies the following structural condition. It must be possible to represent it in the so-called Lindblad form:

$$A(\widehat{\rho}) = \sum_{j\in J} L_j^* L_j \widehat{\rho} + \widehat{\rho} L_j^* L_j - 2L_j \widehat{\rho} L_j^*, \tag{3.12}$$

with some appropriate (but typically not uniquely defined) Lindblad operators L_j, and a finite or infinite index set $J \subset \mathbb{N}$. Furthermore, such models then preserve the mass of the system, i.e. $\mathrm{Tr}\,\widehat{\rho}(t) = \mathrm{Tr}\,\widehat{\rho}_I$, $t \geq 0$.

For the relaxation time Wigner equation we have $A(\widehat{\rho}) = \frac{\widehat{\rho}_0 - \widehat{\rho}}{\tau}$ with some steady state $\widehat{\rho}_0$. Under the natural assumption $\mathrm{Tr}\,\widehat{\rho}_I = \mathrm{Tr}\,\widehat{\rho}_0 = 1$, and if the relaxation time τ is constant, we have (cf. [Arn95b])

$$A(\widehat{\rho}) = \frac{1}{\tau}(\widehat{\rho}_0\,\mathrm{Tr}\,\widehat{\rho} - \widehat{\rho}\,\mathrm{Tr}\,\widehat{\rho}_0) = \sum_{j,k\in\mathbb{N}} L_{jk}^* L_{jk}\widehat{\rho} + \widehat{\rho} L_{jk}^* L_{jk} - 2L_{jk}\widehat{\rho} L_{jk}^*,$$

with the Lindblad operators $L_{jk} = \sqrt{\mu_k/\tau}\ |\varphi_k >< \varphi_j|$. Here, $(\mu_k, \varphi_k)_{k\in\mathbb{N}}$ denotes the eigenpairs of $\widehat{\rho}_0$. Hence, the relaxation time Wigner equation (possibly with a selfconsistent potential) is an admissible open quantum model in Lindblad form.

For the Wigner-Fokker-Planck (WFP) equation with $Q(f)$ from (3.6), the Lindblad condition (3.12) holds iff

$$\begin{pmatrix} \alpha & \gamma + \frac{i\varepsilon}{4}\beta \\ \gamma - \frac{i\varepsilon}{4}\beta & \sigma \end{pmatrix} \geq 0. \tag{3.13}$$

Under this assumption the WFP model is also quantum mechanically correct. Here, the Lindblad operators are linear combinations of x_j and ∂_{x_j} (cf. [AL*04]).

3.3 Analysis of the Wigner-Poisson-Fokker-Planck system

In this section we will sketch the different approaches to the well-posedness analysis for the Cauchy problem of the Wigner-Poisson-Fokker-Planck (WPFP) system in three dimensions:

$$\begin{aligned} \partial_t f + p\cdot\nabla_x f + \theta[V]f &= \beta \mathrm{div}_p(pf) + \sigma\Delta_p f + 2\gamma\mathrm{div}_p(\nabla_x f) + \alpha\Delta_x f,\ t>0, \\ f(x,p,t=0) &= f_I(x,p),\ (x,p)\in\mathbb{R}^6, \\ \Delta_x V(x,t) = n(x,t) &= \int_{\mathbb{R}^3} w(x,p,t)\,dp. \end{aligned} \tag{3.14}$$

For simplicity we set here all physical constants equal to 1, and we chose $C \equiv 0$ as this would not change the subsequent analysis. Also, the Lindblad condition (3.13) is assumed to hold in the sequel.

First we remark that the WPFP model cannot be writen as an equivalent system of countably many Schrödinger equations coupled to the Poisson equation (and this is typical for open quantum systems). Therefore, the approach of [BM91] employed in the well-posedness analysis of the (reversible) Wigner-Poisson system cannot be adapted to WPFP. Hence, there are two remaining frameworks for the analysis: the Wigner function and density matrix formalisms, which we shall both briefly discuss here.

On the quantum kinetic level there are two main analytic difficulties for the nonlinear WPFP system. Since the natural analytic setting for Wigner functions is $f(.,.,t) \in L^2(\mathbb{R}^6)$ we cannot expect that $f(x,.,t) \in L^1(\mathbb{R}^3)$ holds. Hence, the definition of the particle density by $n(x,t) = \int_{\mathbb{R}^3} f(x,p,t)\,dp$ is purely formal. The second key problem is the lack of usable a-priori estimates on the Wigner function which would be needed to prove global-in-time existence of WPFP-solutions: The only useful (and almost trivial) estimate is

$$||f(t)||_{L^2(\mathbb{R}^6)} \leq e^{\frac{3}{2}\beta t}||f_I||_{L^2(\mathbb{R}^6)}, \quad t\geq 0. \tag{3.15}$$

The other physically obvious conservation laws

$$\int f(x,p,t)\,dxdp = const \qquad \text{(mass conservation)},$$

and a simple energy balance involving the kinetic energy $E_{kin} = \frac{1}{2}\int |p|^2 f(x,p,t)\,dxdp$ both include functionals of f that are, a-priori, not necessarily positive and hence not useable on the quantum kinetic level.

Dispersive effects in quantum kinetic equations. Both of the described analytic problems – proper definition of the particle density n (or, equivalently, the electric field $E = \nabla_x V$) and additional a-priori estimates – can be coped with by exploiting dispersive effects of the free-streaming operator jointly with the parabolic regularization of the Fokker-Planck term. Such dispersive techniques for kinetic equations were first developed for the Vlasov-Poisson system (cf. [Per96]) and then adapted to the Vlasov-Poisson-Fokker-Planck equation in [Cas98]. In [ADM04, ADM05] these tools were extended to quantum kinetic theory. They yield first of all an a-priori estimate for the field $E(t)$ in terms of $\|f(t)\|_{L^2(\mathbb{R}^6)}$ only (remember (3.15) !). This estimate allows a *novel definition* of the macroscopic quantities (namely, the self-consistent field, the potential, and the density), which, in contrast to the definition (3.3), (3.4) is now non-local in time. This way, no p-integrability of f is needed.

Next we illustrate these dispersive tools in some more detail. With $G(t) = G(x,p,x',p',t)$ denoting the Green's function of the linear part of (3.14) (cf. [SC*02]), the (linear) WFP equation can be rewritten as

$$\begin{aligned} f(x,p,t) = &\iint G(t) f_I(x',p')dx'dp' \\ &+ \int_0^t \iint G(s)(\theta[V]f)(x',p',t-s)\,dx'dp'ds, \quad t \geq 0. \end{aligned} \tag{3.16}$$

According to the two terms on the r.h.s. we split the electric field

$$E(x,t) = \nabla_x V(x,t) = \frac{x}{4\pi|x|^3} * n(x,t)$$

into $E = E_0 + E_1$ with

$$E_0(x,t) = \frac{x}{4\pi|x|^3} *_x \iiint G(t) f_I(x',p')dx'dp'dp, \tag{3.17}$$

$$E_1(x,t) = \frac{x}{4\pi|x|^3} *_x \int_0^t \iiint G(s)(\theta[V]f)(x',p',t-s)\,dx'dp'dpds.$$

With some tricky reformulation this last equation can be rewritten as

$$\begin{aligned} &(E_1)_j(x,t) \\ &= \frac{1}{4\pi}\sum_{k=1}^{3} \frac{3x_j x_k - \delta_{jk}|x|^2}{|x|^5} *_x \int_0^t \frac{\vartheta(s)}{R(s)^{3/2}} \mathcal{N}\left(\frac{x}{\sqrt{R(s)}}\right) *_x F_k[f](x,t,s)\,ds, \end{aligned} \tag{3.18}$$

$j = 1, 2, 3$; with

$$F_k[f](x,t,s) := \int (\Gamma_k[E_0 + E_1]f)(x - \vartheta(s)p, p, t - s)\, dp, \quad k = 1,2,3,$$

$$\mathcal{N}(x) := (2\pi)^{-3/2} \exp\left(-\frac{|x|^2}{2}\right),$$

$$\vartheta(t) := \frac{1 - e^{-\beta t}}{\beta}; \qquad \vartheta(t) := t, \quad \text{if } \beta = 0,$$

$$R(t) := 2\alpha t + \sigma\left(\frac{4e^{-\beta t} - e^{-2\beta t} + 2\beta t - 3}{\beta^3}\right) + 4\gamma\left(\frac{e^{-\beta t} + \beta t - 1}{\beta^2}\right),$$

and the (vector valued) pseudo-differential operator $\Gamma[E]$ is related to $\theta[V]$ by

$$\theta[V]f(x,p) = \operatorname{div}_p (\Gamma[\nabla_x V]f)(x,p).$$

Notice that (3.18) is a closed equation (more precisely a linear Volterra integral equation of the second kind) for the self-consistent electric field $E_1 \in \mathbb{R}^3$, for any given Wigner trajectory $f \in \mathcal{C}([0,T]; L^2(\mathbb{R}^6))$.

These motivations lead to our new definition of the Hartree-potential:
Definition 3.1 (New definition of mean-field quantities) *To a Wigner trajectory $f \in \mathcal{C}([0,T]; L^2(\mathbb{R}^6))$ we associate*

- *the field $E[f] := E_0 + E_1[f]$, with E_0 given by (3.17), and $E_1[f]$ the unique solution of (3.18),*
- *the potential $V[f] := V_0 + V_1[f]$ with*

$$V_0(x,t) := \frac{1}{4\pi} \sum_{i=1}^{3} \frac{x_i}{|x|^3} *_x (E_0)_i(x,t), \tag{3.19}$$

$$V_1[f](x,t) := \frac{1}{4\pi} \sum_{i=1}^{3} \frac{x_i}{|x|^3} *_x (E_1[f])_i(x,t), \tag{3.20}$$

- *and the position density $n[f] := \operatorname{div} E[f]$ (at least in a distributional sense).*

In contrast to the standard definitions (3.3), (3.4), these new definitions are *non-local in time*. Also, the map $f \mapsto V[f]$ is now *non-linear*. For a given Wigner trajectory these two definitions clearly differ in general. However, they coincide if f is the solution of the WPFP system. These new definitions of the self-consistent field and potential have the advantage that they only require $f \in \mathcal{C}([0,T]; L^2(\mathbb{R}^6))$ and not $f(x,.,t) \in L^1(\mathbb{R}^3)$. If $f(t = 0)$ only lies in $L^2(\mathbb{R}^6)$, the corresponding field and the potential will consequently only be defined for $t > 0$.

The equation (3.18) now easily yields the announced $\|E(t)\|_{L^2(\mathbb{R}^3)}$–estimate for $t \in (0,\infty)$ in terms of f_I and $\|f(t)\|_{L^2(\mathbb{R}^6)}$ only. The fixed-point map

$f \mapsto V[f] \mapsto \tilde{f}$ (where the last steps refers to solving the linear WFP equation (3.16) with given $V(t)$, $t \geq 0$) is now contractive in $\mathcal{C}([0,T]; L^2(\mathbb{R}^6))$ and it yields the global mild solution for the WPFP system (3.14).

Without going into details we briefly list alternative kinetic approaches for the WPFP system that were developed in the last few years: In [ADM04] p-weighted L^2-spaces were used to make the definition of the particle density by $n = \int f\,dp$ meaningful. In [AL*04, CLN04], instead, an L^1-setting is chosen with the same motivation. The pseudo-differential operator $\theta[V]$ is rewritten there as a convolution operator in the p-variable. We remark that such kinetic strategies are valuable as they can, possibly, be extended to the WPFP boundary value problems used for semiconductor device modeling.

The quantum Fokker-Planck system for density matrices. Using the Wigner transforms we first rewrite the WPFP system (3.14) for the integral kernel $\rho(x, x', t)$ from (3.7):

$$\begin{aligned} \rho_t = &-iH_x\rho + iH_{x'}\rho - \frac{\beta}{2}(x - x')\cdot(\nabla_x - \nabla_{x'})\rho \\ &+ \alpha|\nabla_x + \nabla_{x'}|^2\rho - \sigma|x - x'|^2\rho + 2i\gamma(x - x')\cdot(\nabla_x + \nabla_{x'})\rho, \end{aligned} \tag{3.21}$$

coupled to the Poisson equation for V, where $H_{x'}$ is a copy of the Hamiltonian $H = H_x = -\frac{1}{2}\Delta_x + V(x,t)$, but acting on the x'–variable. The corresponding density matrix $\widehat{\rho}$ then satisfies the evolution equation (3.11), augmented with a r.h.s. $iA(\widehat{\rho})$ in Lindblad form (3.12) and coupled to the Poisson equation.

For the whole space case the density matrix formalism provides the most elegant analytic setup. Motivated by the kinetic energy $E_{kin}(\widehat{\rho})$ defined in (3.10) we define the "energy space"

$$\mathcal{E} := \{\widehat{\rho} \in \mathcal{I}_1 \mid \sqrt{1-\Delta}\,\widehat{\rho}\,\sqrt{1-\Delta} \in \mathcal{I}_1\}.$$

For physical quantum states (i.e. $\widehat{\rho} \geq 0$) we then have

$$\|\widehat{\rho}\|_{\mathcal{E}} = \operatorname{Tr}\widehat{\rho} + E_{kin}(\widehat{\rho}).$$

The simple estimate

$$\|n\|_{L^1(\mathbb{R}^3)} \leq \|\widehat{\rho}\|_{\mathcal{I}_1}$$

gives a rigorous meaning (in L^1) to the definition of the particle density (3.9), and the Lieb-Thirring-type estimate (cf. [Arn95a, LP93]) yields

$$\|n\|_{L^3(\mathbb{R}^3)} \leq C\|\widehat{\rho}\|_{\mathcal{E}}.$$

Therefore the nonlinearity $[V(x,t) - V(x',t)]\rho(x,x',t)$ in (3.21) is locally Lipschitz in $\mathcal{E}$. Since the linear part of (3.21) generates a mass conserving semigroup on $\mathcal{E}$, standard semigroup theory yields a unique local-in-time solution to (3.21). A-priori estimates on the mass $\operatorname{Tr}(\widehat{\rho})$ and the total energy $E_{tot}(\widehat{\rho}) = E_{kin}(\widehat{\rho}) + \frac{1}{2}\|\nabla_x V[\widehat{\rho}]\|_{L^2}^2$, due to the energy balance

$$\frac{d}{dt}E_{tot} = 3\sigma \operatorname{Tr} \widehat{\rho}_I - 2\beta E_{kin}(t) - \alpha \|n(t)\|_{L^2}^2,$$

then shows that there exists a unique global mild solution of (3.21) in $\mathcal{C}([0,\infty);\mathcal{E})$ (cf. [AS04]).

4 Macroscopic picture: fluid-type models

The aim of this section is to derive macroscopic quantum models from the following Wigner-Boltzmann equation for the distribution function $f(x,p,t)$:

$$\partial_t f + p \cdot \nabla_x f + \theta[V]f = Q(f), \quad f(x,p,0) = f_I(x,p), \quad (x,p) \in \mathbb{R}^{2d}, \ t > 0. \tag{4.1}$$

Here, (x,p) denotes the position-momentum variables of the phase space, $t > 0$ is the time, $d \geq 1$ the dimension, $Q(f)$ a collision operator, and $\theta[V]f$ is the pseudo-differential operator defined by (3.2). Notice that in the semi-classical limit $\varepsilon \to 0$, the term $\theta[V]f$ converges to $\nabla_x V \cdot \nabla_p f$ and thus, (4.1) reduces to the semi-classical Vlasov equation [MRS90]. The electric potential $V = V(x,t)$ is selfconsistently coupled to the Wigner function f via Poisson's equation

$$\lambda_L^2 \Delta V = \int_{\mathbb{R}^d} f dp - C, \tag{4.2}$$

where λ_L is the scaled Debye length and $C = C(x)$ the doping concentration characterizing the semiconductor device [Jün01].

In classical fluiddynamics, macroscopic models can be derived from the Boltzmann equation by using a moment method. The idea is to multiply the kinetic equation by some monomials $\kappa_i(p)$ and to integrate the equation over the momentum space. This yields the so-called *moment equations*. Usually, not all integrals can be expressed in terms of the moments (which is called the *closure problem*) and an additional procedure is necessary in order to close the equations. Depending on the number of moments which are taken, a variety of fluiddynamical models can be derived [BD96, Lev96]. The aim of this section is to mimic this procedure in the quantum case. Figure 4.1 shows the resulting models arising from special choices of the set of monomials. We will discuss these models in detail in the following subsections. For this, we need to specify the collision operator $Q(f)$ in (4.1). First we introduce in the following subsection the so-called quantum Maxwellian.

4.1 Definition of the quantum Maxwellian

In order to define the quantum Maxwellian, we use the Wigner transform $W(\widehat{\rho})$ of an integral operator $\widehat{\rho}$ on $L^2(\mathbb{R}^d)$ as defined in (3.7), (3.8). Its inverse W^{-1}, also called Weyl quantization, is defined as an operator on $L^2(\mathbb{R}^d)$:

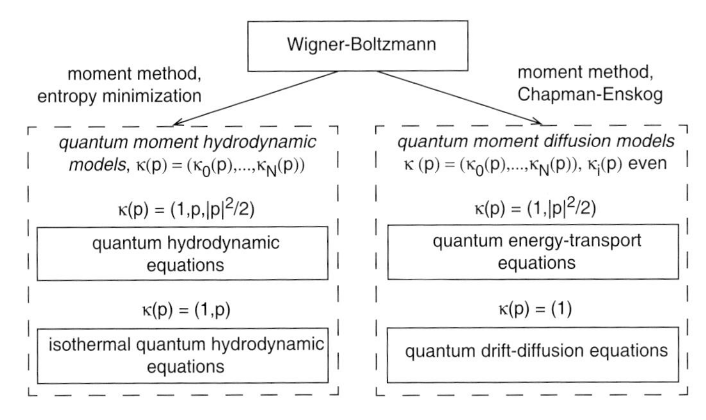

Fig. 4.1. Multiscale hierarchy of macroscopic quantum models.

$$(W^{-1}(f)\phi)(x) = \int_{\mathbb{R}^{2d}} f\Big(\frac{x+y}{2}\Big)\phi(y)e^{ip\cdot(x-y)/\varepsilon}dpdy \quad \text{for all } \phi \in L^2(\mathbb{R}^d).$$

With these definitions we are able to introduce as in [DR03] the *quantum exponential* and the *quantum logarithm* formally by $\operatorname{Exp} f = W(\exp W^{-1}(f))$ and $\operatorname{Log} f = W(\log W^{-1}(f))$, where exp and log are the operator exponential and logarithm, respectively.

Inspired by Levermore's moment method for the classical case [Lev96], Degond and Ringhofer [DR03] have defined the quantum Maxwellian by using the entropy minimization principle. Let a quantum mechanical state be described by the Wigner function f solving (4.1). Then its *relative quantum (von Neumann) entropy* is given by

$$H(f) = \int_{\mathbb{R}^{2d}} f(x,p,\cdot)\Big((\operatorname{Log} f)(x,p,\cdot) - 1 + \frac{|p|^2}{2} + V(x,\cdot)\Big)dxdp.$$

Whereas the classical entropy is a function on the configuration space, the above quantum entropy at given time is a real number, underlining the non-local nature of quantum mechanics.

We define the *quantum Maxwellian* M_f for some given function $f(x,p,t)$ as the solution of the constrained minimization problem

$$H(M_f) = \min\Big\{H(\hat{f}) : \int_{\mathbb{R}^d} \hat{f}(x,p,t)\kappa_i(p)dp = m_i(x,t) \text{ for all } x,\ t,\ i\Big\}, \quad (4.3)$$

where $\kappa_i(p)$ are some monomials in p and $m_i(x,t)$ are the *moments* of f,

$$m_i(x,t) = \langle f(x,p,t)\kappa_i(p)\rangle, \quad i = 0,\ldots,N, \quad (4.4)$$

where we have used the notation $\langle g(p)\rangle = \int g(p)dp$ for functions $g(p)$. The formal solution of this minimization problem (if it exists) is given by $M_f = \operatorname{Exp}(\widetilde{\lambda}\cdot\kappa - \frac{1}{2}|p|^2 - V(x,\cdot))$, where $\kappa = (\kappa_0,\ldots,\kappa_N)$, and $\widetilde{\lambda} = (\widetilde{\lambda}_0,\ldots,\widetilde{\lambda}_N)$ are some Lagrange multipliers. If $\kappa_0(p) = 1$ and $\kappa_2(p) = \frac{1}{2}|p|^2$, setting $\lambda_0 = \widetilde{\lambda}_0 - V$, $\lambda_2 = \widetilde{\lambda}_2 - 1$ and $\lambda_i = \widetilde{\lambda}_i$ otherwise, we can write

$$M_f = \operatorname{Exp}(\lambda \cdot \kappa(p)). \tag{4.5}$$

4.2 Quantum moment hydrodynamic models

In this section we will derive the quantum moment hydrodynamic equations from the Wigner-Boltzmann equation (4.1) with dominant elastic scattering employing a moment method. For a special choice of the moments, the quantum hydrodynamic equations are obtained. If a Fokker-Planck approach is taken for the inelastic collision operator, viscous corrections to the quantum hydrodynamic model are derived.

General quantum moment hydrodynamics. Introducing the hydrodynamic scaling $x' = \alpha x$, $t' = \alpha t$, where $0 < \alpha \ll 1$ measures the typical energy gain or loss during an electron-phonon collision, the Wigner-Boltzmann equation for (f_α, V_α) becomes (omitting the primes)

$$\alpha \partial_t f_\alpha + \alpha(p \cdot \nabla_x f_\alpha + \theta[V_\alpha] f_\alpha) = Q(f_\alpha), \quad (x, p, t) \in \mathbb{R}^{2d} \times (0, \infty), \tag{4.6}$$

together with an initial condition for f_α and the Poisson equation (4.2) for V_α. The collision operator is assumed to split into two parts:

$$Q(f_\alpha) = Q_0(f_\alpha) + \alpha Q_1(f_\alpha),$$

where the first (dominant) part models elastic collisions and the second part models inelastic scattering processes. The operator Q_0 is supposed to satisfy the following properties:

$$\text{(i) If } Q_0(f) = 0 \text{ then } f = M_f, \quad \text{(ii) } \langle Q_0(f) \kappa(p) \rangle = 0, \tag{4.7}$$

where M_f is the quantum Maxwellian introduced in section 4.1 and $\kappa(p)$ is a vector of some monomials $\kappa_i(p)$. If $\kappa(p) = (1, p, |p|^2/2)$, condition (ii) expresses the conservation of mass, momentum, and energy, which is meaningful for elastic collisions. An example fulfilling conditions (i) and (ii) is the BGK-type operator [BGK54] $Q_0(f) = (M_f - f)/\tau$ with the relaxation time $\tau > 0$. An example of an inelastic collision operator Q_1 will be given in the subsection "Viscous quantum hydrodynamic equations" below.

In the following we proceed similarly as in [DR03]. The *moment equations* are obtained by multiplying (4.6) by $\kappa(p)/\alpha$, integrating over the momentum space, and using condition (ii):

$$\partial_t \langle \kappa(p) f_\alpha \rangle + \operatorname{div}_x \langle \kappa(p) p f_\alpha \rangle + \langle \kappa(p) \theta[V_\alpha] f_\alpha \rangle = \langle \kappa(p) Q_1(f_\alpha) \rangle.$$

The second integral on the left-hand side of the moment equations cannot be expressed in terms of the moments; this is called the *closure problem*. We can solve this problem by letting $\alpha \to 0$. Indeed, the formal limit $\alpha \to 0$ in (4.6) gives $Q_0(f) = 0$ where $f = \lim_{\alpha\to 0} f_\alpha$. Hence, by condition (i), $f = M_f$. Then the formal limit $\alpha \to 0$ in the above moment equations yields

$$\partial_t m + \mathrm{div}_x \langle \kappa(p) p M_f \rangle + \langle \kappa(p)\theta[V] M_f \rangle = \langle \kappa(p) Q_1(M_f) \rangle, \tag{4.8}$$

where $m = \langle \kappa(p) M_f \rangle$ are the moments (see (4.4)) and $V = \lim_{\alpha\to 0} V_\alpha$ solves (4.2) with $f = M_f$. The above equations have to be solved for $x \in \mathbb{R}^d$ and $t > 0$, and the initial condition becomes $m(\cdot, 0) = \langle \kappa(p) M_{f_I} \rangle$. In the classical case, Levermore [Lev96] has shown that the moment equations are symmetrizable and hyperbolic. In the present situation, this concept of hyperbolicity cannot be used since (4.8) is not a partial differential equation but a differential equation with non-local operators of the type $\lambda \mapsto \langle \mathrm{Exp}\,(\lambda \cdot \kappa(p)) \rangle$.

The system (4.8) possesses the following (formal) property: If 1 and $\frac{1}{2}|p|^2$ are included in the set of monomials and if the inelastic collision operator conserves mass and dissipates energy, i.e. $\langle Q_1(f) \rangle = 0$ and $\langle \frac{1}{2}|p|^2 Q_1(f) \rangle \le 0$ for all functions f, the total energy

$$E(t) = \int_{\mathbb{R}^{2d}} \left(\left\langle \frac{1}{2}|p|^2 M_f \right\rangle + \frac{\lambda_L^2}{2} |\nabla_x V|^2 \right) dx dp$$

is nonincreasing. To see this, we notice that for all (regular) functions f,

$$\langle \theta[V] f \rangle = 0, \quad \langle p\theta[V] f \rangle = -\langle f \rangle \nabla_x V, \quad \langle \tfrac{1}{2}|p|^2 \theta[V] f \rangle = -\langle p f \rangle \cdot \nabla_x V. \tag{4.9}$$

From the moment equations

$$\partial_t \langle M_f \rangle + \mathrm{div}_x \langle p M_f \rangle = 0, \quad \partial_t \langle \tfrac{1}{2}|p|^2 M_f \rangle + \mathrm{div}_x \langle \tfrac{1}{2}|p|^2 p M_f \rangle \le \langle p M_f \rangle \cdot \nabla_x V$$

and the Poisson equation (4.2) we obtain formally

$$\begin{aligned} \frac{dE}{dt} &\le \int_{\mathbb{R}^d} (\langle p M_f \rangle \cdot \nabla_x V + \lambda_L^2 \nabla_x V \cdot \partial_t \nabla_x V) dx \\ &= \int_{\mathbb{R}^d} (\langle p M_f \rangle \cdot \nabla_x V - V \partial_t \langle M_f \rangle) dx \\ &= \int_{\mathbb{R}^d} (\langle p M_f \rangle \cdot \nabla_x V + V \mathrm{div}_x \langle p M_f \rangle) dx = 0, \end{aligned}$$

proving the monotonicity of the total energy.

In the following section we will specify the choice of the monomials, which enables us to give a more explicit expression of the system (4.8).

Quantum hydrodynamic equations. In classical fluiddynamics, the Euler equations are derived from the Boltzmann equation by using the monomials $\kappa(p) = (1, p, |p|^2/2)$ in the moment equations. In this subsection, we derive the quantum counterpart, the so-called *quantum hydrodynamic (QHD) equations* (see [DR03, JMM05]).

Let $\kappa(p) = (1, p, |p|^2/2)$. The moments $n := m_0$, $nu := m_1$, and $ne := m_2$ are called the *particle, current, and energy densities*, respectively. We also define the *velocity* $u = nu/n$ and the *energy* $e = ne/n$. In this situation, the quantum Maxwellian can be written as $M_f = \mathrm{Exp}\,(\lambda_0 + \lambda_1 \cdot p + \lambda_2|p|^2)$ or, equivalently, as

$$M_f(x,t) = \mathrm{Exp}\Big(A(x,t) - \frac{|p - w(x,t)|^2}{2T(x,t)}\Big), \tag{4.10}$$

where A, w, and T are defined in terms of λ_0, λ_1, and λ_2. In the following we will give a more explicit expression for the quantum moment equations (4.8).

Using (4.9) and observing that the second and third moments can be written as

$$\langle p \otimes pM_f\rangle = P + nu \otimes u, \quad \text{where } P = \langle (p-u)\otimes(p-u)M_f\rangle,$$
$$\langle \tfrac{1}{2}p|p|^2 M_f\rangle = S + (P + neI)u, \quad \text{where } S = \langle \tfrac{1}{2}(p-u)|p-u|^2 M_f\rangle,$$

the quantum moment equations (4.8) become

$$\partial_t n + \mathrm{div}(nu) = \langle Q_1(M_f)\rangle, \tag{4.11}$$
$$\partial_t(nu) + \mathrm{div}(nu\otimes u) + \mathrm{div}\,P - n\nabla V = \langle pQ_1(M_f)\rangle, \tag{4.12}$$
$$\partial_t(ne) + \mathrm{div}\big((P+neI)u\big) + \mathrm{div}\,S - nu\cdot\nabla V = \langle \tfrac{1}{2}|p|^2 Q_1(M_f)\rangle, \tag{4.13}$$

where $u \otimes u$ denotes the matrix with components $u_j u_k$, P is the stress tensor, S the (quantum) heat flux, and I is the identity matrix in $\mathbb{R}^{d\times d}$. The electric potential is given by (4.2) with $f = M_f$ or, in the above notation, by

$$\lambda_L^2 \Delta V = n - C(x). \tag{4.14}$$

The above system, which is solved for $x \in \mathbb{R}^d$ and $t > 0$ with initial conditions for $n(\cdot, 0)$, $nu(\cdot, 0)$, and $ne(\cdot, 0)$, is called the *quantum hydrodynamic equations.* The quantum correction only appears in the terms P and S. We can derive an explicit expression in the $O(\varepsilon^4)$ approximation. For this, we need to expand the quantum Maxwellian M_f in terms of ε^2. As the computations are quite involved, we only sketch the expansion and refer to [JMM05] for details.

The quantum exponential can be expanded in terms of ε^2 yielding $\mathrm{Exp}\,f = e^f - (\varepsilon^2/8)e^f B + O(\varepsilon^4)$, where B is a polynomial in the derivatives of f up to second order. This allows for an expansion of the moments

$$(n, nu, ne) = \int_{\mathbb{R}^d} \mathrm{Exp}\Big(A - \frac{|p-w|^2}{2T}\Big)\Big(1, p, \frac{1}{2}|p|^2\Big)dp,$$

of the stress tensor

$$P = nTI + \frac{\varepsilon^2}{12}n\Big\{\Big(\frac{d}{2}+1\Big)\nabla\log T\otimes\nabla\log T - \nabla\log T\otimes\nabla\log n$$
$$- \nabla\log n\otimes\nabla\log T - (\nabla\otimes\nabla)\log(nT^2) + \frac{R^\top R}{T}\Big\} \tag{4.15}$$
$$+ \frac{\varepsilon^2}{12}T\mathrm{div}\Big(n\frac{\nabla\log T}{T}\Big)I + O(\varepsilon^4), \tag{4.16}$$

and of the quantum heat flux

$$S = -\frac{\varepsilon^2}{12} n\Big\{\Big(\frac{d}{2}+1\Big) R\nabla \log\Big(\frac{n}{T}\Big) + \Big(\frac{d}{2}+2\Big)\operatorname{div} R + \frac{3}{2}\Delta u\Big\}$$
$$+\frac{\varepsilon^2}{12}\left(\frac{d}{2}+1\right) n\Big\{R\nabla \log\Big(\frac{n}{T^2}\Big) + \operatorname{div} R\Big\} + O(\varepsilon^4), \qquad (4.17)$$

where the matrix R with components $R_{ij} = \partial u_i/\partial x_j - \partial u_j/\partial x_i$ is the anti-symmetric part of the velocity gradient and $R^\top$ is the transpose of R. In the semi-classical case $\varepsilon = 0$ the stress tensor reduces to the classical expression $P = nTI$. The term S is purely quantum and vanishes if $\varepsilon = 0$. The energy density is the sum of the thermal, kinetic, and quantum energy,

$$ne = \frac{d}{2}nT + \frac{1}{2}n|u|^2 - \frac{\varepsilon^2}{24}n\Big\{\Delta \log n - \frac{1}{T}\operatorname{tr}(R^\top R) + \frac{d}{2}|\nabla \log T|^2 - \Delta \log T$$
$$- \nabla \log T \cdot \nabla \log n\Big\} + O(\varepsilon^4), \qquad (4.18)$$

where "tr" denotes the trace of a matrix.

A simplified quantum hydrodynamic model up to order $O(\varepsilon^4)$ can be obtained under the assumptions that the inelastic collision part vanishes, $Q_1 = 0$, that the temperature is slowly varying, $\nabla \log T = O(\varepsilon^2)$, and finally, that the vorticity is small, $R = O(\varepsilon^2)$:

$$\partial_t n + \operatorname{div}(nu) = 0, \qquad (4.19)$$

$$\partial_t(nu) + \operatorname{div}(nu \otimes u) + \nabla(nT) - \frac{\varepsilon^2}{12}\operatorname{div}\big(n(\nabla \otimes \nabla)\log n\big) - n\nabla V = 0, \qquad (4.20)$$

$$\partial_t(ne) + \operatorname{div}\big((P + neI)u\big) - \frac{\varepsilon^2}{8}\operatorname{div}(n\Delta u) - nu \cdot \nabla V = 0, \qquad (4.21)$$

with the stress tensor and energy density, respectively,

$$P = nTI - \frac{\varepsilon^2}{12}n(\nabla \otimes \nabla)\log n, \quad ne = \frac{d}{2}nT + \frac{1}{2}n|u|^2 - \frac{\varepsilon^2}{24}n\Delta \log n. \qquad (4.22)$$

We notice that if we choose $\kappa(p) = (1, p)$, we obtain the *isothermal quantum hydrodynamic equations* (4.19)-(4.20) with constant temperature $T = 1$.

The system (4.19)-(4.21) corresponds to Gardner's QHD model except for the dispersive velocity term $(\varepsilon^2/8)\operatorname{div}(n\Delta u)$. The differences between our QHD equations and Gardner's model can be understood as follows. In both approaches, closure is obtained by assuming that the Wigner function f is in equilibrium. However, the notion of "equilibrium" is different. A quantum system, which is characterized by its energy operator $W^{-1}(h)$, with the Weyl quantization W^{-1} and the Hamiltonian $h(p) = |p|^2/2 + V(x)$, attains its minimum of the relative (von Neumann) entropy in the mixed state with Wigner function $f_Q = \operatorname{Exp}(-h/T_0)$. This state represents the *unconstrained* quantum equilibrium. The expansion of f_Q in terms of ε^2 was first given in [Wig32],

$$f_Q(x,p) = \exp(-h(x,p)/T_0)(1+\varepsilon^2 f_2(x,p)) + O(\varepsilon^4)$$

with an appropriate function f_2. As a definition of the quantum equilibrium *with* moment constraints, Gardner employed this expansion of f_Q and modified it mimicking the moment-shift of the Gibbs state in the classical situation:

$$\widetilde{f}_Q(x,p) = n(x)\,\exp\Big(-\frac{h(x,p-u(x))}{T(x)}\Big)\big(1+\varepsilon^2 f_2(x,p-u(x))\big) + O(\varepsilon^4).$$

In contrast to the classical case, $\widetilde{f}_Q$ is *not* the constrained minimizer for the relative von Neumann entropy. On the other hand, the equilibrium state M_f used here is a genuine minimizer of the relative entropy with respect to the given moments. It has been shown in [JMM05] that both approaches coincide if the temperature is constant and if only the particle density is prescribed as a constraint.

Equations (4.19)-(4.22) are of hyperbolic-dispersive type, and the presence of the nonlinear third-order differential operators in (4.20) and (4.21) makes the analysis of the system quite difficult. In particular, it is not clear if the electron density stays positive if it is positive initially. Since the total mass $\int n dx$ and the total energy,

$$\begin{aligned} E(t) = \int_{\mathbb{R}^d} \Big(& \frac{d}{2}nT + \frac{1}{2}n|u|^2 + \frac{\lambda^2}{2}|\nabla V|^2 + \frac{\varepsilon^2}{6}|\nabla\sqrt{n}|^2 + \frac{\varepsilon^2 d}{48}n|\nabla \log T|^2 \\ & + \frac{\varepsilon^2}{24T}n\,\mathrm{tr}(R^\top R)\Big)dx, \end{aligned} \tag{4.23}$$

are conserved quantities of the quantum moment equations (4.11)-(4.18) (if $Q_1 = 0$) [JMM05], this provides some Sobolev estimates. However, this estimate seems to be not strong enough to prove the existence of weak solutions to the system. Indeed, for a special model, a nonexistence result of weak solutions to the QHD equations has been proved [GaJ01]. This result is valid for the one-dimensional isothermal stationary equations, solved in a bounded interval with Dirichlet boundary conditions for the electron density and boundary conditions for the electric potential, the electric field, and the quantum Bohm potential at the left interval point. Moreover, the term Tn_x in (4.20) has been replaced by a more general pressure function $p(n)$ satisfying a growth condition.

The nonexistence result is valid for sufficiently large current densities. On the other hand, for "small" current densities fulfilling a subsonic condition related to classical fluiddynamics, some existence results for the stationary and transient equations have been achieved [HLMO05, Jün98, JL04, JMR02].

The QHD equations contain two parameters: the (scaled) Planck constant ε and the Debye length λ_L. In special regimes of the physical parameters, these constants may be small compared to one, such that the semi-classical limit $\varepsilon \to 0$ or the quasi-neutral limit $\lambda_L \to 0$ may be of interest, leading to simpler models. In fact, the QHD equations reduce in the semi-classical limit

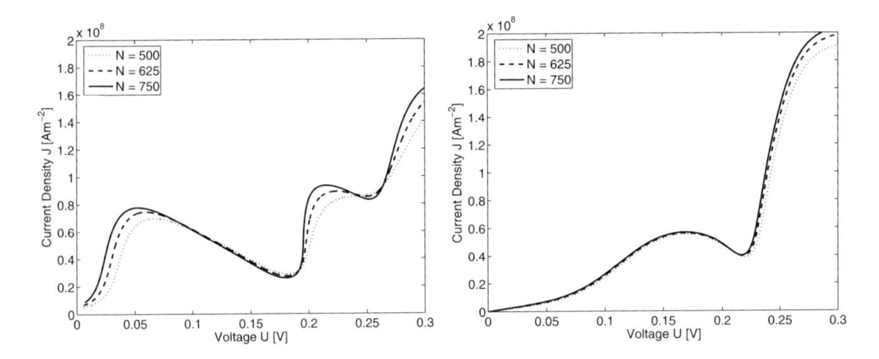

Fig. 4.2. Influence of the number of discretization points N on the current-voltage characteristics for Gardner's QHD equations (left) and for the new QHD model (right).

to the Euler equations. This limit has been proved in the one-dimensional isothermal steady state for sufficiently small current densities in [GyJ00] and for arbitrary large current densities (after adding an ultra-diffusive term in (4.20)) in [GaJ02]. The quasi-neutral limit in the isentropic QHD model has been performed in [LL05], showing that the current density consists, for small Debye length, of a divergence-free vector field connected with the incompressible Euler equations and a highly oscillating gradient vector coming from high electric fields.

The nonisothermal QHD equations have been first solved numerically by Gardner using a finite-difference upwind method, considering the third-order term as a perturbation of the classical Euler equations [Gar94]. However, hyperbolic schemes have the disadvantage that the numerical diffusion may influence the numerical solution considerably [JT06]. This can be seen in Fig. 4.2 (left). The figure shows the current-voltage characteristics of a one-dimensional resonant tunneling diode, computed from the QHD equations (4.19)-(4.21) without the dispersive velocity term but including heat conductivity and relaxation-time terms of Baccarani-Wordeman type. The tunneling diode consists of three regions: the high-doped contact regions and a low-doped channel region. In the channel, a double-potential barrier is included (see [JMM05] for the physical and numerical details).

Due to the numerical viscosity introduced by the upwind method, the solution of Gardner's model strongly depends on the mesh size. On the other hand, the solution to the new QHD equations (4.19)-(4.21) presented in [JMM05] is much less mesh depending (see Fig. 4.2 right).

Notice that the main physical effect of a tunneling diode is that there exists a region in which the current density is decreasing although the voltage is increasing. This effect is called *negative differential resistance* and it is employed, for instance, to devise high-frequency oscillator devices.

The effect of the dispersive velocity term is a "smoothing" of the current-voltage curve. In order to study the influence of this term, we replace the factor $\varepsilon^2/8$ in (4.21) by $\delta^2/8$ and choose various values for δ. Clearly, only $\delta = \varepsilon$ corresponds to the physical situation. Figure 4.3 shows that the characteristics become "smoother" for larger values of δ.

Viscous quantum hydrodynamic equations. We model the inelastic collisions as electron interactions with a heat bath of oscillators in thermal equilibrium (which models the semiconductor crystal). Castella et al. [CE*00] derived for such a situation the collision operator

$$Q_1(w) = \nu \Delta_x w + \nu_1 \Delta_p f + \nu_2 \mathrm{div}_x(\nabla_p f) + \frac{1}{\tau}\mathrm{div}_p(pf).$$

The parameters ν, ν_1, $\nu_2 \geq 0$ constitute the phase-space diffusion matrix, and $\tau > 0$ is a friction parameter, the relaxation time. If $\nu = 0$ and $\nu_2 = 0$, this gives the Caldeira-Leggett operator [CL83]. This model allows to incorporate inelastic scattering in the quantum hydrodynamic equations. Indeed, using the definition of the moments, we compute

$$\langle Q_1(M_f)\rangle = \nu \Delta_x n, \quad \langle p Q_1(M_f)\rangle = \nu \Delta_x(nu) - \nu_2 \nabla_x n - \frac{nu}{\tau},$$
$$\langle \tfrac{1}{2}|p|^2 Q_1(M_f)\rangle = \nu \Delta_x(ne) + d\nu_1 n - \nu_2 \mathrm{div}_x(nu) - \frac{2ne}{\tau}.$$

For simplicity, we suppose in the following that $\nu_1 = \nu_2 = 1/\tau = 0$. Assuming as in the previous section that the temperature gradients and the vorticity are of order ε^2, we obtain the *viscous quantum hydrodynamic equations*:

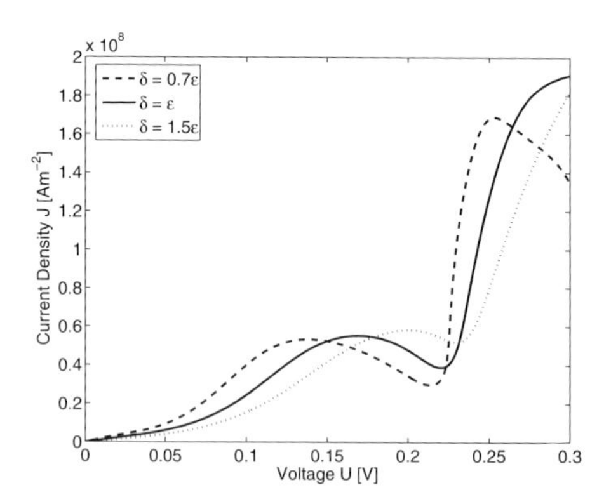

Fig. 4.3. Influence of the dispersive velocity term $(\delta^2/8)(nu_{xx})_x$ on the current-voltage curve.

$$\partial_t n + \operatorname{div}(nu) = \nu \Delta n,$$
$$\partial_t(nu) + \operatorname{div}(nu \otimes u) + \nabla(nT) - \frac{\varepsilon^2}{12}\operatorname{div}\big(n(\nabla \otimes \nabla)\log n\big) - n\nabla V = \nu \Delta(nu),$$
$$\partial_t(ne) + \operatorname{div}\big((P + neI)u\big) - \frac{\varepsilon^2}{8}\operatorname{div}(n\Delta u) - nu \cdot \nabla V = \nu \Delta(ne),$$

where P and ne are defined in (4.22), and V is given by (4.14). Notice that ν is of the same order as ε^2 [GuJ04].

Due to the dissipative terms on the right-hand side of the above system, the total energy

$$E(t) = \int_{\mathbb{R}^d} \Big(\frac{d}{2}nT + \frac{1}{2}n|u|^2 + \frac{\lambda^2}{2}|\nabla V|^2 + \frac{\varepsilon^2}{6}|\nabla\sqrt{n}|^2\Big)dx$$

is no longer conserved but at least bounded:

$$\frac{dE}{dt} + \frac{\nu}{\lambda_L^2}\int_{\mathbb{R}^d} n(n - C)dx = 0.$$

However, it is not clear how to prove the existence of weak solutions or the positivity of the particle density from this equation.

A partial existence result, for sufficiently small current densities in the isothermal stationary model, is presented in [GuJ04]. The main idea is the observation that, in the one-dimensional steady state, we can integrate (4.24) yielding $nu - \nu n_x = J_0$ for some integration constant J_0 which we call the effective current density (since it satisfies $(J_0)_x = 0$). A computation now shows that

$$\Big(\frac{(nu)^2}{n}\Big)_x - \nu(nu)_{xx} = -\nu^2 n\big(n(\log n)_{xx}\big)_x + \Big(\frac{J_0^2}{n}\Big)_x + 2\nu J_0(\log n)_{xx}.$$

Hence, the coefficient of the quantum term becomes $\varepsilon^2/12+\nu^2$, and the viscosity term transforms to $2\nu J_0(\log n)_{xx}$. The smallness condition on the current density is needed in order to control the convective part $(J_0^2/n)_x$. Also in [GuJ04], the inviscid limit $\nu \to 0$ and the semi-classical limit $\varepsilon \to 0$ have been performed.

The isothermal viscous model has been numerically solved in [JMi06, JT06]. The viscosity ν has the effect to "smoothen" the current-voltage characteristics for a tunneling diode, as can be seen from Fig. 4.4 (left). We refer to [JMi06] for details of the employed parameters. The curves are computed from the *isothermal* model. Their behavior is unphysical due to the jump from a low-current to a high-current state. This effect can be explained by the constant temperature assumption. Indeed, in Fig. 4.4 (right) a curve computed from the *nonisothermal* equations is presented. The characteristic shows the correct physical behavior but the viscosity leads to rather small peak-to-valley ratios (ratio of maximal to minimal current density).

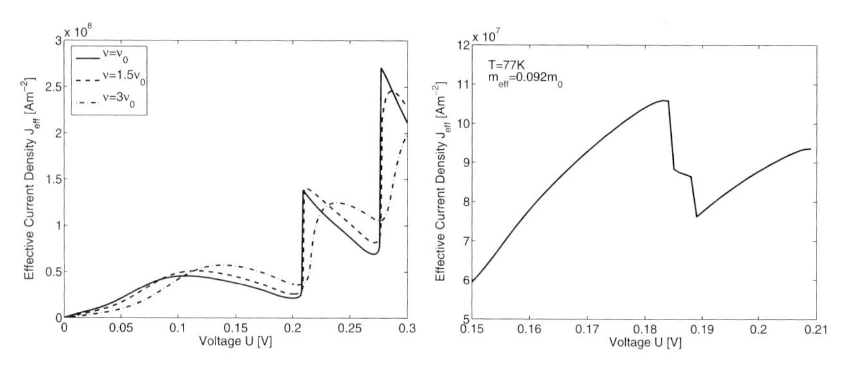

Fig. 4.4. Current voltage characteristics for a tunneling diode for the viscous QHD model. Left: isothermal case for various values of the viscosity; right: non-isothermal model.

4.3 Quantum moment diffusion models

In this section we derive quantum moment diffusion equations from a BGK-type Wigner equation using a Chapman-Enskog method. For special choices of the moments, the quantum energy-transport and the quantum drift-diffusion equations in the $O(\varepsilon^4)$ approximation are obtained.

General quantum moment diffusion equations. We consider the Wigner-Boltzmann equation (4.1) in the diffusion scaling $x' = \alpha x$, $t' = \alpha^2 t$, where $0 < \alpha \ll 1$ is as in the previous section (neglecting the primes):

$$\alpha^2 \partial_t f_\alpha + \alpha(p \cdot \nabla_x f_\alpha + \theta[V_\alpha] f_\alpha) = Q(f_\alpha), \quad (x, p, t) \in \mathbb{R}^{2d} \times (0, \infty). \quad (4.24)$$

Our aim is to perform a Chapman-Enskog expansion in the corresponding moment equations and to perform the formal limit $\alpha \to 0$. For this, we proceed similarly as in [DMR05] using only monomials of even order, for instance $\kappa(p) = (1, \frac{1}{2}|p|^2, \ldots)$. Then the quantum Maxwellian M_f is the formal solution of the constrained minimization problem (4.3) with given moments $m_0(x,t), \ldots, m_N(x,t)$ with respect to the above set of monomials.

We assume that the collision operator can be written as

$$Q(f_\alpha) = Q_0(f_\alpha) + \alpha^2 Q_1(f_\alpha),$$

with $Q_0(f_\alpha)$ modeling the elastic scattering and $Q_1(f_\alpha)$ the inelastic scattering processes. In contrast to the previous section, we assume here that the elastic collisions are modeled by a BGK-type operator [BGK54],

$$Q_0(f) = \frac{1}{\tau}(M_f - f),$$

where $\tau = \tau(x,t) > 0$ is the relaxation time. Then $Q_0(f)$ satisfies the properties (4.7). Concerning inelastic scattering, we suppose only that it preserves the mass, i.e. $\langle Q_1(f) \rangle = 0$ for all functions f.

Multiplying (4.24) by $\kappa(p)/\alpha$, integrating over the momentum space, and using condition (ii) in (4.7), we obtain the moment equations

$$\partial_t\langle\kappa(p)f_\alpha\rangle + \alpha^{-1}(\mathrm{div}_x\langle\kappa(p)pf_\alpha\rangle + \langle\kappa(p)\theta[V]f_\alpha\rangle) = \langle\kappa(p)Q_1(f_\alpha)\rangle.$$

In order to derive the diffusion models, we employ the Chapman-Enskog expansion $f_\alpha = M_{f_\alpha} + \alpha f_\alpha^1$, which defines f_α^1. The formal limit $\alpha \to 0$ in (4.24) gives $Q_0(f) = 0$, where $f = \lim_{\alpha\to 0} f_\alpha$ and hence $f = M_f$, by condition (i) in (4.7). Inserting the Chapman-Enkog expansion in the above moment equations, observing that the integrals $\langle\kappa(p)pM_{f_\alpha}\rangle$ and $\langle\kappa(p)\theta[V_\alpha]M_{f_\alpha}\rangle$ vanish, since $\kappa_i(p)$ is even in p, and performing the limit $\alpha \to 0$, we conclude that

$$\partial_t\langle\kappa(p)M_f\rangle + \mathrm{div}_x\langle\kappa(p)pf^1\rangle + \langle\kappa(p)\theta[V]f^1\rangle = \langle\kappa(p)Q_1(M_f)\rangle, \qquad (4.25)$$

where $f^1 = \lim_{\alpha\to 0} f_\alpha^1$. It remains to determine the limit f^1. Since Q_0 is a BGK-type operator, it holds, using (4.24),

$$f_\alpha^1 = -\frac{\tau}{\alpha}Q_0(f_\alpha) = -\tau\big(\alpha\partial_t f_\alpha + p\cdot\nabla_x f_\alpha + \theta[V_\alpha]f_\alpha - \alpha Q_1(f_\alpha)\big),$$

which implies in the limit $\alpha \to 0$ that $f^1 = -\tau(p\cdot\nabla_x M_f + \theta[V]M_f)$. Inserting this expression for f^1 into (4.25) we obtain the general quantum diffusion equations

$$\begin{aligned}\partial_t m - \mathrm{div}\big(\tau\mathrm{div}\langle p\otimes p\kappa(p)M_f\rangle + \tau\langle\kappa(p)p\theta[V]M_f\rangle\big) + \langle\kappa(p)\theta[V]f^1\rangle \\ = \langle\kappa(p)Q_1(M_f)\rangle,\end{aligned} \qquad (4.26)$$

where we recall that $m = \langle\kappa(p)M_f\rangle$. With the notation (4.5) we see that the expression

$$\begin{aligned}\mathrm{div}(\tau\mathrm{div}\langle p\otimes p\kappa_i(p)M_f\rangle) &= \sum_{j,k,\ell}\frac{\partial}{\partial x_j}\Big(\tau\langle p_jp_k\kappa_i\kappa_\ell\mathrm{Exp}\,(\lambda\cdot\kappa)\rangle\frac{\partial\lambda_\ell}{\partial x_k}\Big) \\ &=: \mathrm{div}(B:\nabla\lambda)\end{aligned}$$

can be interpreted as a diffusion term, and (4.26) can be formulated in a compact form as

$$A\partial_t\lambda - \mathrm{div}(B:\nabla\lambda) = g(\lambda),$$

where $A = \langle\kappa\otimes\kappa M_f\rangle$ and $g(\lambda)$ denotes the lower-order terms in λ. A more explicit expression can be derived in the cases $N = 1$ and $N = 0$ which will be discussed in the following subsections.

Quantum energy-transport equations. Let $N = 1$ and $\kappa(p) = (1, \frac{1}{2}|p|^2)$. For simplicity, we also assume that the relaxation time is constant, $\tau = 1$. Then we can simplify the quantum diffusion equations of the previous subsection. Indeed, employing the formulas (4.9) and

$$\langle\tfrac{1}{2}p|p|^2\theta[V]f\rangle = -(P + neI)\nabla V + \tfrac{\varepsilon^2}{8}n\nabla\Delta V \quad \text{for all functions } f,$$

we obtain from (4.26) the evolution equations for the particle density $m_0 = n$ and the energy density $m_2 = ne$ (see [DMR05]):

$$\partial_t n - \operatorname{div} J_0 = 0, \quad \partial_t(ne) - \operatorname{div} J_2 - J_0 \cdot \nabla V = \langle \tfrac{1}{2}|p|^2 Q_1(f)\rangle, \tag{4.27}$$

$$J_0 = \operatorname{div} P - n\nabla V, \quad J_2 = \operatorname{div} U - (P + neI)\nabla V + \frac{\varepsilon^2}{8} n\nabla\Delta V, \tag{4.28}$$

where $P = \langle p \otimes p M_f\rangle$ is the stress tensor, $U = \langle \frac{1}{2}|p|^2 p \otimes p M_f\rangle$ is a fourth-order moment, and V is given by (4.14). The variables J_0 and J_2 are the particle and energy current densities, respectively. Noticing that the quantum Maxwellian can be written here as

$$M_f(x,t) = \operatorname{Exp}\left(A(x,t) - \frac{|p|^2}{2T(x,t)}\right),$$

one can show that the quantum fluid entropy

$$\eta(t) = \int_{\mathbb{R}^d} M_f(\operatorname{Log} M_f - 1)dxdp = \int_{\mathbb{R}^d} n(A - ne/T + 1)dx$$

is nonincreasing [DMR05].

More explicit equations are obtained in the $O(\varepsilon^4)$ approximation. For this, we need to expand the terms P, U and the energy $ne = \int \frac{1}{2}|p|^2 \operatorname{Exp}(A - |p|^2/2T)dp$ in terms of ε^2. If $\nabla \log T = O(\varepsilon^2)$ and up to order $O(\varepsilon^4)$, some tedious computations lead to the expressions

$$P = nTI - \frac{\varepsilon^2}{12} n(\nabla \otimes \nabla)\log n, \quad ne = \frac{d}{2} nT - \frac{\varepsilon^2}{24} n\Delta \log n, \tag{4.29}$$

$$U = \frac{1}{2}(d+2)nT^2 I - \frac{\varepsilon^2}{24} nT(\Delta \log nI + (d+4)(\nabla \otimes \nabla)\log n),$$

Equations (4.27)-(4.28) with the above constitutive relations for P, U, and ne are called the *quantum energy-transport equations*. Notice that the expressions of P and U differ from those presented in [DMR05]. We expect that this $O(\varepsilon^4)$ model possesses an entropic formulation similar to the classical energy-transport equations [DGJ97] but unfortunately, no entropic structure is currently known.

The quantum drift-diffusion equations. In this subsection we set $N = 0$ and choose $\kappa_0(p) = 1$. Then the quantum Maxwellian reads as $M_f(x,t) = \operatorname{Exp}(A(x,t) - |p|^2/2)$, and similar as in the previous subsection, we obtain

$$\partial_t n - \operatorname{div} J = 0, \quad J = \operatorname{div} P - n\nabla V,$$

where n and P are defined by

$$n = \int_{\mathbb{R}^d} \operatorname{Exp}\left(A - \frac{|p|^2}{2}\right)dp, \quad P = \int_{\mathbb{R}^d} p \otimes p \operatorname{Exp}\left(A - \frac{|p|^2}{2}\right)dp.$$

Again, the electric potential V is given selfconsistently by (4.14). Some analytical properties and numerical results for this nonlocal equation can be found in [GaM06]. In the $O(\varepsilon^4)$ approximation, we can simplify the above model. Indeed, for $T = 1$, we obtain from (4.29) $\mathrm{div}P = \nabla n - \varepsilon^2 n\nabla(\Delta\sqrt{n}/6\sqrt{n}) + O(\varepsilon^4)$, and up to order $O(\varepsilon^4)$ the *quantum drift-diffusion equations*

$$\partial_t n + \frac{\varepsilon^2}{6}\mathrm{div}\Big(n\nabla\Big(\frac{\Delta\sqrt{n}}{\sqrt{n}}\Big)\Big) - \mathrm{div}(\nabla n - n\nabla V) = 0.$$

This fourth-order equation is of parabolic type which simplifies the analysis considerably, in particular compared to the third-order dispersive quantum hydrodynamic equations (4.19)-(4.21). Notice that the quantum term can be written as

$$\mathrm{div}\Big(n\nabla\Big(\frac{\Delta\sqrt{n}}{\sqrt{n}}\Big)\Big) = \frac{1}{2}\mathrm{div}\,\mathrm{div}\big(n(\nabla\otimes\nabla)\log n\big),$$

where $\nabla\otimes\nabla$ denotes the Hessian.

The quantum drift-diffusion equations can be also derived in the relaxation-time limit from the isothermal QHD model including relaxation terms. This limit has been made rigorous in [JLM06], for solutions close to the equilibrium state.

The main mathematical difficulty is to prove the nonnegativity of the solutions. Since the equation is of fourth order, maximum principle arguments cannot be applied. The main idea of the existence analysis in the one-dimensional situation is the observation that the functional $\eta_0(t) = \int(n - \log n)dx$ is nonincreasing [JP00]. More precisely, if the equations are considered on a bounded interval such that $n = 1$ and $n_x = 0$ on the boundary,

$$\frac{d\eta_0}{dt} + \frac{\varepsilon^2}{12}\int_I(\log n)_{xx}^2 dx + \int_I(\log n)_x^2 dx + \frac{1}{\lambda_L^2}\int_I(n - C)\log n dx = 0.$$

By Poincaré's inequality, this provides an H^2 bound (if $C \in L^\infty(I)$) and hence an L^∞ bound for $w = \log n$, showing that $n = e^w$ is nonnegative (we loose positivity due to an approximation procedure). Applying a fixed-point argument, the existence of weak solutions has been proved in [JP00, JV05]. The one-dimensional equations are by now well understood and the regularity, long-time behavior, and numerical approximation of nonnegative weak solutions have been studied [DGJ05, JM06b, JP01, JV05].

Unfortunately, the above idea does not apply in the multi-dimensional case since the functional $\int(n - \log n)dx$ seems not to be nonincreasing anymore. The new idea is to show that the entropy $\eta_1(t) = \int n(\log n - 1)dx$ is bounded,

$$\frac{d\eta_1}{dt} + \frac{\varepsilon^2}{12}\int_{\mathbb{R}^d} n|(\nabla\otimes\nabla)\log n|^2 dx + 4\int_{\mathbb{R}^d}|\nabla\sqrt{n}|^2 dx + \frac{1}{\lambda_L^2}\int_{\mathbb{R}^d}(n - C)n dx = 0,$$

where $\nabla\otimes\nabla$ denotes the Hessian. Since the entropy production integral can be estimated as

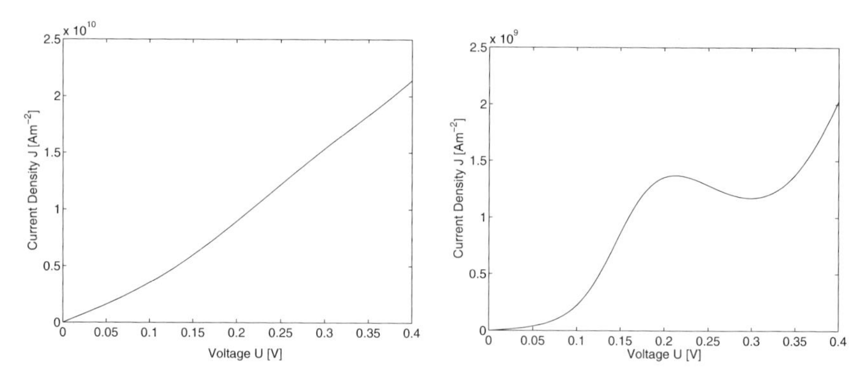

Fig. 4.5. Current voltage characteristics for a tunneling diode from the quantum drift-diffusion model for temperature $T = 300\,\mathrm{K}$ (left) and $T = 77\,\mathrm{K}$ (right).

$$\int_{\mathbb{R}^d} |(\nabla \otimes \nabla)\sqrt{n}|^2 dx \leq c \int_{\mathbb{R}^d} n|(\nabla \otimes \nabla) \log n|^2 dx,$$

for some constant $c > 0$ which depends on the space dimension d, this provides estimates for $\sqrt{n}$ in H^2 and shows that $n = (\sqrt{n})^2 \geq 0$. These estimates allow for a fixed-point argument (see [GST06, JM06b] for a proof in the case of vanishing second-order diffusion and vanishing electric fields).

Concerning the stationary equations, an existence analysis, even in several space dimensions, can be found in [BU98]. When neglecting the second-order diffusion (zero temperature case) and the electric field, we obtain the so-called *Derrida-Lebowitz-Speer-Spohn equation* [DL*91], for which additional nonincreasing functionals have been found [JM06a].

The current-voltage characteristics for a tunneling diode, computed from the one-dimensional quantum drift-diffusion equations, are shown in Fig. 4.5 (left) with the lattice temperature $T = 300\,\mathrm{K}$. We see that the model is not capable to reproduce negative differential differential effects at room temperature. However, when using a smaller lattice temperature, negative differential resistance can be observed (Fig. 4.5 right).

The quantum drift-diffusion model produces good numerical results when coupled to the Schrödinger-Poisson system employed in the channel region [EJ05]. This can be seen from Fig. 4.6 in which the coupled quantum drift-diffusion Schrödinger-Poisson model is compared with the Schrödinger-Poisson system and the coupled drift-diffusion Schrödinger-Poisson equations (see [EJ05] for details).

Acknowlegdments. This work has been supported by the DFG Priority Program 1095 "Analysis, Modeling and Simulation of Multiscale Problems" under Ar 277/3, Ju 359/5. The second author has been partially supported by the DFG, grant Ju 359/3 (Gerhard-Hess Award). The first author thanks Maike Schulte for providing the plots of Sect. 2.

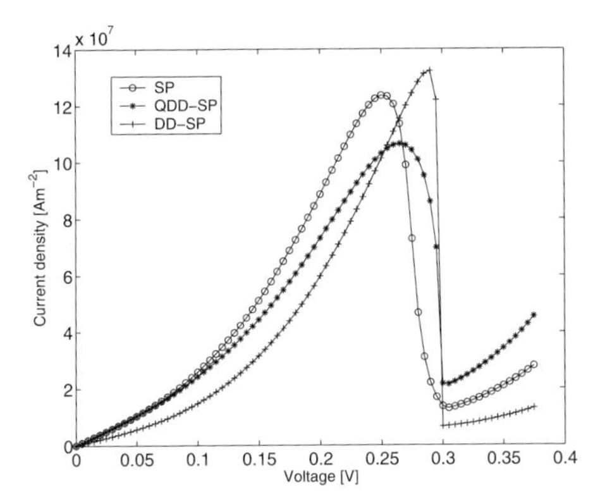

Fig. 4.6. Current voltage characteristics for a tunneling diode from the Schrödinger-Poisson system (SP), the coupled quantum drift-diffusion Schrödinger-Poisson model (QDD-SP), and the coupled drift-diffusion Schrödinger-Poisson model (DD-SP) for temperature $T = 300\,\mathrm{K}$.

References

[AI89] M. Ancona and G. Iafrate. Quantum correction to the equation of state of an electron gas in a semiconductor. *Phys. Rev. B* 39 (1989), 9536–9540.

[AT87] M. Ancona and H. Tiersten. Macroscopic physics of the silicon inversion layer. *Phys. Rev. B* 35 (1987), 7959–7965.

[Arn95a] A. Arnold. Self-consistent relaxation-time models in quantum mechanics. *Comm. PDE* 21(3&4) (1995), 473–506.

[Arn95b] A. Arnold. The relaxation-time von Neumann-Poisson equation. in: Proceedings of ICIAM 95, Hamburg (1995), Oskar Mahrenholtz, Reinhard Mennicken (eds.), *ZAMM* 76-S2 (1996), 293–296.

[Arn98] A. Arnold. Numerically absorbing boundary conditions for quantum evolution equations. *VLSI Design* 6 (1-4) (1998) 313–319.

[ADM04] A. Arnold, E. Dhamo, and C. Manzini. The Wigner-Poisson-Fokker-Planck system: global–in–time solution and dispersive effects, submitted (2005). Technical Report 10/04-N, Angewandte Mathematik, Universität Münster.

[Ar01] A. Arnold. Mathematical concepts of open quantum boundary conditions. *Transp. Theory Stat. Phys.* 30(4-6) (2001) 561–584.

[ADM05] A. Arnold, E. Dhamo, and C. Manzini. Dispersive effects in quantum kinetic equations, submitted (2005). Technical Report 07/05-N, Angewandte Mathematik, Universität Münster.

[AES03] A. Arnold, M. Ehrhardt, I. Sofronov. Approximation, stability and fast calculation of non-local boundary conditions for the Schrödinger equation. *Commun. Mathematical Sciences* 1-3 (2003) 501–556.

[AL*04] A. Arnold, J.L. López, P.A. Markowich and J. Soler. An analysis of quantum Fokker-Planck models: A Wigner function approach. *Rev. Mat. Iberoam.* 20(3) (2004), 771–814.

[AS06] A. Arnold, M. Schulte. Discrete transparent boundary conditions for the Schrödinger equatioon – a compact higher order scheme. in preparation 2006.

[AS04] A. Arnold, C. Sparber. Conservative quantum dynamical semigroups for mean-field quantum diffusion models. *Comm. Math. Phys.* 251(1) (2004), 179–207.
[AR95] A. Arnold, C. Ringhofer. Operator splitting methods applied to spectral discretizations of quantum transport equations. *SIAM J. of Num. Anal.* 32-6 (1995) 1876–1894.
[BD96] N. Ben Abdallah and P. Degond. On a hierarchy of macroscopic models for semiconductors. *J. Math. Phys.* 37 (1996), 3308–3333.
[BMP05] N. Ben Abdallah, F. Méhats, O. Pinaud. On an open transient Schrödinger-Poisson system. *Math. Models Methods Appl. Sci.* 15-5 (2005), 667–688.
[BP06] N. Ben Abdallah, O. Pinaud. Multiscale simulation of transport in an open quantum system: resonances and WKB interpolation. *J. Comput. Phys.* 213-1 (2006) 288–310.
[BU98] N. Ben Abdallah and A. Unterreiter. On the stationary quantum drift-diffusion model. *Z. Angew. Math. Phys.* 49 (1998), 251–275.
[BGK54] P. Bhatnagar, E. Gross, and M. Krook. A model for collision processes in gases. I. Small amplitude processes in charged and neutral one-component systems. *Phys. Review* 94 (1954), 511–525.
[BM91] F. Brezzi, P.A. Markowich. The three-dimensional Wigner-Poisson problem: existence, uniqueness and approximation. *Math. Methods Appl. Sci.* 14(1) (1991), 35–61.
[CL83] A. Caldeira and A. Leggett. Path integral approach to quantum Brownian motion. *Phys. A* 121A (1983), 587–616.
[CLN04] J.A. Cañizo, J.L. López, J. Nieto. Global L^1-theory and regularity for the 3D nonlinear Wigner-Poisson-Fokker-Planck system. *J. Diff. Eq.* 198 (2004), 356–373.
[Cas98] F. Castella. The Vlasov-Poisson-Fokker-Planck System with Infinite Kinetic Energy. *Indiana Univ. Math. J.* 47(3) (1998), 939–964.
[CE*00] F. Castella, L. Erdös, F. Frommlet, and P. Markowich. Fokker-Planck equations as scaling limits of reversible quantum systems. *J. Stat. Phys.* 100 (2000), 543–601.
[DGJ97] P. Degond, S. Génieys, and A. Jüngel. A system of parabolic equations in nonequilibrium thermodynamics including thermal and electrical effects. *J. Math. Pures Appl.* 76 (1997), 991–1015.
[DMR04] P. Degond, F. Méhats, and C. Ringhofer. Quantum hydrodynamic models derived from the entropy principle. *Contemp. Math.* 371 (2005), 107–131.
[DMR05] P. Degond, F. Méhats, and C. Ringhofer. Quantum energy-transport and drift-diffusion models. *J. Stat. Phys.* 118 (2005), 625–665.
[DR03] P. Degond and C. Ringhofer. Quantum moment hydrodynamics and the entropy principle. *J. Stat. Phys.* 112 (2003), 587–628.
[DL*91] B. Derrida, J. L. Lebowitz, E. R. Speer, and H. Spohn. Fluctuations of a stationary nonequilibrium interface. *Phys. Rev. Lett.* 67 (1991), 165–168.
[DA06] E. Dhamo, A. Arnold. An operator splitting method on the periodic Wigner-Poisson-Fokker-Planck system. in preparation, 2006.
[DGJ05] J. Dolbeault, I. Gentil, and A. Jüngel. A nonlinear fourth-order parabolic equation and related logarithmic Sobolev inequalities. To appear in *Commun. Math. Sci.*, 2006.

[EJ05] A. El Ayyadi and A. Jüngel. Semiconductor simulations using a coupled quantum drift-diffusion Schrödinger-Poisson model. *SIAM J. Appl. Math.* 66 (2005), 554–572.

[FZ93] D. Ferry and J.-R. Zhou. Form of the quantum potential for use in hydrodynamic equations for semiconductor device modeling. *Phys. Rev. B* 48 (1993), 7944–7950.

[Fre87] W.R. Frensley. Wigner-function model of a resonant-tunneling semiconductor device *Phys. Rev. B* 36 (1987) 1570–1580.

[GaM06] S. Gallego and F. Méhats. Entropic discretization of a quantum drift-diffusion model. *SIAM J. Numer. Anal.* 43 (2005), 1828-1849.

[GaJ01] I. Gamba and A. Jüngel. Positive solutions of singular equations of second and third order for quantum fluids. *Arch. Rat. Mech. Anal.* 156 (2001), 183–203.

[GaJ02] I. Gamba and A. Jüngel. Asymptotic limits in quantum trajectory models. *Commun. Part. Diff. Eqs.* 27 (2002), 669–691.

[Gar94] C. Gardner. The quantum hydrodynamic model for semiconductor devices. *SIAM J. Appl. Math.* 54 (1994), 409–427.

[GST06] U. Gianazza, G. Savaré, and G. Toscani. A fourth-order nonlinear PDE as gradient flow of the Fisher information in Wasserstein spaces. In preparation, 2006.

[GG*93] H.L. Gruvin, T.R. Govindan, J.P. Kreskovsky, M.A. Stroscio. Transport via the Liouville equation and moments of quantum distribution functions, *Solid State Electr.* 36 (1993), 1697–1709.

[GuJ04] M. Gualdani and A. Jüngel. Analysis of the viscous quantum hydrodynamic equations for semiconductors. *Europ. J. Appl. Math.* 15 (2004), 577–595.

[GyJ00] M. T. Gyi and A. Jüngel. A quantum regularization of the one-dimensional hydrodynamic model for semiconductors. *Adv. Diff. Eqs.* 5 (2000), 773–800.

[HLMO05] F. Huang, H.-L. Li, A. Matsumura, and S. Odanaka. Well-posedness and stability of multi-dimensional quantum hydrodynamics in whole space. Preprint, Osaka University, Japan, 2004.

[Jün98] A. Jüngel. A steady-state quantum Euler-Poisson system for semiconductors. *Commun. Math. Phys.* 194 (1998), 463–479.

[Jün01] A. Jüngel. *Quasi-hydrodynamic Semiconductor Equations.* Birkhäuser, Basel, 2001.

[JL04] A. Jüngel and H.-L. Li. Quantum Euler-Poisson systems: global existence and exponential decay. *Quart. Appl. Math.* 62 (2004), 569–600.

[JLM06] A. Jüngel, H.-L. Li, and A. Matsumura. The relaxation-time limit in the quantum hydrodynamic equations for semiconductors. To appear in *J. Diff. Eqs.*, 2006.

[JMR02] A. Jüngel, M. C. Mariani and D. Rial. Local existence of solutions to the transient quantum hydrodynamic equations. *Math. Models Meth. Appl. Sci.* 12 (2002), 485–495.

[JMa05] A. Jüngel and D. Matthes. A derivation of the isothermal quantum hydrodynamic equations using entropy minimization. *Z. Angew. Math. Mech.* 85 (2005), 806–814.

[JM06a] A. Jüngel and D. Matthes. An algorithmic construction of entropies in higher-order nonlinear PDEs. *Nonlinearity* 19 (2006), 633–659.

[JM06b] A. Jüngel and D. Matthes. The multi-dimensional Derrida-Lebowitz-Speer-Spohn equation. In preparation, 2006.

[JMM05] A. Jüngel, D. Matthes, and J.-P. Milišić. Derivation of new quantum hydrodynamic equations using entropy minimization. Preprint, Universität Mainz, 2005.

[JMi05] A. Jüngel and J.-P. Milišić. Macroscopic quantum models with and without collisions. To appear in *Proceedings of the Sixth International Workshop on Mathematical Aspects of Fluid and Plasma Dynamics*, Kyoto, Japan. *Transp. Theory Stat. Phys.*, 2006.

[JMi06] A. Jüngel and J.-P. Milišić. Numerical approximation of the nonisothermal quantum hydrodynamic equations for semiconductors with viscous terms. In preparation, 2006.

[JP00] A. Jüngel and R. Pinnau. Global non-negative solutions of a nonlinear fourth-oder parabolic equation for quantum systems. *SIAM J. Math. Anal.* 32 (2000), 760–777.

[JP01] A. Jüngel and R. Pinnau. A positivity-preserving numerical scheme for a nonlinear fourth-order parabolic equation. *SIAM J. Num. Anal.* 39 (2001), 385–406.

[JT06] A. Jüngel and S. Tang. Numerical approximation of the viscous quantum hydrodynamic model for semiconductors. To appear in *Appl. Numer. Math.*, 2006.

[JV05] A. Jüngel and I. Violet. The quasineutral limit in the quantum drift-diffusion equations. Preprint, Universität Mainz, Germany, 2005.

[KK*89] N. Kluksdahl, A. M. Kriman, D. K. Ferry, and C. Ringhofer. Self–consistent study of the resonant tunneling diode. *Phys. Rev. B* 39 (1989), 7720–7735.

[KN05] H. Kosina, M. Nedjalkov. Wigner function-based device modeling. in: *Handbook of Theoretical and Computational Nanotechnology* vol. 10 (eds: M. Rieth, W. Schommers), American Scientific Publishers, 2006

[Lev96] C. Levermore. Moment closure hierarchies for kinetic theories. *J. Stat. Phys.* 83 (1996), 1021–1065.

[Lev70] I.B. Levinson. Translational invariance in uniform fields and the equation for the density matrix in the Wigner representation. *Sov. Phys. JETP* 30 (1970) 362–367.

[LL05] H.-L. Li and C.-K. Lin. Zero Debye length asymptotic of the quantum hydrodynamic model for semiconductors. *Commun. Math. Phys.* 256 (2005), 195–212.

[Lin76] G. Lindblad. On the generators of quantum mechanical semigroups. *Comm. Math. Phys.* 48 (1976), 119–130.

[LP93] P.L. Lions. T. Paul. Sur les mesures de Wigner. *Rev. Math. Iberoam.*, 9(3) (1993), 553–561.

[Mad27] E. Madelung. Quantentheorie in hydrodynamischer Form. *Z. Physik* 40 (1927), 322–326.

[MRS90] P. Markowich, C. Ringhofer, and C. Schmeiser. *Semiconductor Equations.* Springer, Vienna, 1990.

[RA05] D.A. Rodrigues, A.D. Armour. Quantum master equation descriptions of a nanomechanical resonator coupled to a single-electron transistor. *New J. Phys.* 7 (2005) 251–272.

[Per96] B. Perthame. Time decay, propagation of low moments and dispersive effects for kinetic equations. *Comm. P.D.E.* 21(1&2) (1996), 659–686.

[Ram02] L. Ramdas Ram-Mohan. *Finite element and boundary emelent applications in quantum mechanics.* Oxford Univ. Press, 2002.

[Rin90] C. Ringhofer. A spectral method for the numerical simulation of quantum tunneling phenomena *SIAM J. Num. Anal.* 27 (1990) 32–50.

[Ris84] H. Risken. *The Fokker-Planck equation.* Springer, 1984.

[SC*02] C. Sparber, J.A. Carrillo, J. Dolbeault, P.A. Markowich. On the long time behavior of the quantum Fokker-Planck equation. *Monatsh. f. Math.* 141(3) (2004), 237–257.

[Str86] M.A. Stroscio. Moment-equation representation of the dissipative quantum Liouville equation. *Superlattices and microstructures* 2 (1986), 83–87.

[Wig32] E. Wigner. On the quantum correction for thermodynamic equilibrium. *Phys. Rev.* 40 (1932), 749–759.

[ZA*05] A. Zisowsky, A. Arnold, M. Ehrhardt, T. Koprucki. Discrete Transparent Boundary Conditions for transient kp-Schrödinger Equations with Application to Quantum-Heterostructures. *ZAMM 85* 11 (2005) 793–805.

Electronic States in Semiconductor Nanostructures and Upscaling to Semi-Classical Models

Thomas Koprucki, Hans-Christoph Kaiser, and Jürgen Fuhrmann

Weierstrass Institute for Applied Analysis and Stochastics, Mohrenstraße 39, 10117 Berlin. koprucki@wias-berlin.de, kaiser@wias-berlin.de, fuhrmann@wias-berlin.de

1 Introduction

Nanostructures are one of the basic features of quantum electronic semiconductor devices. In order to cover quantum effects in semiconductor device simulation, one has to compute the states of the acting electrons and one has to incorporate adequate information about these electronic states into the device simulation tools, which usually operate on a semi-classical level.

In semiconductor devices one basically distinguishes three spatial scales: the atomistic scale of the bulk semiconductor materials (sub-Å), the scale of the interaction zone at the interface between two semiconductor materials together with the scale of the resulting size quantization (nanometer) and the scale of the device itself (micrometer).

At the ab-initio level, the many-body Schrödinger equation for the electrons in the potential of the nuclei gives a complete description of the electronic structure for a semiconductor bulk material. The electrons present in semiconductor materials can be subdivided into two classes: the core electrons and the valence electrons. On the energy scale, the core electrons have much lower levels than the valence electrons. This allows to decouple them from the valence electrons. Together with the nuclei, they form the ionic cores. The valence electrons are responsible for the chemical bonds and for many electronical and optical properties. The electronic states of the valence electrons can be described with high accuracy in the framework of density functional theory. This approach results in a one-particle Schrödinger equation on the *atomistic scale* with a potential given by the ionic cores and the mean field contribution of the interaction between the valence electrons.

The prototype semiconductor nanostructure is a quantum well structure which consists of a stack of different semiconductor materials grown on a substrate, see Fig. 1.2. The thickness of these layers usually ranges from 2 to 20 nm, defining the *nanoscale*. The characteristic feature of the material

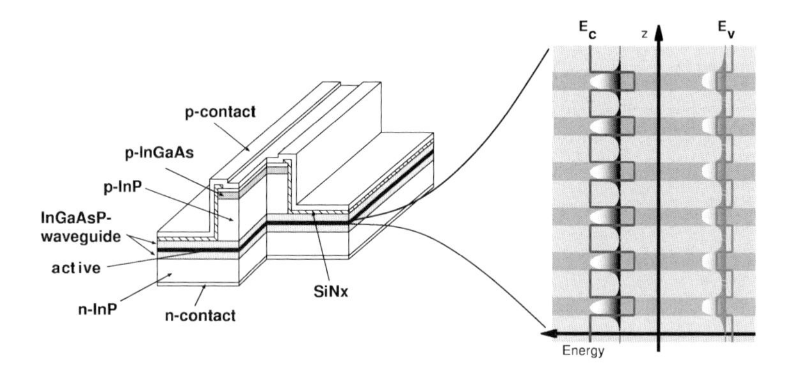

Fig. 1.1. Schema of a SMQW laser diode by HHI, Berlin. Holes injected from the p-contact and electrons injected from the n-contact recombine in the optical active region. The optical active region consisting of six quantum wells is enlarged on the right. E_c and E_v are the band edge profiles for the electrons and holes. The quantum confinement of the electrons and holes within the wells is indicated.

interface between two layers is the abrupt change of the parameters of the crystal (chemical composition, band structure) on a distance of the same order of magnitude as the lattice constant. Basically, this leads to a coupling of the electronic states of the different materials meeting at the interface. The variation of material properties across nanoscale heterostructures induces size-quantization.

The nanostructure is in the focus of the envelope function approximation, which can be understood as a homogenization method with Bloch waves. It leads to $k{\cdot}p$ multi-band Schrödinger-type equations with position-dependent effective mass tensor and band-edge profile, which are state of the art for the description of electronic states in semiconductor nanostructures.

Figure 1.1 shows the essential components of an edge-emitting strained multiple quantum well (SMQW) semiconductor laser diode, the optical active zone of which is a stack of quantum wells separated by barriers from a different material. The transversal simulation of such a laser deals with a cross section measuring up to several microns in diameter. Semi-classical models working on the *device scale* such as drift-diffusion equations are state of the art for the simulation of the electronic behavior of many microelectronic devices. For opto-electronic devices based on nanostructures they can be applied successfully, supposed the constitutive laws in these semi-classical models take into account quantum effects from smaller scale models for the embedded nanostructure [BGK00, BHK03, BK*03, BGH05].

The present paper focuses on the two scale transitions inherent in the hierarchy of scales in the device. In section 2, we start with the description of the band structure of the bulk material by $k{\cdot}p$ Hamiltonians on the atomistic scale. Sect. 3 describes how the envelope function approximation allows to construct kp Schrödinger operators describing the electronic states at the

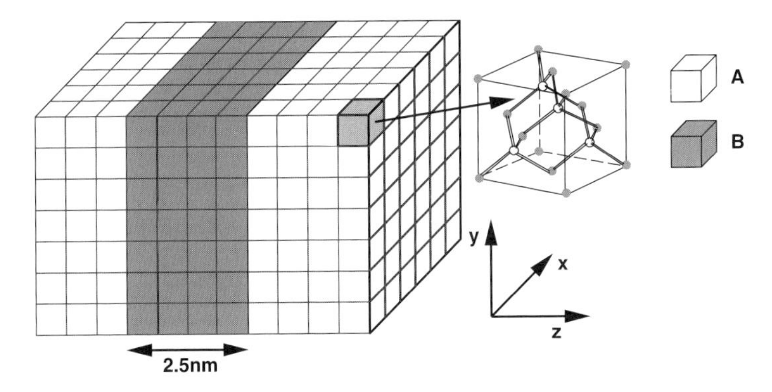

Fig. 1.2. Quantum well: one-dimensional semiconductor nanostructure consisting of two semiconductor materials A and B and two planar material interfaces. The unit cell for a binary semiconductor with zinc-blende crystallographic structure like gallium arsenide is displayed.

nanoscale which are closely related to the $k{\cdot}p$ Hamiltonians from Sect. 2. Special emphasis is placed on the possible existence of spurious modes in the $k{\cdot}p$ Schrödinger model on the nanoscale which are inherited from anomalous band bending on the atomistic scale. Sect. 4 is devoted to the mathematical analysis of these multi-band $k{\cdot}p$ Schrödinger operators. Besides of the confirmation of the main facts about the band structure usually taken for granted, key results are conditions on the coefficients of the $k{\cdot}p$ Schrödinger operator for the nanostructure, which exclude spurious modes and an estimate of the size of the band gap. Sect. 5 gives an overview of properties of the electronic band structure of strained quantum wells. Further, the assumption of flat-band conditions across the nanostructure allows for upscaling of quantum calculations to state equations for semi-classical models. In Sect. 6 we demonstrate this approach for parameters such as the quantum corrected band-edges, the effective density of states, the optical response, and the optical peak gain. Sect. 7 is devoted to the application of the $k{\cdot}p$ Schrödinger theory to low gap quantum wells, a case where a proper rescaling of the optical matrix element is necessary to avoid spurious modes. In Sect. 8 we discuss the application of the $k{\cdot}p$ Schrödinger models to biased quantum wells, the operation mode of electro-optic modulators.

2 Near-band-edge states in semiconductor bulk crystals

The key property of bulk semiconductor materials is, that the atoms form a periodic Bravis lattice defined by its crystallographic unit cell, see Fig. 1.2 and [Car96]. The electronic states of the valence electrons in a semiconductor are essentially given by the solution of the eigenvalue problem for a Schrödinger operator

$$H = -\frac{\hbar^2}{2m_0}\Delta + V_{eff}(r). \tag{2.1}$$

m_0 is the electron rest mass and V_{eff} is the effective potential, consisting of the potential of ionic cores (nuclei and core electrons) and the mean field interaction between the valence electrons. This potential can be given by empirical pseudopotential method (EPM) [CB66, CC76, Car96]. Alternatively, density functional theory allows to obtain the electronic states by solving an effective one-particle Schrödinger equation with a periodic potential.

Due to the translation invariance of the lattice periodic potential, the eigenfunctions of the Schrödinger operator are Bloch waves

$$\Psi(r;k) = e^{ikr}u(r;k), \tag{2.2}$$

defined by a lattice periodic Bloch function $u(r;k)$ depending on the real space vector $r = (x, y, z)$ and parametrically on the wave vector $k = (k_x, k_y, k_z)$, Bloch theorem see [Blo32, Car96]. The Bloch waves as well as the corresponding eigenvalue curves $E(k)$ are periodic in the wave vector k. Thus, it is sufficient (reduced zone scheme) to restrict the considerations to the first *Brillouin zone*, the unit cell of the periodic lattice in the k space, see [Car96].

2.1 $k{\cdot}p$ equation for Bloch waves

The Bloch wave ansatz (2.2) leads to an eigenvalue problem for the Bloch function $u(r;k)$ [Kan66, Kan82]. Including the spin degree of freedom by using a two-component Bloch function one arrives at the at the $k{\cdot}p$ equation. Using the notation of [Bah90] it reads as:

$$Hu_n(r;k) = E_n(k)u_n(r;k) \tag{2.3}$$

with

$$H = H_0 + H_{k\cdot p} + H_k + H_{so} + H_{kso}$$

$$\begin{aligned} H_0 &= -\frac{\hbar^2}{2m_0}\Delta + V_{eff}(x), \quad H_{k\cdot p} = \frac{\hbar}{m_0}k\cdot p, \quad H_k = \frac{\hbar^2k^2}{2m_0}, \\ H_{so} &= \frac{\hbar}{4m_0^2c^2}\big((\nabla V_{eff})\times p\big)\cdot\sigma, \quad H_{kso} = \frac{\hbar^2}{4m_0^2c^2}\big((\nabla V_{eff})\times k\big)\cdot\sigma. \end{aligned}$$

p denotes the quantum mechanical momentum operator defined by $p = -i\hbar\nabla$. $\sigma = (\sigma_x, \sigma_y, \sigma_z)$ is the vector of the Pauli spin matrices

$$\sigma_x = \begin{pmatrix} 0 & 1 \\ 1 & 0 \end{pmatrix}, \quad \sigma_y = \begin{pmatrix} 0 & -i \\ i & 0 \end{pmatrix}, \quad \sigma_z = \begin{pmatrix} 1 & 0 \\ 0 & -1 \end{pmatrix}.$$

H_{so} and H_{kso} describe the spin-orbit interaction.

The eigenvalue curves $E_n(k)$ are the energy bands of the valence electrons in a semiconductor material. Together they form the electronic band structure. The essential property of the band structure in a semiconductor is the

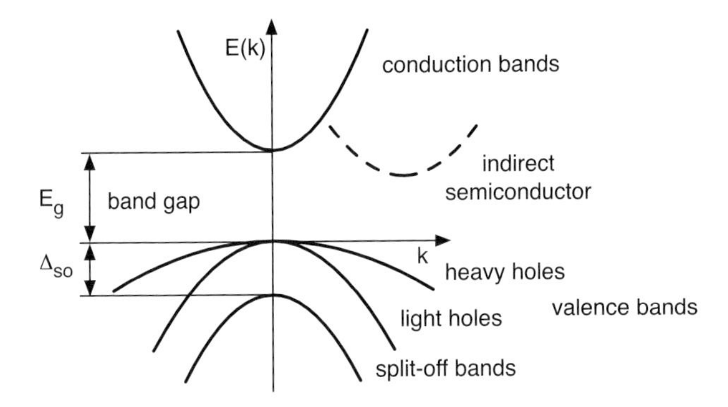

Fig. 2.1. Schematic view of the band structure in a direct semiconductor material (the lowest conduction band and the three topmost valence bands). Due to the Kramers degeneracy (spin degeneracy) each state is doubly degenerate. For indirect semiconductor materials the conduction band minimum is located outside the zone-center $k = 0$ as indicated by the dashed line. Due to crystal symmetry there may exist multiple equivalent band minima in this case, for instance, six in silicon.

existence of a fundamental spectral gap. Energy bands below and above this gap are valence bands and conduction bands, respectively. For the cases of interest, the maximum E_v of the valence band is located at the Γ point, center $k = 0$ of the Brillouin zone. In direct semiconductor materials such as gallium arsenide, the minimum E_c of the conduction bands is also located at the Γ point, whereas for indirect semiconductors such as silicon, it is located outside the zone-center. The band-edges E_c and E_v define the band gap $E_g = E_c - E_v$. Fig. 2.1 shows a schematic band diagram. In thermodynamic equilibrium, the carriers occupy the states near the band extrema. In particular, the conduction band states are occupied by the free roaming electrons and the valence band states are occupied by the positively charged holes. Therefore, for many applications it is sufficient to confine the description of band structure to the near-band-edge states.

2.2 Eight-band $k{\cdot}p$ Hamiltonian for near band-edge states

Basically, any Bloch function $u(r, k)$ can be represented in terms of the zone-center solutions $u_n^\Gamma(r) = u_n(r, k = 0)$:

$$u_n(r, k) = \sum_{n'} C_{n'}(k) u_{n'}^\Gamma(r)$$

The first d near-band-edge states u_n^Γ labeled $n = 1, \ldots, d$ form the set of class-A bands. All remaining states, labeled class-B, are assumed to have only a small influence on the near band-edge states. Typical number of class-A bands are 1, 4, 6 or 8 [Kan82, Bas88, Chu95, Bah90, MGO94]. The class-B bands are far away from the band-edges E_c and E_v and thus have only a small contribution to a state $u_n(r,k)$ of class-A near zone-center. This gives rise to the representation

$$u_n(r,k) \approx \sum_{n'=1}^{d} c_{n'}(k) u_{n'}^{\Gamma}(r), \quad \text{for } n = 1. \cdots, d, \tag{2.4}$$

Taking into account remote band effects arising from the influence of the class-B states by means of Löwdins perturbation scheme one arrives at a nonlinear eigenvalue problem for the class-A bands, see [Kan66, Kan82]. A suitable linearization of this problem provides the corresponding Hamiltonian.

In the following we regard eight class-A bands consisting of the lowest conduction band and the three topmost valence bands, all doubly degenerate, see Fig. 2.1. For materials with diamond or zinc-blende crystallographic structure the space of class-A bands is spanned by $S\uparrow$, $X\uparrow$, $Y\uparrow$, $Z\uparrow$, $S\downarrow$, $X\downarrow$, $Y\downarrow$, $Z\downarrow$, where $\uparrow$ and $\downarrow$ indicate spin up and spin down, respectively. Following the notation of [EW96], the Hamiltonian reads as:

$$H_{8\times 8}(k) = \begin{pmatrix} K(k) + \mathrm{i}\cdot G_z + E & \Gamma \\ \bar{\Gamma} & K(k) - \mathrm{i}\cdot G_z + E \end{pmatrix} \tag{2.5}$$

where the $k{\cdot}p$ matrix is given by

$$K(k) = \begin{pmatrix} A\cdot k^2 & \mathrm{i}P_0 k_x & \mathrm{i}P_0 k_y & \mathrm{i}P_0 k_z \\ -\mathrm{i}P_0 k_x & (L-M)k_x^2 + Mk^2 & Nk_x k_y & Nk_x k_z \\ -\mathrm{i}P_0 k_y & Nk_x k_y & Lk_y^2 + M(k_x^2 + k_z^2) & Nk_y k_z \\ -\mathrm{i}P_0 k_x & Nk_x k_z & Nk_y k_z & Lk_z^2 + M(k_x^2 + k_y^2) \end{pmatrix}$$

$$G_z = \frac{\Delta_{so}}{3}\begin{pmatrix} 0 & 0 & 0 & 0 \\ 0 & 0 & -1 & 0 \\ 0 & 1 & 0 & 0 \\ 0 & 0 & 0 & 0 \end{pmatrix}, \Gamma = \frac{\Delta_{so}}{3}\begin{pmatrix} 0 & 0 & 0 & 0 \\ 0 & 0 & 0 & 1 \\ 0 & 0 & 0 & -\mathrm{i} \\ 0 & -1 & \mathrm{i} & 0 \end{pmatrix}$$

$$E = \begin{pmatrix} E_g & 0 & 0 & 0 \\ 0 & -\Delta_{so}/3 & 0 & 0 \\ 0 & 0 & -\Delta_{so}/3 & 0 \\ 0 & 0 & 0 & -\Delta_{so}/3 \end{pmatrix}$$

As their influence is usually neglected, the matrix elements describing the influence of bulk inversion asymmetry of potential V_{eff} and the influence of k dependent spin-orbit interaction have been omitted, see [Kan82, Bah90].

The parameters of the Hamiltonian are the (parabolic) conduction band mass m_c, the Luttinger parameters $\gamma_1^L, \gamma_2^L, \gamma_3^L$ describing the the heavy hole masses in different crystallographic directions, the band gap energy E_g, the spin-orbit split-off energy Δ_{so}, and the momentum matrix element P_0. These parameters define the coefficients of the Hamiltonian by

$$L = -\frac{\hbar^2}{2m_0}(\gamma_1^L + 4\gamma_2^L) + \frac{P_0^2}{E_g}, M = -\frac{\hbar^2}{2m_0}(\gamma_1^L - 2\gamma_2^L), N = -6\frac{\hbar^2}{2m_0}\gamma_3^L + \frac{P_0^2}{E_g} \tag{2.7}$$

$$A = \frac{\hbar^2}{2m_0}\frac{m_0}{m_c} - P_0^2 \frac{E_g + 2/3\Delta_{so}}{E_g(E_g + \Delta_{so})}. \tag{2.8}$$

P_0 is defined by the Bloch functions S and X

$$P_0 \stackrel{\text{def}}{=} -\mathrm{i}\frac{\hbar}{m_0}\int_{\text{unit cell}} \bar{S}(r)\frac{\hbar}{\mathrm{i}}\frac{\partial}{\partial x}X(r)\,dr, \tag{2.9}$$

and is a measure for the coupling between the conduction bands and the valence bands. It is also known as the optical matrix element, which plays a role in the calculation of the strength of optical transitions. Usually it is given by an energy parameter

$$E_p = \frac{2m_0}{\hbar^2}P_0^2. \tag{2.10}$$

It is possible to obtain the parameters E_g,Δ_{so}, m_c, γ_1^L,γ_2^L, γ_3^L and E_p of the Hamiltonian from experimentally determined properties of the bulk material. There exists a compilation [VM*01] of these band parameters for the 12 major III-V binary semiconductor materials and their ternary and quaternary alloys.

In the literature, also simplified $k{\cdot}p$ Hamiltonians for the valence bands are used. They include the four-band Luttinger-Kohn Hamiltonian for heavy holes and light holes and six-band valence band Hamiltonians [Bas88, Chu95, Car96]. For semiconductors with wurtzite crystallographic structure such as gallium nitride $k{\cdot}p$ Hamiltonians have been established as well [CC96].

2.3 Anomalous band bending

The reciprocal masses A and L on the diagonal of (2.5) depend on the ratio $\zeta = E_p/E_g$. For $\zeta > \zeta_{crit}$ one of them changes the sign. For large wave vectors k parallel to one of the axis in the k space, these diagonal k^2 terms dominate the behavior of the energy bands. Thus, for $\zeta > \zeta_{crit}$ the energy dispersion becomes anomalous in the sense that e.g. for $A < 0$ the bending of the conduction bands becomes negative, see Fig. 3.2. It is known, that this behavior of the band structure can lead to the formation of spurious modes, if $k{\cdot}p$ method is applied to heterostructures, see Sect. 3.4 and [For97, Sol03, MGO94].

2.4 Modification of the band structure by mechanical strain

If a semiconductor material is grown on a substrate with a different lattice constant, e.g. indium gallium arsenide on gallium arsenide, we observe mechanical strain induced by the lattice mismatch between the two crystals. Basically, this mechanical strain leads to a shift of band-edges E_c, E_v resulting in an altered band gap E_g and to a splitting of the heavy hole and light hole bands at the Γ point. Additionally, it has an impact on the anisotropy of the effective mass tensor and the warping of the band structure [CC92].

The influence of strain ε can be included in the $k{\cdot}p$ Hamiltonian by adding the Pikus-Bir deformation interaction matrix $D(\varepsilon)$ defined by deformation potentials [BP74, Bah90, Chu95, EW96].

2.5 Mixed crystal systems

So far, our considerations apply to elementary semiconductors like silicon or binary semiconductors such as gallium arsenide or indium phosphite. Mixed crystal systems are alloys of binary semiconductors. They are of particular interest, because they allow for band gap engineering in the case of ternary compounds like $In_{1-x}Ga_xAs$ and additionally for design of the mechanical strain in the case of quaternary materials like $In_{1-x}Ga_xAs_yP_{1-y}$. Though strictly speaking, the basic assumptions of a periodic crystal lattice do not hold for these materials, they can be treated by similar methods assuming the so-called virtual crystal approximation.

Within the kp framework the band structure of mixed crystal systems can be described by the same Hamiltonian used for the elementary and binary semiconductors. Their parameters can be obtained by interpolation of those of the binary constituents. Often linear interpolation according to the mole fraction x yields a sufficient approximation. For some parameters such as the band gap energy E_g this interpolation has to include a bowing parameter. All the related data can be found in [VM*01].

3 Envelope function approximation for layered semiconductor nanostructures

We have discussed the modeling of the band structure of bulk materials by $k{\cdot}p$ Hamiltonians. Now, this approach is extended to layered semiconductor nanostructures like quantum wells, multiple quantum wells (see Fig. 1.2) and double barrier structures [Sin93], [Bas88], [Chu95]. For these heterostructures, the translation invariance is broken and the microscopic potential V_{eff} cannot be periodic in all space directions. Nevertheless, the heterostructure is a crystalline solid and the atoms form an approximate Bravis lattice. The microscopic potential now consists of a oscillatory part, which corresponds to the potential of the ionic cores, and a globally slowly varying part, which corresponds to the composition of the heterostructure from different bulk materials, see Fig. 3.1.

3.1 Envelope function approximation of the wavefunctions

This obvious two-scale nature of the microscopic potential and of the corresponding microscopic wave function encourages to treat this problem as a *mul-*

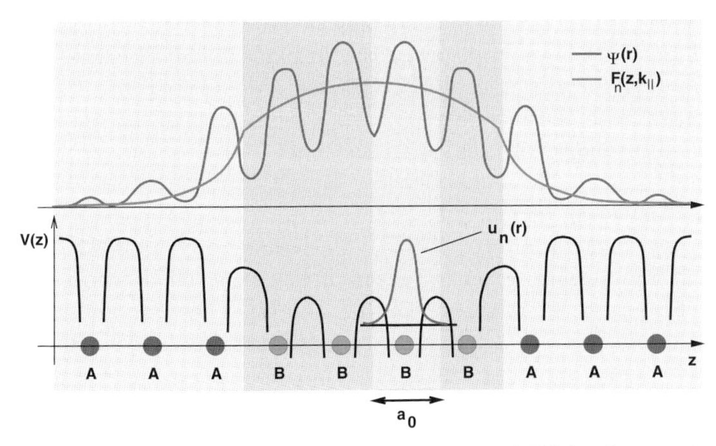

Fig. 3.1. *Schematic view of the microscopic potential $V(z)$, the atomistic wave function Ψ, the Bloch waves $u_n(r)$ and the envelope function $F(z;k_{\|})$ in a one-dimensional heterostructure consisting of atoms of type A and B. The lattice constant is a_0 (typically 0.5 nm). $u_n(r)$ is highly oscillatory on the sub-Å-scale.*

tiscale problem. After first heuristic approaches to this problem, see G. Bastard [Bas88], the first rigourous approach was made by Burt [Bur92, Bur94], who showed, that it is possible to derive a theory for the *slowly-variing* part of the wave functions, the *envelope functions*, in terms of the atomistic scale. This approach leads to a $k{\cdot}p$ Hamiltonian for the envelope functions.

For layered nanostructures, the crystal remains periodic in the in-plane directions $r_{\|} = (x, y)$, perpendicular to the growth direction z. This choice of the coordinate system corresponds to epitaxially grown nanostructures on [001] oriented substrates. As in bulk $k{\cdot}p$ theory, we approximate the electronic state in the nanostructure in the subspace spanned by d lattice periodic, zone-center Bloch functions $u_n^{\Gamma}(r)$, $n = 1, \dots, d$ of class-A. However, we replace the coefficients c_n in (2.4) by the envelope functions $F_n(z;k_{\|})$ depending on the reduced wave vector $k_{\|} = (k_x, k_y)$:

$$\Psi(r;k_{\|}) = \exp(\mathrm{i}k_{\|} \cdot r_{\|}) \sum_{n=1}^{d} F_n(z;k_{\|}) u_n^{\Gamma}(r).$$

3.2 $k{\cdot}p$-Schrödinger operators for layered nanostructures

According to [Bur92],[Bur94],[Bur99] the vector of the envelope functions $F = (F_1, \cdots, F_d)$ for a given reduced wave vector $k_{\|}$ is the solution of an eigenvalue problem

$$H\Big(k_{\|}, -\mathrm{i}\frac{\partial}{\partial z}\Big) F(z;k_{\|}) = E(k_{\|}) F(z;k_{\|}). \tag{3.1}$$

(3.1) is a family iof spatially one-dimensional eigenvalue problems indexed by $k_{\|}$. It provides a description of the in-plane band structure for layered nanostructures, for examples see Sect. 5.

Far away from the interface between two materials, the Schrödinger operator H is formally identical with the bulk Hamiltonian (replacing $k_z \to -i\partial/\partial z$) shifted by the band-edge E_v [Bur92, Bur94]. Therefore, the band-offset ΔE_v between two materials enters as an additional parameter, available for the many material interfaces [VdW89, Kri91, VM*01]. Near the interface between two materials, non-local effects are present which lead to a coupling of the states across the interface [Bur92],[Bur94]. The approximation of these non-local interactions by an interface condition for the envelope functions is sufficient for many applications.

However, one has to be aware that the Burt-Foreman approach relies on the assumption that the Bloch functions used in the approximation are the same for all material layers. As a consequence, the momentum matrix element P_0 has the same value in all materials. However, the experimentally measured value of P_0 given by the optical matrix element E_p (2.10) is varying. Therefore, it is necessary to incorporate this effect in the $k{\cdot}p$ Schrödinger operator by additional interface conditions.

This suggests the following general structure of the Schrödinger operator:

- In each material layer, the Schrödinger operator is derived from a bulk Hamiltonian like (2.5) resulting in a system of d coupled stationary Schrödinger equations, see (4.1).
- At the material interfaces the continuity of the envelope functions and of a flux vector (4.10) is assumed.
- If the optical matrix element P_0 differs only slightly between the materials, one can define an effective value by suitable averaging. Otherwise, by an additional interface condition, one has to define the coupling between the envelope functions related to the Bloch functions S and envelopes related to the Bloch functions X, Y, Z across the interface [Bur99, For97].

The *conventional* $k{\cdot}p$ Schrödinger operators have been derived from the bulk Hamiltonian by replacing $k_z \to -\mathrm{i}\partial/\partial z$, see for instance [Bas88]. This approach also has to be supplemented by coupling conditions at the material interfaces. These interface conditions have been established implicitly using the so-called operator ordering. For the second order terms the BenDaniel-Duke operator ordering $Ak_z^2 \to -\partial/\partial z A(z)\partial/\partial z$ has been used. For the first order terms the heuristic symmetrization rule $Ak_z \to -\frac{\mathrm{i}}{2}[A(z)\partial/\partial z + \partial/\partial z(A(z)\cdot)]$ has been applied [Bas88]. The resulting interface conditions differ from those derived by Burt-Foreman. For a comparison of the different interface conditions see [For93, MGO94].

3.3 Confined states in quantum wells

A quantum well structure consists of a well material layer embedded between two barrier material layers such that band-edge profiles $E_v(z)$ and $E_c(z)$ form a potential well for both holes and electrons, see Fig. 1.1. This potential well leads to the localization of the carriers in the quantum well region. This effect is known as carrier confinement. Another key feature of quantum wells is the appearance of discrete energy levels due to the size quantization induced by the small width of the quantum well, typically ranging from 2-20 nm. Repeated quantum well structures form multiple quantum wells (MQW), see Fig. 1.1.

The effect of carrier confinement is utilized in strained multiple-quantum well (SMQW) laser diodes in order to increase the density of states and the material gain in optical active region of the device. The confined carriers in quantum wells are described by the bounded states of the corresponding $k{\cdot}p$ Schrödinger operator (3.1) which are characterized by the discrete part of the spectrum. The bounded states decay exponentially in the barrier region. Therefore it is possible to use a finite-domain approximation by artificially cutting out a simulation domain Ω and applying homogenous Dirichlet (hard wall) or Neumann (soft wall) boundary conditions.

3.4 Spurious modes

It is known [For97, Sol03] that if the ratio $\zeta = E_p/E_g$, exceeds a critical value ζ_{crit} due to anomalous band bending (see Sect. 2.3) spurious modes occur as eigenfunctions of the $k{\cdot}p$ Schrödinger operator. These spurious modes are concentrated around a large value of k_z, typically outside the first Brillouin zone, and thus spatially oscillatory. The spurious modes lead to a pollution of the spectrum of the operator or even to band gap solutions [For97, Sol03].

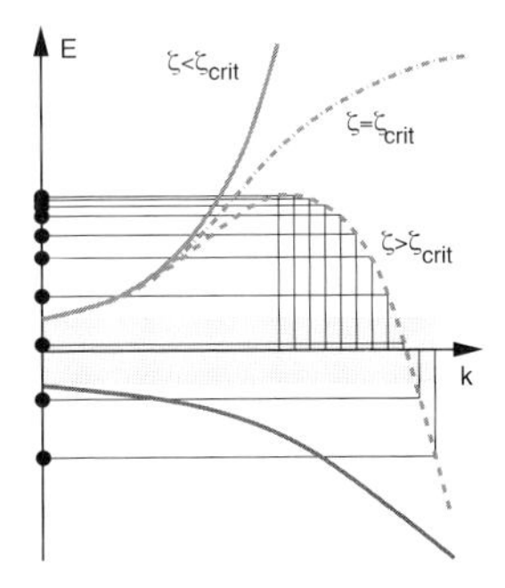

Fig. 3.2. Schematic view of the dependence of the bending of the energy bands on the ratio $\zeta = E_p/E_g$. In the case $\zeta > \zeta_{crit}$ the bending for large k vectors becomes anomalous. In the case of heterostructures this behavior can lead to pollution of the spectrum by spurious modes or even to band gap solutions, as indicated.

This difficulty can be overcome in several ways:

- Appropriate rescaling of the conduction-valence band coupling to achieve $\zeta < \zeta_{crit}$.
- Fitting of a set of bulk band parameters which ensure $\zeta < \zeta_{crit}$.
- Approximation of the eigenfunctions in a function space which guarantees $k_z < k_{crit}$, k_{crit} is the value for which the bending changes its sign.

[Sti01] achieved a conforming set of band parameters by fitting the Hamiltonian to the real bandstructure in a larger part of the Brilluoin zone (≈20%) and not only at the zone center. These effective parameters meet the conditions given by Property 4.2 in Sect. 4 which guarantee that there are no spurious modes.

We have obtained a conforming parameter set by rescaling, see Sect. 7.

4 Mathematical analysis of $k{\cdot}p$ Schrödinger operators

We review the spectral properties of Schrödinger type operators occurring in $k{\cdot}p$ theory of layered semiconductor heterostructures. In this section we denote the growth direction of the layers by x and reduced wave vector $k_{\|}$ by $k = (k_1, k_2) \in \mathbb{C}^2$. The formal structure of these operators is – for d bands $\varphi = (\varphi_1, \ldots, \varphi_d)$:

$$\begin{aligned} -\frac{d}{dx}\Big(m_j \frac{\partial}{\partial x}\varphi_j\Big) &+ \sum_{l=1}^{d}\Big(M_{0\,j\,l}\frac{\partial}{\partial x}\varphi_l - \frac{d}{dx}\big(\overline{M}_{0\,l\,j}\,\varphi_l\big)\Big) \\ &+ \sum_{\alpha=1,2} k_\alpha \sum_{l=1}^{d}\Big(M_{\alpha\,j\,l}\frac{\partial}{\partial x}\varphi_l - \frac{d}{dx}\big(\overline{M}_{\alpha\,l\,j}\,\varphi_l\big)\Big) \\ &+ \sum_{\alpha=1,2} k_\alpha \sum_{l=1}^{d} U_{\alpha\,j\,l}\,\varphi_l + \sum_{\alpha,\beta=1,2} k_\alpha k_\beta \sum_{l=1}^{d} U_{\alpha\,\beta\,j\,l}\,\varphi_l \\ &+ \sum_{l=1}^{d} v_{j\,l}\,\varphi_l + e_j\varphi_j, \quad j = 1, \ldots d \end{aligned} \tag{4.1}$$

We assume the following general properties of the coefficients m_j, $M_{0\,j\,l}$, $M_{\alpha\,j\,l}$, $U_{\alpha\,j\,l}$, $U_{\alpha\,\beta\,j\,l}$, $v_{j\,l}$ and e_j on the space intervall $\Omega = (x_0, x_L)$ of the coordinate of quantization:

Property 4.1. All coefficients are essentially bounded, namely

$$
\begin{aligned}
m_j &\in L^\infty(\Omega,\mathbb{R}), && j=1,\dots,d,\\
e_j &\in L^\infty(\Omega,\mathbb{R}), && j=1,\dots,d,\\
M_\alpha &\in L^\infty(\Omega;\mathcal{B}(\mathbb{C}^d)), && \alpha\in\{0,1,2\},\\
U_\alpha &\in L^\infty(\Omega;\mathcal{B}(\mathbb{C}^d)), && \alpha\in\{1,2\},\\
U_{\alpha\,\beta} &\in L^\infty(\Omega;\mathcal{B}(\mathbb{C}^d)), && \alpha,\,\beta\in\{1,2\},\\
v &\in L^\infty(\Omega;\mathcal{B}(\mathbb{C}^d)),
\end{aligned}
$$

where $\mathcal{B}(\mathbb{C}^d)$ is the Banach space of bounded linear operators on $\mathbb{C}^d$.

Property 4.2. The set of band indices is a disjoint union $\{1,\dots,d\} = D_+ \cup D_-$ of conduction and valence bands that means

$$
\min_{j\in D_+} \operatorname*{vraimin}_{x\in\Omega} m_j(x) > 0, \qquad \max_{j\in D_-} \operatorname*{vraimax}_{x\in\Omega} m_j(x) < 0,
$$

$$
\min_{j\in D_+l} \operatorname*{vraimin}_{x\in\Omega} e_j(x) > 0, \qquad \max_{j\in D_i} \operatorname*{vraimax}_{x\in\Omega} e_j(x) < 0;
$$

D_+ or D_- may be empty. We introduce the conjugation operator Θ on $\mathbb{C}^d$ by

$$
\Theta(c_1,\dots,c_d) = (r_1\,c_1,\dots,r_d\,c_d), \quad r_j = \begin{cases} 1 & \text{if } j\in D_+,\\ -1 & \text{if } j\in D_-. \end{cases}
$$

Property 4.3. For almost all $x\in\Omega$ and all $\alpha,\beta\in\{1,2\}$ the operators $U_\alpha(x)$, $U_{\alpha\,\beta}(x)$, and $v(x)$ are selfadjoint over $\mathbb{C}^d$.

Property 4.4. There is a finite, disjoint partition $x_0 < x_1 < \dots < x_L$ of the interval $\Omega = (x_0, x_L)$ such that the functions $m_j \in \mathbb{R}$, $j=1,\dots,d$, and $M_\alpha \in \mathcal{B}(\mathbb{C}^d)$, $\alpha = 0,1,2$, take exactly one value $\widehat{m}_{j,l}$ and $\widehat{M}_{\alpha,l}$, respectively, on each of the intervals $[x_l, x_{l+1})$.

Following [BK*00] we define parts of the $k{\cdot}p$ Schrödinger operator between $W_0^{1,2} \overset{\text{def}}{=} W_0^{1,2}(\Omega;\mathbb{C}^d)$ and its dual space $W^{-1,2}$, the space of anti-linear forms on $W_0^{1,2}$. For $\varphi,\psi\in W_0^{1,2}$ we set

$$
\langle H\varphi,\psi\rangle = \sum_{j=1}^{d} \int_\Omega m_j\,\frac{\partial}{\partial x}\varphi_j\,\frac{\partial}{\partial x}\overline{\psi}_j\,dx, \tag{4.2}
$$

$$
\begin{aligned}
\langle A_\alpha\varphi,\psi\rangle = \int_\Omega &\Big\langle M_\alpha(x)\,\frac{\partial}{\partial x}\varphi(x),\psi(x)\Big\rangle_{\mathbb{C}^d}\\
&+ \Big\langle M_\alpha^*(x)\,\varphi(x),\frac{\partial}{\partial x}\psi(x)\Big\rangle_{\mathbb{C}^d}\,dx, \quad \alpha=0,1,2,
\end{aligned} \tag{4.3}
$$

$$\langle B_\alpha \varphi, \psi \rangle = \int_\Omega \langle U_\alpha(x)\, \varphi(x), \psi(x) \rangle_{\mathbb{C}^d}\, dx, \qquad \alpha = 1, 2, \tag{4.4}$$

$$\langle B_{\alpha\beta} \varphi, \psi \rangle = \int_\Omega \langle U_{\alpha\beta}(x)\, \varphi(x), \psi(x) \rangle_{\mathbb{C}^d}\, dx, \qquad \alpha, \beta = 1, 2, \tag{4.5}$$

$$\langle V \varphi, \psi \rangle = \int_\Omega \langle v(x)\, \varphi(x), \psi(x) \rangle_{\mathbb{C}^d}\, dx, \tag{4.6}$$

$$\langle E \varphi, \psi \rangle = \sum_{j=1}^{d} \int_\Omega e_j\, \varphi_j(x)\, \overline{\psi}_j(x)\, dx. \tag{4.7}$$

Now we define for each reduced wave vector $k = (k_1, k_2) \in \mathbb{C}^2$ the $k{\cdot}p$ Schrödinger operator

$$H_k : W^{-1,2} \to W^{-1,2} \tag{4.8a}$$

by the sum

$$H_k = H + A_0 + \sum_{\alpha=1,2} k_\alpha (A_\alpha + B_\alpha) + \sum_{\alpha,\beta=1,2} k_\alpha\, k_\beta\, B_{\alpha\beta} + V + E. \tag{4.8b}$$

The terms in (4.8) relate to (2.5): E represents the basic energies of the (class A) bands involved; V contains the spin-orbit interaction and the influence of strain; the operators

$$A_0 + \sum_{\alpha=1,2} k_\alpha B_\alpha \qquad \text{and} \qquad H + \sum_{\alpha=1,2} k_\alpha A_\alpha + \sum_{\alpha,\beta=1,2} k_\alpha\, k_\beta\, B_{\alpha\beta}$$

describe the first and second order $k{\cdot}p$ interactions, respectively. They represent, e.g. for diamond like crystal structures, interband (interaction within conduction bands and valence bands, respectively) and intraband (interaction between conduction and valence bands) coupling, respectively. Making use of the conjugation operator Θ we can split the Schrödinger operator into intraband and interband coupling terms in the following way:

$$H_{k,intra} = \frac{1}{2}(H_k + \Theta H_k \Theta), \quad H_{k,inter} = \frac{1}{2}(H_k - \Theta H_k \Theta).$$

4.1 Spectral properties

We first state spectral properties of the operator (4.8) on the space $W^{-1,2}$.

Theorem 4.5. *See [BK*00]. We assume Properties 4.1–4.4. For any $k \in \mathbb{C}^2$ the operator (4.8) has the same domain as H, namely $W_0^{1,2}$, and all these operators are closed and have a compact resolvent. For any one dimensional complex analytic submanifold $\mathcal{S}$ of $\mathbb{C}^2$ the operator family $\{H_k\}_{\{k\in\mathcal{S}\}}$ is a holomorphic operator family of type (A), see [Kat84, VII.2]. The spectrum of H_k only consists of at most countably many eigenvalues with finite multiplicity, which do not accumulate at any finite point.*

The restriction of the operator (4.8) to the space $L^2 \stackrel{\text{def}}{=} L^2(\Omega; \mathbb{C}^d)$ has the following spectral properties.

Theorem 4.6. *See [BK*00]. We assume Properties 4.1–4.4. The spectra of $H_k|_{L^2}$ and H_k are the same. The resolvent of $H_k|_{L^2}$ is nuclear. For any $k \in \mathbb{C}^2$ the geometric spectral multiplicity is at most d. The domain of $H_k|_{L^2}$ is given by*

$$
\begin{aligned}
\operatorname{dom}(H_k|_{L^2}) = W^{1,2} \cap \Bigg\{ \varphi \;\Bigg|\; & \varphi|_{]x_l, x_{l+1}[} \in W^{2,2}(]x_l, x_{l+1}[), \\
& \widehat{m}_l \lim_{\substack{x \to x_l \\ x > x_l}} \frac{\partial}{\partial x} \varphi(x) - \widehat{m}_{l-1} \lim_{\substack{x \to x_l \\ x < x_l}} \frac{\partial}{\partial x} \varphi(x) + \big(\widehat{M}^*_{0,l+1} - \widehat{M}^*_{0,l}\big) \varphi(x_l) \\
& + \sum_{\alpha = 1,2} k_\alpha \big(\widehat{M}^*_{\alpha,l+1} - \widehat{M}^*_{\alpha,l}\big) \varphi(x_l) = 0,\ l = 0, \ldots, L-1 \Bigg\}.
\end{aligned}
\tag{4.9}
$$

For functions φ from the domain of $H_k|_{L^2}$ the vector

$$
m \frac{\partial}{\partial x} \varphi(x) - \big(M_0^*(x) + \sum_{\alpha=1,2} k_\alpha M_\alpha^*(x)\big) \varphi(x) \tag{4.10}
$$

is continuous across the material interfaces.

Theorem 4.7. *See [BK*00]. We assume Properties 4.1–4.4. If the reduced wave vector $k = (k_1, k_2)$ is from $\mathbb{R}^2$, then the operator $H_k|_{L^2}$ is selfadjoint, has an orthonormal basis of eigenfunctions in L^2, and its geometric and algebraic eigenspaces coincide.*

We now investigate how the spectral properties of the operators $H_k|_{L^2}$ depend on the reduced wave vector k. Unfortunately, the domain of the operators $H_k|_{L^2}$ is not independent of k. Hence, in contrast to Theorem 4.5, the concept of a holomorphic operator familiy of type (A) does not apply anymore. However we can prove, see [BK*00], that for any one dimensional complex analytic submanifold $\mathcal{S} \in \mathbb{C}^2$ the family $\{H_k|_{L^2}\}_{k \in \mathcal{S}}$, is an analytic family of operators in the sense of Kato [Kat84, VII.1.2]. This implies that a closed curve, separating two parts of the spectrum of H_k for $k = k_0$, also separates corresponding parts of the spectrum of H_k for k from a suitable neighbourhood of k_0, see [Kat84, Th. VII.1.7].

4.2 Gap estimate

From the point of view of electronic structure calculation it is of interest for which $k \in \mathbb{R}^2$ the spectral gap between the positive and negative parts of the band-edge operator E can be found in the spectrum of H_k, and how one can estimate the size of the gap in terms of k and the data of the problem. This relates also to the problem of spurious modes in bandstructure calculation.

According to Property 4.2, the lower bound

$$e \stackrel{\text{def}}{=} \min_{j=1,\dots,d} \operatorname*{vraimin}_{x\in\Omega} |e_j(x)| \tag{4.11}$$

of the spectral gap in E is positive, and so is the lowest eigenvalue μ of $|H|$. We define the weighted strength of the band couplings by

$$M \stackrel{\text{def}}{=} \frac{\max_{\alpha=1,2} \|M_\alpha\|^2_{L^\infty(\Omega;\mathcal{B}(\mathbb{C}^d))}}{\min_{j=1,\dots,d} \operatorname*{vraimin}_{x\in\Omega} |m_j(x)|}. \tag{4.12}$$

Moreover, we assume the following property of the coupling matrix M_0:

Property 4.8. $M_0(x)$ *is skewadjoint and* $\Theta M_0(x)$ *is selfadjoint for almost all* $x \in \Omega$.

Property 4.8 implies $M_0 + \Theta M_0 \Theta$ that means that the intra-band part of the operator $M_{0,intra}$, see (4), vanishes. This property is satisfied by the usual eight-band Hamiltonian and its heterostructure equivalent, see (2.5). The skewadjointness of M_0 is a consequence of symmetric operator ordering of the first order terms of conduction-valence band coupling.

Theorem 4.9. *See [BK*00]. In addition to Properties 4.1–4.4 and Property 4.8 we assume that for almost all* $x \in \Omega$

$$\Theta U_\alpha(x) \qquad \textit{are skewadjoint,} \qquad \alpha \in \{1,2\}, \tag{4.13}$$

$$\Theta U_{\alpha\,\alpha}(x) \qquad \textit{are nonnegative operators,} \qquad \alpha \in \{1,2\}, \tag{4.14}$$

and

$$\frac{1}{2} \operatorname*{vraimin}_{\eta\in[0,2\pi[,x\in\Omega} \inf\operatorname{spec}\big(\Theta U_\eta(x) + U_\eta(x)\Theta\big) \stackrel{\text{def}}{=} \nu \geq 0, \tag{4.15}$$

where

$$U_\eta(x) \stackrel{\text{def}}{=} \cos^2\eta U_{11}(x) + \sin^2\eta U_{22}(x) + \sin\eta\cos\eta\big(U_{12}(x) + U_{21}(x)\big).$$

If $|k|$, *and* λ *satisfy the relations*

$$0 \leq \lambda \leq e - \|v\|_{L^\infty([0,x_L];\mathcal{B}(\mathbb{C}^d))}. \tag{4.16a}$$

$$|k| \leq \frac{1}{\delta\sqrt{2}} \tag{4.16b}$$

$$0 < \mu - |k|\sqrt{2}\left(\mu\delta + \frac{M}{\delta}\right) + |k|^2\nu - \|v\|_{L^\infty(\Omega;\mathcal{B}(\mathbb{C}^d))} + e - \lambda, \tag{4.16c}$$

for some $\delta > 0$, *then* λ *belongs to the resolvent set of* H_k.

Thus, we have established a spectral gap $(-\lambda, +\lambda)$ in H_k. In case of $v \equiv 0$ (4.16c) means that H_k at zone center $k = 0$ has the same spectral gap as the band-edges E. Thus, spurious modes, that means band gap solutions, can be excluded in this case.

By properly choosing δ in (4.16) one may obtain sharp estimates of the gap, see [BK*00]. For the optimal choice of $\delta = \delta_{opt}(|k|)$ one obtains the following range of k

$$0 \leq |k|^2 < \frac{e - \|v\|_{L^\infty(\Omega;\mathcal{B}(\mathbb{C}^d))}}{2\,M}. \tag{4.17}$$

for which a spectral gap in H_k persists.

4.3 Remarks

Apart of $k{\cdot}p$ Schrödinger operators with interband coupling there are also operators with a positive or negative definite main part. In the terms of Property 4.2 this means that $D_- = \emptyset$ or $D_+ = \emptyset$, respectively. Important examples are the 4-band Luttinger-Kohn-Hamiltonian [Chu95] and 6-band valence band Hamiltonians, see [Car96, CKI94, Chu95, For93, MGO94, Sin93]. If the operator H from (4.2) is definite, then the operators H_k are semibounded and one obtains more results about the way the eigenvalues and eigenvectors depend on k, see [BK*00].

The whole theory considerably simplifies, if one relaxes Property 4.4 such that the coefficient functions M_α, $\alpha = 0, 1, 2$ are continuous. This allows for a regularization of the operators (4.8). One can prove, see [BK*00], that the resolvents of a regularizing sequence converge in trace class to the operator (4.8) with jumping coefficients M_α.

5 Electronic states in strained quantum wells

The band structure in quantum wells is given by the eigenvalue curves $E_i(k_\|)$ and the vector of eigenfunctions $F_i(z; k_\|) = (F_{i,1}(z; k_\|), \ldots, F_{i,d}(z; k_\|))$ of the $k{\cdot}p$ Schrödinger operator (3.1) depending on the reduced wave vector $k_\| = (k_x, k_y)$. We regard an eight-band kp Hamiltonian of the type (2.5) in the formulation given by [EW96]. Thus, the mathematical results of Sect. 4 apply. The numerical calculations have been performed by means of WIAS-QW [BK`qw`] using a finite volume method.

In particular, the structure under consideration is a $In_{1-x}Ga_xAs_yP_{1-y}$ based strained MQW structure. The stack consists of six 1% compressively strained 7 nm thick quantum wells ($x = 0.239$, $y = 0.826$), which are separated by 10 nm thick 0.3% tensile strained barriers ($x = 0.291$, $y = 0.539$), see [BHK03]. As confirmed by the calculation of the mini-band formation in the MQW structure [BHK03], the barrier width is such that the lowest states in

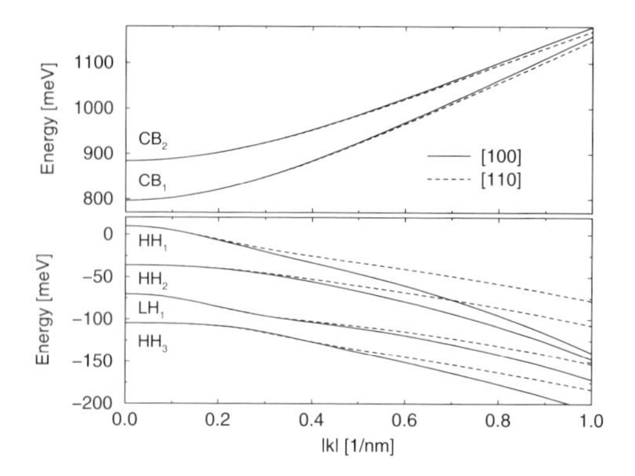

Fig. 5.1. Band structure and warping in a 7nm InGaAsP compressively strained quantum well, calculated with WIAS-QW. Top: conduction bands, bottom: valence bands. $k_{\|}$ in [100]- (solid) and [110]-direction (dashed). We remark that the bands are twice degenerate due to symmetry. Reprinted with permission from [BHK03]. © 2003, IEEE.

the wells decouple, that means, they are strongly localized in the individual quantum wells. This allows to restrict ourselves in the following to single quantum well calculations.

5.1 Band structure

Fig. 5.1 displays the band structure $E_i(k_{\|})$ for the 7 nm InGaAsP quantum well under consideration. Due to their different effective mass, in quantum wells the confinement energy $E_{conf} = E_{v,qw} - E_i(k_{\|} = 0)$ of the heavy hole states is smaller than for light-hole states ($E_{v,qw}$ is the valence band-edge of the well material). Thus, the degeneracy (at $k = 0$) between light and heavy holes in bulk materials, see Fig. 2.1, is lifted. The compressive strain increases the splitting between the heavy hole and light hole states while energetically favoring the heavy holes. Thus, the highest two valence bands correspond to heavy hole states.

We have investigated two different crystallographic directions to demonstrate the angular dependence of the dispersion (warping effect). We observe weak warping for the conduction band, and strong warping and strong non-parbolicities for the valence bands, see Fig. 5.1.

5.2 Momentum matrix elements for interband transitions

Important information on the electronic states within the quantum well is encoded in the intersubband momentum matrix elements. For the transition from the conduction band state F_i to the valence band state F_j these are defined, see [EBWS95], by

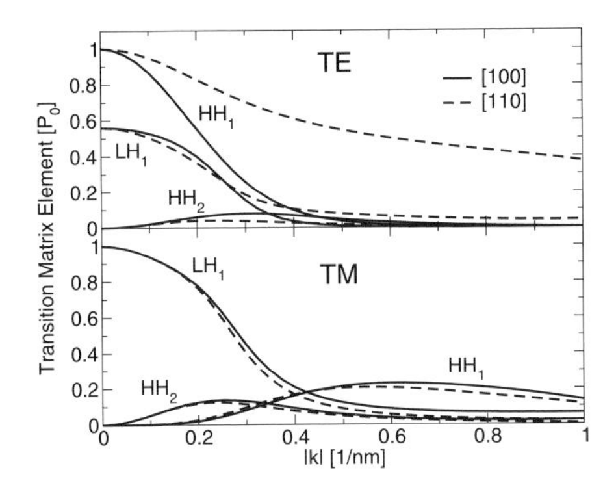

Fig. 5.2. Momentum matrix element dispersion $|ep_{ij}(k_{\|})|^2$ (5.1) for transitions between the lowest conduction band CB_1 shown in Fig. 5.1 top, and the upper valence bands shown in Fig 5.1 bottom. Different polarization directions e are shown. Top: TE-polarization ($p||e_x$), bottom: TM-polarization ($p||e_y$). Note the normalization to the same quantity P_0 in both pictures. Reprinted with permission from [BHK03]. © 2003, IEEE.

$$p_{ij} = \frac{m_0}{\hbar} \sum_{\mu,\nu} \int_{\Omega} \bar{F}_{i,\mu} \left(\nabla_k H_{\mu\nu}(k_{\|}, k_z) \right) \Big|_{ik_z = \frac{\partial}{\partial z}} F_{j,\nu} dz. \tag{5.1}$$

Fig. 5.2 displays the most prominent interband momentum matrix elements for TE and TM polarization. These momentum matrix elements have a distinctive dispersion and warping. In particular, the dominant transition for TE polarization is between the lowest conduction band CB_1 and the topmost valence band HH_1.

6 Upscaling to semi-classical state equations

In the simulation of opto-electronic devices such as strained multiple quantum well (SMQW) laser diodes, semi-classical models turned out to be very successful, provided that information about the optical active zone is derived from smaller scale models. This requires suitable *upscaling schemes* for semi-classical constitutive laws.

The simulation of opto-electronic devices requires at least the description of the flow of electrons and holes, the description of the optical field, and the coupling of these models by the radiative recombination of electrons and holes. In the following, we focus on a semi-classical carrier flow model, and on a specific part of radiative recombination; for models of the optical field, see [BGK00, BGH05, BK*03, BHK03]. In particular, we deal with upscaling schemes for quantities such as the density of states, the optical response, and the optical peak gain, from electronic structure calculations as described in Sect. 5.

6.1 Drift-diffusion equations

The most popular semi-classical models for the carrier flow in a semiconductor device are drift-diffusion models. The basic model of this type is the van Roosbroeck system, see [Gaj93] and the references therein, which describes the flow of electrons and holes in a selfconsistent field due to drift and diffusion. It comprises current-continuity equations for the densities n and p of electrons and holes, respectively, and a Poisson equation for the electrostatic potential φ:

$$q\frac{\partial n}{\partial t} - \nabla \cdot j_n = -qR, \quad q\frac{\partial p}{\partial t} + \nabla \cdot j_p = -qR \tag{6.1}$$

$$\varepsilon_0 \nabla(\varepsilon_s \nabla\varphi) = -q(C + p - n). \tag{6.2}$$

ε_0 is the vacuum dielectric constant, ε_s is the static dielectric constant, q is the elementary charge, and C is the net doping. The recombination rate R in (6.1) involves all non-radiative and radiative recombination processes, and depends at least on n and p. The currents j_n and j_p are driven by the negative gradients of the quasi-Fermi potentials F_n and F_p for electrons and holes, respectively:

$$j_n = -qn\mu_n \nabla F_n, \quad j_p = -qp\mu_p \nabla F_p; \tag{6.3}$$

μ_n and μ_p are the mobilities of electrons and holes.

The current continuity equations describing the motion of electrons and holes have to be completed by laws for the recombination of electrons and holes, and by Fermi-Dirac distributions for the densities of electrons and holes:

$$n = N_c \mathcal{F}_{1/2}\left(\frac{q\varphi - qF_n - E_c}{k_B T}\right), \quad p = N_v \mathcal{F}_{1/2}\left(\frac{E_v + qF_p - q\varphi}{k_B T}\right). \tag{6.4}$$

E_c and E_v denote the conduction and valence band edges, respectively. N_c and N_v are the corresponding densities of states (DOS) given by the expressions

$$N_c = 2\left(\frac{m_c k_B T}{2\pi\hbar^2}\right)^{3/2}, \quad N_v = 2\left(\frac{m_v k_B T}{2\pi\hbar^2}\right)^{3/2}. \tag{6.5}$$

m_c and m_v are the density of state masses, T is the temperature, k_B is Boltzmann's constant, and $\mathcal{F}_{1/2}$ Fermi's integral of the order 1/2:

$$\mathcal{F}_{1/2}(x) = \frac{2}{\sqrt{\pi}} \int_0^\infty \frac{\sqrt{y}}{1 + \exp(y - x)} dy. \tag{6.6}$$

The constitutive equations link the classical drift-diffusion equations to the quantum mechanical model for the electronic structure. Typically one derives parameters in a constitutive law by upscaling of electronic structure information. Examples are the band-edges E_c, E_v and the density of states N_c, N_v.

6.2 Effective band-edges and densities of states

In the flat-band case, $q\varphi - qF_n$ and $q\varphi - qF_p$ are approximately constant across the nanostructure. Therefore, we can define the quasi Fermi levels of the quantum confined electrons and holes by $E_{Fn} = q\varphi - qF_n$ and $E_{Fp} = q\varphi - qF_p$, respectively. SMQW lasers are usually designed such that they operate in this flat-band mode. Assuming thermodynamic equilibrium of confined electrons and holes, respectively, their local density distributions for given quasi Fermi levels E_{Fn} and E_{Fp} are calculated by

$$n_{qw}(z) = \sum_{i \in c} \frac{1}{2\pi^2} \int f\Big(E_i(k_\parallel) - E_{Fn}\Big) \|F_i(z; k_\parallel)\|^2_{\mathbb{C}^d} \, dk_\parallel, \tag{6.7}$$

$$p_{qw}(z) = \sum_{j \in v} \frac{1}{2\pi^2} \int f\Big(E_{Fp} - E_j(k_\parallel)\Big) \|F_j(z; k_\parallel)\|^2_{\mathbb{C}^d} \, dk_\parallel, \tag{6.8}$$

with Fermi's function

$$f(E) = \left(1 + \exp\left(\frac{E}{k_B T}\right)\right)^{-1}. \tag{6.9}$$

We introduce the *average* carrier densities per quantum well by $\bar{n} = \int_\Omega n_{qw}(z) dz / d_{qw}$ and $\bar{p} = \int_\Omega p_{qw}(z) dz / d_{qw}$, where d_{qw} is the thickness of the well. For our example quantum well structure from Sect. 5 the local carrier distributions are plotted in Fig. 6.1 for different values of the sheet concentrations $N = \bar{n}\, d_{qw}$ and $P = \bar{p}\, d_{qw}$.

For a calculated band structure we regard the averaged carrier densities in their dependence on E_{Fn}, E_{Fp} and $k_B T$:

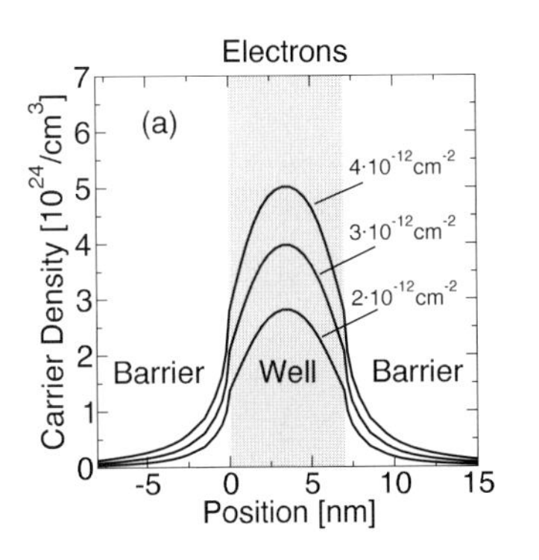

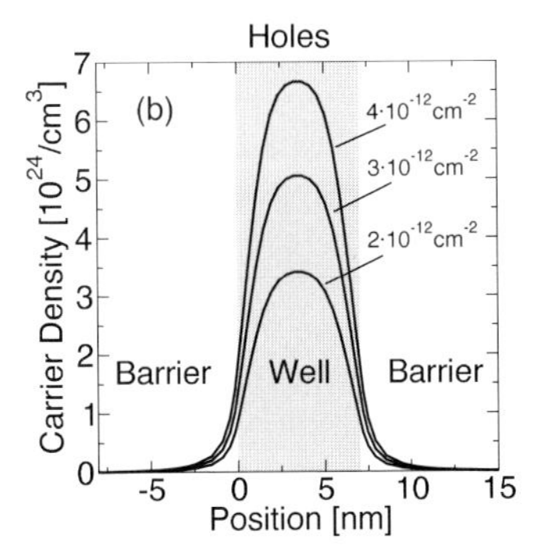

Fig. 6.1. Local carrier density distributions (6.7) and (6.8) for different values of the sheet concentrations $N = P = 2{\cdot}10^{-12} cm^{-2}, 3{\cdot}10^{-12} cm^{-2}, 4{\cdot}10^{-12} cm^{-2}$ for ambient temperature. One observes a stronger confinement of the holes in comparison to the electrons.

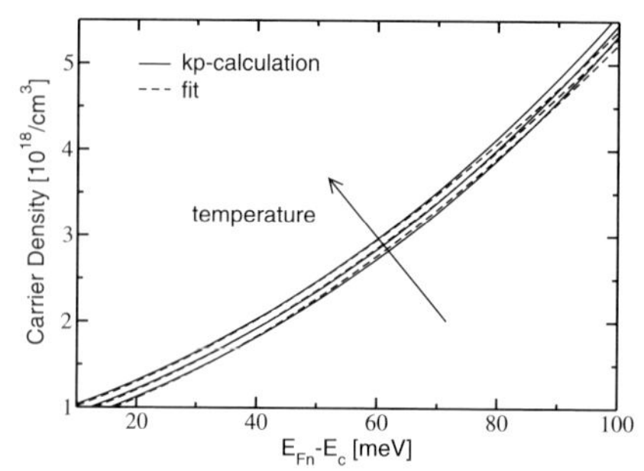

Fig. 6.2. Relation (6.10) (kp-calculation) between electron density n and Fermi level E_{Fn} relative to the net band edge E_c for temperatures T=290K, 315K, 340K. The dashed lines indicate the fit to the macroscopic state-equation (6.11). Reprinted with permission from [BHK03]. © 2003, IEEE.

$$\bar{n} = \bar{n}\left(E_{Fn}, k_B T\right), \quad \bar{p} = \bar{p}\left(E_{Fp}, k_B T\right), \tag{6.10}$$

see Fig. 6.2 and Fig. 6.3 for our example. These relations are fitted to the Fermi-Dirac distributions

$$\bar{n} = N_c \mathcal{F}_{1/2}\left(\frac{E_{Fn} - E_c}{k_B T}\right), \tag{6.11}$$

$$\bar{p} = N_v \mathcal{F}_{1/2}\left(\frac{E_v - E_{Fp}}{k_B T}\right) \tag{6.12}$$

by adjusting the parameters N_c,E_c and N_v, E_v for a specified reference temperature T_0. By (6.5) we obtain the density of state masses m_c and m_v. This procedure provides quantum corrected band-edges E_c, E_v and density of state masses m_c and m_v. Thus, one can treat individual quantum wells as an artificial classical material, whose parameters significantly differ from bulk values of the quantum well materials. This approach has been applied to the the

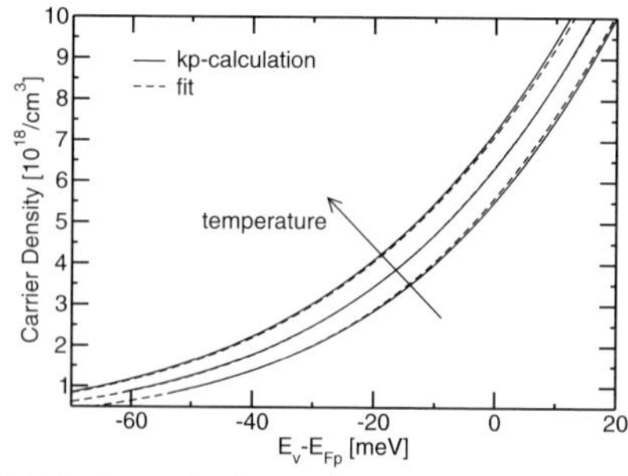

Fig. 6.3. Relation (6.10) (kp-calculation) between hole density p and Fermi level E_{Fp} relative to the net band edge E_v for temperatures T=290K, 315K, 340K. The dashed lines indicate the fit to the macroscopic state-equation (6.12). Reprinted with permission from [BHK03]. © 2003, IEEE.

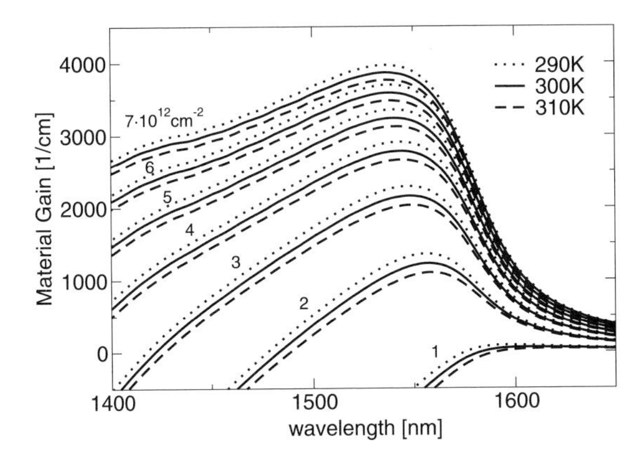

Fig. 6.4. Material gain spectra (TE-polarization) according to (6.13) for different sheet concentrations and temperatures. Reprinted with permission from [BHK03]. © 2003, IEEE.

MQW structure discussed in Sect. 5. Fig. 6.3 and Fig. 6.2 demonstrate the high quality of the fit for this particular structure [BHK03].

6.3 Material gain

Still assuming the flat-band conditions, the band structure and the momentum matrix elements (5.1) enter the expression for the material gain

$$g(\omega) = \frac{\pi\hbar q^2}{\varepsilon_0 m_0^2 n_r c}\frac{1}{d_{qw}}\frac{1}{4\pi^2}\sum_{\substack{i\in c\\ j\in v}}\int\frac{|p_{ij}e|^2}{E_i - E_j} f(E_i - E_{Fn})\,(1 - f(E_j - E_{Fp}))\times$$
$$\left[1 - \exp\left(\frac{\hbar\omega - (E_{Fn} - E_{Fp})}{k_B T}\right)\right]\frac{1}{\pi}\frac{\Gamma}{[(E_i - E_j) - \hbar\omega]^2 + \Gamma^2}\, dk_{\|}, \tag{6.13}$$

where the last factor includes broadening due to collision processes [End97]. The latter have been parametrized by a characteristic intra-band relaxation time τ of 60 fs ($\Gamma = \hbar/\tau$). c is the speed of light and n_r the refractive index.

For the case of an undoped active region and local charge neutrality the calculated gain spectra are drawn in Fig. 6.4 for different excitations and temperatures. The evolution of the corresponding maximum material gain with the carrier density shown in Fig. 6.5 for different temperatures. In [BHK03] it is discussed how to fit the calculated peak gain characteristics to a logarithmic model $g(n) = g_0 \log(n/n_t)$, which is used as a state equation in semi-classical equations.

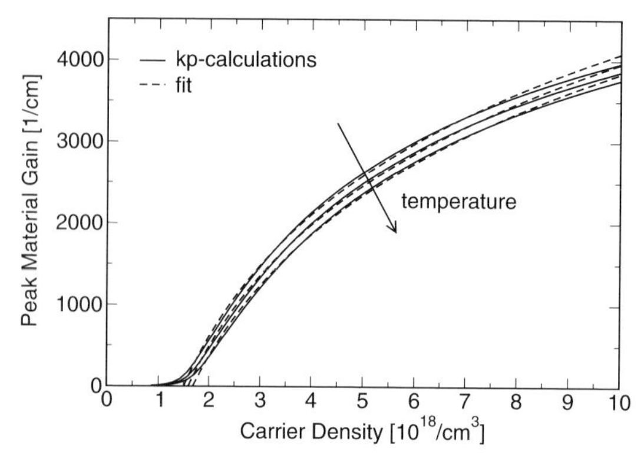

Fig. 6.5. Maximum material gain for different carrier densities and temperatures T=290K, 300K, 310K. Dashed lines indicate the fit to the logarithmic gain model $g(n) = g_0 \log(n/n_t)$. Whereas $g_0 = 2155\text{cm}^{-1}$ is not very sensitive to the temperature and has been kept constant here, the transparency density $n_t = n_t(T)$ roughly linearly increases with the temperature. Reprinted with permission from [BHK03]. © 2003, IEEE.

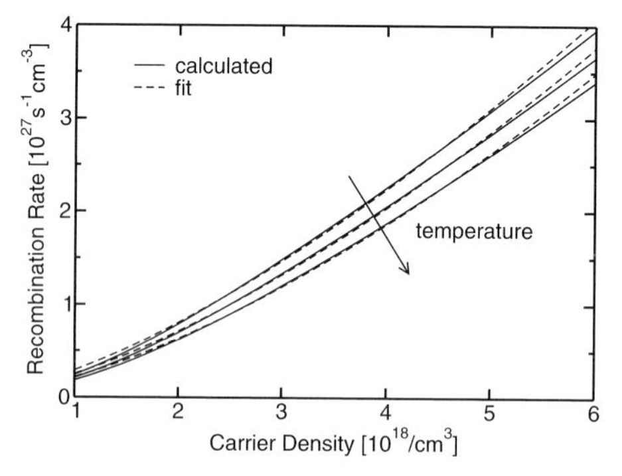

Fig. 6.6. Spontaneous radiative recombination rate (6.14) for different temperatures T=290K, 315K, 340K. Dashed: fit to $R_{rad} = Bn^{\alpha}$, with $\alpha = 1.47\ (290K), 1.51\ (315K), 1.55\ (340K)$. Reprinted with permission from [BHK03]. © 2003, IEEE.

6.4 Spontaneous radiative recombination rate

The spontaneous radiative recombination rate R_{rad} is calculated according to [Wen] by

$$R_{rad} = \frac{n_r q^2}{\pi \hbar^2 c^3 \varepsilon_0 m_0^2} \frac{1}{d_{qw}} \frac{1}{4\pi^2} \sum_{\substack{i \in c \\ j \in v}} \int (E_i - E_j) \left| p_{ij} \right|^2 \times \\ f(E_i - E_{Fn}) f(E_{Fp} - E_j) \, dk_{\|}. \tag{6.14}$$

It is shown in Fig. 6.6 together with the fit to $R_{rad} = Bn^{\alpha}$. The exponent was approximately $\alpha = 1.5$ which differs from the commonly used models corresponding to $\alpha = 2$.

7 Avoiding spurious modes by adjusting E_p

For low band gap semiconductors, the parameters of the eight-band Hamiltonian often cause anomalous band bending, see Sect. 2.3. This may lead to spurious modes of the $k \cdot p$ Schrödinger operator for heterostructures, see Sect. 3.4. However, it is possible to fullfil the condition $\zeta < \zeta_{crit}$ by lowering the value of E_p (2.10):

$$E_p \to E'_p = \alpha E_p, \quad \alpha < 1 \tag{7.1}$$

This adjustment has only a small influence on the band structure for low values of k and therefore leaves the confined states nearly untouched.

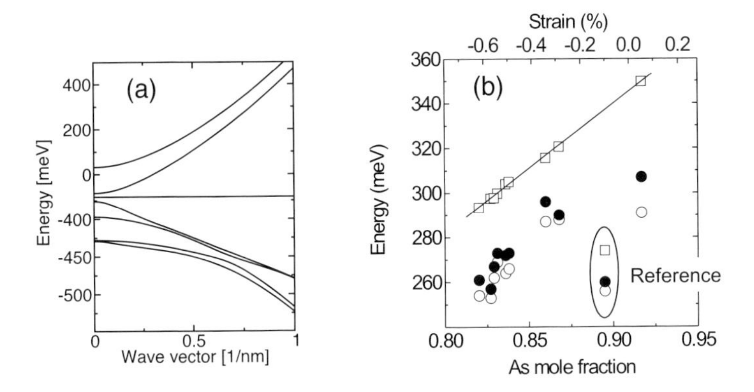

Fig. 7.1. (a) Calculated conduction and valence band structures for a $InAs_xSb_{1-x}/GaSb$ quantum well with the As mole fraction of $x = 0.82$ at ambient temperature, see [KB*05]. (b) Comparison between eight-band $k \cdot p$-band structure calculations and experiments for a set of InAsSb/GaSb multiple quantum well samples with different As mole fractions. Full circles indicate the MQW-edge as determined from transmittance, whereas open circles point to the edge as deduced from PL. Squares mark the edge as obtained from the energy difference of the lowest conduction band (at $k = 0$) and the highest valence band (at $k = 0$) according to 8-band $k \cdot p$-calculations. One observes a systematic IR-shift of the QW-edge data experimentally determined compared to the $k \cdot p$ calculation. This tendency is also present in the $InAs_{0.895}Sb_{0.105}$ bulk-like sample, indicated as 'Reference'. Thus it is not likely to be a residual effect of the $k \cdot p$ calculation. (b) reprinted with permission from [KB*05].

In the limit case $E_p \rightarrow 0$ the conduction bands and the valence bands decouple, and one arrives at the usual six-band Hamiltonian for valence band states. This Hamiltonian still provides a reasonable approximation of the valence band structure [CC92, For93, MGO94, CC96].

The $InAs_xSb_{1-x}$ system is the alloy with the lowest direct band-gap value of all III-V-semiconductor materials. This mixed crystal system is of utmost interest for infrared (IR) optoelectronic applications such as heterostructure-based lasers and detectors [WG*00, GKS00]. In [KB*05] the rescaling approach is applied to the calculation of the electronic states in a set of samples with different mole fractions x. For a comparison of eight-band $k{\cdot}p$ calculation with the experimentally measured properties of the samples, see Fig. 7.1.

8 Biased quantum wells

Typical applications of biased quantum well structures are photonic integrated chips consisting of integrated semiconductor laser/electro-absorption modulator [Bas95, Ch. 6], [Chu95, Ch. 13]. In such devices, the external electric field allows to modulate the absorption or the reflectivity for specific spectral regions.

The applied voltage leads to tilted band edges in the quantum well region, hence, to meta-stable states. If the applied bias is small with respect to the band-edge offsets ΔE_v and ΔE_c of the quantum well to the barrier, then hard wall or soft wall boundary conditions can still be applied for the calculation of the electronic states [SB87, DF93, ASV98]. The calculated eigenvalues $E_i(k_\parallel)$ yield an reasonable approximation of the real parts of the complex eigenvalues corresponding to the meta-stable states of the biased structure.

As an example we discuss the application of this approach to the calculation of the band structure and the local carrier density distributions for a 13 nm thick InGaAsP quantum well, lattice-matched to InP barriers. In Figs. 8.1 and 8.2 we present the results of eight-band $k{\cdot}p$ calculations for various values of the electrical field. Due to the Kramers degeneracy the heavy and light hole bands are double degenerate for an unbiased quantum well, see Fig. 8.1a. This spin degeneracy is lifted if an electric field is applied to the quantum well, see Fig. 8.1b. The spin splitting of the valence bands induced by the external electric field is known as the Rashba effect.

Fig. 8.2 shows titled band edge profiles and carrier density distributions for different applied biases. For the electric field strength $F = 6$ V/μm one observes the onset of accumulation of the hole density in the barrier region, marking the limit of the approach based on the use of hard wall boundary conditions.

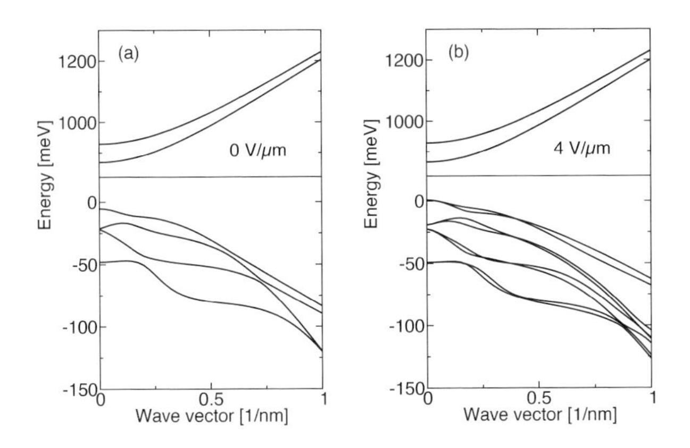

Fig. 8.1. Band structure ([100] direction) in a 13 nm InGaAsP quantum well lattice-matched to InP barriers for two different external electric fields. (a) F=0 V/μm, and (b) F=4 V/μm. At F=0 V/μm, the valence bands are double degenerate. At F=4 V/μm, they split due to the applied external electric field (Rashba effect).

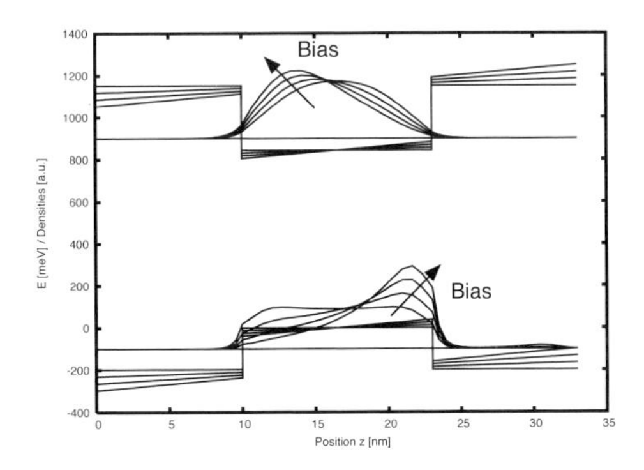

Fig. 8.2. Tilted band edge profiles $E_c(z)$ and $E_v(z)$ and local density distributions for electrons and holes in a 13 nm InGaAsP quantum well lattice-matched to InP barriers for different external electric fields. F=0 V/μm, 2 V/μm, 4 V/μm, 6 V/μm.

References

[ASV98] A. Ahland, D. Schulz, and E. Voges. Efficient modeling of the optical properties of MQW modulators on InGaAsP with absorption edge merging. *IEEE Journal of Quantum Electronics*, 34, 1597–1603, 1998.

[Bah90] T. B. Bahder. Eight-band $k \cdot p$ model of strained zinc-blende crystals. *Physical Review B*, 41(17), 11992–12001, 1990.

[Bas88] G. Bastard. *Wave Mechanics applied to Semiconductor Heterostructures.* Halsted Press, 1988.

[Bas95] M. Bass, editor. *Handbook of optics. 2. Devices, measurement, and properties.* McGraw-Hill, New York, 1995.

[BGH05] U. Bandelow, H. Gajewski, and R. Hünlich. Fabry–Perot Lasers: Thermodynamics-Based Modeling. In J. Piprek, editor, *Optoelectronic Devices.* Springer, 2005.

[BGK00] U. Bandelow, H. Gajewski, and H.-C. Kaiser. Modeling combined effects of carrier injection, photon dynamics and heating in Strained Multi-Quantum Well Lasers. In M. O. Rolf H. Binder, Peter Blood, editor, *Physics and Simulation of Optoelectronic Devices VIII*, volume 3944 of *Proceedings of SPIE*, pages 301–310, August 2000.

[BHK03] U. Bandelow, R. Hünlich, and T. Koprucki. Simulation of Static and Dynamic Properties of Edge-Emitting Multiple-Quantum-Well Lasers. *IEEE Journal of Selected Topics in Quantum Electronics*, 9, 798–806, 2003.

[BKqw] U. Bandelow and T. Koprucki. WIAS-QW. Online: http://www.wias-berlin.de/software/qw.

[BK*00] U. Bandelow, H.-C. Kaiser, T. Koprucki, and J. Rehberg. Spectral properties of $k \cdot p$ Schrödinger operators in one space dimension. *Numerical Functional Analysis and Optimization*, 21, 379–409, 2000.

[BK*03] U. Bandelow, H.-C. Kaiser, T. Koprucki, and J. Rehberg. Modeling and simulation of strained quantum wells in semiconductor lasers. In W. Jäger and H.-J. Krebs, editors, *Mathematics - Key Technology for the Future. Joint Projects Between Universities and Industry*, pages 377–390. Springer Verlag, Berlin Heidelberg, 2003.

[Blo32] F. Bloch. Über die Quantenmechanik der Electronen in Kristallgittern. *Z. Physik*, 52, 555–600, 1932.

[BP74] G. L. Bir and G. E. Pikus. *Symmetry and Strain-Induced Effects in Semiconductors.* John Wiley & Sons, New York, 1974. Übersetzung aus dem Russischen von P. Shelnitz.

[Bur92] M. G. Burt. The justification for applying the effective-mass approximation to microstructures. *J. Physics. Condens. Matter*, 4, 6651–6690, 1992.

[Bur94] M. G. Burt. Direct derivation of effective-mass equations for microstructures with atomically abrupt boundaries. *Physical Review B*, 50(11), 7518–7525, 1994.

[Bur99] M. G. Burt. Fundementals of envelope function theory for electronic states and photonic modes in nanostructures. *J. Physics. Condens. Matter*, 11, R53–R83, 1999.

[Car96] M. Cardona. *Fundamentals of Semiconductors.* Springer, Berlin, 1996.

[CB66] M. L. Cohen and T. K. Bergstresser. Band Structures and Pseudopotential Form Factors for Fourteen Semiconductors of the Diamond and Zinc-blende Structures. *Phys. Rev.*, 141, 789–796, 1966.

[CC76] J. R. Chelikowsky and M. L. Cohen. Nonlocal pseudopotential calculations for the electronic structure of eleven diamond and zinc-blende semiconductors. *Phys. Rev. B*, 14, 556–582, 1976.

[CC92] C. Y.-P. Chao and S. L. Chuang. Spin-orbit-coupling effects on the valence-band structure of strained semiconductor quantum wells. *Physical Review B*, 46(7), 4110–4122, 1992.

[CC96] S. L. Chuang and C. S. Chang. k·p method for strained wurtzite semiconductors. *Phys. Rev. B*, 54, 2491–2504, 1996.

[Chu95] S. L. Chuang. *Physics of optoelectronic Devices.* Wiley & Sons, New York, 1995.

[CKI94] W. W. Chow, S. W. Koch, and M. S. III. *Semiconductor–Laser Physics.* Springer–Verlag, Berlin, 1994.

[DF93] P. Debernardi and P. Fasano. Quantum confined Stark effect in semiconductor quantum wells including valence band mixing and Coulomb effects. *IEEE Journal of Quantum Electronics*, 29, 2741–2755, 1993.

[EBWS95] P. Enders, A. Bärwolff, M. Woerner, and D. Suisky. $k \cdot p$ theory of energy bands, wave functions and optical selection rules in strained tetrahedral semiconductors. *Physical Review B*, 51(23), 16695–16704, 1995.

[End97] P. Enders. Enhancement and spectral shift of optical gain in semiconductors from non–markovian intraband relaxation. *IEEE Journal of Quantum Electronics*, 33(4), 580–588, 1997.

[EW96] P. Enders and M. Woerner. Exact 4×4 block diagonalization of the eight-band $k \cdot p$ Hamiltonian matrix for tetrahedral semiconductors and its application to strained quantum wells. *Semicond. Sci. Technol.*, 11, 983–988, 1996.

[For93] B. A. Foreman. Effective-mass Hamiltonian and boundary conditions for the valence bands of semiconductor microstructures. *Physical Review B*, 48(7), 4964–4967, 1993.

[For97] B. A. Foreman. Elimination of spurious solutions from eight-band $k \cdot p$ theory. *Physical Review B*, 56(20), R12748–R12751, 1997.

[Gaj93] H. Gajewski. Analysis und Numerik von Ladungstransport in Halbleitern (Analysis and numerics of carrier transport in semiconductors). *Mitt. Ges. Angew. Math. Mech.*, 16(1), 35–57, 1993.

[GKS00] H. H. Gao, A. Krier, and V. V. Sherstnev. *Appl. Phys. Lett.*, 77, 872, 2000.

[Kan66] E. O. Kane. The $k \cdot p$ Method. In R. K. Willardson and A. C. Beer, editors, *Semiconductors and Semimetals*, volume 1, chapter 3, pages 75–100. Academic Press, New York and London, 1966.

[Kan82] E. O. Kane. Energy Band Theory. In W. Paul, editor, *Handbook on Semiconductors*, volume 1, chapter 4a, pages 193–217. North-Holland, Amsterdam, New York, Oxford, 1982.

[Kat84] T. Kato. *Pertubation theory for linear operators*, volume 132 of *Grundlehren der mathematischen Wissenschaften.* Springer Verlag, Berlin, 1984.

[KB*05] T. Koprucki, M. Baro, U. Bandelow, T. Tien, F. Weik, J. Tomm, M. Grau, and M.-C. Amann. Electronic structure and optoelectronic properties of strained InAsSb/GaSb multiple quantum wells. *Appl. Phys. Lett.*, 87, 81911/1–181911/3, 2005.

[Kri91] M. P. C. M. Krijn. Heterojunction band offsets and effective masses in III-V quarternary alloys. *Semicond. Sci. Technol.*, 6, 27–31, 1991.

[MGO94] A. T. Meney, B. Gonul, and E. P. O'Reilly. Evaluation of various approximations used in the envelope-function method. *Physical Review B*, 50(15), 10893–10904, 1994.

[SB87] G. D. Sanders and K. K. Bajaj. Electronic properties and optical-absorption spectra of GaAs-$Al_xGa_{1-x}As$ quantum wells in externally appield electric fields. *Phys. Rev. B*, 35, 2308–2320, 1987.

[Sin93] J. Singh. *Physics of semiconductors and their heterostructures.* McGraw–Hill, New York, 1993.

[Sol03] X. C. Soler. *Theoretical Methods for Spintronics in Semiconductors with Applications.* PhD thesis, California Institute of Technology, Pasadena, California, USA, 2003.

[Sti01] O. Stier. *Electronic and Optical Properties of Quantum Dots and Wires.* Dissertation TU Berlin, Germany. Wissenschaft & Technik Verlag, Berlin, 2001.

[VdW89] C. G. Van de Walle. Band lineups and deformation potentials in the model-solid theory. *Phys. Rev. B*, 39, 1871–1883, 1989.

[VM*01] I. Vurgaftman, J. R. Meyer, and L. R. Ram-Mohan. Band parameters for III-V compound semiconductors and their alloys. *J. Appl. Phys.*, 89, 5815–5875, 2001.

[Wen] H. Wenzel. How to use the kp8 programs. Online: http://www.fbh-berlin.de/people/wenzel/kp8.html.

[WG*00] A. Wilk, M. E. Gazouli, M. E. Skouri, P. Cristol, P. Grech, A. N. Baranov, and A. Joullie. *Appl. Phys. Lett.*, 77, 2298, 2000.

Shape Optimization of Biomorphic Ceramics with Microstructures by Homogenization Modeling

Ronald H.W. Hoppe[1,2] and Svetozara I. Petrova[1,3]

[1] Institute of Mathematics, University of Augsburg, 86159 Augsburg.
hoppe@math.uni-augsburg.de, petrova@math.uni-augsburg.de
[2] Department of Mathematics, University of Houston, Houston, TX 77204-3008, USA
[3] Institute for Parallel Processing, Bulgarian Academy of Sciences, 1113 Sofia, Bulgaria

Summary. We consider the modeling, simulation, and optimization of microstructural cellular biomorphic ceramics obtained by biotemplating. This is a process in biomimetics, a recently emerged discipline in materials science where engineers try to mimick or use biological materials for the design of innovative technological devices and systems. In particular, we focus on the shape optimization of microcellular silicon carbide ceramic materials derived from naturally grown wood. The mechanical behavior of the final ceramics is largely determined by the geometry of its microstructure which can be very precisely tuned during the biotemplating process. Our ultimate goal is to determine these microstructural details in such a way that an optimal mechanical performance is achieved with respect to merit criteria depending on the specific application. Within the shape optimization problem the state variables are the displacements subject to the underlying elasticity equations, and the design variables are the geometrical quantities determining the microstructure. Since a resolution of the microstructure is numerically cost-prohibitive, we use the homogenization approach, assuming periodically distributed microcells. Adaptive mesh-refinement techniques based on reliable and efficient a posteriori error estimators are applied in the microstructure to compute the homogenized elasticity coefficients. The shape optimization problem on the macroscopic homogenized model is solved by primal-dual Newton-type interior-point methods. Various numerical experiments are presented and discussed.

1 Introduction

Biomimetics, also called bionics or biomimicry, is a discipline in materials science that has recently attracted a lot of attention. It allows the cost-effective production of high performance technological devices and systems by either mimicking or using naturally grown biological structures (cf., e.g.,

[Eli00, HDL02, GA88]). In contrast to engineering materials, biological structures exhibit a hierarchically built anatomy, developed and optimized in a long-term genetic process. Their inherent cellular, open porous morphology can be used for liquid or gaseous infiltration and subsequently for high temperature reaction processes. A specific example of such a naturally grown biological material is wood which exhibits an anisotropic, porous morphology with excellent strength at low density, high stiffness, elasticity, and damage tolerance. Typical feature of the wood structure is the system of the tracheidal cells which provide the transportation path for water and minerals within the living tree. This open porous system is accessible for infiltration of various metals.

A recent idea in biomimetical applications is to take advantage of naturally grown wood in the production of high performance ceramics to be used as filters and catalysts in chemical processing, heat insulation structures, thermally and mechanically loaded lightweight structures, and medical implants (for instance, for bone substitution). In particular, silicon carbide (SiC) is known as a material that is not only suitable in microelectronical applications due to its bandwidth structure but also useful in mechanical and high temperature applications with regard to its excellent thermomechanical properties. Since wood essentially consists of carbon (C), the idea is to use it as a basic material for the production of highly porous ceramics. Among the large variety of ceramic composites, new biomorphic cellular silicon carbide ceramics from wood were recently produced and investigated, see [GLK98a, GLK98b, OT*95, VSG02]. The conversion of naturally grown wood to highly porous SiC ceramics is done by a process called *biotemplating* which includes two processing steps, illustrated in Fig. 1.1.

Biological porous carbonized preforms (also called $C-$templates) can be derived from different wood structures by drying and high-temperature pyrolysis at temperatures between 800 and 1800^oC and used as templates for infiltrations by gaseous or liquid silicon (Si) to form SiC and SiSiC-ceramics, respectively. During high-temperature processing, the microstructural properties of the bioorganic preforms were retained, so that a one-to-one reproduc-

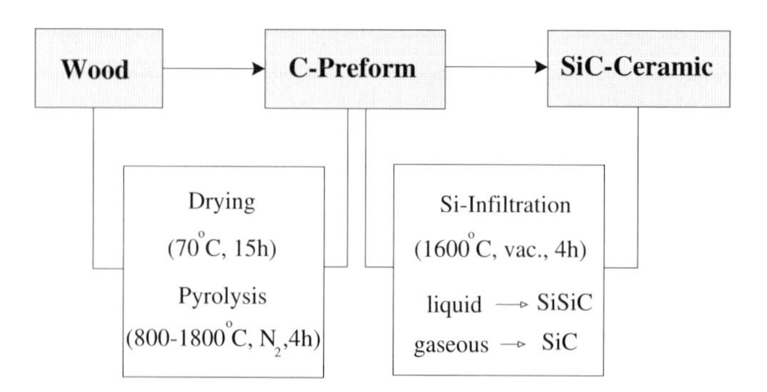

Fig. 1.1. Processing scheme of SiC-ceramics from wood

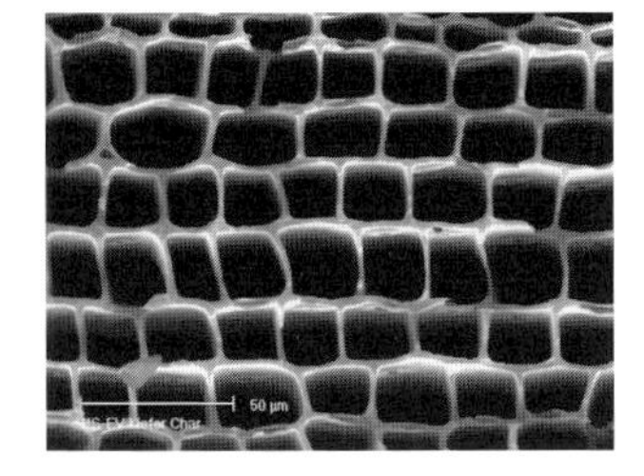
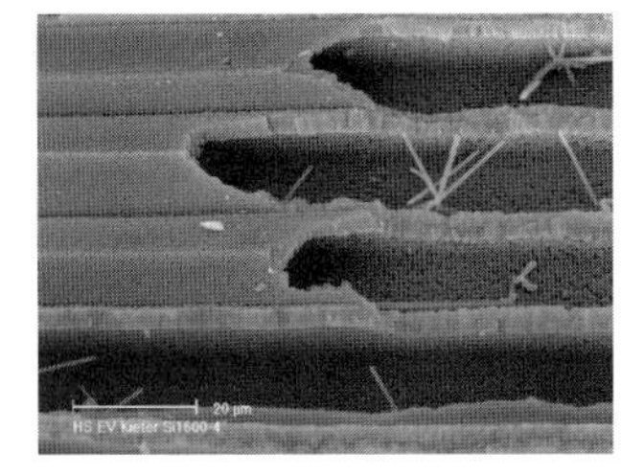

Fig. 1.2. SiC-ceramic derived from pine wood a) radial direction; b) axial direction

tion of the original wood structure was obtained, see Fig. 1.2. The resulting cellular composites show low density, high specific strength, and excellent high temperature stability.

The geometry of the final ceramics, i.e., the widths and lengths of the different layers forming the struts, can be determined very precisely by an appropriate tuning of the process parameters. This raises the question how to choose these microstructural geometrical data in order to achieve an optimal performance with respect to a prespecified merit criterion depending on the specific application. From a mathematical point of view, this issue represents a shape optimization problem where the *state variables* are subject to the underlying elasticity equations and the microstructural data serve as *design variables*. As far as the solution of such a shape optimization problem is concerned, the resolution of the microstructure is cost-prohibitive with respect to both computational time and storage. Therefore, the idea is to perform homogenization, assuming a periodically distributed microstructure, and to apply the optimization to the homogenized model.

In this study, we focus both on the homogenization process and on the application of state-of-the-art optimization strategies for the numerical solution of the shape optimization problem under consideration. The remaining of the paper is organized as follows: In Sect. 2, we describe in detail the homogenization technique that provides a macromechanical model where the components of the homogenized elasticity tensor reflect the microstructural details. Section 3 deals with the setting of our shape optimization problem. In Sect. 4, we present a primal-dual Newton interior-point method and in Sect. 5 we comment on the numerical solution of the condensed primal-dual system. Section 6 concerns adaptive grid-refinement procedures based on a posteriori error estimators. In particular, we use the *Zienkiewicz-Zhu* error estimator proposed in [ZZ87]. Iterative solution techniques for the homogenized elasticity equation and the microcell problem are discussed in Sect. 7. Various numerical results are given in Sect. 8.

2 Homogenized computational model

In this section, we briefly explain the derivation of the homogenized computational model on the macroscale by using the asymptotic homogenization theory. Homogenization has been successfully used in the last three decades for solving multi-scale problems on computational regions occupied by heterogeneous microstructural materials (see, cf., [BP84, BLP78, JKO94, SP80]).

Let $\Omega \subset \mathcal{R}^d$, $d = 2, 3$, be a domain occupied by a heterogeneous material with microstructures of periodically distributed constituents. Suppose that the boundary of Ω, denoted by Γ, consists of a prescribed displacement boundary Γ_D (meas $\Gamma_D > 0$) and a prescribed traction boundary Γ_T, such that $\Gamma = \Gamma_D \cup \Gamma_T$, $\Gamma_D \cap \Gamma_T = \emptyset$. Let $\boldsymbol{b}$ be the body force, $\bar{\boldsymbol{u}}$ be the prescribed displacement on Γ_D, and $\bar{\boldsymbol{t}}$ be the prescribed traction on Γ_T.

The homogenized model for our original heterogeneous material occupying the domain Ω, $\Omega \subset \mathcal{R}^d$, $d = 2, 3$, is illustrated in Fig. 2.1. The main idea of the homogenization is to replace the heterogeneous material by an equivalent homogenized material, extracting the information for the material properties of the various microstructural constituents (or different phases).

The microscopic and macroscopic models are considered simultaneously supposing a strong scale separation, i.e., a large gap in length scale between the macroscopic component and the microstructure. In practical applications the microscopic length scales are orders of magnitude smaller than the physical macroscopic length scale. A main assumption in the homogenization approach is that the original heterogeneous material workpiece is composed of periodically distributed microstructures of various constituents. To couple properly the micro- and macro-scales, we choose a representative volume element (RVE) or a unit microstructure.

Consider a stationary microstructure with a geometrically simple tracheidal periodicity cell $Y = [0, 1]^d$, $d = 2, 3$, (see Fig. 2.2) consisting of an outer layer of carbon (C), interior layer of silicon carbide (SiC), and a

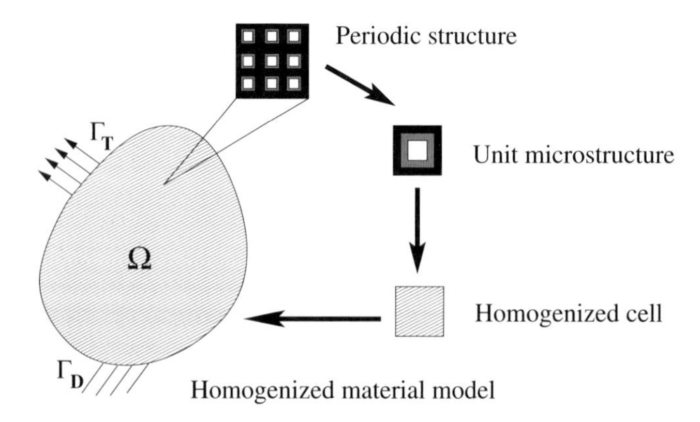

Fig. 2.1. The macroscopic homogenized material model

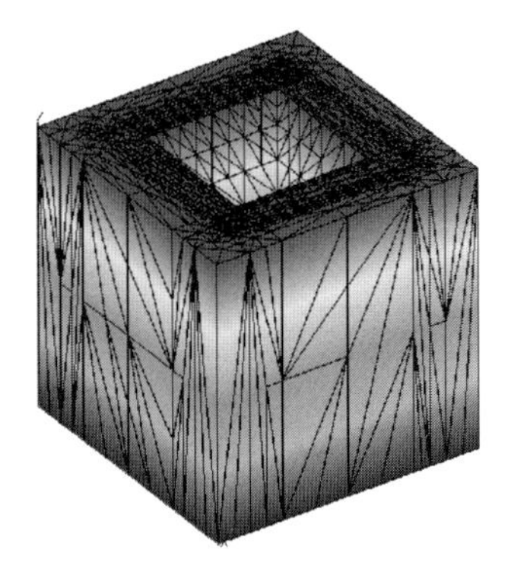

Fig. 2.2. a) 3-D unit periodicity cell Y, b) 2-D unit periodicity cell $Y = P \cup SiC \cup C$

centered pore channel (P, no material). We introduce two space variables $\boldsymbol{x}$ (macroscopic/slow variable) and $\boldsymbol{y}$ (microscopic/fast variable) and denote by ε, $\boldsymbol{y} = \boldsymbol{x}/\varepsilon$, $\varepsilon \ll 1$, the scale parameter (dimensionless number) which, in fact, represents the periodicity under the assumption that ε is very small with respect to the size of Ω, i.e., there exists a large scale gap between the microstructure and the macroscopic component.

The parameter ε allows us to define macrofunctions in terms of the microstructural behavior and vice versa. Thus, for any state function $f(\boldsymbol{y}) := f(\boldsymbol{x}/\varepsilon)$, one can compute the spatial derivatives by using the following differentiation rule

$$\frac{d}{d\boldsymbol{x}} f\left(\boldsymbol{x}, \frac{\boldsymbol{x}}{\varepsilon}\right) = \frac{\partial f(\boldsymbol{x}, \boldsymbol{y})}{\partial \boldsymbol{x}} + \varepsilon^{-1} \frac{\partial f(\boldsymbol{x}, \boldsymbol{y})}{\partial \boldsymbol{y}}.$$

Consider the following elasticity equation defined in the microstructure Y

$$-\nabla \cdot \boldsymbol{\sigma} = \mathbf{F} \qquad \text{in } Y \tag{2.1}$$

with a load vector $\mathbf{F}$. Here, $\boldsymbol{\sigma}$ is the microscopic symmetric stress, $\boldsymbol{u} \in \boldsymbol{H}^1(Y)$, is the corresponding displacement at point $\boldsymbol{y} \in Y$, and $\boldsymbol{e}$ is the microscopic symmetric strain with components

$$e_{ij}(\boldsymbol{u}(\boldsymbol{y})) = \frac{1}{2}\left(\frac{\partial u_i(\boldsymbol{y})}{\partial y_j} + \frac{\partial u_j(\boldsymbol{y})}{\partial y_i}\right). \tag{2.2}$$

The problem (2.1) is subject to periodic boundary conditions on the outer part of ∂Y, Neumann boundary conditions around the pore, and continuity conditions $[\boldsymbol{u}] = 0$ and $[\boldsymbol{\sigma} \cdot \boldsymbol{n}] = 0$ on the interfaces between the different phases, see Fig. 2.2. The symbol [ ] denotes the jump of the function across the corresponding interface with a normal vector $\boldsymbol{n}$ (cf., e.g., [BP84]).

Assuming linearly elastic constituents, the unit microstructure is governed by the Hooke law $\boldsymbol{\sigma} = \boldsymbol{E} : \boldsymbol{e}$ with componentwise $(i, j, k, l = 1, \ldots, d)$ constitutive relations as follows

$$\sigma_{ij}(\boldsymbol{u}) = E_{ijkl}(\boldsymbol{y})\, e_{kl}(\boldsymbol{u}(\boldsymbol{y})). \tag{2.3}$$

Here, we adopt the Einstein convention of a summation on repeated indices. The 4-th order elasticity (also called plain-stress) tensor $\boldsymbol{E}(\boldsymbol{y})$ with components $E_{ijkl}(\boldsymbol{y})$ characterizes the behavior of the material at point $\boldsymbol{y}$ and depends on material constants like Young's modulus and Poisson's ratio. Note that $\boldsymbol{E}(\boldsymbol{y})$ is zero if $\boldsymbol{y}$ is located in the porous subdomain of the microstructure and coincides with the elasticity tensor of the material if $\boldsymbol{y}$ is located in the corresponding microstructural constituent. The elasticity tensor is symmetric in the following sense

$$E_{ijkl} = E_{jikl} = E_{ijlk} = E_{klij} \tag{2.4}$$

and satisfies the following ellipticity conditions

$$E_{ijkl}\, \chi_{ij}\chi_{kl} \geq c\, \chi_{ij}^2, \qquad \forall\, \chi_{ij} = \chi_{ji},$$

for a constant $c > 0$ (cf., e.g., [BP84, BLP78, JKO94]).

Denote by $\boldsymbol{u}_\varepsilon(\boldsymbol{x}) := \boldsymbol{u}(\boldsymbol{x}/\varepsilon)$ the unknown macroscopic displacement vector and consider the following family of elasticity problems

$$-\nabla \cdot \boldsymbol{\sigma}_\varepsilon = \boldsymbol{b} \qquad \text{in } \Omega, \tag{2.5}$$

subject to a macroscopic body force $\boldsymbol{b}$ and a macroscopic surface traction $\boldsymbol{t}$ applied to the portion $\Gamma_T \subset \partial\Omega$. Here, $\boldsymbol{\sigma}_\varepsilon(\boldsymbol{u}_\varepsilon) := \boldsymbol{E}_\varepsilon(\boldsymbol{x})\boldsymbol{e}(\boldsymbol{u}_\varepsilon(\boldsymbol{x}))$ is the stress tensor for $\boldsymbol{x} \in \Omega$ and $\boldsymbol{E}_\varepsilon(\boldsymbol{x}) := \boldsymbol{E}(\boldsymbol{x}/\varepsilon) = \boldsymbol{E}(\boldsymbol{y})$ is the piecewise constant elasticity tensor defined in Y. Following, for instance, [BLP78] for the basic concepts of the homogenization method, the unknown displacement vector is expanded asymptotically as

$$\boldsymbol{u}_\varepsilon(\boldsymbol{x}) = \boldsymbol{u}^{(0)}(\boldsymbol{x},\boldsymbol{y}) + \varepsilon\, \boldsymbol{u}^{(1)}(\boldsymbol{x},\boldsymbol{y}) + \varepsilon^2\, \boldsymbol{u}^{(2)}(\boldsymbol{x},\boldsymbol{y}) + \dots\,, \qquad \boldsymbol{y} = \boldsymbol{x}/\varepsilon, \tag{2.6}$$

where $\boldsymbol{u}^{(i)}(\boldsymbol{x},\boldsymbol{y})$, $i \geq 0$, are $Y-$periodic in $\boldsymbol{y}$, i.e., take equal values on opposite sides of Y. Consider the space $H := \{\boldsymbol{u} | \boldsymbol{u} \in \boldsymbol{H}^1(\Omega),\ \boldsymbol{u} = 0 \text{ on } \Gamma_D\}$. Under the assumptions of symmetry and ellipticity of the elasticity coefficients, it was shown in the homogenization theory that the sequence $\{\boldsymbol{u}_\varepsilon\}$ of solutions of (2.5) tends weakly in H as $\varepsilon \to 0$ to a function $\boldsymbol{u}^{(0)}(\boldsymbol{x}) \in H$, the solution of the following macroscopic homogenized problem with a constant elasticity tensor.

$$-\nabla \cdot \boldsymbol{\sigma} = \boldsymbol{b} \qquad \text{in } \Omega, \tag{2.7}$$

where $\boldsymbol{\sigma} = \boldsymbol{\sigma}(\boldsymbol{u}^{(0)}) := \boldsymbol{E}^H \boldsymbol{e}(\boldsymbol{u}^{(0)}(\boldsymbol{x}))$, $\boldsymbol{x} \in \Omega$, and $\boldsymbol{E}^H$ stands for the homogenized elasticity tensor. Note that $\boldsymbol{u}^{(0)}(\boldsymbol{x})$ depends only on the macroscopic variable $\boldsymbol{x}$ and is independent of the microscopic scale $\boldsymbol{y}$. The leading term $\boldsymbol{u}^{(0)}$ in (2.6) is called a macroscopic displacement and the remaining terms $\boldsymbol{u}^{(i)}$, $i > 0$, are considered as perturbed displacements.

The homogenization method requires to find periodic functions $\boldsymbol{\xi}^{kl}$ satisfying the following problem in a weak formulation to be solved in the microscopic unit cell

$$\int_Y E_{ijpq}(\boldsymbol{y}) \frac{\partial \xi_p^{kl}}{\partial y_q} \frac{\partial \phi_i}{\partial y_j} \, dY = \int_Y E_{ijkl}(\boldsymbol{y}) \frac{\partial \phi_i}{\partial y_j} \, dY, \tag{2.8}$$

where $\boldsymbol{\phi} \in \boldsymbol{H}^1(Y)$ is an arbitrary $Y-$periodic variational function. The function $\boldsymbol{\xi}^{kl}$, also referred to as the characteristic displacement, is found by solving (2.8) in Y with periodic boundary conditions. After computing $\boldsymbol{\xi}^{kl}$, one defines the homogenized coefficients by the following formulas (we refer to [BP84, BLP78, JKO94] for details)

$$E_{ijkl}^H = \frac{1}{|Y|} \int_Y \left(E_{ijkl}(\boldsymbol{y}) - E_{ijpq}(\boldsymbol{y}) \frac{\partial \xi_p^{kl}}{\partial y_q} \right) dY. \tag{2.9}$$

Due to the symmetry conditions (2.4), the 4-th order homogenized elasticity tensor $\boldsymbol{E}^H = (E_{ijkl}^H)$ can be written as a symmetric and usually a dense matrix. In the case $d = 2$ it is a 3×3 matrix and has the form

$$\mathbf{E}^H = \begin{pmatrix} E_{1111}^H & E_{1122}^H & E_{1112}^H \\ & E_{2222}^H & E_{2212}^H \\ \text{SYM} & & E_{1212}^H \end{pmatrix}. \tag{2.10}$$

The 3-d homogenized tensor can be written, respectively, as a 6×6 matrix

$$\mathbf{E}^H = \begin{pmatrix} E_{1111}^H & E_{1122}^H & E_{1133}^H & E_{1112}^H & E_{1123}^H & E_{1113}^H \\ & E_{2222}^H & E_{2233}^H & E_{2212}^H & E_{2223}^H & E_{2213}^H \\ & & E_{3333}^H & E_{3312}^H & E_{3323}^H & E_{3313}^H \\ & & & E_{1212}^H & E_{1223}^H & E_{1213}^H \\ & & & & E_{2323}^H & E_{2313}^H \\ \text{SYM} & & & & & E_{1313}^H \end{pmatrix}. \tag{2.11}$$

The computation of the homogenized elasticity coefficients can be done analytically for some specific geometries as, for instance, layered materials or checkerboard structures. In case of more complicated microstructures, the computation of E_{ijkl}^H has to be done numerically through a suitable microscopic modeling.

Once the constant homogenized coefficients from (2.9) are computed, one comes up with the homogenized macroscopic equation (2.7) given in a weak form as follows

$$\int_\Omega E_{ijkl}^H \frac{\partial u_k^0}{\partial x_l} \frac{\partial v_i}{\partial x_j} d\Omega = \int_\Omega \boldsymbol{b} \cdot \boldsymbol{v} \, d\Omega + \int_{\Gamma_T} \bar{\boldsymbol{t}} \cdot \boldsymbol{v} \, d\Gamma, \qquad \forall \, \boldsymbol{v} \in H, \tag{2.12}$$

where $\boldsymbol{u}^{(0)}(\boldsymbol{x}) := \boldsymbol{u}^{(0)}(\boldsymbol{x}, \boldsymbol{y})$ is the homogenized solution occurring in (2.6).

3 Shape optimization by primal-dual methods

Structural optimization has recently become of increasing interest in computer aided design and optimization of composite structures in materials science (cf., e.g., [Ben95] and the references therein). A typical problem of structural optimization is to minimize a function (called *objective*, *cost* or *criterion* function) over a set of geometrical or behavioral requirements (called *constraints*). The set of structural parameters includes the so-called *state* and *design* parameters, and the problem consists in computing optimal values of the design parameters, such that they minimize the specific objective function. Sizing, shape, and topology optimization problems are different types in structural optimization. Detailed classification of these problems is given, for instance, in [OT83]. In the sizing problems, the goal is to find the optimal thickness distribution of a given material structure. The main difficulty in shape optimization problems arises from the fact that the geometry of a structure is a design variable which means, in particular, that the discretization model associated with the structure has to be changed in the process of optimization, see [All02, SZ92]. In the topology optimization of solid structures we are interested in the determination of the optimal placement of material in space, i.e., one has to determine which points of space are material and which points should remain void (no material). Hence, the main goal of these problems is to find the location of holes and the connectivity of the domain, see [BS03].

Our goal is to optimize mechanical performances of the ceramic composites described in Sect. 1 (such as the compliance or the bending strength) taking into account technological and problem specific constraints on the state and design parameters. Denote the state variables $\boldsymbol{u} = (u_1, ..., u_N)^T$, which are the nodal values of the components of the discrete displacement vector, and the design variables $\boldsymbol{\alpha} = (\alpha_1, ..., \alpha_M)^T$ chosen as the microstructural data determining the geometry of the periodicity cell (widths and lengths of the different materials layers forming the cell walls, see Fig. 2.2). Since the geometrical properties of the final ceramics are not fixed but can be changed and precisely tuned within the processing, we focus on shape optimization of our microcellular SiC ceramic materials. Depending on the specific application, the objective functional $J = J(\boldsymbol{u}, \boldsymbol{\alpha})$ can be chosen according to the following criteria:

- mechanical properties (minimum compliance),
- loading (bending, tension, torsion),
- thermal properties (shock resistance),
- technological properties (minimum weight),
- economical properties (cheapness).

For simplicity, we consider the mean compliance of the structure defined as

$$J(\boldsymbol{u}, \boldsymbol{\alpha}) = \int_{\Omega} \boldsymbol{b} \cdot \boldsymbol{u} \, d\Omega + \int_{\Gamma_T} \bar{\boldsymbol{t}} \cdot \boldsymbol{u} \, d\Gamma, \tag{3.1}$$

Our shape optimization problem reads: Find $(\boldsymbol{u}, \boldsymbol{\alpha}) \in \mathcal{R}^N \times \mathcal{R}^M$ such that

$$J(\boldsymbol{u}, \boldsymbol{\alpha}) = \inf_{\boldsymbol{v}, \boldsymbol{\beta}} J(\boldsymbol{v}, \boldsymbol{\beta}), \tag{3.2}$$

subjected to the following equality and inequality constraints

$$A(\boldsymbol{\alpha})\boldsymbol{u} = \boldsymbol{f}, \qquad g(\boldsymbol{\alpha}) := \sum_{i=1}^{M} \alpha_i = C, \qquad \alpha_{\min}\,\bar{\boldsymbol{e}} \leq \boldsymbol{\alpha} \leq \alpha_{\max}\,\bar{\boldsymbol{e}}, \tag{3.3}$$

where $J(\boldsymbol{u}, \boldsymbol{\alpha})$ is defined by (3.1), $A(\boldsymbol{\alpha})$ is the stiffness matrix corresponding to the homogenized equilibrium equation (2.12), $\boldsymbol{u}$ is the discrete homogenized displacement vector, $\boldsymbol{f}$ is the discrete load vector, and $\bar{\boldsymbol{e}} = (1, 1, \ldots, 1) \in \mathcal{R}^M$. Note that $\alpha_{\min}$ and $\alpha_{\max}$ are technologically motivated lower and upper bounds for the design parameters. In the case of unit microstructure Y, we take the limits $\alpha_{\min} = 0$, $\alpha_{\max} = 0.5$, and $0 < C \leq 0.5$.

4 Primal–dual Newton interior–point method

In the optimization algorithm, we are typically faced with constrained nonconvex nonlinear minimization problems with both equality and inequality constraints on the state variables and design parameters. For the discretized optimization problem we use the primal-dual Newton interior-point methods, recently a topic of intensive research [BHN99, ET*86, FGW02, GOW98, HPS02, HP04b, VS99]. The main idea of these methods is to generate iteratively approximations of the solution which strictly satisfy the inequality constraints. Details are given in this section.

4.1 General nonlinear optimization problem

We consider the following general constrained nonlinear nonconvex programming problem with both equality and inequality constraints

$$\min_{\boldsymbol{x} \in \mathcal{R}^n} f(\boldsymbol{x}), \tag{4.1}$$

subject to

$$\boldsymbol{h}(\boldsymbol{x}) = 0, \qquad \boldsymbol{g}(\boldsymbol{x}) \geq 0, \tag{4.2}$$

where $f : \mathcal{R}^n \to \mathcal{R}$, $\boldsymbol{h} : \mathcal{R}^n \to \mathcal{R}^m, m < n$, and $\boldsymbol{g} : \mathcal{R}^n \to \mathcal{R}^l$ are assumed to be twice Lipschitz continuously differentiable. Note that the constraints (4.2) have to be understood componentwise.

The *Lagrangian function* associated with (4.1)–(4.2) is defined by

$$\mathcal{L}(\boldsymbol{x}, \boldsymbol{y}, \boldsymbol{z}) = f(\boldsymbol{x}) + \boldsymbol{y}^T \boldsymbol{h}(\boldsymbol{x}) - \boldsymbol{z}^T \boldsymbol{g}(\boldsymbol{x}), \tag{4.3}$$

where $\boldsymbol{y} \in \mathcal{R}^m$ and $\boldsymbol{z} \in \mathcal{R}^l$ are the Lagrange multipliers for the equality and inequality constraints, respectively.

The first-order Karush-Kuhn-Tucker (KKT) necessary conditions for optimality of (4.1)–(4.2) read

$$\nabla_{\boldsymbol{x}} \mathcal{L}(\boldsymbol{x}, \boldsymbol{y}, \boldsymbol{z}) = 0, \ \boldsymbol{h}(\boldsymbol{x}) = 0, \ \boldsymbol{g}(\boldsymbol{x}) \geq 0, \ Z\,\boldsymbol{g}(\boldsymbol{x}) = 0, \ \boldsymbol{z} \geq 0,$$

where

$$\nabla_{\boldsymbol{x}} \mathcal{L}(\boldsymbol{x}, \boldsymbol{y}, \boldsymbol{z}) = \nabla f(\boldsymbol{x}) + \sum_{i=1}^{m} y_i \nabla h_i(\boldsymbol{x}) - \sum_{i=1}^{l} z_i \nabla g_i(\boldsymbol{x}) \tag{4.4}$$

is the gradient of the Lagrangian function and Z is the diagonal matrix with a diagonal $\boldsymbol{z}$. We also consider the Hessian of the Lagrangian with respect to $\boldsymbol{x}$ defined by

$$\nabla_x^2 \mathcal{L}(\boldsymbol{x}, \boldsymbol{y}, \boldsymbol{z}) = \nabla^2 f(\boldsymbol{x}) + \sum_{i=1}^{m} y_i \nabla^2 h_i(\boldsymbol{x}) - \sum_{i=1}^{l} z_i \nabla^2 g_i(\boldsymbol{x}), \tag{4.5}$$

where $\nabla^2 f(\boldsymbol{x}), \ \nabla^2 h_i(\boldsymbol{x}), \ 1 \leq i \leq m, \ \nabla^2 g_i(\boldsymbol{x}), 1 \leq i \leq l$ stand for the Hessians of $f(\boldsymbol{x}), h_i(\boldsymbol{x})$, and $g_i(x)$, respectively. Denote by

$$\mathcal{A}(\boldsymbol{x}) = \{i, \ g_i(\boldsymbol{x}) = 0, \ i = 1, \ldots, l\}$$

the set of all indices for which the inequality constraints are equal to zero at $\boldsymbol{x}$. We are interested in finding local minimizers of our optimization problem (4.1)–(4.2). Assume that at least one such point $\boldsymbol{x}^*$ exists satisfying the conditions:

- **Feasibility.** $\boldsymbol{h}(\boldsymbol{x}^*) = 0$ and $\boldsymbol{g}(\boldsymbol{x}^*) \geq 0$.
- **Regularity.** The set $\{\nabla h_1(\boldsymbol{x}^*), \ldots, \nabla h_m(\boldsymbol{x}^*)\} \cup \{\nabla g_i(\boldsymbol{x}^*), i \in \mathcal{A}(\boldsymbol{x}^*)\}$ of gradients of equality and active inequality constraints is linearly independent.
- **Smoothness.** The Hessian matrices $\nabla^2 f(\boldsymbol{x}), \ \nabla^2 h_i(\boldsymbol{x}), \ 1 \leq i \leq m$, and $\nabla^2 g_i(\boldsymbol{x}), \ 1 \leq i \leq l$, exist and are locally Lipschitz continuous at $\boldsymbol{x}^*$.
- **Second-order sufficiency condition.** $\boldsymbol{\eta}^T \nabla_{\boldsymbol{x}}^2 \mathcal{L}(\boldsymbol{x}^*)\boldsymbol{\eta} > 0$ for all vectors $\boldsymbol{\eta} \neq 0$ satisfying $\nabla h_i(\boldsymbol{x}^*)^T \boldsymbol{\eta} = 0, \ 1 \leq i \leq m$, and $\nabla g_i(\boldsymbol{x}^*)^T \eta = 0, \ i \in \mathcal{A}(\boldsymbol{x}^*)$.
- **Strict complementarity.** $z_i^* > 0$ if $g_i(\boldsymbol{x}^*) = 0, \ 1 \leq i \leq l$.

Well-known approaches from the optimization theory for handling problems with inequality constraints are, for instance, the slack variable approach, the active set strategy, and the logarithmic barrier function approach. Each of these approaches results in a nonlinear programming problem with only equality constraints. For example, in the first approach, the constraint $\boldsymbol{g}(\boldsymbol{x}) \geq 0$ can be replaced by $\boldsymbol{g}(\boldsymbol{x}) - \boldsymbol{s} = 0, \boldsymbol{s} \geq 0$ by adding a nonnegative slack variable to each of the inequality constraints. Transformation of the original inequality problem into an equality one, by adding slacks, have been a frequently applied tool in scientific computations in the past twenty years and recently used in cf., [BHN99, ET*86, VS99]. The introduction of slack variables is associated with a small amount of additional work and storage, since they do not enter the objective function and are constrained by simple bounds. The second, active set strategy, approach in nonlinear programming is directly related to the idea of the simplex method in linear programming. At each iterative step from this approach, applying, for example, Newton's method, one has to define which constraints are active at the solution and treat them as equality constraints by ignoring the others. The third approach was used in our practical implementations and we explain it in detail in the next subsection.

4.2 Logarithmic barrier interior-point method

The logarithmic barrier function method was first introduced in [Fri55] and later on popularized by [FM68] in the late sixties of the last century. The basic idea of this method is to replace the optimization problem (4.1)–(4.2) with the following equality constrained optimization problem

$$\min_{\boldsymbol{x} \in \mathcal{R}^n} \beta^{(\rho)}(\boldsymbol{x}), \tag{4.6}$$

subject to

$$\boldsymbol{h}(\boldsymbol{x}) = 0, \tag{4.7}$$

where ρ is a positive scalar, called *barrier parameter*, and

$$\beta^{(\rho)}(\boldsymbol{x}) = f(\boldsymbol{x}) - \rho \sum_{i=1}^{l} \log g_i(\boldsymbol{x}) \tag{4.8}$$

is often referred to as a *barrier function*. To insure existence of the logarithmic terms in (4.8) we implicitly require $g_i(\boldsymbol{x}) > 0,\ 1 \le i \le l$. In such a way, we get a family of subproblems depending on ρ for which it is well-known that under the assumptions conditions from Sect. 4.1 the solution of (4.6)–(4.7) converges to a solution of the original problem (4.1)–(4.2) as ρ decreases to zero (cf., [FM68]). This method obviously is an interior-point method since it keeps the sequence of iterating solutions strictly feasible with respect to the inequality constraints. Note that the logarithmic terms serve as a barrier and result in finding a solution $\boldsymbol{x}^{(\rho)}$ such that $g(\boldsymbol{x}^{(\rho)}) > 0$. The solution points $\boldsymbol{x}^{(\rho)}$ parameterized by ρ define the so-called *central path* or also called *barrier trajectory*.

The gradient of (4.8) is given by

$$\nabla \beta^{(\rho)}(\boldsymbol{x}) = \nabla f(\boldsymbol{x}) - \sum_{i=1}^{l} \frac{\rho}{g_i(\boldsymbol{x})} \nabla g_i(\boldsymbol{x})$$

and the Hessian of $\beta^{(\rho)}(x)$ is defined by

$$\nabla^2 \beta^{(\rho)}(\boldsymbol{x}) = \nabla^2 f(\boldsymbol{x}) - \sum_{i=1}^{l} \frac{\rho}{g_i(\boldsymbol{x})} \nabla^2 g_i(\boldsymbol{x}) + \sum_{i=1}^{l} \frac{\rho}{g_i^2(\boldsymbol{x})} \nabla g_i(\boldsymbol{x}) (\nabla g_i(\boldsymbol{x}))^T. \tag{4.9}$$

The Lagrangian function associated with (4.6)–(4.7) is

$$\mathcal{L}^{(\rho)}(\boldsymbol{x},\boldsymbol{y}) = \beta^{(\rho)}(\boldsymbol{x}) + \boldsymbol{y}^T\boldsymbol{h}(\boldsymbol{x}) = f(\boldsymbol{x}) - \rho\sum_{i=1}^{l}\log g_i(\boldsymbol{x}) + \boldsymbol{y}^T\boldsymbol{h}(\boldsymbol{x})$$

and the gradient of $\mathcal{L}^{(\rho)}(\boldsymbol{x},\boldsymbol{y})$ with respect to $\boldsymbol{x}$ is given by

$$\nabla_{\boldsymbol{x}}\mathcal{L}^{(\rho)}(\boldsymbol{x},\boldsymbol{y}) = \nabla f(\boldsymbol{x}) - \sum_{i=1}^{l}\frac{\rho}{g_i(\boldsymbol{x})}\nabla g_i(\boldsymbol{x}) + \sum_{i=1}^{m} y_i\nabla h_i(\boldsymbol{x}). \tag{4.10}$$

The logarithmic barrier function method consists now of generating a sequence of iterative solutions $\{\boldsymbol{x}\} = \{\boldsymbol{x}^{(\rho)}\}$, local minimizers of the equality constrained subproblems, with $\rho > 0$ decreasing at each iteration. Taking into account the first-order optimality conditions and especially $\nabla_{\boldsymbol{x}}\mathcal{L}^{(\rho)}(\boldsymbol{x},\boldsymbol{y}) = 0$, we see that convergence of $\{\boldsymbol{x}^{(\rho)}\}$ to an optimal solution $\boldsymbol{x}^*$ requires that

$$\lim_{\rho\to 0} y_i^{(\rho)} = y_i^*,\ 1 \le i \le m \qquad \text{and} \qquad \lim_{\rho\to 0}\frac{\rho}{g_i(\boldsymbol{x}^{(\rho)})} = z_i^*,\ 1 \le i \le l, \tag{4.11}$$

where $\{y_i^*\}$ and $\{z_i^*\}$ are the Lagrange multipliers associated with the equality and inequality constraints $g_i(\boldsymbol{x}^{(\rho)}) > 0$, respectively. From $g_i(x^{(\rho)}) \to 0$ and the second relation in (4.11) we get $\rho/g_i^2(\boldsymbol{x}^{(\rho)}) \to \infty$ and hence, the Hessian of the logarithmic barrier function (4.9) would become arbitrarily large. Comparing now relations (4.4) and (4.10), we see that $\rho/g_i(\boldsymbol{x}^{(\rho)})$ serves as a Lagrange multiplier for the inequality constraints. Thus, we can introduce an auxiliary variable $z_i = z_i^{(\rho)} = \rho/g_i(\boldsymbol{x}^{(\rho)}),\ 1 \le i \le l$ which can also be written in the form $z_i^{(\rho)} g_i(\boldsymbol{x}^{(\rho)}) = \rho$. The last relation is usually called *perturbed complementarity* and can be used as a remedy, so that the differentiation will not create ill-conditioning.

We formulate now the perturbed KKT conditions for the logarithmic barrier function problem (4.6)–(4.7), namely

$$\nabla f(\boldsymbol{x}) + \nabla\boldsymbol{h}(\boldsymbol{x})\boldsymbol{y} - \nabla\boldsymbol{g}(\boldsymbol{x})\boldsymbol{z} = 0,\ \boldsymbol{h}(\boldsymbol{x}) = 0,\ Z\boldsymbol{g}(\boldsymbol{x}) = \rho\bar{\boldsymbol{e}},\ \boldsymbol{g}(\boldsymbol{x}) > 0. \tag{4.12}$$

In matrix-vector notations, (4.12) results in the following nonlinear equation with $n + m + l$ components

$$\boldsymbol{F}^{(\rho)}(\boldsymbol{x},\boldsymbol{y},\boldsymbol{z}) = 0 \qquad \text{with} \quad \boldsymbol{F}^{(\rho)}(\boldsymbol{x},\boldsymbol{y},\boldsymbol{z}) = \begin{pmatrix} \boldsymbol{t} + J_{\text{eq}}^T\boldsymbol{y} - J_{\text{in}}^T\boldsymbol{z} \\ \boldsymbol{h} \\ G\boldsymbol{z} - \rho\,\bar{\boldsymbol{e}} \end{pmatrix}, \tag{4.13}$$

where $\boldsymbol{F}^{(\rho)} = \nabla\mathcal{L}^{(\rho)}$ is the gradient of the Lagrangian function with respect to $\boldsymbol{x},\boldsymbol{y}$, and $\boldsymbol{z}$; $\boldsymbol{t} = \nabla f(\boldsymbol{x})$ is the gradient of the objective function, J_{eq} is the Jacobian $m \times n$ matrix of the equality constraints $\boldsymbol{h}(\boldsymbol{x}) = 0$ and J_{in} is the Jacobian $l \times n$ matrix of the inequality constraints $\boldsymbol{g}(\boldsymbol{x}) \ge 0$. In the last equation of (4.13) we have denoted $G = \mathrm{diag}(g_i),\ g_i > 0,\ 1 \le i \le l$ and $\bar{e} = (1, 1, \dots, 1)^T$. Note that at each iteration we have three sets of unknowns: the primal variable $\boldsymbol{x}$, the dual variable $\boldsymbol{y}$, and the perturbed complementarity variable $\boldsymbol{z}$ which we consider independently.

Denote the unknown solution by $\boldsymbol{\Phi} = (\boldsymbol{x}, \boldsymbol{y}, \boldsymbol{z})$ and apply the Newton method to the nonlinear system (4.13) to find the increments $\Delta\boldsymbol{\Phi} = (\Delta\boldsymbol{x}, \Delta\boldsymbol{y}, \Delta\boldsymbol{z})$, namely

$$K \Delta\boldsymbol{\Phi} = -\boldsymbol{F}^{(\rho)}(\boldsymbol{\Phi}), \tag{4.14}$$

which is often referred to as a *primal-dual system*. The vector $\Delta\boldsymbol{\Phi}$ is called *search direction*. The so-called *primal-dual matrix* $K = \left(\boldsymbol{F}^{(\rho)}\right)'(\boldsymbol{\Phi})$ of second derivatives of the Lagrangian function is defined as follows

$$K = \begin{pmatrix} H & J_{\mathrm{eq}}^T & -J_{\mathrm{in}}^T \\ J_{\mathrm{eq}} & 0 & 0 \\ Z J_{\mathrm{in}} & 0 & G \end{pmatrix}, \tag{4.15}$$

where the Hessian of the Lagrangian function $H = \nabla_x^2 \mathcal{L}$ is given by (4.5). Note that the matrix K is sparse, nonsymmetric, independent of ρ, and usually well-conditioned in a sense that its condition number is limited when $\rho \to 0$ (see [Wri98] for more details). In the case of convex optimization (i.e., convex objective function $f(\boldsymbol{x})$, linear equality constraints $\boldsymbol{h}(\boldsymbol{x})$, and concave inequality constraints $\boldsymbol{g}(\boldsymbol{x})$), the Hessian matrix H is positive semidefinite. The properties of the Hessian matrix for inequality constrained optimization problem with logarithmic barrier function method are discussed in [FGW02].

One possible way for solving (4.14) is to symmetrize K taking into account the fact that Z and G are diagonal matrices. This method is proposed in [FGS96] and results in the following symmetric matrix

$$\hat{K} = \begin{pmatrix} H & J_{\mathrm{eq}}^T & -J_{\mathrm{in}}^T \\ J_{\mathrm{eq}} & 0 & 0 \\ -J_{\mathrm{in}} & 0 & -Z^{-1} G \end{pmatrix},$$

which is strongly ill-conditioned with some diagonal elements becoming unbounded as $\rho \to 0$. In particular, for the active inequality constraints, the diagonal entries of $Z^{-1}G$ go to zero, and for the inactive constraints they go to infinity. As the iterates converge, the ill-conditioning of K increases, but it was shown in [FGS96] that the primal-dual solution of the optimization problem is actually independent of the size of the large diagonal elements and can be found by using, for instance, a symmetric indefinite factorization of the primal-dual system.

Another alternative way for solving (4.14) which we use in our practical applications is to eliminate the (1,3) block of (4.15), i.e., due to $\boldsymbol{g}(\boldsymbol{x}) > 0$, we eliminate $\Delta\boldsymbol{z}$ from the third equation of (4.14)

$$\Delta\boldsymbol{z} = -\boldsymbol{z} + G^{-1}(\rho\, \bar{e} - Z J_{\mathrm{in}} \Delta\boldsymbol{x}) \tag{4.16}$$

and replace it in the first equation. This method produces a symmetric linear system with $n+m$ equations of the form

$$\begin{pmatrix} \tilde{H} & J_{\text{eq}}^T \\ J_{\text{eq}} & 0 \end{pmatrix} \begin{pmatrix} \Delta \boldsymbol{x} \\ \Delta \boldsymbol{y} \end{pmatrix} = - \begin{pmatrix} \boldsymbol{t} + J_{\text{eq}}^T \boldsymbol{y} - \rho J_{\text{in}}^T G^{-1} \bar{\boldsymbol{e}} \\ \boldsymbol{h} \end{pmatrix}, \tag{4.17}$$

where $\tilde{H} = H + J_{\text{in}}^T G^{-1} Z J_{\text{in}}$ is often referred to as a *condensed* primal-dual Hessian and (4.17) is called a *condensed* primal-dual system. A detailed analysis of the properties of the condensed primal-dual matrix can be found in [Wri98] where it was shown that the inherent ill-conditioning of the reduced primal-dual matrix is usually benign and does not influence the accuracy of the solution.

Various methods for solving (4.17) and finding the primal-dual steps $(\Delta \boldsymbol{x}, \Delta \boldsymbol{y})$ are proposed in the literature (cg., e.g., [ET*86, GOW98, Wri98]). Note that one needs a reliable and efficient solver of (4.17), since the condensed primal-dual system is solved at every iteration of the optimization loop. In practice, we apply transforming iterations (see [Wit89]) to find the increments. This method will be explained in more detail in Sect. 5.

After finding the solution of (4.17), the algorithm proceeds iteratively from an initial point $(\boldsymbol{x}^{(0)}, \boldsymbol{y}^{(0)}, \boldsymbol{z}^{(0)})$ through a sequence of points determined from the search directions described by (4.16) and (4.17) as follows

$$\boldsymbol{x}^{(k+1)} = \boldsymbol{x}^{(k)} + \alpha_x^{(k)} \Delta \boldsymbol{x}, \ \boldsymbol{y}^{(k+1)} = \boldsymbol{y}^{(k)} + \alpha_y^{(k)} \Delta \boldsymbol{y}, \ \boldsymbol{z}^{(k+1)} = \boldsymbol{z}^{(k)} + \alpha_z^{(k)} \Delta \boldsymbol{z}.$$

The parameters $\alpha_x^{(k)}, \alpha_y^{(k)}, \alpha_z^{(k)} \in (0,1]$ are called *steplengths* and their choice at each iteration is a critical feature of the algorithm to find a local minimizer of the optimization problem.

4.3 Merit functions. Computing the steplengths

In all optimization algorithms it is important to have a reasonable way of deciding whether the new iterate is better than the previous one, i.e., it is essential to measure appropriately the progress in finding a local solution. Merit functions of different types have been a subject of great interest over the past years (see, e.g., [ET*86, GOW98, Wri98]). The main idea of a merit function is to ensure simultaneously a progress toward a local minimizer and toward feasibility. The method of choosing $\alpha^{(k)}$ at each iteration becomes more complicated in general nonlinear programming problems as it is well-known that the Newton method may diverge when the initial estimate of the solution is bad.

Two versions of the Newton method can be applied, namely, the *trust-region* and the *line-search* approach. The first method has recently been applied in, e.g., [BHN99]. Typical for this method is to find a step $\boldsymbol{d}^{(k)}$ which is restricted to a set, called the *trust region*. This set is practically obtained by limiting $|\boldsymbol{d}^{(k)}| \leq r^{(k)}$, where $r^{(k)}$ is the trust region radius. At each iteration, $r^{(k)}$ is updated according to how successful the step has been. For instance, if the a priori chosen merit function M decreases, we accept the step $\boldsymbol{d}^{(k)}$, update the solution $\boldsymbol{\Phi}^{(k+1)} = \boldsymbol{\Phi}^{(k)} + \boldsymbol{d}^{(k)}$ and possibly increase the trust region radius $r^{(k)}$. Otherwise, we decrease $r^{(k)}$ by a damping factor, e.g., $r^{(k)} = r^{(k)}/2$ and compute again the step $\boldsymbol{d}^{(k)}$.

We apply the second variant of the Newton method, the line-search approach. Once the solution $\Delta\boldsymbol{\Phi}^{(k)}$ of (4.14) has been determined, we find a steplength $\alpha^{(k)} > 0$ such that $\boldsymbol{\Phi}^{(k+1)} = \boldsymbol{\Phi}^{(k)} + \alpha^{(k)}\Delta\boldsymbol{\Phi}^{(k)}$ measuring a progress in minimization at each iteration and reducing the merit function in the sense $M(\boldsymbol{\Phi}^{(k+1)}) < M(\boldsymbol{\Phi}^{(k)})$. The ideal value $\alpha^{(k)} = 1$ may not always happen so that various modifications of the basic Newton method have to be implemented. The following basic model algorithm can be considered:

S1. If the conditions for convergence are satisfied, the algorithm terminates with $\boldsymbol{\Phi}^{(k)}$ as the solution.
S2. Compute a search direction $\Delta\boldsymbol{\Phi}^{(k)}$ solving (4.14).
S3. Compute the steplength $\alpha^{(k)} > 0$ for which $M(\boldsymbol{\Phi}^{(k)} + \alpha^{(k)}\Delta\boldsymbol{\Phi}^{(k)}) < M(\boldsymbol{\Phi}^{(k)})$.
S4. Update the estimate for the minimum by $\boldsymbol{\Phi}^{(k+1)} := \boldsymbol{\Phi}^{(k)} + \alpha^{(k)}\Delta\boldsymbol{\Phi}^{(k)}$, $k := k+1$, and go back to step S1.

A standard convergence monitor in nonlinear programming is to choose the Euclidean norm $\|\boldsymbol{F}^{(\rho)}(\boldsymbol{x}, \boldsymbol{y}, \boldsymbol{z})\|$ of the residual produced by the KKT conditions (4.13) as a merit function. However, in many practical implementations, this choice of the merit function is not sufficient, since it does not allow to tell the difference between a local minimizer and a stationary non-minimizing point. The KKT conditions are necessary optimality conditions and hence, the optimization problem (4.1)–(4.2) and the nonlinear problem (4.13) are not equivalent, i.e., the Newton method may find solutions of (4.13) which do not minimize the objective function $f(\boldsymbol{x})$. Therefore, in order to find simultaneously solutions of both problems, a better approach is to rely on a hierarchy of two merit functions (cf., e.g., [GOW98, HPS02]). In general, the choice of merit functions in nonlinear constrained optimization problems is complicated. Several ideas have recently been proposed in the context of primal-dual interior methods (cf., e.g., [BHN99, ET*86, FGW02]). In particular, our *primary merit function* is related to those suggested in [GOW98] and is chosen as a modified augmented Lagrangian incorporating the logarithmic barrier function (4.8) as follows

$$M := M(\boldsymbol{x}, \boldsymbol{y}, \rho, \rho_A) = f(\boldsymbol{x}) - \rho \sum_{i=1}^{l} \log g_i(x) + \boldsymbol{y}^T \boldsymbol{h}(\boldsymbol{x}) + \frac{1}{2}\rho_A\, \boldsymbol{h}(\boldsymbol{x})^T \boldsymbol{h}(\boldsymbol{x}), \tag{4.18}$$

where ρ_A is a positive parameter. Our purpose now is to satisfy the descent conditions and to guarantee a reduction of the merit function in the sense that each iterate should be an improved estimate of the solution of (4.6)–(4.7). Note that a descent is sought only with respect to $\boldsymbol{x}$ taking into account the original optimization problem. A standard way to achieve $M(\boldsymbol{x}+\alpha\Delta\boldsymbol{x},\boldsymbol{y},\rho,\rho_A) < M(\boldsymbol{x},\boldsymbol{y},\rho,\rho_A)$ is to require that $\Delta\boldsymbol{x}$ is a descent direction, i.e., $\Delta\boldsymbol{x}^T\nabla_{\boldsymbol{x}}M < 0$, where $\nabla_{\boldsymbol{x}}M$ is the gradient of the primary merit function with respect to the primal variable $\boldsymbol{x}$. In particular, we have

$$\begin{aligned}\Delta\boldsymbol{x}^T\nabla_{\boldsymbol{x}}M &= \Delta\boldsymbol{x}^T(\boldsymbol{t}-\rho J_{\text{in}}^T G^{-1}\bar{\boldsymbol{e}} + J_{\text{eq}}^T\boldsymbol{y} + \rho_A J_{\text{eq}}^T\boldsymbol{h})\\ &= \Delta\boldsymbol{x}^T(\boldsymbol{t}-\rho J_{\text{in}}^T G^{-1}\bar{\boldsymbol{e}}) - \boldsymbol{h}^T\boldsymbol{y} - \rho_A\boldsymbol{h}^T\boldsymbol{h}, \end{aligned} \tag{4.19}$$

due to $J_{\text{eq}}\Delta\boldsymbol{x} = -\boldsymbol{h}$ from the second equation of (4.17). Hence, $\Delta\boldsymbol{x}^T\nabla_{\boldsymbol{x}}M < 0$ can be satisfied if the augmented Lagrangian parameter ρ_A is sufficiently large, namely

$$\rho_A > \frac{\Delta\boldsymbol{x}^T(\boldsymbol{t}-\rho J_{\text{in}}^T G^{-1}\bar{\boldsymbol{e}}) - \boldsymbol{h}^T\boldsymbol{y}}{\boldsymbol{h}^T\boldsymbol{h}}.$$

Hence, ρ_A can be changed within the optimization loop, if $\Delta\boldsymbol{x}$ is not a descent direction. In our algorithm, we choose

$$\rho_A = \min\left(\frac{5}{\boldsymbol{h}^T\boldsymbol{h}}(\Delta\boldsymbol{x}^T(\boldsymbol{t}-\rho J_{\text{in}}^T G^{-1}\bar{\boldsymbol{e}}) - \boldsymbol{h}^T\boldsymbol{y}), 100\right) \tag{4.20}$$

in the case $\Delta\boldsymbol{x}^T\nabla_{\boldsymbol{x}}M \geq 0$ and continue the loop (see [GOW98, HPS02] for details).

For the *secondary merit function* we choose the l_2- norm of the residual with respect to perturbed KKT-conditions (4.13). We apply the Newton method and choose the steplengths to strictly satisfy the inequality constraints $\boldsymbol{g}(\boldsymbol{x}) > 0$ and the complementarity constraints $\boldsymbol{z} > 0$. Hence, the first requirement for the line-search approach is to insure a strict feasibility. Let $\hat{\alpha}$ and $\hat{\gamma}$ be separate steplengths defined as follows

$$\hat{\alpha} = \max\{\alpha|\boldsymbol{g}(\boldsymbol{x}) + \alpha J_{\text{in}}\Delta\boldsymbol{x} \geq 0\}, \qquad \hat{\gamma} = \max\{\gamma|\boldsymbol{z} + \gamma\Delta\boldsymbol{z} \geq 0\}.$$

Since we maintain interior (i.e., strict feasible) iterates, usually we take a parameter $\tau \in (0,1)$ bounded strongly away from unity and define $\alpha = \min(1,\tau\hat{\alpha})$ and $\gamma = \min(1,\tau\hat{\gamma})$. We use the same steplength γ for the Lagrange multiplier $\boldsymbol{y}$. In practice, both merit functions are used by means of the following strategy: If the steplengths α and γ lead to a reduction of M, they are accepted. If M does not decrease, we check the secondary merit function. If the latter decreases, the steplengths are accepted; otherwise damp the Newton steps by a certain factor and continue the procedure. The barrier parameter $\rho > 0$ is updated by decreasing values until an approximate solution of the nonlinear problem is obtained (cf., e.g., [ET*86, GOW98, HPS02]). We rely on a *watchdog strategy* (see [CL*82]) to ensure progress in finding a local minimizer. If after some fixed number of iterations there is no reduction of M, the augmented Lagrangian parameter ρ_A is chosen sufficiently large in accordance with (4.20).

5 Solving the condensed primal–dual system

The discretized constrained optimization problem (3.2)–(3.3) is solved by the primal-dual interior-point method described in Sect. 4. We consider the diagonal matrices $D_1 := \mathrm{diag}(\alpha_i - \alpha_{\min})$ and $D_2 := \mathrm{diag}(\alpha_{\max} - \alpha_i)$ and introduce $\boldsymbol{z} := \rho D_1^{-1}\bar{\boldsymbol{e}} \geq 0$ and $\boldsymbol{w} = \rho D_2^{-1}\bar{\boldsymbol{e}} \geq 0$ serving as perturbed complementarity. We note that $1 \leq i \leq N$ where N is the number of finite elements in the discretized domain and $\bar{\boldsymbol{e}} = (1, 1, \ldots, 1)^T \in \mathcal{R}^M$. The primal-dual Newton-type interior-point method is applied to three sets of variables: primal feasibility $(\boldsymbol{u}, \boldsymbol{\alpha})$, dual feasibility $(\boldsymbol{\lambda}, \boldsymbol{\eta})$, and perturbed complementarity related to $(\boldsymbol{z}, \boldsymbol{w})$.

Denote the Lagrangian function of (3.2)–(3.3) by

$$\begin{aligned} \mathcal{L}(\boldsymbol{u}, \boldsymbol{\alpha}; \boldsymbol{\lambda}, \eta; \boldsymbol{z}, \boldsymbol{w}) := {} & f(\boldsymbol{u}, \boldsymbol{\alpha}) \\ & + \boldsymbol{\lambda}^T (A(\boldsymbol{\alpha})\, \boldsymbol{u} - \boldsymbol{f}) + \eta\, (g(\boldsymbol{\alpha}) - C) \\ & - \boldsymbol{z}^T (\boldsymbol{\alpha} - \alpha_{\min}\, \bar{\boldsymbol{e}}) - \boldsymbol{w}^T (\alpha_{\max} \bar{\boldsymbol{e}} - \boldsymbol{\alpha}). \end{aligned} \tag{5.1}$$

The Newton method applied to the KKT conditions of (5.1) results in

$$\begin{pmatrix} 0 & \mathcal{L}_{\boldsymbol{u}\boldsymbol{\alpha}} & \mathcal{L}_{\boldsymbol{u}\boldsymbol{\lambda}} & 0 & 0 & 0 \\ \mathcal{L}_{\boldsymbol{\alpha}\boldsymbol{u}} & \mathcal{L}_{\boldsymbol{\alpha}\boldsymbol{\alpha}} & \mathcal{L}_{\boldsymbol{\alpha}\boldsymbol{\lambda}} & \mathcal{L}_{\boldsymbol{\alpha}\eta} & -I & I \\ \mathcal{L}_{\boldsymbol{\lambda}\boldsymbol{u}} & \mathcal{L}_{\boldsymbol{\lambda}\boldsymbol{\alpha}} & 0 & 0 & 0 & 0 \\ 0 & \mathcal{L}_{\eta\boldsymbol{\alpha}} & 0 & 0 & 0 & 0 \\ 0 & Z & 0 & 0 & D_1 & 0 \\ 0 & -W & 0 & 0 & 0 & D_2 \end{pmatrix} \begin{pmatrix} \triangle \boldsymbol{u} \\ \triangle \boldsymbol{\alpha} \\ \triangle \boldsymbol{\lambda} \\ \triangle \eta \\ \triangle \boldsymbol{z} \\ \triangle \boldsymbol{w} \end{pmatrix} = - \begin{pmatrix} \nabla_{\boldsymbol{u}} \mathcal{L} \\ \nabla_{\boldsymbol{\alpha}} \mathcal{L} \\ \nabla_{\boldsymbol{\lambda}} \mathcal{L} \\ \nabla_{\eta} \mathcal{L} \\ \nabla_{\boldsymbol{z}} \mathcal{L} \\ \nabla_{\boldsymbol{w}} \mathcal{L} \end{pmatrix}, \tag{5.2}$$

where I stands for the identity matrix, $Z = \mathrm{diag}(z_i)$ and $W = \mathrm{diag}(w_i)$ are diagonal matrices. Following Sect. 4 we eliminate the increments for $\boldsymbol{z}$ and $\boldsymbol{w}$ from the 5th and 6th rows of (5.2), namely,

$$\triangle \boldsymbol{z} = D_1^{-1}(-\nabla_{\boldsymbol{z}} \mathcal{L} - Z\, \triangle \boldsymbol{\alpha}), \qquad \triangle \boldsymbol{w} = D_2^{-1}(-\nabla_{\boldsymbol{w}} \mathcal{L} + W\, \triangle \boldsymbol{\alpha}) \tag{5.3}$$

and substitute (5.3) in the second row of (5.2). We get the linear system $\tilde{K} \triangle \boldsymbol{\psi} = -\tilde{\boldsymbol{\xi}}$ for the increments of $\boldsymbol{\psi} := (\boldsymbol{u}, \boldsymbol{\alpha}, \boldsymbol{\lambda}, \eta)$, denoted by $\triangle \boldsymbol{\psi} := (\triangle \boldsymbol{u}, \triangle \boldsymbol{\alpha}, \triangle \boldsymbol{\lambda}, \triangle \eta)$ where $\tilde{K}$ is the matrix and $(-\tilde{\boldsymbol{\xi}})$ is the right-hand side of the following condensed primal-dual system

$$\begin{pmatrix} 0 & \mathcal{L}_{\boldsymbol{u}\boldsymbol{\alpha}} & \mathcal{L}_{\boldsymbol{u}\boldsymbol{\lambda}} & 0 \\ \mathcal{L}_{\boldsymbol{\alpha}\boldsymbol{u}} & \tilde{\mathcal{L}}_{\boldsymbol{\alpha}\boldsymbol{\alpha}} & \mathcal{L}_{\boldsymbol{\alpha}\boldsymbol{\lambda}} & \mathcal{L}_{\boldsymbol{\alpha}\eta} \\ \mathcal{L}_{\boldsymbol{\lambda}\boldsymbol{u}} & \mathcal{L}_{\boldsymbol{\lambda}\boldsymbol{\alpha}} & 0 & 0 \\ 0 & \mathcal{L}_{\eta\boldsymbol{\alpha}} & 0 & 0 \end{pmatrix} \begin{pmatrix} \triangle \boldsymbol{u} \\ \triangle \boldsymbol{\alpha} \\ \triangle \boldsymbol{\lambda} \\ \triangle \eta \end{pmatrix} = - \begin{pmatrix} \nabla_{\boldsymbol{u}} \mathcal{L} \\ \tilde{\nabla}_{\boldsymbol{\alpha}} \mathcal{L} \\ \nabla_{\boldsymbol{\lambda}} \mathcal{L} \\ \nabla_{\eta} \mathcal{L} \end{pmatrix}. \tag{5.4}$$

The $\boldsymbol{\alpha}\boldsymbol{\alpha}$-entry of $\tilde{K}$ and the modified entry for the right-hand side are

$$\tilde{\mathcal{L}}_{\boldsymbol{\alpha}\boldsymbol{\alpha}} = \mathcal{L}_{\boldsymbol{\alpha}\boldsymbol{\alpha}} + D_1^{-1}\, Z + D_2^{-1}\, W, \qquad \tilde{\nabla}_{\boldsymbol{\alpha}} \mathcal{L} = \nabla_{\boldsymbol{\alpha}} \mathcal{L} + D_1^{-1}\, \nabla_{\boldsymbol{z}} \mathcal{L} - D_2^{-1}\, \nabla_{\boldsymbol{w}} \mathcal{L}.$$

Direct methods for the solution of (5.4) can be divided into two classes: *range space methods* and *null space methods*. These approaches essentially differ in the grouping of the matrix into a 2×2-block structure. The decomposition of the condensed primal-dual system (5.4) is related to the first approach. In this section, we consider the *null space* decomposition of the condensed primal-dual matrix interchanging the second and the third rows and columns. The resulting matrix can be written according to

$$\tilde{K} = \begin{pmatrix} A_{11} & A_{12} \\ A_{21} & A_{22} \end{pmatrix} = \left(\begin{array}{cc|cc} 0 & \mathcal{L}_{\boldsymbol{u}\boldsymbol{\lambda}} & \mathcal{L}_{\boldsymbol{u}\boldsymbol{\alpha}} & 0 \\ \mathcal{L}_{\boldsymbol{\lambda}\boldsymbol{u}} & 0 & \mathcal{L}_{\boldsymbol{\lambda}\boldsymbol{\alpha}} & 0 \\ \hline \mathcal{L}_{\boldsymbol{\alpha}\boldsymbol{u}} & \mathcal{L}_{\boldsymbol{\alpha}\boldsymbol{\lambda}} & \tilde{\mathcal{L}}_{\boldsymbol{\alpha}\boldsymbol{\alpha}} & \mathcal{L}_{\boldsymbol{\alpha}\eta} \\ 0 & 0 & \mathcal{L}_{\eta\boldsymbol{\alpha}} & 0 \end{array}\right),$$

where the first diagonal block

$$A_{11} = \begin{pmatrix} 0 & \mathcal{L}_{\boldsymbol{u}\boldsymbol{\lambda}} \\ \mathcal{L}_{\boldsymbol{\lambda}\boldsymbol{u}} & 0 \end{pmatrix} \tag{5.5}$$

is now an indefinite but nonsingular matrix. We remind that $\mathcal{L}_{\boldsymbol{\lambda}\boldsymbol{u}} = A(\boldsymbol{\alpha})$ is exactly the stiffness matrix corresponding to the equilibrium equation (2.12). Hence, A_{11}^{-1} exists, and the Schur complement $S := A_{22} - A_{21}A_{11}^{-1}A_{12}$ is defined correctly.

We use the following regular splitting of $\tilde{K}$

$$K^L \tilde{K}^R = M_1 - M_2 \tag{5.6}$$

with left and right factors given below and reasonable matrices M_1 and $M_2 \sim 0$. For solving the system of the form $\tilde{K}\triangle\boldsymbol{\psi} = -\tilde{\boldsymbol{\xi}}$, starting with an initial guess for $\triangle\boldsymbol{\psi} := (\triangle\boldsymbol{u}, \triangle\boldsymbol{\lambda}, \triangle\boldsymbol{\alpha}, \triangle\eta)^T$, the transforming iteration proposed in [Wit89] is described by

$$\triangle\boldsymbol{\psi}^{(\nu+1)} := \triangle\boldsymbol{\psi}^{(\nu)} + K^R M_1^{-1} K^L(-\tilde{\boldsymbol{\xi}} - \tilde{K}\triangle\boldsymbol{\psi}^{(\nu)}), \tag{5.7}$$

where the new iterate $\boldsymbol{\psi}^{(\text{new})}$ is obtained by a line-search in the direction $\triangle\boldsymbol{\psi}$, namely

$$\psi_j^{(\text{new})} = \psi_j^{(\text{old})} + \alpha_j(\triangle\boldsymbol{\psi})_j, \qquad 1 \le j \le 4.$$

The line-search approach and the choice of the steplengths parameters α_j are discussed in Sect. 4.3.

Using an appropriate preconditioner for the stiffness matrix, we approximate the first diagonal block (5.5) as follows

$$A_{11} = \begin{pmatrix} 0 & \mathcal{L}_{\boldsymbol{u}\boldsymbol{\lambda}} \\ \mathcal{L}_{\boldsymbol{\lambda}\boldsymbol{u}} & 0 \end{pmatrix} \sim \begin{pmatrix} 0 & \tilde{\mathcal{L}}_{\boldsymbol{u}\boldsymbol{\lambda}} \\ \tilde{\mathcal{L}}_{\boldsymbol{\lambda}\boldsymbol{u}} & 0 \end{pmatrix} =: \tilde{A}_{11}.$$

Usually, the left and right transformations are of the form

$$K^L = I, \qquad K^R = \begin{pmatrix} I & -\tilde{A}_{11}^{-1}A_{12} \\ 0 & I \end{pmatrix} = \begin{pmatrix} I & 0 & -\tilde{\mathcal{L}}_{\boldsymbol{\lambda}\boldsymbol{u}}^{-1}\mathcal{L}_{\boldsymbol{\lambda}\boldsymbol{\alpha}} & 0 \\ 0 & I & -\tilde{\mathcal{L}}_{\boldsymbol{u}\boldsymbol{\lambda}}^{-1}\mathcal{L}_{\boldsymbol{u}\boldsymbol{\alpha}} & 0 \\ 0 & 0 & I & 0 \\ 0 & 0 & 0 & I \end{pmatrix}.$$

In this case, the regular splitting (5.6) becomes $KK^R = M_1 - M_2$ where

$$M_1 = \begin{pmatrix} 0 & \mathcal{L}_{\boldsymbol{u\lambda}} & 0 & 0 \\ \mathcal{L}_{\boldsymbol{\lambda u}} & 0 & 0 & 0 \\ \mathcal{L}_{\boldsymbol{\alpha u}} & \mathcal{L}_{\boldsymbol{\alpha\lambda}} & \tilde{S} & \mathcal{L}_{\boldsymbol{\alpha}\eta} \\ 0 & 0 & \mathcal{L}_{\eta\boldsymbol{\alpha}} & 0 \end{pmatrix} = \begin{pmatrix} A_{11} & 0 \\ R & Q \end{pmatrix} \tag{5.8}$$

and

$$M_2 = \begin{pmatrix} 0 & 0 & \mathcal{L}_{\boldsymbol{u\alpha}} - \mathcal{L}_{\boldsymbol{u\lambda}}\tilde{\mathcal{L}}_{\boldsymbol{u\lambda}}^{-1}\mathcal{L}_{\boldsymbol{u\alpha}} & 0 \\ 0 & 0 & \mathcal{L}_{\boldsymbol{\lambda\alpha}} - \mathcal{L}_{\boldsymbol{\lambda u}}\tilde{\mathcal{L}}_{\boldsymbol{\lambda u}}^{-1}\mathcal{L}_{\boldsymbol{\lambda\alpha}} & 0 \\ 0 & 0 & 0 & 0 \\ 0 & 0 & 0 & 0 \end{pmatrix}. \tag{5.9}$$

Note that $M_2 \sim 0$ if we have a good preconditioner for the stiffness matrix. In our numerical experiments, we choose a Cholesky decomposition of $\mathcal{L}_{\boldsymbol{u\lambda}}$. The second diagonal block Q in (5.8) is symmetric and indefinite given by

$$Q := \begin{pmatrix} \tilde{S} & \mathcal{L}_{\boldsymbol{\alpha}\eta} \\ \mathcal{L}_{\eta\boldsymbol{\alpha}} & 0 \end{pmatrix} \qquad \text{with} \qquad \tilde{S} := \tilde{\mathcal{L}}_{\boldsymbol{\alpha\alpha}} - \mathcal{L}_{\boldsymbol{\alpha u}}\tilde{\mathcal{L}}_{\boldsymbol{\lambda u}}^{-1}\mathcal{L}_{\boldsymbol{\lambda\alpha}} - \mathcal{L}_{\boldsymbol{\alpha\lambda}}\tilde{\mathcal{L}}_{\boldsymbol{u\lambda}}^{-1}\mathcal{L}_{\boldsymbol{u\alpha}}.$$

We denote the defect in (5.7) by $\boldsymbol{d} = -\tilde{\boldsymbol{\xi}} - \tilde{K}\triangle\psi^{(\nu)}$ and compute the corresponding entries

$$\begin{aligned} \boldsymbol{d_u} &= -\nabla_{\boldsymbol{u}}\mathcal{L} - \mathcal{L}_{\boldsymbol{u\lambda}}\triangle\boldsymbol{\lambda} - \mathcal{L}_{\boldsymbol{u\alpha}}\triangle\boldsymbol{\alpha}, \\ \boldsymbol{d_\lambda} &= -\nabla_{\boldsymbol{\lambda}}\mathcal{L} - \mathcal{L}_{\boldsymbol{\lambda u}}\triangle\boldsymbol{u} - \mathcal{L}_{\boldsymbol{\lambda\alpha}}\triangle\boldsymbol{\alpha}, \\ \boldsymbol{d_\alpha} &= -\tilde{\nabla}_{\boldsymbol{\alpha}}\mathcal{L} - \mathcal{L}_{\boldsymbol{\alpha u}}\triangle\boldsymbol{u} - \mathcal{L}_{\boldsymbol{\alpha\lambda}}\triangle\boldsymbol{\lambda} - \tilde{\mathcal{L}}_{\boldsymbol{\alpha\alpha}}\triangle\boldsymbol{\alpha} - \mathcal{L}_{\boldsymbol{\alpha}\eta}\triangle\eta, \\ \boldsymbol{d}_\eta &= -\nabla_\eta\mathcal{L} - \mathcal{L}_{\eta\boldsymbol{\alpha}}\triangle\boldsymbol{\alpha}. \end{aligned}$$

Taking into account (5.7) one needs to compute $\boldsymbol{\delta} = M_1^{-1}\boldsymbol{d}$, i.e., $M_1\boldsymbol{\delta} = \boldsymbol{d}$. Consequently, we find $\delta_{\boldsymbol{\lambda}} = \tilde{\mathcal{L}}_{\boldsymbol{u\lambda}}^{-1}\boldsymbol{d_u}$ and $\delta_{\boldsymbol{u}} = \tilde{\mathcal{L}}_{\lambda\boldsymbol{u}}^{-1}\boldsymbol{d_\lambda}$. To compute the remaining components of $\boldsymbol{\delta}$ we have to solve systems with an indefinite matrix Q of the form

$$\begin{pmatrix} \tilde{S} & \mathcal{L}_{\boldsymbol{\alpha}\eta} \\ \mathcal{L}_{\eta\boldsymbol{\alpha}} & 0 \end{pmatrix} \begin{pmatrix} \delta_{\boldsymbol{\alpha}} \\ \delta_\eta \end{pmatrix} = \begin{pmatrix} \boldsymbol{d_\alpha} - L_{\boldsymbol{\alpha u}}\boldsymbol{\delta_u} - L_{\boldsymbol{\alpha\lambda}}\boldsymbol{\delta_\lambda} \\ \boldsymbol{d}_\eta \end{pmatrix}.$$

Iterative procedures such as MINRES or Bi-CGSTAB (see [VdV92]) with appropriate stopping criteria can be applied in this case. Compute $K^R\boldsymbol{\delta}$ and find the increments from (5.7) as follows

$$\begin{aligned} \triangle\boldsymbol{u}^{(\text{new})} &= \triangle\boldsymbol{u}^{(\text{old})} + \delta_{\boldsymbol{u}} - \tilde{\mathcal{L}}_{\boldsymbol{\lambda u}}^{-1}\mathcal{L}_{\boldsymbol{\lambda\alpha}}\delta_{\boldsymbol{\alpha}}, & \triangle\boldsymbol{\alpha}^{(\text{new})} &= \triangle\boldsymbol{\alpha}^{(\text{old})} + \delta_{\boldsymbol{\alpha}}, \\ \triangle\boldsymbol{\lambda}^{(\text{new})} &= \triangle\boldsymbol{\lambda}^{(\text{old})} + \delta_{\boldsymbol{\lambda}} - \tilde{\mathcal{L}}_{\boldsymbol{u\lambda}}^{-1}\mathcal{L}_{\boldsymbol{u\alpha}}\delta_{\boldsymbol{\alpha}}, & \triangle\eta^{(\text{new})} &= \triangle\eta^{(\text{old})} + \delta_\eta. \end{aligned}$$

We apply the above algorithm (with a fixed number of iterations) to find the increments of the primal and dual variables $\triangle\boldsymbol{u}, \triangle\boldsymbol{\alpha}, \triangle\boldsymbol{\lambda}, \triangle\eta$ and then use (5.3) to determine the global search direction $\triangle\boldsymbol{\Phi}$.

6 Adaptive grid refinement

Advanced finite element applications in science and engineering provoke the extensive use of adaptive mesh-refinement techniques to optimize the number of degrees of freedom and obtain accurate enough numerical solutions. The adaptive framework requires a locally refined discretization in regions where a better accuracy is necessary.

The computation of the homogenized elasticity coefficients requires the numerical solution of (2.8) with the unit cell as the computational domain. Previous works on shape and topology optimization (cf., e.g., [Ben95, BS03, SZ92]) strongly suggest the use of locally refined grids particularly at material interfaces. In the context of shape optimization such local refinements have been mostly done before the computations relying on a priori geometric informations or in an interactive way (manual remeshing based on computational results). In case of local singularities of the discrete solution, a priori error estimates typically give information about the asymptotic error behavior and thus, are not the best choice to control the mesh. In those parts of the domain where the solution changes rapidly, an automatic grid refinement on the basis of reliable and robust a posteriori error estimators is highly beneficial. In practice, the main goal in adaptive mesh-refinement procedures is to refine the mesh so that the discretization error is within the prescribed tolerance and as possible equidistributed throughout the domain.

In the past twenty years, numerous studies have been devoted to an error control and a mesh-design based on efficient postprocessing procedures (cf., e.g., [CF01, EE*95, HP04a, ZZ87]). A natural requirement for a posteriori error estimates is to be less expensive than the cost of the numerically computed solution. Moreover, appropriate refinement techniques have to be applied to construct the adaptive mesh and implement the adaptive solver. Local reconstruction of the grid is necessary to be done with a computational cost proportional to the number of modified elements.

The a posteriori adaptive strategy can be described as follows:

A1. Start with an initial coarse mesh $\mathcal{T}_0$ fitting the domain geometry. Set $n := 0$.
A2. Compute the discrete solution on $\mathcal{T}_n$.
A3. Use a posteriori error indicator for each element $T \in \mathcal{T}_n$.
A4. If the global error is small enough, then stop. Else refine the marked elements, construct the next mesh $\mathcal{T}_{n+1}$, set $n := n+1$, and go to step A2.

The solution of our linear elasticity equation (2.8) is computed by using adaptive finite element method based on the *Zienkiewicz-Zhu* (referred as ZZ) error estimator. For instance, a recovery technique is analyzed in [ZZ87] for determining the derivatives (stresses) of the finite element solutions at nodes. The main idea of the recovery technique is to develop smoothing procedures which recover more accurate nodal values of derivatives from the original finite element solution.

The necessity of derivative recovering arises from the fact that in the finite element approach the rate of convergence of the derivatives is usually one order less than that of the discrete solution. In particular, the accuracy of the derivatives (stresses) computed by directly differentiating the discrete solution is inferior. Therefore, in many practical problems an improved accuracy of the stresses at nodes is needed.

Denote by $\boldsymbol{\sigma}$ the exact stress, by $\hat{\boldsymbol{\sigma}}$ the discrete finite element discontinuous stress, and by $\boldsymbol{\sigma}^*$ the smoothed continuous *recovered stress.* The computation of $\boldsymbol{\sigma}^*$ was proposed and discussed in [ZZ87] under the assumption that the same basis functions for interpolation of stresses are used as those for the displacements. The recovered stress $\boldsymbol{\sigma}^*$ is computed by smoothing the discontinuous (over the elements) numerical stress $\hat{\boldsymbol{\sigma}}$. The smoothing procedure can be accomplished by nodal averaging method or the L_2-projection technique. Note that the components of $\boldsymbol{\sigma}^*$ are piecewise linear and continuous.

The computational of the global L_2-projection is expensive and the authors of [ZZ87] proposed to use a lumping form of the mass matrix. Thus, the value of the recovered stress $\boldsymbol{\sigma}^*$ at a node P can be computed by averaging the stresses $\hat{\boldsymbol{\sigma}}$ at the elements that share that node. Denote by $Y_P \subset Y$ the neighborhood patch as an union of all triangles/tetrahedra T having node P. Consider

$$\boldsymbol{\sigma}^*(P) = \sum_{T \in Y_P} \omega|_T \, \hat{\boldsymbol{\sigma}}|_T, \quad \omega|_T = \frac{|T|}{|Y_P|}, \quad T \in Y_P, \tag{6.1}$$

i.e., $\boldsymbol{\sigma}^*(P)$ is a weighted average of $\hat{\boldsymbol{\sigma}}$ with weights $\omega|_T$ defined on the elements belonging to Y_P. Least-square technique can also be applied to approximate the stress field at a given node.

It was shown in [ZZ87] that $\boldsymbol{\sigma}^*$ is a better approximation to $\boldsymbol{\sigma}$ than $\hat{\boldsymbol{\sigma}}$ and the following estimate holds

$$\|\boldsymbol{\sigma} - \boldsymbol{\sigma}^*\|_{0,Y} \ll \|\boldsymbol{\sigma} - \hat{\boldsymbol{\sigma}}\|_{0,Y}, \tag{6.2}$$

where Y is the periodicity microcell into consideration. Furthermore, the recovered technique was used in a formulation of a posteriori error estimator by comparing the recovered solution $\boldsymbol{\sigma}^*$ with the finite element solution $\hat{\boldsymbol{\sigma}}$. In particular, the estimate (6.2) allows us to replace the exact (unknown) stress $\boldsymbol{\sigma}$ by $\boldsymbol{\sigma}^*$ and consider $\|\boldsymbol{\sigma}^* - \hat{\boldsymbol{\sigma}}\|_{0,Y}$ as an error estimator.

In many practical implementations reliability and efficiency are highly desirable properties in a posteriori error estimation. It basically means that there exist constants independent of the discrete solution and the mesh which limit the error (in a suitable norm) from below and above. Moreover, technically it is better to use local error estimators which are computationally less expensive. The following local estimator is considered

$$\eta_T := \|\boldsymbol{\sigma}^* - \hat{\boldsymbol{\sigma}}\|_{0,T}. \tag{6.3}$$

The nodal values of the recovered stresses are found locally. The elementwise contributions (6.3) are used further as local error indicators in the adaptive mesh-refinement procedure.

The global ZZ-error estimator is defined by

$$\eta_Y := \left(\sum_{T\in\mathcal{T}_n} \eta_T^2\right)^{1/2}. \tag{6.4}$$

Based on a posteriori processing, the local estimator (6.3) is practically efficient providing recovered values are more accurate, i.e., the quality of the a posteriori error estimator strongly depends on the approximation properties and the accuracy of the recovered solution.

Arbitrary averaging techniques in low order finite element applications for elasticity problems are subject of investigations in [CF01]. In the latter study the authors considered the following global averaging estimator

$$\eta_A := \min_{\boldsymbol{\sigma}^*} \|\boldsymbol{\sigma}^* - \hat{\boldsymbol{\sigma}}\|_{0,Y} \tag{6.5}$$

and proved an equivalence to the error $\|\boldsymbol{\sigma} - \hat{\boldsymbol{\sigma}}\|_{0,Y}$ with lower and upper bounds independent of the shape-regular mesh. Note that in (6.5) $\boldsymbol{\sigma}^*$ is a smoother approximation to $\hat{\boldsymbol{\sigma}}$ obtained by any averaging procedure. In particular, the final error estimate in [CF01] explains the reliability and robustness of the ZZ- a posteriori error estimators in practice.

7 Iterative solution techniques

In this section, we comment on the iterative solvers for the microcell problem (2.8) defined in Y to find the effective coefficients and for the homogenized elasticity equation (2.12) on the global domain Ω. After finite element discretization of the corresponding domain we get the following system of linear equations

$$A\boldsymbol{u} = \boldsymbol{f}, \tag{7.1}$$

where $\boldsymbol{u}$ is the vector of unknown displacements and $\boldsymbol{f}$ is the discrete right-hand side. The stiffness matrix A is symmetric and positive definite but not an M-matrix.

Two typical orderings of the unknowns are often used in practice. In the 3-dimensional case they are presented as follows

$$\left(u_1^{(x)}, u_1^{(y)}, u_1^{(z)}, u_2^{(x)}, u_2^{(y)}, u_2^{(z)}, \ldots, u_N^{(x)}, u_N^{(y)}, u_N^{(z)}\right), \tag{7.2}$$

referred to as a *pointwise displacements ordering* and

$$\left(u_1^{(x)}, u_2^{(x)}, \ldots, u_N^{(x)}, u_1^{(y)}, u_2^{(y)}, \ldots, u_N^{(y)}, u_1^{(z)}, u_2^{(z)}, \ldots, u_N^{(z)}\right), \tag{7.3}$$

called the *separate displacements ordering*. Here, $u_k^{(x)}$, $u_k^{(y)}$, and $u_k^{(z)}$ are the corresponding x,y-, and z- displacement components. For the the first ordering (7.2), the resulting stiffness matrix $A = A^{(point)}$ can be seen as a discretization matrix consisting of elements which are small 3×3 blocks. For the second ordering (7.3), the matrix $A = A^{(block)}$ admits the following 3×3 block decomposition

$$A = \begin{bmatrix} A_{11} & A_{12} & A_{13} \\ A_{21} & A_{22} & A_{23} \\ A_{31} & A_{32} & A_{33} \end{bmatrix}. \tag{7.4}$$

In case of isotropic materials, the diagonal blocks A_{jj}, $j = 1, 2, 3$, in (7.4) are discrete analogs of the following anisotropic Laplacian operators

$$\tilde{D}_1 = a\frac{\partial^2}{\partial x^2} + b\frac{\partial^2}{\partial y^2} + b\frac{\partial^2}{\partial z^2}, \quad \tilde{D}_2 = b\frac{\partial^2}{\partial x^2} + a\frac{\partial^2}{\partial y^2} + b\frac{\partial^2}{\partial z^2}, \quad \tilde{D}_3 = b\frac{\partial^2}{\partial x^2} + b\frac{\partial^2}{\partial y^2} + a\frac{\partial^2}{\partial z^2}$$

with coefficients $a = E(1-\nu)/((1+\nu)(1-2\nu))$ and $b = 0.5E/(1+\nu)$ where E is the Young modulus and ν is the Poisson ratio of the corresponding material. This anisotropy requires a special care to construct an efficient preconditioner for the iterative solution method. Based on Korn's inequality, it can be shown that A and its block diagonal part are spectrally equivalent. The condition number of the preconditioned system depends on the Poisson ratio ν of the materials and the constant in the Korn inequality. For the background of the spectral equivalence approach using block diagonal displacement decomposition preconditioners in linear elasticity problems we refer to [BLA94]. Note that the spectral equivalence estimate will deteriorate for ν close to 0.5 which is not the case in our particular applications.

The PCG method is applied to solve the linear system (7.1). We propose two approaches to construct a preconditioner for A:

(i) construct a preconditioner for $A^{(point)}$

(ii) construct a preconditioner for $A^{(block)}$ of the type $M = \text{diag}(M_{jj})$, where $M_{jj} \sim A_{jj}$, $j = 1, 2, 3$, are "good" approximations to the diagonal blocks of A. In case (i) we have chosen the incomplete Cholesky (IC) factorization of A with an appropriate stopping criterion.

An efficient preconditioner for A_{jj} in case (ii) turns out to be a matrix M_{jj} corresponding to a Laplacian operator $(-\text{div}\,(c\,\text{grad}\,u))$ with a fixed scale factor c. In our case we use, for instance, $c = b/2$ for all three diagonal blocks. *Algebraic MultiGrid* (AMG) method is applied as a "plug-in" solver for A (see [RS86] for details). This method is a purely matrix-based version of the algebraic multilevel approach and has shown in the last decade numerous efficient implementations in solving large sparse unstructured linear systems of equations without any geometric background.

8 Numerical experiments

In this section, we comment on some computational results concerning the microscopic problem to find the homogenized elasticity coefficients and the macroscopic shape optimization problem. For simplicity, we suppose linear elasticity with homogeneous and isotropic constituents in terms of carbon and SiC. The Young modulus E (in GPa) and the Poisson ratio ν of our two materials are, respectively, $E = 10, \ \nu = 0.22$ for carbon and $E = 410, \ \nu = 0.14$ for SiC.

The computation of the characteristic displacement fields $\boldsymbol{\xi}^{kl}$ and the homogenized elasticity coefficients (2.9) requires the solution of linear elastic boundary value problems with the periodicity cell Y as the computational domain. The elasticity equation (2.8) is solved numerically using a conforming finite element discretization of the periodicity cell Y by linear basis functions. Since the periodic displacements $\boldsymbol{\xi}^{kl} = \boldsymbol{\xi}^{lk}$ are symmetric, the equation (2.8) is computed numerically 3 times in the case $d = 2$ and respectively, 6 times in the case $d = 3$. Due to the composite character of our microcell there are material interfaces where the solution changes significantly. Hence, local refinement of the underlying finite element mesh is strongly advised. As discussed in Sect. 6, we use an adaptive grid refinement strategy based on a posteriori error estimator of Zienkiewicz-Zhu type obtained by local averaging of the computed stress tensor. Note that the adaptivity procedure is local and computationally cheap.

Denote the global density of the solid material part in the microstructure by μ, $0 < \mu < 1$. Note that the density of the tracheidal cells of the wood essentially depends on the growth of the tree. If μ is relatively small, we speak about an *early wood* (grown in spring and summer) and respectively, about *late wood* (grown in autumn and winter) for values of μ, close to 1.

We present first some numerical experiments on a plane microstructure ($d = 2$) shown in Fig. 2.2 b). More experiments can be found in [HP04a]. We assume that the material layers in the periodicity cell have equal widths from all sides of the cell. Denote by α_i, $i = 1, 2$, the widths of the carbon and SiC layers, respectively. Figure 8.1 a) illustrates the behavior of the homogenized coefficient E^H_{1212} in case of square hole versus α_1 and α_2 which vary between 0 and 0.5. We compute the effective coefficients E^H_{ijkl} only for a fixed number of values of the design parameters (e.g., 20×20 grid as shown on Fig. 8.1) and then interpolate the values by splines. With regard to the homogenized state equation (2.12), this procedure results in having explicit formulas at hand for the gradients and the Hessian of the Lagrangian function needed in the optimization procedure.

In principal, the hole is located inside the microstructure but we find interesting to demonstrate the behavior, for instance, of E^H_{1212} depending on a rectangular hole $[1 - a] \times [1 - b]$, see Fig. 8.1 b). Note that $a = b = 0$ represents a complete void, $a = b = 1$ realizes a complete solid material, and $0 < a < 1$, $0 < b < 1$ characterize a general porous material. We consider in this example the case when the carbon has completely reacted with the SiC which strongly concerns the so-called *pure biomorphic SiC-ceramics*. Very recently, the chemical experiments have shown that the carbon phase limits the mechanical properties of the composite materials and restricts their high-temperature applications. The final transformation of the original carbonized template to pure ceramic composite requires to offer enough silicon during

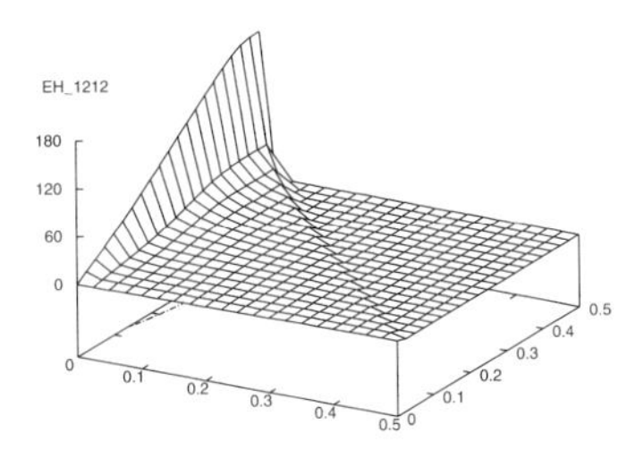

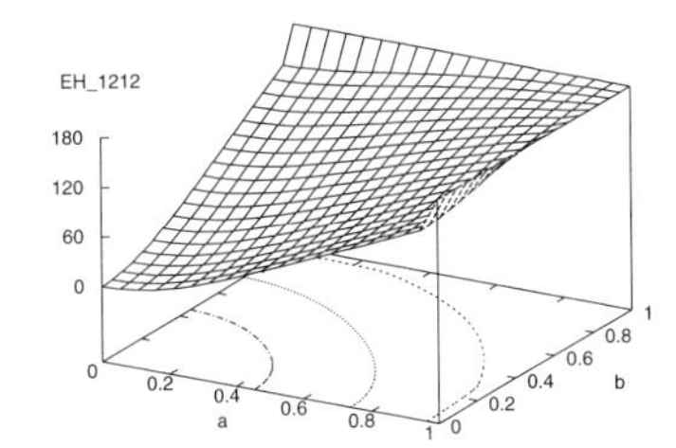

Fig. 8.1. Homogenized coefficient E^H_{1212}: a) w.r.t. the widths of carbon and SiC layers (square hole); b) w.r.t. the sizes $1 - a$ and $1 - b$ of the rectangular hole (pure SiC-ceramic)

the infiltration process and to wait an appropriate time until the carbon is completely consumed by the silicon resulting in a SiC-phase.

Figure 8.2 displays the dependence of the homogenized elasticity coefficients on the density μ of the cell. In particular, we show this behavior versus the width of the SiC layer in case of pure SiC-ceramics. Figure 8.2 a) shows the behavior of the effective coefficients for early wood ($0 \leq \alpha_2 \leq 0.15$, $\mu = 51\%$) and Fig. 8.2 b) demonstrates the coefficients for late wood ($0 \leq \alpha_2 \leq 0.3$, $\mu = 84\%$). One can easily observe from both pictures on this figure a highly nonlinear behavior of the homogenized coefficients.

The mesh-adaptive process is visualized in Fig. 8.3. We see that in case of one material available in the microstructure, an appropriate refinement is done around the corners where the hole with a complete pore is located, see Fig. 8.3 a). In case of more materials, additional mesh-adaptivity is needed across the material interfaces in the microstructure due to the strongly varying material properties in terms of Young's modulus and Poisson's ratio.

In Table 8.1 we give some results for the homogenized elasticity coefficients on the first ten adaptive refinement levels for various values of the density. We report the number of triangles NT and the number of nodes NN on each level when solving problem (2.8). We see from the computed values that the mesh sensitivity on the successive levels is very small. Our adaptive mesh-refinement

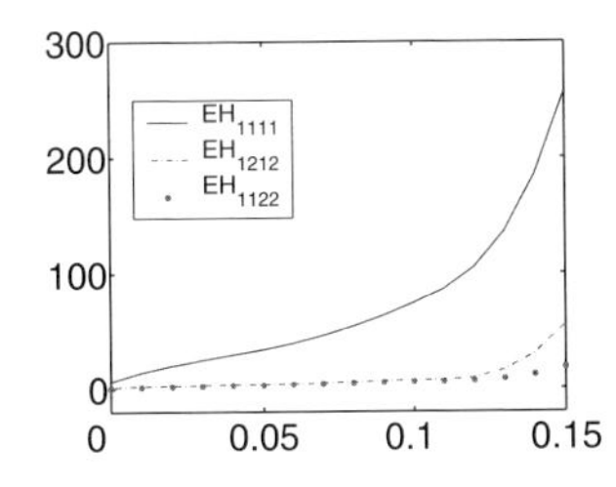

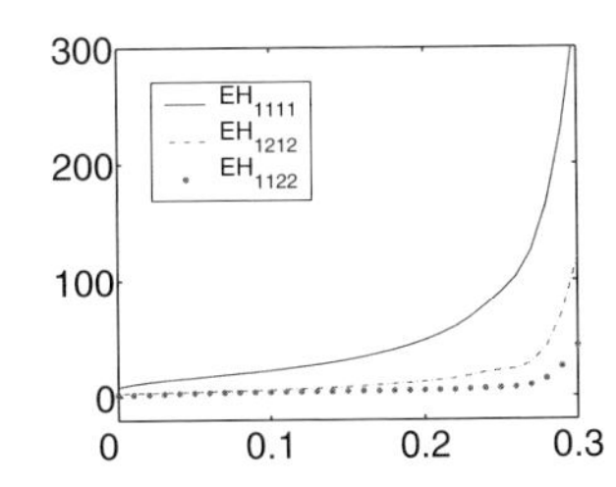

Fig. 8.2. Homogenized coefficients w.r.t. the width α_2 of SiC layer for pure SiC-ceramic: a) early wood, density 51%; b) late wood, density 84%

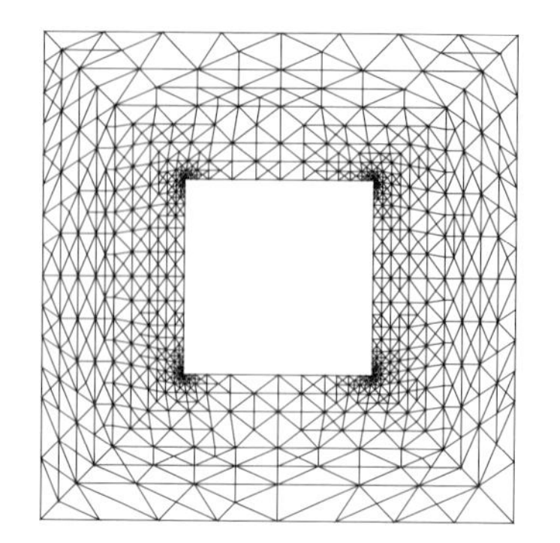
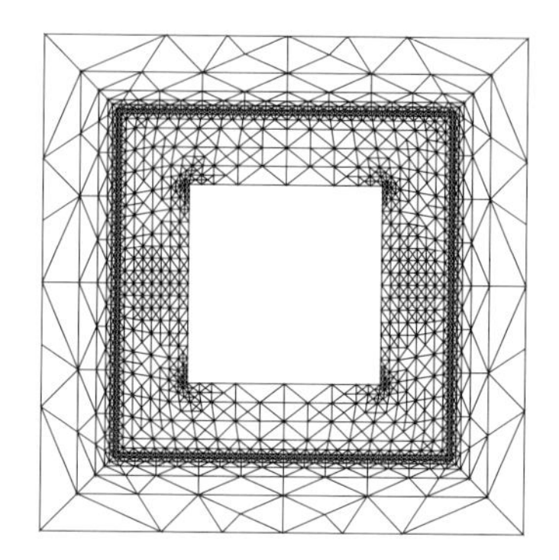

Fig. 8.3. Late wood, density $\mu = 84\%$, 9 adaptive refinement levels: a) SiC, 1527 triangles, 818 nodes; b) carbon and SiC, 3752 triangles, 1916 nodes

procedure stops when a priori given limit for the number of refinement levels is reached.

We are now concerned with the solution of problem (3.2)–(3.3). In Table 8.2 we report some numerical results from running the optimization code varying the constant C with respect to (3.3). Our purpose is to find the optimal widths/lengths of the layers in the composite material and to show the convergence behavior of the optimization algorithm. The domain Ω is chosen to be a circle which corresponds naturally to a cross section of the original wood structure. We have fixed the discretization and vary the initial values for the lengths of the carbon and SiC layers denoted, respectively, by $\alpha_1^{(0)}$ and $\alpha_2^{(0)}$. As before, we report the number of iterations ITER to get convergence, the optimal lengths α_1 and α_2 of the carbon and SiC layers, the last value of the barrier parameter ρ, the final value of the primary merit function M, the l_2-norm of the residual, and the l_2-norm of the complementarity conditions $\boldsymbol{v} = (\boldsymbol{z}, \boldsymbol{w})$ at the last iteration. We see from the experiments that the optimal

Table 8.1. Homogenized coefficients w.r.t. refinement level, a) $\mu = 51\%$, b) $\mu = 84\%$

level	E^H_{1111}	E^H_{1122}	E^H_{1212}	NT	NN
1	64.975	7.664	12.116	168	100
2	63.336	6.642	9.750	220	126
3	58.466	6.682	8.073	288	162
4	56.572	7.012	6.643	484	262
5	54.385	6.245	6.212	712	378
6	52.936	6.091	5.474	1208	630
7	51.914	5.458	5.306	1800	932
8	50.861	4.790	5.217	2809	1444
9	50.455	4.571	5.029	3754	1919
10	49.591	4.359	4.983	5918	3013

level	E^H_{1111}	E^H_{1122}	E^H_{1212}	NT	NN
1	33.430	3.885	9.893	168	100
2	33.064	3.929	9.577	216	126
3	32.844	4.024	9.283	300	168
4	32.291	4.254	8.970	544	296
5	32.144	4.312	8.809	828	438
6	31.909	4.372	8.703	1354	705
7	31.862	4.379	8.526	1892	980
8	31.735	4.399	8.470	2894	1485
9	31.711	4.400	8.373	3752	1916
10	31.487	4.497	8.321	5716	2906

Table 8.2. Convergence results for biomorphic microcellular SiC ceramics

$\alpha_1^{(0)}$	$\alpha_2^{(0)}$	C	ITER	α_1	α_2	ρ	M	$\|\boldsymbol{F}^{(\rho)}\|_2$	$\|\boldsymbol{v}\|_2$
0.05	0.05	0.3	11	3.6e-12	0.3	1.3e-17	1.24	9.63e-6	e-10
0.1	0.1	0.3	11	5.5e-14	0.3	3.0e-21	1.24	1.03e-6	e-12
0.1	0.1	0.4	12	1.6e-16	0.4	1.2e-26	0.85	8.63e-9	e-14
0.2	0.2	0.1	16	5.5e-17	0.1	2.2e-25	7.73	2.23e-8	e-13
0.2	0.2	0.2	13	1.0e-16	0.2	5.3e-26	2.34	1.54e-8	e-14
0.2	0.2	0.3	11	2.5e-16	0.3	6.7e-26	1.24	1.79e-8	e-14
0.24	0.24	0.15	11	5.4e-15	0.15	4.1e-12	3.81	4.99e-7	e-12
0.3	0.1	0.4	11	1.3e-12	0.4	8.5e-19	0.85	5.07e-6	e-10
0.4	0.05	0.1	17	9.8e-15	0.1	6.9e-21	7.73	9.49e-7	e-11

length α_1 of the carbon layer in all the runs is very close to zero, i.e., the solid part of the body is entirely occupied by a silicon carbide layer due to the higher stiffness of this material.

In case of 3-dimensional implementations we decompose the periodic microcell Y first in hexahedra and further we use continuous, piecewise linear finite elements on tetrahedral shape regular meshes. The adaptive grid refinement process is visualized in Fig. 8.4. The mesh adaptivity around the material interfaces has been realized by means of Zienkiewicz-Zhu type a posteriori error which is used heuristically (as an error indicator). One computes the error (6.3) locally for each element and mark for refinement those tetrahedra $\{T\}$ for which

$$\eta_T \geq \gamma \max_{T' \in \mathcal{T}_n} \eta_{T'},$$

where $0 < \gamma < 1$ is a prescribed threshold, for instance, $\gamma = 0.5$. The refinement process is visualized in Fig. 8.4 b) on the cross section of the microstruc-

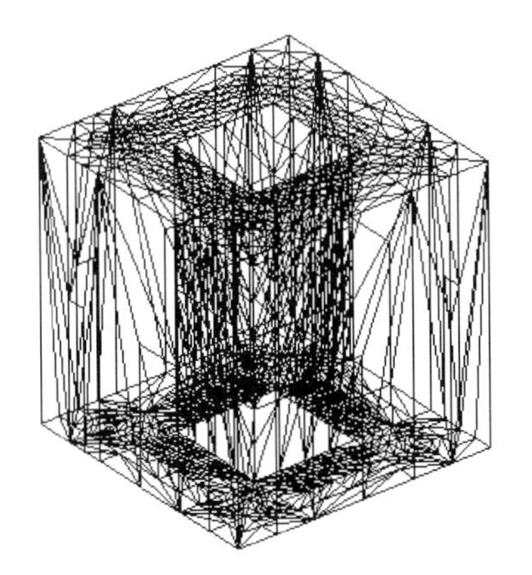
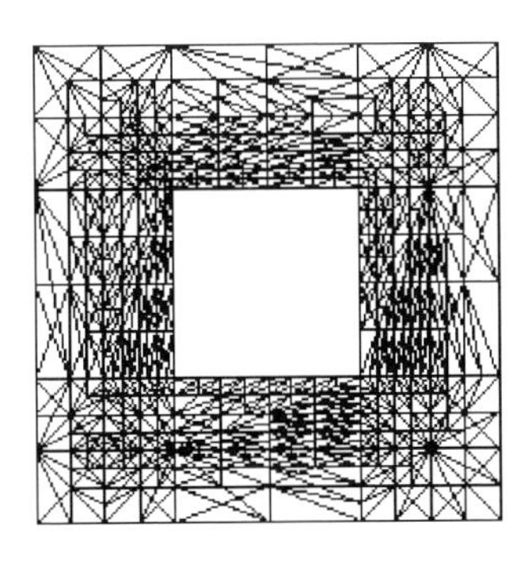

Fig. 8.4. Adaptive refinement a) 3-D unit periodicity cell Y, b) Cross section of Y

Table 8.3. Homogenized coefficients for late wood, density $\mu = 91\%$

level	E^H_{1111}	E^H_{2222}	E^H_{3333}	E^H_{1212}	E^H_{2323}	E^H_{1313}
1	148.35	152.57	153.96	60.22	62.46	59.50
2	154.34	162.64	162.77	69.71	71.31	65.79
3	142.66	148.42	162.79	60.51	65.26	63.23
4	145.84	137.61	161.70	53.91	59.04	62.92
5	127.99	134.32	161.43	49.41	56.19	56.49
6	98.29	111.65	160.71	40.44	46.14	48.45
7	91.79	90.23	158.29	35.70	43.69	46.03
8	82.42	83.00	160.57	30.59	41.03	43.70
9	75.05	75.11	160.22	26.93	39.75	40.97
10	69.66	70.30	159.82	25.47	37.16	39.30

ture Y for widths of the C- and SiC- layers $\alpha_1 = \alpha_2 = 0.15$. Additional adaptive refinement is generated in the stiffer material (SiC) and on the interface between the materials due to the different characteristic constants.

In Table 8.3 we report some values of the computed 3-dimensional homogenized coefficients with respect to the adaptive refinement level for a late wood with density $\mu = 91\%$. More numerical experiments for various values of the density and various number of adaptive levels can be found in [HP06a].

Table 8.4 presents some convergence results for the proposed preconditioners within PCG method. For various values of the density μ of the periodical microstructure we report the number of degrees of freedom NDOF, the number of iterations ITER, and the CPU-time in seconds for the first 11 adaptive refinement levels. One can see from the numerical results a better convergence of AMG-preconditioner compared to IC-factorization. We observe an essential efficiency of AMG for a larger number of unknowns.

Table 8.4. Convergence results with IC and AMG preconditioners, density μ

density	level	1	2	3	4	5	6	7	8	9	10	11
$\mu = 51\%$	NDOF	78	90	126	225	336	579	1185	1908	3360	5598	9987
IC	ITER	9	8	14	23	40	66	105	150	235	269	299
	CPU	e-16	e-16	e-16	0.1	0.2	0.2	0.9	2.4	8.2	20.9	59.1
AMG	ITER	11	13	13	15	18	23	38	57	89	94	99
	CPU	e-16	e-16	e-16	0.2	0.3	0.5	1.5	3	7.6	14.8	23.5
$\mu = 84\%$	NDOF	78	93	150	261	510	1047	2103	3843	6537	10485	18459
IC	ITER	10	11	16	21	44	78	117	171	226	273	301
	CPU	e-16	e-16	0.1	0.1	0.1	0.6	2.4	8.4	24.3	63.7	187.1
AMG	ITER	12	14	14	14	18	31	43	73	69	74	75
	CPU	e-16	e-16	e-16	0.2	0.4	1.1	3	7.5	15.5	25.6	33.8

Acknowlegdments. Research supported by DFG Priority Program 1095 "Analysis, Modeling and Simulation of Multiscale Problems" under Ho 877/5.

References

[All02] G. Allaire. Shape Optimization by the Homogenization Method. Springer, Berlin-Heidelberg-New York, 2002.

[BP84] N. Bakhvalov and G. Panasenko. Averaging Processes in Periodic Media. Nauka, Moscow, 1984.

[Ben95] M.P. Bendsøe. Optimization of Structural Topology, Shape, and Material. Springer, Berlin, 1995.

[BS03] M.P. Bendsøe and O. Sigmund. Topology Optimization: Theory, Methods and Applications. Springer, Berlin-Heidelberg-New York, 2003.

[BLP78] A. Bensoussan, J.L. Lions, and G. Papanicolaou. Asymptotic Analysis for Periodic Structures. Elsevier, North-Holland, Amsterdam, 1978.

[BLA94] R. Blaheta. Displacement decomposition – incomplete factorization preconditioning techniques for linear elasticity problems. *Numer. Linear Algebra Appl.*, 1(2), 107–128, 1994.

[BHN99] R.H. Byrd, M.E. Hribar, and J. Nocedal. An interior point algorithm for large scale nonlinear programming. *SIAM J. Optim.*, 9(4), 877–900, 1999.

[CF01] C. Carstensen and S. A. Funken. Averaging technique for FE–a posteriori error control in elasticity. Part I: Conforming FEM. *Comput. Methods Appl. Mech. Engrg.*, 190, 2483–2498, 2001.

[CL*82] R.M. Chamberlain, C. Lemaréchal, H.C. Pedersen, and M.J.D. Powell. The watchdog technique for forcing convergence in algorithms for constrained optimization. *Math. Progr. Study*, 16, 1–17, 1982.

[Eli00] M. Elices. Structural Biomaterials. Princeton University Press, 2000.

[ET*86] A.S. El-Bakry, R.A. Tapia, T. Tsuchiya, and Y. Zhang. On the formulation and theory of the Newton interior–point method for nonlinear programming. *J. Optim. Theory Appl.*, 89, 507–541, 1996.

[EE*95] K. Eriksson, D. Estep, P. Hansbo, and C. Johnson. Introduction to adaptive methods for differential equations. *Acta Numerica*, 105–158, 1995.

[FM68] A.V. Fiacco and G.P. McCormick. Nonlinear Programming. Sequential Unconstrained Minimization Techniques. John Wiley and Sons, New York, New York, 1968. Republished by SIAM, Philadelphia, Pennsylvania, 1990.

[Fri55] K.R. Frisch. The Logarithmic Potential Method of Convex Programming. Memorandum, University Institute of Economics, Oslo, Norway, 1955.

[FGS96] A. Forsgren, P.E. Gill, and J.R. Shinnerl. Stability of symmetric ill–conditioned systems arising in interior methods for constrained optimization. *SIAM J. Matrix Anal. Appl.*, 17, 187–211, 1996.

[FGW02] A. Forsgren, P.E. Gill, and M.H. Wright. Interior methods for nonlinear optimization. *SIAM Review*, 44, 525–597, 2002.

[GOW98] D.M. Gay, M.L. Overton, and M.H. Wright. A primal–dual interior method for nonconvex nonlinear programming. *Advances in Nonlinear Programming*, (Y. Yuan, ed.), Kluwer, Dordrecht, Holland, 31–56, 1998.

[GA88] L.J. Gibson, M.F. Ashby. Cellular Solids, Structure, and Properties. Pergamon Press, New York, 1988.

[GLK98a] P. Greil, T. Lifka, and A. Kaindl. Biomorphic cellular silicon carbide ceramics from wood: I. Processing and microstructure. *J. Europ. Ceramic Soc.*, 18, 1961–1973, 1998.

[GLK98b] P. Greil, T. Lifka, and A. Kaindl. Biomorphic cellular silicon carbide ceramics from wood: II. Mechanical properties. *J. Europ. Ceramic Soc.*, 18, 1975–1983, 1998.

[HPS02] R.H.W. Hoppe, S.I. Petrova, and V. Schulz, Primal–dual Newton–type interior–point method for topology optimization. *J. Optim. Theory Appl.*, 114(3), 545–571, 2002.

[HP04a] R.H.W. Hoppe and S.I. Petrova. Optimal shape design in biomimetics based on homogenization and adaptivity. *Math. Comput. Simul.*, 65, 257–272, 2004.

[HP04b] R.H.W. Hoppe and S.I. Petrova. Primal–dual Newton interior point methods in shape and topology optimization. *Numer. Linear Algebra Appl.*, 11(5-6), 413–429, 2004.

[HP06a] R.H.W. Hoppe and S.I. Petrova. Efficient solvers for 3-D homogenized elasticity model. In J. Dongarra et al., eds., *Lect. Notes Comput. Sci.*, Springer, 3732, 857–863, 2006.

[HDL02] K.A. Hudgins, A.K. Dillow, and A.M. Lowman. Biomimetic Materials and Design: Biointerfacial Strategies, Tissue Engineering, and Targeted Drug Delivery. Marcel Dekker, New York, 2002.

[JKO94] V.V. Jikov, S.M. Kozlov, and O.A. Oleinik. Homogenization of Differential Operators and Integral Functionals. Springer, 1994.

[OT83] N. Olhoff and J.E. Taylor. On structural optimization. *J. Appl. Mech.*, 50, 1139–1151, 1983.

[OT*95] T. Ota, M. Takahashi, T. Hibi, M. Ozawa, S. Suzuki, Y. Hikichi, and H. Suzuki. Biomimetic process for producing SiC wood. *J. Amer. Ceram. Soc.*, 78, 3409–3411, 1995.

[RS86] J.W. Ruge and K. Stüben. Algebraic multigrid (AMG). In S.F. McCormick, ed., *Multigrid Methods, Frontiers in Applied Mathematics*, volume 5, SIAM, Philadelphia, 1986.

[SP80] E. Sanchez-Palencia, Non–homogeneous Media and Vibration Theory. Lecture Notes in Physics, volume 127, Springer, Berlin-Heidelberg, 1980.

[SZ92] J. Sokolowski and J.-P. Zolésio. Introduction to Shape Optimization. Springer Series in Computational Mathematics, volume 16, Springer, 1992.

[SK91] K. Suzuki and N. Kikuchi. A homogenization method for shape and topology optimization. *Comput. Meth. Appl. Mech. Engrg.*, 93, 291–318, 1991.

[VS99] R.J. Vanderbei and D.F. Shanno. An interior–point algorithm for nonconvex nonlinear programming. *Comput. Optim. Appl.*, 13, 231–252, 1999.

[VdV92] H.A. Van der Vorst. Bi-CGSTAB: A fast and smoothly converging variant of Bi-CG for the solution of nonsymmetric linear systems. *SIAM J. Sci. Stat. Comput.*, 13, 631–644, 1992.

[VSG02] E. Vogli, H. Sieber, P. Greil. Biomorphic SiC-ceramic prepared by Si-gas phase infiltration of wood. *J. Europ. Ceramic Soc.*, 22, 2663–2668, 2002.

[Wit89] G. Wittum. On the convergence of multigrid methods with transforming smoothers. Theory with applications to the Navier–Stokes equations. *Numer. Math.*, 57, 15–38, 1989.

[Wri98] M.H. Wright. Ill–conditioning and computational error in interior methods for nonlinear programming. *SIAM J. Optim.*, 9, 84–111, 1998.

[ZZ87] O.C. Zienkiewicz and J.Z. Zhu. A simple error estimator and adaptive procedure for practical engineering analysis. *Intern. J. Numer. Methods Engrg.*, 24, 337–357, 1987.

Discrete Free Energy Functionals for Elastic Materials with Phase Change

Thomas Blesgen[1], Stephan Luckhaus[2], and Luca Mugnai[2]

[1] Max-Planck-Institute for Mathematics in the Sciences, Inselstrasse 22–26, 04103 Leipzig. blesgen@mis.mpg.de
[2] Mathematisches Institut, Universität Leipzig, Augustusplatz 10–11, 04103 Leipzig. luckhaus@mathematik.uni-leipzig.de, mugnai@mathematik.uni-leipzig.de

Summary. We discuss two different approaches related to Γ-limits of free energy functionals. The first gives an example of how symmetry breaking may occur on the atomistic level, the second aims at deriving a general analytic theory for elasticity on the lattice scale that does not depend on an explicitly chosen reference system.

1 Introduction

The analysis of the mechanical properties of crystals gives rise to internal energies that are connected to the geometry of the considered crystal and are often linked to properties of the atomistic scale as explained in [CK88] and [JF00]. Applications to this theory include among others fatigue phenomena and fracture mechanics. In the past, various attempts were made to develop a mathematically rigid theory. In particular we want to mention [Bal77], [CLL98], [AO05], [OP99], [FT02] and [Tru96]. Nevertheless, up to now, the relationship between macroscopic and atomistic scale is not completely understood.

Here we contribute to this topic. The text is subdivided into two parts. The first gives a simple example where symmetry breaking occurs in the Γ-limit of a one-dimensional monatomic chain when the interatomic distance vanishes. Effects similar to the one presented in this first part may also show to be relevant for numerical approximation schemes where in certain cases a competition between elastic energy and surface energy leads to wrong numerical solutions, see [Ble06].

The second line of investigation is the use of many-body Hamiltonians of Kac type to describe elastic deformations, phase changes and eventually plastic deformations of a domain $\Omega \subset \mathbb{R}^n$ without postulating a reference configuration on the particle level. It is interesting to compare this ansatz to [AO05], where a theory based on algebraic topology is developed.

One of the aims is to make a connection with the existing theory on linear elastic dislocations, see [TK76], [Mer79], [CC*97], [BC05].

2 Phase transitions with symmetry breaking

2.1 The energy functional

For given length $L > 0$, let $\Omega := (0, L) \subset \mathbb{R}$ be a domain that contains a regular monatomic chain.

We suppose that the undeformed discrete reference configuration of Ω is given by a system of $n+1$ atoms with equal distance located at points $R_i^n \in \mathbb{R}$,

$$R_i^n := ih^n \quad 0 \le i \le n.$$

Here, the setting $h^n := L/n$ defines for given number $n \in \mathbb{N}$ the interatomic distance. The limit $n \to \infty$ corresponds to $h^n \searrow 0$. The superscript n is always used to indicate the dependence on the number of subdivisions.

By $\widehat{R}_i^n$, $0 \le i \le n$ we denote the position of atom i after the deformation. Finally, by u_i^n, $0 \le i \le n$ we denote the two-dimensional displacement vector of atom i, given by the relationship

$$u_i^n = \widehat{R}_i^n - R_i^n, \quad 0 \le i \le n.$$

For given deformations $\{u_i^n\}_{0 \le i \le n}$ we introduce the abbreviations

$$p_i^n := \frac{u_{i+1}^n - u_i^n}{h^n}$$

and for shortness the numbers $s_1 := 1$, $s_2 := 2$ and $s_3 := \frac{1}{2}$.

We will study the behaviour of the following energy functional.

$$W^n(u^n) := \begin{cases} +\infty & \text{if } p_i^n = 0 \text{ for some } i, \\ \sum_{k=1}^3 W_k^n(u^n) & \text{else} \end{cases}$$

where

$$W_1^n(u^n) := \sum_{i=0}^{n-2} (h^n)^{-\alpha} \prod_{k=1}^{3} \Big| s_k - \frac{p_{i+1}^n}{p_i^n} \Big|^2, \qquad W_2^n(u^n) := \sum_{i=0}^{n-3} \Big| 1 - \frac{p_{i+2}^n}{p_i^n} \Big|^2,$$

$$W_3^n(u^n) := h^n \sum_{i=0}^{n-2} \Big[\Big(\frac{p_i^n + p_{i+1}^n}{2} - \alpha_1 \Big)^2 \beta_i^n + \Big(\frac{p_i^n + p_{i+1}^n}{2} - \alpha_2 \Big)^2 \gamma_i^n \Big]$$

and

$$\beta_i^n := \Big[1 - (h^n)^{-\alpha} \Big| 1 - \frac{p_{i+1}^n}{p_i^n} \Big|^2 \Big]_+, \gamma_i^n := \Big[1 - (h^n)^{-\alpha} \Big| 2 - \frac{p_{i+1}^n}{p_i^n} \Big|^2 \Big| \frac{1}{2} - \frac{p_{i+1}^n}{p_i^n} \Big|^2 \Big]_+.$$

Here, $0 < \alpha < 1$ and $[x]_+ = x$ for $x \geq 0$ and $[x]_+ = 0$ for $x < 0$.

The concept behind this ansatz is the following. A minimiser of W_1^n either fulfils $p_{i+1}^n \simeq p_i^n$ which specifies one lattice order that is in the sequel referred to as Phase 1, or $p_{i+1}^n \simeq 2p_i^n$ resp. $p_{i+1}^n \simeq \frac{1}{2}p_i^n$ which characterises Phase 2.

W_2^n represents a surface energy. It counts(and limits) the number of transitions between the two phases, as within a phase one asymptotically has $p_{i+2}^n = p_i^n$. Finally, W_3^n represents an elastic energy. We will show below that β_i^n converges in $L^1(\Omega)$ to the indicator function of Phase 1 and γ_i^n to the indicator function of Phase 2 as $n \to \infty$; α_k is the elastic constant to Phase k.

The functional W_1^n represents the electrostatic energy due to interatomic potentials that force the atoms to positions of a certain given lattice order.

For the analysis we extend the discrete deformation values $\{u_i^n\}_i$, to piecewise linear functions u^n in $L^2(\Omega) \cap \mathcal{A}^n$, where $\mathcal{A}^n$ denotes the space of piecewise linear functions, see [BDG99].

2.2 Identification of the Γ-limit for W^n

Now we can state the main result. It characterises the Γ-limit of W^n as n tends to infinity. Let $\chi_1 := \chi$, $\chi_2 := 1 - \chi$. For $u \in H^{1,2}(\Omega)$, $\chi \in BV(\overline{\Omega}, \{0,1\})$ set

$$E(u,\chi) := \frac{1}{4}\int_\Omega |\nabla\chi| + \sum_{k=1}^{2}\int_\Omega \chi_k\,(u' - \alpha_k)^2.$$

Additionally we introduce $W : L^2(\Omega) \to \mathbb{R}$ by

$$W(u) := \begin{cases} \inf_{\chi\in BV(\overline{\Omega},\{0,1\})} E(u,\chi) \text{ if } u \in H^{1,2}(\Omega) \text{ is strictly monotone,} \\ \qquad\qquad +\infty \text{ else.} \end{cases}$$

Theorem 2.1 (Characterisation of the Γ-limit of W^n).
The following statements are valid:
(i) The boundedness of the energy functional $W^n(u^n)$ implies the boundedness of $\left(\int_\Omega |(u^n)'|^2\right)_n$ uniformly in n.
(ii) W is the Γ-limit of W^n as $n \to \infty$ with respect to convergence in $L^2(\Omega)$.

Proof of (i):
Step 1: Construction of the characteristic function χ:
By C we denote various positive constants that may change from line to line.

Let $(u^n) \subset L^2(\Omega)$ be a sequence with $W^n(u^n) \leq C$. We set

$$d_k^i := \left|\frac{p_{i+1}^n}{p_i^n} - s_k\right|, \quad k_0^i := \operatorname{argmin}\Big\{k \mapsto d_k^i \;\Big|\; 1 \leq k \leq 3\Big\}.$$

The boundedness of $W_1^n(u^n)$ implies

$$\sum_{i=0}^{n-1}(h^n)^{-\alpha}\prod_{k=1}^{3}\left(s_k-\frac{p_{i+1}^n}{p_i^n}\right)^2\le C.$$

Therefore there exists a constant $C>0$ such that

$$\sup_i d^i_{k_0^i}\le C(h^n)^{\alpha/2}. \tag{2.1}$$

For n large enough we can thus define an indicator function χ^n to Phase 1 by

$$\chi^n(x):=\begin{cases}0 & \text{if } x\in[ih^n,(i+1)h^n),\ i\le n-2,\ k_0^i\ne 1,\\ 1 & \text{if } x\in[ih^n,(i+1)h^n),\ i\le n-2,\ k_0^i=1,\\ \chi^n(L-2h^n) & \text{if } x\in[L-h^n,L].\end{cases}$$

Next we show that $\chi^n\in BV(\overline{\Omega};\{0,1\})$, i.e.

$$\int_\Omega|\nabla\chi^n|\le C. \tag{2.2}$$

This follows from the boundedness of $W_2^n(u^n)$. Since for large n

$$\frac{p_{i+1}^n}{p_i^n}=s_k+o(1)\quad\text{for some } k\in\{1,2,3\},$$

we see that if $\chi^n(x)$ jumps in $x=(i+1)h^n$ between 0 and 1, then

$$\left(1-\frac{p_{i+2}^n}{p_i^n}\right)^2\ge\frac{1}{4}+o(1)$$

which shows $W_2^n(u^n)\ge\left(\frac{1}{4}+o(1)\right)\int_\Omega|\nabla\chi^n|$ and proves (2.2). Here we adapted the Landau notation and denote by $o(1)$ terms that tend to 0 as $n\to\infty$. With (2.2), well-known compactness results imply the existence of a subsequence (again denoted by) χ^n and a $\chi\in BV(\overline{\Omega},\{0,1\})$ such that $\chi^n\to\chi$ in $L^1(\Omega)$.

Step 2: Convergence of β^n, γ^n in $L^1(\Omega)$:

We extend the discrete quantities $\{\beta_i^n\}_i$, $\{\gamma_i^n\}_i$ to piecewise constant functions in $L^1(\Omega)$ by the definition

$$\beta^n(x):=\begin{cases}\beta_i^n & \text{if } x\in[ih^n,(i+1)h^n) \text{ and } i\le n-2,\\ 0 & \text{if } x\in[L-h^n,L].\end{cases}$$

In the same manner, the extension γ^n of $\{\gamma_i^n\}_i$ is defined.

Straightforward computations show

$$\beta^n\to\chi,\quad \gamma^n\to(1-\chi)\quad\text{in } L^1(\Omega)\text{ for } n\to\infty, \tag{2.3}$$

where the function $\chi \in BV(\overline{\Omega}, \{0,1\})$ is the limit of χ^n found in Step 1.

Step 3: Boundedness of $\int_\Omega |(u^n)'|^2$ uniformly in n:

We choose constants $a \in \mathbb{R}^+$, $b \in \mathbb{R}$ such that

$$\min\{(x-\alpha_1)^2, (x-\alpha_2)^2\} \geq ax^2 - b.$$

Due to the boundedness of $W_3^n(u^n)$ we thus find that there exist constants C_1, $C_2 > 0$ such that

$$C_1 \geq (h^n)^2 C_2 \sum_{i=0}^{n-2} \left(\frac{p_{i+1}^n + p_i^n}{2}\right)^2 (\beta_i^n + \gamma_i^n).$$

Since $p_{i+1}^n = s_k p_i^n + o(1)$ for a $k \in \{1,2,3\}$ and large n we find that

$$\left(\frac{p_{i+1}^n + p_i^n}{2}\right)^2 \geq \left(1 + \frac{1}{2} + o(1)\right)\left(\frac{p_i^n}{2}\right)^2.$$

The term $\left(\beta_i^n + \gamma_i^n\right)$ can for large n be estimated from below by a constant. So we find the existence of a constant $C > 0$ with

$$C \geq (h^n)^2 \sum_{i=0}^{n-2} \left(\frac{p_i^n}{2}\right)^2. \tag{2.4}$$

Due to the estimate $(p_{n-1}^n)^2 \leq (2+o(1))p_{n-2}^n$ the sum in (2.4) can be extended to $i = n-1$ and the estimate still holds.

The sum $\sum_i (p_i^n)^2$ is directly related to $\int_\Omega |(u^n)'|^2$ where u^n is the piecewise affine linear extension of $\{u_i^n\}_i$. With (2.4) extended to $i = n-1$ this yields

$$\sup_n \int_\Omega |(u^n)'|^2 = \sup_n h^n \sum_{i=0}^{n-1} (p_i^n)^2 \leq C. \tag{2.5}$$

Proof of (ii):

Step 4: Lower semicontinuity of W^n:

We have to show: for every sequence $(u^n)_{n\in\mathbb{N}}$ with $u^n \to u$ in $L^2(\Omega)$ there exists a subsequence $(u^{n_k})_{k\in\mathbb{N}}$ with

$$W(u) \leq \liminf_{k\to\infty} W^{n_k}(u^{n_k}).$$

For unbounded $W^n(u^n)$ there is nothing to show. So let $W^n(u^n) \leq C$ for all n. From (2.5) follows u^n, $u \in H^{1,2}(\Omega)$ for all $n \in \mathbb{N}$. Because of the reflexivity of the Hilbert space $H^{1,2}(\Omega)$ we know that there exists a subsequence (again denoted by) u^n such that $u^n \rightharpoonup u$, in $H^{1,2}(\Omega)$ for $n \to \infty$. From Step 2 we know that $\chi^n \to \chi$, $\beta^n \to \chi$, $\gamma^n \to 1-\chi$ in $L^1(\Omega)$ for $n \to \infty$. Because of $\frac{p_{i+1}^n}{p_i^n} \geq \frac{1}{2} + o(1)$, for $n \geq n_0$ we find that u^n is monotone for large n.

Now we estimate $W^n(u^n)$ from below. We claim

$$\liminf_{n\to\infty} W^n(u^n) \geq E(u,\chi) \geq W(u). \tag{2.6}$$

With the help of Theorem 3.4, p.74 in [Dac89], the proof of (2.6) is straightforward, estimating every component of $W^n(u^n)$ separately.

Step 5: Existence of a "recovery sequence":

We have to find a sequence $(u^n) \subset L^2(\Omega)$ converging to u in $L^2(\Omega)$ with

$$W(u) \geq \limsup_{n\to\infty} W^n(u^n).$$

If $W(u) = +\infty$, there is nothing to show. Due to the monotonicity properties of u demonstrated above we know that the functional $\chi \mapsto E(u,\chi)$ is bounded from below in the BV-norm. Using the compactness properties of $BV(\Omega)$ and the coercivity of E, it is clear that $E(u,\cdot)$ attains its minimum, i.e. $W(u) = E(u,\chi)$ for some $\chi \in BV(\overline{\Omega},\{0,1\})$.

Next we show that for piecewise affine, strictly monotone u there exists a sequence u^n with $u^n \to u$ and $W^n(u^n) \to E(u,\chi)$. We start with special cases, then generalise.

Case 1: $u' \equiv a_1 > 0$, $\chi \equiv$ const in Ω:

(a) $\chi \equiv 1$ in Ω: We simply set $u^n := u$ for all n.

(b) $\chi \equiv 0$ in Ω: For $x > 0$ choose u^n such that p_i^n is alternating between $\frac{2}{3}a_1$ and $\frac{4}{3}a_1$. Furthermore u^n satisfies $u^n(x=0) = u(x=0)$.

Case 2: $u' \equiv a_1 > 0$, $\chi \equiv 1$ for $0 \leq x \leq \frac{L}{2}$, $\chi \equiv 0$ for $x > \frac{L}{2}$.

The treatment of this case is more difficult. It is not possible to directly combine the two ansatz functions for u^n of Case 1 because for one index i this would mean $p_i^n = a_1 h^n$ and either $p_{i+1}^n = \frac{2}{3}a_1 h^n$ or $p_{i+1}^n = \frac{4}{3}a_1 h^n$, leading to $\lim_{n\to\infty} W_1^n(u^n) = \infty$.

Therefore we have to introduce a transition layer of width $(h^n)^s$ between the two phases, where $s > 0$ is a small constant to be chosen later. We define

$$\varphi^n(x) := \begin{cases} a_1 & \text{for } 0 \leq x \leq \frac{L}{2}, \\ a_1 + \frac{a_1}{3}(h^n)^{-s}(x - \frac{L}{2}) & \text{for } \frac{L}{2} < x \leq \frac{L}{2} + (h^n)^s, \\ \frac{4}{3}a_1 & \text{for } \frac{L}{2} + (h^n)^s < x \leq L. \end{cases}$$

We set u^n such that $u^n(x=0) = u(x=0)$ and

$$p_i^n := \begin{cases} \varphi^n(ih^n) & \text{for } ih^n \leq \frac{L}{2}, \\ \frac{1}{2}\varphi^n(ih^n),\ \varphi^n(ih^n) \text{ alternating} & \text{for } ih^n > \frac{L}{2}. \end{cases}$$

With this construction, the proof of convergence to 0 of the p_i^n-terms in W_1^n is straightforward. Hence $W_1^n(u^n) \to 0$ as $n \to \infty$.

For the estimation of the functional $W_2^n(u^n)$ we have

$$\left|1 - \frac{p_{i+2}^n}{p_i^n}\right|^2 = \left|1 - \frac{1}{2}\frac{\varphi^n((i+2)h^n)}{\varphi^n(ih^n)}\right|^2 = \left|\frac{1}{2} - \frac{1}{2}\,\frac{\varphi^n(ih^n) - \varphi^n((i+2)h^n)}{\varphi^n(ih^n)}\right|^2.$$

For $I := \frac{\varphi^n(ih^n)-\varphi^n((i+2)h^n)}{\varphi^n(ih^n)}$ simple computations yield

$$I = \begin{cases} 0 & \text{if } (ih^n > \frac{L}{2}) \text{ or } ((i+1)h^n \leq \frac{L}{2}) \\ & \text{or } (\frac{L}{2} < ih^n \leq \frac{L}{2} + (h^n)^s \text{ and } (i+2)h^n > \frac{L}{2} + (h^n)^s), \\ -s_k(h^n)^{1-s} & \text{if } (ih^n > \frac{L}{2} \text{ and } (i+2)h^n \leq \frac{L}{2} + (h^n)^s) \\ & \text{or } (ih^n \leq \frac{L}{2} \text{ and } \frac{L}{2} < (i+1)h^n \leq \frac{L}{2} + (h^n)^s). \end{cases}$$

and for $0 < s < 1$ the convergence of $W_2^n(u^n)$ to $\frac{1}{4}$ can be assured.

For the estimation of $W_3^n(u^n)$, it is clear that outside the strip of width $(h^n)^s$ the summands in $W_3^n(u^n)$ equal $(h^n)^s\big[\chi(a_1-\alpha_1)^2+(1-\chi)(a_1-\alpha_2)^2\big]$. Inside the strip, we have approximately $(h^n)^{s-1}$ summands, where each summand is of the form $(h^n)C$. Thus, the part inside the strip tends to 0 for $n \to \infty$ as long as $s > 0$.

Case 3: General $\chi \in BV(\overline{\Omega}; \{0,1\})$ and piecewise affine, monotone and continuous u: The construction of u^n can be done by iteratively applying the construction given in Case 2.

Case 4: General monotone $u \in H^{1,2}(\Omega)$:

Let u be a generic monotone function in $H^{1,2}(\Omega)$ and let $\{u^n\}$ be a sequence in $\mathcal{A}^n$ such that $u^n \to u$ in $H^{1,2}(\Omega)$. For every n we can apply Case 3 to find a sequence $\{w_l^n\}_l$ such that $w_l^n \to u^n$ in $L^2(\Omega)$ as $n \to \infty$ and $\limsup_l W^l(w_l^n) \leq W(u^n)$. Then we have

$$\limsup_{n\to\infty} \limsup_{l\to\infty} W^l(w_l^n) \leq \limsup_{n\to\infty} W(u^n) = W(u), \tag{2.7}$$

where (2.7) holds because of the strong convergence of u^n to u in $H^{1,2}(\Omega)$. By diagonalisation, we find a sequence $\tilde{u}^n := w_{l(n)}^n$ such that $\tilde{u}^n \to u$ in $L^2(\Omega)$ and $\limsup_{n\to\infty} W^n(\tilde{u}^n) \leq W(u,v)$. □

3 An atomistic model for phase transitions of elastically stressed solids

In this section we present work planned for the last year of support within the priority program. We start with the following Hamiltonian that has been proposed by S. Luckhaus,

$$H(\{x_i\}_{i\in I}) := \int_\Omega \psi(x, \{x_i\}_{i\in I})dx,$$

with

$$\psi(x, \{x_i\}_{i\in I}) = \inf_{A,\tau,\alpha} \Big[\sum_{i\in I} \psi\Big(\frac{x-x_i}{\lambda}\Big) W_\alpha(Ax_i+\tau) + F(A)\Big]. \tag{3.1}$$

Here, I is a finite index set, $\{x_i\}_{i\in I}$ denotes the positions of the atoms, W_α is a periodic, non-negative potential whose zeros are on the unstrained lattice Λ_α corresponding to phase α; F plays the role of an elastic energy, and ψ is a cutoff function, $\lambda \in \mathbb{R}^+$ a scaling parameter. For a spatial point $x \in \Omega$, the infimum in (3.1) is taken with respect to deformations $A = A_x \in \mathrm{GL}(n)$, translations $\tau = \tau_x \in \mathbb{R}^n$ and phase α.

In a suitable way, W_α can be interpreted as a mean field Hamiltonian that is acting on the 'one-particle density'.

This Hamiltonian gives a reasonable description for states which have a lower and upper density close to that of a sheared lattice. One way to incorporate this restriction on the level of the Hamiltonian itself could be to define

$$\tilde{\psi}(x,\{x_i\}_{i\in I}) = \inf_{A,\tau,\alpha}\Big(\sum_{i\in I}\psi\Big(\frac{x-x_i}{\lambda}\Big)W_\alpha(Ax_i+\tau)+F(A)$$
$$+\int_\Omega \psi\Big(\frac{x-y}{\lambda}\Big)\Big[\delta - W_\alpha(Ay+\tau)-\sum_{i\in I}\varphi(y-x_i)\Big]_+ dy\Big)$$

and to set

$$\tilde{h}(\{x_i\}_{i\in I}) := \int_\Omega \tilde{\psi}(x,\{x_i\}_{i\in I}) + \sum_{i\neq j}\varphi(x_i-x_j).$$

In the last line, φ may have compact support or can be a hard core potential, the positive part $[z]_+$ of z is $[z]_+ := z$ for $z \geq 0$ and $[z]_+ := 0$ for $z < 0$.

If λ is large it makes sense to speak of the open connected sets where

$$\tilde{\psi}(x,\{x_i\}_{i\in I}) < o(\lambda^n)$$

as the domains of one elastic phase.

For x in these phase domains we conjecture that the minimal A_x, τ_x, α_x satisfy that (A_x, τ_x) is unique modulo the affine isotropy group of the lattice, and α_x is constant in each domain.

A precise (and hopefully not too restrictive) estimate when this is the case is currently work in progress.

If one assumes the uniqueness of A_x, τ_x and the constancy of α_x in a simply connected subdomain $\tilde{\Omega}$, then one may construct an elastic deformation Φ such that the projection of $\Phi^{-1}(x)$ is τ_x and such that $\nabla(\Phi^{-1})x$ has a projection close to A_x.

Without assuming simple connectedness of $\tilde{\Omega}$ there may be an obstruction to the existence of Φ. Topologically speaking this obstruction is a homomorphism

$$B : \pi_1(\tilde{\Omega}) \to \Lambda_\alpha$$

from the group of affine mappings into the lattice corresponding to phase α. If the linear component is the identity, B coincides with the Burgers vector.

Since the functional – in terms of A – is automatically invariant under the lattice group, it does not make sense to investigate energy minimisers. It is well-known that energy minima do not sustain shear, [FT89].

So, the question is to characterise metastable states. This is completely open at this time.

References

[AO05] Ariza, M.P., Ortiz M.: Discrete crystal elasticity and discrete dislocations in crystals. Arch. Rat. Mech. Anal., **178**, 149–226 (2005)

[Bal77] Ball, J.: Convexity conditions and existence theorems in nonlinear elasticity, *Arch. Rat. Mech. Anal.*, **63**, 337–403 (1977)

[Ble06] Blesgen, T.: On the competition of elastic energy and surface energy in discrete numerical schemes, to appear in *Advances in Computational Mathematics*, (2006)

[BC05] Bonilla, L.L., Carpio, A.: Discrete models for dislocations and their motion in cubic crystals, *Phys. Rev. B*, **12**, 1087–1097 (2005)

[BDG99] Braides, A., Dal Maso, G. Garroni, A.: Variational formulation of softening phenomena in fracture mechanics: the one-dimensional case, *Arch. Rat. Mech. Anal.*, **146**, 23–58 (1999)

[BG99] Braides, A. Gelli, M.S.: Limits of discrete systems without convexity hypothesis, Preprint SISSA (1999)

[CC*97] Carpio, A., Chapman, S.J., Howison, S.D., Ockendon, J.R.: Dynamics of line singularities, *Philosoph. Trans. Royal Soc. London Ser. A*, **1731**, 2013–2024 (1997)

[CLL98] Cato, I., Le Bris, C., Lions, P.-L.: The Mathematical Theory of Thermodynamic Limits: Thomas-Fermi Type Models, Oxford University Press, New York (1998)

[CK88] Chipot, M., Kinderlehrer, D.: Equilibrium configuration of crystals, *Arch. Rat. Mech. Anal.*, **103**, 237–277 (1988)

[Dac89] Dacorogna, B.: *Direct Methods in the Calculus of Variations*, Springer, New York (1989)

[FT89] Fonseca, I, Tartar L.: The displacement problem for elastic crystals. *Proc. Roy. Soc. Edinburgh A*, **113**, 159–180 (1989)

[FT02] Friesecke, G., Theil, F.: Validity and failure of the Cauchy-Born hypothesis in a two-dimensional spring lattice *Journal of Nonlinear Science*, **12**, 445–478 (2002)

[DeG55] E. De Giorgi, Sulla convergenza di alcune successioni di integrali del tipo dell'area, *Rendiconti di Matematica*, **4**, 95-113 (1955)

[DF75] De Giorgi, E., Franzoni, T.: Su un tipo di convergenza variazionali, *Atti Accad. Naz. Lincei*, **8**, 277–294 (1975)

[ILM99] Iosefescu, O, Licht, C., Michaille, G.: The variational limit of a one-dimensional discrete and statistically homogeneous system of material points, Preprint Montpellier (1999)

[JF00] James, R.D., Friesecke, G.: A scheme for the passage from atomistic to continuum theory for thin films, nanotubes and nanorods, *J. Mech. Phys. Sol.*, **48**, 1519–1540 (2000)

[Mer79] Mermin, N.D.: The topological theory of defects in ordered media, *Reviews of Modern Physics*, **51**, 591–648 (1979)
[OP99] Ortiz, M., Phillips, R.: Nanomechanics of defects in solids, *Advances in Applied Mechanics*, **36**, 1–79 (1999)
[TK76] Toulouse, G., Kleman, M.: Principles of classification of defects in ordered media, *Journal de Physique*, **37** L149–L151 (1976)
[Tru96] Truskinovsky, L.: Fracture as a phase transition, *Contemporary Research in the Mechanics and Mathematics of Materials* (Batra R.C., Beatty, M.F., eds.), CIMNE, Barcelona, 322–332 (1996)

Continuum Descriptions for the Dynamics in Discrete Lattices: Derivation and Justification

Johannes Giannoulis[1], Michael Herrmann[2], Alexander Mielke[1,2]

[1] Weierstraß-Institut für Angewandte Analysis und Stochastik, Berlin.
giannoul@wias-berlin.de, mielke@wias-berlin.de

[2] Institut für Mathematik, Humboldt-Universität zu Berlin.
michaelherrmann@math.hu-berlin.de

Summary. The passage from microscopic systems to macroscopic ones is studied by starting from spatially discrete lattice systems and deriving several continuum limits. The lattice system is an infinite-dimensional Hamiltonian system displaying a variety of different dynamical behavior. Depending on the initial conditions one sees quite different behavior like macroscopic elastic deformations associated with acoustic waves or like propagation of optical pulses. We show how on a formal level different macroscopic systems can be derived such as the Korteweg-de Vries equation, the nonlinear Schrödinger equation, Whitham's modulation equation, the three-wave interaction model, or the energy transport equation using the Wigner measure. We also address the question how the microscopic Hamiltonian and the Lagrangian structures transfer to similar structures on the macroscopic level. Finally we discuss rigorous analytical convergence results of the microscopic system to the macroscopic one by either weak-convergence methods or by quantitative error bounds.

1 Introduction

A major topic in the area of multiscale problems is the derivation of macroscopic, continuum models from microscopic, discrete ones. The prototype of a discrete many-particle system is a periodic lattice for modeling a crystal. Starting from the seminal work of Fermi, Pasta, and Ulam ([FPU55]), a lot of interest and work has been attracted to the study of the statical and dynamical behavior of ordered discrete systems. In the dynamical situation one is interested in macroscopic limits that are obtained by choosing well-prepared initial conditions: We choose the initial data in a specified class of functions and want to obtain an evolution equation within this function class, which we call the macroscopic limit problem. This approach is motivated by the theory of modulation equations, which evolved in the late 1960's for problems in fluid mechanics (see e.g. [Mie02] for a survey on this subject). If the linearized model has a space-time periodic solution, one asks how initial modulations of this pattern evolve in time. The modulations occur on much larger spatial and temporal scales; thus the modulation equation is a macroscopic equation.

In mathematically rigorous terms this can be described by studying the following coarse graining diagram:

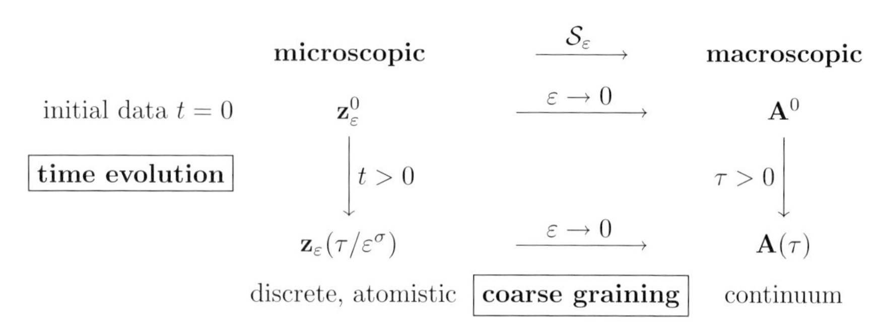

Here $\mathbf{z}_\varepsilon : [0, \tau_*/\varepsilon^\sigma] \to Z_\varepsilon$ denotes the solution of the microscopic model depending on the microscopic time t and $\mathbf{A} : [0, \tau_*] \to Z_0$ is the solution of the macroscopic model. In the best case the diagram commutes, i.e., if the coarse graining $\mathcal{S}_\varepsilon \mathbf{z}_\varepsilon(\tau/\varepsilon^\sigma) \to \mathbf{A}(\tau)$ holds at time $\tau = 0$, then it also holds for all $\tau \in [0, \tau_*]$. Examples of such results will be Theorems 5.1, 5.2, 7.1 and 7.2.

Before establishing these results, we survey methods to derive macroscopic models on the formal level by using a suitable multiscale ansatz and expanding the coefficients of equal powers of the small parameter and of the harmonics of the microscopic fluctuation to 0. The emphasis is to survey the theory and to explain the main techniques and results on simple models like the FPU chain or the Klein-Gordon chain, see Sect. 2.3.

Naturally, our survey can only cover a small part of the rich subject of dynamics in discrete systems. We will totally omit any of the works on static solutions for lattices, see e.g. [FJ00, BG02b, BG02a, Sch05a, Ble05, MBL06]. Moreover, there is a huge body of work concerning the understanding of special solution classes like traveling or standing pulses with or without periodic modulations, see [FW94, MA94, Kon96, FP99, Ioo00, IK00, FM02, FP02, FM03, Jam03, FP04a, FP04b, IJ05, DHM06]. The response of oscillator chains to a simple initial disturbance or to Riemann initial data is studied in [DKV95, DK*96, DK00, BCS01, DHR06], where in particular completely integrable systems like the Toda lattice are of interest. Finally in the framework of non-equilibrium statistical mechanics (cf. for a survey e.g. [Spo91, Bol96]) one is interested in highly disordered systems, where only statistical averages satisfy nice macroscopic equations.

2 The discrete models

In the first subsection we write down the class of systems that can be treated with the methods surveyed below. This includes general polyatomic lattices in any space dimension. The interactions can be general and can occur between several atoms, not just pair potentials, and can have arbitrary range. In the second subsection we treat the linearizations, which simplify a lot and can be treated in particular by Fourier transform methods. There, the central structure are the different dispersion relations, which will be used heavily in the subsequent analysis. Finally we present two simple model problems that represent most of the interesting features. These models will be addressed in most of the following results to illustrate the general results.

2.1 General lattices systems

We model a perfectly period crystal based on a d-dimensional Bravais lattice Γ embedded into $\mathbb{R}^d$. This lattice is homeomorphic to the additive group $\mathbb{Z}^d$ but might have a different metric structure. Each lattice point $\gamma \in \Gamma$ denotes a unit cell in the actual crystal and, hence, the vectors $x_\gamma \in \mathbb{R}^m$ and $\dot{x}_\gamma$ are collections of all the relevant positions and velocities, respectively, of the atoms inside this unit cell. By $(\mathbf{x}, \dot{\mathbf{x}}) \in \ell_2(\Gamma)^m \times \ell_2(\Gamma)^m$ we denote the state of the system, where $\mathbf{x} = (x_\gamma)_{\gamma\in\Gamma}$ and $\dot{\mathbf{x}} = (\dot{x}_\gamma)_{\gamma\in\Gamma}$. By $M \in \mathbb{R}^{m\times m}$ we denote the mass matrix for each cell, which is assumed to be symmetric and positive definite. The total kinetic energy in the crystal is

$$\mathcal{K}(\dot{\mathbf{x}}) = \tfrac{1}{2}\langle\!\langle M\dot{\mathbf{x}}, \dot{\mathbf{x}}\rangle\!\rangle \stackrel{\text{def}}{=} \textstyle\sum_{\gamma\in\Gamma} \tfrac{1}{2}\langle M\dot{x}_\gamma, \dot{x}_\gamma\rangle.$$

The potential energy $\mathcal{V}(\mathbf{x})$ is obtained by adding up all contributions acting on one cell via a single potential $V_{\text{cell}} : \ell_2(\Gamma)^m \to \mathbb{R}$ given the forces of the state $\mathbf{x}$ on the cell at $\gamma = 0$:

$$\mathcal{V}(\mathbf{x}) = \textstyle\sum_{\alpha\in\Gamma} V_{\text{cell}}(T_\alpha \mathbf{x}).$$

Here T_α is the translation operator with $T_\alpha \mathbf{x} = (x_{\alpha+\gamma})_{\gamma\in\Gamma}$. In the case of finite-range interaction the potential V_{cell} only depends on finitely many components, e.g., $V_{\text{cell}}(\mathbf{x}) = V_0(x_0) + \sum_{0<|\gamma|\le R} V_\gamma(x_\gamma - x_0)$ for pair interactions.

The Newtonian equations for this lattice model are given as

$$M\ddot{x}_\gamma = -\mathrm{D}_{x_\gamma}\mathcal{V}(\mathbf{x}) = -\textstyle\sum_{\alpha\in\Gamma} \nabla_{x_{\gamma-\alpha}} V_{\text{cell}}(T_\alpha \mathbf{x}) \quad \text{for } \gamma \in \Gamma. \tag{2.1}$$

Of course this system is invariant under the translations T_α, $\alpha \in \Gamma$, and has the total energy $\mathcal{E}(\mathbf{x}, \dot{\mathbf{x}}) = \mathcal{K}(\dot{\mathbf{x}}) + \mathcal{V}(\mathbf{x})$ as first integral. Moreover, it is a canonical Hamiltonian system with momenta $\mathbf{p} = M\dot{\mathbf{x}}$, Hamiltonian function $\mathcal{H}$, and symplectic form $\boldsymbol{\omega}_{\text{can}}$:

$$\begin{aligned} &\mathcal{H}(\mathbf{x}, \mathbf{p}) = \tfrac{1}{2}\langle\!\langle M^{-1}\mathbf{p}, \mathbf{p}\rangle\!\rangle + \mathcal{V}(\mathbf{x}) \text{ and} \\ &\boldsymbol{\omega}_{\text{can}}\big((\mathbf{v}_1, \mathbf{q}_1), (\mathbf{v}_2, \mathbf{q}_2)\big) = \langle\!\langle \mathbf{v}_1, \mathbf{q}_2\rangle\!\rangle - \langle\!\langle \mathbf{v}_2, \mathbf{q}_1\rangle\!\rangle. \end{aligned} \tag{2.2}$$

Clearly, the Newtonian equations (2.1) are equivalent to the Hamiltonian equations $\dot{\mathbf{x}} = \partial_{\mathbf{p}}\mathcal{H}(\mathbf{x}, \mathbf{p})$, $\dot{\mathbf{p}} = -\partial_{\mathbf{x}}\mathcal{H}(\mathbf{x}, \mathbf{p})$. Moreover, they can be obtained as the Euler-Lagrange equation for the Lagrangian

$$\mathcal{L}(\mathbf{x}, \dot{\mathbf{x}}) = \mathcal{K}(\dot{\mathbf{x}}) - \mathcal{V}(\mathbf{x}). \tag{2.3}$$

2.2 Linear systems and dispersion relations

Linearization leads us to linearized systems, where the potential V is a quadratic form. The linear equation takes the form

$$M\ddot{x}_\gamma = -\sum_{\beta\in\Gamma} A_\beta x_{\gamma+\beta} = \sum_{\alpha\in\Gamma} A_{\gamma-\alpha}x_\alpha \quad \text{for } \gamma\in\Gamma, \tag{2.4}$$

where the interaction matrices satisfy the symmetry condition $A_\beta = A_{-\beta}^\top$ and a decay condition like $\|A_\beta\| \le C\mathrm{e}^{-b|\beta|}$. The quadratic potential energy then reads $\mathcal{V}(\mathbf{x}) = \frac{1}{2}\sum_{\alpha,\gamma\in\Gamma}\langle A_{\gamma-\alpha}x_\alpha, x_\gamma\rangle$.

An essential feature of such harmonic lattices is the presence of many traveling wave solutions in the form of plain waves:

$$x_\gamma(t) = \mathrm{e}^{\mathrm{i}(\theta\cdot\gamma+\omega t)}\Phi \quad \text{where } \theta\in\mathbb{R}^d_* \text{ and } (\mathbb{A}(\theta)-\omega^2 M)\Phi = 0. \tag{2.5}$$

The wave vectors θ are taken from the torus $\mathcal{T}_\Gamma$, which is obtained by factoring $\mathbb{R}^d_* = \mathrm{Lin}(\mathbb{R}^d)$ with respect to the dual lattice. The symbol matrix $\mathbb{A}(\theta)$ reads

$$\mathbb{A}(\theta) = \sum_{\beta\in\Gamma}\mathrm{e}^{\mathrm{i}\theta\cdot\beta}A_\beta \in \mathbb{C}^{m\times m} \quad \text{for } \theta\in\mathcal{T}_\Gamma.$$

Hence, $\mathbb{A}(\theta)$ is Hermitian, and we always impose the basic assumption of stability in the form $\mathbb{A}(\theta)\ge 0$ for all $\theta\in\mathcal{T}_\Gamma$.

Plane-wave solutions as in (2.5) exist if ω and θ satisfy the *dispersion relation*

$$0 = \mathrm{Disp}(\omega,\theta) \stackrel{\text{def}}{=} \det\Big(\omega^2 M - \mathbb{A}(\theta)\Big).$$

Under our stability condition, there are always m non-negative eigenvalue curves

$$\omega = \Omega_k(\theta), \quad k=1,\dots,m,$$

which we order such that $0\le\Omega_1\le\Omega_2\le\dots\le\Omega_m$. The index k is called the band index. Two velocities will be important below, the *phase velocity* c_{ph} and the *group velocity* c_{gr}:

$$c_{\mathrm{ph}} = c_{\mathrm{ph},k}(\theta) \stackrel{\text{def}}{=} \frac{\Omega_k(\theta)}{|\theta|^2}\,\theta \quad \text{and} \quad c_{\mathrm{gr}} = c_{\mathrm{gr},k}(\theta) \stackrel{\text{def}}{=} \nabla\Omega_k(\theta).$$

The dynamics of the linear system is completely determined by M and the symbol matrix $\mathbb{A}:\mathcal{T}_\Gamma\to\mathbb{C}^{m\times m}_{\ge 0}$. This is easily seen by transforming (2.5) into wave vector space. For this define $\mathbf{X}(\theta) = \mathbb{F}\mathbf{x}:\mathcal{T}_\Gamma\to\mathbb{C}^m$ via $\mathbb{F}\mathbf{x}\stackrel{\text{def}}{=}\sum_\gamma \mathrm{e}^{-\mathrm{i}\theta\cdot\gamma}x_\gamma$, then $\mathbf{X}(t)=\mathbb{F}\mathbf{x}(t):\mathcal{T}_\Gamma\to\mathbb{C}^m$ satisfies the equation

$$M\partial_t^2\mathbf{X}(t,\theta) = -\mathbb{A}(\theta)\mathbf{X}(t,\theta) \tag{2.6}$$

if and only if x satisfies (2.5). However, the latter equation is an ODE for each fixed $\theta\in\mathcal{T}_\Gamma$.

For studying the qualitative behavior of the solutions in the subsequent sections, this is not sufficient, and we need to understand the back-transform for large times t. Then, the smoothness properties of the dispersion relations will be important, see Sect. 5.3 and 6.2.

2.3 The chain models of FPU and KG

To illustrate our abstract theory we will frequently refer to the simple scalar and one-dimensional case, viz., $\Gamma = \mathbb{Z} \subset \mathbb{R}$ and $x_j \in \mathbb{R}$. The models have the general form

$$\ddot{x}_j = -V_0'(x_j) + \sum_{k=1}^{K} \big(V_k'(x_{j+k}-x_j) - V_k'(x_j-x_{j-k})\big), \quad j \in \mathbb{Z}. \tag{2.7}$$

Here V_0 is called the on-site potential that couples the atoms to a background field. The interaction is assumed to be pairwise and involves K neighboring atoms.

The *Fermi-Pasta-Ulam chain* (FPU) is obtained by omitting the on-site potential and choosing $K = 1$:

$$\ddot{x}_j = V_1'(x_{j+1}-x_j) - V_1'(x_j-x_{j-1}), \quad j \in \mathbb{Z}. \tag{2.8}$$

The importance of this model is its Galilean invariance, i.e., for all $\xi, c \in \mathbb{R}$ the transformation $(\mathbf{x}, \dot{\mathbf{x}}) \mapsto (x_j+\xi + ct, \dot{x}_j+c)_{j\in\mathbb{Z}}$ leaves (2.8) invariant.

Another simple class is obtained by assuming again $K = 1$ with linear nearest-neighbor interaction and a nonlinear background potential. In analogy to the Klein-Gordon equation this model is called *Klein-Gordon chain* (KG):

$$\ddot{x}_j = x_{j+1} - 2x_j + x_{j-1} - V_0'(x_j), \quad j \in \mathbb{Z}. \tag{2.9}$$

In these two models the dispersion relation has the structure

$$0 = \mathrm{Disp}(\omega, \theta) = \omega^2 - a - 2b(1-\cos\theta) \text{ with } a = V_0''(0) \text{ and } b = V_1''(0),$$

where $a, b \geq 0$ is equivalent to our stability condition. The solution reads

$$\omega = \Omega(\theta) = \big(a + 2b(1-\cos\theta)\big)^{1/2},$$

which is smooth for $a > 0$. For $a = 0$ we find $\Omega(\theta) = \sqrt{b}\,2\,|\sin(\theta/2)|$, which is not differentiable at $\theta = 0$, but the two limits $\pm\sqrt{b}$ of Ω' at $\theta = 0$ are the macroscopic wave speeds.

3 Formal derivation of continuum models

3.1 General multiscale approach

We discuss here the derivation of macroscopic models that appear for solutions having a relatively small amplitude, but we refer to [DHR06] and Sect. 3.7 for results on large amplitude solutions.

The basic ansatz relies on modulations of basic plane waves $\mathrm{e}^{\mathrm{i}(\omega t+\theta\cdot\gamma)}\Phi$ on large spatial scales and suitably chosen slow time scales. We choose $\varepsilon > 0$ to be the small parameter that relates the microscopic and the macroscopic temporal and spatial scales, i.e., we set

$$\tau = \varepsilon^s t \quad \text{and} \quad y = \varepsilon\gamma \in \mathbb{R}^d \quad \text{for } \gamma \in \Gamma \subset \mathbb{R}^d.$$

Of course, there are cases where different scalings in different spatial directions are useful, but for simplicity we restrict ourselves to this case.

We now choose a finite set of wave vectors $\theta_1, \ldots, \theta_N \in \mathcal{T}_\Gamma$ and associated band indices $k_1, \ldots, k_N \in \{1, \ldots, m\}$ and consider the associated plane waves

$$x_\gamma(t) = \mathbf{E}_n(t,\gamma)\boldsymbol{\Phi}_n, \text{ where } \mathbf{E}_n(t,\gamma) \stackrel{\text{def}}{=} \mathrm{e}^{\mathrm{i}(\omega_n t + \theta_n \cdot \gamma)}$$
$$\text{with } \omega_n = \Omega_{k_n}(\theta_n) \text{ and } \boldsymbol{\Phi}_n = \Phi_{k_n}(\theta_n).$$

This may include the case $\theta = 0$ and $\omega = 0$, which relates to the macroscopic limit of solutions without microstructure.

The two-scale method now starts from the ansatz

$$\begin{aligned} &(x_\gamma(t), \dot{x}_\gamma(t)) = R_\varepsilon(\mathbf{A})_\gamma(t), \text{ where } \mathbf{A} = (A_1, \ldots, A_N) \text{ and} \\ &R_\varepsilon(\mathbf{A})_\gamma(t) = \textstyle\sum_{n=1}^N \varepsilon^{\sigma_n} A_n(\varepsilon^s t, \varepsilon\gamma)\mathbf{E}_n(t,\gamma)\boldsymbol{\Phi}_n \\ &\qquad + \textstyle\sum_{n,k=1}^N \varepsilon^{\sigma_n+\sigma_k} \boldsymbol{\Psi}_{n,k}(\varepsilon^s t, \varepsilon\gamma)\mathbf{E}_n\mathbf{E}_k \\ &\qquad + \textstyle\sum_{n,k,l=1}^N \varepsilon^{\sigma_n+\sigma_k+\sigma_l} \boldsymbol{\Psi}_{n,k,l}(\varepsilon^s t, \varepsilon\gamma)\mathbf{E}_n\mathbf{E}_k\mathbf{E}_l + \text{h.o.t.} \end{aligned} \tag{3.1}$$

Here the powers $s, \sigma_1, \ldots, \sigma_N \in \mathbb{R}$ have to be chosen appropriately. We refer to the variety of different models that can be obtained in this way. To obtain real-valued solutions one chooses $A_n = \overline{A}_{N-n}$ and similarly for the higher order terms. In cases with $\theta_n \neq 0$ the functions A_n are the modulating amplitudes of the basic periodic plane wave.

The aim is to derive suitable equations for $A_1, \ldots, A_N$, which make this ansatz (3.1) consistent with the discrete model (2.1). The obtained equations are partial differential equations combined with some algebraic relations. These equations are called the macroscopic equations, because they are posed in terms of the macroscopic variables $\tau = \varepsilon^s t$ and $y = \varepsilon\gamma$. Inserting the ansatz (3.1) into the nonlinear system (2.1) we have to expand both sides in terms of the products $\varepsilon^{\widetilde{q}} \Pi_{n=1}^N \mathbf{E}_n^{q_n}$ with $\widetilde{q} = \sum_{n=1}^N \sigma_n q_n$. Here we have to expand difference quotients $x_{\gamma+\alpha} - x_\gamma$ in terms of spatial derivative of A_n. Moreover, the resonances between the plane waves are important to allow for nontrivial nonlinear interaction. They are characterized by vectors $q \in \mathbb{N}^N$ such that $\Pi_{n=1}^N \mathbf{E}_n^{q_n} \equiv 1$, see Sect. 7.2 for a general theory.

We arrive at a hierarchy of equations that can be parametrized by the multi-index $\mathbf{q} = (q_1, ..., q_N) \in \mathbb{N}^N$. These equations decompose into two groups. If the term $\mathbf{E}_\mathbf{q} = \Pi_{n=1}^N \mathbf{E}_n^{q_n}$ is nonresonant, i.e., different from all the terms $\mathrm{e}^{\mathrm{i}(\omega t + \theta\cdot\gamma)}$ that satisfy the dispersion relation, then the equation for $\Psi_\mathbf{q}(\tau, y)$ is uniquely solvable. The resonant groups associate with the terms $\mathbf{E}_\mathbf{q} = \Pi_{n=1}^N \mathbf{E}_n^{q_n}$ that equal one of the terms $\mathrm{e}^{\mathrm{i}(\Omega_j(\theta)t + \theta\cdot\gamma)}$, which without loss of generality is already in our list, let us say $\mathbf{E}_m$. Naturally, the coefficient $\boldsymbol{\Psi}_\mathbf{q}$ cannot be determined uniquely, because the plane wave $\mathbf{E}_m\boldsymbol{\Phi}_m$ solves the linear problem. Thus, by Fredholm's alternative we obtain a solvability condition for the terms on the left-hand side that contains only lower order terms that are already determined. This gives either a PDE or an algebraic equation on the previously chosen functions. Moreover, the general solution contains a new scalar function $B_{\mathbf{q}}$, namely $\boldsymbol{\Psi}_{\mathbf{q}} = B_{\mathbf{q}} \boldsymbol{\Phi}_m + \boldsymbol{\Psi}^0_{\mathbf{q}}$.

We refer to [Mie02, GM04, GM06] for a more detailed description of this procedure. In fact, without doing any explicit calculation on the specific discrete lattice system (2.1) it is possible to describe the form of the macroscopic equations as follows:

$$\begin{aligned} &\text{If } \omega_k \neq 0: \ \partial_\tau A_k = \textstyle\sum_{\mathbf{q}\in M_k(s)} c_{\mathbf{q}} \Pi_{n=1}^N A_n^{q_n}, \\ &\text{If } \omega_k = 0: \ \partial_\tau^2 A_k = \textstyle\sum_{\mathbf{q}\in M_k(2s)} \widetilde{c}_{\mathbf{q}} \Pi_{n=1}^N A_n^{q_n}, \end{aligned} \tag{3.2}$$

where $M_k(s) \stackrel{\text{def}}{=} \{ \mathbf{q} \,|\, \sigma_k + s = \sum_1^N \sigma_n q_n, \ 0 = \sum_1^N \omega_n q_n, \ \sum_1^N q_n \theta_n = 0 \text{ on } \mathcal{T}_\Gamma \}$. For more details see [Gia06, GMS06] and Sect. 7.2.

The following Sect. 3.2 to 3.7 treat a list of examples, which highlight the generality of the approach.

3.2 The quasilinear wave equation

A simple but important macroscopic model for FPU chains results by the following multiscale ansatz with hyperbolic scaling:

$$x_j(t) = \varepsilon^{-1} X(\varepsilon t, \varepsilon j), \quad \tau = \varepsilon t, \quad y = \varepsilon j. \tag{3.3}$$

Note that here x_j denotes the spatial position of atom j rather than its displacement. We insert the ansatz (3.3) into (2.8) and eliminate the relative displacements by the Taylor expansion $x_{j\pm 1} - x_j \approx \pm \varepsilon \partial_y X(\varepsilon t, \varepsilon j)$. Using $\partial_\tau = \varepsilon \partial_t$ we can identify the macroscopic modulation equations as the nonlinear wave equation

$$\partial_{\tau\tau} X - \partial_y V_1'(\partial_y X) = 0. \tag{3.4}$$

Via $r = \partial_y X$ and $v = \partial_\tau X$ it transforms into the quasilinear first-order system

$$\partial_\tau r - \partial_y v = 0, \quad \partial_\tau v - \partial_y V_1'(r) = 0. \tag{3.5}$$

These equations describe the macroscopic evolution of non-oscillatory solutions of FPU. However, due to the nonlinearity V_1' smooth solutions of (3.5) can form shocks in finite times, and in this case the quasilinear wave equation is not longer an appropriate macroscopic model for FPU. This problem is addressed in [DHR06].

3.3 The Korteweg-de Vries equation

Another example for macroscopic modulation equations, see [SW00, FP99], relies on the KdV-ansatz

$$x_j(t) = \varepsilon\, U\big(\varepsilon^3 t,\, \varepsilon(j{+}ct)\big) \tag{3.6}$$

with scaling $\tau = \varepsilon^3 t$, $y = \varepsilon(j{+}ct)$. We insert the ansatz into (2.8) and use Taylor expansion up to order $O\big(\varepsilon^6\big)$. Comparing the leading order terms we find that c is given by $c^2 = V_1''(0)$. Since the next order terms all cancel, the modulation equation is determined by the terms corresponding to ε^5, and finally we obtain

$$2\,c\,\partial_{\tau y} U - \tfrac{1}{12}\,c^2\,\partial_y U\,\partial_{yy} U - V_1'''(0)\,\partial_{yyyy} U = 0, \tag{3.7}$$

which is a KdV equation for $\partial_y U$.

3.4 The nonlinear Schrödinger equation

We consider the scalar, d-dimensional lattice (2.1) (i.e., $d \in \mathbb{N}$ and $m = 1$)

$$\ddot{x}_\gamma = \textstyle\sum_{0<|\beta|\le R} [V_\beta'(x_{\gamma+\beta}{-}x_\gamma){-}V_\beta'(x_\gamma{-}x_{\gamma-\beta})] - V_0'(x_\gamma), \quad \gamma \in \Gamma, \tag{3.8}$$

and are interested in the macroscopic deformations of a modulated plane wave solution of the linearized system

$$x_\gamma(t) = \varepsilon A(\tau, y)\mathbf{E}(t,\gamma) + \text{c.c.} + O(\varepsilon^2) \quad \text{with} \quad \mathbf{E}(t,\gamma) = \mathrm{e}^{\mathrm{i}(\omega t + \theta\gamma)} \tag{3.9}$$

(c.c.: conjugate complex) for a fixed wave vector $\theta \in \mathcal{T}_\Gamma$ with frequency ω satisfying the dispersion relation $\omega^2 = \Omega^2(\theta) > 0$.

Since the system is dispersive and nonlinear and the amplitude A is weakly scaled by $0 < \varepsilon \ll 1$, we need a slow macroscopic time scale $\tau = \varepsilon^2 t$ comparing to the macroscopic space scale $y = \varepsilon(\gamma{-}c_{\mathrm{gr}} t)$, in order to see the evolution of A as time passes. This is the so called *dispersive scaling.* The choice of y also reflects that we are moving with the pulse at its microscopical group velocity $c_{\mathrm{gr}} = \nabla_\theta \Omega(\theta)$. By this scaling it turns out that the evolution of A is given by the nonlinear Schrödinger equation

$$\mathrm{i}\partial_\tau A = \mathrm{Div}_y\big(\frac{1}{2}\mathrm{D}_\theta^2 \Omega(\theta) \nabla_y A\big) + \rho |A|^2 A. \tag{nlS}$$

For the justification of this equation we refer to Sect. 7.1.

3.5 Three-wave interaction

For the lattice (3.8) we are now interested in a macroscopic description for the evolution of the amplitudes A_n, $n = 1,2,3$, of three nonlinearly interacting modulated plane waves with different wave numbers θ_n and frequencies ω_n, where $\omega_n^2 = \Omega^2(\theta_n)$. Thus, ansatz (3.1) takes the special form

$$x_\gamma(t) = \varepsilon \textstyle\sum_{n=1}^3 A_n(\tau, y)\mathbf{E}_n(t,\gamma) + \text{c.c.} + O(\varepsilon^2) \text{ with } \mathbf{E}_n(t,\gamma) = \mathrm{e}^{\mathrm{i}(\omega_n t + \theta_n \cdot \gamma)}$$

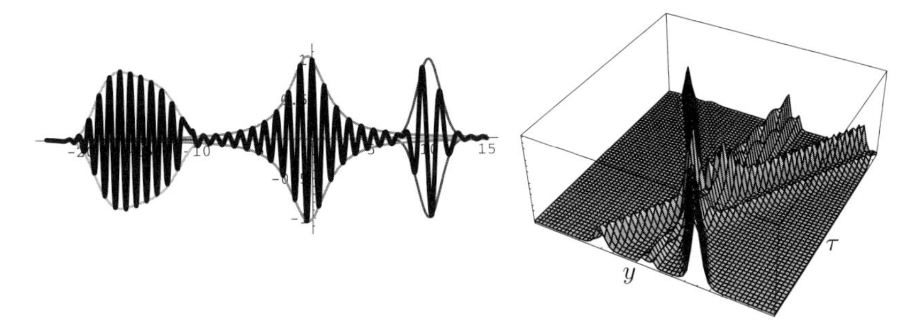

Fig. 3.1. Three-wave interaction. Left: typical initial condition. Right: energy-distribution in space-time.

but now using the hyperbolic scaling $\tau = \varepsilon t$, $y = \varepsilon\gamma$ again. It turns out that, if the wave vectors θ_n and frequencies ω_n are in resonance, viz.,

$$\theta_1 + \theta_2 + \theta_3 = 0 \mod \mathcal{T}_\Gamma \quad \text{and} \quad \omega_1 + \omega_2 + \omega_3 = 0, \tag{3.10}$$

the amplitudes A_n, $n = 1, 2, 3$, satisfy the so called three-wave interaction equations

$$\begin{cases} \omega_1 \partial_\tau A_1 = \omega_1 \nabla_\theta \Omega(\theta_1) \cdot \nabla_y A_1 + c\overline{A}_2 \overline{A}_3, \\ \omega_2 \partial_\tau A_2 = \omega_2 \nabla_\theta \Omega(\theta_2) \cdot \nabla_y A_2 + c\overline{A}_1 \overline{A}_3, \\ \omega_3 \partial_\tau A_3 = \omega_3 \nabla_\theta \Omega(\theta_3) \cdot \nabla_y A_3 + c\overline{A}_1 \overline{A}_2, \end{cases} \tag{3.11}$$

with $c = 2\sum_{0<|\beta|\leq R} V_\beta'''(0) \sum_{n=1}^{3} \sin(\theta_n \cdot \beta) + \mathrm{i}V_0'''(0)$. Each equation consists of a transport part via the group velocity and a nonlinear coupling to the two other modes. Figure 3.1 illustrates the behavior. Without the resonance condition (3.10) being fulfilled, nonlinear terms would not arise and the pulses would just pass through each other. For the justification of this equation we refer to Sect. 7.2.

3.6 Coupled systems

While the two examples above apply to a system with or without background potential V_0, we are now looking at systems with Galilean invariance, where the canonical example is FPU from (2.8). The aim is to understand the coupling between macroscopic deformations and microscopically oscillating pulses. Since in general the macroscopic wave speeds and the microscopic group velocity are different, we use the hyperbolic time scale $\tau = \varepsilon t$. Ansatz (3.1) reduces to

$$x_\gamma(t) = \varepsilon^\alpha X(\tau, y) + \varepsilon^\beta A(\tau, y)\mathbf{E} + \varepsilon^\beta \overline{A}(\tau, y)\overline{\mathbf{E}} + \text{h.o.t.}$$

with $\mathbf{E} = \mathrm{e}^{\mathrm{i}(\omega t + \theta\cdot\gamma)}$ and $\omega = \Omega(\theta)$. Here α and β might be different and depend on the nonlinearities as well as the scaling of the initial data. We treat the case of the FPU chain with $V_1(r) = \frac{a}{2}r^2 + \frac{b}{3}r^3 + \frac{c}{4}r^4$ with $a > 0$ and $b, c \in \mathbb{R}$.

As a first example we consider the case $\alpha = 0$, $\beta = 1$ and find the system

$$\partial_\tau^2 X = c_{\rm m}^2 \partial_\xi^2 X, \quad {\rm i}\partial_\tau A = {\rm i}c_{\rm gr}\partial_y A - \rho_0(\partial_y X)\, A$$

with $c_m := \Omega'(0) = \sqrt{a}$, $c_{\rm gr} := \Omega'(\theta)$ and $\rho_0 := 2b\Omega(\theta)/a$. Since the contributions X and A scale differently, the coupling of X and A takes place only in one equation. We have the two conserved quantities

$$\mathbb{H}(A) = \int_{\mathbb{R}} \omega^2 |A|^2 \,{\rm d}y \quad \text{and} \quad \mathbb{E}(X, X_\tau) = \int_{\mathbb{R}} \tfrac{1}{2} X_\tau^2 + \tfrac{c_{\rm m}^2}{2} X_y^2 \,{\rm d}y.$$

The second example has $\alpha = 0$ and $\beta = 1/2$, which leads to the system

$$X_{\tau\tau} = \left(c_{\rm m}^2 X_y + \rho_1 |A|^2\right)_y, \quad 2{\rm i}\omega A_\tau = {\rm i}\omega c_{\rm gr} A_y - \left(\rho_1 X_y + 2\rho_2 |A|^2\right) A,$$

where $\rho_1 := 2b(1-\cos\theta)$ and $\rho_2 := 3c(1-\cos\theta)^2$. This system is a Lagrangian and Hamiltonian system in the sense to be discussed in Sect. 4. The Lagrangian reads

$$\begin{aligned} &\mathbb{L}(X, A, X_\tau, A_\tau) \\ &= \int_{\mathbb{R}} \omega \, {\rm Im}\left(\overline{A}(2A_\tau - c_{\rm gr} A_y)\right) + \tfrac{1}{2} X_\tau^2 - \tfrac{c_{\rm m}^2}{2} X_y^2 - |A|^2\left(\rho_1 X_y + \rho_2 |A|^2\right) {\rm d}y. \end{aligned}$$

There are two first integrals

$$\begin{aligned} \mathbb{H}(A) &= \int_{\mathbb{R}} \omega^2 |A|^2 \,{\rm d}y, \\ \mathbb{E}(X, A, X_\tau) &= \int_{\mathbb{R}} \omega c_{\rm gr} \,{\rm Im}(\overline{A} A_y) + \tfrac{1}{2} X_\tau^2 + \tfrac{c_m^2}{2} X_y^2 + |A|^2\left(\rho_1 X_y + \rho_2 |A|^2\right) {\rm d}y. \end{aligned}$$

The symplectic structure of the associated Hamiltonian system for (X, A, X_τ) is non-canonical and can easily be deduced as in Sect. 4.2.

3.7 Whitham's modulation equation

In [Whi74] Whitham studies certain nonlinear PDEs and relying on the hyperbolic scaling he develops a theory that is capable to describe the macroscopic evolution of large microscopic oscillations. Here we apply Whitham's approach to three different chain models. We start with KG, cf. (2.7), and make the following multiscale ansatz

$$x_j(t) = \mathfrak{X}\big(\varepsilon t, \, \varepsilon j, \, \varepsilon^{-1} \Theta(\varepsilon t, \, \varepsilon j)\big), \tag{3.12}$$

where $\mathfrak{X}$ is assumed to be 2π-periodic with respect to the phase variable $\phi = \varepsilon^{-1}\Theta$. In this ansatz both the wave number θ and the frequency ω depend on the macroscopic coordinates $(\tau, \, y)$ and are defined by the *modulated phase* Θ via $\theta = \partial_y \Theta$ and $\omega = \partial_\tau \Theta$. It can be shown that to leading order the function $\mathfrak{X}$ must satisfy the following nonlinear advance-delay-differential equation

$$\omega^2 \, \partial_\phi^2 \mathfrak{X} = \nabla_{-\theta} \nabla_{+\theta} \mathfrak{X} - V_0'(\mathfrak{X}) \tag{3.13}$$

with $(\nabla_{\pm\theta}\mathfrak{X})(\phi) = \pm\mathfrak{X}(\phi\pm\theta) \mp \mathfrak{X}(\phi)$. As usual we refer to solutions of this equation as *traveling waves*. The existence problem for solutions of (3.13) with small amplitudes is investigated in [IK00]. For convex potentials V_0 we can provide existence of solutions by adapting an idea from [FV99], compare with the similar problem for FPU in [DHR06]. According to (3.13), the action L of a traveling wave is given by

$$L(\theta,\,\omega) = \tfrac{1}{2\pi}\int_0^{2\pi} \tfrac{\omega^2}{2}\left(\partial_\phi\mathfrak{X}\right)^2 - \left(\nabla_{+\theta}\mathfrak{X}\right)^2 - V_0(\mathfrak{X})\,\mathrm{d}\phi. \tag{3.14}$$

To identify the macroscopic modulation equations it is convenient to use the Lagrangian formalism, see [Whi74] and Section 4.1, because direct expansions in powers of ε turn out to be quite complicate. With some simple averaging the total action of the chain can be expressed by a functional $\mathbb{L}$, which depends on $(\Theta,\,\partial_\tau\Theta)$ only, and we can derive the modulation equations by the principle of least action. It comes out that the modulation equations are equivalent to the following nonlinear system of conservations laws

$$\partial_\tau\theta - \partial_y\omega = 0, \quad \partial_\tau S + \partial_y g = 0, \tag{3.15}$$

where $S = \partial_\omega L$ and $g = \partial_\theta L$. In particular, the system (3.15) is closed by the equation of state (3.14) and the Gibbs equation $\mathrm{d}L = S\mathrm{d}\omega + g\mathrm{d}\theta$.

The modulation theory for FPU, see [DHM06, DHR06] and the references therein, is more complicate than in the KG case due to the nonlinearity of V_1, and the Galilean invariance of (2.8). In particular, we must combine (3.3) and (3.12) as follows

$$x_j(t) = \varepsilon^{-1}X(\varepsilon t,\,\varepsilon j) + \mathfrak{X}\big(\varepsilon t,\,\varepsilon j,\,\varepsilon^{-1}\,\Theta(\varepsilon t,\,\varepsilon j)\big), \tag{3.16}$$

where as before the profile function $\mathfrak{X}$ is assumed to be 2π-periodic with respect to $\phi = \varepsilon^{-1}\Theta$. This ansatz gives rise to four important macroscopic fields, namely the wave number $\theta = \partial_y\Theta$, the frequency $\omega = \partial_\tau\Theta$, the specific length $r = \partial_y X$, and the macroscopic velocity $v = \partial_\tau X$. To leading order, the profile function X must satisfy the traveling wave equation

$$\omega^2\,\partial_\phi^2\,\mathfrak{X} = \nabla_{-\theta}V_1'(\nabla_{+\theta}\,\mathfrak{X}). \tag{3.17}$$

For convex potentials V_1 the existence of solutions can be proved by an convex optimization problem, see [DHR06], and rigorous results without convexity assumptions can be found in [FW94, PP00, Ioo00]. The derivation of the modulation equations for (3.16) again relies on Lagrangian reduction, see for instance [Her04, DHM06], and leads to the following nonlinear system

$$\partial_\tau r - \partial_y v = 0, \quad \partial_\tau v + \partial_y p = 0, \quad \partial_\tau\theta - \partial_y\omega = 0, \quad \partial_\tau S + \partial_y g = 0. \tag{3.18}$$

These equations can be interpreted as the macroscopic conservation laws of mass, momentum, wave number and entropy. As for KG, the constitutive relations for (3.18) result from a careful investigation of the thermodynamic properties of traveling waves. More precisely, it can be shown, at least formally, that (3.17) provides an equation of state $U = U(r, \theta, S)$ as well as the universal Gibbs equation $\mathrm{d}E = \omega \mathrm{d}S - p\mathrm{d}r - g\mathrm{d}\theta + v\mathrm{d}v$, where U and $E = \frac{1}{2}v^2 + U$ denote the internal and total energy, respectively.

The third example is the discrete nonlinear Schrödinger equation

$$-\mathrm{i}\,\dot{a}_j + a_{j+1} - 2\,a_j + a_{j-1} + \varrho\,|a_j|^2\,a_j = 0, \quad j \in \mathbb{Z}, \tag{3.19}$$

with complex valued a_j and real parameter ϱ. This equation has exact solutions (traveling waves) of the form $a_j = B\,\mathrm{e}^{\mathrm{i}(\theta j + \omega t)}$ with real amplitude B if ω obeys the nonlinear dispersion relation $\omega + \cos\theta - 2 + \varrho\,B^2 = 0$. The modulation theory for several variants of (3.19) was studied in [HLM94] and bases on the multiscale ansatz $a_j(t) = B(\varepsilon t, \varepsilon j)\,\mathrm{e}^{\mathrm{i}\,\Theta(\varepsilon t, \varepsilon j)/\varepsilon}$, where as before we set $\theta = \partial_y \Theta$ and $\omega = \partial_\tau \Theta$. One obtains the macroscopic balance laws

$$\partial_\tau\big(A^2\big) - \partial_y\big(2\,A^2\,\sin\theta\big) = 0, \quad \partial_\tau\theta + \partial_y\big(\varrho\,A^2 + \cos\theta\big) = 0, \tag{3.20}$$

where the second evolution equation is equivalent to $\partial_\tau\theta - \partial_y\omega = 0$. We mention that (3.20) can also be derived by means of Hamiltonian or Lagrangian reduction discussed in the next section, see [HLM94] for the details.

4 Hamiltonian and Lagrangian structures

The derivation of macroscopic equations for discrete models (or continuous models with microstructure) can be seen as a kind of reduction of the infinite dimensional system to a simpler subclass. If we choose well ordered initial conditions, we hope that the solution will stay in this order and evolve according to a slow evolution with macroscopic effects only. We may interprete this as a kind of (approximate) invariant manifold, and the macroscopic equation describes the evolution on this manifold, the functions $A_1, ..., A_N$ defining kind of coordinates. For such a reduction procedure it is a natural question how the original Hamiltonian and Lagrangian structures, as described in Sect. 2.1, "reduce" to the macroscopic equation. Here we just survey the main ideas and some examples and refer to [GHM06] for the full details.

Before addressing this question we first address the exact reduction of a Hamiltonian and Lagrangian systems to exactly invariant manifolds (cf. e.g. [Mie91]). First consider the Lagrangian setting for $\mathcal{L}$ defined on $\mathrm{T}Q$. Assume that we have an invariant manifold $\mathbb{M} \subset \mathrm{T}Q$ given in the form

$$\mathbb{M} = \{\,(q, \dot{q}) = S(p, \dot{p}) \in \mathrm{T}Q \mid (p, \dot{p}) \in \mathrm{T}P\,\}.$$

Then, we may define the reduced Lagrangian $L^{\mathrm{red}} = \mathcal{L} \circ S : \mathrm{T}P \to \mathbb{R}$. An easy calculation proves that any solution p of the reduced Lagrangian system

$$0 = -\tfrac{\mathrm{d}}{\mathrm{d}t}\big(\partial_{\dot{p}} L^{\mathrm{red}}(p, \dot{p})\big) + \partial_p L^{\mathrm{red}}(p, \dot{p})$$

leads to a solution $(q, \dot{q}) = S(p, \dot{p})$ of the original Lagrangian system. Vice versa, any solution of the latter system that also lies in $\mathbb{M}$ solves the reduced Lagrangian system.

In the Hamiltonian case the tangent bundle structure of $Z = \mathrm{T}Q$ is generalized to a general symplectic structure $\boldsymbol{\omega}$ on the state space Z. Together with the Hamiltonian $\mathcal{H}$ the Hamiltonian system reads

$$\boldsymbol{\Omega}(z)\dot{z} = \mathrm{D}\mathcal{H}(z) \quad \text{or} \quad \dot{z} = \mathbf{J}(z)\mathrm{D}\mathcal{H}(z),$$

where $\boldsymbol{\omega}_z(v_1, v_2) = \langle \boldsymbol{\Omega}(z)v_1, v_2 \rangle$ and $\mathbf{J}(z) = \boldsymbol{\Omega}(z)^{-1} : \mathrm{T}_z^* Z \to \mathrm{T}_z Z$. For a symplectic, flow-invariant submanifold $\mathbb{M} = \{ z = R(y) \in Z \mid y \in Y \}$ we define the reduced symplectic structure $\boldsymbol{\Omega}^{\mathrm{red}}$ and the reduced Hamiltonian H^{red} via

$$\boldsymbol{\Omega}^{\mathrm{red}}(y) = \mathrm{D}R(y)^* \boldsymbol{\Omega}(R(y))\mathrm{D}R(y) \quad \text{and} \quad H^{\mathrm{red}}(y) = \mathcal{H}(R(y)).$$

Using the flow-invariance of $\mathbb{M}$ it is easy to see that any solution of the reduced Hamiltonian system $\boldsymbol{\Omega}^{\mathrm{red}}(y)\dot{y} = \mathrm{D}H^{\mathrm{red}}(y)$ solves the original system and vice versa if starting on $\mathbb{M}$.

Our applications will of course use the ansatz R_ε from (3.1) for the reduction, which can be seen as an approximation of an invariant manifold.

4.1 Lagrangian reduction

The multiscale ansatz (3.1) discussed above was chosen such that it is formally consistent and in many cases it is possible to justify the ansatz by a rigorous error analysis, as surveyed in Sections 6.1 and 7. Hence, we consider the multiscale ansatz as a parametrization of an (approximate) invariant manifold. Inserting the ansatz (3.1) into the Lagrangian $\mathcal{L}$ defined in (2.3) we obtain a reduced Lagrangian in the form

$$L^{\mathrm{red}}(\varepsilon, \mathbf{A}, \partial_\tau \mathbf{A}) = \varepsilon^\rho \mathbb{L}(\mathbf{A}, \partial_\tau \mathbf{A}) + O(\varepsilon^{\rho+1}), \ \text{where } \mathbf{A} = (A_1, \ldots, A_N).$$

Here L^{red} is still an infinite sum over $\gamma \in \Gamma$. However, when expanding in powers of ε, the multiscale ansatz leads to a limit that is an integral over the macroscopic space variable $y \in \mathbb{R}^d$. The infinite sum can be considered as a Riemann sum for the spatial integral.

Since L^{red} is independent of τ, the solutions of the reduced Euler-Lagrange equation conserve the associated energy $\mathbb{E}$ obtain as

$$\mathbb{E}(\mathbf{A}, \mathbf{A}_\tau) = \langle\!\langle \partial_\tau \mathbf{A}, \partial_{\mathbf{A}_\tau} \mathbb{L}(\mathbf{A}, \mathbf{A}_\tau) \rangle\!\rangle - \mathbb{L}(\mathbf{A}, \mathbf{A}_\tau).$$

It is proved in [GHM06] that the Lagrangian equation for $\mathbf{A}$ associated with the lowest order term $\mathbb{L}$ of the reduced Lagrangian $L^{\mathrm{red}}(\varepsilon, \cdot)$ really provides exactly the macroscopic equation (3.2) derived in Sect. 3.1.

Here we illustrate this result using a simple example based on the Klein-Gordon chain (2.9) with the potential $V_0(x) = \frac{a}{2}x^2 + \frac{b}{4}x^4$. We consider a single modulated pulse in the form

$$x_j(t) = \varepsilon^{1/2} A(\varepsilon t, \varepsilon j)\mathbf{E} + \varepsilon^{1/2}\overline{A}(\varepsilon t, \varepsilon j)\overline{\mathbf{E}} \quad \text{with } \mathbf{E} = \mathrm{e}^{\mathrm{i}(\omega t + \theta j)}, \tag{4.1}$$

where $\omega = \Omega(\theta)$. Inserting this ansatz into $\mathcal{L}$ and using $\vartheta = \mathrm{e}^{\mathrm{i}\theta} - 1$ we find

$$\begin{aligned} L^{\mathrm{red}}(\varepsilon, A, A_\tau) = & \textstyle\sum_{\mathbb{Z}} \Big(\frac{\varepsilon}{2}\omega^2 |A\mathbf{E} - \overline{A}\,\overline{\mathbf{E}}|^2 + \varepsilon^2 \mathrm{i}\omega (A\mathbf{E} - \overline{A}\,\overline{\mathbf{E}})\big(A_\tau \mathbf{E} + \overline{A}_\tau \overline{\mathbf{E}}\big) \\ & - \tfrac{\varepsilon}{2} |A\vartheta\mathbf{E} + \overline{A}\,\overline{\vartheta}\,\overline{\mathbf{E}}|^2 - \varepsilon^2 \big(A\vartheta\mathbf{E} + \overline{A}\,\overline{\vartheta}\,\overline{\mathbf{E}}\big)\big(A_y \mathbf{E} + \overline{A}_y \overline{\mathbf{E}}\big) \\ & - \tfrac{\varepsilon a}{2} |A\mathbf{E} + \overline{A}\,\overline{\mathbf{E}}|^2 - \tfrac{\varepsilon^2 b}{4} |A\mathbf{E} + \overline{A}\,\overline{\mathbf{E}}|^4 + O(\varepsilon^3) \Big) \\ = & \, \varepsilon \mathbb{L}(A, A_\tau) + O(\varepsilon^2) \quad \text{with} \\ \mathbb{L}(A, A_\tau) = & \textstyle\int_{\mathbb{R}} \mathrm{i}\omega \big(A\overline{A}_\tau - \overline{A}A_\tau\big) - \big(\vartheta A \overline{A}_y + \overline{\vartheta}\,\overline{A} A_y\big) - \frac{3b}{2}|A|^4 \,\mathrm{d}y. \end{aligned}$$

The important observation for this calculation is that the lowest order terms cancel, which can be seen as a manifestation of equipartition of kinetic and potential energy in the plane waves. Moreover, the terms involving $\mathbf{E}^k$ with $k \neq 0$ also drop out by periodicity. This averaging is a formal procedure here, but we will see in the next subsection that in a two-scale setting with an extra phase variable it can be made exact.

Using $\vartheta - \overline{\vartheta} = 2\mathrm{i}\sin\theta = 2\mathrm{i}\omega(\theta)\omega'(\theta)$ the Euler-Lagrange equation reads

$$0 = -\partial_\tau\big(\partial_{\overline{A}_\tau}\mathbb{L}\big) - \partial_y\big(\partial_{\overline{A}_y}\mathbb{L}\big) + \partial_{\overline{A}}\mathbb{L} = -2\mathrm{i}\omega A_\tau + 2\mathrm{i}\omega\omega' A_y - 3b|A|^2 A. \tag{4.2}$$

Of course, this is exactly the desired macroscopic modulation equation, which can be obtained as in Sect. 3.1. Moreover, because of invariance in τ, there is a first integral, namely the associated energy

$$\mathbb{E}(A, \partial_\tau A) = \textstyle\int_{\mathbb{R}} \mathrm{i}\omega\omega' \big(\overline{A}A_y - A\overline{A}_y\big) + \frac{3b}{2}|A|^4 \,\mathrm{d}y.$$

4.2 Hamiltonian reduction

In the Hamiltonian setting we might also try to derive the reduced Hamiltonian by inserting the multiscale ansatz (3.1) into the Hamiltonian $\mathcal{H}$ defined in (2.2). We obtain

$$\widetilde{H}(\varepsilon, \mathbf{A}, \partial_\tau \mathbf{A}) = \varepsilon^{\varrho}\mathbb{H}(\mathbf{A}, \partial_\tau \mathbf{A}) + O(\varepsilon^{\varrho+1}).$$

In the example of the previous subsection we immediately find $\varrho = 0 < \rho = 1$ and $\mathbb{H}(\mathbf{A}, \partial_\tau \mathbf{A}) = \int_{\mathbb{R}} 2\omega^2 |A|^2 \,\mathrm{d}y$. Moreover, the symplectic form can be reduced and we obtain

$$\boldsymbol{\Omega}_\varepsilon^{\mathrm{red}} = \boldsymbol{\Omega}_0 + O(\varepsilon) \quad \text{with } \boldsymbol{\Omega}_0 = 2\mathrm{i}\omega.$$

It is easy to see that the function $\mathbb{H}$ is also a first integral of the macroscopic system (4.2). However, it is not the desired energy $\mathbb{E}$, and the flow associated with the Hamiltonian system $\boldsymbol{\Omega}_0 \partial_\tau A = \mathrm{D}\mathbb{H}(A)$ is the phase translation $A(0, \cdot) \mapsto \mathrm{e}^{-2\mathrm{i}\omega t} A(t, \cdot)$. The discrepancy is easily understood, because in $\mathcal{H}$ the leading terms of the kinetic and potential theory are added while they cancel in $\mathcal{L}$. Note that $\mathbb{H}$ is associated with the phase symmetry of (4.2) that is not present in the original discrete system. It is introduced into the problem via the multiscale ansatz and it manifests itself only in the limit.

Thus, to treat the Hamiltonian limit correctly it is suitable to embed the discrete Hamiltonian system into a continuous one that has the corresponding symmetries. In this systems we can compensate for drifts in the phases via the phase velocity and for drifts with the group velocities by going into suitably moving frames. On the level of Hamiltonians this leads to a subtraction of the corresponding first integrals. The terms balance in exactly the right way such that the same cancellations occur as in the Lagrangian setting. This is the content of the following classical result in the theory of Hamiltonian systems with symmetry.

Proposition 4.1. *Let $(Z, \mathcal{H}, \boldsymbol{\Omega})$ be a Hamiltonian system, which is equivariant with respect to the one-parameter symmetry group $(T_\alpha)_{\alpha\in\mathbb{R}}$ with associated first integral $\mathcal{I}$. Then $z : [0,T] \to Z$ solves $\boldsymbol{\Omega}\dot{z} = \mathrm{D}\mathcal{H}(z)$ if and only if $\widetilde{z} : t \mapsto T_{ct}z(t)$ solves $\boldsymbol{\Omega}\dot{\widetilde{z}} = \mathrm{D}\widetilde{\mathcal{H}}_c(\widetilde{z})$, where $\widetilde{\mathcal{H}}_{c,\omega} = \mathcal{H} - c\mathcal{I}$.*

We illustrate the idea in the pulse propagation problem treated in the previous subsection. The continuous Hamiltonian system is defined on the cylinder space-phase $\Xi = \mathbb{R} \times \mathbb{S}^1$ and has the configuration space $\mathrm{L}^2(\Xi)$. For functions $u \in \mathrm{L}^2(\Xi)$ we consider the system

$$\begin{aligned} &\partial_t^2 u = \Delta_{(1,0)}u - au - bu^3 \quad \text{with } a > 0 \text{ and} \\ &\Delta_{(\varepsilon,\delta)}u(\eta,\phi) = u(\eta{+}\varepsilon, \phi{+}\delta) - 2u(\eta,\phi) + u(\eta{-}\varepsilon, \phi{-}\delta). \end{aligned} \tag{4.3}$$

Introducing $p = \partial_\tau u$ this is a canonical Hamiltonian system with

$$H^{\mathrm{cont}}(u,p) = \int_\Xi \tfrac{1}{2}p^2 + \tfrac{1}{2}\big(\nabla_{(1,0)}u\big)^2 + \tfrac{a}{2}u^2 + \tfrac{b}{4}u^4 \,\mathrm{d}\eta\,\mathrm{d}\phi. \tag{4.4}$$

Here the important fact is that this system contains the KG chain exactly, because the system decouples completely into an uncountable family of KG chains just displaced by $(\eta,\phi) \in [0,1) \times \mathbb{S}^1$. Moreover, (4.3) is invariant under translations in the spatial direction η as well as in the phase direction ϕ. This leads to the two first integrals

$$I^{\mathrm{sp}}(u,p) = \int_\Xi p\,\partial_\eta u \,\mathrm{d}\eta\,\mathrm{d}\phi \quad \text{and} \quad I^{\mathrm{ph}}(u,p) = \int_\Xi p\,\partial_\phi u \,\mathrm{d}\eta\,\mathrm{d}\phi. \tag{4.5}$$

The flows associated with the canonical symplectic structure and with one of these first integral leads to the transport along the corresponding direction with constant speed one.

Using the symmetry of H^{cond} we can go into a frame moving with the phase speed $c_{\mathrm{ph}} = \omega/\theta$. According to Proposition 4.1 the corresponding Hamiltonian is $H^{\mathrm{ph}}(u,p) = H^{\mathrm{cont}}(u,p) - \omega I^{\mathrm{ph}}(u,p)$. Into this Hamiltonian we insert the suitably adjusted multiscale ansatz (4.1), namely

$$\begin{aligned} u(t,\eta,\phi) &= \varepsilon^{1/2}A(\varepsilon t,\varepsilon\eta)\mathbf{E}_{\mathrm{ph}} + \varepsilon^{1/2}\overline{A}(\varepsilon t,\varepsilon\eta)\overline{\mathbf{E}}_{\mathrm{ph}}, \\ p(t,\eta,\phi) &= \varepsilon^{1/2}\mathrm{i}\omega\big(A(\varepsilon t,\varepsilon\eta)\mathbf{E}_{\mathrm{ph}} - \overline{A}(\varepsilon t,\varepsilon\eta)\overline{\mathbf{E}}_{\mathrm{ph}}\big) \\ &\quad + \varepsilon^{3/2}\big(\partial_\tau A(\varepsilon t,\varepsilon\eta)\mathbf{E}_{\mathrm{ph}} + \partial_\tau\overline{A}(\varepsilon t,\varepsilon\eta)\overline{\mathbf{E}}_{\mathrm{ph}}\big), \end{aligned}$$

where $\mathbf{E}_{\mathrm{ph}} = \mathrm{e}^{\mathrm{i}(\phi+\theta\varepsilon)}$ does no longer depend on time. Through the subtraction of the properly chosen multiple of the corresponding first integral we exactly obtain the cancellation of the leading terms. Moreover, integration over $\phi \in \mathbb{S}^1$ makes all terms $\mathbf{E}_{\mathrm{ph}}^k$ with $k \neq 0$ exactly 0. Hence, the resulting reduced Hamiltonian has the expansion

$$H_\varepsilon^{\mathrm{red}}(A, \partial_\tau A) = \varepsilon \mathbb{E}(A) + O(\varepsilon^2)$$

with $\mathbb{E}$ from above. A simple calculation shows that $\boldsymbol{\Omega}_0 \partial_\tau A = \mathrm{D}\mathbb{E}(A)$ is exactly the macroscopic equation (4.2).

4.3 Derivation of KdV from the FPU chain

Here we apply both the Lagrangian and Hamiltonian reduction from above to the FPU chain with KdV-multiscale ansatz, see (3.6). For simplicity we restrict to the infinite chain with $V_1(0) = V_1'(0) = 0$, and we always assume that all arising integrals do exist.

Following the idea in [BP06] we embed the discrete system into a continuous one. For this example we choose the continuous configuration space Q to be $\mathrm{L}^2(\mathbb{R})$ and identify each discrete configuration $(x_j)_{j\in\mathbb{Z}}$ with an piecewise linear function $w = w(\eta) \in \mathrm{L}^2(\mathbb{R})$ defined by $x_j = w(j)$. Since (3.6), i.e. $w(t, \eta) = \varepsilon U(\varepsilon^3 t, \varepsilon(\eta+ct))$, describes slow macroscopic modulations without fast oscillations, there is no need for introding phase variables. The Lagrangian $\mathcal{L}$ of the continuous system is given by $\mathcal{L}(w, \dot{w}) = \mathcal{K}(\dot{w}) - \mathcal{V}(w)$, with

$$\mathcal{V}(w) = \textstyle\int_{\mathbb{R}} V_1(\nabla_+ w)\,\mathrm{d}\eta, \quad \mathcal{K}(\dot{w}) = \int_{\mathbb{R}} \dot{w}^2\,\mathrm{d}\eta \tag{4.6}$$

with $(\nabla_+ w)(\eta) = w(\eta+1) - w(\eta)$. The continuous system is invariant under the group of translations, and this gives rise to a further conserved quantity $\mathcal{I}$. Exploiting Noether's theorem we find the first integral $\mathcal{I}(w, \dot{w}) = \int_{\mathbb{R}} \dot{w}\,\partial_\eta w\,\mathrm{d}\eta$, which has no counterpart in the discrete microscopic FPU chain.

Inserting the ansatz (3.6) into the energies and using $\int_{\mathbb{R}} \partial_y U \partial_{yy} U\,\mathrm{d}y = 0$, $\int_{\mathbb{R}} \partial_y U \partial_{yyy} U\,\mathrm{d}y = -\int_{\mathbb{R}} (\partial_{yy} U)^2 \mathrm{d}y$, and $c^2 = V_1''(0)$ we find

$$\begin{aligned}
\mathcal{K}(\dot{w}) &= \varepsilon^3\,\tfrac{1}{2}\,\mathbb{H}(U) + \varepsilon^5\,\mathbb{I}(U, \partial_\tau U) + O(\varepsilon^7),\\
\mathcal{V}(w) &= \varepsilon^3\,\tfrac{1}{2}\,\mathbb{H}(U) + \varepsilon^5\,\mathbb{E}(U) + O(\varepsilon^7),\\
\mathcal{I}(w, \dot{w}) &= \varepsilon^3\,c^{-1}\,\mathbb{H}(U) + \varepsilon^5\,c^{-1}\,\mathbb{I}(U, \partial_\tau U) + O(\varepsilon^7),
\end{aligned}$$

where

$$\begin{aligned}
\mathbb{H}(U) &= c^2 \textstyle\int_{\mathbb{R}} (\partial_y U)^2\,\mathrm{d}y, \quad \mathbb{I}(U, \partial_\tau U) = c \int_{\mathbb{R}} \partial_\tau U\,\partial_y U\,\mathrm{d}y,\\
\mathbb{E}(U) &= -\tfrac{1}{24}\,c^2 \textstyle\int_{\mathbb{R}} (\partial_{yy} U)^2 \mathrm{d}y + \tfrac{1}{6}\,V_1'''(0) \int_{\mathbb{R}} (\partial_y U)^3 \mathrm{d}y.
\end{aligned}$$

Consequently, with $\mathbb{L} = \mathbb{I} - \mathbb{E}$ we find

$$\begin{aligned}\mathcal{L}(w,\,\dot{w}) &= \varepsilon^5\,\mathbb{L}(U,\,\partial_\tau U) + O(\varepsilon^7),\\ \mathcal{H}(w,\,\dot{w}) &= \varepsilon^3\,\mathbb{H}(U) + \varepsilon^5\,\mathbb{I}(U,\,\partial_\tau U) + \varepsilon^5\,\mathbb{E}(U) + O(\varepsilon^7),\\ \mathcal{H}(w,\,\dot{w}) - c\,\mathcal{I}(w,\,\dot{w}) &= \varepsilon^5\,\mathbb{E}(U) + O(\varepsilon^7),\end{aligned}$$

and it follows that the reduced Lagrangian equation equals (3.7).

In the next step we reduce the Hamiltonian structure. For the microscopic continuous system the canonical momentum is given by $p = \dot{w}$ with Hamiltonian $\mathcal{H}(w,p) = \mathcal{K}(p) + \mathcal{V}(w)$. For (w,p) the multiscale ansatz (3.6) means

$$(w,p) = R_\varepsilon(U)(\eta) = \big(\varepsilon U(\varepsilon\eta),\, \varepsilon^4\,\partial_\tau U(\varepsilon\eta) + \varepsilon^2\,c\,\partial_y U(\varepsilon\eta)\big),$$

where the last term is due to the frame moving with speed c. Reduction of the canonical symplectic form $\boldsymbol{\Omega}$ with $\langle \boldsymbol{\Omega}\,(w,\,p),\,(\tilde{w},\,\tilde{p})\rangle = \int_{\mathbb{R}} w\tilde{p} - \tilde{w}p\,\mathrm{d}\eta$ leads to

$$\begin{aligned}&\langle \boldsymbol{\Omega}\,R_\varepsilon(U), R_\varepsilon(\tilde{U})\rangle = \varepsilon^2\langle \boldsymbol{\Omega}^{\mathrm{red}}U,\tilde{U}\rangle + O\big(\varepsilon^4\big) \text{ with}\\ &\langle \boldsymbol{\Omega}^{\mathrm{red}}U,\tilde{U}\rangle = c\int_{\mathbb{R}} \big(U\partial_y\tilde{U} - \tilde{U}\partial_y U\big)\,\mathrm{d}\eta = -2\,c\int_{\mathbb{R}} \partial_y U\,\tilde{U}\,\mathrm{d}\eta.\end{aligned}$$

From this we conclude $\boldsymbol{\Omega}^{\mathrm{red}} = -2\,c\,\partial_y$. Note that $\boldsymbol{\Omega}^{\mathrm{red}}$ is defined on $\mathrm{L}^2(\mathbb{R})$, whereas $\boldsymbol{\Omega}$ lives on $\mathrm{L}^2(\mathbb{R})\times\mathrm{L}^2(\mathbb{R})$. This dimension reduction is natural, because the multiscale ansatz (3.6) yields a coupling of w and p in leading order. Finally it follows immediately that the reduced Hamiltonian equation $\boldsymbol{\Omega}^{\mathrm{red}}U_\tau = \mathrm{D}\mathbb{E}(U)$ is again equivalent to (3.7).

4.4 Derivation of nlS from the KG chain

We consider the KG chain (2.9) with $V_0(x) = \frac{a}{2}x^2 + \frac{b}{4}x^4$. The sum of the kinetic and potential energy gives the Hamiltonian

$$H(\mathbf{x},\dot{\mathbf{x}}) = \sum_{j\in\mathbb{Z}} \big(\tfrac{1}{2}\dot{x}_j^2 + \tfrac{1}{2}(x_{j+1}-x_j)^2 + \tfrac{a}{2}x_j^2 + \tfrac{b}{4}x_j^4\big)\,.$$

Since we are interested in modulated pulses, we proceed as in Section 4.2 and embed the discrete chain on $\mathbb{Z}$ into the cylinder $\Xi = \mathbb{R}\times\mathbb{S}^1$ leading to the continuous Hamiltonian system (4.3) with Hamiltonian H^{cont} in (4.4).

Again we have the two symmetries of spatial translations T^{sp} and phase translations T^{ph} leading to the two first integrals I^{sp} and I^{ph} given in (4.5). However, we proceed differently, because we are interested in a dispersive ansatz $u(t,\eta) = \varepsilon A(\varepsilon^2 t, \varepsilon(\eta + ct))\mathbf{E}$ + c.c. + h.o.t., where $c = c_{\mathrm{gr}}$, cf. (3.9). Thus, we apply Proposition 4.1 using the symmetry transformation

$$(\widetilde{u},\widetilde{p}) = T^{\mathrm{sp}}_{ct}T^{\mathrm{ph}}_{(\omega-c\theta)t}(u,p), \qquad \widetilde{\mathcal{H}} = \mathcal{H} - cI^{\mathrm{sp}} - (\omega - c\theta)I^{\mathrm{ph}}.$$

The associated canonical Hamiltonian system $\boldsymbol{\Omega}^{\mathrm{can}}(\widetilde{u},\widetilde{p}) = \mathrm{D}\widetilde{\mathcal{H}}(\widetilde{u},\widetilde{p})$ on $\mathrm{L}(\Xi)^2$ is still fully equivalent to a family of uncoupled KG chains.

Inserting the scaling exposes the macroscopic behavior. For this define

$$(u(\eta,\phi), p(\eta,\phi)) = (\varepsilon U(\varepsilon\eta, \phi - \theta\eta), \varepsilon P(\varepsilon\eta, \phi - \theta\eta)),$$

which keeps the canonical structure, if we move a factor the ε, which arises from the transformation rule $\mathrm{d}y = \varepsilon\,\mathrm{d}\eta$, into a the time parametrization $\tau = \varepsilon^2 t$. We obtain the new Hamiltonian

$$\mathcal{H}_\varepsilon(U,P) = \int_\Xi \tfrac{1}{2\varepsilon^2}\Big(\big[P-\omega U_\phi-\varepsilon c U_y\big]^2 + \big(\nabla_{(\varepsilon,\theta)}U\big)^2 \\ +aU^2 - \big[\omega P U_\phi + \varepsilon c P U_y\big]^2\Big) + \tfrac{b}{4}U^4\,\mathrm{d}y\,\mathrm{d}\phi,$$

where $\nabla_{(\varepsilon,\theta)}U(y,\phi) = U(y+\varepsilon,\phi+\theta) - U(y,\phi)$. Now we see that the suitably transformed version of the modulational ansatz (3.9), viz.,

$$(U(y,\phi),P(y,\phi)) = R_\varepsilon(A)(y,\phi) = (\mathrm{Re}\,A(y)\mathrm{e}^{\mathrm{i}\phi}, \omega\,\mathrm{Re}\,A(y)\mathrm{e}^{\mathrm{i}\phi}) + O(\varepsilon),$$

leads to the expansion

$$\mathcal{H}_\varepsilon(R_\varepsilon(A)) = \mathbb{H}_{\mathrm{nlS}}(A) + O(\varepsilon) \text{ with } \mathbb{H}_{\mathrm{nlS}}(A) = \int_{\mathbb{R}} \omega\omega''|A_y|^2 + \tfrac{3b}{8}|A|^4\,\mathrm{d}y$$

and the reduced symplectic structure $\boldsymbol{\Omega}^{\mathrm{red}} = 2\mathrm{i}\omega$. Thus, we recover the one-dimensional version of nlS given in Sect. 3.4.

5 Weak convergence methods

For static problems there is a rich literature concerning the Γ-convergence of potential energy functionals of discrete models to continuum models (cf. [FJ00, FT02, BG02a, BG02b, MBL06]). Here we want to summarize some first results for dynamic problems that rely on weak convergence.

5.1 An abstract weak convergence result

In [Mie06a] it was shown that linear elastodynamics can be derived from a general linear lattice model as described in Sect. 2. However, this result used exact periodicity and linearity in an essential way. The abstract approach presented here will be discussed in [Mie06b] in full details. Its main advantage lies in the flexibility, which allows for applications in nonlinear and macroscopically heterogeneous settings.

We consider a family of Hamiltonian systems parametrized by $\varepsilon \in [0,1]$,

$$\boldsymbol{\Omega}_\varepsilon(z)\dot{z} = \mathrm{D}\mathcal{H}_\varepsilon(z), \tag{5.1}$$

and we are interested in the limit behavior for $\varepsilon \to 0$. Again, ε measures the ratio between the microscopic and the macroscopic spatial scales, viz., $y = \varepsilon\gamma$.

We consider the situation that all $\mathcal{H}_\varepsilon$ are defined on one reflexive Banach space Z, but may take the value $+\infty$ outside the subspace Z_ε. It is a question of general interest to characterize the further conditions on the convergence of $\mathcal{H}_\varepsilon$ to $\mathcal{H}_0$ and of $\boldsymbol{\Omega}_\varepsilon$ to $\boldsymbol{\Omega}_0$ such that suitable limits z of solutions z_ε of

(5.1) are solutions of the limit problem (5.1) for $\varepsilon = 0$. A first guess would be that $\mathcal{H}_0$ is the Γ-limit of $\mathcal{H}_\varepsilon$, i.e.

$$\begin{aligned}
&\text{(G1)} \quad z_\varepsilon \rightharpoonup z \quad \Longrightarrow \quad \mathcal{H}_0(z) \le \liminf_{\varepsilon\to 0} \mathcal{H}_\varepsilon(z_\varepsilon),\\
&\text{(G2)} \quad \forall\, z \in Z\ \exists(\widetilde{z}_\varepsilon)_{\varepsilon\in(0,1)} : \ \widetilde{z}_\varepsilon \rightharpoonup z \text{ and } \mathcal{H}_0(z) = \lim_{\varepsilon\to 0} \mathcal{H}_\varepsilon(\widetilde{z}_\varepsilon).
\end{aligned}$$

However, we will see below that it cannot be expected in general.

We assume that the subspaces $Z_\varepsilon \subset Z$ are closed and that $\mathcal{H}_\varepsilon \in \mathrm{C}^1(Z_\varepsilon, \mathbb{R})$ for $\varepsilon \in [0,1]$. Moreover, there exist mappings $G_\varepsilon \in \mathrm{Lin}(Z_0, Z_\varepsilon)$ such that we have

$$Z_\varepsilon \ni z_\varepsilon \rightharpoonup z \in Z_0 \quad \Longrightarrow \quad G_\varepsilon^* \mathrm{D}\mathcal{H}_\varepsilon(z_\varepsilon) \rightharpoonup \mathrm{D}\mathcal{H}_0(z) \text{ in } Z_0^*. \tag{5.2}$$

Finally we assume that the symplectic operators $\boldsymbol{\Omega}_\varepsilon$ are independent of $z \in Z$ and that there exists a larger Banach space W such that Z embeds continuously and densely into W such that $\boldsymbol{\Omega}_\varepsilon : W \to Z^*$ has an inverse operator for all $\varepsilon \in [0,1]$ with the norm bounded independently of ε. For the convergence we ask the condition

$$Z_\varepsilon \ni z_\varepsilon \rightharpoonup z \in Z_0 \quad \Longrightarrow \quad G_\varepsilon^* \boldsymbol{\Omega}_\varepsilon z_\varepsilon \rightharpoonup \boldsymbol{\Omega}_0 z \text{ in } Z^*. \tag{5.3}$$

Now we use the fact that solutions z_ε of (5.1) also solve the weak equation

$$\int_0^T \langle \mathrm{D}\mathcal{H}_\varepsilon(z_\varepsilon(t)), \varphi_\varepsilon(t)\rangle + \langle \boldsymbol{\Omega}_\varepsilon z_\varepsilon(t), \dot{\varphi}_\varepsilon(t)\rangle \,\mathrm{d}t - \langle \boldsymbol{\Omega}_\varepsilon z_\varepsilon, \varphi_\varepsilon\rangle\big|_0^T = 0 \tag{5.4}$$

for all $\varphi_\varepsilon \in \mathrm{C}^1([0,T], Z_\varepsilon)$. Choosing $\varphi_\varepsilon(t) = G_\varepsilon \varphi(t)$ for some $\varphi \in \mathrm{C}^1([0,T], Z_0)$ and using suitable a priori bounds on z_ε in $\mathrm{C}^0([0,T], Z) \cap \mathrm{C}^1([0,T], W)$ it is possible to extract a weakly convergent subsequence with $z_\varepsilon(t) \rightharpoonup z(t)$ for some $z \in \mathrm{C}^0([0,T], Z_\mathrm{w}) \cap \mathrm{L}^\infty([0,T], W)$. By the assumptions (5.2) and (5.3) we pass to the limit in (5.4) and obtain

$$\int_0^T \langle \mathrm{D}\mathcal{H}_0(z), \varphi\rangle + \langle \boldsymbol{\Omega}_0 z, \dot{\varphi}\rangle \,\mathrm{d}t - \langle \boldsymbol{\Omega}_0 z, \varphi\rangle\big|_0^T = 0.$$

Under suitable assumptions it then follows that z solves (5.1) for $\varepsilon = 0$.

5.2 Elastodynamics

The program described in the previous subsection can be applied to polyatomic Klein-Gordon chains, which we also allow to have large-scale variations in the stiffness and masses. The KG chains under consideration are assumed to have a periodicity of N on the microscopic level, and all quantities may change also on the macroscopic scale $y = \varepsilon j$. For $k \in \mathbb{Z}_N = \{\, j \bmod N \,|\, j \in \mathbb{Z}\,\}$ we have given functions $m_k, a_k, b_k, c_k \in \mathrm{L}^\infty(\mathbb{R})$, which are all bounded from below by a positive constant. The KG chain is then given by the canonical Hamiltonian system on $\ell^2 \times \ell^2$

$$\begin{aligned}
\mathcal{H}_\varepsilon^{\mathrm{discr}}(\mathbf{x}, \mathbf{p}) = \sum_{j\in\mathbb{Z}} \Big(&\frac{p_j^2}{2m_{[j]}(\varepsilon j)} + \frac{a_{[j]}(\varepsilon j)}{2}(x_{j+1} - x_j)^2 \\
&+ \frac{\varepsilon^2 b_{[j]}(\varepsilon j)}{2} x_j^2 + \frac{\varepsilon^2 c_{[j]}(\varepsilon j)}{4} x_j^4 \Big),
\end{aligned} \tag{5.5}$$

where $[j] = j \bmod N$. To derive a suitable continuum model we embed $\ell^2 \times \ell^2$ into $Z = Z_0 = \mathrm{H}^1(\mathbb{R}) \times \mathrm{L}^2(\mathbb{R})$ via

$$\begin{aligned} &Z_\varepsilon = \{\, (u,v) \in Z \mid u|_{[\varepsilon j, \varepsilon j+\varepsilon]} \text{ affine, } v|_{(\varepsilon j-\varepsilon/2,\ \varepsilon j+\varepsilon/2)} \text{ constant} \,\}\text{and} \\ &(u,v) = E_\varepsilon(\mathbf{x},\mathbf{p}) \text{ with } (u(\varepsilon j), v(\varepsilon j)) = (x_j, p_j) \text{ for all } j \in \mathbb{Z}. \end{aligned} \tag{5.6}$$

The associated Hamiltonian $\mathcal{H}_\varepsilon$ coincides with $\mathcal{H}_\varepsilon^{\mathrm{discr}}$ up to a factor ε, which relates to the time rescaling, namely $\mathcal{H}_\varepsilon(u,v) =$

$$\int_{\mathbb{R}} \tfrac{v(y)^2}{2M(y,y/\varepsilon)} + \tfrac{A(y,y/\varepsilon)}{2} u'(y)^2 \,\mathrm{d}y + \sum_{j\in\mathbb{Z}} \varepsilon\big(\tfrac{B(\varepsilon j,j)}{2} u(\varepsilon j)^2 + \tfrac{C(\varepsilon j,j)}{4} u(\varepsilon j)^4\big),$$

where $M(y,z) = m_{[k]}(y)$ for $z \in (k-1/2,\, k+1/2)$, $A(y,z) = a_{[k]}(y)$ for $z \in (k,\, k+1)$ for $k \in \mathbb{Z}$, with similar formulas for B and C.

The important step in the analysis is the construction of the operator $G_\varepsilon \colon Z_0 \to Z_\varepsilon$. We define $(u_\varepsilon, v_\varepsilon) = G_\varepsilon(u,v)$ via $v_\varepsilon(y) = \frac{M(y,y/\varepsilon)}{M^*(y)} v(y)$ and

$$\int_{\mathbb{R}} A(y,y/\varepsilon) u_\varepsilon'(y)\widetilde{u}'(y) + u_\varepsilon(y)\widetilde{u}(y) \,\mathrm{d}y = \int_{\mathbb{R}} A^*(y) u'\widetilde{u}' + u\widetilde{u} \,\mathrm{d}y$$

for all $\widetilde{u}$ with $(\widetilde{u}, 0) \in Z_\varepsilon$, see (5.6). Here A^* is the averaged stiffness and M^* the averaged masses

$$A^*(y) = \big(\tfrac{1}{N}\int_0^N A(y,z)^{-1}\,\mathrm{d}z\big)^{-1} \quad \text{and} \quad M^*(y) = \tfrac{1}{N}\int_0^N M(y,z)\,\mathrm{d}z.$$

It is then possible to prove the abstract conditions 5.2 and 5.3, which leads to the following results, cf. [Mie06b].

Theorem 5.1. *Let $E_\varepsilon : \ell^2 \times \ell^2 \to Z = \mathrm{H}^1(\mathbb{R}) \times \mathrm{L}^2(\mathbb{R})$ be the embedding in (5.6). Let $(x^\varepsilon, p^\varepsilon) : [0,\, T/\varepsilon] \to \ell^2 \times \ell^2$ be solutions of the canonical Hamiltonian system associated with $\mathcal{H}_\varepsilon^{\mathrm{discr}}$ in (5.5). If for $\tau = 0$ we have*

$$\begin{pmatrix} I & 0 \\ 0 & M(\cdot,\cdot/\varepsilon) \end{pmatrix} E_\varepsilon \begin{pmatrix} \mathbf{x}^\varepsilon(\tau/\varepsilon) \\ \varepsilon\mathbf{p}^\varepsilon(\tau/\varepsilon) \end{pmatrix} \rightharpoonup \begin{pmatrix} u(\tau) \\ M^*(\cdot)v(\tau) \end{pmatrix} \quad \text{in } Z,$$

then this convergence holds for all $\tau \in [0,T]$, where $(u,v) : [0,T] \to Z$ is a solution of the macroscopic wave equation arising from the canonical Hamiltonian system with

$$\mathcal{H}_0(u,v) = \int_{\mathbb{R}} \tfrac{1}{2M^*(y)} v^2 + \tfrac{A^*(y)}{2}(u')^2 + \tfrac{B^*(y)}{2}u^2 + \tfrac{C^*(y)}{4}u^4 \,\mathrm{d}y,$$

where $B^*(y) = \frac{1}{N}\int_0^N B(y,z)\,\mathrm{d}z$ and $C^*(y) = \frac{1}{N}\int_0^N c(y,z)\,\mathrm{d}z$.

It should be noted that $\mathcal{H}_0$ is not the Γ-limit of $\mathcal{H}_\varepsilon$ when using canonical variables. However, if we use the Lagrangian coordinates $(u^\varepsilon, \dot{u}^\varepsilon) = (u^\varepsilon, M(\cdot,\cdot/\varepsilon)^{-1}p^\varepsilon)$, then it is the Γ-limit.

5.3 Energy transport via Wigner-Husimi measures

Waves in dispersive media travel with a speed that depends on their wave length. We now discuss this for the general linear model introduced in Sect. 2.2. Wave propagation is driven by the group velocity $c_{\mathrm{gr}} = \nabla\Omega_j(\theta)$, which depends on the wave vector $\theta \in \mathcal{T}_\Gamma$ and the band number $j \in \{1,\dots,m\}$. Thus, at each macroscopic point $y \in \mathbb{R}^d$ we need to know how much energy is located in which band and in which wave-vector regime.

The relevant mathematical tool is the Wigner measure or the Husimi measure, which was used in [Gér91, LP93, MMP94, GMMP97, TP04] to study transport of oscillations (relating to energy, density, or other physical quantities). The case of discrete lattices is analyzed in detail in [Mac04, Mie06a]. For this we rewrite (2.6) into diagonal and rescaled form

$$\frac{\partial}{\partial\tau}U^\varepsilon(\tau,\theta) = \mathbb{B}(\varepsilon,\theta)U^\varepsilon(\tau,\theta) \text{ with } \mathbb{B}(\varepsilon,\theta) = \frac{\mathrm{i}}{\varepsilon}\mathrm{diag}(\Omega_1(\theta),...,\Omega_m(\theta)). \quad (5.7)$$

The Wigner transform $W^\varepsilon[u^\varepsilon]$ of $u^\varepsilon = \mathbb{F}^{-1}U^\varepsilon$ is now defined as a matrix-valued distribution on $\mathbb{R}\times\mathcal{T}_\Gamma$. For the diagonal entries it is possible to pass to the limit $\varepsilon \to 0$ and one finds the Wigner measure $\mu_j^{\mathrm{W}} = \lim_{\varepsilon\to 0}(W^\varepsilon[u^3])_{jj}$. More precisely, we have the following result, see [Mie06a].

Theorem 5.2. *Let $u^\varepsilon : [0,T] \to \mathrm{L}^2(\mathcal{T}_\Gamma,\mathbb{C}^m)$ be a family of solutions for* (5.7) *with $\|u^\varepsilon(0)\|_{\mathrm{L}^2} \le C$. Let $j \in \{1,...,m\}$ and $S_j \subset \mathcal{T}_\Gamma$ be given such that $\Omega_j \in \mathrm{C}^1(\mathcal{T}_\Gamma\backslash S_j)$. If for $\tau = 0$ we have*

$$\lim_{\varepsilon\to 0}(W^\varepsilon[u^\varepsilon](\tau))_{jj} = \mu_j^{\mathrm{W}}(\tau) \text{ in } \mathcal{D}(\mathbb{R}^d\times\mathcal{T}_\Gamma) \text{ and } \mu_j^{\mathrm{W}}(0,\mathbb{R}^d\times S_j) = 0,$$

then this convergence holds for all $\tau \in [0,T]$, where $\mu_j^{\mathrm{W}} : [0,T] \to \mathcal{M}(\mathbb{R}^d\times\mathcal{T}_\Gamma)$ is a solution of the energy-transport equation

$$\partial_\tau\mu_j^{\mathrm{W}} = \nabla\Omega_j(\theta)\cdot\partial_y\mu_j^{\mathrm{W}} \quad \text{on } [0,T]\times\mathbb{R}^d\times\mathcal{T}_\Gamma.$$

Using this result it is possible to obtain the energy distribution by integration over θ, namely

$$e(\tau,y)\mathrm{d}y = \sum_{j=1}^m \int_{\theta\in\mathcal{T}_\Gamma} \mu_j^{\mathrm{W}}(0,\, y-\nabla\Omega_j(\theta)\tau, \mathrm{d}\theta).$$

The above theorem is restricted to the case that no mass concentrates on the singular set S_j, where the dispersion relation is not smooth and, hence, the group velocity is not defined. However, using the Husimi measure as developed in [Mie06a] it is possible to treat this case also in some cases.

6 Quantitative estimates via Gronwall estimates

Another technique for the justification of continuum models uses quantitative estimates to control the error between the macroscopic equation and the

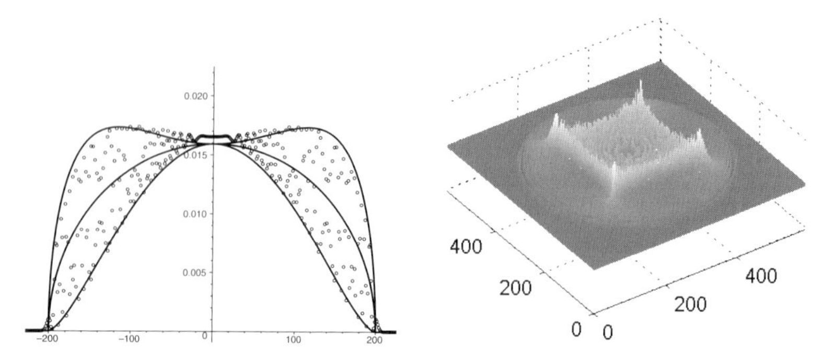

Fig. 5.1. Left: energy distribution at $t = 200$ for the linear chain $\ddot{x}_j = x_{j+1} - 2x_j + x_{j-1}$ with initial data $x_j(0) = \delta_j$ and $\dot{x}_j(0) = 0$. Right: displacement for the square lattice $\mathbb{Z}^2$ with simple nearest-neighbor interaction at time $t = 120$, cf. [Pat06].

microscopic equation. We present the abstract idea in Sect. 6.1 and apply in Sect. 7. This method can also be used to prove dispersive stability results as discussed Sect. 6.2.

We work totally in the original microscopic lattice model

$$\dot{z} = Lz + \mathcal{N}(z), \tag{6.1}$$

where Z is the Banach space for the state $z(t)$, and $L: Z \to Z$ is the linear part, which is assumed to generate a bounded semigroup $(\mathrm{e}^{Lt})_{t\geq 0}$, i.e.

$$\exists\, C_L > 0\ \forall\, t \geq 0\ \forall\, z \in Z: \quad \|\mathrm{e}^{Lt} z\| \leq C_L \|z\|. \tag{6.2}$$

We also rely on our standard assumption that the solution $z = 0$ is energetically stable, as the Hamiltonian energy is conserved. Here it means

$$\exists\, C_E > 0\ \forall \text{ sln. } z \text{ of (6.1) } \forall\, t \geq 0: \quad \|z(t)\| \leq C_E \|z(0)\|. \tag{6.3}$$

The nonlinearity $\mathcal{N}: Z \to Z$ is assumed to be locally Lipschitz. However, the essential features have to be addressed by using additional Banach spaces Y and W such $Y \subset Z \subset W$ with continuous embeddings and

$$\text{(i) } \forall\, z \in Z\text{: } \|z\|_W \leq \|z\| \qquad \text{(ii) } \forall\, \widetilde{z} \in Y\text{: } \|\widetilde{z}\| \leq \|\widetilde{z}\|_Y. \tag{6.4}$$

In applications to lattices we have in mind

$$Y = \ell_1(\Gamma, \mathbb{R}^k)^2, \quad Z = \ell_2(\Gamma, \mathbb{R}^k)^2, \quad W = \ell_\infty(\Gamma, \mathbb{R}^k)^2. \tag{6.5}$$

In Sect. 6.1 the importance is that $\mathcal{N}$ satisfies

$$\begin{aligned} &\exists\, C > 0\ \exists\, \nu > 0\ \forall\, z_1, z_2 \in Z \text{ with } \|z_1\|_W, \|z_2\|_W \leq 1: \\ &\|\mathcal{N}(z_1) - \mathcal{N}(z_2)\| \leq C_{\mathcal{N}} \big(\|z_1\|_W + \|z_2\|_W\big)^\nu \|z_1 - z_2\|. \end{aligned} \tag{6.6}$$

In Sect. 6.2 the importance of Y is the *dispersive decay estimate*

$$\exists\,\kappa\in(0,1)\;\exists\,C_W>0\;\forall\,z\in Y\;\forall\,t>0: \quad \|\mathrm{e}^{Lt}z\|_W\le\frac{C_W}{(1+t)^\kappa}\|z\|_Y. \tag{6.7}$$

of the linear semigroup. For the nonlinearity we then use

$$\exists\,C_{\mathcal N}>0\;\exists\,\alpha,\nu>0\;\forall\,z\in Z: \quad \|\mathcal N(z)\|_Y\le C_{\mathcal N}\|z\|_W^\nu\|z\|^\alpha. \tag{6.8}$$

With Y, Z and W as in (6.5) a standard nonlinearity $\mathcal N((x_\gamma)_{\gamma\in\Gamma})=(n(x_\gamma))_{\gamma\in\Gamma}$ with $|n(\xi_1)-n(\xi_2)|\le C(|\xi_1|+|\xi_2|)^\beta|\xi_1-\xi_2|$ will satisfy (6.6) with $\nu=\beta$ and (6.8) with $\nu=\beta-1$ and $\alpha=2$.

6.1 Error control for approximate solutions

The basic idea is to construct an approximate solution z_{app}, which in fact will be given in the form $z_{\mathrm{app}}=R_\varepsilon(\mathbf A)$, and to derive an estimate for the associated error. For any $z\in \mathrm C^1([0,T],Z)$ we define the *residual* via

$$\mathrm{Res}(z)(t)=\dot z(t)-Lz(t)-\mathcal N(z(t)). \tag{6.9}$$

The following result shows that the smallness of the residual together with the stability condition (6.2) implies that the error between z_{app} and an exact solution is small.

Theorem 6.1. *Assume that the conditions* (6.2), (6.4i) *and* (6.6) *hold. Moreover, let* C_R, C_A, τ_*, σ, α, $\varrho>0$ *be given as well as a family* $(z_{\mathrm{app}}^\varepsilon)_{\varepsilon\in(0,1)}$ *of approximate solutions* $z^\varepsilon_{\mathrm{app}}\in\mathrm C^1([0,\tau_*/\varepsilon^\sigma],Z)$ *satisfying*

$$\|z^\varepsilon_{\mathrm{app}}(t)\|_W\le C_A\varepsilon^\alpha \quad \textit{and} \quad \|\mathrm{Res}(z^\varepsilon_{\mathrm{app}})(t)\|\le C_R\varepsilon^\varrho \tag{6.10}$$

for all $t\in[0,\tau_*/\varepsilon^\sigma]$. *Moreover, assume*

$$\varrho>\alpha+\sigma \quad \textit{and} \quad \nu\alpha\ge\sigma. \tag{6.11}$$

Then, for each $d>0$ *there exist* $\varepsilon_0\in(0,1)$ *and* $D>0$ *such that for all* $\varepsilon\in(0,\varepsilon_0]$ *any exact solution* z *of* (6.1) *with* $\|z(0)-z^\varepsilon_{\mathrm{app}}(0)\|\le d\varepsilon^{\varrho-\sigma}$ *satisfies*

$$\|z(t)-z^\varepsilon_{\mathrm{app}}(t)\|\le D\varepsilon^{\varrho-\sigma} \textit{ for } t\in[0,\tau_*/\varepsilon^\sigma]. \tag{6.12}$$

In (6.11) the case $\nu\alpha>\sigma$ is not really interesting, as in this regime the nonlinearity is not really active. In the first inequality ϱ may be as big as we like, what improves the order of approximation in (6.12) but does not allow us to extend the length of the time interval, i.e, to make σ bigger, because it is restricted by the second inequality.

Proof. For the construction of ε and D we define $C_1=C_L(d+C_R\tau_*)$ and $C_2=C_LC_{\mathcal N}(3C_A)^\nu$ and let $D=2C_1\mathrm e^{C_2\tau_*}$ and $\varepsilon_0=\min\{1,(C_A/D)^\delta\}$, where $\delta=1/(\varrho-\alpha-\sigma)>0$.

We write the exact solution z of (6.1) in the form $z(t) = z^{\varepsilon}_{\mathrm{app}}(t) + \varepsilon^{\beta} R(t)$ with $\beta = \varrho - \sigma$. Clearly $\|R(0)\| \le d$ and we have to show $\|R(t)\| \le D$ for all $t \in [0, \tau_*/\varepsilon^{\sigma}]$. Inserting this ansatz into (6.1) and applying the variation-of-constants formula we find

$$R(t) = \mathrm{e}^{Lt} R(0) + \int_0^t \frac{\mathrm{e}^{L(t-s)}}{\varepsilon^{\alpha}} \big(\mathcal{N}(z^{\varepsilon}_{\mathrm{app}}(s) + \varepsilon^{\beta} R(s)) - \mathcal{N}(z^{\varepsilon}_{\mathrm{app}}(s)) - \mathrm{Res}(z^{\varepsilon}_{\mathrm{app}})(s)\big)\,\mathrm{d}s.$$

Defining $r(t) = \|R(t)\|$ and using the available estimates give

$$r(t) \le C_L d + \int_0^t C_L \big(C_{\mathcal{N}}[C_A\varepsilon^{\alpha} + C_A\varepsilon^{\alpha} + \varepsilon^{\beta} D]^{\nu} r(s) + C_R \varepsilon^{\varrho-\beta}\big)\,\mathrm{d}s,$$

where we assumed $r(s) \le D$ on $[0, t_D]$ and $t \le t_D$. Note that $d < D$ and r is continuous, which implies $t_D > 0$. We will show that $t_D = \tau_*/\varepsilon^{\sigma}$.

Assuming $\varepsilon \in (0, \varepsilon_0]$ we arrive at $r(t) \le C_L d + C_L C_R \varepsilon^{\sigma} t + C_2 \varepsilon^{\alpha\nu} \int_0^t r(s)\,\mathrm{d}s$. Because of $\varepsilon^{\sigma} t \le \tau_*$ we find $r(t) \le C_1 + \varepsilon^{\sigma} C_2 \int_0^t r(s)\,\mathrm{d}s$ and Gronwall's lemma gives $r(t) \le C_1 \mathrm{e}^{C_2 \varepsilon^{\sigma} t} \le C_1 \mathrm{e}^{C_2 \tau_*} = D/2$ for all $t \in [0, t_D]$. However, this shows that $r(t)$ cannot reach D. As a consequence we may choose $t_D = \tau_*/\varepsilon^{\sigma}$ and we are done. ■

6.2 Dispersive stability

Here we present conditions which guarantee that the dispersive decay estimate (6.7) for the linear semigroup can be transfered to the full nonlinear problem. We follow ideas from [Sch96, MSU01] and refer to [Pat06] for more satisfactory results.

Theorem 6.2. *Assume that* (6.3), (6.7), *and* (6.8) *hold with* $\nu\kappa > 1$. *Then, there exist* $C, \eta > 0$ *such that all solutions* z *of* (6.1) *with* $\|z(0)\|_Y \le \eta$ *satisfy*

$$\|z(t)\|_W \le \frac{C}{(1+t)^{\kappa}} \|z(0)\|_Y \quad \textit{for all} \quad t > 0. \tag{6.13}$$

Proof. We follow the ideas in [MSU01] Lemma 3 and adapt it to the more general case. We rely on $0 < \kappa < 1 < \nu\kappa$, which yield the estimate

$$\int_0^t \frac{\mathrm{d}s}{(1+s)^{\kappa\nu}(1+t-s)^{\kappa}} \le \frac{c_{\nu,\kappa}}{(1+t)^{\kappa}} \text{ with } c_{\nu,\kappa} = \Big(\frac{2^{\kappa}}{\kappa\nu-1} + \frac{2^{\kappa\nu}}{1-\kappa}\Big). \tag{6.14}$$

This is easily obtained by estimating $\int_0^{t/2}$ and $\int_{t/2}^t$ separately. Using the variation-of-constants formula together with the available estimates we find

$$\|z(t)\|_W \le \frac{C_W}{(1+t)^{\kappa}} \|z(0)\|_Y + \int_0^t \frac{C_W}{(1+t-s)^{\kappa}} C_{\mathcal{N}} \|z(s)\|_W^{\nu} \|z(s)\|^{\alpha}\,\mathrm{d}s.$$

With $r(t) = \max\{\,(1+s)^{\kappa}\|z(s)\|_W \mid s \in [0,t]\,\}$ and $\delta = \|z(0)\|_Y$ we obtain

$$(1+t)^{\kappa}\|z(t)\| \le C_W \delta + \int_0^t \frac{C_W C_{\mathcal{N}} C_E^{\alpha} r(t)^{\nu} \delta^{\alpha}}{(1+t-s)^{\kappa}(1+s)^{\kappa\nu}}\,\mathrm{d}s.$$

Employing (6.13) and using that r is nondecreasing we find

$$r(t) \leq C_W\delta + C_*\delta^\alpha r(t)^\nu \text{ for all } t \geq 0, \text{ where } C_* = c_{\nu,\kappa}C_W C_\mathcal{N} C_E^\alpha.$$

We now choose η such that $C_*\eta^\alpha(3C_W\eta)^\nu \leq C_W\eta$ and claim that $r(t)$ remains less than $3C_W\delta$ if $\|z(0)\|_Y = \delta \leq \eta$, i.e., the desired assertion holds with $C = 3C_W$. Let $t_W = \sup\{\, t \geq 0 \mid \forall s \in [0,t] : \; r(s) \leq 3C_W\delta\,\}$, then for $t \in [0, t_W]$ and $0 < \delta \leq \eta$ we have

$$r(t) = C_W\delta + C_*\delta^\alpha(3C_W\delta)^\nu \leq 2C_W\delta < 3C_W\delta.$$

Since r is also continuous, we conclude $t_W = \infty$. ∎

The typical application of the above result involves the spaces $Y = \ell_1$ and $Z = \ell_2$ and $W = \ell_\infty$. Hence, for a nonlinearity with $\mathcal{N}(\mathbf{x}) = (n(x_j))_{j\in\Gamma}$ and $|n(x_j)| \leq C_n|x_j|^\beta$ we have (6.8) with $\alpha = 2$ and $\nu = \beta-2$. Moreover, the theory in [Pat06] provides explicity values of κ, which can be determined directly for the properties of the dispersion relations $\omega = \Omega_m(\theta)$ discussed in Sect. 2.2. For this note that e^{Lt} can be written as a discrete convolution

$$\mathrm{e}^{Lt}(\mathbf{x},\dot{\mathbf{x}}) = \big(\textstyle\sum_{\alpha\in\Gamma} G_{\gamma-\alpha}(t)(x_\alpha,\dot{x}_\alpha)\big)_{\gamma\in\Gamma},$$

where the Green's functions $G_\gamma(t) \in \mathbb{R}^{2m\times 2m}$ satisfy $G_0(0) = \mathrm{id}$ and $G_\gamma(0) = 0$ for $\gamma \neq 0$. Each component of each $G_\gamma(t)$ can be calculated via oscillatory integrals of the type

$$\int_{\theta\in\mathcal{T}_\Gamma} \mathrm{e}^{\mathrm{i}(\Omega_k(\theta)t+\theta\cdot\gamma)} g(\theta)\,\mathrm{d}\theta$$

with given smooth functions g. Uniform decay properties in $\gamma \in \Gamma$ for such integrals strongly depend on the non-degeneracy of $\mathrm{D}^2\Omega_k(\theta)$. Integrating over balls in $\mathcal{T}_\Gamma$, where $\det \mathrm{D}^2\Omega_k(\theta)$ is bounded away from 0, we easily obtain a decay like $t^{-d/2}$. However, due to periodicity, degeneracies must occur, and the uniform decay is always worse.

For instance, the one-dimensional FPU and the KG chains from Section 2.3 lead to $\kappa = 1/3$, because $\Omega : \mathbb{S}^1 \to \mathbb{R}$ has turning points and the third derivative is nonzero in these points. As a consequence the above method leads to the following very preliminary dispersive decay result.

Proposition 6.3. *Consider the KG chain (2.9) with V_0 of the form $V_0'(x) = ax + O(|x|^\beta)$ for $|x| \to 0$ with $a > 0$ and $\beta > 5$. Then, there exists $\delta > 0$ and $C > 0$ such that for each initial condition $(\mathbf{x}(0), \dot{\mathbf{x}}(0))$ we have*

$$\|(\mathbf{x}(0),\dot{\mathbf{x}}(0))\|_{\ell_1} \leq \delta \implies \|(\mathbf{x}(t),\dot{\mathbf{x}}(t))\|_{\ell_\infty} \leq \frac{C\|(\mathbf{x}(0),\dot{\mathbf{x}}(0))\|_{\ell_1}}{(1+t)^{1/3}} \quad \text{for all } t \geq 0.$$

This result is still very weak in terms of the restriction on β, and we refer to [Pat06] for improved results . See also [Zua05, IZ05] for related dispersive decay results in discrete approximations of PDEs.

7 Justification of modulation equations

In this section we provide rigorous justification results for two examples. In contrast to Sect. 5 we will use the quantitative estimates provided in Sect. 6.1. The ideas are based on the justification theory developed for general modulation equations, see [KSM92, Sch94, Sch98] and the surveys [MSU01, Mie02]. In particular, we mention the papers [Sch95, Sch05b], which contain examples, where the modulation equations, derived formally as in Sect. 3, *fail to predict* the dynamics of the microscopic system correctly. Thus, the justification results are needed to validate the formally obtained macroscopic equations.

To explain the main ideas and still stay sufficiently simple we consider for both subsequent examples the d-dimensional, scalar model (3.8). The main observation about the multiscale ansatz $x_\gamma^{A,\varepsilon} = \varepsilon^\sigma A(\varepsilon\gamma)\mathbf{E} + \text{c.c.}$ is that it satisfies the estimates

$$\|(x_\gamma^{A,\varepsilon})_{\gamma\in\Gamma}\|_{\ell_2} \leq C_s\varepsilon^{\sigma-d/2}\|A\|_{\mathrm{H}^s} \text{ and } \|(x_\gamma^{A,\varepsilon})_{\gamma\in\Gamma}\|_{\ell_\infty} \leq C_s\varepsilon^{\sigma}\|A\|_{\mathrm{H}^s},$$

for any $s > d/2$. Thus, our solutions $z = (\mathbf{x}, \dot{\mathbf{x}})$ will be small only in $W = \ell_\infty(\Gamma)^2$ but may be large in $Z = \ell_2(\Gamma)^2$. However, for using the abstract approach provided in Theorem 6.1 we need to make the residual of the approximate solution $z_{\text{app}} = R_\varepsilon(\mathbf{A})$ small in Z. This means that the order of approximation of the formal ansatz R_ε in (3.1) has to be taken sufficiently high depending on the dimension d.

7.1 Nonlinear Schrödinger equation

We want to justify the nonlinear Schrödinger equation

$$\mathrm{i}\partial_\tau A = \mathrm{div}_y(\frac{1}{2}\mathrm{D}_\theta^2\Omega(\theta)\nabla_y A) + \rho|A|^2 A \tag{nlS}$$

as a macroscopic modulation equation for the microscopic lattice system (3.8), for the formal derivation see Sect. 3.4. We use the dispersive scaling $\tau = \varepsilon^2 t$ and $y = \varepsilon(t - c_{\text{gr}}t)$ for the basic periodic pattern $\mathbf{E} = \mathrm{e}^{\mathrm{i}(\omega t+\theta\cdot\gamma)}$, where $\omega = \Omega(\theta)$ and $c_{\text{gr}} = \Omega'(\theta)$. To derive an evolution equation for the macroscopic modulation amplitude $A : [0,\infty)\times\mathbb{R}^d \to \mathbb{C}$ we have to use the improved ansatz

$$x_\gamma(t) = R_\varepsilon^K(A)_\gamma(t) := \textstyle\sum_{k=1}^K \varepsilon^k \sum_{n=-k}^k A_{k,n}(\tau, y)\mathbf{E}^n,$$

where all the coefficient functions $A_{k,n}$ can be calculated formally if the non-resonance condition of order K holds, namely

$$n^2\Omega(\theta)^2 \neq \Omega(n\theta)^2 \text{ for } n = 0, 2, 3, ..., K. \tag{7.1}$$

Of course, we have $A = A_{1,1}$, where A satisfies (nlS). The other coefficient functions satisfy $A_{k,-n} = \overline{A}_{k,n}$ and are either algebraic expressions of functions $\partial_\tau^r\partial_y^s A_{p,n}^q$ with $r+2|s|+pq = k$, $p \leq k-1$ or (for $n = 1$, where the non-resonance condition fails) they satisfy some linear inhomogeneous Schrödinger-type equations.

Since all coefficients of the terms $\varepsilon^k \mathbf{E}^n$ with $k = 1, ..., K$ are equated to 0, the residual of the ansatz $z_{\text{app}} = (R_\varepsilon^K(A), \frac{\mathrm{d}}{\mathrm{d}t} R_\varepsilon^K(A)) : [0, \tau_*/\varepsilon^2] \to Z = \ell(\Gamma)^2$ satisfies

$$\|\text{Res}(z_{\text{app}})(t)\|_{\ell_\infty} \leq C\varepsilon^{K+1}\|A\|_{\mathrm{H}^s} \text{ and } \|\text{Res}(z_{\text{app}})(t)\|_{\ell_2} \leq C\varepsilon^{K+1-d/2}\|A\|_{\mathrm{H}^s}$$

for any suitable $s > K+2+d/2$. Thus, we have all the ingredients to apply Theorem 6.1. However, we note that the dispersive time scale $\tau = \varepsilon^2 t$ needs $\sigma = 2$, while the amplitude $\|z_{\text{app}}(t)\|_{\ell_\infty} \sim \varepsilon^\alpha$ with $\alpha = 1$. Now condition (6.11) only holds for $\nu \geq 2$. Thus, the nonlinearity $\mathcal{N}$ needs to be cubic (cf. (6.6)). The following result realizes this condition by assuming $V_\beta'''(0) = 0$, see [GM04] for the case $d = 1$.

Theorem 7.1. *Let $K \in \mathbb{N}$ with $K > 2+d/2$ and assume that the scalar d-dimensional lattice model* (3.8) *has potentials $V_\beta \in \mathrm{C}^{K+2}(\mathbb{R})$ with $V_\beta(0) = V_\beta'(0) = V_\beta'''(0) = 0$. Choose a wave vector $\theta \in \mathcal{T}_\Gamma$ satisfying the non-resonance conditions* (7.1). *Let $A \in \mathrm{C}([0,\tau_*], \mathrm{H}^{K+3}(\mathbb{R}^d, \mathbb{C})) \cap \mathrm{C}^1([0,\tau_*], \mathrm{H}^{K+1}(\mathbb{R}^d, \mathbb{C}))$ be an arbitrary solution of (nlS). Then, for each $d > 0$ there exist $\varepsilon_0 \in (0,1)$ and $D > 0$ such that for all $\varepsilon \in (0, \varepsilon_0]$ any exact solution $\mathbf{x}$ of* (3.8) *with*

$$\|(\mathbf{x}(0), \dot{\mathbf{x}}(0)) - (R_\varepsilon^{K-2}(A)(0), \dot{R}_\varepsilon^{K-2}(A)(0))\|_{\ell_2} \leq d\varepsilon^{K-1-d/2}$$

satisfies, for all $t \in [0, \tau_*/\varepsilon^2]$,

$$\|(\mathbf{x}(t), \dot{\mathbf{x}}(t)) - (R_\varepsilon^{K-2}(A)(t), \dot{R}_\varepsilon^{K-2}(A)(t))\|_{\ell_2} \leq D\varepsilon^{K-1-d/2}.$$

The condition $V_\beta'''(0) = 0$ allows us to apply the simple abstract result of Sect. 6.1. However, this condition is not necessary. In the case of nonlinearities that also have a quadratic part it is still possible to derive a similar result if we impose more restrictive non-resonance conditions. To treat that case one uses ideas from the theory of normal forms to transform the system via a near identity transform into a system that has the same linear part but no quadratic part in the nonlinearity. We refer to [Sch98, GM06] for positive results and mention also [Sch05b] for an example, where the result fails due to fact that the more restrictive non-resonance condition is violated.

7.2 Interaction of several modulated pulses

We report on results in [Gia06] and consider the scalar d-dimensional model (3.8) for which we want to show how the three-wave interaction equations (3.11) can be justified in terms of explicit error estimates. Given are three wave vectors $\theta_n \in \mathcal{T}_\Gamma$ and associated frequencies ω_n with $\omega_n^2 = \Omega^2(\theta_n)$, which are in resonance, namely

$$\theta_1 + \theta_2 + \theta_3 = 0 \quad \text{in } \mathcal{T}_\Gamma, \qquad \omega_1 + \omega_2 + \omega_3 = 0. \tag{7.2}$$

Following [Gia06, GMS06] we use the following type of non-resonance condition for other combinations of these wave vectors. We set $\theta_{-n} := -\theta_n$ and $\omega_{-n} := -\omega_n$ and say that the mode system $\{(\theta_n, \omega_n) : n = 1, 2, 3\}$ is *closed of order* K, if for all $k \in \{1, ..., K\}$ and all $n_1, ..., n_k \in \widetilde{N} = \{-3, -2, -1, 1, 2, 3\}$ the following holds:

$$\big(\textstyle\sum_1^k \omega_{n_l}\big)^2 = \Omega\big(\textstyle\sum_1^k \theta_{n_l}\big)^2 \iff \begin{cases} \exists\, n_* \in \widetilde{N}\colon\ \theta_{n_*} = \sum_1^k \theta_{n_l} \\ \qquad\quad \text{and } \omega_{n_*} = \sum_1^k \omega_{n_l}. \end{cases} \tag{7.3}$$

Here we use the hyperbolic scaling $\tau = \varepsilon t$ and $y = \varepsilon\gamma$ and, as explained at the beginning of Sect. 7, we need the improved multiscale ansatz

$$\mathbf{x}(t) = R_\varepsilon^K(\mathbf{A})(t) = \textstyle\sum_{k=1}^K \varepsilon^k \sum_{n_1,\dots,n_k \in \widetilde{N}} B_{n_1,\dots,n_k}(\tau, y)\mathbf{E}_{n_1} \dots \mathbf{E}_{n_k} \tag{7.4}$$

with $\mathbf{A} = (A_1, A_2, A_3)$, $\mathbf{E}_n = \mathrm{e}^{\mathrm{i}(\omega_n t + \theta_n \cdot \gamma)}$, $B_n = A_n$ and $\overline{B}_{n_1,\dots,n_k} = B_{-n_1,\dots,-n_k}$. Thus, to leading order we have three wave packets, which we expect to travel with their group velocities and to have interactions with the other wave packets.

As explained in Sect. 3.1 it is possible to determine the coefficient functions $B_{n_1,\dots,n_k}$ in such a way that the approximate solution $z_{\mathrm{app}} = (R_\varepsilon^K(\mathbf{A})(t), \dot{R}_\varepsilon^K(\mathbf{A})(t))$ and the residual $\mathrm{Res}(z_{\mathrm{app}})$ satisfy

$$\|z_{\mathrm{app}}(t)\|_{\ell_\infty} \le C\varepsilon^\alpha \text{ with } \alpha = 1 \quad\text{and}\quad \|\mathrm{Res}(z_{\mathrm{app}})(t)\|_{\ell_2} \le C\varepsilon^{K+1-d/2}$$

if the triple $\mathbf{A} = (A_1, A_2, A_3) : [0, \tau_*] \to \mathrm{L}^2(\mathbb{R}^d, \mathbb{C})^3$ is a sufficiently smooth solution of the three-wave interaction equation (3.11). Since $\tau = \varepsilon^\sigma t$ with $\sigma = 1$, we may apply Theorem 6.1 with $\nu = 1$, which means that nonlinearities with quadratic parts are allowed.

The precise statement from [Gia06] reads as follows.

Theorem 7.2. *Let* $K \in \mathbb{N}$ *with* $K > 1 + d/2$ *and assume that the scalar, d-dimensional lattice model* (3.8) *has potentials* $V_\beta \in \mathrm{C}^{K+2}(\mathbb{R})$ *with* $V_\beta(0) = V_\beta'(0) = 0$ *for* $|\beta| < R$. *Assume that the mode system* $\{(\theta_n, \omega_n) : n = 1, 2, 3\}$ *satisfies the resonance condition* (7.2) *and is closed of order* K *(cf.* (7.3)*). Let* $\mathbf{A} \in \mathrm{C}([0, \tau_*], \mathrm{H}^{K+2}(\mathbb{R}^d; \mathbb{C})) \cap \mathrm{C}^{K+1}([0, \tau_*], \mathrm{H}^1(\mathbb{R}^d; \mathbb{C}))$ *be an arbitrary solution of* (3.11). *Then, for each* $d > 0$ *there exist* $\varepsilon_0 \in (0, 1)$ *and* $D > 0$ *such that for all* $\varepsilon \in (0, \varepsilon_0]$ *any exact solution* $\mathbf{x}$ *of* (3.8) *with*

$$\|(\mathbf{x}(0), \dot{\mathbf{x}}(0)) - (R_\varepsilon^{K-1}(\mathbf{A})(0), \dot{R}_\varepsilon^{K-1}(\mathbf{A})(0))\|_{\ell_2} \le d\varepsilon^{K-d/2}$$

satisfies, for all $t \in [0, \tau_*/\varepsilon]$,

$$\|(\mathbf{x}(t), \dot{\mathbf{x}}(t)) - (R_\varepsilon^{K-1}(\mathbf{A})(t), \dot{R}_\varepsilon^{K-1}(\mathbf{A})(t))\|_{\ell_2} \le D\varepsilon^{K-d/2}.$$

The whole theory can be generalized in several aspects. First we may consider mode systems with N different wave vectors, where $N \ge 4$. Then, we obtain a system of N equations for $A_1, ..., A_N$, where only those quadratic terms

$\overline{A}_{n_2}\overline{A}_{n_3}$ occur in the equation for $\partial_\tau A_{n_1}$ if the three modes $(\theta_{n_l}, \omega_{n_l})_{l=1,2,3}$ satisfy the resonance condition (7.2). Other triple interactions do not matter on this time scale either because the frequencies or the wave vectors do not resonate. Quadruple or higher interactions are too small in amplitude to influence the macroscopic behavior (cf. [Gia06]).

Second it is possible to do the very same analysis for systems rather than for a scalar problem. Of course, then we have to pay attention to the different frequency bands. We also refer to [GMS06], where multipulse interactions are treated for nonlinear Schrödinger equations with periodic potentials, see [CMS04, Spa06].

Similar phenomena arise in such different subjects as phonon collisions (cf. [Spo05]) and in surface water waves (cf. [SW03]).

Acknowlegdments. This work was substantially supported by the DFG Priority Program SPP 1095 *Analysis, Modeling and Simulation of Multiscale Problems* under Mi 459/3. The authors are grateful to W. Dreyer, G. Friesecke, T. Kriecherbauer, C. Patz, G. Schneider, C. Sparber, S. Teufel, F. Theil and H. Uecker for fruitful discussions.

References

[BCS01] A. M. Balk, A. V. Cherkaev, and L. I. Slepyan. Dynamics of chains with non-monotone stress-strain relations. I. Model and numerical experiments. II. Nonlinear waves and waves of phase transition. *J. Mech. Phys. Solids*, 49(1), 131–148, 149–171, 2001.

[BG02a] A. Braides and M. S. Gelli. Continuum limits of discrete systems without convexity hypotheses. *Math. Mech. Solids*, 7, 41–66, 2002.

[BG02b] A. Braides and M. S. Gelli. Limits of discrete systems with long-range interactions. *J. Convex Anal.*, 9, 363–399, 2002. Special issue on optimization (Montpellier, 2000).

[Ble05] T. Blesgen. Two-phase structures as singular limit of a one-dimensional discrete model. *Cubo*, 7(2), 69–79, 2005.

[Bol96] C. Boldrighini. Macroscopic limits of microscopic systems. *Rend. Mat. Appl. (7)*, 16, 1–107, 1996.

[BP06] D. Bambusi and A. Ponno. On metastability in FPU. *Commun. Math. Physics*, 264, 539–561, 2006.

[CMS04] R. Carles, P. A. Markowich, and C. Sparber. Semiclassical asymptotics for weakly nonlinear Bloch waves. *J. Statist. Phys.*, 117(1-2), 343–375, 2004.

[DHM06] W. Dreyer, M. Herrmann, and A. Mielke. Micro–macro transition for the atomic chain via Whitham's modulation equation. *Nonlinearity*, 19, 471–500, 2006.

[DHR06] W. Dreyer, M. Herrmann, and J. Rademacher. Wave trains, solitons and modulation theory in FPU chains. In "Analysis, Modeling and Simulation of Multiscale Problems, A. Mielke (edr), Springer-Verlag, 2006.".

[DK00] W. Dreyer and M. Kunik. Cold, thermal and oscillator closure of the atomic chain. *J. Phys. A*, 33(10), 2097–2129, 2000. (Corrigendum: J. Phys. A 33 (2000) 2458).

[DK*96] P. Deift, S. Kamvissis, T. Kriecherbauer, and X. Zhou. The Toda rarefaction problem. *Comm. Pure Appl. Math.*, 49(1), 35–83, 1996.

[DKV95] P. Deift, T. Kriecherbauer, and S. Venakides. Forced lattice vibrations. I, II. *Comm. Pure Appl. Math.*, 48, 1187–1249, 1251–1298, 1995.

[FJ00] G. Friesecke and R. D. James. A scheme for the passage from atomic to continuum theory for thin films, nanotubes and nanorods. *J. Mech. Phys. Solids*, 48(6-7), 1519–1540, 2000.

[FM02] G. Friesecke and K. Matthies. Atomic-scale localization of high-energy solitary waves on lattices. *Physica D*, 171, 211–220, 2002.

[FM03] G. Friesecke and K. Matthies. Geometric solitary waves in a 2D mass-spring lattice. *Discrete Contin. Dyn. Syst. Ser. B*, 3, 105–114, 2003.

[FP99] G. Friesecke and R. L. Pego. Solitary waves on FPU lattices: I. Qualitative properties, renormalization and continuum limit. *Nonlinearity*, 12, 1601–1627, 1999.

[FP02] G. Friesecke and R. L. Pego. Solitary waves on FPU lattices: II. Linear implies nonlinear stability. *Nonlinearity*, 15, 1343–1359, 2002.

[FP04a] G. Friesecke and R. L. Pego. Solitary waves on FPU lattices: III. Howland-type Floquet theory. *Nonlinearity*, 17, 207–227, 2004.

[FP04b] G. Friesecke and R. L. Pego. Solitary waves on FPU lattices: IV. Proof of stability at low energy. *Nonlinearity*, 17, 229–251, 2004.

[FPU55] E. Fermi, J. Pasta, and S. Ulam. Studies of nonlinear problems. Technical report, Los Alamos Scientific Laboratory of the University of California, May 1955. Report LA-1940.

[FT02] G. Friesecke and F. Theil. Validity and failure of the Cauchy–Born hypothesis in a two–dimensional mass–spring lattice. *J. Nonlin. Science*, 12, 445–478, 2002.

[FV99] A.-M. Filip and S. Venakides. Existence and modulation of traveling waves in particle chains. *Comm. Pure Appl. Math.*, 52(6), 693–735, 1999.

[FW94] G. Friesecke and J. A. D. Wattis. Existence theorem for solitary waves on lattices. *Comm. Math. Phys.*, 161(2), 391–418, 1994.

[Gér91] P. Gérard. Microlocal defect measures. *Comm. Partial Differential Equations*, 16(11), 1761–1794, 1991.

[GHM06] J. Giannoulis, M. Herrmann, and A. Mielke. Lagrangian and Hamiltonian redcution for lattice systems. *In preparation*, 2006.

[Gia06] J. Giannoulis. On the interaction of modulated pulses in discrete lattices. *In preparation*, 2006.

[GM04] J. Giannoulis and A. Mielke. The nonlinear Schrödinger equation as a macroscopic limit for an oscillator chain with cubic nonlinearities. *Nonlinearity*, 17, 551–565, 2004.

[GM06] J. Giannoulis and A. Mielke. Dispersive evolution of pulses in oscillator chains with general interaction potentials. *Discr. Cont. Dynam. Systems Ser. B*, 6, 493–523, 2006.

[GMMP97] P. Gérard, P. A. Markowich, N. J. Mauser, and F. Poupaud. Homogenization limits and Wigner transforms. *Comm. Pure Appl. Math.*, 50, 323–379, 1997. Erratum: Comm. Pure Appl. Math. 53 (2000) 280–281.

[GMS06] J. GIANNOULIS, A. MIELKE, and C. SPARBER. Interaction of modulated pulses in the nonlinear Schrödinger equation with periodic potential. *In preparation*, 2006.
[Her04] M. HERRMANN. *Ein Mikro-Makro-Übergang für die nichtlineare atomare Kette mit Temperatur.* PhD thesis, Institut für Mathematik, Humboldt-Universität zu Berlin, December 2004.
[HLM94] M. H. HAYS, C. D. LEVERMORE, and P. D. MILLER. Macroscopic lattice dynamics. *Physica D*, 79(1), 1–15, 1994.
[IJ05] G. IOOSS and G. JAMES. Localized waves in nonlinear oscillator chains. *Chaos*, 15, ?? 15 pp., 2005.
[IK00] G. IOOSS and K. KIRCHGÄSSNER. Travelling waves in a chain of coupled nonlinear oscillators. *Comm. Math. Phys.*, 211(2), 439–464, 2000.
[Ioo00] G. IOOSS. Travelling waves in the Fermi-Pasta-Ulam lattice. *Nonlinearity*, 13(3), 849–866, 2000.
[IZ05] L. IGNAT and E. ZUAZUA. Dispersive properties of a viscous numerical scheme for the Schrödinger equation. *C. R. Math. Acad. Sci. Paris*, 340, 529–534, 2005.
[Jam03] G. JAMES. Centre manifold reduction for quasilinear discrete systems. *J. Nonlinear Sci.*, 13, 27–63, 2003.
[Kon96] V. KONOTOP. Small-amplitude envelope solitons in nonlinear lattices. *Phys. Rev. E*, 53, 2843–2858, 1996.
[KSM92] P. KIRRMANN, G. SCHNEIDER, and A. MIELKE. The validity of modulation equations for extended systems with cubic nonlinearities. *Proc. Roy. Soc. Edinburgh Sect. A*, 122, 85–91, 1992.
[LP93] P.-L. LIONS and T. PAUL. Sur les mesures de Wigner. *Rev. Mat. Iberoamericana*, 9, 553–618, 1993.
[MA94] R. S. MACKAY and S. AUBRY. Proof of existence of breathers for time-reversible or Hamiltonian networks of weakly coupled oscillators. *Nonlinearity*, 7(6), 1623–1643, 1994.
[Mac04] F. MACIÀ. Wigner measures in the discrete setting: high-frequency analysis of sampling and reconstruction operators. *SIAM J. Math. Anal.*, 36(2), 347–383, 2004.
[MBL06] L. MUGNAI, T. BLESGEN, and S. LUCKHAUS. On gamma-limits of discrete free energy functionals. In "Analysis, Modeling and Simulation of Multiscale Problems, A. Mielke (edr), Springer-Verlag, 2006."
[Mie91] A. MIELKE. *Hamiltonian and Lagrangian flows on center manifolds.* With applications to elliptic variational problems. Lecture Notes in Mathematics Vol. 1489. Springer-Verlag, Berlin, 1991.
[Mie02] A. MIELKE. The Ginzburg–Landau equation in its role as a modulation equation. In B. Fiedler, editor, *Handbook of Dynamical Systems II*, pages 759–834. Elsevier Science B.V., 2002.
[Mie06a] A. MIELKE. Macroscopic behavior of microscopic oscillations in harmonic lattices using Wigner-Husimi measures. *Arch. Rat. Mech. Analysis*, 181, 401–448, 2006.
[Mie06b] A. MIELKE. Weak convergence methods for the micro-macro transition from from disrcete to continuous systems. *In preparation*, 2006.
[MMP94] P. A. MARKOWICH, N. J. MAUSER, and F. POUPAUD. A Wigner-function approach to (semi)classical limits: electrons in a periodic potential. *J. Math. Phys.*, 35, 1066–1094, 1994.

[MSU01] A. Mielke, G. Schneider, and H. Uecker. Stability and diffusive dynamics on extended domains. In B. Fiedler, editor, *Ergodic Theory, Analysis, and Efficient Simulation of Dynamical Systems*, pages 563–583. Springer–Verlag, 2001.

[Pat06] C. Patz. *Dispersive behavior in discrete lattices.* PhD thesis, In preparation, 2006.

[PP00] A. A. Pankov and K. Pflüger. Traveling waves in lattice dynamical systems. *Math. Meth. Appl. Sci.*, 23, 1223–1235, 2000.

[Sch94] G. Schneider. Error estimates for the Ginzburg-Landau approximation. *Z. angew. Math. Phys.*, 45(3), 433–457, 1994.

[Sch95] G. Schneider. Validity and limitation of the Newell-Whitehead equation. *Math. Nachr.*, 176, 249–263, 1995.

[Sch96] G. Schneider. Diffusive stability of spatial periodic solutions of the Swift-Hohenberg equation. *Comm. Math. Phys.*, 178(3), 679–702, 1996.

[Sch98] G. Schneider. Justification of modulation equations for hyperbolic systems via normal forms. *NoDEA Nonlinear Differential Equations Appl.*, 5(1), 69–82, 1998.

[Sch05a] A. Schlömerkemper. Mathematical derivation of the continuum limit of the magnetic force between two parts of a rigid crystalline material. *Arch. Ration. Mech. Anal.*, 176, 227–269, 2005.

[Sch05b] G. Schneider. Justification and failure of the nonlinear Schrödinger equation in case of non-trivial quadratic resonances. *J. Diff. Eqns.*, 216, 354–386, 2005.

[Spa06] C. Sparber. Effective mass theorems for nonlinear schrödinger equations. *SIAM J. Appl. Math.*, 66, 820–842, 2006.

[Spo91] H. Spohn. *Large scale dynamics of interacting particles.* Springer-Verlag, New York, 1991.

[Spo05] H. Spohn. The phonon Boltzmann equation, properties and link to weakly anharmonic lattice dynamics. *J. Stat. Physics*, 2005. To appear (arXiv:math-ph/0505025).

[SW00] G. Schneider and C. E. Wayne. Counter-propagating waves on fluid surfaces and the continuum limit of the Fermi-Pasta-Ulam model. In B. Fiedler, K. Gröger, and J. Sprekels, editors, *International Conference on Differential Equations*, volume 1, pages 390–404. World Scientific, 2000.

[SW03] G. Schneider and C. E. Wayne. Estimates for the three-wave interaction of surface water waves. *European J. Appl. Math.*, 14, 547–570, 2003.

[TP04] S. Teufel and G. Panati. Propagation of Wigner functions for the Schrödinger equation with a perturbed periodic potential. In *Multiscale methods in quantum mechanics*, Trends Math., pages 207–220. Birkhäuser Boston, 2004.

[Whi74] G. B. Whitham. *Linear and nonlinear waves.* Wiley-Interscience [John Wiley & Sons], New York, 1974. Pure and Applied Mathematics.

[Zua05] E. Zuazua. Propagation, observation, control and numerical approximation of waves approximated by finite difference method. *SIAM Review*, 47, 197–243, 2005.

Wave Trains, Solitons and Modulation Theory in FPU Chains

Wolfgang Dreyer[1], Michael Herrmann[2], and Jens D.M. Rademacher[1]

[1] Weierstraß-Institut für Angewandte Analysis und Stochastik, Berlin.
dreyer@wias-berlin.de, rademach@wias-berlin.de
[2] Institut für Mathematik, Humboldt-Universität zu Berlin.
michaelherrmann@math.hu-berlin.de

Summary. We present an overview of recent results concerning wave trains, solitons and their modulation in FPU chains. We take a thermodynamic perspective and use hyperbolic scaling of particle index and time in order to pass to a macroscopic continuum limit. While strong convergence yields the well-known p-system of mass and momentum conservation, we generally obtain a weak form of it in terms of Young measures. The modulation approach accounts for microscopic oscillations, which we interpret as temperature, causing convergence only in a weak, average sense. We present the arising Whitham modulation equations in a thermodynamic form, as well as analytic and numerical tools for the resolution of the modulated wave trains. As a prototype for the occurrence of temperature from oscillation-free initial data, we discuss various Riemann problems, and the arising dispersive shock fans, which replace Lax-shocks. We predict scaling and jump conditions assuming a generic soliton at the shock front.

1 Introduction

We consider chains of N identical particles as plotted in Fig. 1.1, nearest neighbor coupled in a nonlinear potential $\Phi : \mathbb{R} \to \mathbb{R}$ by Newton's equations

$$\ddot{x}_\alpha = \Phi'(x_{\alpha+1} - x_\alpha) - \Phi'(x_\alpha - x_{\alpha-1}), \tag{1.1}$$

where $\dot{} = \frac{\mathrm{d}}{\mathrm{d}t}$ is the time derivative, $x_\alpha(t)$ the atomic position, and $\alpha = 1, \ldots, N$ the particle index. Since the work of Fermi, Pasta and Ulam [FPU55] one usually refers to (1.1) as FPU chains.

We mainly consider general, convex potentials Φ. While our focus lies on nonlinear Φ', the harmonic potential with linear forces is an instructive, completely integrable example. A nonlinear example, but still completely integrable, is the famous Toda chain, see [Tod70, DK*74, Hén74] with potential

$$\Phi(r) = \exp(1 - r) - (1 - r). \tag{1.2}$$

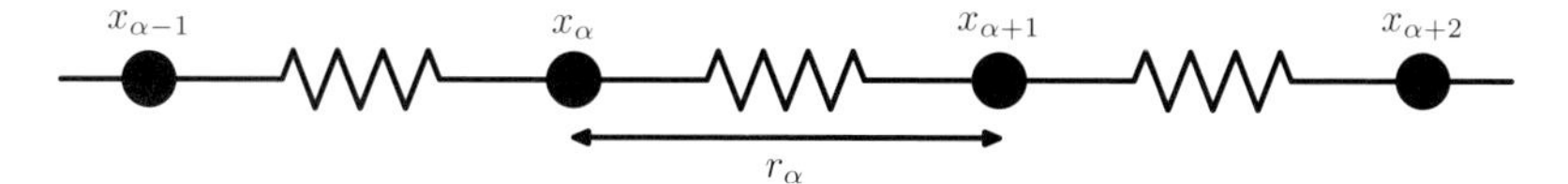

Fig. 1.1. The atomic chain with nearest neighbour interaction.

For our purposes it is convenient to use the atomic distances $r_\alpha = x_{\alpha+1} - x_\alpha$ and velocities $v_\alpha = \dot{x}_\alpha$ as the basic variables, changing (1.1) to the system

$$\dot{r}_\alpha = v_{\alpha+1} - v_\alpha\,, \qquad \dot{v}_\alpha = \Phi'(r_\alpha) - \Phi'(r_{\alpha-1}). \tag{1.3}$$

Rather than investigating solutions of (1.3) for finite N, we focus on the thermodynamic limit $\varepsilon = 1/N \to 0$ in the *hyperbolic* scaling of the *microscopic* coordinates t and α, which is defined by the *macroscopic* time $\overline{t} = \varepsilon t$ and particle index $\overline{\alpha} = \varepsilon\alpha$. It is natural to scale the atomic positions in the same way, i.e. $\overline{x} = \varepsilon x$, which leaves atomic distances and velocities scale invariant. For a survey on other reasonable scaling we refer to Sect. 2 and [GHM06a].

Our main goal is to derive a micro-macro transition for the atomic chain, i.e. we aim to replace the high dimensional ODE (1.1) by a few macroscopic PDEs. The derivation of such a continuum limit is simple as long as the atomic data vary on the macroscopic scale only, see (1.9) below. If, however, the atomic data oscillate on the microscopic scale, the problem is tremendously complicated, because then distances and velocities *do not* converge to macroscopic functions. Instead, at each point in the macroscopic space-time, the local (r, v)-distribution converges to a nontrivial Young-measures, see Sect. 5. We interpret the microscopic oscillations as a form of temperature in the chain, see Sect. 2, and refer to oscillation-free limits as *cold*.

The main problem in the case of temperature is to find an appropriate description for the structure and evolution of the oscillations. Even if we are interested in the macroscopic behavior of averaged quantities only, the microscopic oscillations determine the evolution of the *internal energy*, that is, the amount of energy which is stored in purely microscopic motion. In other words, any reasonable macroscopic limit for oscillatory solutions needs to describe thermodynamic effects, such as creation of temperature and transport of heat. Unfortunately, no rigorous theory is known that applies without further assumptions.

Numerical simulations as discussed in Sect. 5 and Sect. 6, as well as rigorous results for the Toda chain, cf. [HFM81, Kam91, VDO91, DK*96, DM98], suggest that for certain solutions of (1.1), the arising microscopic oscillations can be described by *modulated traveling waves*. Traveling waves are highly oscillatory exact solutions of (1.1). The most relevant kind for our purposes are *wave trains* which are periodic functions of a single phase variable, depending on four characteristic parameters. Modulated wave trains arise when these parameters vary on the macroscopic scale.

A characteristic property of wave trains is that the arising Young measures are supported on closed curves. As a consequence, they correspond to a very special kind of temperature which is not related to our usual notion of thermalization. However, they give rise to a thermodynamically consistent macroscopic theory involving temperature, entropy, and so on. Moreover, if cold initial data form macroscopic shocks, then Newton's equations self-organize into microscopic oscillations in form of modulated wave trains, and in this sense our notion of temperature turns out to be generic.

Some aspects of the problems addressed in this article have much in common with certain *zero dispersion limits*, which we will briefly discuss next to illustrate our point of view. This is not to be confused with so called zero diffusive-dispersive limits, where diffusive effects prevail, cf. e.g. [Sch82, KL02]. The most prominent example is Burger's equation

$$\partial_{\overline{t}}\, u + u\, \partial_{\overline{\alpha}}\, u = 0, \tag{1.4}$$

which, on a formal level, is the zero dispersion limit of the KdV equation

$$\partial_{\overline{t}}\, u + u\, \partial_{\overline{\alpha}}\, u + \varepsilon\, \partial_{\overline{\alpha}}^{3}\, u = 0. \tag{1.5}$$

The main question is under which conditions the solutions of (1.5) converge as $\varepsilon \to 0$ to (weak) solutions of (1.4). The rigorous theory for this problem was developed by Lax and Levermore in [LL83, Ven85] by relying on the complete integrability of (1.5).

It is well known that generic initial data $u_{\rm ini}$ yield a critical time $\overline{t}_{\rm crit}$ such that (1.4) has a unique smooth solution for $0 < \overline{t} < \overline{t}_{\rm crit}$, but for $\overline{t} > \overline{t}_{\rm crit}$ solutions exist in a weak sense only, having at least one discontinuity at some $\overline{\alpha}_{\rm crit}$, and satisfying $\partial_{\overline{t}}\, u + \frac{1}{2}\partial_{\overline{\alpha}}\left(u^2\right) = 0$ in the sense of distributions.

Imposing the same initial datum $u_{\rm ini}$ for KdV, the typical behavior for $\varepsilon \to 0$ is as follows, see for instance the surveys in [Lax86, Lax91, LLV93]. For $0 < \overline{t} < \overline{t}_{\rm crit}$ the solutions u_ε of (1.5) converge strongly to the unique smooth solution of (1.4). However, for $\overline{t} > \overline{t}_{\rm crit}$ the KdV-solutions become highly oscillatory in a neighborhood of $\overline{\alpha}_{\rm crit}$ with wavelength $1/\sqrt{\varepsilon}$, and converge to a weak limit $\langle u \rangle$ only. The main point is that $\langle u \rangle$ does *not* satisfy Burgers equation, i.e. $\partial_{\overline{t}}\, \langle u \rangle + \frac{1}{2}\partial_{\overline{\alpha}}\, \langle u \rangle^2 \neq 0$, because $\langle u \rangle^2 \neq \langle u^2 \rangle$.

A discrete zero dispersion limit was studied in [GL88], replacing (1.4) by

$$\dot{u}_\alpha + \tfrac{1}{2}\, u_\alpha \left(u_{\alpha+1} - u_{\alpha-1}\right) = 0. \tag{1.6}$$

This scheme is equivalent to a dispersive spatial discretization of (1.4), because the identification $u_\alpha(t) = u(\varepsilon t,\, \varepsilon\alpha)$ transforms (1.6) into $\partial_{\overline{t}}\, u + u\, \nabla^{\pm\varepsilon} u = 0$ with $\nabla^{\pm\varepsilon} u(\overline{t}, \overline{\alpha}) = \big(u(\overline{t}, \overline{\alpha} + \varepsilon) - u(\overline{t}, \overline{\alpha} - \varepsilon)\big)/2\varepsilon$, The numerical study in [GL88] provides evidences for the same qualitative limiting behavior as for KdV. Further examples for numerical schemes with interesting zero dispersion limit can be found in [HL91] and [LL96].

Towards modulation theory, [GL88] found a simple description for *modulated binary oscillations*, which provides an *approximate* solution of (1.6) satisfying $u_\alpha \approx v_\alpha + (-1)^\alpha w_\alpha$, where v_α and w_α vary on the macroscopic scale only. The modulation equations for binary oscillations read

$$\partial_{\overline{t}} a + a\, \partial_{\overline{\alpha}} b = 0, \quad \partial_{\overline{t}} b + b\, \partial_{\overline{\alpha}} a = 0. \tag{1.7}$$

where $a = v + w$ and $b = v - w$. This system is strictly hyperbolic if and only if a and b have the same sign, and conservation laws for $\ln a$ and $\ln b$ imply that strictly positive initial data stay positive for all times.

Let $a = a(\overline{t}, \overline{\alpha})$ and $b = b(\overline{t}, \overline{\alpha})$ be a smooth solution of (1.7) defined until $\overline{t}_{\mathrm{crit}}$, and denote the corresponding modulated binary oscillations by

$$u_\alpha^{\mathrm{mod}}(t) = v(\varepsilon t,\, \varepsilon\alpha) + (-1)^\alpha\, w(\varepsilon t,\, \varepsilon\alpha). \tag{1.8}$$

It is proven in [GL88] that (1.8) indeed yields approximate solutions of the microscopic system for $t < \varepsilon^{-1}\overline{t}_{\mathrm{crit}}$ in the sense that $u_\alpha^{\mathrm{mod}}(t) - u_\alpha(t)$ converges to zero as $\varepsilon \to 0$ for $\overline{t} < \overline{t}_{\mathrm{crit}}$ if $u_\alpha(0) = u_\alpha^{\mathrm{mod}}(0)$. For larger times we expect that modulated binary oscillations are not longer close to an exact solution.

Returning to the atomic chain (1.1), we next derive the macroscopic evolution of *cold* data, i.e. we assume macroscopic fields $r(\overline{t}, \overline{\alpha})$ and $v(\overline{t}, \overline{\alpha})$ such that $r_\alpha(t) = r(\varepsilon t,\, \varepsilon\alpha)$, $v_\alpha(t) = v(\varepsilon t,\, \varepsilon\alpha)$. Substitution into (1.3) yields

$$\partial_{\overline{t}} r - \nabla^{+\varepsilon} v \; = \; 0, \quad \partial_{\overline{t}} v - \nabla^{-\varepsilon} \Phi'(r) \; = \; 0, \tag{1.9}$$

In the limit $\varepsilon \to 0$ we formally obtain the so called *p-system* consisting of the macroscopic conservation laws for mass and momentum given by

$$\partial_{\overline{t}} r - \partial_{\overline{\alpha}} v = 0, \quad \partial_{\overline{t}} v - \partial_{\overline{\alpha}} \Phi'(r) = 0. \tag{1.10}$$

It is well known that, for convex Φ, the p-system is hyperbolic and that for smooth solutions the energy is conserved according to

$$\partial_{\overline{t}} \left(\tfrac{1}{2} v^2 + \Phi(r)\right) - \partial_{\overline{\alpha}} \left(v\, \Phi'(r)\right) = 0. \tag{1.11}$$

In analogy to the previous discussion, the p-system describes the thermodynamic limit for cold atomic data as long as these data are smooth on the macroscopic scale. However, we will show next that, if the nonlinearity forms a shock, then the p-system is no longer a thermodynamically consistent model for the macroscopic evolution. For simplicity, we assume that the flux function Φ' is convex so that all eigenvalues of (1.10) are genuinely nonlinear. According to the Lax theory of hyperbolic system, cf. [Smo94, Daf00, LeF02], a shock wave propagates with a constant shock speed c so that r and v satisfy the Rankine-Hugoniot jump conditions across the shock given by

$$-c[\![r]\!] - [\![v]\!] = 0, \quad -c[\![v]\!] - [\![\Phi'(r)]\!] = 0, \tag{1.12}$$

where $[\![\cdot]\!]$ denotes the jump. However, (1.12) implies that the jump condition for the energy must be violated, i.e. for shocks with (1.12) we have

$$-c[\![\frac{1}{2}v^2 + \Phi(r)]\!] - [\![v\,\Phi'(r)]\!] \neq 0.$$

Consequently, the p-system predicts some production for the macroscopic energy (the Lax criterion selects only shocks with negative production).

In contrast, Newton's equations always conserve mass, momentum *and* energy. Therefore, the p-system cannot describe the thermodynamic limit beyond the shock at which the atomic data start to oscillate. Indeed, some amount of energy is dissipated into *internal energy* leading to a *dispersive shock fan.* It is one of the merits of modulation theory that it can describe the microscopic oscillations emerging from cold shocks as discussed in Sect. 6.

The article is organized as follows. In Sect. 2: we briefly sketch the thermodynamical framework. We survey some existence and approximation results of *wave trains* and *solitons* in Sect. 3, including multi-phase wave trains, thermodynamic properties and new a priori estimates. Section 4 gives a brief overview on Whitham's modulation theory applied to FPU chains, leading to a system of four conservation laws for wave train parameters. In Sect. 5 we briefly summarize some aspects of numerical justification by evaluating the aforementioned Young measures and testing assumptions of modulation theory. The shock problem for cold Riemann data is discussed in Sect. 6, and we characterize the behavior of all macroscopic fields at the shock front by assuming that is consists of a generic soliton.

2 Thermodynamic framework

Thermodynamics describes the evolution of deformation and heat on the macroscopic scale in a body, which may be isolated from the surroundings or is subjected to external supply of mechanical forces and heat. In the following we will illustrate the strategy of thermodynamics for a macroscopic body in one space dimension, that is microscopically constituted by an atomic chain. To this end thermodynamics considers, at any *Lagrangian space-time point* $(\bar{t}, \bar{\alpha})$ a certain number of *specific densities* $u_j(\bar{t}, \bar{\alpha})$, $j = 1...M$, and determines these fields by means of a system of PDEs for given initial and boundary data. The most important densities in 1D are the specific volume (mean distance) r, the macroscopic velocity v, and the specific total energy $E = \frac{1}{2}v^2 + U$, uniquely decomposed into kinetic energy and specific internal energy U.

The PDE system relies on M equations of balance that read in regular points

$$\partial_{\bar{t}} u_j + \partial_{\bar{\alpha}} F_j = P_j, \quad j \in \{1, 2, \ldots, M\}, \tag{2.1}$$

where f_j and P_j are called *fluxes* and *productions*, respectively. The fundamental balance equations are the conservation laws for mass, momentum and energy given by

$$\partial_{\bar{t}} r - \partial_{\bar{\alpha}} v = 0\,, \quad \partial_{\bar{t}} v + \partial_{\bar{\alpha}} p = 0\,, \quad \partial_{\bar{t}} E + \partial_{\bar{\alpha}} f = 0, \tag{2.2}$$

where p denotes the pressure and f is the energy flux, satisfying $f = pv + q$ with heat flux q. Further conservation laws are possible, but those are material and process dependent.

In order that (2.1) becomes a closed system for the variables, thermodynamics has to model constitutive equations that relate, in a material dependent manner, the fluxes and productions to the densities themselves and/or their time and space derivatives. The generality of the constitutive functions is restricted by universal principles like Galileian invariance and the *entropy principle*, and by material dependent symmetry principles. The entropy principle consists of several parts, see [MR98] for more details.

1. There exists an entropy pair (S, g), given by (material dependent) constitutive functions in terms of the densities u_j, so that the entropy density S is a concave function.
2. The constitutive laws closing (2.1) yield a further balance equation

$$\partial_{\bar{t}} S + \partial_{\bar{\alpha}} g = \Sigma \geq 0 \quad \text{with} \quad \Sigma = 0 \Longleftrightarrow P_j = 0, \tag{2.3}$$

   where Σ denotes the non-negative *entropy production.*
3. The definition of (absolute) temperature T_{macro} is given by

$$T_{\text{macro}} = \frac{\partial U}{\partial S}. \tag{2.4}$$

   Note that this phenomenological definition is a priori unrelated to any microscopic model.
4. The law of Clausius-Duhem holds, i.e. $T_{\text{macro}} g = q$.

This abstract framework is the basic paradigm of Rational Thermodynamics and assumed to hold in all cases. However, the constitutive laws depend on the chosen macroscopic scaling and are generally unknown. Nevertheless, the scaling predicts the way in which the fluxes (and productions) can and cannot depend on the densities. For instance, in the hyperbolic scaling, the macroscopic equations must be invariant (to leading order) under $(\bar{t}, \bar{\alpha}) \mapsto (\lambda\bar{t}, \lambda\bar{\alpha})$, whereas the parabolic scaling $(\bar{t}, \bar{\alpha}) = (\varepsilon^2 t, \varepsilon\alpha)$ implies macroscopic invariance under $(\bar{t}, \bar{\alpha}) \mapsto (\lambda^2\bar{t}, \lambda\bar{\alpha})$.

Therefore, in the hyperbolic scaling all constitutive relations for the fluxes must be local, i.e. F_j depends pointwise on the densities u_j, so that (2.2) is a first order system. We mention that hyperbolicity of (2.1) is guaranteed if the entropy S is a concave function, see [MR98]. Generally, for the hyperbolic scaling, all constitutive relations can be encoded in a Gibbs equation with a single thermodynamic potential. On the other hand, for parabolic scaling, we expect that the fluxes depend (mainly) on the spatial derivatives of the densities. In the simplest case the energy flux f is given by Fourier's law, i.e. $f = \partial_{\bar{\alpha}} U$, so that the energy balance leads to the heat equation.

In conclusion, we sketch the macroscopic thermodynamics for the atomic chain (1.1) as it results from modulated wave trains. It turns out that the macroscopic system (2.1) consists of the three fundamental and a fourth equation, the *conservation of wave number* $\partial_{\bar{t}} k - \partial_{\bar{\alpha}} \omega = 0$, with wave number k and frequency ω. In addition, there is a fifth conservation law for the entropy S, i.e. $\Sigma = 0$ in (2.3), and all fluxes are given by the thermodynamic potential $U = U(r, k, S)$ through the Gibbs equation

$$\mathrm{d}E = \omega\,\mathrm{d}S - p\,\mathrm{d}r - g\,\mathrm{d}k + v\,\mathrm{d}v. \tag{2.5}$$

Note that the equation of state depends on the chain, i.e. on the potential Φ, whereas (2.5) is universal. From (2.4) and (2.5), we infer that the macroscopic temperature T_{macro} equals the wave train frequency ω. Interestingly, here there is a difference between T_{macro} and the kinetic temperature defined as the mean kinetic energy of the atoms, see Sect. 3. However, it turns out that the Clausius-Duhem law is satisfied, i.e. we find $q = \omega g$.

3 Traveling waves

Traveling waves are exact solutions of the infinite chain (1.1) for $N = \infty$ of the form $x_\alpha(t) = x(\alpha - ct)$ depending on a single phase variable ϕ and traveling with a constant speed c. In the context of the macroscopic limits that we consider, relevant traveling waves are solitons, which vanish as $\phi \to \pm\infty$, and wave trains, which are periodic in ϕ. Due to Galilean invariance, we can allow additional drift in space-time of the form

$$x_\alpha(t) = r\alpha + vt + y_\alpha(t),$$

where the profile $y_\alpha(t)$ solves the modified lattice equations

$$\ddot{y}_\alpha = \Phi'(r + y_{\alpha+1} - y_\alpha) - \Phi'(r + y_\alpha - y_{\alpha-1}) \tag{3.1}$$

and traveling waves $y_\alpha(t) = \mathbb{Y}(\alpha - ct)$ solve the second order advance-delay differential equation

$$c^2 \partial_{\phi\phi} \mathbb{Y}(\phi) = \Phi'(r + \mathbb{Y}(\phi + 1) - \mathbb{Y}(\phi)) - \Phi'(r + \mathbb{Y}(\phi) - \mathbb{Y}(\phi - 1)). \tag{3.2}$$

3.1 Wave trains

Normalizing the period of wave trains to 1 and using $c = \omega/k$ with wave number k and frequency ω, we obtain the form

$$x_\alpha(t) = r\alpha + vt + \mathbb{X}(k\alpha + \omega t), \tag{3.3}$$

where $\mathbb{X}(\varphi)$ is the 1-periodic *wave profile* function. There are unique choices for the average distance r and the average velocity v such that $\int_0^1 \mathbb{X}(\varphi)\mathrm{d}\varphi = 0$. Upon substitution into Newton's equations, we obtain the analogue of (3.2)

$$\omega^2 \partial_{\varphi\varphi}\mathbb{X} = \Phi'(r + \mathbb{X}(\varphi + k) - \mathbb{X}(\varphi)) - \Phi'(r + \mathbb{X}(\varphi) - \mathbb{X}(\varphi - k)), \tag{3.4}$$

with the three parameters r, k, ω. Another useful formulation is the fixed point equation, or nonlinear eigenvalue problem, for $\mathbb{V} = \partial_\varphi \mathbb{X}$

$$\omega^2 \mathbb{V} = \mathcal{F}(\mathbb{V}) := \hat{A}_k \partial\Phi\left(r + \hat{A}_k \mathbb{V}\right), \tag{3.5}$$

where the operator $\hat{A}_k$ and the Nemyckii operator $\partial\Phi$ are defined by

$$(\hat{A}_k\mathbb{V})(\varphi) := A_k\mathbb{V}(\varphi) - k\int_0^1 \mathbb{V}(s)\mathrm{d}s\,, \quad A_k\mathbb{V}(\varphi) := \int_{\varphi-k/2}^{\varphi+k/2} \mathbb{V}(\tau)\mathrm{d}\tau$$

$$\partial\Phi(\mathbb{V})(\varphi) := \Phi'(\mathbb{V}(\varphi)).$$

Distances and velocities of the microscopic wave trains are then

$$r_\alpha(t) = r + A_k\mathbb{V}(k\alpha + \omega t + k/2)\,, \quad v_\alpha(t) = v + \omega\mathbb{V}(k\alpha + \omega t). \tag{3.6}$$

Existence and approximation of wave trains

We give an overview of the variational approach to wave train existence and approximation by numerical schemes that are based on maximizing

$$\mathcal{W}(\mathbb{V}) = \int_0^1 \Phi\left(r + \hat{A}_k\mathbb{V}\right)\mathrm{d}\varphi, \ \mathbb{V} \in H_\gamma := \left\{\mathbb{V} \in L^2([0,1]) : \frac{1}{2}\|\mathbb{V}\|_{L^2}^2 \le \gamma\right\}.$$

Problem 3.1. For given parameters r, k and $\gamma > 0$ we seek maximizers of $\mathcal{W}$ in H_γ, i.e. we solve $W(r, k, \gamma) = \max_{\mathbb{V}\in H_\gamma} \mathcal{W}(\mathbb{V})$.

Theorem 3.2. *Problem 3.1 always has a solution. In particular, there exists a maximizer* $\mathbb{V}$ *with* $\|\mathbb{V}\|_{L^2} = \sqrt{2\gamma}$ *together with a positive Lagrangian multiplier* $\omega_1^2 > 0$ *such that* $\mathbb{V}$ *and* ω^2 *solve* (3.5).

Scheme 3.3. *Let any parameter set for problem 3.1 be given, and let* $\mathbb{V}_0 \in H_\gamma$ *be an arbitrary initial datum with* $\hat{A}_k\mathbb{V}_0 \neq 0$. *Then we define inductively two sequences* $(\mathbb{V}_n)_{n\in\mathbb{N}} \subset H_\gamma$ *and* $(\omega_n)_{n\in\mathbb{N}}$ *by the following iteration step*

$$\mathbb{V}_{n+1} = \frac{1}{f_n}\mathbb{W}_n\,, \quad \mathbb{W}_n = \mathcal{F}\mathbb{V}_n\,, \quad f_n = \frac{\|\mathbb{W}_n\|_{L^2}}{\sqrt{2\gamma}}\,, \quad \omega_{n+1} = \sqrt{f_n}.$$

In [Her04] it is proved that this scheme is compact, and numerical simulations indicate that Scheme 3.3 converges.

Remark 3.4. In fact, Theorem 3.3.2 in [Her04] proves that every closed cone of functions that is invariant under $\mathcal{F}$ contains at least one traveling wave. For the cone of even monotone functions used below, this was also shown in [FV99], Theorem 2.14.

By means of Scheme 3.3 we can compute wave trains with prescribed parameter $\gamma = \frac{1}{2} \parallel \mathbb{V} \parallel_{L^2}^2$. There are variants of 3.3 which allow to prescribe either the entropy S or the temperature T of a wave train (for the definition of S and T see Sect. 3.1 below). Hence, wave trains are parametrized (at least) by (k, r, γ), as well as trivially by v; the latter is relevant for the modulation equations discussed in Sect. 4. On may view the parameter ω of the wave train equation (3.4) depending on (r, k, γ) via a 'dispersion relation', here expressed as the Lagrange multiplier. We emphasize, that it is not known whether the set of wave trains is a smooth three-dimensional manifold of orbits; note that phase shifts $\mathbb{V}(\cdot + s)$ trivially give rise to an (at least) one-dimensional kernel of the linearization $\omega^2 - D\mathcal{F}(\mathbb{V})$ of (3.5) spanned by $\partial_\varphi \mathbb{V}$. Moreover, for given parameters there is a discrete multiplicity of solutions, because solutions for mk, $m \in \mathbb{N}$, are solutions for k as well, though these do not have minimal period 1. We conjecture that wave trains are unique in cones defined by monotonicity properties of $\mathbb{V}$ as discussed below.

Existence and approximation of multi-phase wave trains

We present new results concerning the existence of multi-phase waves, which will be published with full details elsewhere. As before, our variational approach is essentially restricted to convex interaction potentials Φ, but allows for arbitrary large amplitudes.

For simplicity we consider only two-phase wave trains having two wave numbers k_1 and k_2. However, all results can easily be generalized to other multi-phase wave trains. Moreover, to avoid technicalities we always suppose that Φ is defined on the whole real axis with bounded and continuous second derivative Φ''.

A *two-phase wave train* is an exact solution of Newton's equations satisfying

$$x_\alpha(t) = r\alpha + vt + \mathbb{X}(k_1\alpha + \omega_1 t,\, k_2\alpha + \omega_2 t). \tag{3.7}$$

Here r, v, k_1, k_2, ω_1 and ω_2 are given parameters, and the profile function $\mathbb{X}$ is assumed to have zero average and be 1-periodic with respect to each phase variable $\varphi_i = k_i\alpha + \omega_i t$. The ansatz (3.7) gives rise to the advance-delay differential equation

$$\big(\omega_1^2\, \partial_{\varphi_1}^2 + \omega_2^2\, \partial_{\varphi_2}^2\big)\mathbb{X} = \nabla^- \partial\Phi\big(r + \nabla^+\mathbb{X}\big) \tag{3.8}$$

where $\nabla^\pm$ are difference operators defined by

$$\big(\nabla^\pm\mathbb{X}\big)(\varphi_1, \varphi_2) := \pm\mathbb{X}(\varphi_1 \pm k_1, \varphi_2 \pm k_2) \mp \mathbb{X}(\varphi_1, \varphi_2)$$

Our aim is to identify an optimization problem with a single scalar constraint such that (3.8) is equivalent to the corresponding Euler-Lagrange equation with multiplier ω_1^2. Consequently, we regard the ratio $\beta = \omega_2/\omega_1$ as parameter of this problem.

Let $\mathcal{T}_2 \cong [0,1] \times [0,1]$ be the two dimensional torus with its canonic Lebesgue measure, and let all function spaces which follow be defined on $\mathcal{T}_2$. We consider the Sobolev space

$$H_0^1 = \left\{ X \in H^1 : \int_{\mathcal{T}_2} X = 0 \right\}, \quad \| X \|_{H_0^1}^2 := \int_{\mathcal{T}_2} (\partial_{\varphi_1} \mathbb{X})^2 + \beta^2 (\partial_{\varphi_2} \mathbb{X})^2. \quad (3.9)$$

This norm is equivalent to the standard norm on H_0^1 as long as $\beta \neq 0$. Let E be the canonic embedding $E : H_0^1 \to L^2$, and E^* its adjoint operator $E^* : L^2 \to H^{-1} = \left(H_0^1\right)^*$. Note that E is compact, and that here we have identified L^2 with its dual L^{2*}. By $\triangle$ we denote the Laplace operator corresponding to (3.9), i.e.

$$\triangle := \partial_{\varphi_1}^2 + \beta^2 \, \partial_{\varphi_2}^2 .$$

Recall that $-\triangle : H_0^1 \to H^{-1}$ is an isometric isomorphism between Hilbert spaces, and that the difference operators $\nabla^{\pm} : L^2 \to L^2$ are continuous with $\left(\nabla^+\right)^* = -\nabla^-$. Moreover, our assumptions on Φ imply that the convex functional $\mathbb{X} \mapsto \int_{\mathcal{T}_2} \Phi(\mathbb{X})$ is well defined and continuous on L^2.

The spaces and operators from above allow to regard the wave train equation as an equation in H^{-1}. In particular, (3.8) is equivalent to

$$-\omega_1^2 \triangle X = E^* \left(\nabla^+\right)^* \partial\Phi\left(r + \nabla^+ E \, \mathbb{X}\right), \quad (3.10)$$

where the Nemyckii operator $\partial\Phi : L^2 \to L^2$ with $\mathbb{X} \mapsto \Phi'(\mathbb{X})$ is the Gateaux differential of the functional (3.11). For fixed $\gamma > 0$ we define the closed convex set $H_\gamma \in L^2$ and the convex Gateaux differentiable functional $\mathcal{W}$ as follows

$$H_\gamma = \left\{ X \in H_0^1 \; : \; \tfrac{1}{2} \, \| X \|_{H_0^1}^2 \leq \gamma \right\}, \quad \mathcal{W}(\mathbb{X}) := \int_{\mathcal{T}_2} \Phi\left(\nabla^+ E \, \mathbb{X}\right). \quad (3.11)$$

Now (3.10) yields the following constrained optimization problem.

Problem 3.5. For given parameters r, k_1, k_2, $\gamma > 0$ and $\beta \neq 0$ we seek maximizers of $\mathcal{W}$ in H_γ, i.e. we solve

$$W(r, k_1, k_2, \beta, \gamma) = \max_{\mathbb{X} \in H_\gamma} \mathcal{W}(\mathbb{X}).$$

Theorem 3.6. *Problem 3.5 always has a solution. In particular, there exists a maximizer* $\mathbb{X}$ *with* $\| \mathbb{X} \|_{H_0^1} = \sqrt{2\gamma}$ *together with a positive Lagrangian multiplier* $\omega_1^2 > 0$ *such that* $\mathbb{X}$ *and* ω_1^2 *solve* (3.10).

Remark 3.7. By construction, (3.10) is an identity in H^{-1}. However, $\mathbb{X} \in H^1$ implies that the right hand side of (3.10) is again an element of H^1, and the theory of elliptic regularity provides $\mathbb{X} \in H^3$. Moreover, we can prove further regularity by exploiting Sobolev's embedding theorems.

In analogy to the single-phase wave trains, we can solve the optimization problem 3.5 iteratively using an adapted abstract approximation scheme.

Geometry and phase velocity of wave trains

Since the modulation equations for the macroscopic limit of the chain depend on wave trains, it is essential to understand properties of wave trains and their parameter variation. Motivated by numerical simulations, we investigate geometric properties of wave trains in the phase space of distances and velocities. With the shock problem in mind, see Sect. 6, we are also interested in the transition to solitons as the wave number tends to zero.

From (3.5) we infer that if $(\mathbb{V}, \omega, k, r)$ is a solution to (3.5), then $(-\mathbb{V}, \omega, 1-k, r)$ is also a solution, and vice versa: $\mathbb{V}(\varphi; k) = -\mathbb{V}(\varphi + 1/2; 1-k)$. In case of a binary oscillation, $k = 1/2$, the symmetry implies that (3.4) reduces to the planar Hamiltonian ODE

$$\omega^2 \partial_{\varphi\varphi}\mathbb{X} = \Phi'(r - 2\mathbb{X}) - \Phi'(r + 2\mathbb{X}). \tag{3.12}$$

More generally, for rational $k = n/m$ equation (3.4) can be written as an m-dimensional second order Hamiltonian ODE with components $\mathbb{X}_j = \mathbb{X}(\cdot + jn/m)$, $j = 0, \dots m-1$. This system is equivariant under the $\mathbb{Z}_m$ action $\mathbb{X}_j \to \mathbb{X}_{j+1}$, where indices are taken modulo m, and $\mathbb{Z}_m$ lies in the isotropy subgroup of wave trains.

The microscopic phase space of distances and velocities is in fact the phase space of the ODE (3.12) for $k = 1/2$. Therefore, the orbits

$$Q := \{(r + A_k\mathbb{V}(\varphi + k/2), v + \omega\mathbb{V}(\varphi)) \,|\, 0 \le \varphi < 1\} \tag{3.13}$$

are convex, non self-intersecting curves and nested for different ω with fixed (r, k).

We can prove some of these properties for general wave number k, see also Fig. 5.4, and define the positive cones

$$\begin{aligned}\mathcal{M}_\pm := \{\mathbb{V}(1+\varphi) = \mathbb{V}(\varphi),\ \int_0^1 \mathbb{V}(s)\mathrm{d}s = 0,\ \mathbb{V}(\varphi) = \mathbb{V}(-\varphi),\\ \operatorname{sgn}(\mathbb{V}(\varphi_1) - \mathbb{V}(\varphi_2)) = \pm 1,\ 0 < \varphi_1 < \varphi_2 < 1/2\},\end{aligned}$$

so that $\mathbb{W} \in \mathcal{M}_\pm$ has unique global extrema at $\varphi = 0$ and $\varphi = 1/2$, and $\mathbb{W}(\varphi_1) = \mathbb{W}(\varphi_2)$ is equivalent to $\varphi_1 \in \{\varphi_2, -\varphi_2, 1 - \varphi_2\}$. By symmetry $\mathbb{W} \in \mathcal{M}_\pm$, implies $\mathbb{W}(\cdot + 1/2) \in \mathcal{M}_\mp$. The basis of the following results is the observation $\mathcal{F} : \overline{\mathcal{M}_\pm} \to \mathcal{M}_\pm$, which was noted in [FV99]. Throughout this article, we are only interested in wave trains in $\mathcal{M}_+ \cup \mathcal{M}_-$, and conjecture that wave trains are unique within these cones.

Let $Q_\Phi = \{\Phi''(r + A_k\mathbb{V}(\varphi)) \mid 0 < \varphi \leq 1\}$. We will estimate the phase velocity $c_{\rm ph} := \omega/k$ of wave trains and the size of Q in terms of

$$M := \max Q_\Phi(\mathbb{V})\,, \quad m := \min Q_\Phi(\mathbb{V}).$$

Remark 3.8. Note that $\sqrt{m}$, $\sqrt{M}$ are the characteristic velocities of the p-system. Applying Theorem 3.10 below for monotone Φ'', these values are attained at $\varphi = 0, 1/2$, respectively.

Next, we report our main, new results concerning the general geometry and phase velocity of wave trains; full proofs will appear elsewhere.

Theorem 3.9. *Assume $\Phi'' > 0$. Then (3.5) has solutions $(\omega, \mathbb{V}) \in \mathbb{R}_0 \times \mathcal{M}_+$ for any $(\gamma, r, k) \in \mathbb{R}^2 \times (0, 1/2)$ such that $k \to 0 \Leftrightarrow \omega \to 0$.*

More precisely, the phase velocity of these wave trains satisfies

$$b(k)\sqrt{m} \;\leq\; \frac{|\omega|}{k} \;\leq\; \sqrt{M}, \tag{3.14}$$

where $b(k) \in (0, 1/2)$, $b(k) = 1/2$ for $k \leq 2\varphi_*$ with $\mathbb{V}(\varphi_*) = 0$ the unique root of $\mathbb{V}$. For $2\varphi_* < k < 1/2$, we can take

$$b(k) = \frac{1}{4} - \frac{\int_{-k/4}^{k/4} |\mathbb{V}(\varphi)|^2 + |\mathbb{V}(\varphi + 1/2)|^2 \mathrm{d}\varphi}{4\|\mathbb{V}\|_2^2}.$$

Theorem 3.9 states that the lower phase velocity bound of wave trains is estimated by a correction of the p-system characteristic. Indeed, small amplitude wave trains have $m \sim \Phi''(r)$, so that the harmonic phase velocities $\sqrt{m}\sin(\pi k)/\pi k$ apply, which, being *smaller* than $\sqrt{m}$, necessitate a correction such as $b(k)$.

Theorem 3.10. *Assume $\Phi'' > 0$ and consider solutions $\mathbb{V} \in \mathcal{M}_+$ to (3.5) for given (r, k, γ). Then the curve Q is smooth, closed, convex and non self-intersecting. Its unique extrema in r-direction lie at $\varphi = -k/2, (1-k)/2$ and in v-direction at $\varphi = 0, 1/2$, and it is bounded by $|\omega| \leq k\sqrt{M}$ and*

$$\frac{1}{2}k|\mathbb{V}(\varphi)| \;\leq\; |A_k\mathbb{V}(\varphi)| \;\leq\; \sqrt{k}\|\mathbb{V}\|_2. \tag{3.15}$$

If Φ'' is monotone, and ω independent of (r, k), then ω is a strictly monotone function of γ and the curves Q are nested near the extrema in r- and v-directions for fixed (r, k). The sign of monotonicity is that of ω.

While the unique points on Q with vertical and horizontal slope lie at $\varphi = (1-k)/2, 1/2$ and $\varphi = 0, -k/2$, the limiting profile for $k \to 0$ is not necessarily parametrized by φ. Indeed, for a limiting soliton, we expect that only one of these pairs converges to the point (r, v) of the soliton's background state as $k \to 0$.

The harmonic dispersion relation (3.18) renders ω a function of k, so that the last part of Theorem 3.10 does not apply. Indeed, in this case $\mathbb{V}$ is independent of k and ω, see (3.17).

Remark 3.11. The estimates on the size of Q imply that a nontrivial solitary limit as $k \to 0$ requires unbounded norm parameter γ, growing at least like $1/k$. Since Theorem 3.9 also implies $\omega \to 0$ as $k \to 0$, we expect that the monotonicity of γ in ω holds in general for small ω.

Nestedness of Q near the extrema in r and v directions for fixed (r, k) is a biproduct of our approach. However, it seems difficult to prove the numerical observation that the entire phase plots are nested.

Thermodynamics of wave trains

Wave trains represent exact solutions of Newton's equations which are highly oscillating on the microscopic scale. However, on the macroscopic scale we cannot resolve the microscopic oscillations but must pass to a thermodynamic description involving energy, pressure, temperature and the like. It turns out that for each wave train all these thermodynamic quantities are constant on the macroscopic scale. As a consequence, we can interpret each exact wave train solution of Newton's equation as a "thermodynamic state" of the chain. This idea turns out to be fruitful in modulation theory discussed in Sect. 4, where we allow for macroscopic modulations of the thermodynamic states.

Most thermodynamic quantities are defined as mean values of the oscillating atomic data in a wave train solution:

$$
\begin{aligned}
W &= \int_0^1 \Phi(r + A_k\mathbb{V}(\varphi))\,\mathrm{d}\varphi && \text{specific internal potential energy density,}\\
p &= -\int_0^1 \Phi'(r + A_k\mathbb{V}(\varphi))\,\mathrm{d}\varphi && \text{pressure = negative specific force density,}\\
K &= \frac{\omega^2}{2}\int_0^1 \mathbb{V}(\varphi)^2\,\mathrm{d}\varphi && \text{specific internal kinetic energy density,}
\end{aligned}
$$

and further

$$
\begin{aligned}
T &= 2K && \text{kinetic temperature,}\\
F &= K - W && \text{specific internal action density,}\\
U &= K + W && \text{specific internal energy density,}\\
E &= \frac{1}{2}v^2 + U && \text{specific total energy density.}
\end{aligned}
$$

Note that $K = \omega^2\gamma$, where γ is norm parameter used above. All these quantities are constants for exact wave trains. However, in modulation theory they become fields in $\overline{t}$ and $\overline{\alpha}$ whose macroscopic evolution is described by the modulation equations. In particular, although all quantities are defined by integrals over the phase variable φ, in modulation theory they represent *specific densities.*

There are other important thermodynamic quantities which are not directly related to mean values of the atomic data. It turns out that

$$S := \omega \int_0^1 \mathbb{V}(\varphi)^2 \, \mathrm{d}\varphi \,, \quad g := - \int_0^1 \mathbb{V}(\varphi) \, \Phi'(r + A_k \mathbb{V}(\varphi)) \, \mathrm{d}\varphi \tag{3.16}$$

can be interpreted as the macroscopic *entropy density* and *entropy flux*, respectively. It is proven in [Her04, DHM06] that any smooth family of wave trains provides an equation of state together with a corresponding Gibbs equation.

independent variables	*thermodynamic potential*	*Gibbs equation*
(r, k, γ)	$W = W(r, k, \gamma)$	$\mathrm{d}W = \omega^2 \, \mathrm{d}\gamma - p \, \mathrm{d}r - g \, \mathrm{d}k$
(r, k, ω)	$F = F(r, k, \omega)$	$\mathrm{d}F = S \, \mathrm{d}\omega + p \, \mathrm{d}r + g \, \mathrm{d}k$
(r, k, S)	$U = U(r, k, S)$	$\mathrm{d}U = \omega \, \mathrm{d}S - p \, \mathrm{d}r - g \, \mathrm{d}k$

The different variants of equations of state and Gibbs equations are all equivalent as long as the respective changes of coordinates are well defined.

The Gibbs equation becomes very important in modulation theory, where it provides the closure for the modulation equations. In particular, if the equation of state is known, then all other constitutive relations are determined by the Gibbs equation.

Examples for wave trains

For a few specific potential, explicit expressions are known for both the profile functions and the equation of state. The following examples are taken from [Her04, DHM06].

The harmonic chain with interaction potential $\Phi(r) = c_0 + c_1 \, r + \frac{c_2}{2} \, r^2$. Here the linearity of Φ' implies that (3.5) may be solved by means of Fourier transform. Some simple calculations yield the following family of traveling waves, parameterized by (r, k, γ),

$$\mathbb{V}(\varphi) = 2\sqrt{\gamma} \sin(2\pi\varphi), \qquad A_k \mathbb{V}(\varphi + k/2) = (\sin(\pi k)/\pi) \, \mathbb{V}(\varphi). \tag{3.17}$$

Here $\mathbb{V}$ is independent of (r, k) and $A_k \mathbb{V}$ independent of r. Degeneracy of the harmonic chain is also reflected by the *harmonic dispersion relation*

$$\omega(k) = \sqrt{c_2} \sin(\pi k)/\pi, \tag{3.18}$$

which provides the frequency ω as function of k, and does not depend on r or γ. Consequently, for the harmonic chain we cannot choose r, k and ω as set of independent variables. From (3.17) we infer that the equation of state reads

$$W(r,\,k,\,\gamma) = c_0 + c_1\,r + \frac{1}{2}\,c_2\,r^2 + \omega(k)^2\gamma,$$

which implies $g(r,\,k,\,\gamma) = -c_2\,\sin{(2\pi k)}\gamma$ and $S(r,\,k,\,\gamma) = 2\omega(k)\gamma$. Moreover, we can replace γ by S, and obtain

$$U(r,\,k,\,S) = c_0 + c_1 r + \frac{1}{2}c_2 r^2 + \omega(k)S. \tag{3.19}$$

The hard sphere model with interaction radius r_0. Here all atomic interaction are modeled as elastic collision between hard spheres with radius r_0. This gives rise to an interaction potential Φ with

$$\Phi(r) = +\infty \quad \text{for} \quad r < r_0, \quad \Phi(r) = 0 \quad \text{for} \quad r \geq r_0.$$

Although this potential is not smooth the notion of traveling waves may be generalized to this case, and again we are able to derive explicit expressions for wave trains. Some basic arguments lead to the following family of traveling waves, parameterized by $(r,\,k,\,\omega)$,

$$\mathbb{V}(r,\,k,\,\varphi) = \begin{cases} -(r-r_0)/k & \text{if } 0 \leq \varphi < k, \\ +(r-r_0)/(1-k) & \text{if } k \leq \varphi < 1. \end{cases}$$

Note that here the frequency $\omega > 0$ is a free parameter and may be chosen independently of r and k. The corresponding equation of state reads

$$U(r,\,k,\,S) = \tfrac{1}{2}\,(r-r_0)^{-2}\,S^2\,k\,(1-k). \tag{3.20}$$

We mention that the hard sphere model describes the high energy limit for certain potentials, see [Tod81] for the Toda potential, and [FM02], as well as Sect. 3.2, for Lennard-Jones potentials.

The third example is the *small amplitude limit*, where the amplitude δ of $\mathbb{V}$ is defined as the first fourier coefficient, i.e. for odd $\mathbb{V}$ we find $\delta = \int_0^1 \mathbb{V}(\varphi)\,\cos{(2\pi\varphi)}\,\mathrm{d}\varphi$. To identify the leading order terms we expand the nonlinear interaction potential Φ around the mean distance r up to fourth order. To leading order the frequency ω must satisfy the harmonic dispersion relation which now reads $\Omega_0(r,\,k) = \sqrt{\Phi''(r)}\sin{(k\,\pi)}/\pi$. According to [DHM06], the amplitude δ and the action F can be expressed in powers of $\omega - \Omega_0(r,\,k)$ as follows

$$\begin{aligned}\delta^2 &= \frac{\omega^2 - \Omega_0(r,\,k)^2}{2\,\Omega_0(r,\,k)^2}\,G(r,\,k),\\ F(r,\,k,\,\omega) &= -\Phi(r) + G(r,\,k)(\omega - \Omega_0(r,\,k))^2 + \mathcal{O}\big((\omega - \Omega_0(r,\,k))^3\big),\end{aligned} \tag{3.21}$$

where $G(r,\,k)$ is given by

$$G(r,\,k) = \frac{\Phi''(r)^2}{2\,\pi^2\,\Phi''(r)\,\Phi^{(4)}(r)\,\big(1-\cos{(2\pi k)}\big) + \big(\Phi'''(r)\big)^2\big(1+\cos{(2\pi k)}\big)}.$$

3.2 Solitary waves

Homoclinic orbits in ODEs are typically accompanied by large wave length periodic orbits in the sense that there exists a parameter curve of periodic orbits converging to the homoclinic orbit as the period tends to infinity [VF92]. For the lattices we consider, the situation is similar: wave trains exist for arbitrarily larger wave number and limit to solitons as the wave number tends to zero. This was proven for certain monotone waves and potentials under growth assumptions in [PP00] by a mountain pass approach.

We thank Karsten Matthies (University of Bath) for providing notes on which the remaining part of this section is based. We report some of his joint work with Gero Friesecke, mainly concerning solitons in (1.1) in the form (3.1) for a large class of possibly non-convex potentials Φ.

A prototype of physically realistic interaction is given by the standard Lennard-Jones potentials

$$\Phi(r) = a\Big(r^{-m} - r_*^{-m}\Big)^2 \text{ for } r > 0, \quad a > 0,\ m \in \mathbb{N}.$$

where Φ is minimized when neighbouring particles are placed at some specific equilibrium distance $r = r_* > 0$, and tends to infinity as the neighbour distance tends to zero.

Since the particle positions x_α corresponding to displacements y_α are $x_\alpha = r_*\alpha + y_\alpha$, this means that $\Phi(r)$ must have a minimum at $r = r_*$ and that $\Phi(r) \to \infty$ as $r \to 0$. More precisely we assume:

(H1) (Minimum at r_*) $\Phi \in C^3(0,\infty)$, $\Phi(r_*) = \Phi'(r_*) = 0$, $\Phi''(r_*) > 0$
(H2) (Growth) $\Phi(r) \geq c_0 r^{-1}$ for some $c_0 > 0$ and all r close to 0 and $\Phi(r) = \infty$ for $r \leq 0$.
(H3) (Hardening) $\Phi'''(r) < 0$ in $(0, r_*]$, $\Phi(r_* + r) < \Phi(r_* - r)$ in $(0, r_*)$.

Here we seek solitons whose profile $\mathbb{Y}(\phi)$ solves (3.2) with $r = r_*$. The construction in [FW94] for the existence of solitons is based on the variational problem

$$\begin{aligned} &\text{Minimize } \gamma_*(\mathbb{Y}) := \frac{1}{2}\int_{\mathbb{R}} \partial_\phi \mathbb{Y}(\phi)^2 d\phi \text{ among } \mathbb{Y} \in W^{1,2}_{\mathrm{loc}}(\mathbb{R}) \text{ satisfying} \\ &\partial_\varphi \mathbb{Y} \in L^2(\mathbb{R}),\ \mathcal{W}_*(\mathbb{Y}) := \int_{\mathbb{R}} \Phi(r + \mathbb{Y}(\phi+1) - \mathbb{Y}(\phi))\mathrm{d}\phi. \end{aligned} \tag{3.22}$$

Remark 3.12. It is instructive to compare this ansatz with the one used for wave trains in Sect. 3.1, where the real line is replaced by the unit interval and $\mathcal{W}$ maximized for fixed norm parameter γ. This lead to a relatively simple convex maximization problem for convex potentials. In contrast, (3.22) is a kind of dual problem, where $\mathcal{W}_*$ is fixed and the norm parameter γ_* minimized; a more challenging formulation that allows for non-convex potentials.

The goal is to determine the Γ-limit of the variational problem and the limiting profile in the high-energy regime. Since this regime is highly discrete and involves strong forces, neither classical continuum approximations nor weak coupling approximations are possible.

The limiting profile for $\mathcal{W}_*(\mathbb{Y}) \to \infty$ was derived in [FM02]. Here we recover this as a corollary of the following Γ-convergence result. We let

$$H^* := \{\mathbb{Y} \in W^{1,2}_{\text{loc}}(\mathbb{R}) | \mathbb{Y}(0) = 0, \partial_\phi \mathbb{Y} \in L^2(\mathbb{R})\},$$

and for every displacement profile $\mathbb{Y}$ we denote the relative displacement profile by $r(\phi) = \mathbb{Y}(\phi+1) - \mathbb{Y}(\phi)$. As in (3.22) we consider the functional γ_* on

$$\begin{aligned} H^*_K &= \{\mathbb{Y} \in H^* \,|\, \mathcal{W}_*(\mathbb{Y}) = K\} \\ H^*_\infty &= \{\mathbb{Y} \in H^* \,|\, r(\phi) \geq -r_*; \forall \phi \in \mathbb{R} \\ &\qquad \exists \text{ compact nonempty set } S_\mathbb{Y} \subset \mathbb{R} \text{ with } r_{|S_\mathbb{Y}} = -r_*\}. \end{aligned}$$

Theorem 3.13. (Γ-convergence) *Assume that the interaction potential satisfies (H1), (H2). Then the problem 'Minimize $\gamma_*(\mathbb{Y})$ for $\mathbb{Y} \in H^*_K$' Γ-converges to the problem 'Minimize $\gamma_*(\mathbb{Y})$ for $\mathbb{Y} \in H^*_\infty$' in the sense that*

1. (lim-inf-inequality) *If $\mathbb{Y}^{(K)} \rightharpoonup \mathbb{Y}$ in H^* with $\mathbb{Y}^{(K)} \in H^*_K$, $\mathbb{Y}^{(K)}$ translation normalized (i.e. $r^{(K)}(0) = \min_{\phi\in\mathbb{R}} r^{(K)}(\phi)$), then $\mathbb{Y} \in H^*_\infty$ and $\gamma_*(\mathbb{Y}) \leq \liminf_{K\to\infty} \gamma_*(\mathbb{Y}^{(K)})$,*
2. (Existence of recovery sequence) *For all $\mathbb{Y} \in H^*_\infty$ there exists a sequence $\mathbb{Y}^{(K)} \in H^*_K$ with $\mathbb{Y}^{(K)} \rightharpoonup \mathbb{Y}$ in H^* and $\gamma_*(\mathbb{Y}^{(K)}) \to \gamma_*(\mathbb{Y})$.*

A consequence is the following piecewise linear asymptotic displacement profile, corresponding to piecewise constant velocity profile.

Corollary 3.14. (*Asymptotic shape of minimizers) Every translation normalized sequence $\mathbb{Y}^{(K)}$ of minimizers of γ_* on H^*_K converges in H weakly to the up to translation unique minimizer $\mathbb{Y}_\infty$ of the limit problem, where*

$$\mathbb{Y}_\infty(\phi) = \begin{cases} 0,\ \phi \leq 0 \\ -r_*\phi,\ \phi \in [0,1] \\ -r_*,\ \phi \geq 1. \end{cases}$$

In a mechanical interpretation, this is a compression wave localized on a single atomic spacing. The limiting dynamics are hard-sphere dynamics like in a Boltzmann gas, see Fig. 3.1. In particular the work shows that dispersionless transport of energy is not restricted to the long-wave regime.

We mention that Friesecke and Matthies analyse a two dimensional counterpart of (3.1) in [FM03], see Fig. 3.1. The existence of longitudinal solitary waves along one of the lattice directions was shown for typical potentials under some mild nondegeneracy assumptions. These traveling waves are unique,

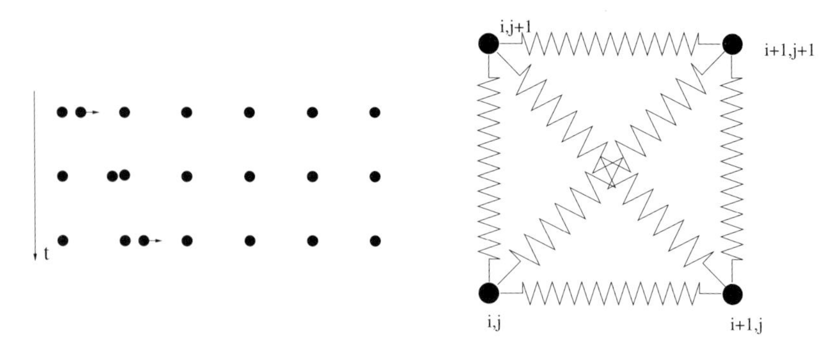

Fig. 3.1. Left: Hard-sphere soliton. Right: cell of springs in 2D lattice.

i.e. there are no other localized traveling wave in the same direction, e.g. there do not exist transversal traveling waves. It is surprising that purely harmonic springs are included here, because solitary waves do not occur in harmonic chains.

4 Modulation Theory

4.1 Macroscopic evolution of data with temperature

In this section we use the theory of Young measures, see for instance [Tay96, War99, Daf00], and derive some restrictions for any thermodynamic limit of the chain.

Let $\Omega = \{(\bar{t}, \bar{\alpha}) : 0 \leq \bar{t} \leq \bar{t}_{\text{fin}}, \bar{\alpha} \in [0, 1]\}$, and let $(N_i)_i$ be a sequence with $N_i \to \infty$. Moreover, for any i let $Q_\alpha^{(i)}(t) = (r_\alpha^{(i)}(t)\, v_\alpha^{(i)}(t))$, $0 \leq t \leq N_i \bar{t}_{\text{fin}}$ and $\alpha = 1, ..., N_i$, be a solution of Newton's equation, and suppose that the total energy of the initial data is proportional to N_i, i.e.

$$\sum_{\alpha=1}^{N_i} \left(\frac{1}{2} \left(v_\alpha^{(i)}(0) \right)^2 + \Phi\left(r_\alpha^{(i)}(0) \right) \right) = \mathcal{O}(N_i). \tag{4.1}$$

Under some suitable assumptions on the potential (say boundedness of Φ'' for simplicity) the functions $Q_\alpha^{(i)}(t)$ are compact with respect to the convergence of Young measures in the following sense. There is at least a subsequence, still denoted by $(N_i)_i$, and a family of probability measures $(\bar{t}, \bar{\alpha}) \mapsto \mu(\bar{t}, \bar{\alpha}, dQ)$ such that for any continuous observable $\psi = \psi(Q) = \psi(r, v)$ the following convergence is satisfied

$$\int_\Omega \psi\left(Q^{(i)}(N_i \bar{t}, N_i \bar{\alpha}) \right) \xi(\bar{t}, \bar{\alpha})\, \mathrm{d}\bar{t}\, \mathrm{d}\bar{\alpha} \xrightarrow{i \to \infty} \int_\Omega \langle \psi \rangle (\bar{t}, \bar{\alpha})\, \xi(\bar{t}, \bar{\alpha})\, \mathrm{d}\bar{t}\, \mathrm{d}\bar{\alpha}. \tag{4.2}$$

Here ξ is an smooth test function, and $\langle\psi\rangle(\bar{t},\,\bar{\alpha})$ is given by

$$\langle\psi\rangle(\bar{t},\,\bar{\alpha}) = \int_{\mathbb{R}^2} \psi(Q)\,\mu(\bar{t},\,\bar{\alpha},\,dQ). \tag{4.3}$$

For fixed $(\bar{t},\,\bar{\alpha}) \in \Omega$, the probability measure $\mu(\bar{t},\,\bar{\alpha},\,dQ)$ describes the microscopic oscillations in the vicinity of $(\bar{t},\,\bar{\alpha})$, and for any observable Ψ the number $\langle\Psi\rangle(\bar{t},\,\bar{\alpha})$ gives the *local mean value* of Ψ.

Here we consider the *common* probability distribution of distance and velocity instead of their separate statistics so that any measure $\mu(\bar{t},\,\bar{\alpha},\,dQ)$ can be interpreted as a weight function defined on the *microscopic state space* which is spanned by distance and velocity.

In Sect. 1 we have seen that Newton's equations are equivalent to the two *microscopic conservation laws* (1.3), from which one can derive the microscopic conservation of energy $\dot{e}_\alpha(t) = -f_\alpha(t) + f_{\alpha+1}(t)$ with $e_\alpha(t) = \frac{1}{2}v_{\alpha+1}^2(t) + \Phi(r_\alpha(t))$ and $f_\alpha(t) = -v_\alpha(t)\,\Phi'(r_\alpha(t))$. As a direct consequence, *every* Young measure limit must satisfy the following *macroscopic* conservation laws of mass, momentum and energy

$$\begin{aligned} \partial_{\bar{t}}\,\langle r\rangle - \partial_{\bar{\alpha}}\,\langle v\rangle &= 0, \\ \partial_{\bar{t}}\,\langle v\rangle - \partial_{\bar{\alpha}}\,\langle\Phi'(r)\rangle &= 0, \\ \partial_{\bar{t}}\left\langle \tfrac{1}{2}v^2 + \Phi(r)\right\rangle - \partial_{\bar{\alpha}}\,\langle v\,\Phi'(r)\rangle &= 0. \end{aligned} \tag{4.4}$$

This system of PDEs gives some restrictions for any young measure limit of the atomic chain. However, in general we can not express the fluxes in terms of the densities, and hence the system (4.4) is not closed, i.e. it does not determine the macroscopic evolution completely. We mention that (4.4) shows that any Young measure limit is a *measure-valued* solution of the p-system in the sense of DiPerna, see [Hör97, Daf00]. In addition, it is a measure-valued solution of the energy equation (1.11).

Within modulation theory we will start with some assumptions concerning the structure of the microscopic oscillations in the chain. Afterwards we will identify further macroscopic evolution laws extending (4.4), and constitutive relations that close the extended system.

4.2 Whitham modulation equations for wave trains

Here we describe Whitham's modulation theory for the atomic chain with hyperbolic scaling. For further examples concerning modulation theories of discrete system we refer to [HLM94, SW00, FP99, DK00, GM04, GM06], and to [GHM06a, GHM06b] for an overview.

A *modulated traveling wave* is an approximate solution of Newton's equation (1.1) satisfying

$$x_\alpha(t) = \frac{1}{\varepsilon}X(\varepsilon t,\,\varepsilon\alpha) + \tilde{\mathbb{X}}\left(\varepsilon t,\,\varepsilon\alpha,\,\frac{1}{\varepsilon}\Theta(\varepsilon t,\,\varepsilon\alpha)\right), \tag{4.5}$$

where X and Θ are macroscopic functions. The generic traveling wave parameters $(r,\, v,\, k,\, \omega)$ now are macroscopic fields depending on $(\overline{t},\, \overline{\alpha})$, and read

$$\omega = \partial_{\overline{t}}\,\Theta, \quad k = \partial_{\overline{\alpha}}\,\Theta, \quad v = \partial_{\overline{t}}\,X \quad r = \partial_{\overline{\alpha}}\,X. \tag{4.6}$$

The microscopic oscillations are described by

$$\tilde{\mathbb{X}}(\overline{t},\, \overline{\alpha},\, \varphi) = \mathbb{X}\big(r(\overline{t},\, \overline{\alpha}),\, v(\overline{t},\, \overline{\alpha}),\, k(\overline{t},\, \overline{\alpha}),\, \omega(\overline{t},\, \overline{\alpha}),\, a(\overline{t},\, \overline{\alpha}),\, \varphi\big), \tag{4.7}$$

where $\mathbb{X}(r,\, v,\, k,\, \omega,\, a,\, \varphi)$ is a smooth family of 1-periodic wave trains depending on the parameters $\mathbf{u} = (r,\, v,\, k,\, \omega,\, a)$ as well as on the phase φ. We use an additional parameter a, which might be the entropy S or the parameter γ. However, in any case we impose an abstract dispersion relation

$$\omega = \Omega(r,\, k,\, a). \tag{4.8}$$

The modulation equations are PDEs which describe the macroscopic evolution of the modulated parameter $(r,\, v,\, k,\, \omega,\, a)$, and ensure that (4.5) indeed provides approximate solutions. For their formal derivation we use Whitham's variational approach, see [Whi74, FV99, DHM06], and [GHM06a, GHM06b] for a more general setting.

In a first step we insert the ansatz (4.5) into the expression for the total action in the atomic chain, and expand all arising terms in powers of ε. This gives rise to the *reduced action integral*

$$\text{total action} = \mathbb{L}(X,\, \Theta,\, a) = \int_0^{\overline{t}_{\text{fin}}} \int_0^1 L\big(\mathbf{u}(\overline{t},\, \overline{\alpha})\big)\, \mathrm{d}\overline{\alpha}\, \mathrm{d}\overline{t}, \tag{4.9}$$

with $L(\mathbf{u}) = \mathcal{L}(\mathbf{u},\, \mathbb{X}(\mathbf{u},\, \cdot))$ and

$$\mathcal{L}(\mathbf{u},\, \mathbb{X}) = \int_0^1 \left(\frac{1}{2}(v + \omega\, \partial_\varphi \mathbb{X})^2 - \Phi(r + \nabla_k \mathbb{X}) \right) \mathrm{d}\varphi,$$

where $(\nabla_k \mathbb{X})(\varphi) = \mathbb{X}(\varphi + k)$. In a second step we apply the principle of least action to (4.9). The variation with respect to a gives $\partial_a L = 0$, which recovers the dispersion relation (4.8), and the variations with respect to X and Θ yield

$$\partial_{\overline{t}}\, \partial_v L + \partial_{\overline{\alpha}}\, \partial_r L = 0 \quad \text{and} \quad \partial_{\overline{t}}\, \partial_\omega L + \partial_{\overline{\alpha}}\, \partial_k L = 0, \tag{4.10}$$

respectively. Moreover, the definitions (4.6) imply two further evolution equations, namely $\partial_{\overline{t}}\, r - \partial_{\overline{\alpha}}\, v = 0$ and $\partial_{\overline{t}}\, k - \partial_{\overline{\alpha}}\, \omega = 0$.

In the last step we reformulate all macroscopic identities by using the thermodynamic definitions from Sect. 3, and as a consequence we find that the modulation equations take the form

$$\partial_{\overline{t}} \Big(r,\, v,\, k,\, S\Big) + \partial_{\overline{\alpha}} \Big(-v,\, +p,\, -\omega,\, +g\Big) = 0. \tag{4.11}$$

These equations represent the macroscopic conservation laws for mass, momentum, wave number and entropy. Moreover, they imply the conservation of energy via

$$\partial_{\bar{t}} E + \partial_{\bar{\alpha}} \left(pv + g\omega\right) = 0. \tag{4.12}$$

and thus we can regard the system (4.11) as an extension of (4.4). Recall that the closure for (4.11) and (4.12) is provided by the equation of state $E = \frac{1}{2}v^2 + U(r, k, S)$ and the Gibbs equation (2.5). However, for almost all interaction potential Φ we lack explicit expressions for the equations of state, and therefore we cannot characterize the properties of (4.11).

Finally, we display the modulation equations for the harmonic chain

$$\partial_{\bar{t}} \Big(r,\, v,\, k,\, S\Big) - \partial_{\bar{\alpha}} \Big(v,\, c_2\, r,\, \omega(k),\, \omega'(k)\, S\Big) = 0, \tag{4.13}$$

which follow from (4.11) by means of the equation of state (3.19), and the harmonic dispersion relation (3.18).

4.3 The justification problem

So far, there is no known rigorous derivation of the modulation equations for the nonlinear case. For this reason we formulate a conjecture, following similar results for partial differential equations [KSM92, Sch98, Mie02]. We assume that the potential Φ is sufficiently smooth, and that a smooth family of traveling waves $\mathbb{X}(\mathbf{u}, \varphi)$ with independent parameters $\mathbf{u} = (r, v, k, \omega)$ is given. Moreover, we assume that the following set $\mathcal{M}$ is open

$$\mathcal{M} = \left\{ \mathbf{u} = (r,\, v,\, k,\, \omega) \;\middle|\; \begin{array}{l} \text{the system (4.11) is strictly hyperbolic in } \mathbf{u}, \\ \text{the traveling wave } \mathbb{X}(\mathbf{u},\, \cdot) \text{ is linearly stable} \end{array} \right\}.$$

For a given solution $\widetilde{\mathbf{u}} = \widetilde{\mathbf{u}}(\bar{t}, \bar{\alpha})$ of (4.11) we define

$$M_\alpha^\varepsilon(t) = \begin{pmatrix} \widetilde{r}(\varepsilon t,\, \varepsilon\alpha) + (\hat{A}_k \mathbb{V})\Big(\widetilde{\mathbf{u}}(\varepsilon t,\, \varepsilon\alpha),\, \frac{1}{\varepsilon}\widetilde{\Theta}(\varepsilon t,\, \varepsilon\alpha) + \frac{1}{2}\widetilde{k}(\varepsilon t,\, \varepsilon\alpha)\Big) \\ \widetilde{v}(\varepsilon t,\, \varepsilon\alpha) + \widetilde{\omega}(\varepsilon t,\, \varepsilon\alpha)\, \mathbb{V}\Big(\widetilde{\mathbf{u}}(\varepsilon t,\, \varepsilon\alpha),\, \frac{1}{\varepsilon}\widetilde{\Theta}(\varepsilon t,\, \varepsilon\alpha)\Big) \end{pmatrix},$$

where $\mathbb{V}$ abbreviates $\partial_\varphi \mathbb{X}$, and the modulated phase $\widetilde{\Theta}$ is given by (4.6). We believe that the following conjecture is in the heart of the matter.

Conjecture 4.1. Let $\widetilde{\mathbf{u}}$ be a sufficiently smooth solution of Whitham's equation defined for $\bar{t} \in [0, \bar{t}_{\text{fin}}]$, and suppose that $\widetilde{\mathbf{u}}$ takes values in $\mathcal{M}$. Then there exists a suitable Banach space Υ_ε, and some exponent $\kappa > 0$ such that

$$\|Q^\varepsilon(t) - M^\varepsilon(t)\|_{\Upsilon_\varepsilon} = \mathcal{O}(\varepsilon^\kappa), \quad \|Q^\varepsilon(0)\|_{\Upsilon_\varepsilon} = \mathcal{O}(1) \tag{4.14}$$

for all ε, and all t with $0 \le \varepsilon t \le \bar{t}_{\text{fin}}$.

At the moment we are far from being able to prove this conjecture in this general form. However, it does hold rigorously for the harmonic chain and the hard sphere model.

The proof for the harmonic chain essentially relies on the linearity of Newton's equations, which allows to control the residuum, see [DHM06]. In addition, there is further rigorous derivation of (4.11) in the context of Wigner measures. For the details we refer to [Mie06], and for similar results to [Mac02, Mac04]. The rigorous justification for the hard sphere model is based on the observation that both the microscopic dynamics and the macroscopic equations become much simpler in the Eulerian representation of thermodynamics, cf. [Her04]

On a formal level we expect a close relation between stability of wave trains and hyperbolicity of modulation equations; if Whitham's equations (4.11) are not hyperbolic, then the corresponding initial value problem is ill-posed, which indicates that traveling waves are unstable due to a Benjamin-Feir instability, see e.g. [Whi74, BM95]. However, for arbitrary interaction potential Φ, neither stability criterions nor hyperbolicity conditions are available up to now. Having linearly degenerate eigenvalues, the harmonic chain and hard sphere model are not prototypical and do not provide further insight. Only the small amplitude limit gives some criteria for the hyperbolicity of (4.11). Starting with the equation of state (3.21) we can compute the characteristic speeds for (4.11), see [DHM06], and end up with the following criterion. The system (4.11) has four real eigenvalues, and is thus hyperbolic, if

$$\begin{aligned}\hat{N}(r,\,k) =& \Big(\Phi'''(r)\Big)^2 \Big(7 - 8\cos(2\pi k) + \cos(4\pi k)\Big) + \\ & \Phi''(r)\,\Phi^{(4)}(r)\,\Big(4\cos(2\pi k) - 3 - \cos(4\pi k)\Big)\end{aligned}$$

is positive, but has two imaginary eigenvalues for $\hat{N}(r,\,k) < 0$. For $k = 1/2$ the corresponding formula was already given in [Fla96].

5 Numerical justification of modulation theory

Although there is no rigorous justification for the modulation equations (4.11), numerical simulations strongly indicate that they provide the right thermodynamic description for a wide class of initial value problems for the atomic chain. We refer to [DH06] which gives a detailed thermodynamic interpretation of several numerical experiments. The main results can be summarized as follows.

(*i*) If all macroscopic fields are smooth, then the arising oscillations in the atomic data can be described in terms of modulated traveling waves, and the macroscopic dynamics is governed by the modulation system (4.11).

(*ii*) Modulated traveling waves describe the microscopic oscillations emerging when cold data form shocks.

(iii) If the shocks emerge from data with temperature, then usually the microscopic oscillations exhibit a more complicated structure, and (single-phase) modulation theory fails in this case.

Concerning the last item, results for the Toda chain suggest a hierarchy of modulation models, enumerated by the number of phases, where shocks on a lower level require the model of the next level, see e.g. the review [LLV93] and the references therein.

Note that these numerical observation are valid only if the interaction potential Φ is convex, the macroscopic scale results from the hyperbolic scaling, and the microscopic initial data are given by modulated traveling waves.

In this section we give an brief survey on the numerical justification from [DH06], and present a typical example with periodic boundary conditions and smooth macroscopic fields. Moreover, in Sect. 6 we study the numerical solution of a Riemann problem with cold initial data, and give an improved discussion of its macroscopic limit.

In order to study the macroscopic behavior of the atomic chain for large N we must evaluate the thermodynamic properties of the numerical data which are the macroscopic fields of the local mean values, and the local distribution functions of the atomic data. The computation of both mean values and distribution functions relies on *mesoscopic space-time windows.* In what follows let $\mathcal{F} = I_{\mathrm{T}}^{\mathcal{F}} \times I_{\mathrm{P}}^{\mathcal{F}}$ be such a window where I_{T} and I_{P} are sets of time steps and particle indices, respectively. The window F is mesoscopic if and only if it is very small on the macroscopic scale, but contains a lot of particles as well as time steps, i.e. $\sharp I_{\mathrm{T}}, \sharp I_{\mathrm{P}} \sim N^{\kappa}$ for some exponent κ with $0 < \kappa < 1$. In particular, any F describes the microscopic vicinity of a certain macroscopic point $Z_{\mathcal{F}} = (\overline{t}_{\mathcal{F}}, \overline{\alpha}_{\mathcal{F}})$.

For any atomic observable ψ we can easily compute the mean value $\langle\psi\rangle_{\mathcal{F}}$ of ψ with respect to each window $\mathcal{F}$ by a simple averaging formula. Note that there is a close relation to the notion of Young measures. In particular, if the atomic data converge for $N \to \infty$ in the sense of Young measures, then $\langle\psi\rangle_{\mathcal{F}}$ is a good approximation for $\langle\psi\rangle(\overline{t}_{\mathcal{F}}, \overline{\alpha}_{\mathcal{F}})$ from (4.3). Moreover, by means of $\mathcal{F}$ we can compute the complete measure $\mu\big(\overline{t}_{\mathcal{F}}, \overline{\alpha}_{\mathcal{F}}, dQ\big)$, see [DH06] for the details.

The micro-macro transition of modulation theory relies on the hypothesis that all atomic oscillations can be described by modulated wave trains. If this is right, then the microscopic distributions functions within any space-time window $\mathcal{F}$ must be equivalent to an exact wave train. Of course, the parameters of this wave train may depend on $\mathcal{F}$. In order to justify this hypothesis for given $\mathcal{F}$, we have to identify four wave train parameters, namely the specific length $r_{\mathcal{F}}$, the mean velocity $v_{\mathcal{F}}$, the wave number $k_{\mathcal{F}}$ and a fourth parameter which might be the parameter $\gamma_{\mathcal{F}}$, the frequency $\omega_{\mathcal{F}}$, the entropy $S_{\mathcal{F}}$, or the temperature $T_{\mathcal{F}}$.

The values of $r_{\mathcal{F}}$, $v_{\mathcal{F}}$ and $T_{\mathcal{F}}$ are given by mean values of microscopic observables. This reads $r_{\mathcal{F}} = \langle r \rangle_{\mathcal{F}}$, $v_{\mathcal{F}} = \langle v \rangle_{\mathcal{F}}$, and $T_{\mathcal{F}} = \langle v^2 \rangle_{\mathcal{F}} - \langle v \rangle_{\mathcal{F}}^2$. Determining $k_{\mathcal{F}}$ and $\omega_{\mathcal{F}}$ is not so obvious, because they have no immediate physical interpretation on the microscopic scale. To overcome this problem we introduce *auxiliary observables* Ψ_k and Ψ_ω, see [DH06] for their definitions, and set

$$k_{\mathcal{F}} := \langle \Psi_k \rangle_{\mathcal{F}}, \qquad \omega_{\mathcal{F}} := \langle \Psi_\omega \rangle_{\mathcal{F}}.$$

In the next step we start a numerical scheme similar to (3.3), which allows to prescribe the values $r_{\mathcal{F}}$, $v_{\mathcal{F}}$, $k_{\mathcal{F}}$ and $T_{\mathcal{F}}$, see [DH05] for details, and compute an exact wave train with these parameters. For any $\mathcal{F}$, the scheme yields a profile function $\mathbb{V}_{\mathcal{F}}$ as well as a frequency $\omega_{\mathcal{F}}^{\mathrm{TW}}$ which does not result from the auxiliary observable Ψ_ω but satisfies a dispersion relation.

Finally, we compare the microscopic distribution functions from the numerical data with their *macroscopic predictions* which can be expressed in terms of $\mathbb{V}_{\mathcal{F}}$. In particular, according to modulation theory, the support of the microscopic distribution functions must equal the curve

$$\varphi \mapsto Q^{\mathrm{TW}}(\varphi) = \Big(r_{\mathcal{F}} + \hat{A}_k \mathbb{V}_{\mathcal{F}}(\varphi + k_{\mathcal{F}}/2),\; v_{\mathcal{F}} + \omega_{\mathcal{F}} \mathbb{V}_{\mathcal{F}}(\varphi) \Big). \tag{5.1}$$

This rather strong prediction can be check for given numerical data.

Smoothly modulated initial data

We study the evolution of data with temperature by imposing initial data in form of smoothly modulated binary oscillations, i.e. we set

$$r_\alpha(0) = \begin{cases} r^{\,\mathrm{odd}}(\varepsilon\alpha) & \text{if } \alpha \text{ is odd,} \\ r^{\,\mathrm{even}}(\varepsilon\alpha) & \text{if } \alpha \text{ is even,} \end{cases} \qquad v_\alpha(0) = \begin{cases} v^{\,\mathrm{odd}}(\varepsilon\alpha) & \text{if } \alpha \text{ is odd,} \\ v^{\,\mathrm{even}}(\varepsilon\alpha) & \text{if } \alpha \text{ is even,} \end{cases}$$

where $r^{\,\mathrm{odd}}$, $r^{\,\mathrm{even}}$, $v^{\,\mathrm{odd}}$ and $v^{\,\mathrm{even}}$ may be read off from Fig. 5.1. We solved Newton's equation for the Toda chain with $N = 4000$ up to the macroscopic time $\bar{t}_{\mathrm{fin}} = 0.4$ by means of the Verlet scheme, see [SYS97, HLW02].

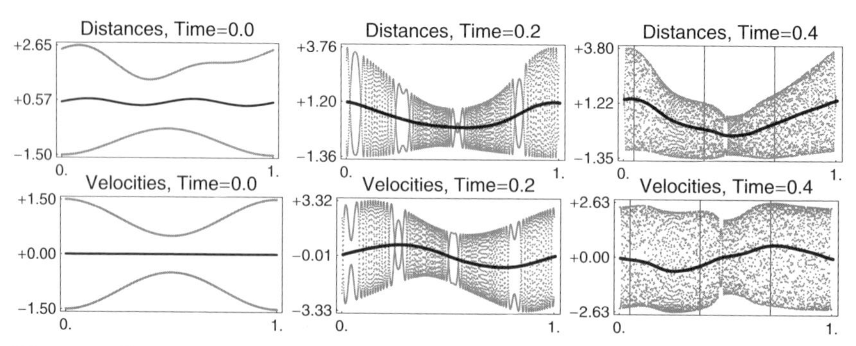

Fig. 5.1. Snapshots of the atomic distances and velocities at several macroscopic times. The vertical lines at $\bar{t} = 0.4$ mark the space-time windows for Fig. 5.4.

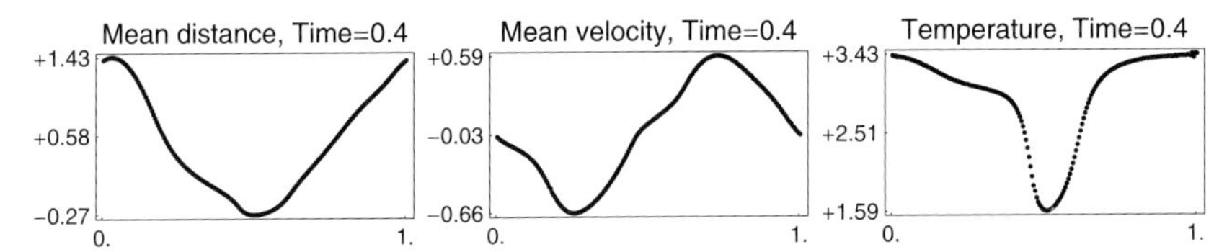

Fig. 5.2. Selected macroscopic fields as functions of $\overline{\alpha}$ for $\overline{t} = 0.4$.

Figure 5.1 contains snapshots of the atomic data for several macroscopic times, where the black colored curves represent the local mean values, and Figure 5.2 shows the profiles for some macroscopic fields at time $\overline{t} = 0.4$. We observe that the atomic data are highly oscillating on the microscopic scale so that any appropriate mathematical descriptions of the limit $N \to \infty$ must rely on measures. The computation of wave number and frequency is illustrated in Fig. 5.3, showing the oscillating values of the auxiliary observables Ψ_k and Ψ_Ω as well as their macroscopic mean values.

In Fig. 5.4 we compare the microscopic distribution functions with their macroscopic predictions from modulation theory for six mesoscopic space-time windows at $\overline{t} = 0.4$. For each of these windows we represent the distribution function of microscopic data by a density plot with high (Gray) and low (White) probability for finding a particle. Note that the support of every distribution functions is contained in closed curves, and that the distribution functions vary on the macroscopic scale.

The black dots in Fig. 5.4 represent the macroscopic predictions: we project 20 points $Q_i = Q^{\mathrm{TW}}(i/20)$ of the curve (5.1), into the density plots. Figure 5.4 reveals that the curve (3.13) coincides with the support of the microscopic distribution functions, and that the distance between Q_{i+1} and Q_i is related to the gray level of the microscopic distribution functions. In conclusion, we can describe the microscopic oscillations within any window $\mathcal{F}$ by a periodic wave train. Moreover, we can conclude that the macroscopic evolution of the thermodynamic fields is indeed governed by the modulation equations (4.11), see the discussion in [DH06].

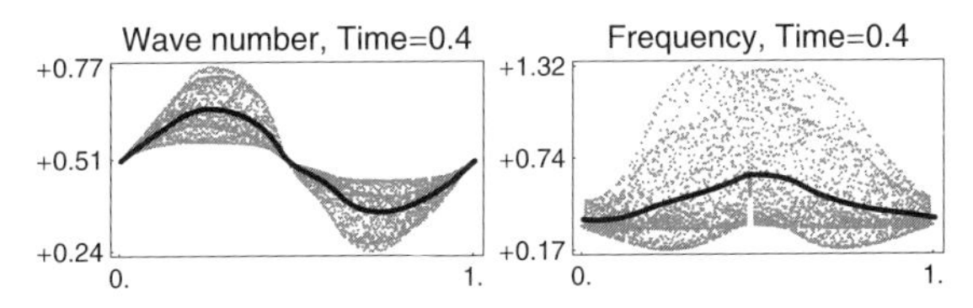

Fig. 5.3. Wave number and frequency: oscillating auxiliary variables and macroscopic mean values.

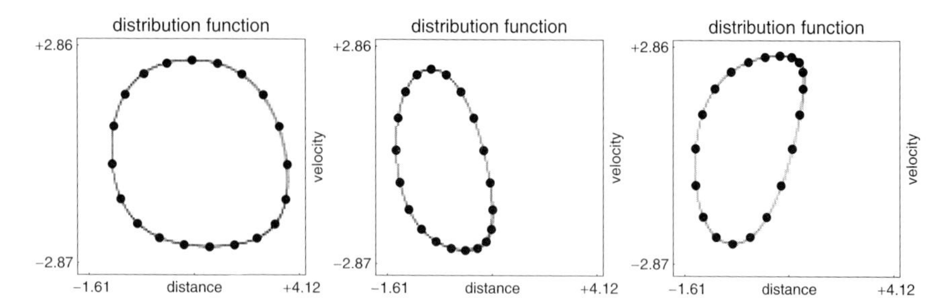

Fig. 5.4. Distribution functions of the atomic data in three selected space-time windows at $\overline{t} = 0.4$; for the $\overline{\alpha}$-coordinates see Figure 5.1. Each picture contains a density plot of the atomic data (White and Gray) together with an illustration of the macroscopic prediction (Black).

6 The shock problem

Since we expect a hyperbolic system describing the macroscopic limit, it is natural to investigate Riemann problems and interpret the results in terms of hyperbolic theory. A goal of this is to indicate selection principles for Riemann solvers that account for the macroscopic limit of atomic chains.

We would naively expect to find rarefaction fans, shocks and possibly contact discontinuities, that are selected by characteristic curves and entropy conditions and whose velocities are determined by characteristic velocities and Rankine-Hugoniot conditions.

It turns out that this picture is invalid when microscopic oscillations occur, leading to modulated wave trains as mentioned in Sect. 4. Instead, we find a situation very similar to the zero dispersion limit of the KdV equation mentioned in Sect. 1, where *dispersive shock fans* replace Lax-shock, and where velocities are not given by characteristic velocities of the limiting Burger's equation, corresponding to the p-system in our case. Faced with a large number of publications on this matter, we restrict references here to [LLV93, LP05, El05] and the bibliographies therein.

We focus on cold initial data, i.e. constant displacements and velocities with a single jump at some $\bar{\alpha}_*$, i.e.

$$(r,v)(\bar{\alpha},0) = (r_-,v_-)\,,\ \bar{\alpha} \le \bar{\alpha}_*\,,\ \text{and}\ (r,v)(\bar{\alpha},0) = (r_+,v_+)\,,\ \bar{\alpha} > \bar{\alpha}_*.$$

The macroscopic limit of the harmonic potential for such Riemann problems is cold and described by (4.13). It is therefore described by the corresponding p-system, which is a linear 1D wave equation, whose dynamics can be understood directly from the d'Alembert solution form, so there are only contact discontinuities.

For general nonlinear potentials, there is numerical evidence that dispersive shocks appear for initial data leading to Lax-shocks of the p-system, while rarefaction data leads to cold macroscopic limits described by the p-system.

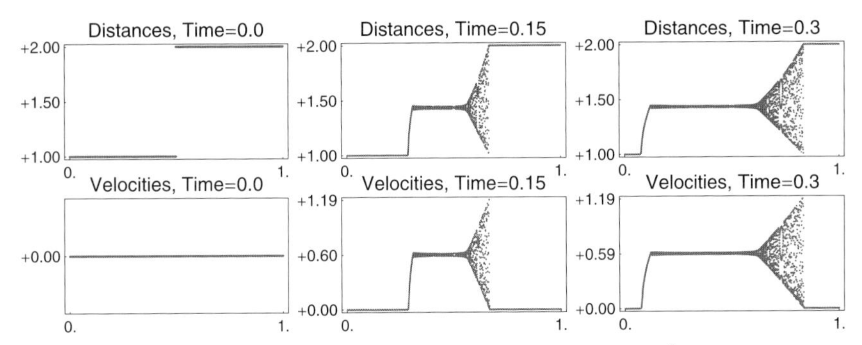

Fig. 6.1. Riemann problem with $N = 4000$ and $\Phi(r) = (r-1)^2/2 - \cos(2(r-1))/4$ with one rarefaction wave and one dispersive shock: snapshots of atomic distances and velocities for $\bar{t} = 0.0$, $\bar{t} = 0.15$, and $\bar{t} = 0.3$.

In Fig. 6.1 we plot a typical situation for illustration, and sketch a dispersive shock fan in Fig. 6.3. We are particularly interested in the transition of the Whitham modulation at its front.

Remark 6.1. For convex flux, i.e. $\Phi''' > 0$, the p-system can be solved uniquely in terms of at most two rarefaction or shock waves [Smo94]. For non-convex flux the situation is more complicated, and the entropy conditions for the p-system no longer agree, because eigenvalues are no longer genuinely nonlinear [KS97, LeF02]. A specific choice of a convex-concave potential for (1.1), numerically yields a macroscopically cold, strong shock, connecting states with equal characteristic velocities, and traveling with a different Rankine-Hugoniot velocity. In particular, it is not a contact discontinuity or Lax shock, but a (fast) undercompressive shock. Details on this phenomenon will be published elsewhere.

The macroscopic dynamics in space-time for Riemann data appear to be self-similar, hence reducible to a macroscopic velocity variable $c = \alpha/t = \bar{\alpha}/\bar{t}$. More formally, we assume that the Young measure $\mu(c)$ arising in the macroscopic limit for (initially cold) Riemann problems at each c is either a point measure or supported on a closed curve, corresponding to a wave train, so that from the modulation ansatz (4.5) we obtain $\tilde{\mathbb{X}}(c, \varphi)$ and analogously, from Sect. 4.3, an expression $M^\varepsilon(c, \varphi)$ for the vector of modulated distances and velocities. We use the phase variable φ to parametrize the support of $\mu(c)$. In case $\mu(c)$ is a point measure, we obtain a strong limit where $\tilde{\mathbb{X}}(c, \varphi) \equiv 0$.

A dispersive shock spans a range of speeds from the shock back velocity, c_b, to the shock front velocity, c_f. To ease notation, we assume $0 < c_\mathrm{b} < c_\mathrm{f}$, and that the constant states to the left and right of the dispersive shock are (r_-, v_-) and (r_+, v_+) as sketched in Fig. 6.3.

It is instructive to view the modulation of wave trains in a dispersive shock as the selection of a curve in the set of wave trains $\tilde{\mathbb{X}}(c, \varphi)$ parametrized by $c_\mathrm{b} < c < c_\mathrm{f}$ in terms of the wave train parameters $(r(c), k(c), \omega(c))$. This curve

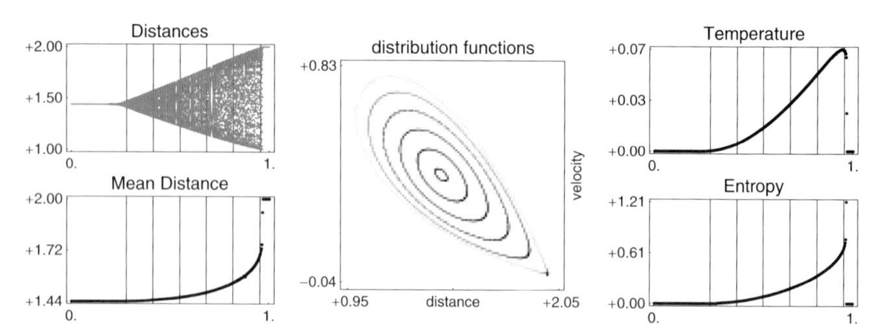

Fig. 6.2. Example for dispersive shocks. Left: snapshots of atomic distances and their local mean values. Center: superposition of several distribution functions within the shock; positions of the space-time windows are marked by vertical lines. Right: snapshots of temperature and entropy.

bridges the energy jump between the constant states (r_-, v_-) and (r_+, v_+), and the wave trains become singular at c_b and c_f. Based on numerical evidence [HFM81] and results for the Toda chain [VDO91, Kam91], we assume that $M^\varepsilon(c_\mathrm{b}) \equiv (r_-, v_-)$ has zero amplitude, and the shock front $M^\varepsilon(c_\mathrm{f})$ corresponds to a *soliton* with background state (r_+, v_+), where $k(c_\mathrm{f}) = \omega(c_\mathrm{f}) = 0$. Note that this is a singular limit of (3.4) and that Theorem 3.10 implies infinite kinetic energy $\gamma(c_\mathrm{f})$. We plot wave trains and fields within a dispersive shock in Fig. 6.2.

More precisely, the shock front is assumed to be a homoclinic orbit

$$\mathcal{H}(s) := \lim_{c \to c_\mathrm{f}} M^\varepsilon(c_\mathrm{f})$$

in the phase scaling $\varphi = \omega s$ with asymptotic state $\lim_{s \to \pm\infty} \mathcal{H}(s) = (r_+, v_+)$. We expect the convergence to the asymptotic state is exponential in s, thus L_p-norms of $[\mathcal{H}(s) - (r_+, v_+)]$ are finite. In terms of the wave train profile $\mathbb{X}$ and $\mathbb{V} = \partial_\varphi \mathbb{X}$ we can write the second component $\mathcal{H}_2$ of $\mathcal{H}$ as

$$\mathbb{H}(s) := \mathcal{H}_2(s) - v_+ = \lim_{c \to c_\mathrm{f}} \omega(c)\mathbb{V}(c, \omega(c)s) = \lim_{c \to c_\mathrm{f}} \frac{\mathrm{d}}{\mathrm{d}s}\mathbb{X}(c, \omega(c)s).$$

Both the vanishing amplitude at c_b and sinusoidal oscillations, and the homoclinic orbit at c_f are natural codimension-1 singularity along a curve of wave trains viewed as periodic orbits.

Assuming a soliton at the shock front means in particular that the modulation system does not have a strong shock, which is challenging to confirm numerically as discussed below. Instead, we conjecture that at the shock front the entropy S jumps and (r, v, k, ω) are continuous with unbounded derivative. Heuristically, the excess energy at the jump in the initial data cannot be dissipated by the conservative system, but is transported dispersively via oscillations with two new degrees of freedom, frequency and wave number.

Properties at and near the soliton

We predict the scaling of temperature and related quantities assuming the scaling in generic or conservative homoclinic bifurcations of ODEs [VF92], where the unfolding parameter, here c, is exponentially small in the period, here $1/\omega$. We thus expect $c_{\rm f} - c \sim {\rm e}^{-\kappa/\omega}$, for some $\kappa > 0$, and so

$$\omega(c) \sim k(c) \sim 1/\log(c_{\rm f} - c),$$

because Theorem 3.9 implies the same scaling in k. Indeed, this scaling could be confirmed for the case of Toda potential using the explicit solutions in [DKV95], and also appears in the formal derivations in [El05].

Temperature, entropy, entropy flux. The definition $T = \omega^2 \int_0^1 \mathbb{V}(\varphi)^2 {\rm d}\varphi$ of the temperature of a wave train yields

$$T(c) = \int_0^1 \left[\omega(c)\mathbb{V}(c,\varphi)\right]^2 {\rm d}\varphi = \omega(c) \int_0^{1/\omega(c)} \left[\omega(c)\mathbb{V}(c,\omega(c)s)\right]^2 {\rm d}s$$

and thus (assuming smoothness) the limiting temperature of the soliton

$$\lim_{c\to c_{\rm f}} T(c) = \lim_{c\to c_{\rm f}} \omega(c) \int_0^{1/\omega(c)} \left[\omega(c)\mathbb{V}(c,\omega(c)s)\right]^2 {\rm d}s = \int_0^{\infty} \mathbb{H}(s)^2 {\rm d}s \lim_{c\to c_{\rm f}} \omega(c) = 0,$$

because the L^2 norm of $\mathbb{H}$ is finite. Then the scalings of temperature T, entropy S and entropy flux g, see (3.16), are given by

$$T(c) \sim \left(\log(c_{\rm f} - c)\right)^{-1}, \quad g(c) \sim S(c) = T(c)/\omega \sim 1,$$

where we used $cS' = g'$. We thus predict that the temperature is continuous for all c and decays to zero like $1/\log$. Entropy and entropy flux vanish in cold regions, but continuously approach a finite, non-zero value and jump to zero beyond the shock front.

Since the temperature also decays towards the shock back, we expect a unimodal curve $T(c)$ with a unique maximum, as is the case in e.g. a planar ODE where the interior of a homoclinic orbit is filled with periodic orbits and an elliptic equilibrium.

However, these scalings and limiting values are difficult to confirm numerically, because the $1/\log$ decay is hard to resolve, and the shock front could not be simulated in isolation from the rest of the modulation region due to boundary effects.

Norm parameter γ. On account of Theorem 3.10, the norm parameter γ grows at least like $1/k$, so that $\gamma(c) \geq -C\log(c_{\rm f} - c)$, for a constant $C > 0$. This agrees with the prediction from the above entropy scaling, because

$$\gamma(c) = \frac{S}{2\omega} = \frac{T}{2\omega^2} \sim \frac{1}{\omega} \sim \log(c_{\rm f} - c).$$

Mean distance and velocity. Assuming that c unfolds the homoclinicity as a generic (or Hamiltonian) ODE, the flow time through a fixed small region near (r_+, v_+) grows logarithmically in c and thus for the average values we obtain the scaling

$$r(c) = r_+ - r_1/\log(c_\mathrm{f} - c) + h.o.t. \quad v(c) = v_+ - v_1/\log(c_\mathrm{f} - c) + h.o.t.,$$

with some constants r_1, v_2, since the limiting values are those of the corresponding Riemann data.

Note that the first equation in (4.4) implies $-cr' = v'$ in the sense of distributions, where $' = \mathrm{d}/\mathrm{d}c$. Therefore, $-c_\mathrm{f} r_1 = v_1$, and so

$$c_\mathrm{f} = -v_1/r_1 \tag{6.1}$$

replaces the Rankine-Hugoniot jump condition.

Propagation speeds. The modulation equations yield five equations for the propagation speed of the shock front; four in term of leading order expansions such as $-cr' = v' \to -c = \mathrm{d}v/\mathrm{d}r$ above, and one jump condition $c[\![S]\!] = [\![g]\!]$. Indeed, in numerical simulations of dispersive shocks all these velocities are close to that obtained from the slope of the shock front in space-time.

The conservation of wave number implies $-ck' = \omega'$ and thus throughout the dispersive shock we have $-c = c_\mathrm{g} := \mathrm{d}\omega/\mathrm{d}k$, which is the group velocity and not the phase velocity $c_\mathrm{ph} := -\omega/k$ of wave trains. Note that here, $-c$ is the expected propagation velocity due to the choide of sign for ω in (3.3) and (4.7).

In particular, the shock front should move with the limiting group velocity, while the soliton speed naturally is the limiting phase velocity. However, in the solitary limit, phase and group velocity typically coincide, because for $L = 1/k$ we have the identity

$$c_\mathrm{g} = c_\mathrm{ph} - L\frac{\mathrm{d}c_\mathrm{ph}}{\mathrm{d}L},$$

where $\frac{\mathrm{d}c_\mathrm{ph}}{\mathrm{d}L}$ is exponentially small for generic and conservative homoclinic bifurcations in ODE [VF92]; the identity follows from $\frac{\mathrm{d}c_\mathrm{ph}}{\mathrm{d}k} = \frac{c_\mathrm{g} - c_\mathrm{ph}}{k}$.

Recall that the phase velocities of wave trains were estimated in (3.14) and rigorously imply that the soliton velocity is bounded (essentially) by p-system characteristics velocities c_-, c_- of the left and right states r_-, r_+. However, in numerical simulations, the shock front velocity c_f never exhausted these bounds, but was strictly between c_- and c_+.

On the other hand, the shock velocity of the p-system is given by the Rankine-Hugoniot condition $c_\mathrm{rh} = \sqrt{(\Phi'(r_-) - \Phi'(r_+))/(r_- - r_+)}$, and in all cases (for Φ' monotone) we numerically found the velocity ordering sketched in Fig. 6.3, that is,

$$c_\mathrm{b} < c_+ < c_\mathrm{rh} < c_\mathrm{f} < c_-,$$

where $c_\mathrm{rh} - c_\mathrm{f} \sim 5\%$. Characteristics point into the dispersive shock fan, and indeed, we seem to find dispersive shocks only if $c_- < c_\mathrm{rh} < c_+$, see also Remark 6.1.

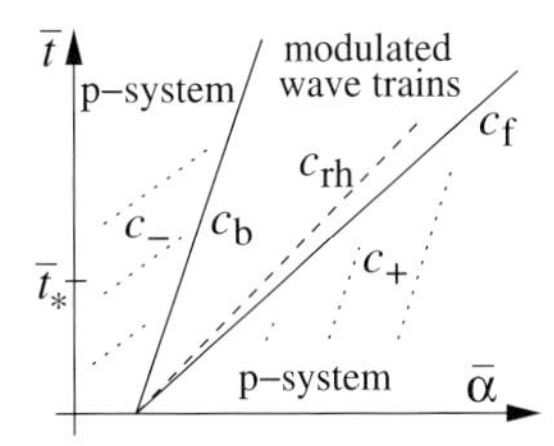

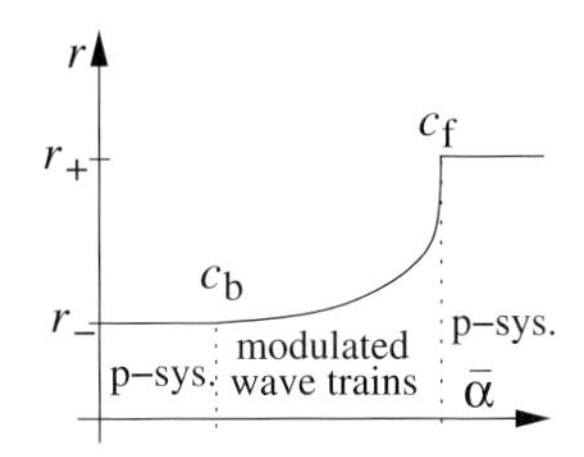

Fig. 6.3. Left: Sketch of a dispersive shock for the macroscopic limit of a shock problem in (1.1). Dashed line is the p-system Lax shock with speed c_{rh}, dotted the p-system characteristic velocities $c_{\pm}$ of left and right states $(r_{\pm}, v_{\pm})$. Right: sketch of the r-modulation at some time $\bar{t}_* > 0$ with $1/\log$ scaling at c_{f}.

Finally, we mention that the velocity c_{b} of the shock back, where wave trains have small amplitude, numerically agrees with the prediction from harmonic modulation equations, i.e. $c_{\mathrm{b}} = \sqrt{\Phi''}(r_-)\sin(\pi k(c_{\mathrm{b}}))/\pi k_{\mathrm{b}}$.

Remarks and open problems

The occurrence of dispersive shocks has only been proven rigorously for some completely integrable cases, in particular the Toda chain [VDO91, Kam91]. Unfortunately, the literature on this issue is not easily accessible to non-specialists, and we found it inconclusive concerning the rigorous justification of a hyperbolic system of Whitham modulation equations. In fact, neither the observation that the shock front is a soliton, nor the scaling at the shock front, nor the velocity of the shock back seem to be worked out.

Similarly, to our knowledge, the selection mechanism for the soliton has not been formulated in terms of initial values for the Riemann problem (though the shock front velocity for the Toda shock problem can be computed explicitly [VDO91]). An observation towards a selection principle could be that in numerical experiments for vanishing initial velocities, the dispersive shock exhausts precisely the range between the initial jump in the r-component. We also observe that the dispersive shock in (c, r, v)-space is a graph over the plane $(0, r, v)$. In other words, the modulation parameter curve $(r, v, k, \omega)(c)$ appears to be selected in such a way that wave train orbits Q are (nested) level sets of an unknown function.

We hope that a study of the explicit solutions for the case of Toda potential, and results for zero dispersion limits mentioned in Sect. 1, provide more insight into the general situation, in particular the prediction of dispersive shocks and the shock front velocity.

Acknowlegdments. This work has been supported by the DFG Priority Program 1095 "Analysis, Modeling and Simulation of Multiscale Problems". We gratefully acknowledge notes by Karsten Matthies on which Sect. 3.2 is based, and fruitful discussions with Alexander Mielke and Thomas Kriecherbauer.

References

[BM95] T. Bridges and A. Mielke. A proof of the Benjamin-Feir instability. *Archive Rational Mech. Analysis*, 133, 145–198, 1995.

[Daf00] C. Dafermos. *Hyperbolic Conservation Laws in Continuum Physics*, volume 325 of *Grundlehren d. mathem. Wissenschaften.* Springer, Berlin, 2000.

[DH05] W. Dreyer and M. Herrmann. On the approximation of periodic traveling waves for the atomic chain. WIAS preprint 1030, 2005.

[DH06] W. Dreyer and M. Herrmann. Numerical experiments on modulation theory for the nonlinear atomic chain. WIAS preprint 1031, revised version, 2006.

[DHM06] W. Dreyer, M. Herrmann, and A. Mielke. Micro-macro transition for the atomic chain via Whitham's modulation equation. *Nonlinearity*, 19(2), 471–500, 2006.

[DK00] W. Dreyer and M. Kunik. Cold, thermal and oscillator closure of the atomic chain. *J. Phys. A: Math. Gen.*, 33, 2097–2129, 2000.

[DK*74] P. Deift, S. Kamvissis, T. Kriecherbauer, and X. Zhou. The Toda lattice ii. existence of integrals. *Phys. Rev. B*, 9(4), 1924–1925, 1974.

[DK*96] P. Deift, S. Kamvissis, T. Kriecherbauer, and X. Zhou. The Toda rarefaction problem. *Comm. Pure Appl. Math*, 49(1), 35–83, 1996.

[DKV95] P. Deift, T. Kriecherbauer, and S. Venakides. Forced lattice vibrations, Part I. *Comm. Pure Appl. Math.*, 48, 1187–1250, 1995.

[DM98] P. Deift and T. T.-R. McLaughlin. *A continuum limit of the Toda lattice*, volume 131/624 of *Mem. Americ. Math. Soc.* Americican Mathematical Society, 1998.

[El05] G. El. Resolution of a shock in hyperbolic systems modified by weak dispersion. *Chaos*, 15, 2005.

[Fla96] S. Flach. Tangent bifurcation of band edge plane waves, dynamical symmetry breaking and vibrational localization. *Physica D*, 91, 223–243, 1996.

[FM02] G. Friesecke and K. Matthies. Atomic-scale localization of high-energy solitary waves on lattices. *Physica D*, 171, 211–220, 2002.

[FM03] G. Friesecke and K. Matthies. Geometric solitary waves in a 2d mass spring lattice. *Discrete Contin. Dyn. Syst. Ser. B*, 3, 105–114, 2003.

[FP99] G. Friesecke and R. L. Pego. Solitary waves on FPU lattices. I. Qualitative properties, renormalization and continuum limit. *Comm. Math. Phys.*, 12(6), 1601–1627, 1999.

[FPU55] E. Fermi, J. Pasta, and S. Ulam. Studis on nonlinear problems. *Los Alamos Scientific Laboraty Report*, LA–1940, 1955.

[FV99] A.-M. Filip and S. Venakides. Existence and modulation of traveling waves in particle chains. *Comm. Pure Appl. Math.*, 51(6), 693–735, 1999.

[FW94] G. Friesecke and J. A. D. Wattis. Existence theorem for solitary waves on lattices. *Comm. Math. Phys.*, 161(2), 391–418, 1994.

[GHM06a] J. Giannoulis, M. Herrmann, and A. Mielke. Continuum descriptions for the dynamics in discrete lattices: derivation and justification. In "Analysis, Modeling and Simulation of Multiscale Problems, A. Mielke (edr), Springer-Verlag, 2006.".

[GHM06b] J. Giannoulis, M. Herrmann, and A. Mielke. Lagrangian and Hamiltonain redcution for lattice systems. In preparation, 2006.

[GL88] J. Goodman and P. Lax. On dispersive difference schemes. *Comm. Pure. Appl. Math.*, 41, 591–613, 1988.

[GM04] J. Giannoulis and A. Mielke. The nonlinear Schrödinger equation as a macroscopic limit for an oscillator chain with cubic nonlinearities. *Nonlinearity*, 17, 551–565, 2004.
[GM06] J. Giannoulis and A. Mielke. Dispersive evolution of pulses in oscillator chains with general interaction potentials. *Discr. Cont. Dynam. Systems Ser. B*, 6, 493–523, 2006.
[Hén74] M. Hénon. Integrals of the Toda lattice. *Phys. Rev. B*, 9(4), 1921–1923, 1974.
[Her04] M. Herrmann. *Ein Mikro-Makro-Übergang für die nichtlineare atomare Kette mit Temperatur.* Phd thesis, Humboldt-Universität zu Berlin, 2004.
[HFM81] B. Holian, H. Flaschka, and D. McLaughlin. Shock waves in the toda lattice: Analysis. *Phys. Rev. A*, 24(5), 2595–2623, 1981.
[HL91] T. Y. Hou and P. Lax. Dispersive approximations in fluid dynamics. *Comm. Pure Appl. Math.*, 44, 1–40, 1991.
[HLM94] M. Hays, C. D. Levermore, and P. Miller. Macroscopic lattice dynamics. *Physica D*, 79(1), 1–15, 1994.
[HLW02] E. Hairer, C. Lubich, and G. Wanner. *Geometric Numerical Integration*, volume 31 of *Springer Series in Comp. Mathem.* Springer, Berlin, 2002.
[Hör97] L. Hörmander. *Lectures on Nonlinear Hyperbolic Differential Equations*, volume 26 of *Mathématiques & Applications.* Springer, Paris, 1997.
[Kam91] S. Kamvissis. *On the long time behavior of the double infinite Toda chain under shock initial data.* PhD thesis, New York University, 1991.
[KL02] C. Kondo and P. LeFloch. Zero diffusion-dispersion limits for hyperbolic conservation laws. *SIAM Math. Anal.*, 22, 1320–1329, 2002.
[KS97] P. Krejči and I. Straškraba. A uniqueness criterion for the Riemann problem. *Hiros. Math. J.*, 27, 307–346, 1997.
[KSM92] P. Kirrmann, G. Schneider, and A. Mielke. The validity of modulation equations for extended systems with cubic nonlinearities. *Proc. Roy. Soc. Edinburgh Sect. A*, 122, 85–91, 1992.
[Lax86] P. Lax. On dispersive difference schemes. *Physica D*, 18, 250–254, 11986.
[Lax91] P. Lax. The zero dispersion limit, a deterministic analogue of turbulence. *Comm. Pure Appl. Math.*, 44, 1047–1056, 1991.
[LeF02] P. G. LeFloch. *Hyperbolic Conservation Laws.* Lectures in Mathematics: ETH Zürich. Birkhäser, Basel, 2002.
[LL83] P. Lax and C. Levermore. The small dispersion limit of the Korteweg-de Vries equation, I, II, III. *Comm. Pure Appl. Math.*, 36, no.3, 253–290; no.5, 571–593; no.6, 809–830, 1983.
[LL96] C. Levermore and J. Liu. Large oscillations arising in a dispersive numerical scheme. *Physica D*, 99, 191–216, 1996.
[LLV93] P. D. Lax, C. D. Levermore, and S. Venakides. The generation and propagation of oscillations in dispersive initial value problems and their limiting behavior. In A. Fokas and V. Zakharov, editors, *Important developments in soliton theory*, pages 205–241. Springer, Berlin, 1993.
[LP05] P. Lorenzoni and S. Paleari. Metastability and dispersive shock waves in fermi-pasta-ulam system. oai:arXiv.org:nlin/0511026, 2005.
[Mac02] F. Maciá. *Propagación y control de vibraciones en medios discretos y continous.* PhD thesis, Universidad Complutense de Madrid, 2002.
[Mac04] F. Maciá. Wigner measures in the discrete setting: high-frequency analysis of sampling and reconstruction operators. *SIMA*, 36(2), 347–383, 2004.

[Mie02] A. Mielke. The Ginzburg–Landau equation in its role as a modulation equation. In B. Fiedler, editor, *Handbook of Dynamical Systems II*, pages 759–834. Elsevier Science B.V., 2002.

[Mie06] A. Mielke. Macroscopic behavior of microscopic oscillations in harmonic lattices using wigner-husimi measures. *Arch. Rat. Mech. Analysis*, 181, 401–448, 2006.

[MR98] I. Müller and T. Ruggeri. *Rational Extended Thermodynamics*, volume 37 of *Springer Tracts in Natural Philosophy*. Springer, 2. edition, 1998.

[PP00] A. A. Pankov and K. Pflüger. Traveling waves in lattice dynamical systems. *Math. Meth. Appl. Sci.*, 23, 1223–1235, 2000.

[Sch82] M. Schonbek. Convergence of solutions to nonlinear dispersive equations. *Comm. Part. Diff. Equa.*, 7, 959–1000, 1982.

[Sch98] G. Schneider. Justification of modulation equations for hyperbolic systems via normal forms. *NoDEA Nonlinear Differential Equations Appl.*, 5(1), 69–82, 1998.

[Smo94] J. Smoller. *Shock Waves and Reaction-Diffusion Equations*, volume 258 of *Grundlehren d. mathem. Wissenschaften*. Springer, 2. edition, 1994.

[SW00] G. Schneider and C. E. Wayne. Counter-propagating waves on fluid surfaces and the continuum limit for the Fermi-Pasta-Ulam model. In K. Gröger, B. Fiedler, and J. Sprekels, editors, *International Conference on Differential Equations*, pages 390–403. World Scientific, 2000.

[SYS97] R. D. Skeel, G. Yang, and T. Schlick. A family of symplectic integrators: Stability, accuracy, and molecular dynamics Applications. *SIAM J. Sci. Comput.*, 18, 203–222, 1997.

[Tay96] M. E. Taylor. *Partial Differential Equations III: Nonlinear Theory*, volume 117 of *Applied Mathematical Sciences*. Springer, New York, 1996.

[Tod70] M. Toda. Waves in nonlinear lattices. *Prog. Theor. Phys.*, 45, 174–200, 1970.

[Tod81] M. Toda. *Theory of nonlinear lattices*, volume 20 of *Springer Series in Solid-State Sci.* Springer, Berlin, 1981.

[VDO91] S. Venakides, P. Deift, and R. Oba. The Toda shock problem. *Comm. Pure Appl. Math.*, 44, 1171–1242, 1991.

[Ven85] S. Venakides. The generation of modulated wavetrains in the solution of the Korteweg-de Vries equation. *Comm. Pure Appl. Math.*, 38, 883–909, 1985.

[VF92] A. Vanderbauwhede and B. Fiedler. Homoclinic period blow-up in reversible and conservative systems. *Z. Angew. Math. Phys.*, 43, 292–318, 1992.

[War99] G. Warnecke. *Analytische Methoden in der Theorie der Erhaltungsgleichungen*, volume 138 of *Teubner Texte z. Mathem.* B.G. Teubner, 1999.

[Whi74] G. B. Whitham. *Linear and Nonlinear Waves*, volume 1237 of *Pure And Applied Mathematics*. Wiley Interscience, New York, 1974.

On the Effect of Correlations, Fluctuations and Collisions in Ostwald Ripening

Barbara Niethammer[1], Felix Otto[2], and Juan J.L. Velázquez[3]

[1] Institut für Mathematik, Humboldt-Universität zu Berlin, Unter den Linden 6, 10099 Berlin. niethamm@mathematik.hu-berlin.de
[2] Institut für Angewandte Mathematik, Rheinische Friedrich-Wilhelms-Universität Bonn, Wegelerstraße 10, 53115 Bonn. otto@iam.uni-bonn.de
[3] Departamento de Matemática Aplicada, Facultad de Matemáticas, Universidad Complutense, 28040 Madrid, Spain. JJ_Velazquez@mat.ucm.es

1 Introduction

A phenomenon of fundamental importance in materials science is the late stage coarsening in a phase transformation in two-phase mixtures. Phase separation occurs for example during heat treatment, where the mixture is brought into a thermodynamically unstable state, in which two different phases are energetically favored. If the volume fraction of one phase is small, phase separation, triggered by thermal fluctuations, results in a polydisperse mixture of spherical particles of the minority phase, immersed in a background phase, the "matrix". Due to the large surface energy of the fine mixture the system is still far from equilibrium. In the last stage of the phase transformation the particles compete to reduce the total surface area by transferring mass from smaller to larger particles. As a consequence, large particles grow at the expense of the smaller ones which eventually disappear. This particular form of coarsening is also known as Ostwald ripening.

The process is well-described by a class of free boundary problems (cf. Sect. 2.1), which are however too complex to allow for predictions on the dynamics of large particle ensembles. A major advance was made in the late 50's by Lifshitz and Slyozov [LS61] and independently by Wagner [Wag61], in their nowadays classical LSW-theory for Ostwald ripening. They derived a mean-field model in the regime of vanishing volume fraction of particle phase. Within this mean-field theory they predicted universal self-similar large time asymptotics, which enabled them to make quantitative predictions on the coarsening statistics. In particular, this analysis implies universal growth rates for the mean particle size, which serves as a measure for the coarsening rate, and a universal particle size distribution.

More than 25 years after the theory was developed, a lively discussion started in the physics and metallurgical community [Bro89, Bro90a,

Bro90b, HHR92, HH*89, Hoy90] whether universal large-time behavior within the LSW model is indeed strictly valid or whether the asymptotics depend on the initial data. This issue has by now been resolved by careful numerical simulations and also by a rigorous mathematical analysis [GMS98, Mee99, NP99, NP01]. In fact, it turns out that the LSW model alone does not display universal self-similar large-time behavior if no additional effects are taken into account.

While the latter disadvantage of the LSW theory was not generally accepted, the limited regime of validity of the LSW model due to the mean-field assumption, which neglects local interactions between particles, was recognized early as a significant shortcoming. Several approaches were taken to develop theories which take into account finite volume fraction of the new phase (see e.g. [AV94, BW79, HV88, KET86, Mar87, MR84, TE93, TK84, TKE87, VG84, YE*93]). A key question in this topic is whether and how correlations between particles develop during the evolution and how they influence the coarsening rate.

In this article we review recent progress in the mathematical analysis of fundamental issues in models of Ostwald ripening. We will describe the regime of validity of zero-order mean-field models and their failure in predictive power due to weak selection of self-similar asymptotic states. To overcome these drawbacks we consider extensions of the mean-field model in the sense of a perturbative theory which accounts for finite volume fraction of particles. A main part of this article will be devoted to a review of recent results determining the scaling and precise form of the corrective terms.

First we will clarify why the first order corrections have different relative size in finite and infinite systems respectively. Second, as a main part of this review, we describe the main ideas and results of a perturbative theory which takes correlations between particles into account. Under the assumption that particles are weakly correlated we derive a closed system of equations for the one- and two-particle distribution function. By our method we recover a theory proposed by Marder [Mar87] in the applied literature, but under a more natural closure hypothesis. However, it appears, that the assumption of small correlations is destroyed for the largest particles by the evolution. Here fluctuations are relevant and we present the results of a corresponding recent theory [NV06a] which takes this effect into account. We also briefly discuss the possible role of collisions between particles. This effect seems on first glance negligible, but on closer inspection it turns out, that it might have an important effect on the long-time behavior.

The problem of Ostwald ripening is not only a basic phenomenon in the aging of materials. The structure of the model also poses interesting and challenging mathematical problems and can be viewed as a paradigm of a many-particle system with long range interactions through a slowly decaying field. Similar problems and questions arise for example in sedimentation problems.

We limit our discussion to the issues related to the mathematical analysis of basic models for Ostwald ripening. Further aspects, which are relevant for applications, such as the role of mechanical stresses, but seem presently much less accessible by a mathematical analysis, are discussed for example in the review papers [FPL99, Voo85, Voo92] and the references therein.

2 Models

2.1 The Mullins-Sekerka evolution

The Cahn-Hilliard model [Peg89] describes nucleation, growth and the early stages of coarsening. We are interested in the case where an initially uniform mixture is unstable which means that the free energy E favors two phases. Each phase is characterized by a specific value of the concentration and they are separated by an interfacial layer of characteristic width and of minimal area. The dynamics, driven by the negative gradient $-\nabla u$ of the chemical potential u and limited by the diffusion of each component, reduces the free energy while the spatial average of the relative concentration is preserved.

Coarsening in the Cahn-Hilliard model is well-described by the Mullins-Sekerka sharp-interface model [ABC94, Peg89]. It describes the position of the interfacial layer by the boundary ∂D of a region $D \subset \Omega \subset \mathbb{R}^3$. It is based on the assumption that diffusion in the bulk takes place on a much faster time scale than evolution of the interface. Thus, the potential relaxes at each time instantaneously to equilibrium. The model is given in dimensionless variables by

$$\Delta u = 0 \quad \text{in the bulk } \Omega \backslash \partial D, \tag{2.1}$$

$$u = \kappa \quad \text{at the interface } \partial D\,, \tag{2.2}$$

$$V = [\nabla u \cdot \mathbf{n}] \qquad \text{at the interface } \partial D. \tag{2.3}$$

Here κ denotes the mean curvature of the interface, $\mathbf{n}$ its normal, $[\nabla u \cdot \mathbf{n}]$ the jump of the normal component of the gradient $\nabla u \cdot \mathbf{n}$ across the interface and V its normal velocity. Equation (2.2) is the well known Gibbs-Thomson law, which accounts for surface tension and (2.3) is the Stefan condition, which reflects local conservation of energy. The main global features of this interfacial motion are that it reduces the area of ∂D while it preserves the volume of D, if (2.1), (2.2) are supplemented with periodic or Neumann boundary conditions on $\partial\Omega$.

Some of the upcoming analysis is motivated by the fact that the Mullins-Sekerka evolution has an interpretation as a gradient flow. It is the gradient flow of the surface energy on the manifold of phase distributions with constant volume where the metric tensor is given by the H^{-1} norm in the bulk.

2.2 Small volume fraction

In the following we consider the regime where the volume fraction covered by D is very small. Once the interfacial regime emerges, D consists of many disconnected "particles", which quickly become approximately radially symmetric, and whose centers essentially do not move. Then it is legitimate to replace (2.3) by taking the average over each particle. The particles are then characterized by their immovable center X_i and their radii $R_i(t)$ and (2.3) reduces to

$$\dot{R}_i = \frac{1}{4\pi R_i^2} \int_{\partial P_i} [\nabla u \cdot \mathbf{n}] \, dS, \tag{2.4}$$

where $P_i := B(X_i, R_i(t))$, and (2.2) gives

$$u = \frac{1}{R_i} \qquad \text{on } \partial P_i. \tag{2.5}$$

It has been established by a rigorous analysis in [AF99, AF03] that the Mullins-Sekerka problem is in the regime of small volume fraction well approximated by (2.1), (2.5) and (2.4).

It is not difficult to establish that the system (2.1), (2.4) and (2.5) is locally well-posed if initially particles do not overlap. Such a local in time solution can be extended up to a time when a particle vanishes or when two particles touch. The first event is not critical; we just extend the evolution up to the time when a particle disappears, remove the particle and continue with the remaining ones. In this way we obtain a continuous in time, piecewise smooth solution. The second phenomenon, that particles touch, is problematic. In particular so, as the simplification to spherical shape is not a good approximation if particles are close. In principle one has to go back to the full Mullins-Sekerka evolution (2.1)-(2.3), or one constructs an ad-hoc regularization by merging the two particles into one with the same volume. Since we are interested in the dynamics of a large set of particles with small volume fraction, we expect that such an event is rare if it occurs at all and does not have an influence on the global behavior of the system. The latter is certainly true to leading order, but we will in particular also be interested in higher order terms, where the effect of collisions might have to be taken into account. We will address this issue in the end of this article.

During the subsequent evolution, the large particles grow at the expense of small particles, which eventually vanish. The state of the system is now completely described by the radii $\{R_i\}_i$ of the particles and it is tempting to try to identify the evolution law for $\{R_i\}_i$. Lifshitz, Slyozov and Wagner [LS61, Wag61] have done this by formal arguments and found that this evolution reduces to leading order solely to an evolution of the distribution of the $\{R_i\}_i$. We now investigate in more detail the formal argument of LSW and discuss its crucial assumptions.

2.3 Leading order approximation

In the following we denote by

- ϕ the volume fraction of the particles,
- ρ the density of particles.

The typical distance $\langle d\rangle$ between neighboring particles is given by $\rho^{-1/3}$, whereas the typical particle radius $\langle R\rangle$ is given by

$$\frac{4\pi}{3}\langle R\rangle^3 = \rho\,\phi.$$

We are always interested in the regime of small volume fraction, i.e.

$$\phi \ll 1 \tag{2.6}$$

which in particular implies that

$$\langle R\rangle \ll \langle d\rangle. \tag{2.7}$$

According to (2.1), the chemical potential u is a harmonic function outside the particles. Thanks to the separation of particles expressed in (2.7) there exists a function $\bar{u}$ such that

- $\bar{u}$ is "slowly varying" in the sense that the length scale of variations of $\bar{u}$ is much larger than $\langle d\rangle$,
- $\bar{u} \approx u$ at distances from the particles which are much larger than $\langle R\rangle$.

This function $\bar{u}$ is called the mean field.

Now recall that u is harmonic outside the particles and satisfies the Gibbs-Thomson law (2.5) on the boundary of the particles. Since $u \approx \bar{u}$ away from the particles and since $\bar{u}$ is approximately constant for $|x - X_i| \ll \langle d\rangle$, we find that

$$u \begin{cases} = \frac{1}{R_i} & \text{for } R_i \geq |x - X_i| \\ \approx (1 - R_i\,\bar{u})\,\frac{1}{|x-X_i|} + \bar{u} & \text{for } R_i \leq |x - X_i| \ll \langle d\rangle \end{cases}$$

and thus according to (2.4)

$$\dot{R}_i = [\nabla u \cdot \mathbf{n}] \approx \frac{1}{R_i^2}\,(R_i\,\bar{u} - 1). \tag{2.8}$$

The mean-field $\bar{u}$ has still to be determined. It turns out that the length scale over which $\bar{u}$ is varying is intrinsically determined by the screening effect between particles. This effect is crucial in the theory of Ostwald ripening and we will address is separately in the next section.

2.4 The screening length

The screening length ξ describes the effective range of particle interactions. It is analogous to the classical Debye-Hückel screening length in electrostatics and can heuristically be derived as follows.

Consider a point charge at $X_0 \in \mathbb{R}^3$ surrounded by conducting balls $B(R_i, X_i)$ which are uniformly distributed according to a number density ρ and have volume fraction $\phi \ll 1$. The point charge at X_0 creates an electric field and a corresponding potential G and thus induces a negative charge on $\partial B(R_i, X_i)$. This induced charge roughly equals $-4\pi R_i G(X_i)$, where $4\pi R_i$ is the capacity of a single ball in $\mathbb{R}^3$. In a dilute system capacity is approximately additive which implies that the total negative charge is approximately given by $-4\pi\langle R\rangle\rho G$. Hence the effective electric potential satisfies

$$-\Delta G = \delta_{X_0} - 4\pi\langle R\rangle\rho\, G \qquad \text{in } \mathbb{R}^3,$$

and thus

$$G(x) = \frac{1}{4\pi|x - X_0|} e^{-\frac{|x-X_0|}{\xi}}, \tag{2.9}$$

where

$$\xi = \frac{1}{\sqrt{4\pi\langle R\rangle\rho}} \tag{2.10}$$

is the screening length.

Formula (2.9) shows that the presence of the balls has the effect that the effective range of the electric potential is limited to ξ, whereas the electric potential in a system without balls is just $\frac{1}{4\pi|x-X_0|}$ and decays only slowly.

2.5 The classical LSW model

The classical theory by Lifshitz, Slyozov and Wagner (LSW) [LS61, Wag61] is based on (2.8) and the assumption that the interaction range between particles is infinite. Hence, the mean-field $\bar{u}$ is constant in space and for each time determined by the constraint that the volume fraction of particles is conserved. This leads to the set of equations

$$\begin{aligned} \dot{R}_i &= \frac{1}{R_i^2}\left(R_i\bar{u} - 1\right), \\ \bar{u}(t) &= \frac{\sum_i 1}{\sum_i R_i(t)}, \end{aligned} \tag{2.11}$$

where the sum extends over the particles with positive volume. Notice that the critical radius is given by $1/\bar{u}$, in other words, the critical radius is just the mean radius in the systems. Particles larger than the critical radius grow, smaller ones shrink. However, the critical radius typically increases over time, so that more and more particles start to shrink and finally disappear.

The system (2.11) now translates without further approximation into the following evolution law for the one-particle number density $f_1 = f_1(R,t)$:

$$\partial_t f_1 + \partial_R\Big(\frac{1}{R^2}\left(R\bar{u}(t) - 1\right) f_1\Big) = 0 \tag{2.12}$$

with

$$\bar{u}(t) = \frac{\int f_1(R,t)\, dR}{\int R f_1(R,t)\, dR}. \tag{2.13}$$

It is interesting to note that the LSW-model has an interpretation as a gradient flow, a structure which is inherited from the Mullins-Sekerka model (see also [FSF03] for a direct derivation from thermodynamic extremal principles). This structure is most easily described for a finite collection of particles $\mathbf{R} := \{R_i\}_i$, which corresponds to a discrete size distribution. The corresponding manifold of a set of particles with fixed volume is characterized by

$$\mathcal{M} := \Big\{ \mathbf{R} = \{R_i\}_i \mid \sum_i R_i^3 = const. \Big\} \tag{2.14}$$

and the tangent space by

$$T_{\mathbf{R}}\mathcal{M} := \Big\{ \mathbf{V} = \{V_i\}_i \mid \sum_i R_i^2 V_i = 0 \Big\}.$$

The energy is given by the surface energy

$$E(\mathbf{R}) = \tfrac{1}{2} \sum_i R_i^2. \tag{2.15}$$

On $\mathcal{M}$ we introduce a Riemannian metric via the metric tensor g given by

$$g_{\mathbf{R}}(\mathbf{V}, \mathbf{W}) = \sum_i R_i^3 V_i W_i \qquad \text{for all } \mathbf{V}, \mathbf{W} \in T_{\mathbf{R}}\mathcal{M}. \tag{2.16}$$

For a system to be a gradient flow, it means that the direction of the evolution at each time, here represented by $\dot{\mathbf{R}} = \{\dot{R}_i\}$, is in the direction of steepest descent on the energy landscape. This implies that for every time the equation

$$g_{\mathbf{R}}(\dot{\mathbf{R}}, \mathbf{V}) = -dE(\mathbf{R})\mathbf{V} \qquad \text{for all } \mathbf{V} \in T_{\mathbf{R}}\mathcal{M}$$

must be satisfied. In view of (2.15) and (2.16) this is equivalent to

$$\dot{R}_i = -\frac{1}{R_i^2} + \frac{\lambda}{R_i}$$

for a Lagrange-multiplier λ, which is defined by volume conservation. One easily finds that $\lambda = \sum_i 1 / \sum_i R_i$, i.e. $\lambda = \bar{u}$ as defined in (2.11).

A recent proof of well-posedness of the LSW model and variants of it [NP05] makes significant use of the corresponding dissipation identity (see also [CG00, Lau01, Lau02, NP00] for further results on well-posedness).

Of main interest is whether the LSW model exhibits universal dynamic scaling. In fact, the model has the scale invariance $R \sim t^{1/3}$. Therefore, it is not surprising that it has one smooth self-similar solution. Folklore suggests that this self-similar solution characterizes the long-time behavior of all solutions, thereby capturing the statistical self-similarity of Ostwald ripening. Based on this, LSW have predicted the following growth law for the expected particle radius $\langle R \rangle$:

$$\langle R \rangle \approx \left(\frac{4}{9}t\right)^{1/3} \tag{2.17}$$

and a specific unique shape for the rescaled size distribution. Unfortunately, the classical LSW-theory has two serious shortcomings which will be discussed in the next section.

3 Validity and failure of classical mean-field models

3.1 Long-time dynamics and weak selection

In fact, (2.12) has not only one but a one-parameter family of self-similar solutions. All members of this family have compact support and can be characterized by their behavior at the end of the support, that is their behavior for large particles. More precisely, for any $\alpha \in (-1, \infty]$ there exists a self-similar solution behaving like a polynomial with power α (resp. like an exponential for $\alpha = \infty$) at the end of its support. While Wagner [Wag61] seems not to recognize the existence of self-similar solutions for $\alpha < \infty$, Lifshitz and Slyozov [LS61] find all self-similar solutions, but argue that only the one with $\alpha = \infty$ is stable. Their stability argument takes so called "encounters" between particles into account, which are not represented in the LSW model (2.12). However, it seems that the latter argument has not been taken into account in the controversial discussion in the applied literature [Bro89, Bro90a, Bro90b, HHR92, HH*89, Hoy90] and the question remained, whether or not, universal self-similar behavior in the LSW model (2.12). Only several years later it has been recognized in [GMS98, MS96] and shown by a rigorous analysis in [NP99, NP01] that the long-time behavior of the LSW-equation is not at all universal but depends very sensitively on the data.

Let us describe the main features of the long-time behavior in more detail: the first mathematically rigorous results were obtained in [NP99] (and in [NP01] for a related problem; see also [CG04] for simulations) for data with compact support. The main result gives a necessary condition for convergence towards any self-similar solution with $\alpha \in (-1, \infty)$. It turns out that the data have to be regularly varying at the end of their support with the same power α. This result in particular also implies that for a large class of data - the ones which are not regularly varying - no self-similar behavior occurs. The results of [NP99] were improved in [NV06b], where it is shown that the condition of regular variation is also sufficient for convergence for data which are small perturbations of the self-similar solution. Convergence for general regularly varying data has been established for the case that α is not too large. Furthermore, a kind of stability result is provided, which is perhaps the most interesting from the point of view of applications. It states – roughly speaking – that for data which are not regularly varying, but whose variation is bounded, the system still coarsens with the expected rates. In contrast to [KO02], where weak upper bounds for coarsening rates within the

Cahn-Hilliard model are established (see also [CNO06] for the dependence of coarsening rates on volume fraction), the analysis in [NV06b] concerns the simpler mean-field model but provides much stronger results, namely pointwise upper and lower bounds.

All the results [NP99, NP01, NV06b] are for self-similar solutions with finite α. In [NV06c] the case $\alpha = \infty$, that is the case of the LSW solution, is considered. A necessary and locally sufficient criterion on the data for convergence to this self-similar solution is established. In a certain sense the domain of attraction is much larger, since the corresponding condition is much less stringent than regular variation. We also refer to [Car06, CP98], where analogous results for a simplified LSW model have been obtained.

While the case $\alpha = \infty$ is already more involved than for finite α, the analysis of the long-time behavior of the LSW-model for data with noncompact support is again much harder. Partial results have been obtained in [Vel06a, Vel06b], which show that also in this case the long-time behavior is extremely sensitive to the initial data.

Let us also mention here an ad-hoc approach to overcome the unphysical "weak selection problem" in the LSW model. was added in (2.12) In [Mee99] and [Vel98] a second-order term was added to (2.12) accounting for fluctuations from the nucleation regime. The main role of this term is to create an infinite tail for the radius distribution and it is argued that then the only possible limit is the LSW-solution. However, this regularization is derived on an ad-hoc basis and the physical relevance seems not clear at this stage. In Sect. 5 below, we will present results from a recent analysis [NV06a], where a second order term is derived from the Mullins-Sekerka evolution, which accounts for pair-interaction between particles.

3.2 Narrow range of validity of the classical LSW-theory

The second problem within the classical LSW-theory is that all self-similar profiles and the corresponding coarsening rates do not agree with experiments. It is common belief that this discrepancy is due to the finiteness of the volume fraction of particles. As mentioned previously, one can find a large variety of approaches in the physics literature to predict the first-order correction to the zero order theory (e.g. [AV94, BW79, HV88, KET86, Mar87, MR84, TE93, TK84, TKE87, VG84, YE*93]). One main goal of this paper is to summarize how to access the first-order correction analytically.

Before we start let us briefly summarize the by now fairly complete results on the rigorous derivation of zero-order models. The first rigorous derivation of the classical LSW-model in the case that the total system size is much smaller than the screening length was obtained in [Nie99, Nie00]. In this regime it has been shown in [Vel00] that stochastic effects through noise in the data cannot regularize the weak selection problem in the LSW theory over the relevant time scales (times over which still a sufficiently large number of particles is present). In the case that the system is of the order of the screening length or larger one obtains in fact an inhomogeneous extension of the LSW model. In this regime the evolution for $\{R_i\}_i$ reduces only to an evolution of the joint distribution of particle radii and particle centers $f_1(R, X, t)$. We have again (2.8), but (2.13) generalizes to

$$-\Delta_X \bar{u} + 4\pi \left(\bar{u} \int R\, f_1\, dR - \int f_1\, dR \right) = 0. \tag{3.1}$$

The kinetic equation (2.12) now has X as a parameter. Again, this equation clearly demonstrates effective screening in an arrangement of charged particles, since the Green's function of the operator $-\Delta u + \mu u = 0$, where $\mu = 4\pi \int Rf\, dR$, decays exponentially fast over length scales of the order of ξ. Since in view of (2.10) we have $\xi \sim \left(\frac{\langle d\rangle^3}{\langle R\rangle}\right)^{1/2} \sim \langle d\rangle \phi^{-1/6}$ we have in the regime of small volume fraction that $\xi \gg \langle d\rangle$ and consequently the interaction range of one particle still contains a large number of particles.

In the case that the system is of the order of the screening length equation (3.1) coupled with (2.12) has been rigorously derived in [NO01b] (see also [NO01a] for the more difficult two-dimensional case), a construction of correctors and deterministic error estimates is provided in [GN02].

The work [NV04a, NV04b] handles the most relevant situation where the system size is much larger than the screening length. This implies that when rescaling the system with respect to the natural length scale, the screening length, one obtains a homogenization problem in an unbounded domain. One important step in the analysis is well-posedness for the limit problem in unbounded domains [NV04c] and the result [NV06d] which establishes that the fundamental solution of the microscopic problem decays exponentially w.r.t. the screening length. This allows to "localize" the homogenization procedure in [NV04a, NV04b]. Special care is taken to have sufficiently weak assumptions on the particle distribution which include an initially random distribution of particles. First, it is shown that the evolution preserves such a statistically uniform distribution over time, which allows a proper definition of "the" screening length. Second, it is established that only a very small fraction of particles can overlap and that this does not affect the macroscopic evolution law for the remaining particles. Thus, this result also rules out corrections on the zero order level due to a stochastic nature of the data.

4 First-order corrections

We have seen that the LSW mean-field model is in a certain sense degenerate since it allows for a family of self-similar solutions, where none of them is in any sense distinguished.

Experiments on Ostwald ripening on the other hand only partly confirm the predictions of the LSW-theory: The growth exponent in (2.17) is confirmed but the rate is considerably larger, see [AS*99] for a recent experiment. The general belief is that the deviation is due to the finiteness of ϕ. The zero order theory treats the spatial arrangement as if the particles were infinitely far away. Hence it tends to overestimate the distance over which particles have to diffuse and thus the constant in (2.17) underestimates the rate of coarsening. The magnitude of the expected correction will depend on the spatial correlation between growing and shrinking particles (sinks and sources for the diffusion field u).

4.1 Finite systems

If one tries to analytically access the scaling of the lowest order correction to LSW, one finds that the analysis for large but finite clusters is quite different from that for infinite clusters and yields a different prediction of the scaling. This has lead to a controversy about the relevance of either analysis. More precisely, there seems to be a contradiction between the analysis for infinite systems, which predicts the scaling $\phi^{1/2}$ of the correction term and the analysis for finite systems, which predicts the scaling $\phi^{1/3}$. Based on "snapshot" numerical simulations, Fradkov, Glicksman & Marsh [FGM96], see also [MG*98], predict a cross-over from $\phi^{1/3}$ to $\phi^{1/2}$ with increasing system size. We will now explain the results as well as the snapshot perspective in more detail.

4.1.1 Heuristic arguments

We will give a heuristic argument for a cross-over of the order of the correction to the LSW theory in finite systems. This is most easily explained by the monopole approximation of (2.1)-(2.2), which seems to go back to Weins and Cahn [WC73]: In this approximation, the growth rates $\{B_i\}_i$ of the particle volumes, that is $-B_i := \frac{d}{dt}[\frac{4\pi}{3} R_i^3] = 4\pi R_i^2 \frac{dR_i}{dt}$, are obtained by solving the linear system of equations

$$\frac{1}{R_i} = \bar{u} + \frac{B_i}{R_i} + \sum_{j\neq i} \frac{B_j}{d_{ij}} \tag{4.1}$$

and

$$\sum_i B_i = 0, \tag{4.2}$$

where $d_{ij} := |X_i - X_j|$ is the distance between particle centers. Observe that $\bar{u}$ can be interpreted as a Lagrange multiplier for (4.2). It has been argued that the error coming from the monopole approximation is of higher order in ϕ than the first-order correction to the LSW theory. Indeed, the error is of order $\phi^{2/3}$ as can be deduced e.g. from equation (2.42) of [AV94].

We consider a periodic setting with n particles in the box $\Omega_n := (-\frac{1}{2}(\frac{n}{\rho})^{1/3}, \frac{1}{2}(\frac{n}{\rho})^{1/3})^3$. Then, solving the monopole approximation (4.1) means to invert the matrix $Id - \mathbf{g}$, where $\mathbf{g} = g_{ij} = -\frac{R_i}{d_{ij}}$. This is possible if $\mathbf{g}$ is small enough in an appropriate sense. If we take for instance the matrix norm corresponding to $\max_i |B_i|$ we find

$$\|\mathbf{g}\| = \sup_i \sum_j |g_{ij}| \sim \langle R \rangle \rho \Big(\frac{n}{\rho}\Big)^{2/3}.$$

In view of (2.10) this gives $\|\mathbf{g}\| \sim \Big(\frac{\text{system size}}{\text{screening length}}\Big)^2$. Thus, $\mathbf{g}$ is invertible if the system size is much smaller than the screening length. In this case, the system (4.1) can be solved by an asymptotic expansion using the Neumann series, which leads to

$$B_i = (1 - R_i \bar{u}) + \sum_j g_{ij}(1 - R_j \bar{u}) + \sum_j \sum_k g_{ij} g_{jk}(1 - R_k \bar{u}) + \dots$$

and $\bar{u}$ has to be determined such that $\sum_i B_i = 0$ holds to the desired order. Since the entries of the matrix scale as $\phi^{1/3}$, the ratio between $\langle R \rangle$ and $\langle d \rangle$, this yields an expansion for B_i in the parameter $\phi^{1/3}$.

4.1.2 Numerical simulations

The starting point in the numerical simulations of Fradkov, Glicksman and Marsh [FGM96, MG*98] is a fixed joint distribution of particle centers $\{X_i\}_i$ in the sphere of volume $\frac{n}{\rho}$ and particle radii $\{R_i\}_i$ with the following properties

- $\{X_i\}_i$ has number density ρ in the sphere of volume $\frac{n}{\rho}$.
- R_i is distributed according to the self-similar LSW distribution with mean volume ϕ.
- R_i and R_j are independent for $i \neq j$.
- $\{X_i\}_i$ and $\{R_i\}_i$ are independent.

The parameters to be varied are the average number $n \gg 1$ and the volume fraction $\phi \ll 1$ of particles within Ω_n.

"Snapshot" analysis means the study of the joint distribution of $\{X_i, R_i, B_i\}_i$, where the $\{B_i\}_i$ are determined according to (4.1). In particular, one is interested in the deviation of $\{B_i\}_i$ from the LSW growth rates $\{B_i^{LSW}\}_i$, implicitly given by the obvious truncation of (4.1)

$$\frac{1}{R_i} = \bar{u}^{LSW} + \frac{B_i^{LSW}}{R_i} \quad \text{and} \quad \sum_i B_i^{LSW} = 0. \tag{4.3}$$

In [FGM96, MG*98], the deviation is measured by considering the rate of change of particle radius $\frac{dR_i}{dt} = -\frac{1}{4\pi R_i^2} B_i$. It is found by numerical simulations for $n \in [100, 3162]$ [FGM96, MG*98] that the quantity

$$\frac{\langle \frac{1}{4\pi R_i^2} (B_i - B_i^{LSW}) \rangle}{\langle \frac{1}{4\pi R_i^2} B_i^{LSW} \rangle}$$

shows a cross-over between $\phi^{1/3}$ and $\phi^{1/2}$ when the sphere becomes larger than the screening length. In view of (2.10) this means when

$$n \sim \phi^{-1/2}. \tag{4.4}$$

4.1.3 Rigorous results

To our knowledge, the numerically observed cross-over has first been unambiguously reproduced by analytical tools in [HNO05b]. This is done under assumptions used in the numerical simulation [FGM96], that is, the monopole approximation and the snapshot perspective.

We have argued that the cross-over in the correction term occurs when the system size becomes of the order of the screening length, in other words, when (4.4) holds. We call in the following particle systems "subcritical systems" or "supercritical systems" if they are smaller or larger than the screening length, respectively.

Heuristic arguments and the numerical simulations in [FGM96, MG*98] suggest that the correction in a supercritical system should scale as:

$$\frac{\langle \frac{1}{4\pi R_i^2} (B_i - B_i^{LSW}) \rangle}{\langle \frac{1}{4\pi R_i^2} B_i^{LSW} \rangle} \sim \phi^{1/2} \quad \text{for} \quad n \gg \phi^{-1/2}. \tag{4.5}$$

In view of (2.10), the quantity $\phi^{1/2}$ is just the ratio between the relevant length scales $\langle R \rangle$ and ξ. Furthermore, we gave some heuristic arguments that the correction term for subcritical systems should scale as $\phi^{1/3}$. Therefore, both scalings coincide at the cross-over $n \sim \phi^{-1/2}$ when the correction term in a subcritical system scales as

$$\frac{\langle \frac{1}{n} \sum_i \frac{1}{R_i^2} (B_i - B_i^{LSW}) \rangle}{\langle \frac{1}{n} \sum_i \frac{1}{R_i^2} B_i^{LSW} \rangle} \sim n^{-1/3} \phi^{1/3} \quad \text{for} \quad n \ll \phi^{-1/2}. \tag{4.6}$$

Instead of considering the expected relative deviation in the rate of change of the mean radius we will investigate the relative deviation in the rate of change of energy. This idea is motivated by the gradient flow structure of the Mullins-Sekerka evolution. More precisely we consider

$$\frac{\dot{E}^{LSW} - \dot{E}}{|\langle \dot{E}^{LSW} \rangle|}, \tag{4.7}$$

where E is the interfacial energy of the particles, i.e.

$$E = \frac{1}{2n} \sum_i R_i^2,$$

and its rate of change is

$$\dot{E} = -\frac{1}{n}\sum_i \frac{B_i}{R_i},$$

while

$$\dot{E}^{LSW} = -\frac{1}{n}\sum_i \frac{B_i^{LSW}}{R_i},$$

with

$$B_i^{LSW} = 1 - R_i \bar{u}^{LSW}, \qquad \bar{u}^{LSW} = \frac{\sum_i 1}{\sum_i R_i}.$$

Since the energy is decreasing, $\dot{E}$ is always negative. Likewise $\dot{E}^{LSW}$ is always negative, but we expect the difference in (4.7) to be positive for most realizations, since the LSW theory should underestimate the coarsening rate. The reason for measuring the correction in terms of (4.7) is that because of the gradient flow structure of the evolution, the quantity (4.7) can be expressed variationally (see (4.9)).

The main result in [HNO05b] is that with high probability

$$\frac{\dot{E}^{LSW} - \dot{E}}{|\langle \dot{E}^{LSW} \rangle|} \sim \left\{ \begin{array}{ll} n^{-1/3}\,\phi^{1/3} & \text{for} \quad n \ll \phi^{-1/2} \\ \phi^{1/2} & \text{for} \quad n \gg \phi^{-1/2} \end{array} \right\}. \tag{4.8}$$

Observe that this result is somewhat stronger than (4.5) and (4.6) in the sense that we make a qualitative statement about the entire distribution, not just its expected value.

The result is derived under some regularity assumptions on the distribution of the particle centers, which are satisfied with high probability for independently distributed centers (see Sect. 2.4 of [HNO05b] for details). Furthermore it is assumed that the radii are independently and identically distributed with respect to a probability density with compact support. In the following $\langle \cdot \rangle$ denotes the expected value with respect to the joint probability measure P of the variables $\{R_i\}_i$.

Henceforth we say that a stochastic variable F satisfies $F \leq C$ with high probability if for any $\varepsilon > 0$ there exists $C(\varepsilon)$ such that $P\big(\{F \leq C\}^c\big) \leq \varepsilon$.

Theorem 4.1. *([HNO05b], Thm. 2.1) (The sub-critical regime)*
If $n \ll \phi^{-1/2}$ we have with high probability that

$$\frac{\dot{E} - \dot{E}^{LSW}}{|\langle \dot{E}^{LSW} \rangle|} \geq -C\, n^{-1/3}\, \phi^{1/3}.$$

Furthermore

$$\frac{\langle \dot{E} - \dot{E}^{LSW} \rangle}{|\langle \dot{E}^{LSW} \rangle|} \leq -\frac{1}{C}\, n^{-1/3} \phi^{1/2}.$$

Theorem 4.2. *([HNO05b], Thm. 2.2) (The super-critical regime)*
If $n \ll \phi^{-1/2}$ and $\phi \le \phi_0$ we have with high probability that

$$-C\,\phi^{1/2} \le \frac{\dot{E} - \dot{E}^{LSW}}{|\langle \dot{E}^{LSW} \rangle|} \le -\frac{1}{C}\,\phi^{1/2}.$$

The proof of these estimates relies on the fact that the underlying evolution has the structure of a gradient flow and thus (4.7) has a variational formulation. To see this, note that a solution of (4.1) can be characterized as a solution of

$$\min_{\{\tilde{B}_i\}_i;\sum_i \tilde{B}_i=0} \left\{ \frac{1}{n}\sum_i \frac{1}{2R_i}\tilde{B}_i^2 + \frac{1}{n}\sum_i\sum_{j\neq i} \frac{\tilde{B}_i\tilde{B}_j}{2d_{ij}} - \frac{1}{n}\sum_i \frac{\tilde{B}_i}{R_i} \right\}.$$

For the solution B_i we have

$$\frac{1}{n}\sum_i \frac{1}{2R_i}B_i^2 + \frac{1}{n}\sum_i\sum_{j\neq i}\frac{B_iB_j}{2d_{ij}} - \frac{1}{n}\sum_i\frac{B_i}{R_i} = -\frac{1}{n}\sum_i \frac{B_i}{2R_i} = \frac{1}{2}\dot{E}.$$

Therefore we can write the deviation in the rate of change of energy from the LSW result in the form:

$$\begin{aligned}\dot{E} - \dot{E}^{LSW} = &\min_{\{\tilde{B}_i\}_i,\sum_i \tilde{B}_i=0} \left\{ \frac{1}{n}\sum_i \frac{1}{R_i}\tilde{B}_i^2 \right.\\ &\left. + \frac{1}{n}\sum_i\sum_{j\neq i}\frac{\tilde{B}_i\tilde{B}_j}{d_{ij}} - \frac{1}{n}\sum_i \frac{2\tilde{B}_i}{R_i} + \frac{1}{n}\sum \frac{B_i^{LSW}}{R_i}\right\}.\end{aligned}$$

We recall that $B_i^{LSW} = 1 - \frac{R_i}{\overline{R}}$, where $\overline{R} = \frac{1}{n}\sum_i R_i$, and use $\sum_i \tilde{B}_i = 0$ to find

$$\begin{aligned}\sum_i \frac{1}{R_i}\left(\tilde{B}_i - \left(1 - \frac{R_i}{\overline{R}}\right)\right)^2 &= \sum_i \frac{1}{R_i}\tilde{B}_i^2 - \sum_i \frac{2}{R_i}\tilde{B}_i\left(1-\frac{R_i}{\overline{R}}\right)\\ &\quad + \sum_i \frac{1}{R_i}\left(1 - \frac{R_i}{\overline{R}}\right)^2\\ &= \sum_i \frac{1}{R_i}\tilde{B}_i^2 - \sum_i \frac{2\tilde{B}_i^2}{R_i} + \sum_i \left(\frac{1}{R_i} - \frac{1}{\overline{R}}\right).\end{aligned}$$

Thus we can express the deviation of the rate of change of energy in the compact form:

$$\dot{E} - \dot{E}^{LSW} = \min_{\{\tilde{B}_i\}_i;\sum_i \tilde{B}_i} \left\{ \frac{1}{n}\sum_i \frac{1}{R_i}\left(\tilde{B}_i - B_i^{LSW}\right)^2 + \frac{1}{n}\sum_i\sum_{j\neq i}\frac{\tilde{B}_i\tilde{B}_j}{d_{ij}} \right\}. \tag{4.9}$$

The variational formulation has the advantage that one can obtain a useful upper bound by finding a suitable trial field $\tilde{B}_i$. The construction of a proper trial field in the super-critical case is guided by the intuition that due to the screening effect the system separates into independent subsystems of the size of the screening length. Indeed, the LSW construction in subsystems of the size of order of the screening length provides a suitable upper bound. The mathematically most interesting part is the lower bound, which is established in Prop. 3.3 of [HNO05b] by the use of Fourier analysis and a suitable splitting of the interaction field in near- and far-field respectively.

Remark: Notice that in the sub-critical regimes we only succeed to derive a lower bound, whereas we obtain an upper bound only for the expected value. It is not surprising, that subcritical systems have less good self-averaging properties than supercritical systems and, in fact, a recent rigorous result [CH*06] shows, that for any $M > 0$ there is a finite probability $\rho_M > 0$ such that $(\dot{E} - \dot{E}^{LSW})/|\langle \dot{E}^{LSW} \rangle| > M$.

4.2 Infinite systems

We are now interested in the precise form of the expected growth rate of a particle up to order $\phi^{1/2}$ in large (or infinite) systems and under natural assumptions on the statistics of the particles.

We start by reviewing some results from the physics literature. Marqusee and Ross [MR84] derive the evolution of the one-point statistics under the assumption of independently and identically distributed particles. They do this by manipulating the non-convergent series in the monopole approximation, which they interpret as a multiple scattering series.

However, it is obvious that the assumption that $\{(R_i, X_i)\}_i$ are statistically independent is not preserved by the evolution: A medium sized particle in the neighborhood of a large particle will shrink faster than in an average environment. Hence a large particle eventually influences the statistics of $\{(R_i, X_i)\}_i$ within the screening length. This in turn will influence the evolution of that large particle.

Marder [Mar87] takes this effect into account and derives the evolution of the two-point statistics up to an error $o(\phi^{1/2})$. Starting from the monopole approximation he generates a hierarchy of equations for the expectation value of B_1 conditioned on the position and radius $(R_1, X_1), \cdots, (R_k, X_k)$ of a finite number of particles. He truncates the hierarchy on the level of two-particle statistics by a closure hypothesis.

In the second part of the paper, Marder performs an analysis of the evolution for the two-point statistics. He assumes that particles are initially independently distributed and then linearizes around the Marqusee-Ross theory. The resulting equations are solved numerically. As an effect of correlations, Marder's theory predicts a significantly stronger broadening of the self-similar particle size distribution than the Marqusee-Ross theory.

Yet a different calculus has been developed in Tokuhama, Enomoto and Kawasaki [KET86, TE93, TK84, TKE87]. They also start from the monopole approximation (4.1) but allow for arbitrary correlations. By splitting the matrix in the monopole approximation into a deterministic and a fluctuating part they obtain a first-order correction in $\phi^{1/2}$ which resemble the theory of Marqusee-Ross and in addition contains a term coming from correlations which is however not explicit.

4.3 A perturbative theory to capture correlations

A new method to identify the conditional expectations of particle growth rates has recently been proposed in [HNO05a]. We will now summarize the method and results of this work, which recovers Marder's theory but under a more natural closure assumption. As a byproduct, one also obtains a simple derivation of Marqusee-Ross's evolution for the one-particle statistics.

The main step is to rederive the expected value of the growth rate of particle 1 conditioned on particles 1 and 2, denoted by $\langle B_1|(R_1, X_1), (R_2, X_2)\rangle$. Motivated by a cluster expansion, we assume that the joint probability distribution of $\{(R_i, X_i)\}_{i\geq 1}$ has a special form which only depends on the one- and two-particle statistics. To express $\langle B_1|(R_1, X_1), (R_2, X_2)\rangle$ in terms of these one- and two-particle statistics we employ a method which allows us to separate screening and correlation effects. The idea is to relate the system with all particles $\{(R_i, X_i)\}_{i\geq 1}$ to the system $\{(R_i, X_i)\}_{i\geq k+1}$ where k particles have been removed. This amounts to one step in the Schwarz alternating method, a deterministic argument which captures the screening effects. If $\{(R_i, X_i)\}_{i\geq 1}$ are independent, $\{(R_i, X_i)\}_{i\geq 1}$ and $\{(R_i, X_i)\}_{i\geq k}$ are statistically equivalent in an infinite system. Hence expectations conditioned on the removed particles $\{(R_i, X_i)\}_{i\leq k}$ can be replaced by unconditioned expectations. This allows to derive closed equations for conditional expectations.

4.3.1 Statistical setup

As before we consider a periodic setting with n particles in the box Ω_n, but think now of the limit $n \uparrow \infty$. In particular we always assume $n^2\phi \gg 1$, i.e. the box size is much larger than the screening length.

We make the following assumptions on the statistics of our system. The distribution of $\{(R_i, X_i)\}_{i\geq 1}$ is defined by a probability distribution

$$p_n(R_1, X_1, \ldots, R_n, X_n)\, dR_1\, d^3X_1 \ldots dR_n\, d^3X_n =: p_n(1, \ldots, n)\, d(1) \ldots d(n).$$

It is natural to assume that the distribution is invariant under particle exchange, that is

$$p_n(\sigma(1), \ldots, \sigma(n)) = p_n(1, \ldots, n) \tag{4.10}$$

for all permutations σ, and invariant under translation, that is

$$p_n(R_1, X_1 - x, \ldots, R_n, X_n - x) = p_n(R_1, X_1, \ldots, R_n, X_n) \tag{4.11}$$

for all $x \in \mathbb{R}^3$. The probability distribution of the k-particle statistics $\{(R_i, X_i)\}_{1 \le i \le k}$ is then given by

$$p_k(1, \ldots, k) = \int p_n(1, \ldots, n)\, d(k+1) \ldots d(n).$$

Conditional expectations of a random variable $v = v(1, \ldots, n)$ are given by

$$\langle v(1, \ldots, n) \,|\, 1, \ldots, k \rangle = \int v(1, \ldots, n) \frac{p_n(1, \ldots, n)}{p_k(1, \ldots, k)}\, d(k+1) \ldots d(n).$$

In the limit $n \to \infty$ it is more convenient to work with number densities. The one-particle density is given by

$$f_1(R) \overset{(4.11)}{=} f_1(R, X) = \langle \sum_{i=1}^{n} \delta(R - R_i)\delta(X - X_i) \rangle \overset{(4.10)}{=} n p_1(R, X). \tag{4.12}$$

The two-particle density is given by

$$\begin{aligned} f_2(R, \tilde{R}, X - \tilde{X}) &\overset{(4.11)}{=} f_2(R, X, \tilde{R}, \tilde{X}) \\ &= \langle \sum_{i=1}^{n} \sum_{j \neq i}^{n} \delta(R - R_i)\delta(X - X_i)\delta(\tilde{R} - R_j)\delta(\tilde{X} - X_j) \rangle \\ &\overset{(4.10)}{=} n(n-1) \langle \delta(R - R_1)\delta(X - X_1)\delta(\tilde{R} - R_2)\delta(\tilde{X} - X_2) \rangle \\ &= n(n-1) p_2(R, X, \tilde{R}, \tilde{X}) \end{aligned}$$

and so on. The volume fraction ϕ is given by the one-particle density via

$$\phi = \int \frac{4\,\pi}{3}\, R^3 f_1(R)\, dR\,,$$

the number density by

$$\rho = \int f_1(R)\, dR\,,$$

and the capacity density, which defines the screening length, via

$$\frac{1}{\xi^2} = \int 4\,\pi\, R\, f_1(R)\, dR\,. \tag{4.13}$$

We are ultimately interested in the expected values of the growth rates $\langle B_1 \,|\, 1 \rangle$, $\langle B_1 \,|\, 1, 2 \rangle$, $\langle B_1 \,|\, 1, 2, 3 \rangle$ etc., which appear in the evolution equations for the one-, two- and three-particle statistics. Indeed, we have for $f_1 = f_1(R_1, t)$ and $f_2 = f_2(R_1, X_1, R_2, X_2, t)$ that

$$\begin{aligned} \frac{\partial f_1}{\partial t} &= \frac{\partial}{\partial R_1} \Big(\frac{1}{R_1^2} \langle B_1 \,|\, 1 \rangle\, f_1 \Big), \\ \frac{\partial f_2}{\partial t} &= \frac{\partial}{\partial R_1} \Big(\frac{1}{R_1^2} \langle B_1 \,|1, 2 \rangle\, f_2 \Big) + \frac{\partial}{\partial R_2} \Big(\frac{1}{R_2^2} \langle B_2 \,|1, 2 \rangle\, f_2 \Big), \quad \text{etc.} \end{aligned} \tag{4.14}$$

4.3.2 Statistical assumptions

As pointed out before, we will derive the growth rates $\langle B_1 \,|\, 1\rangle$ etc. under different assumptions on the distribution of particles.

Marqusee-Ross assumed that at any time the system described by $\{(X_i, R_i)\}_{i\geq 1}$ is statistically homogeneous and uncorrelated. More precisely, this means in terms of the probability distribution that

$$p_n(1,\ldots,n) = \Pi_{i=1}^{n} p_1(i). \tag{4.15}$$

Our main goal however is to allow also for correlations between particles. Pair-, triple- and higher correlations in the particle distribution are given by

$$\begin{aligned} q_2(1,2) &= p_2(1,2) - p_1(1)p_1(2), \\ q_3(1,2,3) &= p_3(1,2,3) - \big(p_1(1)p_1(2)p_1(3) \\ &\qquad + q_2(1,2)p_1(3) + q_2(2,3)p_1(1) + q_2(1,3)p_1(2)\big), \end{aligned}$$

etc..

In the following we only retain pair correlations, that is we postulate

$$q_k \equiv 0 \qquad \text{for } k \geq 3.$$

Moreover, we neglect products of p_2 in the cluster expansion, such that we arrive at

$$p_n(1,\ldots,n) = \Pi_{i=1}^{n} p_1(i) + \sum_{i=1}^{n} \sum_{j>i} q_2(i,j) \Pi_{k\neq i,j} p_1(k). \tag{4.16}$$

In the large system limit it is convenient to work with number density based quantities:

$$\begin{aligned} g_2(1,2) &:= n(n-1)q_2(1,2) = f_2(1,2) - \tfrac{n-1}{n} f_1(1)f_2(2) \\ g_3(1,2,3) &:= f_3(1,2,3) - \tfrac{(n-1)(n-2)}{n^2} f_1(1)f_1(2)f_1(3) \\ &\qquad - \tfrac{(n-2)}{n}\big(g_2(1,2)f_1(3) + g_2(2,3)f_1(1) + g_2(1,3)f_1(2)\big). \end{aligned}$$

We assume that pair correlations are small, that is

$$\frac{g_2(i,j)}{f_1(i)f_1(j)} = O(\phi^{1/2}), \tag{4.17}$$

and vanish for large distances, that is

$$\frac{g_2(i,j)}{f_1(i)f_1(j)} = o(\phi^{1/2}) \qquad \text{for } \xi \ll |X_i - X_j| \ll \left(\frac{n}{\rho}\right)^{1/3}. \tag{4.18}$$

Due to the good ergodicity properties enforced by (4.18) the spatial average of u equals the ensemble average, that is

$$⨍_{\Omega_n} u(x)\, d^3x = \langle ⨍_{\Omega_n} u(x)\, d^3x\rangle =: \bar{u} \qquad \text{in the limit } n \to \infty. \tag{4.19}$$

4.3.3 On the expected value of the Green's function

As we have seen in Sect. 2.4, one way to express the screening effect is to establish that the effective operator of the Laplace operator in domains with holes is the Helmholtz operator (see e.g. [CM97] and the references therein).

We will present here the main ideas of another derivation of this fact, which will also illustrate our method by a simplified problem. The link is made via the expected value of the Green's function which agree up to order $O(\phi^{1/2})$.

We denote by $G_j^{(1,\ldots,k)}(x)$ the Green's function for $\mathbb{R}^3 \setminus \cup_{i\geq k+1} P_i$ with singularity in X_j, $j \in \{1, \ldots, k\}$: i.e.

$$\begin{aligned} -\Delta_x G_j^{(1,\ldots,k)} &= 4\pi\delta_{X_j} \qquad \text{in } \mathbb{R}^3 \setminus \cup_{i\geq k+1} P_i, \\ G_j^{(1,\ldots,k)} &= 0 \qquad \text{in } \cup_{i\geq k+1} P_i. \end{aligned} \tag{4.20}$$

Lemma 4.3. *([HNO05a], Lemma 3.3)*
Under the assumptions (4.16), (4.17) *and* (4.18) *we obtain in the infinite volume limit*

$$\langle G_1^{(1)}(x) \,|\, 1 \rangle - \frac{1}{|x - X_1|} e^{-\frac{|x-X_1|}{\xi}} = O(\phi^{1/2}) \min \left\{ \frac{1}{\xi}, \frac{1}{|x - X_1|} \right\},$$

for all $x \in \mathbb{R}^3 \setminus P_1$.

We indicate the main idea of the proof. Roughly speaking, our claim is that $\langle G_1^{(1)}(x) \,|\, 1 \rangle$ is an approximate solution of

$$-\Delta \langle G_1^{(1)}(x) \,|\, 1 \rangle + \frac{1}{\xi^2} \langle G_1^{(1)}(x) \,|\, 1 \rangle = 4\pi\delta(x - X_1). \tag{4.21}$$

We give here a sketch of the argument; the control of the error terms can be found in the proof of Lemma 4.4 in [HNO05a].

To this purpose we introduce the charges of $G_1^{(1)}$ on $\{\partial P_i\}_{i\geq 2}$ by

$$B_{1,i}^{(1)} := \frac{1}{4\pi} \int_{\partial P_i} \frac{\partial G_1^{(1)}}{\partial \mathbf{n}}, \tag{4.22}$$

so that up to dipolar terms

$$-\Delta G_1^{(1)}(x) = 4\pi\delta(x - X_1) + \sum_{i\geq 2} B_{1,i}^{(1)} \, 4\pi\delta(x - X_i). \tag{4.23}$$

As a first approximation the Green's function for the system $\{(R_i, X_i)\}_{i\geq 2}$ can be approximated by the Green's function for the reduced system $\{(R_i, X_i)\}_{i\geq 3}$ as follows:

$$G_1^{(1)}(x) \approx G_1^{(1,2)}(x) - G_1^{(1,2)}(X_2) \, R_2 \, G_2^{(1,2)}(x),$$

which in view of (4.22) leads to

$$B_{1,2}^{(1)} \approx -R_2 G_1^{(1,2)}(X_2).$$

Inserting this into (4.23) (with particle 2 replaced by i) already yields a form similar to (4.21):

$$\begin{aligned} &-\Delta G_1^{(1)}(x) + \sum_{i\geq 2} 4\pi R_i\, \delta(x-X_i)\, G_1^{(1,i)}(x) \\ &= -\Delta G_1^{(1)}(x) + \sum_{i\geq 2} 4\pi R_i\, \delta(x-X_i)\, G_1^{(1,i)}(X_i) \\ &\approx 4\pi\delta(x-X_1). \end{aligned} \tag{4.24}$$

We now take conditional expectations:

$$\begin{aligned} &-\Delta\langle G_1^{(1)}(x)\,|\,1\rangle + \Big\langle \sum_{i\geq 2} 4\pi R_i\, \delta(x-X_i)\, \langle G_1^{(1,2)}(x)\,|\,1,2\rangle\,\Big|\,1\Big\rangle \\ &\overset{(4.10)}{=} -\Delta\langle G_1^{(1)}(x)\,|\,1\rangle + \Big\langle \sum_{i\geq 2} 4\pi R_i\, \delta(x-X_i)\, \langle G_1^{(1,i)}(x)\,|\,1,i\rangle\,\Big|\,1\Big\rangle \\ &= -\Delta\langle G_1^{(1)}(x)\,|\,1\rangle + \Big\langle \sum_{i\geq 2} 4\pi R_i\, \delta(x-X_i)\, G_1^{(1,i)}(x)\,\Big|\,1\Big\rangle \\ &\approx 4\pi\delta(x-X_1). \end{aligned} \tag{4.25}$$

Since our system is nearly decorrelated, we expect

$$\langle G_1^{(1,2)}(x)\,|\,1,2\rangle \approx \langle G_1^{(1,2)}(x)\,|\,1\rangle\,. \tag{4.26}$$

This allows us to appeal to the following argument: In the infinite volume limit the removal of one particle is immaterial:

$$\langle G_1^{(1,2)}(x)\,|\,1\rangle \approx \langle G_1^{(1)}(x)\,|\,1\rangle\,. \tag{4.27}$$

Inserting (4.26) and (4.27) into (4.25) yields

$$-\Delta\langle G_1^{(1)}(x)\,|\,1\rangle + \langle \sum_{i\geq 2} 4\pi R_i\, \delta(x-X_i)\,|\,1\rangle\langle G_1^{(1)}(x)\,|\,1\rangle \approx 4\pi\delta(x-X_1). \tag{4.28}$$

Since our system is nearly decorrelated, we expect

$$\Big\langle \sum_{i\geq 2} 4\pi R_i\, \delta(x-X_i)\,\Big|\,1\Big\rangle \approx \Big\langle \sum_{i\geq 2} 4\pi R_i\, \delta(x-X_i)\Big\rangle. \tag{4.29}$$

In the infinite volume limit, we have

$$\begin{aligned} \Big\langle \sum_{i\geq 2} 4\pi R_i\, \delta(x-X_i)\Big\rangle &\approx \Big\langle \sum_{i\geq 1} 4\pi R_i\, \delta(x-X_i)\Big\rangle \\ &\overset{(4.12)}{=} \int 4\pi R_1 \delta(x-X_1) f_1(R_1)\, dR_1\, d^2X_1 \\ &= \int 4\pi R_1\, f_1(R_1)\, dR_1 \\ &\overset{(4.13)}{=} \frac{1}{\xi^2}\,. \end{aligned} \tag{4.30}$$

Inserting (4.29) and (4.30) into (4.28) yields (4.21).

4.3.4 Separation of screening and correlations

Using the same strategy as in Lemma 4.3 one can also relate the growth rate of particle 1 conditioned on a finite number of particles to the system where this finite number of particles has been removed.

In analogy to the last section we denote by $u^{(1,\dots,k)}$ the solution of the elliptic boundary value problem (2.2), (2.3) in the system where particles $1,\dots,k$ have been removed. Our crucial result in the derivation of first order corrections to LSW is the following hierarchy for the expected growth rates.

Lemma 4.4. *([HNO05a], Lemma 3.5) Under the assumptions* (4.16), (4.17) *and* (4.18) *we obtain*

$$\langle B_1 \,|\, 1\rangle = \Big(1 + \frac{R_1}{\xi}\Big)\big(1 - R_1 \langle u^{(1)}(X_1) \,|\, 1\rangle\big) + o(\phi^{1/2}), \tag{4.31}$$

$$\begin{pmatrix} \langle B_1 \,|\, 1,2\rangle \\ \langle B_2 \,|\, 1,2\rangle \end{pmatrix} = \begin{pmatrix} 1 + \frac{R_1}{\xi} & -\frac{R_1}{d_{12}} e^{-\frac{d_{12}}{\xi}} \\ -\frac{R_2}{d_{12}} e^{-\frac{d_{12}}{\xi}} & 1 + \frac{R_2}{\xi} \end{pmatrix} \cdot \begin{pmatrix} 1 - R_1 \langle u^{(1,2)}(X_1) \,|\, 1,2\rangle \\ 1 - R_2 \langle u^{(1,2)}(X_2) \,|\, 1,2\rangle \end{pmatrix} + o(\phi^{1/2}), \tag{4.32}$$

$$\begin{pmatrix} \langle B_1 \,|\, 1,2,3\rangle \\ \langle B_2 \,|\, 1,2,3\rangle \\ \langle B_3 \,|\, 1,2,3\rangle \end{pmatrix} = \begin{pmatrix} 1 + \frac{R_1}{\xi} & -\frac{R_1}{d_{12}} e^{-\frac{d_{12}}{\xi}} & -\frac{R_1}{d_{13}} e^{-\frac{d_{13}}{\xi}} \\ -\frac{R_2}{d_{12}} e^{-\frac{d_{12}}{\xi}} & 1 + \frac{R_2}{\xi} & -\frac{R_2}{d_{23}} e^{-\frac{d_{23}}{\xi}} \\ -\frac{R_3}{d_{13}} e^{-\frac{d_{13}}{\xi}} & -\frac{R_3}{d_{23}} e^{-\frac{d_{23}}{\xi}} & 1 + \frac{R_3}{\xi} \end{pmatrix} \cdot \begin{pmatrix} 1 - R_1 \langle u^{(1,2,3)}(X_1) \,|\, 1,2,3\rangle \\ 1 - R_2 \langle u^{(1,2,3)}(X_2) \,|\, 1,2,3\rangle \\ 1 - R_3 \langle u^{(1,2,3)}(X_3) \,|\, 1,2,3\rangle \end{pmatrix} + o(\phi^{1/2}). \tag{4.33}$$

Like in LSW, that is $B_1 = 1 - R_1 \bar{u}$, the formulas in Lemma 4.4 relate the conditional particle growth rates to mean-fields $\langle u^{(1)}(X_1) \,|\, 1\rangle$, etc. The new elements are the factors

$$\Big(1 + \frac{R_1}{\xi}\Big), \quad \begin{pmatrix} 1 + \frac{R_1}{\xi} & -\frac{R_1}{d_{12}} e^{-\frac{d_{12}}{\xi}} \\ -\frac{R_2}{d_{12}} e^{-\frac{d_{12}}{\xi}} & 1 + \frac{R_2}{\xi} \end{pmatrix}, \quad \dots \quad ,$$

which capture screening. As opposed to the LSW-theory, which overestimates the distance between particles, these screening factors reflect the fact that the interaction range is finite and contributes as an amplification factor in the growth rates.

The effect of correlations in the particle distribution are contained in the expressions for the mean-field $\langle u^{(1)}(X_1) \,|\, 1\rangle$, $\langle u^{(1,2)}(X_1) \,|\, 1,2\rangle$ etc..

4.3.5 Independent particles

The results of the previous section now allow an easy derivation of the theory for independently distributed particles which was first derived by Marqusee-Ross. Our approach avoids the non-converging series in the monopole approximation, instead we directly calculate finite particle statistics. In fact, our hierarchy of formulas, Lemma 4.4, allows us to make efficient use of the assumption of statistical independence (4.15).

Proposition 4.5. *([HNO05a], Proposition 3.7) Under the assumption* (4.10), (4.11) *and* (4.15) *we find in the infinite volume limit*

$$\langle B_1 \rangle = 0, \tag{4.34}$$

$$\langle B_1 \,|\, 1 \rangle = \Big(1 + \frac{R_1}{\xi}\Big)(1 - R_1 \bar{u}) + o(\phi^{1/2}). \tag{4.35}$$

It is straightforward to derive (4.35) from formula (4.31) in Lemma 4.4. Indeed, since $u^{(1)}$ does not depend on particle 1, we have $\langle u^{(1)}(x) \,|\, 1\rangle = \langle u^{(1)}(x)\rangle$. Now, since particles are statistically independent, we find $\langle u^{(1)}(x)\rangle = \langle u(x)\rangle$. However, the infinite system $\{(R_i, X_i)\}_{i\geq 1}$ is statistically equivalent to $\{(R_i, X_i)\}_{i\geq 2}$. Hence, due to translation invariance we find that $\langle u(x)\rangle$ is constant and recalling (4.19) we finally have $\langle u^{(1)}(x) \,|\, 1\rangle = \bar{u}$. Since particles are identically distributed we obtain (4.34) directly from (4.2). The mean field $\bar{u}$ is determined by (4.34) & (4.35) and given by

$$\bar{u} \;=\; \frac{1 + \frac{\langle R_1 \rangle}{\xi}}{\langle R_1 \rangle + \frac{\langle R_1^2 \rangle}{\xi}} + \frac{1}{\langle R_1 \rangle} o(\phi^{1/2}). \tag{4.36}$$

The system (4.14), (4.35) and (4.36) is precisely the Marqusee-Ross theory.

4.3.6 Weakly correlated particles

In this section, we rederive the Marder theory under assumptions (4.16), (4.17) and (4.18) on the statistics of the particles.

Proposition 4.6. *([HNO05a], Proposition 3.9)*
Under the assumptions (4.10), (4.11), (4.16), (4.17) *and* (4.18) *we find in the infinite volume limit*

$$\langle B_1 \rangle = 0,$$

$$\langle B_1 \,|1 \rangle = \Big(1 + \frac{R_1}{\xi}\Big)(1 - R_1(\bar{u} + \delta u_1)) + o(\phi^{1/2}), \tag{4.37}$$

$$\langle B_1 \,|1,2 \rangle = \Big(1 + \frac{R_1}{\xi}\Big)(1 - R_1(\bar{u} + \delta u_1 + \delta u_2)) - \frac{R_1}{d_{12}} e^{-\frac{d_{12}}{\xi}} (1 - R_2 \bar{u}) + o(\phi^{1/2}) , \tag{4.38}$$

where for $i = 1, 2$

$$\delta u_i = \int \frac{e^{-\frac{|y-X_1|}{\xi}}}{|y-X_1|}(1 - R\bar{u})\frac{g_2(R_i, X_i, R, y)}{f_1(R_i)}\, dR\, d^3y = O(\phi^{1/2}). \tag{4.39}$$

The mean field $\bar{u}$ is as in the Marqusee-Ross theory implicitly determined by $\langle B_1 \rangle = 0$.

The last term in (4.38) quantifies how a large particle 2 will negatively affect the growth rate $-B_1$ of particle 1: Particle 1 will grow below average. Hence the large particle 2 over the course of time affects the particle cloud in its neighborhood, as described by (4.14). The quantity $g_2(R_2, X_2, R, y)$ keeps book of this impact. which leads to the deviation δu_2 in the mesoscopic mean field from its average value $\bar{u}$ as described by (4.39). The equation (4.37) (with particle 1 replaced by particle 2) shows how this in turn influences the growth rate of particle 2.

Up to an implicit term of order $O(\phi)$, (4.37), (4.38) is indeed identical with Marder's result [Mar87, (2.31)]. Our derivation differs however from Marder's in the initial assumption. Marder postulates, cf. [Mar87, (2.25)], the following relation between conditional expected charge distributions

$$\langle \sum_{i\geq 1} B_i \delta(x - X_i) \,|\, 1, 2 \rangle = \langle \sum_{i\geq 1} B_i \delta(x - X_i) \,|\, 1 \rangle + \langle \sum_{i\geq 1} B_i \delta(x - X_i) \,|\, 2 \rangle.$$

This assumption seems unsatisfactory, since it is an assumption on the solution $\{B_i\}_{i\geq 1}$. We replace this assumption on the solution by the assumption on the cluster expansion (4.16), which is an assumption on the data $\{(R_i, X_i)\}_{i\geq 1}$.

4.3.7 Is Marder's model self-consistent?

The model derived in Sect. 4.3 has still a snapshot perspective in the sense that we derive the results under the assumptions (4.16) and (4.17) which are a priori not preserved under the evolution. One can show (cf. [HNO05a], Ch. 3.6) that the assumptions are self-consistent for times of order $\langle R \rangle^3$ (the natural time scale) for the "bulk" of particles. However, it is argued in [NV06a] that the assumptions are not self-consistent in a boundary layer near the largest particles in the system.

The basic reasoning goes as follows. Suppose one solves the model (4.14), (4.37), (4.38) for uncorrelated initial data, where $f_1(R_1, 0)$ has compact support. Consequently, the support of $f_2 = f_2(R_1, R_2, X_1 - X_2, 0)$ is also compact in R_1 and R_2. However, the evolution of R_1 and R_2, determined by (4.38), depends on space due to the term $e^{-d_{12}/\xi}/d_{12}$ in (4.38). Therefore, particles R_1 and R_2 which are at a distance smaller than ξ evolve differently than particles R_1 and R_2 which are at a distance much larger than ξ. As a consequence also the support of f_2 in R_1 and R_2 varies in space. It is not too difficult to see (cf. [NV06a]), that regions in the variables (R_1, R_2, X) develop where f_2 identically vanished but where $f_1(R_1)$ and $f_1(R_2)$ not. In these regions g_2 is of the order $f_1(R_1)f_1(R_2)$, such that (4.17) is not satisfied.

5 Fluctuations of largest particles

For the reasons described in the previous section one cannot assume that correlations are small around the largest particles. Consequently, a different kind of correction has to be derived. The onset of large correlations for the largest particles is a kind of hydrodynamic regime which has to be described by a suitable boundary layer. Such a model has been derived in [NV06a]. A main ingredient is a closure relation which expresses f_2 by evaluating f_1 at a certain shift in the variable R_1. The second main task is then to explicitly compute the shift term to leading order. Similarly to the derivation of the results of Sect. 4.3, a main idea in the analysis is to express relevant quantities in a system of particles through the ones in a system where a particle has been removed. The final step consists in a Taylor expansion of f_1 that leads to a second order equation for the evolution of f_1.

The resulting model has the following form:

$$\frac{\partial f_1(R_1,t)}{\partial_t} - \frac{\partial}{\partial R_1}\left(\left(\frac{1}{(R_1)^2} - \frac{1}{R_1\langle R\rangle}\right) f_1(R_1,t)\right) \tag{5.1}$$
$$= \frac{\partial}{\partial R_1}\left(\left[\frac{1}{R_1}\int_0^t \frac{\int_{\mathbb{R}^3} W(s,t,x)\frac{1}{4\pi|x|}dx}{R_L(s,t,R_1)\frac{\partial R_L(s,t,R_1)}{\partial R}}ds\right]\frac{\partial f_1(R_1,t)}{\partial R_1}\right)$$

where the function W satisfies the integral equation

$$W(s,t,x) + \frac{4\pi}{\langle R\rangle(s)}\int_0^s K(s,\tau,\bar{t})\left(\int_{\mathbb{R}^3} G(x-y,t)W(\tau,t,y)\,dy\right)d\tau \tag{5.2}$$
$$= 4\pi G(x,t)\int\left(1 - \frac{R_L(s,t,R)}{\langle R\rangle(s)}\right)\left(1 - \frac{R}{\langle R\rangle(t)}\right) f_1(R,t)\,dR$$

and the kernel K is given by

$$K(s,\tau,\bar{t}) = \int_{\{R\,:\,R_L(s,t,R)>0\}} \frac{\frac{\partial R_L(s,t,R)}{\partial R}}{R_L(\tau,t,R)\frac{\partial R_L(\tau,t,R)}{\partial R}} f_1(R,\bar{t})\,dR \tag{5.3}$$

Here G is as in (2.9) (with $X_0 = 0$) and $R_L = R_L(s,t,R)$ satisfies the LSW equation, that is $\partial_s R_L = \frac{1}{R_L^2}(R_L/\langle R\rangle - 1))$, with $R_L(t,t,R) = R$.

If we investigate the order of terms, one finds that (5.2) implies that W is of order $\frac{1}{\xi}$. This in turn gives that the "diffusion coefficient", the term in front of $\frac{\partial f_1}{\partial R_1}$ is also of order $\frac{1}{\xi}$. Compared with the terms on the left-hand side, we find that the relative size of the diffusive term if indeed of order

$\frac{\langle R \rangle}{\xi} = O(\phi^{1/2})$. However, close to the end of the support, where f_1 is small, e.g. like a power law or an exponential, the term $\frac{\partial f_1}{\partial R_1}$ becomes large to f_1 and the second order term is of the same order as f_1 in a boundary layer of size $\phi^{1/4}$ at the end of the support. This effect is exactly the same as in other well-known situations where boundary layer develop, such as in the Navier-Stokes equation with no-slip boundary conditions.

In [NV06a] also self-similar solutions to (5.1)-(5.3) are investigated. By asymptotic expansions a self-similar solution is constructed, which is a perturbation of the LSW self-similar solution. The perturbation has the shape of a Gaussian tail and induces correction to the mean particle size of order $\phi^{1/4}$. We find

6 Collisions versus Correlations and Fluctuations

We already pointed out in Sect. 2.2 that particles may collide during their evolution and that in a discussion of corrections to the LSW model such events have to be taken into account. At first glance, this effect seems to be of higher order than correlations, since the number of particles per unit volume which are involved in collisions is of order $\rho\phi$, hence the correction of the LSW model due to collisions should have relative size of order ϕ.

However, the situation is more subtle, because collision of particles of medium size produces a large particle which might dominate the long-time behavior. In fact, ideas along these lines have already been formulated by Lifshitz and Slyozov in their original paper [LS61]. They suggest that a coagulation term with a kernel which is additive in the volume variable should be added on the right hand side of the LSW model. It is predicted that the corresponding correction to the self-similar LSW solution is of order $\ln \frac{1}{\phi}$. The main reason for this large correction is due to the kinetic character of collisions. This implies that the fraction of particles which are transported to the super-critical regime is of order ϕ, whereas the diffusive correction due to fluctuations in (5.1) transports only the largest particles to a supercritical regime. This effect is exponentially small in ϕ.

In a forthcoming article [NV06e] we will investigate the derivation and analysis of the "coagulation model" by a mathematical analysis and compare the size of the effects with the ones induced by (4.38) and (5.1) respectively.

This discussion highlights that there are many analogies in the analysis of Ostwald ripening to the mathematical analysis of gas dynamics. In gas dynamics there exist two relevant limits to obtain kinetic equations for the distribution of particle positions and velocities, namely the Boltzmann and the Bogoliubov [AP81] limit. The assumptions in the Boltzmann limit are that the range of interactions between particles is small compared to their average distance. In this case it is possible to reduce the dynamics to an equation for the one-particle distribution function, the well-known Boltzmann equation. In the second limit, the range of interactions between particles is large compared with their average distance. The dynamics can be simplified under the assumption of that the ratio between average potential energy and kinetic energy is small. The resulting model are equations for the one- and two-particle distribution function. In Ostwald ripening, collisions between particles play a role which is analogous to collisions of gas particles in the Boltzmann equation, whereas the results in [Mar87, HNO05a] might be considered as a generalization of the Bogoliubov method.

Acknowlegdments. This work has been supported by the DFG Priority Program 1095 "Analysis, Modeling and Simulation of Multiscale Problems" under Ni 505/2.
Juan Velázquez also gratefully acknowledges support through the DGES Grant MTM2004-05634, a Humboldt prize, awarded by the Alexander von Humboldt foundation, and the hospitality of the Max-Planck Institute for Mathematics in the Sciences and the Humboldt-University Berlin.

References

[ABC94] N. Alikakos, P. Bates, and X. Chen. Convergence of the Cahn–Hilliard equation to the Hele–Shaw model. *Arch. Rat. Mech. Anal.*, 128, 165–205, 1994.

[AF99] N. Alikakos and G. Fusco. The equations of Ostwald ripening for dilute systems. *J. Stat. Phys.*, 95, 5/6, 851–866, 1999.

[AF03] N. Alikakos and G. Fusco. Ostwald ripening for dilute systems under quasistationary dynamics. *Comm. Math. Phys.*, 238, 3, 429–479, 2003.

[AP81] A. I. Akhiezer and S. V. Peletminskiĭ. *Methods of statistical physics*, volume 104 of *International Series in Natural Philosophy*. Pergamon Press, Oxford, 1981. With a foreword by N. N. Bogoliubov [N. N. Bogolyubov], Translated from the Russian by M. Schukin.

[AS*99] J. Alkemper, V. A. Snyder, N. Akaiwa, and P. W. Voorhees. Dynamics of late–stage phase separation: A test of theory. *Phys. Rev. Let.*, 82, 2725–2729, 1999.

[AV94] N. Akaiwa and P. W. Voorhees. Late–stage phase separations: Dynamics, spatial correlations and structure functions. *Phys. Rev.E*, 49, 3860–3880, 1994.

[Bro89] L. C. Brown. A new examination of classical Coarsening Theory. *Acta Metall.*, 37, 71–77, 1989.

[Bro90a] L. C. Brown. Reply to comments by Hillert... *Scripta Metall.*, 24, 963–966, 1990.

[Bro90b] L. C. Brown. Reply to comments by Hoyt... *Scripta Metall.*, 24, 2231–2234, 1990.

[BW79] A. D. Brailsford and P. Wynblatt. The dependence of Ostwald ripening kinetics on particle volume fraction. *Acta met.*, 27, 489–497, 1979.

[Car06] J. Carr. Stability of solutions in a simplified LSW model. 2006. preprint.

[CG00] J.-F. Collet and T. Goudon. On solutions to the Lifshitz–Slyozov model. *Nonlinearity*, 13, 1239–1262, 2000.

[CG04] J. Carrillo and T. Goudon. A numerical study on large–time asymptotics of the Lifshitz–Slyozov system. *J. Sc. Comp.*, 20 1, 69–113, 2004.

[CH*06] S. Conti, A. Hönig, B. Niethammer, and F. Otto. Non-universality in low volume fraction Ostwald Ripening. 2006. Preprint.

[CM97] D. Cioranescu and F. Murat. A strange term coming from nowhere. In A. Cherkaev and R. Kohn, editors, *Topics in the Mathematical Modelling of Composite Materials*, pages 45–94. Birkhäuser, 1997.

[CNO06] S. Conti, B. Niethammer, and F. Otto. Coarsening rates in off-critical mixtures. *SIAM J. Math. Anal.*, 37,6, 1732–1741, 2006.

[CP98] J. Carr and O. Penrose. Asymptotic behaviour in a simplified Lifshitz–Slyozov equation. *Physica D*, 124, 166–176, 1998.

[FGM96] V. E. Fradkov, M. E. Glicksman, and S. P. Marsh. Coarsening kinetics in finite clusters. *Phys. Rev. E*, 53, 3925–3932, 1996.

[FPL99] P. Fratzl, O. Penrose, and J. Lebowitz. Modeling of phase separation in alloys with coherent elastic misfit. *J. Statist. Phys.*, 95(5-6), 1429–1503, 1999.

[FSF03] F. D. Fischer, J. Svoboda, and P. Fratzl. A thermodynamic approach to grain growth and coarsening. *Phil. Mag.*, 83, 9, 1075–1093, 2003.

[GMS98] B. Giron, B. Meerson, and P. V. Sasorov. Weak selection and stability of localized distributions in Ostwald ripening. *Phys. Rev. E*, 58, 4213–6, 1998.

[GN02] A. Garroni and B. Niethammer. Correctors and error–estimates in the homogenization of a Mullins–Sekerka problem. *IHP Analyse nonlinéaire*, 19 4, 371–393, 2002.

[HHR92] M. Hillert, O. Hunderi, and N. Ryum. Instability of distribution functions in particle coarsening. *Scripta metall.*, 26, 1933–1938, 1992.

[HH*89] M. Hillert, O. Hunderi, N. Ryum, and T. Saetre. A comment on the Lifshitz-Slyozov-Wagner theory of particle coarsening. *Scripta metall.*, 23, 1979–1982, 1989.

[HNO05a] A. Hönig, B. Niethammer, and F. Otto. On first–order corrections to the LSW theory I: infinite systems. *J. Stat. Phys.*, 119 1/2, 61–122, 2005.

[HNO05b] A. Hönig, B. Niethammer, and F. Otto. On first–order corrections to the LSW theory II: finite systems. *J. Stat. Phys.*, 119 1/2, 123–164, 2005.

[Hoy90] J. J. Hoyt. On the steady-state particle size distribution during coarsening. *Scripta metall.*, 24, 163–166, 1990.

[HV88] S. Hardy and P. Voorhees. Ostwald Ripening in a System with a High Volume Fraction of Coarsening Phase. *Met. Trans. A*, 19 A, 2713–2721, 1988.

[KET86] K. Kawasaki, Y. Enomoto, and M. Tokuhama. Elementary derivation of kinetic equation for Ostwald ripening. *Physica A*, 135, 426–445, 1986.

[KO02] R. V. Kohn and F. Otto. Upper bounds for coarsening rates. *Comm. Math. Phys.*, 229, 375–395, 2002.

[Lau01] P. Laurençot. Weak solutions to the Lifshitz–Slyozov–Wagner equation. *Indiana Univ. Math. J.*, 50, 3, 1319–1346, 2001.

[Lau02] P. Laurençot. The Lifshitz–Slyozov–Wagner equation with total conserved volume. *SIAM J. Math. Anal.*, 34, 2, 257–272, 2002.

[LS61] I. M. Lifshitz and V. V. Slyozov. The kinetics of precipitation from supersaturated solid solutions. *J. Phys. Chem. Solids*, 19, 35–50, 1961.

[Mar87] M. Marder. Correlations and Ostwald ripening. *Phys. Rev. A*, 36, 858–874, 1987.

[Mee99] B. Meerson. Fluctuations provide strong selection in Ostwald ripening. *Phys. Rev. E*, 60, 3, 3072–5, 1999.

[MG*98] H. Mandyam, M. E. Glicksman, J. Helsing, and S. P. Marsh. Statistical simulations of diffusional coarsening in finite clusters. *Phys. Rev. E*, 58,2, 2119–2130, 1998.

[MR84] J. A. Marqusee and J. Ross. Theory of Ostwald ripening: Competitive growth and its dependence on volume fraction. *J. Chem. Phys.*, 80, 536–543, 1984.

[MS96] B. Meerson and P. V. Sasorov. Domain stability, competition, growth and selection in globally constrained bistable systems. *Phys. Rev. E*, 53, 3491–4, 1996.

[Nie99] B. Niethammer. Derivation of the LSW theory for Ostwald ripening by homogenization methods. *Arch. Rat. Mech. Anal.*, 147, 2, 119–178, 1999.

[Nie00] B. Niethammer. The LSW–model for Ostwald Ripening with kinetic undercooling. *Proc. Roy. Soc. Edinb.*, 130 A, 1337–1361, 2000.

[NO01a] B. Niethammer and F. Otto. Domain coarsening in thin films. *Comm. Pure Appl. Math.*, 54, 361–384, 2001.

[NO01b] B. Niethammer and F. Otto. Ostwald Ripening: The screening length revisited. *Calc. Var. and PDE*, 13 1, 33–68, 2001.

[NP99] B. Niethammer and R. L. Pego. Non–self–similar behavior in the LSW theory of Ostwald ripening. *J. Stat. Phys.*, 95, 5/6, 867–902, 1999.

[NP00] B. Niethammer and R. L. Pego. On the initial-value problem in the Lifshitz-Slyozov-Wagner theory of Ostwald ripening. *SIAM J. Math. Anal.*, 31,3, 457–485, 2000.

[NP01] B. Niethammer and R. L. Pego. The LSW model for domain coarsening: Asymptotic behavior for total conserved mass. *J. Stat. Phys.*, 104, 5/6, 1113–1144, 2001.

[NP05] B. Niethammer and R. L. Pego. Well-posedness for measure transport in a family of nonlocal domain coarsening models. *Indiana Univ. Math. J.*, 54, 2, 499–530, 2005.

[NV04a] B. Niethammer and J. J. L. Velázquez. Homogenization in coarsening systems I: deterministic case. *Math. Meth. Mod. Appl. Sc.*, 14,8, 1211–1233, 2004.

[NV04b] B. Niethammer and J. J. L. Velázquez. Homogenization in coarsening systems II: stochastic case. *Math. Meth. Mod. Appl. Sc.*, 14,9, 2004.

[NV04c] B. Niethammer and J. J. L. Velázquez. Well–posedness for an inhomogeneous LSW–model in unbounded domains. *Math. Annalen*, 328 3, 481–501, 2004.

[NV06a] B. Niethammer and J. J. L. Velázquez. A model accounting for fluctuations in Ostwald Ripening. 2006. Preprint.

[NV06b] B. Niethammer and J. J. L. Velázquez. Global stability and bounds for coarsening rates within the LSW mean-field theory. *Comm. in PDE*, 2006. to appear.

[NV06c] B. Niethammer and J. J. L. Velázquez. On the convergence towards the smooth self-similar solution in the LSW model. *Indiana Univ. Math. J.*, 55, 761–794, 2006.

[NV06d] B. Niethammer and J. J. L. Velázquez. Screening in interacting particle systems. *Arch. Rat. Mech. Anal.*, 180 3, 493–506, 2006.

[NV06e] B. Niethammer and J. J. L. Velázquez. Correlations versus particle collisions in Ostwald Ripening. In preparation.

[Peg89] R. L. Pego. Front migration in the nonlinear Cahn–Hilliard equation. *Proc. R. Soc. Lond.* **A**, 422, 261–278, 1989.

[TE93] M. Tokuhama and Y. Enomoto. Theory of phase-separation in quenched dynamic binary mixtures. *Phys. Rev. E*, 47, 2, 1156–1179, 1993.

[TK84] M. Tokuhama and K. Kawasaki. Statistical–mechanical theory of coarsening of spherical droplets. *Physica A*, 123, 386–411, 1984.

[TKE87] M. Tokuhama, K. Kawasakii, and Y. Enomoto. Kinetic equation for Ostwald Ripening. *Physica A*, 134, 183–209, 1987.

[Vel98] J. J. L. Velázquez. The Becker–Döring equations and the Lifshitz–Slyozov theory of coarsening. *J. Stat. Phys.*, 92, 195–236, 1998.

[Vel00] J. J. L. Velázquez. On the effect of stochastic fluctuations in the dynamics of the Lifshitz–Slyozov–Wagner model. *J. Stat. Phys.*, 99, 57–113, 2000.

[Vel06a] J. J. L. Velázquez. On the dynamics of the characteristic curves for the LSW model. 2006. Preprint.

[Vel06b] J. J. L. Velázquez. Self-similar behavior for noncompactly supported solutions of the LSW model. 2006. Preprint.

[VG84] P. W. Voorhees and M. E. Glicksman. Solution to the multi–particle diffusion problem with application to Ostwald Ripening – I. Theory. *Acta met.*, 32, 2001–2011, 1984.

[Voo85] P. W. Voorhees. The theory of Ostwald ripening. *J. Stat. Phys.*, 38, 231–252, 1985.

[Voo92] P. W. Voorhees. Ostwald ripening of two–phase mixtures. *Ann. Rev. Mater. Sc*, 22, 197–215, 1992.

[Wag61] C. Wagner. Theorie der Alterung von Niederschlägen durch Umlösen. *Z. Elektrochemie*, 65, 581–594, 1961.

[WC73] J. Weins and J. W. Cahn. The effect of size and distribution of second phase particles and voids on sintering. In E. G. C. Kuczynski, editor, *Sintering and related phenomena*, page 151. Plenum, London, 1973.

[YE*93] J. H. Yao, K. R. Elder, H. Guo, and M. Grant. Theory and simulation of Ostwald ripening. *Phys. Rev. B*, 47, 14110–14125, 1993.

Radiative Friction for Charges Interacting with the Radiation Field: Classical Many-Particle Systems

Sebastian Bauer[1] and Markus Kunze[1]

Universität Duisburg-Essen, Fachbereich Mathematik, 45117 Essen.
sbauer@ing-math.uni-essen.de, mkunze@ing-math.uni-essen.de

Summary. We consider an ensemble of classical particles modelled by means of a continuity equation for a distribution function which is coupled back to the self-induced fields. For such infinite dimensional systems simpler effective equations are derived in the limit $c \to \infty$. A main emphasis is on higher order approximations, in particular on the first post-Newtonian order where radiation starts to play a role.

1 Introduction

The most precise existing theory of gravitation, the theory of general relativity, predicts that certain astrophysical systems, such as colliding black holes or neutron stars, will give rise to gravitational radiation. There is a major international effort under way to detect these gravitational waves [Bra04]. In order to relate the general theory to predictions of what the detectors will see it is necessary to use approximation methods since the exact theory is too complicated. The mathematical status of these approximations remains unclear and only very partial results exist. Therefore it is useful to start with model problems. One option is the relativistic Vlasov-Maxwell system which plays an important role in plasma physics. Although the field part of this system is electromagnetic and spacetime is flat, such a model is already fairly difficult. It is often used to gain a better understanding of the mathematical structures involved in more realistic gravity models. A further option for a simplified model is the scalar theory of gravitation, as described by the Vlasov-Nordström theory [Cal03]. It has already been considered as a model problem for numerical relativity in [ST93].

Among the approximation methods used to study gravitational radiation those which are most accessible mathematically are the post-Newtonian approximations. Some information on these has been obtained in [Ren94, Ren92] but further rigorous progress seems difficult at this point. For that reason it seems to be useful to investigate (from the viewpoint of approximation methods) the two systems presented above, i.e., Vlasov matter coupled to the Maxwell fields and Vlasov matter coupled to a scalar gravitational field governed by the Nordström equation. We shall explain the post-Newtonian expansion of the Vlasov-Maxwell system in some detail in Sect. 2 and sketch our results concerning the Vlasov-Nordström system in Sect. 3.

Remark: Our contribution is one part of the research project jointly with G. Panati, H. Spohn, and S. Teufel within the Schwerpunkt. The second part will be covered in the separate contribution [PST06].

2 The Vlasov-Maxwell system

The Vlasov-Maxwell system from kinetic theory models the evolution of a plasma or gas composed of many collisionless particles which move under the influence of their self-generated electromagnetic field. For the sake of simplicity we assume that there are only two different species of particles with mass normalized to unity and charge normalized to plus unity and minus unity, respectively. The particle distributions on phase space are modelled through the nonnegative distribution functions f^+ and f^-, $f^\pm = f^\pm(t,x,p)$, depending on time $t \in \mathbb{R}$, position $x \in \mathbb{R}^3$, and momentum $p \in \mathbb{R}^3$. It is assumed that collisions between single particles are sufficiently rare such that they can be neglected. Therefore all forces between the particles are mediated by the electromagnetic fields. The dynamics are governed by

$$\left.\begin{aligned} &\partial_t f^\pm + \hat{p}\cdot\nabla_x f^\pm \pm (E + c^{-1}\hat{p}\times B)\cdot\nabla_p f^\pm = 0, \\ &c\nabla\times E = -\partial_t B, \qquad c\nabla\times B = \partial_t E + 4\pi j \\ &\nabla\cdot E = 4\pi\rho, \qquad \nabla\cdot B = 0, \\ &\rho := \int (f^+ - f^-)\,dp, \qquad j := \int \hat{p}\,(f^+ - f^-)\,dp \end{aligned}\right\} \quad (\mathrm{RVM}_c)$$

Here

$$\hat{p} = (1 + c^{-2}p^2)^{-1/2} p \in \mathbb{R}^3 \tag{2.1}$$

is the relativistic velocity associated to p. The Lorentz force $E + c^{-1}\hat{p}\times B$ realizes the coupling of the Maxwell fields $E = E(t,x) \in \mathbb{R}^3$ and $B = B(t,x) \in \mathbb{R}^3$ to the Vlasov equation, and conversely the density functions $f^\pm$ enter the field equations via the scalar charge density $\rho = \rho(t,x)$ and the current density $j = j(t,x) \in \mathbb{R}^3$, which act as source terms for the Maxwell equations. The parameter c denotes the speed of light for given units of time and space of the physical system. In order to state the Cauchy problem for (RVM_c) initial data for the densities and for the fields have to be prescribed,

$$f^\pm(0,x,p) = f^{\circ,\pm}(x,p), \quad E(0,x) = E^\circ(x), \quad B(0,x) = B^\circ(x). \tag{2.2}$$

Henceforth we treat the speed of light c as a parameter and study the behavior of the system as $c \to \infty$. It will be explained below that after a suitable rescaling the fields are slowly varying in their space and time variables. Thus the limit $c \to \infty$ corresponds to an adiabatic limit and to slowly moving particles. Our general goal is to establish conditions under which the solutions of (RVM$_c$) converge to a solution of an effective system. For the Vlasov-Maxwell system the first result in this direction was obtained in [Sch86], where it has been shown that as $c \to \infty$ the solutions of (RVM$_c$) approach a solution of the Vlasov-Poisson system at the rate $\mathcal{O}(c^{-1})$; see [AU86, Deg86] for similar results and [Lee04] for the case of two spatial dimensions. The respective Newtonian limits of other related systems are derived in [Ren94, CL04].

It was one aim of this project to replace the Vlasov-Poisson system by other effective equations to achieve higher order convergence and more precise approximations. In [BK05] this led to an effective system whose solutions stay as close as $\mathcal{O}(c^{-3})$ to a solution of the full Vlasov-Maxwell system. In the context of individual particles, this post-Newtonian (PN) order of approximation is usually called the Darwin order; see [KS00, Spo04] and the references therein. We also mention that weak convergence properties of other kinds of Darwin approximations for the Vlasov-Maxwell system have been studied in [DR92, BF*03].

In the next order, and in analogy to the case of individual particles [KS01], radiation effects play a role for the first time. Therefore the well-known problems related to the existence of unphysical solutions, usually called "run-away solutions", are expected to turn up [Jac99]. For individual particles these problems can be resolved rigorously by restricting the dynamics to a suitable center-like manifold in the phase space; see [KS01, Spo04]. Since the phase space of the Vlasov-Maxwell system is infinite dimensional (as densities are considered), it is clear that several new mathematical difficulties have to be surrounded in this step. In [Bau06b] we determined effective equations for the Vlasov-Maxwell system on the center manifold, which led to a slightly dissipative Vlasov-like effective equation, free of "run-away" solutions, and we proved that solutions of these equations stay as close as $\mathcal{O}(c^{-4})$ to a solution of the full Vlasov-Maxwell system.

Compared to systems of coupled individual particles, for the Vlasov-Maxwell system one immediately is faced with the fact that so far only the existence of local solutions is known in general. These solutions are global under additional conditions, for instance if a suitable a priori bound on the velocities is available; see [GS86]. This means that from the onset we will have to restrict ourselves to solutions of (RVM$_c$) which are only defined on some time interval $[0, T]$ that may be very small. On the other hand, in [Sch86] it has been shown that such a time interval can be found which is uniform in $c \geq 1$, so it seems reasonable to accept this limit.

In Sect. 2.1 we first carry out a formal expansion of (RVM$_c$) in c^{-1} as $c \to \infty$. It turns out that up to the Darwin order this approximation yields the correct effective system. However, for the next order this naive procedure does not yield the optimal result. To get a clue on how an improved approximation may look like we derive the leading order dipole radiation term in

Sect. 2.2, and thereafter we establish an improved system (called the radiation approximation) which gives a better approximation to the full Vlasov-Maxwell system. More details on this are elaborated in Sect. 2.3. Finally a comparison of the continuous models to the individual particle models is done in Sect. 2.4.

2.1 The naive post-Newtonian expansion

We adopt the definition of a post-Newtonian approximation from [KR01b]; see also [Ren92] for the Einstein case. Thus the matter and the fields are described by a one-parameter family $(f^{\pm}(c), E(c), B(c))$ of solutions to (RVM$_c$), depending on the parameter $c \in [c_0, \infty[$. This means that $(f^{\pm}(c), E(c), B(c))$ describes a family of solutions of physical systems which are represented in parameter-dependent units, where the numerical value of the speed of light is given by c. A more conventional physical description of the post-Newtonian expansion would say that in a fixed system of units the occurring velocities are small compared to the speed of light. Taking this viewpoint means that we consider (RVM$_c$) at a fixed c (say $c = 1$) by rescaling the prescribed non-negative initial densities $f^{\circ,\pm}$, for which we suppose that $f^{\circ,\pm} \in C^{\infty}(\mathbb{R}^3 \times \mathbb{R}^3)$ have compact support. To be more precise, denote $\bar{v} = \int\int \hat{p} f^{\circ,\pm}(x,p)\, dx\, dp$, where $\hat{p}$ is taken for $c = \varepsilon^{-1/2}$; cf. (2.1). Then $\bar{v}$ is viewed as an average velocity of the system. Then we introduce $f^{\circ,\pm,\varepsilon}(x,p) = \varepsilon^{3/2} f^{\circ,\pm}(\varepsilon x, \varepsilon^{-1/2} p)$ and consider $f^{\circ,\pm,\varepsilon}$ for $c = 1$. It follows that

$$\bar{v}^{\varepsilon} = \int\int \hat{p} f^{\circ,\pm,\varepsilon}(x,p)\, dx\, dp = \sqrt{\varepsilon} \int\int \hat{p} f^{\circ}(x,p)\, dx\, dp = \sqrt{\varepsilon}\, \bar{v},$$

i.e., the systems with initial distribution functions $f^{\circ,\pm,\varepsilon}$ have small velocities compared to the systems associated to $f^{\circ,\pm}$. Under this scaling the masses remain unchanged, as $\int\int f^{\circ,\pm,\varepsilon}(x,p)\, dx\, dp = \int\int f^{\circ,\pm}(x,p)\, dx\, dp$. Next observe that $(f^{\pm}, E, B)$ is a solution of (RVM$_c$) with $c = \varepsilon^{-1/2}$ if and only if

$$\begin{aligned} f^{\pm,\varepsilon}(t,x,p) &= \varepsilon^{3/2} f^{\pm}(\varepsilon^{3/2} t, \varepsilon x, \varepsilon^{-1/2} p), \\ E^{\varepsilon}(t,x) &= \varepsilon^2 E(\varepsilon^{3/2} t, \varepsilon x), \\ B^{\varepsilon}(t,x) &= \varepsilon^2 B(\varepsilon^{3/2} t, \varepsilon x), \end{aligned} \tag{2.3}$$

is a solution of (RVM$_c$) with $c = 1$. By definition of the rescaled fields these fields are slowly varying in their space and time variables. Thus the limit $c \to \infty$ corresponds to an adiabatic limit. Henceforth we will return to the original formulation and consider the system as $c \to \infty$, treating the speed of light c as a parameter. However, due to the rescaling outlined above all theorems can also be formulated in a parameter independent fashion. In that case the value of c is fixed, say $c = 1$, and the initial data have to be modified according to (2.3); see [BK05] for details.

We start with a formal expansion of all quantities occurring in (RVM$_c$) in powers of c^{-1},

$$
\begin{aligned}
f^{\pm} &= f_0^{\pm} + c^{-1} f_1^{\pm} + c^{-2} f_2^{\pm} + c^{-3} f_3^{\pm} + \dots, \\
E &= E_0 + c^{-1} E_1 + c^{-2} E_2 + c^{-3} E_3 + \dots, \\
B &= B_0 + c^{-1} B_1 + c^{-2} B_2 + c^{-3} B_3 + \dots, \\
\rho &= \rho_0 + c^{-1} \rho_1 + c^{-2} \rho_2 + c^{-3} \rho_3 + \dots, \\
j &= j_0 + c^{-1} j_1 + c^{-2} j_2 + c^{-3} j_3 + \dots .
\end{aligned}
\tag{2.4}
$$

In addition, also the initial densities are assumed to allow an expansion as $f^{\circ,\pm} = f_0^{\circ,\pm} + c^{-1} f_1^{\circ,\pm} + \dots$. Finally $\hat{p} = p - (c^{-2}/2) p^2 p + \dots$ by (2.1), where $p^2 = |p|^2$. These expansions can be substituted into (RVM$_c$). Comparing coefficients at every order gives a hierarchy of equations for the coefficients. The equations at order k will be addressed as the $k/2$ PN equations, and

$$
f^{\pm, k/2\,\mathrm{PN}} = \sum_{j=0}^{k} c^{-j} f_j^{\pm}, \quad E^{k/2\,\mathrm{PN}} = \sum_{j=0}^{k} c^{-j} E_j, \quad B^{k/2\,\mathrm{PN}} = \sum_{j=0}^{k} c^{-j} B_j,
$$

is the $k/2$ PN approximation. This notation is used due to the fact that in the context of general relativity post-Newtonian approximations are usually counted in orders of c^{-2}.

At order zero the well known Vlasov-Poisson system of plasma physics is obtained,

$$
\left.
\begin{aligned}
\partial_t f_0^{\pm} + p \cdot \nabla_x f_0^{\pm} \pm E_0 \cdot \nabla_p f_0^{\pm} = 0, \\
E_0(t,x) = -\int |z|^{-2} \bar{z}\, \rho_0(t, x+z)\, dz, \\
\rho_0 = \int (f_0^{+} - f_0^{-})\, dp, \\
f_0^{\pm}(0,x,p) = f_0^{\circ,\pm}(x,p),
\end{aligned}
\right\}
\tag{VP$_{\mathrm{plasma}}$}
$$

where $\bar{z} = |z|^{-1} z$. Note that the degrees of freedom of the electromagnetic fields up to this order are lost, reflecting that the limit $c \to \infty$ is singular and the hyperbolic field equations become elliptic. As mentioned above, this 0 PN approximation is made rigorous in [Sch86].

Concerning a general k we assume that the lower order coefficients have already been computed. Then the fields at order k have to solve

$$
\begin{aligned}
\nabla \times E_k &= -\partial_t B_{k-1}, & \nabla \cdot E_k &= 4\pi \rho_k, \\
\nabla \times B_k &= \partial_t E_{k-1} + 4\pi j_{k-1}, & \nabla \cdot B_k &= 0.
\end{aligned}
$$

The Vlasov equation to that order is

$$
\partial_t f_k^{\pm} + p \cdot \nabla_x f_k^{\pm} \pm E_0 \cdot \nabla_p f_k^{\pm} = \mp E_k \cdot \nabla_p f_0^{\pm} + R_k^{\pm},
$$

where the $R_k^{\pm}$ can be calculated from the known quantities $f_j^{\pm}$, $\nabla_x f_j^{\pm}$, $\nabla_p f_j^{\pm}$, E_j, and B_j for $j = 0, \dots, k-1$. A special feature of this hierarchy is as follows.

If we assume for the initial data that $f_k^{\circ,\pm} = 0$ for all odd k, then using the explicit form of R_k it can be shown that we can set

$$f_{2l+1}^{\pm} = 0, \quad E_{2l+1} = 0, \quad B_{2l} = 0 \tag{2.5}$$

consistently, for $l = 0, 1, 2, \ldots$. This simplification will be employed throughout. To solve the equations for (f_k, E_k, B_k), we observe that once E_k is known, then $f_k^{\pm}$ can be calculated using characteristics. Note that for all orders k the characteristic flow is determined by the vector field $(p, \pm E_0)$. On the other hand, if the $f_k^{\pm}$ are known, then ρ_k and j_k are fixed. Using the vector identity $-\nabla \times \nabla \times + \nabla\nabla\cdot = \Delta$, we can rewrite the field equations as

$$\begin{aligned} E_{2k} &= 4\pi\Delta^{-1}(\nabla\rho_{2k} + \partial_t j_{2k-2}) + \Delta^{-1}(\partial_t^2 E_{2k-2}), \\ B_{2k+1} &= \Delta^{-1}(\partial_t^2 B_{2k-1}) - 4\pi\Delta^{-1}(\nabla \times j_{2k}), \end{aligned} \tag{2.6}$$

where quantities carrying a negative index are understood to be zero. Assuming that all densities are compactly supported we can solve these field equations. Of course without boundary conditions the solutions are not unique, and at least for higher orders they will not vanish at infinity. Nevertheless, if we take those fields, then the coupled equations can be solved by a fixed-point iteration for E_k. Thus on a formal level the (naive) PN approximation scheme is well defined.

According to this scheme, B_1 is given by

$$B_1(t,x) = \int |z|^{-2}\bar{z} \times j_0(t, x+z)\, dz, \tag{2.7}$$

where $j_0 = \int p\,(f_0^+ - f_0^-)\, dp$. The couple $(f_2^{\pm}, E_2)$ is the solution to

$$\left.\begin{aligned} \partial_t f_2^{\pm} + p\cdot\nabla_x f_2^{\pm} \pm E_0\cdot\nabla_p f_2^{\pm} & \\ = \tfrac{1}{2}p^2\, p\cdot\nabla_x f_0^{\pm} \mp (E_2 + p\times B_1)\cdot\nabla_p f_0^{\pm}&, \\ E_2(t,x) = \frac{1}{2}\int \bar{z}\,\partial_t^2\rho_0(t,x+z)\,dz - \int |z|^{-1}\partial_t j_0(t,x+z)\,dz & \\ - \int |z|^{-2}\bar{z}\,\rho_2(t,x+z)\,dz&, \\ \rho_2 = \int (f_2^+ - f_2^-)\,dp&, \\ f_2(0,x,p) = f_2^{\circ,\pm}(x,p)&. \end{aligned}\right\} \tag{LVP$_{\text{plasma}}$}$$

Thus the 1 PN approximation (Darwin approximation) corresponding to $k = 2$ is

$$f^{\pm,1\,\mathrm{PN}} = f_0^{\pm} + c^{-2}f_2^{\pm}, \quad E^{1\,\mathrm{PN}} = E_0 + c^{-2}E_2, \quad B^{1\,\mathrm{PN}} = c^{-1}B_1. \tag{2.8}$$

This Darwin system is Hamiltonian in the following sense. If the conserved energy

$$\mathcal{E} = \int\int \sqrt{1 + c^{-2}p^2/2}\,(f^+ + f^-)\,dx\,dp + \frac{1}{8\pi}\int (E^2 + B^2)\,dx$$

of (RVM_c) is expanded according to (2.4), then the Darwin energy $\mathcal{E}_{\mathrm{D}} = \mathcal{E}_{\mathrm{D,\,kin}} + \mathcal{E}_{\mathrm{D,\,pot}}$ is obtained. Explicitly its kinetic energy and potential energy parts are given by

$$\mathcal{E}_{\mathrm{D,\,kin}} = \int\int [(p^2/2 - c^{-2}p^4/8)(f_0^+ + f_0^-) + c^{-2}p^2/2(f_2^+ + f_2^-)]\,dx\,dp,$$

$$\mathcal{E}_{\mathrm{D,\,pot}} = \frac{1}{8\pi}\int [E_0^2 + 2c^{-2}E_0 \cdot E_2 + c^{-2}B_1^2]\,dx.$$

It can be checked that $\mathcal{E}_{\mathrm{D}}$ is conserved along solutions of the 1 PN approximation. The approximation properties of the Darwin system w.r. to solutions of the full Vlasov-Maxwell system are investigated in [BK05]. In this paper it is shown that if we adapt the initial data (2.2) to suit the initial data of the 1 PN approximation, then the solutions are tracked down with an error of order c^{-3}. Hence the naive post-Newtonian expansion is valid up to this order.

For the next level ($k = 3$ or 1.5 PN), $B_3 = \Delta^{-1}(\partial_t^2 B_1) - 4\pi\Delta^{-1}(\nabla \times j_2)$ and also $B_1 = -4\pi\Delta^{-1}(\nabla \times j_0)$ by (2.6). Thus $B_3 = -4\pi\Delta^{-2}(\partial_t^2\nabla \times j_0) - 4\pi\Delta^{-1}(\nabla \times j_2)$ allows for the solution

$$B_3(t,x) = \frac{1}{2}\int |z|\,\partial_t^2\nabla \times j_0(t, x+z)\,dz + \int |z|^{-1}\nabla \times j_2(t, x+z)\,dz, \quad (2.9)$$

where

$$j_2 = \int [p\,(f_2^+ - f_2^-) - (p^2/2)p\,(f_0^+ - f_0^-)]\,dp.$$

It follows that $f^{\pm,1.5\,\mathrm{PN}} = f_0^\pm + c^{-2}f_2^\pm + c^{-3}f_3^\pm = f^{\pm,1\,\mathrm{PN}}$, $E^{1.5\,\mathrm{PN}} = E_0 + c^{-2}E_2 + c^{-3}E_3 = E^{1\,\mathrm{PN}}$, and $B^{1.5\,\mathrm{PN}} = c^{-1}B_1 + c^{-3}B_3 = B^{1\,\mathrm{PN}} + c^{-3}B_3$, due to (2.5) and (2.8). Therefore the energy $\mathcal{E}_{\mathrm{D}}$ from above does not have to be changed in comparison to the 1 PN order. Hence at the 1.5 PN order (corresponding to c^{-3}) we would obtain a Hamiltonian system with no effects due to radiative friction visible. This suggests that the naive post-Newtonian approximation has to be improved in order to resolve such effects which are indeed present in the system. In order to get a clue on how such a refinement should look like it is useful to study the energy which is radiated to future null infinity by the full Vlasov-Maxwell system.

2.2 Dipole Radiation

The starting point for the following calculation is the local energy conservation $\partial_t e + \nabla\cdot\mathcal{P} = 0$ for classical solutions of the Vlasov-Maxwell system. The energy density e and the momentum density $\mathcal{P}$ are given by

$$e(t,x) = c^2\int \sqrt{1 + c^{-2}p^2}\,(f^+ + f^-)(t,x,p)\,dp + \frac{1}{8\pi}\left(|E(t,x)|^2 + |B(t,x)|^2\right),$$

$$\mathcal{P}(t,x) = c^2\int p(f^+ + f^-)(t,x,p)\,dp + \frac{c}{4\pi}\,E(t,x) \times B(t,x).$$

Defining the local energy in the ball of radius $r > 0$ as $\mathcal{E}_r(t) = \int_{|x|\le r} e(t,x)\,dx$, this conservation law and the divergence theorem imply that $\frac{d}{dt}\,\mathcal{E}_r(t) = -\int_{|x|=r} \bar{x}\cdot\mathcal{P}(t,x)\,d\sigma(x)$, where $\bar{x} = |x|^{-1}x$ denotes the outer unit normal. Our assumptions on the support of the distribution functions are such that the contribution of $\int p\,(f^+ + f^-)\,dp$ to $\mathcal{P}$ vanishes for $|x| = r$ large. Hence we arrive at

$$\frac{d}{dt}\,\mathcal{E}_r(t) = \frac{c}{4\pi}\int_{|x|=r} \bar{x}\cdot(B\times E)(t,x)\,d\sigma(x).$$

Therefore the energy flux radiated to null infinity at time t is obtained as

$$\lim_{r\to\infty}\frac{c}{4\pi}\int_{|x|=r} \bar{x}\cdot(B\times E)(t+c^{-1}r,x)\,d\sigma(x),$$

where $t + c^{-1}r$ is the advanced time. In [BL*06, Thm. 1.4] it is shown that for suitable solutions of the Vlasov-Maxwell system which are isolated from incoming radiation in the limit $c\to\infty$ the total amount of radiated energy is given by

$$\frac{2}{3c^3}\,|\ddot{D}(t)|^2, \tag{2.10}$$

where D is the dipole moment of the Newtonian limit ($\mathrm{VP}_{\mathrm{plasma}}$), defined as $D(t) = \int x\rho_0(t,x)\,dx$; see [BL*06, Thm. 1.4] for the exact statement and the remarks below for the retarded system (${}_{\mathrm{ret}}\mathrm{RVM}_c$) which models "no incoming radiation". This result yields a mathematical formulation and a rigorous proof of the Larmor formula in the case of Vlasov matter. Returning to the approximations, we should therefore introduce into the effective equation a radiation reaction force causing this loss of energy. As already suggested in [KR01a, KR01b] we thus modify the Vlasov equation of the Newtonian distribution by incorporating a small correction into the force term as

$$\partial_t f_0^\pm + p\cdot\nabla_x f_0^\pm \pm \left(E_0 + \frac{2}{3c^3}\dddot{D}\right)\cdot\nabla_p f_0^\pm = 0. \tag{2.11}$$

The additional term is the generalization of the radiation reaction force used in particle models; see [Jac99, (16.8)]. We also note that for this system the quantity

$$\begin{aligned}\mathcal{E}_{\mathrm{S}}(t) = \frac{1}{2}\int\!\!\int p^2(f_0^+ + f_0^-)(t,x,p)\,dx\,dp + \frac{1}{8\pi}\int |E_0(t,x)|^2\,dx \\ - \frac{2}{3c^3}\,\dot{D}(t)\cdot\ddot{D}(t)\end{aligned} \tag{2.12}$$

is decreasing. More precisely one obtains $\frac{d}{dt}\,\mathcal{E}_{\mathrm{S}}(t) = -\frac{2}{3c^3}|\ddot{D}(t)|^2$, cf. (2.10). The subscript S refers to Schott who considered similar quantities for particle models; see [Spo04]. Although $\mathcal{E}_{\mathrm{S}}$ has no definite sign, its decrease can be attributed to the effect of radiation damping. If instead of $\mathcal{E}_{\mathrm{S}}$ the usual positive energy $\mathcal{E}_{\mathrm{VP}} = \frac{1}{2}\int\int p^2(f_0^+ + f_0^-)\,dx\,dp + \frac{1}{8\pi}\int |E_0|^2\,dx$ of the Vlasov-Poisson system is considered, then along solutions of (2.11) the relation $\frac{d}{dt}\mathcal{E}_{\rm VP} = \frac{2}{3c^3}(\ddot{D}\cdot\dddot{D} - |\ddot{D}|^2)$ is obtained which has no straightforward interpretation.

Thus we consider (2.11) to be a promising candidate of an effective equation for the relativistic Vlasov-Maxwell system. However, one immediately runs into the problem that initial data have to be supplied for $D(0)$, $\dot{D}(0)$, and $\ddot{D}(0)$, as third derivatives of D occur in the equations. Since only $D(0)$ and $\dot{D}(0)$ are determined by $f_0^{\circ,\pm}$ and since there is no obvious way to extract the missing information from the approximation scheme, additional degrees of freedom seem to be generated. This phenomenon is also known from the theory of accelerated single charges and leads to a multitude of unphysical (so-called run-away) solutions. In [KS01] it has been observed that in the particle model this problem has the structure of a geometric singular perturbation problem, and the "physical" dynamics are obtained on a center-like manifold of the full dynamics.

In order to adopt this language to the model under consideration, we assume that we are supplied with a (local in time) classical solution $(f_0^\pm, E_0)$ of (2.11) and assume that the support of $f_0^\pm(t,\cdot,\cdot)$ remains compact for all t in the interval of existence of the solution. We define the bare mass by $M(t) = \int\int (f_0^+ + f_0^-)(t,x,p)\,dx\,dp$. Then (2.11) yields mass conservation $\partial_t M = 0$ as well as charge conservation for both species $\partial_t \rho_0^\pm + \nabla\cdot j_0^\pm = 0$, where $j_0^\pm = \int p f_0^\pm\,dp$ and $\rho_0^\pm = \int f_0^\pm\,dp$. From $D = \int x\rho_0\,dx$ and (2.11) we then find that $\dot{D} = \int\int p\,(f_0^+ - f_0^-)\,dx\,dp$ and $\ddot{D} = D^{[2]} + \frac{2}{3c^3}M\dddot{D}$, where

$$D^{[2]}(t) = \int\int E_0(t,x)(f_0^+ + f_0^-)(t,x,p)\,dx\,dp. \tag{2.13}$$

Defining $y = \ddot{D}$ and $\eta = \frac{2}{3c^3}M$, this can be rewritten as $y = D^{[2]} + \eta\dddot{D}$. Putting

$$\begin{aligned} F^\pm(f_0^\pm, y) &= -p\cdot\nabla_x f_0^\pm \mp (E_0 + M^{-1}(y - D^{[2]}))\cdot\nabla_p f_0^\pm, \\ G(f_0^\pm, y) &= y - D^{[2]}, \end{aligned}$$

it follows that (2.11) can be recast as the singular perturbation problem

$$\left.\begin{aligned} \dot{f}_0^\pm &= F^\pm(f_0^\pm, y) \\ \eta\dot{y} &= G(f_0^\pm, y) \end{aligned}\right\} \tag{SGPP$_\eta$}$$

In contrast to [KS01] we are dealing with a phase space of infinite dimension. Thus the proof of the existence of an invariant manifold is hard. We shall return to that question in a forthcoming paper. For the moment we shall take the existence of a smooth invariant manifold for granted and assume that it is given as a smooth graph $h_\eta = h_\eta(f_0^\circ)$, where h_η is defined on (a subset of) $C_0^\infty(\mathbb{R}^3\times\mathbb{R}^3)\times C_0^\infty(\mathbb{R}^3\times\mathbb{R}^3)$ and takes values in $\mathbb{R}^3$. For the moment f_0° denotes $(f_0^{\circ,+}, f_0^{\circ,-})$, and similarly $f_0 = (f_0^+, f_0^-)$ and $F = (F^+, F^-)$. The manifold $\mathcal{M}_\eta = \{(f_0^\circ, h_\eta(f_0^\circ))\}$ is invariant under the flow of (SGPP$_\eta$) if the solution of (SGPP$_\eta$) subject to the initial conditions $(f_0(0), y(0)) = (f_0^\circ, h_\eta(f_0^\circ))$ satisfies

$$y(t) = h_\eta(f_0(t,\cdot,\cdot)). \tag{2.14}$$

We want to determine a system of Vlasov-Poisson type which is a good approximation of the dynamics on the manifold. For this reason we assume that we can expand h_η in η about 0 as $h_\eta = h_0 + \eta h_1 + \mathcal{O}(\eta^2)$. Setting $\eta = 0$ in the second relation of (SGPP$_\eta$), $0 = G(f_0, h_0(f_0)) = h_0(f_0) - D^{[2]}$ is obtained, so that $h_0 = D^{[2]}$; note that $D^{[2]}$ depends on f_0. Thus

$$\begin{aligned}\eta\dot{y} &= G(f_0, h_0(f_0) + \eta h_1(f_0) + \mathcal{O}(\eta^2)) = h_0(f_0) + \eta h_1(f_0) + \mathcal{O}(\eta^2) - D^{[2]}\\ &= \eta h_1(f_0) + \mathcal{O}(\eta^2).\end{aligned}$$

On the other hand, differentiating (2.14) yields

$$\eta\dot{y} = \eta\,\langle h_\eta'(f_0), \dot{f}_0\rangle = \eta\,\langle h_0'(f_0), \dot{f}_0\rangle + \mathcal{O}(\eta^2),$$

and consequently $h_1(f_0) = \langle h_0'(f_0), F(f_0, h_0(f_0))\rangle$ by (SGPP$_\eta$). Explicitly,

$$\begin{aligned}&\langle h_0'(f_0), F(f_0, h_0(f_0))\rangle\\ &\quad = h_0'(f_0)\cdot\int\!\!\int\Big(-p\cdot\nabla_x(f_0^+ - f_0^-) - E_0\cdot\nabla_p(f_0^+ + f_0^-)\Big)(\cdot,x,p)\,dx\,dp.\end{aligned}$$

From (2.14) and the above it follows that $y = h_\eta(f_0) = h_0(f_0) + \eta h_1(f_0) + \mathcal{O}(\eta^2) = D^{[2]} + \eta\,\langle D^{[2]\prime}(f_0), F(f_0, D^{[2]}(f_0))\rangle + \mathcal{O}(\eta^2)$. After a straightforward computation we obtain

$$\langle D^{[2]\prime}(f_0), F(f_0, D^{[2]}(f_0))\rangle = D^{[3]}, \tag{2.15}$$

where

$$D^{[3]}(t) = 2\int\Big(H^+(t,x)j_0^-(t,x) - H^-(t,x)j_0^+(t,x)\Big)\,dx, \tag{2.16}$$

defining

$$H^\pm(t,x) := \oint |z|^{-3}(-3\bar{z}\otimes\bar{z} + \mathrm{id})\rho_0^\pm(t,x+z)\,dz \in \mathbb{R}^{3\times 3}. \tag{2.17}$$

Note that $H(z) = -3(\bar{z}\otimes\bar{z}) + \mathrm{id}$ is bounded on $\mathbb{R}^3\setminus\{0\}$, homogeneous of degree zero, and satisfies $\int_{|z|=1} H(z)\,d\sigma(z) = 0$. Formally, $D^{[3]}$ is close to $\dddot{D}$, since $\dddot{D} = \frac{d}{dt}(D^{[2]} + \frac{2}{3c^3}M\ddot{D}) = \frac{d}{dt}D^{[2]} + \mathcal{O}(c^{-3}) = D^{[3]} + \mathcal{O}(c^{-3})$ by (2.15).

We introduce the "reduced radiating Vlasov-Poisson system" as

$$\left.\begin{aligned}\partial_t f_0^\pm + p\cdot\nabla_x f_0^\pm \pm\Big(E_0 + \frac{2}{3c^3}\,D^{[3]}\Big)\cdot\nabla_p f_0^\pm &= 0,\\ E_0(t,x) &= -\int |z|^{-2}\bar{z}\,\rho_0(t,x+z)\,dz,\\ \rho_0 &= \int (f_0^+ - f_0^-)\,dp,\\ f_0^\pm(0,x,p) &= f_0^{\circ,\pm}(x,p),\end{aligned}\right\} \tag{rrVP$_c$}$$

where $D^{[3]}$ is defined by (2.16) and (2.17). The next proposition addresses the existence and uniqueness of local classical solutions of (rrVP$_c$). Furthermore, it provides some useful estimates. For the initial data we assume

$$\begin{aligned} &f_0^{\circ,\pm} \in C_0^\infty(\mathbb{R}^3 \times \mathbb{R}^3), \quad f_0^{\circ,\pm} \geq 0, \\ &f_0^{\circ,\pm}(x,p) = 0 \text{ for } |x| \geq r_0 \text{ or } |p| \geq r_0, \quad \|f_0^{\circ,\pm}\|_{W^{4,\infty}} \leq S_0, \end{aligned} \tag{2.18}$$

with some $r_0, S_0 > 0$ fixed.

Proposition 2.1. *If $f_0^{\circ,\pm}$ satisfies the above hypotheses, then there exists $0 < \tilde{T} \leq \infty$ such that the following holds for $c \geq 1$.*

(a) There is a unique classical solution $(f_0^\pm, E_0)$ of (rrVP$_c$) existing on a time interval $[0, T_c[$ with $\tilde{T} \leq T_c \leq \infty$. In addition, $\frac{d}{dt}D^{[2]} = D^{[3]}$, where $D^{[2]}$ is defined by (2.13).

(b) For every $T < \tilde{T}$ there is a constant $M_1(T) > 0$ such that for all $0 \leq t \leq T$,

$$f_0^\pm(t,x,p) = 0 \quad \textit{if} \quad |x| \geq M_1(T) \quad \textit{or} \quad |p| \geq M_1(T).$$

(c) Even $f_0^\pm \in C^\infty$ holds, and for every $T < \tilde{T}$ there is constant $M_2(T) > 0$ such that for all $0 \leq t \leq T$,

$$|\partial^\alpha f_0^\pm(t,x,p)| + |\partial_t^\beta E_0(t,x)| + |\partial_t^\gamma D^{[3]}(t)| \leq M_2(T)$$

for every $x \in \mathbb{R}^3$, $p \in \mathbb{R}^3$, $|\alpha| \leq 4$, $\beta \leq 2$, and $\gamma \leq 1$.

See [Bau06a] for the proof. Note that the constants $\tilde{T}$, $M_1(T)$, and $M_2(T)$ do only depend on the "basic" constants r_0 and S_0. In particular $\tilde{T}$, $M_1(T)$, and $M_2(T)$ are independent of c. Since the second moment $\int\int p^2(f_0^+ + f_0^-)\,dx\,dp$ cannot be bounded a priori by using energy conservation, it seems difficult to prove global existence of classical solutions to (rrVP$_c$). Recall that both methods yielding global existence of Vlasov-Poisson type systems essentially relied on such an a bound; see [Pfa92, Sch91, LP91].

By means of (rrVP$_c$) the approximations are improved by replacing at order zero the solution to (VP$_{\text{plasma}}$) by the solution to (rrVP$_c$). Thus let (f_0, E_0) be the solution of (rrVP$_c$); note that this solutions depends on c, as opposed to the solution of (VP$_{\text{plasma}}$). Next define B_1, $(f_2^\pm, E_2)$, and B_3 according to (2.7), (LVP$_{\text{plasma}}$), and (2.9), respectively. We remark that solutions to (LVP$_{\text{plasma}}$) do exist on $[0, T_c[$ (where T_c is from Proposition 2.1) and enjoy the usual properties, provided that both $f_0^{\circ,\pm}$ and $f_2^{\circ,\pm}$ satisfy the assumptions (2.18); see [Bau06a].

In the following section we are going to explain that

$$\begin{aligned} f^{\pm,\mathrm{R}} &= f_0^\pm + c^{-2} f_2^\pm, \\ E^{\mathrm{R}} &= E_0 + c^{-2}E_2 + (2/3)c^{-3}D^{[3]}, \\ B^{\mathrm{R}} &= c^{-1}B_1 + c^{-3}B_3, \end{aligned} \tag{2.19}$$

yields a higher order pointwise approximation of (RVM$_c$) than the Vlasov-Poisson or the Darwin system considered in [BK05]. We call (2.19) the radiation approximation. In the terminology of post-Newtonian approximations it is the 1.5 PN approximation.

Using the Vlasov equation and integration by parts the following formulas are found.

Proposition 2.2. *The fields E^{R} and B^{R} can be written as*

$$\begin{aligned}
E^{\mathrm{R}}(t,x) = &-\int |z|^{-2}\bar{z}(\rho_0 + c^{-2}\rho_2)(t,x+z)\,dz \\
&+\frac{1}{2}c^{-2}\int\int |z|^{-2}\Big\{3(\bar{z}\cdot p)^2\bar{z} - p^2\bar{z}\Big\}(f_0^+ - f_0^-)(t,x+z,p)\,dz\,dp \\
&-c^{-2}\int\int |z|^{-1}\Big\{\bar{z}\otimes\bar{z} + 1\Big\} \\
&\quad\Big(E_0(t,x+z) + (2/3)c^{-3}D^{[3]}(t)\Big)(f_0^+ + f_0^-)(t,x+z,p)\,dz\,dp \\
&+\frac{2}{3}c^{-3}D^{[3]}(t),
\end{aligned} \tag{2.20a}$$

and

$$\begin{aligned}
B^{\mathrm{R}}(t,x) = &\,c^{-1}\int\int |z|^{-2}(\bar{z}\wedge p)(f^{+,\mathrm{R}} - f^{-,\mathrm{R}})(t,x+z,p)\,dz\,dp \\
&-\frac{3}{2}c^{-3}\int\int |z|^{-2}(\bar{z}\cdot p)^2(\bar{z}\wedge p)(f_0^+ - f_0^-)(t,x+z,p)\,dz\,dp \\
&+\frac{1}{2}c^{-3}\int\int |z|^{-1}\Big\{(\bar{z}\wedge p)\otimes\bar{z} + (\bar{z}\cdot p)\bar{z}\wedge(\cdots)\Big\} \\
&\quad\Big(E_0(t,x+z) + (2/3)c^{-3}D^{[3]}(t)\Big)(f_0^+ + f_0^-)(t,x+z,p)\,dz\,dp \\
&-c^{-3}\int \bar{z}\wedge\Big(H^+(t,x+z)j_0^-(t,x+z) \\
&\qquad - H^-(t,x+z)j_0^+(t,x+z)\Big)\,dz.
\end{aligned} \tag{2.20b}$$

At the end of this section we want to mention that there is another variant of a damped Vlasov Poisson type system, considered in [KR01a, KR01b]. While for that system a global solution theory is at hand, the authors did not compare approximations based on their solutions to solutions of the full system.

2.3 1.5 PN comparison dynamics

For the 1.5 PN comparison dynamics, as outlined in Sect. 2.2, the initial data $(f_0^{\circ,\pm}, f_2^{\circ,\pm})$ are given. The field quantities are to be computed from the resulting densities $(f_0^\pm, f_2^\pm)$ by means of (2.20a), (2.20b), $f^{\pm,\mathrm{R}} = f_0^\pm +$

$c^{-2}f_2^{\pm}$, and (2.17). In comparison to that, for the Cauchy problem of (RVM$_c$) the initial fields E° and B° also have to be specified; see (2.2). Thus it is the question for which choice of initial data (2.19) yields a good comparison dynamics and for which not. Certainly it is possible to choose initial data for the fields such that the densities of the two dynamics evolve in a completely different way.

In the first part of this section we therefore fix the initial fields for the Vlasov-Maxwell dynamics from the comparison dynamics; see formula (IC) below. From a mathematical viewpoint this procedure comes with the additional advantage that results on the existence and uniqueness of local-in-time solutions for both the full dynamics and the comparison dynamics have been established; see [GS86, Sch86, Bau06a]. This way it is possible in Theorem 2.4 to obtain a pointwise approximation up to the order $\mathcal{O}(c^{-4})$. It should be mentioned that in [Sch86, Deg86, BK05] the fields are adapted in the same way up to the relevant orders.

There are two drawbacks of this method. In essence post-Newtonian expansion is an expansion of the relativistic velocity $\hat{p}$ and the retarded time $t - c^{-1}|x-y|$. It is clear that assuming localized sources the expansion of the retarded time is only a good approximation in the near zone of the source where $|x - y| \ll c$. This is reflected in the fact that the estimates for the fields in Theorem 2.4 and Theorem 2.5 are only local in the space variable x. Thus also the adapted initial fields are only reliable in the near zone, as is underlined by the fact that they have infinite energy. From a more physical point of view it is moreover questionable to use the Cauchy problem at all. For post-Newtonian expansions the main interest lies in localized systems which are isolated from the rest of the world and which have already evolved for a long time with small velocities. Therefore the Cauchy problem might not be the right formulation since it is not clear how to incorporate these properties into the initial fields.

In physics textbooks isolated systems are characterized by the absence of incoming radiation, i.e., there is no energy coming into the system from past null infinity; see [Cal04]. Here past null infinity is that region of spacetime which is reached along backward light cones. In case that the sources are given, fields which are free of incoming radiation are usually calculated by means of retarded potentials. In the second part of this section we consider a family of solutions of (RVM$_c$), parameterized by c, passing through $f^{\circ,\pm}$ at time $t = 0$; in fact the initial data may also depend on c according to (IC), but this dependence is suppressed in our notation. Then in contrast to the Cauchy problem for (RVM$_c$) the electromagnetic fields are just computed by means of the retarded potentials; see ($_{\mathrm{ret}}$RVM$_c$). The underlying physical picture is that in the absence of incoming radiation every solution of (RVM$_c$) will approach a solution of ($_{\mathrm{ret}}$RVM$_c$), i.e., solutions of ($_{\mathrm{ret}}$RVM$_c$) form a kind of initial layer. Since it is our goal to model slow systems we assume that the momenta are bounded uniformly in $c \geq 1$ and time $t \in \mathbb{R}$. It is beyond the scope of the present survey paper to investigate the existence of solutions with mathematical rigor. Instead we simply introduce Assumption (A) which summarizes all the properties needed. Note however that in [Cal04] the existence of such global solutions is proved for small $f^{\circ,\pm}$ and also uniqueness is discussed.

Vlasov-Maxwell dynamics with adapted initial data

To achieve the improved approximation accuracy we match the initial data of (RVM_c) by the data for the radiation system. For prescribed initial densities $f_0^{\circ,\pm}$ and $f_2^{\circ,\pm}$ we determine $(f_0^{\pm}, E_0)$, B_1, $(f_2^{\pm}, E_2)$, and B_3 according to what has been explained above. Then we consider the Cauchy problem for (RVM_c), where the initial data are taken as

$$\left.\begin{aligned} f^{\pm}(0,x,p) &= f^{\circ,\pm}(x,p) = f_0^{\circ,\pm}(x,p) + c^{-2} f_2^{\circ,\pm}(x,p) + c^{-4} f_{c,\,\mathrm{free}}^{\circ,\pm}(x,p), \\ E(0,x) &= E^{\circ}(x) = E_0(0,x) + c^{-2}E_2(0,x) + (2/3)c^{-3}D^{[3]}(0) \\ &\qquad + c^{-4}E^{\circ}_{c,\,\mathrm{free}}(x), \\ B(0,x) &= B^{\circ}(x) = c^{-1}B_1(0,x) + c^{-3}B_3(0,x) + c^{-4}B^{\circ}_{c,\,\mathrm{free}}(x). \end{aligned}\right\} \tag{IC}$$

In contrast to the contributions at orders 0 to 3, which are fixed by the values of the approximations, $(f_{c,\,\mathrm{free}}^{\circ,\pm}, E^{\circ}_{c,\,\mathrm{free}}, B^{\circ}_{c,\,\mathrm{free}})$ can be chosen freely. They are only subjected to the constraints $\nabla\cdot E^{\circ}_{c,\,\mathrm{free}} = 4\pi\int (f^{\circ,+}_{c,\,\mathrm{free}} - f^{\circ,-}_{c,\,\mathrm{free}})\,dp$ and $\nabla\cdot B^{\circ}_{c,\,\mathrm{free}} = 0$. Note that the constraint equations at the lower orders are satisfied by fiat. Furthermore, we shall assume that

$$\begin{gathered} f^{\circ,\pm}_{c,\,\mathrm{free}} \in C^{\infty}(\mathbb{R}^3\times\mathbb{R}^3), \quad E^{\circ}_{c,\,\mathrm{free}}, B^{\circ}_{c,\,\mathrm{free}} \in C_0^{\infty}(\mathbb{R}^3), \\ f^{\circ,\pm}_{c,\,\mathrm{free}} = 0 \quad \text{if} \quad |x| \ge r_0 \quad \text{or} \quad |p| \ge r_0, \\ \|f^{\circ,\pm}_{c,\,\mathrm{free}}\|_{L^{\infty}} \le S_0, \quad \|E^{\circ}_{c,\,\mathrm{free}}\|_{W^{1,\infty}} + \|B^{\circ}_{c,\,\mathrm{free}}\|_{W^{1,\infty}} \le S_0, \end{gathered} \tag{2.21}$$

holds uniformly in c, with some $r_0, S_0 > 0$ fixed. Before we formulate the approximation result, let us recall that solutions of (RVM_c) with initial data (IC) exist at least on some time interval $[0,\hat{T}[$ which is independent of $c \ge 1$; see [Sch86, Thm. 1].

Proposition 2.3. *Assume that $f_0^{\circ,\pm}$ and $f_2^{\circ,\pm}$ satisfy (2.18). If $f^{\circ,\pm}$, E°, and B° are defined according to (IC), then there exists $0 < \hat{T} \le \infty$ (independent of c) such that for all $c \ge 1$ there is a unique smooth solution $(f^{\pm}, E, B)$ of (RVM_c) with initial data (IC) on the time interval $[0,\hat{T}[$. In addition, for every $T < \hat{T}$ there are constants $M_3(T), M_4(T) > 0$ such that for all $0 \le t \le T$,*

$$\begin{gathered} f^{\pm}(t,x,p) = 0 \quad \text{if} \quad |x| \ge M_3(T) \quad \text{or} \quad |p| \ge M_3(T), \\ |f^{\pm}(t,x,p)| + |E(t,x)| + |B(t,x)| \le M_4(T), \end{gathered}$$

for every $x, p \in \mathbb{R}^3$ and $c \ge 1$.

Actually in [Sch86, Thm. 1] E° and B° do not depend on c, but an inspection of the proof shows that the assertions remain valid for initial fields as defined by (IC).

The first main approximation result at 1.5 PN is as follows; see [Bau06b].

Theorem 2.4. *Assume that* $f_0^{\circ,\pm}$ *and* $f_2^{\circ,\pm}$ *satisfy (2.18). Then calculate* $(f_0^\pm, E_0)$, B_1, $(f_2^\pm, E_2)$, *and* B_3 *by means of* ($rrVP_c$), *(2.7)*, (LVP_{plasma}), *and (2.9), respectively. Thereafter choose the initial data* $(f^{\circ,\pm}, E^\circ, B^\circ)$ *for* (RVM_c) *according to (IC) and (2.21). Let* (f, E, B) *denote the solution of* (RVM_c) *with initial data (IC) and let* $(f^{\pm,\mathrm{R}}, E^{\mathrm{R}}, B^{\mathrm{R}})$ *be defined by (2.19). Then for every* $T < \min\{\tilde{T}, \hat{T}\}$ *and* $r > 0$ *there are constants* $M(T) > 0$ *and* $M(T,r) > 0$ *such that for all* $0 \le t \le T$,

$$\begin{aligned} |f^\pm(t,x,p) - f^{\pm,\mathrm{R}}(t,x,p)| &\le M(T)c^{-4} \quad (x \in \mathbb{R}^3), \\ |E(t,x) - E^{\mathrm{R}}(t,x)| &\le M(T,r)c^{-4} \quad (|x| \le r), \\ |B(t,x) - B^{\mathrm{R}}(t,x)| &\le M(T,r)c^{-4} \quad (|x| \le r), \end{aligned}$$

for every $p \in \mathbb{R}^3$ *and* $c \ge 1$.

The constants $M(T)$ and $M(T,r)$ are independent of $c \ge 1$, but they do depend on the basic constants r_0 and S_0. Note that if (RVM_c) is compared to the Vlasov-Poisson system ($\mathrm{VP}_{\mathrm{plasma}}$) only, one obtains the Newtonian approximation

$$|f^\pm(t,x,p) - f_0^\pm(t,x,p)| + |E(t,x) - E_0(t,x)| + |B(t,x)| \le M(T)c^{-1},$$

see [Sch86, Thm. 2B]. If it is compared to the Darwin system the estimates

$$\begin{aligned} &|f^\pm(t,x,p) - f^{\pm,1\,\mathrm{PN}}(t,x,p)| + |B(t,x) - B^{1\,\mathrm{PN}}(t,x)| \le M(T)c^{-3}, \\ &|E(t,x) - E^{1\,\mathrm{PN}}(t,x)| \le M(T,r)c^{-3}, \end{aligned}$$

are found; see [BK05, Thm. 1.1] and recall (2.8). At first glance it could seem it is a strong limitation to Theorem 2.4 that the time interval $[0,T] \subset [0, \min\{\tilde{T}, \hat{T}\}[$ might be very small, as $\tilde{T}$ and $\hat{T}$ might be very small. Regarding this point we remind the rescaling from Sect. 2.1 which allows to reformulate the result in an ε-dependent fashion on the time interval $[0, \varepsilon^{-3/2}T]$ as $\varepsilon \to 0$, i.e., for long times.

The retarded Vlasov-Maxwell dynamics

Following [Cal04] we introduce the retarded relativistic Vlasov-Maxwell system as

$$\left.\begin{aligned} \partial_t f^\pm + \hat{p} \cdot \nabla_x f^\pm \pm (E + c^{-1}\hat{p} \times B) \cdot \nabla_p f^\pm = 0, \\ E(t,x) = -\int \frac{dy}{|x-y|} (\nabla\rho + c^{-2}\partial_t j)(t - c^{-1}|x-y|, y), \\ B(t,x) = c^{-1} \int \frac{dy}{|x-y|} \nabla \times j(t - c^{-1}|x-y|, y), \\ \rho = \int (f^+ - f^-)\, dp, \qquad j = \int \hat{p}\,(f^+ - f^-)\, dp. \end{aligned}\right\} \qquad ({}_{\mathrm{ret}}\mathrm{RVM}_c)$$

If we assume that $(f^{\pm}, E, B)$ is a smooth solution of $({}_{\mathrm{ret}}\mathrm{RVM}_c)$, then ρ and j satisfy the continuity equation $\partial_t \rho + \nabla \cdot j = 0$. Therefore the retarded fields are a solution of the Maxwell equations, i.e., (f, E, B) also solves (RVM_c). Note that it is necessary to know the densities for all times $]-\infty, t]$ in order to compute the fields at time t. Hence there is no sense to the notation of a local solution of this system. As in the case of the Cauchy problem every solution of $({}_{\mathrm{ret}}\mathrm{RVM}_c)$ satisfies $f^{\pm}(t,x,p) = f^{\pm}(0, X^{\pm}(0;t,x,p), P^{\pm}(0;t,x,p))$, where $s \mapsto (X^{\pm}(s;t,x,p), P^{\pm}(s;t,x,p))$ solves the characteristic system

$$\dot{X} = \hat{P}, \quad \dot{P} = \pm(E + c^{-1}\hat{P} \times B), \tag{2.23}$$

with initial data $X^{\pm}(t;t,x,p) = x$ and $P^{\pm}(t;t,x,p) = p$. Thus $0 \le f^{\pm}(t,x,p) \le \|f^{\pm}(0,\cdot,\cdot)\|_{\infty}$ holds.

As before let $f_0^{\circ,\pm}$, $f_2^{\circ,\pm}$, and $f_{c,\,\mathrm{free}}^{\circ,\pm}$ satisfy (2.18) with some $r_0, S_0 > 0$. Put

$$f^{\circ,\pm} = f_0^{\circ,\pm} + c^{-2} f_2^{\circ,\pm} + c^{-4} f_{c,\,\mathrm{free}}^{\circ,\pm}.$$

We make the following **Assumption (A):**

(a) For every $c \ge 1$ there is a global solution $f^{\pm} \in C^4(\mathbb{R} \times \mathbb{R}^3 \times \mathbb{R}^3)$ of $({}_{\mathrm{ret}}\mathrm{RVM}_c)$ passing through $f^{\circ,\pm}$ at time $t = 0$, i.e., $f^{\pm}(0,x,p) = f^{\circ,\pm}(x,p)$ for $x, p \in \mathbb{R}^3$.
(b) There is a constant $P_1 > 0$ such that $f^{\pm}(t,x,p) = 0$ for $|p| \ge P_1$ and $c \ge 1$. In particular, $f^{\pm}(t,x,p) = 0$ for $|x| \ge r_0 + P_1|t|$ by (2.23).
(c) For every $T > 0$, $R > 0$, and $P > 0$ there is a constant $M_5(T,R,P) > 0$ such that

$$|\partial_t^{\alpha+1} f^{\pm}(t,x,p)| + |\partial_t^{\alpha} \nabla_x f^{\pm}(t,x,p)| \le M_5(T,R,P)$$

for $|t| \le T$, $|x| \le R$, $|p| \le P$, and $|\alpha| \le 3$, uniformly in $c \ge 1$.

Our second main approximation result at 1.5 PN is taken from [Bau06b].

Theorem 2.5. *Assume that $(f^{\pm}, E, B)$ is a family of solutions of $({}_{\mathrm{ret}}RVM_c)$ satisfying Assumption (A) with constants P_1 and $M_5(T,R,P)$. Take $\tilde{T} > 0$ from Proposition 2.1. Then for every $T < \tilde{T}$ and $r > 0$ there are constants $M(T) > 0$ and $M(T,r) > 0$ such that for all $0 \le t \le T$,*

$$\begin{aligned} |f^{\pm}(t,x,p) - f^{\pm,\mathrm{R}}(t,x,p)| &\le M(T)c^{-4} \quad (x \in \mathbb{R}^3), \\ |E(t,x) - E^{\mathrm{R}}(t,x)| &\le M(T,r)c^{-4} \quad (|x| \le r), \\ |B(t,x) - B^{\mathrm{R}}(t,x)| &\le M(T,r)c^{-4} \quad (|x| \le r), \end{aligned}$$

for every $p \in \mathbb{R}^3$ and $c \ge 2P_1$. The constants $M(T)$ and $M(T,r)$ do only depend on r_0, S_0, P_1, and $M_5(\cdot,\cdot,\cdot)$. In particular they are independent of $c \ge 2P_1$.

2.4 Comparison to the particle model

We shall compare our results for the continuous density Vlasov models to the corresponding results for individual particle models. Usually the latter are denoted the Abraham-Lorentz system; see [KS00, KS01, Spo04]. Both systems are quite similar for the Hamiltonian approximations up to 1 PN. In addition, dissipative corrections at 1.5 PN make it necessary to have a closer look at the underlying phase space. The true comparison dynamics lives on a center-like manifold in an extended phase space. In addition, the dynamics on this manifold can be approximated by a modified Vlasov-Poisson system coupled to a second order equation, as in (SGPP$_\eta$). In [KR01b, Sect. 3] it is argued that the force term of the 1.5 PN approximation used in that paper can be obtained formally in the limit "number of particles $\to \infty$" from the individual particle models. In comparison to [KS00, KS01] the main difference to the present paper lies in the treatment of the initial data. For the individual particle model the initial data for the fields are supposed to be of "charged soliton" type. One can think of these fields as generated by charges forced to move freely for $-\infty < t \leq 0$ with their initial velocities. For the approximation this leads to an initial time slip t_0 which the charges need to "forget" their initial data. The initial data for the approximation are then fixed by matching the data of the full system at time t_0. Therefore the initial data for the approximation are given only implicitly, since first one has to compute a solution of the full system over $[0, t_0]$. Regarding the Cauchy problem for the Vlasov-Maxwell system, the matching is done the other way round. For a given initial density one computes the fields of the approximations and imposes their values at $t = 0$ as initial data for the fields of the full system. Hence these initial data are given more explicitly. In fact it is possible to calculate them by fixing only $f_0^{\circ,\pm}$ and $f_2^{\circ,\pm}$; see (2.20a) and (2.20b). Also Theorem 2.4 and Theorem 2.5 seem to be stronger than the results obtained for the particle model, as the passage from 1 PN to 1.5 PN did improve the approximation only in certain directions [KS01, (3.21), (3.32)]. It seems reasonable to expect that a matching of the initial data at $t = 0$ as described above could also improve the earlier results on the Abraham-Lorentz system.

3 The Vlasov-Nordström system

Recently there has been some interest in a simplified but still relativistic model of gravitation in which Vlasov matter is coupled to a scalar theory of gravitation; the latter essentially goes back to Nordström [Nor13]. The metric tensor used in General Relativity is replaced by a scalar function and the Einstein equations are replaced by a wave equation. In [CR03] the following system (VN$_c$) has been considered.

$$S(f) - \Big(S(\varphi)p + \gamma c^2 \nabla_x \varphi\Big) \cdot \nabla_p f = 4S(\varphi)f, \quad -\partial_t^2 \varphi + c^2 \Delta_x \varphi = 4\pi\mu,$$

where $\mu = \int \gamma f \, dp$. Once again $f = f(t, x, p)$ denotes the density to find a particle at time t at position x with momentum p, where $t \in \mathbb{R}$, $x \in \mathbb{R}^3$, and $p \in \mathbb{R}^3$. The scalar gravitational potential $\varphi = \varphi(t, x)$ is generated by the particles via the source μ, and c denotes the speed of light. In addition,

$$p^2 = |p|^2, \quad \gamma = (1 + c^{-2}p^2)^{-1/2}, \quad \hat{p} = \gamma p, \quad \text{and} \quad S = \partial_t + \hat{p} \cdot \nabla_x.$$

The initial data are

$$f(0, x, p) = f^\circ(x, p), \quad \varphi(0, x) = \varphi^0(x), \quad \partial_t \varphi(0, x) = \varphi^1(x).$$

For a physical interpretation and a derivation of this system see [CL04]. In this formulation (VN_c) exhibits many similarities to the relativistic Vlasov-Maxwell system. Thus it is not surprising that many techniques developed for the Vlasov-Maxwell system also apply to the Vlasov-Nordström system. However, regarding basic questions like existence and uniqueness of solutions (VN_c) is by far better understood than (RVM_c). Global existence of classical solutions for unrestricted data is proved in [Cal06], also see [Lee05] for the 2D case.

Again we are concerned with the non-relativistic limit $c \to \infty$ of (VN_c). Under certain circumstances it has been made rigorous in [Ren94] that the (gravitational) Vlasov-Poisson system is the non-relativistic limit of the full Einstein-Vlasov system. In [CL04] it has been shown that as $c \to \infty$ also solutions of (VN_c) converge to a solution of a Vlasov-Poisson system with an error of the order $\mathcal{O}(c^{-1})$. These facts support the belief that (VN_c) may serve well as a model problem for Einstein-Vlasov.

To derive the higher order (post-Newtonian) approximations we follow the naive approach from Sect. 2.1 and first expand all relevant quantities in powers of c^{-1},

$$\begin{aligned} f &= f_0 + c^{-1} f_1 + c^{-2} f_2 + \ldots, \\ \mu &= \mu_0 + c^{-1} \mu_1 + c^{-2} \mu_2 + \ldots, \\ \varphi &= \varphi_0 + c^{-1} \varphi_1 + c^{-2} \varphi_2 + \ldots. \end{aligned}$$

Comparing coefficients yields the equations $-\Delta_x \varphi_0 = 0$ and $-\Delta_x \varphi_1 = 0$. Thus we set $\varphi_0 = \varphi_1 = 0$. As mentioned above, at order zero the gravitational Vlasov-Poisson system ($\mathrm{VP}_{\mathrm{grav}}$)

$$\begin{aligned} &\partial_t f_0 + p \cdot \nabla_x f_0 - \nabla_x \varphi_2 \cdot \nabla_p f_0 = 0, \quad \mu_0 = \int f_0 \, dp, \\ &\varphi_2(t, x) = -\int |z|^{-1} \mu_0(t, x + z) \, dz, \quad f_0(0, x, p) = f^\circ(x, p), \end{aligned}$$

is obtained; see [CL04] for a proof including the necessary error estimates. At the first order the linearized Vlasov-Poisson system

$$\left.\begin{aligned}\partial_t f_1 + p\cdot\nabla_x f_1 - \nabla_x\varphi_3\cdot\nabla_p f_0 - \nabla_x\varphi_2\cdot\nabla_p f_1 = 0,\\ \mu_1 = \int f_1\,dp,\quad \Delta_x\varphi_3 = 4\pi\mu_1 + \partial_t^2\varphi_1,\end{aligned}\right\}$$

appears. Hence if we suppose that $f_1(0,x,p) = 0$, then we can set $f_1 = 0$ and $\varphi_3 = 0$, which also yields $\mu_1 = 0$. For the second order one derives an inhomogeneous Vlasov equation coupled to a Poisson equation (LVP$_{\text{grav}}$),

$$\begin{aligned}&\partial_t f_2 + p\cdot\nabla_x f_2 - \nabla_x\varphi_2\cdot\nabla_p f_2 - \nabla_x\varphi_4\cdot\nabla_p f_0\\ &\quad = 4f_0\tilde{S}(\varphi_2) + (p^2/2)\,p\cdot\nabla_x f_0 + \Big(\tilde{S}(\varphi_2)p - (p^2/2)\nabla_x\varphi_2\Big)\cdot\nabla_p f_0,\\ &\mu_2 = \int (f_2 - (p^2/2)f_0)\,dp,\quad \Delta_x\varphi_4 = 4\pi\mu_2 + \partial_t^2\varphi_2,\end{aligned}$$

where $\tilde{S} = \partial_t + p\cdot\nabla_x$. For (LVP$_{\text{grav}}$) we choose homogeneous initial data $f_2(0,x,p) = 0$.

We now follow the route of adapted initial data as above. Similarly as before it would also be possible to approximate solutions to the retarded Vlasov-Nordström system, the latter being defined analogously to ($_{\text{ret}}$RVM$_c$). For the retarded Vlasov-Nordström system the leading order radiation contribution can be determined explicitly. It is due to monopole radiation and does not vanish for spherically symmetric solutions [BL*06, ST93].

For the adapted initial data let f° be given. Then we calculate (f_0,φ_2) and (f_2,φ_4) according to (VP$_{\text{grav}}$) and (LVP$_{\text{grav}}$). Now we consider (VN$_c$) with initial data

$$\begin{aligned}f(0,x,p) &= f^\circ(x,p),\\ \varphi^0(x) = \varphi(0,x) &= c^{-2}\varphi_2(0,x) + c^{-4}\varphi_4(0,x) + c^{-6}\varphi^0_{\text{free}}(x),\\ \varphi^1(x) = \partial_t\varphi(0,x) &= c^{-2}\partial_t\varphi_2(0,x) + c^{-4}\partial_t\varphi_4(0,x) + c^{-6}\varphi^1_{\text{free}}(x),\end{aligned}$$

where $\varphi^0_{\text{free}}, \varphi^1_{\text{free}} \in C_0^\infty(\mathbb{R}^3)$. The following approximation theorem from [Bau05] shows that the Darwin approximation

$$f^{\mathrm{D}} = f_0 + c^{-2}f_2,\quad \varphi^{\mathrm{D}} = c^{-2}\varphi_2 + c^{-4}\varphi_4,$$

yields a higher order pointwise approximation of the Vlasov-Nordström (VN$_c$). Before we formulate this result let us recall that solutions of (VN$_c$) with matched initial data do exist at least on some time interval $[0,T]$ which is independent of $c \geq 1$; see [CL04, Thm. 3].

Theorem 3.1. *Assume that $f^\circ \in C^\infty(\mathbb{R}^3\times\mathbb{R}^3)$ is nonnegative and compactly supported. From f° calculate (f_0,φ_2) and (f_2,φ_4). Thereafter introduce the matched initial data for (VN$_c$) as above. Let (f,φ) denote the corresponding solution of (VN$_c$) and let $(f^{\mathrm{D}},\varphi^{\mathrm{D}})$ be defined as the Darwin approximation. Then there is a constant $M(T) > 0$, and for every $r > 0$ one can select $M(T,r) > 0$, such that for all $0 \leq t \leq T$,*

$$\begin{aligned}
|f(t,x,p) - f^{\mathrm{D}}(t,x,p)| &\le M(T)c^{-4} && (x \in \mathbb{R}^3),\\
|\varphi(t,x) - \varphi^{\mathrm{D}}(t,x)| &\le M(T,r)c^{-4} && (|x| \le r),\\
|\partial_t\varphi(t,x) - \partial_t\varphi^{\mathrm{D}}(t,x)| &\le M(T)c^{-4} && (x \in \mathbb{R}^3),\\
|\nabla_x\varphi(t,x) - \nabla_x\varphi^{\mathrm{D}}(t,x)| &\le M(T,r)c^{-6} && (|x| \le r),
\end{aligned}$$

for every $p \in \mathbb{R}^3$ *and* $c \ge 1$. *The constants* $M(T)$ *and* $M(T,r)$ *are independent of* $c \ge 1$.

Acknowlegdments. This work has been supported by the DFG Priority Program 1095 "Analysis, Modeling and Simulation of Multiscale Problems" under the grant Ku 1022/3.

References

[AU86] K. Asano and S. Ukai. On the Vlasov-Poisson limit of the Vlasov-Maxwell equation. In T. Nishida, M. Mimura M., and H. Fujii, editors, *Patterns and Waves*, Stud. Math. Appl. Volume 18, pages 369–383, Amsterdam, 1986, North-Holland.

[Bau05] S. Bauer. Post-Newtonian approximation of the Vlasov-Nordström system. *Comm. Partial Differential Equations*, 30:957–985, 2005.

[Bau06a] S. Bauer. The reduced Vlasov-Poisson system with radiation damping. In preparation, 2006.

[Bau06b] S. Bauer. Post-Newtonian dynamics at order 1.5 in the Vlasov-Maxwell system. Preprint `arXiv math-ph/0602031`, 2006.

[BK05] S. Bauer and M. Kunze. The Darwin approximation of the relativistic Vlasov-Maxwell system. *Ann. Henri Poincaré*, 6:283–308, 2005.

[BL*06] S. Bauer, M. Kunze, G. Rein, and A.D. Rendall. Multipole radiation in a collisionless gas coupled to electromagnetism or scalar gravitation. Preprint `arXiv math-ph/0508057`. To appear in *Comm. Math. Phys.*, 2006.

[BF*03] S. Benachour, F. Filbet, Ph. Laurençot, and E. Sonnendrücker. Global existence for the Vlasov-Darwin system in $\mathbb{R}^3$ for small initial data. *Math. Methods Appl. Sci.*, 26:297–319, 2003.

[Bra04] C. Bradaschia, editor, Proceedings of the 5th Edoardo Amaldi Conference on Gravitational Waves. In *Classical Quantum Gravity*, 21:S377–S1263, 2004.

[Cal03] S. Calogero. Spherically symmetric steady states of galactic dynamics in scalar gravity. *Classical Quantum Gravity*, 20:1729–1742, 2003.

[Cal04] S. Calogero. Global small solutions of the Vlasov-Maxwell system in the absence of incoming radiation. *Indiana Univ. Math. J.*, 53:1331–1364, 2004.

[Cal06] S. Calogero. Global classical solutions to the 3D Nordström-Vlasov system. Preprint `arXiv math-ph/0507030`, 2006.

[CL04] S. Calogero and H. Lee. The non-relativistic limit of the Nordström-Vlasov system. *Comm. Math. Sci.*, 2:19–34, 2004.

[CR03] S. Calogero and G. Rein. On classical solutions of the Nordström-Vlasov system. *Comm. Partial Differential Equations*, 28:1–29, 2003.

[Deg86] P. Degond. Local existence of solutions of the Vlasov-Maxwell equations and convergence to the Vlasov-Poisson equation for infinite light velocity. *Math. Methods Appl. Sci.*, 8:533–558, 1986.

[DR92] P. Degond and P. Raviart. An analysis of the Darwin model of approximation to Maxwell's equations. *Forum Math.*, 4:13–44, 1992.

[GS86] R.T. Glassey and W. Strauss. Singularity formation in a collisionless plasma could occur only at high velocities. *Arch. Rat. Mech. Anal.*, 92:59–90, 1986.

[Jac99] J.D. Jackson. *Classical Electrodynamics.* 3rd edition, Wiley, New York, 1999.

[KR01a] M. Kunze and A.D. Rendall. The Vlasov-Poisson system with radiation damping. *Ann. Henri Poincaré*, 2:857–886, 2001.

[KR01b] M. Kunze and A.D. Rendall. Simplified models of electromagnetic and gravitational radiation damping *Classical Quantum Gravity*, 18:3573–3587, 2001.

[KS00] M. Kunze and H. Spohn. Slow motion of charges interacting through the Maxwell field. *Comm. Math. Phys.*, 212:437–467, 2000.

[KS01] M. Kunze and H. Spohn. Post-Coulombian dynamics at order c^{-3}. *J. Nonlinear Science*, 11:321–396, 2001.

[Lee04] H. Lee. The classical limit of the relativistic Vlasov-Maxwell system in two space dimensions. *Math. Methods Appl. Sci.*, 27:249–287, 2004.

[Lee05] H. Lee. Global existence of solutions of the Nordström-Vlasov system in two space dimensions. *Comm. Partial Differential Equations*, 30:663–687, 2005. Erratum: *Comm. Partial Differential Equations*, 30:1261-1262, 2005.

[LP91] P.-L. Lions and B. Perthame. Propagation of moments and regularity for the 3-dimensional Vlasov-Poisson system. *Invent. Math.*, 105:415–430, 1991.

[Nor13] G. Nordström. Zur Theorie der Gravitation vom Standpunkt des Relativitätsprinzips. *Ann. Phys. Lpz.*, 42:533, 1913.

[PST06] G. Panati, H. Spohn, and S. Teufel. Motion of electrons in adiabatically perturbed periodic structures, in this volume.

[Pfa92] K. Pfaffelmoser. Global classical solutions of the Vlasov-Poisson system in three dimensions for general initial data. *J. Differential Equations*, 95:281–303, 1992.

[Ren92] A.D. Rendall. On the definition of post-Newtonian approximation. *Proc. R. Soc. Lond. A*, 438:341–360, 1992.

[Ren94] A.D. Rendall. The Newtonian limit for asymptotically flat solutions of the Vlasov-Einstein system. *Comm. Math. Phys.*, 163:89-112, 1994.

[Sch86] J. Schaeffer. The classical limit of the relativistic Vlasov-Maxwell system. *Comm. Math. Phys.*, 104:403–421, 1986.

[Sch91] J. Schaeffer. Global existence of smooth solutions to the Vlasov-Poisson system in three dimensions. *Commun. Partial Differ. Equations*, 16:1313–1335, 1991.

[ST93] S.L. Shapiro and S.A. Teukolsky. Scalar gravitation: A laboratory for numerical relativity. *Phys. Rev. D*, 47:1529–1540, 1993.

[Spo04] H. Spohn. *Dynamics of Charged Particles and Their Radiation Field.* Cambridge University Press, Cambridge, 2004.

Numerical Integrators for Highly Oscillatory Hamiltonian Systems: A Review

David Cohen[1], Tobias Jahnke[2], Katina Lorenz[1], and Christian Lubich[1]

[1] Mathematisches Institut, Univiversität Tübingen, 72076 Tübingen. {Cohen, Lorenz, Lubich}@na.uni-tuebingen.de

[2] Freie Universität Berlin, Institut für Mathematik II, BioComputing Group, Arnimallee 2–6, 14195 Berlin. Tobias.Jahnke@math.fu-berlin.de

Summary. Numerical methods for oscillatory, multi-scale Hamiltonian systems are reviewed. The construction principles are described, and the algorithmic and analytical distinction between problems with nearly constant high frequencies and with time- or state-dependent frequencies is emphasized. Trigonometric integrators for the first case and adiabatic integrators for the second case are discussed in more detail.

1 Introduction

Hamiltonian systems with oscillatory solution behaviour are ubiquitous in classical and quantum mechanics. Molecular dynamics, in particular, has motivated many of the new numerical developments in oscillatory Hamiltonian systems in the last decade, though the potential range of their applications goes much farther into oscillatory multi-scale problems of physics and engineering.

Since the publication of the last review article on the numerical solution of oscillatory differential equations by Petzold, Jay & Yen [PJY97] in 1997, algorithms and their theoretical understanding have developed substantially. This fact, together with the pleasure of presenting a final report after six years of

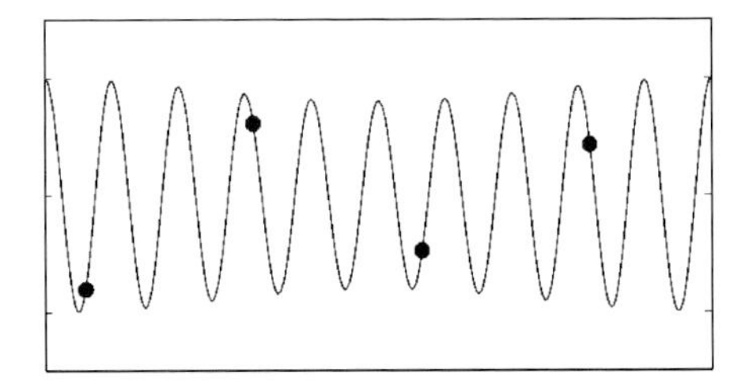

Fig. 1.1. Oscillations and long time steps

funding by the DFG Priority Research Program 1095 on multiscale systems, have incited us to write the present review, which concentrates on Hamiltonian systems. A considerably more detailed (and therefore much longer) account than given here, appears in the second edition of the book by Hairer, Lubich & Wanner [HLW06, pp. 471–565]. Numerical methods for oscillatory Hamiltonian systems are also treated in the book by Leimkuhler & Reich [LR04, pp. 257–286], with a different bias from ours.

The outline of this review is as follows. Sect. 2 describes some classes of oscillatory, multi-scale Hamiltonian systems, with the basic distinction between problems with nearly constant and with varying high frequencies. Sect. 3 shows the building blocks with which integrators for oscillatory systems have been constructed. As is illustrated in Fig. 1.1, the aim is to have methods that can take large step sizes, evaluating computationally expensive parts of the system more rarely than a standard numerical integrator which would resolve the oscillations with many small time steps per quasi-period. Sect. 4 deals with trigonometric integrators suited for problems with almost-constant high frequencies, and Sect. 5 with adiabatic integrators for problems with time- or solution-dependent frequencies.

2 Highly oscillatory Hamiltonian systems

We describe some problem classes, given in each case by a Hamiltonian function H depending on positions q and momenta p (and possibly on time t). The canonical equations of motion $\dot{p} = -\nabla_q H$, $\dot{q} = \nabla_p H$ are to be integrated numerically.

2.1 Nearly constant high frequencies

The simplest example is, of course, the harmonic oscillator given by the Hamiltonian function $H(p,q) = \frac{1}{2}p^2 + \frac{1}{2}\omega^2 q^2$, with the second-order equation of motion $\ddot{q} = -\omega^2 q$. This is trivially solved exactly, a fact that can be exploited for constructing methods for problems with Hamiltonian

$$H(p,q) = \frac{1}{2}p^T M^{-1} p + \frac{1}{2}q^T A q + U(q) \tag{2.1}$$

with a positive semi-definite constant stiffness matrix A of large norm, with a positive definite constant mass matrix M (subsequently taken as the identity matrix for convenience), and with a smooth potential U having moderately bounded derivatives.

The chain of particles illustrated in Fig. 2.1 with equal harmonic stiff springs is an example of a system with a single high frequency $1/\varepsilon$. With the mid-points and elongations of the stiff springs as position coordinates, we have

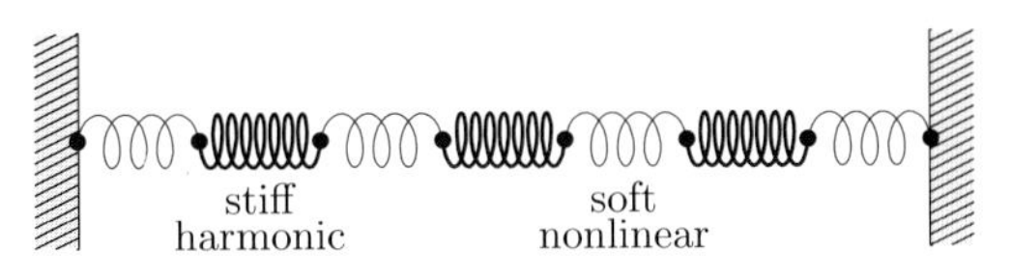

Fig. 2.1. Chain with alternating soft nonlinear and stiff linear springs

$$A = \frac{1}{\varepsilon^2}\begin{pmatrix} 0 & 0 \\ 0 & I \end{pmatrix}, \qquad 0 < \varepsilon \ll 1. \tag{2.2}$$

Other systems have several high frequencies as in

$$A = \frac{1}{\varepsilon^2}\,\mathrm{diag}(0, \omega_1, \ldots, \omega_m), \qquad 0 < \varepsilon \ll 1, \tag{2.3}$$

with $1 \le \omega_1 \le \cdots \le \omega_m$, or a wide range of low to high frequencies without gap as in spatial discretizations of semilinear wave equations.

In order to have near-constant high frequencies, the mass matrix need not necessarily be constant. Various applications lead to Hamiltonians of the form studied by Cohen [Coh06] (with partitions $p = (p_0, p_1)$ and $q = (q_0, q_1)$)

$$H(p,q) = \frac{1}{2} p_0^T M_0(q)^{-1} p_0 + \frac{1}{2} p_1^T M_1^{-1} p_1 + \frac{1}{2} p^T R(q) p + \frac{1}{2\varepsilon^2} q_1^T A_1 q_1 + U(q) \tag{2.4}$$

with a symmetric positive definite matrix $M_0(q)$, constant symmetric positive definite matrices M_1 and A_1, a symmetric matrix $R(q)$ with $R(q_0, 0) = 0$, and a potential $U(q)$. All the functions are assumed to depend smoothly on q. Bounded energy then requires $q_1 = O(\varepsilon)$, so that $p^T R(q) p = O(\varepsilon)$, but the derivative of this term with respect to q_1 is $O(1)$. A simple example of (2.4) is given by a triatomic (water) molecule as illustrated in Fig. 2.2, with strong linear forces that approximately keep the distances and the angle fixed.

2.2 Explicitly time-dependent high frequencies

Here the prototype model is the harmonic oscillator with time-dependent frequency, $H(p,q,t) = \frac{1}{2}p^2 + \frac{1}{2}\varepsilon^{-2}\omega(t)^2 q^2$, with $\omega(t)$ and $\dot\omega(t)$ of magnitude ~ 1 and $\varepsilon \ll 1$. Solutions of the equation of motion $\ddot q = -\varepsilon^{-2}\omega(t)^2 q$ oscillate with a quasi-period $\sim \varepsilon$, but the frequencies change on the slower time scale ~ 1. The action (energy divided by frequency) $I(t) = H(p(t), q(t))/\omega(t)$ is an almost-conserved quantity, called an adiabatic invariant; see, e.g., Henrard [Hen93].

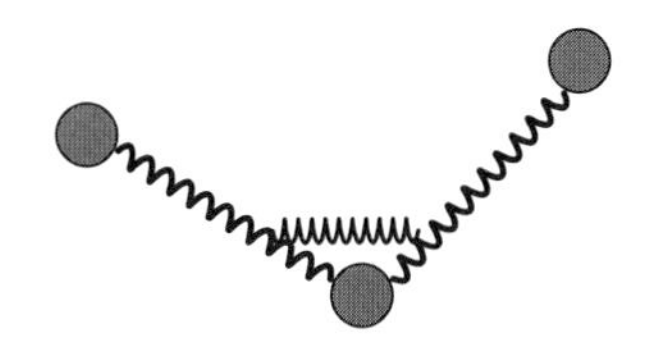

Fig. 2.2. Triatomic molecule

Numerical methods designed for problems with nearly constant frequencies (and, more importantly, nearly constant eigenspaces) behave poorly on this problem, or on its higher-dimensional extension

$$H(p,q,t) = \frac{1}{2}p^T M(t)^{-1} p + \frac{1}{2\varepsilon^2} q^T A(t) q + U(q,t), \tag{2.5}$$

which describes oscillations in a mechanical system undergoing a slow driven motion. Here $M(t)$ is a positive definite mass matrix, $A(t)$ is a positive semi-definite stiffness matrix, and $U(q,t)$ is a potential, all of which are assumed to be smooth with derivatives bounded independently of the small parameter ε. This problem again has adiabatic invariants associated with each of its high frequencies as long as the frequencies remain separated. However, on small time intervals where eigenvalues almost cross, rapid non-adiabatic transitions may occur, leading to further numerical challenges.

2.3 State-dependent high frequencies

Similar difficulties are present, and related numerical approaches have recently been developed, in problems where the high frequencies depend on the position, as in the problem class studied analytically by Rubin & Ungar [RU57], Takens [Tak80], and Bornemann [Bor98]:

$$H(p,q) = \frac{1}{2}p^T M(q)^{-1} p + \frac{1}{\varepsilon^2} V(q) + U(q), \tag{2.6}$$

with a constraining potential $V(q)$ that takes its minimum on a manifold and grows quadratically in non-tangential directions, thus penalizing motions away from the manifold. In appropriate coordinates we have

$$V(q) = \frac{1}{2} q_1^T A(q_0) q_1 \quad \text{for} \quad q = (q_0, q_1)$$

with a positive definite matrix $A(q_0)$.

A multiple spring pendulum with stiff springs as illustrated in Fig. 2.3 is a simple example, with angles as slow variables q_0 and elongations of stiff springs as fast variables q_1. In contrast to the triatomic molecule of Fig. 2.2, where also the angle is kept approximately constant, here the frequencies of the high

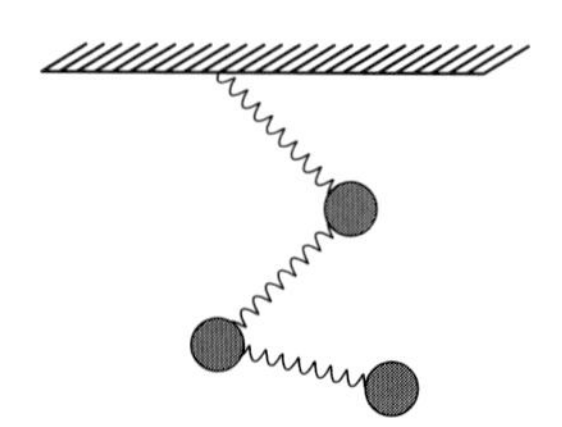

Fig. 2.3. Triple pendulum with stiff springs

oscillations depend on the angles which change during the motion. Different phenomena occur, and different numerical approaches are appropriate for the two different situations.

As in the case of time-dependent frequencies, difficulties (numerical and analytical) arise when eigenfrequencies cross or come close, which here can lead to an indeterminacy of the slow motion in the limit $\varepsilon \to 0$ (Takens chaos).

2.4 Almost-adiabatic quantum dynamics and mixed quantum-classical molecular dynamics

A variety of new developments in the numerics of oscillatory problems within the last decade were spurred by problems from quantum dynamics; see, e.g., [BN*96, DS03, FL06, HL99b, HL99c, HL03, Jah03, Jah04, JL03, LT05, NR99, NS99, Rei99]. Though these problems can formally be viewed as belonging to the classes treated above, it is worthwhile to state them separately: time-dependent quantum dynamics close to the adiabatic limit is described by an equation

$$i\varepsilon\dot{\psi} = H(t)\psi \tag{2.7}$$

with a finite-dimensional hermitian matrix $H(t)$ with derivatives of magnitude ~ 1 representing the quantum Hamiltonian. This is a complex Hamiltonian system with the time-dependent Hamiltonian function $\frac{1}{2}\psi^* H(t)\psi$ (consider the real and imaginary parts of ψ as conjugate variables, and take an ε^{-1}-scaled canonical bracket).

A widely used (though disputable) model of mixed quantum-classical mechanics is the Ehrenfest model

$$\begin{aligned} \ddot{q} &= -\nabla_q\Big(\psi^* H(q)\psi\Big) \\ i\varepsilon\dot{\psi} &= H(q)\psi \end{aligned} \tag{2.8}$$

with a hermitian matrix $H(q)$ depending on the classical positions q. This corresponds to the Hamiltonian function $\frac{1}{2}p^T p + \frac{1}{2}\psi^* H(q)\psi$. The small parameter ε here corresponds to the square root of the mass ratio of light (quantum) and heavy (classical) particles. While this is indeed small for electrons and nuclei, it is less so for protons and heavy nuclei. In the latter case an adiabatic reduction to just a few eigenstates is not reasonable, and then one has to deal with a quantum Hamiltonian which is a discretization of a Laplacian plus a potential operator that depends on the classical position. Both cases show oscillatory behaviour, but the appropriate numerical treatment is more closely related to that in Sects. 2.3 and 2.1 in the first and second case, respectively. Irrespective of its actual physical modeling qualities, the Ehrenfest model is an excellent model problem for studying numerical approaches and phenomena for nonlinearly coupled slow and fast, oscillatory motion.

3 Building-blocks of long-time-step methods: averaging, splitting, linearizing, corotating

We are interested in numerical methods that can attain good accuracy with step sizes whose product with the highest frequency in the system need not be small; see Fig. 1.1. A large variety of numerical methods to that purpose has been proposed in the last decade, and a smaller variety among them has also been carefully analysed. All these long-time-step and multiscale methods are essentially based on a handful of construction principles, combined in different ways. In addition to those described in the following, time-symmetry of the method has proven to be extremely useful, whereas symplecticity appears to play no essential role in long-time-step methods.

3.1 Averages

A basic principle underlying all long-time-step methods for oscillatory differential equations is the requirement to avoid isolated pointwise evaluations of oscillatory functions, but instead to rely on averaged quantities.

Following [HLW06, Sect. VIII.4], we illustrate this for a method for second-order differential equations such as those appearing in the previous section,

$$\ddot{q} = f(q), \qquad f(q) = f^{[\mathrm{slow}]}(q) + f^{[\mathrm{fast}]}(q). \tag{3.1}$$

The classical Störmer-Verlet method with step size h uses a pointwise evaluation of f,

$$q_{n+1} - 2q_n + q_{n-1} = h^2 f(q_n), \tag{3.2}$$

whereas the exact solution satisfies

$$q(t+h) - 2q(t) + q(t-h) = h^2 \int_{-1}^{1} (1-|\theta|)\, f\big(q(t+\theta h)\big)\, d\theta\,. \tag{3.3}$$

The integral on the right-hand side represents a weighted average of the force along the solution, which will now be approximated. At $t = t_n$, we replace

$$f\big(q(t_n + \theta h)\big) \approx f^{[\mathrm{slow}]}(q_n) + f^{[\mathrm{fast}]}\big(u(\theta h)\big)$$

where $u(\tau)$ is a solution of the differential equation

$$\ddot{u} = f^{[\mathrm{slow}]}(q_n) + f^{[\mathrm{fast}]}(u)\,. \tag{3.4}$$

We then have

$$h^2 \int_{-1}^{1} (1-|\theta|)\Big(f^{[\mathrm{slow}]}(q_n) + f^{[\mathrm{fast}]}\big(u(\theta h)\big)\Big) d\theta = u(h) - 2u(0) + u(-h)\,. \tag{3.5}$$

For the differential equation (3.4) we assume the initial values $u(0) = q_n$ and $\dot{u}(0) = \dot{q}_n$ or simply $\dot{u}(0) = 0$. This initial value problem is solved numerically, e.g., by the Störmer-Verlet method with a micro-step size $\pm h/N$ with

$N \gg 1$ on the interval $[-h, h]$, yielding numerical approximations $u^N(\pm h)$ and $\dot{u}^N(\pm h)$ to $u(\pm h)$ and $\dot{u}(\pm h)$, respectively. No further evaluations of $f^{[\text{slow}]}$ are needed for the computation of $u^N(\pm h)$ and $\dot{u}^N(\pm h)$. This finally gives the symmetric two-step method of Hochbruck & Lubich [HL99a],

$$q_{n+1} - 2q_n + q_{n-1} = u^N(h) - 2u^N(0) + u^N(-h)\,. \qquad (3.6)$$

The method can also be given a one-step formulation, see [HLW06, Sect. VIII.4]. Further symmetric schemes using averaged forces were studied by Hochbruck & Lubich [HL99c] and Leimkuhler & Reich [LR01].

The above method is efficient if solving the fast equation (3.4) over the whole interval $[-h, h]$ is computationally less expensive than evaluating the slow force $f^{[\text{slow}]}$. Otherwise, to reduce the number of function evaluations we can replace the average in (3.5) by an average with smaller support,

$$q_{n+1} - 2q_n + q_{n-1} = h^2 \int_{-\delta}^{\delta} K(\theta)\Big(f^{[\text{slow}]}(q_n) + f^{[\text{fast}]}\big(u(\theta h)\big)\Big)\,d\theta \qquad (3.7)$$

with $\delta \ll 1$ and an averaging kernel $K(\theta)$ with integral equal to 1. This is further approximated by a quadrature sum involving the values $f^{[\text{fast}]}\big(u^N(mh/N)\big)$ with $|m| \le M$ and $1 \ll M \ll N$. The resulting method is an example of a *heterogeneous multiscale method* as proposed by E [E03] and Engquist & Tsai [ET05], with macro-step h and micro-step h/N. Method (3.7) is in between the Störmer-Verlet method (3.2) ($\delta = 0$) and the averaged-force method (3.6) ($\delta = 1$).

In the above methods, the slow force is evaluated, somewhat arbitrarily, at the particular value q_n approximating the oscillatory solution $q(t)$. Instead, one might evaluate $f^{[\text{slow}]}$ at an averaged position $\overline{q}_n$, defined by solving approximately an approximate equation

$$\ddot{u} = f^{[\text{fast}]}(u), \quad u(0) = q_n,\ \dot{u}(0) = 0, \qquad \text{and setting} \quad \overline{q}_n = \int_{-\delta}^{\delta} \widetilde{K}(\theta)\, u(\theta h)\, d\theta,$$

with another averaging kernel $\widetilde{K}(\theta)$ having integral 1. Such an approach was first studied by García-Archilla, Sanz-Serna & Skeel [GSS99] for the impulse method (see below), and subsequently in [HL99a] for the averaged-force method, in order to reduce the sensitivity to step size resonances in the numerical solution. For that purpose, it turned out that taking $\delta = 1$ (or an integer) is essential.

3.2 Splitting

The Störmer-Verlet method (see [HLW03]) can be interpreted as approximating the flow φ_h^H of the system with Hamiltonian $H(p,q) = T(p) + V(q)$ with $T(p) = \frac{1}{2}p^T p$ by the symmetric splitting

$$\varphi_{h/2}^{V} \circ \varphi_h^{T} \circ \varphi_{h/2}^{V} .$$

In the situation of a potential $V = V^{[\text{fast}]} + V^{[\text{slow}]}$, we may instead use a different splitting of $H = (T + V^{[\text{fast}]}) + V^{[\text{slow}]}$ and approximate the flow φ_h^H of the system by

$$\varphi_{h/2}^{V^{[\text{slow}]}} \circ \varphi_h^{T+V^{[\text{fast}]}} \circ \varphi_{h/2}^{V^{[\text{slow}]}} .$$

This is the *impulse method* that was proposed in the context of molecular dynamics by Grubmüller, Heller, Windemuth & Schulten [GH*91] and Tuckerman, Berne & Martyna [TBM92]:

$$\begin{aligned} &\text{1. kick: set } p_n^+ = p_n - \tfrac{1}{2}h\,\nabla V^{[\text{slow}]}(q_n) \\ &\text{2. oscillate: solve } \ddot{q} = -\nabla V^{[\text{fast}]}(q) \text{ with initial values } (q_n, p_n^+) \\ &\qquad\qquad\quad \text{over a time step } h \text{ to obtain } (q_{n+1}, p_{n+1}^-) \\ &\text{3. kick: set } p_{n+1} = p_{n+1}^- - \tfrac{1}{2}h\nabla V^{[\text{slow}]}(q_{n+1}) \ . \end{aligned} \tag{3.8}$$

Step 2 must in general be computed approximately by a numerical integrator with a smaller time step. If the inner integrator is symplectic and symmetric, as it would be for the Störmer-Verlet method, then also the overall method is symplectic and symmetric.

García-Archilla, Sanz-Serna & Skeel [GSS99] mollify the impulse method by replacing the slow potential $V^{[\text{slow}]}(q)$ by a modified potential $V^{[\text{slow}]}(\overline{q})$, where $\overline{q}$ represents a local average as considered above.

3.3 Variation of constants formula

A particular situation arises when the fast forces are linear, as in

$$\ddot{q} = -Ax + g(q) \tag{3.9}$$

with a symmetric positive semi-definite matrix A of large norm. With $\Omega = A^{1/2}$, the exact solution satisfies

$$\begin{aligned} \begin{pmatrix} q(t) \\ \dot{q}(t) \end{pmatrix} = &\begin{pmatrix} \cos t\Omega & \Omega^{-1}\sin t\Omega \\ -\Omega \sin t\Omega & \cos t\Omega \end{pmatrix} \begin{pmatrix} q_0 \\ \dot{q}_0 \end{pmatrix} \\ &+ \int_0^t \begin{pmatrix} \Omega^{-1}\sin(t-s)\Omega \\ \cos(t-s)\Omega \end{pmatrix} g\big(q(s)\big)\, ds \ . \end{aligned} \tag{3.10}$$

Discretizing the integral in different ways gives rise to various numerical schemes proposed in the literature for treating (3.9) (the earliest references are Hersch [Her58] and Gautschi [Gau61]). This also gives reinterpretations of the methods discussed above when they are applied to (3.9). We consider a class of *trigonometric integrators* that reduces to the Störmer-Verlet method for $A = 0$ and gives the exact solution for $g = 0$ [HLW06, Chap. XIII]:

$$q_{n+1} = \cos h\Omega \, q_n + \Omega^{-1} \sin h\Omega \, \dot{q}_n + \frac{1}{2} h^2 \Psi \, g(\Phi q_n) \tag{3.11}$$

$$\dot{q}_{n+1} = -\Omega \sin h\Omega \, q_n + \cos h\Omega \, \dot{q}_n + \frac{1}{2} h \big(\Psi_0 \, g(\Phi q_n) + \Psi_1 \, g(\Phi q_{n+1}) \big) \, . \tag{3.12}$$

Here $\Psi = \psi(h\Omega)$ and $\Phi = \phi(h\Omega)$, where the *filter functions* ψ and ϕ are smooth, bounded, real-valued functions with $\psi(0) = \phi(0) = 1$. Moreover, we have $\Psi_0 = \psi_0(h\Omega)$, $\Psi_1 = \psi_1(h\Omega)$ with even functions ψ_0, ψ_1 satisfying $\psi_0(0) = \psi_1(0) = 1$. The method is symmetric if and only if

$$\psi(\xi) = \operatorname{sinc}(\xi) \, \psi_1(\xi) \, , \qquad \psi_0(\xi) = \cos(\xi) \, \psi_1(\xi) \ , \tag{3.13}$$

where $\operatorname{sinc}(\xi) = \sin(\xi)/\xi$. In addition, the method is symplectic (for $g = -\nabla U$) if and only if

$$\psi(\xi) = \operatorname{sinc}(\xi) \, \phi(\xi) \ . \tag{3.14}$$

The two-step form of the method reads

$$q_{n+1} - 2\cos(h\Omega) \, q_n + q_{n-1} \;=\; h^2 \Psi g(\Phi q_n) \ . \tag{3.15}$$

Various methods of Sects. 3.1 and 3.2 can be written in this way, with different filters Ψ and Φ, when they are applied to (3.9):

$\psi(\xi) = \operatorname{sinc}^2(\frac{1}{2}\xi)$	$\phi(\xi) = 1$	Gautschi [Gau61] and averaged method (3.6)
$\psi(\xi) = \operatorname{sinc}(\xi)$	$\phi(\xi) = 1$	Deuflhard [Deu79] and impulse method (3.8)
$\psi(\xi) = \operatorname{sinc}^2(\xi)$	$\phi(\xi) = \operatorname{sinc}(\xi)$	García-Archilla & al. [GSS99]: mollified i.m.
$\psi(\xi) = \operatorname{sinc}^2(\xi)$	$\phi(\xi) = 1$	Hairer & Lubich [HL00]
$\psi(\xi) = \operatorname{sinc}^3(\xi)$	$\phi(\xi) = \operatorname{sinc}(\xi)$	Grimm & Hochbruck [GH06]

As will be seen in Sect. 4, the choice of the filter functions has a substantial influence on the long-time properties of the method.

3.4 Transformation to corotating variables

For problems where the high frequencies and the corresponding eigenspaces depend on time or on the solution, as in (2.5)–(2.8), it is useful to transform to corotating variables in the numerical treatment.

We illustrate the basic procedure for Schrödinger-type equations (2.7) with a time-dependent real symmetric matrix $H(t)$ changing on a time scale ~ 1, for which the solutions are oscillatory with almost-period $\sim \varepsilon$. A time-dependent linear transformation $\eta(t) = T_\varepsilon(t)\psi(t)$ takes the system to the form

$$\dot{\eta}(t) = S_\varepsilon(t) \, \eta(t) \quad \text{ with } \quad S_\varepsilon = \dot{T}_\varepsilon T_\varepsilon^{-1} - \frac{i}{\varepsilon} \, T_\varepsilon H T_\varepsilon^{-1}. \tag{3.16}$$

A first approach is to freeze $H(t) \approx H_*$ over a time step and to choose the transformation

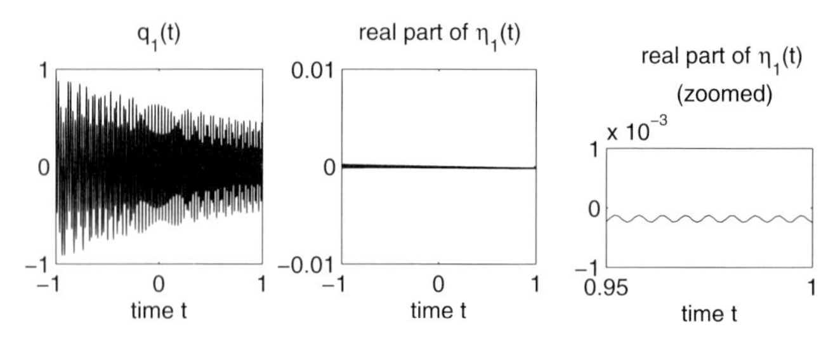

Fig. 3.1. Oscillatory solution component and adiabatic variable as functions of time

$$T_\varepsilon(t) = \exp\Big(\frac{it}{\varepsilon} H_*\Big)$$

yielding a matrix function $S_\varepsilon(t)$ that is highly oscillatory and bounded in norm by $O(h/\varepsilon)$ for $|t - t_0| \le h$, if $H_* = H(t_0 + h/2)$. Numerical integrators using this transformation together with an appropriate treatment of the oscillatory integrals, are studied by Hochbruck & Lubich [HL99c], Iserles [Ise02, Ise04], and Degani & Schiff [DS03]. Step sizes are still restricted by $h = O(\varepsilon)$ in general, but can be chosen larger in the special case when the derivatives of $\frac{1}{\varepsilon}H(t)$ are moderately bounded.

A uniformly bounded matrix $S_\varepsilon(t)$ in (3.16) is obtained if we diagonalize

$$H(t) = Q(t)\,\Lambda(t)\,Q(t)^T$$

with a real diagonal matrix $\Lambda(t) = \mathrm{diag}\,(\lambda_j(t))$ and an orthogonal matrix $Q(t)$ of eigenvectors depending smoothly on t (possibly except where eigenvalues cross). We define $\eta(t)$ by the unitary *adiabatic transformation*

$$\eta(t) = \exp\Big(\frac{i}{\varepsilon}\Phi(t)\Big) Q(t)^T \psi(t) \quad \text{with} \quad \Phi(t) = \mathrm{diag}\,(\phi_j(t)) = \int_0^t \Lambda(s)\,ds, \tag{3.17}$$

which represents the solution in a rotating frame of eigenvectors. Such transformations have been in use in quantum mechanics since the work of Born & Fock [BF28] on the adiabatic invariants $I_j(t) = |\eta_j(t)|^2$ in Schrödinger equations. Figure 3.1 illustrates the effect of this transformation, showing solution components in the original and in the adiabatic variables.

The transformation (3.17) to adiabatic variables yields a differential equation where the ε-independent skew-symmetric matrix

$$W(t) = \dot{Q}(t)^T Q(t)$$

is framed by oscillatory diagonal matrices:

$$\dot{\eta}(t) = \exp\Big(\frac{i}{\varepsilon}\Phi(t)\Big)\, W(t)\, \exp\Big(-\frac{i}{\varepsilon}\Phi(t)\Big)\, \eta(t). \tag{3.18}$$

Numerical integrators for (2.7) based on the transformation to the differential equation (3.18) are given by Jahnke & Lubich [JL03] and Jahnke [Jah04]. The simplest of these methods freezes the slow variables $\eta(t)$ and $W(t)$ at the midpoint of the time step, makes a piecewise linear approximation to the phase $\Phi(t)$, and then integrates the resulting system exactly over the time step. This gives the following *adiabatic integrator*:

$$\eta_{n+1} = \eta_n + hB(t_{n+1/2})\,\frac{1}{2}(\eta_n + \eta_{n+1}) \quad \text{with} \tag{3.19}$$

$$B(t) = \Big(\exp\Big(-\frac{i}{\varepsilon}(\phi_j(t) - \phi_k(t))\Big)\ \mathrm{sinc}\Big(\frac{h}{2\varepsilon}(\lambda_j(t) - \lambda_k(t))\Big)\, w_{jk}(t)\Big)_{j,k}.$$

More involved – and substantially more accurate – methods use a Neumann or Magnus expansion in (3.18) and a quadratic phase approximation. Numerical challenges arise near avoided crossings of eigenvalues, where $\eta(t)$ remains no longer nearly constant and a careful choice of step size selection strategy is needed in order to follow the non-adiabatic transitions; see [JL03] and [HLW06, Chap. XIV].

The extension of this approach to (2.5), (2.6), and (2.8) is discussed in Sect. 5. The transformation to adiabatic variables is also a useful theoretical tool for analysing the error behaviour of multiple time-stepping methods applied to these problems in the original coordinates, such as the impulse and mollified impulse methods considered in Sect. 3.2; see [HLW06, Chap. XIV].

4 Trigonometric integrators for problems with nearly constant frequencies

A good understanding of the behaviour of numerical long-time-step methods over several time scales has been gained for Hamiltonian systems with almost-constant high frequencies as considered in Sect. 2.1. We here review results for single-frequency systems (2.1) with (2.2) (and $M = I$) from Hairer & Lubich [HL00] and [HLW06, Chap. XIII], with the particle chain of Fig. 2.1 serving as a concrete example. The variables are split as $q = (q_0, q_1)$ according to the blocks in (2.2). We consider initial conditions for which the total energy $H(p, q)$ is bounded independently of ε,

$$H(p(0), q(0)) \le \text{Const.}$$

The principal theoretical tool is a *modulated Fourier expansion* of both the exact and the numerical solution,

$$q(t) = \textstyle\sum_k z^k(t)\, e^{ikt/\varepsilon}, \tag{4.1}$$

an asymptotic multiscale expansion with coefficient functions $z^k(t)$ changing on the slow time scale 1, which multiply exponentials that oscillate with frequency $1/\varepsilon$. The system determining the coefficient functions turns out to have a Hamilton-type structure with formal invariants close to the total and oscillatory energies.

The results on the behaviour of trigonometric integrators (3.15) on different time scales have been extended from single- to multi-frequency systems (possibly with resonant frequencies) by Cohen, Hairer & Lubich [CHL05], and to systems (2.4) with non-constant mass matrix by Cohen [Coh04, Coh06].

4.1 Time scale ε

On this time scale the system (2.1) with (2.2) only shows near-harmonic oscillations with frequency $1/\varepsilon$ and amplitude $O(\varepsilon)$ in the fast variables q_1, which are well reproduced by just any numerical integrator.

4.2 Time scale ε^0

This is the time scale of motion of the slow variables q_0 under the influence of the potential $U(q)$. Here it is of interest to have an error in the numerical methods which is small in the step size h and uniform in the product of the step size with the high frequency $1/\varepsilon$. The availability of such uniform error bounds depends on the behaviour of the filter functions ψ and ϕ in (3.15) at integral multiples of π. Under the conditions

$$\psi(2k\pi) = \psi'(2k\pi) = 0, \quad \psi((2k-1)\pi) = 0, \quad \phi(2k\pi) = 0 \tag{4.2}$$

for $k = 1, 2, 3, \ldots$, it is shown in [HLW06, Chap. XIII.4] that the error after n time steps is bounded by

$$\|q_n - q(nh)\| \le C\,h^2, \quad \|\dot{q}_n - \dot{q}(nh)\| \le C\,h \quad \text{for} \quad nh \le \text{Const.}, \tag{4.3}$$

with C independent of h/ε and of bounds of derivatives of the highly oscillatory solution.

Error bounds without restriction of the product of the step size with the frequencies are given for general positive semi-definite matrices A in (2.1) by García-Archilla, Sanz-Serna & Skeel [GSS99] for the mollified impulse method ($\psi(\xi) = \operatorname{sinc}^2(\xi)$, $\phi(\xi) = \operatorname{sinc}(\xi)$), by Hochbruck & Lubich [HL99a] and Grimm [Gri05a] for Gautschi-type methods ($\psi(\xi) = \operatorname{sinc}^2(\xi/2)$ and suitable ϕ), and most recently by Grimm & Hochbruck [GH06] for general A and general classes of filter functions ψ and ϕ.

4.3 Time scale ε^{-1}

An energy exchange between the stiff springs in the particle chain takes place on the slower time scale ε^{-1}. To describe this in mathematical terms, let $q_{1,j}$ be the jth component of the fast position variables q_1, and consider

$$I_j = \frac{1}{2}\,\dot{q}_{1,j}^2 + \frac{1}{2\varepsilon^2}\,q_{1,j}^2,$$

which in the example represents the harmonic energy in the jth stiff spring. The quantities I_j change on the time scale ε^{-1}. To leading order in ε, their change is described by a differential equation that determines the coefficient of $e^{it/\varepsilon}$ in the modulated Fourier expansion (4.1). It turns out that for a trigonometric method (3.15), the differential equation for the corresponding coefficient in the modulated Fourier expansion of the numerical solution is consistent with that of the exact solution for all step sizes if and only if

$$\psi(\xi)\,\phi(\xi) = \mathrm{sinc}(\xi) \qquad \text{for all} \;\; \xi \geq 0.$$

It is interesting to note that this condition for correct numerical energy exchange is in contradiction with the condition (3.14) of symplecticity of the method, with the only exception of the impulse method, given by $\psi = \mathrm{sinc}$, $\phi = 1$. That method, however, does not satisfy (4.2) and is in fact extremely sensitive to near-resonances between frequency and step size (h/ε near even multiples of π). A way out of these difficulties is to consider trigonometric methods with more than one force evaluation per time step, as is shown in [HLW06, Chap. XIII].

In Fig. 4.1 we show the energy exchange of three numerical methods for the particle chain of Fig. 2.1, with $\varepsilon = 0.02$ and with potential and initial data as in [HLW06, p. 22]. At $t = 0$ only the first stiff spring is elongated, the other two being at rest position. The harmonic energies I_1, I_2, I_3, their sum $I = I_1 + I_2 + I_3$, and the total energy H (actually $H - 0.8$ for graphical reasons) are plotted along the numerical solutions of the following methods, with step sizes $h = 0.015$ and $h = 0.03$:

(A) impulse method ($\psi = \mathrm{sinc}$, $\phi = 1$)
(B) mollified impulse method ($\psi = \mathrm{sinc}^2$, $\phi = \mathrm{sinc}$)
(C) heterogeneous multiscale method (3.7) with $\delta = \sqrt{\varepsilon}$.

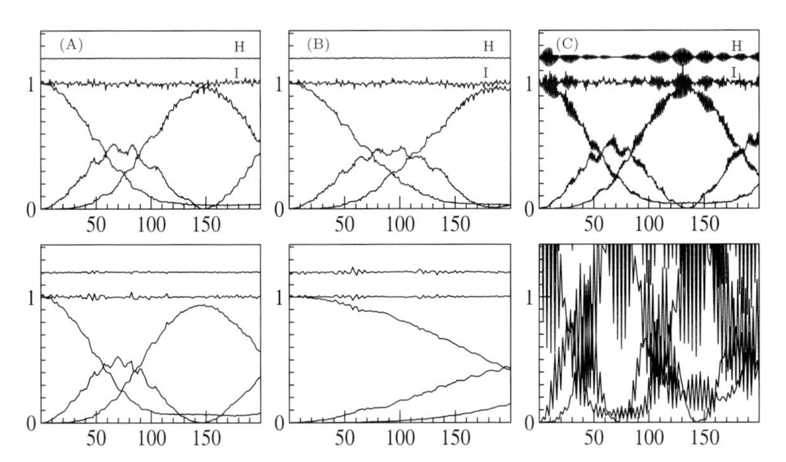

Fig. 4.1. Energy exchange between the stiff springs for methods (A)-(C), with $h = 0.015$ (upper) and $h = 0.03$ (lower), for $\varepsilon = 0.02$

For this problem with linear fast forces, the averaging integrals of Sect. 3.1 and the solution of (3.8) are computed exactly. We notice that for the larger step size only method (A) reproduces the energy exchange in a correct way. Method (C) behaves very similarly to the Störmer-Verlet method.

4.4 Time scales ε^{-N} with $N \geq 2$

In Fig. 4.1 it is seen that the total oscillatory energy I remains approximately conserved over long times. Along the exact solution of the problem, I is in fact conserved up to $O(\varepsilon)$ over exponentially long times $t \leq e^{c/\varepsilon}$; see Benettin, Galgani & Giorgilli [BGG87] and Cohen, Hairer & Lubich [CHL03] for different proofs based on canonical transformations of Hamiltonian perturbation theory and modulated Fourier expansions (4.1), respectively. Along the numerical solution by a trigonometric integrator (3.15), near-conservation of the oscillatory energy I and the total energy H are shown in [HL00] and [HLW06, Chap. XIII] over times $t \leq \varepsilon^{-N}$ under a non-resonance condition between the frequency and the step-size:

$$\left| \sin\left(\frac{kh}{2\varepsilon}\right) \right| \geq c\sqrt{h} \quad \text{for} \quad k = 1, \ldots, N.$$

It is known from [HL00] that the condition

$$\psi(\xi) = \operatorname{sinc}^2(\xi)\, \phi(\xi)$$

is necessary to have long-time conservation of the total energy uniformly for all values of h/ε, and numerical experiments indicate that this condition on the filter functions may also be sufficient. Otherwise, energy conservation is lost at least when h/ε is close to an even multiple of π.

Figure 4.2 shows, for the same data and methods as before, the maximum deviations of H and I on the interval $[0, 1000]$ as functions of h/ε.

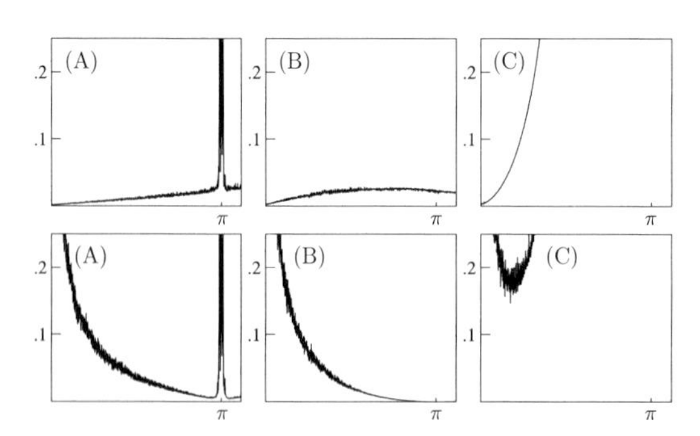

Fig. 4.2. Maximum deviation of total energy (upper) and oscillatory energy (lower) as functions of h/ε (for step size $h = 0.02$)

5 Adiabatic integrators for problems with varying frequencies

Adiabatic integrators are a novel class of numerical integrators that have been devised in [JL03, Jah04, Jah03, LJL05, Lor06, HLW06] for various kinds of oscillatory problems with time- or solution-dependent high frequencies, including (2.5)–(2.8). These integrators have in common that the oscillatory part of the problem is transformed to adiabatic variables (cf. Sect. 3.4) and the arising oscillatory integrals are computed analytically or approximated by an appropriate expansion. The methods allow to integrate these oscillatory differential equations with large time steps in the adiabatic regime of well-separated frequencies and follow non-adiabatic transitions with adaptively refined step sizes.

5.1 Adiabatic integrators for quantum-classical molecular dynamics

Following Jahnke [Jah03], we sketch how to construct a symmetric long-time-step method for problem (2.8), which couples in a nonlinear way slow motion and fast oscillations with frequencies depending on the slow variables.

Proceeding as in Sect. 3.4, the quantum system is transformed to adiabatic variables by

$$\eta(t) = \exp\left(\frac{i}{\varepsilon}\Phi(t)\right) Q\big(q(t)\big)^T \psi(t)\,, \tag{5.1}$$

where $H(q) = Q(q)\Lambda(q)Q(q)^T$ is a smooth eigendecomposition of the Hamiltonian and

$$\Phi(t) = \int_{t_0}^{t} \Lambda\big(q(s)\big)\, ds, \qquad \Phi = \mathrm{diag}(\phi_j) \tag{5.2}$$

is the phase matrix, a diagonal matrix containing the time integrals over the eigenvalues $\lambda_j(q(t))$ along the classical trajectory. Inserting (5.1) into (2.8) yields the new equations of motion

$$\ddot{q} = -\eta^* \exp\left(\frac{i}{\varepsilon}\Phi\right) K(q) \exp\left(-\frac{i}{\varepsilon}\Phi\right)\eta, \tag{5.3}$$

$$\dot{\eta} = \exp\left(\frac{i}{\varepsilon}\Phi\right) W(q,\dot{q}) \exp\left(-\frac{i}{\varepsilon}\Phi\right)\eta \tag{5.4}$$

with the tensor $K(q)$ and skew-symmetric matrix $W(q,\dot{q})$ given as

$$\begin{aligned} K(q) &= Q(q)^T \nabla_q H(q) Q(q), \\ W(q,\dot{q}) &= \left(\frac{d}{dt}Q(q)\right)^T Q(q) \;=\; \big(\nabla_q Q(q)\,\dot{q}\big)^T Q(q). \end{aligned} \tag{5.5}$$

Equations (5.3) and (5.4) can be restated as

$$\ddot{q} = -\eta^* \Big(E(\Phi) \bullet K(q) \Big) \eta, \tag{5.6}$$

$$\dot{\eta} = \Big(E(\Phi) \bullet W(q,\dot{q}) \Big) \eta. \tag{5.7}$$

where $\bullet$ means entrywise multiplication, and $E(\Phi)$ denotes the matrix

$$E(\Phi) = \Big(e_{jk}(\Phi) \Big), \qquad e_{jk}(\Phi) = \exp\left(\frac{i}{\varepsilon}(\phi_j - \phi_k) \right). \tag{5.8}$$

The classical equation can be integrated using the averaging technique from Sect. 3. We insert (5.6) into (3.3) and, in order to approximate the integral, keep the smooth variables $\eta(t)$ and $K(q(t))$ fixed at the midpoint t_n of the interval $[t_{n-1}, t_{n+1}]$. Since more care is necessary for the oscillating exponentials, we replace Φ by the linear approximation

$$\Phi(t_n + \theta h) \approx \Phi(t_n) + \theta h \Lambda\big(q(t_n)\big). \tag{5.9}$$

These modifications yield the *averaged Störmer-Verlet method* of [Jah03]:

$$q_{n+1} - 2q_n + q_{n-1} = -h^2 \, \eta_n^* \Big(E(\Phi_n) \bullet \mathcal{I}(q_n) \bullet K(q_n) \Big) \eta_n \tag{5.10}$$

$$\text{with} \quad \mathcal{I}(q_n) = \int_{-1}^{1} (1 - |\theta|) \, E\big(\theta h \Lambda(q_n)\big) \, d\theta.$$

The matrix of oscillatory integrals $\mathcal{I}(q_n)$ can be computed analytically: its entries $\mathcal{I}_{jk}(q_n)$ are given by

$$\mathcal{I}_{jk}(q_n) = \int_{-1}^{1} (1 - |\theta|) \exp(i\theta\xi_{jk}) \, d\theta = \operatorname{sinc}^2(\tfrac{1}{2}\xi_{jk})$$

$$\text{with} \quad \xi_{jk} = \frac{h}{\varepsilon}\big(\lambda_j(q_n) - \lambda_k(q_n)\big). \tag{5.11}$$

Note that in the (computationally uninteresting) small-time-step limit $h/\varepsilon \to 0$ the integrator (5.10) converges to the Störmer-Verlet method.

The easiest way to approximate the quantum vector $\eta(t)$ is to keep $\eta(t) \equiv \eta(0)$ simply constant. According to the quantum adiabatic theorem [BF28] the resulting error is only $O(\varepsilon)$ as long as the eigenvalues of $H(q(t))$ are well separated and the eigendecomposition remains smooth. A more reliable method, which in its variable-time-step version follows non-adiabatic transitions in η occurring near avoided crossings of eigenvalues, is obtained by integrating Eq. (5.7) from $t_n - h$ to $t_n + h$, using the linear approximation (5.9) for $\Phi(t)$, and freezing the slow coupling matrix $W(q(t), \dot{q}(t))$ at the midpoint t_n. This yields the adiabatic integrator from [Jah03],

$$\eta_{n+1} - \eta_{n-1} = 2h\Big(E(\Phi_n) \bullet \mathcal{J}(q_n) \bullet W_n\Big)\eta_n \tag{5.12}$$

$$\text{with } \mathcal{J}(q_n) = \frac{1}{2}\int_{-1}^{1} E\big(\theta h\Lambda(q_n)\big)\, d\theta.$$

The (j,k)-entry of the matrix of oscillatory integrals $\mathcal{J}(q_n)$ is simply

$$\mathcal{J}_{jk}(q_n) = \frac{1}{2}\int_{-1}^{1} \exp(i\theta\xi_{jk})\, d\theta = \text{ sinc } \xi_{jk}.$$

The explicit midpoint rule is recovered in the limit $h/\varepsilon \to 0$. The derivative contained in (5.5) and the integral in (5.2) are not known explicitly but can be approximated by the corresponding symmetric difference quotient and the trapezoidal rule, respectively. These approximations are denoted by W_n and Φ_n in the above formulas.

The approximation properties of method (5.10), (5.12) for large step sizes up to $h \le \sqrt{\varepsilon}$ are analysed in [Jah03]. A discrete quantum-adiabatic theorem is established, which plays an important role in the error analysis.

5.2 Adiabatic integrators for problems with time-dependent frequencies

Adiabatic integrators for mechanical systems with a time-dependent multi-scale Hamiltonian (2.5) are presented in [HLW06, Chap. XIV] and [Lor06], following up on previous work by Lorenz, Jahnke & Lubich [LJL05] for systems (2.5) with $M(t) \equiv I$ and $A(t)$ a symmetric positive definite matrix. To simplify the presentation, we ignore in the following the slow potential and set $U \equiv 0$.

The approach is based on approximately separating the fast and slow time scales by a series of time-dependent canonical linear coordinate transformations, which are done numerically by standard numerical linear algebra routines. The procedure can be sketched as follows:

- The Cholesky decomposition $M(t) = C(t)^{-T}C(t)$ and the transformation $q \mapsto C(t)q$ change the Hamiltonian in such a way that the new mass matrix is the identity.
- The eigendecomposition

$$A(t) = Q(t)\begin{pmatrix} 0 & 0 \\ 0 & \Omega(t)^2 \end{pmatrix} Q(t)^T, \qquad \Omega(t) = \text{diag}(\omega_j(t))$$

  of the symmetric stiffness matrix $A(t)$ allows to split the positions $q = (q_0, q_1)$ and momenta $p = (p_0, p_1)$ into slow and fast variables q_0, p_0 and q_1, p_1, respectively.
- The fast positions and momenta are rescaled by $\varepsilon^{-1/2}\Omega(t)^{1/2}$ and by $\varepsilon^{1/2}\Omega(t)^{-1/2}$, respectively.

- The previous transforms produce a non-separable term $q^T K(t)p$ in the Hamiltonian. One block of the matrix $K(t)$ is of order $O(\varepsilon^{-1/2})$ and has to be removed by one more canonical transformation.

The Hamiltonian in the new coordinates $p = (p_0, p_1)$ and $q = (q_0, q_1)$ then takes the form

$$H(p,q,t) = \frac{1}{2} p_0^T p_0 + \frac{1}{2\varepsilon} p_1^T \Omega(t) p_1 + \frac{1}{2\varepsilon} q_1^T \Omega(t) q_1 + q^T L(t) p + \frac{1}{2} q^T S(t) q$$

with a lower block-triangular matrix L and a symmetric matrix S of the form

$$L = \begin{pmatrix} L_{00} & 0 \\ \varepsilon^{1/2} L_{10} & L_{11} \end{pmatrix}, \qquad S = \begin{pmatrix} S_{00} & \varepsilon^{1/2} S_{01} \\ \varepsilon^{1/2} S_{10} & \varepsilon S_{11} \end{pmatrix}.$$

Under the condition of bounded energy, the fast variables q_1 and p_1 are now of order $O(\varepsilon^{1/2})$. The equations of motion read

$$\begin{aligned} \dot p_0 &= f_0(p,q,t) \\ \dot q_0 &= p_0 + g_0(q,t) \\ \begin{pmatrix} \dot p_1 \\ \dot q_1 \end{pmatrix} &= \frac{1}{\varepsilon} \begin{pmatrix} 0 & -\Omega(t) \\ \Omega(t) & 0 \end{pmatrix} \begin{pmatrix} p_1 \\ q_1 \end{pmatrix} + \begin{pmatrix} f_1(p,q,t) \\ g_1(q,t) \end{pmatrix} \end{aligned}$$

with functions

$$\begin{pmatrix} f_0 \\ f_1 \end{pmatrix} = -L(t)p - S(t)q, \quad \begin{pmatrix} g_0 \\ g_1 \end{pmatrix} = L(t)^T q,$$

which are bounded uniformly in ε. The oscillatory part now takes the form of a skew-symmetric matrix multiplied by $1/\varepsilon$, similar to (2.7). We diagonalize this matrix and define the diagonal phase matrix Φ as before:

$$\begin{pmatrix} 0 & -\Omega(t) \\ \Omega(t) & 0 \end{pmatrix} = \Gamma i \Lambda(t) \Gamma^*, \qquad \Gamma = \frac{1}{\sqrt{2}} \begin{pmatrix} I & I \\ -iI & iI \end{pmatrix}, \tag{5.13}$$

$$\Lambda(t) = \begin{pmatrix} \Omega(t) & 0 \\ 0 & -\Omega(t) \end{pmatrix}, \qquad \Phi(t) = \int_{t_0}^{t} \Lambda(s)\, ds. \tag{5.14}$$

The transformation to adiabatic variables is now taken as

$$\eta = \varepsilon^{-1/2} \exp\Big(-\frac{i}{\varepsilon}\Phi(t)\Big)\, \Gamma^* \begin{pmatrix} p_1 \\ q_1 \end{pmatrix} \tag{5.15}$$

with the factor $\varepsilon^{-1/2}$ introduced such that $\eta = O(1)$. The equations of motion become

$$\begin{aligned} \dot p_0 &= -L_{00} p_0 - S_{00} q_0 - \varepsilon S_{01} Q_1 \eta \\ \dot q_0 &= p_0 + L_{00}^T q_0 + \varepsilon L_{10}^T Q_1 \eta \end{aligned} \tag{5.16}$$

for the slow variables, and

$$\dot{\eta} = \exp\Bigl(-\frac{i}{\varepsilon}\Phi\Bigr)\, W\, \exp\Bigl(\frac{i}{\varepsilon}\Phi\Bigr)\, \eta - P_1^*\left(L_{10}p_0 + S_{10}q_0\right) \tag{5.17}$$

for the adiabatic variables, where

$$W = \Gamma^* \begin{pmatrix} -L_{11} & -\varepsilon S_{11} \\ 0 & L_{11}^T \end{pmatrix} \Gamma, \qquad \begin{pmatrix} P_1 \\ Q_1 \end{pmatrix} = \Gamma \exp\Bigl(\frac{i}{\varepsilon}\Phi\Bigr).$$

Slow and fast degrees of freedom are only weakly coupled, because in the slow equations (5.16) the fast variable η always appears with a factor ε. The oscillatory part has the familiar form of a coupling matrix framed by oscillatory exponentials, cf. (3.18) and (5.4). Under a separation condition for the frequencies $\omega_j(t)$, the fact that the diagonal of W is of size $O(\varepsilon)$ implies that the expressions $I_j = |\eta_j|^2$ are adiabatic invariants. I_j is the action (energy divided by frequency)

$$I_j = \frac{1}{\omega_j}\left(\frac{1}{2}\, p_{1,j}^2 + \frac{\omega_j^2}{2\varepsilon^2}\, q_{1,j}^2\right).$$

An adiabatic integrator for (2.5) is obtained by the following splitting (for details see [HLW06, Chap. XIV] and Lorenz [Lor06]):

1. Propagate the slow variables (p_0, q_0) with a half-step of the symplectic Euler method. For the oscillatory function $Q_1(t)$, replace the evaluation at $t_{n+1/2} = t_n + h/2$ by the average

$$Q_1^- \approx \frac{2}{h}\int_{t_n}^{t_{n+1/2}} Q_1(t)\, dt,$$

   obtained with a linear approximation of the phase $\Phi(t)$ and analytic computation of the integral.
2. Propagate the adiabatic variable η with a full step of a method of type (3.19) for (5.17).
3. Propagate the slow variables (p_0, q_0) with a half-step of the adjoint symplectic Euler method, with an appropriate average of $Q_1(t)$.

The approximation properties of this method are analyzed in [HLW06, Chap. XIV], where it is shown that the error over bounded time intervals, in the original variables of (2.5), is of order $O(h^2)$ in the positions and $O(h)$ in the momenta, uniformly in ε for $h \le \sqrt{\varepsilon}$. Numerical comparisons with other methods illustrate remarkable benefits of this approach [Lor06, LJL05].

We present numerical illustrations from Lorenz [Lor06] for the time-dependent Hamiltonian (2.5) with $M(t) \equiv I$ and

$$A(t) = \begin{pmatrix} t+3 & \delta \\ \delta & 2t+3 \end{pmatrix}^2$$

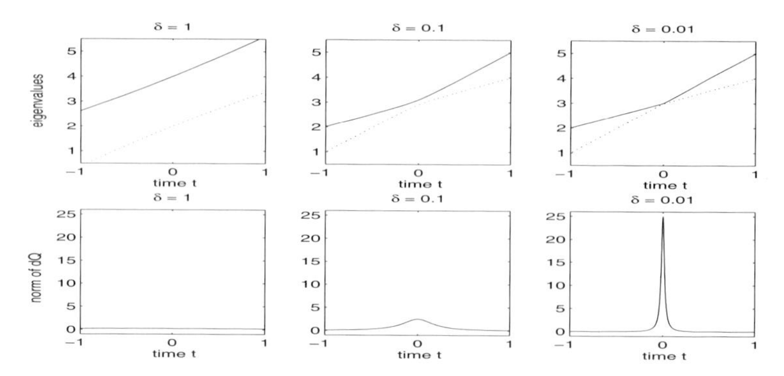

Fig. 5.1. Frequencies ω_j (upper) and $\|\dot{Q}\|$ (lower) for $\delta = 1,\ 0.1,\ 0.01$

on the time interval $[-1, 1]$. The behaviour of the components of the solution $q(t)$ and the adiabatic variable $\eta(t)$ are as in Fig. 3.1 for $\delta = 1$ and $\eta = 0.01$.

Figure 5.1 shows the frequencies and the norm of the time derivative of the matrix $Q(t)$ that diagonalizes $A(t)$ for various values of the parameter δ. For small values of δ, the frequencies approach each other to $O(\delta)$ at $t = 0$, and $\|\dot{Q}(0)\| \sim \delta^{-1}$. This behaviour affects the adiabatic variables $\eta_j(t)$, as is shown in Fig. 5.2.

For $\delta \sim \varepsilon^{1/2}$, there appears an $O(1)$ change in η in an $O(\delta)$ neighbourhood of $t = 0$, and for smaller values of δ the components of η essentially exchange their values; cf. Zener [Zen32] for the analogous situation in Schrödinger-type equations (2.7). Small step sizes are needed near $t = 0$ to resolve this behaviour. Figure 5.3 shows the step sizes chosen by a symmetric adaptive step selection algorithm described in [HLW06, Chap. XIV] for different values of δ. Errors of similar size are obtained in each case.

5.3 Integrators for motion under a strong constraining force

The methods and techniques of the previous subsection can be extended to problems (2.6) with solution-dependent high frequencies [HLW06, Lor06]. The

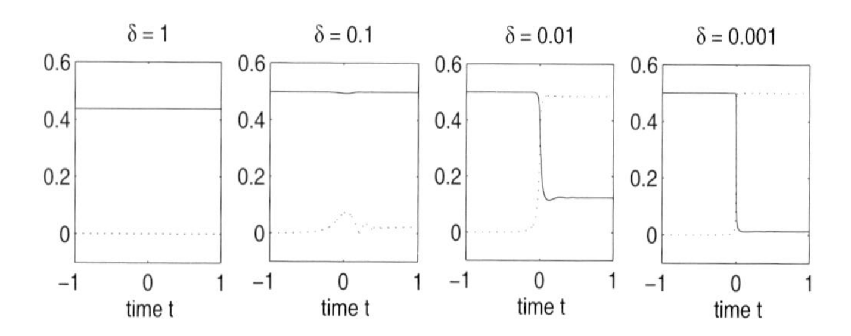

Fig. 5.2. Adiabatic variables η_j as functions of time

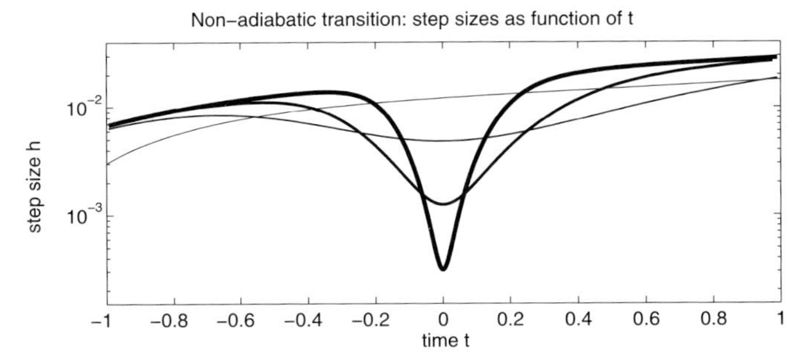

Fig. 5.3. Step size vs. time for $\delta = 2^0,\ 2^{-2},\ 2^{-4},\ 2^{-6}$ (increasingly thicker lines)

procedure is again to apply a series of canonical coordinate transformations to transform (numerically) to a system with nearly-separated slow and fast components:

$$\begin{aligned}
\dot p_0 &= -\nabla_{q_0}\Big(\frac{1}{2}\,p_0^T M_0(q_0)^{-1}p_0 + U(q_0,0)\Big)\\
&\quad -\nabla_{q_0}\Big(\frac{1}{2\varepsilon}\,p_1^T\Omega(q_0)p_1 + \frac{1}{2\varepsilon}\,q_1^T\Omega(q_0)q_1\Big) + f_0(p,q)\\
\dot q_0 &= M_0(q_0)^{-1}p_0 + g_0(p,q) \qquad\qquad (5.18)\\
\begin{pmatrix}\dot p_1\\ \dot q_1\end{pmatrix} &= \frac{1}{\varepsilon}\begin{pmatrix}0 & -\Omega(q_0)\\ \Omega(q_0) & 0\end{pmatrix}\begin{pmatrix}p_1\\ q_1\end{pmatrix} + \begin{pmatrix}f_1(p,q)\\ g_1(p,q)\end{pmatrix}
\end{aligned}$$

with the diagonal matrix $\Omega(q_0)$ of frequencies $\omega_j(q_0)$ and smooth functions of magnitude $f_0 = O(\varepsilon)$, $g_0 = O(\varepsilon)$ and $f_1 = O(\varepsilon^{1/2})$, $g_1 = O(\varepsilon^{1/2})$ in the case of bounded energy and well-separated frequencies. The fast motion of (p_1, q_1) is followed numerically in the adiabatic variables η, which are again defined by (3.17). In these coordinates, the integrator used is then similar to those of Sects. 5.1 and 5.2. Alternatively, the system is integrated in the original coordinates by multiple time-stepping methods such as the (mollified) impulse method of Sect. 3.2, and the above-mentioned transformations are only used as a theoretical tool for analysing the numerical method. We refer to [HLW06, Chap. XIV] for more details.

The actions $I_j = |\eta_j|^2$ are again adiabatic invariants in the case of well-separated frequencies, remaining $O(\varepsilon^{1/(m+1)})$ close to their initial values over bounded time intervals if additionally all the expressions $\omega_j \pm \omega_k \pm \omega_l$ have zeros of multiplicity at most m. It is worthwhile to note that the oscillatory energy appearing in (5.18) can be written as

$$\frac{1}{2\varepsilon}\,p_1^T\Omega(q_0)p_1 + \frac{1}{2\varepsilon}\,q_1^T\Omega(q_0)q_1 = \textstyle\sum_j I_j\,\omega_j(q_0)$$

so that the limit equation of the slow variables for $\varepsilon \to 0$ takes the form first discovered by Rubin & Ungar [RU57],

$$\dot{p}_0 = -\nabla_{q_0}\Big(\frac{1}{2}p_0^T M_0(q_0)^{-1}p_0 + U(q_0,0) + \sum_j I_j\,\omega_j(q_0)\Big)$$
$$\dot{q}_0 = M_0(q_0)^{-1}p_0$$

with the oscillatory energy acting as an extra potential. However, as was noted by Takens [Tak80], the slow motion can become indeterminate in the limit $\varepsilon \to 0$ when the frequencies do not remain separated; see also Bornemann [Bor98]. In contrast to the integration of the slow limit system with constant actions I_j, the numerical integration of the full oscillatory system by an adiabatic integrator with adaptive time steps detects changes in the actions. Moreover, it can follow an almost-solution (having small defect in the differential equation) that passes through a non-adiabatic transition.

Acknowlegdments. This work has been supported by the DFG Priority Program 1095 "Analysis, Modeling and Simulation of Multiscale Problems" under LU 532/3-3. In addition to the discussions with participants of this program, we particularly acknowledge those with Assyr Abdulle, Ernst Hairer, and Gerhard Wanner.

References

[BGG87] G. Benettin, L. Galgani & A. Giorgilli, *Realization of holonomic constraints and freezing of high frequency degrees of freedom in the light of classical perturbation theory. Part I*, Comm. Math. Phys. 113 (1987) 87–103.

[BF28] M. Born, V. Fock, *Beweis des Adiabatensatzes*, Zs. Physik 51 (1928) 165-180.

[Bor98] F. Bornemann, *Homogenization in Time of Singularly Perturbed Mechanical Systems*, Springer LNM 1687 (1998).

[BN*96] F. A. Bornemann, P. Nettesheim, B. Schmidt, & C. Schütte, *An explicit and symplectic integrator for quantum-classical molecular dynamics,* Chem. Phys. Lett., 256 (1996) 581–588.

[BS99] F. A. Bornemann, C. Schütte, *On the singular limit of the quantum-classical molecular dynamics model,* SIAM J. Appl. Math., 59 (1999) 1208–1224.

[Coh04] D. Cohen, *Analysis and numerical treatment of highly oscillatory differential equations*, Doctoral Thesis, Univ. de Genève (2004).

[Coh06] D. Cohen, *Conservation properties of numerical integrators for highly oscillatory Hamiltonian systems*, IMA J. Numer. Anal. 26 (2006) 34–59.

[CHL03] D. Cohen, E. Hairer & C. Lubich, *Modulated Fourier expansions of highly oscillatory differential equations*, Found. Comput. Math. 3 (2003) 327–345.

[CHL05] D. Cohen, E. Hairer & C. Lubich, *Numerical energy conservation for multi-frequency oscillatory differential equations*, BIT 45 (2005) 287–305.

[DS03] I. Degani & J. Schiff, *RCMS: Right correction Magnus series approach for integration of linear ordinary differential equations with highly oscillatory solution*, Report, Weizmann Inst. Science, Rehovot, 2003.

[Deu79] P. Deuflhard, *A study of extrapolation methods based on multistep schemes without parasitic solutions*, Z. angew. Math. Phys. 30 (1979) 177–189.

[E03] W. E, *Analysis of the heterogeneous multiscale method for ordinary differential equations*, Comm. Math. Sci. 1 (2003) 423–436.

[ET05] B. Engquist & Y. Tsai, *Heterogeneous multiscale methods for stiff ordinary differential equations*, Math. Comp. 74 (2005) 1707–1742.

[FL06] E. Faou & C. Lubich, *A Poisson integrator for Gaussian wavepacket dynamics*, Report, 2004. To appear in Comp. Vis. Sci.

[GSS99] B. García-Archilla, J. Sanz-Serna, R. Skeel, *Long-time-step methods for oscillatory differential equations*, SIAM J. Sci. Comput. 20 (1999) 930-963.

[Gau61] W. Gautschi, *Numerical integration of ordinary differential equations based on trigonometric polynomials*, Numer. Math. 3 (1961) 381–397.

[Gri05a] V. Grimm, *On error bounds for the Gautschi-type exponential integrator applied to oscillatory second-order differential equations*, Numer. Math. 100 (2005) 71–89.

[Gri05b] V. Grimm, *A note on the Gautschi-type method for oscillatory second-order differential equations*, Numer. Math. 102 (2005) 61–66.

[GH06] V. Grimm & M. Hochbruck, *Error analysis of exponential integrators for oscillatory second-order differential equations*, J. Phys. A 39 (2006)

[GH*91] H. Grubmüller, H. Heller, A. Windemuth & K. Schulten, *Generalized Verlet algorithm for efficient molecular dynamics simulations with long-range interactions*, Mol. Sim. 6 (1991) 121–142.

[HL00] E. Hairer, C. Lubich, *Long-time energy conservation of numerical methods for oscillatory differential equations.* SIAM J. Num. Anal. 38 (2000) 414-441.

[HLW06] E. Hairer, C. Lubich & G. Wanner, *Geometric Numerical Integration. Structure-Preserving Algorithms for Ordinary Differential Equations.* Springer Series in Computational Mathematics 31. 2nd ed., 2006.

[HLW03] E. Hairer, C. Lubich & G. Wanner, *Geometric numerical integration illustrated by the Störmer–Verlet method*, Acta Numerica (2003) 399–450.

[Hen93] J. Henrard, *The adiabatic invariant in classical mechanics*, Dynamics reported, New series. Vol. 2, Springer, Berlin (1993) 117–235.

[Her58] J. Hersch, *Contribution à la méthode aux différences*, Z. angew. Math. Phys. 9a (1958) 129–180.

[HL99a] M. Hochbruck & C. Lubich, *A Gautschi-type method for oscillatory second-order differential equations*, Numer. Math. 83 (1999) 403–426.

[HL99b] M. Hochbruck & C. Lubich, *A bunch of time integrators for quantum/classical molecular dynamics*, in P. Deuflhard et al. (eds.), Computational Molecular Dynamics: Challenges, Methods, Ideas, Springer, Berlin 1999, 421–432.

[HL99c] M. Hochbruck & C. Lubich, *Exponential integrators for quantum-classical molecular dynamics*, BIT 39 (1999) 620–645.

[HL03] M. Hochbruck & C. Lubich, *On Magnus integrators for time-dependent Schrödinger equations*, SIAM J. Numer. Anal. 41 (2003) 945–963.

[Ise02] A. Iserles, *On the global error of discretization methods for highly-oscillatory ordinary differential equations*, BIT 42 (2002) 561–599.

[Ise04] A. Iserles, *On the method of Neumann series for highly oscillatory equations*, BIT 44 (2004) 473–488.

[Jah03] T. Jahnke, *Numerische Verfahren für fast adiabatische Quantendynamik*, Doctoral Thesis, Univ. Tübingen (2003).

[Jah04] T. Jahnke, *Long-time-step integrators for almost-adiabatic quantum dynamics*, SIAM J. Sci. Comput. 25 (2004) 2145–2164.

[JL03] T. Jahnke & C. Lubich, *Numerical integrators for quantum dynamics close to the adiabatic limit*, Numer. Math. 94 (2003) 289–314.

[LT05] C. Lasser & S. Teufel, *Propagation through conical crossings: an asymptotic semigroup*, Comm. Pure Appl. Math. 58 (2005) 1188–1230.

[LR01] B. Leimkuhler & S. Reich, *A reversible averaging integrator for multiple time-scale dynamics*, J. Comput. Phys. 171 (2001) 95–114.

[LR04] B. Leimkuhler & S. Reich, *Simulating Hamiltonian Dynamics*, Cambridge Monographs on Applied and Computational Mathematics **14**, Cambridge University Press, Cambridge, 2004.

[Lor06] K. Lorenz, *Adiabatische Integratoren für hochoszillatorische mechanische Systeme*, Doctoral thesis, Univ. Tübingen (2006).

[LJL05] K. Lorenz, T. Jahnke & C. Lubich, *Adiabatic integrators for highly oscillatory second order linear differential equations with time-varying eigendecomposition*, BIT 45 (2005) 91–115.

[Net00] P. Nettesheim, *Mixed quantum-classical dynamics: a unified approach to mathematical modeling and numerical simulation,* Thesis FU Berlin (2000).

[NR99] P. Nettesheim & S. Reich, *Symplectic multiple-time-stepping integrators for quantum-classical molecular dynamics*, in P. Deuflhard et al. (eds.), Computational Molecular Dynamics: Challenges, Methods, Ideas, Springer, Berlin (1999) 412–420.

[NS99] P. Nettesheim & C. Schütte, *Numerical integrators for quantum-classical molecular dynamics*, in P. Deuflhard et al. (eds.), Computational Molecular Dynamics: Challenges, Methods, Ideas, Springer, Berlin (1999) 412–420.

[PJY97] L. R. Petzold, L. O. Jay & J. Yen, *Numerical solution of highly oscillatory ordinary differential equations*, Acta Numerica 7 (1997) 437–483.

[Rei99] S. Reich, *Multiple time scales in classical and quantum-classical molecular dynamics,* J. Comput. Phys. 151 (1999) 49–73.

[RU57] H. Rubin & P. Ungar, *Motion under a strong constraining force*, Comm. Pure Appl. Math. 10 (1957) 65–87.

[Tak80] F. Takens, *Motion under the influence of a strong constraining force*, Global theory of dynamical systems, Proc. Int. Conf., Evanston/Ill. 1979, Springer LNM 819 (1980) 425–445.

[TBM92] M. Tuckerman, B.J. Berne & G.J. Martyna, *Reversible multiple time scale molecular dynamics*, J. Chem. Phys. 97 (1992) 1990–2001.

[Zen32] C. Zener, *Non-adiabatic crossing of energy levels*, Proc. Royal Soc. London, Ser. A 137 (1932) 696–702.

Energy Level Crossings in Molecular Dynamics

Folkmar Bornemann[1], Caroline Lasser[2], and Torben Swart[2]

[1] Zentrum Mathematik, Technische Universität München, 85747 Garching bei München. bornemann@ma.tum.de

[2] Fachbereich Mathematik und Informatik, Freie Universität Berlin, Arnimallee 2-6, 14195 Berlin. lasser@mi.fu-berlin.de, swart@mi.fu-berlin.de

Summary. Energy level crossings are the landmarks that separate classical from quantum mechanical modeling of molecular systems. They induce non-adiabatic transitions between the otherwise adiabatically decoupled electronic level spaces. This review covers results on the analysis of propagation through level crossings of codimension two, a mathematical justification of surface hopping algorithms, and a spectral study of a linear isotropic system.

1 Introduction

Molecular systems are a prime example of a multiscale problem. The light electrons move rapidly, in a highly oscillatory fashion, while the nuclei, as the heavier parts of the molecule, move much slower. T his separation of mass and subsequently time and space scales is at the heart of Born-Oppenheimer approximation. It allows for a drastic reduction of problem size when dealing with molecular systems.

A quantum mechanical, non-relativistic description of a molecule is given by the molecular Schrödinger operator

$$
\begin{aligned}
H_{\mathrm{mol}} = & -\sum_{j=1}^{N} \tfrac{1}{2M_j} \Delta_{q_j} - \sum_{j=1}^{n} \tfrac{1}{2} \Delta_{x_j} \\
& + \sum_{j<k} |x_j - x_k|^{-1} + \sum_{j<k} Z_j Z_k |q_j - q_k|^{-1} - \sum_{j,k} Z_k |x_j - q_k|^{-1},
\end{aligned}
$$

where the vectors q and x denote the positions of N nuclei and n electrons, and M_j, Z_j denote mass and charge of the jth nucleus. For the simplicity of notation, one assumes that the nuclei have identical mass M and introduces a scale parameter

$$0 < \epsilon = \sqrt{1/M} \ll 1,$$

which is of the order 10^{-2}, typically. One rewrites the operator as

$$H_{\mathrm{mol}} = -\frac{\epsilon^2}{2}\Delta_q + H_{\mathrm{el}}(q),$$

where the electronic Hamiltonian $H_{\mathrm{el}}(q)$ acts, for a fixed nuclear configuration q, on the electronic degrees of freedom only.

The first step of Born-Oppenheimer approximation consists in solving the electronic eigenvalue problem for all nuclear configurations,

$$\forall q: \; H_{\mathrm{el}}(q)\chi(q,x) = E(q)\chi(q,x).$$

For the second step, one assumes that the electronic energy levels of interest are uniformly separated from the remainder of the electronic spectrum. That is, if one is interested in two levels $E^-(q)$ and $E^+(q)$,

$$\forall q: \; \mathrm{dist}(\{E^+(q), E^-(q)\}, \sigma(H_{\mathrm{el}}(q)) \setminus \{E^+(q), E^-(q)\}) > \delta$$

for some $\delta > 0$. Then, one looks for a diabatic basis $\{\widetilde{\chi}^{\pm}(q,x)\}$ of the electronic subspace $\mathrm{span}\{\chi^{\pm}(q,x)\}$, such that the mapping $q \mapsto \widetilde{\chi}^{\pm}(q,x)$ is smooth. If one replaces the Coulomb interactions inbetween nuclei, and between nuclei and electrons, by a mollified charge distribution, then the electronic Hamiltonian depends smoothly on X, which guarantees existence of a diabatic basis. A diabatic basis $\{\widetilde{\chi}^{\pm}(q,x)\}$ is expected to be different from the adiabatic basis $\{\chi^{\pm}(q,x)\}$ if the electron energy levels $E^{\pm}(q)$ have the same symmetry, see [LS06]. Given a diabatic basis, one builds a hermitian matrix

$$V(q) = \begin{pmatrix} V_{--}(q) & V_{-+}(q) \\ V_{+-}(q) & V_{++}(q) \end{pmatrix},$$

whose entries consists of the expectation values of the electronic Hamiltonian with respect to the diabatic basis functions,

$$V_{kl}(q) = \langle \widetilde{\chi}^k(q,\cdot), H_{\mathrm{el}}(q)\widetilde{\chi}^l(q,\cdot)\rangle_{L^2_{el}}, \qquad k,l \in \{-,+\}.$$

The Born-Oppenheimer Hamiltonian is then given as

$$H_{\mathrm{BO}} = -\frac{\epsilon^2}{2}\Delta_q + V(q).$$

It is a two-level Schrödinger operator acting only on the nucleonic degrees of freedom. Let $\widetilde{\chi}(q,x) = (\widetilde{\chi}^+(q,x), \widetilde{\chi}^-(q,x))^t$. If $\psi(q,t)$ is a solution of the time-dependent Born-Oppenheimer problem

$$i\epsilon\partial_t\psi = H_{\mathrm{BO}}\psi, \qquad \psi(q,0) = \psi_0(q), \tag{1.1}$$

then $\psi(q,t)\cdot\widetilde{\chi}(q,x)$ is an approximate solution of the full molecular problem

$$i\epsilon\partial_t\Psi = H_{\mathrm{mol}}\Psi, \qquad \Psi(q,x,0) = \psi_0(q)\cdot\widetilde{\chi}(q,x)$$

by an error of order ϵ as $\epsilon \to 0$, see [ST01].

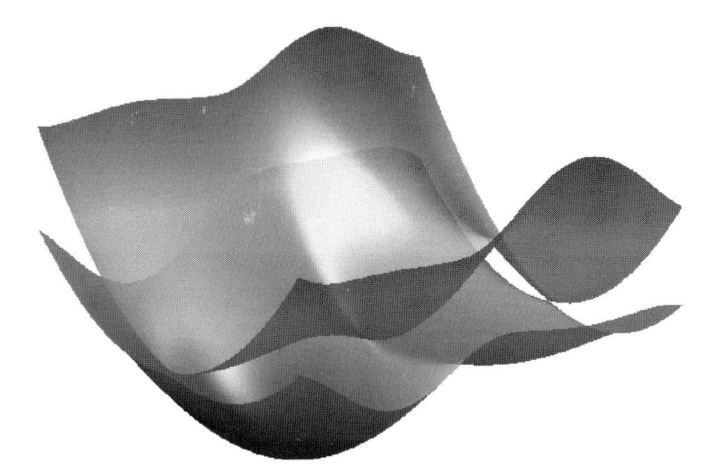

Fig. 1.1. The energy levels of the model for the cis-trans isomerization of retinal in rhodopsin of [HS00]. They cross, when the angular variable ϕ is approximately $\frac{\pi}{2}$ or $\frac{3\pi}{2}$, and the collective coordinate vanishes. In the plot, the abscissa corresponds to ϕ, the ordinate to y. The two local minima of the lower energy level are associated with the cis and the trans configuration of the molecule. (See page 694 for a colored version of the figure.)

For polyatomic molecules, which consist of more than two nuclei, one has to expect the crossing of electron energy levels, that is the existence of nucleonic configurations q with $E^+(q) = E^-(q)$. These crossings, or more precisely, those with a crossing manifold of codimension two or higher,

$$\operatorname{codim}\{q \in \mathbb{R}^d;\, E^+(q) = E^-(q)\} \geq 2,$$

induce non-adiabatic transitions that are of leading order in ϵ as $\epsilon \to 0$: Denote the orthogonal eigenprojectors of the matrix $V(q)$ by $\Pi^\pm(q)$. There is a large set of initial data ψ_0 with

$$\Pi^\pm \psi_0 = 0 \qquad \text{and} \qquad \exists t > 0 : \Pi^\pm(e^{-iH_{\mathrm{BO}}t/\epsilon}\psi_0) = O(1), \quad \epsilon \to 0.$$

That is, the solution of the two-level system performs a non-adiabatic transition from the eigenspace associated with the eigenvalue $E^\mp(q)$ to the one associated with $E^\pm(q)$.

Non-adiabatic transitions are typically linked with ultrafast isomerization processes, radiationless decay, or molecular collisions. The most spectacular example of a femtosecond isomerization modelled by an electron level crossing, the cis-trans isomerization of retinal in rhodopsin, is the first step of vision, see [HS00]. Rhodopsin is the light-absorbing pigment of the rods, which are responsible for the acute, but coarse colorless vision. The model incorporates two electronic levels $E^-(q)$ and $E^+(q)$ and considers two nucleonic degrees of freedom $q = (\phi, y)$, an angular variable $\phi \in \mathbb{T}$ and a collective coordinate

$y \in \mathbb{R}$. The levels $E^{\pm}(q)$ cross twice, when $\phi \approx \frac{\pi}{2}$ or $\phi \approx \frac{3\pi}{2}$ and $y = 0$, see Fig. 1.1. The two local minima of $E^{-}(q)$ are associated with the cis and the trans configuration of the molecule; the lower steeper one represents the stable cis configuration and the higher flatter one the unstable trans configuration. Initially, the wave function is localized in the cis minimum. After photons excited the wave function vertically to the upper level $E^{+}(q)$, it runs down to approach the two crossing points and performs a non-adiabatic transition there, down to the trans minimum of the lower level $E^{-}(q)$. This isomerization is considered the first step in a chain of events that culminate in a change of the impulse pattern sent along the optic nerve.

Generally, physical models incorporate more than two nucleonic degrees of freedom. The numerical solution of Schrödinger equations with high dimensional configuration spaces, however, is a challenging task. A grid based representation of the wave function scales exponentially in the number of space dimensions. Since the wave function is highly oscillatory, with oscillations in space and time of about the order ϵ, a full grid based discretization in four space dimensions is still considered at the borderline of current computer power.

On the other hand, the wave function itself does not have any direct physical interpretation. Meaningful are quadratic quantities of the wave function like the position density or expectation values. A suitable vehicle for encoding quadratic quantities of a wave function $\psi \in L^2(\mathbb{R}^d, \mathbb{C}^2)$ is the associated Wigner function

$$W(\psi)(q,p) = (2\pi)^{-d} \int_{\mathbb{R}^d} e^{i\,y\cdot p} \psi(q - \tfrac{\epsilon}{2}y) \otimes \overline{\psi}(q + \tfrac{\epsilon}{2}y) dy, \qquad (q,p) \in \mathbb{R}^{2d},$$

which is a function on phase space $\mathbb{R}^{2d}$ taking values in the space of hermitian 2×2-matrices. We refer to Chapter 1.8 in [Fol89] for an exposition of the basic properties. Here, we will restrict ourselves to mentioning marginal distributions and the relation to expectation values.

Integration of the Wigner function with respect to momentum and position space result in position and momentum density, respectively,

$$\int_{\mathbb{R}^d} \mathrm{tr}\,(W(\psi)(q,p))\, dp = |\psi(q)|^2, \quad \int_{\mathbb{R}^d} \mathrm{tr}\,(W(\psi)(q,p))\, dq = (2\pi\epsilon)^{-d}\, |\widehat{\psi}(p/\epsilon)|^2.$$

Recall that Weyl quantization associates with a smooth, compactly supported function on phase space, $a \in C_0^\infty(\mathbb{R}^{2d}, \mathbb{C}^{2\times 2})$, a bounded operator $a(q, -i\epsilon\nabla_q)$ on the Hilbert space $L^2(\mathbb{R}^d, \mathbb{C}^2)$, whose action is defined by

$$a(q, -i\epsilon\nabla_q)\psi(q) = \int_{\mathbb{R}^{2d}} e^{i(q-y)\cdot p} a(\tfrac{1}{2}(q+y), \epsilon p)\psi(y) dy dp.$$

The expectation values of a Weyl quantized operator is encoded by the Wigner function, too:

$$\int_{\mathbb{R}^{2d}} \operatorname{tr}\left(W(\psi)(q,p)a(q,p)\right) dqdp = \langle \psi, a(q, -i\epsilon\nabla_q)\psi\rangle_{L^2}.$$

The last identity, combined with the theorem of Calderón-Vaillancourt that asserts that the operator norm of $a(q, -i\epsilon\nabla_q)$ is bounded by a finite sum of sup-norms of derivatives of a, allows one to view the Wigner function as a distribution,

$$W(\psi) : \ a \mapsto \int_{\mathbb{R}^{2d}} \operatorname{tr}\left(W(\psi)(q,p)a(q,p)\right) dqdp.$$

Let ψ be the solution of the two-level system (1.1). If one considers the Wigner matrix $W(\psi)$ with respect to an eigenbasis of the eigenvalues $E^{\pm}(q)$, the diagonals of the matrix $W(\psi)$ show a much more favorable behavior as compared to the highly oscillatory wave function ψ. In the semiclassical limit $\epsilon \to 0$ the diagonals can approximately be described by classical transport and a non-adiabatic transfer of weight between them, see Theorem 2.2 below. Hence, level populations, that is

$$\|\Pi^{\pm}\psi(t)\|_{L^2}^2 = \int_{\mathbb{R}^{2d}} \Pi^{\pm}(q)W(\psi(t))(q,p)\Pi^{\pm}(q)dqdp$$

or other quadratic quantities related to the projected wave functions $\Pi^{\pm}\psi(t)$, can be computed efficiently, even in high dimensional situations.

2 Analysis of the dynamics

The mathematical analysis of time-dependent two-level Schrödinger systems

$$i\epsilon\partial_t\psi^\epsilon = \left(-\tfrac{\epsilon^2}{2}\Delta_q + V(q)\right)\psi^\epsilon, \qquad \psi^\epsilon(0) = \psi_0^\epsilon \in L^2(\mathbb{R}^d, \mathbb{C}^2) \tag{2.1}$$

with crossing eigenvalues has been pioneered by Hagedorn [Hag94]. For time-reversible molecular systems the potential matrix is real-symmetric,

$$V(q) = w(q)\mathrm{Id} + \begin{pmatrix} v_1(q) & v_2(q) \\ v_2(q) & -v_1(q) \end{pmatrix}, \tag{2.2}$$

where $w, v_1, v_2 \in C^\infty(\mathbb{R}^d, \mathbb{R})$ are smooth, real-valued functions with decay properties guaranteeing the essential self-adjointness of the Hamilton operator

$$H = -\tfrac{\epsilon^2}{2}\Delta_q + V(q).$$

Denoting $v(q) = (v_1(q), v_2(q))^t$, the matrix $V(q)$ has the eigenvalues

$$w(q) \pm \sqrt{v_1(q)^2 + v_2(q)^2} = w(q) \pm |v(q)|.$$

The crossing manifold $\{q \in \mathbb{R}^d;\ v(q) = 0\}$ of coinciding eigenvalues has codimension two if one assumes

$$v(q) = 0 \ \Rightarrow \ \operatorname{rank} Dv(q) = 2,$$

where $Dv(q) = (\nabla_q v_1(q), \nabla_q v_2(q))$. These crossings are called conical intersections in the chemical physics literature, see [DYK04]. They seem to be the most commonly studied type of level crossings.

The mathematical results on the dynamics near conical intersections rely on the smallness of the semiclassical parameter $0 < \epsilon \ll 1$, giving dynamical descriptions that are asymptotic with respect to $\epsilon \to 0$. They fall into three groups: the propagation of Gaussian wave packets [Hag94], of two-scale Wigner measures [FG02, FG03, Fer03, FL03], and of Wigner functions [LT05, FL06]. Resolving the non-adiabatic transitions at the crossing manifold, all approaches rely on some kind of normal form, which in its essence is the Landau-Zener operator [Zen32]

$$-i\epsilon\partial_s + \begin{pmatrix} s & \gamma \\ \gamma & -s \end{pmatrix}, \qquad \gamma \in \mathbb{R}.$$

In the framework of microlocal analysis, that is, locally in phase space, normal forms for generic level crossings have been derived by Fermanian Kammerer and Gérard [FG02, FG03]. T he microlocal normal forms of Colin de Verdière [CdV03, CdV04] even allow for superpolynomial error estimates.

2.1 Heuristics

An intuitive, but *non-rigorous* argument that shows the Landau-Zener operator to essentially gear the dynamics through conical intersections has been given for the linear isotropic potential

$$V(q) = \begin{pmatrix} q_1 & q_2 \\ q_2 & -q_1 \end{pmatrix} \tag{2.3}$$

by Teufel and the second author [LT05]. We will adapt these heuristics to general conical intersections in the following. We note, that they result in an explicit formula for the gap γ, which will be a crucial ingredient of the effective asymptotics and the subsequent numerical algorithm that we aim for.

The Schrödinger operator H is the semiclassical Weyl quantization of the matrix-valued symbol

$$\mathbb{R}^{2d} \to \mathbb{R}^{2\times 2}, \quad (q,p) \mapsto \tfrac{1}{2}|p|^2 + V(q).$$

Given its eigenvalues

$$\mathbb{R}^{2d} \to \mathbb{R}, \quad (q,p) \mapsto \tfrac{1}{2}|p|^2 + w(q) \pm |v(q)|$$

one associates the two Hamiltonian systems

$$\dot{q} = p, \qquad \dot{p} = -\nabla_q w(q) \mp \frac{Dv(q)v(q)}{|v(q)|} \tag{2.4}$$

and the corresponding classical flows $\Phi^t_\pm$, which are well-defined away from the crossing manifold.

As a first step, one formally inserts a classical trajectory $(q(t), p(t))$ of one of the Hamiltonian systems into the trace-free part of the symbol of the full operator, obtaining the purely time-dependent problem

$$i\epsilon\dot{\phi}(t) = \begin{pmatrix} v_1(q(t)) & v_2(q(t)) \\ v_2(q(t)) & -v_1(q(t)) \end{pmatrix} \phi(t). \tag{2.5}$$

Such systems show non-adiabatic transitions in the region, where the gap between the eigenvalues is minimal (see e. g. [Bor98, HJ04, BT05]). A necessary condition for the gap between the eigenvalues to become minimal along the chosen trajectory is

$$\tfrac{d}{dt}|v(q(t))|^2 = 0.$$

This condition is satisfied if the trajectory passes the hypersurface

$$S = \left\{(q,p) \in \mathbb{R}^{2d};\ Dv(q)p \cdot v(q) = 0\right\}.$$

Let $\alpha \in [-1, 1]$ be an angle to be determined later. A conjugation by the half-angle rotation matrix

$$\begin{pmatrix} \cos\frac{\alpha}{2} & -\sin\frac{\alpha}{2} \\ \sin\frac{\alpha}{2} & \cos\frac{\alpha}{2} \end{pmatrix}$$

transforms problem (2.5) to

$$i\epsilon\dot{\phi}(t) = \begin{pmatrix} (\cos\alpha, \sin\alpha)^t \cdot v(q(t)) & (\cos\alpha, \sin\alpha)^t \wedge v(q(t)) \\ (\cos\alpha, \sin\alpha)^t \wedge v(q(t)) & -(\cos\alpha, \sin\alpha)^t \cdot v(q(t)) \end{pmatrix} \phi(t),$$

where $x \wedge y = x_1y_2 - x_2y_1$ denotes the symplectic product of the two vectors $x, y \in \mathbb{R}^2$. Assuming that the chosen trajectory passes the hypersurface of minimal gap through the point (q_*, p_*) at time $t = 0$, one linearizes,

$$v(q(t)) = v(q_*) + tDv(q_*)p_* + O(t^2).$$

Aiming at a Landau-Zener problem with the diagonals linearly depending on time and the off-diagonals constant, one chooses the rotation angle as

$$\alpha = \arccos\left(\frac{\nabla_q v_1(q_*) \cdot p_*}{|Dv(q_*)p_*|}\right) = \arcsin\left(\frac{\nabla_q v_2(q_*) \cdot p_*}{|Dv(q_*)p_*|}\right).$$

Since $Dv(q_*)p_* \cdot v(q_*) = 0$, the linearized system then reads as

$$i\epsilon\dot{\phi}(t) = \begin{pmatrix} t|Dv(q_*)p_*| & \frac{Dv(q_*)p_* \wedge v(q_*)}{|Dv(q_*)p_*|} \\ \frac{Dv(q_*)p_* \wedge v(q_*)}{|Dv(q_*)p_*|} & -t|Dv(q_*)p_*| \end{pmatrix} \phi(t).$$

If $|Dv(q_*)p_*| \gg 0$, one sets a new semiclassical parameter $\tilde{\epsilon} = \epsilon/|Dv(q_*)p_*|$, and obtains the Landau-Zener problem

$$i\tilde{\epsilon}\dot{\phi}(t) = \begin{pmatrix} t & \gamma \\ \gamma & -t \end{pmatrix} \phi(t)$$

with gap

$$\gamma = \frac{Dv(q_*)p_* \wedge v(q_*)}{|Dv(q_*)p_*|^2}.$$

Let $\phi^{\pm}(t)$ denote the components of the vector $\phi(t)$ with respect to the eigenbasis of the Landau-Zener matrix, and put $\phi^{\pm}(\pm\infty) = \lim_{t\to\pm\infty} \phi^{\pm}(t)$. If

$$\begin{pmatrix} \phi^+(-\infty) \\ \phi^-(-\infty) \end{pmatrix} = \begin{pmatrix} 1 \\ 0 \end{pmatrix} \qquad \text{or} \qquad \begin{pmatrix} \phi^+(-\infty) \\ \phi^-(-\infty) \end{pmatrix} = \begin{pmatrix} 0 \\ 1 \end{pmatrix},$$

then

$$\begin{pmatrix} |\phi^+(+\infty)|^2 \\ |\phi^-(+\infty)|^2 \end{pmatrix} = \begin{pmatrix} 1-T & T \\ T & 1-T \end{pmatrix} \begin{pmatrix} |\phi^+(-\infty)|^2 \\ |\phi^-(-\infty)|^2 \end{pmatrix}$$

with a transition rate

$$T \sim \exp\Big(-\frac{\pi}{\tilde{\epsilon}}\gamma^2\Big), \qquad \tilde{\epsilon} \to 0.$$

In particular, the Landau-Zener rate T exhibits that the non-adiabatic transitions are of leading order in $\tilde{\epsilon}$ if the chosen trajectory experiences a gap γ that is of order $\sqrt{\tilde{\epsilon}}$.

Remark 2.1. We note that even though the time-dependent problem (2.5) we started with is formulated just in terms of the position coordinate $q(t)$, the resulting Landau-Zener problem has a gap γ that depends on phase space information, namely the point (q_*, p_*) at which the trajectory attains the minimal gap between the eigenvalues $w(q) \pm |v(q)|$.

2.2 Branching process

The heuristics motivates the following definition of random trajectories and a corresponding Markov process for effectively describing the dynamics through conical intersections.

One attaches to points $(q,p) \in \mathbb{R}^{2d}$ in phase space a label -1 or $+1$, indicating reference to the eigenvalue $w(q) - |v(q)|$ or $w(q) + |v(q)|$. Moreover, one chooses a positive number $R > 0$, defining the set

$$\{(q,p) \in \mathbb{R}^{2d};\ |v(q)| \le R\sqrt{\epsilon}\}$$

as a distinguished tubular neighborhood of the crossing manifold. For labelled phase space points $(q,p,j) \in \mathbb{R}^{2d} \times \{-1,+1\}$, one sets

$$\mathcal{T}^{(q,p,j)} : [0,+\infty) \to \mathbb{R}^{2d} \times \{-1,+1\}$$

such that $\mathcal{T}^{(q,p,j)}(t) = (\Phi_j^t(q,p), j)$ as long as

$$|v(q^j(t))| > R\sqrt{\epsilon} \qquad \text{or} \qquad Dv(q^j(t))p^j(t) \cdot v(q^j(t)) \neq 0.$$

A jump from j to $-j$ occurs with probability

$$T(q_*, p_*) = \exp\left(-\frac{\pi}{\epsilon} \frac{(Dv(q_*)p_* \wedge v(q_*))^2}{|Dv(q_*)p_*|^3} \right) \tag{2.6}$$

whenever $\Phi_j^t(q,p)$ hits the manifold of minimal gap

$$S = \left\{ (q,p) \in \mathbb{R}^{2d};\ Dv(q)p \cdot v(q) = 0 \right\}.$$

at time t in a point $(q_*, p_*) \in S$ with $|v(q_*)| \leq R\sqrt{\epsilon}$.

The randomized evolution $\mathcal{T}^{(q,p,j)}(t)$ defines a Markov process. The associated backwards semi-group $\mathcal{L}^t$ is given by its action on a class of continuous scalar-valued functions $a = a(q,p,j)$ satisfying suitable boundary conditions at the manifold of minimal gap S,

$$\mathcal{L}^t a(q,p,j) := \mathbf{E}^{(q,p,j)} a\big(\mathcal{T}^{(q,p,j)}(t)\big),$$

see [LT05, FL06]. This definition naturally extends to matrix-valued functions of the form $a = a^+ \Pi^+ + a^- \Pi^-$ with $a^\pm \in C_0^\infty(\mathbb{R}^{2d} \setminus S, \mathbb{C})$, that is, to functions that commute with the potential matrix V. By duality, the semigroup acts on Wigner functions also,

$$\mathcal{L}^t W(\psi) : a \mapsto \int_{\mathbb{R}^{2d}} \operatorname{tr}\big(W(\psi)(q,p)(\mathcal{L}^t a)(q,p) \big)\, dq dp.$$

Theorem 2.2 ([LT05, FL06]). *Let $(\psi_0^\epsilon)_{\epsilon > 0}$ be a bounded sequence in $L^2(\mathbb{R}^d, \mathbb{C}^2)$ such that $\Pi^- \psi_0^\epsilon = 0$ and there exists $\delta > 0$ with*

$$\lim_{\epsilon \to 0} \int_{S_\delta} |W(\psi_0^\epsilon)(q,p)|\, dq dp = 0,$$

where $S_\delta = \{(q,p) \in \mathbb{R}^{2d};\ |v(q)|, |Dv(q)p \cdot v(q)| \leq \delta\}$. Let V be a matrix with conically intersecting eigenvalues as given in (2.2). Suppose, that $q \mapsto |v(q)|$ is convex, and that $Dv(q)\nabla w(q) \cdot v(q) \leq 0$ for all $q \in \mathbb{R}^d$.

Then, for all $T > 0$, the solution $\psi^\epsilon(t)$ of the Schrödinger equation (2.1) with initial data $\psi^\epsilon(0) = \psi_0^\epsilon$ satisfies

$$\sup_{t \in [0,T]} \int_{\mathbb{R}^{2d}} \big(W(\psi^\epsilon(t)) - \mathcal{L}^t W(\psi_0^\epsilon) \big)\, a(q,p)\, dq dp$$
$$= O(R^{-1}) + O(R^3\sqrt{\epsilon}) + O(\sqrt{\epsilon}\,|\ln \epsilon|)$$

for all $a = a^+ \Pi^+ + a^- \Pi^-$ with $a^\pm \in C_0^\infty(\mathbb{R}^{2d} \setminus S, \mathbb{C})$.

Remark 2.3. Note, that the error is minimal as $R = \epsilon^{-1/8}$ and is of order $\epsilon^{1/8}$.

Remark 2.4. The assumptions on the potential $V(q)$ guarantee that minus-trajectories issued from the upper level never meet the crossing again. Hence, the system dealt with does not show any interlevel interferences, which could not be resolved by merley working on the diagonal of the Wigner matrix.

Remark 2.5. The assumptions on the initial data $(\psi_0^\epsilon)_{\epsilon>0}$ ensure that the wave function does not localize near the manifold of minimal gap and the crossing manifold initially. This is due to the semigroup encorporating only an effective treatment of the non-adiabatic transitions, which becomes valid when the solution has passed by the crossing and the manifold of minimal gap.

In [LT05], Theorem 2.2 has been proven for the linear isotropic potential (2.3) with an error of $o(1)$ as $\epsilon \to 0$. The proof for general potentials providing the error bound with respect to ϵ and R is given in [FL06]. It falls into two parts: Away from the crossing, there is only classical transport. One shows by semiclassical Weyl calculus that the error of propagation is of order $O(R^{-1}) + O(\sqrt{\epsilon})$. Near the crossing, non-adiabatic transitions become relevant. For proving the correctness of the asymptotic transition rates one resorts to a refined version of the microlocal normal form of [CdV03].

2.3 Surface hopping algorithm

The semigroup $\mathcal{L}^t$ suggests a numerical algorithm that can be seen as a rigorous counterpart to the surface hopping algorithms of chemical physics. Such algorithms have been introduced by Tully and Preston in [TP71] for studies of molecular collisions. To our knowledge Theorem 2.2 is the first mathematically rigorous justification for such an approach. For high dimensional problems in photochemistry surface hopping seems to be one of the most popular algorithms employed. We refer to the review [ST05] as a pointer to the vast chemical literature on algorithms of this type.

A numerical realization of the semigroup $\mathcal{L}^t$ is achieved by the following steps: one projects the initial wave function to the energy levels and computes the associated Wigner functions. After sampling the Wigner functions, one propagates along the classical trajectories of the Hamiltonian systems (2.4) and opens up a new trajectory on the other level whenever a trajectory passes through the hypersurface of minimal gap. When splitting up, the weight associated with the trajectories is distributed according to the Landau-Zener transition coefficient (2.6).

[LST06] thoroughly validates this algorithm for systems with linear isotropic potentials. Figure 2.1 shows the relative error of level populations for the following test case: the initial data are Gaussian wave packets associated with the upper level localized at a distance of $5\sqrt{\epsilon}$ from the crossing with average momentum of order one. The time evolution stops at time $t = 10\sqrt{\epsilon}$.

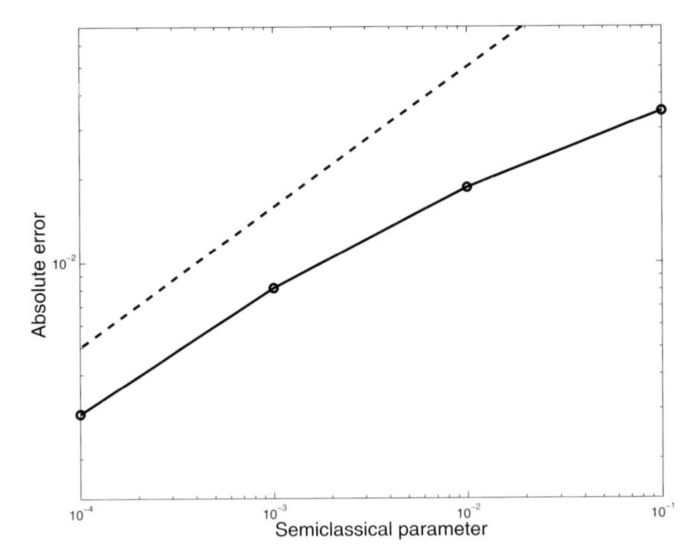

Fig. 2.1. The absolute error of the surface hopping algorithm versus a Strang splitting scheme for level populations with respect to the semiclassical parameter ϵ. The dashed line is the function $\frac{1}{2}\sqrt{\epsilon}$, while the solid line is the absolute error for the final populations on the two levels.

The level populations $\|\Pi^{\pm}\psi(t)\|_{L^2}^2$ are computed by the rigorous surface hopping algorithm as well as by a numerically converged Strang splitting scheme. As a function of the semiclassical parameter, the resulting absolute error is bounded by $\frac{1}{2}\sqrt{\epsilon}$. All the other experiments of [LST06] show a comparable error of order $\sqrt{\epsilon}$, indicating, that the $\epsilon^{1/8}$ error bound of Theorem 2.2 is not sharp.

3 Spectral study

Mathematically rigorous spectral studies of operators with crossing eigenvalues have aimed at resolvent estimates [Jec0303, Jec05] and bounds on the number of resonances [Néd96, Néd01, Néd03]. More explicit, quantitative investigations showing a clear spectral fingerprint of non-adiabatic origin have been undertaken by Avron and Gordon in the zero energy regime [AG00a, AG00b]. Complementarily to these results, the joint work of the second author with Fujiié and Nédélec [FLN06] deals with Bohr-Sommerfeld conditions for energies bounded away from zero.

The common model operator of [AG00a, AG00b, FLN06] has a linear isotropic potential matrix with conically intersecting eigenvalues,

$$H = -\epsilon^2 \Delta_q + V(q) = -\epsilon^2 \Delta_q + \begin{pmatrix} q_1 & q_2 \\ q_2 & -q_1 \end{pmatrix}.$$

Its scalar counterparts are the one-level operators

$$H^{\pm} = -\epsilon^2 \Delta_q \pm |q|.$$

The upper level operator H^+ has a confining potential, which is bounded from below by zero and increases to infinity as $|q| \to \infty$. Hence, H^+ has purely discrete spectrum with strictly positive eigenvalues (see Theorems XIII.47 and XIII.67 in [RS78]). The lower level operator H^- has a repulsive potential. Its commutator with the generator of dilations $D = \frac{1}{2i}(q \cdot \nabla_q + \nabla_q \cdot q)$ is positive,

$$[H^-, iD] = -2\Delta_q + |q|,$$

and yields a global Mourre estimate. Hence, H^- has purely absolutely continuous spectrum (see Corollary 4.10 in [CF*87]). The full operator H, however, inherits the purely absolutely continuous spectrum of H^-, while echoing the discrete spectrum of H^+ with resonances close to the real axis.

The resonances of the operator H are defined by complex dilation (see Theorem 2.1 in [Néd96]). They are the eigenvalues $E \in \mathbb{C}$ of the complex scaled Hamiltonian

$$H_\theta = -\epsilon^2 e^{-2i\theta} \Delta_q + e^{i\theta} V(q),$$

which is a non-selfadjoint operator with discrete spectrum in the lower half-plane independent of the dilation parameter $\theta \in]0, \frac{\pi}{3}[$.

Remark 3.1. For a large class of scalar Schrödinger operators, resonances defined by complex dilation have been identified with the poles of a suitable continuation of the resolvent or of the scattering matrix. The underlying physical picture is that of a slowly decaying state, whose life-time is set by the imaginary part of the resonance. We refer to Chapter 8 in [CF*87], Chapter 16 in [HS96], or the review [Zwo99] as introductory reading for the theory of resonances.

A resonance E of H is determined by solving $H\psi = E\psi$ in the distributional sense and validating decay and regularity properties of the dilated resonant state $q \mapsto \psi(e^{-i\theta}q)$. One uses the radial symmetry in $V(q)$ to reduce the partial differential operator H to a direct sum of ordinary differential systems: the operator P is unitarily equivalent to

$$\bigoplus_{\nu \in \mathbb{Z}+\frac{1}{2}} H_\nu(r, -i\epsilon\partial_r; \epsilon), \qquad H_\nu(r, \rho; \epsilon) = \begin{pmatrix} r^2 - \rho & \epsilon\nu/r \\ \epsilon\nu/r & r^2 + \rho \end{pmatrix}, \tag{3.1}$$

where $(r, \rho) \in \mathbb{R}^+ \times \mathbb{R}$. Nédélec [Néd96] has derived this equivalence by a Fourier transformation, a change to polar coordinates $(r, \phi) \in \mathbb{R}^+ \times \mathbb{T}$, a conjugation by the half-angle rotation matrix

$$\begin{pmatrix} \cos\frac{\phi}{2} & -\sin\frac{\phi}{2} \\ \sin\frac{\phi}{2} & \cos\frac{\phi}{2} \end{pmatrix},$$

and a final Fourier series ansatz in the angular variable ϕ. A similar decomposition labelled by half-integers has been obtained by Avron and Gordon [AG00a, AG00b], using the commutation relation of H with an angular momentum operator L,

$$[H, L] = 0, \qquad L = q \wedge (-i\epsilon\nabla_q) + \tfrac{1}{2}\begin{pmatrix} 0 & i \\ -i & 0 \end{pmatrix}.$$

Both decompositions share the fact, that the half-integer labelling turns the conical intersection $q \mapsto \pm|q|$ into a family of avoided crossings. However, while in [AG00a, AG00b] the ordinary differential systems are solved in the zero energy regime in terms of generalized hypergeometric functions, the aim of [FLN06] is an explicit asymptotic analysis ($\epsilon \to 0$) of non-zero energies by means of an exact WKB method.

According to the decomposition (3.1) one associates with a resonance E of H an angular momentum number $\nu \in \mathbb{Z} + \frac{1}{2}$ if E corresponds to a distributional solution of the ordinary differential problem

$$H_\nu(r, -i\epsilon\partial_r; \epsilon)u = Eu,$$

such that $r \mapsto u(e^{-i\theta}r)$ satisfies appropriate boundary conditions as $r \to 0$ and $r \to +\infty$. Consider $E \in]a, b[$ for positive numbers $0 < a < b$. If $\epsilon > 0$ is sufficiently small, the energy surface

$$\big\{(r, \rho) \in \mathbb{R}^+ \times \mathbb{R};\ \det\left(H_\nu(r, \rho; \epsilon) - E\right) = 0\big\}$$

consists of two connected curves, a closed simple one and one being unbounded. Let $A_\nu(E, \epsilon)$ be the action associated with the closed curve,

$$A_\nu(E, \epsilon) = 2\int_{r_0}^{r_1} \sqrt{\det(H_\nu(r, 0; \epsilon) - E)}\, dr,$$

where $0 < r_0 < r_1$ are the first and second positive zero of the mapping $r \mapsto \det(H_\nu(r, 0; \epsilon) - E)$, and the square root is taken positive. As a function of E, the action $A_\nu(E, \epsilon)$ is extended analytically into a complex neighborhood of the interval $]a, b[$. The first result is the following Bohr-Sommerfeld type quantization condition of resonances with fixed angular momentum.

Theorem 3.2 ([FLN06]). *Let $E_0 > 0$ and $\nu \in \mathbb{Z} + \frac{1}{2}$ be given. Then there exist $\delta > 0$, $\epsilon_0 > 0$, and a function $c(E, \epsilon) : \{(E, \epsilon) \in \mathbb{C}\times]0, \epsilon_0[;\ |E - E_0| < \delta\} \to \mathbb{C}$ with $c(E, \epsilon) \to 0$ uniformly in E as $\epsilon \to 0$, such that E is a resonance of $H = -\epsilon^2\Delta_q + V(q)$ with angular momentum ν if and only if (E, ϵ) satisfies the following quantization condition:*

$$\sqrt{\frac{\pi\epsilon}{2}}\, \nu\, e^{-i\pi/4}\, E^{-3/4}\, e^{iA_\nu(E,\epsilon)/\epsilon} + 1 = c(E, \epsilon). \tag{3.2}$$

To our knowledge, Theorem 3.2 is the first Bohr-Sommerfeld quantization condition for a Schrödinger system with crossing eigenvalues. The prefactor before the exponential of the action carrying the second scale $\sqrt{\epsilon}$ is a clear signature of non-adiabaticity, stemming from a connection formula involving the Landau-Zener problem

$$i\epsilon\partial_r u = \begin{pmatrix} r & -\gamma \\ \overline{\gamma} & -r \end{pmatrix} u, \qquad \gamma = \epsilon \tfrac{\nu}{\sqrt{2}} E^{-3/4} + O(\epsilon^2). \tag{3.3}$$

For the proof of Theorem 3.2, the exact WKB method of Gérard and Grigis [GG88] is extended from scalar Schrödinger equations to a class of 2×2 first order differential systems, covering the case $H_\nu(r, -i\epsilon\partial_r; \epsilon)u = Eu$. The exact WKB solutions are of the form

$$u(r) = e^{\pm iz(r)/\epsilon} w(r), \qquad z(r) = \int_{r_*}^{r} \sqrt{\det(H_\nu(s, 0; \epsilon) - E)}\, ds.$$

They are locally defined away from turning points, which are the zeros of the mapping $r \mapsto \det(H_\nu(r, 0; \epsilon) - E)$. For $E \in]a, b[$, there are three positive turning points $r_0 < r_1 < r_2$, where r_0 tends to zero, while r_1 and r_2 coalesce at $\sqrt{E}$ as $\epsilon \to 0$.

For obtaining global solutions, one connects the exact WKB solutions at the turning points, using the good ϵ-asymptotics of the amplitude vector w. More precisely, one constructs an exact solution, which vanishes at the origin, and represents it after several connection procedures as a linear combination of solutions with controlled behavior at infinity. For $r \to +\infty$, there is a fundamental system of solutions $u_\infty^\pm$, such that $r \mapsto u_\infty^+(e^{-i\theta}r)$ is exponentially growing and $r \mapsto u_\infty^-(e^{-i\theta}r)$ exponentially decaying. The quantization formula (3.2) corresponds to the condition that the connection coefficient of the exponentially growing solution u_∞^+ vanishes.

The origin is a regular singular point of the equation $H_\nu(r, -i\epsilon\partial_r; \epsilon)u = Eu$ with indices $\pm\nu$. Moreover, the first turning point r_0 tends to zero as $\epsilon \to 0$. One constructs an exact WKB solution in a small complex neighborhood of the origin, which corresponds to the index $+\nu$. Studying the two parameter asymptotics of this solution as $(r, \epsilon) \to (0, 0)$, one encounters the same difficulties as in the context of the Langer modification for the radial Schrödinger equation, see also [FR00]. The $o(1)$ error estimate in Theorem 3.2 originates just from here. The rest of the proof gives better control on the convergence rate.

At $r = \sqrt{E}$, the second and third turning points r_1 and r_2 coalesce as $\epsilon \to 0$. The connection formula at this double turning point is calculated using a microlocal reduction to a normal form. Microlocally near $(r, \rho) = (\sqrt{E}, 0)$, the equation $H_\nu(r, -i\epsilon\partial_r; \epsilon)u = Eu$ looks like the Landau-Zener problem (3.3). A further reduction step leads to the saddle-point problem

$$r(-i\epsilon\partial_r)u = \tfrac{|\gamma|^2}{2} u.$$

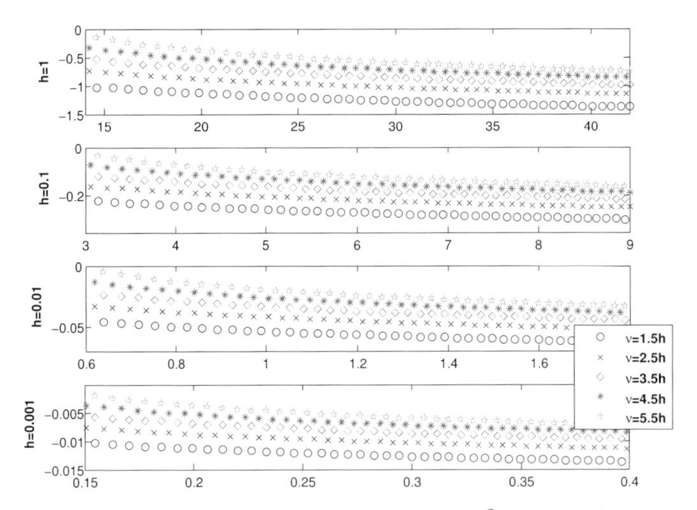

Fig. 3.1. Resonances of the model operator $H = -\epsilon^2 \Delta_q + V(q)$. The parameter k lies in $\{11, 12, \ldots, 60\}$, while ν is chosen in $\{1.5, 2.5, \ldots, 5.5\}$. The semiclassical parameter ϵ varies from 10^{-3} to 1. (See page 694 for a colored version of the figure.)

For this problem exact microlocal connection formulas are known [CP94, Ram96] and one has to lift them to the exact WKB solutions. The second scale $\sqrt{\epsilon}$ in the quantization condition (3.2) originates from this connection.

An asymptotic study (Proposition 8.2 in [FLN06]) of the action yields

$$A_\nu(E, \epsilon) = \tfrac{4}{3} E^{3/2} + \pi\nu\epsilon + O(\epsilon^2 |\ln \epsilon|) \qquad (\epsilon \to 0).$$

Plugging this expansion into the Bohr-Sommerfeld condition (3.2), one is motivated to define the following family of almost horizontal sequences:

$$\Gamma_\nu(a, b; \epsilon) = \Big\{ \lambda \in \mathbb{C};\ \lambda = \lambda_{k\nu}\epsilon - i\,\tfrac{3}{8}\left(\epsilon \ln \tfrac{1}{\epsilon} - \epsilon \ln \tfrac{\pi\nu^2}{2\lambda_{k\nu}\epsilon}\right), \\ k \in \mathbb{Z} \text{ s.t. } \lambda_{k\nu}\epsilon \in]a, b[\Big\}$$

with

$$\lambda_{k\nu} = \tfrac{3\pi}{16}(8k - 4\nu + 5) \qquad (k \in \mathbb{Z},\ \nu \in \mathbb{Z} + \tfrac{1}{2}).$$

Since $\lambda_{k\nu}\epsilon \in]a, b[$, the second term of the imaginary part of $\lambda \in \Gamma_\nu(a, b; \epsilon)$ is of $O(\epsilon)$ and smaller than the first term $-\frac{3}{8}\epsilon \ln \frac{1}{\epsilon}$. Hence, $\Gamma_\nu(a, b; \epsilon)$ is an almost horizontal sequence in the lower half-plane with distance of $O(\epsilon \ln \frac{1}{\epsilon})$ from the real axis. The asymptotic distribution of resonances with real part in the positive interval $]a, b[$ reads as follows:

Theorem 3.3 ([FLN06]). *Let $N \in \mathbb{N}$ and $0 < a < b$ be given. Then there is $\epsilon_0 > 0$ and a positive function $c :]0, \epsilon_0[\to \mathbb{R}^+$ with $c(\epsilon) = o(\epsilon)$ as $\epsilon \to 0$ such that for each $\lambda \in \bigcup_{\nu \le N} \Gamma_\nu(a, b; \epsilon)$ there exists one and only one resonance of $H = -\epsilon^2 \Delta_q + V(q)$ within the set $\{E \in \mathbb{C};\ |E - \lambda^{2/3}| < c(\epsilon)\}$.*

The plots in Fig. 3.1 illustrate the distorted lattice of resonances given by Theorem 3.3. The larger the angular momentum number ν is taken, the closer the resonance is to the real axis and the longer the life time of the corresponding resonant states will be. This observation is in wonderful agreement with the dynamical properties of the operator P. The one-level operators $H^{\pm}$ induce Hamiltonian systems conserving angular momentum $q \wedge p = q_1 p_2 - q_2 p_1$, which also encodes how close classical trajectories arrive near the crossing manifold $\{q = 0\}$. On the one hand, a high angular momentum number ν of a resonance mirrors a periodic orbit of the upper level with high angular momentum. Such orbits in turn imply existence of localized quasimodes and long-living resonant states. On the other hand, small angular momentum numbers ν correspond to orbits close to the crossing manifold. Close to the crossing, non-adiabatic transitions to the unbounded motion of the minus-system are possible. In this regime shorter life-times and resonances far away from the real axis have to be expected.

References

[AG00a] J. Avron, A. Gordon. Born-Oppenheimer wave function near level crossing. *Phys. Rev. A*, 62: 062504-1–062504-9, 2000.

[AG00b] J. Avron, A. Gordon. Born-Oppenheimer approximation near level crossing. *Phys. Rev. Lett.*, 85(1): 34–37, 2000.

[BT05] V. Betz, S. Teufel. Precise coupling terms in adiabatic quantum evolution: The generic case. *Comm. Math. Phys.* 260: 481–509, 2005.

[Bor98] F. Bornemann. Homogenization in time of singulary perturbed mechanical systems. *Lecture Notes in Mathematics* 1687, Spinger-Verlag, 1998.

[CdV03] Y. Colin de Verdière. The level crossing problem in semi-classical analysis. I. The symmetric case. *Ann. Inst. Fourier (Grenoble)* 53(4): 1023–1054, 2003.

[CdV04] Y. Colin de Verdière. The level crossing problem in semi-classical analysis. II. The hermitian case. *Ann. Inst. Fourier (Grenoble)*, 54(5): 1423–1441, 2004.

[CP94] Y. Colin de Verdière, B. Parisse. Équilibre instable en régime semi-classique. I. Concentration microlocale. *Comm. Par. Diff. Eq.*, 19(9&10): 1535–1563, 1994.

[CF*87] H. Cycon, R. Froese, W. Kirsch, B. Simon. *Schrödinger operators with application to quantum mechanics and global geometry.* Texts and monographs in physics, Springer-Verlag, 1987.

[DYK04] W. Domcke, D. Yarkony, H. Köppel (eds.). *Conical intersections.* World Scientific Publishing, Advanced Series in Physical Chemistry, Vol. 15, 2004.

[Fer03] C. Fermanian Kammerer. Wigner measures and molecular propagation through generic energy level crossings. *Rev. Math. Phys.* 15: 1285–1317, 2003.

[FG02] C. Fermanian Kammerer, P. Gérard. Mesures semi-classiques et croisements de modes. *Bull. Soc. math. France*, 130(1): 123–168, 2002.

[FG03] C. Fermanian Kammerer, P. Gérard. A Landau-Zener formula for non-degenerated involutive codimension 3 crossings. *Ann. Henri Poincaré* 4: 513–552, 2003.

[FL03] C. Fermanian Kammerer, C. Lasser. Wigner measures and codimension two crossings. *Jour. Math. Phys.* 44(2): 507–527, 2003.

[FL06] C. Fermanian Kammerer, C. Lasser. Modeling of molecular propagation through conical intersections: an asymptotic description. *In preparation.*

[Fol89] G. Folland. *Harmonic analysis in phase sapce.* Princeton University Press, 1989.

[FLN06] S. Fujiié, C. Lasser, L. Nédélec. Semiclassical resonances for two-level Schrödinger operator with a conical intersection. Preprint 194 of the DFG-priority program 1095, 2006.

[FR00] S. Fujiié, T. Ramond. Exact WKB analysis and the Langer modification with application to barrier top resonances. C. Howls (ed.), Toward the exact WKB analysis of differential equations, linear or non-linear, Kyoto University Press, 15–31, 2000.

[GG88] C. Gérard, A. Grigis. Precise estimates of tunneling and eigenvalues near a potential barrier. *J. Differ. Eq.*, 72(1): 149–177, 1988.

[Hag94] G. Hagedorn. Molecular propagation through electron energy level crossings. *Mem. A. M. S.*, 111(536), 1994.

[HJ04] G. Hagedorn, A. Joye. Time development of exponentially small non-adiabatic transitions. *Comm. Math. Phys.* 250(2): 393–413, 2004.

[HS96] P. Hislop, I. Sigal. *Introduction to spectral theory with applications to Schödinger operators.* Springer-Verlag, 1996.

[HS00] S. Hahn, G. Stock. Quantum-mechanical modeling of the femtosecond isomerization in rhodopsin. *J. Chem. Phy. B* 104: 1146-1149, 2000.

[Jec0303] T. Jecko. Semiclassical resolvent estimates for Schrödinger matrix operators with eigenvalues crossings. *Math. Nachr.*, 257(1): 36–54, 2003.

[Jec05] T. Jecko. Non-trapping condition for semiclassical Schrödinger operators with matrix-valued potentials. *Math. Phys. El. J.*, 11(2), 2005.

[LS06] C. Lasser, T. Swart. The non-crossing rule for electronic energy levels. *In preparation.*

[LST06] C. Lasser, T. Swart, S. Teufel. Propagation through conical crossings: a surface hopping algorithm. *In preparation.*

[LT05] C. Lasser, S. Teufel. Propagation through conical crossings: an asymptotic semigroup. *Comm. Pure Appl. Math.* 58(9): 1188–1230, 2005.

[Néd96] L. Nédélec. Résonances semi-classique pour l'opérateur de Schrödinger matriciel en dimension deux. *Ann. Inst. Henri Poincaré Phys. Théor.*, 65(2): 129–162, 1996.

[Néd01] L. Nédélec. Resonances for matrix Schrödinger operators. *Duke Math. J.* 106(2): 209–236, 2001.

[Néd03] L. Nédélec. Existence of resonances for matrix Schrödinger operators. *Asympt. Anal.* 35(3-4): 301–324, 2003.

[Ram96] T. Ramond. Semiclassical study of quantum scattering on the line. *Commun. Math. Phys.*, 177: 221–254, 1996.

[RS78] M. Reed, B. Simon. Methods of modern mathematical physics IV: Analysis of operators. *Academic Press*, 1978.

[ST01] H. Spohn, S. Teufel. Adiabatic decoupling and time-dependent Born-Oppenheimer theory. *Comm. Math. Phys.* 224: 113–132, 2001.

[ST05] G. Stock, M. Thoss. Classical description of nonadiabatic quantum dynamics. *Adv. Chem. Phys.* 131: 243–375, 2005.

[TP71] J. Tully, R. Preston. Trajectory surface hopping approach to nonadiabatic molecular collisions: the reaction of H^+ with D_2. *J. Chem. Phys.* 55(2): 562–572, 1971.

[Zen32] C. Zener. Non-adiabatic crossing of energy levels. *Proc. Roy. Soc. Lond.* 137:696–702, 1932.

[Zwo99] M. Zworski. Resonances in physics and geometry. *Notices Amer. Math. Soc.*, 46(3): 319–328, 1999.

Motion of Electrons in Adiabatically Perturbed Periodic Structures

Gianluca Panati[1], Herbert Spohn[1], and Stefan Teufel[2]

[1] Zentrum Mathematik, TU München, Boltzmannstraße 3, 85747 Garching. panati@ma.tum.de, spohn@ma.tum.de
[2] Mathematisches Institut, Universität Tübingen, Auf der Morgenstelle 10, 72076 Tübingen. stefan.teufel@uni-tuebingen.de

Summary. We study the motion of electrons in a periodic background potential (usually resulting from a crystalline solid). For small velocities one would use either the non-magnetic or the magnetic Bloch hamiltonian, while in the relativistic regime one would use the Dirac equation with a periodic potential. The dynamics, with the background potential included, is perturbed either through slowly varying external electromagnetic potentials or through a slow deformation of the crystal. In either case we discuss how the Hilbert space of states decouples into almost invariant subspaces and explain the effective dynamics within such a subspace.

1 Introduction

In a crystalline solid the conduction electrons move in the potential created by the ions and the core electrons. Somewhat mysteriously and linked to the Pauli exclusion principle, the Coulomb repulsion between conduction electrons may be ignored, within a good approximation. Thereby one arrives at a fundamental model of solid state physics, namely an ideal Fermi gas of electrons subject to a periodic crystal potential. Let Γ be the periodicity lattice. It is a Bravais lattice and generated through the basis $\{\gamma_1, \gamma_2, \gamma_3\}$, $\gamma_j \in \mathbb{R}^3$, as

$$\Gamma = \{\gamma = \sum_{j=1}^{3} \alpha_j \gamma_j \quad \text{with } \alpha \in \mathbb{Z}^3\}. \tag{1.1}$$

The crystal potential V_Γ is then Γ-periodic, i.e., $V_\Gamma : \mathbb{R}^3 \to \mathbb{R}$ and $V_\Gamma(x + \gamma) = V_\Gamma(x)$ for all $\gamma \in \Gamma$, and the electrons are governed by the one-particle hamiltonian

$$H_{\mathrm{SB}} = -\frac{1}{2}\Delta_x + V_\Gamma \,. \tag{1.2}$$

H_{SB} is the (Schrödinger)-Bloch hamiltonian. A wave function $\psi_t \in L^2(\mathbb{R}^3)$ evolves in time according to the Schrödinger equation

$$i\frac{\partial}{\partial t}\psi_t = H_{\mathrm{SB}}\psi_t\,. \tag{1.3}$$

We have chosen units such that the mass of an electron $m_{\mathrm{e}} = 1$ and $\hbar = 1$. The electron charge, e, is absorbed in V_Γ. Since V_Γ is periodic, electrons move ballistically with an effective dispersion relation given by the Bloch energy bands E_n, see below for a precise definition. E_n is periodic with respect to the lattice Γ^* dual to Γ, $E_n(k+\gamma^*) = E_n(k)$ for all $\gamma^* \in \Gamma^*$, $k \in \mathbb{R}^3$. This feature makes the dynamical properties of a Bloch electron very different from a massive particle with dispersion $E_{\mathrm{free}}(k) = \frac{1}{2}k^2$ valid in case $V_\Gamma = 0$.

The thermodynamics of the electron gas is studied taking H_{SB} as a starting point. Dynamically, however, one wants to probe the response of the electrons to external forces which very crudely come in two varieties.

(i) *External electromagnetic potentials.* Electrostatic potentials manufactured in a lab have a slow variation on the scale of the lattice Γ. Therefore we set $V_{\mathrm{ext}}(x) = e\phi(\varepsilon x)$, e the charge of the electron, with ε a dimensionless parameter and ϕ independent of ε. $\varepsilon \ll 1$ means that the potential V_{ext} has a slow variation when measured with respect to the lattice spacing of Γ. Note that the electrostatic force is $\mathcal{O}(\varepsilon)$ and thus weak. External magnetic fields on the other hand can be so strong that the radius of gyration is comparable to the lattice spacing. It then makes sense to split the vector potential as $A_0 + A_{\mathrm{ext}}$, where $A_0(x) = -\frac{1}{2}B_0 \wedge x$ with $B_0 \in \mathbb{R}^3$ a constant magnetic field. Included in H_{SB}, this yields the magnetic Bloch hamiltonian

$$H_{\mathrm{MB}} = \frac{1}{2}(-i\nabla_x - A_0)^2 + V_\Gamma\,. \tag{1.4}$$

A_{ext} is a probing vector potential in addition to A_0. A_{ext} is slowly varying on the scale of the lattice, $A_{\mathrm{ext}}(x) = A(\varepsilon x)$ with A independent of ε, and the corresponding magnetic field is small of order ε. Including all electromagnetic potentials, for simplicity with the electric charge absorbed into A and ϕ, the hamiltonian becomes

$$H = \frac{1}{2}\big(-i\nabla_x - A_0(x) - A(\varepsilon x)\big)^2 + V_\Gamma(x) + \phi(\varepsilon x)\,. \tag{1.5}$$

(ii) *Mechanical forces.* The crystal lattice can be deformed through external pressure and shear. Thereby an electric polarization is induced, an effect which is known as piezoelectricity. If charges are allowed to flow, in this way mechanical pressure can be transformed into electric currents. The mechanical forces are time-dependent but slow on the typical time-scale of the electrons. Therefore in (1.2) $V_\Gamma(x)$ is replaced by $V_{\Gamma(\varepsilon t)}(x, \varepsilon t)$. $\Gamma(\varepsilon t)$ is the instantaneous periodicity lattice and is defined as in (1.1). $V_{\Gamma(t)}$ is space-periodic, i.e. $V_{\Gamma(t)}(x+\gamma, t) = V(x,t)$ for all $\gamma \in \Gamma(t)$. The special case of a time-independent lattice, $\Gamma(t) = \Gamma$, but a still slowly in time varying crystal potential is also of interest. For example, one may imagine a unit cell with two nuclei. If the two nuclei are displaced relative to each other, then Γ remains fixed while the crystal potential in the unit cell changes with time. The resulting piezoelectric hamiltonian reads

$$H_{\mathrm{PE}}(t) = -\frac{1}{2}\Delta_x + V_{\Gamma(\varepsilon t)}(x, \varepsilon t)\,. \tag{1.6}$$

Our general goal is to understand, in each case, the structure of the solution of the time-dependent Schrödinger equation for small ε. Obviously, H in (1.5) is a space-adiabatic problem, while (1.6) corresponds to a time-adiabatic problem. However in the latter case it turns out to be profitable to transform to a time-independent lattice, say $\Gamma(0)$. Then also terms varying slowly in space are generated. Thus, in the general case the full power of the space-adiabatic perturbation theory [PST03a, Teu03] will be needed. A word of caution must be issued here for the magnetic Bloch hamiltonian. To use the methods from [PST03a] in this context, the unperturbed Hamiltonian must be periodic, which is the case only if the magnetic flux per unit cell is rational. One can then define an enlarged magnetic unit cell such that H_{MB} is invariant with respect to magnetic translations. If the magnetic flux is not rational, the crutch is to include in A_0 a nearby rational flux part of the magnetic field, with a small denominator, and to treat the remainder as A_{ext}.

To achieve our goal, depending on the context we use one of the periodic hamiltonians as backbone. The periodic hamiltonian is denoted by H_{per} with H_{per} either H_{SB}, or H_{MB}, or H_{PE} at fixed t, or H_{LS} from (1.8), or H_{DB} from (1.9). As explained below, the Hilbert space $\mathcal{H} = L^2(\mathbb{R}^3)$ then splits as $\mathcal{H} = \bigoplus_{n=0}^{\infty} \mathcal{H}_n$, where n is the band index. Each subspace $\mathcal{H}_n$ is invariant under $\exp[-itH_{\mathrm{per}}]$ and H_{per} restricted to $\mathcal{H}_n$ is unitarily equivalent to multiplication by $E_n(k)$. $E_n(k)$ is the effective hamiltonian associated to the n-th band. The complexity of the full problem has been reduced substantially, since only a single band dynamics has still to be studied. Modifying H_{per} such that it becomes slowly varying in space-time is, vaguely speaking, a small perturbation. Thus one would expect that the invariant subspace $\mathcal{H}_n$ is to be substituted by a slightly tilted subspace. On this subspace $E_n(k)$ will turn into a more complicated effective hamiltonian. The difficulty is that, while the dynamics generated by H_{per} can be computed by solving a purely spectral problem, none of the perturbed hamiltonians can be understood in this way. In particular, one has to spell out carefully over which time scale the slightly tilted subspace associated to $\mathcal{H}_n$ remains approximately invariant and in what sense the dynamics generated by the effective hamiltonian approximates the true time evolution.

To lowest order the effective hamiltonian can be guessed from elementary considerations and belongs to a standard tool of solid state physics [AM76]. The guess provides however little hint on the validity of the approximation. There one needs a mathematical theorem which states precise conditions on the initial wave function and provides an error bound, from which the time scale for validity can be read off.

Under the header "geometric phase" physicists and quantum chemists have realized over the past twenty years, say, that the first order correction to the effective hamiltonian carries a lot of interesting physics, see [BM*03] for a recent comprehensive overview. For the magnetic Bloch hamiltonian the first order correction yields a Hall current proportional to the Chern number of the magnetic Bloch vector bundle. Similarly, the modern theory of piezoelectricity, expresses the piezocurrent as an integral of the Berry connection over the Brillouin zone, see King-Smith, Vanderbilt [KSV93] and Resta [Res94]. First order effective Hamiltonians are no longer guessed so easily and it is convenient to have the systematic scheme [PST03a] available.

In nature electrons are spin $\frac{1}{2}$ particles. The wave function is thus $\mathbb{C}^2$-valued and the hamiltonian in (1.5) is modified to

$$H = \frac{1}{2}\big(-i\nabla_x - A_0(x) - A(\varepsilon x)\big)^2 + V_\Gamma(x) + \phi(\varepsilon x) - \frac{1}{2}\sigma\cdot\big(B_0 + \varepsilon B(\varepsilon x)\big) \quad (1.7)$$

with $B = \nabla \wedge A$ for the slowly varying part of the magnetic field. Here $\sigma = (\sigma_1, \sigma_2, \sigma_3)$ is the 3-vector of Pauli spin matrices. Besides the term proportional to the uniform magnetic field B_0, H acquires a subleading term of order ε. More accurately one may want to include the spin-orbit coupling. The periodic piece of the hamiltonian reads then

$$H_{\mathrm{LS}} = -\frac{1}{2}\Delta_x + V_\Gamma(x) + \frac{1}{4}\sigma\cdot\big(\nabla V_\Gamma(x) \wedge (-i\nabla_x)\big) \quad (1.8)$$

and the slowly varying potential is added as in (1.7) with the additional subleading term $\varepsilon\frac{1}{4}\sigma\cdot(\nabla\phi(\varepsilon x)\wedge(-i\nabla_x))$.

Depending on the crystalline solid, the conduction electrons can move so fast that relativistic corrections become important. On the one-particle level an obvious choice is then the Dirac equation with a periodic potential V_Γ. Wave functions are $\mathbb{C}^4$-valued and the hamiltonian reads

$$H_{\mathrm{DB}} = \beta m_{\mathrm{e}} c^2 + c\alpha\cdot p + V_\Gamma\,, \quad p = -i\nabla_x\,. \quad (1.9)$$

We introduced here the mass, m_{e}, of the electron and the speed of light, c. The 4×4 matrices $\beta, \alpha_1, \alpha_2, \alpha_3$ are standard and defined in [Tha94, Ynd96], for example. Note that the Lorentz frame is fixed by the solid, i.e. by V_Γ.

In fact, the non-relativistic limit for H_{DB} yields the spin-orbit hamiltonian H_{LS} [Tha94, Ynd96]. If $\|V_\Gamma\|$ is bounded, for sufficiently large c, the Dirac hamiltonian H_{DB} has a spectral gap, which widens as $c\to\infty$. Projecting onto the electron subspace, to leading order in $1/c$ one obtains the Pauli-Bloch hamiltonian $-(1/2m_{\mathrm{e}})\Delta_x + V_\Gamma$ with the spin-orbit coupling in (1.8) as a correction of strength $1/(m_{\mathrm{e}}c)^2$. In addition the crystal potential is corrected by $-\Delta_x V_\Gamma(x)/8(m_{\mathrm{e}}c)^2$.

In our contribution we will provide some background on how to establish, including error bounds, the validity of the approximate dynamics as generated by an effective hamiltonian, including order ε corrections, for most of the models mentioned in the introduction. For this purpose it is necessary to briefly recall the spectral theory for the periodic hamiltonian, which is done in the following section. In the subsequent sections we deal with particular cases in more detail. We start with the non-magnetic Bloch hamiltonian, see (1.5) with $B_0 = 0$. For the magnetic Bloch hamiltonian we explain how $B_0 \to 0$ and $B_0 \to \infty$ may be viewed as particular adiabatic limits. Piezoelectricity is discussed in the last section.

Remark. Our contribution is one part of the research project jointly with S. Bauer and M. Kunze within the Schwerpunkt. Their part will be covered in [BK06]. Both contributions appear now as almost disjoint, which only reflects that we wanted to present a coherent story. The unifying aspect is an adiabatic limit for wave-type evolution equations. In this contribution we stay on the level of effective hamiltonians while in [BK06] one pushes the scheme to the first dissipative correction.

2 The periodic hamiltonians

We consider a general dimension, d, and assume that the periodicity lattice Γ is represented as

$$\Gamma = \left\{ x \in \mathbb{R}^d : x = \textstyle\sum_{j=1}^{d} \alpha_j \gamma_j \text{ for some } \alpha \in \mathbb{Z}^d \right\}, \tag{2.1}$$

where $\{\gamma_1, \ldots, \gamma_d\}$ are vectors spanning $\mathbb{R}^d$. We denote by Γ^* the dual lattice of Γ with respect to the standard inner product in $\mathbb{R}^d$, i.e. the lattice generated by the dual basis $\{\gamma_1^*, \ldots, \gamma_d^*\}$ determined through the conditions $\gamma_i^* \cdot \gamma_j = 2\pi\delta_{ij}$, $i, j \in \{1, \ldots, d\}$. The centered fundamental domain M of Γ is defined by

$$M = \left\{ x \in \mathbb{R}^d : x = \textstyle\sum_{j=1}^{d} \beta_j \gamma_j \text{ for } \beta_j \in [-\frac{1}{2}, \frac{1}{2}] \right\}, \tag{2.2}$$

and analogously the centered fundamental domain M^* of Γ^*. The set M^* is the *first Brillouin zone* in the physics parlance.

Assumption 1. *The crystal potential $V_\Gamma : \mathbb{R}^d \to \mathbb{R}$ satisfies $V_\Gamma(x + \gamma) = V_\Gamma(x)$ for all $\gamma \in \Gamma$, $x \in \mathbb{R}^d$. V_Γ is infinitesimally bounded with respect to $-\Delta$.*

It follows from Assumption 1 that the periodic hamiltonians discussed below are self-adjoint on the domain of $-\Delta$.

2.1 The Bloch hamiltonian

We consider

$$H = -\frac{1}{2}\Delta + V_\Gamma \,. \tag{2.3}$$

The periodicity of H is exploited through the Bloch-Floquet-Zak transform, or just the Zak transform for sake of brevity [Zak68]. The advantage of such a variant is that the fiber at k of the transformed Hamiltonian operator has a domain which does not depend on k.

The Zak transform is defined as

$$(\mathcal{U}_{\mathrm{Z}}\psi)(k,x) := \sum_{\gamma\in\Gamma} \mathrm{e}^{-\mathrm{i}k\cdot(x+\gamma)}\psi(x+\gamma)\,, \qquad (k,x)\in\mathbb{R}^{2d}, \tag{2.4}$$

initially for a fast-decreasing function $\psi \in \mathcal{S}(\mathbb{R}^d)$. One directly reads off from (2.4) the following periodicity properties

$$\big(\mathcal{U}_{\mathrm{Z}}\psi\big)(k,y+\gamma) = \big(\mathcal{U}_{\mathrm{Z}}\psi\big)(k,y) \quad \text{for all} \quad \gamma\in\Gamma\,, \tag{2.5}$$

$$\big(\mathcal{U}_{\mathrm{Z}}\psi\big)(k+\lambda,y) = \mathrm{e}^{-\mathrm{i}y\cdot\lambda}\,\big(\mathcal{U}_{\mathrm{Z}}\psi\big)(k,y) \quad \text{for all} \quad \lambda\in\Gamma^*\,. \tag{2.6}$$

From (2.5) it follows that, for any fixed $k\in\mathbb{R}^d$, $\big(\mathcal{U}_{\mathrm{Z}}\psi\big)(k,\cdot)$ is a Γ-periodic function and can thus be regarded as an element of $\mathcal{H}_{\mathrm{f}} = L^2(M)$. $M = \mathbb{R}^d/\Gamma$ and it has the topology of the d-dimensional torus $\mathbb{T}^d$. On the other side, Equation (2.6) involves a unitary representation of the group of lattice translations on Γ^* (isomorphic to Γ^* and denoted as Λ), given by

$$\tau:\Lambda\to\mathcal{U}(\mathcal{H}_{\mathrm{f}})\,,\quad \lambda\mapsto\tau(\lambda)\,,\quad \big(\tau(\lambda)\varphi\big)(y) = \mathrm{e}^{\mathrm{i}y\cdot\lambda}\varphi(y). \tag{2.7}$$

It is then convenient to introduce the Hilbert space

$$\begin{aligned}\mathcal{H}_\tau &= \big\{\psi\in L^2_{\mathrm{loc}}(\mathbb{R}^d,\mathcal{H}_{\mathrm{f}}):\ \psi(k-\lambda) = \tau(\lambda)\,\psi(k) \qquad \text{for all } \lambda\in\Lambda\big\}\\ &= L^2_\tau(\mathbb{R}^d\,,\ \mathcal{H}_{\mathrm{f}})\,,\end{aligned} \tag{2.8}$$

equipped with the inner product

$$\langle\psi,\,\varphi\rangle_{\mathcal{H}_\tau} = \int_{M^*} dk\,\langle\psi(k),\,\varphi(k)\rangle_{\mathcal{H}_{\mathrm{f}}}\,. \tag{2.9}$$

Obviously, there is a natural isomorphism between $\mathcal{H}_\tau$ and $L^2(M^*,\mathcal{H}_{\mathrm{f}})$ given by restriction from $\mathbb{R}^d$ to M^*, and with inverse given by τ-equivariant continuation, as suggested by (2.6). Equipped with these definitions, one checks that the map in (2.4) extends to a unitary operator

$$\mathcal{U}_{\mathrm{Z}}: L^2(\mathbb{R}^d)\to\mathcal{H}_\tau\cong L^2(M^*,L^2(M)), \tag{2.10}$$

with inverse given by

$$(\mathcal{U}_{\mathrm{Z}}^{-1}\varphi)(x) = \int_{M^*} dk\,\mathrm{e}^{\mathrm{i}k\cdot x}\varphi(k,[x]), \tag{2.11}$$

where $[\,\cdot\,]$ refers to the a.e. unique decomposition $x=\gamma_x+[x]$, with $\gamma_x\in\Gamma$ and $[x]\in M$.

As already mentioned, the advantage of this construction is that the transformed hamiltonian is a fibered operator over M^*. Indeed, for the Zak transform of the hamiltonian operator (2.3) one finds

$$\mathcal{U}_{\mathrm{Z}} H \mathcal{U}_{\mathrm{Z}}^{-1} = \int_{M^*}^{\oplus} dk\, H(k) \tag{2.12}$$

with fiber operator

$$H(k) = \frac{1}{2}\big(-\mathrm{i}\nabla_y + k\big)^2 + V_\Gamma(y)\,, \quad k \in M^*\,. \tag{2.13}$$

By Assumption 1, for fixed $k \in M^*$, the operator $H(k)$ acts on $L^2(M)$ with the Sobolev space $H^2(M)$ as domain independently of $k \in M^*$. Each fiber operator $H(k)$ has pure point spectrum accumulating at infinity. For definiteness the eigenvalues are enumerated according to their magnitude $E_0(k) \leq E_1(k) \leq E_2(k) \leq \ldots$ and repeated according to their multiplicity. $E_n : M^* \to \mathbb{R}$ is the n-th energy band function. It is continuous on M^* when viewed as a d-torus. Generically the eigenvalues $E_n(k)$ are non-degenerate. Of course, there may be particular points in k-space where particular energy bands touch each other and the corresponding eigenvalue becomes degenerate. The normalized eigenfunction corresponding to $E_n(k)$ is the Bloch function and denoted by $\varphi_n(k) \in H^2(M)$. It is determined only up to a k-dependent phase factor. A further arbitrariness comes from points where energy bands touch. We denote by $P_n(k)$ the projection along $\varphi_n(k)$ and set

$$P_n = \int_{M^*}^{\oplus} dk P(k)\,, \qquad \mathcal{H}_n = P_n L^2(\mathbb{R}^d)\,. \tag{2.14}$$

Through the Zak transform we have achieved the product structure

$$\mathcal{H} = \mathcal{H}_{\mathrm{s}} \otimes \mathcal{H}_{\mathrm{f}}\,, \qquad \mathcal{H}_{\mathrm{s}} = L^2(M^*)\,,\ \mathcal{H}_{\mathrm{f}} = L^2(M)\,. \tag{2.15}$$

$\psi \in \mathcal{H}_{\mathrm{n}}$ is of the form $\phi(k)\varphi_n(k,y)$. The band index n fixes the local pattern of the wave function ψ while $\phi(k)$ provides the slow variation. Therefore $L^2(M^*)$ is the Hilbert space of the slow degrees of freedom. On the other hand for fixed k, one has oscillations in time determined by the eigenvalues $E_n(k)$. On long time scales, these become fast oscillations. Therefore $\mathcal{H}_{\mathrm{f}} = L^2(M)$ is the Hilbert space of the fast degrees of freedom.

Since $[P_n, H] = 0$, the subspaces $\mathcal{H}_n$ are invariant under $\mathrm{e}^{-\mathrm{i}Ht}$. $P_n \mathrm{e}^{-\mathrm{i}Ht} P_n$ is unitarily equivalent to multiplication by $\mathrm{e}^{-\mathrm{i}E_n(k)t}$ on $L^2(M^*)$. Note that, in general, $\mathcal{H}_n$ is not a spectral subspace for H. The band functions generically have overlapping ranges. Therefore, if slowly varying terms are added to the hamiltonian, the dynamics can no longer be captured so easily by a spectral analysis of the perturbed hamiltonian.

2.2 The magnetic Bloch hamiltonian

We consider $d = 3$. The hamiltonian reads

$$H = \frac{1}{2}\big(- i\nabla_x - A(x)\big)^2 + V_\Gamma(x)\,, \quad x \in \mathbb{R}^3\,, \tag{2.16}$$

with $A(x) = -\frac{1}{2}B \wedge x$, $B \in \mathbb{R}^3$. Physically the most relevant case is $d = 2$. It is included here by setting $x = (x_1, x_2, 0)$ and $B = (0, 0, B_0)$. Following Zak [Zak64], see also [DGR02], one introduces the magnetic translations

$$(T_\alpha\psi)(x) = \big(e^{-i\alpha\cdot(-i\nabla_x + A(x))}\psi\big)(x) = e^{i\alpha A(x)}\psi(x - \alpha) \tag{2.17}$$

with $\alpha \in \mathbb{R}^3$. They satisfy the Weyl relations

$$T_\alpha T_\beta = e^{-\frac{i}{2}B\cdot(\alpha\wedge\beta)}T_{\alpha+\beta} = e^{-iB\cdot(\alpha\wedge\beta)}T_\beta T_\alpha\,. \tag{2.18}$$

To have a commuting subfamily we need

Assumption 2. *The magnetic field B is such that $B \cdot (\gamma \wedge \gamma') \in 2\pi\mathbb{Q}$ for all $\gamma, \gamma' \in \Gamma$.*

In the two-dimensional case our assumption requires that the magnetic flux per unit cell, $B_0 \cdot (\gamma_1 \wedge \gamma_2)$, is a rational multiple of 2π.

Under the Assumption 2 there exists a sublattice $\Gamma_0 \subset \Gamma$ such that $B\cdot(\gamma \wedge \gamma') \in 2\pi\mathbb{Z}$ for every $\gamma, \gamma' \in \Gamma_0$. Γ_0 is not unique. The set $\{T_\alpha\}_{\alpha\in\Gamma_0}$ is a family of commuting operators, which commute with H. Since $T_\alpha T_\beta = \pm T_{\alpha+\beta}$, the magnetic translations still form only a projective group. It becomes a group by an even smaller sublattice $\Gamma_1 \subset \Gamma_0$ such that $B \cdot (\gamma \wedge \gamma') \in 4\pi\mathbb{Z}$ for all $\gamma, \gamma' \in \Gamma_1$. Another common choice is a further modification of the phase through

$$\mathcal{T}_\alpha = e^{-\frac{i}{2}\varphi(\alpha)}T_\alpha \tag{2.19}$$

with $\varphi(\alpha) = B_1\alpha_2\alpha_3 + B_3\alpha_1\alpha_2 - B_2\alpha_1\alpha_3$. Then $\mathcal{T}_\alpha\mathcal{T}_\beta = \mathcal{T}_{\alpha+\beta}$ for all $\alpha, \beta \in \Gamma_0$.

We can now proceed as in the non-magnetic case. The Zak transform becomes

$$(\mathcal{U}_\mathrm{Z}\psi)(k, x) = \sum_{\gamma\in\Gamma_0} e^{-ik\cdot(x+\gamma)}\mathcal{T}_\gamma\psi(x)\,, \quad (k, x) \in \mathbb{R}^6\,. \tag{2.20}$$

The properties of $\mathcal{U}_\mathrm{Z}\psi$ are as in (2.2), (2.3) provided Γ is replaced by Γ_0, and $\mathcal{H}_\tau$ is replaced by $\mathcal{H}_\tau^B = \{u \in L^2_\mathrm{loc}(\mathbb{R}^d, \mathcal{H}_\mathrm{f}) : (2.23) \text{ below holds true}\}$. In particular, H of (2.16) admits the fiber decomposition

$$\mathcal{U}_\mathrm{Z} H \mathcal{U}_\mathrm{Z}^{-1} = \int_{M^*}^{\oplus} dk H(k) \tag{2.21}$$

with M^* the first Brillouin zone of Γ_0^* and with the fiber operator

$$H(k) = \frac{1}{2}(-i\nabla_y + \frac{1}{2}B \wedge y + k)^2 + V_\Gamma(y)\,. \tag{2.22}$$

The domain of $H(k)$ is independent of k but, in contrast to $H(k)$ from (2.13), a function u in the domain has to satisfy the more complicated boundary condition

$$e^{-\frac{i}{2}y\cdot(\alpha\wedge B)}u(y-\alpha)=u(y)\,. \tag{2.23}$$

2.3 Dirac hamiltonian, spin-orbit coupling

The Dirac hamiltonian with periodic potential reads

$$H=\beta-i\alpha\cdot\nabla_x+V_\Gamma(x)\,, \tag{2.24}$$

where we have set $m_e=1$, $c=1$. As for the Bloch hamiltonian, H admits the fiber decompositon

$$H=\int_{M^*}^{\oplus}dkH(k) \tag{2.25}$$

with fiber hamiltonian

$$H(k)=\beta+\alpha\cdot(-i\nabla_y+k)+V_\Gamma(y)\,. \tag{2.26}$$

$H(k)$ acts on $L^2(M,\mathbb{C}^4)$ with periodic boundary conditions (2.5). The free Dirac operator has a spectral gap of size 2, in our units, between the electron and positron subspace. If we assume $\|V_\Gamma\|<1$, then this gap persists and the eigenvalues can be labelled as $E_0(k)\le E_1(k)\le\dots$ in the electron subspace and as $E_{-1}(k)\ge E_{-2}(k)\ge\dots$ in the positron subspace. One has $E_{-1}(k)<E_0(k)$ for all $k\in M^*$. (In fact the labelling can be achieved without a restriction on $\|V_\Gamma\|$, see [Mau03]).

For $V_\Gamma=0$, the eigenvalue $E(k)$ is two-fold degenerate reflecting the spin $\frac{1}{2}$ of the electron, resp. positron. This degeneracy persists if the periodic potential is inversion symmetric, see [Mau03] for details.

Proposition 2.1. *Let H be given by (2.24) with $\|V_\Gamma\|<1$. Let there exist $a\in\mathbb{R}^3$ such that $V_\Gamma(x+a)=V_\Gamma(-x+a)$. Then each $E_n(k)$ is at least two-fold degenerate.*

Proof. Without loss of generality we may assume $a=0$. We use the standard basis for the α-matrices, see [Tha94]. In this basis time-reversal symmetry is implemented by the anti-unitary operator

$$T\psi(y)=-i\alpha_3\alpha_1\psi^*(y)\,, \tag{2.27}$$

where the complex conjugation is understood component-wise. Using that $\alpha_\ell\alpha_3\alpha_1=-\alpha_3\alpha_1\overline{\alpha}_\ell$, $\ell=1,2,3$, where $\overline{\alpha}_\ell$ refers to matrix element-wise complex conjugation, one checks that

$$-i\nabla_y\alpha_\ell T=-iT\nabla_y\alpha_\ell\,,\quad k\alpha_\ell T=-Tk\alpha_\ell \tag{2.28}$$

and therefore

$$T^{-1}H(k)T = H(-k)\,. \tag{2.29}$$

Secondly we use space inversion as

$$R\psi(y) = \beta\psi(-y)\,. \tag{2.30}$$

Then

$$R^{-1}H(k)R = H(-k)\,. \tag{2.31}$$

Combining both symmetries implies

$$T^{-1}R^{-1}H(k)RT = H(k)\,. \tag{2.32}$$

If $H(k)\psi = E\psi$, then also $RT\psi$ is an eigenfunction with the same eigenvalue. Thus our claim follows from $\langle\psi, RT\psi\rangle = 0$. To verify this identity we note that $-i\alpha_3\alpha_1 = \mathrm{diag}\,(\sigma_2, \sigma_2)$ and $\langle\chi, R\sigma_2\chi^*\rangle = 0$ for an arbitrary two-spinor χ.

Corollary 2.2. *The eigenvalue $E_n(0)$ of $H(0)$ is at least two-fold degenerate.*

Proof. Since $T^*H(0)T = H(0)$ by (2.29) and $\langle\psi, (-i\alpha_3\alpha_1)\psi^*\rangle = 0$, the claim follows.

If V_Γ is not inversion symmetric, generically an energy band is two-fold degenerate at $k = 0$ and then splits into two non-degenerate bands. Note that a non-degenerate eigenvalue $E_n(k)$ has an associated eigenvector with a definite spin orientation.

The Pauli equation with spin-orbit coupling has the hamiltonian

$$H = -\frac{1}{2}\Delta_x + V_\Gamma(x) + \frac{1}{4}\sigma\cdot\big(\nabla V_\Gamma(x)\wedge(-i\nabla_x)\big)\,. \tag{2.33}$$

After Zak transform the corresponding fiber hamiltonian becomes

$$H(k) = \frac{1}{2}(-i\nabla_y + k)^2 + V_\Gamma(y) + \frac{1}{4}\sigma\cdot\big(\nabla V_\Gamma(y)\wedge(-i\nabla_y + k)\big) \tag{2.34}$$

with periodic boundary conditions. H of (2.33) is bounded from below. But otherwise the band structure is similar to the periodic Dirac operator. Proposition 2.1 and Corollary 2.25 hold as stated. In the proof one only has to use the appropriate time-reversal operator, which is $T\psi = \sigma_2\psi^*$ in the σ_3-eigenbasis.

2.4 Gap condition and smoothness

Let us consider one of the periodic hamiltonians, H_{per}, with fiber decomposition $H(k)$. H_{per} is adiabatically perturbed to H^ε. Very crudely the corresponding unitary groups should be close. To make such a notion quantitative a gap condition must be imposed. We denote by $\sigma(H)$ the spectrum of the self-adjoint operator H.

Gap condition: *We distinguish a family of m physically relevant energy bands $\{E_j(k),\ n \le j \le n+m-1\} = \sigma_0(k)$. This family satisfies the gap condition if*

$$\mathrm{dist}\{\sigma_0(k)\,,\ \sigma(H(k)) \setminus \sigma_0(k)\} \ge g > 0 \quad \text{for all } k \in M^*\,. \tag{2.35}$$

We repeat that the gap condition is not a spectral condition for H_{per}. Let us set $P^0 = \bigoplus_{j=n}^{n+m-1} P_j$.

Under the gap condition the projector $P(k)$ depends smoothly, in many cases even (real) analytically, on k. $\mathrm{Ran}\, P^0(k)$ is spanned by the basis $\{\varphi_j(k)\}_{j=n,\dots,n+m-1}$. If the m relevant energy bands have no crossings amongst each other, then φ_j is necessarily an eigenvector of $H(k)$ satisfying $H(k)\varphi_j(k) = E_j(k)\varphi_j(k)$. But if there are band crossings, it can be convenient not to insist on $\varphi_j(k)$ being an eigenvector of $H(k)$. Thus, while $P^0(k)$ is unique, the spanning basis is not. In applications it is of importance to know whether there is at least some choice of $\varphi_j(k)$, $j = n, \dots, n+m-1$, such that they have a smooth k-dependence. Locally, this can be achieved. However, since M^* has the topology of a torus, a global extension might be impossible. In fact this will generically happen for the magnetic Bloch hamiltonian, see [DN80, Nov81, Lys85] for examples. Somewhat surprisingly, a reasonably general answer has been provided only recently [Pan06]. For the case of the Bloch hamiltonian, analyticity has been proved before in cases $d = 1$, m arbitrary, and d arbitrary, $m = 1$, see Nenciu [Nen83, Nen91] and Helffer, Sjöstrand [HS89]. They rely on analytical techniques. In [Pan06] topological methods are developed.

Proposition 2.3. *In case of the non-magnetic Bloch hamiltonian let either $d \le 3$, $m \in \mathbb{N}$ or $d \ge 4$, $m = 1$. Then there exists a collection of smooth maps $\mathbb{R}^d \ni k \mapsto \varphi_j(k) \in L^2(M)$, $j = n, \dots, n+m-1$, with the following properties*

(i) the family $\{\varphi_j(k)\}_{j=n,\dots,n+m-1}$ is orthonormal and spans the range of $P^0(k)$.

(ii) each map is equivariant in the sense that

$$\varphi_j(k) = \tau(\lambda)\varphi(k+\lambda) \quad \textit{for all } k \in \mathbb{R}^d\,,\ \lambda \in \Lambda\,, \tag{2.36}$$

where $\tau(\lambda)$ is multiplication by $e^{i\lambda\cdot y}$. The same property holds for the non-magnetic periodic Dirac operator and Pauli operator with spin-orbit coupling.

Remark. The proof uses the first Chern class of the vector bundle whose fiber at k is the span of the family $\{\varphi_j(k)\}_{j=n,\dots,n+m-1}$ i.e. $\mathrm{Ran}\, P^0(k)$. To establish continuity, and thus smoothness, this first Chern class has to vanish, a property, which does not hold for a magnetic Bloch hamiltonian except for some particular energy bands.

If ε is small, excitations across the energy gap are difficult to achieve. More precisely to P^0 one can associate a projection operator Π^ε such that for arbitrary $\ell, \ell' \in \mathbb{N}$, $\tau \in \mathbb{R}_+$, it holds

$$\|(1 - \Pi^\varepsilon)e^{-iH^\varepsilon t}\Pi^\varepsilon \psi\| \le c_{\ell,\ell'}(\tau)\varepsilon^\ell \|\psi\| \tag{2.37}$$

for $0 \le t \le \varepsilon^{-\ell'}\tau$ with suitable constants $c_{\ell,\ell'}(\tau)$ independent of ε. In other words that the subspaces $\Pi^\varepsilon\mathcal{H}$ and $(1 - \Pi^\varepsilon)\mathcal{H}$ almost decouple, i.e. decouple at any prescribed level of precision and over any polynomial length of the time span under consideration. For the specific case of the Bloch hamiltonian more quantitative details on the decoupling are provided in Sect. 3.

If the gap condition is not satisfied, the dynamical properties are much more model dependent. Firstly the gap condition can be violated in various ways. In our context, since $H(k)$ has discrete spectrum, the violation comes through band crossings. The behavior close to a band crossing has to be studied separately [LT05, BT05]. In other models the energy band sits at the bottom of the continuous spectrum of $H(k)$ without gap [Teu02]. Then an assertion like Equation (2.37) holds only under a suitable restriction to small ℓ, ℓ', usually $\ell, \ell' = 1$ or perhaps $\ell = 2$, $\ell' = 1$.

The inequality (2.37) makes no assertion about the dynamics inside the almost invariant subspace $\Pi^\varepsilon\mathcal{H}$. While there is a general theory available [PST03a], we prefer to discuss the examples separately in the subsequent sections.

3 Nonmagnetic Bloch hamiltonians: Peierls substitution and geometric phase corrections

We discuss in more detail the effective dynamics for the Schrödinger equation with a periodic potential. For concreteness we fix the spatial dimension to be 3. Under Zak transform the nonmagnetic Bloch hamiltonian becomes

$$\mathcal{U}_Z\Big(\frac{1}{2}\big(-i\nabla_x - A(\varepsilon x)\big)^2 + V_\Gamma(x) + \phi(\varepsilon x)\Big)\mathcal{U}_Z^{-1} = H_Z^\varepsilon \tag{3.1}$$

with

$$H_Z^\varepsilon = \frac{1}{2}\big(-i\nabla_y + k - A(i\varepsilon\nabla_k^\tau)\big)^2 + V_\Gamma(y) + \phi(i\varepsilon\nabla_k^\tau)\,. \tag{3.2}$$

Here ∇_k^τ is differentation with respect to k and satisfying the y-dependent boundary conditions (2.6). H_Z^ε is a self-adjoint operator on $L^2_\tau(\mathbb{R}^3, H^2(M))$, compare with (2.8).

In (3.2) we observe that the external potentials couple the fibers. To emphasize this feature we think of (3.2) as being obtained through Weyl quantization from the operator valued function

$$H_0(k, r) = \frac{1}{2}\big(-i\nabla_y + k - A(r)\big)^2 + V_\Gamma(y) + \phi(r) \tag{3.3}$$

as defined on $(r, k) \in \mathbb{R}^6$ and acting on $\mathcal{H}_\mathrm{f}$ with fixed domain $H^2(M)$, see [PST03b] for details. In this form one is reminded of the Weyl quantization of the classical hamiltonian function $h_{\text{cl}}(q,p) = \frac{1}{2}p^2 + V(q)$ which yields the semiclassical hamiltonian

$$H_{\text{sc}} = \frac{1}{2}(-i\varepsilon\nabla_x)^2 + V(x) \tag{3.4}$$

acting in $L^2(\mathbb{R}^3)$. The analysis of (3.4) yields that on the time-scale $\varepsilon^{-1}t$ the wave packet dynamics governed by H_{sc} well approximates the flow generated by h_{cl}. In contrast, the adiabatic analysis deals with operator valued symbols, as in (3.3), and has as a goal to establish that the dynamics decouples into almost invariant subspaces and to determine the approximate dynamics within each such subspace.

To be specific, let us then fix throughout one band index n and let us assume that the band energy E_n is nondegenerate and satisfies the gap condition. Therefore we know that $E_n : M^* \to \mathbb{R}$ is smooth and we can choose the family of Bloch functions $\varphi_n(k)$, with $H(k)\varphi_n(k) = E_n(k)\varphi_n(k)$, such that φ_n depends smoothly on k. For each $\ell \in \mathbb{N} = \{0, 1, \ldots\}$ there exists then an orthogonal projection Π_ℓ^ε on $\mathcal{H}_\tau$ such that

$$\|[H_Z^\varepsilon\,,\, \Pi_\ell^\varepsilon]\| \le c_\ell \varepsilon^{\ell+1} \tag{3.5}$$

for some constants c_ℓ. Integrating in time one concludes that the subspaces $\Pi_\ell^\varepsilon \mathcal{H}_\tau$ are almost invariant in the sense that

$$\|(1 - \Pi_\ell^\varepsilon) e^{-i\varepsilon^{-\ell'} t H_Z^\varepsilon} \Pi_\ell^\varepsilon \psi\| \le \|\psi\|(1 + |t|) c_\ell\, \varepsilon^{\ell+1} \varepsilon^{-\ell'} \tag{3.6}$$

for any $\ell, \ell' \in \mathbb{N}$. Note that the adiabatic time scale, order $\varepsilon^{-\ell'}$, can have any power law increase, at the expense of choosing the order of the projection Π_ℓ^ε sufficiently large. Only for times of order $e^{1/\varepsilon}$ one observes transitions away from the almost invariant subspace. The zeroth order projection is attached to the n-th band under consideration, while the higher orders are successively smaller corrections to Π_0^ε. To construct Π_0^ε one considers the projection onto the n-th band, $|\varphi_n(k)\rangle\langle\varphi_n(k)|$, as an operator valued function with values in $B(\mathcal{H}_{\text{f}})$. From it we obtain the minimally substituted projection $|\varphi_n(k - A(r))\rangle\langle\varphi_n(k - A(r))|$. Its Weyl quantization is ε-close to the orthogonal projection Π_0^ε.

The second task is to determine the approximate time-evolution on $\Pi_\ell^\varepsilon \mathcal{H}_\tau$. Since the subspace changes with ε, it is more convenient to unitarily map $\Pi_\ell^\varepsilon \mathcal{H}_\tau$ to an ε-independent reference Hilbert space, which in our case is simply $L^2(M^*)$. The dynamics on $L^2(M^*)$ is governed by an effective hamiltonian. It is written down most easily in terms of a hamiltonian function $h_\ell^\varepsilon : M^* \times \mathbb{R}^3 \to \mathbb{R}$. h_ℓ^ε is a smooth function. We also may regard it as defined on $\mathbb{R}^3 \times \mathbb{R}^3$ and Γ^*-periodic in the first argument. h_ℓ^ε admits the power series

$$h_\ell^\varepsilon = \sum_{j=0}^{\ell} \varepsilon^j h_j \tag{3.7}$$

with ε-independent functions h_j. The effective quantum hamiltonian is obtained from h_ℓ^ε through Weyl quantization. The index ℓ regulates the time scale over which the approximation is valid and the size of the allowed error.

In [PST03b] we provide an iterative algorithm to compute h_j. In practice only h_0 and h_1 can be obtained, at best h_2 under simplifying assumptions. While this may look very restrictive, it turns out that already h_1 yields novel physical effects as compared to h_0. Even higher order corrections seem to be less significant.

To lowest order one obtains

$$h_0(k,r) = E_n(k - A(r)) + \phi(r)\,, \tag{3.8}$$

which Weyl-quantizes to

$$\mathcal{W}^\varepsilon[h_0] = E_n(k - A(i\varepsilon\nabla_k)) + \phi(i\varepsilon\nabla_k) \tag{3.9}$$

acting on $L^2(M^*)$, where $i\nabla_k$ is the operator of differentiation with periodic boundary conditions. (The twisted boundary conditions appearing in (3.2) are absorbed into the unitary map of $\Pi_0^\varepsilon\mathcal{H}$ to $L^2(M^*)$.) In solid state physics the Weyl quantization (3.9) is referred to as *Peierls substitution.* (3.8), (3.9) have a familiar form. The periodic potential merely changes the kinetic energy $\frac{1}{2}k^2$ of a free particle to $E_n(k)$. The main distinctive feature is the periodicity of the kinetic energy. For example, in presence of a linear potential ϕ, $\phi(x) = -E\cdot x$, an electron, initially at rest, will start to accelerate along E but then turns back because of periodicity in k.

To first order the effective hamiltonian reads

$$h_1(k,r) = \big(\nabla\phi(r) - \nabla E_n(\widetilde{k}) \wedge B(r)\big)\cdot \mathcal{A}_n(\widetilde{k}) - B(r)\cdot\mathcal{M}_n(\widetilde{k})\,, \tag{3.10}$$

with the kinetic wave number $\widetilde{k} = k - A(r)$. The coefficients $\mathcal{A}_n$ and $\mathcal{M}_n$ are the geometric phases. They carry information on the Bloch functions $\varphi_n(k)$. $\mathcal{A}_n$ is the Berry connection given through

$$\mathcal{A}_n(k) = i\langle\varphi_n(k)\,,\ \nabla_k\varphi_n(k)\rangle_{\mathcal{H}_f} \tag{3.11}$$

and $\mathcal{M}_n$ is the Rammal-Wilkinson phase given trough

$$\mathcal{M}_n(k) = \frac{1}{2} i\langle\nabla_k\varphi_n(k), \wedge(H(k) - E_n(k))\nabla_k\varphi_n(k)\rangle_{\mathcal{H}_f}\,. \tag{3.12}$$

The Bloch functions φ_n are only determined up to a smooth phase $\alpha(k)$, i.e. instead of $\varphi_n(k)$ one might as well use $e^{-i\alpha(k)}\varphi_n(k)$ with smooth $\alpha : M^* \to \mathbb{R}$. Clearly $\mathcal{M}_n$ is independent of the gauge field α. On the other hand, $\mathcal{A}$ is gauge-dependent while its curl

$$\Omega_n = \nabla \wedge \mathcal{A}_n \tag{3.13}$$

is gauge independent. From time-reversal one concludes that

$$\Omega_n(-k) = -\Omega_n(k)\,. \tag{3.14}$$

In particular, in dimension $d = 2$ for the first Chern number of the Bloch vector bundle one obtains

$$\int_{M^*} dk \Omega_n(k) = 0\,. \tag{3.15}$$

For the magnetic Bloch hamiltonian, (3.14) is violated in general, see Sect. 4. The integral in (3.15) can take only integer values (in the appropriate units) and the first Chern number may be different from zero. Physically this leads to the quantization of the Hall current [PST03b, SN99].

We still owe the reader precise a statement on the error in the approximation. At the moment we work in the representation space at precision level $\ell = 1$. Let H_{eff} be the Weyl quantization of $h_0 + \varepsilon h_1$, see (3.8) and (3.10). There is then a unitary $U^\varepsilon : \Pi_1^\varepsilon \mathcal{H}_\tau \to L^2(M^*)$ such that for all $\psi \in \mathcal{H}_\tau$

$$\|\big(e^{-iH_Z^\varepsilon t} - U^{\varepsilon *} e^{-iH_{\mathrm{eff}} t} U^\varepsilon\big)\Pi_1^\varepsilon \psi\| \le c\|\psi\|(1+|\tau|)\varepsilon^2 \tag{3.16}$$

with $|t| \le \varepsilon^{-1}\tau$ and some constant c independent of $\|\psi\|$, τ, and ε.

4 Magnetic Bloch hamiltonians: the Hofstadter butterfly

We turn to a magnetic Bloch hamiltonian in the form (1.4), in dimension $d = 2$ and with a transverse constant magnetic field B_0. We want to explain how the limits $B_0 \to \infty$ and $B_0 \to 0$ can be understood with adiabatic methods. As a remark, it is worthwhile to recall that, when the physical constants are restored, the dimensionless parameter B_0 is given by

$$B_0 = \frac{\mathcal{B}_0 S}{2\pi\hbar c/e}\,, \tag{4.1}$$

where S is the area of the fundamental cell of Γ and $\mathcal{B}_0$ the strength of the magnetic field, both expressed in their dimensional units. Thus B_0 corresponds physically to the magnetic flux per unit cell divided by hc/e, as the fundamental quantum of magnetic flux. This section is based essentially on [FP06], which elaborates on previous related results [Bel86, HS89].

Adiabatic limits are always related to separation of time-scales. In the present case, one indeed expects that as $B_0 \to \infty$ the cyclotron motion induced by B_0 is faster than the motion induced by V_Γ, while in the limit $B_0 \to 0$ the microscopic variations of the wave function induced by V_Γ are expected to be faster than the cyclotron motion.

Let us focus first on the Landau regime $B_0 \to \infty$. In order to make quantitative the previous claim, one introduces the operators

$$\begin{cases} L_1 = \frac{1}{\sqrt{B_0}}\left(p_1 + \frac{1}{2}B_0\, x_2\right)\,, \\ L_2 = \frac{1}{\sqrt{B_0}}\left(p_2 - \frac{1}{2}B_0\, x_1\right)\,, \end{cases} \qquad [L_1, L_2] = i\mathbf{1}, \tag{4.2}$$

and the complementary pair of operators

$$\begin{cases} G_1 = \frac{1}{B_0}\left(p_1 - \frac{1}{2}B_0\, x_2\right), \\ G_2 = \frac{1}{B_0}\left(p_2 + \frac{1}{2}B_0\, x_1\right), \end{cases} \qquad [G_1, G_2] = \tfrac{i}{B_0}\mathbf{1}, \tag{4.3}$$

where the relative sign is chosen such that $[L_i, G_j] = 0$, for $i, j = 1, 2$.

If $V_\Gamma = 0$, then H_{MB} describes a harmonic oscillator, with eigenfunctions localized on a scale $|B_0|^{-1/2}$; this corresponds to the cyclotron motion in classical mechanics. Since $[G_i, H_{\mathrm{MB}}] = 0$, the operators G_1 and G_2 describe conserved quantities, which semiclassically correspond to the coordinates of the center of the cyclotron motion.

If $V_\Gamma \neq 0$, but the energy scale $\|V_\Gamma\|$ is smaller than the cyclotron energy $\approx B_0$, then the operators G_i have a non-trivial but slow dynamics. By introducing the adiabatic parameter $\eta = 1/B_0$ the hamiltonian reads

$$H_{\mathrm{MB}} = \frac{1}{2\eta}\left(L_1^2 + L_2^2\right) + V_\Gamma\left(G_2 - \sqrt{\eta}L_2, -G_1 + \sqrt{\eta}L_1\right). \tag{4.4}$$

In view of the commutator $[G_1, G_2] = i\eta\mathbf{1}$, one can regard $\eta\, H_{\mathrm{MB}}$ as the η-Weyl quantization (in the sense of the mapping $(q, p) \mapsto (G_1, G_2)$) of the operator-valued symbol

$$h_{\mathrm{MB}}(q, p) = \frac{1}{2}\left(L_1^2 + L_2^2\right) + \eta\, V_\Gamma\left(p - \sqrt{\eta}L_2, -q + \sqrt{\eta}L_1\right). \tag{4.5}$$

For each fixed $(q, p) \in \mathbb{R}^2$, $h_{\mathrm{MB}}(q, p)$ is an operator acting in the Hilbert space $\mathcal{H}_{\mathrm{f}} \cong L^2(\mathbb{R})$ corresponding to the fast degrees of freedom. If $\|V_\Gamma\|_{\mathcal{B}(\mathcal{H})} < \infty$, then $h_{\mathrm{MB}}(q, p)$ has purely discrete spectrum, with eigenvalues

$$\lambda_{n,\eta}(q, p) = (n + \frac{1}{2}) + \eta V_\Gamma(p, -q) + \mathcal{O}(\eta^{3/2}), \qquad n \in \mathbb{N},$$

as $\eta \downarrow 0$. The index $n \in \mathbb{N}$ labels the *Landau levels.* For η small enough, each eigenvalue band is separated from the rest of the spectrum by an uniform gap. Thus we can apply space-adiabatic perturbation theory to show that the band corresponds to an almost-invariant subspace $\Pi_{n,\eta}L^2(\mathbb{R}^2)$. Let us focus on a specific $n \in \mathbb{N}$. One can prove that the dynamics inside $\mathrm{Ran}\,\Pi_{n,\eta}L^2(\mathbb{R}^2)$ is described by an effective hamiltonian, which at the first order of approximation in η reads

$$h_1^\eta = (n + \frac{1}{2}) + \eta V_\Gamma(G_1, -G_2). \tag{4.6}$$

The first term in (4.6) is a multiple of the identity, and as such does not contribute to the dynamics as far as the expectation values of observables are concerned. Leading-order dynamics is thus described by the second term, which does not depend on the Landau level $n \in \mathbb{N}$. Since V_Γ is a biperiodic function and (G_1, G_2) a canonical pair, the second term is a Harper-like operator. The spectrum of such operators exhibit a complex fractal behavior (*Hofstadter butterfly*) sensitively depending on the diophantine properties of $\alpha = \frac{B_0}{2\pi}$ (notice that $V_\Gamma(G_1, G_2)$ depends on α through the commutator $[G_1, G_2] = iB_0^{-1}\mathbf{1}$). The Cantor structure of the spectrum was proven first in [BS82] for the case $V_\Gamma(x_1, x_2) = \lambda \cos x_1 + \cos x_2$ (Harper model), for a dense set of the parameter values. Later Helffer and Sjöstrand accomplished a detailed semiclassical analysis of the Harper operator [HS89]. As a final step the Cantor spectrum has been proven by Puig ($\lambda \neq 0$, α Diophantine) [Pui04], and by Avila and Jitomirskaya [AJ05] for all the conjectured values of the parameters: $\lambda \neq 0$, α irrational (the *Ten Martini conjecture*, as baptized by B. Simon).

Secondly we turn to the opposite limit $B_0 \to 0$, where the slow part of the dynamics is still described by the magnetic momentum operators $\widetilde{L}_j = \sqrt{B_0} L_j$ ($j = 1, 2$), with commutator of order $\mathcal{O}(B_0)$. However the decomposition given by (4.2) and (4.3) is no longer convenient.

Since A_0 is a linear function, $A_0(\varepsilon x) = \frac{1}{2}\varepsilon B_0 \wedge x$, the slow variation limit $\varepsilon \to 0$ agrees with the weak field limit $B_0 \to 0$. We then pose $\varepsilon = B_0$ and we regard H_{MB} in (1.4) as an adiabatic perturbation of the periodic hamiltonian (2.3). Thus we are reduced to the situation described in Sect. 3: to each isolated Bloch band of the unperturbed hamiltonian there corresponds a subspace $\Pi_{n,\varepsilon} L^2(\mathbb{R}^2)$ which is approximately invariant under the dynamics as $\varepsilon \downarrow 0$. The dynamics inside this subspace is described by Peierls substitution (3.9), which now reads

$$\mathcal{W}^\varepsilon[h_0] = E_n(k - \frac{1}{2} e_3 \wedge (i\varepsilon \nabla_k)), \tag{4.7}$$

as an operator acting in $L^2(\mathbb{T}^2, dk)$. Here $B_0 = (0, 0, \varepsilon)$ and e_i is the unit vector in the i-th direction.

Formula (4.7) shows that the leading order effective hamiltonian depends only on the operators $(K_1, K_2) = K$,

$$K = k - \frac{1}{2} e_3 \wedge (i\varepsilon \nabla_k),$$

which roughly speaking are the Fourier transform of the pair $(\widetilde{L}_1, \widetilde{L}_2)$, and not on the complementary pair of operators. The same property holds true for the effective hamiltonian h_ℓ^ε, at any order of approximation $\ell \in \mathbb{N}$, see [FP06], with important consequences on the splitting of magnetic subbands at small but finite B_0.

An operator in the form (4.7), shortly written $E_n(K_1, K_2)$, is *isospectral* to an Harper-like operator, namely $E_n(G_1, G_2)$ acting in $L^2(\mathbb{R})$. Indeed the first numerical evidence of the butterfly-like Cantor structure of the spectrum of Harper-like operators appeared when Hofstadter investigated the spectrum of $\cos K_1 + \cos K_2$ as a function of ε [Hof76]. On the other side, an operator of the form $E_n(K_1, K_2)$ is not *unitarily equivalent* to the Harper operator $E_n(G_1, G_2)$. The important geometric and physical consequences of this fact are developed in [FP06].

Having explained the two extreme cases, $B_0 \to 0$ and $B_0 \to \infty$, the reader may wonder about the intermediate values of the magnetic field, $B_0 \approx 1$. As explained already in Sect. 2.3 it is convenient to introduce the magnetic translations

$$\mathcal{T}_\alpha = e^{-\frac{i}{2}\varphi(\alpha)} \exp(iB_0\, \alpha \cdot G), \qquad \alpha \in \Gamma_0,$$

see (2.17) and (2.19). If B_0 satisfy Assumption 2, then $\{\mathcal{T}_\alpha\}$ is a commutative group, thus leading to the magnetic Zak transform (2.20). H_{MB} is then a fibered operator over the magnetic Bloch momentum $\kappa \in \mathbb{T}^2$. At each κ the spectrum of $H_{\mathrm{MB}}(\kappa)$ is pure point and the corresponding eigenvalues $\mathcal{E}_n^{B_0}$ are the *magnetic Bloch bands.*

In view of this structure, one might argue that the adiabatic perturbation of the hamiltonian which includes, on top of the constant magnetic field B_0, a slowly varying magnetic potential $A(\varepsilon x)$ as in (1.5) can be treated with the methods of Sect. 3. There is however one crucial element missing. Indeed one can still associate to each magnetic Bloch band $\mathcal{E}_n^{B_0}$, isolated from the rest of the spectrum, an almost-invariant subspace $\mathrm{Ran}\, \Pi_n^{B_0}$. On the other side the construction of the effective hamiltonian relies on smoothness which may be impeded for topological reasons. Indeed the analogue of Proposition 2.3 is generically false for magnetic Bloch hamiltonians, as well-understood [DN80, Nov81, Lys85]. In geometric terminology this fact is rephrased by saying that the magnetic Bloch bundle is generically non-trivial (in technical sense). This important fact has sometimes been overlooked. For example, Assumption B in [DGR04] is equivalent to the triviality of the magnetic Bloch bundle. Under this assumption the magnetic case is already covered by the results in [PST03b]. Thus the problem of adiabatic perturbation of a generic magnetic Bloch hamiltonian appears to be an open, in our view challenging, problem for the future.

5 Piezoelectricity

In the year 1880 the brothers Jacques and Pierre Curie discovered that some crystalline solids (like quartz, tourmaline, topaz, ...) exhibit a macroscopic polarization if the sample is strained.

It turns out that also this effect can be understood in the framework of adiabatically perturbed periodic hamiltonians, cf. [PST06, Lei05]. The perturbation is now slowly in time,

$$H_{\mathrm{PE}}(t) = -\frac{1}{2}\Delta_x + V_{\Gamma(\varepsilon t)}(x, \varepsilon t)\,. \tag{5.1}$$

If the potential $V_\Gamma(x, \varepsilon t)$ has no center of inversion, i.e. there is no point with respect to which the potential has space-reflection symmetry, then the slow variation of the periodic potential is expected to generate a non-zero current and can be shown to do so for particular examples [ABL97]. By translation invariance this current if averaged over a unit cell is everywhere the same and we denote the average current by $J^\varepsilon(t)$. For the following discussion we assume that V_Γ varies only for times in the finite interval $[0,T]$. Integrating the current per volume over the relevant time interval yields the average polarization,

$$\Delta \mathbf{P}^\varepsilon = \int_0^T \mathrm{d}t\, J^\varepsilon(t)\,.$$

In this section we discuss results that relate the current $J^\varepsilon(t)$ directly to the quantum mechanics of non-interacting particles governed by the hamiltonian (5.1), without the detour via the semiclassical model. For this we need to solve the Schrödinger equation with initial state $\rho(0) = P(0)$ being the spectral projection of $H_{\mathrm{PE}}(0)$ below the Fermi energy $E(0)$. Since the piezo effect occurs only for insulators, we can assume that $E(0)$ lies in a gap of the spectrum of $H_{\mathrm{PE}}(0)$ and, in order to simplify the discussion, we also assume that this gap does not close in the course of time. Hence there is a continuous function $E : [0,T] \to \mathbb{R}$ such that $E(t)$ lies in a gap of $H_{\mathrm{PE}}(t)$ for all t. The state at time t is given by

$$\rho^\varepsilon(t) = U^\varepsilon(t,0)\,P(0)\,U^\varepsilon(t,0)^*\,,$$

where the unitary propagator $U^\varepsilon(t,0)$ is the solution of the time-dependent Schrödinger equation

$$\mathrm{i}\varepsilon\frac{\mathrm{d}}{\mathrm{d}t}U^\varepsilon(t,0) = H_{\mathrm{PE}}(t)\,U^\varepsilon(t,0) \qquad \text{with} \quad U^\varepsilon(0,0) = \mathbf{1}\,. \tag{5.2}$$

With the current operator given by

$$j^\varepsilon := \frac{\mathrm{i}}{\varepsilon}\,[H(t),x] = -\frac{\mathrm{i}}{\varepsilon}\nabla_x\,, \tag{5.3}$$

and the trace per volume defined as

$$\mathcal{T}(A) := \lim_{\Lambda_n\to\mathbb{R}^3}\frac{1}{|\Lambda_n|}\mathrm{Re}\,\mathrm{Tr}(\mathbf{1}_{\Lambda_n}A)\,, \tag{5.4}$$

with $\mathbf{1}_{\Lambda_n}$ being the characteristic function of a 3-dimensional box Λ_n with finite volume $|\Lambda_n|$, the average current in the state $\rho^\varepsilon(t)$ is

$$J^\varepsilon(t) = \mathcal{T}(\rho^\varepsilon(t)\,j^\varepsilon)\,.$$

Finally the average polarization is

$$\Delta \mathbf{P}^\varepsilon = \int_0^T \mathrm{d}t\ \mathcal{T}(\rho^\varepsilon(t)\,J^\varepsilon)\,, \tag{5.5}$$

which is the main quantity of physical interest. The given framework allows us to describe the macroscopic polarization of a solid by a pure *bulk property*, i.e. independently of the shape of the sample.

In the simplest but most important case (see Paragraph (ii) in Sect. 1 for a discussion of the model), the periodic potential $V_\Gamma(x, \varepsilon t)$ is periodic with respect to a time-*independent* lattice Γ. For this case King-Smith and Vanderbilt [KSV93] derived a formula for $\Delta\mathbf{P}$ based on linear response theory, which turned out to make accurate predictions for the polarization of many materials. Their by now widely applied formula reads

$$\Delta\mathbf{P} = \frac{1}{(2\pi)^3} \sum_{n=0}^{N_c} \int_{M^*} \mathrm{d}k \, \big(\mathcal{A}_n(k,T) - \mathcal{A}_n(k,0)\big), \tag{5.6}$$

where the sum runs over all the occupied Bloch bands and $\mathcal{A}_n(k,t)$ is the Berry connection coefficient for the n-th Bloch band at time $t \in \mathbb{R}$,

$$\mathcal{A}_n(k,t) = \mathrm{i}\langle \varphi_n(k,t), \nabla_k \varphi_n(k,t)\rangle_{L^2(M)} \, .$$

Although $\mathcal{A}_n$ depends on the choice of the Bloch function φ_n, the average polarization (5.6) defines a gauge invariant quantity, i.e. it is independent of the choice of Bloch functions.

In [PST06] we show that $\Delta\mathbf{P}^\varepsilon$ defined in (5.5) approaches $\Delta\mathbf{P}$ as given by the King-Smith and Vanderbilt formula (5.6) with errors smaller than any power of ε, whenever the latter is well defined. More precisely we show that under suitable technical conditions on $V_\Gamma(t)$ the average polarization is well defined and that for any $N \in \mathbb{N}$

$$\Delta\mathbf{P}^\varepsilon = -\frac{1}{(2\pi)^d} \int_0^T \mathrm{d}t \int_{M^*} \mathrm{d}k \; \Theta(k,t) + \mathcal{O}(\varepsilon^N) \, , \tag{5.7}$$

where

$$\Theta(k,t) := -\mathrm{i} \, \mathrm{tr}\left(P(k,t)\left[\partial_t P(k,t), \, \nabla_k P(k,t)\right]\right) \, , \tag{5.8}$$

and $P(k,t)$ is the Bloch-Floquet fiber decomposition of the spectral projector $P(t) = \mathbf{1}_{(-\infty,E(t)]}(H_{\mathrm{PE}}(t))$. Whenever all Bloch bands within $\mathrm{Ran}P(k,t)$ are isolated, the explicit term in (5.7) agrees with (5.6). Note however that (5.7) is more general, since it can be applied also to situations where band crossings occur within the set of occupied bands.

From the point of view of adiabatic approximation, this result is actually quite simple, since one just needs the standard time-adiabatic theory. At time $t = 0$ the state $\rho(0)$ is just the projection $P(0)$ onto the subspace of the isolated group of occupied bands. Since these bands remain isolated during time evolution, this subspace is adiabatically preserved according to the original adiabatic theorem of Kato [Kat50], i.e.

$$\rho^\varepsilon(t) = P(t) + \mathcal{O}(\varepsilon) \, ,$$

and one can compute the higher order corrections to $\rho^\varepsilon(t)$ using the higher order time-adiabatic approximation due to Nenciu [Nen93]. In order to get explicit results, one has to do the adiabatic approximation for each fixed

$k \in M^*$ separately. This is possible since $H_{\text{PE}}(t,k)$ is still fibered in k, due to translation invariance with respect to a time-independent lattice. However, since we need to differentiate with respect to k in order to compute the current, as suggested by formula (5.8), the expansion needs to be done uniformly on spaces of suitable equivariant functions. This makes the proof more technical than expected at first sight.

Alternatively one can derive also for $H_{\text{PE}}(t)$ the semiclassical equations of motion including first order corrections:

$$\begin{cases} \dot{q} = \nabla_k E_n(k,t) - \varepsilon\, \Theta_n(k,t), \\ \dot{k} = 0\,. \end{cases} \tag{5.9}$$

And again averaging the velocity over the first Brillouin zone yields the correct quantum mechanical average current that is the contribution from the n-th band.

Note the striking similarity between the semiclassical corrections in (5.8) and the electromagnetic field. If we define the geometric vector potential

$$\mathcal{A}_n(k,t) = \mathrm{i}\langle \varphi_n(k,t), \nabla_k \varphi_n(k,t)\rangle_{L^2(M)},$$

and the geometric scalar potential

$$\phi_n(k,t) = -\mathrm{i}\langle \varphi_n(k,t), \partial_t \varphi_n(k,t)\rangle_{L^2(M)},$$

in terms of the Bloch function $\varphi_n(k,t)$ of some isolated band, then in complete analogy to the electromagnetic fields we have

$$\Theta_n(k,t) = -\partial_t \mathcal{A}_n(k,t) - \nabla_k \phi_n(k,t), \tag{5.10}$$

and

$$\Omega_n(k,t) = \nabla_k \wedge \mathcal{A}_n(k,t)\,. \tag{5.11}$$

Time-dependent deformations of a crystal generically also lead to a time-dependent periodicity lattice $\Gamma(t)$, see (5.1). This more general situation is considered in [Lei05, LP06]. Now the lattice momentum k is no longer a conserved quantity and the full space-adiabatic perturbation theory is required in order to compute the corresponding piezoelectric current. As a result an additional term appears in the semiclassical equations of motion, reflecting the deformation of the lattice of periodicity.

Acknowlegdments. We thank Ulrich Mauthner, Max Lein, and Christof Sparber for most informative discussions. This work has been supported by the DFG Priority Program 1095 "Analysis, Modeling and Simulation of Multiscale Problems" under Sp 181/16-3.

References

[AM76] N.W. Ashcroft and N.D. Mermin. *Solid State Physics*. Saunders, New York, 1976.

[AJ05] A. Avila and S. Jitomirskaya. Solving the Ten Martini Problem. Lecture Notes in Physics 690, pages 5–16, 2006.

[ABL97] J.E. Avron, J. Berger, and Y. Last. Piezoelectricity: quantized charge transport driven by adiabatic deformations. *Phys. Rev. Lett.*, 78, 511–514, 1997.

[BK06] S. Bauer and M. Kunze. Radiative friction for particles interacting with the radiation field: classical many-particle systems. In "Analysis, Modeling and Simulation of Multiscale Problems, A. Mielke (edr), Springer-Verlag, 2006.".

[Bel86] J. Bellissard. K-theory of C^*-algebras in solid-state physics. Lecture Notes in Physics, volume 257, pages 99–156, 1986.

[BS82] J. Bellissard and B. Simon. Cantor spectrum for the almost Mathieu equation. *J. Funct. Anal.*, 48, 408–423, 1982.

[BT05] V. Betz and S. Teufel. Precise coupling terms in adiabatic quantum evolution: the generic case. *Comm. Math. Phys.*, 260, 481–509, 2005.

[BM*03] A. Bohm, A. Mostafazadeh, H. Koizumi, Q. Niu, and J. Zwanziger. *The Geometric Phase in Quantum Systems*. Springer-Verlag, Berlin, 2003.

[DGR02] M. Dimassi, J.C. Guillot, and J. Ralston. Semiclassical asymptotics in magnetic Bloch bands. *J. Phys. A*, 35, 7597–7605, 2002.

[DGR04] M. Dimassi, J.-C. Guillot, and J. Ralston. On Effective hamiltonians for adiabatic perturbations of magnetic Schrödinger operators. *Asymptotic Analysis*, 40, 137–146, 2004.

[DN80] B.A. Dubrovin and S.P. Novikov. Ground state of a two-dimensional electron in a peridodic magnetic field. *Zh. Eksp. Teo. Fiz*, 79, 1006–1016, translated in *Sov. Phys. JETP*, 52 vol. 3, 511–516, 1980.

[FP06] F. Faure and G. Panati. Peierls substitution, Hofdstadter butterfly and deformations of bundles. In preparation.

[HS89] B. Helffer and J. Sjöstrand. Analyse semi-classique pour l'équation de Harper I-III. *Mem. Soc. Math. France* (N.S), 34, tome 116, 1989, and 39, tome 117, 1990, and 40, tome 118, 1990.

[HS89] B. Helffer and J. Sjöstrand. Equation de Schrödinger avec champ magnétique et équation de Harper. *Schrödinger Operators,* Lecture Notes in Physics, volume 345, pages 118–197, 1989.

[Hof76] D.R. Hofstadter. Energy levels and wave functions of Bloch electrons in rational and irrational magnetic fields. *Phys. Rev. B*, 14, 2239–2249, 1976.

[Kat50] T. Kato. On the adiabatic theorem of quantum mechanics. *Phys. Soc. Jap.*, 5, 435–439, 1950.

[KSV93] R.D. King-Smith and D. Vanderbilt. Theory of polarization in crystalline solids. *Phys. Rev. B*, 47, 1651–1654, 1993.

[LT05] C. Lasser and S. Teufel. Propagation through conical crossings: an asymptotic semigroup. *Comm. Pure Appl. Math.*, 58, 1188–1230, 2005.

[Lei05] M. Lein. A dynamical approach to piezoelectricity. Diplomarbeit, Physik Department, TU München, 2005.

[LP06] M. Lein and G. Panati. Piezoelectricity: beyond the fixed lattice approximation, in preparation.

[Lys85] A.S. Lyskova. Topological characteristic of the spectrum of the Schrödinger operator in a magnetic field and in a weak potential. *Theor. Math. Phys.*, 65, 1218–1225, 1985.

[Mau03] U. Mauthner. Ph.D. thesis, TU München. In preparation.

[Nen83] G. Nenciu. Existence of the exponentially localised Wannier function. *Comm. Math. Phys.*, 91, 81–85, 1983.

[Nen91] G. Nenciu. Dynamics of band electrons in electric and magnetic fields: rigorous justification of the effective Hamiltonian. *Rev. Mod. Phys.*, 63, 91–127, 1991.

[Nen93] G. Nenciu. Linear adiabatic theory. Exponential estimates. *Comm. Math. Phys.* 152, 479–496, 1993.

[Nov81] S.P. Novikov. Magnetic Bloch functions and vector bundles. Typical dispersion law and quantum numbers. Sov. Math. Dokl., 23, ??, 1981.

[Pan06] G. Panati. Triviality of Bloch and Bloch-Dirac bundles. arXiv:math-ph/0601034, 2006.

[PST03a] G. Panati, H. Spohn, and S. Teufel. Space-adiabatic perturbation theory. *Adv. Theor. Math. Phys.*, 7, 145–204, 2003.

[PST03b] G. Panati, H. Spohn, and S. Teufel. Effective dynamics for Bloch electrons: Peierls substitution and beyond. *Comm. Math. Phys.*, 242, 547–578, 2003.

[PST06] G. Panati, C. Sparber, and S. Teufel. A simple semiclassical description of piezoelectricity. In preparation.

[Pui04] J. Puig. Cantor spectrum for the almost Mathieu operator. *Comm. Math. Phys.*, 244, 297–309, 2004.

[Res94] R. Resta. Macroscopic polarization in crystalline dielectrics, the geometric phase approach. *Rev. Mod. Physics*, 66, 899–915, 1994.

[SN99] G. Sundaram and Q. Niu. Wave-packet dynamics in slowly perturbed crystals, gradient corrections and Berry-phase effects. *Phys. Rev. B*, 59, 14195–14925, 1999.

[Teu02] S. Teufel. Effective N-body dynamics for the massless Nelson model and adiabatic decoupling without spectral gap. *Ann. Henri Poincaré*, 3, 939–965, 2002.

[Teu03] S. Teufel. *Adiabatic Perturbation Theory in Quantum Dynamics.* Lecture Notes in Mathematics Vol. 1821, Springer-Verlag, Berlin, 2003.

[Tha94] B. Thaller. *The Dirac Equation.* Springer-Verlag, Heidelberg, 1992.

[Ynd96] F.J. Yndurain. *Relativistic Quantum Mechanics and Introduction to Field Theory.* Springer-Verlag, Berlin, 1996.

[Zak64] J. Zak. Magnetic translation group. *Phys. Rev. A*, 134, 1602–1606, 1964.

[Zak68] J. Zak. Dynamics of electrons in solid in external fields. *Phys. Rev.*, 168, 686–695, 1968.

Graph Algorithms for Dynamical Systems

Michael Dellnitz[1], Mirko Hessel-von Molo[1], Philipp Metzner[2], Robert Preis[1], and Christof Schütte[2]

[1] Institute for Mathematics, University of Paderborn, 33095 Paderborn. dellnitz@uni-paderborn.de, mirkoh@uni-paderborn.de, robsy@uni-paderborn.de
[2] Department of Mathematics, Freie Universität Berlin, Arnimallee 14, 14195 Berlin. metzner@math.fu-berlin.de, schuette@math.fu-berlin.de

1 Introduction

This article is concerned with the numerical analysis of dynamical systems using methods that are based on a discretized description of the system as a graph. The graph-based description provides a unifying framework to approach a wide and diverse variety of dynamical systems, from time-discrete maps via ordinary differential equations to stochastic differential equations describing e. g. diffusion in a potential landscape.

Within this variety, this article focusses on those dynamical systems that can possess a 'multiscale structure' in the sense that they exhibit interesting dynamical behavior on more than one timescale. We will explain what we mean with this phrase by means of some examples. Consider in Fig. 1 one trajectory of Chua's circuit, that is described by the well-known three-dimensional ordinary differential equation

$$\begin{aligned} \dot{x} &= \alpha(y - m_0 x - \frac{1}{3} m_1 x^3) \\ \dot{y} &= x - y + z \\ \dot{z} &= -\beta y \end{aligned}$$

(see e. g. [HP*96]). It is clearly visible that relatively long parts of the whole trajectory are contained in two 'leaves' within which the trajectory shows a spiralling motion, with only some quick 'jumps' between the two leaves.

Similar phenomena can be observed in systems of quite different mathematical type. As an example, consider a stochastic process in $\mathbf{R}^n$ describing diffusion within a potential given by a function $V : \mathbf{R}^n \to \mathbf{R}$. The system is given by the Smoluchovski equation

$$\dot{X}(t) = -\nabla V(X(t)) + \varepsilon \dot{W}(t)$$

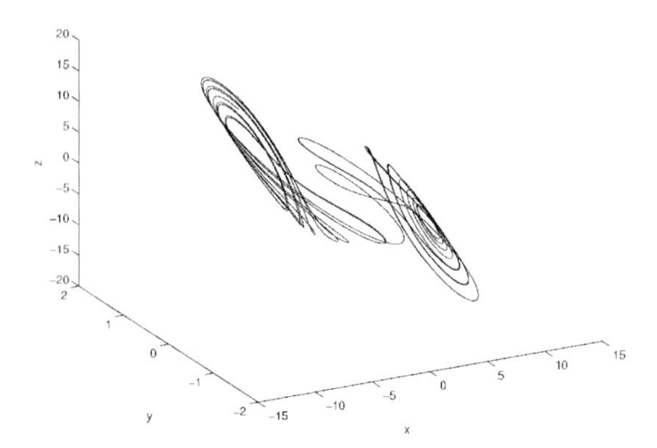

Fig. 1.1. A trajectory of Chua's circuit that switches relatively rarely between two almost invariant sets.

with W_t being a standard n-dimensional Wiener process and ε a small parameter. (A variation of this example is described in more detail in Sect. 4.) If we assume $V(x) \to \infty$ for $\|x\| \to \infty$ (in order to avoid sample paths drifting off to infinity), then any sample path will spend most of the time in the vicinity of the local minima of V, with transitions between the minima being rare events.

The common feature of the Chua circuit example and of the diffusion example is the existence of subsets of the state space that are, although not being invariant under the dynamics considered, nevertheless *almost* invariant in the sense that on a short timescale, a change of a trajectory between the sets is an event rarely encountered. This suggests to analyze such systems on the *short* timescale as if those almost invariant sets were indeed invariant, concentrating on features of the dynamics within the sets, and neglecting outside interactions. On the *long* timescale, on the contrary, the dynamics of such systems can be considered as some kind of 'flipping process' between several almost invariant 'superstates'. In this view, the first step of an analysis that separates different timescales is the identification of almost invariant sets in the dynamics, which forms the prime motivation for the work presented in this article.

As was already alluded to in the beginning, we choose the approximation of continuous dynamics through discrete Markov chains as the unifying approach to various kinds of dynamical systems. Reading a transition matrix as the adjacency matrix of a graph naturally transforms the situation into a graph theoretic framework. The problem of identifying almost invariant sets thus becomes the problem of finding partitions of a graph that are optimal with respect to certain cost functions, for which a plethora of solution or approximation methods is at hand.

The remainder of this article is organized as follows. In Sect. 2, together with some notation we introduce the basic concepts used in this work, in particular the concept of almost invariant sets that is central for the contents of this article. We formulate the problem of identifying almost invariant partitions and then reformulate it first as a discrete optimization problem and then as a graph-theoretic problem. Sect. 3 takes up the last formulation and introduces algorithmic possibilities graph theory offers for the solution of the problem. We pay particular attention to the concept of the congestion of a graph with its connection to dynamical systems concepts. Sect. 4 illustrates the use of shortest-path-algorithms for the detection of dynamically meaningful transition paths between almost invariant sets. Here the crucial point is the appropriate choice of edge weights in the graph, for which two particular examples are presented.

2 Numerical Analysis of Dynamical Systems

In this section we introduce the concept of almost invariant sets of a dynamical system, and we describe a standard framework for their numerical analysis using hierarchical set oriented methods.

2.1 Dynamical Systems and Invariant Measures

A map $f : X \to X$ on a compact subset $X \subset \mathbb{R}^n$ defines a discrete-time dynamical system with state space X. Trajectories of the system are sequences of points in X of the form

$$x_{k+1} = f(x_k),\ k = 0, 1, \dots .$$

A particularly important class of such maps f is that of time-T maps of an ordinary differential equation. In this case, under mild assumptions on the ODE (local existence and uniqueness) f is even a diffeomorphism; in the following we will assume this to be the case. Note that the state space X need not be the maximal domain of the map f. In many cases it is more appropriate to consider the dynamical system on some (invariant) subset of the maximal domain, e. g. an attractor, an ergodic component, the set of non-wandering points or the chain recurrent set.

In this work, we are interested in questions about the *global* dynamical behavior of the dynamical system $f : X \to X$. A powerful approach to these questions is to use the *transfer operator* (or *Perron-Frobenius operator*) associated with f, which, instead of generating single trajectories of points in X, describes the evolution of sets or, more generally, of (signed) measures on X. More precisely, the transfer operator associated with f is the linear operator $P : \mathcal{M} \to \mathcal{M}$,

$$(P\nu)(S) = \nu(f^{-1}(S)), \quad S \text{ measurable},$$

on the space $\mathcal{M}$ of signed measures on the Borel σ-algebra over X.

In the following we will assume that μ is an invariant measure for f, that is, the probability measure μ satisfies

$$\mu(S) = \mu(f^{-1}(S)) = (P\mu)(S) \text{ for all measurable } S \subset X ,$$

and thus is a fixed point of the transfer operator. Moreover we assume that μ is a unique so-called SRB-measure (*Sinai-Ruelle-Bowen*) in the sense that this is the only invariant measure which is robust under small random perturbations, in other words the only physically relevant invariant measure for the dynamical system f.

2.2 Almost Invariant Sets

For two measurable sets $S_1, S_2 \subset X$ we define the *transition probability* ρ from S_1 to S_2 as

$$\rho(S_1, S_2) := \frac{\mu(S_1 \cap f^{-1}(S_2))}{\mu(S_1)},$$

whenever $\mu(S_1) \neq 0$. The transition probability $\rho(S) := \rho(S, S)$ from a set $S \subset X$ into itself is called the *invariance ratio* of S. If for a number $\delta \in [0, 1]$ the relation

$$\rho(S) \geq \delta$$

holds, S is called an δ-*almost invariant set.* In practice, we will be interested in numbers $\delta = 1 - \varepsilon$ with $0 < \varepsilon << 1$. When no precise bound δ on the invariance ratio is important, we will also simply speak of almost invariant sets.

The following observation will be crucial for the rest of this article. Let S be an δ-almost invariant set, with $\delta = 1 - \varepsilon$. From $\mu(S) = \mu(f^{-1}(S))$ one has on the one hand that

$$\mu(S) = \mu(f^{-1}(S)) = \mu(S \cap f^-1(S)) + \mu(X \setminus S \cap f^{-1}(S)). \tag{2.1}$$

On the other hand,

$$\mu(X \setminus S) = \mu(X \setminus S \cap f^{-1}(S)) + \mu(X \setminus S \cap f^{-1}(X \setminus S)). \tag{2.2}$$

As S is δ-almost invariant, (2.1) means that

$$\mu(X \setminus S \cap f^{-1}(S)) \leq \varepsilon\mu(S) \tag{2.3}$$

which together with (2.2) implies that

$$\begin{aligned}\mu(X \setminus S \cap f^{-1}(X \setminus S)) &\geq \mu(X \setminus S) - \varepsilon\mu(S)\\ &= \left(1 - \varepsilon\frac{\mu(S)}{\mu(X \setminus S)}\right) \cdot \mu(X \setminus S),\end{aligned}$$

and thus

$$\rho(X \setminus S) \geq \left(1 - \varepsilon \frac{\mu(S)}{\mu(X \setminus S)}\right).$$

In short, this means that the complement of an almost invariant set is also almost invariant, with the respective invariance ratios being the more similar the closer the ratio $\frac{\mu(S)}{\mu(X \setminus S)}$ is to one.

This observation naturally motivates the problem of determining a *partition* of X consisting of almost invariant sets of roughly equal weight. For the rest of this article we will be concerned with this problem.

Although it may seem obvious, it is important to note that unlike e. g. the decomposition into ergodic components, which is unique for any given system, the decomposition of the state space into almost invariant sets will in general not be unique. In fact, any small (with respect to e. g. Lebesgue measure) variation of an almost invariant set will also be an almost invariant set, probably with a slightly different invariance ratio.

We now formally define the problem of finding a partition of almost invariant sets in the spatially continuous setting we have been considering so far. It will be reformulated twice in the course of this article, first for a spatially discretized setting and later in the language of graph theory.

Problem 2.1. For some fixed $p \in \mathbb{N}^+$ find a collection of pairwise disjoint sets $\mathcal{S} = \{S_1, \ldots, S_p\}$ with $\bigcup_{1 \leq i \leq p} S_i = X$ and $\mu(S_i) > 0$, $1 \leq i \leq p$, such that

$$\rho(\mathcal{S}) := \frac{1}{p} \sum_{i=1}^{p} \rho(S_i) \rightarrow \max .$$

2.3 Discretization of the Transfer Operator

For the detection and approximation of almost invariant sets we need to explicitly deal with the transfer operator. Since an analytical expression for it will only be derivable for none but the most simple systems, we need to derive a finite-dimensional approximation to it. The following description is based on results from e.g. [DH*97, DJ99, DFJ01, DJ02].

The basic idea for the discretization is to construct a sufficiently fine covering of the state space of the system consisting of *boxes*, i. e. generalized rectangles, by means of a *subdivision algorithm* as described in [DH97]. The basic principle of the subdivision algorithm is as follows. One starts with a box $Q \supset X$ containing the state space. Setting $\mathcal{B}_0 = \{Q\}$, a sequence $(\mathcal{B}_n)_{n \in \mathbf{N}}$ is iteratively constructed, with each iteration step $i \rightarrow i + 1$ of the iteration consisting of two parts. In the first part, from the collection $\mathcal{B}_i$ a new collection $\tilde{\mathcal{B}}_{i+1}$ is constructed by subdividing each box $B \in \mathcal{B}_i$ along a prescribed coordinate axis into two new boxes. In the second part, $\mathcal{B}_{i+1}$ is constructed as the collection of those boxes that do intersect with X, i. e.

$$\mathcal{B}_{i+1} = \left\{ B \in \tilde{\mathcal{B}}_{i+1} \mid B \cap X \neq \emptyset \right\}$$

There are several modifications of this scheme, in particular regarding the choice of boxes to be subdivided. While in the simple scheme every box is subdivided in every step, one can reduce the numerical effort by introducing an additional selection criterion that decides which boxes to subdivide and which ones to leave at the present level. More detailed expositions of the subdivision scheme can be found e. g. in [DH97, DJ99, Jun01].

Of course, in practice one cannot infinitly go on with the construction of an arbitrarily fine box covering, but will have to stop the process at some level, which results in a covering of the state space X by a finite collection $\mathcal{B} = \{B_1, \ldots, B_b\}$ of boxes, i. e.

$$X \subset \bigcup_{i=1}^{b} B_i \quad \text{with} \quad m(B_i \cap B_j) = 0 \quad \text{for } i \neq j\ .$$

Here m denotes Lebesgue measure.

To discretize the transfer operator, we replace the space $\mathcal{M}$ of signed measures over the Borel σ-algebra by the finite-dimensional space $\mathcal{M}_{\mathcal{B}}$ of signed measures on the σ-algebra that is given by the set of arbitrary unions of boxes in $\mathcal{B}$. The standard basis for this vector space is given by those measures that assign the weight 1 to precisely one box in $\mathcal{B}$ and 0 to all other boxes.

With respect to this basis, the discretized transfer operator $P_{\mathcal{B}} : \mathcal{M}_{\mathcal{B}} \to \mathcal{M}_{\mathcal{B}}$ is represented by the matrix of transition probabilities

$$P_{\mathcal{B}} = (p_{ij}), \quad \text{where} \quad p_{ij} = \frac{m(f^{-1}(B_i) \cap B_j)}{m(B_j)}, \quad 1 \leq i, j \leq b. \tag{2.4}$$

In the compution of the transition probabilities p_{ij}, the denominator poses no problem, as the boxes B_i are generalized rectangles. For the computation of $m(f^{-1}(B_i) \cap B_j)$, that is, the measure of the subset of B_j that is mapped into B_i, there are several possibilities described e. g. in [DFJ01]. A method that is often used is the Monte Carlo approach as described in [Hun94]:

$$m(f^{-1}(B_i) \cap B_j) \approx \frac{1}{K} \sum_{k=1}^{K} \chi_{B_i}(f(x_k)),$$

where the x_k's are selected at random in B_j from a uniform distribution. Evaluation of $\chi_{B_i}(f(x_k))$ only means that we have to check whether or not the point $f(x_k)$ is contained in B_i. There are efficient ways to perform this check based on a hierarchical construction and storage of the collection $\mathcal{B}$ (see [DH97, DH*97]).

Note that once we have computed an approximation $P_{\mathcal{B}}$ of the transfer operator we can obtain a discretized version of the natural invariant measure μ of the box covering $\mathcal{B}$ of A as the eigenvector to the eigenvalue 1 of $P_{\mathcal{B}}$.

As described in the beginning of this section, a region will be of interest if it is almost invariant in the sense that typical points are mapped into the region itself with high probability. Evidently the infinite dimensional optimization problem 2.1 needs to be discretized in order to be treated numerically. To this end we again restrict ourselves to subsets that are unions of elements of the partition $\mathcal{B}$. Consider the *transition matrix* $P_{\mathcal{B}}$ from (2.4). Then, our goal in the discretized setting is to solve the following problem.

Problem 2.2 (Boxes). For some $p \in \mathbb{N}^+$ find a collection of pairwise disjoint sets $\mathcal{S} = \{S_1, \ldots, S_p\}$ with $\bigcup_{1 \le i \le p} S_i = \mathcal{B}$ and $\mu(S_k) > 0$, $1 \le i \le p$, such that

$$\rho(\mathcal{S}) = \frac{1}{p} \sum_{k=1}^{p} \rho(S_k) = \frac{1}{p} \sum_{k=1}^{p} \frac{\sum_{B_i, B_j \subset S_k} p_{ij} \cdot \mu(B_j)}{\sum_{B_j \subset S_k} \mu(B_j)} \to \max \ .$$

2.4 Graph Formulation

In this section, we go one step further with reformulating the problem of finding almost invariant sets of a dynamical system. As it turns out, the optimization problem 2.2 can be translated into the problem of finding an optimal cut in a graph. To see this, we first show how the matrix describing the discretized transfer operator can also be understood as a matrix describing a directed graph, and then show that the quantity $\rho(\mathcal{S})$ can be naturally expressed in terms of edge and vertex weights of the graph.

As in the previous section, let $\mathcal{B}$ be a box covering of X. Let $G = (V, E)$ be a graph with vertex set $V = \mathcal{B}$ and directed edge set

$$E = E(\mathcal{B}) = \{(B_1, B_2) \in \mathcal{B} \times \mathcal{B} : f(B_1) \cap B_2 \neq \emptyset\} \ .$$

The function $vw : V \to \mathbb{R}$ with $vw(B_i) = \mu(B_i)$ assigns a weight to the vertices and the function $ew : E \to \mathbb{R}$ with $ew((B_i, B_j)) = \mu(B_i) p_{ji}$ assigns a weight to the edges. Furthermore, let

$$\bar{E} = \bar{E}(\mathcal{B}) = \{\{B_1, B_2\} \subset \mathcal{B} : (f(B_1) \cap B_2) \cup (f(B_2) \cap B_1) \neq \emptyset\} \ .$$

This defines an undirected graph $\bar{G} = (V, \bar{E})$ with a weight function $\bar{ew} : \bar{E} \to \mathbb{R}$ with $\bar{ew}(\{B_i, B_j\}) = \mu(B_j) p_{ij} + \mu(B_i) p_{ji}$ on the edges. The difference between the graphs G and $\bar{G}$ is that in $\bar{G}$ the edge weight between two vertices is the sum of the edge weights of the two directed edges between the same vertices in G. Thus, the total edge weights of both graphs are identical.

To formulate the problem of partitioning the state space into almost invariant sets, we will define two cost functions that describe how much weight remains within a certain set on the one hand, and how much weight changes the set of a partition on the other hand. In order to so, we first need some more notation and write $\mu(S) = \sum_{i \in S} \mu(B_i)$ for $S \subset V$, and with $\bar{S} = V \setminus S$ we further denote

$$E_{S,S} = \sum_{i,j \in S} \mu(B_i) p_{ij} \quad \text{and} \quad E_{S,\bar{S}} = \frac{1}{2} \sum_{i \in S, j \in \bar{S}} \mu(B_i) p_{ij} + \mu(B_j) p_{ji} .$$

Definition 2.3. *For a set $S \subset V$ we define*

$$C_{int}(S) = \frac{E_{S,S}}{\mu(S)}$$

as the ***internal cost*** *of S, and*

$$C_{ext}(S) = \frac{E_{S,\bar{S}}}{\mu(S) \cdot \mu(\bar{S})}$$

as the ***external cost*** *of S.*

Note that both cost functions are independent from the choice between the directed graph G or the undirected graph $\bar{G}$. We can therefore choose the simpler undirected graph, and we will do that in the following.

Definition 2.4. *For a partition $\mathcal{S} = \{S_1, \ldots, S_p\}$ of V we define*

$$C_{int}(\mathcal{S}) = \frac{1}{p} \sum_{i=1}^{p} C_{int}(S_i) \tag{2.5}$$

as the ***internal cost*** *of $\mathcal{S}$, and*

$$C_{ext}(\mathcal{S}) = \frac{\sum_{1 \leq i < j \leq p} E_{S_i, S_j}}{\prod_{i=1}^{p} \mu(S_i)} \tag{2.6}$$

as the ***external cost*** *of $\mathcal{S}$.*

Intuitively, optimal almost invariant partitions have maximal internal cost and minimal external cost. Therefore, both the internal and external costs are useful cost functions for the problem of computing almost invariant sets. However, the minimization of the external cost is not equivalent to the maximization of the internal cost. In fact, the maximization of the internal cost favors in general parts that are on average very loosely coupled to the rest of the system. However, the size of these parts can in principle become very small. On the other hand the minimization of the external cost favors balanced weighting of the components.

It is an easy task to check that $\rho(\mathcal{S}) = C_{\text{int}}(\mathcal{S})$. Thus, the optimization of problem 2.2 is identical to the optimization of the internal costs of the partition $\mathcal{S}$ in equation (2.5) written in graph notation. Therefore, we have established the following graph partitioning problem.

Problem 2.5 (Graph). For some fixed $p \in \mathbb{N}^+$ find a collection of pairwise disjoint sets $\mathcal{S} = \{S_1, \ldots, S_p\}$ with $\bigcup_{1 \leq i \leq p} S_i = V$ and $vw(S_i) > 0$, $1 \leq i \leq p$, such that

$$C_{\text{int}}(\mathcal{S}) \to \max \ . \tag{2.7}$$

One can also consider the analogous problem for the external cost function.

Problem 2.6 (Graph). For some fixed $p \in \mathbb{N}^+$ find a collection of pairwise disjoint sets $\mathcal{S} = \{S_1, \ldots, S_p\}$ with $\bigcup_{1 \leq i \leq p} S_i = V$ and $vw(S_i) > 0$, $1 \leq i \leq p$, such that

$$C_{\text{ext}}(\mathcal{S}) \to \min \ . \tag{2.8}$$

3 Computation of Almost Invariant Sets as a Graph Partitioning Problem

In this section we show how existing graph partitioning methods and tools can be applied to compute almost invariant sets. We first describe some state of the art graph partitioning heuristics. Then, we introduce the notion of *congestion* in a graph and its use in the analysis of dynamical systems, in particular in view of the problem of finding a partition of almost invariant sets. Finally, we explain how the congestion can be used as a criterion to decide on the number of almost invariant sets.

3.1 Graph Partitioning Heuristics

In this section we briefly review existing approaches and algorithms for partitioning the vertex set of a graph. As most variations of the partitioning problem – including those we are interested in in this article – are **NP**-complete, the algorithms we present are approximation algorithms that are often based on some heuristic for obtaining good partitionings.

The existing methods and tools for graph partitioning do not exactly optimize the cost functions we introduced in the previous section. We therefore give an overview of the most successful graph partitioning methods and implementations and point out the necessary modifications to such tools.

For the remainder of this section we assume that we aim to partition a the vertex set of a graph into a known number of p parts. In Sect. 3.3 we will present a way to identify this number.

We want to calculate a partition of the vertex set V of a graph $G = (V, E)$ into p parts $V = S_1 \cup \cdots \cup S_p$ such that one of our cost functions of equations (2.5) or (2.6) is optimized. However, the calculation of an optimal solution of both cost functions is **NP**-complete. Another widely discussed partitioning problem is to minimize the cut size $cut(\pi) = \sum_{1 \leq i < j \leq p} E_{S_i,S_j}$ of the partition π under the constraint that all parts have an equal (or almost equal) number of vertices. This problem is sometimes called *Balanced Partitioning Problem* and is **NP**-complete, even in the simplest case when a graph with constant vertex and edge weights is to be partitioned into two parts [GJ79].

Efficient graph partitioning strategies have been developed for a number of different applications. Efficiency and generalizations of graph partitioning methods strongly depend on specific implementations. There are several software libraries, each of which provides a range of different methods. Examples are CHACO [HL94], JOSTLE [Wal00], METIS [KK98a], SCOTCH [Pel96] or PARTY [Pre98]. The goal of the libraries is to both provide efficient implementations and to offer a flexible and universal graph partitioning interface to applications. These libraries are designed to create solutions to the balanced partitioning problem.

The tool PARTY has been developed by one of the authors and we have used it for partitioning the graphs in this paper. PARTY, like other graph partitioning tools, follows the Multilevel Paradigm. The multilevel graph partitioning strategies have been proven to be very powerful approaches to efficient graph-partitioning [Bou98, Gup97, HL95, KK98b, KK99, MPD00, PM*94, Pre00]. The efficiency of this paradigm is dominated by two parts: graph coarsening and local improvement.

The graph is coarsened down in several levels until a graph with a sufficiently small number of vertices is constructed. A single coarsening step between two levels can be performed by the use of graph matchings. A matching of a graph is a subset of the edges such that each vertex is incident to at most one matching edge. A matching of the graph is calculated and the vertices incident to a matching edge are contracted to a super-vertex. Experimental results reveal that it is important to contract those vertices which are connected via an edge of a high weight, because it is very likely that this edge does not cross between parts in a partition with a low weight of crossing edges. PARTY uses a fast approximation algorithm which is able to calculate a good matching in linear time [Pre00].

PARTY stops the coarsening process when the number of vertices is equal to the desired number of parts. Thus, each vertex of the coarse graph is one part of the partition. However, it is also possible to stop the coarsening process as soon as the number of vertices is sufficiently small. Then, any standard graph partitioning method can be used to calculate a partition of the coarse graph.

Finally, the partition of the smallest graph is projected back level-by-level to the initial graph. The partition is locally refined on each level. Standard methods for local improvement are Kernighan/Lin [KL70] type of algorithms with improvement ideas from Fiduccia/Mattheyses [FM82]. An alternative local improvement heuristic is the Helpful-Set method [DMP95] which is derived from a constructive proof of upper bounds on the bisection width of regular graphs [HM92, MD97, MP01].

As mentioned above, the tools are designed for solving the balanced graph partitioning problem. Thus, the optimization criterion is different from our cost functions of Sect. 2.4. The coarsening step of the multilevel approach does not consider the balancing of the weights of the super-vertices. It is the local refinement step which not only improves the partition locally but also balances the weights of the parts. Thus, we have to modify the local improvement part of the multilevel approach. We therefore modified the Kernighan/Lin implementation in PARTY such that it optimizes the cost-function C_{int}. Overall, we use the algorithm of Fig. 3.1 to calculate almost invariant sets.

As an example to illustrate the partitioning we consider a graph that was obtained as the discretization of the dynamics of a pentane molecule that is considered in detail in [DH*00]. This molecule has two dihedral angles which are used as state space coordinates. The left plot of Fig. 3.2 shows the box collection and all transitions between boxes. As we will see in sect. 3.3, it is adequate to partition the graph into five or seven parts. These are shown in the center and right plots of Fig. 3.2.

Partition graph $G_0 = (V_0, E_0)$ into p parts

$i = 0$;
WHILE ($|V_i| > p$)
  calculate a graph matching $M_i \subset E_i$;
  use M_i to coarse graph $G_i = (V_i, E_i)$ to a graph $G_{i+1} = (V_{i+1}, E_{i+1})$;
  $i := i + 1$;
END WHILE
let π_i be a p-partition of G_i such that each vertex is one part;
WHILE ($i > 0$)
  $i := i - 1$;
  use M_i to project π_{i+1} to a p-partition π_i of G_i;
  modify the partition π_i on G_i locally to optimize $C_{\text{int}}(\pi_i)$;
END WHILE
output π_0.

Fig. 3.1. Computing a graph partitioning with the multilevel approach.

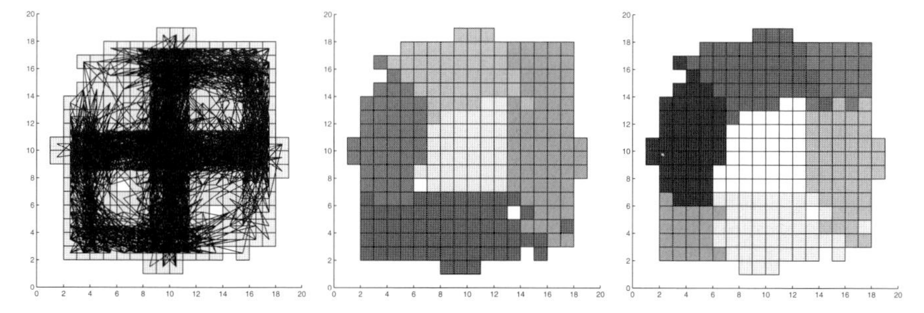

Fig. 3.2. Partition of graph describing the dynamics Pentane300. Left: The graph with all edges. Center: Partition consisting of five parts with $C_{\text{int}} = 0.980$. Right: Partition consisting of seven parts with $C_{\text{int}} = 0.963$. (See page 695 for a colored version of the figure.)

3.2 Congestion

Standard graph partitioning methods partition the graph into a predefined number of parts. However, if we do not know the resulting number of almost invariant sets a priori, we need to find mechanisms which help us to decide on a natural number of parts. As we will see, such a mechanism can be devised on the basis of the concept of *congestion* of a graph.

Intuitively, the congestion of a graph is a quantity that can be used to identify 'bottlenecks' in the graph, i. e. edges that connect subgraphs which have relatively many internal and relatively few external edges. This description already explains the relevance of the congestion for the problem of finding almost invariant sets of a dynamical system. The concept is based on the idea of so-called multi-commodity flows on the graph (see e.g. [Lei92, Sin93]).

In the following, we first formally define the congestion. As it is often not feasibly to precisely compute this quantity for a given graph, we will then shortly discuss heuristics for an approximation of the congestion. These heuristics will produce upper bounds on the congestion, which we can use in a lower bound on the external cost of a bisection of a graph that is discussed immediately afterwards.

The first concept we need for the definition of the congestion is that of a *single-commodity flow* in a graph. Such a flow may be imagined as a way of describing the transport of a certain quantity c of some good from a source s to a target t through a network of roads that is given by the graph.

Definition 3.1. *Let $G = (V, E)$ be an undirected graph with the vertex set $V = \{1, \ldots, d\}$. Let $s, t \in V$ be vertices of G with $s \neq t$, let $c \in \mathbf{R}$. A* ***single-commodity flow*** *f of the commodity c from s to t on G is a function $f : V \times V \to \mathbf{R}$ such that*

i. $f(v, w) = 0$ *for all* $\{v, w\} \notin E$ *(flow on edges only),*
ii. $f(v, w) = -f(w, v)$ *for all* $v, w \in V$ *(symmetry),*
iii. $\sum_{w \in V} f(v, w) = 0$ *for all* $v \in V \backslash \{s, t\}$ *(flow conservation) and*
iv. $\sum_{w \in V} f(s, w) = \sum_{w \in V} f(w, t) = c.$

As the next step, we will define *multi-commodity flows* which generalize single-commodity flows. Intuitively, a multi-commodity flow describes the transport of certain commodities from every vertex to every vertex of the graph. The formal definition is as follows.

Definition 3.2. *Let $c_{s,t} \in \mathbf{R}$ for $1 \leq s, t \leq d$. A* **multi-commodity flow** *F of the commodities $c_{s,t}$ on G is a function $F : V \times V \times V \times V \to \mathbf{R}$ such that for each pair $(s, t) \in V \times V$, the function $F(s, t, \cdot, \cdot)$ is a single-commodity flow of the commodity $c_{s,t}$ from s to t on G.*

With these concepts, we are in a position that allows us to introduce the *congestion* of a graph as we are using it in this work.

Definition 3.3. *Let $G = (V, E)$ be an undirected graph with the vertex set $V = \{1, \ldots, d\}$, vertex weights μ_i^d for $i \in V$, and edge weights A_{ij} for $\{i, j\} \in E$. For $s, t \in V$, let $c_{s,t} = \mu_s^d \cdot \mu_t^d$. Denote by $\mathcal{F}$ the set of all multi-commodity flows of the commodities $c_{s,t}$ on G. For an edge $\{v, w\} \in E$ and $F \in \mathcal{F}$, the* ***edge congestion of*** *$\{v, w\}$* ***in*** *F is*

$$\mathrm{cong}(\{v, w\}, F) = \frac{\sum_{1 \leq s,t \leq d} |F(s, t, v, w)|}{A_{vw}} . \tag{3.1}$$

The **flow congestion of** *F on G is*

$$\mathrm{cong}(F) = \max_{\{v,w\} \in E} \mathrm{cong}(\{v, w\}, F) , \tag{3.2}$$

and finally the **congestion** *of the graph G is*

$$\mathrm{cong}(G) = \min_{F \in \mathcal{F}} \mathrm{cong}(F) \ . \tag{3.3}$$

In principle, other choices of commodities $c_{s,t}$ than those used in this definition are also possible. However, the choice we made here seems the most appropriate for the use we will make of the congestion in this paper, as will become clear in the following section.

Approximation of the congestion

The computation of the congestion $\mathrm{cong}(G)$ of a graph can be costly and is often infeasible. We will see in the next section that the congestion can be used to bound the external cost of a partition. However, that bound holds for the congestion $\mathrm{cong}(F)$ of any multi-commodity flow F. Thus, a sub-optimal flow produces a sub-optimal, but still valid bound. In this section we discuss heuristics for calculating a flow with a small flow congestion.

There are some hints of how to construct a flow with a small flow congestion. Clearly, cycles in the flow should be avoided. Furthermore, it is easy to imagine that a low-congestion flow should - at least primarily - go along shortest paths between the pairs of vertices. Here, the length of a path is the sum of the reciprocal values of the edge weights along the path.

A straightforward method is to send the flow along shortest paths only. If more than one shortest paths exist, the flow value can be split among them. If the edge weights are constant, all shortest paths can be calculated in time $O(|V| \cdot |E|)$. This can be done by $|V|$ independent Breath-First searches in time $O(|E|)$ each. If the edge weights are non-negative, all shortest paths can be calculated in time $O(|V| \cdot (|V| \cdot \log|V| + |E|))$, e.g. with $|V|$ runs of the single-source shortest path Dijkstra algorithm using Fibonacci heaps. We refer to [CLR90] for a deeper discussion of shortest paths algorithms.

A different method is to consider n commodities at a time. For each vertex v_s, $1 \leq s \leq n$, consider the commodities $c_{s,t}$, $1 \leq t \leq n$, i.e. all commodities with v_s as the source. For each source v_s we calculate a flow F_s which transports all commodities from v_s to all other vertices, i.e. it replaces n single-commodity flows such that $F_s(v, w) = \sum_{t=1}^{n} F(s, t, v, w)$. F_s is a single-source, multiple-destination commodity flow. Definitions (i.) and (ii.) for the single-commodity flow (Definition 3.1) remain unchanged whereas the definitions of (iii.) and (iv.) are replaced by

v. $\sum_{w \in V} F_s(v_t, w) = -c_{s,t}$ for all $t \neq s$ (from source s to target t) and

vi. $\sum_{w \in V} F_s(v_s, w) = \sum_{t=1}^{n} c_{s,t} - c_{s,s}$ (from source s to all targets t except to source s itself).

It is left to show how we calculate a single-source flow F_s. We use algorithms from diffusion load balancing on distributed processor networks for this task, see e.g. [DFM99, EF*99, EMP00]. Here, the problem is to balance the work load in a distributed processor network such that the volume of data movement is as low as possible. We use these algorithms in the setting that the vertex v_s models a processor with load $\sum_{t=1}^{n} c_{s,t}$ and all other vertices model a processor with no load. Furthermore, the processors are heterogeneous with a capacity of $c_{s,t}$ for processor v_t [EMP00]. The diffusion algorithms calculate a balancing flow such that each vertex/processor v_t gets a load of $c_{s,t}$. That is exactly what we need in our context. The resulting balancing flow has a nice property: it is minimal in the l_2-norm, i.e. the diffusion algorithms minimize the value $\sqrt{\sum_{1 \le v,w \le n} |F_s(v,w)|}$. This ensures that there are no cycles in any flow F_s. Furthermore, the flows are not restricted along shortest paths and can avoid high traffic along shortest paths. However, the flows are still favored to be along reasonably short paths. Thus, it is expected that the overall edge congestion of the resulting flow is reasonably small and that the flow congestion is close to the congestion of the graph.

The PARTY library includes efficient code of a variety of diffusion algorithms. We use them to calculate the single-source, multiple-destination flows for each source s and then add up the values to get the multi-commodity flow. Numerical experiments indicate that the resulting flow is indeed very small.

An Example: Pentane

To illustrate the meaning of the congestion in the context of dynamical systems, we again consider as example the graph describing the dynamics of a pentane molecule from [DH*00]. The graph corresponding to this dynamical system is shown in the left part of Fig. 3.3. The middle and the right part of Fig. 3.3 show the edges with low and with high congestion, respectively. The coloring of the boxes indicates the partition into seven almost invariant sets.

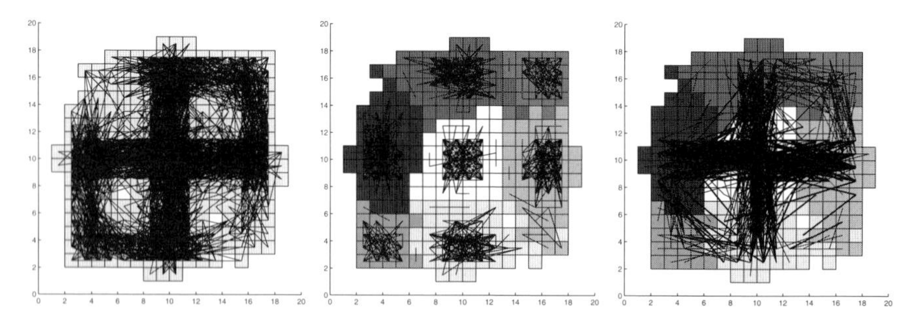

Fig. 3.3. Congestion of the Pentane. Left: all transitions. Center: only transitions with a low congestion. Right: only transitions with a high congestion. (See page 695 for a colored version of the figure.)

It can be observed that – as expected – edges with low congestion can mainly be found inside the almost invariant sets. On the other hand edges between different almost invariant sets have a large congestion. Thus, a high congestion indicates that there are at least two regions in the phase space which are only loosely coupled. As we will see in Sect. 3.3, this observation is the basis for using the congestion as an identifier for the number of almost invariant sets which have to be approximated.

The congestion bound on C_{ext}

We will now see how the concept of congestion of a graph can be used for the analysis of dynamical systems, in particular for the problem of finding a partition into almost invariant sets.

The congestion can be used to derive a lower bound on $C_{ext}(S)$ for any $S \subset I$. As before, we use multi-commodities $c_{s,t} = vw(v_s) \cdot vw(v_t)$ for each source $v_s \in V$ and each destination $v_t \in V$. On the one hand side, for any multi-commodity flow with commodities $c_{s,t}$ as given above, at least $\mu(S)\cdot\mu(\bar{S})$ 'units' have to cross the cut between S and $\bar{S}$, and as many in the opposite direction. On the other hand, with $E_{S,\bar{S}}$ being the sum of edge weights of edges crossing the cut, by definition of the congestion, at most $\text{cong}(G) \cdot E_{S,\bar{S}}$ units can cross the cut. Therefore we have $2 \cdot \mu(S) \cdot \mu(\bar{S}) \leq \text{cong}(G) \cdot E_{S,\bar{S}}$, which at once gives the important inequality

$$C_{\text{ext}}(S) = \frac{E_{S,\bar{S}}}{\mu(S) \cdot \mu(\bar{S})} \geq \frac{2}{\text{cong}(G)}. \tag{3.4}$$

Obviously, a high and tight lower bound can only be achieved with a small congestion. Although the congestion can be computed in polynomial time, it remains to be very costly. Nevertheless, the congestion can be approximated by the congestion of any flow. Heuristics for calculating a small congestion were discussed above. Further discussion of lower bounds based on different variations of multi-commodity flows can be found in [Sen01].

3.3 Identification of the Number of Almost Invariant Sets

We now discuss the problem of identifying an appropriate number of almost invariant sets a given space should be partitioned into.

Informally, we want to determine a number $p \in \mathbf{N}$ such that there is a partition of V consisting of p parts $V = S_1 \cup \dots \cup S_p$ with a high internal cost. As the internal cost is monotonically decreasing for the optimal partitions with an increasing number of parts p, we are looking for a number p such that an (almost) optimal partition into $p-1$ parts has an only slightly higher internal cost than an (almost) optimal partition consisting of p parts while (almost) optimal partitions into $p+1$ parts have a substantially lower internal cost. The idea is that if we try to split a *compact set* (one with a small congestion), the internal cost will drop substantially. Thus, our strategy is to start with the whole vertex set as the initial set and keep on bisecting the sets until they become compact sets. This leads us to a strategy of how to determine the number of compact parts. It can be phrased as a general method:

Recursively bisect the vertices of the graph until all parts are compact.

Recursive bisection is a widely used technique in graph partitioning. Although there are many partitioning methods which directly partition the vertices of a graph into a number of parts, we cannot apply them here, because we do not know the number of parts a priora priori. Aditionally, the graph bisection methods are often much more efficient than their generalized counterparts.

We have seen in the previous section that the congestion of a graph can be used to derive a lower bound on the external cost of a set bisection, i.e. a large congestion indicates a large external cost and, therefore, also a small internal cost. We use the congestion in order to decide whether a set is compact or not. In our experiments we use a threshhold of 5 and say that if the congestion is larger than 5 than the set has at least one bottleneck and is not compact. Thus, our strategy is to subdivide the parts until all parts have a congestion of at most 5.

One needs to solve two tasks in order to follow the recursive bisection strategy and we described both in the previous sections. We use the methods described in Sect. 3.1 to recursively calculate bisections of a graph. Furthermore, we use the congestion in order to indicate whether a part is compact or not.

Figure 3.4 illustrates the recursive bisection process in the partitioning of the graph in the pentane molecule example from [DH*00] which we already used before. From the top left to the bottom right picture, the levels of the recursive procedure are shown rowwise. As we can observe in the first picture, the first bisection results in one part of 43 boxes and a very low congestion of 0.88. However, the other part consisting of 212 boxes has a high congestion of 168.82. We continue to bisect parts with a congestion value higher than 5. Thus, after a total of 4 bisection levels we get a partition into 7 parts and the highest congestion of any part is 3.67 .

4 Short Paths

Broadly speaking, the previous section has been concerned with the use of graph partitioning algorithms to obtain information about almost invariant sets of a dynamical system. In this section, we will consider the use of another class of graph algorithms, namely that of shortest-path algorithms, in the context of dynamical systems. We will see that such algorithms can be used to compute discrete approximations to transition paths of a dynamical system. The crucial question for this undertaking is the choice of a weight function that defines the *length* of an edge. We will present two such functions for different

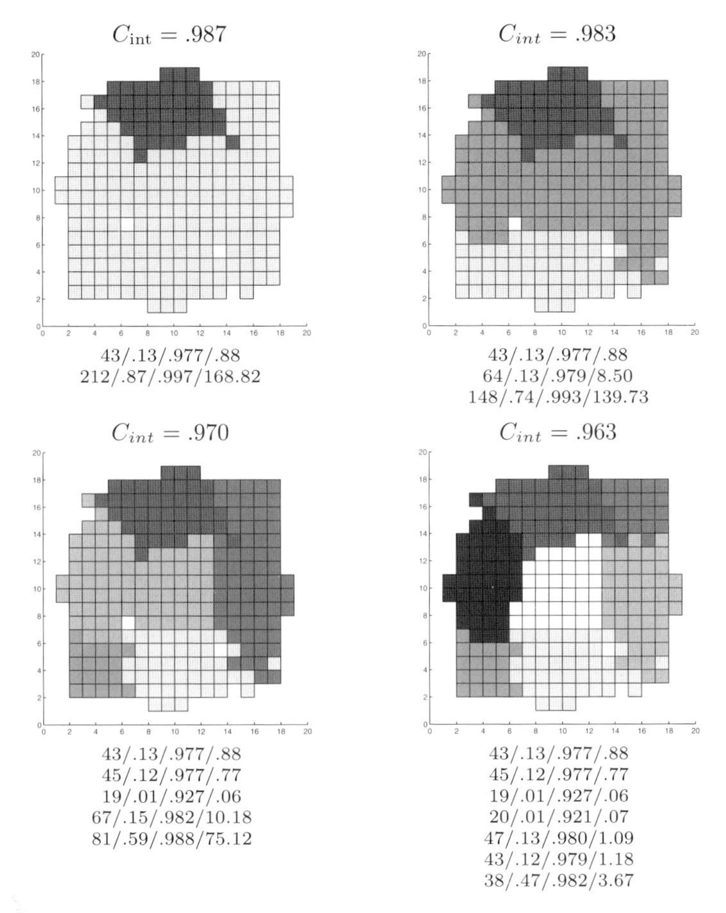

Fig. 3.4. Recursive bisection of the graph for the pentane example. The values indicate: number of boxes / invariant measure / internal cost C_{int} / congestion of subgraph. The graph has 255 vertices and a congestion of 139.67. (See page 696 for a colored version of the figure.)

types of dynamics. Both have a natural motivation, and we will compare the results of both approaches.

Definition 4.1. *Let $G = (V, E)$ be any graph with edge weights given by a function $ew : E \to \mathbf{R}$. A sequence $[(v_1, v_2), (v_2, v_3), \dots (v_i, v_{i+1})]$ of edges $(v_j, v_{j+1}) \in E$, $1 \le j \le i$, is called a* ***path*** *from vertex v_1 to vertex v_{i+1} of size i and of length $l = \sum_{j=1}^{i} ew((v_j, v_{j+1}))$. A* **shortest path** *from a vertex v_s to a vertex v_d is a path of minimum length from v_s to v_d. The* **distance** $\text{dist}(v_s, v_d)$ *from v_s to v_d is the length of a shortest path from v_s to v_d.*

The standard algorithm used for computing shortest paths in graphs is the Dijkstra algorithm. It solves the so called *Single Source Shortest Path Problem* where the shortest paths from one source vertex $v_s \in V$ to all other vertices $v \in V$ have to be determined. The *Single Source, Single Destination Shortest Path Problem* is a special case in which only one path from v_s to a designated destination vertex v_d has to be determined. In both cases the runtime of the Dijkstra algorithm is $O(|V|\log(|V|) + |E|)$. For a profound discussion of this standard algorithm we refer to e. g. [CLR90], in the following we will only roughly sketch its basic principle.

Given a vertex v_s as starting vertex, the algorithm maintains a list of distances to v_s assigned to every other vertex that is initialised with the value ∞ and in the end contains the lengths of the shortest paths from v_s to any vertex. In the first step, the distances of all neighbors of v_s are set to the weight of the edge connecting them to v_s. These vertices form the initial *halo set*, i.e. they are the vertices for which one path from v_s is known but it is not known whether this path is a shortest path. In the main loop of the algorithm, it removes a vertex v_{min} with the minimum known distance from the halo set, and considers all neighbors of v_{min}. If a neighbor is also in the halo set, the algorithm checks whether a path through v_{min} would result in a distance from v_s less than the current known distance. If a neighbor is not yet in the halo set, it is added to it, with its distance value being the sum of the distance of v_{min} and the length of the edge connecting the neighbor to v_{min}. The algorithm terminates when a prescribed target vertex is reached or when the halo set becomes empty.

By two slight modifications, the Dijkstra algorithm can be generalized to find a shortest path from any vertex of a source set $V_s \subset V$ to any vertex of a destination set $V_d \subset V$. The first modification is that in the initialization step all vertices of V_s are assigned the distance value 0, and that all neighbors of vertices from V_s that do not themselves belong to V_s form the initial halo set. The second modification is that in the main loop, every time a vertex v is removed from the halo set, it is checked whether $v \in V_d$.

4.1 Several Short Paths

For the purposes of this article, the purely graph theoretic consideration of shortest paths we have seen until now has to be extended by some ideas related to the specialized setting of graphs describing (temporal and/or spatial) discretizations of continuous dynamical systems. In particular we have in mind the fact that the numerical realizations of these graphs necessarily come with a discretization error which makes it doubtful whether the notion of *the* shortest path between two vertices v_s and v_d is really a meaningful quantity in our applications – even leaving out the possible existence of several shortest paths. Therefore, we are not only interested in one (or all) precisely shortest paths, but we are also interested in all paths which are only slightly longer than a path with the shortest length.

For this reason, we want to calculate all paths from a vertex v_s to a vertex v_d which have a length of at most $(1 + \epsilon)\,\mathrm{dist}(v_s, v_d)$. In order to do so, we need to apply the Dijkstra algorithm only two times. Firstly, we calculate all distances from v_s to all other vertices and denote these distances by $\mathrm{dist}_1(v)$ for all vertices $v \in V$. Among all distances this also includes the distance between v_s and v_d. Secondly, we consider a new graph $G_r = (V, F)$ where F consists of the edges in E with direction reversed. Then, we calculate all distances from v_d to all other vertices in G_r, and denote these distances by $\mathrm{dist}_2(v)$ for all vertices $v \in V$. Note that $dist_2(v)$ is also the distance from v to v_d in G for any vertex $v \in V$.

It is now simple to decide whether or not an edge (v_i, v_j) lies on a path between v_s and v_d of length at most $\mathrm{dist}(v_s, v_d)(1+\epsilon)$. Such a path has to consist of three parts: a path from v_s to v_i, the edge (v_i, v_j) itself and a path from v_j to v_d. The shortest length for the first part is $\mathrm{dist}_1(v_i)$ and the shortest length of the last part is $\mathrm{dist}_2(v_j)$. Thus, an edge (v_i, v_j) lies on a path between v_s and v_d of length at most $(1+\epsilon)\,\mathrm{dist}(v_s, v_d)$ if and only if

$$\mathrm{dist}_1(v_i) + ew((v_i, v_j)) + \mathrm{dist}_2(v_j) \leq (1+\epsilon)\,\mathrm{dist}(v_s, v_d)\ .$$

The result is a subset $E_{sp} \subset E$ of edges belonging to the short paths.

4.2 Choices of edge weights

Until now, we have considered graphs with edge weights $ew((v_i, v_j)) = \mu_i^d P_{ji}$ that where introduced in Sect. 2.4. While this weighting is appropriate for graph partitioning algorithms which aim to minimize the internal cost of a partition, it is less useful for shortest path algorithms.

Instead, we want to use an edge weight such that the length of a path $((v_1, v_2), (v_2, v_3), \ldots, (v_i, v_{i+1}))$ from a vertex v_1 to a vertex v_{i+1} reflects the product of the probabilities to choose the next edge along the path, i.e. $\prod_{j=1}^{i} P_{j+1,j}$. Thus, a high probability to go along this path should be reflected by a short path and vice versa.

We can do this by using shortest paths algorithms on the graph with edge weights

$$ew(v_i, v_j) := \frac{1}{\log(P_{ji})} = -\log(P_{ji})\ .$$

Then the length of a path $((v_1, v_2), (v_2, v_3), \ldots, (v_i, v_{i+1}))$ is

$$l = \sum_{j=1}^{i} ew((v_j, v_{j+1})) = -\log(\prod_{j=1}^{i} P_{j+1,j})\ .$$

Note that the product of probabilities on the right hand side of this equation is the probability that the Markov chain described by the matrix P produces the considered sample path when started in v_1.

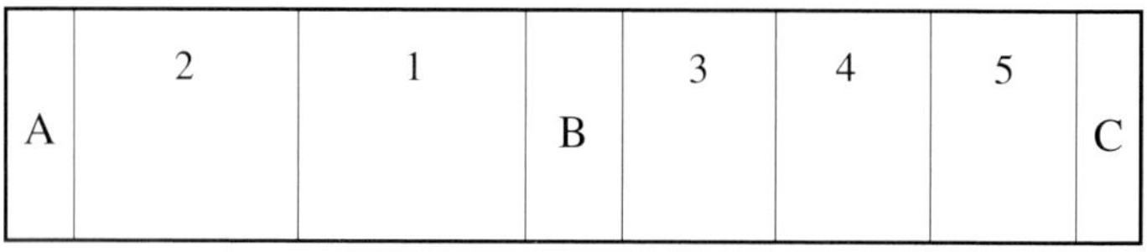

A	2	1	B	3	4	5	C

Fig. 4.1. Schematic representation of the example. The rectangular domain of the pure diffusion with a reflecting boundary is discretized into 8 boxes.

Motivational example

As an example for a type of dynamics for which the edge weight introduced in the previous section seems inappropriate, we consider diffusion in a flat potential landscape (i. e. with $V \equiv 0$, see below). We choose a rectangular domain and apply reflecting boundary conditions. In Fig. 4.1 we give a schematic picture of the situation. Suppose, we start the process in box B. From the symmetry of the domain and the nature of diffusion, it is clear that the probability to end up in box A is the same as to reach the box C, namely 0.5. But the particular decomposition of the domain, with the boxes 3, 4 and 5 having only two thirds of the width of the boxes 1 and2, implies that the transition probabilities between boxes 1 and 2 on the one hand side and between boxes 3 and 4, and 4 and 5 on the other hand are all equal. This means that the discrete path $(B, 3, 4, 5, C)$ is less probable than the path $(B, 1, 2, A)$, in contradiction to the continuous picture.

Free Energy

Another important quantity to characterize the transition behavior of a dynamics in a complex system is the free energy barrier which the dynamics has to overcome on its way between two almost invariant sets. Suppose we consider the Smoluchowsky dynamics generated by the stochastic differential equation

$$\dot{X}(t) = -\nabla V(X(t)) + \sqrt{2\beta^{-1}}\dot{W}(t) \tag{4.1}$$

where $X(t) \in \mathbf{R}^n$, W is a standard Browian motion, $V : \mathbf{R}^n \to \mathbf{R}$ is a potential and β is a parameter that is referred to as the inverse temperature. The probability to find the equilibrated system in a certain region, say $C \subset \mathbf{R}^n$, is given by

$$\mu(C) = Z^{-1} \int_C \exp(-\beta V(x)) \mathrm{d}x \tag{4.2}$$

where Z is the normalization factor. The traditional way to define the free energy is by means of the marginal density with respect to a given reaction coordinate $\xi : \mathbf{R}^n \mapsto \mathbf{R}$

$$Z(q) = Z^{-1} \int_{\mathbf{R}^n} \exp(-\beta V(x)) \delta(\xi(x) - q) \mathrm{d}x. \tag{4.3}$$

Then the free energy is given by the logarithm of the partition sum $Z(q)$:

$$F(q) = -\beta^{-1} \ln Z(q). \tag{4.4}$$

Discrete free energy

Now consider a reversible Markov process on a finite state space $S = \{s_1, \ldots, s_n\}$ and let $\pi = (\pi_1, \ldots, \pi_n)$ its unique stationary distribution. Analogously to the continuous case, we define the free energy in terms of a probability distribution

$$F(i) = -\ln \pi_i > 0, \quad i \in S. \tag{4.5}$$

New weight

Given two disjoint sets $A, B \subset S$, we are interested in the state space path which crosses the lowest free energy barriers on its way from A to B. To this end, we introduce the new edge weights

$$w(i,j) = |F_j - F_i|. \tag{4.6}$$

Let $p = (i_1, \ldots, i_s)$ be a path such that

$$F_{i_j} \leq F_{i_{j+1}} \Leftrightarrow \pi_{i_j} \geq \pi_{i_{j+1}}, \quad j = 1, \ldots, s-1 \tag{4.7}$$

then the length of the path is

$$l(p) = \sum_{j=1}^{s-1} w_{i_j, i_{j+1}} = F_{i_s} - F_{i_1}. \tag{4.8}$$

This means that the weight of such a path is simply given by the free energy difference between the last and the first state of the path. Moreover, if we fix the states i_1 and i_s, then all paths connecting these two states and satisfying (4.7), have the same length. Next consider a path $p = (i_1, \ldots, i_n)$ which can be decomposed into two parts $p_1 = (i_1, \ldots, i_s)$ and $p_2 = (i_s, \ldots, i_n)$ such that

$$\begin{cases} F_{i_j} \leq F_{i_{j+1}}, & j = 1, \ldots, s-1 \\ F_{i_j} \geq F_{i_{j+1}}, & j = s, \ldots, n-1 \end{cases}. \tag{4.9}$$

One immediately verifies that the length of such a path is given by

$$l(p) = 2F_{i_s} - (F_{i_1} + F_{i_n}) \geq 0. \tag{4.10}$$

Again, the length of the path depends only on free energy differences, namely the barriers $F_{i_s} - F_{i_1}$ and $F_{i_s} - F_{i_n}$. Consequently, if we fix the states i_1 and i_n then the shortest path between i_1 and i_n w.r.t. to the weights (4.6) is the one which crosses the lowest free energy barriers.

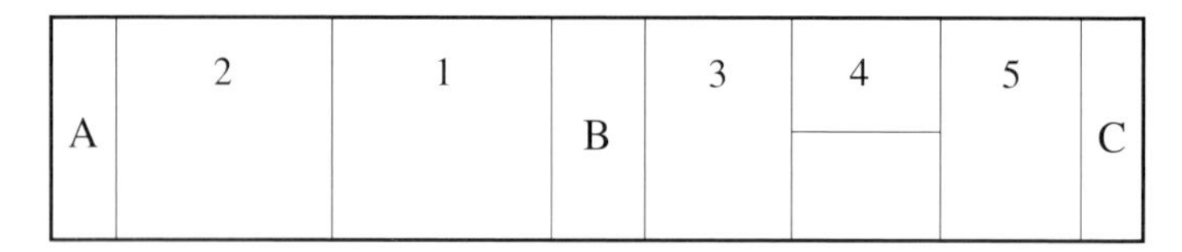

Fig. 4.2. Schematic representation of the modified example.

Interpretation

The first observation is that the new weight allows to find barriers between invariant sets. Furthermore, it is more insensitive with respect to the underlying discretization. To explain this issue, let us go back to example in 4.2. The probabilities to find the equilibrated dynamics in the boxes 1 and 2 are equal, and so are the probabilities of the boxes 3, 4 and 5. That means that there are no free energy barriers for the dynamics on its way to the box A or C, respectively, conditioned on starting in the box B and thus the paths are equal. But this exactly results from the new weight: The lengths of both paths are equal,

$$l(B, 1, 2, A) = F_A - F_B = F_C - F_B = l(B, 3, 4, 5, C) = const.$$

In the previous example the volumes of the boxes are equal. What happens if the volumes of the boxes differ? Suppose we decompose the box 4 into two boxes with equal volume. In Fig. 4.2 we give a schematic representation of the modified example. For this discretization both weights would tell that the path $(B, 1, 2, A)$ is the preferred one since both the transition probabilities w.r.t. the box 4 and its stationary distribution decrease. But nevertheless, the new weight is more insensitive to the underlying discretization because the length of a path does not depend on the *entire* path but only on the barriers which the path overcomes.

4.3 Modified update step in the Dijkstra algorithm

The twofold contribution of a barrier can be seen as reflecting the reversibility of the process. If it is necessary to know the value of the sum of barriers only in one direction then this can be done by modifying the update step in the Dijkstra algorithm. Let v be the current node in the main loop of the Dijkstra algorithm and let k be a neighbor which has to be updated. Instead of using the weight $w(v, k)$ we propose to use the weight $\tilde{w}(v, k) = \max\{0, F_k - F_v\}$ for updating the distance of the node k. Doing so, the bidirectional Dijkstra algorithm with the modified update-step computes the same paths as the unmodified one, but the length of a path only depends on the barrier in one direction.

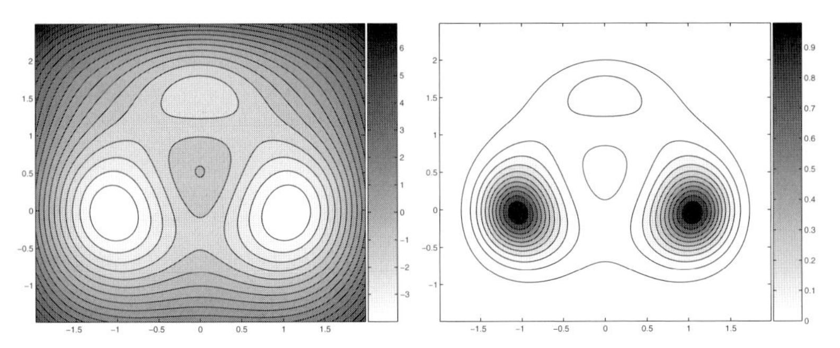

Fig. 4.3. Left :Contour plot of the three-hole potential. Right: Equilibrium distribution at $\beta = 1.67$.

4.4 Illustrative example: diffusion in a potential landscape

In the following example we study the behavior of the bidirectional Dijkstra-Algorithm in the presents of two possible transition channels. We compare the families of transition paths resulting from the probability weight and the free energy weight. For this purpose we choose the three-well potential

$$
\begin{aligned}
V(x,y) = 3e^{-x^2-(y-\frac{1}{3})^2} - 3e^{-x^2-(y-\frac{5}{3})^2} \\
-5e^{-(x-1)^2-y^2} - 5e^{-(x+1)^2-y^2}
\end{aligned}
$$

which already has been investigated in [PS*03]. As one can see in the left picture of Fig. 4.3 the two deep minima at $(-1,0)$ and $(1,0)$ are connected by an upper and a lower channel. We choose the inverse temperature $\beta = 1.67$ such that despite the dominance of the two deep minima there is still a little probability to find the dynamics in the shallow minimum around $(0,\frac{5}{3})$. The dynamical bottlenecks in the upper channel are two saddle points with equal potential energy whereas the dynamics in the lower channel only has to overcome one saddle point with potential energy higher than that of the upper ones.

The following experiments are based on a discrete realization of the dynamics given in (4.1) for the inverse temperature $\beta = 1.67$. To be more precise, we use the Euler-Maruyama-scheme

$$
x_{n+1} = x_n - \nabla V(x_n)\tau + \sqrt{2\beta^{-1}\tau}\,\eta_n, \quad n = 0,\dots,N-1 \tag{4.11}
$$

to discretize the SDE (4.1) in time, where $x_n \in \mathbf{R}^2$, τ is the time step and η_n denotes a realization of a gaussian random variable with mean zero and variance one. We choose the time step $\tau = 10^{-3}$ and generate a trajectory of total length τN with $N = 10^6$.

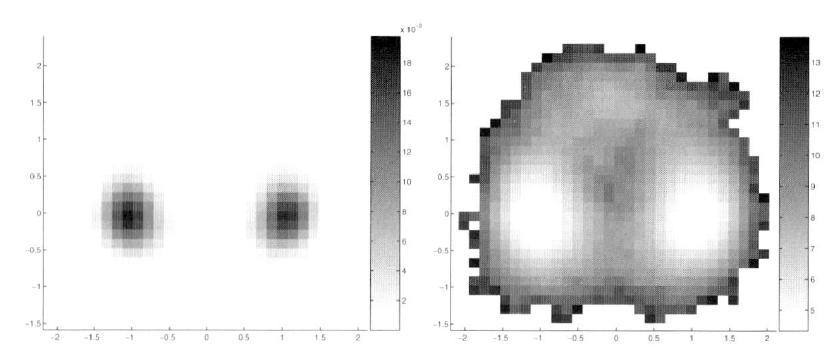

Fig. 4.4. Left: Stationary distribution of the transition matrix. Right: The free energy of the boxes.

Equidistant discretization

Next we decompose the rectangular domain into 30×30 equidistantly spaced boxes. In the left picture of Fig. 4.4 we depict the stationary distribution of the reversible transition matrix and in the right picture we plot the corresponding free energy.

In Fig. 4.5 we illustrate the results of bidirectional Dijkstra for both weights, the weights which are based on the transition probability and the weights (4.6) incorporating the free energy. For all computations we use the same sets A and B which consists of boxes covering the two deep minima, respectively. In the two columns of Fig. 4.5 we draw the edges which belong to the family of shortest paths between the sets A and B. From top to bottom we increase the parameter ϵ which results in increasing number of edges. The left column shows the edges of the most probable paths, whereas in the right column we draw the edges of paths which crosses the lowest free energy barriers.

As one can see, both methods detect for small ϵ the lower transition channel as the preferred one. But with increasing ϵ the families of transition paths differ. The family of transition paths resulting from transition probability weight for $\epsilon = 0.6$ includes paths which overcome the big barrier in the middle of the potential. Since the probability that the dynamics leaves the basin of attraction scales exponentially with the barrier which has to be overcame, the paths over the big barrier make no sense. Although the dynamics could get trapped in the upper shallow minima, the two lower saddle points rather should allow the dynamics to make transition than to go over the big barrier in the middle. This behavior is reflected by the free energy weight as can be seen in the last picture of the right column.

Acknowlegdments. This work has been supported by the DFG Priority Program 1095 "Analysis, Modeling and Simulation of Multiscale Problems" under grants number De 448/8 and Schu 1368/2.

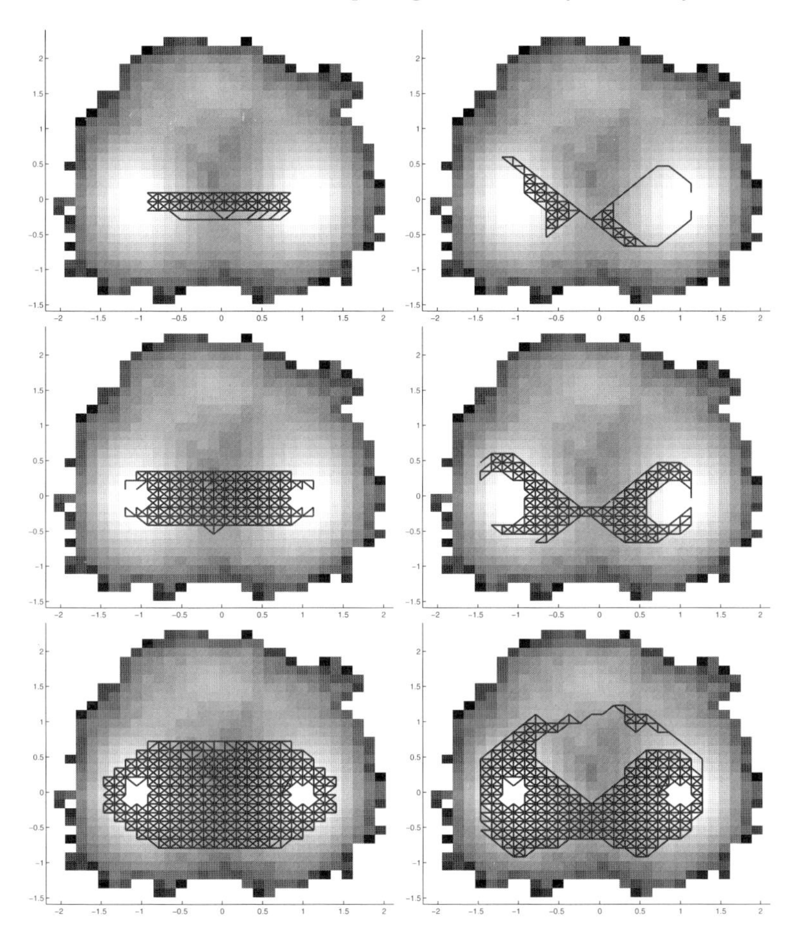

Fig. 4.5. Left column: Family of most probable paths between the set A (left minimum) and the set B (right minimum). From the top to the bottom we choose $\epsilon = 0.1, \epsilon = 0.3$ and $\epsilon = 0.6$. Right column: Family of paths which crosses the lowest free energy barriers. From the top to the bottom we choose $\epsilon = 0.01, \epsilon = 0.05$ and $\epsilon = 0.13$. (See page 697 for a colored version of the figure.)

References

[Bou98] N. Bouhmala. Impact of different graph coarsening schemes on the quality of the partitions. Technical Report RT98/05-01, University of Neuchatel, Department of Computer Science, 1998.

[CLR90] T. Cormen, C. Leiserson, and R. Rivest. *Introduction to Algorithms.* MIT Press, 1990.

[DFJ01] M. Dellnitz, G. Froyland, and O. Junge. The algorithms behind GAIO – set oriented numerical methods for dynamical systems. In *Ergodic Theory, Analysis, and Efficient Simulation of Dynamical Systems*, pages 145–174, 2001.

[DFM99] R. Diekmann, A. Frommer, and B. Monien. Efficient schemes for nearest neighbor load balancing. *Parallel Computing*, 25(7), 789–812, 1999.

[DH97] M. Dellnitz and A. Hohmann. A subdivision algorithm for the computation of unstable manifolds and global attractors. *Numerische Mathematik*, 75, 293–317, 1997.

[DH*00] P. Deuflhard, W. Huisinga, A. Fischer, and C. Schütte. Identification of almost invariant aggregates in reversible nearly uncoupled Markov chains. *Lin. Alg. Appl.*, 315, 39–59, 2000.

[DH*97] M. Dellnitz, A. Hohmann, O. Junge, and M. Rumpf. Exploring invariant sets and invariant measures. *Chaos*, 7(2), 221–228, 1997.

[DJ99] M. Dellnitz and O. Junge. On the approximation of complicated dynamical behavior. *SIAM J. Numer. Anal.*, 36(2), 491–515, 1999.

[DJ02] M. Dellnitz and O. Junge. Set oriented numerical methods for dynamical systems. In B. Fiedler, G. Iooss, and N. Kopell, editors, *Handbook of Dynamical Systems II: Towards Applications*, pages 221–264. World Scientific, 2002.

[DMP95] R. Diekmann, B. Monien, and R. Preis. Using helpful sets to improve graph bisections. In D. Hsu, A. Rosenberg, and D. Sotteau, editors, *Interconnection Networks and Mapping and Scheduling Parallel Computations*, volume 21 of *DIMACS Series in Discrete Mathematics and Theoretical Computer Science*, pages 57–73. AMS, 1995.

[EF*99] R. Elsässer, A. Frommer, B. Monien, and R. Preis. Optimal and alternating-direction loadbalancing schemes. In P. A. et al., editor, *Euro-Par'99 Parallel Processing*, LNCS 1685, pages 280–290, 1999.

[EMP00] R. Elsässer, B. Monien, and R. Preis. Diffusive load balancing schemes on heterogeneous networks. In *12th ACM Symp. on Parallel Algorithms and Architectures (SPAA)*, pages 30–38, 2000.

[FM82] C. Fiduccia and R. Mattheyses. A linear-time heuristic for improving network partitions. In *Proc. IEEE Design Automation Conf.*, pages 175–181, 1982.

[GJ79] M. Garey and D. Johnson. *COMPUTERS AND INTRACTABILITY - A Guide to the Theory of NP-Completeness.* Freemann, 1979.

[Gup97] A. Gupta. Fast and effective algorithms for graph partitioning and sparse matrix reordering. *IBM J. of Research and Development*, 41, 171–183, 1997.

[HL94] B. Hendrickson and R. Leland. The chaco user's guide: Version 2.0. Technical Report SAND94-2692, Sandia National Laboratories, Albuquerque, NM, 1994.

[HL95] B. Hendrickson and R. Leland. A multilevel algorithm for partitioning graphs. In *Proc. Supercomputing '95.* ACM, 1995.

[HM92] J. Hromkovič and B. Monien. The bisection problem for graphs of degree 4 (configuring transputer systems). In Buchmann, Ganzinger, and Paul, editors, *Festschrift zum 60. Geburtstag von Günter Hotz*, pages 215–234. Teubner, 1992.

[HP*96] A. Huang, L. Pivka, C. Wu, and M. Franz. Chua's equation with cubic nonlinearity. *Int. J. Bif. Chaos*, 6, 1996.

[Hun94] F. Hunt. A monte carlo approach to the approximation of invariant measures. *Random Comput. Dynam.*, 2, 111–133, 1994.

[Jun01] O. Junge. An adaptive subdivision technique for the approximation of attractors and invariant measures: Proof of convergence. *Dynamical Systems*, 16(3), 213–222, 2001.

[KK98a] G. Karypis and V. Kumar. *METIS Manual, Version 4.0.* University of Minnesota, Department of Computer Science, 1998.

[KK98b] G. Karypis and V. Kumar. Multilevel k-way partitioning scheme for irregular graphs. *J. of Parallel and Distributed Computing*, 48, 96–129, 1998.

[KK99] G. Karypis and V. Kumar. A fast and high quality multilevel scheme for partitioning irregular graphs. *SIAM J. on Scientific Computing*, 20(1), 1999.

[KL70] B. Kernighan and S. Lin. An effective heuristic procedure for partitioning graphs. *The Bell Systems Technical J.*, pages 291–307, 1970.

[Lei92] F. Leighton. *Introduction to Parallel Algorithms and Architectures: Arrays, Trees, Hypercubes.* Morgan Kaufmann Publishers, 1992.

[MD97] B. Monien and R. Diekmann. A local graph partitioning heuristic meeting bisection bounds. In *8th SIAM Conf. on Parallel Processing for Scientific Computing*, 1997.

[MP01] B. Monien and R. Preis. Bisection width of 3- and 4-regular graphs. In *26th International Symposium on Mathematical Foundations of Computer Science (MFCS)*, LNCS 2136, pages 524–536, 2001.

[MPD00] B. Monien, R. Preis, and R. Diekmann. Quality matching and local improvement for multilevel graph-partitioning. *Parallel Computing*, 26(12), 1609–1634, 2000.

[Pel96] F. Pellegrini. SCOTCH 3.1 user's guide. Technical Report 1137-96, LaBRI, University of Bordeaux, 1996.

[PM*94] R. Ponnusamy, N. Mansour, A. Choudhary, and G. Fox. Graph contraction for mapping data on parallel computers: A quality-cost tradeoff. *Scientific Programming*, 3, 73–82, 1994.

[Pre98] R. Preis. *The PARTY Graphpartitioning-Library, User Manual - Version 1.99.* Universität Paderborn, Germany, 1998.

[Pre00] R. Preis. *Analyses and Design of Efficient Graph Partitioning Methods.* Heinz Nixdorf Institut Verlagsschriftenreihe, 2000. Dissertation, Universität Paderborn, Germany.

[PS*03] S. Park, M. Sener, D. Lu, and K. Schulten. Reaction paths based on mean first-passage times. *J. Chem. Phys.*, 2003.

[Sen01] N. Sensen. Lower bounds and exact algorithms for the graph partitioning problem using multicommodity flows. In *Proc. European Symposium on Algorithms*, 2001.

[Sin93] A. Sinclair. *Algorithms for Random Generation & Counting: A Markov Chain Approach.* Progress in Theoretical Computer Science. Birkhäuser, 1993.

[Wal00] C. Walshaw. *The Jostle user manual: Version 2.2.* University of Greenwich, 2000.

Conditional Averaging for Diffusive Fast-Slow Systems: A Sketch for Derivation

Jessika Walter[1] and Christof Schütte[2]

[1] Department of Mathematics, Ecole Polytechnique Fédérale de Lausanne, 1015 Lausanne, Switzerland. jessika.walter@epfl.ch

[2] Department of Mathematics and Computer Science, Free University Berlin. schuette@math.fu-berlin.de

Summary. This article is concerned with stochastic differential equations with disparate temporal scales. We consider cases where the fast mode of the system rarely switches from one almost invariant set in its state space to another one such that the time scale of the switching is as slow as the slow modes of the system. In such cases descriptions for the effective dynamics cannot be derived by means of standard averaging schemes. Instead a generalization of averaging, called conditional averaging, allows to describe the effective dynamics appropriately. The basic idea of conditional averaging is that the fast process can be decomposed into several 'almost irreducible' sub-processes, each of which can be treated by standard averaging and corresponds to one metastable or almost invariant state. Rare transitions between these states are taken into account by means of an appropriate Markov jump process that describes the transitions between the states. The article gives a derivation of conditional averaging for a class of systems where the fast process is a diffusion in a double well potential.

1 Introduction

In complex system modeling, one often finds mathematical models that consist of differential equations with different temporal and spatial scales. As a consequence, mathematical techniques for the elimination of some of the smallest scales have achieved considerable attention in the last years; the derivation of reduced models by means of averaging techniques [FW84, AKN93, SV85, Kif02, Fre78, Kif01, Kif92, BLP78], homogenization techniques [BS97, Bor98, BS99], or stochastic modelling [MTV01, Mor65, Zwa73, MTV02] may serve as typical links to this discussion.

This article is concerned with stochastic differential equations where the fast mode of the system rarely switches from one almost invariant set in its state space to another one such that the time scale of the switching is as slow as the slow modes of the system. The basic idea is that the fast process then can be decomposed into several 'almost irreducible' subprocesses, each of which corresponds to one metastable or almost invariant state. To quantify this principle, the rare transitions between these states are described by means of the expected exit times that can be used to parametrize a Markov chain model mimicking the transitions between the states.

The Averaging Principle

Let $V : \mathbf{R}^m \times \mathbf{R}^n \to \mathbf{R}$ and consider the SDE

$$\dot{x}^\epsilon = -D_x V(x^\epsilon, y^\epsilon) + \sigma \dot{W}_1 \tag{1.1}$$

$$\dot{y}^\epsilon = -\frac{1}{\epsilon} D_y V(x^\epsilon, y^\epsilon) + \frac{\varsigma}{\sqrt{\epsilon}} \dot{W}_2, \tag{1.2}$$

with $\epsilon > 0$ and W_j $(j = 1, 2)$ standard Brownian motions. If we assume $\sigma = \varsigma$, the above SDE is well-known as the Smoluchowski equation. For $\epsilon \ll 1$, this system consists of a fast variable, y, and a slow one, x. Under suitable conditions on V (cf. [FW84]), averaging completely characterizes the limit x^0 of the slow dynamics x^ϵ for $\epsilon \to 0$ by an averaged SDE

$$\dot{x}^0 = - \int_{\mathbf{R}^n} D_x V(x^0, y)\, \mu_{x^0}(y) \mathrm{d}y + \sigma \dot{W}_1, \tag{1.3}$$

where μ_x denotes the invariant density of the fast dynamics for fixed x:

$$\mu_x(y) = \frac{1}{Z_x} \exp(-\frac{2}{\varsigma^2} V(x, y)), \quad Z_x = \int_{\mathbf{R}^n} \exp(-\frac{2}{\varsigma^2} V(x, y))\, \mathrm{d}y, \tag{1.4}$$

which is assumed for each x to be the unique invariant density.

Metastable Fast Dynamics & Exit Times

Let us now assume that the fast dynamics exhibit *metastable states* , i.e., that the effective dynamics in the fast degrees of freedom (DOF) can be described by (rare) jumps between these sets, while in between the jumps the dynamics remains within one of these metastable subsets. Under this condition averaging may fail to reproduce the effective dynamics of the original system, mainly for the following reason: The averaging principle is based on the fact that the fast DOF completely explore the accessible state space before any change in the slow DOF happen; this can fail to hold if metastability is observed in the fast dynamics; in particular there is some subset of the accessible state space from which the fast motion will most probably exit only on some scale of order ord(1) or even larger. Let us make this rigorous by introducing the *mean exit time* for the process y_x^ϵ from one of the metastable subsets, where y_x^ϵ is governed by the SDE (1.2) for fixed x. If we assume the existence of two metastable sets $\mathbf{R}^n = B_x \cup B_x^c$ with $B_x \cap B_x^c = \emptyset$, the mean exit time $\bar{\tau}_x^\epsilon(y)$ from B_x is the expected value of the first exit time $\tau_x^\epsilon(y)$ of the process y_x^ϵ from B_x started at $y_x^\epsilon(t = 0) = y$, which is defined by

$$\tau_D^\epsilon(y) = \inf\left\{t \in \mathbf{R}^+ : \int_0^t \mathbf{1}_{D^c}(y_x^\epsilon(s))\,\mathrm{d}s > 0,\; y_x^\epsilon(0) = y\right\} \tag{1.5}$$
$$\tau_x^\epsilon(y) = \tau_{B_x}^\epsilon(y)$$

where D^c denotes the complement of the set D.

Although we would expect that exit times depend on the starting point, i.e., $y_x^\epsilon(0) = y$, it can be shown that there do exist subsets D, for which the exit time is basically independent for all states $y \in D$. Especially for a metastable collection of sets D_i of the Smoluchowski dynamics, in the limit of vanishing noise intensity we are able to assign a first exit time $\bar{\tau}^\epsilon$ to an entire subset D_i rather than to single points $y \in D_i$, see [HMS02, SH02, SH00]. The question of the asymptotic behaviour of the mean exit time for vanishing noise term ς has been discussed in detail by, for example, FREIDLIN and WENTZELL in [FW84], from which the following result is taken (up to some slight modifications tailored to (1.1)&(1.2)) :

Theorem 1.1 ([FW84, Thm. 4.1 of Chap. 4], [SH00]). *Let the potential $V(x,\cdot)$ be twice continuously differentiable, let $y_{\min}$ be one of its local minima, and B_x a metastable subset with sufficiently smooth boundary ∂B_x containing $y_{\min}$ in its interior, but containing no other local minimum of $V(x,\cdot)$ within its interior. Without loss of generality we may assume that $V(x, y_{\min}) = 0$. Suppose that y_0 is the unique point on the boundary ∂B_x with*

$$V_{\text{bar}}^x = V(x, y_0) = \min\{V(x,y) : y \in \partial B_x\}.$$

The mean exit time $\bar{\tau}_x^\epsilon$ for the process y_x^ϵ with $y_x^\epsilon(0) \in B_x$ then satisfies

$$\lim_{\varsigma \to 0} \varsigma^2 \ln \frac{\bar{\tau}_x^\epsilon}{\epsilon} = 2\,V_{\text{bar}}^x.$$

As we are interested in the case where the averaging principle fails, let us have a closer look on the relation between the time scale of the fast motion and the exit times from metastable subsets in the fast DOF. The result of the above theorem tells us two things: First, rapid mixing of the fast DOF ($\bar{\tau}_x^\epsilon \ll 1$) can be realized by fixing ς and the potential energy function; then we are always able to find an ϵ small enough such that averaging yields a good approximation of the effective dynamcis. Second, if we decrease ς or increase the potential energy barrier, the smallness parameter ϵ has to be chosen exponentially small such that the averaged system still is a satisfactory approximation. If we want to study the effect of metastabilities in the fast motion, it is natural to relate $V_{\text{bar}}^x/\varsigma^2$ to ϵ so that the exit times from metastable sets vary on a timescale of order ord(1), that is, so that

$$\bar{\tau}_x^\epsilon \simeq C(x)\,\epsilon\,\exp\!\left(\frac{2}{\varsigma^2} V_{\text{bar}}^x\right) = \text{ord}(1), \tag{1.6}$$

where $C(x)$ denotes the subexponential pre-factor that necessarily depends on x. Subsequently, the relation symbol $\simeq$ denotes asymptotic equality and ord is used to indicate comparison to the same order.

Conditional Averaging

The scaling assumption (1.6) on ς represents a modeling step which will lead towards the derivation of the *principle of conditional averaging* that may yield an appropriate reduced model in cases where the ordinary averaging scheme fails: Since we observe rapid sampling of the invariant density μ_x in each of the metastable subsets, we propose to average over each of these sets alone and to couple the resulting systems by a Markovian switching process which describes the flipping behaviour between the metastable sets. Then, in the case of (at most) two metastable subsets $B_x^{(1)}$ and $B_x^{(2)}$ for fixed x, the conditionally averaged limit dynamics has the form

$$\dot{x}^0 = -\int D_x V(x^0, y)\, \mu_{x^0}^{(\tilde{I}(t,x^0))}(y)\, \mathrm{d}y + \sigma \dot{W}_1, \tag{1.7}$$

$$\mu_x^{(1)}(y) = \frac{1}{\mu_x(B_x^{(1)})}\mu_x(y)\mathbf{1}_{B_x^{(1)}}(y), \quad \mu_x^{(2)}(y) = \frac{1}{\mu_x(B_x^{(2)})}\mu_x(y)\mathbf{1}_{B_x^{(2)}}(y), \tag{1.8}$$

with $\tilde{I}(t,x)$ denoting the Markov chain model with state space $\mathbf{S} = \{1,2\}$, where the rates of the jumps reproduce the transition rates of the original system. In [SW*03] explicit values for the generating rate matrix are obtained by using the most dominant eigenvalue $\lambda_1^\epsilon(x) < 0$ of the generator of the fast dynamics (1.2) together with the weights $\mu_x(B_x^{(i)})$ of the metastable states on the fiber of the fast state space.

Approach

The authors of [SW*03] derived the limit dynamics (1.7) in terms of multiscale analysis of the Fokker-Planck equation, but there is no rigorous proof. The goal of this paper is to obtain a deeper insight into the nature of the conditionally averaged system.

Subsequently we consider the SDE

$$\dot{x}^\epsilon = -D_x V(x^\epsilon, y^\epsilon) + \sigma \dot{W}_1 \tag{1.9}$$

$$\dot{y}^\epsilon = -\frac{1}{\epsilon} D_y V(x^\epsilon, y^\epsilon) + \frac{\varsigma}{\sqrt{\epsilon}}\dot{W}_2, \tag{1.10}$$

with $\epsilon > 0$ and W_j $(j = 1,2)$ standard Brownian motions. We assume the fast dynamics (1.10) to exhibit metastable states $B_x^{(1)}$ and $B_x^{(2)}$ so that the exit times from the metastable subsets happen on a time scale of order ord(1) or even larger.

Under these assumptions, we may take advantage of the methodology employed to extract the effective dynamics (1.7). This result (each metastable subset of the fast dynamics is connected to one averaged equation) motivates the idea to construct a system of fast-slow equations which allows for the incorporation of temporal fast scale effects in a natural way: the fast motion *within* one metastable subset is approximated by an irreducible subprocess that corresponds to a stochastic differential equation. The result is quantified by the parametrization of a Markov chain model $I(t,x)$ that controls the switches from one (sub)process to the other according to the transition rates between the metastable subsets of the original dynamics. We thus obtain a stochastic process where the slow variable at each instance is coupled to one of two fast variables but where a stochastic switching process controls the switches from one fast variable to the other. Then, under appropriate assumptions on the potential V and for ς small, a good approximation of the original dynamics (1.9)&(1.10) may be given by

$$\dot{x}^\epsilon = -D_x V(x^\epsilon, y^\epsilon) + \sigma \dot{W}_1 \tag{1.11}$$

$$\dot{y}^\epsilon = -\frac{1}{\epsilon}\,\omega^{(I(t,x^\epsilon))}(x^\epsilon)\big(y^\epsilon - m^{(I(t,x^\epsilon))}(x^\epsilon)\big) + \frac{\varsigma}{\sqrt{\epsilon}}\dot{W}_2, \tag{1.12}$$

where $\omega^{(i)}(x)$ denotes curvature of $V(x,\cdot)$ in the potential minima of the metastable subsets $B_x^{(i)}$ for $i=1,2$, and $m^{(i)}(x)$ the respective minima.

A reduced system in the slow variable solely is then obtained by applying the well-known averaging results from [Pap76, Kur73, FW84] to each of these stochastic differential equations. Denoting $\mu_x^{\mathrm{OU}(i)}$ the (unique) invariant density of the process defined by (1.12) for fixed x and $I(t,x)=i$, the averaged system then has the form

$$\dot{x}^0 = -\int D_x V(x^0, y)\,\mu_{x^0}^{\mathrm{OU}(I(t,x^0))}(y)\,\mathrm{d}y + \sigma\,\dot{W}_1, \tag{1.13}$$

where the $\mu_x^{\mathrm{OU}(i)}$ denote the invariant densities of the Ornstein-Uhlenbeck (OU) processes (1.12) (for these we have explicit expressions).

That is, we derive a description of the effective dynamics in two steps. In a first step we replace the fast dynamics in each of the metastable subsets by appropriate OU processes which are coupled to each other by a Markovian switching process that reproduces the transition times between $B_x^{(1)}$ and $B_x^{(2)}$ of the original process. In a second step we simply use the invariant density of the OU processes in order to obtain the reduced system (1.13) by means of standard averaging. Recalling the conditionally averaged system

$$\dot{x}^0 = -\int D_x V(x^0, y)\,\mu_{x^0}^{(\tilde{I}(t,x^0))}(y)\,\mathrm{d}y + \sigma\dot{W}_1, \tag{1.14}$$

with $\mu_x^{(i)}$, $i=1,2$ defined by (1.8), it is of considerable interest to compare the effective dynamics obtained by the two different approaches, namely (1.13) on the one hand and (1.14) on the other. Note that the jump processes I that corresponds to (1.13) will be derived in a different way than the jump process $\tilde{I}$ of (1.14). However, we will see that I and $\tilde{I}$ are comparable in a certain way.

2 System under Consideration

Subsequently we study the SDE (1.1)&(1.2), where the following basic assumptions about the potential $V = V(x, y)$ are made:

Assumption 2.1 (i) $V \in \mathcal{C}^\infty(\mathbf{R}^{m+1})$;
(ii) $V(x,\cdot)$ *is a double-well potential for all* $x \in \mathbf{R}^m$ *with two local minima at* $y = m^{(1)}, m^{(2)}$ *and one local maximum at* $y = y_0$ *with* $m^{(1)} < y_0 < m^{(2)}$*; to point out the dependence on* x *we will also write* $m^{(i)}(x)$, $i = 1, 2$.
(iii) *the position of the saddle point does not depend on* x*, without loss of generality we may assume* $y_0(x) = 0$ *for every* x*;*
(iv) *the extrema are non-degenerate uniformly in* x*, i.e., for* $i = 1, 2$

$$D_{yy}V(x, m^{(i)}) = \omega^{(i)}(x) \geq \tilde{\omega}^{(i)} > 0, \; D_{yy}V(x, y_0) = -\omega_0(x) \leq -\tilde{\omega}_0 < 0.$$

Therefore, for fixed x, the particle spends a 'long time' in one basin (=potential well), then quickly undergoes a transition into the other basin, in which it spends another 'long time', and so on. The condition $y_0(x) = 0$ implies that for every $x \in \mathbf{R}^m$ the locations of the two basins do not depend on x such that the natural decomposition of the entire state space into metastable subsets is simply given by $B^{(1)} \cup B^{(2)}$, where[3]

$$B^{(1)} = \{(x, y) \in \mathbf{R}^{m+1} \mid y < 0\}, \quad B^{(2)} = \{(x, y) \in \mathbf{R}^{m+1} \mid y > 0\}. \tag{2.1}$$

The double-well potentials may serve as toy models mimicking a larger system whose potential energy surface presents several basins corresponding to metastable states.

As outlined in the introduction, we proceed in two steps to derive a reduced model for the effective slow variable dynamics. The key point for the first step is rooted in the design of $V(x, \cdot)$ which already suggests that an averaging procedure should incorporate metastabilities in the fast dynamics that are induced by the double-well structure: If the noise level in the fast equation is small, the diffusion sample paths of the fast process are located near the local minima of the potential wells, and transitions between the two potential wells can be considered as rare events. Then, the diffusion can be decomposed into two sub-processes

$$(x^\epsilon(t), y^\epsilon_{(i)}(t)) = (x^\epsilon(t), y^\epsilon(t))\, \mathbf{1}_{B^{(i)}}(x^\epsilon(t), y^\epsilon(t)), \qquad i = 1, 2,$$

and a two-state Markov chain $I(t, x)$ mimicking the transitions between $B^{(1)}$ and $B^{(2)}$ which happen along the y dynamics and thus depend on the position of the slow one. Our approach is based on a quantification of the rates at

[3] In [SW*03], the metastable decomposition for fixed x is defined by the zero z of the second eigenfunction $u_1(x, \cdot)$ of the fast dynamics generator. It is shown in [Wal05] that the zero z of $u_1(x, \cdot)$ actually is approximated by the saddle point of the potential $V(x, \cdot)$.

which the fast process moves between the two subsets on the one hand and, on the other, on an appropriate replacement of the almost irreducible fast (sub)processes by appropriately chosen OU-processes evolving independently of each other. Thus, the most basic questions we have to address concern the fast process (1.2) for fixed slow variable x, which is done in Sect. 3. In so doing, we basically have to decompose the fast process into the *intra-well* small fluctuations of the diffusion around the potential minima and the *inter-well* dynamics of the diffusion. For both parts we then obtain by means of small noise asymptotics basic results that are then picked up in order to assemble in Sect. 4 the full dynamics approximation (1.11)&(1.12) including the slow variables motion.

The second step of the approach relies on the small noise approximation and is based upon averaging results that can be found in a vast number of articles. A simple application of a theorem in [FW84] then provides us in Theorem 4.1 with the reduced dynamics (1.13). In the Appendix we show how the averaged dynamics can be derived by using multiscale asymptotics of the Fokker-Planck equation corresponding to the small noise approximation (1.11)&(1.12). In Sect. 5 we compare the averaged dynamics (1.13) to the conditionally averaged system (1.7).

Another important concern of the approach is the relationship between the noise level ς in the fast diffusion and the smallness parameter ϵ: The fast diffusions inter-well and intra-well approximations are justified for vanishing noise ς, so that we suggest $\varsigma \to 0$ to zero as $\epsilon \to 0$. Our considerations will result in a coupling rule for ς and ϵ that incorporates the asymmetry of the double-well potential. In Lemma 3.5 the choice of the noise level ς is coupled to ϵ as well as the slow variable x so that the exit times from the metastable subsets of the fast dynamics vary on a time scale of ord(1) or larger resulting in (1.6).

Biomolecules operate at ambient temperature and solvent condition, and most biomolecular processes can only be understood in a thermodynamical context. Therefore, most experiments on biomolecular systems are performed under the equilibrium conditions of constant temperature T, particle number, and volume. Statistical mechanics tells us that statistical ensembles of molecular systems with internal energy V under these circumstances should be modelled by the equilibrium density $\exp(-\beta V)$. For the Smoluchowski system (1.1)&(1.2) this means to enforce $\sigma = \varsigma$ (for fixed ϵ), such that experiments can be arranged with inverse temperature $\beta = -2/\varsigma^2$. However, if the noise intensity ς depends on the slow variable x, it is hardly possible to interpret the model system in the context of equilibrium ensembles. Therefore our later Lemma 3.5 is not satisfactory as its application removes the system under consideration far away from the mathematical modeling of biological processes. Thus, the investigations have to be extended to situations where ς depends on ϵ solely. This will be done in Appendix B.

3 Basic Results on Fast Process

Let Assumption 2.1 be valid in all of the following. For small noise intensity ς, the process y^ϵ corresponding to the Smoluchowski equation (1.2) for fixed x is almost decomposable into two subprocesses $y^\epsilon_{(1)}$, $y^\epsilon_{(2)}$, each attracted to a minimum $m^{(i)}(x)$, $i = 1, 2$ of the function $V(x, \cdot)$.

Thus, we consider the fast motion $y^\epsilon_x(t)$ for fixed slow variables $x \in \mathbf{R}^m$:

$$\dot{y}^\epsilon_x = -\frac{1}{\epsilon} D_y V(x, y^\epsilon_x) + \frac{\varsigma}{\sqrt{\epsilon}} \dot{W}_2, \tag{3.1}$$

and distinguish between the two different regions of attraction $O^{(1)}_x$ and $O^{(2)}_x$ where $O^{(i)}_x$ is an open subset of $B^{(i)}_x$ with $m^{(i)}(x)$ in its interior. The subsets $B^{(1)}_x$ and $B^{(2)}_x$ are defined by the potential energy barrier:

$$B^{(1)}_x = \{y \in \mathbf{R} : y < y_0(x)\}, \quad \text{and} \quad B^{(2)}_x = \{y : y > y_0(x)\}, \tag{3.2}$$

with $y_0(x) = 0$ denoting the saddle point of the potential $V(x, \cdot)$.

In the limit of small noise level ς, Theorem 3.1 below will provide us for small ς with an approximation of the fast dynamics (3.1) restricted to a single metastable set by a simple Ornstein-Uhlenbeck (OU) process mimicking the rapid mixing in each of these subsets prior to exiting. There is no information in these stationary limits about the possible jumps from the branch $y = m^{(1)}(x)$ to the branch $y = m^{(2)}(x)$, or conversely. To address the question of the overall behaviour of the stationary state, we will consider in Theorem 3.3 and Corollary 3.4 below the new discrete-space process on $\{m^{(1)}(x), m^{(2)}(x)\}$ assigning information about the inter-well dynamics.

3.1 Approximation of Intra-well Dynamics

For vanishing noise intensity ς, in each of the subsets $O^{(i)}_x$ the fast diffusion will consist of small fluctuations around the potential minima $m^{(1)}(x)$ and $m^{(2)}(x)$, respectively. The drift term in (1.2) can now be expanded in a Taylor series with respect to y. Taylor-expansion of $D_y V(x, \cdot)$ around $m^{(i)}(x)$, $i = 1, 2$ gives

$$D_y V(x, y) = D_{yy} V(x, m^{(i)}(x)) \,(y - m^{(i)}(x)) + \mathcal{O}(|y - m^{(i)}(x)|^2), \tag{3.3}$$

where we have used $D_y V(x, m^{(i)}(x)) = 0$. For y sufficiently close to $m^{(i)}(x)$, this provides us with an approximation of the SDE (3.1):

$$\dot{y}^\epsilon_{\mathrm{OU}(i)} = -\frac{1}{\epsilon} \omega^{(i)}(x) \,(y^\epsilon_{\mathrm{OU}(i)} - m^{(i)}(x)) + \frac{\varsigma}{\sqrt{\epsilon}} \dot{W}_2, \tag{3.4}$$

with $\omega^{(i)}(x)$, $i = 1, 2$ denoting the curvature of $V(x, \cdot)$ in $m^{(i)}(x)$, see Assumption 2.1. The solution of the stochastic differential equation (3.4) is known as a process of Ornstein-Uhlenbeck type, or OU process for short. To distinguish it from the 'decoupled' processes $y^\epsilon_{(i)} \in O^{(i)}_x$, $i = 1, 2$ that originate from (3.1) we denote it $y^\epsilon_{\mathrm{OU}(i)}$ for $i = 1, 2$. We omit the index for the fixed variable x.

The quality of the approximation will depend on how close the original motion stays in the vicinity of the minima $m^{(i)}(x)$, $i = 1, 2$. This can be made more precise by applying the small noise expansion method for stochastic differential equations. The basic assumption of asymptotically expanding the solution process $y^\epsilon_{(i)}$ for $i = 1, 2$ into powers of the noise intensity ς leads to a reduction of the equation (3.1) into a sequence of time-dependent OU processes. Mostly the first order is quite adequate and amounts to a linearisation of the original equation about the deterministic solution. The reader may refer to [Gar85], where it is shown that the procedure yields a convergent power series of ς. Tailored to the approach (3.4), the procedure yields a power series

$$y^\epsilon_x = y^\epsilon_{\mathrm{OU}(i)} + \varsigma^2 R(\varsigma),$$

where the remainder $R(\varsigma, t)$ is the solution of an SDE and stochastically converges to $r(0, t)$. That is, it exists a limiting SDE with solution $R(0, t)$ such that for all $T \in \mathbf{R}^+$

$$\operatorname{st-}\lim_{\varsigma \to 0} \{ \sup_{t \in [0,T]} |R(\varsigma, t) - R(0, t)| \} = 0,$$

where $\operatorname{st-}\lim_{n\to\infty} \xi_n = \xi$ denotes $\lim_{n\to\infty} \mathbf{P}\{|\xi_n - \xi| \geq \delta\} = 0$ for every $\delta > 0$ and a sequence $\{\xi_n\}$ of random variables.

Theorem 3.1 ([Gar85, Chapters 6.2, 4.3.7]). *Let y^ϵ_x be given by the SDE (3.1) where ϵ and x are chosen arbitrary but fixed. Suppose that the process starts for some $i = 1, 2$ in an open subset $O^{(i)}_x$ of $B^{(i)}_x$ containing $m^{(i)}(x)$ in its interior, and let $y^\epsilon_{\mathrm{OU}(i)}$ be the solution of (3.4). Then we have for all $T \in \mathbf{R}^+$*

$$|y^\epsilon_x(t) - y^\epsilon_{\mathrm{OU}(i)}(t)| = \mathcal{O}(\varsigma^2),$$

where $\mathcal{O}$ is understood as being satisfied with respect to stochastic convergence uniformly in $t \in [0, T]$ (as $\varsigma \to 0$).

Remark 3.2. As the OU process (3.4) is ergodic, the stationary density $\mu_x^{\mathrm{OU}(i)}$ is simply given by the Gaussian with mean $m^{(i)}(x)$ and variance $\varsigma^2/(2\omega^{(i)}(x))$. Aiming at a comparison of $\mu_x^{\mathrm{OU}(i)}$ and $\mu_x^{(i)}$, as defined by (1.8) and (1.4), it is shown in [Wal05] that

$$\lim_{\varsigma \to 0} (\mu_x^{\mathrm{OU}(i)} - \mu_x^{(i)}) = 0 \qquad \text{in } L^1(\mathbf{R}).$$

Note that we do not get convergence in L^∞.

3.2 Asymptotics of Inter-well Dynamics

To give a picture of the essential dynamics in the fast state space, we consider the statistics of the exit times from the metastable sets and approximate the transition events of the diffusion by jump times of an associated continuous-time, finite state-space Markov chain (the double-well potential implies a two-state Markov chain). In principle, one can compute the exit times via direct numerical simulation. The approximated exit times can then be used to construct a transition rate matrix $\mathcal{Q}$ that generates stochastic matrices $\exp(t\mathcal{Q})$ for all times $t > 0$. However, the computational effort of estimating the expected exit times can be avoided by resorting to the rich literature on the derivation of asymptotic formula for the jump times that are strongly connected to the dominant spectrum of the corresponding generator, see e.g. [Pav02, BGK02, BE*02, HKN04]. Whereas the first papers only gave the asymptotic behaviour of the logarithm of expected exit times (cf. Theorem 1.1), in [Pav02, BGK02, BE*02] one also finds estimates for the prefactor.

For small noise intensity ς, transitions between the potential wells occur at Kramers' time that is given up to exponential order by $\exp((2/\varsigma^2)\Delta V)$, where ΔV is the potential barrier that the process must cross to reach the other potential well. The first exit time of the Markov process $y_x^\epsilon(t)$ from D started at $y_x^\epsilon(0) = y$ as defined in (1.5) measures only exits that happen for some non-null time interval and depends on the realization of the Markov process.

We are interested in the transition times between the metastable subsets $B_x^{(1)}$ and $B_x^{(2)}$. If the noise intensity does not vanish, they are not identical to the exit times $\tau^\epsilon_{B_x^{(i)}}$, $i = 1, 2$. Instead we have to modify the metastable subsets slightly such that a (small) neighbourhood around the saddle point is included, i.e., we consider $B_x^{(1)} + \delta = (-\infty, \delta)$ and $B_x^{(2)} - \delta = (-\delta, \infty)$ instead with $\delta > 0$ being a small parameter. Recall that $O_x^{(i)} \subset B_x^{(i)}$, $i = 1, 2$ are some regions of attraction (excluding a neighbourhood around the saddle point and including the potential minima, that is, $m^{(i)}(x) \in O_x^{(i)}$ for $i = 1, 2$). Then, the first exit times from $B_x^{(1)} + \delta$ and $B_x^{(2)} - \delta$ are basically independent for all starting points $y \in O_x^{(1)}$ and $y \in O_x^{(2)}$, respectively. This enables us to assign the expected exit times from $B_x^{(1)} + \delta$ and $B_x^{(2)} - \delta$ to the entire subsets $O_x^{(1)}$ and $O_x^{(2)}$ rather than to single points.

In the next theorem, we denote the expected transition times from $B_x^{(i)}$ to $B_x^{(j)}$ with $i \neq j$ by $\mathcal{T}^\epsilon_{i \to j}(x)$, $i = 1, 2$.

Theorem 3.3 ([Pav02, BGK02]). *The metastable inter-well transitions of the dynamics (3.1) are given by the following precise asymptotic estimates*[4] *as* $\varsigma \to 0$*:*

---

[4] We emphasize again that in the following we will speak of (metastable) transition times between $B_x^{(1)}$ and $B_x^{(2)}$ or *metastable* exit times instead of exit times from

$$\mathcal{T}^{\epsilon}_{1\to 2}(x) = \mathbf{E}_{y\in O_x^{(1)}}[\tau^{\epsilon}_{B_x^{(1)}+\delta}] \simeq \epsilon \frac{2\pi}{\sqrt{\omega^{(1)}(x)\,\omega_0(x)}} \exp\big(\frac{2}{\varsigma^2}V^{(1)}_{\text{bar}}(x)\big), \quad (3.5)$$

$$\mathcal{T}^{\epsilon}_{2\to 1}(x) = \mathbf{E}_{y\in O_x^{(2)}}[\tau^{\epsilon}_{B_x^{(2)}-\delta}] \simeq \epsilon \frac{2\pi}{\sqrt{\omega^{(2)}(x)\,\omega_0(x)}} \exp\big(\frac{2}{\varsigma^2}V^{(2)}_{\text{bar}}(x)\big), \quad (3.6)$$

where $V^{(1)}_{\text{bar}}(x) = V(x,y_0) - V(x,m^{(1)}(x))$ *and* $V^{(2)}_{\text{bar}}(x) = V(x,y_0) - V(x,m^{(2)}(x))$ *denote the left and right potential barriers.*

In [Pav02] the result is obtained in terms of the largest eigenvalue $\neq 0$ of the associated infinitesimal generator, which corresponds (apart from suitable weights) to the inverse of the expected diffusion exit times. The connection will be discussed in the next corollary.

Our goal is to build a two-state Markov chain and view inter-well transitions of the diffusion as simple jumps of this chain. Correspondence between the diffusion and the chain will be established by exploiting that exit times are asymptotically almost exponential random variables which is shown in [SH02, HMS02]. Relying on this fact, we define the jump rates for the reducing Markov-chain as the reciprocal of the expected exit times $\mathcal{T}^{\epsilon}_{i\to j}(x)$, $i \neq j$, which provides us with the rate matrix $\mathcal{Q}^{\epsilon}_x$ being defined by

$$\mathcal{Q}^{\epsilon}_x := \begin{pmatrix} -1/\mathcal{T}^{\epsilon}_{1\to 2}(x) & 1/\mathcal{T}^{\epsilon}_{1\to 2}(x) \\ 1/\mathcal{T}^{\epsilon}_{2\to 1}(x) & -1/\mathcal{T}^{\epsilon}_{2\to 1}(x) \end{pmatrix}, \qquad \mathcal{Q}^{\epsilon}_x \begin{pmatrix} 1 \\ 1 \end{pmatrix} = 0. \qquad (3.7)$$

The following corollary shows that the invariant density of the reducing Markov chain is asymptotically given by the weights over the potential wells.

Corollary 3.4. *Let us denote the (assumed positive and unique) invariant density of* $\mathcal{Q}^{\epsilon}_x$ *by* $\psi(x) = (\psi_1(x), \psi_2(x))$, *that is,*

$$\psi(x)\,\mathcal{Q}^{\epsilon}_x = 0 \qquad \text{with} \quad \psi_1(x) + \psi_2(x) = 1.$$

Then we find that $\psi(x)$ *is given asymptotically as* $\varsigma \to 0$ *by* $(\mu_x(B_x^{(1)}), \mu_x(B_x^{(2)}))$, *explicitly,*

$$\psi_i(x) \simeq \mu_x(B_x^{(i)}), \qquad i = 1, 2. \qquad (3.8)$$

The rate matrix $\mathcal{Q}^{\epsilon}_x$ *can be expressed in terms of the invariant density by introducing the second eigenvalue* $\lambda^{\epsilon}_1(x)$ *of the infinitesimal generator that corresponds to the diffusion (3.1). In so doing, we asymptotically obtain*

$$\mathcal{T}^{\epsilon}_{1\to 2}(x) \simeq \frac{1}{|\lambda^{\epsilon}_1(x)|\mu_x(B_x^{(2)})} \qquad \text{and} \qquad \mathcal{T}^{\epsilon}_{2\to 1}(x) \simeq \frac{1}{|\lambda^{\epsilon}_1(x)|\mu_x(B_x^{(1)})},$$

and, conclusively,

$B_x^{(i)}$, $i = 1, 2$, for the asymptotic estimates are given for the mean values of the first exit times from $B_x^{(1)} + \delta$ and $B_x^{(2)} - \delta$ with $\delta > 0$, where the precise choice of the parameter δ is not important.

$$\mathcal{Q}_x^\epsilon \simeq |\lambda^\epsilon(x)| \begin{pmatrix} -\mu_x(B_x^{(2)}) & \mu_x(B_x^{(2)}) \\ \mu_x(B_x^{(1)}) & -\mu_x(B_x^{(1)}) \end{pmatrix}. \tag{3.9}$$

Proof. To establish (3.8), we simply have to verify

$$\frac{\mu_x(B_x^{(1)})}{\mu_x(B_x^{(2)})} \simeq \frac{\mathcal{T}_{1\to 2}^\epsilon(x)}{\mathcal{T}_{2\to 1}^\epsilon(x)}. \tag{3.10}$$

To this end, we may apply Laplace's method of asymptotic evaluation of integrals depending on the parameter ς. According to Laplace , we easily get the asymptotic estimates in the small noise limit

$$\frac{\mu_x(B_x^{(1)})}{\mu_x(B_x^{(2)})} = \tag{3.11}$$
$$\sqrt{\frac{\omega^{(2)}(x)}{\omega^{(1)}(x)}} \exp\Big(- \frac{2}{\varsigma^2}\big(V(x, m^{(1)}(x)) - V(x, m^{(2)}(x))\big)\Big) \big(1 + \mathcal{O}(\varsigma)\big),$$

and, by using $V(x, m^{(1)}(x)) - V(x, m^{(2)}(x)) = -(V_{\text{bar}}^{(1)}(x) - V_{\text{bar}}^{(2)}(x))$ together with (3.5)&(3.6), we end up with (3.10). The informations about the behaviour of $\lambda_1^\epsilon(x)$ are again based on the results of PAVLYUKEVICH in [Pav02] who derived the asymptotic formula of $\lambda_1^\epsilon(x)$ in the small noise limit by expanding λ_1^ϵ into a power series. For *asymmetric* double-well potential this gives the accurate asymptotics for $\lambda_1^\epsilon(x)$ in terms of quantities concerning the *shallow* well of the potential:

$$|\lambda_1^\epsilon(x)| = \frac{1}{\epsilon} \frac{\sqrt{\omega^{(1)}(x)\omega_0(x)}}{2\pi} \exp\big(- \frac{2}{\varsigma^2} V_{\text{bar}}^{(1)}(x)\big) \big(1 + \mathcal{O}(\varsigma)\big),$$

where we assume without loss of generality

$$V_{\text{bar}}^{(1)}(x) = \min\{V_{\text{bar}}^{(1)}(x), V_{\text{bar}}^{(2)}(x)\}.$$

This result has been derived for asymmetric double-well potentials, such that the weight on the deep well is approximately 1, that is, $\mu_x(B_x^{(2)}) \approx 1$. This obviously is fulfilled for small values of ς due to $\mu_x(B_x^{(2)}) \to 1$ as $\varsigma \to 0$. However, to include the case of symmetric double-well potentials (then we have $\mu_x(B_x^{(2)}) = \mu_x(B_x^{(1)}) = 0.5$) we prefer to rewrite the asymptotics of λ_1^ϵ according to

$$|\lambda_1^\epsilon(x)|\, \mu_x(B_x^{(2)}) = \frac{1}{\epsilon} \frac{\sqrt{\omega^{(1)}(x)\omega_0(x)}}{2\pi} \exp\big(- \frac{2}{\varsigma^2} V_{\text{bar}}^{(1)}(x)\big) \big(1 + \mathcal{O}(\varsigma)\big), \tag{3.12}$$

which allows us due to (3.5) to express the transition rate $1 \to 2$ asymptotically by $|\lambda_1^\epsilon|\mu_x(B_x^{(2)})$. Using the asymptotic estimates (3.11) and (3.12) provides us with an alternative formulation for the asymptotics of $\lambda_1^\epsilon(x)$ by using the curvature in the deep well (and the weight over the shallow well):

$$|\lambda_1^\epsilon(x)|\,\mu_x(B_x^{(1)}) = \frac{1}{\epsilon}\frac{\sqrt{\omega^{(2)}(x)\omega_0(x)}}{2\pi}\,\exp\big(-\frac{2}{\varsigma^2}V_{\mathrm{bar}}^{(2)}(x)\big)\,\big(1+\mathcal{O}(\varsigma)\big).$$

3.3 Freezing Metastable Transitions

We complete the analysis of the fast process (3.1) with establishing a relationship between the smallness parameter ϵ and the noise level ς such that the scaling assumption (1.6) is fulfilled. According to Theorem 3.3 this can explicitly be realized only if $\exp(-(2/\varsigma^2)\Delta V)$ scales like ϵ. Here, ΔV denotes the barrier that has to be crossed, that is, $\Delta V = V_{\mathrm{bar}}^{(1)}(x)$ or $\Delta V = V_{\mathrm{bar}}^{(2)}(x)$. A natural way of realizing (1.6) was to rescale the potential energy barrier in an appropriate manner (see [Wal05]).

However, due to the asymptotic investigations in Theorem 3.3 we leave the potential untouched and rescale the diffusion ς instead. An easy calculation leads to the following lemma.

Lemma 3.5. *To freeze the metastable transition times on a time scale* $t \geq \mathrm{ord}(1)$ *for every* x *as* $\epsilon \to 0$ *it is convenient to set*

$$\varsigma = \varsigma(\epsilon, x) = \left(\frac{2\min\{V_{\mathrm{bar}}^{(i)}(x) \mid i=1,2\}}{\ln(K/\epsilon)}\right)^{1/2}, \qquad K > 0. \tag{3.13}$$

Remark 3.6. In Lemma 3.5 we actually have to use the minimum of the two barriers $V_{\mathrm{bar}}^{(1)}$, $V_{\mathrm{bar}}^{(2)}$: Replacing $\min\{V_{\mathrm{bar}}^{(i)}(x)\}$ by $V_{\mathrm{bar}}^{(2)} = (1+\delta)V_{\mathrm{bar}}^{(1)}$ for $\delta > 0$ would lead to $\mathcal{T}_{1\to 2}^\epsilon = \mathrm{ord}(\epsilon^\delta)$. According to Corollary 3.4, the need for using the minimal barrier is equally expressed by demanding that the second eigenvalue $\lambda_1^\epsilon(x)$ asymptotically is part of the dominant spectrum.

As outlined in Sect. 2, it is of considerable interest to study how to avoid coupling of the diffusion ς to x and still obtain large time conformational changes in the asymptotic limit $\epsilon \to 0$. Based upon Lemma 3.5, the following considerations will lead to meaningful conclusions (a short description is given in Appendix B) that are strongly connected to results obtained by the approach via multiscale asymptotics with disparate transition scales (see [Wal05, Chapter 3]): Depending on the noise intensity σ in the slow variable dynamics (1.1), the x trajectory will stay with overwhelming probability in a bounded domain $D(\sigma)$ of its state space[5]; if we choose $V_{\mathrm{bar}}^{\mathrm{small}}$ according to the rule

[5] Note that σ is not related to ς, and we do not demand for small values of σ. However, it should be clear that a choice of $V_{\mathrm{bar}}^{\mathrm{small}}$ could depend on the actual size of σ. This becomes more clear by considering Figures 6.5 and B.1. Therefore, we write $D = D(\sigma)$ for the bounded region.

$$V_{\text{bar}}^{\text{small}} = \min\{V_{\text{bar}}^{(i)}(x) \mid x \in D(\sigma),\ i = 1, 2\},$$

and set

$$\varsigma(\epsilon) = \left(\frac{2V_{\text{bar}}^{\text{small}}}{\ln(K/\epsilon)}\right)^{1/2}, \qquad K > 0, \tag{3.14}$$

we expect the metastable transitions to happen on a time scale $t \geq \text{ord}(1)$. If the potential energy barriers outside the domain $D(\sigma)$ are smaller than $V_{\text{bar}}^{\text{small}}$, the particle will for very small ϵ instantly jump over the barrier once it has reached the complement of $D(\sigma)$. Then, the time of the metastable transitions will be somehow connected to the expected exit time of the x dynamics from $D(\sigma)$. The above idea is justified by rigorously examining the asymptotics of the metastable transition times considered in the entire state space. Tailored to exemplary situations we outline the procedure in Appendix B.

4 Derivation of Reduced Dynamics

We return to the dynamics (1.1)&(1.2) and use the results of the preceding section for the design of a small noise approximation of the original process. The approximated system is then used in Theorem 4.1 as the basic system for the application of standard averaging theorems resulting in the reduced slow variable dynamics.

Small Noise Approximation

In all of the following let (x^ϵ, y^ϵ) be the solution of the SDE (1.1)&(1.2).

Exact jump process.

As a first step let us introduce the process $\hat{I}$ that describes the jumps between the metastable sets in y-direction as given by the original dynamics:

$$\hat{I}(t) = 1 + \mathbf{1}_{B^{(2)}_{x^\epsilon(t)}}(y^\epsilon(t)). \tag{4.1}$$

With this defined, let us denote by $(\hat{x}^\epsilon_{\text{OU}}, \hat{y}^\epsilon_{\text{OU}})$ the random process determined by

$$\dot{x} = -D_x V(x, y) + \sigma \dot{W}_1 \tag{4.2}$$

$$\dot{y} = -\frac{1}{\epsilon}\,\omega^{(\hat{I}(t))}(x)\big(y - m^{(\hat{I}(t))}(x)\big) + \frac{\varsigma}{\sqrt{\epsilon}}\dot{W}_2, \tag{4.3}$$

Now, suppose that the initial points are chosen such that $(x^\epsilon(0), y^\epsilon(0)) = (\hat{x}^\epsilon_{\text{OU}}(0), \hat{y}^\epsilon_{\text{OU}}(0)) = (x, y)$.

According to Theorem 3.1 we obtain for any $x, y, T > 0$ and $\epsilon > 0$ that the process $(x^\epsilon(t), y^\epsilon(t))$, $t \in [0, T]$ of the original dynamics and the random process $(\hat{x}^\epsilon_{\mathrm{OU}}(t), \hat{y}^\epsilon_{\mathrm{OU}}(t))$ get arbitrarily close to each other for $\varsigma \to 0$.

However, we will not concentrate on the rigorous mathematical justification of this result, mainly for reasons given in the next paragraph.

Approximate jump process.

This result may be very nice. However, it has the crucial disadvantage that we will never have the process $\hat{I}$ without knowing the actual solution of the original process. Therefore, we replace the jump process $\hat{I}$ by its approximate version I as constructed above. Obviously, this will prevent us from being able to construct any kind of *pathwise* convergence. However, it will finally allow to construct an approximate dynamics that is explicit in the sense that it does not depend on any knowledge about the original process. To this end, we denote by $(x^\epsilon_{\mathrm{OU}}, y^\epsilon_{\mathrm{OU}})$ the random process determined by

$$\dot{x} = -D_x V(x, y) + \sigma \dot{W}_1 \tag{4.4}$$

$$\dot{y} = -\frac{1}{\epsilon}\, \omega^{(I(t,x))}(x)\big(y - m^{(I(t,x))}(x)\big) + \frac{\varsigma}{\sqrt{\epsilon}} \dot{W}_2, \tag{4.5}$$

with $I(t, x) \in \mathbf{S} = \{1, 2\}$ denoting the x-dependent Markov chain model whose transition rate matrix $\mathcal{Q}^\epsilon_x = (q_{ij})_{i,j}$ is given by its entries

$$\begin{aligned} q_{11}(x) &= -q_{12}(x), \qquad q_{22}(x) = -q_{21}(x), \\ q_{12}(x) &= \frac{1}{\epsilon} \frac{\sqrt{\omega^{(1)}(x)\omega_0(x)}}{2\pi} \exp\!\Big(-\frac{2}{\varsigma^2} V^{(1)}_{\mathrm{bar}}(x)\Big), \\ q_{21}(x) &= \frac{1}{\epsilon} \frac{\sqrt{\omega^{(2)}(x)\omega_0(x)}}{2\pi} \exp\!\Big(-\frac{2}{\varsigma^2} V^{(2)}_{\mathrm{bar}}(x)\Big). \end{aligned} \tag{4.6}$$

Again, suppose that the initial points are chosen such that $(x^\epsilon(0), y^\epsilon(0)) = (x^\epsilon_{\mathrm{OU}}(0), y^\epsilon_{\mathrm{OU}}(0)) = (x, y)$ and $I(t = 0, x) = i$ for $(x, y) \in B^{(i)}$.

According to Theorems 3.1 and 3.3 and Corollary 3.4 we can expect for any $x, y, T > 0$ and $\epsilon > 0$ to obtain a good approximation of $(x^\epsilon(t), y^\epsilon(t))$, $t \in [0, T]$ by the random process $(x^\epsilon_{\mathrm{OU}}(t), y^\epsilon_{\mathrm{OU}}(t))$ whenever the noise level ς in the fast equation is small enough.
We will call the dynamics (4.4)&(4.5) in the following *small noise approximation* or *OU approximated dynamics.*

Averaging

In Theorem 4.1 we finally arrive at the reduced slow variable system by applying standard averaging theorems to the small noise approximation (4.4)&(4.5) where the transition rates of the jump process $I(t, x)$ are given by (4.6).

Theorem 4.1 ([FW84, Chapter 7]). *Let $(x^\epsilon_{\mathrm{OU}}, y^\epsilon_{\mathrm{OU}})$ be given by (4.4)&(4.5) and denote by $x^0(t)$ the solution of the differential equation*

$$\dot{x} = -\int D_x V(x,y)\, \mu_x^{\mathrm{OU}(I(t,x))}(y)\, \mathrm{d}y \, + \sigma \dot{W}_1, \tag{4.7}$$

where $\mu_x^{\mathrm{OU}(i)}$ is the (unique) invariant density of the process defined by (3.4) for fixed x. Then for any $T > 0$ and $\varsigma > 0$ we have

$$\operatorname{st-lim}_{\epsilon \to 0} \{ \sup_{t \in [0,T]} |x^\epsilon_{\mathrm{OU}}(t) - x^0(t)| \} = 0.$$

Subsequently, we refer to the slow variable dynamics (4.7) as the *OU averaged dynamics*. In Appendix A we use multiscale asymptotics of the corresponding Fokker-Planck equation to derive the OU averaged dynamics from the small noise approximation.

5 Comparison to Conditional Averaging

To complete the discussion and re-establish reference to the conditionally averaged system (1.7) we finally examine its closeness to the OU averaged dynamics (4.7). In so doing, we basically compare the behaviour in the asymptotic limit $\varsigma \to 0$ of

1. the drift term in (4.7) and (1.7) for fixed $\tilde{I}(t,x) = I(t,x) = i$;
2. the corresponding transition chains $\tilde{I}$ and I that control the switches between $i = 1$ and $i = 2$.

First, let us consider the transition chain $\tilde{I}(t,x)$ of the conditionally averaged system (1.7) as given in [SW*03]. There, the transition rates of the jump process $\tilde{I}$ are defined by the rate matrix

$$\tilde{Q}_x = |\lambda_1^\epsilon(x)| \begin{pmatrix} -\mu_x(B_x^{(2)}) & \mu_x(B_x^{(2)}) \\ \mu_x(B_x^{(1)}) & -\mu_x(B_x^{(1)}) \end{pmatrix}, \tag{5.1}$$

where $\lambda_1^\epsilon(x)$ is the second eigenvalue of the infinitesimal generator of the diffusion (1.2) that is assumed to be of order ord(1). We compare the entries in (5.1) to the transition rates q_{ij} of the jump process $I(t,x)$ corresponding to the OU averaged dynamics that are defined in (4.6). Exploiting the asymptotic results of Corollary 3.4 and under a certain additional assumption, the transitions rates q_{ij} are asymptotically equal to the rates of $\tilde{Q}_x$. The additional assumption that has to be fulfilled concerns the metastable decomposition as derived by applying conditional averaging: In [SW*03], the limit dynamics are derived by *projecting the ensemble dynamics of the original system onto the subspace spanned by the dominant spectrum* of the infinitesimal generator $\mathcal{L}_x$ of (1.2). Then, the metastable decomposition $B_x^{(1)}$ and $B_x^{(2)}$ will be defined in

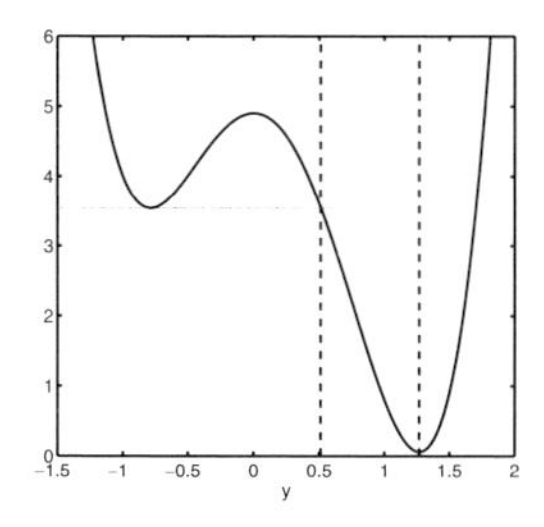

Fig. 5.1. Asymmetric double-well potential. Any decomposition defined by a point z_* that is situated between the dashed lines leads to asymptotically wrong results if the decomposition is used for the conditionally averaged dynamics.

terms of the second eigenfunction $u_1(x,\cdot)$ of $\mathcal{L}_x$ and not by the saddle point of the potential $V(x,\cdot)$ as is done in (3.2). Thus, for the definition of $\tilde{Q}$ we have to use

$$B_x^{(1)} = \{y : u_1(x,y) < 0\} \quad \text{and} \quad B_x^{(2)} = \{y : u_1(x,y) > 0\},$$

where we can assume that $B_x^{(1)}$ is the left subset. It should be clear that the zero z of $u_1(x,\cdot)$ must be somewhere between the two potential minima $m^{(1)}$ and $m^{(2)}$, and in fact, it is only a small step from using results in [Pav02] to show that z asymptotically (as $\varsigma \to 0$) approaches the saddle point $y_0(x)$, cf. [Wal05]. The attentive reader may convince himself that it the result is of crucial importance, as other choices of the zero z between the potential minima may lead to fatal approximation errors (not only for the transition rates but also for the drift term), compare illustration in Fig. 5.1.

Having obtained the asymptotic equality as $\varsigma \to 0$ of the jump rates of $\tilde{I}$ and I, we still have to compare the drift terms in (4.7) and (1.7) for fixed $\tilde{I}(t,x) = I(t,x) = i$. The terms vary for fixed x in the probability density that is used to obtain the averaged force on the slow variable x. We apply standard Laplacian asymptotics in the limit of vanishing noise $\varsigma \to 0$, which provides us for $i = 1,2$ with the precise estimates

$$\begin{aligned}\int D_x V(x,y)\,\mu_x^{\mathrm{OU}(i)}(y)\,\mathrm{d}y &= D_x V(x,m^{(i)}(x))\,\big(1+\mathcal{O}(\varsigma)\big), \\ \int D_x V(x,y)\,\mu_x^{(i)}(y)\,\mathrm{d}y &= D_x V(x,m^{(i)}(x))\,\big(1+\mathcal{O}(\varsigma)\big),\end{aligned} \tag{5.2}$$

where the derivative $D_x V(x,m^{(i)}(x))$ is taken wrt. the first component solely.

Conclusively, let us suppose that $\varsigma = \varsigma(\epsilon)$ is coupled to ϵ by using (3.14). Replacing ς by $\varsigma(\epsilon)$ in the fast equation (1.2) of the original process will lead to a time scale separation of the fast dynamics in y and the metastable transitions between the potential wells. Then, application of the ordinary averaging procedure will destroy the information about slow mixing between the two branches and the result becomes inappropriate to render the effective dynamics. By contrast, application of Theorem 4.1 does not require to fix ς: Even if $\varsigma = \varsigma(\epsilon)$ due to (3.14), the reduced model (4.7) will represent the effective dynamics of (4.4)&(4.5). We get this result because the averaging procedure does not affect the Markov chain $I(t,x)$ that stores the distributional information of the metastable transitions. Therefore, by examining the averaged system (4.7) as $\varsigma \to 0$, we will obtain the differential equation

$$\dot{x} = D_x V(x, m^{(I(t,x))}(x)) + \sigma \dot{W},$$

that is considered as the final limit SDE of the original process (1.1)&(1.2) with $\varsigma = \varsigma(\epsilon)$ given by (3.14) as $\epsilon \to 0$.

6 Numerical Experiments

In this section we illustrate the results from the preceding section by numerical experiments with an appropriate test example.

We consider the Smoluchowski equation (1.1)&(1.2) where the potential for the numerical analysis is given by:

$$V(x,y) = 2.5\,(y^2 - 1)^2 \; - \; 0.8\,x\,y^3 \; + \; 0.005\,x^4 \; + \; 1.6, \tag{6.1}$$

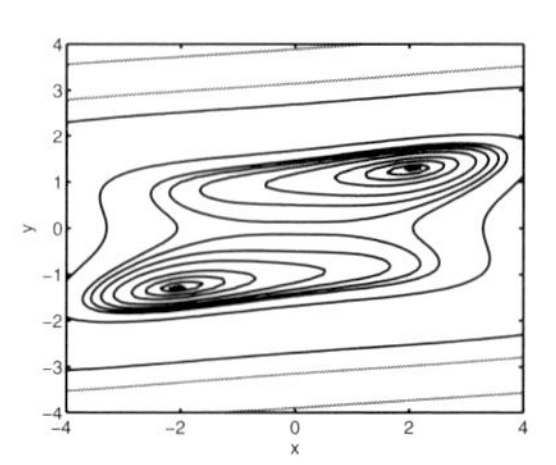

Fig. 6.1. Full potential V.

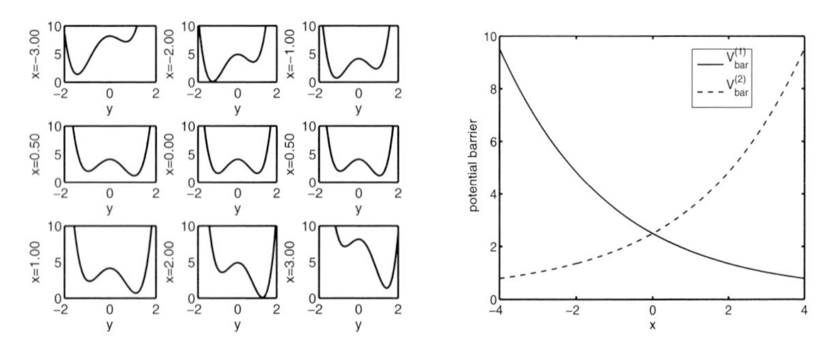

Fig. 6.2. Left: Potentials $V(x,\cdot)$ in y for different values of x. Right: potential barriers $V^{(1)}_{\text{bar}}(x)$ (full line) and $V^{(2)}_{\text{bar}}(x)$ (dashed line).

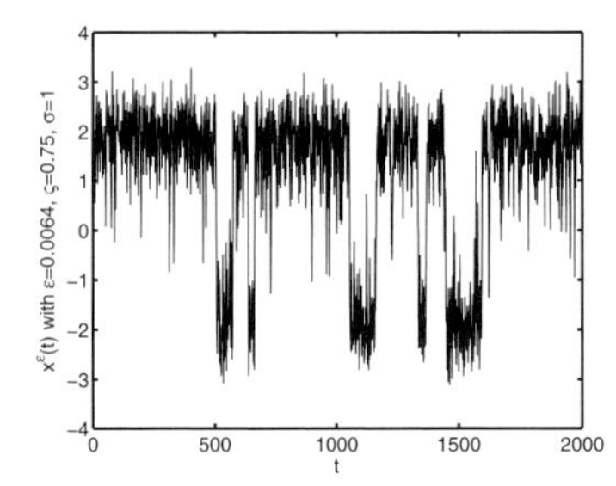

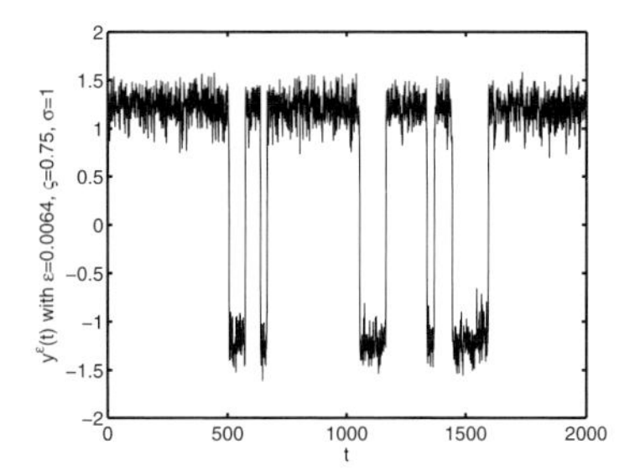

Fig. 6.3. Typical realization of the original dynamics for $\sigma = 1.0$, $\varsigma = 0.75$ and $\epsilon = 0.0064$. Left: trajectories x coordinate; right: trajectories y coordinate.

which clearly satisfies Assumption 2.1. The potential energy surface is shown in Fig. 6.1. At the left hand side of Fig. 6.2 we illustrate the double-well potentials $V(x, \cdot)$ for different values of x. The saddle point always is $y_0(x) = 0$ and takes the value $V(x, 0) = 4.1 + 0.005\,x^4$, the potential minima are

$$m^{(i)}(x) = 0.12\,x + (-1)^i \sqrt{0.0576\,x^2 + 4}, \qquad i = 1, 2.$$

The right side of Fig. 6.2 shows the potential barriers $V_{\text{bar}}^{(1)}(x)$ (the left barrier) and $V_{\text{bar}}^{(2)}(x)$ (the right barrier) as functions of x.

In Fig. 6.3 we show a typical realization of the dynamics (1.1)&(1.2) with $\sigma = 1.0$, $\varsigma = 0.75$ and $\epsilon = 0.0064$. For the generation of the trajectories we use the Euler-Maryuana scheme with internal time step $dt = \epsilon/100$. We clearly observe that jumps between the metastable decomposition $B^{(1)} = \{(x, y) \,|\, y < 0\}$ and $B^{(2)} = \{(x, y) \,|\, y > 0\}$ induce metastable transitions in the x dynamics between $x < 0$ and $x > 0$. Comparison with the averaged trajectory in Fig. 6.4 reveals inappropriateness of the standard averaging procedure (1.3). In Fig. 6.4 right we illustrate the averaged potential $\overline{V}$ (known as Fixman potential or conformational free energy landscape)that is associated with the realization at the left:

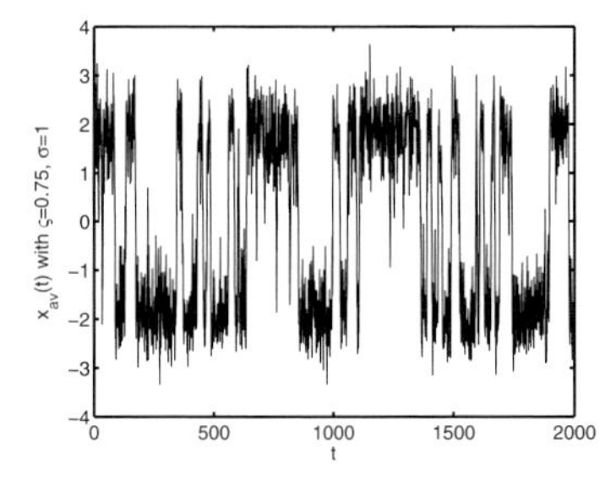

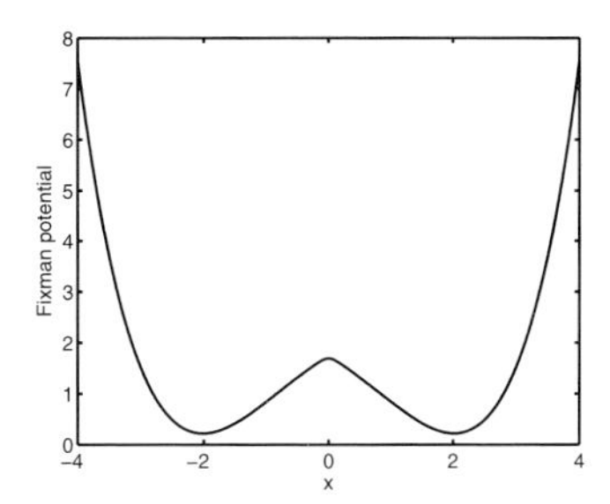

Fig. 6.4. Left: Typical realization of the simply averaged dynamics (1.3) for $\sigma = 1.0$, $\varsigma = 0.75$. Right: Fixman potential that corresponds to the trajectory at the left.

$$\overline{V}(\varsigma, x) = -\frac{\varsigma^2}{2} \ln \int \exp\big(-\frac{2}{\varsigma^2} V(x,y)\big)\, \mathrm{d}y.$$

Using standard Laplace asymptotics provides us with the potential in the limit $\varsigma \to 0$ of vanishing fast diffusion

$$\overline{V}(x) = \min\big\{V\big(x, m^{(1)}(x)\big), V\big(x, m^{(1)}(x)\big)\big\}.$$

In Fig. 6.4 we additionally plotted $\overline{V}(x)$, which graphically is completely identical to $\overline{V}(\varsigma = 0.75, x)$.

Fig. 6.3&6.4 explicitly visualize the simply averaged dynamics to be inappropriate to render the effective dynamical behaviour of $x^\epsilon(t)$ as $\epsilon \to 0$. For small ϵ diffusion in y is very fast compared to diffusion in x. However, the important (and only) barriers of the potential are barriers in y direction. Thus, for fixed ϵ, decreasing the noise intensity ς in the fast equation increases the metastability in y. Consequently, by choosing different ς one can analyze the effect of increasing metastability on averaging. To this end, it is convenient to use the x averaged values of the expected transition rates $1/\mathcal{T}^\epsilon_{i\to j}(x)$. As detailed in Appendix B this provides us in the asymptotic limit $\varsigma \to 0$ with the expected transition times $\overline{\mathcal{T}}^\epsilon_{1\to 2}$ between the metastable decomposition $B^{(1)} \cup B^{(2)}$ in the (x, y) state space.

We generated $N = 2000$ realizations of the original dynamics for $\epsilon = 0.0064$, $\sigma = 1.0$ and $\varsigma = 0.75, 0.7, 0.65, 0.60$, and waited for the first exit times from $B^{(1)}$. The top row in Fig. 6.5 illustrates the location of the trajectories x-coordinate right before the transitions occured; the pictures at the bottom display the function under the integral in (B.4) (normalized to 1) and nicely illustrate that the major contribution to the integral in (B.4) will move rightwards as $\varsigma \to 0$, for $V^{(1)}_{\mathrm{bar}}(x) \to 0$ as $x \to \infty$. Comparison of the upper and

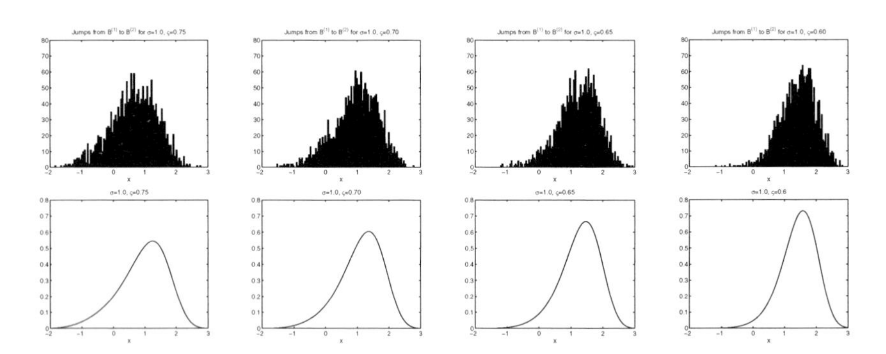

Fig. 6.5. Top: Transition location (from $B^{(1)}$ to $B^{(2)}$) of the trajectories x-coordinate computed by means of $N = 2000$ realizations of the conditionally averaged dynamics for $\sigma = 1.0$, $\epsilon = 0.0064$ fixed and $\varsigma = 0.75, 0.7, 0.65, 0.6$. Bottom: Function under the integral in (B.4) normalized to 1 by using the same parameters as above.

Table 6.1. Expectation values of transition times from $B^{(1)}$ to $B^{(2)}$ corresponding to Fig. 6.5. The values for $\overline{\mathcal{T}}^{\epsilon}_{1\to 2}$ are obtained by using (B.2) or (B.4).

Mean transition times	$\varsigma = 0.75$	$\varsigma = 0.7$	$\varsigma = 0.65$	$\varsigma = 0.60$
mean value from 2000 real.	213	462	1240	4285
averaged value $\overline{\mathcal{T}}^{\epsilon}_{1\to 2}$	119	323	1037	4157

the lower pictures reveals almost coincidence between the contribution to the integral in (B.4) and the actual location in the x space of the jumps from $B^{(1)}$ to $B^{(2)}$. Finally, we compare in Table 6.1 the averaged values of the transition times to the numerically obtained values by means of the $N = 2000$ realizations. We observe that ς has to be chosen small to get closeness.

Discretization

The pathwise simulation of the dynamics consisting of the two state Markov jump process $I(t,x)$ is developed by using a specific stochastic particle method ([HSS01]). To this end, recall the infinitesimal generator $\mathcal{Q}^{\epsilon}_x = (q^{\epsilon}_{ij}(x))_{i,j}$ that allows to calculate the hopping probabilities between the states $\mathbf{S} = \{1,2\}$. The transition matrix $P^{\epsilon}_{\tau}(x) = (p^{\epsilon}_{ij}(\tau,x))$ at time τ is then obtained by

$$P^{\epsilon}_{\tau}(x) = \exp(\tau \mathcal{Q}^{\epsilon}_x).$$

A straightforward calculation reveals

$$p^{\epsilon}_{12}(\tau,x) = \frac{q^{\epsilon}_{12}(x)}{q^{\epsilon}_{12}(x) + q^{\epsilon}_{21}(x)} \left(1 - e^{-\tau(q^{\epsilon}_{12}(x)+q^{\epsilon}_{21}(x))}\right), \tag{6.2}$$

$$p^{\epsilon}_{21}(\tau,x) = \frac{q^{\epsilon}_{21}(x)}{q^{\epsilon}_{12}(x) + q^{\epsilon}_{21}(x)} \left(1 - e^{-\tau(q^{\epsilon}_{12}(x)+q^{\epsilon}_{21}(x))}\right). \tag{6.3}$$

The entries of $\mathcal{Q}^{\epsilon}_x$ are given in (4.6) by the inverse of the precise estimates of the expected transition times over the potential energy barrier in y direction.

The stochastic particle method requires two steps. We shortly demonstrate it for the OU averaged dynamics (4.7).

Step 1: Transport. The first step consists of determining an updated position $x(t + \mathrm{d}t)$ by solving

$$\dot{x} = -\int D_x V(x,y)\, \mu_x^{\mathrm{OU}(i)}(y)\, \mathrm{d}y + \sigma \dot{W}_1,$$

over $[0, \mathrm{d}t]$ with initial point $x(t)$.

Step 2: Exchange. The second step models the exchange between the states $I(t,x) = 1$ and $I(t,x) = 2$. Thus, if $i = 1$, we set $i = 2$ with hopping probability $p_{1\to 2} = p^{\epsilon}_{12}(\mathrm{d}t, x(t + \mathrm{d}t))$ and remain at $i = 1$ with probability $1 - p_{1\to 2}$. Vice versa, if $i = 2$, we set $i = 1$ with hopping probability $p_{2\to 1} = p^{\epsilon}_{21}(\mathrm{d}t, x(t + \mathrm{d}t))$ and remain at $i = 2$ with probability $1 - p_{2\to 1}$. Return to step 1 by setting $x(t) = x(t + \mathrm{d}t)$.

Parameter Choice

Subsequently, we choose the noise intensity in the slow equation $\sigma = 1$ and the smallness parameter is $\epsilon = 0.0064$. Trajectories are illustrated with $\varsigma = 0.75$, whereas for comparison of exit times we use different values of ς.

Recalling coupling ς to ϵ according to (3.14), some words seem to be necessary concerning the comparison of the full dynamics to the OU approximated ones: Without loss of generality we can choose ς arbitrary without considering the coupling, for the experiments are performed for a fixed value of ϵ. Therefore, for fixed $\epsilon = \epsilon^*$ and fixed $\varsigma = \varsigma^*$ we can always find a constant $K = K^*$ (or a barrier $V_{\text{bar}}^{\text{small}} = V_{\text{bar}}^{\text{small}*}$) such that $\varsigma(\epsilon^*) = \varsigma^*$ under (3.14). Even if we take (3.13) as the basis of our computation, we can desist from the coupling rule, for the constant K then can be chosen dependent of x, such that we still arrive at $\varsigma(\epsilon^*, x) = \varsigma^*$. Actually, the postulation of relating ς to ϵ only serves as a formal justification of the OU approximation. For the numerical implementation only the size of ς by its own is of importance, not its relation to ϵ.

The motivation to choose $\sigma = 1.0$ and not $\sigma = \varsigma$ can be infered from Fig. 6.5. In case of smaller values of σ, say $\sigma = 0.75$, the x-coordinate of the trajectory will hardly reach the region where the jumps mostly happen. Then we had to choose ς larger, which on its part would result in a worse approximation of the intra-well fast dynamics. We will come back to this problem below.

6.1 Comparison Between Original Dynamics and Small Noise Approximation

Here, we carry out numerical studies in order to compare the Smoluchowski dynamics (1.1)&(1.2) with those governed by system (4.4)&(4.5) with fast OU processes and transition chain $I(t, x)$ that controls the switches between the two OU processes.

Typical realizations of both the original dynamics and the OU approximated ones are shown in Fig. 6.6. The trajectories have been generated using the Euler-Maruyama scheme with time step $\mathrm{d}t = \epsilon/100$ for both systems. Apparently, the transition rates between $B^{(1)}$ and $B^{(2)}$ coincide to some extend and the oscillating motion (around the potential minima in y) inbetween the transitions seems to be well approximated by using OU processes in the fast equation. We clearly observe that jumps induce metastable transitions in the x dynamics between $x < 0$ and $x > 0$. However, for the trajectories being in $B^{(1)}$ we observe that the x-coordinate of the original dynamics spreads considerably further rightwards than the x trajectory of the OU approximated system (and for the trajectories in $B^{(2)}$ the original dynamics x-coordinate spreads further leftwards).

The above observation suggests that the original dynamics have noticeable smaller transition times between $B^{(1)}$ and $B^{(2)}$, as the original dynamics more

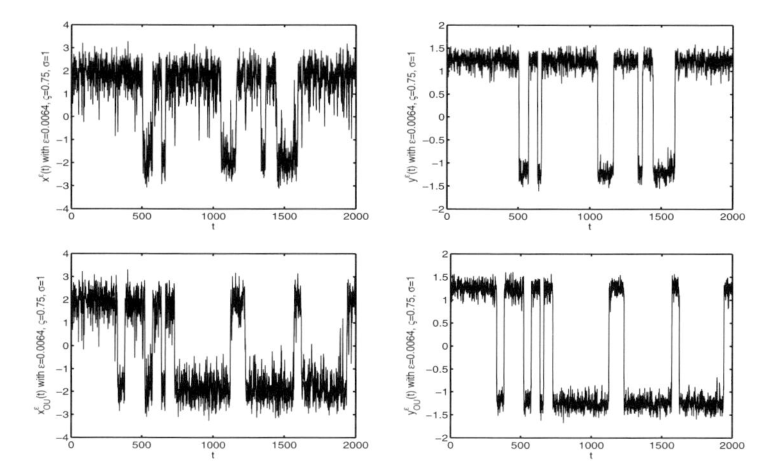

Fig. 6.6. Typical realization of the original dynamics (top) and the approximated dynamics with fast OU processes (bottom). At the left we see the x, at the right the y coordinate. The realizations have been computed for the same realization of the white noise (in the slow and in the fast equation).

often reaches a domain where the potential barriers (in y direction) are small. This is confirmed by Table 6.2, where we computed the expected transition times from $B^{(1)}$ to $B^{(2)}$ by means of $N = 2000$ realizations for two different values of ς and $\sigma = 1.0$, $\epsilon = 0.0064$ fixed. We come back to this problem in the next section where we include the averaged dynamics into the numerical examinations. Actually, it will turn out that ς has to be chosen very small to get perfect coincidence of both the original and the OU approximated system.

6.2 Results Including Averaged Dynamics

We now demonstrate pre-eminence of the OU averaged dynamics (4.7). To complete the representation we include the conditionally averaged dynamics (1.7).

In Fig. 6.7 we compare realizations of the averaged to the full dynamics' x-coordinate. Every trajectory has been computed with the same realization of white noise $\dot{W}_1$, $\dot{W}_2$, such that the internal time step has been set to $\mathrm{d}t = \epsilon/100$

Table 6.2. Exit times from the set $B^{(1)}$ for the original dynamics and the OU approximated system.

dynamical model	$\varsigma = 0.8$	$\varsigma = 0.75$
original dynamics	113	210
OU approximated dynamics	136	265

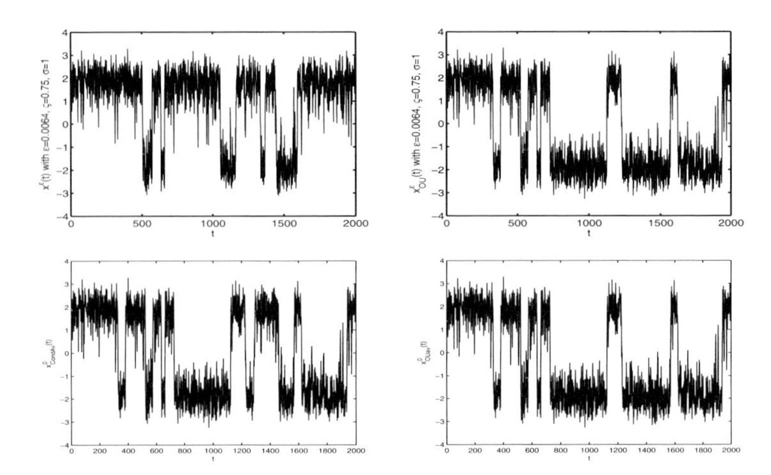

Fig. 6.7. Realizations of the original dynamics x coordinate (top, left), the x coordinate of the (full) OU approximated dynamics (top, right), the OU averaged dynamics (bottom, right), and the conditionally averaged system (bottom, left).

even for the averaged dynamics. The Markov jump process $I(t,x)$ is realized by using the same realization of random numbers for every concerned system. Concerning the systems with OU processes (full and OU averaged), we observe pathwise convergence of the x trajectories, whereas comparison of the original dynamics with the conditionally averaged system reveals distributional coincidence.

In order to present numbers instead of pictures we want to compute the expectation values of the metastable transition times from $x < 0$ to $x > 0$ for different values of ς. It is natural to expect that this is realized by computing the first exit times from the set $S + \delta$ with $S = \{x \in \mathbf{R} \mid x < 0\}$, where $\delta > 0$ has to be large enough to guarantee that the process effectively reaches some (small) region of attraction in the complement of S. But Fig. 6.5 nicely shows that the x-coordinate can spread far into the positive region even when it is restricted to the metastable set S. Thus, we suggest to define the stopping time as the first exit from $B^{(1)}$ instead, respectively the first jump from $I(t,x) = 1$ to $I(t,x) = 2$. At least for $\varsigma \leq 7.5$ (compare Fig. 6.6) this is equivalent to the metastable transitions from $x < 0$ to $x > 0$. From $N = 2000$ realizations for $\epsilon = 0.0064$ and $\sigma = 1.0$ we get a very good agreement between the OU approximated dynamics and the OU averaged dynamics, and a good agreement between the original and the conditionally averaged dynamics.

However, there still remains the problem of difference between the OU averaged and the conditionally averaged dynamics. To overcome the problem, we illustrate in Fig. 6.8 the potentials that correspond to the respective trajectories for $\varsigma = 0.75, 0.60, 0.3$. For $I(t,x) = i \in \{1,2\}$ and ς fixed, the conditionally averaged potential $\overline{V}^{(i)}(\varsigma, x)$ and the OU averaged potential $\overline{V}^{\mathrm{OU}(i)}(\varsigma, x)$ are

Table 6.3. Comparison of exit times from the metastable set $S = \{x \in \mathbf{R} \mid x < 0\}$. For $\varsigma = 0.65$, it was not possible to compute the exit times of the full dynamics' motion within a reasonable period of time.

dynamical model	$\varsigma = 0.8$	$\varsigma = 0.75$	$\varsigma = 0.65$
original dynamics	113	210	$--$
conditinally averaged dynamics	105	213	1240
OU approximated dynamics	136	265	$--$
OU averaged dynamics	135	265	1537

defined implicitly by

$$D_x \overline{V}^{\mathrm{OU}(i)}(\varsigma, x) = \int D_x V(x,y)\, \mu_x^{\mathrm{OU}(i)}(y)\, \mathrm{d}y,$$

$$D_x \overline{V}^{(i)}(\varsigma, x) = \int D_x V(x,y)\, \mu_x^{(i)}(y)\, \mathrm{d}y,$$

and we easily show that

$$\overline{V}^{(i)}(\varsigma, x) = -\frac{\varsigma^2}{2} \ln \int_{B_x^{(i)}} \exp\big(-\frac{2}{\varsigma^2} V(x,y)\big)\, \mathrm{d}y.$$

Exploiting the estimation method of Laplace we obtain asymptotical identity of both potentials:

$$\lim_{\varsigma \to 0} \overline{V}^{(i)}(\varsigma, x) = \lim_{\varsigma \to 0} \overline{V}^{\mathrm{OU}(i)}(\varsigma, x) = V(x, m^{(i)}(x)).$$

Fig. 6.8 reveals $V^{(i)}(\varsigma, x) \approx \overline{V}^{\mathrm{OU}(i)}(\varsigma, x)$ for $\varsigma \leq 0.3$, whereas they differ visibly for $\varsigma \geq 0.60$ mainly in that region where the jumps from $i = 1$ to $i = 2$ mostly happen. This perfectly explains the significant difference concerning the transition times in Tables 6.2&6.3.

A Derivation of Reduced System by Multiscale Analysis

Here, we show how the averaged system (4.7) can be derived from the system (4.4)&(4.5) with fast OU processes by using multiscale asymptotics wrt.

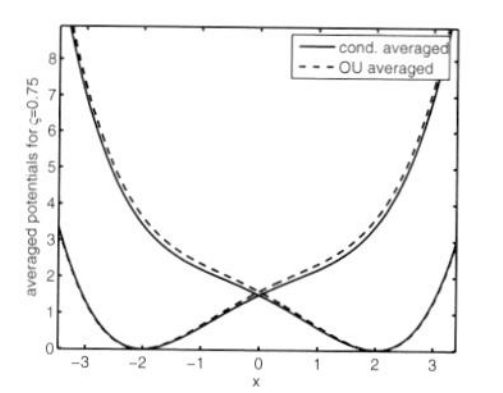

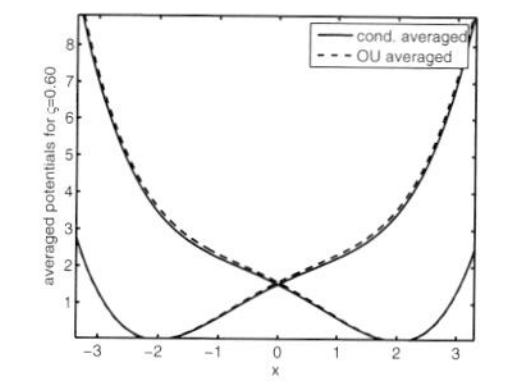

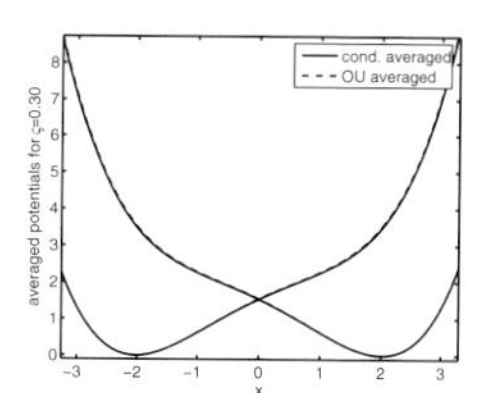

Fig. 6.8. Comparison of conditinally averaged (full line) to OU averaged potentials (dashed line). From left to right: $\varsigma = 0.75, 0.60, 0.3$.

the smallness parameter ϵ. As the method is applied to ensemble instead of single dynamics we have to set up before the necessary requirements concerning the evolution of probability densities.

Let us extend the fast-slow system with two OU processes to a finite number of OU processes. Thus, we consider the process $(x^{\epsilon}_{\mathrm{OU}}, y^{\epsilon}_{\mathrm{OU}}) \in \mathbf{R}^m \times \mathbf{R}$ that is determined by the following SDE:

$$\dot{x}(t) = -D_x V(x, y) + \sigma \dot{W}_1, \tag{A.1}$$

$$\dot{y}(t) = -\frac{1}{\epsilon}\, \omega^{(I(t,x))}(x)\left(y - m^{(I(t,x))}(x)\right) + \frac{\varsigma(x)}{\sqrt{\epsilon}} \dot{W}_2, \tag{A.2}$$

where $I(t, x)$ is a right-continuous Markov chain on a probability space taking values in a finite state space $\mathbf{S} = \{1, 2, ..., N\}$ and $\omega^{(i)}(x)$ takes values in $\mathbf{R}^+$ for all $i \in \mathbf{S}$. The noise intensity of the fast diffusion may depend on x, but is assumed to be strictly positive, that is $\varsigma(x) \geq c > 0$. To simplify notation we perform the asymptotic procedure without a possible dependence of ς on the Markov chain $I(t, x)$; a generalization in this direction had no effect on the computation. The generator $\mathcal{Q}_x = (q_{ij}(x))_{N\times N}$ of the switching chain $I(t, x)$ depends on the slow variable x and contains the transition rates $q_{ij} = q_{ij}(x) > 0$ from i to j if $i \neq j$ while

$$q_{ii}(x) = -\sum_{i\neq j} q_{ij}(x). \tag{A.3}$$

For fixed $x \in \mathbf{R}^m$ and $i \in \mathbf{S}$ the diffusion dynamics (A.2) is known as OU process and consequently ergodic. The (unique) stationary density $\mu_x^{\mathrm{OU}(i)}$ is given by

$$\mu_x^{\mathrm{OU}(i)}(y) = \frac{1}{\varsigma(x)}\sqrt{\frac{\omega^{(i)}(x)}{\pi}}\, \exp\left(-\,\omega^{(i)}(x)\frac{\left(y - m^{(i)}(x)\right)^2}{\varsigma(x)^2}\right), \tag{A.4}$$

which is a Gaussian with mean $m^{(i)}(x)$ and variance $\varsigma(x)^2/(2\omega^{(i)}(x))$, and thus independent of ϵ.

The evolution of probability densities $p^{\epsilon} \in L^1(\mathbf{R}^{m+1} \times \mathbf{S})$ under the dynamics given by (A.1)&(A.2) is described by the forward *Fokker-Planck equation*. Here, we are working in unweighted function spaces, that is, the density p^{ϵ} gives the physical probability to find the system in state (x, y) at time t. For later use it may be helpful to slightly change notation for the densities p^{ϵ}: The agreement $p^{\epsilon}_{(i)}(t, x, y) := p^{\epsilon}(t, x, y, i)$ enables us to represent p^{ϵ} as an N-dimensional vector according to $p^{\epsilon} = (p^{\epsilon}_{(1)}, ..., p^{\epsilon}_{(N)})$ with $p^{\epsilon}_{(i)} \in L^1(\mathbf{R}^{m+1})$. Now, the Fokker-Planck equation is regarded on some suitable subspace of $L^1(\mathbf{R}^{m+1} \times \mathbf{S})$, and reads

$$\partial_t p^{\epsilon} = \mathcal{A}^{\epsilon} p^{\epsilon}, \qquad \mathcal{A}^{\epsilon} = \frac{1}{\epsilon}\mathcal{A}_x + \mathcal{A}_y + \mathcal{Q}^T, \tag{A.5}$$

$$\mathcal{A}_x = \begin{pmatrix} \mathcal{A}_x^{(1)} & 0 & 0 & 0 \\ 0 & \mathcal{A}_x^{(2)} & 0 & 0 \\ 0 & 0 & \ddots & 0 \\ 0 & 0 & 0 & \mathcal{A}_x^{(N)} \end{pmatrix}, \ \mathcal{A}_y = \begin{pmatrix} \mathcal{A}_y^{(1)} & 0 & 0 & 0 \\ 0 & \mathcal{A}_y^{(2)} & 0 & 0 \\ 0 & 0 & \ddots & 0 \\ 0 & 0 & 0 & \mathcal{A}_y^{(N)} \end{pmatrix}$$

where $\mathcal{A}_x^{(i)}$ and $\mathcal{A}_y^{(i)}$ are given for $f \in L^1(\mathbf{R}^{m+1})$ by

$$\mathcal{A}_x^{(i)} f(x,y) = \frac{\varsigma(x)^2}{2} \Delta_y f(x,y) + D_y \Big(\omega^{(i)}(x) \big(y - m^{(i)}(x) \big) f(x,y) \Big)$$
$$\mathcal{A}_y^{(i)} f(x,y) = \frac{\sigma^2}{2} \Delta_x f + D_x \big(D_x V(x,y) f(x,y) \big).$$

Note that we actually have to use $\mathcal{Q}^T$ in (A.5), for the rate matrix $\mathcal{Q}$ is basically considered to be part of the backward Chapman-Kolmogorov equation, that is, it describes the evolution of the expectations of functions of the state of the system. Consequently, the probability to be in state (x,y) is given by

$$\langle p^\epsilon(t,x,y), \mathbf{1} \rangle_{\mathbf{S}} = \sum_{i \in \mathbf{S}} p^\epsilon_{(i)}(t,x,y),$$

$\langle \cdot, \cdot \rangle_{\mathbf{S}}$ denoting the Euclidean inner product in $\mathbf{R}^N$.

Our aim is to average with respect to the fast variable y and obtain an averaged equation for the slow variable x alone. To this end, we will use multiscale analysis.

Projection Operator

We would like to derive an equation for the distribution function in x:

$$\int \langle p^\epsilon(t,x,y), \mathbf{1} \rangle_{\mathbf{S}} \, \mathrm{d}y = \sum_{i \in \mathbf{S}} \int p^\epsilon_{(i)}(t,x,y) \, \mathrm{d}y,$$

which would be valid in the limit where ϵ becomes very small. To this end, we introduce the vector $\overline{p}^\epsilon(t,x) = (\overline{p}^\epsilon_{(1)}, ..., \overline{p}^\epsilon_{(N)})^T$ with densities $\overline{p}^\epsilon_{(i)} \in L^1(\mathbf{R}^m)$ defined by

$$\overline{p}^\epsilon_{(i)}(t,x) = \int p^\epsilon_{(i)}(t,x,y) \, \mathrm{d}y.$$

It is expected that an approximate solution of the full dynamics would be obtained by multiplying each $\overline{p}^\epsilon_{(i)}(t,x)$ by the stationary distribution $\mu_x^{\mathrm{OU}(i)}$ of the SDE (A.2) for fixed $I(t,x) = i$. We formalize this by defining a *projection operator* $\Pi = \mathrm{diag}(\Pi^{(1)}, ..., \Pi^{(N)})$ acting on functions $f = (f_1, ..., f_N)^T \in L^1(\mathbf{R}^{m+1} \times \mathbf{S})$ by

$$(\Pi f)(x,y) = \mathrm{diag}(\mu_x^{\mathrm{OU}(1)}, ..., \mu_x^{\mathrm{OU}(N)}) \int f(x,y) \, \mathrm{d}y.$$

It is obvious that Π projects any function into the subspace of all functions which can be written in the form

$$f = (f_1, ..., f_N)^T, \quad f_i(x,y) = \overline{f}_i(x)\,\mu_x^{\mathrm{OU}(i)}(y), \tag{A.6}$$

where $\overline{f}_i$ is an arbitrary function of $L^1(\mathbf{R}^m)$, thus $\overline{f} = (\overline{f}_1, ..., \overline{f}_N)^T \in L^1(\mathbf{R}^m \times \mathbf{S})$. In the following we study the case where the initial condition $p^\epsilon(t=0,x,y)$ can be expressed by

$$p^\epsilon(t=0,x,y) = (\Pi p^\epsilon(t=0))(x,y).$$

However, functions f of the form (A.6) are all solutions of

$$\mathcal{A}_x f = 0,$$

that is, the space into which Π projects is the kernel or nullspace of $\mathcal{A}_x$ expressed by $\mathcal{A}_x\Pi = 0$. Due to the properties of $\mathcal{A}_x^{(i)}$ considered as an operator acting on functions g in y, that is $g = g(y) \in L^1(\mathbf{R})$, we furthermore have:

$$\Pi\mathcal{A}_x = 0 = \mathcal{A}_x\Pi. \tag{A.7}$$

This is easily seen by introducing the formal adjoint $\mathcal{T}_x^{(i)}$ of $\mathcal{A}_x^{(i)}$, i.e., a differential operator such that for all $u \in L^1(\mathbf{R})$, $v \in L^\infty$ (or $u, v \in L^2(\mathbf{R})$) we have

$$\langle \mathcal{A}_x^{(i)} u, v\rangle_{L^2} = \langle u, \mathcal{T}_x^{(i)} v\rangle_{L^2}, \qquad \langle u, v\rangle_{L^2} := \int u(y)\,\overline{v(y)}\,\mathrm{d}y.$$

If we consider $\Pi^{(i)}$ – for fixed x – as an operator acting on functions in y, we can rewrite it by

$$\Pi^{(i)} u = \langle u, \mathbf{1}\rangle_{L^2}\,\mu_x^{\mathrm{OU}(i)}.$$

Together with the well-known fact that $\mathcal{T}_x^{(i)}\mathbf{1} = 0$ (see, e.g., [SHD01, Hui01]) we finally get the desired result (A.7).

Multiscale Analysis

We now make the following ansatz for the solution of the Fokker-Planck equation with the initial conditions described above:

$$p^\epsilon = p^0 + \epsilon\, p^1 + \epsilon^2 p^2 + ...$$

This ansatz is inserted into the Fokker-Planck equation (A.5) and then, by comparison of coefficients of different powers of ϵ we get:

$$\epsilon^{-1}: \quad \mathcal{A}_x\, p^0 = 0 \tag{A.8}$$

$$\epsilon^{0}: \quad \mathcal{A}_x\, p^1 + (\mathcal{A}_y + \mathcal{Q}^T)\, p^0 = \partial_t p^0 \tag{A.9}$$

$$\epsilon^{1}: \quad \mathcal{A}_x\, p^2 + (\mathcal{A}_y + \mathcal{Q}^T)\, p^1 = \partial_t p^1 \tag{A.10}$$

Step 1: (A.8) immediately yields that $p^0 \in \mathcal{N}(\mathcal{A}_x)$, i.e.,

$$\begin{aligned} \Pi p^0 &= p^0, \quad \text{equivalently} \\ p^0(t,x,y) &= \operatorname{diag}(\mu_x^{\mathrm{OU}(1)},...,\mu_x^{\mathrm{OU}(N)})\,\overline{p}^0(t,x), \end{aligned} \tag{A.11}$$

for a function $\overline{p}^0 \in L^1(\mathbf{R}^m \times \mathbf{S})$ depending only on x.

Step 2: Let Π act on (A.9) and use (A.7). This time we get:

$$\Pi(\mathcal{A}_y + \mathcal{Q}^T)\Pi p^0 = \partial_t \Pi p^0. \tag{A.12}$$

By using (A.11) simple calculations reveal for $\overline{p}^0 = (\overline{p}^0_{(1)},...,\overline{p}^0_{(N)})^T$:

$$\partial_t \overline{p}^0 = (\overline{\mathcal{A}} + \mathcal{Q}^T)\,\overline{p}^0, \tag{A.13}$$

with

$$\overline{\mathcal{A}} = \begin{pmatrix} \overline{\mathcal{A}}^{(1)} & 0 & 0 & 0 \\ 0 & \overline{\mathcal{A}}^{(2)} & 0 & 0 \\ 0 & 0 & \ddots & 0 \\ 0 & 0 & 0 & \overline{\mathcal{A}}^{(N)} \end{pmatrix},$$

$$\overline{\mathcal{A}}^{(i)} = \frac{\sigma^2}{2}\Delta_x + D_x\Big(\int D_x V(x,y)\,\mu_x^{\mathrm{OU}(i)}(y)\,\mathrm{d}y\,\cdot\,\Big),$$

$\overline{\mathcal{A}}$ acting on $L^1(\mathbf{R}^m \times \mathbf{S})$. Thus $\overline{p}^0$ is determined by a Fokker-Planck equation, and its solution gives us p^ϵ up to error $\mathcal{O}(\epsilon)$. The associated SDE is given by

$$\dot{x} = -\int D_x V(x,y)\,\mu_x^{\mathrm{OU}(I(t,x))}(y)\,\mathrm{d}y + \sigma \dot{W}_1, \tag{A.14}$$

with solution process $x^0(t)$ where $I(t,x) \in \mathbf{S}$ controls the switches between the different OU processes due to the rate matrix $\mathcal{Q} = \mathcal{Q}_x$. The SDE (A.14) describes the limit dynamics of (A.1)&(A.2) in the sense that its solution satisfies $x^\epsilon_{\mathrm{OU}} \to x^0$ as $\epsilon \to 0$ either pathwise [FW84], or in the distributional sense [Kur73, MTV99].

B Asymptotics of Transition Times

Here, we come back to the problem addressed in Sect. 3.3. In order to avoid coupling ς to the slow variable dynamics x we relax the postulation $\mathcal{T}^\epsilon_{i\to j}(x) \geq \mathrm{ord}(1)$, $i \neq j$ in Lemma 3.5 that is required for every x and $i = 1, 2$. Instead of considering the transition times on every fibre of the fast state space for fixed x, we introduce the expected transition times $\overline{\mathcal{T}}^\epsilon_{1\to 2}$ and $\overline{\mathcal{T}}^\epsilon_{2\to 1}$ between the metastable decomposition $B^{(1)} \cup B^{(2)}$ in the entire (x, y) state space. This enables us to identify large time conformational changes with the stipulation

$$\overline{\mathcal{T}}^{\epsilon}_{1\to 2}, \overline{\mathcal{T}}^{\epsilon}_{2\to 1} \geq \mathrm{ord}(1).$$

We obviously have

$$\mathcal{T}^{\epsilon}_{i\to j}(x) \geq \mathrm{ord}(1) \qquad \Longrightarrow \qquad \overline{\mathcal{T}}^{\epsilon}_{1\to 2}, \overline{\mathcal{T}}^{\epsilon}_{2\to 1} \geq \mathrm{ord}(1),$$

whereas the other direction need not to be valid.

With these preparations we claim the following: If we define the relationship between ς and ϵ by

$$\varsigma(\epsilon) = \left(\frac{2V^{\mathrm{small}}_{\mathrm{bar}}}{\ln(K/\epsilon)}\right)^{1/2}, \tag{B.1}$$
$$V^{\mathrm{small}}_{\mathrm{bar}} = \min\{V^{(i)}_{\mathrm{bar}}(x) \mid x \in D(\sigma),\, i = 1, 2\},$$

where $D(\sigma)$ is some appropriately chosen bounded connected domain[6] of the slow variable state space, the metastable transitions $\overline{\mathcal{T}}^{\epsilon}_{1\to 2}, \overline{\mathcal{T}}^{\epsilon}_{2\to 1}$ are of order one or even larger.

In what follows we show how to compute $\overline{\mathcal{T}}^{\epsilon}_{i\to j}$ which is strongly connected to the asymptotic order of the transition times $\mathcal{T}^{\epsilon}_{i\to j}(x)$ on every fiber. We will consider two possible situations that are exemplary for the different approaches. We first examine the consequences of the asymptotic order of $\mathcal{T}^{\epsilon}_{i\to j}(x)$ in general, and afterwards relate the results to the functions $V^{(i)}_{\mathrm{bar}}(x)$ and a coupling $\varsigma = \varsigma(\epsilon)$ given by (B.1).

In Theorem B.1 below we assume the transition times $\mathcal{T}^{\epsilon}_{i\to j}(x)$ to asymptotically go to infinity, where we do not specify wherefrom the asymptotic investigations come from, that is, we leave open which parameter causes the asymptotic behaviour. Thus, possible (and reasonable) choices were $\varsigma \to 0$ and ϵ fixed, $\epsilon \to 0$ so that $\varsigma(\epsilon) \to 0$, or, not less supposable, we could assume a scaling of the potential barrier. The next result becomes apparent in [Wal05, Chapter 3], where the approach is justified by means of multiscale analysis for distinguished time scales. There the metastable transitions are assumed to happen on the longest time scale, which requires the averaging of the metastable transition rates (represented by the second eigenvalue of the corresponding generator) for fixed x wrt. the invariant density of the conditionally averaged potentials.

Theorem B.1 ([Wal05, Chapter 3.3.3]). *Suppose $\mathcal{T}^{\epsilon}_{i\to j}(x) \to \infty$ almost everywhere for $i, j = 1, 2$ and $i \neq j$. Then the metastable transition times $\overline{\mathcal{T}}^{\epsilon}_{i\to j}$ are basically independent of the starting point and are asymptotically derived by means of averaging the x-dependent transition rates[7] against the invariant*

[6] See explanation in Sect. 3.3.

[7] Note, that we actually have to average the transition rates and *not* the transition times.

probability distribution of the x dynamics conditioned upon remaining within the metastable set $B^{(i)}$ *and taking the inverse of the averaged transition rates, that is,*

$$\overline{\mathcal{T}}^{\epsilon}_{i\to j} \simeq \frac{1}{\mathbf{E}_{\bar{\mu}^{(i)}}[1/\mathcal{T}^{\epsilon}_{i\to j}(x)]},$$

where the quantity $\mathbf{E}_{\bar{\mu}^{(i)}}[1/\mathcal{T}^{\epsilon}_{i\to j}(x)]$ is given by

$$\mathbf{E}_{\bar{\mu}^{(i)}}[1/\mathcal{T}^{\epsilon}_{i\to j}(x)] = \int 1/\mathcal{T}^{\epsilon}_{i\to j}(x)\,\bar{\mu}^{(i)}(x)\,\mathrm{d}x, \tag{B.2}$$

$$\bar{\mu}^{(i)}(x) = \frac{1}{Z^{(i)}}\exp\Big(-\frac{2}{\sigma^2}\Big(-\frac{\varsigma^2}{2}\ln\int_{B^{(i)}_x}\exp\big(-\frac{2}{\varsigma^2}V(x,y)\big)\mathrm{d}y\Big)\Big). \tag{B.3}$$

Here, $Z^{(i)}$ denotes the normalization constant and depends on ς as well. Define the jump process $\overline{I}(t)$ by its transition rates $1/\overline{\mathcal{T}}^{\epsilon}_{i\to j}$. Then we find that the random process $(x^{\epsilon}_{\mathrm{OU}}, y^{\epsilon}_{\mathrm{OU}})$ determined by system (4.4)&(4.5) is asymptotically given by the SDE

$$\begin{aligned}\dot{x} &= -D_xV(x,y) + \sigma\dot{W}_1\\ \dot{y} &= -\frac{1}{\epsilon}\,\omega^{(\overline{I}(t))}(x)\big(y - m^{(\overline{I}(t))}(x)\big) + \frac{\varsigma}{\sqrt{\epsilon}}\dot{W}_2.\end{aligned}$$

In the limit of small noise ς the evaluation of the expression (B.2) asymptotically reduces to

$$\mathbf{E}_{\bar{\mu}^{(i)}}[1/\mathcal{T}^{\epsilon}_{i\to j}(x)] \simeq \tag{B.4}$$
$$\frac{1}{\epsilon}\frac{1}{Z^{(i)}}\int\frac{\sqrt{\omega^{(i)}(x)\omega_0(x)}}{2\pi}\exp\Big(-\frac{2}{\varsigma^2}V^{(i)}_{\mathrm{bar}}(x)\Big)\exp\Big(-\frac{2}{\sigma^2}V(x,m^{(i)}(x))\Big)\,\mathrm{d}x.$$

Proof. We only have to show (B.4). The rest is verified in [Wal05]. First, we consider the averaged density $\bar{\mu}^{(i)}(x)$: Using standard Laplacian asymptotics, we get for ς small

$$\int_{B^{(i)}_x}\exp\big(-\frac{2}{\varsigma^2}V(x,y)\big)\mathrm{d}y = \varsigma\sqrt{\frac{\pi}{\omega^{(i)}(x)}}\exp\big(-\frac{2}{\varsigma^2}V(x,m^{(i)}(x))\big)\big(1+\mathcal{O}(\varsigma)\big),$$

and, exploiting $(\varsigma^2/2)\ln(\varsigma\sqrt{\pi/\omega^{(i)}(x)}) \to 0$ as $\varsigma\to 0$, we end up with the asymptotic limit (from (B.3))

$$\bar{\mu}^{(i)}(x) \simeq \frac{1}{Z^{(i)}}\exp\big(-\frac{2}{\sigma^2}V(x,m^{(i)}(x))\big).$$

Together with Theorem 3.3 we immediately obtain (B.4).

Remark B.2. If we consider the asymptotics of the transition times for vanishing ς and ϵ fixed, it is easily seen that the assumptions of Theorem B.1 are fulfilled. However, if we consider the asymptotic limit for $\epsilon\to 0$ and $\varsigma = \varsigma(\epsilon)$ as given by (B.1), the behaviour of $\mathcal{T}^{\epsilon}_{i\to j}(x)$ will depend on the course of the functions $V^{(i)}_{\mathrm{bar}}(x)$, $i = 1, 2$.

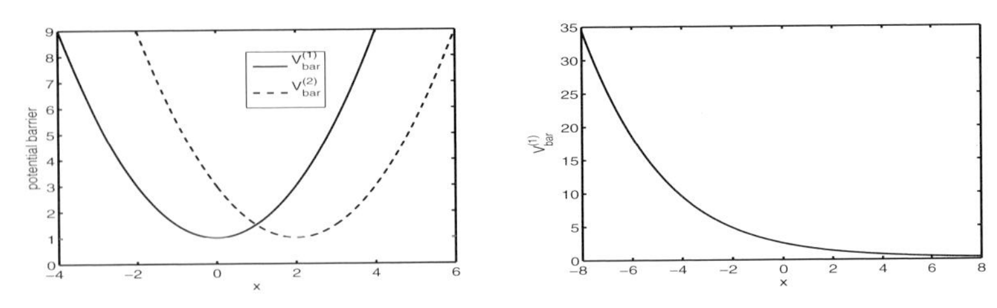

Fig. B.1. Exemplary possibilities for the functions $V_{\text{bar}}^{(i)}(x)$, $i = 1, 2$. In the left picture we find that the left and right potential barriers are bounded away from zero by a positive constant. This is prevented in the picture at the right, for $V_{\text{bar}}^{(1)}$ converges to zero.

In the situation illustrated in the left panel of Fig. B.1 we may apply the approach in Theorem for sure, if the relationship between ς and ϵ is defined in an appropriate way. This is formulated in the next corollary.

Corollary B.3. *Suppose that* $\inf\{V_{\text{bar}}^{(i)}(x) \mid x \in \mathbf{R},\ i = 1, 2\} = V_{\text{bar}}^{\text{small}} > 0$ *and* $V_{\text{bar}}^{(i)} > V_{\text{bar}}^{\text{small}}$ *almost everywhere. Let us define the small noise intensity* ς *by*

$$\varsigma(\epsilon) = \left(\frac{2V_{\text{bar}}^{\text{small}}}{\ln(K/\epsilon)} \right)^{1/2}, \qquad K > 0. \tag{B.5}$$

Then, we obtain in the asymptotic limit $\epsilon \to 0$

$$\overline{\mathcal{T}}_{i\to j}^{\epsilon} \simeq \left(\int 1/\mathcal{T}_{i\to j}^{\epsilon}(x)\, \bar{\mu}^{(i)}(x)\, \mathrm{d}x \right)^{-1}$$

Assume in addition that $V_{\text{bar}}^{(1)}$ *attains its smallest value and let the minimum of* $V_{\text{bar}}^{(1)}$ *occur at, say,* x_0. *Moreover, we assume* $V_{\text{bar}}^{(1)} > V_{\text{bar}}^{(1)}(x_0) \geq V_{\text{bar}}^{\text{small}}$ *for all* $x \neq x_0$ *and* $D_{xx}V_{\text{bar}}^{(1)}(x_0) \neq 0$. *Then*

$$\mathbf{E}_{\bar{\mu}^{(1)}}[1/\mathcal{T}_{1\to 2}^{\epsilon}(x)] \simeq \tag{B.6}$$
$$\sqrt{\frac{\omega^{(1)}(x_0)\omega_0(x_0)}{4\,\pi\,\partial_x^2 V_{\text{bar}}^{(1)}(x_0)}}\ \frac{1}{Z^{(1)}}\ \exp(-\frac{2}{\sigma^2}V(x_0, m^{(1)}(x_0)))\ \frac{\varsigma}{\epsilon}\ \exp(-\frac{2}{\varsigma^2}V_{\text{bar}}^{(1)}(x_0)).$$

For $\delta \geq 0$ *so that* $V_{\text{bar}}^{(1)}(x_0) = (1+\delta)V_{\text{bar}}^{\text{small}}$, *we finally obtain*

$$\overline{\mathcal{T}}_{1\to 2}^{\epsilon} = \text{ord}(\epsilon^{-\delta}\sqrt{\ln(1/\epsilon)}). \tag{B.7}$$

Proof. The first part immediately follows from Theorem B.1, for we have almost everywhere

$$V_{\text{bar}}^{(i)}(x) = (1+\delta_x^{(i)})V_{\text{bar}}^{\text{small}}, \quad \delta_x^{(i)} > 0 \quad \Longrightarrow \quad \mathcal{T}_{i\to j}^{\epsilon}(x) = \text{ord}(\epsilon^{-\delta_x^{(i)}}).$$

(B.6) follows from (B.4) by using Laplace's method in the limit of vanishing noise ς. With the assumed coupling of ς according to (B.5), we then obtain from (B.6)

$$\mathbf{E}_{\bar{\mu}^{(1)}}[1/\mathcal{T}^{\epsilon}_{1\to 2}(x)] = \mathrm{ord}(\epsilon^{-\delta}\sqrt{\ln(1/\epsilon)}).$$

One could also contemplate a situation, such as that schematically indicated at the right-hand side of Fig. B.1. Here, $V^{(1)}_{\mathrm{bar}}(x) \to 0$ as $x \to \infty$, and there is no local minimum $V^{\mathrm{small}}_{\mathrm{bar}}$ such that $V^{\mathrm{small}}_{\mathrm{bar}} \leq V^{(1)}_{\mathrm{bar}}(x)$ for all x. Access to this problem is established in the next proposition where the argumentation has to be carried out rather intuitively. As outlined in Remark B.2, in Proposition B.4 it is coercive to consider the asymptotic behaviour as $\epsilon \to 0$ together with a reasonable coupling of $V^{(i)}_{\mathrm{bar}}(x)/\varsigma^2$ that is not yet specified.

Proposition B.4. *Suppose that* $\min\{\mathcal{T}^{\epsilon}_{1\to 2}(x), \mathcal{T}^{\epsilon}_{2\to 1}(x)\} \to 0$ *asymptotically for* $x \in D$ *where* D *is some subset of positive Lebesgue measure. We define a decomposition of* $D = D_1 \cup D_2$ *by*

$$D_1 = \{x \in D \mid \min\{\mathcal{T}^{\epsilon}_{1\to 2}(x), \mathcal{T}^{\epsilon}_{2\to 1}(x)\} = \mathcal{T}^{\epsilon}_{1\to 2}(x)\},$$
$$D_2 = \{x \in D \mid \min\{\mathcal{T}^{\epsilon}_{1\to 2}(x), \mathcal{T}^{\epsilon}_{2\to 1}(x)\} = \mathcal{T}^{\epsilon}_{2\to 1}(x)\}.$$

To simplify argumentation, we assume that D_i, $i = 1, 2$ *are connected subsets of* D *and* $(D_1 \cap D_2)\backslash\partial(D_1 \cap D_2) = \emptyset$. *Moreover, we restrict to the case where for* $x \in D^c$ *with* D^c *denoting the complement of* D *we have* $\min\{\mathcal{T}^{\epsilon}_{1\to 2}(x), \mathcal{T}^{\epsilon}_{2\to 1}(x)\} \to \infty$. *Now, the following is satisfied: The metastable transition times* $\overline{\mathcal{T}}^{\epsilon}_{i\to j}$ *from* $B^{(i)}$ *to* $B^{(j)}$ *will depend on the starting point* $x_0 = x^{\epsilon}(0)$ *and we write* $\overline{\mathcal{T}}^{\epsilon}_{i\to j}[x_0]$. *For* $(x^{\epsilon}(0), y^{\epsilon}(0)) \in B^{(1)}$ *with* $x^{\epsilon}(0) = x_0$ *we asymptotically obtain*

$$\overline{\mathcal{T}}^{\epsilon}_{1\to 2}[x_0] \simeq \mathbf{E}_{x_0}[\tau_{D^c\cup D_2}(x^{\epsilon}(t))], \tag{B.8}$$

where $\tau_{D^c\cup D^2}(x^{\epsilon}(t))$ *denotes first exit time of the process* $x^{\epsilon}(t)$ *from the set* $D^c \cup D^2$. *Instead of considering the exit times of the process* $x^{\epsilon}(t)$, *we can equally well consider the exit times of the conditionally averaged dynamics (1.7) with* $I(t, x) = 1$ *fixed. In the limit of* $\varsigma \to 0$ *we will be allowed to replace* x^{ϵ} *in (B.8) by the small noise approximation or the OU averaged dynamics as defined in Sect. 4 and still obtain the correct asymptotics. And, conclusively, by using (5.2), we arrive for vanishing* ς *at*

$$\overline{\mathcal{T}}^{\epsilon}_{1\to 2}[x_0] \simeq \mathbf{E}_{x_0}[\tau_{D^c\cup D_2}(x^0(t))] = \mathrm{ord}(1)$$

where $x^0(t)$ *is determined by*

$$\dot{x} = -D_x V(x, m^{(1)}) + \sigma \dot{W}_1.$$

In exact the same way we obtain asymptotics for $\overline{\mathcal{T}}^{\epsilon}_{2\to 1}$.

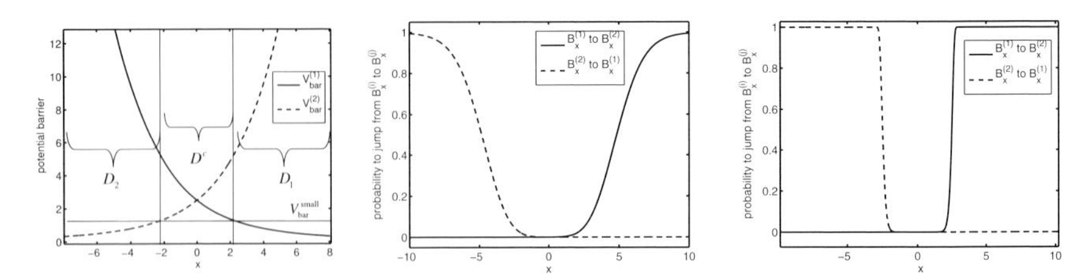

Fig. B.2. left: Illustration of $V^{(i)}_{\text{bar}}$, $i = 1, 2$ and $V^{\text{small}}_{\text{bar}}$ and the resulting subsets D_1, D_2, D^c; middle: transition probabilities (time step $\text{d}t = 0.01$) corresponding to (B.1) with $V^{\text{small}}_{\text{bar}} < V^{(1)}_{\text{bar}}(m)$ and $\epsilon = 10^{-3}$; right: transition probabilities with $\epsilon = 10^{-12}$.

Proof. A careful inspection of the transition probabilities $p^{\epsilon}_{12}(t, x)$ as defined in (6.2) and (6.3) with $q^{\epsilon}_{ij}(x) = 1/\mathcal{T}^{\epsilon}_{i\to j}$ reveals for each time step $\text{d}t$ pointwise convergence for almost every x

$$\lim_{\epsilon\to 0} p^{\epsilon}_{12}(\text{d}t, x) = 0, \quad x \in D_2 \cup D^c, \qquad \lim_{\epsilon\to 0} p^{\epsilon}_{12}(\text{d}t, x) = 1, \quad x \in D_1.$$

This shows that for ϵ small enough, the particle in $B^{(1)}$ will instantly jump over the barrier once it has reached D_1 and as long as it stays in $D_2 \cup D^c$ nothing will happen.

Example B.5. Let $V^{(i)}_{\text{bar}}$ be given as illustrated in the left picture[8] of Fig. B.2, that is, $V^{(1)}_{\text{bar}}$ is strictly monotonically decreasing with $V^{(1)}_{\text{bar}}(x) \to 0$ as $x \to \infty$, and $V^{(2)}_{\text{bar}}$ is strictly monotonically increasing with $V^{(2)}_{\text{bar}}(x) \to 0$ as $x \to -\infty$. Then there exists an intersection point m such that $V^{(1)}_{\text{bar}}(m) = V^{(2)}_{\text{bar}}(m)$. Now, choose $V^{\text{small}}_{\text{bar}}$ such that $V^{\text{small}}_{\text{bar}} < V^{(i)}_{\text{bar}}(m)$ and define the relation between ϵ and ς according to (B.1). The resulting subsets D_1, D_2 and $D^c = (D_1 \cup D_2)^c$ are shown in Fig. B.2. The picture in the middle shows the transition probabilities $p_{1\to 2} = p^{\epsilon}_{12}(\text{d}t, x)$, $p_{2\to 1} = p^{\epsilon}_{21}(\text{d}t, x)$ to jump over the barrier for moderately chosen $\epsilon = 10^{-3}$ and time step $\text{d}t = 1/100$. At the right we illustrate the transition probabilities for very small $\epsilon = 10^{-12}$. We clearly observe that for vanishing ϵ the particle will jump over the barrier once it has reached D_1 and D_2, respectively.

Example B.6. Let the assumptions be given as in Example B.5, but this time we choose $V^{\text{small}}_{\text{bar}} > V^{(i)}_{\text{bar}}(m)$. In this case, $D^c = \emptyset$ and the state space is decomposed into the sets D_1 and D_2 that are separated by the point m with $V^{(1)}_{\text{bar}}(m) = V^{(2)}_{\text{bar}}(m)$. An illustration is given in Fig. B.3. Again, the transition probabilities for $\epsilon = 10^{-12}$ at the right-hand side reveal

$$p_{1\to 2} = p^{\epsilon}_{12}(\text{d}t, x) \approx 0 \;\text{ for } x \in D_2, \quad p_{2\to 1} = p^{\epsilon}_{12}(\text{d}t, x) \approx 1 \;\text{ for } x \in D_1.$$

[8] We have chosen the potential from Sect. 6.

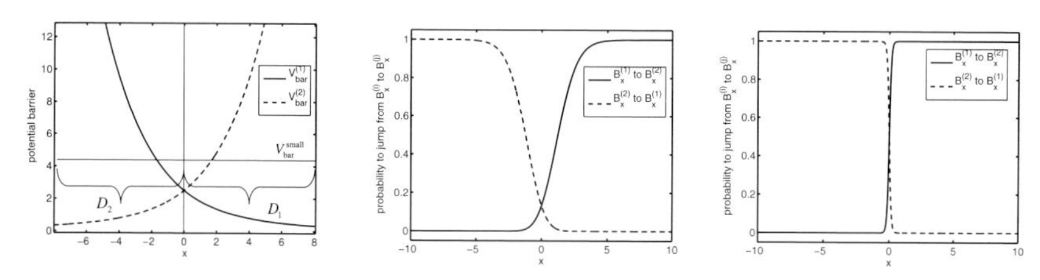

Fig. B.3. Same as Fig. B.2, but this time $V_{\text{bar}}^{\text{small}} > V_{\text{bar}}^{(1)}(m)$.

References

[AKN93] V. I. Arnold, V. V. Kozlov, and A. I. Neishtadt. *Mathematical Aspects of Classical and Celestial Mechanics.* Springer, Berlin, 1993.

[BLP78] A. Bensoussan, J. L. Lions, G. Papanicolaou. *Asymptotic Analysis for Periodic Structures.* Elsevier Science & Technology Books, July 1978.

[Bor98] F. A. Bornemann. *Homogenization in Time of Singularly Perturbed Mechanical Systems.* Number 1687 in Lecture Notes in Mathematics. Springer-Verlag, 1998.

[BS97] F. A. Bornemann and C. Schütte. Homogenization of Hamiltonian systems with a strong constraining potential. *Physica D*, 102:57–77, 1997.

[BS99] F. A. Bornemann and C. Schütte. On the singular limit of the quantum-classical molecular dynamics model. *1999*, 59:1208–1224, SIAM J. Appl. Math.

[BE*02] A. Bovier, M. Eckhoff, V. Gayrard, and M. Klein. Metastability in reversible diffusion processes I: Sharp estimates for capacities and exit times. *J. Eur. Math. Soc.*, 6:399–424, 2004.

[BGK02] A. Bovier, V. Gayrard, and M. Klein. Metastability in reversible diffusion processes II: Precise estimates for small eigenvalues. *J. Eur. Math. Soc.*, 7:69–99, 2005.

[Fre78] M. Freidlin. The averaging principle and theorems on large deviations. *Russ. Math. Surv.*, 33(5):107–160, 1978.

[FW84] M. Freidlin and A. Wentzell. *Random Perturbations of Dynamical Systems.* Springer, New York, 1984. Series in Comprehensive Studies in Mathematics.

[Gar85] C. W. Gardiner. *Handbook of Stochastic Methods.* Springer, Berlin, 2nd enlarged edition edition, 1985.

[HKN04] B. Helffer, M. Klein, and F. Nier. Quantitative analysis of metastability in reversible diffusion processes via a witten complex approach. preprint, 2004.

[HSS01] I. Horenko, B. Schmidt, and C. Schütte. A theoretical model for molecules interacting with intense laser pulses: The Floquet–based quantum–classical Liouville equation. *J. Chem. Phys.*, 115(13):5733–5743, 2001.

[Hui01] W. Huisinga. *Metastability of Markovian systems: A transfer operator based approach in application to molecular dynamics.* PhD thesis, Free University Berlin, 2001.

[HMS02] W. Huisinga, S. Meyn, and C. Schütte. Phase transitions & metastability in Markovian and molecular systems. accepted in Ann. Appl. Probab., 2002.

[Kif92] Y. Kifer. Averaging in dynamical systems and large deviations. *Invent. Math.*, 110:337–370, 1992.

[Kif01] Y. Kifer. Stochastic versions of Anosov's and Neistadt's theorems on averaging. *SD*, 1:1–21, 2001.

[Kif02] Y. Kifer. L^2–diffusion approximation for slow motion in averaging. preprint, 2002.

[Kur73] T. G. Kurtz. A limit theorem for perturbed operator semigroups with applications to random evolutions. *J. Funct. Anal.*, 12:55–67, 1973.

[MTV01] A. J. Majda, I. Timofeyev, and E. Vanden–Eijnden. A mathematical framework for stochastic climate models. *Comm. Pure Applied Math.*, 54:891–974, 2001.

[MTV99] A. J. Majda, I. Timofeyev, and E. Vanden-Eijnden. Models for stochastic climate prediction. *Proc. Natl. Acad. Sci.*, 96(26):14687–14691, 1999.

[MTV02] A. J. Majda, I. Timofeyev, and E. Vanden-Eijnden. A priori tests of a stochastic mode reduction strategy. *Physica D*, 170:206–252, 2002.

[Mor65] H. Mori. Transport collective motion and Brownian motion. *Prog. Th. Phys. Supp.*, 33:423–455, 1965.

[Pap76] G. Papanicolaou. Some probabilistic problems and methods in singular perturbation. *Rocky Mountain Math. J.*, 6:653–674, 1976.

[Pav02] I. Pavlyukevich. *Stochastic Resonance*. Logos, Berlin, 2002.

[SV85] J. Sanders and F. Verhulst. *Averaging Methods in Nonlinear Dynamical Systems*. Springer, New York, 1985.

[SH00] C. Schütte and W. Huisinga. On conformational dynamics induced by Langevin processes. In B. Fiedler, K. Gröger, and J. Sprekels, editors, *EQUADIFF 99 - International Conference on Differential Equations*, volume 2, pages 1247–1262, Singapore, 2000. World Scientific.

[SH02] C. Schütte and W. Huisinga. Biomolecular conformations can be identified as metastable sets of molecular dynamics. In P. G. Ciaret and J.-L. Lions, editors, *Handbook of Numerical Analysis*, volume Computational Chemistry. North–Holland, 2002. in press.

[SHD01] C. Schütte, W. Huisinga, and P. Deuflhard. Transfer operator approach to conformational dynamics in biomolecular systems. In B. Fiedler, editor, *Ergodic Theory, Analysis, and Efficient Simulation of Dynamical Systems*, pages 191–223. Springer, 2001.

[SW*03] C. Schütte, J. Walter, C. Hartmann, and W. Huisinga. An averaging principle for fast degrees of freedom exhibiting long-term correlations. *SIAM Multiscale Modeling and Simulation*, submitted. Presently available via www.math.fu-berlin.de/∼biocomp, 2003.

[Wal05] J. Walter. *Averaging for Diffusive Fast-Slow Systems with Metastability in the Fast Variable*. PhD thesis, Free University Berlin, 2005.

[Zwa73] R. Zwanzig. Nonlinear generalized Langevin equations. *J. Stat. Phys.*, 9:215–220, 1973.

Colored Plates

B. Nestler, F. Wendler: Simulations of Complex Microstructure Formations.

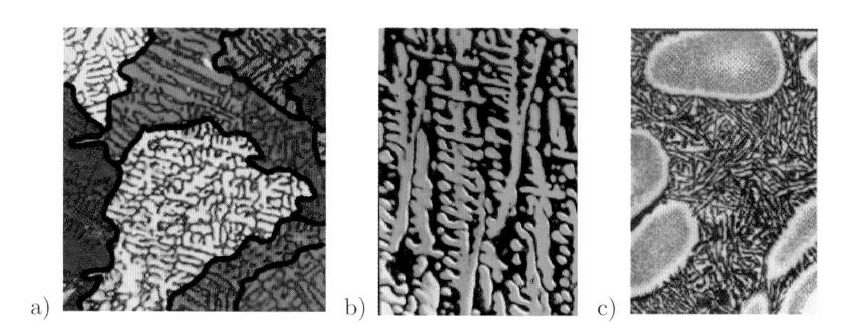

Fig. 1.1. Experimental micrographs of $Al - Si$ alloy samples: a) Grain structure with different crystal orientations, b) network of primary Al dendrites and c) interdendritic eutectic microstructure of two distinguished solid phases in the regions between the primary phase dendrites. (This figure is displayed in the text on page 114.)

Fig. 2.1. Left image: Schematic drawing of a domain separation by four different phase regions; Middle image: Polycrystalline grain structure in Al-Si; Right image: Multiphase solification microstructure with dendrites and a eutectic structure. (This figure is displayed in the text on page 117.)

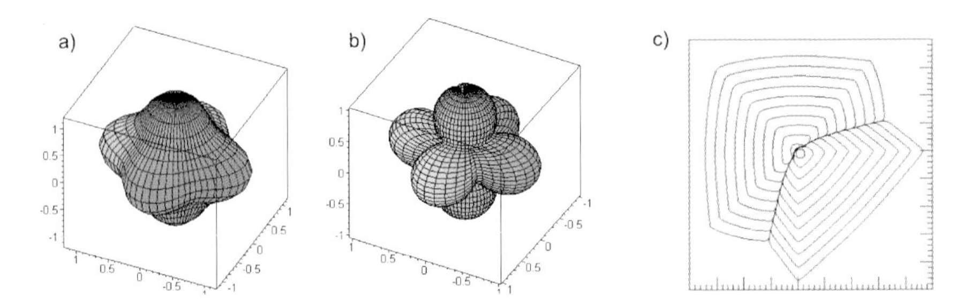

Fig. 2.2. 3D surface plot of a) a smooth and b) a facetted cubic anisotropy. c) contour plots of two adjacent growing, 45° misoriented cubic crystals applying the smooth anisotropy formulation in Eq. (2.5) with $\delta = 0.2$. (This figure is displayed in the text on page 120.)

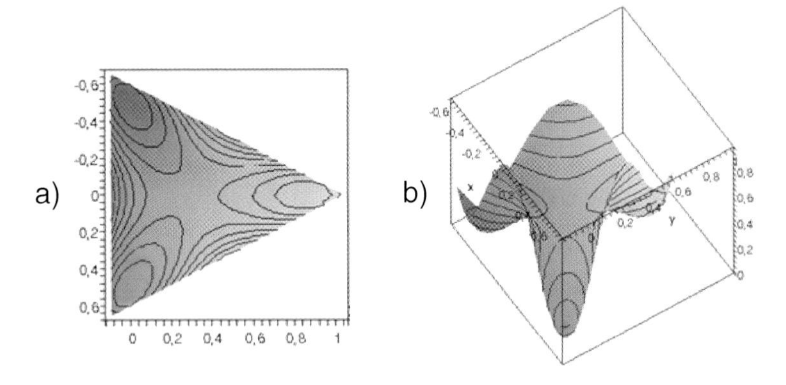

Fig. 2.3. Plot of the multi-well potential $w_{st}(\varphi)$ for $N = 3$ and equal surface entropy densities $\gamma_{\alpha\beta}$. (This figure is displayed in the text on page 121.)

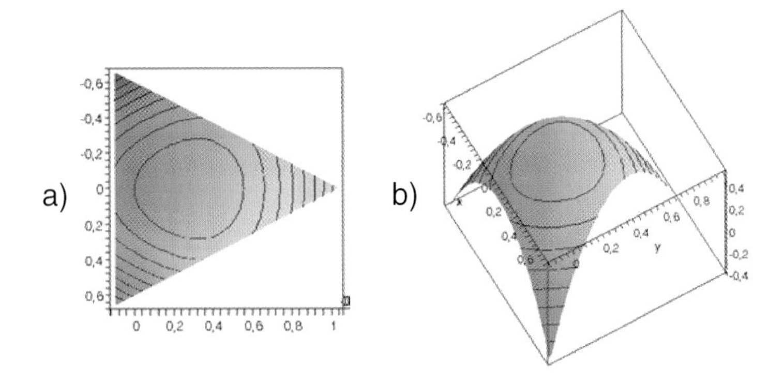

Fig. 2.4. Plot of the multi-obstacle potential $w_{ob}(\varphi)$ for $N = 3$ and equal surface entropy densities $\gamma_{\alpha\beta}$. (This figure is displayed in the text on page 122.)

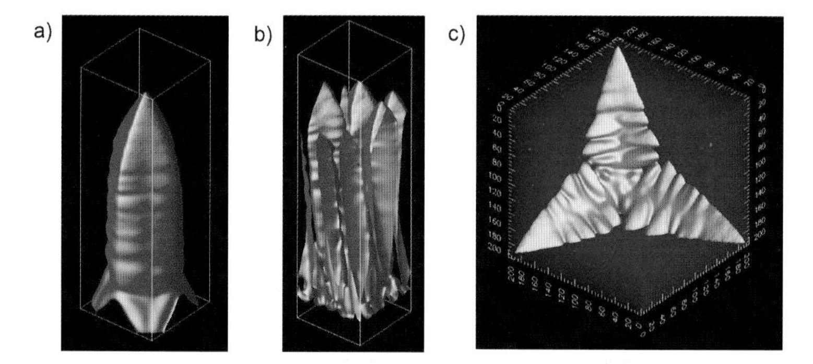

Fig. 4.5. 3D Ni dendritic growth at an undercooling of $\Delta = 0.6$. a) single channel dendrite, b) dendritic array with an orientation inclination of 15° with rsp. to normal, c) equiaxial dendrite. (This figure is displayed in the text on page 135.)

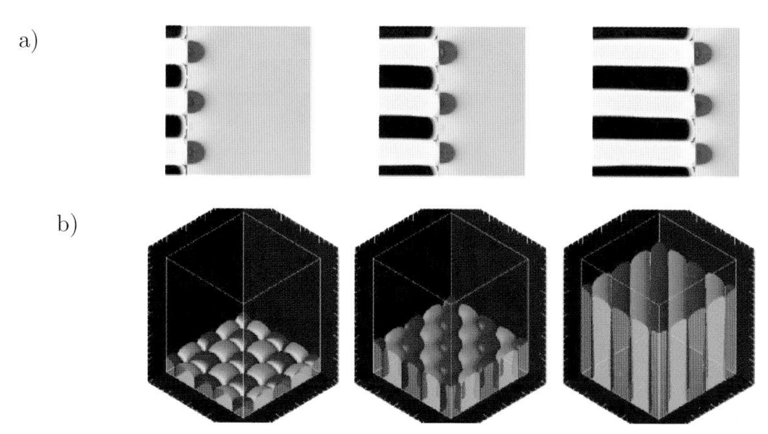

Fig. 4.8. Establishment of regular lamellar solidification at the eutectic composition in 2D (a) and 3D (b). (This figure is displayed in the text on page 138.)

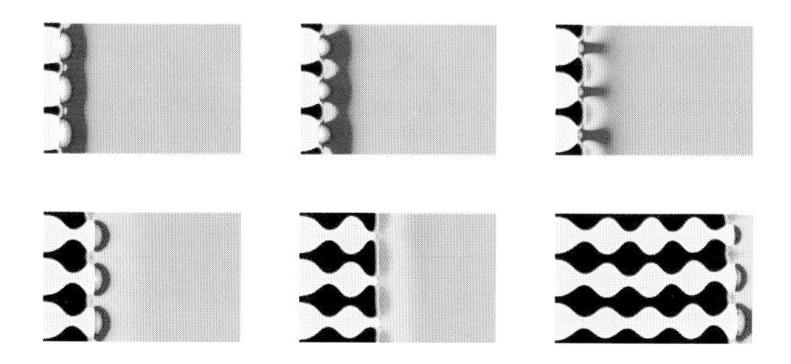

Fig. 4.9. Regular oscillations along the solid-solid interface driven by the motion of the triple junction/triple line in 2D. (This figure is displayed in the text on page 138.)

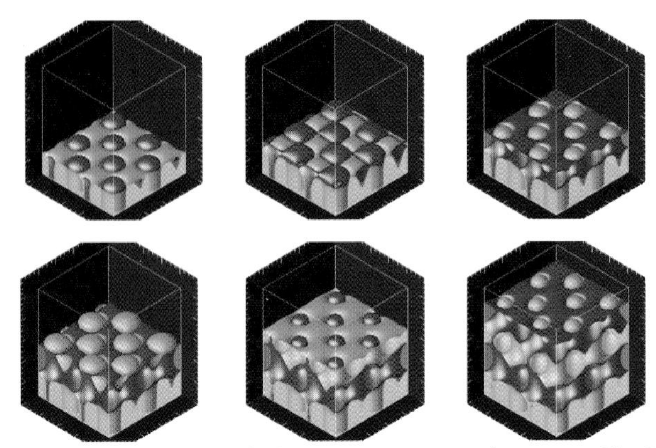

Fig. 4.10. Topological change of the microstructure due to oscillations along the solid-solid interface in 3D. (This figure is displayed in the text on page 139.)

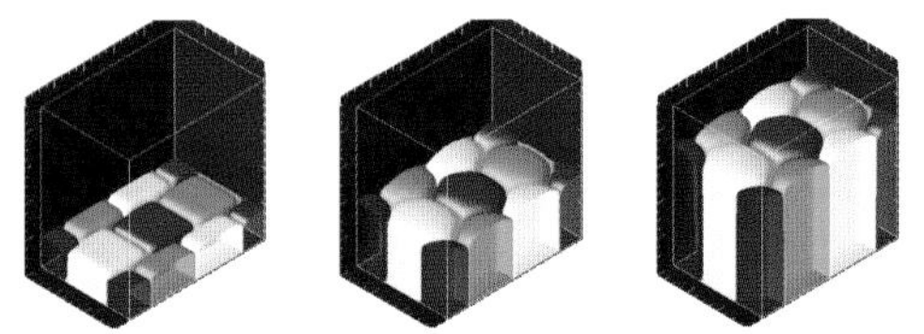

Fig. 4.12. Formation of a 3D hexagonal rod-like structure in a ternary eutectic system with isotropic surface energies and three different solid phases α, β and γ. (This figure is displayed in the text on page 139.)

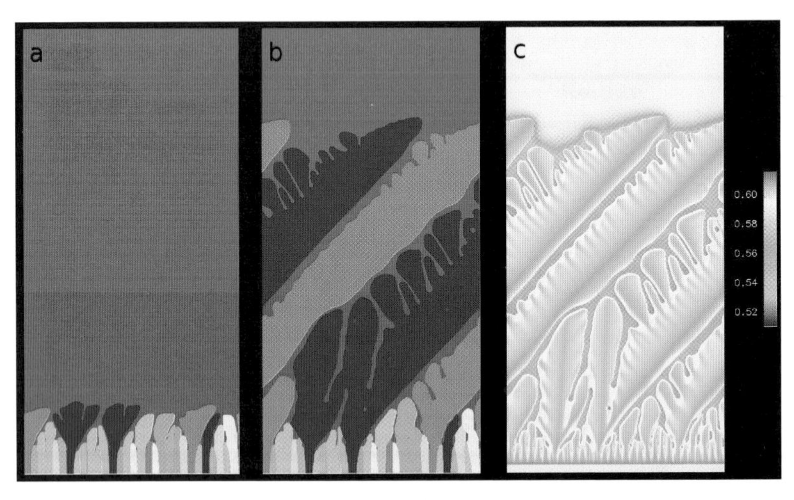

Fig. 4.15. Selection process in a polycrystalline dendritic front: The colours in a) and b) indicate the orientations of the dendrites for two different time steps, whereas in c) the Ni concentration is shown. (This figure is displayed in the text on page 141.)

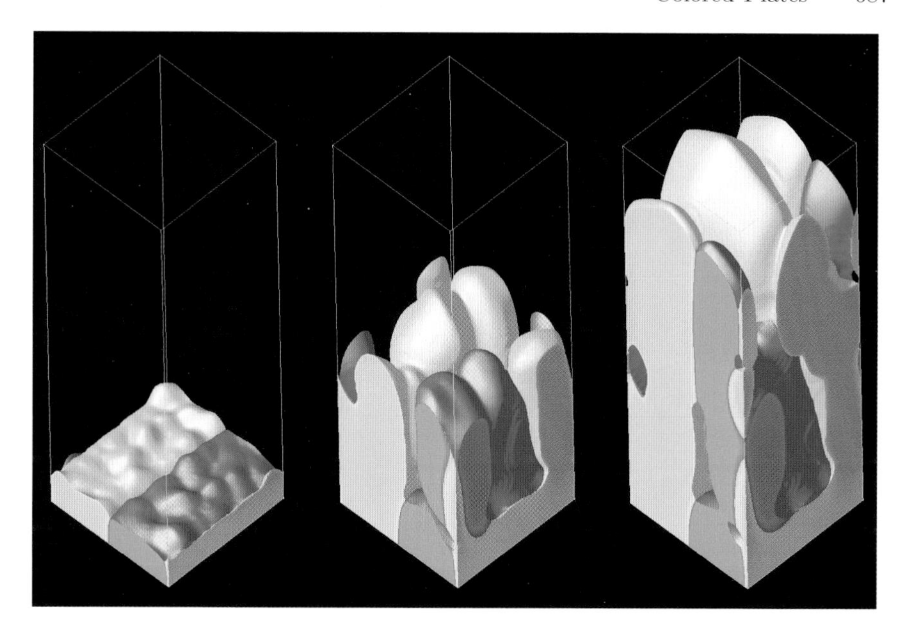

Fig. 4.16. Three time steps of two misaligned NiCu grains in 3D starting from a rough planar initial state (lateral periodic boundary conditions). (This figure is displayed in the text on page 142.)

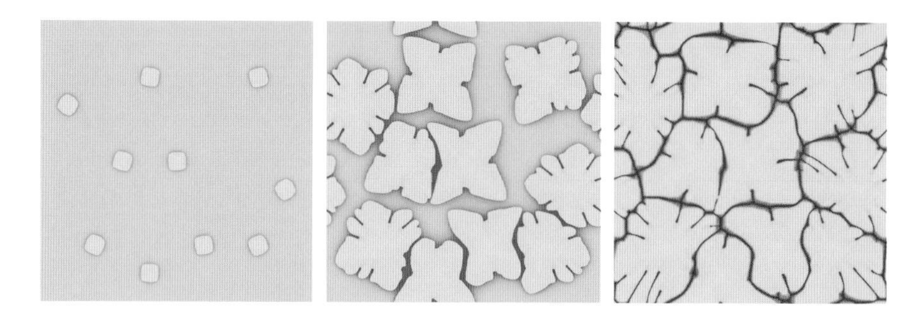

Fig. 4.17. Growth of dendritic NiCu grains into a 20 K undercooled melt illustrated by the Ni concentration (range: 0.41−0.62). The complete solidification (right image) is reached after further reducing the temperature by 15 K in a second step. (This figure is displayed in the text on page 143.)

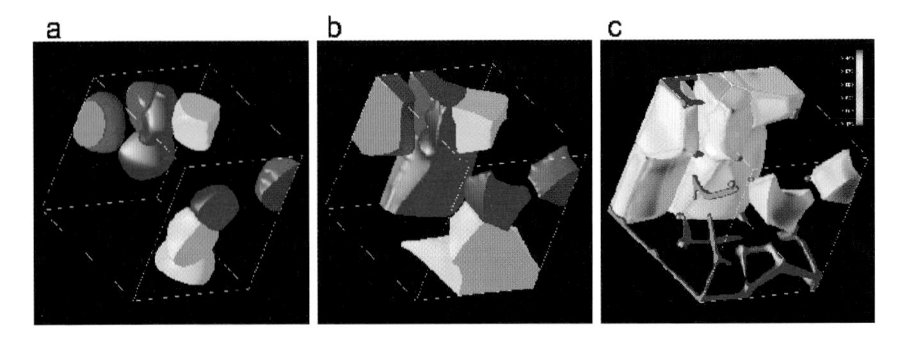

Fig. 4.18. a) and b): Growth of a polycrystalline NiCu structure with 30 grains on a domain of size $100 \times 100 \times 100$. The isosurfaces of selected grains for two time steps are displayed. c): Heat treatment with partial melting along the grain vertices (from [WN06]). (This figure is displayed in the text on page 143.)

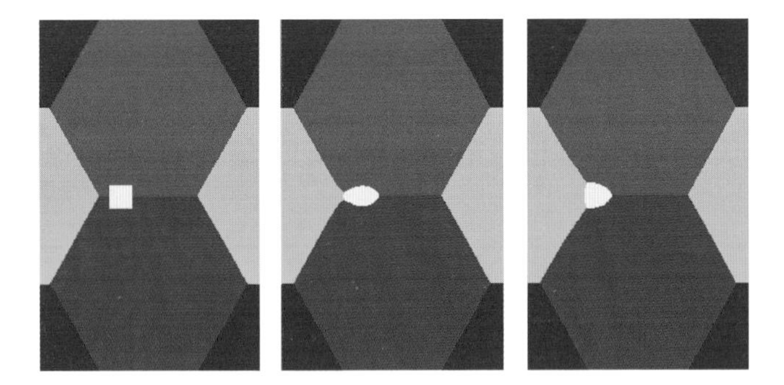

Fig. 4.21. Periodic test geometry with an inserted melt inclusion (light grey) at the horizontal boundary, shown for three time steps (from left to right). (This figure is displayed in the text on page 146.)

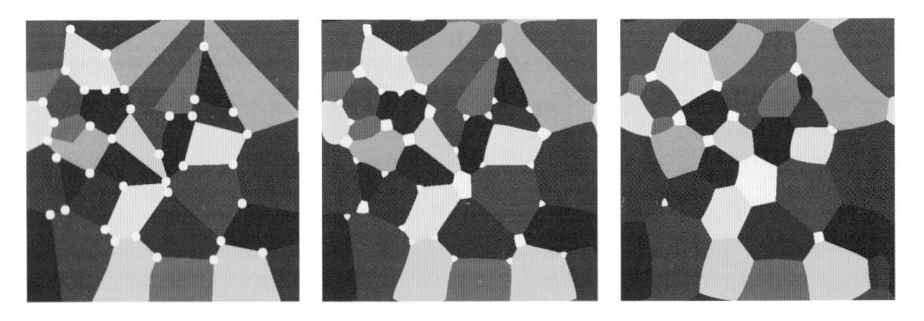

Fig. 4.22. Coarsening of a grain structure with fluid inclusions (light grey) at the triple junctions for three time steps and for a ratio of the surface energies $\gamma_{ss}/\gamma_{sl} = 1.2$. (This figure is displayed in the text on page 146.)

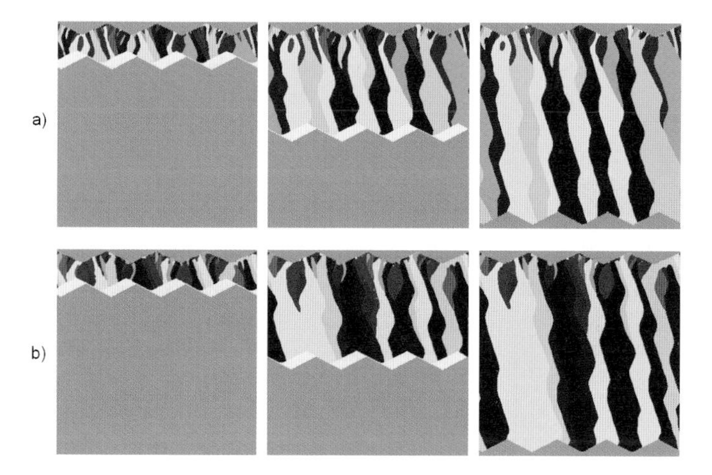

Fig. 4.24. Polycrystalline grain growth in a crack-seal process with facetted anisotropy of the surface energies of the grain boundaries. a) and b) show three time steps of two simulation runs with different starting grain distribution. (This figure is displayed in the text on page 148.)

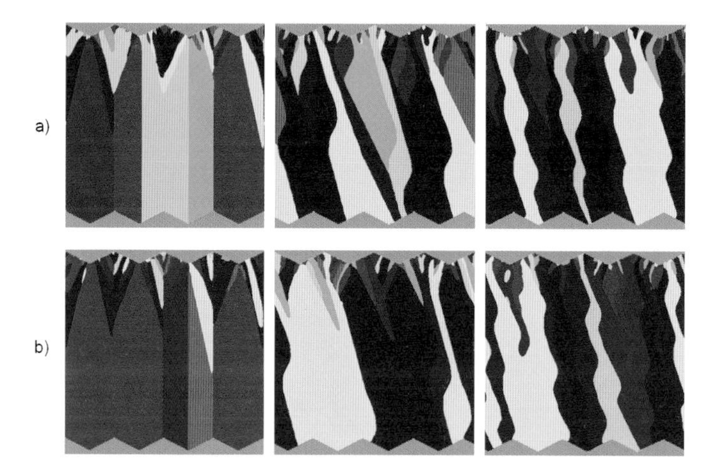

Fig. 4.25. Effect of different shear rates on the resulting morpholgy. From left to right: no shear, $\Delta x = 10$ cells, $\Delta x = 20$ cells. Two simulation runs a) and b) with different grain distributions are displayed. (This figure is displayed in the text on page 149.)

Garcke, Lenz, Niethammer, Rumpf, Weikard: Multiple Scales in Phase Separating Systems with Elastic Misfit.

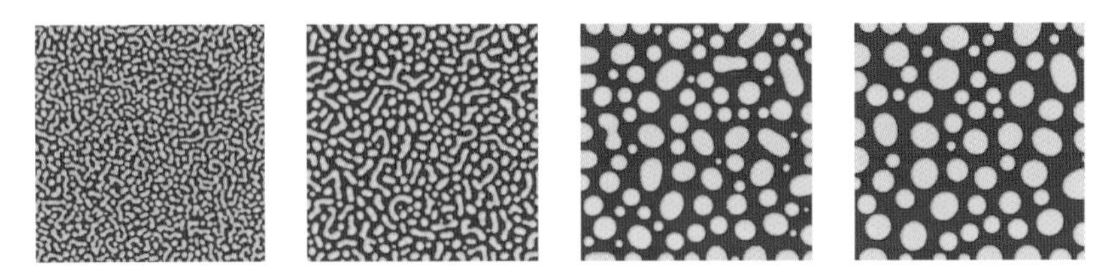

Fig. 1.1. Evolution starting from a perturbation of a uniform state. (This figure is displayed in the text on page 154.)

Fig. 1.2. Alignment of interfaces driven by homogeneous, anisotropic elasticity. (This figure is displayed in the text on page 154.)

Fig. 4.2. Effects of inhomogeneous elasticity: On the left side the green phase is the elastically harder one, the blue phase is softer. On the right side it is vice versa. The volume fraction of both phases are the same. (This figure is displayed in the text on page 166.)

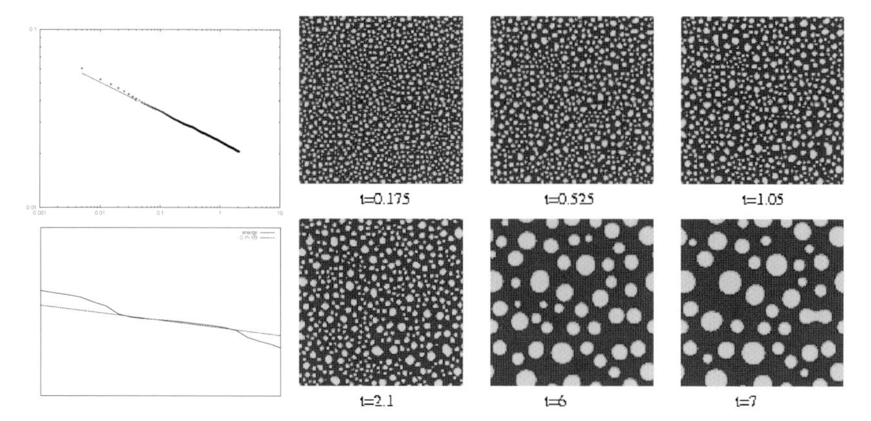

Fig. 5.1. Graph of the energy at an early and a very late stage of the evolution (two graphs on the left side), different time steps of the evolution (on the right side). (This figure is displayed in the text on page 167.)

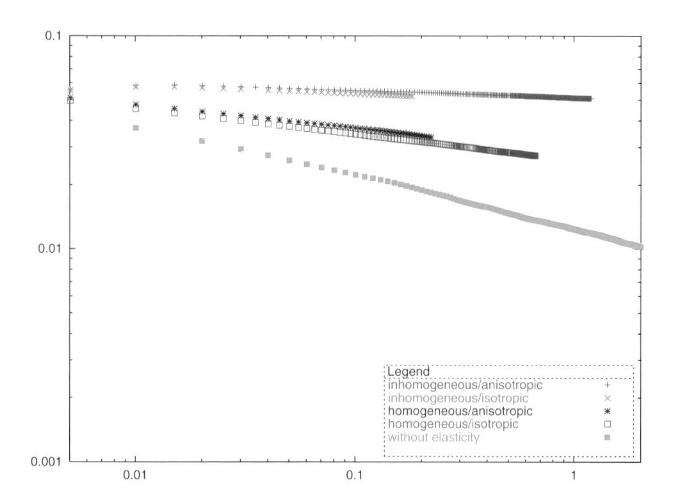

Fig. 5.2. Graph of the nonelastic part of the energy. (This figure is displayed in the text on page 168.)

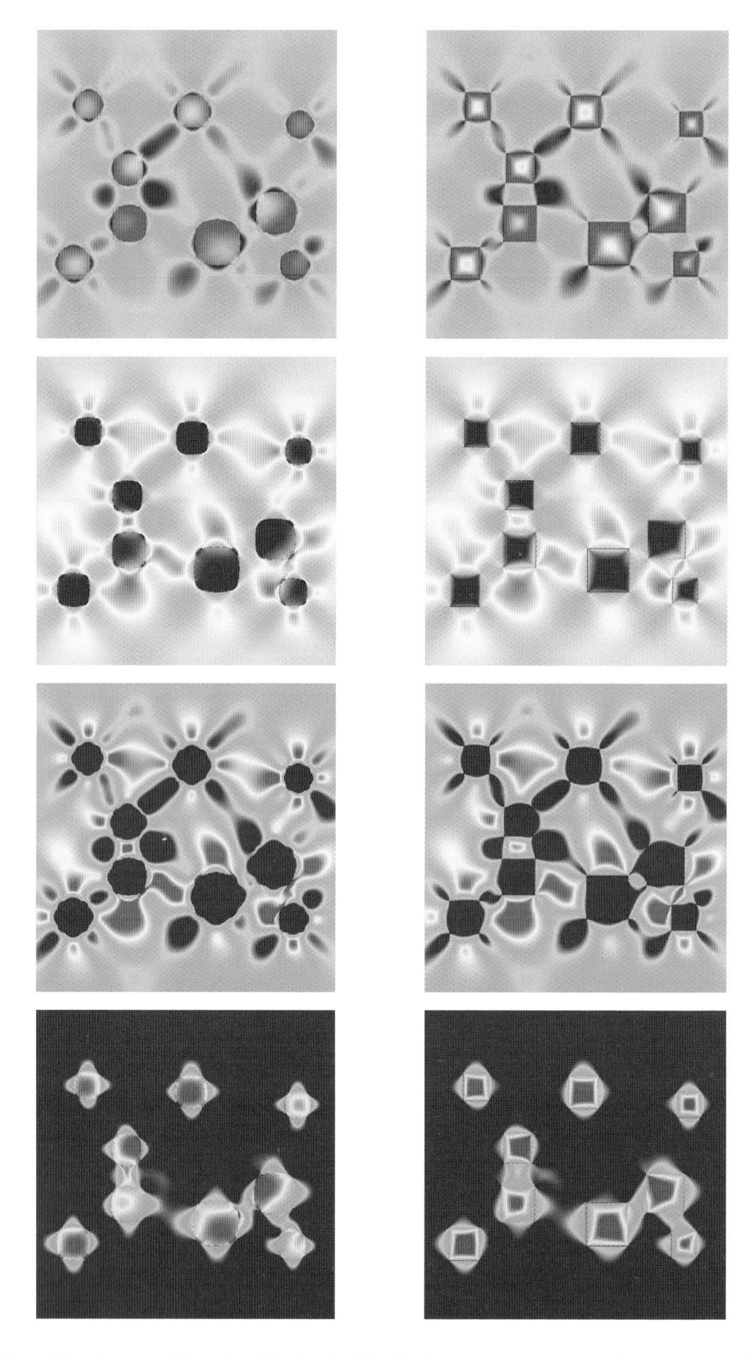

Fig. 7.4. The trace of the elastic strain (both top rows, compared to zero strain and eigenstrain), the trace of the stress (middle) and energy density (bottom) for the Mullins–Sekerka (left) and the reduced model (right), in the initial configuration. (This figure is displayed in the text on page 175.)

Bartels, Carstensen, Conti, Hackl, Hoppe, Orlando: Relaxation and the Computation of Effective Energies and Microstructurcs in Solid Mcchanics.

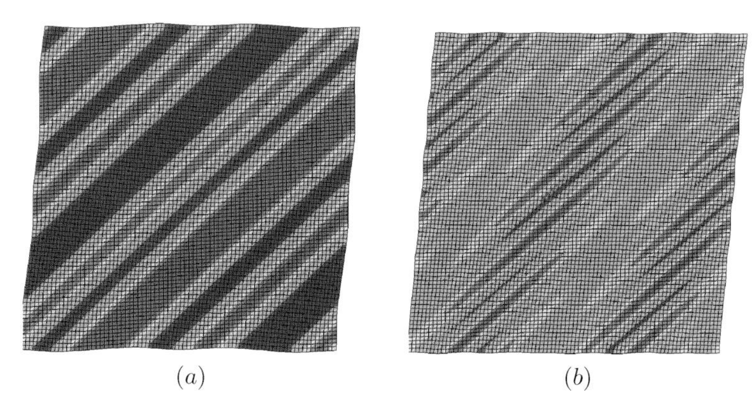

Fig. 5.1. Single-slip plasticity (a) First order laminates and (b) Second order laminates as from (3.12) and (3.16) respectively, assuming periodic boundary conditions. (This figure is displayed in the text on page 218.)

S. Conti, G. Dolzmann: Derivation of Elastic Theories for Thin Sheets and the constraint of incompressibility.

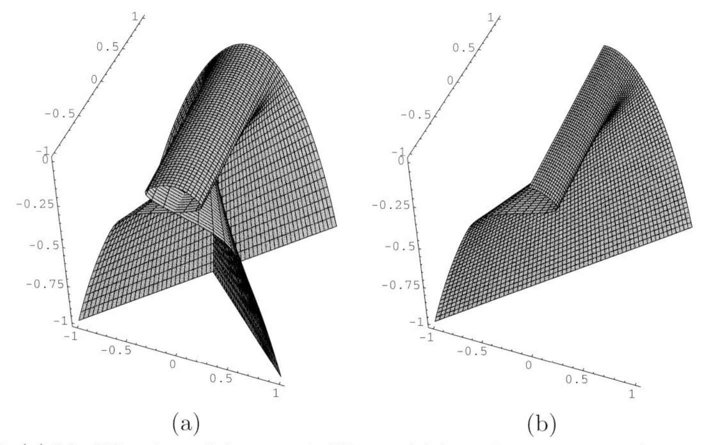

Fig. 4.2. (a) Modification of the map in Figure 4.1 in order to remove the singularity. (b) Plot of one half of surface in (a). This view illustrates that the modification is a smooth surface in $\mathbb{R}^3$ for which the tangent plane in each point has full rank. (This figure is displayed in the text on page 235.)

F. Bornemann, C. Lasser, T. Swart: Energy Level Crossings in Molecular Dynamics.

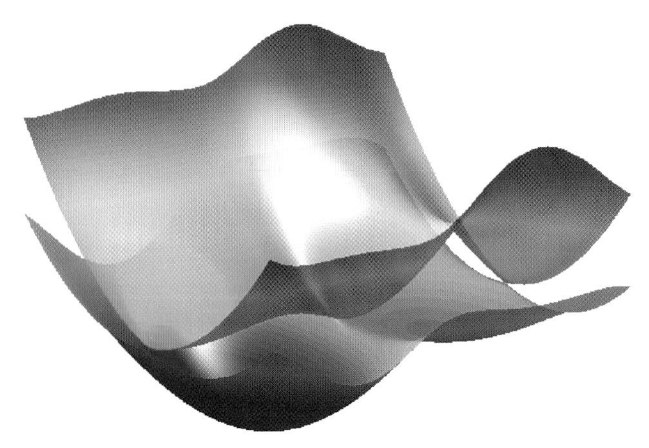

Fig. 1.1. The energy levels of the model for the cis-trans isomerization of retinal in rhodopsin of [HS00]. They cross, when the angular variable ϕ is approximately $\frac{\pi}{2}$ or $\frac{3\pi}{2}$, and the collective coordinate vanishes. In the plot, the abscissa corresponds to ϕ, the ordinate to y. The two local minima of the lower energy level are associated with the cis and the trans configuration of the molecule. (This figure is displayed in the text on page 579.)

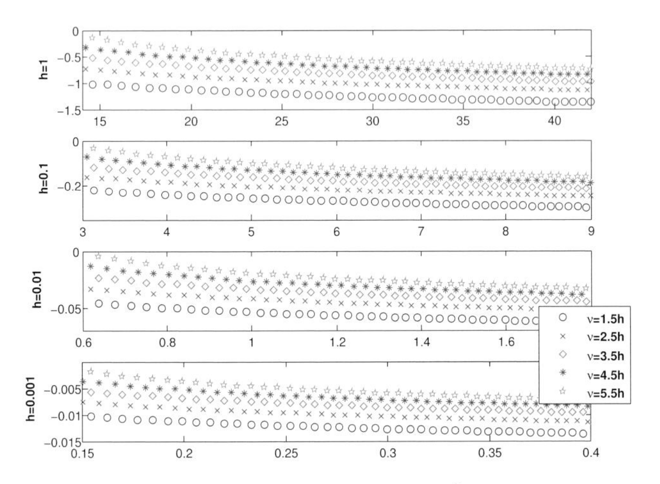

Fig. 3.1. Resonances of the model operator $H = -\epsilon^2 \Delta_q + V(q)$. The parameter k lies in $\{11, 12, \ldots, 60\}$, while ν is chosen in $\{1.5, 2.5, \ldots, 5.5\}$. The semiclassical parameter ϵ varies from 10^{-3} to 1. (This figure is displayed in the text on page 591.)

Dellnitz, Hessel-von Molo, Metzner, Preis, Schütte: Graph Algorithms for Dynamical Systems.

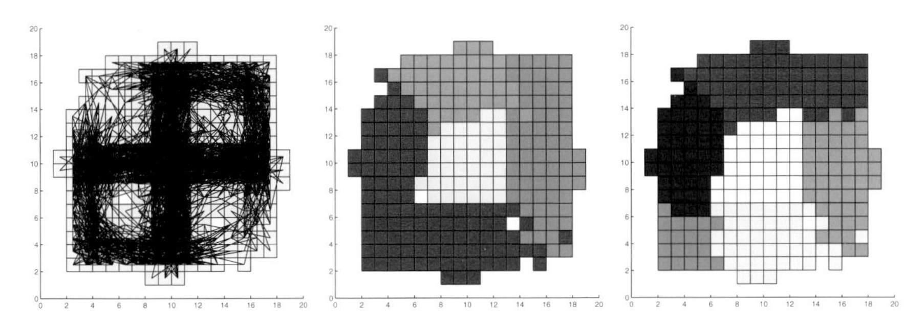

Fig. 3.2. Partition of graph describing the dynamics Pentane300. Left: The graph with all edges. Center: Partition consisting of five parts with $C_{\text{int}} = 0.980$. Right: Partition consisting of seven parts with $C_{\text{int}} = 0.963$. (This figure is displayed in the text on page 629.)

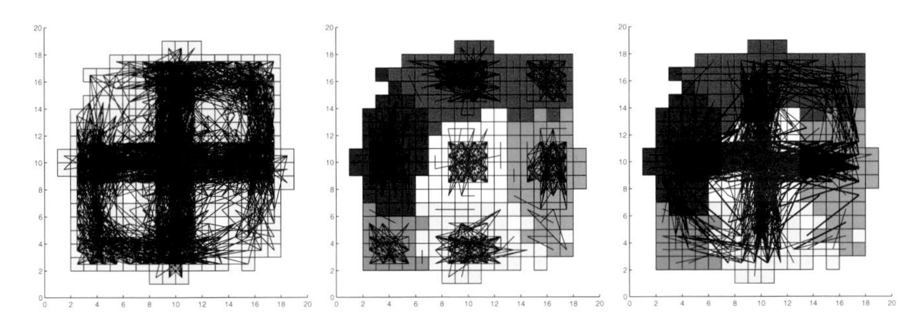

Fig. 3.3. Congestion of the Pentane. Left: all transitions. Center: only transitions with a low congestion. Right: only transitions with a high congestion. (This figure is displayed in the text on page 632.)

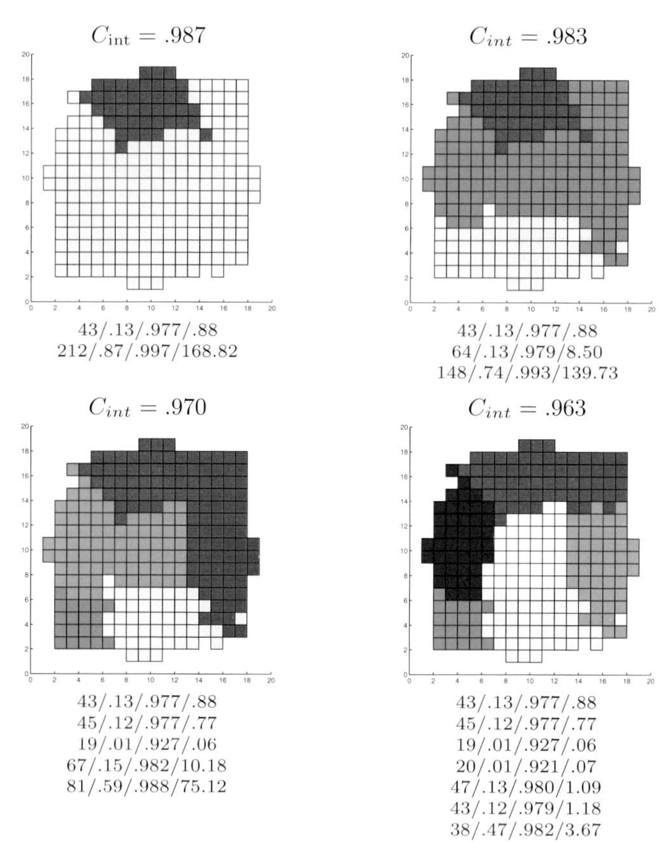

Fig. 3.4. Recursive bisection of the graph for the pentane example. The values indicate: number of boxes / invariant measure / internal cost C_{int} / congestion of subgraph. The graph has 255 vertices and a congestion of 139.67. (This figure is displayed in the text on page 635.)

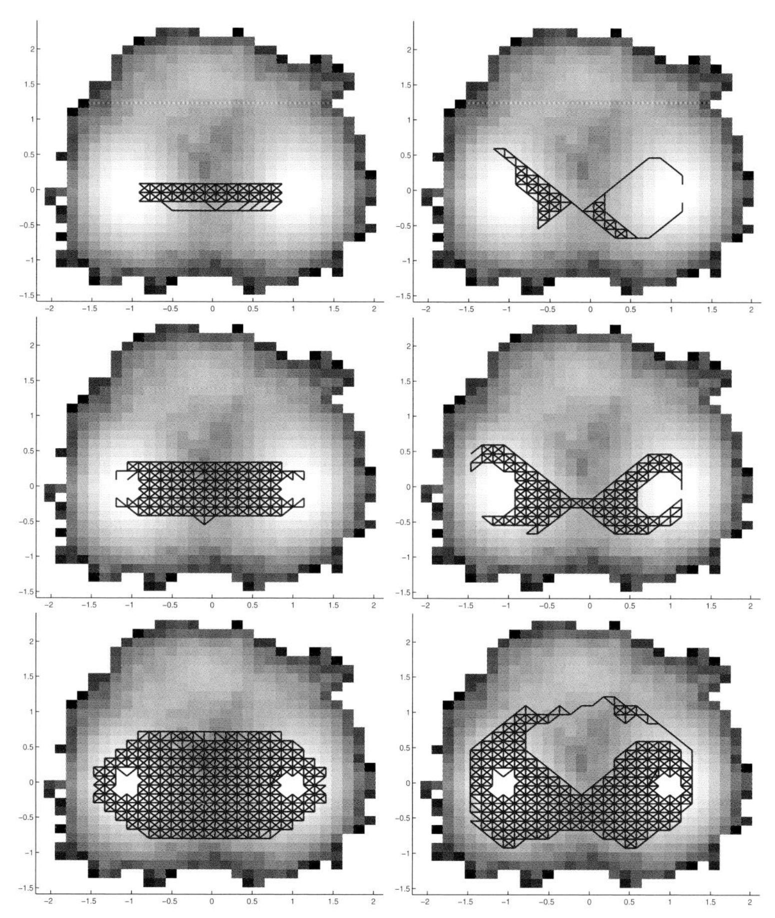

Fig. 4.5. Left column: Family of most probable paths between the set A (left minimum) and the set B (right minimum). From the top to the bottom we choose $\epsilon = 0.1, \epsilon = 0.3$ and $\epsilon = 0.6$. Right column: Family of paths which crosses the lowest free energy barriers. From the top to the bottom we choose $\epsilon = 0.01, \epsilon = 0.05$ and $\epsilon = 0.13$. (This figure is displayed in the text on page 643.)

Printing: Krips bv, Meppel
Binding: Stürtz, Würzburg